Handbook of Energy Technology
Trends and Perspectives

Other Books by V. Daniel Hunt

- **ENERGY DICTIONARY,** 1979, Van Nostrand Reinhold, New York, N.Y.
- **ENERGY ISSUES IN HEALTH CARE,** 1979, The Energy Institute, Washington, D.C.
- **WINDPOWER—A HANDBOOK ON WIND ENERGY CONVERSION SYSTEMS,** 1981, Van Nostrand Reinhold, New York, N.Y.
- **HANDBOOK OF CONSERVATION AND SOLAR ENERGY**—TRENDS AND PERSPECTIVES, 1982, Van Nostrand Reinhold, New York, N.Y.

Handbook of Energy Technology

Trends and Perspectives

V. Daniel Hunt
Director
The Energy Institute

VAN NOSTRAND REINHOLD COMPANY
NEW YORK   CINCINNATI   TORONTO   LONDON   MELBOURNE

Van Nostrand Reinhold Company Regional Offices:
New York   Cincinnati

Van Nostrand Reinhold Company International Offices:
London   Toronto   Melbourne

Copyright © 1982 by Van Nostrand Reinhold Company

Library of Congress Catalog Card Number: 80-22195
ISBN: 0-442-22555-5

All rights reserved. No part of this work covered by the copyright hereon may be reproduced or used in any form or by any means—graphic, electronic, or mechanical, including photocopying, recording, taping, or information storage and retrieval systems—without permission of the publisher.

Manufactured in the United States of America

Published by Van Nostrand Reinhold Company
135 West 50th Street, New York, N.Y. 10020

Published simultaneously in Canada by Van Nostrand Reinhold Ltd.

15 14 13 12 11 10 9 8 7 6 5 4 3 2 1

Library of Congress Cataloging in Publication Data
Hunt, V. Daniel.
  Handbook of energy technology.
  Includes index.
  1. Power resources—Handbooks, manuals, etc.
2. Power (Mechanics)—Handbooks, manuals, etc.
I. Title.
TJ163.2.H874   621.042   80-22195
ISBN 0-442-22555-5

Preface

Energy is the lifeblood of our economy. Without adequate energy supplies now and in the future, the security of Americans, their jobs, and their ability to provide for their families will be threatened and their standard of living will be lowered. Every American is painfully aware that our national energy situation has deteriorated over the past several years. Gasoline prices have more than doubled. Our oil import bill has risen 96 percent. Our energy supplies have become increasingly vulnerable because U.S. continental oil production is below 1973 levels. The threat of sudden shortages, curtailments, and gas lines has unfortunately become a recurring reality due to poor government planning and management.

This steady deterioration has not only compounded our economic problems of inflation, recession, and dollar weakness; but also, even more importantly, it has affected our confidence as a nation. Energy shortages, spiraling costs, and increasing insecurity are beginning to darken our hopes and expectations.

A few years ago, President Carter declared energy the "moral equivalent of war" and sent Congress hundreds of recommendations for action, including the creation of the Department of Energy. Since then, the budget for the Federal government's energy bureaucracy has grown to about $10 billion per year. The Congress has joined in the stampede, taking action on several hundred more energy bills since 1977. As a result, the federal bureaucracy was engaged in projects nationwide: allocating gasoline, setting building temperatures, printing rationing coupons, and readying standby plans to close factories, and distributing "no drive day" stickers to American motorists—all the time saying, "we must make do with less."

This disappointing cycle of shrinking energy prospects and expanding government regulation and meddling is unnecessary. The proven America values of individual enterprise can help solve our energy problems.

We need an alternative strategy of aggressively boosting the nation's energy supplies; stimulating new energy technology and more efficient energy use; restoring choice and freedom in the marketplace for energy consumers and producers alike; and eliminating energy shortages and disruptions that are a roadblock to renewed national economic growth and rising living standards.

The United States must proceed on a steady and orderly path toward energy self-sufficiency. But, in the interim, our pressing need for insurance against supply disruption should not be made hostage to the whims of foreign governments.

In order to increase domestic production of energy, the decontrol of the price at the well head of oil was needed. Decontrol of prices on all oil products and an end to government authority to allocate petroleum supplies except in a national emergency should be stressed. The market restrictions on the use of natural gas should be eliminated.

Coal, our most abundant energy resource, can bridge the gap between our other energy sources and renewable energy sources. In

1977, President Carter promised to double coal production by 1985. Instead, coal production has increased by only 11 percent to date and future prospects are dim.

To effectively use our coal resources, coal transportation systems must be upgraded and the government controls on them revised. Government regulations regarding the mining and use of coal must be simplified.

Coal, and nuclear energy offer the most realistic intermediate solutions to America's energy needs. We must support accelerated use of nuclear energy through technologies that have been demonstrated to be efficient and safe. The safe operation, as well as design, of nuclear power plants should have a high priority to ensure the continued availability of this important energy source. The licensing process should be streamlined through consolidation of the present process and use of standardized reactor designs.

Nuclear power development requires sound plans for nuclear waste disposal and storage, and reprocessing of spent fuel. Technical solutions to these problems exist, and decisive federal action to choose and implement solutions is essential. Regional away-from-reactor storage of spent fuel should be implemented. The development of permanent storage facilities for nuclear wastes must be expedited. Since waste disposal is a national responsibility, no state should bear an unacceptable share of this responsibility.

We must continue to support the development of new synthetic fuel technologies for liquid, gaseous, and solid hydrocarbons which can be derived from coal, oil shale, and tar sands.

Gasohol is an important, immediately available source of transportation fuel that is helping to extend our petroleum reserves. We must encourage development of the domestic gasohol industry.

The government must continue supporting research to expedite new energy technologies, including solar energy, geothermal, wind, magnetic fusion, and biomass.

Inefficient energy use results from government subsidization of imported oil and holding the price of natural gas substantially below its market value. When the price of energy is held artifically low, there is no incentive for conservation. This kind of energy consumption stems not from the excesses of the public, but from the policy distortions of government.

A policy of decontrol, development of our domestic energy resources, and incentives for new supply technology and conservation technologies will substantially reduce our dependence on imported oil.

Virtually all major environmental legislation in the past decade reflected a bipartisan concern over the need to maintain a clean and healthful environment. Although the new environmental policies have resulted in improved air quality, cleaner waters, and more careful analysis of toxic chemicals, the price paid has far exceeded the direct cost of designing and installing new environmental control technology. In the energy area, the increased complexity of regulations, together with continual changes in the standards imposed, have resulted in tremendous delays in the planning and construction of new facilities ranging from electric power plants to oil refineries, pipelines, and synthetic fuel plants.

An effective balance between energy and environmental goals must be achieved. Government requirements should be firmly based on the best scientific evidence available, they must be enforced evenhandedly and predictably, and the process of their development and enforcement must have finality.

In order to solve our energy problem, we must increase our domestic energy production capability. In the short term, therefore, the nation must move forward on all fronts simultaneously, including oil and gas, coal, alcohol fuels, and nuclear energy. Renewable resources must be brought on-line to replace conventional sources. Finally, in conjunction with this new production initiative, we must strive to maximize conservation and the efficient use of energy.

The return to the traditions that gave vitality and strength to this nation is urgent. The free world—indeed western civilization—needs a strong United States. That strength requires a prospering economy with a vigorous domestic energy industry. That vigor can only be achieved in an atmosphere of freedom—one that encourages individual initiatives and personal resourcefulness.

V. Daniel Hunt
The Energy Institute
Fairfax Station, Virginia

Notice

This handbook was prepared as an account of work sponsored by Van Nostrand Reinhold Company. Neither Van Nostrand Reinhold nor The Energy Institute, nor any of their employees, make any warranty, express or implied, or assume any legal liability or responsibility for the accuracy, completeness or usefulness of any information, apparatus, product or process disclosed, or represent that its use would not infringe on privately owned rights.

The view, opinions and conclusions in this book are those of the author and do not necessarily represent those of the United States Government or the United States Department of Energy.

Public domain information and those documents abstracted, edited or otherwise utilized are noted in the acknowledgements or on specific illustrations.

Acknowledgments

I served as Project Manager of a forty-five member team of professional staff members of TRW's Energy Systems Group who were responsible for the development of the following Department of Energy Program Summary Documents:

- *Energy Technology Program Summary Document*
- *Fossil Energy Program Summary Document*
- *Solar, Geothermal, Electric and Storage Systems Program Summary Document*
- *Fission Energy Program Summary Document*
- *Nuclear Waste Management Program Summary Document*
- *Magnetic Fusion Energy Program Summary Document*

The Department of Energy Program Manager for this effort was Mr. Ted D. Tarr. Mr. Tarr provided enlightened management direction and was responsible for the scope and direction of the above series of reports.

In addition to Mr. Tarr, I would like to express my appreciation to TRW for their indirect support in preparing this handbook, and in particular Mr. Robert T. McWhinney Jr., Operations Manager of the Systems Planning Office.

I acknowledge the support and assistance of Mr. Timothy F. O'Leary, Manager of the Energy Technology program summary material utilized in Part I of this handbook. Participation by Mr. Gerald E. Stock, Manager of the Solar, Geothermal, Electric and Storage Systems information; Mr. Vic Trebules who served as the DOE Fission Energy Coordinator; Mr. Don Frazier, Nuclear Waste Management report manager; and especially Dr. Harlan Watson, who developed the Magnetic Fusion Program Summary material. I also extend thanks to all those Department of Energy and TRW staff members who developed or reviewed various versions of this material.

The material in this book is based on the Program Summary Documents which are now in the public domain and a thorough revision and editing by the author. The author accepts responsibility for all errors.

A book of this magnitude is dependent upon excellent staff, and I have been fortunate. Judith A. Anderson was an excellent coordinator of this project as well as technical editor. Special thanks are extended to Janet C. Hunt, Anne Potter, and Michelle M. Donahue for the typing of the manuscript.

The majority of art, graphs and photographs were provided through the courtesy of the Department of Energy.

Contents

Preface v

Part I
ENERGY TECHNOLOGY OVERVIEW

1. U.S. Energy Policy 1
2. Role of Energy Technology 7
3. Energy Technology Programs 20

Part II
FOSSIL ENERGY

1. U.S. Energy Policy 57
2. FY 1980 Fossil Energy Overview 60
3. Activity Description—Coal 82
4. Activity Description—Petroleum 279
5. Activity Description—Gas 320
6. Field Activities 335
7. Commercialization 348
8. Environmental and Socioeconomic Implications 352
9. International Programs 358
10. University Activities 368

Part III
SOLAR, GEOTHERMAL, ELECTRIC AND STORAGE SYSTEMS

1. Introduction 373
2. Solar, Geothermal, Electric and Storage Systems Overview 376
3. Solar Technology 400
4. Geothermal 450
5. Electric Energy Systems 474
6. Energy Storage Systems 488
7. Regional Activities 520
8. Commercialization 527
9. Environmental Aspects 536
10. Socioeconomic Aspects 544
11. International 549

Part IV
FISSION ENERGY

1. The National Energy Plan and Fission Energy Policy	567
2. Fission Energy Program Management	583
3. Converter Reactor Systems	589
4. Breeder Reactor Systems	627
5. Advanced Nuclear Systems	703
6. Deployment Considerations	737

Part V
NUCLEAR WASTE MANAGEMENT

1. U.S. Energy Policy	751
2. FY 1980 Nuclear Waste Management Overview	755
3. Subactivity Descriptions	778
4. Field Activities	845
5. Commercialization	849
6. Environmental Implications	851
7. Socioeconomic Issues	856
8. Regional Activities	859
9. International Activities	861

Part VI
MAGNETIC FUSION ENERGY

1. U.S. Energy Policy	867
2. FY 1980 Overview—Activities of the Office of Fusion Energy	870
3. Subactivity Descriptions	893
4. Field Activities	938
5. Commercialization	944
6. Environmental Implications	948
7. Regional Activities	953
8. International Programs	956

Part VII
REFERENCE INFORMATION

Glossary	965
Definition of RD and D Scale-up Phases	975
Abbreviations and Acronyms	977
References	982
Energy Units and Conversion Factors	987
Milestone Symbols	992
Index	993

Handbook of Energy Technology

Trends and Perspectives

Part I

Energy Technology Overview

1.
U.S. Energy Policy

BACKGROUND

Energy has been relatively inexpensive in the United States. The nation has abundant reserves of coal, oil, and natural gas. Americans used these resources liberally to raise their living standards; the supply or price of energy was not a major concern. Consequently, the country produced homes, cars, appliances, and factory equipment that use energy wastefully by today's standards.

In the past, economic growth and energy consumption have been very closely related. Between 1960 and 1977, the U.S. economy grew 81 percent from $737 billion to $1.3 trillion in constant 1972 dollars. U.S. energy consumption grew 70 percent, from 45 quads to 76 quads during the same period. Before the energy crisis in late 1973, the relationship was even closer. Between 1960 and 1972, GNP grew at an average annual rate of 3.9 percent. Energy consumption grew at an average annual rate of 4.0 percent during those years. (See Figure I 1-1.)

The complacency about energy ended rather abruptly in 1973 with the oil embargo. The embargo, and the quadrupling of imported oil prices, showed Americans the implications of depending on imported oil.

Rising energy costs have been one cause of inflation. A higher price for this basic resource is quickly felt throughout the economy. These higher costs have made it more difficult to reach full employment by rendering uneconomic a portion of the country's stock of capital equipment. The high cost of imported energy has undermined the value of the dollar abroad, creating more inflationary pressure, and could weaken American foreign policy.

Although the link between energy consumption and economic growth will persist, the tie between them now should not be as close. Energy consumption is likely to grow far more slowly than the economy. Many European countries have provided their people a high living standard with much lower energy consumption than the United States. Energy prices may induce the U.S. economy to grow more energy efficient in regard to energy use.

Even though U.S. energy consumption should grow less than the gross national product and well below historical growth rates, energy demand may still surpass 100 quads by 1990. The mix of energy sources will proba-

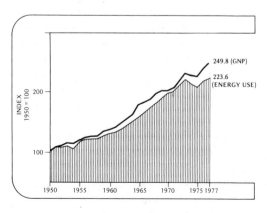

Figure I-1-1. Growth of GNP and energy use.

bly change in meeting this demand. Oil and gas may provide a smaller share of the total supply. Coal and nuclear power may increase their shares while the share provided by other energy sources could be relatively constant, although the actual amount of energy they provide may rise. By the year 2000, however, alternative energy sources could provide a large and rapidly rising share of U.S. energy.

Since 1945, petroleum and natural gas have provided more energy in the United States than any other source. Abundant supplies from easily accessible domestic fields induced the nation to rely less on coal and more on convenient fuels. During the late 1940s, the United States began to augment its domestic supplies of oil with imports to satisfy rising demand. By 1960, imported oil amounted to 18 percent of U.S. petroleum consumption, and imported gas amounted to 1 percent of demand for that product. By 1977, these figures had risen to 45 percent and 5 percent, respectively.

Despite new oil and gas supplies from Alaska, offshore drilling, and synthetics, the country's dependence on imports could still grow. Some projections indicate that imports could account for 51 percent of U.S. oil consumption and 17 percent of gas consumption by 1990. Other studies indicate that world oil demand may begin to outrun supply as early as 1985. Recent discoveries of oil and gas, however, could push those dates into the more distant future. (See Figure I 1-2.)

The absolute worldwide exhaustion of oil and gas resources will probably never occur. As the demand for petroleum rises and the supply fails to keep pace, its price will rise. Energy users will switch away from petroleum, and producers will find it economic to use more expensive methods of recovering oil. As the process continues, the use of petroleum will become more and more restricted.

Because of this, the United States is searching for alternative energy sources such as solar or geothermal energy. The nation is also seeking ways to conserve energy. This would allow the economy to expand without the danger of having its energy lifeline cut.

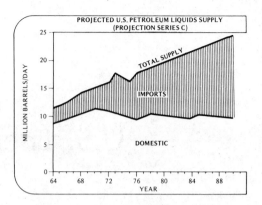

Figure I-1-2. U.S. Petroleum liquids supply.

The energy situation is one of the most difficult problems facing the United States. The supply of energy and its price concern almost everyone. Society is being forced to adapt very quickly to new relationships and to adopt a more frugal way of life. The transition is not easy. It involves giving up habits acquired during a lifetime and lowering expectations for the future.

NATIONAL ENERGY ACTS

The National Energy Acts, recently passed into law, provide many tools with which the country's energy goals can be achieved. They may save 2.4 to 3.0 million barrels of oil a day by 1985. The laws encourage energy conservation so that the nation's resources may be used more efficiently. The Acts also promote the use of more abundant domestic natural resources such as coal. Finally, they encourage increased production of domestic energy.

After the acts were passed, President Carter stated, "We have declared to ourselves and the world our intent to control our use of energy, and thereby to control our own destiny as a nation."

The National Energy Acts consist of five pieces of legislation:

- The National Energy Conservation Policy Act of 1978
- The Powerplant and Industrial Fuel Use Act of 1978
- The Energy Tax Act of 1978

- The Natural Gas Policy Act of 1978
- The Public Utility Regulatory Policies Act of 1978

National Energy Conservation Policy Act of 1978

The intent of this act is to reduce the growth in energy use and conserve nonrenewable energy resources without inhibiting economic growth. It provides a variety of incentives for homeowners to make their dwellings more energy efficient. The act also mandates efficiency standards for home appliances and efficiency labeling of industrial equipment. Some of its major provisions are as follows:

- A utility conservation program for residences
- Weatherization grants for low-income families
- A solar energy loan program for solar heating and cooling equipment in residences
- An energy conservation loan program
- A grant program for schools and hospitals to improve their energy efficiency
- Energy audits for public buildings
- Appliance efficiency standards
- Civil penalties for violating automobile fuel economy standards

Powerplant and Industrial Fuel Use Act of 1978

This law seeks to induce certain electric power plants and major fuel burning installations (MFBI) to use proportionately more domestic coal or other fuels instead of natural gas or petroleum. Major provisions of the act include:

- A prohibition against new facilities using oil or natural gas without a Federal exemption
- Department of Energy authority to require facilities to use coal
- Restrictions on the use of natural gas as a boiler fuel
- Loans for air pollution equipment for electric utilities

Energy Tax Act of 1978

The purpose of this Act is to induce individuals and companies to conserve energy and convert to new energy technologies. This Act provides:

- Residential insulation and conservation tax credits
- Tax credits for insulation and solar or wind equipment
- A tax exemption for gasohol
- A tax on automobiles that fall below Federal mileage standards
- A tax credit for development of geothermal energy
- Expensing of intangible drilling costs for geothermal energy
- An additional 10-percent investment tax credit for investment in industrial energy conservation or alternative energy sources
- Denial of tax credits for new gas or oil boilers

Public Utility Regulatory Policies Act of 1978

This law gives the Department of Energy (DOE) and the Federal Energy Regulatory Commission (FERC) major roles, along with the states, in reshaping electric and natural gas utility practices. The purposes of the act are to conserve energy and capital, to make the most efficient use of facilities and resources, and to assure equitable rates to consumers. The act contains:

- Federal rate design standards
- A requirement that states consider adopting the Federal Standards
- A requirement that gas utilities consider adopting standards on advertising and service termination
- FERC rules favoring cogeneration
- FERC authority to order interconnection of electric power facilities and wheeling of electricity
- Funding to help state and consumer representation in proceedings
- Loans to help development of small hydroelectric projects

- Legislation to speed crude oil transportation systems

Natural Gas Policy Act of 1978

This law estblishes equal prices for interstate and intrastate natural gas. It also deregulates the price of natural gas in stages. Unless a new law is passed, all controls will be off beginning in 1989. The main provisions of the law are as follows:

- A series of maximum prices for various categories of natural gas
- Deregulation of new gas and certain intrastate gas on January 1, 1985
- A system of incremental pricing so that large industrial users bear the burden of higher prices until controls are removed
- Presidential power to allocate intrastate gas during an emergency
- A prohibition on curtailing interstate supplies of natural gas unless the gas is needed by residences or small business, or if curtailment would endanger life

Table I 1-1 shows the expected imported energy saving in 1985 generated by the National Energy Acts.

Table I 1-1. 1985 Imported energy savings.

ACTION	SAVING IN 1985 (THOUSANDS OF BARRELS OF OIL PER DAY)
Conservation	
Building conservation programs/appliances	410
Auto and truck standards	265
Utility rate reform	0–160
Natural gas pricing	1,000–1,400
Coal conversion	300
Energy taxes	
Residential tax credits	225
Gas guzzler tax	80
Business energy tax credits	110
TOTAL	2,390–2,950

2.
Role of Energy Technology

ENERGY TECHNOLOGY'S IMPLEMENTATION OF POLICY

Program Overview

Former Secretary Schlesinger stated "the purpose of the National Energy Act is to put in place a policy framework for decreasing oil imports by:

- replacing oil and gas with abundant domestic fuels in industry and electric utilities,
- reducing energy demand through improved efficiency,
- increasing production of more conventional sources of domestic energy through more rational pricing policies, and
- building a base for the development of solar and renewable energy sources."

Energy technology is providing the technical foundation for developing new energy sources and improved ways of using energy. In supporting the goals of the National Energy Acts as well as previous legislation, Energy Technology is concentrating its efforts in accord with two criteria.

The first is near-term commercial use. Priority is being given to technologies that can be developed quickly and moved into the economy. This can rapidly affect energy production and consumption.

The second is concentration on local and regional needs. No national solution to the energy problem can be expected. An approach that works well in one part of the country could be a failure somewhere else. In the past, options were overlooked because they are not applicable over wide areas; but they may provide the ideal solution to a local problem.

Different regions of the country have different energy needs and will require different technologies to meet them. The various energy technologies complement each other; they do not exclude each other. Decreasing energy use through conservation is the natural helpmate to raising energy supply. People in various regions have different tastes, customs, and traditions that can be satisfied only by a variety of technologies. People can select the energy path that brings them to their own destinations. Pursuing many technologies gives people the freedom to choose.

Along with these criteria, Energy Technology is pursuing a long-term research and development program. The goal is to provide technology products for a virtually inexhaustible energy supply.

The key theme of all the programs is to develop products, not just technologies. It is insufficient to do basic engineering for difficulties which might appear in the distant future. Energy technology is taking a businesslike approach to the energy problem. Technology is developed to serve a specific purpose and to compete in the commercial marketplace. It must be the right tool to do the job at a price that industry and consumers can afford.

The principal mechanisms in achieving this objective are not strictly technological. They are social, economic, environmental, and legal as well. The National Energy Acts recognize this by the variety of incentives, penalties, and regulations that they provide. Although technology is only one of many paths toward solving the energy problem, significant progress can be made if the nation maintains its commitment to research and development. Technologies can be brought to commercialization. However, the major objective is not just to do research and wait for something to happen. The objective is to make things happen.

Making things happen can be difficult. There are many problems in bringing a technology from the drawing board to the marketplace.

First, it is expensive. Literally billions of dollars can be spent bringing a technology to commercialization.

Second, it takes time. Research and development often must continue for decades before a technology can be used.

Third, there must be a balance in the projects being worked on. Concentrating on just one or two greatly raises the risk of failure. There must be a reasonable mix of efforts in fossil fuel, solar energy, hydropower, nuclear energy, geothermal energy, and magnetic fusion energy.

Fourth, the public must be involved. Citizens and industry must accept a program for it to have any positive result. The public and industry should be involved from the very beginning. Their participation will not only make the work easier, but will also promote greater success. The wider the spectrum of participation in a project, the more aware people will be of the possibilities of conservation and alternative energy sources. And they will be more willing to change the way they use energy.

Finally, social, environmental, and institutional problems must be identified and resolved early. It is a waste of time and resources to resolve problems late in the development process that could have been addressed earlier. Generally, the more the process has been advanced, the more difficult, expensive, and time consuming it is to resolve problems.

Program Priorities. The highest priority in the development of technologies is to reduce the U.S. dependence on imported oil. Several ways of achieving this goal are immediately evident:

- Convert coal to liquid fuels
- Reduce consumption by conservation
- Increase the domestic oil supply through such methods as developing tertiary recovery techniques or developing alternate resources such as shale oil
- Develop alternative sources of energy

All options are being pursued, but the highest priority is being given to developing new liquid fuel supply technologies. Emphasis is also being placed on short-term and mid-term solutions that could lessen U.S. demand for petroleum in the critical 1985 to 1990 period. Some technologies are approaching a stage where commercialization efforts can begin. Until these are established in the marketplace, they will have a very high priority compared to long-range solutions to the energy problem.

Projects Ready for Commercialization. The effort to develop usable energy products has been successful in many areas. Recently, a DOE task force evaluated the commercialization potential of 23 energy technologies. Some were already being used, while others were ready to begin their entry into the economy.

Hydrothermal–geothermal energy sources are already in operation in some areas and could make significant additional contributions to energy supplies.

Small wind machines are already used in special circumstances, although work remains to be done on large machines before they are ready for the open market.

In situ oil shale processing is now commercial, as are enhanced oil recovery methods. These two concepts to increase domestic oil production will directly reduce our dependence on foreign energy.

Likewise, low- and medium-Btu coal gasifiers and enhanced gas recovery methods, all of which are now ready for commercialization, will reduce dependence on foreign sources.

Commercialization efforts can now begin on "industrial" applications of atmospheric fluidized bed combustors. These combustors burn coal more cleanly. Their lower environmental impact could induce some industries now using gas or oil to convert to coal as their basic energy source.

Technology Not Ready for Commercialization. Some projects are far along the research and development path but are not quite ready to begin the commercialization process. These projects will receive the highest R&D priority. They have the greatest potential for a quick payoff in terms of alternatives to oil and gas.

Coal liquefaction. Coal liquefaction will be one of the major priority items in Energy Technology's efforts in the immediate future. The technology is fairly well advanced and holds promise of a fast return in lower oil imports.

As shown in Figure I 2-1, products derived from coal liquefaction could replace petroleum products in two markets. The first uses low-ash, low-sulfur boiler fuels to produce electric power or industrial steam. The other market uses high-grade fuels such as gasoline, heating oil, and chemical feedstocks. The development of coal liquefaction will provide increased domestic production of clean fossil fuel and feedstocks. Because these fuels could be used with existing equipment, they would not require expensive refitting of industrial equipment.

The strategy in developing liquefaction is to support a number of processes simultaneously, but only the most promising will go to the pilot plant stage. The need for several techniques is made necessary by the variety of coals that must be processed and the requirements for a range of fuels.

High-Btu Gasification. The need to develop high-Btu coal gasification processes is based on the same reasoning as in coal liquefaction. They would serve to offset the need to import gas and save the capital invested in existing industrial equipment by eliminating the need to switch to alternate fuels. (See Figure I 2-2.)

The thrust of the program will be to develop the first generation plants to a point where commercialization can begin. Work will also be performed to identify and develop second- and third-generation processes which promise more economic or environmentally acceptable technologies.

Again, because of the various types of coal used as feedstock, several processes are being developed. Data are being gathered to provide information on the economic and environmental aspects of each process.

The heating value of this gas is about 1,000 Btu per cubic foot. It is essentially free of sulfur and other pollutants and should not pose an environmental hazard.

Fluidized Bed Combustion for Utilities. While the industrial size combustors are ready to be-

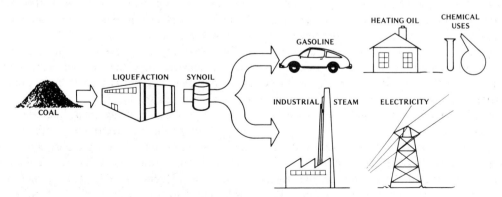

Figure I-2-1. Products of coal liquefaction.

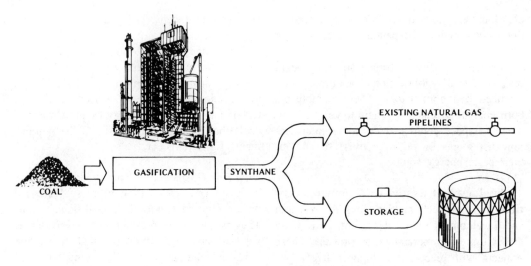

Figure I-2-2. Coal gasification.

gin their entry into the marketplace, the larger ones need more research and development before they could gain acceptance. Bringing the very large scale combustors that the utilities need to a point where they can begin commercialization will be a major part of energy technology's effort. Problems to be overcome include:

- The development of combustion systems capable of directly burning high-sulfer coal and other coals in a manner that does not degrade the environment
- The development of higher efficiency combustion systems
- The development of more durable components
- The reliable operation of combined cycle systems

Large Wind Machines. More research is needed to tap the sun's power through large wind machines. Small wind machines are ready for commercialization, but large machines face additional problems. Small machines have the possibility of being mass produced at relatively low cost. Large machines, on the other hand, must be tailored to their location.

The rotor is another problem. As a rule, the rotor accounts for half the cost of the machine. Doubling the diameter of the rotor quadruples the power of the machine. Research is being done to determine the fatigue load of the large blades and the risk of blade failure. Until this information is available, commercialization is risky.

Since large wind machines have the potential for delivering significant amounts of power in New England, the Pacific Northwest, and the High Plains, research and development of this concept deserves priority. This is especially true because the technology is approaching a point where commercialization might become feasible.

Photovoltaics. Photovoltaic devices convert sunlight directly into electricity. Intuitively, this is one of the more desirable means of producing electricity because it is clean, has no moving parts, does not require combustion, and is infinitely renewable. The devices are already being used in space satellites and in remote applications where small amounts of power are needed.

The technology, however, is still in its infancy. Consequently, the cost of generating electricity this way is too high in most applications. (See Figure I 2-3.)

Rapid progress is being made in reducing the cost of producing electricity in this manner. Projections indicate that the cost could be brought down from about $1 per kWh to about 10 to 15 cents in 1982, 5 to 8 cents in 1986, and 4 to 6 cents in 1990. Photovoltaics

Role of Energy Technology 11

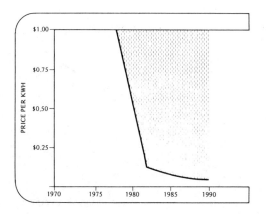

Figure I-2-3. Projected cost of photovoltaic electricity.

could conceivably supply a significant amount of electric demand as early as 1985.

Nuclear waste. The handling of nuclear waste is another priority item. Nuclear power is already providing nearly 4 percent of the U.S. energy. It is expected to rise to nearly 10 percent by 1990. Some nuclear wastes are long-lived and are perceived by the public as being dangerous. Ways must be found to handle the problems of nuclear waste, if nuclear power is to fulfill its potential.

Specific objectives of the research and development program in nuclear wastes are to:

- Identify a site (or sites) and construct a repository for spent nuclear fuel and contaminated materials
- Develop safe packaging technologies for nuclear wastes
- Provide more storage space for the spent fuel that will accumulate under the policy of nuclear nonproliferation

Other priorities. These priorities are not an exhaustive list of the efforts that can be made. They represent only the most urgent programs or the ones that hold promise of short-term, high-impact results.

Some other priority items are to encourage the production of alcohol or methane from wood, corn, and cane products; to promote the more widespread use of forest residues for direct combustion in boilers as a substitute for oil or natural gas; and to develop an improved battery system to fit in with the energy storage requirements of solar technologies. The range of energy technology's efforts is shown in Figure I 2-4.

Along with the priority given to moving technologies into the marketplace, long-range

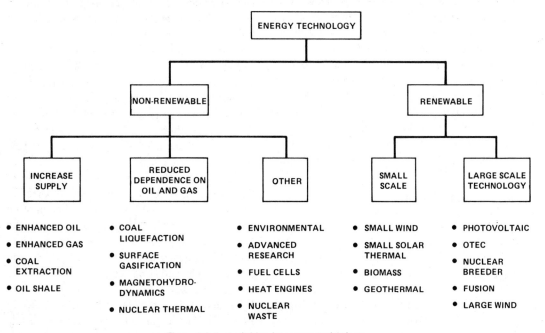

Figure I-2-4. Activities in energy technology.

research and development will continue to be supported. Long-range efforts are conducted to increase knowledge and to bring new products to the marketplace after the turn of the century.

The Magnetic Fusion Program is a long-term effort that may provide almost unlimited energy in the next century. The concept is still highly experimental, and two basic research paths are being followed. One is magnetic confinement and the other is inertial confinement. Although each of the options has attractive features, it is still far too early to follow one path at the expense of the other.

Likewise, the Magnetohydrodynamics Program is a long-term effort to provide energy. In this activity, coal is burned directly to produce electricity. Experiments have shown that this method is a very efficient producer of electricity and a very low producer of pollution. This concept, too, is still in the experimental stage, with activities being carried out in cooperation with the Soviet Union.

Long-term projects hold the key to U.S. energy supply in the twenty-first century and beyond. The energy problem will never be solved completely. The nation must search continually for new energy sources or more efficient ways of using energy. New technologies must continually be brought to the market, and planning for the next 100 years must be done today.

There will not be a single solution to U.S. energy problems. The resolution will not come from nuclear power, fossil energy, wind, or biomass. Rather, the solution will be found in a combination of them. The country must bring the entire range of its technologies and resources to bear on its energy problems.

Supply and Demand Energy

Projections of energy supply and demand done by the Energy Information Administration (EIA) show that domestic energy consumption could surpass 100 quads by the end of the 1980s. This compares to about 76 quads consumed in 1977. As shown in Figure I 2-5, the long-term growth rate in energy consumption slowed drastically in the

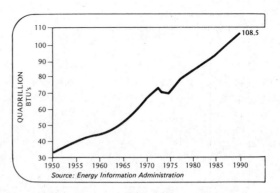

Figure I-2-5. U.S. Domestic energy consumption.

mid-1970s due to the oil embargo, sharply higher energy prices, and the worst recession since the 1930s.

Through the end of this decade, energy consumption is projected to grow a bit faster, but will taper off in the 1980s.

Within this growth in energy consumption, the mix of energy sources is expected to change considerably. The change is shown in Figure I 2-6. Oil now comprises about 49 percent of the energy used in this country, but it will fall to 47 percent by 1985 and 45 percent by 1990. Likewise, gas now accounts for 25 percent of all energy consumed in the United States. By 1985 its importance should drop to 20 percent, and it should decline still further to 18 percent by 1990.

Coal, on the other hand, will increase its share from 19 percent now to 23 percent in 1990. Nuclear power is projected to rise from providing only 3.6 percent of domestic energy now to nearly 10 percent by 1990. Hydrothermal and geothermal energy are expected to raise their share from 3.3 percent to 4.6 percent in the same period.

New technologies play a very minor role in the initial EIA projections. The projection does not consider new initiatives that may be undertaken to bring new technologies into the marketplace. Synthetic fuels are projected to supply between 0.2 and 0.4 quad in 1985 and between 0.7 and 1.35 quads in 1990. Solar will supply between 0.08 and 0.24 quad in 1985 and between 0.26 and 0.95 quad in 1990. The total for all new technologies is projected to be between 0.58 and 1.24 quads

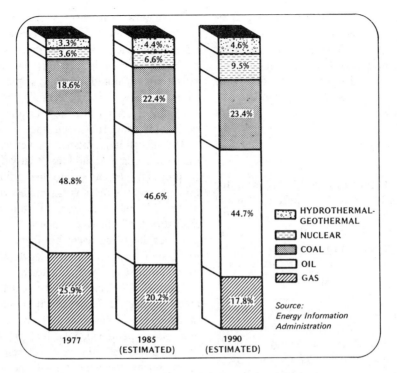

Figure I-2-6. Percent U.S. energy by source.

in 1985 and between 1.66 and 3.6 quads in 1990.

Energy Saving from New Technology

By the year 2000, the energy saving from new technologies could be increased considerably from the 1990 projections. According to the Domestic Policy Review, solar energy could supply as much as 18.1 quads, or 13 percent of projected energy consumption. At $25 per barrel of oil, solar technologies might supply more than 9 quads of energy.

In the fossil fuel area, heat engines and waste heat recovery could be expected to yield about 1.5 quads in 1985 and 10 quads by 2000. Advanced combustors, such as fluidized beds, could save 4 quads by 1985.

Advanced techniques in recovering energy resources could add greatly to reserves. It is estimated that 300 quads of methane exist in conjunction with coal and may be recoverable. In addition to gas that could be produced with existing methods, between 226 and 3,340 trillion cubic feet of natural gas may be available through enhanced recovery techniques. Likewise, one trillion additional barrels of oil may be available through enhanced recovery and shale.

Research and development in coal liquefaction and gasification can provide new energy from an old source. But its contribution will be difficult to measure until the extent of a market is determined.

Similarly, R&D on pollution control measures may not lead to direct energy saving, but it opens new energy sources. Coal that was difficult to use can become an economic resource. The nation will be able to use the energy it has without harmful side effects. Oxides of sulfur and nitrogen need not be spewed into the air, or toxic metals spread over the earth. The country can have both a clean environment and abundant energy. Again, the impact is difficult to quantify until the techniques are developed and a market surveyed.

Obviously, these estimates contain wide margins for error. Foretelling the immediate future is difficult; foretelling 10 to 20 years into the future is almost impossible. Similarities between long-range forecasts and actual events, if not aberrant, are at least unusual. The projections give a range of possibilities of an order of magnitude. They should not be taken as absolute standards by which to judge success or failure.

BUDGET

The Energy Technology budget request for fiscal year 1980 totals $3.5 billion. This is about $6 million above fiscal year 1979. Funding for the Solar, Geothermal, Electric and Storage System programs has risen by nearly $61 million, and the budget for Nuclear Waste Management is expected to be up by nearly $132 million. The Magnetic Fusion Program may have $8 million more available in the next fiscal year. The Fission Energy Program, on the other hand, may have its budget reduced by about $185 million. Fossil Energy could be cut by $12 million. Details of budget funding are provided in Table I 2-1.

SIGNIFICANT ISSUES

Commercialization

A goal in energy technology is to bring new technologies into general use, with the cooperation of industry and the public. Commercialization is the last stage of development, when a technology is adopted by the private sector. For a technology to become commercial, technological and engineering problems must be resolved and it must be economic. Additional conditions must be satisfied as well:

- The public must accept the technology as an effective way of meeting a need
- The public must believe that action is being taken to guard against possible adverse effects
- Industry must believe that the technology can be implemented without serious problems and that it can be operated profitably

There is a difference of opinion over how far the Government legitimately should pursue commercialization and how much of the cost it should bear. It is generally agreed that the Government should provide some incentives to move new energy technologies into the marketplace. After all, they have an influence on a wide range of social goals.

The question remains, however, whether the Government should be relatively passive in providing incentives to the private sector through such devices as tax credits or whether it should intervene directly through strict regulation of private industry. Both approaches are being used, but there is disagreement over whether the balance should tip toward incentives or regulations.

There is additional disagreement about how much capital the Government should provide in bringing a technology to commercialization. Some people believe that private industry should bear the cost of research and development, because it will be using the technology. Others say that only the Government should be willing to undertake such long-range, high-risk efforts.

Socioeconomics

Socioeconomic issues are very closely related to issues in commercialization. They revolve around the question of how far the Government should go to influence the marketplace. Some believe that the Government should exert as little influence as possible in the marketplace. They feel that the relative prices of oil, gas, coal, or other energy sources will lead the economy to choose the most efficient mix. Government intervention, they believe, will lead to an inefficient allocation of resources and a lower standard of living than what otherwise might have been obtained.

On the other hand, proponents of Government intervention believe that Government regulation is the surest way of promoting a "fair" distribution of benefits. Intervention would reduce the time necessary for a technology to be widely accepted, raising the benefits derived from it. Government action might also reduce unemployment or rising

Table I-2-1. Budget for FY 1978-1980.

PROGRAMS	FY 1978 BA ACTUAL ($MILLION)	FY 1979 BA ESTIMATE ($MILLION)	FY 1980 BA ESTIMATE ($MILLION)	FY 1980 BA INCREASE (DECREASE) ($MILLION)
Fossil Energy Program				
Coal				
Mining R&D	63	76	60	(16)
Coal Liquefaction	127	206	122	(84)
Surface Coal Gasification	196	160	169	9
Heat Engines and Heat Recovery	42	58	46	(12)
Combustion Systems	66	59	57	(2)
Fuel Cells	35	41	20	(21)
Magnetohydrodynamics	72	80	72	(8)
Other	58	1	117	116
Subtotal	659	681	663	(18)
Petroleum				
Enhanced Oil Recovery	46	54	21	(33)
Oil Shale	29	49	28	(21)
Other	4	5	8	3
Subtotal	79	108	57	(51)
Gas				
Enhanced Gas Recovery	27	34	28	(6)
Subtotal	27	34	28	(6)
General Reduction	0	63	0	0
TOTAL	765	760	748	(11)
Solar, Geothermal, Electric and Storage Program				
Solar Technology				
Solar Electric	240	303	353	50
Solar Thermal Electric	104	100	121	21
Photovoltaics	64	104	130	26
Wind Systems	37	61	67	6
Ocean Systems	36	38	35	(3)
Biomass	21	43	58	15
Other	9	12	30	18
Subtotal	270	358	441	83
Geothermal	108	158	141	(17)
Electric Energy Systems	42	41	27	(14)
Energy Storage Systems	51	50	66	7
TOTAL	471	617	676	59
Fission Energy Program				
Converter Reactor Systems	87	87	96	(19)
Breeder Reactor Systems	736	742	590	(152)
Special Nuclear Systems	75	56	40	(15)
Naval Reactors	255	297	278	(19)
TOTAL	1,151	1,191	968	(203)
Nuclear Waste Management Program				
Commercial Waste Management	184	191	199	8
Defense Waste Management	294	257	372	115
Spent Fuel Disposition	5	11	20	9
TOTAL	483	459	592	133
Magnetic Fusion Program				
Confinement Systems	109	141	137	(4)
Development and Technology	56	50	58	(1)
Applied Plasma Physics	53	65	76	11
Reactor Projects	113	88	91	3
Other	1	3	2	(1)
TOTAL	332	356	364	8
TOTAL ENERGY TECHNOLOGY	3,204	3,383	3,367	(16)

Note: Detail may not add to total due to rounding; initial figures may be subject to change.

prices that might accompany a change in energy sources.

The new technologies are "infant industries" that may have difficulty becoming established, but hold the promise of quickly becoming strong enough to stand alone. Many people believe that Government should aid such new endeavors.

Only the Government, interventionists believe, can take "externalities" into account; that is, the benefits derived by society as a whole are considered, not just the benefits enjoyed by an individual or a single company. Thus, the Government would take account of an improved balance of payments, a stronger dollar, and possibly lower inflation in making investment decisions, while a business would not.

Environment

The environment is not just flora and fauna; it encompasses health, safety, social and economic issues as well. The potential impacts of Energy Technology's programs range from severe to trivial. The burning of fossil fuels, the more widespread use of nuclear energy, and the disposal of nuclear waste have potentially more serious environmental implications. The environmental effect of solar and wind technologies, on the other hand, is very slight.

Fossil Energy. Concern for the physical environment and the health of the population are two of the biggest problems in the use of fossil fuels. Switching from using oil and gas to a greater reliance on coal is being hampered in large part by the increased pollution that this would cause. Coal burning emits sulfur dioxide, nitrogen oxides, and particles to the atmosphere, as well as leaving a residue of solid waste. Coal mining, especially strip mining, can be very unsightly. Abandoned mines have marred the landscape, and working mines create noise and dust. Coal liquefaction and gasification require large amounts of water. The most likely candidate for these processes, western coal, is located in an area that is chronically short of water. Many people contend that the water should be reserved for agriculture. Increasing the use of coal means increasing its potential for damaging the environment.

One of the major portions of the Fossil Energy effort is to find solutions to these environmental problems. But the environmental effort does not stop there. Activities are directed toward developing knowledge of fossil energy chemistry, materials, technology, processing, engineering, analysis, sampling, worker management and protection, and property management and protection.

The environmental effects of programs are considered from the very beginning. To ensure their acceptability, Fossil Energy sets these objectives:

- Incorporation of environmental activities into funding proposals
- Full consideration of environmental values when projects are selected
- Public participation in decisions
- Compliance with Federal and state environmental laws
- Collection of environmental data for future regulatory decisions

Solar, Geothermal, Electric and Storage Systems. Even the clean, renewable, and inexhaustible energy sources pose some environmental problems.

Solar energy can pollute through the release of optical coating and cleaning materials and the manufacture of photovoltaic cell material. Such pollution, however, would be an extremely minor problem.

Wind energy could cause a minor problem of interference with television reception and noise generation. Wind machines could release minor amounts of oil and other lubricants. Large wind machines could also be an aesthetic problem due to their size.

Using biomass for fuel creates pollution through smoke, dust, and discharges from machinery. It could also create a problem if it becomes so popular that it results in extensive removal of vegetation.

Geothermal energy causes pollution, in that drilling creates noise and dust, and operation can release gases to the atmosphere or contaminate local water. Taking water out of the ground can cause the land to sink, while injecting water into the ground might cause earthquakes.

Electric energy production and transmission generate pollutants through the basic fuel that is transformed into electric current and the electric field created by high voltage transmission.

Energy storage can hurt the environment through leakage from batteries, leaching salt from aquifers in pumped hydroelectric storage, and increased seismic activity.

Nuclear Energy. In normal operation, nuclear power plants release waste heat to the atmosphere or nearby rivers and streams.

Accidents, however, could release radioactivity. A release of radioactive material into the atmosphere could be serious because the wind could carry it into heavily populated areas.

Nuclear reactors also generate radioactive waste material, which creates long-term storage problems with their own environmental dangers.

Nuclear Waste. The waste generated by nuclear facilities must be reprocessed, stored or disposed of. The first option has been foreclosed due to its potential for being improperly used to generate material for nuclear weapons.

Storing and disposing of nuclear waste has one major environmental danger—that some of the material might escape. Nuclear waste must be stored for extremely long periods of time for the radioactive material to decay to safe levels. There is a danger that containers will rupture and contaminate the environment with radioactive material. Transporting the material carries the same danger.

Magnetic Fusion. Because magnetic fusion is still highly experimental, its environmental impact is still largely unknown. It appears to have less risk of spreading nuclear materials than does nuclear fission energy. On the other hand, workers may be exposed to magnetic fields with unknown long-term results. In addition, scarce resources, such as beryllium, may be used in large quantities.

Scientific and Technological Issues

The scientific and technological issues in the area of research are generally the questions asked about any new product. Will it work? How much does it cost? When will it be ready? What are the alternatives?

The specific issues, of course, depend on the program. The issues in fossil energy are different from the ones in solar or nuclear energy, but there are some common threads.

Questions about pollution and other environmental effects are asked about all new technologies. The cost of a new technology compared to other ways of operating is always a major issue. The dependability and effectiveness of a new technology must always be considered.

Regional Needs

Regional energy needs must be met as well as national ones. There is no single solution to the nation's energy problem. There are many solutions, and some that work well in one area may not work well in another. Geothermal energy, for instance, works much better in California than in Pennsylvania due to the structure of the earth's crust; wind energy would work better in New England than in the South due to the generally stronger winds in the Northeast.

Instead of relying on just one or two energy sources, the United States should draw upon a range of resources, even if some of them will work only in a limited area. There is no need for the Federal Government to neglect useful energy sources because they will supply just one region. The Government decided this question long ago when it created the Tennessee Valley Authority.

Approach to Technology

The same type of reasoning applies to the controversy over whether the United States should follow the "hard path" or the "soft path" in seeking new energy sources. The "hard path" is the use of large, centralized, nonrenewable energy sources such as coal, oil, and fission. The "soft path" is the use of small, decentralized, inexhaustible energy such as solar or wind power.

The nation can travel along many paths to a greater energy supply. Solar power and power from coal and oil can all be used. The use of nuclear power does not automatically exclude the use of wind power. The nation must make use of all its energy resources, not just some of them.

MANAGEMENT APPROACH

The management approach is goal-oriented in that all activities are designed ultimately to achieve the aims of the U.S. energy policy. Program objectives are set with national energy goals in mind; milestones are established to judge progress. Programs and projects are subject to periodic evaluation, during which their progress and potential are considered. A decision is then made whether to continue or terminate them.

The experience, knowledge and skills of the technical staff, its major contractor installations, and component consultants with specialized backgrounds are used in the planning and execution of the program. Significant problems are analyzed periodically by these groups or by special committees such as the Domestic Policy Review in Solar Energy or the Interagency Review Group in Nuclear Waste. Alternatives are considered, trade-offs analyzed, and resource requirements forecast. The results of study projects and meetings are distributed widely. In this way Energy Technology can draw on a wide spectrum of professional expertise to determine the direction of its programs.

Organization

As shown in Figure I 2-7, the Energy Technology consists of five areas of responsibility:

- Solar, Geothermal, Electric and Storage Systems—planning development, and implementation of programs in solar energy, geothermal energy, energy storage systems, and electric energy systems
- Fossil Energy Programs—planning, development, and execution of advanced research and technology development in coal, oil, gas, improved conversion efficiency and environmental control
- Nuclear Energy—planning, development, and implementation of programs in nuclear energy
- Nuclear Waste Management—planning, development, and execution of processing and isolation; spent fuel storage and transfer; transportation of nuclear waste; and decommissioning and decontaminating Energy Technology nuclear facilities
- Magnetic Fusion Energy—planning, development, and implementation of research and development in magnetic fusion energy

The other offices support the line offices with program analyses, strategies, and pri-

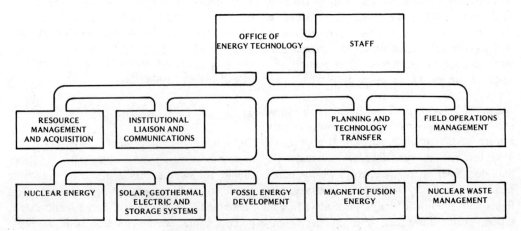

Figure I-2-7. Energy technology organization.

orities; market and issue analyses; budget formulation, execution, and review; and other support activities.

Decentralization

The tasks performed at the Department of Energy Headquarters properly deal with broad issues and policy. The job is of such a magnitude that it tends to crowd out the day-to-day concerns of managing specific projects. It is DOE policy, therefore, to manage projects where they are being carried out. Project management is being moved from Headquarters to the field.

In October 1977, the detailed management of projects began moving to the field. Schedule, performance, and resource administration are now the responsibilities of field personnel. A Project Charter, approved by Headquarters, sets forth the responsibility, authority, and accountability for the work to be performed.

Figure I 2-8 shows how project management has changed. In fiscal year 1977, Headquarters controlled 53 percent of project funds, and field operations controlled 47 percent. This year, Headquarters controls only 33 percent of the funds, while field management controls 67 percent. By fiscal year 1980 field control should rise to 73 percent, and Headquarters control should drop to 27 percent. It would then be considered completely decentralized.

Personnel problems have been the greatest difficulty yet encountered. Technical management has been transferred out of Headquarters, but there is still need for procurement, legal, administrative, and other support personnel.

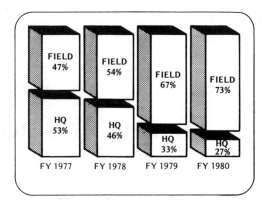

Figure I-2-8. Percent of funds managed at headquarters and the field.

Positions have been identified that could be moved from Headquarters to the field. Such positions totaled nearly 80 in fiscal year 1978, and about 50 were transferred. This generally involved moving jobs rather than people. Job slots and responsibilities were moved to the field where they could be filled.

Decentralization is being implemented at three levels: Programs, Major Segments of Programs, and Projects. Operations offices and Technology Centers are being restructured to execute programs, Lead Labs are being restructured to execute major segments of programs, and field offices are being reshaped to manage projects effectively.

As noted earier, decentralization is well underway. It is difficult and time consuming, but it is being done without retarding technological advance. The plan is monitored continuously to ensure that problems are quickly identified and resolved.

3.
Energy Technology Programs

FOSSIL ENERGY

Program Overview

Oil and gas account for only about 11 percent of known U.S. fossil energy reserves, but they supply about 75 percent of our fossil energy demands. Coal, on the other hand, makes up about 77 percent of national recoverable reserves, but it accounts for less than 20 percent of fossil energy consumed. National energy policy is designed to shift energy consumption away from scarce resources and toward those that are more plentiful.

The three major areas in Fossil Energy research and development are coal, petroleum, and gas. Research and development will concentrate on technologies that:

- Increase domestic production of coal, oil and gas
- Ensure that coal-burning facilities use coal economically and in an environmentally acceptable manner
- Transform coal into liquids and gas
- Improve the efficiency and economics of fossil fuel use

Research and development is particularly needed in liquid fuels. Domestic supplies are dwindling and large amounts are needed for transportation and industrial feedstocks. Alternatives to petroleum-derived liquids (fermentation ethanol) are only beginning to become available, and construction of large-scale production facilities are expensive.

Likewise, research and development programs will be required before coal can significantly expand its role as an energy provider. Environmental, health, and safety problems must be overcome, and new coal technologies must be economically attractive. A major objective of the Fossil Energy Program is to meet these challenges, smoothing the transition to a more coal-based economy. Coal research programs are directed toward developing more efficient technologies and making coal more economically competitive and environmentally acceptable.

The transition to a more coal-based economy will be easier if coal is transformed into liquids or gases that can be used in existing equipment. Economical coal liquids and gases would find a ready market and would quickly reduce America's dependence on imported energy. Enhanced recovery of oil and gas would also substitute domestic energy for imports. Primary production leaves most of the oil and gas in the ground. Recovering oil and gas from shale would further reduce the U.S. dependence on foreign energy.

Issues. Concerns in fossil energy development generally revolve around commercialization, environmental, and socioeconomic issues. Commercialization issues generally focus on whether a technology is sufficiently developed and can compete in the marketplace. If the technology shows promise of being able to compete once it becomes established, initial Government support may be indicated. Even if a technology is fully competitive, Government controls, tax policies, and regulations may be altered to encourage further development.

Environmental issues must also be faced before a fossil technology can gain widespread acceptance. The effect on air and water, as well as worker health and safety, must be considered. In addition, the effect of emerging laws, standards, and regulations on the technology must be examined before it can be used on a wide scale.

Socioeconomic issues refer to the effects that large-scale development may have on the social and economic life of an area. A technology's influence on industry size, location, and rate of development must be considered. Likewise, the effect on a regional or national economy can be an issue. Government may have a role in alleviating the adverse social and economic effects of large-scale energy development.

Technology Transfer. The Fossil Energy R&D program is continually transferring technology to the private sector. Potential users are kept informed as research and development progresses. Cost-sharing of pilot and demonstration plants transfers knowledge quickly and ensures that the private sector is interested in the technology. Emphasis is placed on commercialization planning, because this is the ultimate objective of research and development. When technology transfer and commercialization become the principal concerns, however, a program may be shifted.

Fossil Energy Programs

Coal. Coal is the most abundant fossil energy resource in the United States. Moreover, it is versatile; it can be burned directly or transformed into liquid, gas, or feedstock. But its full potential cannot be realized until a number of problems are solved. Coal is bulky—it is more difficult to transport and burn than liquid or gaseous fuels. Coal burning can be a source of pollution in the form of sulfur dioxide, nitrogen oxides, particulates, and solid waste. The processes for making liquids and gas from coal have yet to be fully developed. The coal program is structured to help surmount these problems and allow the nation to tap the vast potential of its coal reserves.

The program is designed to produce technologies that extract, process, and use coal in an environmentally acceptable manner. The industrial sector is given special attention because it has many options for using coal but is presently a major consumer of oil and gas. Many industries should be able to switch to coal without technical difficulty. The coal program also includes development of heat engines, heat recovery systems, combustion systems, and fuel cells.

The ultimate objective of the coal program is to decrease the nation's dependence on imported oil and increase its use of coal. (See Figure I 3-1.) Progress toward this objective has important economic and national security benefits. To attain this goal, the strategy sets four specific objectives:

- Ensure that facilities now burning coal can continue to do so under environmental standards—this would make coal combustion a more attractive option for new facilities during the next decade
- Demonstrate the capability to produce synthetic liquids and gas from coal by the mid-to-late 1980s so that significant capacity could be built in the 1990s, if needed
- Develop systems to use coal in a more efficient and environmentally acceptable manner in the 1990s and beyond
- Improve the technology base for economic, efficient, and environmentally acceptable use of coal

To accomplish these goals, DOE supports a range of projects, because coal must serve a variety of needs. It may be transformed into solid, liquid, or gaseous fuel or used to generate electricity. Potential users of technologies will employ different types of coal and will face a range of environmental constraints. In

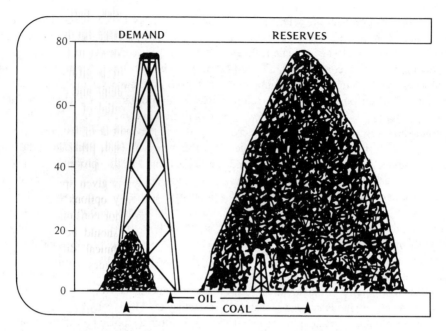

Figure I-3-1. Coal and oil demand/reserves as percent of U.S. total.

addition, most users want systems specifically designed to meet their needs.

Government funding must be selective and balanced among the various programs. Because pilot and demonstration facilities are very expensive, the Government will demonstrate only the most promising. Funds should be concentrated to ensure success in each key program.

In selecting large-scale demonstration projects for Government support, the coal strategy relies upon three criteria:

- Projects that demonstrate processes with a wide range of applications should be favored over those that have limited applicability.
- Projects should be selected that comple-

- Finally, projects should be chosen according to their potential for environmental acceptability and for technical and economic success. Government-supported projects should lead to commercial technologies with significant potential for increasing coal use.

The Coal Program is composed of research, development and demonstration in the following areas:

- Mining Research and Development
- Coal Liquefaction
- Coal Gasification
- Advanced Research and Technology Development
- Advanced Environmental Control Technology
- Heat Engines and Heat Recovery
- Combustion Systems
- Fuel Cells
- Magnetohydrodynamics

Mining research and development. Mining research and development is being carried out on equipment and systems for underground mining, surface mining, and coal preparation. The objectives of the program are to improve mining economics, maximize resource recovery, improve worker health and safety, and protect the environment by transferring new technologies to private industry.

Coal liquefaction. Coal liquefaction could provide energy from coal to replace petroleum in uses such as boiler fuel, gasoline, heating oil, and chemical feedstock. The objective of the liquefaction program is to facilitate the beginning of a synthetic liquid-fuel industry.

Four liquefaction processes—Solvent Refined Coal-I and -II, H-Coal, and the Exxon

Donor Solvent (EDS) process—are in advanced development. But all need further development to improve their operation and economics.

The Solvent Refined Coal processes convert high-sulfur, high-ash coals into low-sulfur and nearly ash-free fuels. Technical demonstration plants are being designed to evaluate both of the SRC processes.

The H-Coal process converts high-sulfur coal into either a boiler fuel or a synthetic crude oil. Construction of an H-Coal pilot plant began in FY 1977 and was completed late in FY 1979. Plant operation is scheduled to begin in FY 1980.

The Donor Solvent Process liquefies coal by transferring hydrogen to it. The liquefied coal may then be refined to obtain more usable products. A pilot plant is scheduled to begin operating in FY 1980.

Coal Gasification. Coal gasification could enable coal to supplement the declining reserves of natural gas. The many gasification processes under development, both surface and "in situ" methods, have the same objective: to supply an economic and environmentally acceptable synthetic gas which can substitute for natural gas.

Surface Coal Gasification is made up of High-Btu Gasification, which produces pipeline-quality gas; Low-Btu Gasification, which produces gas for industrial and utility use; and Third Generation Gasification, which advances the technologies for both High- and Low-Btu Gasification. Two High-Btu gasification methods and several Low-Btu methods are currently under investigation. Six contractors are evaluating small commercial gasifiers. These units can supply gas for a single industrial plant. Figure I 3-2 shows a coal gasification pilot plant.

"In situ" coal gasification refers to the transformation of coal into gas while it remains underground. Underground gasification could put to use some of the 1,800 billion tons of unmineable coal in the United States. Low- and medium-Btu gas can be produced for industrial use or for chemical feedstock. The gas may also be upgraded to pipeline-quality and put directly into the existing natural gas transmission system.

Figure I-3-2. Coal gasification pilot plant.

Advanced environmental control technology. Environmental standards pose the greatest constraint on increasing coal use. Unless new environmentally acceptable technologies are available, many facilities that might use coal may have to remain on other fuels.

Advanced Environmental Control seeks ways to ensure that facilities will be able to use coal without endangering the environment. The program is divided into Flue Gas Cleanup, Gas Stream Cleanup, and Technology Support.

Flue Gas Cleanup investigates ways of cleaning the stack gases of conventional combustion units so that emissions can meet environmental standards. Gas Stream Cleanup refers to removing pollutants during or prior to combustion. Technology Support develops technologies for waste management, instrumentation, process controls, economic comparisons, and other activities that help make coal more environmentally benign.

Heat engines and heat recovery. The goal of this program is to develop heat engine and heat recovery technologies that will maximize energy saving in the utility, industrial, commercial, and residential sectors. This can be done by:

- Increasing utilities' direct use of coal by developing integrated coal systems that can displace oil and gas
- Improving energy efficiency with cogeneration systems that can supply both electric and heat energy with the same amount of fuel used before to generate electricity
- Developing technologies to produce useful work from waste heat

The Heat Engines and Heat Recovery Program is divided into three areas:

- Integrated Coal Conversion and Utilization Systems
- Advanced Cogeneration Systems
- Heat Recovery Component Technology

Integrated coal conversion and utilization systems use advanced combustors or combined coal conversion end-use systems to generate electricity. They could find widespread use, as coal and coal-derived fuels are likely to be the major fossil energy sources for electric utilities in the future. Gas turbine systems appear to be the most promising at the moment.

Present heat systems, such as boilers in electric utilities, waste a significant portion of their fuel energy. Such waste heat is a large, potential source of useful energy.

Advanced cogeneration systems produce both electricity and process heat on an industrial or commercial site. The waste heat derived from generating electricity can be used for space heating, hot water heating, or even cooling. Industrial cogeneration can supply electrical needs as well as heat for various processes. Such systems can convert 75 to 85 percent of their fuel energy into useful forms. (See Figure I 3-3.)

Heat recovery component technology can provide the means to recover waste heat and put it to work. If even a fraction of waste heat were recovered, it could significantly reduce the dependence of the United States on imported energy.

Combustion systems. About 90 percent of the fuel used in this country is burned, and this process is likely to remain dominant in the foreseeable future. The Combustion Systems Program seeks to improve combustion efficiency by:

- Developing fluidized-bed combustion systems capable of burning high-sulfur and other coals in an environmentally acceptable manner
- Substituting coal for oil in combustors capable of burning coal-oil mixtures
- Improving the reliability and efficiency of present boilers and furnaces

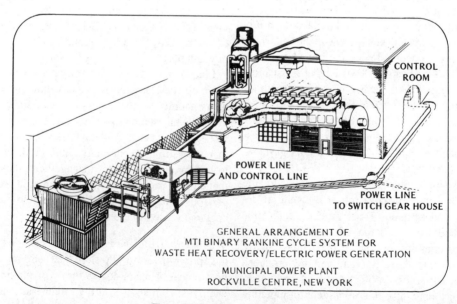

Figure I-3-3. Waste heat recovery.

Fluidized-bed combustors generate less air pollution than conventional coal burners. The technology shows promise of meeting forthcoming stringent source environmental standards. Several fluidized-bed combustors are being tested, and industrial boilers have been judged ready for commercialization.

Engines are being improved. They evolved during an era of plentiful and inexpensive energy. Despite recent advances, their fuel efficiency could be raised by 20 to 30 percent.

Likewise, most industrial furnaces and boilers were designed when energy costs were low. They were built to operate at low cost rather than be energy efficient. Laboratory and design studies are now in progress to improve this equipment.

One way to extend fossil energy resources would be to burn coal-oil mixtures instead of just oil. Existing boilers, heaters, and furnaces are being studied to see which can be converted to this fuel mixture.

In addition, alternative fuels—such as agricultural and forest residues, industrial waste, and fossil liquids—are being examined as substitutes for coal or petroleum distillates.

Fuel cells. Fuel cells convert fuel into electricity through an electrochemical process shown in Figure I 3-4. Because of their unique features, they can be factory built and located close to the point of demand, reducing the energy lost in transmission. Their modular construction permits a size flexibility ranging from kilowatts to megawatts, and their non-polluting, quiet operation enables them to be located on site even in crowded residential areas. Fuel cells may be used in a number of applications:

- Industrial cogeneration where their high electrical and thermal efficiencies can be used
- Integrated energy systems for residential and commercial buildings
- Electric utility generation

Magnetohydrodynamics. Magnetohydrodynamics converts heat directly into electricity by passing a conducting fluid through a magnetic field. Such a system produces electricity very efficiently with little pollution. At least 50 percent more electricity is generated from a given quantity of heat than in conventional generators, and sulfur is removed as part of the generating process.

The system is still experimental, however. Magnetohydrodynamics shifted to experimental engineering only in 1976, although progress since then has exceeded expectations.

Petroleum. Domestic production of sufficient liquid fuels is now the primary energy supply

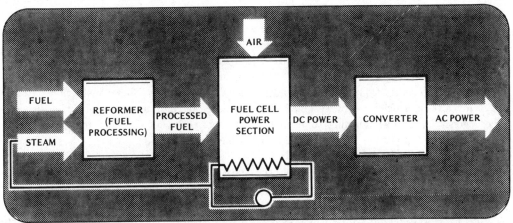

1. *THE REFORMER SECTION PROCESSES HYDROCARBON FUEL FOR FUEL CELL USE*
2. *THE POWER SECTION CONVERTS PROCESSED FUEL AND AIR INTO DC POWER*
3. *THE INVERTER PRODUCES USEABLE AC POWER TO MEET CUSTOMER REQUIREMENTS*

Figure I-3-4. Major sections of a fuel cell power plant.

problem in the United States. United States oil reserves as of January 1978 were 29.5 billion barrels. Domestic crude oil production this year will be about 3 billion barrels, down from the high of 3.4 billion in 1970. Oil imports are reaching 50 percent of U.S. consumption. The resulting balance of payments deficits and the possibility of an embargo or other supply disruption pose serious threats to the U.S. economy. A basic tenet of the national energy policy is that "the United States must reduce its vulnerability to potentially devastating supply interruptions." The goal is to reduce oil imports to less than 6 million barrels per day by 1985. Conservation alone cannot achieve such a large reduction, and massive switching to other fuels is impractical now, especially in the transportation sector. Therefore, domestic oil production must be increased.

The Petroleum Program is working toward this goal through raising the yield from known reservoirs, deriving oil from shale, and improving drilling technology.

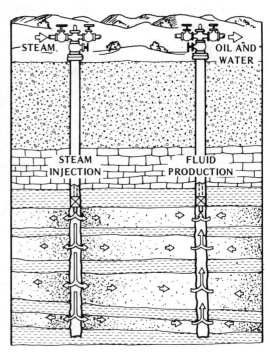

Figure I-3-5. Enhanced oil recovery—steam displacement.

Enhanced oil recovery. Oil produced by natural reservoir pressure, lift pumps, or other means of primary production recovers about 20 percent of the oil in place. By pumping water or gas into the reservoir to increase or maintain the pressure (secondary production), the yield can be increased to more than 30 percent. More advanced methods of improving the recovery rates (tertiary recovery) are called enhanced oil recovery. The target of enhanced oil recovery is the recovery of some of the nearly 70 percent of oil remaining in place.

Twenty-two cost-shared efforts are underway to test enhanced recovery methods. These generally involve injecting some material—steam (as shown in Figure I 3-5) or water, for instance—into the ground. This increases oil pressure and makes it possible to raise production. Many tests are needed, because each field or deposit has unique characteristics, and technologies must be developed for specific situations.

Oil shale. Oil shale is another potentially significant domestic source of oil. Substantial oil shale deposits are distributed across the country, but the most economic resources are in a relatively small region of western Colorado and eastern Utah. Several firms have shown interest in producing oil from these rich deposits for around $20 per barrel. Oil shale may thus, be our least expensive alternative source of petroleum, especially considering that refined shale oil is quite similar to a light crude oil. Although several above-ground and in situ processes are under advanced development, production may be limited to environmental problems. These restrictions, which include air emissions, spent-shale disposal, and protection of aquifers may limit large-scale development.

The program's short-term objective is to provide the technology for commercial oil shale operation before 1985. Over the longer term, the objective is to improve the economics of shale oil and to lessen the environmental impact of producing it.

In situ methods, those which produce oil without mining the shale, appear to be very promising. They require less water and manpower, and avoid the problem of waste disposal. The Anvil Points Oil Shale Facility is pictured in Figure I 3-6.

Advanced process technology. The Advanced Process Technology Program develops the technology to extract and use oil and gas resources. The elements of the program are:

- Advanced Exploratory Research (for oil, gas, shale, and tar sands)
- Shale Oil Refining and Utilization
- Product Characterization and Utilization

The program extends knowledge of petroleum resources, its characteristics, and methods by which it can be exploited. Advanced Exploratory Research concentrates on technologies that could improve resource extraction. Shale Oil Refining and Utilization is concerned with refining shale oil, determining its nature, and developing its technologies. Finally, Product Characterization and Utilization maintains and improves the data on crude oil and its products.

Drilling and Offshore Technology. This program seeks to improve drilling rates and to extend drilling to very deep and hostile areas. Its major goals are as follows:

Figure I-3-6. Anvil points oil shale facility.

- Increase drilling speed by 30 percent by 1985
- Develop technologies to drill below 30,000 feet in deep offshore waters and in the Arctic
- Reduce drilling and production costs
- Provide drilling technology to other DOE programs

Gas. Natural gas has occupied a significant role in the domestic energy market because of its convenience, cleanliness, and low cost. This role will continue in the near future because of the large capital investment in transmission pipelines and distribution systems that serve industrial, commercial, and residential users. Gas currently supplies about one-fourth of the total energy used in the United States, although it constitutes less than 5 percent of the total domestic conventional energy reserves. Demand has grown steadily, but additional reserves from new discoveries have not kept pace.

The objective of the natural gas recovery program is to obtain significant production of natural gas from unconventional sources, reducing our dependence on foreign energy. Stimulating such production depends on changing the production cost structure to increase production at any given gas price; accelerating investment in research and development and hastening the application of improved technology; and undertaking high-risk, high-return research and development in areas not covered by industry.

The strategy is to identify and assess new sources of natural gas and to develop technologies to extract it. This includes sharing costs with the private sector to enhance technology transfer and commercialization. The strategy is designed to ensure that gas will be available as the price of imports, liquefied natural gas, and oil products rise. Gas extraction technologies will be available as they become economic. Exploration and development risks will than be low enough to attract private development.

The Enhanced Gas Recovery Program is concentrating on gas shale in the eastern United States, sandstone gas reservoirs in the Rocky Mountain region, and the free methane in coal seams. The program also provides environmental and other support activities to ensure that each project meets environmental, health, and safety requirements.

Eastern gas shale. This program seeks to determine the extent of the resource and to develop techniques for locating fractured reservoirs (see Figure I 3-7). It is also developing techniques to extract the gas once it is located. Technologies will be evaluated through field experiments, which will be conducted in two phases. Phase I will be the drilling of three or more "screening" wells at 20 locations. Phase II will be the selection of sites for commercial demonstrations.

Western gas sands. This project also seeks to determine the extent of the gas resource and to develop efficient production techniques. It is also trying to identify the relationship between the physical and chemical properties of the reservoirs. Research and development now is concentrated on developing the tools and techniques needed to fracture sandstone sufficiently to recover an economic quantity of gas. As many as four major field tests may begin by FY 1980.

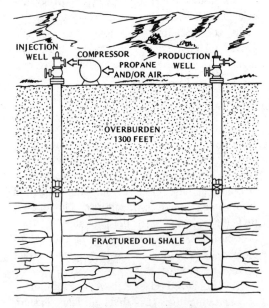

Figure I-3-7. Gas shale production.

Methane from coal. An estimated 800 trillion cubic feet of methane is trapped in coal seams and adjacent rock. Some 300 trillion cubic feet of this gas may be readily recoverable. Methane is now removed from mines as a safety measure, with about 250 million cubic feet per day emitted into the atmosphere.

The methane-from-coal project seeks to develop economic methods to extract methane from active coal mines and from "unmineable" coal beds. Most current techniques for removing coal-bed methane were developed by mine operators. Further research, development, and demonstration are necessary to make the techniques economic for recovering methane as well as removing it. Five major research and development projects will continue in FY 1980.

SOLAR, GEOTHERMAL, ELECTRIC AND STORAGE SYSTEMS

Program Overview

Exhaustible energy supplies in the United States are becoming more expensive as they are being depleted. The country is increasingly dependent on costly and unreliable foreign sources of oil. Energy derived from the sun and the earth, however, can provide the United States with a clean and secure source of power.

In these areas, program objectives are designed to improve both the near-term and long-term U.S. energy posture. The specific objectives of the program are:

- To develop technologies that use solar and geothermal energy to displace imported and exhaustible fuels
- To develop electric transmission, distribution, and energy storage systems which integrate intermittent energy sources into the national grid; and to make the most efficient use of both traditional and alternative energy sources
- To test and demonstrate technologies that are energy-efficient or that use renewable energy, so they can compete in commercial marketplaces by the

mid-1980s and mature as major energy sources within this century
- To obtain several quads of energy annually from these sources by the late 1980s

A key element of the program is to develop usable technology products. The Solar Technology Activities, for instance, are aimed at developing methods of deriving energy from the sun. Along with this, the Division of Electric Energy Systems is working on ways to integrate the new energy into the Nation's power system, determine its impact on peak power supplies, and provide cheap, reliable equipment to convert the direct current into usable alternating current. In Energy Storage, work is under way to find better means to store the energy so it can be used even if the sun is not shining and the wind is not blowing. Energy Technology pursues all aspects of a developing technology to ensure that a complete energy system can enter the market.

The programs are conducted with the cooperation of eventual users and manufacturers of the technologies. This increases their chance for successful commercialization. For example, the geothermal energy program includes participation by local communities, regional interests, and the States.

For most of the technologies, the major problem is cost reduction. Products must also be reliable and easily maintained before they can capture a significant share of the market, but these problems are relatively minor compared to the high current cost of the technologies. A technological breakthrough, however, can have a dramatic impact on costs. This has happened many times in the past. The manufacture of electronic components and products such as radios and calculators, is a case in point. The same thing could happen in solar technology, as manufacturers gain experience, costs and prices could fall sharply.

The problems are not just technological. Mass marketing of the new technologies may require a change in the way certain items are bought. While the technologies can greatly reduce the need to buy fuel, they often cost more initially. The higher initial cost can be offset in the long run by fuel saving, but decisions generally are not made on the total capital and operating costs of a system. Items are usually compared on the basis of initial cost. The "life cycle cost" should be the standard of comparison.

Requirements and Considerations. Each technology in Solar, Geothermal, Electric and Storage Systems has its own requirements. They can be grouped in four general categories:

- Scientific and technological requirements
- Commercialization considerations
- Environmental impact
- Socioeconomic concerns

The specific requirements and considerations relevant to each of the programs are summarized in Table I 3-1.

Programs

Solar Technology. Solar Technology encompasses the development of power-generating systems (other than solar heating and cooling, and agricultural and industrial process heat) which derive energy directly or indirectly from the sun. The program is divided into two major parts: Solar Electric Applications, which is developing solar electric systems, and Biomass Systems, which is developing systems that use organic materials to produce energy.

Solar electric applications. The Solar Electric Applications Activity is developing solar electric systems that will contribute significantly to national energy supplies. The objectives of this program are to:

- Develop and demonstrate solar systems at commercially competitive costs by the mid-1980s.
- Ensure that solar energy systems are compatible with existing utility systems and operations
- Establish solar energy's environmental acceptability and operational safety

Table I-3-1. Requirements and considerations in energy technology activity areas.

PROGRAM AREAS / ISSUES	SOLAR					GEOTHERMAL			ELECTRIC		STORAGE			
	Solar Thermal Power	Photovoltaic Energy Conversion	Wind Energy Conversion	Ocean Systems	Fuels from Biomass	Hydrothermal Resources	Geopressured Resources	Geothermal Technology Development	Power Supply Integration	Power Delivery	Batteries and Electrochemical	Thermal and Hydrogen	Mechanical and Magnetic	Technical and Economic Analysis
SCIENTIFIC AND TECHNOLOGICAL														
EXTENT OF ENERGY RESOURCE					•	•	•							
IMPROVED MATERIALS	•	•	•	•							•	•	•	•
OPTIMAL WORKING FLUIDS	•			•										
SYSTEM EFFICIENCY	•	•	•		•	•		•		•		•	•	•
IMPROVED ENERGY CONVERSION	•	•		•	•	•		•				•	•	
IMPROVED ENERGY DENSITY								•				•		
TECHNICAL FEASIBILITY				•				•						
COMMERCIALIZATION														
PROCESS ECONOMICS	•	•	•	•	•	•	•		•	•	•	•	•	•
UTILITY INTERFACE		•	•			•	•		•					
USER ACCEPTANCE (INDUSTRIAL)	•			•		•	•		•	•	•			
USER ACCEPTANCE		•							•					
ENVIRONMENTAL														
PUBLIC HEALTH									•		•	•	•	
OCCUPATIONAL SAFETY				•		•	•							
AIR					•	•	•							
WATER						•	•	•						
LAND	•	•	•	•		•	•							
SOCIOECONOMIC														
LOCAL ECONOMIC EFFECTS	•	•		•		•	•	•						
OWNERSHIP OF ENERGY SOURCE	•	•	•			•	•	•						
LEGAL PROBLEMS	•	•	•											
UTILITY INTERFACE	•	•	•							•	•			
LAND/WATER USE	•		•	•				•		•				

- Clarify and resolve legal and institutional issues
- Help create the necessary institutional infra-structure.

The major components of the Solar Electric Applications Program are:

- Solar Thermal Power Systems
- Photovoltaic Electric Conversion
- Wind Energy Conversion Systems
- Ocean Systems

Solar thermal technologies concentrate the heat from the sun to heat water or some other fluid to drive a turbogenerator. The primary objective is to provide an alternative to scarce oil in electric generating plants. The activity also includes the use of waste heat after electricity has been produced. Research and development is being carried out on both large and small power systems.

Large power systems focus on plants in the 10-MW to 300-MW range. Several concepts are being pursued. Storage-coupled plants, solar/hybrid plants, and solar repowering of existing plants are all being considered.

In storage-coupled systems, the sun's energy is concentrated and used to heat a fluid which, in turn, drives a turbine generator. The heated fluid may also be stored to provide electricity when the sun is not shining.

Solar repowering refers to an existing electric generating unit that is provided with a solar-fired boiler. When the sun is shining, solar energy can be used to generate electricity. The solar/hybrid concept is similar except that an entirely new plant is developed instead of using existing equipment. Both technologies have a backup conventional boiler for use when the sun is not shining.

A central receiver system has been built in Albuquerque. Its 5-MW capacity was achieved in 1978, and construction of a 10-MW experiment is in progress. Construction of another facility near Barstow, California, should be completed in 1981. The cost of this system is being shared with Southern California utilities.

Small Solar Thermal Power Systems are generally less than 10-MW and are usually located fairly close to energy users. The energy generated by the small solar system can be used as process heat by industry, for space heating and cooling, or to generate electricity. The small units can be used in remote areas, as power systems for small communities or for industrial energy.

Studies being done on large solar power systems are being extended to provide information on small systems. Total Energy System experiments are being conducted in Albuquerque, New Mexico, Ft. Hood, Texas, and Mississippi County Community College in Shenandoah, Georgia. Data from the large-scale experiments, Total Energy Systems, and early small-scale system experiments will be combined to select promising systems for further development.

Photovoltaic devices convert light directly into electricity. When light energy strikes certain materials called semiconductors, electricity is created. In devices specially designed for the purpose, wires may be attached and electricity drawn off. The basic scientific principles are well known, and the technology for specialized uses is relatively advanced.

The major component in all photovoltaic systems is the "array" that collects the sunlight and converts it to electricity. The array is composed of a number of electrically interconnected sealed panels, each of which contains many solar cells. The array now accounts for half of the total system cost. Other system components—called the balance of system—include power conditioners, storage elements (batteries), controls, and structural members.

Arrays of photovoltaic cells can be sized for a number of different applications. Relatively small arrays can provide electricity for individual residences. Intermediate-sized arrays could serve facilities such as industrial plants and residential communities. Finally, large arrays could be connected in a central power station to provide electricity for a utility grid.

On November 4, 1978, the President of the United States signed the Solar Photovoltaics Act. The Act provides for accelerated research, development, and demonstration of photovoltaic technologies to speed their commercialization. Specifically, the goal is to reduce the cost of photovoltaic electricity from $7 per peak watt now to $1 (1975 dollars) in 1988. The cost has already fallen sharply from the $22 per peak watt in 1976.

There is already an established market for photovoltaic systems where small amounts of power are needed at remote locations. For instance, photovoltaic cells have provided power for spacecraft and railroad signal systems. However, except for specialized uses, photovoltaic energy is not yet economically competitive with conventional electricity.

The market for photovoltaic cells will not expand significantly until costs decline. The main objective of the Photovoltaic Program, then, is to reduce these costs. This can be done by aggressively pursuing advanced research and technology development, conducting real-world testing, and accelerating market growth through selected Federal Government purchases. At the same time, the Program will address the technical, environmental, institutional, and social issues involved in fostering the widespread use of photovoltaic power systems.

The overall program is composed of four key elements:

- Advanced research and development—to

establish the technical feasibility of new or improved photovoltaic arrays
- Technology development—to develop low-cost manufacturing technology
- Systems engineering and standards—to integrate photovoltaic components into energy systems
- Tests and applications—to experiment with various photovoltaic applications

Wind Energy Conversion Systems

From the standpoint of early application, one of the most promising ways to tap the sun's energy is through the use of wind systems. Wind energy is derived from the sun, making this resource continually renewable and nonpolluting.

Wind energy systems have been in use for many centuries, traditionally for irrigation and milling. Since the turn of this century, wind energy systems have been used to generate electric power. The systems were generally small, ranging in output from a few watts to a few kilowatts. Such machines almost vanished when the Rural Electrification Administration brought centralized power to the countryside in the 1930s.

The objective of the Wind Energy Conversion Systems Program is to accelerate the development of reliable and economically competitive wind energy systems. To achieve this objective, it will be necessary to advance wind technology, develop a sound industrial base, and address nontechnical issues which may delay the introduction of new wind machines. The development of rugged, economical wind systems capable of providing up to 30 years of reliable, relatively maintenance-free service remains the program's primary challenge. The wind machine at Sandusky, Ohio, is shown in Figure I 3-8.

Wind systems of three relatively discrete sizes are being explored: small machines of up to 100 kW, for use at dispersed sites where loads are light; an intermediate size of nominal 100 to 200 kW rating for a variety of applications; and megawatt-scale units for large loads, including central power grids.

The small-scale systems development program has two major components: the small

Figure I-3-8. 100-Kilowatt wind machine.

wind systems development, testing, and evaluation managed by DOE's Rocky Flats Wind Energy Test Center; and the farm and rural applications testing by the USDA's Agricultural Research Service. Two intermediate systems are operating at Clayton, New Mexico, and Culebra, Puerto Rico; and two more were installed at Block Island, Rhode Island, and Oahu, Hawaii.

Because of the relatively advanced state of aerodynamics technology, basic research is not included in the activity. Instead, the program centers on mission analyses, on the development of more effective machines, and on the accumulation and analysis of site-specific data.

A study of remote and isolated area markets was made during fiscal year 1978, and a marketing study of high potential wind systems was initiated. Other studies are addressing the operational and economic questions associated with using wind power mixed along with conventional power systems.

Ocean Systems

Ocean Systems may provide an inexhaustible and renewable energy source for generating base-load electricity and producing energy-intensive products. Systems currently under study include:

- Ocean Thermal Energy Conversion
- Salinity gradients
- Ocean currents
- Wave energy

The Ocean Thermal concept uses the temperature difference between warm surface waters and cold water from the depths to operate a heat cycle and generate electricity. (See Figure I 3-9.)

Power may be extracted from salinity gradients by using the energy potential that exists across a membrane between two solutions of differing salinity.

Ocean currents are another source of energy that may be exploited. In this application, large diameter hydroturbines are held below the ocean surface by tension moorings and driven by the current.

Wave energy systems extract mechanical energy from ocean waves and convert it into electricity. This concept is still highly experimental.

Figure I-3-9. Ocean thermal energy.

An engineering test facility, OTEC–1, is currently planned to give the earliest ocean-based performance of an Ocean Thermal Energy system. Test data on the cold water pipe, biofouling, corrosion, and cleaning will be collected beginning in early 1980. This system will be used for scaling experiments in an ocean environment. The results of experiments at this facility will be used for the selection of power-cycle subsystems.

At the same time, advanced research and development will concentrate on advanced heat exchangers, cold water pipes, biofouling resistance, and corrosion resistance. The results will be used to design improved systems.

In other areas, methods to harness the energy of ocean currents, wave energy, and salinity will be explored.

Biomass energy systems. Biomass is organic material such as plants, manure, and agricultural and forestry residues. Residues are leaves, stalks, branch, bark, or other material left over after harvesting or processing. This material contains stored chemical energy produced by plants. Biomass can be collected and burned directly or converted to other useful energy forms. It is used principally as a source of heat in industry, commerce, and residences; but it can also be converted into alcohol, synthetic natural gas, and petrochemical feedstocks.

The primary objective of the Biomass Energy Systems Program is to develop the means of converting biomass into usable fuels or energy-intensive products. As biomass supply and conversion technologies become cost competitive, detailed plans can be made for their commercial introduction.

Program strategy is based on the time when technologies are expected to be commercially available. In the near term (1985), efforts are directed toward process improvement and, where necessary, demonstrations of commercial-scale applications. Projects for the mid term (1985–2000) are concentrating on laboratory-scale studies, process economics studies, and engineering design studies. Projects expected to be commercial after the year 2000 are being developed through applied research and exploratory development. The Biomass Systems Program cooperates with USDA and other Government agencies in its research.

Direct combustion of wood and wood residues is expected to have the greatest near-term impact as a source of industrial process heat, electricity, and residential heat. Fermentation of starches and sugars to grain alcohol will also have a near-term effect. Anaerobic digestion of manures is expected to have an impact on the commercial market by 1985. Concepts being developed for mid-term commercial impact include energy farms to provide a reliable biomass supply, and the production of medium-Btu gas and fuel oil by thermochemical conversion. Technologies now in the applied and exploratory research phase include aquatic energy forms, development of exotic species, biophotolysis, and other advanced conversion technologies.

Strategy. The overall Federal strategy for solar technologies is:

- To discover opportunities for displacing critical exhaustible resources, such as oil and natural gas, and to provide a market environment favorable to solar systems
- To bring forth, through research and development, market-ready solar systems
- To reduce costs to parity with conventional systems
- To accelerate market development through procurement that stimulates the solar industry
- To study issues in national policy for inexhaustible energy resources and systems
- To promote rapid commercialization through technology transfer

Some of the most promising near-term technologies being developed by Solar Technology are in its Biomass Program.

In Wind Energy, an extensive procurement program for small (less than 40 kW) machines is planned, because they can be used in various dispersed applications. Larger wind systems (greater than 1-MW) also are being developed for mid-term applications.

Ocean thermal energy conversion, the most advanced ocean energy systems concept, will be demonstrated at a 1-MW test facility to determine scale-up costs and potential.

In addition, research and development of more cost-effective photovoltaic cells is being emphasized. The commercialization of photovoltaic systems will be accelerated by the Federal Government purchases ordered by the Solar Photovoltaic RD&D Act of 1978 and the National Energy Act.

The Solar Technology Program also is evaluating the long-range potential of new solar technologies with uncertain technological and economic characteristics.

In summary, the Solar Technology strategy

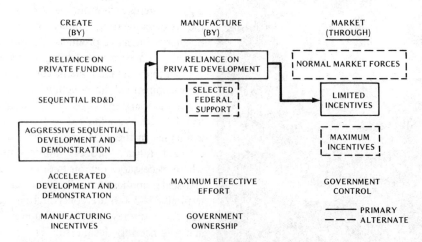

Figure I-3-10. Government role options.

Table I-3-2. Impact of solar energy (Quads).

	1978 ACTUAL	2000 BASE CASE ($25/BBL)	MAXIMUM PRACTICAL CASE
HYDRO	3.0	3.4	3.8
BIOMASS	1.8	3.1	5.5
INDUSTRIAL PROCESS HEAT		1.0	2.6
HOT WATER, HEATING & COOLING		0.9	2.0
WIND		0.6	1.7
PASSIVE HEATING		0.2	1.0
PHOTOVOLTAICS	—	0.1	1.0
SOLAR THERMAL ELECTRIC	*	0.1	0.4
OTEC	—	0	0.1
SATELLITE POWER	—	0	0
TOTAL	4.8	9.4	18.1

Source: Domestic Policy/Review

is threefold: to foster product development and improvement, by supporting research and development; to enlarge manufacturing capacity and improve production techniques primarily by engaging in early market "buys"; and to stimulate the formation of market channels and institutions through a program of selected tests and applications coupled with other incentives (see Figure 3-10).

The Domestic Policy Review (DPR) estimated the possible effect all solar technologies could have by the end of this century. DPR totals give an indication of how solar technologies could help to solve the United States' energy problems. The estimates are shown in Table I 3-2.

Geothermal Energy. Geothermal energy is the heat retained in the Earth's crust. In some areas the heat is close enough to the surface that it is possible to tap it as a source of power for industrial use or the generation of electricity.

There are three general types of geothermal energy that may prove useful before the year 2000 in resolving the United States' energy problems:

- Vapor- and liquid-dominated hydrothermal energy
- Geopressurized hydrothermal energy
- Hot dry rock

Vapor- and liquid-dominated geothermal energy are hot water and steam that have been produced by contact with relatively shallow masses of hot rock. The liquid or vapor is trapped below ground in fractured rocks or porous sediment. Vapor reservoirs are relatively rare, but have been used as a source of power for many years. Larderello in Italy has been operating since 1904 and The Geysers in California (shown in Figure I 3-11) has been operating since 1960. Hot-water dominated reservoirs are about 20 times as numerous as ones dominated by steam. In using these, the liquid is passed through a heat exchanger to vaporize a hydrocarbon fluid and the vapor is used to drive a turbine. The spent geothermal fluid is then injected back into the ground. In some cases it is possible to extract valuable minerals from the fluid before its disposal.

In addition to generating electricity, geothermal energy can be used for industrial and residential space heating and cooling. It can also be used for industrial process heat and in agricultural applications.

Geopressurized hydrothermal resources are hot-water reservoirs containing methane which are trapped under high pressure. They are located mostly along the Gulf Coast of the United States. This resource offers three types of energy: thermal, kinetic, and dissolved methane. Although the total resource is believed to be large, the economics of recovery are unknown.

Hot dry rock, as the name implies, is rock that has a high heat content but little or no

Figure I-3-11. Ventilating geothermal wells and gathering pipes at the geysers, a geothermal steam field in California.

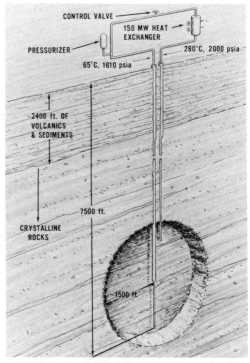

Figure I-3-12. Hot dry rock.

water. It requires the introduction of water to supply usable power. (See Figure I 3-12.) This resource is also estimated to be very large, but significant use of it is probably several decades away.

The overall purpose of the Geothermal Energy Program is to stimulate commercial development. The program has the following objectives and time frame:

(present–1985)
- Confirm the magnitude and longevity of hydrothermal resources
- Determine the extent and recoverability of geopressured resources
- Establish the feasibility of using geothermal heat for electric power generation and direct heat applications with available technology

(1985–2000)
- Develop advanced technology for the discovery, confirmation, production, and application of geothermal energy
- Define the hot dry rock resource base
- Develop an extraction technology for hot dry rock resources

(2000–2020)
- Develop an economical technology for exploiting normal gradient geothermal heat

Some activities currently underway include:

- Confirmation of geothermal-hydrothermal resources
- Research in direct heat applications
- Operation of experimental facilities to produce electric power
- Design of a demonstration plant
- Development of new drill bits

The goal is to raise the commercial use of geothermal energy to amounts shown in Table I 3-3.

Electric Energy Systems. The Electric Energy Systems Program is concerned with the power supply, power delivery, and management of the nation's 4.3 million mile electric system. As new energy technologies are developed to supply power, the electric system must be capable of accommodating the energy in an efficient, flexible, and environmentally acceptable manner. The Division of Electric Energy Systems develops technologies relevant to the systems aspects of U.S. electric networks.

The Electric Energy Systems research, development, and demonstration efforts are done in cooperation with the electric utilities, the

Table I-3-3. Geothermal energy contribution (Quads).

	QUADS/YEAR NEAR TERM 1985	MID-TERM 2000	LONG TERM 2020
ELECTRICAL APPLICATIONS	0.2-0.3	1.5-3.0	5.0-10.0
DIRECT THERMAL APPLICATIONS	0.1-0.2	0.5-2.0	6.0-8.0
GEOPRESSURED METHANE	0 - .02	2.0-4.0	5.0-10.0
TOTAL	0.3-0.5	4.0-9.0	16.0-28.0

Electric Power Research Institute, equipment manufacturers, and other Federal agencies. The program expedites the development of high-risk, long-term, high-payback technologies and the development of technologies which have a regional or national impact. The Electric Energy Systems Program is divided into two major activities, Power Supply Integration and Power Delivery. Its objectives are to:

- Ensure that all of the elements of future electric energy systems (fossil, nuclear, solar, etc.) are integrated to form an efficient system (see Figure I 3-13)
- Ensure the system can deliver electrical energy from the source to the user
- Develop system techniques to maintain reliability as the system increases in size and complexity
- Help implement the National Energy Acts

The successful RD&D and industry implementation of the Electric Energy Systems Program is projected to save 1.0 million barrels of oil per day equivalent by 1985 and 5.0 million by the year 2000. Beyond these direct energy savings, the systems approach toward electric network RD&D will save by reducing capital investment requirements, conserving limited energy resources, lowering cooling water requirements, and reducing land requirements.

The Power Supply Integration component of the Electric Energy Systems Activity is

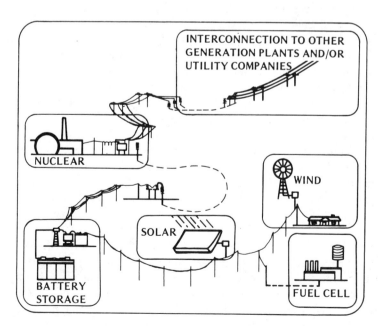

Figure I-3-13. Linking sources and uses of energy.

divided into three sections: Load Management, Solar Generic/Integration of New Source Technologies, and System Control and Development.

Load Management refers to altering the pattern of electricity use to obtain the best results from the generation, transmission, and distribution capabilities of the electric power system. This is expected to produce near-term benefits, since components are either being field tested or actively promoted by utility companies. Moreover, the automated distribution and control components will enable dispersed solar technologies to be used within a system originally designed for central generation.

Solar Generic/Integration of New Source Technologies helps remove power system restraints on new technologies. It is seeking ways to integrate new, dispersed electricity sources into the power system. Modeling and verification will ensure that the consumer benefits from new technologies and improved efficiency.

System Control and Development provides design procedures to control an increasingly complex electric power system. As new technologies become a significant part of the electric power system, they may require new controls and control strategies. This program, unfunded for FY 1980, is being reevaluated in the light of the National Energy Acts. At this point, its future is uncertain.

The second major division of the Electric Energy Systems Program is Power Delivery. While Power Supply Integration was primarily concerned with developing the "software" to analyze problems within the electric system, the Power Delivery Program is mainly concerned with developing the "hardware" that makes the system work. The Program's major components are the Electric Field Effects of high-voltage transmission, and Underground Cables and Compact Stations.

The Electric Field Effects part of the program is concerned with the environmental effects of high-voltage electricity transmission. Studies have indicated that strong electric fields may alter some biological processes. Furthermore, studies are under way to determine the biological effect of noise from transmission lines.

The Underground Cables and Compact Stations Subprogram is working on underground superconducting cables and compressed gas insulated cables. Underground compact stations are intended to route power to specific places in a complex grid. Work in this area is being reduced pending an analysis of spending priorities.

Energy Storage Systems. Improved methods of storing energy will make it possible for intermittent sources of energy to provide power continuously. Wind machines supply power only when the wind blows, and solar technologies provide energy only when the sun shines. When the air is calm, at night, or on cloudy days, solar or wind devices would require a backup energy source unless their energy can be stored.

Energy storage systems could significantly reduce the need for expensive backup energy sources. Energy can be produced and stored until needed, lessening the need for imported oil and saving the capital that would otherwise have gone into a secondary energy system. The lower cost of a solar or wind system that could provide continuous power would speed the commercialization of inexhaustible energy sources.

Storage could also promote conservation through recovering waste heat from utilities and industries. The heat could be used during periods of high energy demand. Operations could be made more efficient.

Energy storage technologies can be used in space heating and cooling, transportation, industrial processes and utilities. The broad range of energy storage technologies is shown in Figure I 3-14. The three types of energy stored are heat, electric, and mechanical energy. Energy sources are shown at the left of the diagram; they include solar energy, hydroelectric power utilities, waste heat from power plants, and industrial processes. Storage technologies appropriate to each energy source are shown in the center of the figure. For instance, electricity produced by hydroelectric, nuclear, or fossil fuel power plants may be

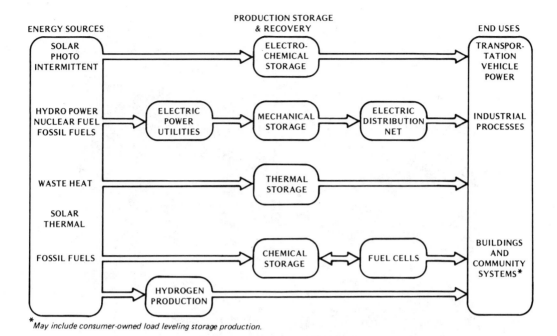

Figure I-3-14. Uses of storage systems.

stored using electrochemical storage, mechanical storage, or thermal storage.

The program is divided into two major sections: Battery Storage, and Thermal and Mechanical Storage.

The Battery Storage Program is developing inexpensive, long-life, high-performance batteries for use in transportation, solar energy systems, and electric networks. The program consists of Near-Term Batteries, Advanced Battery Development, Solar Applications, Electrochemical Systems Research, Supporting Research, Electrolyte Technology, and Dispersed Battery Applications.

The Thermal and Mechanical Storage Program seeks to provide reliable, inexpensive, high-performance energy storage technologies for a variety of purposes. The six areas within this program are Thermal Storage, Chemical/Hydrogen Storage, Mechanical Storage, Magnetic Storage, Utility Applications, and Technical and Economic Analysis.

One obvious application of advanced energy storage systems is their use in vehicles. This would permit stationary energy sources such as coal or nuclear power, to be used in mobile applications, thereby displacing oil and gas. Batteries and flywheels are being developed for use in electric and hybrid vehicles.

Storage systems can also be used to heat or cool buildings. Aquifer storage, for instance, can supply cool water in the summer and hot water in the winter.

Utilities could store energy to generate peak loads. Such an application would substitute coal, nuclear, or solar energy for oil and gas.

FISSION ENERGY

Program Overview

Commercialization of light water reactors (LWR's) has been underway for more than a decade, and fission now supplies about 13 percent of the nation's electricity (see Figure 3-15).

Despite nuclear power's impressive growth, the industry has faced some problems. In the past few years, projections of nuclear capacity have declined sharply. Construction delays, deferrals, and cencellations have all taken

Figure I-3-15. Brown's ferry power plant.

their toll. The most frequently mentioned reasons for the lack of orders are lower projections of electricity demand; utility financing and economics; uncertainty in Government nuclear waste policy; and delays in licensing, environmental approval and public acceptance.

Government nuclear policy now defers the reprocessing of spent fuel and has suspended breeder reactor demonstrations. In this context, the Fission Energy Program is designed to:

- Enhance LWR potential for deployment by improving technology and finding ways to extend uranium resources
- Defer reprocessing and recycling of spent fuel until all options can be assessed
- Maintain the breeder option through a strong research and development program

The Fission Energy Program seeks to develop advanced nuclear technologies that minimize the risk of proliferation, while deferring further demonstration efforts in technologies based on plutonium. A major system evaluation program is underway wherein a large number of alternative reactor concepts, alternative fuel cycles, and combinations of the two are being assessed. They are being judged on their proliferation resistance and their capability to extend the energy obtainable from uranium beyond that provided by the once-through fuel cycle. Concurrently, base-technology work on advanced fission options is continuing.

Thus, the immediate goal of the advanced fission programs is to assemble all the information required for a decision on the best approach to follow in using fission energy. Appropriate options may then be pursued to develop an essentially inexhaustible fission resource or to bridge the gap until other inexhaustible resource technologies can be commercially realized.

The Fission Energy Program is structured to continue the development of nuclear fission technology in three general areas:

- Converter Reactor Systems—To improve the efficiency of the use of uranium and

moderately extend the uranium resource base
- Breeder Reactor Systems—Programs to extend the uranium resource base
- Advanced Nuclear Systems—Programs to study and apply nuclear nonproliferation technologies, evaluate advanced systems, and support space and national security activities

Converter Reactor Systems

Thermal Reactor Technology. During the period of FY 1980 through 1983, the course of the Fission Energy Program may be dictated, in large measure, by the need to ensure that uranium could continue to share the burden of meeting the Nation's future power requirements. The objectives of the program are:

- To improve the utilization of uranium fuel in existing LWR's
- To reduce the radiation exposure of LWR plant personnel
- To resolve technical problems which can cause deterioration of plant availability
- To enhance the safety of LWR's
- To support international cooperation in the nuclear field

These objectives are accomplished through the development, demonstration, and widespread commercial application of improved technologies. This program includes direct involvement and cost sharing by the electric utility industry and its equipment and service suppliers.

At the end of FY 1977, 22 cost-sharing or no-fee contracts, together worth $102 million, had been negotiated and signed. DOE's share is $44 million (or 43 percent) and industry's share is $58 million. The work includes participation by nine major utilities, which have a combined installed nuclear operating capacity of 16,200-MWe; all the LWR vendors and fuel suppliers; and other industry equipment and service suppliers.

Advanced Isotope Separation. The Converter Reactor Systems' objective is to conserve uranium by developing separation technologies that will permit economic production of uranium from the tails of current enrichment facilities and from the existing tails stockpile.

The Advanced Isotope Separation effort involves the development of three techniques—two laser techniques and a plasma technique. The devices should demonstrate the components and scientific principles required to design an engineering demonstration facility. The equipment should be operating by 1981, at which time they can be evaluated. A decision can then be made whether to proceed to engineering development.

In addition, technologies are assessed for their proliferation potential as well as their ability to enrich uranium. Such assessments constantly seek to apply separation technologies to energy-related problems

Gas-Cooled Thermal Reactors. The main thrust of this Program is to assist the utility industry to determine the requirements, timing, design characteristics, and type of development programs which could best serve utility interests in gas-cooled thermal reactors.

This type of reactor uses fuel more efficiently than do light water reactors. More electricity can be generated per unit of uranium, lessening the environmental impact of hot-water or nuclear wastes.

The Philadelphia Electric Company completed a demonstration project in 1974, and the Public Service Company of Colorado is beginning another. The Colorado plant has operated at 68 percent of fuel power.

Advanced Reactor Systems. This new program develops technologies to reduce the risk of nuclear proliferation. Such technologies must preserve the attractive qualities of nuclear energy without risking the spread of nuclear weapons. To achieve these goals, the Advanced Reactor Systems Program:

- Develops and demonstrates technologies which permit research reactors to use fuels that cannot be used in weapons but meet the operator's needs
- Develops reactor features to maximize

the energy obtained from uranium and minimize access to weapons-grade material

Breeder Reactor Systems

Previously, the Liquid Metal Fast Breeder Reactor had been assigned highest priority for development and demonstration. Subsequent to the change in nuclear policy, the long-term objectives of this program have been replaced by interim goals focused on nonproliferation, including alternative fuels and safety issues.

The interim goals of the Breeder Reactor Activity are as follows:

- To maintain the technical and engineering basis of the breeder reactor, and develop information for selection of a breeder system consistent with U.S. nonproliferation objectives
- To provide technical support to the Administration's foreign policy initiatives on nuclear nonproliferation

The breeder program is designed to produce information and resolve questions associated with the meeting of the interim goals of the program by late 1979 or 1980. By that time, the overall fission program studies should have been completed, and the National Uranium Resource Evaluation Program should be well underway. These studies, along with a continuing R&D program, could permit decisions on the future direction of fission power. In the meantime, the fast breeder R&D program should center around the Fast Flux Test Facility and include testing of reference and alternative fuels for nonproliferation reactor systems; testing of materials applicable to breeder reactors; and learning from the experience in construction, operation, and maintenance of a liquid metal reactor system.

Liquid Metal Fast Breeder Reactor (LMFBR). Beginning in FY 1979, an LMFBR conceptual design study is planned to be conducted during approximately 30 months. The study should identify plant design, cost and scheduling requirements, and minimize the risk of proliferation. It should fill the urgent need for focusing efforts in the many technologies involved.

The objective of the breeder technology program is to provide engineering information and technology support for future plants and facilities. These efforts will increase safety and reliability, reduce costs, and permit easier and faster licensing.

The program during the 1978–1981 period is planned to permit a decision on the LMFBR in 1981 and to allow flexibility in the choice of fuel and fuel cycle, if a decision is made to go ahead with the program.

Water-Cooled Breeder Reactors. The purpose of this program is to improve significantly the use of nuclear fuel for the generation of electrical energy. As part of the Breeder Reactor Program, the Light Water Breeder Reactor (LWBR) Program has as its immediate objective to confirm that breeding can be achieved in light water reactors.

An LWBR core has been developed, designed, fabricated, and installed, and it is now operating in the DOE-owned pressurized water reactor plant of the Shippingport Atomic Power Station at Shippingport, Pennsylvania. It is expected that the core could be operated for 3 to 4 years. At the end of this period, the core will be removed, and the spent fuel shipped to the Naval Reactors Expended Core Facility in Idaho for a detailed examination. It is expected that the expended core could contain about 1 percent more fissionable material than the initial loading.

In addition, work is underway in the Advanced Water Breeder Applications Program to develop additional technical information to assist industry to evaluate the Water-cooled Breeder concept.

Gas-cooled Fast Breeder Reactor. Another breeder concept, historically being developed as a backup to the LMFBR and therefore on a lower priority, is the gas-cooled fast breeder reactor.

The program seeks to develop gas-cooled fast breeders as a long-term option. It also provides technology and other support to an international cooperative effort. Key projects and development tasks are being planned to correspond with a 30-year commercial deployment schedule. Due to the large technology base existing in the United States and abroad, the program requires a relatively modest development program.

Fuel Cycle. On April 7, 1977, President Carter announced a new U.S. nuclear energy policy, which deferred indefinitely the reprocessing of commercial spent fuel to allow time to investigate proliferation-resistant fuel cycle alternatives. The nuclear fuel cycle program had been directed toward the commercialization of fuel reprocessing and plutonium recycling. As the result of the new policy, the program was reoriented to the investigation of a full range of alternative fuel cycles. Fuel cycles under consideration involve a variety of fuel materials containing combinations of uranium, plutonium, thorium, and fission products. In FY 1980, a generic approach will be taken to establish a fuel-processing technology base, emphasizing fast breeder reactor fuels but which may also apply to converter reactors. (See Figure I 3-16.)

The major goal of the fuel cycle program is to develop safe, protected processes which could enhance the contribution of nuclear fission to the nation's energy economy. The program seeks to develop technologies for the recovery and reuse of fuel and to develop and improve technologies for the recovery and reuse of fuel and to develop and improve the safeguards of fuel processing systems.

The specific objectives of the program are to:

- Develop proliferation-resistant fuel cycles that enable nuclear energy to help meet the country's needs
- Provide analytical support to the LWBR Proof-of-Breeding experiment

Advanced Nuclear Systems

Advanced Systems Evaluation. One of the primary objectives of the advanced fission development program is to assess all advanced fission options for their relative nonprolifera-

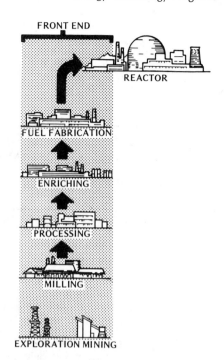

Figure I-3-16. Fuel-Cycle—Front End.

tion, economic, technical, and institutional merits.

In pursuit of this objective, two major system evaluation programs, the domestic Nonproliferation Alternative Systems Assessment Program (NASAP) and the International Nuclear Fuel Cycle Evaluation Program (INFCE), have been established to assess the nonproliferation advantages and other characteristics of a wide variety of advanced reactor concepts and fuel cycles. (See Figure I 3-17.)

NASAP is reevaluating nuclear energy systems, with emphasis on nonproliferation characteristics and on meeting nonproliferation objectives. During FY 1978, candidate systems were selected and methodologies were developed for assessing proliferation resistance, resource availability and utilization, commercial potential, economics, technical feasibility, safety, environmental and licensing impacts, and international acceptability. All remaining NASAP activities were completed in FY 1979.

The INFCE Program studies ways to minimize the danger of nuclear weapons prolifera-

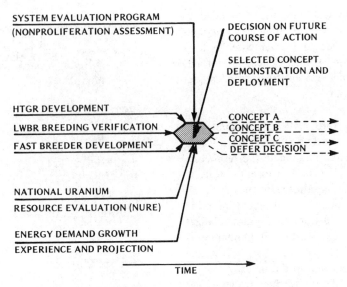

Figure I-3-17. System evaluation.

tion without jeopardizing the peaceful use of nuclear energy. The program is helping to build an international consensus on ways to use nuclear energy and prevent misuse.

The first drafts of the working groups reports were available in May 1979, and the overall report was published in February 1980.

The Economics Evaluation Program maintains and updates information on the technology, capital costs, fuel costs, and operating and maintenance costs of nuclear power.

The Advanced Technology Systems Assessment Program provides information for decisions on the development or deployment of nuclear systems. The scope of new approaches or new ideas is not restricted; emphasis is given to preliminary evaluation of concepts to see if more detailed study is warranted.

Space and Terrestrial Systems. The unique characteristics of nuclear-powered electric generators—compact size, light weight, and long life—enable operation of the sensing, analytical, and communication systems of spacecraft, satellites, and other remotely located devices for long time periods without relying on external sources of energy. Thus, the Space and Terrestrial Systems Program, by its technical initiatives, supports the national security as well as the civilian scientific exploration of space. By applying the technology developed for space applications, the program also fosters beneficial terrestrial uses of nuclear energy.

The principal objectives of the Space and Terrestrial Systems Program are to:

- Deliver nuclear-electric systems to other Federal agencies for space missions
- Provide power units for scientific experiments, telemetry, navigation, communication, or surveillance purposes
- Develop "nuclear batteries" for use in remote or undersea sites
- Investigate various isotopes and their utilization, using the skills developed for the handling and applications of plutonium-238

NAVAL REACTORS DEVELOPMENT

The Naval Nuclear Propulsion Program is a joint Department of Energy and Department of the Navy effort that designs and develops nuclear propulsion plants for installation in Naval ships of various sizes to meet Navy needs. The Division of Naval Reactors is responsible for the design, development, construction, testing, operation, and maintenance of nuclear propulsion plants and reactors for naval ships and their land prototypes.

Submarine Propulsion Reactors

The Submarine Propulsion Reactors Program designs, develops, and tests improved reactor cores and nuclear propulsion plants for a wide variety of submarine applications to meet the needs of the Navy, and provides support for operation, refueling, overhaul, and maintenance of the operating submarine fleet.

Surface Ship Propulsion Reactors

Design, development, and testing of improved reactor cores and nuclear propulsion plants for various sizes and types of Navy surface ships is carried out under the Surface Ship Propulsion Reactors Program. It also includes providing technical support for operation, refueling, and maintenance of the operating surface-ship fleet.

Supporting Research and Development

Supporting research and development for the Naval Reactors program is provided at the Expended Core Facility, and the High Temperature Test Facility. Development of advanced-concept methods and techniques that can be used in the Naval Reactors programs is also pursued.

NUCLEAR WASTE MANAGEMENT

Program Overview

Policy. The U.S. policy is to store or dispose of spent fuel discharged from nuclear reactors without reprocessing it. This is commonly called the "once-through fuel cycle." In 1977, the United States deferred indefinitely all commercial reprocessing of nuclear fuel until all aspects of proliferation could be considered. The change in policy halted the development of reprocessing techniques that industry had been considering.

Eventually, the present storage space for nuclear fuel could be filled. Unless additional storage space is found, reactors may be unable to discharge spent fuel. In the first half of the 1980s, spent fuel might have to be shipped from reactors whose storage area is filled to those with additional space. During the second half of that decade, reactors could lose a portion of their discharge capability. In the 1990s, reactors may no longer be able to discharge spent fuel, forcing them to shut down.

To forestall this eventuality, the Government has announced a new fuel storage policy. The Government will take title to spent fuel and store it for a one-time fee. This will provide temporary storage until a decision is made on reprocessing. The Government may also store spent fuel from foreign reactors.

The Terminal Storage Program was developed to meet the need to dispose of radioactive waste. Permanent disposal of nuclear waste will be necessary whether or not the United States begins to reprocess fuel.

Background. The Government has been handling nuclear waste since the days of the Manhattan Project. The methods used then are still used today:

- Shallow land-burial of low-level waste
- Storage of high-level liquid waste in shielded subsurface tanks (Figure I 3-18)
- Storage of spent fuel and other materials in shielded water basins

Some waste has also been stored in vaults and on surface pads, dumped in the ocean, held in settling ponds, and disposed of as grout in deep shale formations.

High-level waste. High-level waste is the product of reactor fuel reprocessing. It has resulted mostly from processing of military reactor fuels. At the Hanford and Savannah River storage sites, the waste is held in large tanks, (Figure I 3-18) and at the Idaho reservation, the liquid waste is converted to a solid. Several leaks have occurred in older tanks releasing radioactive fluid. While these leaks have not contaminated underground aquifers or resulted in any injury, they have cast doubt on the tanks as a method of long-term storage. Figure I 3-19 shows samples of encapsulated high-level waste.

Serious work began in the 1960s to develop

Figure I-3-18. High-level waste storage tanks.

technologies for disposing of high-level waste. Efforts were focused on solidifying and disposing of liquid waste in stable geological formations. Salt formations seemed to be the most promising. A project in the Lyons, Kansas area in the late 1960s and early 1970s was generally successful, but had to be terminated due to political objections and some engineering problems.

With nuclear reactors running out of storage area, the geological program has been given new life. The Office of Waste Isolation is identifying potential repository sites. This broad-based effort includes repository design studies, equipment development, waste migration studies, and detailed planning for licensing, construction, and operation of a repository by 1985.

Low-level waste. Low-level waste is loaded into drums and buried in trenches, but opposition to this disposal method has grown in the past few years. In 1970, the Government decided that long-lived transuranic waste would no longer be buried but would be stored on the surface until a deep geological depository is ready.

Waste Management

Long-term management is provided for nuclear materials held by the Federal Government. It also provides the technology for converting waste into forms which can be contained for long periods. Planning for nuclear

Figure I-3-19. Encapsulated high-level waste.

waste management has been divided into plans for defense waste management, commercial waste management, and spent fuel disposition.

Defense Waste Management. The Defense Waste Management Program consists of the following subprograms:

- Interim Waste Operations
- Long-term Waste Management Technology
- Terminal Storage
- Transportation
- Decontamination and Decommissioning

Interim waste operations. The Interim Waste Operations Subprogram is continuing present practices for safe handling and storage of radioactive wastes pending long-term waste management programs. The confinement of existing and currently generated waste products will be continually improved. Treatment and storage facilities for radioactive waste will be constructed and operated so that new techniques can be implemented quickly. (See Figure I 3-20.) Multiple containment, now used, minimizes the potential for release of radioactive material and minimizes the consequences of accidents which might occur.

Upgrading shallow land-burial for low-level waste will focus on improved technology and stabilization of sites when they are no longer needed. Standards will be developed through assessment studies; stabilization will minimize long-term maintenance and surveillance. The Government will discourage waste generation. Alternatives to shallow land-burial will be pursued. Intermediate-level liquid wastes generated at Oak Ridge National Laboratory (ORNL) will be disposed of by shale fracture.

Transuranium-contaminated solid wastes will continue to be placed into retrievable storage until treatment facilities are available. The construction of a treatment facility will proceed at the earliest practical time.

Long-term waste management. This activity covers R&D on:

- Long-term management of high-level waste
- Contaminated solid waste
- Airborne waste

Figure I-3-20. Waste separation plant.

An early demonstration will be made of immobilization, packaging, and emplacement of defense high-level waste at a suitable repository. Waste treatment and handling technology will be developed based on health, safety, environmental, and economic considerations.

Terminal storage. The strategy is to design, construct, and operate the Waste Isolation Pilot Plant (WIPP). WIPP will be a geologic repository located in a deep salt formation in southwestern New Mexico. It will be used for the permanent isolation of defense transuranium waste and to conduct experiments with various forms of high-level waste in a geologic repository environment.

Transportation. This subprogram will ensure the availability of systems to transport radioactive wastes. As much care must be taken in moving waste material as in storing it. Sometimes it must be carried through heavily populated areas, where leakage could be dangerous.

Decontaminating and decommissioning. A subprogram of decontaminating and decommissioning (D&D) radioactive facilities will be developed and pursued when such facilities are declared excess. R&D will be conducted where required to support specific projects.

Commercial Waste Management. The Commercial Waste Management Program consists of the following subprograms:

- Terminal Isolation
- Waste Treatment Technology
- Supporting Studies
- Solidification Process Demonstration
- Decontamination and Decommissioning
- Low-level Waste Management Operations
- West Valley Nuclear Center Activities

Terminal isolation. Two sites for the terminal isolation of commercial high-level waste and transuranium waste will be developed in different regions of the country. The location for the first repository will probably be in salt. The location for the second, in a different region, will be in either salt or crystalline rock.

Waste treatment technology. The objective of the Waste Treatment Technology Subprogram is to develop and demonstrate waste handling and treatment technology that will provide waste forms which meet repository criteria. Preliminary design specifications will be developed to immobilize each waste type.

Supporting studies. Evaluations of health and safety, economics, risk, quality assurance, and environmental impact will be developed to support the waste management program. The Committee on Radioactive Waste Management of the National Academy of Science, national experts who evaluate and report on all major waste management programs at all plant operating sites, is working in this area.

Solidification process demonstration. This project is intended to demonstrate the application of an existing process for immobilizing commercial wastes. The project will be completed in FY 1979.

Decontamination and decommissioning. The objective of this portion of the Commercial Waste Program is to provide industry with DOE expertise in decontamination and decommissioning (D&D) through information centers, data dissemination, advice, and consulation.

Low-level waste management operations. DOE does not have any direct responsibility for commercial burial grounds at this time. If Congress modifies the law, DOE may become responsible for the operation of those facilities.

West Valley Nuclear Center activities. DOE does not have any direct responsibility in this area. Studies are in progress that may identify a role for DOE in some future capacity.

Spent Fuel Disposition. On October 18, 1977, the Department of Energy announced its policy to accept and take title for spent fuel from U.S. nuclear reactors and a limited quantity from reactors located in other countries. The spent fuel program will provide the necessary storage facilities and supporting activities to carry out this commitment.

The decision to defer indefinitely the reprocessing of spent fuel in the U.S. and the United States' urging that other nations join in this deferral require retrievable storage of spent fuel for an indefinite period of time pending a decision either to recover and reuse that fuel or to dispose of it permanently. The handling of spent fuel is shown in Figure I 3-21.

The objective of the spent fuel storage program is to provide for the safe storage of U.S. and some foreign spent fuel, consistent with national nuclear energy goals and U.S. nonproliferation objectives.

This program should remove the uncertainty faced by utilities about the disposition of spent fuel discharged from reactors, and it should reduce foreign pressures to reprocess spent fuel because of full storage basins.

Issues

Public opposition to nuclear waste disposal is the most visible issue. Before the program can be successful, the public must be convinced that the disposal method is safe. Regulatory requirements are also issues that must be resolved. Regulations have yet to be formulated and the uncertainty of operating standards could retard the program.

The criteria for selecting a site, the geological media for disposal, and the form that the waste should take have to be decided. Likewise, packaging spent fuel requires further research and development.

The risk involved in disposing of materials that will be radioactive for thousands of years is unknown. It is not inconceivable that civilization could collapse and all knowledge of the storage sites be forgotten before the material becomes safe. The site must be safe even if untouched for millennia.

The issue of public versus private responsibility for nuclear waste disposal has not yet been decided.

MAGNETIC FUSION

Program Overview

Magnetic Fusion is still highly experimental but has the potential of providing virtually unlimited energy. Scientific research is oriented toward demonstrating the scientific feasibility of magnetic fusion, although its

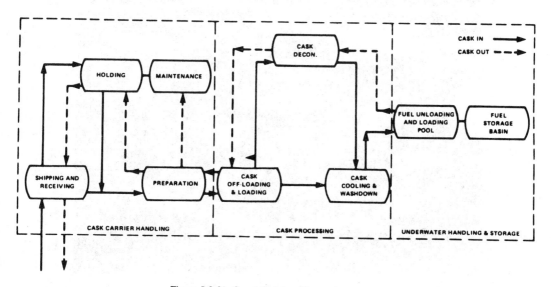

Figure I-3-21. Spent fuel handling and storage.

economic feasibility is still unknown. Magnetic Fusion may prove to have social or environmental problems which have not been identified. Nevertheless, it holds promise of being a clean, efficient, and prolific source of energy.

The use of fusion to produce electric power seems to be the most promising use of this energy source, but other applications are being studied. These include the direct production of hydrogen or synthetic fuels, the production of industrial chemicals and primary metals, and the breeding of materials for use in fission reactors. Two mechanisms for providing energy from magnetic fusion are shown in Figures I 3-22 and I 3-23.

The near-term objectives of the Magnetic Fusion Energy Program emphasize:

- Demonstration of magnetic fusion's scientific feasibility
- Maintenance and development of a scientific and technological base
- Establishment of an engineering development base
- Encouragement of research in alternative concepts

Two different paths are being followed to develop fusion energy: magnetic confinement and inertial confinement. In magnetic confinement, the hot gases which will undergo fusion reactions are contained in carefully shaped magnetic fields. In inertial confinement, the fusion fuel is contained within a small pellet. A large amount of energy is deposited in the pellet by a laser beam, for example, and fusion reactions are ignited. Each of these approaches comprises different schemes for achieving a controlled thermonuclear reaction. Several have attractive features and have achieved promising results, but narrowing of options at this time would be premature.

Four strategic objectives have been set for fusion research:

- To develop a scientific, technological, and engineering base so that the viability of various applications can be assessed
- To construct and operate scaling experiments which will provide data for the design of energy-producing prototype reactors
- To perform design studies of energy-producing systems, guiding the R&D program toward a final product
- To construct and operate energy-producing experimental fusion facilities to de-

Figure I-3-22. Fusion research—Elmo bumpy torus.

Energy Technology Programs 51

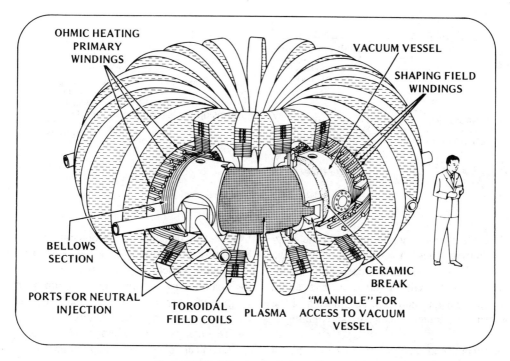

Figure I-3-23. Princeton large torus diagram.

termine the potential of fusion energy for commercialization

Scientific feasibility has not yet been achieved. That is, fusion reactions have so far absorbed more energy than they have released. Based on the program's steady advance over the past few years, however, it may be achieved within the next five years or so.

Following the development of scientific feasibility, the program will move from applied research into a development phase. Engineering test facilities will be constructed for the most promising techniques in each confinement system.

Demonstration projects will come next. They will involve the operation of an Engineering Prototype Reactor for the most promising engineering test facilities. They will use the elements of the successful engineering test facility in a pilot plant. The demonstration phase will be completed with the operation of one or more commercial demonstration reactors in which power is produced at a cost that makes it economically attractive to the private sector.

The program is shown schematically in Figure I 3-24.

Current Activity. The Office of Fusion Energy (OFE) is responsible for magnetic fusion research and development within the Department of Energy. The work has been broken down into four interrelated areas:

- Confinement systems—experimental activities to demonstrate and refine methods of heating and containing high-temperature plasmas
- Development and technology—the development of technological and engineering bases to design, construct and operate larger, more complex fusion experiments and facilities
- Applied plasma physics—experimental and theoretical studies of fusion plasma phenomena and the study of new confinement concepts as alternatives to the two major current approaches
- Reactor projects—support for the construction of major new test facilities

Confinement systems. In this subactivity, work will continue on the two approaches to mag-

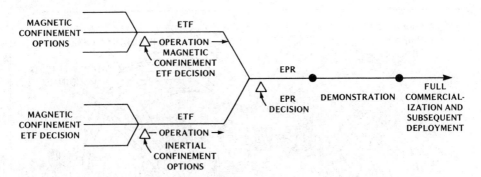

Figure I-3-24. DOE fusion program logic.

netic confinement—the tokamak and the magnetic mirror. The tokamak systems have brought fusion closer to reactorlike conditions. Research will be done in five areas: heating, transport and sealing, plasma shape optimization, impurity control, and fueling.

Development and technology. In Development and Technology, studies are continuing in six areas: plasma heating, magnetic systems, reactor materials, fusion systems engineering, environment and safety, and fusion energy applications.

Applied plasma physics. Studies in this area are developing the data base in atomic and molecular physics necessary for impurity control, exploring new plasma confinement concepts, and investigating new plasma heating and production methods.

Reactor projects. Construction continues on the Tokamak Fusion Test Reactor and several neutron sources needed for materials testing.

Issues

There are six major scientific and technological issues in Magnetic Fusion:

- Prototype development—fusion power concepts require more experiments to evaluate their feasibility as prototypes
- Materials behavior—potential structural materials must be tested for long periods in a fusion reactor environment
- Plasma behavior—development of the ability to predict plasma behavior under a wide variety of conditions
- Scaling laws—ability to scale up from present experiments to larger experiments and prototype reactors
- Technology development—overcoming technical obstacles which limit fusion advances
- Engineering issues—developing facilities that are simple, reliable, and inexpensive

Nonscientific issues include:

- Control of radioactivity
- Safe design of reactors
- Occupational safety of reactor personnel
- Siting of reactors
- Resource requirements
- Private-sector participation

Need for Fusion R&D

Domestic reserves of oil and gas are dwindling. Advanced recovery techniques can extend the reserves but cannot stretch them infinitely. Coal, while plentiful now, could be rapidly depleted if the nation's energy consumption rises in the future as it has in the past. Eventually, exhaustible resources will be consumed or at least made so scarce and expensive that their use will be severely restricted.

There is a need to develop an inexhaustible energy source. The three most likely candidates for the job are solar energy, fission breeder reactors, and fusion energy. All of the technologies require more development before they will contribute substantially to America's energy needs. Magnetic fusion has the farthest to go, but its potential rewards are great. The Country cannot rely on just one or two sources of energy; it should bring the full range of technologies to bear on energy problems. The fusion option should not be foreclosed because others also look promising.

Part II
Fossil Energy

1.
U. S. Energy Policy

On April 29, 1977, the Administration submitted to Congress the National Energy Plan (NEP) and accompanying legislative proposals that are designed to produce a coherent policy concerning energy for the United States. The National Energy Act (NEA), passed by Congress in mid-October 1978, allows implementation of vital parts of the NEP.

The National Energy Plan recognizes that petroleum and natural gas are resources which cannot continue indefinitely to support the present high rates of growth in energy consumption. To meet the demand for oil, the United States has turned increasingly to imports. This has resulted in a progressively larger dependence on an uncertain future world oil market and, therefore, a growing vulnerability to supply interruptions. With this serious situation confronting us, the NEP establishes three overriding energy objectives:

> ... as an immediate objective that will become even more important in the future, to reduce dependence on foreign oil and vulnerability to supply interruptions.
>
> ... in the mid-term, to keep U.S. imports sufficiently low to weather the period when world oil production approaches its capacity limitation; and
>
> ... in the long-term, to have renewable, essentially inexhaustible sources of energy for sustained economic growth. (NEP, p. 6)

The strategy of the NEP contains three major components to achieve these objectives:

First, by carrying out an effective conservation program in all sectors of energy use, through reform of utility rate structures and by making energy prices reflect true replacement costs, the nation should reduce the annual rate of growth of demand to less than 2 percent. That reduction would help achieve both the near-term and the mid-term goals. It would reduce vulnerability and prepare the nation's stock of capital goods for the time when world oil production will approach capacity.

Second, industries and utilities using oil and natural gas should convert to coal and other abundant fuels. Substitution of other fuels for oil and gas would reduce imports and make gas more widely available for household use. An effective conversion program would thus contribute to meeting both the immediate and medium-term goals.

Third, the nation should pursue a vigorous research and development program to provide renewable and other resources to meet U.S. energy needs in the next century. The Federal Government should support a variety of energy alternatives in their early stages, and continue support through the development demonstration stage for technologies that are technically, economically, and environmentally most promising. (NEP, p. 32)

THE FOSSIL ENERGY PROGRAM

The United States relies upon fossil energy for more than 90 percent of its energy requirements. Three-fourths of this is supplied by petroleum and natural gas, which constitute only 11 percent of proven U.S. reserves. Coal provides another 20 percent, yet accounts for about 77 percent of proven U.S. fossil energy reserves. Oil shale, which constitutes an enormous potential resource, is untapped. To support the goals of the U.S. energy policy, Fossil Energy will develop technology that will:

- Increase domestic production of coal, oil, and gas
- Assure that current and new facilities that burn coal can do so in an economically viable and environmentally acceptable manner
- Provide a capability to convert coal to liquid and gaseous fuels
- Allow more efficient and more economically attractive utilization of fossil energy resources

FOSSIL ENERGY RESOURCE STRATEGIES

The Fossil Energy Program is concerned with solids, liquids, and gases. The associated resource strategies are complementary in that each is designed to help achieve the fossil energy goals previously mentioned. Within each strategy are programs which, while placing major emphasis on that strategy, also are applicable to other strategies.

Coal Resource Strategy (Solid Fuels)

The DOE coal strategy is organized around the major applications of coal—direct combustion and conversion to liquids and gases. The industrial sector is given special attention because it has significant opportunities for using coal in some areas where oil and gas are presently used. To decrease the nation's dependence on imported oil and increase its use of coal, the strategy sets four specific objectives.

The first objective is to ensure that stationary facilities now burning coal can continue to do so under applicable environmental standards and that coal combustion can become an increasingly viable option for new facilities in the industrial and utility sectors during the next decade.

This objective will be supported by Fossil Energy RD&D programs to provide the following:

- Support of coal preparation and physical coal cleaning processes to increase the amount of coal that can be burned directly in existing facilities or used with flue gas desulfurization systems (FGD) in new and existing facilities
- Development of chemical coal cleaning technology to assist in meeting strict environmental standards and to provide additional options for new facilities
- Development of both regenerative and nonregenerative FGD systems, emphasizing improvements in operating performance and reliability
- Development of atmospheric fluidized bed (AFB) technologies, emphasizing commercialization of small industrial AFB's in the near term with development of utility-scale systems drawing from experience with industrial AFB's

The second objective is to demonstrate the capability for producing synthetic liquids and gas from coal by the mid- to late-1980s so that a significant capacity could be built in the 1990s if oil prices rise.

This objective will be supported by the following:

- Demonstration of the capability to produce heavy liquids from coal to displace residual fuel oil and other products
- Limited investment in alternative methods of producing lighter liquids such as gasoline and distillate fuels from coal as a longer run option for the transportation sector
- Completion of detail design for the demonstration of an advanced gasification process in a full-scale module to show ability to use a broader range of coals at lower costs
- Completion of Phase 1 design for possible future demonstration of technology

for medium-Btu gas in an industrial application
- Accelerated R&D on highly advanced process concepts which offer potential for significant cost reductions
- Characterization of the environmental issues associated with the production and utilization of coal-derived liquids and gas

The third objective is to develop systems that will use coal in a more economic, efficient, and environmentally acceptable manner for the 1990s and beyond.

Because higher priority is given to the first two objectives, this objective is supported by more moderately paced programs for the following:

- Magnetohydrodynamics (MHD) electric generation
- Fuel cells based on coal-derived fuels for either urban central station or dispersed electric generation
- High-temperature turbines for improved efficiency as well as turbines that allow combustion of heavier and dirtier fuels
- Other advanced thermodynamic cycles and heat recovery systems

The last objective provides fundamental improvements in the technology base for economic, efficient, and environmentally acceptable use of coal and its products. This objective will be achieved by a broad range of basic and applied research activities to provide the basis for a technically diverse coal program. Development of a strong technology base to keep new concepts flowing will be combined with rigorous selection procedures for choosing candidates for further development. Since process scaleup is costly and demands continuing resource commitments, an analytical selection process will be applied to key transition points in the technology development process.

Petroleum Resource Strategy (Liquid Fuels)

Because a liquid fossil fuels shortage is the major incentive to the national program, all fossil energy goals are applicable. Some of the objectives of this strategy are as follows:

- Improve existing and develop new exploration, development, and enhanced recovery technologies
- Develop technology to recover liquid fuels from oil shale that can be processed into petroleum substitutes
- Develop equipment and processes capable of using a wide variety of synthetic fuels in an environmentally acceptable manner

This RD&D program is designed to stimulate private industry participation for those situations in which economic and technical risks are currently inhibiting development.

Gas Resource Strategy (Gaseous Fuels)

The objective of the natural gas recovery program is to obtain significant production of natural gas from unconventional sources to reduce our dependence on foreign sources of gas or oil.

The strategy is to identify and assess new sources of natural gas that have significant production potential and to develop technologies to extract this resource at acceptable costs and in an environmentally and socially acceptable manner. These sources include Devonian shales, tight sands, methane from coal seams, and geopressured aquifers. (Geopressured methane is covered under the Geothermal Program.)

This strategy involves:

- Determining characteristics and production potential of the different resources
- Developing reliable exploratory techniques for the significant resources
- Developing, demonstrating, and transferring effective extraction technologies

The strategy includes cost-sharing projects with the private sector to enhance timely transfer and commercialization of the technology. It is designed to ensure that economical and environmentally acceptable sources of supplementary gas will be available.

2.
FY 1980 Fossil Energy Overview

PROGRAM STRUCTURE AND OBJECTIVES

Activity Structure and Budget

Fossil energy activities are associated with three main resources: coal, petroleum, and gas. DOE and its predecessor organizations have assessed requirements in keeping with stated objectives, analyzed potential payoffs and risks associated with selected technologies and processes, and structured financial requirements accordingly. Total annual funding has increased from $58 million in FY 1973 to $881 million in FY 1979. In FY 1980, the Department of Energy is requesting Congress to appropriate a lesser amount than it did for FY 1979. Of a total FY 1980 request of $747.6 million, $662.7 million is planned for Coal Resources, $57.4 million for Petroleum, and $27.6 million for Gas.

The Office of Fossil Energy has been reorganized to better align related tasks to management areas. The budget structure accommodates this organizational change and ensuing functional redistribution while retaining general similarity to last year's structure. Figure II 2-1 compares FY 1979 and FY 1980 funding levels.

A summarized history of funding levels comparing FY 1973 to FY 1977 budget authorities is shown in Table II 2-1. The changes in emphasis in various categories can be observed. More detail on the recent FY 1978, 1979, and 1980 budgets is provided in Table II 2-2. This table is a fundamental part of this section because it summarizes, by functional area, allocation of funds to projects through technical development phases toward possible commercialization.

Scientific and Technological Issues

A sound scientific and technological base is essential for reliable, efficient operation of existing processes being tested and for the invention and improvement of new and significantly improved processes.

There is a need for better information on the following:

- Rock fracture and drilling mechanics
- Chemical structure of coal, oil shale, and mechanism of liquefaction, gasification, and combustion
- Essential catalytic properties
- Coal beneficiation
- Fluid bed combustion
- Metals and ceramics applications
- Pollution control

Commercialization Issues

The issues to be addressed in determining the commercial viability of a given technology include the following:

- Technological and Resource Factors—Is the resource of sufficient magnitude and the technology sufficiently proven?

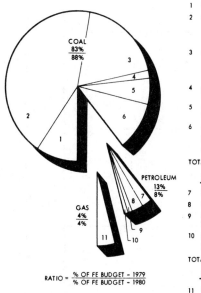

		FY 79	FY 80	DIFFERENCE	%-1979/%-1980
COAL					
1	COAL MINING	76,136	60,350	-15,786	9/8
2	COAL PROCESSING	381,024	301,656	-79,368	47/40
	• COAL LIQUEFACTION	206,426	122,306	-84,120	26/17
	• COAL GASIFICATION	159,598	169,350	+ 9,752	20/23
	• IN SITU GASIFICATION	15,000	10,000	- 5,000	2/1
3	COAL UTILIZATION	211,272	217,600	+ 6,328	26/29
	• COMBUSTION	58,901	57,400	- 1,501	7/8
	• FUEL CELLS	41,000	20,000	-21,000	5/3
	• HEAT ENGINES AND HEAT RECOVERY	58,000	46,000	-12,000	7/6
4	ADVANCED ENVIRONMENTAL CONTROL TECHNOLOGY	7,000	43,250	+36,250	1/6
5	ADVANCED RESEARCH AND TECHNOLOGY DEVELOPMENT	46,371	50,950	+ 4,579	6/7
6	MAGNETOHYDRODYNAMICS (MHD)	80,000	72,000	- 8,000	10/10
	PROGRAM DIRECTION	10,255	11,057	+ 702	1/1
	UNOBLIGATED PRIOR YEAR FUNDS	-78,021			
	TOTAL COAL RESOURCE	680,666	662,663	-18,003	83/88
PETROLEUM					
7	ENHANCED OIL RECOVERY	54,072	21,400	-32,672	7/3
8	OIL SHALE	48,600	28,200	-20,400	6/4
9	DRILLING AND OFFSHORE TECHNOLOGY	2,600	3,000	+ 400	1/1
10	ADVANCED PROCESS TECHNOLOGY	2,200	4,000	+1,800	1/1
	PROGRAM DIRECTION	679	786	+ 107	1/1
	TOTAL PETROLEUM RESOURCE	108,151	57,386	-50,765	13/8
GAS					
11	ENHANCED GAS RECOVERY	33,863	27,593	-6,270	4/4
	SUBTOTAL FOSSIL ENERGY	822,680	747,642	-75,038	100/100
	GENERAL REDUCTION	-63,000			
	TOTAL FOSSIL ENERGY	759,680	747,642	-12,038	

(Figures in Thousands of Dollars)

Figure II-2-1. Comparison of FY 1979 and 1980 funding.

- Institutional Factors—Will Federal controls, tax policies, and regulations or state and local institutional problems hamper development?
- Economic Factors—Is the technology competitive on its own or will it need Government support? If so, for how long? What are the reasons, if any, that have dissuaded industry from large-scale participation? What is the cost/benefit analysis?
- Environmental and Social Factors—Is the technology environmentally acceptable? If not, are waivers to air or water pollution standards justified? What are the socioeconomic consequences of commercialization? What are the environmental costs versus social/economic benefits?

Environmental Issues

Implementation of the Fossil Energy Program will require that factors affecting the physical and biological environment, worker health and safety, and property management and protection be addressed. The following factors will be considered:

- The effect on the human environment of each major fossil energy action
- The impact of emerging laws, standards, and regulations on Fossil Energy programs, technologies, and prospects for commercialization
- Waste disposal from plant operations
- The need for additional monitoring and characterization activities and the impact of Fossil Energy technologies
- Water resource utilization and quality maintenance
- Necessary actions to aid technology development in areas where naturally occurring background emissions exceed Federal pollution standards

Socioeconomic Issues

The development of major new technologies will have a social and economic impact on the

Table II-2-1. Fossil energy development FY 1973 - FY 1977 total funding.

RESOURCE/ACTIVITY	FY 1973	FY 1974	FY 1975	FY 1976 AND TQ	FY 1977
COAL					
Mining Research and Development[2]	$ —	$ —	$ —	$ —	$ 51.9 [3]
Liquefaction	11.4	45.5	107.7 [3]	164.4 [3]	111.4 [3]
Surface Gasification	36.6	53.7	109.8	117.6	135.2
In Situ Gasification	0	1.7	6.6	7.8	8.2
Advanced Research and Technology Development	0.8	4.8	23.3	44.2	44.1
Heat Engines and Heat Recovery	0	0	4.1	13.5	25.9
Combustion Systems	0.5	15.5	35.9	59.6	56.9
Fuel Cells [4]	—	—	—	—	19.1
Magnetohydrodynamics	3.5	7.5	14.3	41.3	40.0
TOTAL COAL RESOURCE	52.8	128.7	301.7	448.4	492.7
PETROLEUM					
Enhanced Oil Recovery	3.1	8.7	26.4	30.5	23.8
Oil Shale	2.5	3.2	4.9	17.4	19.3
Drilling and Offshore Technology	0 [5]	0 [5]	0	2.9	2.4
Advanced Process Technology	—	—	1.8	2.2	1.8
TOTAL PETROLEUM RESOURCE	5.6	11.9	33.1	53.0	47.3
GAS					
Enhanced Gas Recovery	[5]	[5]	[5]	16.9	18.6
TOTAL GAS RESOURCE	0	0	0	16.9	18.6
TOTAL FOSSIL ENERGY	$ 58.4	$ 140.6	$ 334.8	$ 518.3	$ 558.6

1. Based on FY 1980 budget structure. Includes both Operating Expenses and Capital Acquisition funds.
2. Funded by Department of the Interior prior to FY 1978.
3. Includes $78.021 million in unobligated funds for the Clean Boiler Fuel project which was used to offset the FY 1979 appropriation for the coal program.
4. Funded under Conservation prior to FY 1978.
5. Funds included under Enhanced Oil Recovery total.

Table II-2-2. Fossil energy development FY 1980 budget estimates summary table by technologies budget authority (dollars in thousands).

	COAL RESOURCE	ACTUAL FY 1978	APPN FY 1979	ESTIMATE FY 1980
I.	**MINING RESEARCH AND DEVELOPMENT**			
	a. Underground Coal Mining			
	1. Mine Planning and Development	$ 7,000	$ 13,970	$ 8,130
	2. Production Mining	25,300	21,500	21,500
	3. Transport and Other Services	7,600	8,700	2,540
	4. Studies and Support	6,385	9,260	5,830
	TOTAL UNDERGROUND COAL MINING	46,285	53,430	38,000
	b. Surface Coal Mining			
	1. Mine Planning and Development	400	500	500
	2. Production Mining	10,100	11,000	6,385
	3. Studies and Support	975	1,115	1,115
	TOTAL SURFACE COAL MINING	11,475	12,615	8,000
	c. Coal Preparation and Analysis			
	1. Physical Coal Preparation	825	1,825	1,825
	2. Chemical Coal Cleaning	705	2,110	5,205
	3. Studies and Support	3,490	5,936	7,120
	TOTAL COAL PREPARATION AND ANALYSIS	5,020	9,871	14,150
	Capital Equipment	0	220	200
	Construction — Mining and energy related technologies education center/coal utilization facility	0	570	0
	TOTAL MINING RESEARCH AND DEVELOPMENT	$ 62,780	$ 76,136	$ 60,350
II.	**COAL LIQUEFACTION**			
	a. Direct Hydrogenation			
	1. Ebullated–Bed (H–Coal) Pilot Plant	$ 26,000	$ 21,000	$ 35,000
	2. Fixed–Bed Hydrogenation (Synthoil) PDU	11,000	2,889	0
	3. Multistage Liquefaction Process	1,000	0	0
	TOTAL DIRECT HYDROGENATION	38,000	23,889	35,000
	b. Solvent Extraction			
	1. Solvent Refined Coal (SRC) Pilot Plant	15,000	13,500	15,000
	2. Solvent Refined Coal (SRC) Utilization	1,000	0	0
	3. Chemical from Coal Project	400	0	0
	4. Donor Solvent Liquefaction (EDS) Pilot Plant	30,300	34,600	30,000
	5. CO–Steam Process	500	540	0
	TOTAL SOLVENT EXTRACTION	47,200	48,640	45,000
	c. Third Generation Processes			
	1. Process Research/Technology Development	5,300	8,700	10,000
	2. Zinc Chloride Catalyst PDU	2,000	2,100	2,500
	3. Disposable Catalyst Hydrogenation	0	1,011	9,500
	4. Flash Liquefaction Process	1,500	2,000	3,000
	TOTAL THIRD GENERATION PROCESSES	8,800	13,811	25,000

Table II-2-2. (Continued).

COAL RESOURCE	ACTUAL FY 1978	APPN FY 1979	ESTIMATE FY 1980
d. Support Studies and Engineering Evaluations			
1. Cresap Liquefaction Test Facility	$ 12,120	$ 10,000	$ 0
2. Coal Liquids Refining	1,500	0	0
3. Environmental Studies and Other Support	2,500	4,286	10,006
TOTAL SUPPORT STUDIES AND ENGINEERING EVALUATIONS	16,120	14,286	10,006
e. Liquefaction Demonstration Plants			
1. Solvent Refined Coal (SRC) Demonstration Plants	12,000	104,000	7,000
Operating Expenses	(12,000)	(14,000)	(7,000)
Construction	0	(90,000)	0
2. Design and Technical Support	4,000	1,600	0
TOTAL LIQUEFACTION DEMONSTRATION PLANTS	16,000	105,600	7,000
Capital Equipment	400	200	300
TOTAL COAL LIQUEFACTION	$126,520	$206,426	$122,306
III. SURFACE COAL GASIFICATION			
a. High−Btu Gasification			
1. Bi−Gas Pilot Plant	$ 3,000	$ 7,900	$ 0
2. Self−Agglomerating Burner PDU	1,500	0	0
3. Synthane Pilot Plant	14,500	2,620	0
4. Peat Gasification	0	4,500	0
5. Steam Iron Pilot Plant	5,700	0	0
TOTAL HIGH−BTU GASIFICATION	24,700	15,020	0
b. Low−Btu Gasification			
1. Fixed−Bed Gasification	4,550	4,190	6,000
2. Fluidized−Bed Gasification	5,000	4,000	8,000
3. Entrained−Bed Gasification, Atmospheric Process	0	1,490	0
4. Hydrogen−from−Coal Facility	0	0	0
5. Ash Agglomeration Process	1,300	0	0
6. Gasifiers in Industry	8,880	2,000	500
TOTAL LOW−BTU GASIFICATION	19,730	11,680	14,500
c. Third Generation Processes			
1. Hydrogasification	3,500	6,147	11,000
2. Catalytic Gasification Pilot Plant	3,500	7,228	12,000
3. Process Research/Technology Development	6,600	9,700	4,000
TOTAL THIRD GENERATION PROCESSES	13,600	23,075	27,000
d. Special Projects and Support Studies			
1. Hot Gas Cleanup	3,600	1,000	0
2. Gasification Test Facility	20,000	10,000	15,000
TOTAL SPECIAL PROJECTS AND SUPPORT STUDIES	23,600	11,000	15,000
e. Technical Support			
1. Engineering Evaluations	3,000	2,100	0
2. Equipment, Materials and Process Development	4,100	2,100	0
3. Environmental Studies and Other Support	7,150	8,328	0
4. Design and Technology Support	14,400	6,495	12,350
TOTAL TECHNICAL SUPPORT	28,650	19,023	12,350

Table II-2-2. (Continued).

COAL RESOURCE	ACTUAL FY 1978	APPN FY 1979	ESTIMATE FY 1980
f. Gasification Demonstration Plants			
1. High–Btu Pipeline Gas Demonstration Plant A	$ 34,000	$ 42,000	$ 71,000
Operating Expenses	(16,000)	(2,000)	(16,000)
Construction	(18,000)	(40,000)	(55,000)
2. High–Btu Pipeline Gas Demonstration Plant B	0	10,000	14,000
Operating Expenses		(8,000)	(14,000)
Construction		(2,000)	(0)
3. Low/Medium–Btu Fuel Gas Demo Plant, Industrial	40,500	17,000	0
Operating Expenses	(4,500)	(3,000)	(1,000)
Construction	(36,000)	(14,000)	(0)
4. Low–Btu Fuel Gas Demonstration Plant, Utility	3,000	0	0
Operating Expenses	(3,000)	(0)	(0)
Construction	(0)	(0)	(0)
5. Low–Btu Fuel Gas Demo Plant, Small Industrial	7,000	10,300	15,000
Operating Expenses	(4,000)	(1,300)	(5,000)
Construction	(3,000)	(9,000)	(10,000)
TOTAL GASIFICATION DEMONSTRATION PLANTS	84,500	79,300	100,000
Capital Equipment	800	500	500
TOTAL SURFACE COAL GASIFICATION	$195,580	$159,598	$169,350
IV. IN SITU COAL GASIFICATION			
a. Western Low–Btu Gasification (Linked Vertical Wells)	$ 4,200	$ 4,300	$ 2,700
b. Western Medium–Btu Gasification (Packed Bed)	2,200	2,450	3,200
c. Eastern Coal Technology	1,200	0	0
d. Steeply Dipping Bed	1,100	4,550	3,000
e. Environmental Support	1,200	900	600
f. Supporting Research	1,100	2,100	400
Capital Equipment	1,600	700	100
TOTAL IN SITU COAL GASIFICATION	$ 12,600	$ 15,000	$ 10,000
V. ADVANCED RESEARCH AND TECHNOLOGY DEVELOPMENT			
a. Processes	$ 5,201	$ 7,850	$ 14,800
b. Direct Utilization	5,972	9,450	7,100
c. Materials and Components	7,564	9,290	8,900
d. Program Development and Coordination	8,363	13,056	7,500
Capital Equipment	300	375	400
Construction	9,600	6,350	12,050
TOTAL ADVANCED RESEARCH AND TECHNOLOGY	$ 37,000	$ 46,371	$ 50,950

Table II-2-2. (Continued).

COAL RESOURCE	ACTUAL FY 1978	APPN FY 1979	ESTIMATE FY 1980
VI. ADVANCED ENVIRONMENTAL CONTROL TECHNOLOGY			
a. Flue Gas Cleanup			
1. Lime/Limestone Reliability	$ 0	$ 800	$ 11,050
2. Advanced Flue Gas Desulfurization	0	600	8,300
3. Advanced Flue Gas Cleanup	0	1,300	5,700
TOTAL FLUE GAS CLEANUP	0	2,700	25,050
b. Gas Stream Cleanup			
1. Turbine Cleanup Systems	0	1,400	7,000
2. Molten Carbonate Cleanup Systems	0	0	1,400
3. Process Modification Technology	0	1,000	2,000
TOTAL GAS STREAM CLEANUP	0	2,400	10,400
c. Technology Support	0	1,900	7,000
Capital Equipment	0	0	800
TOTAL ADVANCED ENVIRONMENTAL CONTROL TECHNOLOGY	$ 0	$ 7,000	$ 43,250
VII. HEAT ENGINES AND HEAT RECOVERY			
a. Integrated Coal Conversion Utilization Systems	$ 25,500	$ 26,700	$ 24,800
b. Advanced Cogeneration Systems			
1. Directly Fired Heat Cycles	4,578	11,920	9,200
2. Externally Fired Heat Cycles	1,217	2,000	3,580
3. Prototype Systems Assessment	2,830	1,000	220
TOTAL ADVANCED COGENERATION SYSTEMS	8,625	14,920	13,000
c. Heat Recovery Component Technology			
1. Low Grade Heat Recovery	2,750	3,435	4,000
2. High Grade Heat Recovery	4,075	9,825	4,000
3. Heat Exchanger Technology	1,150	2,920	0
TOTAL HEAT RECOVERY COMPONENT TECHNOLOGY	7,975	16,180	8,000
Capital Equipment	235	200	200
TOTAL HEAT ENGINES AND HEAT RECOVERY	$ 42,335	$ 58,000	$ 46,000
VIII. COMBUSTION SYSTEMS			
a. Atmospheric Fluidized Beds (AFB)			
1. AFB Boiler, 30 MW	$ 4,000	$ 8,700	$ 7,000
2. AFB Industrial Applications	10,000	0	7,300
3. AFB Turbine Test Unit (TTU)	2,000	914	800
4. AFB Component Test and Integration Unit (CTIU)	8,500	11,800	9,800
5. Anthracite Applications	0	2,186	1,000
TOTAL ATMOSPHERIC FLUIDIZED BEDS	24,500	23,600	25,900

Table II-2-2. (Continued).

COAL RESOURCE	ACTUAL FY 1978	APPN FY 1979	ESTIMATE FY 1980
b. Pressurized Fluidized Beds (PFB)			
1. PFB Combined Cycle Pilot Plant	$ 7,500	$ 2,500	$ 3,500
2. PFB Component Test & Integration Facility (CTIU)	4,729	3,800	3,000
3. Hot Gas Cleanup	1,500	3,000	0
4. International Energy Agency	2,000	4,934	3,500
5. PFB Utility EDS Plant	0	0	10,000
TOTAL PRESSURIZED FLUIDIZED BEDS	15,229	14,234	20,000
c. Advanced Combustion Technology			
1. Engine Combustion Technology	2,160	2,150	2,000
2. Improved Oil and Gas Burners	405	500	800
3. Coal Combustion Support and Engineering Eval.	10,471	4,692	3,700
TOTAL ADVANCED COMBUSTION TECHNOLOGY	13,036	7,342	6,500
d. Alternative Fuel Utilization			
1. Coal Oil Mixtures	−2,000	5,450	2,500
2. Alternative Fuel Combustion	1,215	1,950	0
3. Coalbed Methane Utilization	2,700	2,000	0
TOTAL ALTERNATIVE FUEL UTILIZATION	1,915	9,400	2,500
e. Combustion Systems Demonstration Plants			
1. Fluid–Bed Demonstration Plants	7,000	2,000	0
Operating Expenses			0
Construction			0
2. Design and Technical Support	4,000	1,600	2,500
TOTAL COMBUSTION SYSTEMS DEMO PLANTS	11,000	3,600	2,500
Capital Equipment	465	725	0
TOTAL COMBUSTION SYSTEMS	$ 66,145	$ 58,901	$ 57,400
IX. FUEL CELLS			
a. 4.8 MW Electric Utility Power Plant Development	$ 16,500	$ 5,500	$ 0
b. Phosphoric Acid Systems Development	13,000	14,000	5,000
c. Molten Carbonate Systems Development	3,000	17,000	14,500
d. Fuel Cell Applied Research	2,700	4,000	0
Capital Equipment	200	500	500
TOTAL FUEL CELLS	$ 35,400	$ 41,000	$ 20,000
X. MAGNETOHYDRODYNAMICS			
a. Open–Cycle Plasma Systems			
1. Component Development and Integration Facility (CDIF)	$ 21,000	$ 21,889	$ 20,600
Operating Expenses			
Construction			
2. Engineering Development	27,300	32,171	26,900
3. Engineering Test Facility	2,941	400	800
4. Systems Engineering	6,754	10,486	8,100
5. Supporting Research	9,300	13,554	13,600
TOTAL OPEN–CYCLE PLASMA SYSTEMS	67,800	78,500	70,000

Table II-2-2. (Continued).

COAL RESOURCE	ACTUAL FY 1978	APPN FY 1979	ESTIMATE FY 1980
b. Closed–Cycle Systems	$ 3,200	$ 1,000	$ 2,000
Capital Equipment	500	500	0
TOTAL MAGNETOHYDRODYNAMICS	$ 71,500	$ 80,000	$ 72,000
XI. PROGRAM DIRECTION	$ 9,455	$ 10,255	$ 11,057
SUBTOTAL COAL RESOURCES	659,315	758,687	662,663
UNOBLIGATED PRIOR YEAR FUNDS		–78,021	
TOTAL COAL RESOURCES	$659,315	$680,666	$662,663
PETROLEUM RESOURCE			
I. ENHANCED OIL RECOVERY			
a. Micellar–Polymer Flooding	$ 12,500	$ 10,360	$ 5,200
b. Carbon Dioxide Flooding	6,600	6,720	3,200
c. Thermal Recovery	12,000	19,430	4,900
d. Improved Waterflooding	3,500	4,680	0
e. Environmental Studies and Other Support	11,340	12,172	7,600
Capital Equipment	300	700	500
TOTAL ENHANCED OIL RECOVERY	$ 46,240	$ 54,072	$ 21,400
II. OIL SHALE			
a. Shale Oil	$ 24,032	$ 43,350	$ 23,500
b. Shale Gas	3,850	4,350	3,500
Capital Equipment	900	900	1,200
TOTAL OIL SHALE	$ 28,782	$ 48,600	$ 28,200
III. DRILLING AND OFFSHORE PRODUCTION TECH.			
a. Drilling	$ 1,000	$ 1,509	$ 1,500
b. Offshore Technology	0	585	1,000
c. Environment and Support	600	506	500
Capital Equipment	0	0	0
TOTAL DRILLING AND OFFSHORE TECHNOLOGY	$ 1,600	$ 2,600	$ 3,000
IV. ADVANCED PROCESS TECHNOLOGY			
a. Advanced Exploratory Research	$ 0	$ 0	$ 1,700
b. Shale Oil Refining and Utilization	0	1,000	500
c. Product Characterization and Utilization	1,400	1,200	1,300
Capital Equipment	0	0	500
TOTAL ADVANCED PROCESS TECHNOLOGY	$ 1,400	$ 2,200	$ 4,000
V. PROGRAM DIRECTION	$ 595	$ 679	$ 786
TOTAL PETROLEUM RESOURCES	$ 78,617	$108,151	$ 57,386

Table II-2-2. (Continued).

GAS RESOURCE	ACTUAL FY 1978	APPN FY 1979	ESTIMATE FY 1980
I. ENHANCED GAS RECOVERY			
a. Eastern Gas Shale Project	$ 14,000	$ 18,000	$ 9,000
b. Western Gas Sands Project	4,300	7,500	8,800
c. Methane from Coal	2,000	4,000	5,000
d. Environment and Support	4,985	3,902	4,110
e. Geopressure Aquifers	1,500	0	0
Capital Equipment	300	200	400
TOTAL ENHANCED GAS RECOVERY	$ 27,085	$ 33,602	$ 27,310
II. PROGRAM DIRECTION	316	261	283
TOTAL GAS RESOURCE	$ 27,401	$ 33,863	$ 27,593
SUBTOTAL FOSSIL ENERGY	765,333	822,680	747,642
GENERAL REDUCTION		-63,000	
TOTAL FOSSIL ENERGY	$765,333	$759,680	$747,642

region in which they are established. Every effort will be made to minimize the negative effect of such development and to enhance the positive aspects. Major social and economic factors addressed include:

- The effects of industry size, location, and rate of development
- The impact on the regional economies and the national economy of the implementation of various Fossil Energy technologies
- The possible lifestyle changes caused by large-scale energy development; e.g., boom-towns
- The Federal Government's role in ameliorating adverse social and economic impacts on the sparsely settled regions where large-scale energy development is likely to occur.

FOSSIL ENERGY STRATEGY

Coal

Background: Coal as an Energy Source. Coal is an abundant resource—it constitutes about 77 percent of U.S. conventional energy reserves, but currently supplies less than 20 percent of energy consumption. Moreover, coal can supply all demand sections since it can be burned directly or transformed into liquid, gas, or feedstock. But the full potential of coal cannot be realized until a number of problems are solved. Coal is bulky—it is more difficult to transport and burn than liquids and gaseous fuels and the transportation system for increased quantities is inadequate. Coal burning can be a major source of pollution, in the form of sulfur dioxide, nitrogen oxides, particulates, and solid waste. The processes for making liquids and gas from coal have yet to be fully developed. The DOE coal strategy is designed to help surmount these problems and allow the nation to utilize the potential of its coal resources.

The DOE coal strategy is organized around the extraction, processing, and utilization of coal in an environmentally acceptable manner. The industrial sector is given special attention because it has options for using coal and is presently a major consumer of oil and gas.

Coal strategy. The ultimate objective of the coal strategy is to decrease the nation's dependence on imported oil and increase the use of more abundant energy resources such as coal in an environmentally safe manner. Progress toward this objective has important economic and national security benefits. To attain this goal, the strategy sets the following objectives:

- Ensure that stationary facilities now burning coal can continue to do so under applicable environmental standards, so that coal combustion can become an increasingly viable option for new facilities in the utility and industrial sectors during the next decade
- Demonstrate the capability for producing synthetic liquids and gas from coal by the mid- to late-1980s so that significant capacity could be built in the 1990s, if required
- Develop systems that will use coal in a more efficient and environmentally acceptable manner for the 1990s and beyond
- Achieve fundamental improvements in the technology base for economic, efficient, and environmentally acceptable use of coal in all time periods

Rationale for Current Program Priorities. To accomplish these goals, the DOE strategy selects a range of projects that should receive Government support, recognizing a variety of needs that must be served. Potential users of technologies will want to employ different types of coal and will face a range of environmental constraints. They will also want different types of product fuels—solids, liquids, gases, or electric power. In addition, most users will want systems specifically designed to meet individual needs.

The Government must be selective in its funding strategy. Because pilot and demonstration facilities involve major financial commitments, the Government cannot hope to support the demonstration of all of the possible combinations of processes, goals, and products. Funds should be concentrated to provide the "critical mass" that is necessary for success in each key program.

In selecting large-scale demonstration projects for Government support, the coal strategy generally relies on the following three criteria:

- Projects that demonstrate processes with a wide range of applications should be favored over those that have limited applicability. For example, some processes can produce synthesis gas for feedstock use as well as hydrogen for direct coal liquefaction. Technologies like these, which serve a number of functions, should be preferred.
- Projects should be selected that complement each other in order to maximize the return on Federal investment. For example, a basic coal gasification process can be a key component for the production of medium-Btu gas, high-Btu gas, chemicals from coal, and coal liquids. By coordinating the choice of projects, the impact of any one can be increased several times.
- Projects should be chosen according to their potential for environmental acceptability and technical and economic success. Government supported projects should lead to commercially viable technologies, with significant potential for increasing coal use.

Strategic Planning and Alternative Scenarios for Coal. Once it is decided that a technology merits Government assistance, a different set of factors should determine the appropriate method and pace of Government support. A major element of the coal strategy is to involve the private sector in all phases of technology development—conceptualizaion, research, development, demonstration, and commercialization. The DOE coal strategy looks at two key criteria in deciding when and what type of Government assistance should be provided: the level of risk/benefit associated with the technology and the pace of development that is required.

The Federal role in providing assistance is tied to the level of risks and benefits associated with each technology. Projects that depend on higher future energy prices in order to be economic are currently not attractive to the private sector. These projects may require substantial Federal cost-sharing through demonstration-scale facilities. Other technologies that are nearly competitive may only need economic incentives such as tax credits or a stable Federal policy. In some cases, the Federal Government can share risks on major projects through loan guarantees or favorable tariff treatment. In each case, the Federal Government needs to determine the most appropriate policy instruments to advance the technologies that have the greatest potential.

The pace of development will depend on the status of the technology, the lead time for development, and the length of time before the technology will become economic. The major determinant of the competitiveness of most new technologies will be the world price of oil. The pace of development, along with the type of support provided, will determine the size of budget outlays in any given year.

Because mining, direct combustion, liquid, and gas technologies involve different risks and benefits, the appropriate pace and method of Government support varies for each application of coal extraction, processing, and utilization technology.

Need for Coal R&D Programs. The needs for coal extraction, processing, and utilization programs are diverse and are dependent on complex issues. The basis for each of these programs is addressed below.

Mining research and development. From 1945 to 1968, the mining industry experienced increasing productivity; beginning in 1969, this trend underwent a significant reversal in both underground and surface mining—productivity in underground mining dropped by 50 percent. This decline in productivity coincided with an increasing demand for coal and is attributed mostly to more stringent safety, health, and environmental restrictions.

No simple answers are forthcoming to these complex problems. The Mining R&D Program necessarily includes many technological approaches to cover the spectrum of requirements for different equipment, techniques, and applications. Of roughly 6,200 producing mines in the United States, only 11 companies can be considered large; these 11 each produced in excess of 10 million of the 665 million tons of total U.S. output in 1976. Equipment manufacturers are generally unable to find mine operators who are willing to accept the risk associated with trying new equipment. In addition, 90 percent of the equipment manufacturers are too small to afford extensive R&D. Labor unrest, social change, regulatory (health, safety, and environmental) constraints, and miner skills, attitudes, and motivations add to the complexities of developing mining systems.

Liquids. Petroleum liquids provide almost half of U.S. energy needs but constitute less than 8 percent of domestic energy reserves. The demand for oil is projected to grow at over 4 percent a year (conservative estimate representative of range of estimates from various series in EIA, Administrator's 1978 Report to the Congress, Volume 2, Table 64, p. 133). Because domestic production is declining (except for a small jump caused by Alaskan production), oil imports are increasing at over twice that rate. The consequences of these trends for the nation are serious.

Increased dependence on foreign oil jeopardizes national security and weakens the value of the dollar. Reliance on imports also leaves the economy vulnerable to rises in world oil prices.

Supplemental gas. Natural gas supplied 26 percent (or 19 quads) of the total domestic energy consumption in 1977. Demand for dependable supplies of natural gas exceeds supplies available from conventional domestic sources. Although natural gas pricing legislation is expected to encourage additional production of conventional natural gas, shortfalls will eventually result. At this time, coal-derived synthetic gas can help fill this gap in the form of high-Btu (pipeline quality) gas or medium- or low-Btu gas for industrial applications. Development of this capability is important as a backstop to high world energy prices in the 1990s and beyond.

Direct combustion. The demand for coal in 1985 is projected to range from 1.0 billion to 1.2 billion tons (conservative estimate representative of range of estimates from various series in EIA, Administrator's 1978 Report to the Congress, Volume 2, Table 64, p. 133). Because close to 90 percent of the coal consumed in this country is and will continue to be burned directly, any program to increase coal utilization should place primary emphasis on promoting direct coal combustion.

Industrial feedstocks. Industrial consumption of natural gas and oil for chemical feedstocks (1.4 MMBD oil equivalent in 1976) could be reduced significantly if coal-based synthesis gas were used. Moreover, technology for producing synthesis gas from coal is already well known. Three commercial processes for making synthesis gas have been widely used abroad (Lurgi, Koppers-Totzek, and Winkler) and two improved processes are inactive private sector development (Texaco and Shell-Koppers). The petrochemical industry is showing active interest in coal-base synthesis gas because it is nearly economical at today's prices, and the economics will improve as natural gas and oil become more costly. Initial application could be stimulated with modest financial or regulatory incentives.

Increased efficiency. Typically, heat engines extract approximately 30 percent of the available energy from the fuel. Additionally, many power-generating facilities are designed to burn oil and gas. Many technologies are available to extract energy from coal more efficiently, but they are in early stages of development, and the economics are such that the private sector cannot develop these technologies without Federal participation.

Environmental requirements. Environmental standards pose the greatest constraint on increasing direct coal use. Unless new technologies are available for meeting these standards, many facilities that might use coal will turn to other fuels.

The environmental performance required for both utility and industrial coal-burning facilities depends largely on the age and location of the facility. Most existing facilities meet the emission standards of the Clean Air Act. New facilities on which construction began before proposal of the EPA's revised New Source Performance Standards (NSPS) must meet the current NSPS. However, future coal use will be increasingly affected by the stricter standards in upcoming NSPS revisions.

Development of new environmental control technologies will be closely linked with development of more efficient heat engines.

Petroleum

Background: Petroleum as an Energy Source. The level of imports of liquid fuels is now the primary energy supply problem in the United States and will continue to be so for the next 10 to 20 years. With oil imports reaching 50 percent of the U.S. consumption, the outflow of dollars and the possibility of an embargo or other supply disruption poses serious threats to the U.S. economy. A basic principle of the National Energy Plan is that "the United States must reduce its vulnerability to potentially devastating supply interruptions." The National Energy Plan established a goal of reducing oil imports to less than 6 million barrels a day by 1985. This goal cannot be met by reducing demand only. Conservation alone cannot achieve such a large reduction, and massive switching to fuels is largely impractical, especially in the transportation sector. Therefore, domestic production of liquid fuels must be increased. Offshore production, oil shale, synthetic liquids from coal, and enhanced recovery are possible sources of additional oil.

United States oil reserves as of January 1978 were 29.5 billion barrels. Domestic crude oil production in 1979 was about 3 billion barrels, still declining from the high reached in the early 1970s.

Petroleum liquids provide almost half of U.S. energy needs but constitute less than 8 percent of domestic energy reserves. The demand for oil is projected to grow at least 4 percent a year. Because domestic production is declining, oil imports are increasing at over twice that rate. The consequences of these trends for the nation are serious.

Although total demand is increasing, the product slate demanded is becoming generally lighter, mostly due to increased demand for transportation fuels. At the same time, the crude oil input to produce these products is becoming heavier for U.S. crudes, but is remaining fairly constant for the world.

A liquids strategy should address the demand for an increasingly light product state. If the nation continues to rely on existing supplies of petroleum products, two options will be available. The United States can import a higher portion of light crude, or refinery adjustments can be made to transform the increasingly heavy domestic and potential synthetic barrel into the increasingly lighter product state. The cost of converting heavy crudes to lighter products will place an upper boundary on any price for light crudes. Of the available technological options, enhanced oil recovery has the highest probability of near-term impact with the lowest cost premium.

Today, crude petroleum is refined to produce several kinds of end-products. The transportation sector, which uses over 60 percent of all liquids, requires almost entirely light liquids such as gasoline and middle distillates. Liquid use by other sectors is distributed fairly evenly among the residential/commercial, industrial, and utility sectors. The following facts derived from the National Energy Goals should be considered in determining current program priorities:

- Domestic production of oil and gas began to decline in the early 1970s, and this trend is continuing with no sign of abatement.
- Extensive proven reserves still exist in the 48 continental states. Costs of production in marginal reservoirs have, however, resulted in decisions either not to produce or to stop production.

- Additional proven reserves lie in the "frontier areas." These have been defined as (1) undeveloped areas in Alaska, both onshore and offshore; (2) offshore in waters more than 600 feet deep; and (3) reserves at depths greater than those currently worked in commercial oil fields.
- If technologies could be developed that would make oil production less costly and if the costs and associated risks of "wildcatting" could be reduced, more wells would be drilled in unproven fields.

The Petroleum Program is designed to encourage and support industrial participation in developing and demonstrating techniques to increase the rate of production; to recover more oil from existing reservoirs; to recover refinable oil from shale; to provide technology for more efficient drilling; and to develop the technology base relating to the production and use of petroleum and shale oil.

Strategic Planning for Petroleum. Oil produced by natural reservoir pressure or by lift pumps, known as primary production, recovers about 20 percent of the oil-in-place. By pumping water or gas into the reservoir to increase or maintain the pressure (secondary production), the yield can be increased to over 30 percent. More advanced methods of improving the recovery rates (tertiary recovery) have come to be known as enhanced oil recovery (EOR). The potential target for EOR is estimated at 334 billion barrels of oil not economically recoverable with conventional technology. This represents about two-thirds of the original oil in place. EOR technology under development has the potential to recover 5 to 10 percent of this currently unrecoverable resource.

Oil shale constitutes another potentially significant domestic source of liquids. Substantial oil shale deposits are distributed across the country, but the resources now considered most economic are located in a relatively small region of western Colorado and eastern Utah. Several firms have indicated substantial interest in producing shale oil from these rich deposits at net costs of around $20/bbl. Shale oil may thus be our least expensive alternative source of liquids, especially considering that refined shale oil is quite similar to a sweet, light crude oil. Although resources are large and several above-ground and in situ processes are under advanced development, potential production of shale oil from the most economic Western reserves is likely to be limited by environmental constraints. These restrictions, which include air emissions from mining and processing, spent shale disposal, and protection of aquifers, could limit large-scale developments in remote Western areas.

The Advanced Process Technology (APT) Program will undertake and maintain a broad technology base to provide a fundamental understanding of mechanisms required to extract and utilize unconventional liquid and gas resources. The elements of the APT Program are as follows:

- Advanced Exploratory Research (for oil, gas, shale, and tar sands)
- Shale Oil Refining and Utilization
- Product Characterization and Utilization

The Drilling and Offshore Technology Program efforts are aimed at improving drilling penetration rates and extending domestic drilling operations to very deep and hostile frontier areas. This program can have significant application to other fossil fuel programs such as enhanced oil and gas recovery, in situ coal gasification, drilling in geopressured reservoirs, coal bed methane drainage, and drilling for geothermal wells.

Need for Petroleum R&D Programs. The Economic Regulatory Administration of DOE has published regulations that will allow oil produced by EOR techniques to receive the world oil price. This is expected to encourage the private sector to increase their efforts in EOR. However, some of the more promising EOR techniques are technically immature and costly. Industry is working with these more advanced techniques, but at a rate commensurate with their degree of risk and, thus, too slowly to have much near-term impact. The DOE EOR program, therefore, will sponsor research and development on the more promising of those advanced technologies so that they can be accelerated to a state of commercial readiness and made available to a greater number of participants, including the smaller independent producers who might otherwise be excluded from such research. If the response of the small producers to the new prices is not sufficient, additional incentives may be needed to support this special part of the petroleum industry.

Liquids from oil shale offer great promise for both near-term and long-term supplies of liquid hydrocarbons from a major domestic resource. Although several technologies (some developed with Federal funds) have been in existence for many years to extract raw shale oil from the Green River Formation of Colorado, Wyoming, and Utah (the "Western Shales"), the economics of the various processes so far have not been promising enough for commercialization. This situation, with partially developed and proven technology combined with the environmental concerns now facing oil shale development, requires significant new Federal initiatives if shale oil is to make an effective entry into the liquid fuels market. The Oil Shale Program is predicated upon the following assumptions:

- Production of domestic crude oil will continue to decline.
- Liquids from shale and coal offer the major areas of opportunity to establish a domestic source of liquid hydrocarbons to offset increasing imports of crude and the ensuing economic effects, particularly impacts on the balance of payments.
- First generation technologies, primarily surface retorting, have been developed to a level where only minimal DOE support is required. Such support will be directed toward selection or development of an optimum surface-retorting system to support modified in situ mining and retorting.
- Modified in situ technologies offer the best opportunity for an environmentally acceptable second generation system to retort liquids from Western oil shales.

The APT Program is directly related to the Enhanced Oil Recovery, Gas Recovery, and Oil Shale Programs. Its basic objective is to develop more advanced or novel technologies that can be brought to commercial readiness by those related programs; undertake and maintain a broad technology base to provide an understanding of extraction mechanics common to those programs; and characterize products obtained from both conventional and unconventional energy resources.

Unlike technology programs in which new and innovative processes are brought through successive scaleup developments, the technological activities of the Drilling and Offshore Technology Program are those of continuing improvement of existing technologies. Evolutionary technologies have extended drilling capabilities from a little more than 600 feet to 5 miles or more. The frontiers of drilling have been extended to deep offshore regions and into the Land Arctic. Offshore drilling in the Arctic involves better knowledge of shore-fast and pack-ice dynamics and under-ice oceanography and bottom conditions. These considerations, still largely unknown, require considerably more research before drilling operations in the Ocean Arctic become routine. Development of such technologies, particularly in an accelerated time scale, is both expensive and technologically risky. If paced by the economics of the marketplace, it is highly unlikely that such technologies will be developed in the near-term. Thus, Government leadership is essential if accelerated development of such processes is to occur.

Gas

Background: Gas as an Energy Souce. Natural gas has occupied a significant role in the domestic energy market because of its convenience, cleanliness, and low cost. Strong pressures exist for a continued significant role in the near future as a result of the large capital investment already in transmission pipelines and distributions systems, to serve industrial and commercial/residential users. The gap between unrestricted demand and supply will have to be filled by new natural gas sources if gas is to maintain its share of the market in the near-term, while the national economy makes its conversion to long-term alternatives.

Natural gas currently supplies about one-fourth of the total energy used in the United States while it constitutes less than 5 percent of the total domestic conventional energy reserves. Demand for gas has grown steadily, but additional reserves from new discoveries have not kept pace. Since 1973, gas production has declined significantly, resulting in delivery curtailments and fuel switching in industry. The continuing supply shortfall culminated in severe industrial disruption in the winter of 1976–77. Although this decline has slowed, the trend continues today, and current projections indicate that by 1985 gas from known reserves will be able to satisfy only 65 to 73 percent of current rates of consumption of natural gas. The shortfall in domestic natural gas supply is highlighted by six important supply indicators:

- Total proved reserves have declined by 27 percent since 1970, from 283 Tcf to 207 Tcf (32 Tcf are in Alaska).
- The ratio of proved reserves to production is at an all-time low, less than 11 to 1 (about 9.5 to 1 when Alaska is excluded).
- New additions to supply, from new discoveries and extensions of known fields, have replaced only 1 Tcf for every 2.5 Tcf consumed over the last 6 years.
- Although exploration in 1976 was three times that in 1970, stimulated largely by significant price increases, additions made to reserves per completed exploratory well have declined from 4 Bcf per well in 1970 to 1 Bcf per well in 1976.
- New frontiers (Alaska, deeper waters, deeper wells) could contribute significantly, but finding, producing, and delivering this gas will be costly compared to past performance and will require long lead times.
- Developmental drilling, which supplies 80 percent of additions to reserves, has more than doubled in the past 7 years.

However, the resource base for future developmental drilling is not being replenished because of low exploratory success.

Rationale for Current Program Priorities. National policy set forth in the 1977 National Energy Plan (NEP) calls for substantially increased domestic energy supplies at economically acceptable costs, and identifies unconventional gas resources as potential increased natural gas supply targets. As a result, the DOE Gas Recovery Research, Development, and Demonstration (RD&D) Program has been established. Six large, unconventional gas resources have been identified as having significant potential for development and positive impact on future supplies:

- The low-permeability (tight) lenticular gas sandstones of the western Rocky Mountain region
- The tight, blanket-type sands of the Southern and Southwestern regions of the United States
- The tight, low-pressure, shallow gas formations of the Northern Great Plains
- The gas-bearing Devonian and Mississippian shales of the Eastern United States in Appalachia and the Midwest
- The natural gas present within coal seams and associated strata
- The high-temperature, high-pressure (geopressured) aquifers of the Gulf Coast region

These unconventional gas resources have received relatively little attention from private industry to date due to technological problems associated with achieving economical production. Eastern shales, Western sands, and coal formations are all characterized by low permeability to gas flow and low porosity. Development of deep, hot, geopressured aquifers could require highly sophisticated production systems. There is limited ongoing industrial activity in the Eastern shales and tight gas basins:

- An estimated 1 Tcf/year is produced currently from these sources.

- This production is limited to the formations and areas that are geologically favorable.

Industry is expected to defer or reject development beyond these levels due to the following conditions:

- All the resources are marked by geological circumstances that make production costly and risky.
- Current levels of measurement capability and understanding of the basic gas-generation and gas-trapping mechanisms are inadequate to characterize the resource fully and/or to identify the most promising segments; thus, optional recovery systems cannot be designed at present.
- The full extent of the potential is unknown, with much of the resource in untested areas.
- The requisite technology is only in the experimental phase, requiring tests and analyses to determine optimal well stimulation and field development technologies in light of specific geologic conditions.
- The resource characterization and technology R&D required to overcome these barriers are marked by high costs and high risks.
- The costs of conducting R&D cannot be recouped through production benefits to individual producers. The benefits primarily accrue to other producers and to consumers.
- In light of these difficulties, industries prefer alternative investment opportunities. Moreover, higher prices or other financial inducements provide no direct incentives to perform R&D, so additional development of only the most favorable geologic targets may be undertaken.

The intent of the Gas Recovery Program is to help overcome the above barriers by:

- Shifting the price-supply curve through R&D aimed at higher recovery, lower cost technology
- Accelerating resource characterization and development of new technology
- Reducing the costs and risks of R&D in high-potential/high-risk resource areas
- Accelerating deployment of new technologies toward full-scale commercialization

Strategic Planning and Alternative Scenarios for Gas. Stimulation of incremental gas production from unconventional sources relies on three strategic objectives:

- Shift the price-supply curve through more efficient technology, thus providing increased production at any given gas price
- Accelerate the timing of R&D investments and subsequent application of the improved technology
- Undertake high-risk/potentially high-return R&D in areas currently deferred by industry.

The strategic objectives of the Gas Recovery Program are aimed at increasing the near-term (before 1985) and longer-term (1985 to 2000) production of natural gas from the unconventional sources by developing economical recovery techniques and transferring the technology to the gas industry for early commercialization. Multiple technologies must be developed due to large variations in geologic and reservoir conditions and must be tested in the field. Technology development and testing will be closely supported by resource characterization studies (delineation of the size, geographic extent, and intrinsic physical/chemical properties of each resource), and research pertaining to mechanisms of recovery. Environmental impact assessments will be performed to assure environmental acceptability of developing technologies.

Supplemental pipeline gas supplies include imports of natural gas via LNG or pipeline from Canada or Mexico, special domestic supplies including gas from Alaska and from unconventional geologic formations, and pipeline gas made from coal. The cost of some supplemental sources is strongly linked to world oil prices, while other sources are largely independent of the world oil price.

At present, gas imports, liquefied natural gas (LNG), Mexican imports, and industrial fuel oil are generally cheaper than domestic supplementary gas from either "unconventional sources" or from coal. However, domestic supplements will become more economical than LNG as the world price of oil rises. Technologies for domestic supplemental gas supplies should be developed so they will be available as they become economic.

The wide ranges in gas price estimates available reflect lack of knowledge concerning the resource base as well as production technology and costs. Although the feasibility and amounts of production available at various costs are not well understood, some of the unconventional sources are likely to be competitive with imported gas distillate at increased world oil prices. Characterization of the size and cost of these unconventional gas resources is a priority activity.

Synthetic pipeline gas will become available from coal conversion processes in the future.

Need for Gas R&D Programs. The economy can adjust to the shortfall of domestic natural gas in different ways. First, the residential/commercial and industrial sectors can use electricity, but it is generally more costly. Alternatively, distillate fuel oil can substitute for natural gas chiefly in the industrial sector, an adjustment which increases oil imports. Finally, a variety of supplemental sources, including imports and unconventional gas, could be developed to meet expected demand. As world oil prices rise, these sources may become increasingly attractive as alternative supplies of pipeline gas.

It is anticipated that ultimate development and commercialization of the unconventional gas resources will be accomplished by private industry. Technology developments from the Gas Recovery Program are intended to sufficiently lower the exploration and development risks associated with these gas resources to levels which represent attractive prospects for development by the gas industry. It is expected that, upon completion of the program, gas production by private industry will continue with no further Government support. The Gas Recovery RD&D projects are cost-shared with industry to as great an extent possible in order to promote industry involvement and effect more complete technology transfer. (However, industry enthusiasm for such cooperative ventures tends to be dampened by the same concerns that have prevented extensive development of the unconventional sources.) In addition to major economic and technological uncertainties addressed by this program, other potential constraints such as restrictions, potential environmental problems, price regulations, tax requirements, the possibility of antitrust legislation, and the capacity of the drilling and stimulation service industries to perform the work must be addressed before full communication of these technologies can be achieved.

MANAGEMENT APPROACH

Overview

Fossil Energy (FE) has devised a management approach that is designed to make FE responsive to the mission of DOE. The new organization, shown in Figure II 2-2, describes the program flow and management of major fossil energy research and development, and demonstration projects. The basic FE responsibilities are divided into two major functional roles, policy and program development and implementation. The division of responsibility under the new management approach will be for Headquarters to handle policy and program development and for the field activities to assume the implementation responsibility.

Program Management—Headquarters

Program management, policy initiatives, and governmental interface with the Congress, Office of Management and Budget (OMB), and other agencies and industry will be performed by Headquarters. Specific responsibilities of headquarters include the following:

- Prescribe policy for the Fossil Energy Program

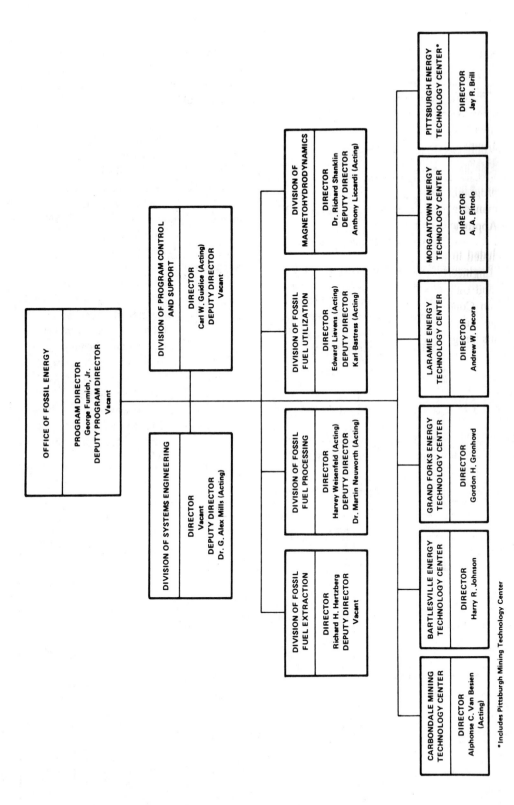

Figure II-2-2. DOE fossil energy organization.

*Includes Pittsburgh Mining Technology Center

- Establish those technologies in which the program will be developed
- Establish and maintain liaison with other Assistant Secretaries
- Develop program plans for each assigned area
- Develop and justify budgets
- Interface with the Congress and other groups that influence the program
- Measure work progress in the various program/project areas and inform DOE management of program results
- Approve field procurement plans

Included in the Headquarters staff is a Systems Engineering Group to develop the options and trade-offs for FE and a Division of Program Control and Support to develop and manage the various programs. Table II 2-3 presents programmatic responsibility by division.

Project Management—Field Activities

The various field activities will assume responsibility for implementation aspect of the FE Program, including contracting authority to a certain extent. Lead activities have been assigned for the various subprograms and projects as follows:

Bartlesville	- Enhanced Oil Recovery
	- Internal Combustion Engines
Carbondale	- Mining R&D
Grand Forks	- "Applications Center" for Low-Rank Coals
Laramie	- Oil Shale
	- Non-Shale In Situ Processes
Morgantown	- Unconventional Gas Recovery (Western Tight Gas Sands has not yet been assigned)
	- Fluidized-Bed Combustion
	- Gas Steam Cleanup
	- Coal Feeding Systems
Pittsburgh	- Coal Liquefaction
	- Synthetic Liquids Characterization
	- Coal Mixtures
	- Direct Coal Combustion
	- MHD

Table II-2-3. Program management.

FY 1980 PROGRAM STRUCTURE	FOSSIL FUEL EXTRACTION	FOSSIL FUEL PROCESSING	FOSSIL FUEL UTILIZATION	MHD	PROGRAM CONTROL & SUPPORT	SYSTEMS ENGRG
Mining Research and Development	X					
Coal Liquefaction		X				
Surface Coal Gasification		X				
In Situ Coal Gasification	X					
Advanced Research and Technology Development		X	X		X	X
Advanced Environmental Control Technology			X			
Heat Engines and Heat Recovery			X			
Combustion Systems			X			
Fuel Cells			X			
Magnetohydrodynamics				X		
Enhanced Oil Recovery	X					
Oil Shale	X					
Drilling and Offshore Recovery	X					
Advanced Process Technology	X					
Enhanced Gas Recovery	X					

Chicago Operations - Fuel Gas Demonstration
Idaho Operations - MHD/Butte
Oak Ridge - SRC-I and -II

Project management responsibility for the various university projects is assigned to the appropriate field activity.

The field activities will be responsible for the following:

- Development of integrated and individual plans for subprogram and projects
- Development of procurement plans
- Chairing source evaluation boards and technical advisory committees
- Management of projects consistent with milestones and costs
- Maintaining and reporting obligations/costs on project and program bases
- Development and maintenance of adequate technology bases for each program area

Industry

In addition to the management accomplished at the Headquarters and field activity levels, DOE and industry work together as follows.

Industry provides technical support such as analysis and quick reaction solutions to management problems, and plant operation and management for demonstration projects and some smaller installations.

Technical support is procured using normal competitive practices from manufacturing firms as well as from high technology companies.

Plant operations and management is established in a variety of ways: direct contracting, cost sharing, contract services for Government facilities and others.

3.
Activity Description—Coal

COAL RESOURCE OVERVIEW

Coal is the nation's most plentiful fuel resource, with enough reserves to last several centuries. It offers the greatest potential in the 1980s for relieving or minimizing the current U.S. energy shortfall. For example, the United States has reserves estimated at 3.2 trillion tons of unmined coal, whereas supplies of oil and natural gas are becoming rapidly depleted. Major efforts of the Coal Research, Development, and Demonstration Program are aimed at tapping this abundant source of energy for direct use as a fuel in an efficient and environmentally acceptable manner, or for processing into synthetic fuels to provide a substitute for limited oil and natural gas resources.

The principal objectives of the Coal Resource Program are: to accelerate the development of environmentally acceptable technology for extracting coal, converting coal to liquid and gaseous fuels by both surface and in situ processing; to stimulate improved methods for the direct combustion of coal; to foster rapid development of advanced power conversion systems, including fuel cells and magnetohydrodynamics, for generating electricity from coal; to validate technical, economic, and environmental acceptability of second-generation coal processes by design, construction, and operation of near-commercial scale modules; to expand the technological base and the supply of technically trained manpower; to support the evolving coal conversion and utilization technologies; and to stimulate technology in the area of greater heat conversion efficiency from fossil fuel products.

The basic objectives of the Coal Resource Program will be met by implementing activities in the areas of coal conversion and improved direct coal utilization. The activity areas to be discussed include:

- Mining R&D
- Coal Liquefaction
- Surface Coal Gasification
- In Situ Coal Gasification
- Advanced Research and Technology Development
- Advanced Environmental Control Technology
- Heat Engines and Heat Recovery
- Combustion Systems
- Fuel Cells
- Magnetohydrodynamics

The Mining Research and Development Activity supports the substitution of abundant energy sources (i.e., coal and oil shale) for less abundant sources (i.e., oil and natural gas) by developing technologies that will improve the economics of production; maximize resource recovery; protect the health and safety of mine workers; and protect the environment. Based upon the National Energy Plan and the major requirements of mining technologies identified above, the functions of the Mining Research and Development Activity are to conduct research in the areas of underground coal mining, surface coal mining, and coal preparation.

Coal conversion activities develop processes to convert coal into products that substitute for those derived from oil and natural gas. These substitutes include crude oil, fuel oil, and distillates; chemical feedstock, pipeline quality (high-Btu) and fuel (low- and intermediate-Btu) gas; and other byproducts, such as char, that may be useful in energy production. The Liquefaction and Gasification Activities address the development of these products and their use in the market.

Significant advances have been made in the coal conversion and utilization technology. The current effort in this area is directed towards demonstrating second-generation technology on a near-commercial scale in the early 1980s. Four coal liquefaction processes (SRC-I, SRC-II, H-Coal and EDS) are now in advanced stages of development. Surface coal gasification processes have been commercially available for many years, but are being further developed to improve process efficiencies and to reduce associated costs. In situ coal conversion has been demonstrated by highly successful subpilot scale field tests. Government and industry are jointly participating with a near-term goal to develop at least one commercial underground conversion process during 1985–1987.

The Advanced Research and Technology Development projects in the DOE Energy Technology Centers, national laboratories, and in conjunction with other Government agencies, private industry, and universities will develop the basic and applied technology and data upon which present and future fuel conversion and utilization processes depend. Included under this activity are program planning and development efforts essential to the formulation of fossil energy plans and policy.

The Advanced Environmental Control Activity has received new emphasis based on a transfer of several projects from EPA to DOE in FY 1979. The activity is directed toward developing technology to minimize the environmental impacts of new or existing fossil fuel utilization systems. The desired goals will be achieved through research and development of technology in the areas of flue gas cleanup and gas stream cleanup.

The Heat Engines and Heat Recovery Activity will develop technology to maximize energy savings in the utility, industrial and residential/commercial market sectors. Three major objectives for achieving the goal are to increase the direct utilization of coal in the utilities by developing advanced, highly efficient, integrated coal conversion/utilization systems; to develop advanced cogeneration systems that can supply both electric and thermal needs for the equivalent fuel energy now being used to generate basic electrical requirements; and to develop advanced heat recovery component technology for producing useful work from waste heat and minimizing heat rejection by the utilities, industrial, and residential/commercial sectors.

The objectives of the Combustion Systems Activity are to develop fluidized-bed coal combustion systems capable of directly burning high-sulfur and other coals of all ranks and quality in an environmentally acceptable and economic manner; develop technology to substitute coal for a substantial portion of oil in existing combustors capable of firing coal-oil mixtures; and improve the reliability and efficiency of present boilers and furnaces.

Fuel cells represent a new energy-conversion alternative which is highly efficient, flexible, modular, economically attractive, and environmentally acceptable. These characteristics permit location of power plants in large load centers and even on site in urban, rural, and residential areas. The near-term (3-6 years) objective of the DOE's Fuel Cell Activity is to establish the commercial feasibility of fuel cell power plants for electric utility applications, industrial cogeneration, and for building total energy systems.

The magnetohydrodynamics (MHD) system of power production from coal is considered a potential future source of low cost, central station, electric power. Low NO_x emissions, inherent sulfur removal, and high efficiency of the MHD system make it environmentally attractive. Commercialization is expected by the late 1980s or early 1990s.

Table II 3-1 summarizes the funding levels by activity for the FY 1978 to 1980 period.

MINING RESEARCH AND DEVELOPMENT

The mission of the Mining Research and Development (MR&D) Activity is to develop

Table II-3-1. Funding levels for coal resource activities for FY 1978 - 1980.

COAL RESOURCE ACTIVITIES	BUDGET AUTHORITY (DOLLARS IN THOUSANDS)		
	ACTUAL FY 1978	APPROPRIATION FY 1979	ESTIMATE FY 1980
Mining Research and Development	$ 62,780	$ 76,136	$ 60,350
Coal Liquefaction	126,520	206,426	122,306
Surface Coal Gasification	195,580	159,598	169,350
In Situ Coal Gasification	12,600	15,000	10,000
Advanced Research and Technology Development	37,000	46,371	50,950
Advanced Environmental Control Technology	0	7,000	43,250
Heat Engines and Heat Recovery	42,335	58,000	46,000
Combustion Systems	66,145	58,901	57,400
Fuel Cells	35,400	41,000	20,000
Magnetohydrodynamics	71,500	80,000	72,000
Program Direction	9,455	10,255	11,057
Unobligated Prior Year Funds		−78,021	
TOTAL	$659,315	$680,666	$662,663

and transfer to industry the technologies (both equipment and techniques) required to supply solid fuels at acceptable economic and social costs.

Activity objectives are to improve systems technology; to develop cost-effective equipment and techniques; to develop, test, demonstrate, and transfer to industry new and innovative mining concepts; and to develop economically competitive preparation technologies required to assure that coal is an environmentally acceptable fuel.

These objectives will be met by implementing efforts in three subactivities: Underground Coal Mining, Surface Coal Mining, and Coal Preparation.

The Underground Coal Mining Subactivity centers around coal mine planning and development, room-and-pillar mining, panel mining, novel systems, and haulage. Subactivity emphasis is to improve existing technology; to develop novel systems for thick, multiple, and pitching seam mining; and to assure the lowest cost coal by integrating mining operations into a total system.

The Surface Coal Mining Subactivity centers around equipment and systems for open cast and contour mining. Emphasis is on more effective use of standard equipment and improvement of present mining systems. However, newly developed equipment that can improve performance and be readily integrated into the mining cycle is also included.

The Coal Preparation Subactivity is aimed at developing improved technologies for the classification and upgrading of coal, including desulfurization and removal of trace fugitive elements in raw coal. Subactivity emphasis is on the development and demonstration of improved coal preparation technology to ensure an adequate supply of environmentally acceptable coal for utilities and industrial consumers.

The MR&D Activity provides the technology for producing the nation's most plentiful fuel by satisfying four basic requirements:

- Improve the economics of production through increasing capital and labor productivity by developing more efficient cutting and haulage systems and mine development methods; by increasing automation; by adopting labor-saving techniques; and by increasing machine reliability
- Maximize resource recovery by improving the efficiency of pillar recovery in room-and-pillar operations; by increasing the effectiveness of panel mining systems; and by developing techniques for

the mining of thick, multiple, and pitching coal seams
- Protect the health and safety of mineworkers by providing remotely operated systems and automated roof support systems; by implementing improved ventilation and degasification techniques; and by providing improved monitoring/warning systems
- Protect the environment by improving the quality of coal preparation; by controlling subsidence; by reducing acid mine drainage; by minimizing the problems of waste disposal; by improving over-burden handling; and by developing improved, cost-effective restoration techniques.

Funding levels by subactivity for the FY 1978 to FY 1980 period are listed in Table II 3-2.

Throughout the following discussions of the MR&D Activity, both brief descriptions and current status assessments are presented for the individual tasks that comprise the Underground Coal Mining, Surface Coal Mining, and Coal Preparation Research Subactivities.

Underground Coal Mining

Underground coal mining tasks can be divided into the following four general areas:

- Mine planning and development
- Production mining
- Transport and other services
- Studies and support

Mine planning is the key to efficient coal production. Adequate preparation at this stage can assure economical production and minimize undesirable side effects, not only during mining, but also long after mining has ceased. Proper planning can lead to the minimization of adverse socioeconomic impacts; selection of the right mining/preparation method; minimization of inflows of gas and water; minimization of environmental impacts associated with mining; and maximization of resource recovery. Coal mine development, both shaft and in-mine, involves the cutting of access openings necessary to allow a production mining system to be applied. These access openings provide for ventilation, man and material transportation, and coal transportation. The speed of mine development determines when the mine reaches full production. Coal costs are decreased by quicker development since the time that the initial invested capital sits idle is reduced.

At the present time the majority of underground coal production in the United States results from two production methods, room-and-pillar mining and panel mining. Room-and-pillar, the source of 95 percent of our

Table II-3-2. Funding levels for MR&D subactivities for FY 1978 - 1980.

MINING RESEARCH AND DEVELOPMENT	BUDGET AUTHORITY (DOLLARS IN THOUSANDS)		
SUBACTIVITIES	ACTUAL FY 1978	APPROPRIATION FY 1979	ESTIMATE FY 1980
Underground Coal Mining	$46,285	$53,430	$38,000
Surface Coal Mining	11,475	12,615	8,000
Coal Preparation and Analysis	5,020	9,871	14,150
Capital Equipment	0	220	200
Construction — Mining and energy related technologies education central/coal utilization facility	0	570	0
TOTAL	$62,780	$75,836	$60,350

underground coal, is by far the predominant method. Panel mining, the source of the remaining 5 percent, offers the potential for increased productivity, worker safety, and decreased environmental impacts, and is therefore being given R&D focus in the program. In addition to concentrating on the two current dominant methods, efforts aimed at developing and evaluating novel mining systems are also underway.

Once the initial mining operation begins, effective haulage and other supporting systems are needed. Haulage systems must, for example, move men, supplies, and mined coal safely, rapidly, and economically between the mine working section and the surface. Haulage can be a major bottleneck to efficient production.

The studies and support task provides engineering services and program and cost management needed to ensure a balanced program. This task also includes studies and seminars concerning advanced technologies and the operation of the Surface Test Facility. This facility will allow testing of equipment and systems under simulated mining conditions.

Mining Planning and Development

Mine Planning. Mine planning has become an important phase to the overall development of an underground coal mine. Major concerns of mine planning are the prediction and control of hazardous gases such as methane, prediction and control of subsidence, and mine waste disposal.

In FY 1979, data was collected and analyzed to develop a methodology which will allow for the prediction and control of strata gases (methane) and fluids. Based upon these analyses, techniques and equipment presently used to control these phenomena will be optimized. This effort is in addition to the ongoing methane drainage program, which encompasses vertical, directional, and horizontal drilling for methane degasification.

A major problem with underground mining is subsidence. Subsidence is not only hazardous to a miner's health and safety, but may also be environmentally unacceptable. To guarantee that all the alternatives for subsidence prediction and control have been investigated, a review and assessment of foreign technology was conducted in FY 1979. Based upon these assessments, development of an empirical prediction model for subsidence was also initiated in FY 1979. In addition to developing the formula, a monitoring system was initiated. Once the reliability of the formula and monitoring system has been demonstrated, procedures to minimize the damage caused by subsidence will be perfected. A model to identify the conditions related to subsidence and the development of procedures to counteract subsidence will be developed.

Mine waste disposal has been another problem area which is amenable to planning. Waste disposal can impede a mine's progress and increase the environmental degradation associated with the mine. In addition to waste disposal techniques, recycling methods will also be investigated. In FY 1979, the various properties of mine waste were identified and potential uses for the waste are still under evaluation. Beginning in FY 1980, further investigations will be conducted concerning waste utilization. Also, various underground disposal procedures will be developed for the unusable mine waste.

Effluents from the mine are another concern of mine planning. During FY 1979 and 1980, various mechanisms were developed which will identify potential sources of water pollution within the mine. Also, mine closure procedures, to minimize the leaching of these pollutants, will be developed. Beginning in FY 1980, these findings will be incorporated into the development of procedures for effective waste management. Once the procedures have been determined, in-mine pollution controls will be developed.

Major milestones for the Underground Coal Mining Subactivity are given in Figure II 3-1.

Shaft Development. A key element in underground mining is the preparatory excavation required to construct ventilation shafts, man and material access shafts, and in-mine roadways (entries) requisite for coal production. These mine development activities affect the time required to open or expand mines and thus, the cost of coal.

Activity Description—Coal 87

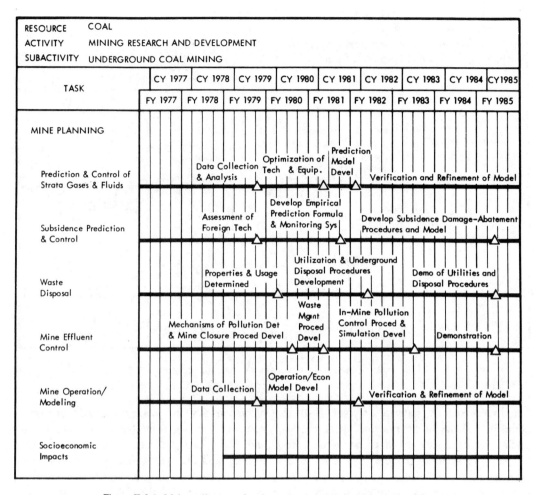

Figure II-3-1. Major milestones for the underground coal mining subactivity.

The present method of coal mine shaft sinking, which only achieves an average rate of 40 feet per month, uses a drill-blast-muck cycle. In this labor-intensive operation, crews must constantly be transported in and out of the shaft to accommodate blasting, they must remain idle as the shaft is cleared of dust and fumes, and a number of precautionary measures must be taken to ensure health and safety. In addition, the limited pool of skilled labor for conventional shaft sinking contributes to the method's high cost.

A program for the development of mechanized shaft-sinking systems to reduce the time and cost associated with current methods was initiated in 1975. This effort emphasizes development of (1) a down-hole-powered (power supply at the cutting head) system for construction of man and material shafts, exceeding 20 feet in diameter, commensurate with the needs of major new mines; and (2) a surface-powered system for construction of shafts up to 20 feet in diameter (a practical upper limit for such systems) for ventilation needs and lesser man and material requirements. Both systems have the design capability to sink the shafts blind; i.e., without having to depend on previous development in the mine itself.

Blind shaft borer. The blind shaft borer (BSB) is a system with the design capability of sinking greater than 20-foot-diameter shafts to depths of 2,000 feet, the minimum required to accommodate projected production capacities and increased depth of reserves. Fabrication of the BSB and related support systems is in process. The major BSB support systems

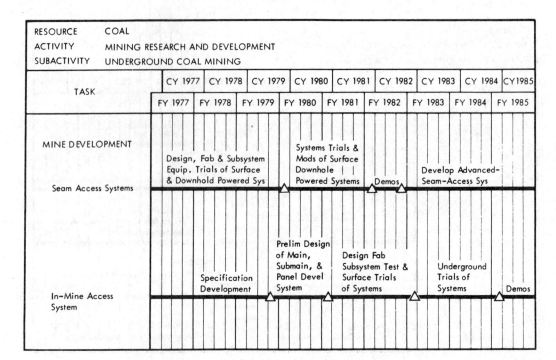

Figure II-3-2. Major milestones for mine development.

are the secondary haulage (top of BSB to surface) and the lining systems (presently 12-inch-thick concrete). Even using state-of-the-art support systems for the first field trials, shaft construction rates of 36 feet per day are projected, an order of magnitude improvement over current rates.

The first equipment trials incorporating state-of-the-art support systems were begun during FY 1979. The construction of an 1,150-foot deep shaft at the Oakgrove Mine in Alabama will be completed in FY 1980. This will be followed by an evaluation and redesign phase. A second trial period during FY 1980 incorporating newly designed support systems will concentrate on overall system performance and compatibility of support systems.

Rotary drilling system. In FY 1979, fabrication continued on a surface-powered rotary drilling system capable of excavating up to 20-foot-diameter ventilation shafts and remotely placing a 12-inch-thick concrete lining. This method uses scaled-up oil well drilling technology where men stay safely on the surface during the drilling operation. The lining placement system, which is part of this development, borrows from the civil construction and electronics industries, combining slip-forming and remote sensing and controls. The rate of excavation is expected to be comparable to that of the BSB system. Trials will begin in FY 1980.

In-mine access. Development of new areas for coal production typically involves the use of a continuous miner advancing a heading in sets of five roadways (entries). Ten to fifteen entries are generally required on a main heading. The large number of entries that are driven is largely due to ventilation and haulage demand and restrictions on entry dimensions. Continuous miners are not designed to cut roof or floor rocks; therefore, entry height is limited to coal seam thickness, and width is limited by roof stability. In addition, crosscuts between entries must be driven at intervals of 100 feet or less to meet current regulations (Refer to Figure II-3-2).

The large number of entries and cross-cuts required, combined with roof support and ventilation considerations, dictates that the majority of time be spent in moving the continuous miner from one location to another and in preparing to cut coal. This is particularly true in development operations, because the coal pillars left between entries must be large in order to provide support for the life of the mine. The result is an extremely slow rate of development.

Despite the relatively large pillars left, ground control problems persist because of the number of entries involved. Continuous maintenance is required, therefore, and disruption of services results.

Research activities to alleviate the in-mine development problems have been done in part under the production-related research; i.e., the development of a better continuous miner and the integration of bolting. Meanwhile, in the development area, other basic equipment, technology, and approaches have been examined.

Evaluations to date have focused on the potential of three basic applications: (1) slope access from the surface and main entry development (tunnel boring); (2) submain and panel development involving both the cutting of coal and roof and/or floor strata (mixed face); and (3) submain and panel development restricted to the coal seam (in-seam).

Tunnel Boring System

As a step toward alleviating development problems, a cost-sharing contract was undertaken to examine the technical and economic feasibility of applying tunnel boring machines (TBM's) to mine development.

TBM's (shown in Figure II 3-3) came into their own in the early 1970s in the construction industry, finding wide use in the construction of transportation tunnels, sewer systems, and other underground operations involving large diameter openings in both hard and soft strata.

TBM's show potential for mine development because of the possible speed of excavation, often hundreds of feet per day in civil works projects; the stability of the circular

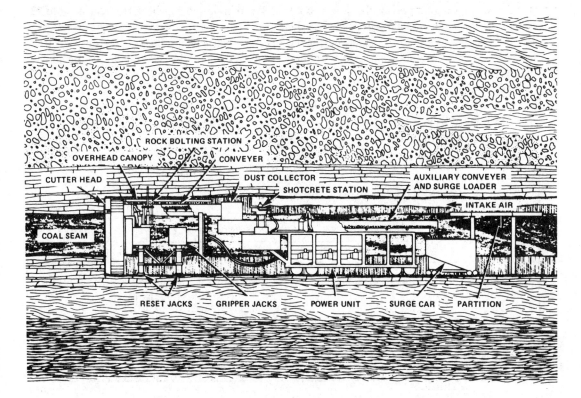

Figure II-3-3. Tunnel boring machine.

opening; and the fact that large openings could be driven independent of strata considerations. As to the latter, it was determined that, from a ventilation standpoint, the 18-foot-diameter TBM chosen for the cost-sharing project provided the equivalent of five entries in the particular mine involved. Stability of opening is important from a maintenance (and safety) standpoint, and speed is essential in addressing the basic development problem.

Under this joint effort, a 6,000-foot-long tunnel is being bored off the East Mains of the Federal No. 2 Mine to connect existing mine workings with an air shaft and to open the area for longwall mining. Technical and economic feasibility is being evaluated in view of the constraints relative to the hazardous environment and requirements imposed by Federal and State mining laws. Another major objective is to obtain the data requisite to developing specifications for a TBM system optimally designed for the coal industry.

The system has now operated safely for about 4,000 feet and remains in an operational state in a region previously shown to be too gassy for conventional development. In FY 1979, development of an advanced tunnel boring system was initiated, incorporating data from previous efforts.

Once access to the seam has been attained, the process of accessing areas located further into the mine begins. In-mine roadways must be driven to provide conduits for ventilation as well as men and material passages that are needed to set up the many production sections. Since this is a very slow process, specifications for more rapid in-mine access systems were developed during FY 1979. This effort includes the development systems. In FY 1981, the improved systems will enter final design and fabrication (Figure II-3-4).

Production Mining.

Room-and-pillar mining. The initial production stage of a room-and-pillar operation involves the driving of a set of parallel interconnected entries (tunnels). Mining is accomplished by the cyclic performance of operations in cutting, loading, and roof support. The interconnection of the entries is necessitated by the requirement for closed loops for ventilation and produces a checkerboard pattern of pillars. The second production stage of the room-and-pillar mining method involves partial mining of the remaining pillars of coal.

The continuous miners used in room-and-pillar mining are capable of producing 6 to 12 tons per minute. However, the typical average is only half that production for only 20 percent of the shift because of the nature of the method as currently practiced. The thrust of the room-and-pillar effort is to develop systems that will permit more truly continuous mining throughout the shift. The initial goal is

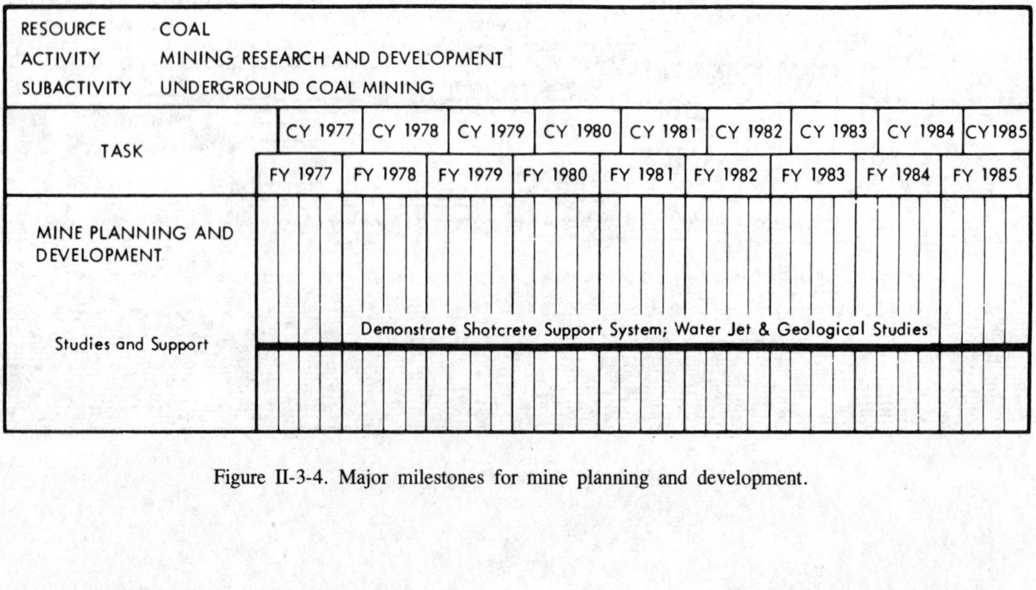

RESOURCE	COAL									
ACTIVITY	MINING RESEARCH AND DEVELOPMENT									
SUBACTIVITY	UNDERGROUND COAL MINING									
TASK		CY 1977	CY 1978	CY 1979	CY 1980	CY 1981	CY 1982	CY 1983	CY 1984	CY 1985
		FY 1977	FY 1978	FY 1979	FY 1980	FY 1981	FY 1982	FY 1983	FY 1984	FY 1985
MINE PLANNING AND DEVELOPMENT										
Studies and Support				Demonstrate Shotcrete Support System; Water Jet & Geological Studies						

Figure II-3-4. Major milestones for mine planning and development.

to reduce lost production time through increased system reliability and improved combinations/cycling of machine functions. Once this is accomplished, emphasis will shift to raising the average production rate closer to the potential of the basic machinery.

Major milestones for room-and-pillar mining are presented in Figure II 3-5.

Continuous miner/bolters. The continuous miner/bolter effort, an integral part of room-and-pillar mining, deals with better permanent and temporary roof support methods. Focusing on roof support constraints, two parallel approaches are being undertaken: total system development and improved component development. These systems use continuous miners to which a roof bolting capability has been added (miner/bolters). (See Figure II 3-6.) Laboratory testing is underway, and underground testing was completed in FY 1979. Based upon the initial results, designs of other new systems will begin in FY 1980. These new systems will begin preliminary testing also in FY 1980.

Meanwhile, major improvements are being made in the roof support components used on these new complex systems. The component development work includes contracts for longer-than-seam height drills, high-speed water jet drills, and bolter modules. Each of these elements is being perfected through a series of laboratory and underground tests by which design and reliability problems can be isolated and solved. This parallel effort allows a methodical development plan for the com-

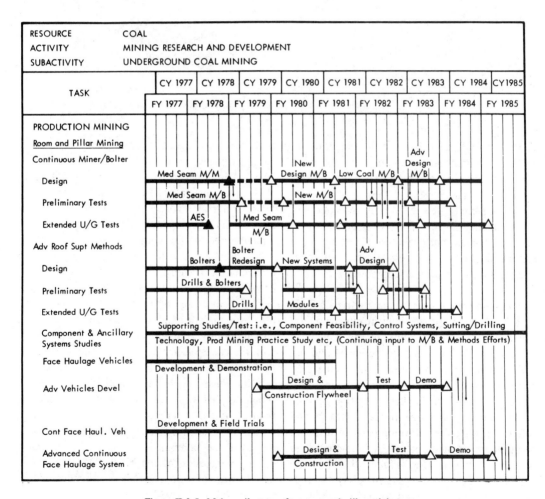

Figure II-3-5. Major milestones for room-and-pillar mining.

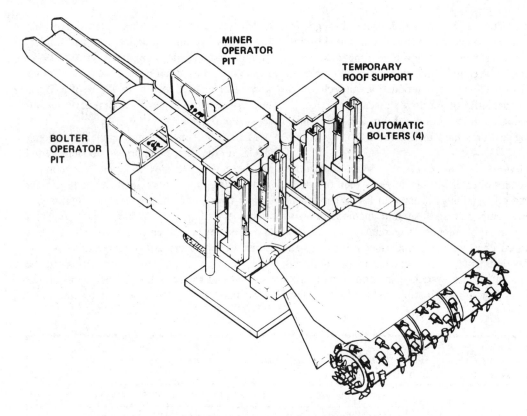

Figure II-3-6. Artist concept of a miner/bolter.

ponents simultaneously with investigation of system problems on the miner/bolters. As the components reach the high reliability needed for new complex systems, they will be combined to form a new advanced generation of miner/bolters. During FY 1979, development of components and reliability testing were continued. During FY 1980, tests will continue on these improved components.

Remote-controlled continuous miner. Even with hardware improvements in miner/bolters, actual production falls short of the machine's potential. This gap between potential and actual production is primarily due to machine operator variability. To eliminate decision-making constraints in the operation of the miner and to maximize operator health and safety, miner/bolter development efforts are being concentrated on a remote-controlled continuous miner. (See Figure II 3-7.) The automated extraction system (AES), a full-face programmable continuous miner with roof bolting capability and self-advancing ventilation, was tested underground during FY 1979. Additional AES component needs include coal interface detectors, guidance systems, and maintenance management systems. Roof support component development needs include studies for longer-than-seam height drills, high-speed water jet drills, bolter modules control systems, and work to improve component reliability. All of these elements are being perfected through a series of laboratory and underground tests by which design and reliability problems can be isolated and solved. This parallel effort will allow a methodical development plan for the components simultaneously with investigation of system problems for roof supports and the miner/bolters. As the components reach the high reliability needed for new complex systems, they will be combined to form a new, advanced generation of machines. Development and testing of this ancillary equipment will continue into FY 1980.

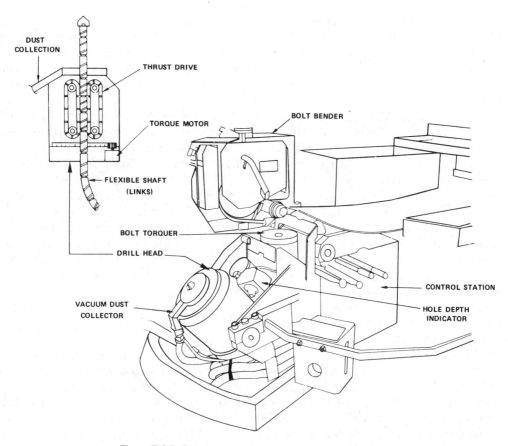

Figure II-3-7. Remote operated, longer-than-seam-height drill.

Face haulage vehicles. In the United States, nearly 90 percent of underground room-and-pillar mines use shuttle cars to transport mined coal from the working face to a secondary haulage system. The majority of these units are electric-powered and, to transmit power to the vehicle, use a trailing cable from the reel on the shuttle car to a stationary tie-off point. This umbilical cord connection is a safety hazard, restricts underground travel, and limits the system to two operating vehicles. Application of a self-contained type of power source, eliminating the trailing cable restriction, would reduce costs, enhance miner safety, and permit use of more than two vehicles. Current battery-powered shuttle cars use large, heavy, lead-acid battery packs, which take up a significant portion of the shuttle car and leave less space available for a payload of coal. An iron-nickel battery and charger system, which is lighter and occupies less volume than current systems, is being developed for use in a coal mine hauler. Surface field trials have been started and upon completion will be followed by an underground test demonstration.

Another technology currently available to power shuttle cars is the energy accumulator and storage system of the flywheel. The flywheel can be recharged while the car is unloading and has the potential of producing three to five times the watt hours per pound available from lead-acid batteries. If development is successful, the flywheel-powered shuttle car will have inherent capabilities advantageous in the face haulage function. The powerpack will be smaller and lighter than existing self-contained units; it will have fast recharge and high density; and it will be fumeless, clean, safe, and reliable. Evaluation of a high-strength, low-mass flywheel in tandem with a hydraulic or electric drive system is currently in process. This will be followed by design and fabrication through FY 1979 and FY 1980, then surface test and demonstration.

Continuous face haulage. The introduction of continuous mining machines prompted the mining industry to reassess shuttle car batch-type haulage. Between location changes, the continuous miner can cut and load coal on a nearly continuous basis; however, after loading a shuttle car, the high-capacity machine must wait while the second shuttle car is spotted under the miner tailboom. Delays in the haulage system, usually due to waiting for shuttle cars, take 25 percent of the continuous miner cycle-time. Computer simulations of a face operation indicate that up to a 45 percent increase in coal production could be realized through the use of a continuous haulage system associated with a secondary haulage system to handle the increased load.

Several types of continuous haulage systems are under study and development. These include a floor-mounted automated bridge conveyor train, a monorail-mounted bridge conveyor system, extendable belt bridge conveyors, a flexible belt conveyor train, coal injectors for hydraulic pipeline transport of coal from the mine face, and a multiple-unit continuous haulage system. The latter is a series of self-propelled, four-wheeled steerable cars with conveyors mounted in such a way that one conveyor discharges onto a secondary conveyor in a continuous train. Underground field trials are continuing during FY 1979 and FY 1980. Fabrication and surface testing of the monorail bridge conveyor, automated bridge conveyor train, and the coal injectors for hydraulic transportation of coal were completed in FY 1979. Underground field trials will commence in FY 1980. In addition, design and fabrication of an advanced continuous face haulage system will commence in FY 1980.

Panel Mining

Panel mining is a high production approach to coal mining that is widely used in Europe but has found only recent acceptance in the United States. To date, this method accounts for only 4 percent of the coal mined by underground methods. Basically, large rectangular blocks of coal are defined and extracted by successively slicing one of the sides. For example, the predominant method now used in the United States is retreat longwall panel mining. The complete block of coal to be mined is usually 300 to 500 feet wide and 3,000 to 6000 feet long. Up to 4000 tons per shift have been extracted from a 7-foot seam. Typically, however, only 700 to 900 tons per shift are realized because of equipment problems and disruptive geological conditions. High capital requirement for the equipment is another constraint to the use of longwall mining.

Major milestones are given in Figure II 3-8.

High production retreat panel mining. High production retreat panel mining is aimed at improving the performance of the basic longwall system. Development of advanced technology components were initiated in FY 1979. In FY 1980, the production mine for the advance technology face (ATF) will begin development. During this period, the production mine will become available for trials of improved components such as shearers, conveyors, and longwall roof supports. Additionally, in FY 1980 a cost-benefit study on the use of wider shearer drums will be completed. This study will result in the procurement of equipment in FY 1980 for the underground trials of a wide web extraction technique, which is scheduled to commence in FY 1981. Since health and safety problems are expected to worsen with increased production, a study is planned in FY 1980 to determine the production constraints associated with dust and methane. Also, conceptual designs of equipment and techniques to minimize these problems will be initiated. Studies to increase the shearer haulage speed will result in the design of prototype hardware in FY 1980. Component studies in the area of innovative cutter loader systems and conveyor systems will be undertaken in FY 1980 with subsequent pro-

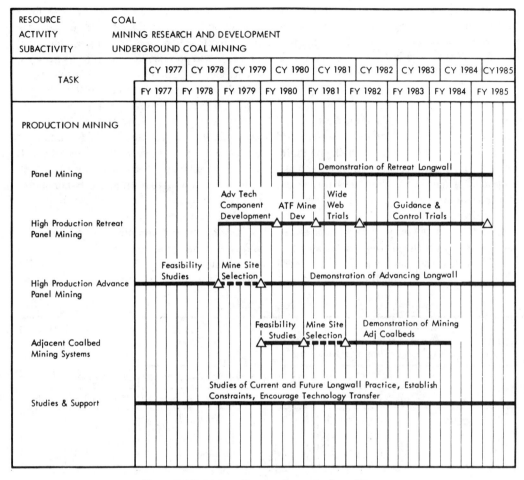

Figure II-3-8. Major milestones for production mining.

curement and trials of prototype equipment. (See Figure II 3-8 for major milestones.)

High production advance panel mining. A high production advance panel mining effort is aimed at the application and improvement of advancing longwall technology in the United States. There are various legal and economic problems which must be considered in applying this technology. At the present time, feasibility and mine site studies are underway. In FY 1980, if the feasibility has been demonstrated, a request for proposal (RFP) will be issued for a demonstration. (See Figure II 3-8 for major milestones.)

Adjacent coal bed mining systems. Feasibility studies will be conducted on adjacent coal bed mining systems in FY 1980. Most U.S. experience has been with mining virgin coal beds with no previously mined areas above the seam. Considerable European experience in adjacent coal recovery has led to the wide application of longwall mining. The feasibility studies will assess the applicability of the existing technology to U.S. geologic conditions. (See Figure II 3-8 for major milestones.)

Panel mining support studies. Support studies will be conducted which will benefit the entire panel mining subtask. One such study, longwall system availability, is planned for FY 1980. This study will provide a high quality, statistically significant data base on the performance of longwall mines in the United States. Other studies include current and future longwall practices and constraints, assessment of new technology, and the identification and solution of specific geologic problems. (Major milestones are shown in Figure II 3-8.)

Novel systems. In the area of novel systems, attention is focused on the modification and application of existing mining systems to unique mining situations (i.e., thick, multiple, and steeply pitching seams (TMSP)) and development of advanced mining systems which offer economic advantages over current systems.

Since the vast majority of U.S. coal mined underground is produced from coal seams less than 10 feet thick, the coal industry techniques and equipment which have been developed are quite naturally better suited to these thinner seams. In mining thick seam coal (10 to 40 feet) with present techniques, there has to be a compromise in recovery, safety, or profit; therefore, good recovery is usually sacrificed. Recovery percentages below 30 percent might be expected when present room-and-pillar techniques are applied to a coal seam 20 or more feet thick; the remaining 70 percent or more of the coal in place is lost. The effort to find a better way to mine thick coal seams is important from the standpoint of increased production of coal and improved resource recovery.

The objective of the advanced mining systems effort in underground mining is to explore, develop, and evaluate innovative concepts for producing coal and reducing the

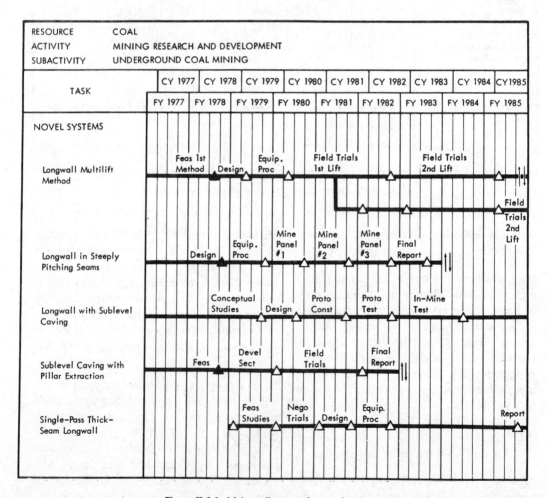

Figure II-3-9. Major milestones for novel systems.

cost per ton. Surveys of state-of-the-art, critical experiments, prototype development and testing, full-scale development and testing, and, finally, in-mine demonstrations are performed to prove each concept.

Major milestones for novel systems are shown in Figure II 3-9.

Longwall multilift method. During FY 1979, site selection was completed and a detailed mine design plan was finalized for the first U.S. application tests of the longwall multilift method for extracting thick underground coal. One method consists of working a panel of a longwall mining face starting at the top of the thick coal deposit and working down in horizontal slices. (See Figure II 3-10.) For example, in a 20-foot seam the first pass extracts the upper half of the seam and lays a screen mesh down on top of the remaining part of the seam. This screen acts as an artificial roof when the bottom half of the seam is mined. Resource recovery using the longwall multilift method will approach 60 percent, twice that of conventional room-and-pillar methods. During FY 1979, equipment was procured;

field trials of the first lift will commence in FY 1980.

Longwall in steeply pitching seam. Efforts will continue to develop methods to mine the large amounts of coal in steeply pitching seams, previously unrecoverable because the steep operating angle (greater than 25°) precludes use of continuous mining methods. The alternative of longwall mining will be tested in Colorado. Planned development of the first pitching seam panel for actual operations began in FY 1979. Mining of the first panel will occur during FY 1979 and FY 1980. Mining of the second panel will start in FY 1980.

Longwall in sublevel caving. Development, construction, and field testing of a prototype longwall sublevel caving technique will continue through FY 1981, the first application of such a system in the United States. The method consists of working a longwall face along the bottom of the thick seam and drawing or recovering the top coal from the rubble as it caves into the extracted cut (gob).

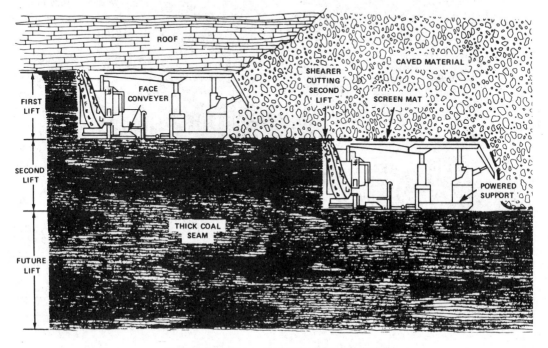

Figure II-3-10. Longwall multilift mining method.

Gobbed coal is recovered on a conveyor either by gravitational flow or mechanically stimulated methods. (See Figure II 3-11.) Projected resource recovery for this method will approach 80 to 85 percent.

Sublevel caving with pillar extraction. The sublevel caving with pillar extraction (SCPE) method is a modification of the normal method of pillar mining, with the addition of top coal recovery. Mine openings will be driven against the bottom of the seam using room-and-pillar techniques. Top coal recovery is accomplished on retreat by drilling and blasting the top coal in increments and loading out the fallen top coal with a continuous miner.

In FY 1979, an SCPE technique was tested for process verification. Field trials will be initiated in FY 1980 and completed in 1982. SCPE uses the same equipment and methods that are used in room-and-pillar continuous mining operations. The method is designed to provide for a 75-percent resource recovery rate of coal and is applicable where erratic seam thickness occurs.

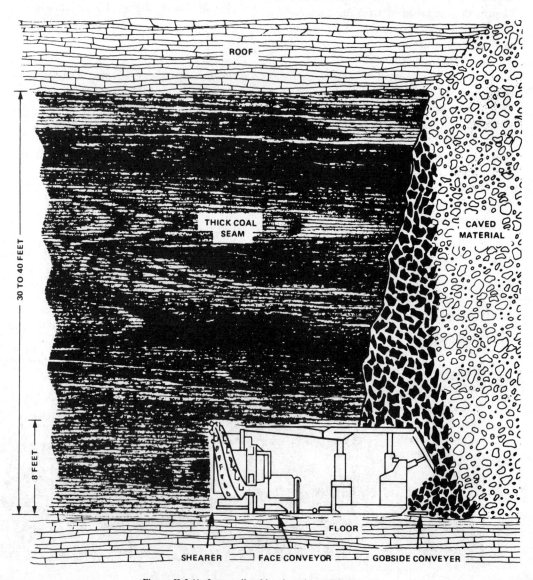

Figure II-3-11. Longwall sublevel caving mining method.

Advanced Mining Systems

Efforts in high-volume hydraulic mining address the problems of recovering coal that lies in moderate to thick seams (equal to or greater than 6 feet) having pitch angles greater than 4 degrees. Most technology for hydraulic mining has been used in other countries. The purpose of this effort is to explore the technique's applicability to U.S. conditions. The design phase for a practical system will be followed by construction of a prototype hydraulic coal mining system and a field demonstration.

Development work is underway on high-energy, low-volume hydraulic jet augmentation of mechanical mining systems. Use of cavitating, percussive, and continuous water jets is being investigated. This effort is focusing on developing the best combination of a low-volume jetting system with a boom-type miner to determine how much the jet increases production in coal and rock.

A variable wall miner being developed for panel mining uses a side cutter auger to cut and convey the coal, all with a single implement. A bread-board prototype was designed, constructed, and tested in the Surface Test Facility in FY 1979. Testing will commence in FY 1980.

A borehole mining system is being evaluated to provide a concept of an underground system operated at the surface. The prototype apparatus operates on a 16-inch diameter borehole drilled through the coal seam. A high-pressure water line, nozzle, and slurry pump combine to erode the coal, form a slurry, and pump the slurry to the surface. To date, coal has been produced at a rate of 10 tons/hour from a depth of 90 feet in a pitching coal seam. During FY 1979, a full-scale borehole mining system was designed and constructed; tests were begun in FY 1980.

Transport and Other Services. Improved haulage systems to effectively move men, supplies, and mined coal between the mine working section and the surface are of utmost importance in overall development of a cost-effective underground coal mining system. The major project being studied is the automated rail haulage system. Other development and testing efforts include a low coal conveyor clearance system and a hydraulic transport system. Figure II 3-12 shows major milestones for this area.

Automated rail haulage. An automated rail haulage system uses unit trains which circulate within the mine, without operator assistance, from various loading sites to the unloading point and back to the same or alternative site. The system requires a locomotive designed specifically for automation, with controls and sensors in the entire system necessary for unmanned operation. The overall system eliminates workers from the most hazardous portions of the rail haulage operation.

The automated rail haulage system is expected to reduce rail haulage costs 7 to 12 percent, to increase overall in-mine productivity 3 to 6.5 percent, and to reduce haulage accidents by 50 percent. Design and fabrication of components of the automated rail haulage system were completed during FY 1979. A production prototype will undergo surface testing in FY 1980.

Studies and Support

Surface test facility. Surface testing of new and modified underground coal mining equipment is currently limited, in most cases, to functional checkout of major systems with no load applied. As a result, the first real tests of equipment performance under load are now conducted in underground production operations. The Surface Test Facility will provide capability to conduct and evaluate research and development efforts above ground on mine equipment and systems in a simulated underground mining environment. This will expedite the introducton of safer, more efficient technology in the underground mining operation. The test facility was operational during FY 1979. Planned typical tasks for the facility include:

- The longwall cutter/loader will be evaluated for pitch, roll, yaw, and haulage parameters.
- The continuous miner will be evaluated

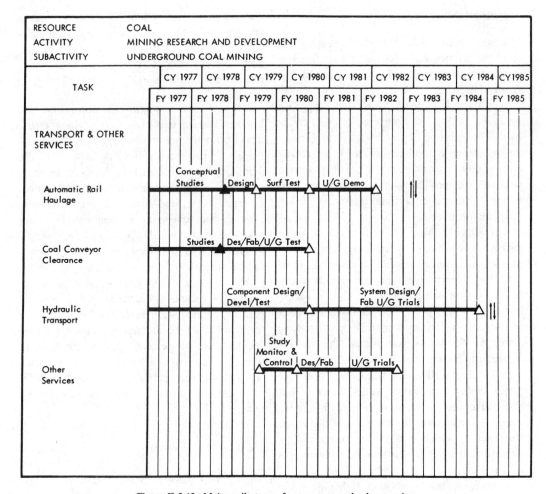

Figure II-3-12. Major milestones for transport and other services.

for stability of controls and for reliability and effectiveness of the interface detectors and guidance control components.
- The variable wall miner will be evaluated to determine its cutting and conveying capability.

Major milestones for the Surface Test Facility are shown in Figure II 3-13.

Surface Coal Mining

Coal production from surface mines has steadily increased during the past decade. It appears that this trend will continue in the immediate future because surface mining production costs are less than underground mining costs. In comparison to underground mining, surface mining also has the added advantages of a generally higher resource recovery rate and less health and safety problems. But even with these advantages, there are some limitations: the economics of production, with presently available surface mining equipment, are very sensitive to the ratio of overburden thickness to seam thickness; there are practical upper limits to the absolute depths of single pass mining; and as practiced in the past, surface mining has negatively affected the nation's environmental quality.

In a general sense, there are three types of problems facing surface mining. First, there is a need for better equipment and systems to extend the capabilities of surface mining. Past improvements have been in the direction of ever larger equipment, and there are strong

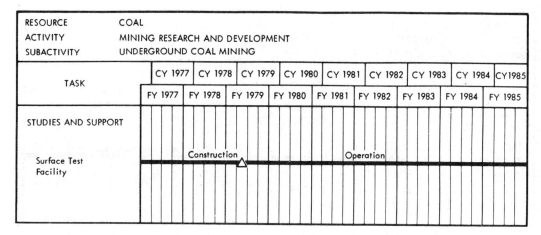

Figure II-3-13. Major milestones for the surface test facility.

indications that this approach is reaching, or has reached, its limits. Second, there is a need for new surface mining equipment that will meet environmental protection regulations at the lowest possible cost. This is necessary to avoid a significant productivity decline similar to the underground coal mining productivity decay which followed the passage of health and safety legislation. Third, there is a need for a new technology which, by design, assures environmental and worker protection and improved efficiency. To meet these needs, modifications to conventional mining equipment and new technologies must be developed.

In order to address the RD&D requirements of surface coal mining, the following program structure has been developed:

- Mine Planning and Development (MP&D)
- Production Mining (PM)
- Studies and Support

Mine Planning and Development. Federal and State surface mining regulations have precipitated an increasing awareness of mine planning. In order to conform with statutory environmental requirements, and to increase productivity within these constraints, a systematic approach for removing the overburden, mining the coal, and reclaiming the land is needed. The mine planning and development task needs to emphasize improvements in equipment combinations and in design of layout and haulage configurations. To ensure that proper procedures are instituted or standardized, studies that reviewed the various mine planning and development techniques were conducted in FY 1978. Based upon the results of these studies, design criteria for a surface mine site are being developed, and published for industry starting in FY 1980. Major milestones for mine planning and development are presented in Figure II 3-14.

Production Mining

Open cast mining. Open cast mining is a general term used to describe surface mining methods employed in the interior and Western coal provinces. This method generally involves developing an initial box cut through the overburden to expose the coal seam. The coal is extracted and a second cut is made through the overburden. As each succeeding cut is made, the overburden or spoil is directly or indirectly cast by the primary stripping (e.g., dragline, shovel, etc.) unit into the cut previously excavated.

Open cast mining methods can be grouped into two categories—area mining and modified open pit mining. Area mining is characterized by the block-wise stripping of the coal seam and by a single bench. Modified open pit mining refers to the removal of over-

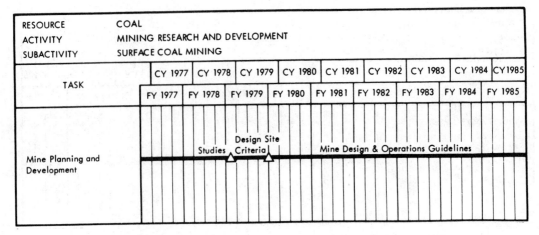

Figure II-3-14. Major milestones for mine planning and development.

burden/coal in a series of benches that are oriented in a geometric manner.

Area mining is used primarily for mining surface coal reserves west of the Mississippi River. Even though geographic and geologic conditions are diverse, two basic excavation procedures are used: dragline casting, and shovel and truck mining. Many mines use or plan to use draglines with bucket capacities of 60 to 75 cubic yards and booms of about 300 feet. While draglines of this size are capable of digging overburden depths of 150 to 200 feet in one lift, their effectiveness is constrained by dumping radius limits and the resultant inability to cast a substantial proportion of the material without rehandling. Present dragline methods require approximately 10 to 40 percent of the overburden to be rehandled, which adds substantially to the cost. These dragline casting limitations have led to the selection of shovel and haul trucks for mining of coal and stripping of overburden benches in deeper mines. Although these truck and shovel operations provide a flexible mining system, overburden handling can be three to four times more costly than dragline casting. Although dozers and scrapers are sometimes used in surface operations, deeper surface mines generally cannot use this equipment combination because of high operating costs. In FY 1978, analyses of conventional mining equipment effectiveness in handling overburden were conducted. Limitations of existing equipment were determined by this analysis. In FY 1979, to overcome these limitations, fabrication of new and modified systems was begun. Initial testing of the systems will commence in FY 1980.

The casting limitations and high rehandle percentages associated with dragline area mining systems have prompted the development of modified open pit mining methods. A modified open pit mine configuration consists of a geometric sequence of benches for excavation, haulage, and spoil disposition. The modified open pit concept permits total, direct placement of overburden a dragline operation. The equipment utilized by a modified open pit system typically consists of mobil units for overburden removal, spoil handling, materials, thereby eliminating the spoil rehandling problems associated with coal extraction, and haulage. Since this type of equipment is readily available, a new mine can be brought into production in a relatively short period of time as compared to the long lead time required for the construction and delivery of a dragline.

Continuous mining systems need to be developed for both overburden removal and coal excavation in the modified open pit. These systems will increase productivity by eliminating the cyclic nature of the excavation process. Depending upon the nature of the primary machine configuration involved, these continuous mining machines can be classified into four categories: (1) boom-type bucket wheel excavators (BWE); (2) integral-wheel bucket wheel excavators; (3) drum shearer and auger-cutter type excavators; and (4) vertical arm belt loading excavators. Several of the machines could be used for both coal and overburden extraction. Systems development and procedures were investigated and developed beginning in FY 1979. Near the end of FY 1980, the most promising systems will enter detailed design and fabrication. Once the fabrication has been completed, field trials and demonstrations will start in FY 1982.

Major milestones are pictured in Figure II 3-15.

Contour mining. Contour mining is practiced in the mountainous, hilly terrain of Appalachia. The two basic methods employed are haulback mining and mountaintop removal. In haulback mining, overburden removal commences at an outcrop of coal and proceeds around the contour of the mountain until the economic stripping limit is reached. The drilled and blasted overburden is removed primarily by front-end loaders and dozers, placed in haulage vehicles, taken from the active pit area, and deposited in the pit where the coal has been removed. The mountaintop removal method allows for the economical recovery of coal located near the top of mountains, ridges, or knolls. This method generally involves making many cuts across a mountaintop. Although similar to other hilltop stripping processes, the slopes are generally steeper and topographic relief is greater. The mountaintop is not restored to its original contour, but rather it is flattened or restored to gently rolling hills. Additionally, construction equipment is often used instead of draglines for the removal of overburden.

The present method of haulback mining (Figure II 3-16) requires material back hauled by trucks, which is labor intensive. The installation of conveyors is needed to reduce the need for trucks, thereby reducing pit congestion and labor requirements. The main advantages in developing cost-effective conveyor systems are that cost per ton will be reduced, the mines will become safer, and reclamation practices in steep slope areas will be improved.

In order to comply with the reclamation

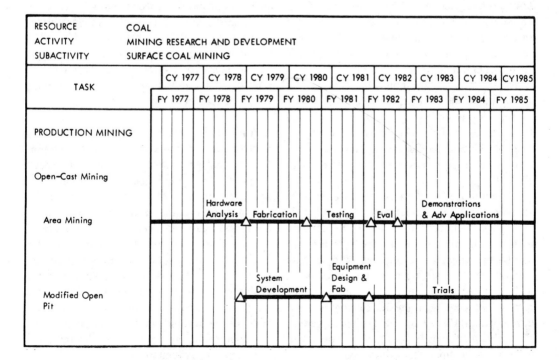

Figure II-3-15. Major milestones for open-cast mining.

Figure II-3-16. Longwall conveyor haulage.

laws promulgated by Federal and State Governments, haulback mining has been instituted in contour mining. The conventional haulback method depends greatly upon trucks. In FY 1978, design and fabrication of a low wall conveyor system was initiated. This system will improve the economics of contour mining by reducing truck haulage requirements. Field trials began in FY 1979, after the conveyor system had been fabricated. Evaluation of the conveyor's effectiveness will be conducted in FY 1981.

An additional method being developed to recover coal in mountainous terrain is crossridge mountaintop removal. In contrast to haulback mining—where the cuts proceed around the mountain at the coal outcrop, this method involves slicing a mountaintop like a loaf of bread. Mine designs for the efficient removal of coal near the mountaintops were developed in FY 1978. Field trials began in FY 1979. Additionally, designs for conveyors to be used in mountaintop removal are being studied. In FY 1979, conveyors entered equipment fabrication. The conveyors will be incorporated into the trials of crossridge mountaintop removal method in FY 1981. When the trials are completed, an evaluation of the conveyor's effectiveness will be conducted. Major milestones for contour mining are presented in Figure II 3-17.

Studies and Support. Today, spoil leveling and reclamation make up a substantial proportion of coal production costs. Prior to enactment of Federal and State reclamation laws, little, if anything, was done to reduce pollution and restore the land. Increasing public awareness and knowledge of the effects of mine pollution led to the enactment of the Surface Mining Control and Reclamation Act of 1977 (P.L. 95–87). To keep pace with the evolution of mining techniques, innovative reclamation methods must be developed. Without innovations in reclamation methods and equipment, the threat of accelerated costs and reduced production exists.

The mining industry's response has been to

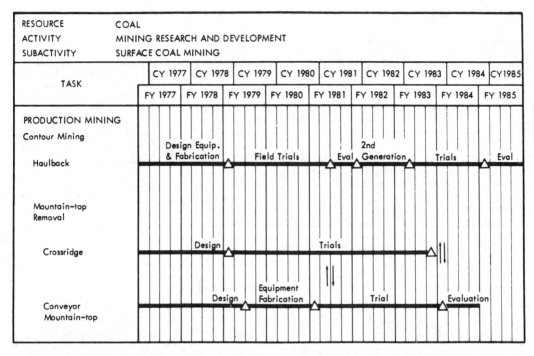

Figure II-3-17. Major milestones for contour mining.

rely on available equipment, dozers, scrapers, graders, front-end loaders, and in some instances, draglines to meet reclamation needs. This is an expensive compromise for many operations, since the equipment was not designed or developed to effectively handle the specific requirements of earth movement for contouring spoil piles. At the present time, an estimated 70 percent of central and western mines use tractor dozers for spoil leveling, with reclamation costs varying from $1,500 to over $20,000 per acre.

Thus, the primary emphasis in studies and support is in the area of reclamation studies. The studies being conducted include the hydrology and water quality of watersheds subjected to surface mining, and the adverse effects of noise, fugitive dust, and vibrations from blasting. Additionally, equipment to increase the effectiveness of land reclamation is currently being investigated. These studies will continue for an extended period of time. (See Figure II 3-18.) The techniques and equipment will be evaluated and improved continuously throughout the period.

Coal Preparation and Analysis

The primary function of a coal preparation facility is to deliver a finished coal product that has been custom prepared in terms of quality. With modern extraction techniques, the amount of impurities has increased, mandating that a greater percentage of coal be upgraded. Powerplants and other consumers are using furnaces and transport systems that demand very uniform feedstocks. In addition, fine-size coal must be prepared for shipment to avoid windage losses and freezing in the winter. As a result, coal preparation has become a very important element in encouraging the increased utilization of coal.

The purpose of coal preparation research is to support the increased use of coal in the United States by cleaning coal to meet environmental standards, metallurgical requirements, and feedstock specifications for both conventional combustion and advanced coal conversion processes. The Clean Air Act Amendments of 1977 mandate that large coal-fired industries and utilities meet the more

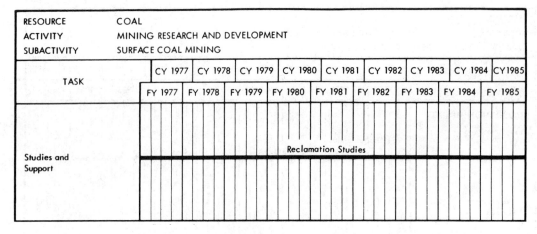

Figure II-3-18. Milestones for studies and support.

stringent sulfur oxides, particulates, and nitrogen oxides standards. The uncertainty concerning further promulgation of more stringent air quality and significant deterioration standards has inhibited industry from investing in coal preparation technologies. These standards can impede the use of coal if the cost of compliance is prohibitive.

Physical coal preparation describes the process of crushing, sizing, and cleaning the coal on the basis of its physical properties (e.g., specific gravity), which is typical of most of today's coal preparation for industrial and commercial users. Physical coal preparation (PCP) technology is used at most modern coal mines. There are about 400 such plants representing an investment of approximately $1 billion (replacement cost is about $4 billion). In addition to its traditional role of providing a high thermal value fuel at minimum cost, physical cleaning is being increasingly relied upon to remove sulfur from coal. Despite inherent limitations for PCP, studies have shown that physical cleaning, followed by stack gas cleaning to remove additional sulfur, is almost always cheaper than cleanup of the effluent gas alone.

Novel coal cleaning technology such as chemical coal cleaning (CCC) holds promise for meeting more stringent air quality regulations as well as providing a customized coal feedstock for a range of coal conversion technologies. Chemical cleaning can potentially remove a major portion of the total (inorganic and organic) sulfur in coals. These technologies also offer potential benefits for coal conversion. Unlike PCP, which produces a coal similar to the parent feedstock, CCC can be made to produce a product whose physical and chemical properties are modified from those of the feed. Conceptually, coking properties or chemical balances can be modified in the chemical reaction to produce the ideal feed for a given conversion process. There are presently no commercial chemical coal cleaning techniques, and much remains to be done to develop commercial technology, which is far more revolutionary than PCP technology.

To effectively develop optimal coal cleaning technologies, a number of other requirements must be addressed. These requirements include the characterization of coal properties; economic assessment of competing beneficiation processes; the measurement and characterization of trace element materials in coal; characterization of chemically bonded sulfur in coal; and on-line instrumentation for coal prep plant management.

Therefore, to systematically approach the requirements associated with coal preparation, the following program structure has been developed:

- Physical Coal Preparation
- Chemical Coal Cleaning
- Studies and Support

Physical Coal Preparation. Physical removal of coal impurities is the least expensive and most environmentally acceptable preparation technique. Since the technology is fairly well developed for coarse coal cleaning, the task for developing physical cleaning processes is directed principally toward upgrading coal finer than one-half inch.

High-gradient magnetic separation (HGMS) is a new physical cleaning technique that provides a practical means for separating small, weakly magnetic pyrite particles from coal. Oil agglomeration is a technique used to separate ash minerals from coal by selectively coating the coal particles with a thick oil film. The technical feasibility of both techniques needs to be demonstrated. Once technically demonstrated, the economics of the technique needs to be explored. Also, preliminary studies indicate that wet HGMS and oil agglomeration can be combined to produce an enhanced coal product. This combined technique needs further development.

Bench-scale experiments will be continued to establish the technical feasibility of wet HGMS. A 5-TPH evaluation unit was operated in FY 1979 to demonstrate the technical feasibility of oil agglomeration. Results will be evaluated in FY 1980. Based upon the evaluation, designs will be initiated on a 20-TPH evaluation unit that will combine wet HGMS and oil agglomeration processes into a single operation. This evaluation unit will be constructed in FY 1981, operated in FY 1982, and evaluated in the following year.

Dry HGMS equipment is also being evaluated. During FY 1980, an evaluation unit will be designed. Construction will begin in FY 1981 with operations commencing in FY 1982.

Froth flotation is a technique used to separate ash and pyrite from coal. Although it has widespread application in the coal preparation industry, problems persist with oxidized coal flotation, reagent feed control, and flotation quality control. A froth flotation evaluation unit, which was scheduled for construction in FY 1979 as part of the Bruceton coal preparation test facility, will commence operations in FY 1980. It is anticipated that the evaluation unit will operate for three years, examining both pyrite flotation and oxidized coal flotation. Special emphasis will be focused on partially oxidizing pyrite surfaces to enhance flotation response.

The burning of Western coals, particularly lignites, is hindered by a high sodium and water content. An evaluation unit, constructed in FY 1979, tested different lignite upgrading techniques in FY 1979 and FY 1980. A 10-TPH pilot plant will be designed in FY 1981 and constructed in FY 1982. An evaluation unit, using fiberglass vessels and tanks, will be constructed as part of the Bruceton facility in FY 1980 to test sodium removal techniques. If successful, a dryer will be designed in FY 1982 and a 5-TPH evaluation unit will be constructed in FY 1983.

Advanced gravity separation studies on the performance of cyclones, Richert cones, and spirals are also underway. The heavy density cyclone studies will focus on the use of magnetite as a medium. Evaluation of first studies was conducted in FY 1979, and these studies will continue in FY 1980 with dissemination of results to industry in FY 1981.

Major milestones are presented in Figure II 3-19.

Chemical Coal Cleaning. Longer term, high-risk applications for coal beneficiations need to be directed toward the development of new chemical desulfurization technologies. Laboratory tests have already shown that six basic processes remove substantially all of the pyrite sulfur, and varying percentages of the organic sulfur in raw coal. These processes, and variations to these processes, are listed in Table II 3-3.

A plan consisting of four major elements is necessary to develop the remaining chemical desulfurization processes for commercial operations. These are laboratory R&D, engineering R&D, pilot plant operations, and demonstration plant operations. The first two elements are being performed in parallel over the FY 1978–82 time period, utilizing DOE resources at PERC, Ames, Pittsburgh Mining Operations, and contractor laboratories. Preliminary laboratory evaluations have indicated that some of these processes may not be com-

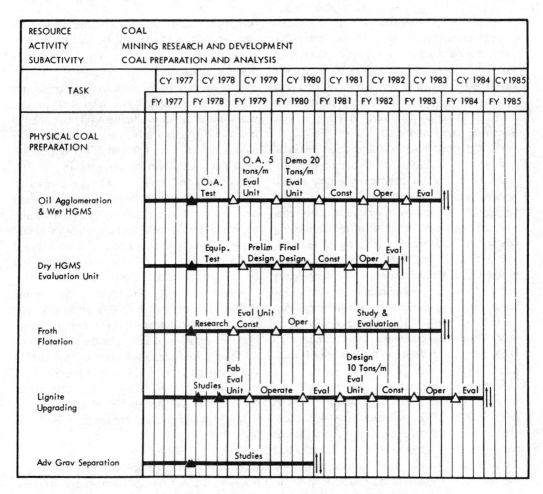

Figure II-3-19. Major milestones for coal preparation and analysis.

mercially feasible. A substantial portion of the additional R&D effort needs to be devoted to studies in laboratory batch and small-scale continuous flow units, and to engineering development of critical process subsystems. Pilot plant and commercial plant cost estimates also need to be performed.

Several processes utilize the oxydesulfurization approach (air or O_2/H_2O), some of which add a basic reagent to improve organic sulfur removal under less severe operating conditions. The oxydesulfurization approach shows a reasonable chance of commercialization, although the various processes require an in-depth evaluation of their economic feasibility. The oxydesulfurization research effort could result in a commercial operation within the FY 1986-87 time frame. Beginning in FY 1979, a study evaluating the three major oxydesulfurization alternatives was initiated. Based upon the results of this study, design of an oxydesulfurization alternatives will be initiated. Based upon the results of this study, design of an oxydesulfurization evaluation unit will be initiated in FY 1980. This unit will be operational for two years and will then be redesigned in FY 1983 or FY 1984. It is planned to operate the redesigned unit for an additional two years with the results then being transferred to industry.

Other processes, listed in the table above, which remove organic sulfur are unlikely to reach commercial operation before FY 1987 unless greatly accelerated through development or solution of those major technical and economic issues now retarding such develop-

Table II-3-3. Chemical coal cleaning processes and variations.

PROCESS	TREATMENT	% SULFUR REMOVAL PYRITIC	ORGANIC	DEVELOPMENT STATUS
1. Magnex	FE(CO$_5$) Magnetic	0-90+	0	1 TPD pilot units
2. KVB	NO$_x$/H$_2$O/NaOH	0-900-90+	0-40	Preliminary bench scale
3. Battelle	NaOH/CA(OH)$_2$/H$_2$O	0-90+	0-70	Small pilot unit
4. Meyers	FE(SO$_4$)$_3$/H$_2$O/O$_2$/Naphtha	0-90+	0	8 TPD pilot plant
5. Ledgemont	O$_2$/H$_2$O	0-90+	0	Bench-scale
	O$_2$/H$_2$O/NH$_3$	0-90+	0-25	Bench-scale
6. Bureau of Mines/ERDA	Air/H$_2$O	0-90+	0-40	1-4 lb/hr pilot unit
7. Ames	Air or O$_2$/H$_2$O/Na$_2$CO$_3$	0-90+	0-25+	Bench-scale
8. Chlorinolysis	Cl$_2$/H$_2$O	0-90+	20-70	Bench-scale
9. Mixed Leachant (Soviet Union)	Fe$_2$(SO$_4$)$_3$/HNO$_3$	0-90+	0	5-10 TPD pilot plant
10. General Electric	Microwave/Na$_2$CO$_3$	0-90+	0-70	Preliminary bench-scale

ment. Except for the chlorinolysis and KVB processes being developed, present knowledge does not justify such acceleration. These techniques were evaluated in FY 1979 and FY 1980. Designs for evaluation units will be initiated in FY 1981 with operations commencing in FY 1982. Modifications will be incorporated into the unit(s) in 1984.

Major milestones for chemical coal cleaning are given in Figure II 3-20.

Studies and Support. Coal cleaning processes are the heart of a coal preparation plant; however, many ancillary unit operations also occur in coal preparation. Such unit operations include size reduction, sizing, watering, pelletizing, and handling. With the exception of pelletizing, all of these unit operations are in common use today. Nonetheless, further developments in these areas are needed to reduce costs and abate pollution.

Fine-size coal is associated with advanced preparation processes. An evaluation unit to pelletize subbituminous coal will be constructed in FY 1980 to avoid windage losses from railroad cars. Studies will also be continued and results published on additives to prevent coal freeze-up in railroad cars.

Sizing studies will continue in FY 1980, concentrating on a wet environment. Sintering and coking studies will also continue. During FY 1980, this effort will consist of bench-scale experiments to develop a low-sulfur, low-ash fuel that can be burned in household and other small furnaces without the use of particulate and smoke suppressors. Designs will be developed in FY 1981 and an evaluation unit constructed in FY 1982 to demonstrate the technology to produce a high-quality smokeless fuel.

In FY 1980, a national survey and evaluation will be accelerated to locate existing ponds, determine their volume and coal content, and project the grade and potential amount of coal that can be recovered from each pond. Results of the survey will be published the following year, with an annual update thereafter.

Efforts are underway in controlling fine-size wastes. In FY 1980, efforts will be accelerated to develop a filtering process to obviate the need for tailing ponds. An evaluation unit using the filtering process will be designed in FY 1981 and constructed in FY 1982.

Techniques are being explored to recover fine coal already in sludge ponds. A report on the use of spirals was published in FY 1979. From FY 1980 through 1982, agglomeration techniques will be explored.

In FY 1980, coarse refuse material will be evaluated for Btu recovery potential, as well as for uses in the construction industry. Designs will be initiated in FY 1981 on a fluidized-bed combustor (FBC) to utilize the Btu content of wastes.

Additionally, a small study will be initiated in FY 1980 to determine the amount of water pollution resulting from gob piles, sludge ponds, and other wastes from cleaning plants. In FY 1981, the study will be expanded to

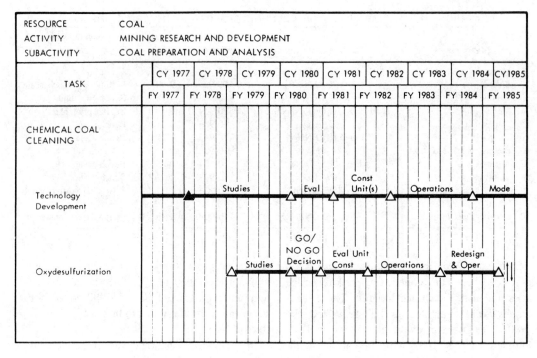

Figure II-3-20. Major milestones for chemical coal cleaning.

include methods of minimizing the discharge of spent reagents.

To better understand the characteristics and properties associated with the clean burning of coal, studies will be continued to identify and characterize coal material, carbon/sulfur compounds, and trace elements found in various coal seams in the United States. Washability studies will be continued until samples from most of the major seams have been tested. Long-term studies on the surface chemistry of coal, pyrite, and ash will also be continued.

Economic feasibility studies are also underway. During FY 1979, the economic feasibility of oxydesulfurization, HGMS, and oil agglomeration was analyzed; lignite drying will be assessed in FY 1980 and 1981. In FY 1982, the effort will be accelerated to assess the competing chemical desulfurization processes that are still under consideration. Additionally, ongoing efforts to determine the economic feasibility of cleaning in conjunction with flue gas desulfurization (FGD) and fluidized-bed combustors (FBC) are being accelerated. FGD results were published in FY 1980 and FBC results will be published in FY 1982.

In the area of instrumentation and automation, development was initiated in FY 1979 on prototype ash, sulfur, and water meters. In FY 1980, efforts will be concentrated on the design of on-line meters for coal preparation plants. The meters will be assembled in FY 1982 and installed and operated the following year. Additionally, a modest effort will be initiated in FY 1980 on applications of computer simulation in coal preparation facilities.

Major milestones for studies and support are shown in Figure II 3-21.

COAL LIQUEFACTION

The development of commercial coal liquefaction processes will provide a route, within environmental guidelines, to increase domestic production of clean fossil fuels and feedstocks. Synthetic fuels derived from coal liquefaction processes can replace petroleum-refined products most effectively in two distinct markets. One market uses low-ash, low-sulfur boiler fuels suitable for clean generation of electric power and industrial steam. The other market uses high-grade fuels such

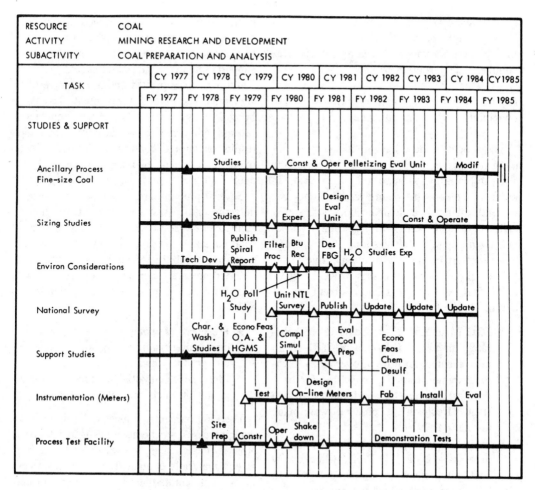

Figure II-3-21. Major milestones for studies and support.

as gasoline, heating oil, and chemical feedstocks.

The objective of the Liquefaction Program is to facilitate the establishment of a synthetic liquid fuels industry. Specifically, the program plans are to:

- Accelerate commercialization by defining and reducing the technical, institutional, socioeconomic, and environmental risks of the liquefaction process through technical development programs
- Develop and demonstrate, in partnership with industry, selected improved processes to convert U.S. coals into environmentally acceptable liquid fuels
- Perform laboratory studies and process development on advanced liquefaction processes

The general strategy of the Liquefaction Program is to support several liquefaction processes in parallel from laboratory scale, through process development unit (PDU). Only the most promising candidates will be selected for advancement to the pilot plant stage and only those processes identified as meeting commercial market needs will be considered for technical demonstration. This approach is based on a number of considerations. First, market requirements for a wide range of fuels produced from a variety of coals, coupled with the need for a large number of plants to produce significant quantities, may require demonstration of several liquefaction processes. This variety of processes is evident in today's crude-oil-based liquid fuels industry. By the same token, the future synthetic liquid fuels industry, which will replace a large share of this fuel market cannot rely on any single process configuration. Second, the processes under development provide complementary support to each other in many ways. Third, all processes under development require supporting research to improve overall operability, increase efficiency, and reduce costs.

The processes described in this section implement the liquefaction strategy of producing a clean-burning synthetic oil from coal, in order to release petroleum and natural gas presently being fired in steam boilers for other end uses. This clean-burning may subsequently be upgraded to gasoline, fuel oil, and other fuels if economically competitive.

Four liquefaction processes (SRC-I, SRC-II, H-Coal, and Exxon Donor Solvent) are in advanced development stages:

- SRC-I and SRC-II are being tested separately in a 6 T/D and a 50 T/D (coal) pilot plant, both now in operation. Technical demonstration plants (one for each process) are in the preliminary design stage. Decision to proceed was made during FY 1979.
- H-Coal is undergoing continued development in a 2-1/2 T/D PDU, while a 600 T/D pilot plant is under construction with operation scheduled for early FY 1980.
- The Exxon Donor Solvent (EDS) is undergoing continued development in a small PDU, while a 250 T/D pilot plant is under construction with operation scheduled to begin during the first half of FY 1980.

The Solvent Refined Coal (SRC) processes and EDS processes are included in the Solvent Extraction Subactivity of the Coal Liquefaction Program. The H-Coal process is included in the Direct Hydrogenation Subactivity.

Also included in the Coal Liquefaction Program are support research and development activities that provide backup for advanced liquefaction processes and improvements in liquefaction technology in general. In addition, indirect liquefaction (liquid fuels via syngas from coal) is under active evaluation and one or more process configurations may be selected for further development. These support research and development activities could lead to improved catalysts and result in significant economic improvement. The 1 ton/hr reactor system of the flash liquefaction process will operate at capacity. Laboratory results encourage continued efforts in developing an efficient two-stage process to produce distillate fuels.

Several problem areas common to most liquefaction processes are under study, for example, liquid-solid separation, utilization of solid-containing bottoms for hydrogen production, and durability of auxiliary equipments.

Environmental concerns associated with coal liquefaction processes are primarily as follows:

- Presence of potentially toxic and carcinogenic organic compounds in coal liquid products and residues
- Presence of potentially toxic trace metals in the products and by-products

The environmental impact of coal liquefaction processes are being investigated under a program designed to analyze/characterize hazardous materials contained in various process streams or emitted from operating installations. Studies are also being conducted at pilot plant level to develop adequate occupational safety and health procedures for worker protection.

Table II 3-4 provides the funding levels by subactivity for the period FY 1978 through FY 1980.

Direct Hydrogenation

In FY 1980, this subactivity includes one active project, the H-Coal Process. From the process standpoint, H-Coal is similar to third generation processes; i.e., zinc chloride catalyst, and disposable catalyst hydrogenation, in that all catalytically add hydrogen to coal for liquefaction to oil while removing coal sulfur as gaseous hydrogen sulfide, which may further be converted to elemental sulfur for sale

Table II-3-4. Funding for coal liquefaction subactivities.

COAL LIQUEFACTION SUBACTIVITIES	BUDGET AUTHORITY (DOLLARS IN THOUSANDS)		
	ACTUAL FY 1978	APPROPRIATION FY 1979	ESTIMATE FY 1980
Direct Hydrogenation	$ 38,000	$ 23,889	$ 35,000
Solvent Extraction	47,200	48,640	45,000
Third Generation Processes	8,800	13,811	25,000
Support Studies and Engineering Evaluations	16,120	14,286	10,006
Liquefaction Demonstration Plants	16,000	105,600	7,000
Capital Equipment	400	200	300
TOTAL	$126,520	$206,426	$122,306

or storage. The major differences lie in the mechanics of the reactor and/or the type of catalyst used.

Specific objectives of the Direct Hydrogenation Subactivity are to determine/improve operational reliability and process economics.

Ebullated-Bed (H-Coal) Pilot Plant. The H-Coal process (Figure II 3-22) is a catalytic hydroliquefaction process that converts high-sulfur coal to either a boiler fuel that will meet sulfur emission regulations or to a refinery syncrude. Coal is dried and crushed to minus 40 mesh, then slurried with recycled oil and pumped to a pressure of 200 atm. Compressed hydrogen is added to the slurry, and the mixture is preheated and charged continuously to the bottom of the ebullated-bed catalytic reactor. Upward passage of the internally recycled reaction mixture maintains the catalyst in a fluidized state (catalyst activity is maintained by the semi-continuous addition of fresh catalyst and the withdrawal of spent catalyst). The temperature of the ebullated-bed catalytic reactor is controlled by adjusting the temperature of reactants entering the unit.

The H-Coal process was developed by Hydrocarbon Research, Inc. (HRI) as a further application of the ebullated-bed processing technology originally used to convert heavy oil petroleum residues into lighter fractions (H-oil process). Early development of the H-Coal process involved research with a bench-scale unit and a PDU, as well as preparation of a conceptual process design. Independent evaluations in 1963 and in 1976 confirmed the technology and feasibility of the H-Coal process.

Since 1964, HRI has been developing the H-Coal process continuously under the mixed sponsorship of ERDA (now DOE) and a private industry consortium.

Based on the experimental results obtained from a bench-scale and a w-1/2 tons/day PDU operating under both syncrude and boiler fuel modes, final detailed design of a 600 T/D pilot plant was completed in December 1977.

The pilot plant site was selected (Catlettsburg, Kentucky) by Fossil Energy at the recommendation of HRI, from sites offered by the private partners. A completed environmental assessment shows that the plant will not significantly affect the environment of the locality. Badger Plants, Inc., is responsible for construction of the pilot plant. Operation of the pilot plant will be the responsibility of Ashland Synthetic Fuels with technical support by subcontract with HRI.

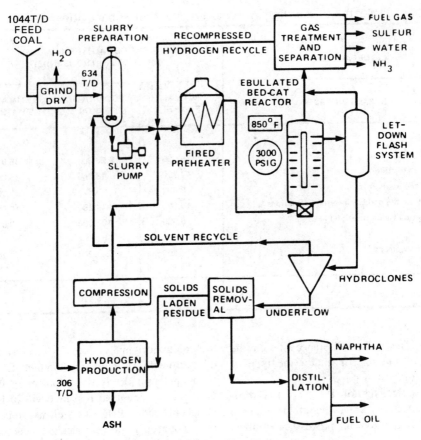

Figure II-3-22. The H-coal process.

The industry consortium sponsoring the H-Coal project is presently composed of the Electric Power Research Institute (EPRI), Ashland Oil, Inc., Standard Oil of Indiana, Conoco Coal Development Company, Mobil Oil Company, and the Commonwealth of Kentucky. DOE is providing 80 percent of the funds, and the consortium is providing 20 percent.

Construction of the pilot plant started in January 1977, with completion scheduled for late FY 1979. In FY 1980, plant startup and operability tests will be performed in both syncrude and boiler fuel modes. Material balance data around all process units will be used to establish process yield data as a function of coal type and operating conditions.

Major milestones for the H-Coal process are presented in Figure II 3-23.

Fixed-bed Hydrogenation (Synthoil) PDU. Construction of the 10 T/D coal liquefaction process development unit located at the Pittsburgh Energy Technology Center is nearly complete. This PDU was originally designed to develop the synthoil process. Subsequent evaluation and testing indicated further development should not be pursued. The PDU will be maintained in a standby condition for possible use in developing advanced processes. This project is to be phased out in FY 1979.

Solvent Extraction

Solvent extraction projects involve the use of a solvent to obtain a liquefied product from coal. The solvent acts as the agent which transfers hydrogen to the coal to promote liquefaction.

Solvent Refined Coal (SRC) Pilot Plant. The SRC process converts high-sulfur, high-ash coals to a nearly ash-free, low-sulfur fuel. Figure II 3-24 shows a schematic of the SRC-

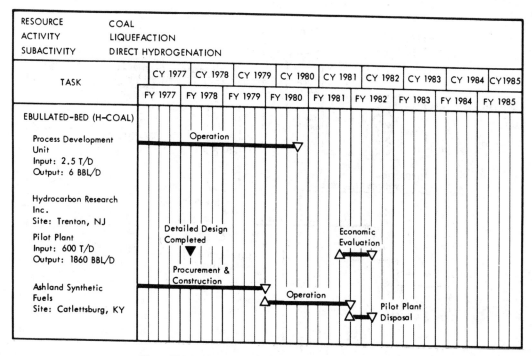

Figure II-3-23. Major milestones for the H-coal process.

I process. Feed is first pulverized and mixed with a coal-derived solvent in a slurry mix tank. The slurry is then combined with hydrogen produced from other steps in the process, and is pumped through a fired preheater. The heated mixture is then fed into a dissolver where about 90 percent of the coal (on a moisture and ash-free basis) is dissolved. In the dissolver, several reactions occur simultaneously: coal is depolymerized and hydrogenated, resulting in an overall decrease in molecular weight; the solvent is hydrocracked to form lower-molecular-weight hydrocarbons that range from methane to light oil; and much of the organic sulfur is removed by hydrogenation in the form of hydrogen sulfide.

From the dissolver, the mixture is passed on to a separator where the raw product gas is separated from the slurry of undissolved solids and coal solution. Raw gas is sent to a hydrogen recovery and a gas desulfurization unit. Hydrogen recovered is recycled with the slurry coming from the slurry mix tank. Hydrocarbon gases are recovered and hydrogen sulfide is converted to elemental sulfur.

Undissolved solids and the coal solution in the slurry are separated in a solid-liquid separation unit. In a commercial plant, the solids could be sent to a gasifier-converter where they react with steam and oxygen to produce hydrogen for use in the process. Process solvent is recovered from the cool solution by distillation and recycled to slurry the coal feed. The residue which remains is the final product, solvent refined coal (SRC-I process).

Typical properties of the SRC-I product obtained from a Western Kentucky bituminous coal feed having 7.1 percent ash and 3.4 percent sulfur are 0.1 percent ash and 0.8 percent sulfur. It has a solidification point of around 350°F to 400°F and a heating value of about 16,000 Btu/lb.

A modification of the SRC-I process produced distillate liquid products (SRC-II) instead of solid solvent refined coal (see Figure II 3-25). In this modification, part of the product slurry is recycled as solvent for the pulverized coal feed instead of 450°F-plus boiling-range distillate. As a result of increased severity of reaction conditions, hydrogenation reaction is greater and a major part

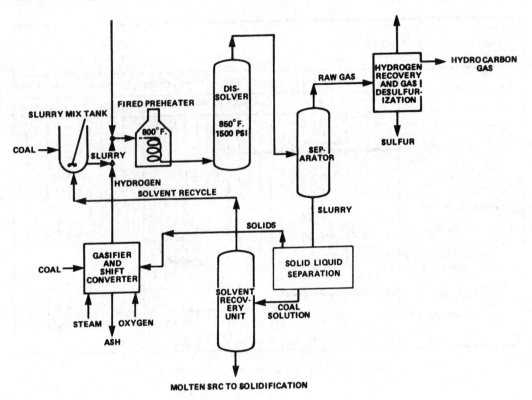

Figure II-3-24. SRC (solid) process — SRC-I.

of the coal is converted to a liquid-distillate product. The quantity of unconverted coal and vacuum residue is controlled so it is in balance with the requirements for gasifier feed to produce the process hydrogen requirements. This eliminates the solid/liquid separation step required for production of fuel in solid form.

The SRC project was begun in 1962, when Spencer Chemical Company was awarded a research contract to study the technical feasibility of a coal deashing process. In 1965, the process was successfully demonstrated in a 50 lb/hr continuous flow unit, and work on the original contract scope was completed. During the term of the contract, Gulf Oil Corporation acquired Spencer Chemical Company. After reorganization, the contract was extended and was assigned to the research department of Pittsburg and Midway (P&M) Coal Mining Company, and another subsidiary of Gulf Oil Corporation. Current efforts are underway at the Ft. Lewis, Washington, pilot plant and the Wilsonville, Alabama, PDU. A combustion test of the SRC solid product performed by the Southern Company Services in a 22.5-MW utility boiler demonstrated its capability to meet current emission standards for sulfur and nitrogen oxides. Similarly, a combustion test of the SRC distillate fuel oil (replacing No. 6 boiler fuel oil) in a utility boiler located at the 74th St. Station of Consolidated Edison Co. in New York City successfully demonstrated that emissions will comply with EPA standards for sulfur and nitrogen oxides.

Technical demonstration plants are being designed to evaluate large-scale operation of both the SRC-I and SRC-II processes.

The objective of the 50 T/D pilot plant located at Ft. Lewis, Washington, is to carry out essential process studies and provide continued technical support to the demonstration projects. Construction was completed by Rust Engineering in FY 1975. Operation of the pilot plant is under the direction of P&M Coal Mining Company. Feed rates exceeding design have been satisfactorily attained. Opera-

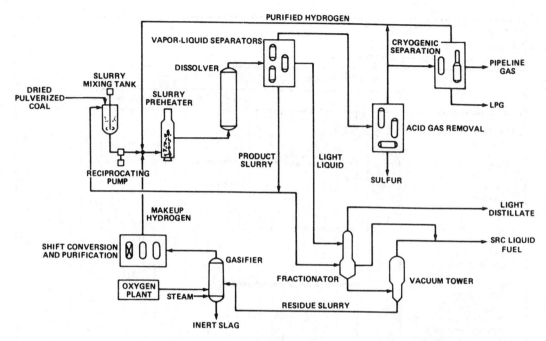

Figure II-3-25. SRC (liquid) process — SRC-II.

tions to produce 3,000 tons of solid product for full-scale combustion tests were completed during FY 1977. Pilot plant modifications, which permit the recycling of unconverted coal, were completed during FY 1977. This modified pilot plant produced over 5,000 barrels of distillate fuel oil for combustion testing during FY 1977 and FY 1978. An improved filter was evaluated during FY 1978. A study of solvent deashing and other process investigations will continue during FY 1979 and FY 1980 in support of the demonstration plants. Operations will be conducted in both the SRC-I (solid) and SRC-II (distillate) modes.

Ft. Lewis pilot plant operation includes both on-site and off-site environmental monitoring program in order to document the quality of the working and neighboring plant environment. In addition, toxicity of process liquids and solids is being studied.

Operation of the 6 T/D SRC PDU at Wilsonville, Alabama, was originally industry-financed through EPRI. Since 1976, it has been supported jointly by EPRI and DOE to supplement efforts of the Ft. Lewis pilot plant. Wilsonville provides a flexible test unit for screening additional coals and for selection of various process improvement options for additional testing at Fort Lewis. This cooperative effort will explore further improvements in production of solid SRC fuel, including solid-liquid separation methods, during FY 1979 and 1980.

Major milestones for SRC are given in Figure II 3-26.

Donor Solvent Liquefaction (EDS) Pilot Plant. The Donor Solvent process (Figure II 3-27) is a noncatalytic process that liquefies coal by the use of a hydrogen donor solvent obtained from coal-derived distillate. The donor solvent transfers hydrogen to the coal, thus promoting liquefaction of the coal. The process is less complex than SRC and utilizes engineering and design technology similar to that practiced in the petroleum industry. It is sufficiently flexible to allow for different varieties of coal feed and produces a wide spectrum of liquid products depending on market demand.

Coal is ground and slurried with the recycled donor solvent. The slurry is heated by a fired heater, and pre-heated gaseous hydro-

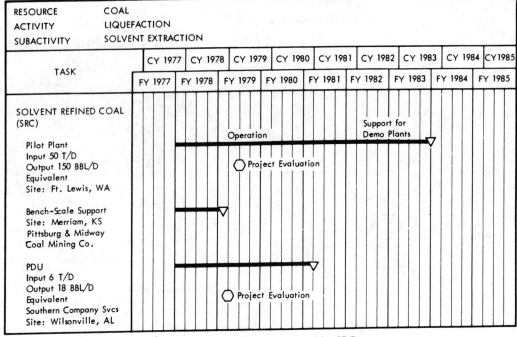

Figure II-3-26. Major milestones for SRC.

gen is added. The reaction is carried out in a tubular reactor without internals. Products from the liquefaction reactor are sent to several stages of separation units for recovery of gas, naphtha, middle distillate, and bottoms comprised primarily of unreacted coal and mineral matter. Distillation is the means for solid-liquid product separation.

The heavy bottoms from distillation are sent to a flexicoker to produce additional liquids and low-Btu gas for in-plant fuel use. Hydrogen for in-plant use is provided by steam reforming of C_1–C_2 gases produced in the process. The hydrogen is recycled to the liquefaction and solvent hydrogenation sections.

A portion of the middle distillate product is sent to the solvent hydrogenation step, using a catalytic fixed-bed reactor to produce donor solvent to be recycled to the slurry preparation step. Depending on the ultimate product utilization, the primary liquid products may be further refined.

The plant is balanced in that it is self-sufficient in both process fuel and H_2 requirements. The process gives high yields of low-sulfur liquids from either bituminous, or sub-bituminous coals or lignites. For Illinois bituminous coal, the liquid yield is determined to be 2.6 barrels of C_4/1000°F liquids per ton of dry coal feed.

The project was started in 1966 with Exxon Research and Engineering Company (ER&E) funding. The developmental program is structured into five phases: predevelopment, planning and design, detailed engineering, construction, and operation. Through 1975, cumulative ER&E funds of $32 million were expended on Phases I and II, resulting in the development and demonstration of the process in laboratory-scale reactors of up to one-half ton/day.

In August 1975, ER&E submitted an unsolicited proposal to ERDA (now DOE) to cost-share an estimated $268 million future-development cost on a 50/50 basis. Because the project cost was substantial, ER&E also sought financial participation from other in-

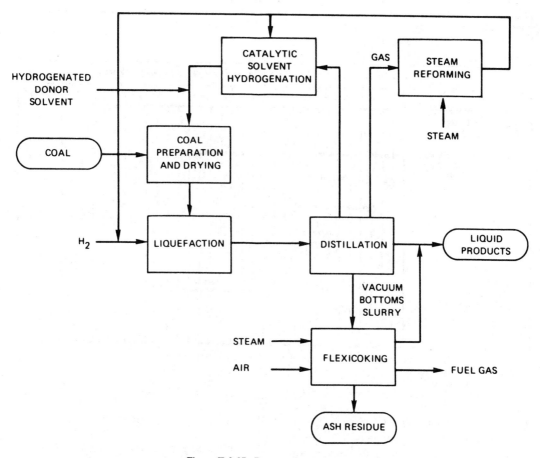

Figure II-3-27. Donor solvent process.

dustrial partners. Because industrial cost-sharing participants were not firmed up at the time ER&E could start Phase III, it was divided into two separate phases to permit discrete funding.

- Phase III–A: laboratory and engineering research and development
- Phase III–B: detailed engineering.

Phase III–A spanned the period between January 1976 and mid-1977. Total cost of this phase was $13 million, shared between ERDA and ER&E on a 50/50 basis.

Phases III–B to V will span the period between mid-1977 and mid-1982. Total cost is expected to be $227 million plus contingency that will also be cost-shared on a 50/50 basis by the Federal government and private industry.

During FY 1980, the following work was performed:

- Complete construction of the pilot plant
- Begin pilot plant operations
- Complete revamp and begin operation of the prototype flexicoker
- Investigate liquefaction/bottoms processing interactions
- Evaluate process improvements
- Prepare conceptual design study for a pioneer plant.

The major effort in FY 1980 will be to complete construction and commence operation of the 250 T/D pilot plant and prototype flexicoker.

Major milestones for Donor Solvent Liquefaction are shown in Figure II 3-28.

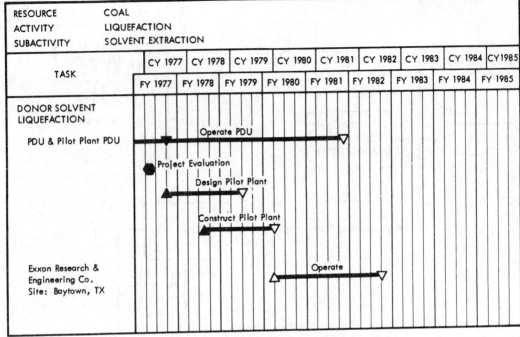

Figure II-3-28. Major milestones for donor solvent liquefaction.

CO-Steam Process. The CO-Steam process, developed by Pittsburg Energy Technology Center (PETC), is designed to convert low-rank coal such as lignite into a low-sulfur fuel oil by the noncatalytic reaction of coal with carbon monoxide or synthetis gas in an oil slurry. Grand Forks Energy Technology Center (GFETC) is investigating this process to produce clean liquid fuel from inexpensive raw material.

During FY 1979, the CO-Steam project was completed. In FY 1980, use of the facility will be redirected with the following technical objectives:

- Determine the catalytic effects of mineral matter contained in various low-rank coal
- Study process derived liquid fractions as recycle solvents
- Investigate reaction parameters.

This effort will be included under the support studies and Engineering Evaluations subactivity in FY 1980, and was funded as a separate project.

Third Generation Processes

This subactivity includes research and technology development for third generation processes that offer potentially significant advantages over second generation processes now in technical demonstration or pilot plant phases of development. These advanced processes are in various developmental phases, from the laboratory and bench scale to larger continuous process units, including PDU's. Processes which are selected for support offer significant improvements in one or more of the following areas:

- Hydrogen utilization
- Operability
- Product quality and yield
- Plant investment and operating costs
- Types of coal utilized.

Future development of these advanced processes beyond the laboratory or "process research" scale are expected to benefit from the availability of many existing PDU's and pilot plants, since these facilities can be modified to serve the purpose of scale-up experimentation and demonstration. Furthermore, the experience accumulated from development of second generation processes can be used as guidelines for technical direction.

Zinc Chloride Catalyst PDU. The $ZnCl_2$ catalyst process is designed to convert bituminous and subbituminous coal into distillates (in the gasoline range) by severe catalytic hydrocracking. The process may be applied either to coal as a one-step process or to coal extract as a two-step process. The process configuration will be set by economic considerations related primarily to the extent of catalyst recovery.

In the process (Figure II 3-29), coal is dried and pulverized before introduction to a feed tank where it is slurried with a process-derived recycle oil. The slurry feed proceeds to the hydrocracking reactor where it is mixed with hydrogen and the $ZnCl_2$. The reactor operates between 675°F and 825°F and between 1,500 and 3,500 psig. In the reactor, coal is cracked to distillates primarily in the gasoline range. All products go to a receiver where gas is separated from the liquid which is further processed to produce gasoline. The gasoline and light fuel oil distillates are essentially solid-free.

From the hydrocracking reactor, the solid discharge containing spent catalyst, nitrogen, sulfur compounds, ash, and carbonaceous residue is fed to a fluidized-bed combustor which operates at 1700°F and 2 psig. $ZnCl_2$ is separated from the rest of the residue as a vapor, condensed, and recycled back to the hydrocracking reactor together with fresh $ZnCl_2$.

Work being performed is aimed at developing a process which selectively produced high-octane gasoline, while providing high reaction rate, constant catalyst activity, and high catalyst recovery by regeneration.

During the period 1975 to 1977, major research activities were conducted on a 100-lb/hour bench-scale unit to study the conversion of subbituminous coal into distillate fuels, and to investigate zinc chloride catalyst regeneration. This work was successfully completed. Primary products from the bench-scale experiments have been characterized as 90 Research Octane Number (RON) gasoline and a low-sulfur, low-nitrogen content fuel oil. Primary and secondary zinc chloride catalyst regeneration achieved a minimum catalyst recovery of 99.5 percent. Other accomplishments included the development of a mathematical model describing coal reaction kinetics in the hydrocracking reactor. In June 1978, construction of a 1 T/D PDU was completed. Shakedown of the hydrocracking section is complete, and material balance operation feeding SRC-I product from the Takoma (Ft. Lewis) Pilot Plant is now underway. Shakedown of the catalyst regeneration section is in progress. Operation of the catalyst regeneration feeding spent zinc chloride melt from the hydrocracking operation was scheduled for the second quarter of FY 1979 after sufficient spent melt can be collected.

Based on the SRC operation and the results of economic analyses now in progress, a decision will be made whether to proceed with subbituminous coal runs in the PDU during the third quarter of FY 1979. The purpose of the proposed work is two-fold: demonstration of plant operability using subbituminous coals and evaluation of test data results will allow a decision to be made in FY 1980 whether to conduct tests with bituminous coal.

Major milestones for the zinc chloride catalyst are given in Figure II 3-30.

Disposable Catalyst Hydrogenation. This process uses inexpensive catalysts that avoid the extra complexity/cost of catalyst recovery and regeneration. During a single pass with coal through the hydrogenation reactor, the catalyst converts coal to sulfur-free fuel oil. (See Figure II 3-31.)

Coal is dried and mixed with a process-derived oil and the disposable catalyst. The paste is compressed to 2,000-4,000 psi and preheated. Recycle gas and make-up hydrogen are added to the paste before hydrogenation reactions take place in a pressure vessel in liquid phase. Gaseous and sludge discharge from the reactor are sent to a gas separator and letdown system. Part of the gas is recycled. The gaseous fuel product is composed

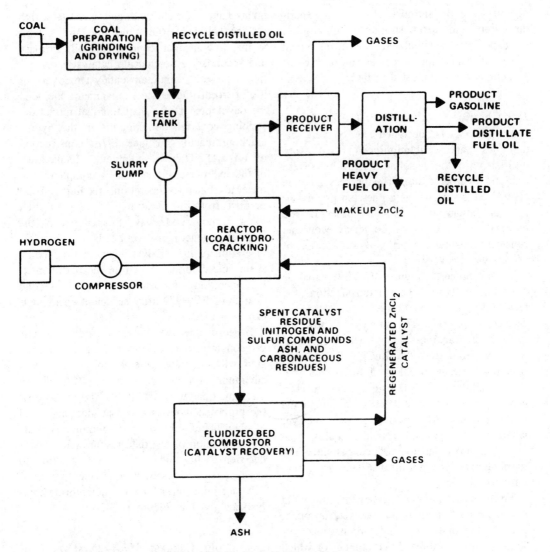

Figure II-3-29. Zinc chloride catalyst process.

of light hydrocarbons. The oil is distilled to obtain liquid fuel and pasting oil, while the residue and heavy oil are processed to recover light fractions, paste oil, and heavy liquid products.

During FY 1979, a major coordinated program was carried out at the Pittsburgh Energy Technology Center (PETC) and at industrial facilities with encouraging results. Catalytically active species have been identified in coal mineral matter. The mineral matter has been technically characterized and techniques developed to activate the most powerful species. Inexpensive ores have been evaluated in laboratory units.

During FY 1980, promising leads were further pursued in the laboratory and larger scale testing will be intensified. A major effort will involve extended operations of a 1,200 lb/day coal liquefaction facility at the PETC. The most promising catalyst will be tested in an existing DOE pilot plant. An independent engineering evaluation of this process will be conducted and the process compared with other processes already in the pilot plant stage.

Activity Description—Coal 123

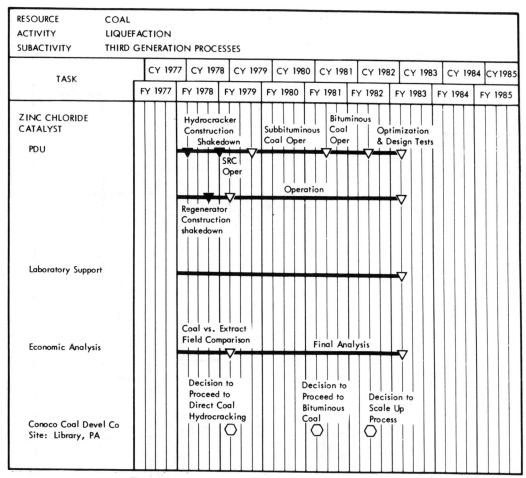

Figure II-3-30. Major milestones for the zinc chloride catalyst.

Major milestones for disposable catalyst hydrogenation are shown in Figure II 3-32.

Flash Liquefaction Process. The flash liquefaction process is being developed to evaluate the concept of direct hydrogenation of pulverized coal injected into an entrained-bed reactor using heated hydrogen (1500°F) as the transport gas. (See Figure II 3-33.) If required, some oxygen will also be added to sustain heat balance for the endothermic reaction. Rapid mixing and reaction of coal with hydrogen take place in the reactor at about 1800°F and about 1,000 psi. The reaction time is of the order of 10 to 100 milliseconds. The product mixture is then quenched. Char and tar are separated from the vapor phase. The vapor is condensed, and oil product and water phase are separated.

During 1976, methods for pneumatic transport and injection of the pulverized coal into the entrained-bed reactor were developed through the use of cold-flow tests to determine flow behavior. Upon satisfactory completion of this evaluation, further testing then moved on to a 1/4 ton/hr reactor system. The objectives were to evaluate the effects of reactor configuration and operating conditions. Based on the experimental findings, a 1 ton/hr reactor was designed, built and tested to determine process parameters. In addition to these contractually funded activities, research efforts to provide supporting information have been sponsored by Rocketdyne. During 1978,

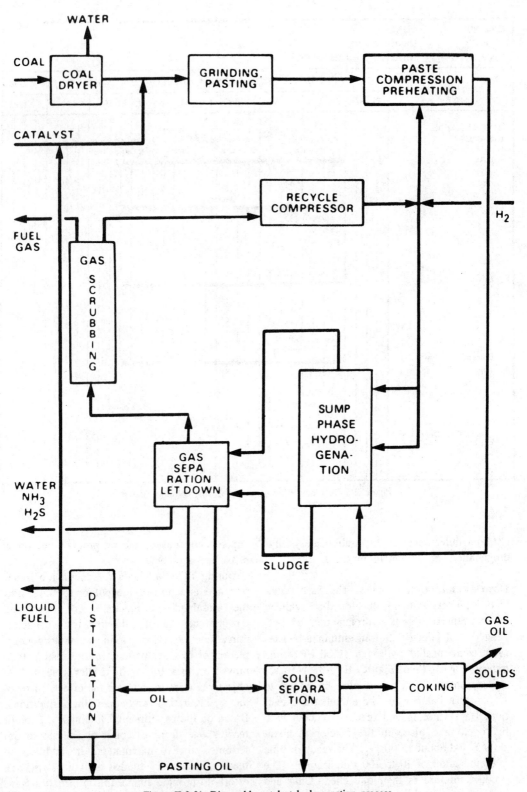

Figure II-3-31. Disposable catalyst hydrogenation process.

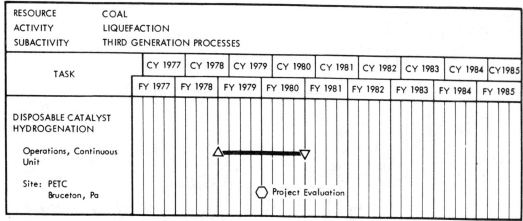

Figure II-3-32. Major milestones for disposable catalyst hydrogenation.

tests were conducted with the 1 ton/hr reactor system to further develop the process. Presently, the reactor system is being expanded to a full process development unit with testing beginning in FY 1979. Operations will continue in FY 1980 with additional equipment to recover the benzene, toluene, and xylene liquids. Further design and economic studies will be conducted to permit evaluation of scaled-up operations. Data from extended operations will be provided for pilot plant design.

Figure II 3-34 shows major milestones for flash liquefaction.

Process Research/Technology Development. The FY 1981 request for this research area will provide for the acceleration of ongoing projects and for initiating new third generation coal liquefaction processes. Among the candidate processes to be studied is Two-Stage Liquefaction.

Two-stage liquefaction uses a short contact time solvent extraction reactor in the first stage to liquefy the coal after which the product is deashed and sent to a second stage expanded bed hydrocracker for upgrading to distillable fuel oil products. Recent laboratory results have shown this integrated process consumes substantially less hydrogen and produces a cleaner liquid product. The fuel oil is also lower in nitrogen content than the oils produced in other processes.

In FY 1979, process studies were conducted using short residence time and additional upgrading using coal extract from the Wilsonville Pilot Plant. New innovative catalysts for the upgrading step are being tested and an economic evaluation of the process is being made to guide further research.

During FY 1980, laboratory efforts will be supplemented with extensive operations of a 1 T/D process development unit. A more definitive, and independent, economic evaluation will be made based on product yields and hydrogen consumption data obtained from continuous operations in an integrated fashion with optimum operating conditions for each stage. Work under this activity will be conducted at the Pittsburgh Energy Technology Center and at other sites to be selected.

Among the other processes to be considered for further development are several indirect liquefaction routes and a multistage direct catalytic liquefaction process.

Support Studies and Engineering Evaluations

This subactivity includes mechanical component testing, environmental studies, engineering evaluations, architect/engineering and design services, characterization of liquid fuels from coal, upgrading and refining of raw products, and support program planning and reporting.

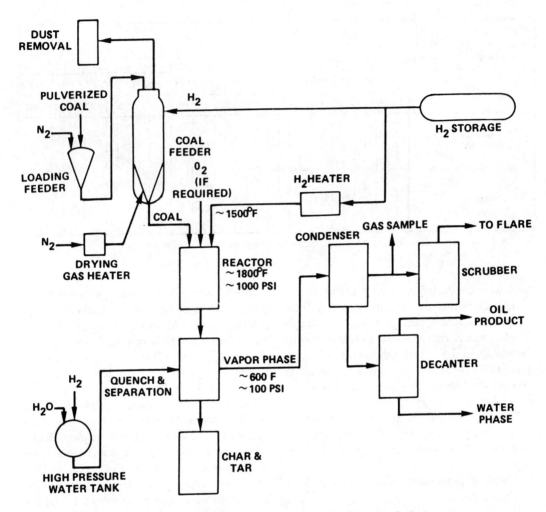

Figure II-3-33. Partial liquefaction by direct hydrogenation (flash liquefaction) process.

Cresap Liquefaction Test Facility. This facility, located at Cresap, West Virginia, was originally designed and built as a coal liquefaction pilot plant for the development of the Consolidated Synthetic Fuel (CSF) Process. It was inactivated in 1970 due to numerous mechanical and operating problems. In early 1978, Fluor Engineers and Constructors, Inc., under a contract to DOE, completed renovation of the plant to serve as a major support center for the development, testing and evaluation of key process equipments critical to most coal liquefaction processes. Other planned utilization of the refurbished installation include (1) demonstration of integrated plant operation, initially under conditions specified in the CSF process; and (2) development of third generation advanced coal liquefaction processes.

In the CSF Process, coal is crushed to 100-mesh size and combined with coal-derived solvent in a mixer. The slurry is pressurized and preheated, then pumped to an extractor where coal liquefaction reactions occur. Discharge from the extractor reactor is subsequently treated in a liquid/solid separation unit to remove unreacted carbonaceous material. The rejected solid residue is fed to a carbonizer for conversion to char, oil and gas. Oil, thus produced, is recycled to the solid/liquid separation unit while char and gas are recovered. The solids-free oil leaving the liq-

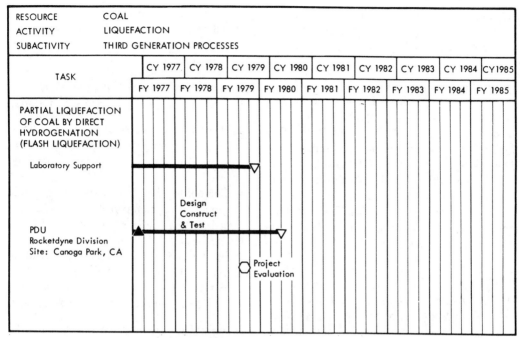

Figure II-3-34. Major milestones for flash liquefaction.

uid/solid separation unit is heated and fed into a flash still where the heavy components of the coal extract are separated from the light liquids. The light liquids are further fractionated into light and middle distillates. The middle distillate may either be recycled as a component of coal solvent or recovered as product fuel oil. The heavier components of the coal extract are heated in a fired preheater, hydrogenated, and fractionated into three distillate fractions. The middle distillate is used as donor solvent, and the heavy distillate is recovered as fuel oil product.

Reactivation of the Cresap facility was initiated in May 1974. Due to extensive plant deterioration, elaborate modifications were incorporated in the renovation effort to meet current design standards of piping, electrical, and mechanical codes. Refurbishment of the plant was completed in early FY 1978.

An important milestone in the Cresap project was the achievement of totally integrated operation (CSF Process Mode) in March 1978, producing a low-ash fuel oil. Solvent de-ashing was the method used for liquid/solid separation and appeared to be a viable approach. During the integrated operation, high-pressure slurry pumps were found to be particularly troublesome because of material erosion and seal failure. The results have been used as a basis for instituting future pump improvements. Other operating data obtained included the performance of valves, heaters, heat exchangers, and in-line meters. All the collected information will be used as the basis for equipment evaluation. As part of an environmental assessment task, a hygienic subtask has been initiated to monitor and minimize operators' contact with coal-derived liquids. Currently, this subtask involves periodic skin tests and complete physical checkups.

Because of schedule delays at Cresap, other large pilot plant facilities can be used as alternative sources for component performance and data. In FY 1979, operations are continuing at a reduced funding level in order to plan for an orderly shutdown by FY 1980. Shutdown plans include placing the plant in a standby condition so that it can be restarted in the future, if required, at minimum expense.

Figure II 3-35 gives major milestones for the Cresap Test Facility.

Coal Liquids Refining. The objective of this project is to determine the applicability of commercial universal oil products (UOP) conversion processes for all distillate coal liquids. Studies on upgrading residue containing coal liquid were completed in 1976. Distillate coal liquids from the H-Coal and Exxon Donor Solvent processes are now being investigated. The results from upgrading distillate coal liquids have been very encouraging to date. This project was completed in FY 1979.

Environmental Studies and Other Support. The objective of engineering evaluations is to maintain a continuous review of the technical status of processes that are probable candidates for future commercialization, in order to ensure cost-effective application of research funds.

Preliminary reviews are being performed for all coal liquefaction projects. Engineering studies by independent third parties will be performed for the processes that appear to be more probable candidates for further development.

Major environmental and support projects include participation in the Fossil Energy Program's activities regarding environmental impact statements and monitoring and improving the environmental performance of specific plants and processes.

Two-stage liquefaction (Cities Service). The objectives of the TSL project are as follows:

- Determine the expanded bed conditions (of an LC-firing unit) required to hydrotreat SRC extract at different severity levels
- Upgrade a production quantity of SRC extract using commercial-scale expanded bed hydroprocessing equipment
- Demonstrate that a first stage SRC process (to produce coal extract) can be combined with a second stage expanded bed hydroprocessing step to economically provide a valuable liquid fuel from coal

Successful operations have provided results indicating that the two-stage liquefaction process does utilize hydrogen more efficiently, which could result in lower costs for fuel oil production. Improved nitrogen removal has also been demonstrated.

During FY 1979, advanced research was conducted on conversion and denitrification of various coal extracts, effect of diffusional resistance on catalyst performance and screening new catalysts. This research will continue in FY 1980.

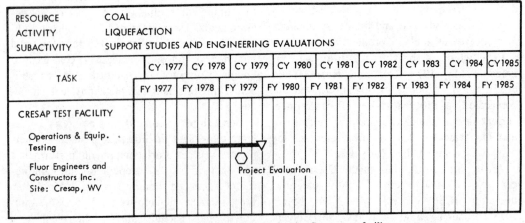

Figure II-3-35. Major milestones for the Cresap test facility.

Gasification of residual coal liquefaction products (Texaco, Inc.). The objective of this project is to determine the suitability of residual carbonaceous material from coal liquefaction processes as feed stocks for the Texaco gasifier to produce hydrogen or synthesis gas, thus significantly improving process thermal efficiency.

The project has successfully demonstrated that high-ash, coal-derived residues can be gasified directly using the Texaco gasification. Operations will continue in FY 1980, evaluating operating parameters and residues from various liquefaction processes. Data will be provided for commercial plant designs.

Liquefaction Demonstration Plants

Solvent Refined Coal (SRC) Demonstration Plants. The history and description of this process are described earlier.

The SRC process was selected for two demonstration projects in FY 1978 because: (1) the technology is the most mature of the liquefaction processes; (2) the Wilsonville and Fort Lewis pilot plants have been operating successfully, producing clean burning solid and liquid synthetic fuels from high-sulfur coals; and (3) preliminary economic evaluations indicate that this process is cost competitive with other liquefaction processes. Planned operation of the SRC demonstration plants will:

- Provide an accurate basis for determining investment and operation costs for commercial-sized applications of this technology
- Ascertain the environmental acceptability of this fuel product for direct use in electric power generation
- Assess the marketability of a range of fuel products obtainable from a given coal
- Determine the acceptability and degree of burner retrofit required of this fuel product by utilities and industry
- Demonstrate the technical viability of certain process steps such as solid/liquid separation, hydrogen generation and residue extraction systems. Other technical aspects such as materials of construction, equipment design, and fabrication technologies needed for commercial-scale applications will also be demonstrated.

The SRC Demonstration Program was initiated in July 1978 with the award of multiphased contracts to the Southern Company Services Inc. for demonstration of the Solid SRC (SRC-I) process and to the Pittsburg and Midway Coal Mining Company for demonstration of the Liquid SRC (SRC-II) process. Both demonstration plants would be full-scale "first modules" of multimodule commercial plants to be completed by the sponsoring industrial partners.

Both contractors are currently at work on a phase zero preliminary design. In Phase Zero, the contractors are preparing preliminary demonstration plant designs and cost estimates long lead procurement plans, market and economic assessments, detailed management plans for subsequent phases, and cost sharing proposals and environmental analyses for their projects. Each contractor is expected to have completed phase zero by the end of the third quarter FY 1979.

The administration has decided that both projects are to be continued as currently planned. After the conclusion of phase zero for both projects a decision will be made as to which of the two plants will be built. The remaining funds from the terminated project will be reprogrammed for use in the surviving project.

As a result of the decision at the conclusion of Phase Zero, the contractors initiated Phase I of the projects in FY 1979.

A multiphased contract for the SRC-I demonstration plant was signed by the Secretary on July 10, 1978, with Southern Company Services, Inc. Principal subcontractors are Air Products and Chemicals, Inc., Wheelabrator-Frye, Inc., Catalytic, Inc., and Rust Engineering, Inc. The proposed site is near Neuman, Kentucky. The plant will be designed to process Western Kentucky (#9,14) coal to produce a low-ash, low-sulfur, environmentally acceptable solid fuel that readily substitutes in conventional coal transportation and handling facilities and burns in coal-burning boilers without significant modification. The principal market for solid SRC-I is the coal-burning utilities in the Midwest and Southwest that will be required to install scrubbers or convert to oil or gas to meet environmental standards. The Phase Zero effort was concluded on July 19, 1979.

A multiphased contract for the SRC-II demonstration plant was signed by the Secretary on July 10, 1978, with the Pittsburg and Midway Coal Mining Company, a subsidiary of the Gulf Oil Corporation. The proposed site is near Morgantown, West Virginia. The plant will be designed to process Powhatten (Pittsburgh seam) coal to produce a virtually ash-free, low-sulfur, environmentally acceptable liquid fuel that readily substitutes in conventional liquid fuel transportation and handling facilities and burns in oil fired utility boilers without significant modification. The principal market for liquid SRC-II is the oil burning utilities in the urban Northeast that are currently burning imported oil. The Phase Zero effort was concluded on April 30, 1979.

Major milestones for SRC are presented in Figure II 3-36.

Design and Technical Support. The Design and Technology Support project for Liquefaction Demonstration Plants covers the following:

- Equipment development and component

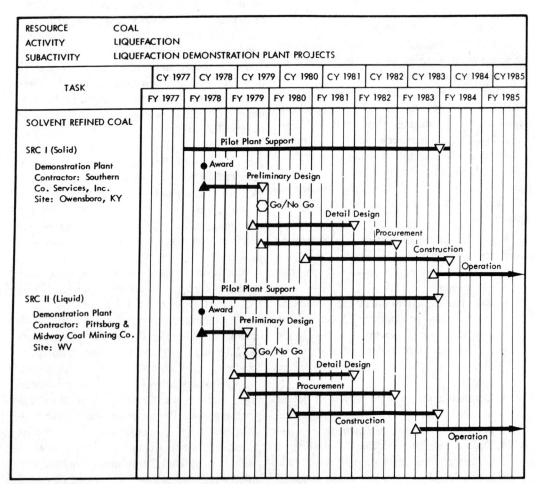

Figure II-3-36. Major milestones for SRC.

testing: Successful operation of large-scale equipment and components is essential for commercial-plant success. There is an array of such equipment (i.e., large valves and coal feeders) that is generic to the operation of all demonstration plants and must therefore, be developed and tested hand-in-hand with the balance of the plant. Funding is largely for continuation of work previously started.
- Engineering support: Project technical support will be provided by the national laboratories, Energy Technology Centers (ETC's), and various private engineering firms.
- Environmental and socioeconomic studies: Analysis of the environmental impact of large coal liquefaction pilot/demonstration plants and future commercial installations will continue. Potential social, ecological, aesthetic, and economic impacts and methods for achieving their equilibrium are studied.
- Planning and program support: These support tasks are similarly a continuation of established activities. They include modeling and system studies, economic and financial analyses, technical reports, and management support.

SURFACE COAL GASIFICATION

The Surface Coal Gasification Activity is composed of three subactivities: High-Btu Gasification for clean pipeline quality gas, Low-Btu Gasification for clean industrial and utility fuel gas, and Third-Generation Gasification for advancing the state-of-the-art of both high- and low-Btu technologies. These subactivities are supported by special projects, studies, and demonstration projects. Table II 3-5 summarizes the funding levels by subactivity for the FY 1978 to FY 1980 period.

High-Btu Gasification

National patterns of growth in population and economic activity require increasing consumption of natural gas. At the same time, proven reserves have decreased, despite intensified drilling activity. These forces have, in the short-term, caused local and regional shortages. In the long term, the country is faced with ultimate resource depletion. Development of alternative domestic fuel resources of a quality and convenience equal to natural gas is a high national priority. The High-Btu Gasification Subactivity is meeting that priority by:

Table II-3-5. Funding for surface coal gasification for FY 1978 - 1980.

SURFACE COAL GASIFICATION	BUDGET AUTHORITY (DOLLARS IN THOUSANDS)		
SUBACTIVITIES	ACTUAL FY 1978	APPROPRIATION FY 1979	ESTIMATE FY 1980
High—Btu Gasification	$ 24,700	$ 15,020	$ 0
Low—Btu Gasification	19,730	11,680	14,500
Third Generation Processes	13,600	23,075	27,000
Special Projects and Support Studies	23,600	11,000	15,000
Technical Support	28,650	19,023	12,350
Gasification Demonstration Plants	84,500	79,300	100,000
Capital Equipment	800	500	500
TOTAL	$195,580	$159,598	$169,350

- Identifying and accelerating the development of third-generation technology to improve process economics on a commercial scale for the 1985–2000 period
- Developing and demonstrating, in cooperation with industry, new and improved second-generation gasification technology necessary for the construction of commercial-scale plants

The Government-supported High-Btu Gasification Activity includes four major second-generation projects for development of improved processes to manufacture Substitute Natural Gas (SNG), with the following specifications:

- *Heating Value:* 950 to 1,000 Btu/scf
- *Purity level:* essentially free of sulfur and other potential pollutants and hazardous impurities
- *Process gas exit pressure:* approximately 1,000 psi
- *Combustion characteristics:* similar to natural gas; exhaust gases should meet EPA and other regulations

Although many gasification processes under development have the same objectives, each is characterized by important differences that warrant concurrent development. These differences include reaction conditions, coal pretreatment, method of feed, reactor configuration, heat supply, gas purification, and methanation process. The Lurgi Gasification process has been used as a benchmark for measuring process economics of these processes currently under development. Improvements to present technology will be measured primarily in terms of comparative capital and operating costs, as well as ability to operate successfully with U.S. caking-type coals. Development of each candidate process will continue to a point where industry can construct, with an acceptable degree of risk, a commercial plant based on one or more of the concepts.

DOE's High-Btu Coal Gasification effort encompasses the development of processes to convert, in an environmentally acceptable manner, coals of various qualities to a clean-burning natural gas substitute. Since each of the high-Btu processes employs different process schemes, the effluent and emission characteristics of these processes are also expected to vary. Process characterization studies are currently underway to provide adequate and representative data on the effluents and emissions of pilot plants. It is expected that these data will be useful to regulatory agencies for future development of environmental and occupational standards. The water resources requirements associated with commercial-size facilities are also of concern. Thus, studies are underway to determine the adequacy of water supplies, the feasibility of using low-quality water for process applications, and the economics of recycle.

Processes currently being investigated or previously tested on a pilot-plant scale under High-Btu Gasification include: HYGAS, Bi-Gas and Synthane.

Bi-gas Pilot Plant. During FY 1979, the plant will be operated to obtain additional data, and will then be deactivated and placed in a mothballed status. Some of the reasons for not continuing the project into FY 1980 are as follows:

- The Gas Research Institute's withdrawal of support
- Operating and other technical difficulties, which have prevented operation of the plant as originally planned

Synthane Pilot Plant. The synthane process (see Figure II 3-37) was developed by the Bureau of Mines in the early 1960s. It is now under further development at Pittsburgh Energy Technology Center (PETC), Bruceton, Pennsylvania. A preliminary design contract was awarded to M.W. Kellogg Company in 1970 to determine whether there were sufficient data to design a pilot plant. In June 1971, Lummus was awarded a contract to design a 75 T/D Synthane pilot plant, which was built by Rust Engineering in 1975. Lummus is now operating the pilot plant to ac-

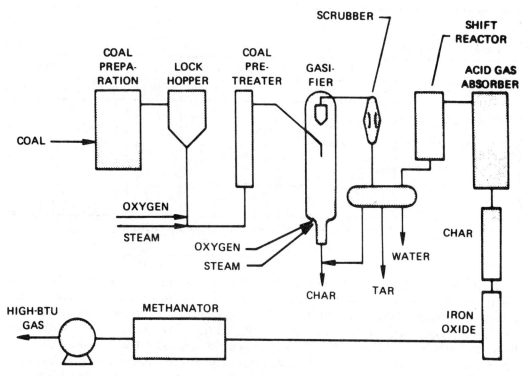

Figure II-3-37. Synthane process.

quire data to evaluate technical and operational feasibility of the process and to provide information required for scale-up.

Construction of the pilot plant was completed in March 1975. Commissioning and pre-startup testing of the process units through the gasification system were completed in June 1976. The first gasification run was made in July 1976. The pilot plant was successfully operated on Rosebud coal in August 1977. The data obtained were considered satisfactory to provide a reasonable basis for extrapolation to a large-scale facility. In September 1977, after the plant was converted to using an Eastern caking coal, pretreater and gasification tests were conducted. In August 1978, successful operation was achieved with an extended four-day run in the integrated system.

During FY 1979, a DOE evaluation team reviewed the synthane project. It was their conclusion that additional investment to modify the pilot plant as well as possibly two years further operation would be required to achieve a confidence level in the process that would justify scale-up to larger component (commercial) development is not warranted by the estimated technical and economic potential of the process. Weighed against other candidates, it has been decided to deactivate the plant in FY 1979 and place it in standby status pending further evaluation of its use in the catalytic Gasification Program.

Peat Gasification. Funds provided in the FY 1979 budget will support processed development into FY 1980, when a go/no-go decision on the construction of a pilot plant will be made.

The objective of this work is to generate engineering and kinetic data to be used in developing a process to convert peat into pipeline quality gas. Reaction kinetics are obtained from three reactor bench-scale studies. A preliminary economic evaluation is being performed to obtain an estimate for the cost of SNG from peat. Physical studies are underway on size reduction, fluidization, and

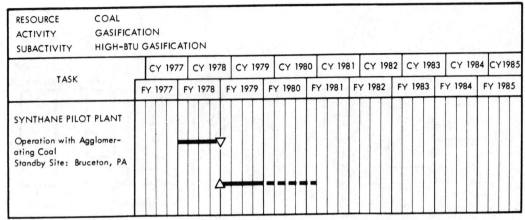

Figure II-3-38. Major milestones for the synthane pilot plant.

slurrying characteristics of peat. Studies on gasification of peat with H_2 and H_2O mixtures have been completed. During FY 1979, reaction performance (rates, conversion, selectivity, etc.) will be studied in coiled-tube and lift-line reactors. A steam-oxygen char (peat) gasifier also is being studied with emphasis on obtaining sintering profiles.

Low-Btu Gasification

The potential market for clean-burning low-Btu gas as a fuel for electric utilities and industrial users and as a chemical feedstock has grown as a result of increasing regional uncertainties of fuel supply, the decrease in natural gas reserves, and more stringent environmental regulations.

The objectives of the Low-Btu Gasification Activity are as follows:

- To develop and demonstrate, in cooperation with industry, the technology for construction of commercial-scale plants to convert coal to environmentally acceptable gaseous fuels with a heating value of 100 to 500 Btu/scf
- To establish the required techniques for using these coal-derived gaseous fuels for chemical feedstocks, industrial fuels, and combined-cycle systems for electric power generation
- To provide incentives to production and utilization of low-Btu gas by industry

Low-Btu gasification is a promising method for using coal as a utility and industrial fuel in an economic and environmentally acceptable manner. The generation of low-Btu gas from coal requires a less complex plant than that required for production of high-Btu gas. As a result, capital and operating costs are lower per equivalent Btu. For very efficient generation of electric power, low-Btu gas is a potential fuel for the gas turbine/steam turbine combined cycle. Low-Btu gas also can be used in industrial plants normally dependent on natural gas, oil, or coal. In all cases, the lower heating value of low-Btu gas compared to natural gas requires that the low-Btu gas be used at or near the production site.

Coal-derived medium-Btu gas has recently been identified as an attractive fuel for conversion of existing facilities, and for new plants designed for using natural gas or fuel oil. This gas has a heating value of 240 to 300 Btu/scf and can be produced at lower cost than SNG. A major effort is underway to identify the potential of gasifiers currently in use or under development for conversion to the production of medium-Btu gas. The primary activity of this program is the economical integration of gasifiers, gas cleanup systems, and end-use systems.

For industrial use of low-Btu gas, a Program Opportunity Notice (PON) was released for the Gasifiers in Industry Project. Six pro-industrial and institutional uses for low-Btu gas from coal.

Concurrently, promising new concepts for improved gasification techniques are being actively researched at the PDU and pilot plant level. These include fixed-bed, entrained-bed, fluidized-bed, and molten salt systems. Improved gas cleanup methods that will operate at elevated temperature and pressure are under development. These will enhance the economic potential of low-Btu gas as a clean industrial and public utility fuel. These new techniques will be tested as components of integrated systems combining gas production, gas cleanup, and gas utilization units to establish their operability and economic potential.

Four systems are currently under study: (1) an atmospheric pressure entrained-bed was constructed, started up, and operated through FY 1979; (2) successful, continuous operation of the devolatilization and gasifier sections of the two-stage fluidized-bed PDU has demonstrated that the unit can devolatilize caking coals without pretreatment and can successfully agglomerate a variety of coal and char feed materials; (3) a molten salt gasification PDU was built at a site in California, and recently began operation; and (4) design data and a full plant definition are nearing completion for the pressurized entrained-bed pilot plant.

Although low-Btu coal gasification produces an environmentally acceptable product, environmental impacts are associated with the processing and conversion of coal itself. A major concern is that air, water, and solid waste emissions may exceed local environmental standards for pilot plant facilities. This concern is being dealt with by the Office of Fossil Energy, in cooperation with the Environmental Protection Agency. Environmental monitoring and research studies of future low-Btu facilities are being coordinated. Environmental reviews and evaluations of proposed low-Btu sites are being made prior to selection, in order to anticipate potential environmental and socioeconomic problems. Modern environmental control techniques are being used in the pilot plant designs, and their acceptability is being evaluated. New techniques for trace element control, water reuse and cleanup, and control of very fine particulates are being researched to lessen the environmental impact of low-Btu gas utilization systems.

Fixed-bed Gasification

Stirred process. The objective of this project is to produce low-Btu gas for testing, cleanup, and combustion systems. The current system consists of three major sections: gasification, particulate removal, and sulfur removal.

As shown in Figure II 3-39, unsized coal is fed from lock hoppers into the top of the gasifier, which operates at pressures between 100 psi and 300 psi. Steam and air are introduced below the grate at the bottom. The nominal steam/coal feed ratio is 1 to 2, and the air/coal ratio is 3 to 1. Gasifier temperatures range from 2400°F just above the grate to 800° to 1200° at the gas exit. A water-cooled stirrer is used to control caking and to eliminate voids. The product gas has a heating value of 130 to 170 Btu/scf.

From the gasifier, the gas is passed through a cyclone and a tar condenser to remove particulates and tar. It is then scrubbed with water in a venturi scrubber, a disengaging chamber, and a direct cooler/scrubber for final cleanup and cooling. H_2S is removed and converted to sulfur in a high pressure Stretford plant. The gas is passed through a pressure letdown valve and vented.

The low-Btu gasification (moving fixed-bed) pilot plant has been operated at the Morgantown Energy Technology Center (METC) for several years. The system was modified in 1976 to provide improved methods for coal feeding, ash removal, stirring, lock hoppers, and data collection.

During FY 1979, in an attempt to obtain data on an integrated system (including devices such as humidifiers for tar separation, a novel high-pressure Stretford system, etc.), the METC Fixed-Bed (Stirred) Gasification/Full Flow Cleanup System was tested on Pittsburgh seam coal. Front end testing of the cleanup system was successful. Several integrated pilot plant runs have been completed and preliminary data reduction indicate satisfactory operation.

The pilot plant test program will continue in FY 1980, and a tar combustor and novel waste water evaporator will be added. The products from the tar combustor will be reinjected into the producer; the waste water evaporator precludes the need for a conventional water treatment system.

Major milestones are presented in Figure II 3-40.

Slagging Process. Coal, periodically charged to the lock hopper, moves by gravity flow into the gas generator. Some cooled product gas is compressed and recycled through the coal lock to prevent steam and tar vapors from entering the coal lock and condensing on the incoming coal. The coal is continuously gasified by an oxygen-steam mixture introduced through four water-cooled towers at the bottom of the gasifier. Molten slag is formed at the hearth and flows through a central one-inch diameter taphole into a water quench bath. Gas from the high-temperature reaction zone can be drawn through the taphole to aid slag flow. This gas is cooled and metered in a separate circuit.

The product gas leaving the gasifier contains water vapor and tar. It is scrubbed in a spray cooler with condensed recycle liquor. The gas liquor is periodically discharged from the spray cooler to a settling tank. The washed gas is then cooled to 60°F in an indirect cooler before being sampled, metered, and flared. (See Figure II 3-41.) The heating value of this gas is 350 Btu/scf.

This project was initiated in FY 1975 at the Grand Forks Energy Technology Center and is using a 25 T/D PDU-scale, slagging ash, fixed-bed coal gasification facility that had been constructed and operated in the late 1950s and early 1960s and shut down in 1966. The results from that earlier effort had been very favorable in terms of capacity and steam requirements.

The reactivation of the gasifier was completed in early FY 1976. In FY 1977 and FY 1978, the task of comparing effluents from various Western coals was continued, and efforts directed to those techniques and methods by which the quantity of effluents and

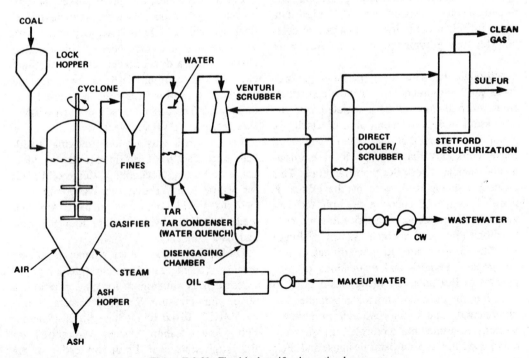

Figure II-3-39. Fixed-bed gasification, stirred process.

their composition altered to reduce the need for treatment facilities. During FY 1979 the gasifier was modified so that caking as well as noncaking coals can be gasified. Operation is planned on caking coals into FY 1980 and FY 1981.

Major milestones for the slagging process are presented in Figure II 3-42.

Fluidized-bed Gasification (Pressurized)

Two-stage Pressurized Process. This project focuses on the development of a low-Btu gasification process that can use caking coals without pretreatment. As shown in Figure II 3-43, the conceptual process consists of two principal process units: a pressurized fluid-bed devolatilizer and a fluidized-bed gasifier-combustor.

Crushed coal is dried with recycled fuel in a fluidized-bed dryer and introduced to the devolatilizer unit through a central draft tube. The coal and recirculating char are carried upward by hot gas flowing from the gasifier-combustor. This gas provides most of the heat to the unit; it devolatilizes and partially hydrogasifies the coal. The recycled solids, descending in the annular downcomer, are needed to dilute the feed coal and to temper the hot inlet gases. A high ratio (up to 100:1) of recycle solids to coal feed is maintained to prevent or control agglomeration of the coal as it passes through its plastic phase. A dense, dry char is withdrawn at the top of the draft tube.

Char from the devolatilizer and fines removed from the product gas are burned with air in the lower leg of the gasifier-combustor at 2100°F to provide gasification heat. Steam is injected into the same area for temperature control and to provide hydrogen for gasification. Ash from fines combustion agglomerates on the char ash and segregates in the lower leg for removal. Gasification occurs in a fluidized bed in the upper portion of the gasifier-combustor.

The hot low-Btu from the gasifier-combustor improves in heating value as it passes through the devolatilizer. The product gas

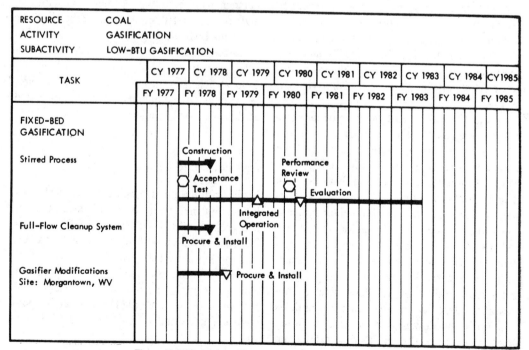

Figure II-3-40. Major milestones for fixed-bed gasification.

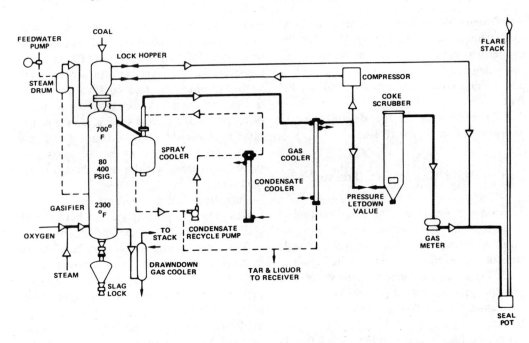

Figure II-3-41. Fixed-bed gasifier, slagging process.

from the devolatilizer is stripped of fines and ash in cyclones, cooled, and then scrubbed with water. The product gas heating value is 120 to 150 Btu/scf.

Initial development of the concept required the design and construction of a process development unit with a coal feed design capacity of 1,200 lb/hr. The PDU and its auxiliary subsystems were completed at the Westinghouse laboratory in Waltz Mill, Pennsylvania, in 1975.

During CY 1978, tests of the gasifier-combustor were run to obtain further data on operations, with direct feeding of coal. Two tests were run, in which highly caking Pittsburgh seam coal was gasified with air and steam. Gasification time on coal in these tests were 112 and 160 hours with voluntary shutdown in both cases. The PDU was then modified to allow the use of oxygen. After a shakedown run, two tests were accomplished, gasifying Pittsburgh seam coal with oxygen and steam. While the gasifier operated successfully, deposits of dry carbon particles built up in the downstream cyclone requiring shutdown after 50 hours. During fabrication of a modified cyclone, tests with Indiana and Kentucky coals were successfully run with oxygen and steam. Medium-Btu gas with heating values as high as 270 Btu/scf has been produced in the oxygen blown testing. During CY 1979, the complete coal gasification system will be operated in an integrated manner with both the gasifier, devolatilizer, and all associated equipment. Tests will be run with both air and oxygen.

Current planning is to terminate the three-stage pressurized fluidized project in February 1980, when the current contract will expire. No further developmental effort is anticipated at this time. (See Figure II 3-44.)

Fast Fluid-Bed Process

In the fast fluid-bed process, -20 mesh coal is fed into the lower section of the fast fluid-bed gasifier. (See Figure II 3-45.) The incoming coal is mixed with char fed from a companion slow fluid-bed gasifier at a rate of ten-parts char to one-part feed coal. The coal and char react with air and steam fed into the bottom of the generator. (The gasifier is operated at approximately 2000°F to 2400°F and 10 atm.)

The gas/solids from the gasifier pass through a primary cyclone to remove practically all solids, which are then discharged

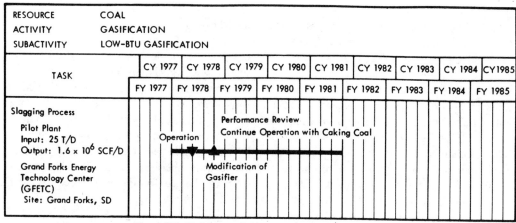

Figure II-3-42. Major milestones for the slagging process.

into the nitrogen-gas charged fluidized-bed reactor directly below. The char from the slow bed reactor is then fed into the fast fluid-bed reactor via a transfer leg. Ash from the fast fluid-bed reactor drops to the bottom and is discharged. Gas and particulates from the primary cyclone are passed on to a secondary cyclone for further separation.

The objective of this effort is to determine operating parameters, feasibility, and operability. Design parameters of the PDU are as follows:

- Coal feed rate: 600 lbs/hr
- Solids density in fast bed: 7 lbs/scf
- Average gas velocity in fast bed: 10 ft/sec
- Average temperature in fast bed: 1700°F
- Pressure: 150 psig

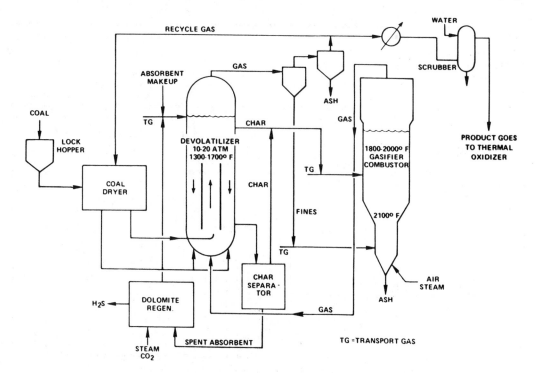

Figure II-3-43. Fluidized-bed gasification, two-stage pressurized process.

Gas/solid contact in the fast fluidization mode of operation occurs at high gas velocities in the range of 7.0 to 20.0 feet per second and high solids loading in the range of 10.0 to 20.0 pounds per cubic foot. This mode of operation can be considered an optimum region of operation between the (standard) fluidized bed that operates in the range of 0.5 to 5.0 feet per second gas velocity and at solid loadings of 20.0 to 40.0 pounds per cubic foot and the entrained or dilute-phase transport mode that operates in the range of 30.0 to 60.0 feet per second gas velocity with a solid loading in the range of 0.1 to 1.0 pounds per cubic foot.

The potential advantages for the fast fluid-bed coal gasification process include: (1) higher gasifier capacity by an order of magnitude over the standardized fluid bed; (2) no tar formation; (3) higher turn down capacity with minimum loss in efficiency; (4) lower operating temperature than the entrained bed mode, allowing increased flexibility in the materials of construction; (5) decreased potential for the formation of explosive mixtures; and (6) potential elimination of pretreatment of caking coals.

The Fast Fluidized-Bed process is under development at facilities of Hydrocarbon Research, Inc. (HRI), in Trenton, New Jersey. The current PDU program was initiated in July 1976. During FY 1978, the PDU was operated, using Eastern bituminous coal at 1700°F (no ash formed). A Phase II program is being conducted during FY 1979. Higher temperature operation (2200°F) using various coals is planned for FY 1980. (See Figure II 3-46.)

Entrained-bed Gasification, Atmospheric Process. This project originated with a co-funded study by Combustion Engineering to evaluate entrained bed gasification modes for use in the utility industry. Following the study, the decision was made in 1973 to select the atmospheric, two-stage slagging mode.

The current contract was undertaken in 1974 for the design, construction, and operation of a 5 TPD pilot plant at Windsor, Connecticut. The plant was completed, essentially

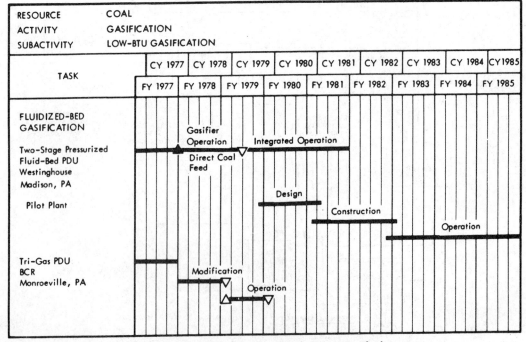

Figure II-3-44. Major milestones for the two-stage pressurized process.

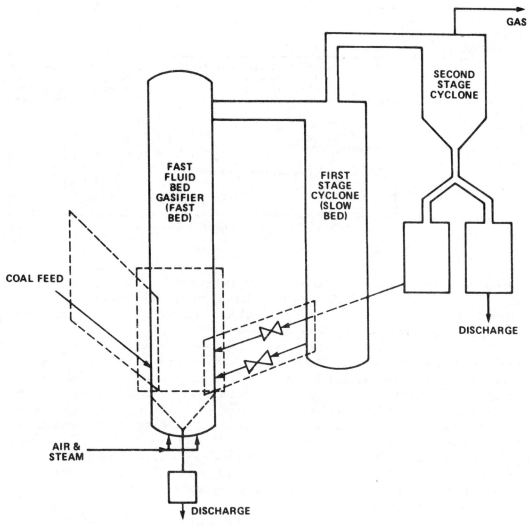

Figure II-3-45. Fast fluid bed process.

on schedule, and has undergone startup and parametric testing, using Pittsburgh seam coal. Operation to date has been successful.

In FY 1980, no funding has been requested, because the project will have been completed in FY 1979 as originally planned and scheduled. The test program using Pittsburgh seam coal was completed in June 1979. Operation of the plant beyond this date is not anticipated.

Gasifiers in Industry. This program promotes the use of low-Btu gas in one or more industrial processes, for steam boilers, for direct process heat, and in industrial dryers. Its objective is to demonstrate the integration of a low-Btu gasification system with industrial end-uses, using present day state-of-the-art components.

Coal will be gasified at the industrial plant, and purified, as required for the specific application. (See Figure II 3-47.) Transportation costs dictate that low-Btu gas be used near its point of production. Therefore, factors controlling the connection of the gasifier to an end-user of the gas must be evaluated. These factors will determine acceptability of coming advanced gasification techniques.

Small-sized individual gasifiers are commercially available. Under this program, they will be the forerunners of synthetic fuel conversion from coal in industry. The Federal

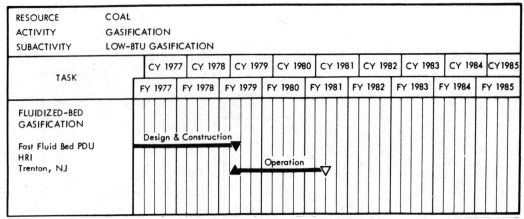

Figure II-3-46. Major milestones for fast fluid bed process.

program is of short duration. It focuses on the use of caking coals, of two-stage gasifiers which have not yet been demonstrated on U.S. coals, and on the evaluation of environmental acceptability. A recent PON called for proposals to achieve these objectives. Six contractors were selected; projects are discussed individually below; significant milestones are also shown in Figure II 3-48.

1. *The Glen Gery Brick Company* (with Acurex-Aerotherm) has instrumented an existing Wellman-Galusha gasifier. This gasifier is producing low-Btu gas from anthracite coal, and the gas is being used to fire a brick kiln at Glen Gery's plant in York, Pennsylvania. Valuable information has been obtained on selection and operation of instruments for measuring gasifier performance. Operational data, including material and energy balances are being collected. Completion of the operation of this gasifier under the DOE agreement was scheduled for the second quarter of FY 1979. This represents additional data collected. A complete report on the operation will be available shortly thereafter. Total estimated cost is $1.47 million. The DOE share is $0.72 million (50 percent cost-sharing).

2. *University of Minnesota* (with Foster Wheeler) will use a low-sulfur Wyoming bituminous coal to produce low-Btu gas as fuel for an existing natural gas-fired boiler. A two-stage STOIC gasifier will be used to produce a gas low in particulate matter. The heavy oil contained in the top gas will be removed by an electrostatic precipitator and collected. This oil has properties similar to No. 6 oil and will be stored for use during the coldest months.

The design effort on this project has shown that substantial modifications of existing foreign designs are required to provide reliable, environmentally acceptable operation in the United States. Construction is continuing on this project and startup will be completed in early CY 1979. The first year of continuous operation and detailed data gathering is planned for FY 1979. Three additional years' operating and maintenance data will be collected.

DOE is planning an environmental monitoring program for this project which will define and characterize emissions and their effects on the environment. This program is being coordinated with EPA and NIOSH. Total estimated cost is $5.5 million. The DOE share is $2.5 million (46 percent cost-sharing).

3. *Pike County, Kentucky* (with Mason and Hanger and the Kentucky Center for Energy Research) will produce gas from a low-sulfur Kentucky bituminous coal, a caking variety which requires a gasifier with an agitator. Thus, Wellman-Galusha gasifier equipped with particulate removal devices will be used. Gas produced will fire a steam boiler, used for heating and cooling several facilities within the Douglas site, including housing, a school, a nursing home, a fire station, and commercial facilities.

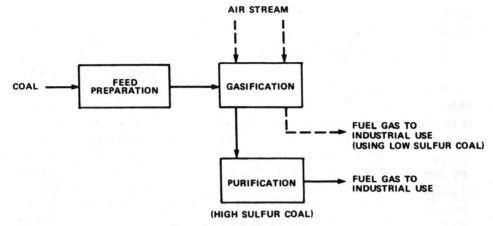

Figure II-3-47. Gasifiers in industry.

Detailed design and procurement of long lead-time items is currently underway. Design was completed in early FY 1979 and construction was begun during the first quarter of FY 1979. Activities during FY 1979 included completion of construction. Startup and operation will occur in FY 1980.

The environmental aspects of this installation will also be monitored in the joint DOE, EPA, National Institute of Occupational Safety and Health (NIOSH) program. Details are currently being developed.

Pike County envisions a second stage of this project. It would include addition of a sulfur removal plant and use of the clean, desulfurized low-Btu gas as a fuel for industrial clients in an adjacent industrial park. DOE is not currently participating in this phase.

Estimated cost of this project is $5.8 mil-

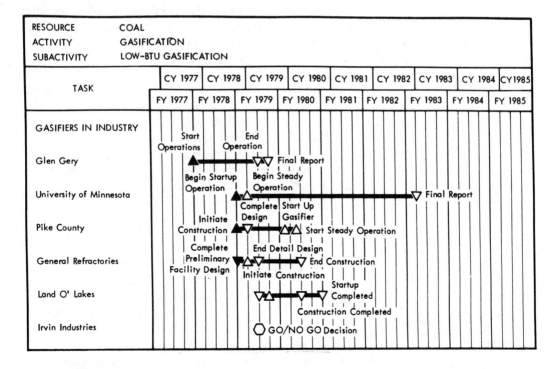

Figure II-3-48. Major milestones for gasifiers in industry.

lion. DOE's share is $2.9 million (50 percent cost-sharing).

4. *General Refractories* (with Holly, Kenney, Schott, and Woodall-Duckham) will use gas to fuel tunnel kilns, periodic kilns, and dryers at Hitchins, Kentucky, plant. A Woodall-Duckham two-stage gasifier will be used. Particulates will be removed, but most of the tar will be retained in the gas.

Process and preliminary designs, as well as the cost estimate, were completed during FY 1978. FY 1979 work included completion of detailed design and the start of plant construction which will be completed in FY 1980.

Estimated project cost is $4.2 million, of which the DOE share is $2.1 million (50 percent cost-sharing).

5. *Land O'Lakes* will use low-Btu gas to fire a boiler and a spray dryer at their Perham, Minnesota, plant. Using low-Btu gas to spray-dry dairy products is a unique use and requires careful design and extensive testing.

A two-stage gasifier will be used. The design will include extensive particulate removal for the stream to be used in spray drying. Particulates and condensibles will be removed from the remainder of the gas. A low-sulfur, nonagglomerating bituminous coal from Utah will be used. Design was completed in FY 1979 and construction begun. Construction and plant startup is scheduled for FY 1980.

Total estimated cost is $6.5 million. DOE's share is $3.25 million (50 percent cost-sharing).

6. *Irvin Industries* (with Mason and Hanger and the Kentucky Center for Energy Research) had proposed to use gas as fuel for a 172-acre industrial park at Georgetown, Kentucky. DOE, however, has evaluated the project and will not continue it into FY 1980.

Third-generation Processes

In addition to the gasification projects described in the preceding paragraphs, a number of projects are classified as third generation.

Hydrogasification. Hydrogasification produces high-Btu gas by the direct reaction of hydrogen with coal. From a thermodynamics standpoint, this method has excellent potential for maximizing thermal efficiency and producing the least costly synthetic pipeline gas. Elliott and Von Fredersdorff, in the coal gasification chapter of Lowry's book *Chemistry of Coal Utilization*, state, "The hydrogasification of coal to produce methane has an inherent thermochemical advantage in comparison with the catalytic methanation of synthesis gas produced by gasifying coal with oxygen and steam." This advantage was exploited in the early hydropyrolysis experimentation at the Bureau of Mines, which led to the development of the two-stage Hydrane process. As this process matured to pilot plant status, two developments occurred independently. In the first, Cities Service demonstrated a laboratory-scale, entrained-bed, single-stage hydrogasifier with short-residence-time characteristics. In the second, Rocketdyne showed potential for using rocket combustor technology to achieve improved mixing with high mass flux (lbs/hr ft^3) through-put in a hydrogen/coal reactor. It was determined that the two-stage Hydrane reactor was too complex for commercial application, and that a single-stage, short-residence-time hydrogasifier based on rocket principles was a more attractive option. A conceptual flowsheet for the overall process is shown in Figure II 3-49.

Hydrogasification has been under development at the Pittsburgh Energy Technology Center since the early 1960s. A principal result was the Hydrane reactor. Current contractual efforts are geared to development of an advanced reactor concept with superior commercial potential. Strategy involves testing and development to generate data for use in commercial-scale plant design.

Plans are as follows:

- FY 1977–1978: Hydrogasification feasibility study
- FY 1978–1981: advanced development and preliminary design
- FY 1981–1986: design, construction, and operation of very large pilot or demonstration

Exploratory development work under Contract EX–77–C–01–2518 was completed in

August 1978. Work on this phase was performed at two levels of operation: 2–4 lbs/hr and 1/4 ton/hr. A new contract (ET–78–C–01–3125) was signed with Rockwell International for advanced development studies. Tests will initially be performed at 1/4 ton/hr and then at 4 tons/hr. The contract was effective September 30, 1978, and will run for 3 years.

Figure II 3-50 gives major milestones for hydrogasification.

Catalytic Gasification Pilot Plant. In the Exxon Catalytic Coal Gasification process, carbon monoxide and hydrogen are recycled to the reactor to keep the CO/H_2 content as high as possible, thus forcing the net products of gasification reactions to be CO_2 and CH_4. The recycle rate is set such that there is no yield of CO and H_2 in the gasifier.

As shown in Figure II 3-51, coal entering the system is impregnated with a recovered make-up catalyst prior to entering the reaction vessel, where the coal is gasified with steam at 1200° to 1300°F in the presence of equilibrium concentrations of CO and H_2. The gasifier product gas exchanges heat with incoming water, generating steam. The steam and recycled CO/H_2 are preheated to about 150°F above gasification temperatures prior to injection to balance system heat loss.

Char/ash residue containing catalyst is removed from the gasifier. Catalyst is recovered by water through a countercurrent leaching operation. It is estimated that 50 to 75 percent of the carbonate may be reclaimed in this manner. Some catalyst reacts with coal-ash to form an insoluble potassium aluminosilicate, with about 5 percent weight of coal feed estimated lost in the insoluble form. Recovery of up to 90 percent of the potassium by routes such as acid wash of char has been demonstrated and will undergo further investigation.

After cooling the product gas and generating the steam required for the process, the gas is sent to a series of separation steps to produce CO_2 (which is vented), produce methane, and recycle CO/H_2.

Advantages of the catalytic SNG process are:

- Pretreatment is not required for caking coals
- The need for oxygen for other means of providing high-level heat directly in the gasifier is eliminated

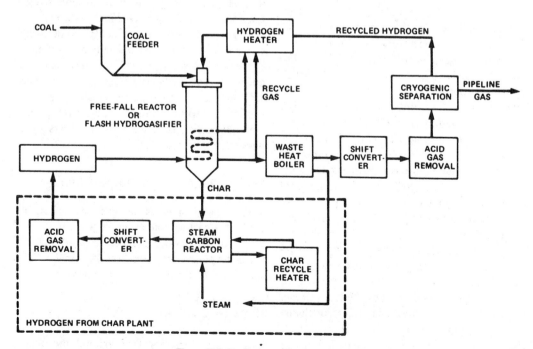

Figure II-3-49. Hydrogasification process.

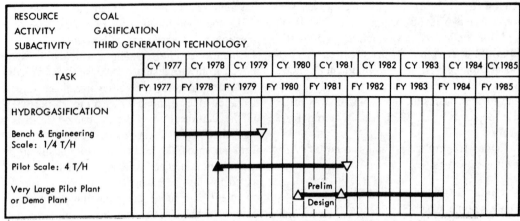

Figure II-3-50. Major milestones for hydrogasification.

- Gasifier temperaures are reduced
- Shift and methanation steps are eliminated
- Potentially higher thermal efficiency than that of thermal coal gasification processes is possible because of reduced need for high-level heat input and greatly heating and cooling of gas streams.

The Exxon Catalytic Coal Gasification process is under development at Exxon's research facilities in Baytown, Texas, with DOE support. Benchwork on catalytic gasification has been in progress since 1968. The current plan is to continue bench-scale development to obtain scale-up information for design and operation of a pilot plant. The PDU was constructed during FY 1978. Operation will continue to mid-FY 1980. Bench-scale research and development and engineering will be continuous through this period. A Request for Proposal (RFP) for the transfer of technology from Exxon to a third party process development company (possible licensor) is underway.

A feasibility study, performed by the Exxon Research and Engineering Company (October, 1979), examined the alternatives of testing this process in a converted pilot plant vis-a-vis a new facility. The study concluded that the Synthane pilot plant was most suitable for conversion. Basic and detailed design followed during FY 1979, with retrofit construction to begin in FY 1980 and operation planned from FY 1982 through FY 1984.

Major milestones are shown in Figure II 3-52.

Process Research/Technology Development. The FY 1980 request will provide for the continued development of the Hi Mass Flux gasification process. During FY 1979, the Hi Mass Flux process was evaluated using a Coal/Steam-Oxygen mixture as opposed to previous work using a coal/air mixture.

During FY 1980, experimental work using coal/steam-oxygen mixtures will be continued by Bell Aerospace Corp. In these studies, operating parameters will be optimized and scale-up factors investigated. Analysis and design efforts will be initiated toward the possible construction of a scaled-up Hi Mass Flux pilot plant.

Special Projects and Support Studies

This program includes hot gas cleanup methods and a gasification system test facility. The former activity was concluded during FY 1979 when developmental work on the Molten Salt and Iron Oxide processes was completed.

Hot Gas Cleanup

Molten Salt Process. To maximize the overall efficiency of combined-cycle power

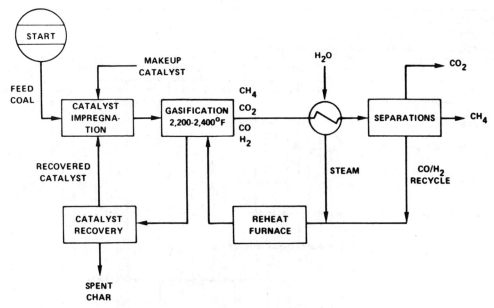

Figure II-3-51. Catalytic gasification process.

production, the sensible heat in low-Btu gas must be conserved. Also, the gas must be cleaned to remove particles and sulfur compounds before it enters the gas turbines. Existing cleanup technologies require the gas to be cooled, thus wasting its sensible heat. The Molten Salt Cleanup process pictured in Figure II 3-53 will operate at much higher temperatures, allowing more efficient use of high-sulfur coals for production of power.

Under contract with DOE, Battelle Northwest Laboratory operates a PDU at Richland, Washington. The objective of this project is to demonstrate process feasibility in the continuous operating mode.

Upon completion of the operation of the PDU in FY 1979, the derived data will be used to conduct cost analyses for the design and construction of a hot-gas cleanup facility.

Iron Oxide Process. The objective of this project, performed for DOE by Air Products, was to develop a solid absorbent for H_2S removal from hot (1000° to 1500°F) low-Btu gases at near-atmospheric pressure. This will improve the overall thermal efficiency of low-Btu gasification/electric generation systems.

The conceptual design of an iron oxide hot-gas cleanup system consists of four steps: particulate removal, H_2S removal, sorbent regeneration, and sulfur recovery.

Particulates in a hot (1000° to 1500°F) raw gas stream for a gasifier are removed in a cyclone (Figure II 3-54). The gas is then passed through an absorber containing an iron oxide fly-ash sorbent where the iron oxide is converted to iron sulfides.

When the sorbent becomes saturated, as indicated by an increase in the H_2S concentration in the product gas, the gas flow is shifted to the second absorber. The first absorber is then regenerated by oxidation with air, converting the sorbent back to iron oxide and releasing SO_2. Steam and nitrogen also can be introduced during the regeneration to moderate the absorber temperature.

The off-gases from the regenerator are reduced in a carbon-bed catalytic reactor, producing elemental sulfur from sulfur dioxide. It has been demonstrated that 96 percent of the sulfur in the raw gas can be collected as elemental sulfur.

At present, solid absorbents for H_2S cleanup are being developed and tested at laboratory scale (30 to 50 scf/hr feed-gas rate). A bench-scale unit (10,000 scf/hr) has been constructed at METC, and operations are continuing. (See Figure II 3-55.)

Elemental sulfur production during SO_2 re-

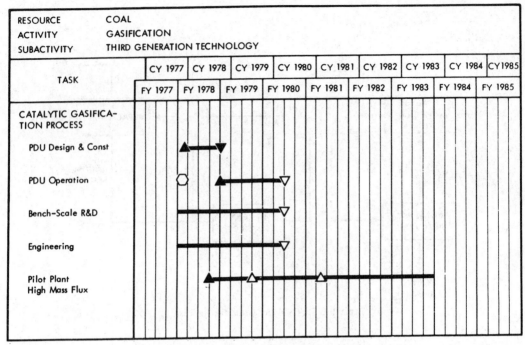

Figure II-3-52. Major milestones for catalytic gasification process.

generation of Fe_2O_3 sorbent-bed is being investigated. A prototype of the alkali monitor developed for METC by Ames National Laboratory is being tested.

A gasification simulator is being designed to produce major gas components, heteroatomic compounds, tars, and particulates at gasification temperatures in lab gasification studies.

Gasification Test Facility. The Gasification Systems Multi-test Facility will be a central site wherein all components of a coal gasification system can be tested and evaluated on a commercial scale. The facility will be constructed in stages as required to meet the programmatic priorities of the Gasification Development Programs. The initial phase will include a state-of-the-art, and one or two second-generation gasifiers. The state-of-the-art gasifier will serve two roles: (1) to evaluate the particular technology selected; and (2) to provide a continuing source of synthesis/fuel gas to support tests or other downstream equipment or processes. The facility will be constructed with maximum flexibility, to permit changes in the test program, the addition of new concepts, and changes in programmatic emphasis.

In subsequent phases, more advanced gasifiers (i.e., third generation) as well as clean-up systems, methanator test materials, and new equipment will be installed and included as part of the test program. The coordination of this test program will be the responsibility of a contractor yet to be selected.

First operation of this facility in FY 1980 will require funds to complete the conceptual designs, define the program in detail, complete the design of site support facilities, and specify and order long lead time equipment.

Upon completion, the facility will be used to test and evaluate the concepts of different gasification systems as they evolve from the overall gasification program. This project will be managed by Energy Technology and will be totally funded by DOE. Although several sites are being considered for this facility, a specific location has not been selected at this time.

Major milestones for the facility are given in Figure II 3-56.

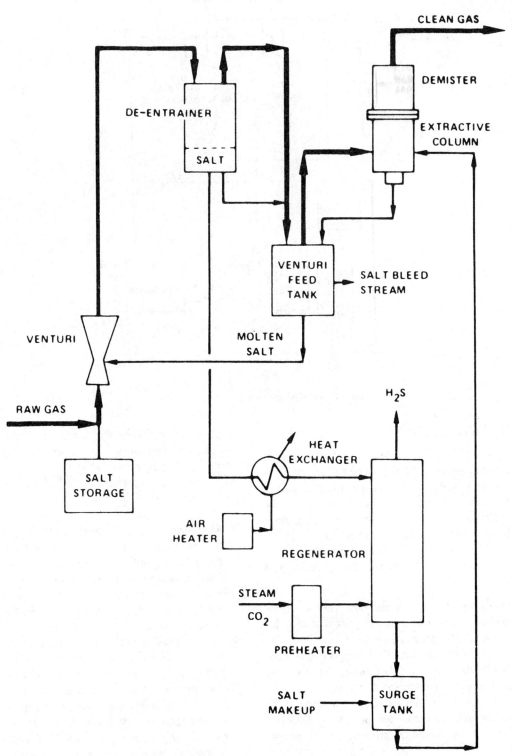

Figure II-3-53. Hot-gas cleanup, molten salt process.

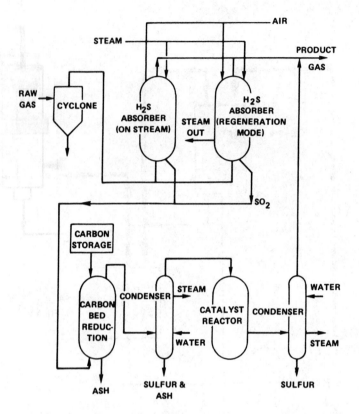

Figure II-3-54. Hot-gas cleanup, iron oxide process.

Technical Support

This subactivity provides for technical support efforts essential to the overall Surface Gasification Activity. Studies including engineering evaluations; equipment and material evaluations; process development; and generic environmental studies (rather than process specific) have been integrated into this subactivity under the Design and Technical Support Task in FY 1980.

Engineering Evaluations. The Gas Research Institute (GRI) has initiated work that complements DOE effort on gasification. The Engineering Evaluations Task supports the joint DOE/GRI High-Btu Gasification Activity. The objectives of this program are (1) to develop each process to its maximum potential, (2) gather data essential for process evaluation, (3) identify candidate processes or combination of component processing steps that will have the greatest commercial potential, (4) carry out commercial plant concept designs, and (5) identify knowledge gaps; i.e., data, materials, and equipment required to construct a commercial gasification facility.

The status of engineering evaluations of high-Btu processes is as follows:

- Monitoring and evaluation of the Bi-gas Steam Oxygen, CO_2 Acceptor, and Synthane pilot plants have been completed. This monitoring and evaluation activity will continue for any new processes that will be a part of the combined DOE/GRI Joint Program.
- A conceptual commercial design using a Western subbituminous coal was completed for the Bi-Gas, HYGAS Steam-Oxygen and Steam-Iron, Synthane, and the CO_2 Acceptor processes in order to compare the economics with a conventional Lurgi process. A parallel design was also prepared using Pittsburgh seam coal for the same foregoing processes except for CO_2 Acceptor, which is not

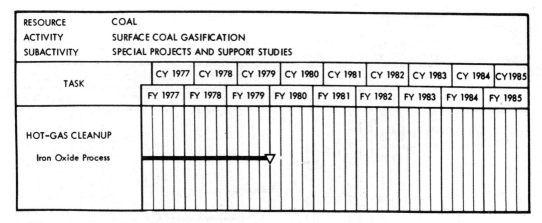

Figure II-3-55. Milestone for the iron oxide process.

suitable for bituminous coal. In addition, a conceptual commercial design was completed which compares the CO_2 Acceptor process against a conventional Lurgi process using lignite as the feed coal
- Process studies have been completed for the following unit operations: coal preparation and handling, coal pre-treatment, coal feed injection, gas purification, shift conversion, methanation, and effluent treatment
- Mechanical development studies for commercial-size components have been completed. These studies reaffirm and/or modify those problem areas where mechanical development and/or engineering studies are required to meet the anticipated needs of demonstration and/or commercial gasification facilities for commercial-size equipment
- Engineering studies were completed in the areas of: safety assurance of high-Btu coal gasification pilot plants, materials of construction for high-Btu coal gasification plants, and material development studies for high Btu-gasification.

The FY 1980 effort will include continuation of scientific and engineering support on the entire gasification program, evaluation of processes, development, review and evaluation of equipment design, and economic analyses of gasification process plants.

Equipment, Material, and Process Development

New Equipment Development. The objectives for this program are to generate new concepts for specialized equipment, to modify the most suitable commercial equipment or, if unavailable, to design new equipment that will be needed in a commercial-scale facility. Requirements are reviewed and project needs defined. Studies will then be implemented for the development of suitable grinders, valves, separators, and power recovery turbines.

Lock Hopper Valve Development. The objective of this program is to eliminate problems of gas leakage, sealing, erosion, and tar deposits by developing valves for lock hopper systems to feed dry, crushed coal. The approach is to complete design of components, construct and install the system, purchase valves, utilize lock hopper valve system, and report on the valves tested.

Catalytic Methanation. The objective of this program is to develop new and improved technology for catalytic conversion of synthesis gas to high-Btu gas. The approach is to improve catalyst life and productivity, increase rate and reduce cost of Raney nickel coatings, and minimize cost of catalytic methanation.

Catalyst Development. The objective of this

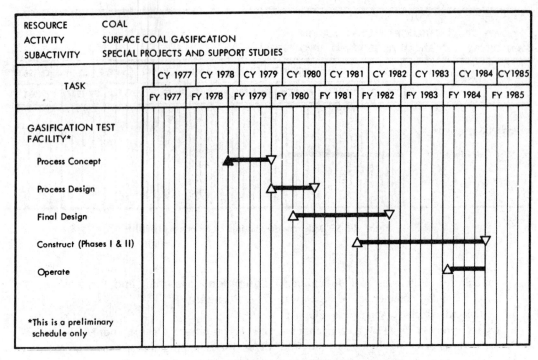

Figure II-3-56. Major milestones for the gasification test facility.

program is to determine factors that affect life and activity of tube-wall reactor catalyst. The approach is to identify factors related to catalyst deactivation, prepare deactivation resistant catalysts, and test longevity of new catalysts.

The FY 1980 effort will include field management of the coal feeder development program and conduct of final development tests. An effort will be initiated for design, development, and testing of cryogenic equipment for catalytic and hydrogasification processes.

Environmental Studies and Other Support. During FY 1979, technical support continued for the High Btu Pipeline Gas, Low/Medium Btu Fuel Gas, and Gasification Test Facility. Support included review and evaluation of Environmental Report, initiation of preparation of EIA and technical review of Industrial Partner's activities. The FY 1980 effort represents a continuation of activities on these three major gasification demonstration projects as well as review and assessment of major pilot plants and gasifiers in industry activities.

Design and Technical Support. In addition to the aforementioned tasks, which will, in FY 1980, be included under this task, Design and Technical Support includes supporting research, laboratory studies on novel approaches to coal gasification, and support for project evaluation and trade-off analysis. In FY 1979, cost effectiveness of process systems was studied, including the impact of novel materials and equipment. Studies at the Institute of Gas Technology include the chemistry, the fluid dynamics, and the thermodynamics of coal gasification processes and basic technical information supporting the pilot plant work. Studies were made on peat gasification in several existing PDU facilities. The "Coal Conversion Systems Technical Data Book" was continued in FY 1979 to compile a comprehensive source of information on coal properties and conversion processes for the entire Fossil Energy Program. The use of advanced techniques for computer modeling of coal conversion reactions was applied to the Synthane concept in FY 1979 and to the Westinghouse gasifier which for the first time took the caking tendencies of Eastern coals into account while modeling the combined mixing and agglomeration steps of the reaction process.

Gasification Demonstration Plants

High-Btu Demonstration Plants. This program includes the following demonstration plant projects for producing pipeline-quality synthetic gas:

- Project A: The Continental Oil Company (CONOCO) contract employs the British Gas/Lurgi slagging gasifier as the basic coal gasification technique, and demonstrated downstream processes, such as Lurgi technology, for gas purification and water clean-up
- Project B: The Illinois Coal Gasification Group (ICGG) contract, employs the COED/COGAS process
- HYGAS (Conceptual Design Only): The contract with Procon, Inc., employs technology developed by the Institute of Gas Technology (IGT).

Congress has authorized the initial engineering of Projects A and B (CONOCO and ICGG), and detailed engineering design, construction, and operation of one demonstration plant. A selection will, therefore, be made when adequate design and other relevant technical and economic information have been generated to permit an equitable and viable choice between the two. The project not selected may be considered for continuation, at a much slower pace, as a candidate for a second demonstration plant.

Additionally, a contract was awarded to Procon, Inc., for the conceptual designs of a HYGAS commercial plant and a demonstration plant. The contract has been continued with IGT for the HYGAS pilot plant operation to validate the process design parameters and technology.

The primary goal of these projects is production of synthetic pipeline quality gas (approximately 2000 Btu/SCF) to be used as a substitute for natural gas.

CONOCO and ICGG projects resulted from an RFP issued by ERDA in October 1975 for the design, construction, and operation of a demonstration plant. Five proposals were received, and CONOCO and ICGG were selected, contracts awarded, and work began in mid-1977. Both contracts provided for the use of Appalachian, Eastern, Midwestern agglomerating (caking) coals and operation of existing pilot plants in technical support programs to confirm technical adaptability of essential process features.

The CONOCO proposal is based upon the Slagging Lurgi Fixed-Bed Gasifier developed by British Gas at Westfield, Scotland; this project has been identified as the "High-Btu Pipeline Gas A" project.

The ICGG proposal—"High-Btu Pipeline Gas B"—is based upon the COGAS process (the COGAS pilot plant is located at Leatherhead, England), but includes the COGAS Development Company's COED coal pyrolysis process developed in Princeton, New Jersey, at a large pilot plant during the early 1970s.

Both CONOCO and ICGG have prepared conceptual designs of commercial-scale plants, established capacity and general concepts for the demonstration facility, performed considerable site-specific evaluation of the demo plant in the locations proposed, and completed the confirmatory technical support programs to validate basic data; and have developed certain business considerations for the government-contractor, cost-shared plant. An evaluation of this information as the basis for a selection between CONOCO and ICGG was undertaken in November 1978.

Management of the project is broken down into three separate phases for both contract and control purposes:

- Phase I includes both conceptual design for a commercial process and a demonstration plant for that process.
- Phase II includes plant construction and all supporting activities, equipment purchases, etc.
- Phase III includes operation of the plant and final process evaluation.

The schedule shown in Figure II 3-57 rep-

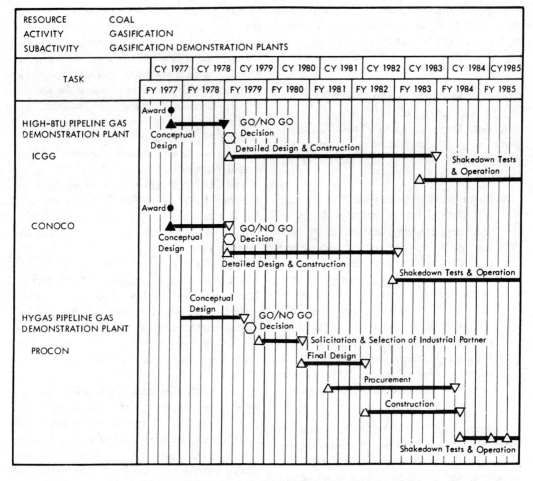

Figure II-3-57. Major milestones for gasification demonstration plants.

resents activities for both the CONOCO and ICGG projects, as planned by the contractors.

CONOCO (Project A) and ICGG (Project B) will be evaluated together as part of the high-Btu pipeline gas program. The main features of the two projects are outlined below:

- CONOCO (Project A)
  — Feed coal washed, crushed and sized ¼" to 1-½"
  — Four gasifiers in parallel to 445 psig and 2000°F, with gas cooling
  — Shift conversion
  — Gas clean-up and sulfur recovery; tars recycled to produce gas
  — Phenol extraction and ammonia recovery from waste water Methanation
  — Gas drying and compression to 850 psig
- ICGG (COGAS process)—COGAS gasifier
  — Feed coal is dried and crushed
  — Pyrolyzed in four fluidized beds in successive stages, oil condensed from pyrolysis volatiles—COED Process
  — Resultant char gasified in two-part gasifier (char combustion with air feed to supply heat for steam gasification in other vessel)—COGAS Process
  — Phenol extraction and ammonia recovery from waste water
  — Gas clean-up, sulfur recovered and converted to sulfuric acid

— Combined shift/methanation
— Gas drying and compression to 1000 psig
— Tars and oils are hydrocracked using syngas hydrogen

British Gas/Slagging Lurgi Process. The objective of the CONOCO project is to verify the technical, economic, and environmental acceptability of the British Gas/Lurgi process by building, operating, and evaluating a demonstration plant.

The slagging process, shown in Figure II 3-58, is a modification on the existing commercial Lurgi process in that it will accept caking coals and operate at higher temperatures. The process reduces the carbon content of the molten ash stream and thereby increases the conversion of carbon in coal to gas. Carbon conversion to products is better than 99.9 percent. The process also produces less tar, phenol, and heavy hydrocarbons than the older type commercial Lurgi Coal Gasification units. The tars recycled to the gasifier produce a higher yield of gas.

Gases leaving the gasifiers are cooled by quenching with a gas liquor (recycled water) spray. The H_2/CO ratio as required for downstream methanation is adjusted through the converter while the remainder bypasses the converter. In the purification step, naptha and water are separated by condensation, while hydrogen sulfide and carbon dioxide are removed by the Lurgi Rectisol process. In the methanation step, methane is produced from carbon monoxide and hydrogen over a fixed-bed nickel catalyst.

In the gas liquor separator, coal fines, tar, and tar oil are removed from water. The fines with tar and oil are recycled to the extinction in the gasifiers. The Lurgi Phenosolvan process extracts phenols from the water. Ammonia is removed from the gas by the CLL process to produce aqueous ammonia. An air separation plant will produce the oxygen used in the gasifier, a requirement for making pipeline quality gas.

In addition to tar precipitation, oil separation, and phenol extraction processes, the plant will include all the service sections required, such as steam production, water purification, air separation plant, incineration, waste treatment, as well as product loading, tankage, and buildings.

British Gas Corporation, as support for this project, ran extensive coal gasification test pe-

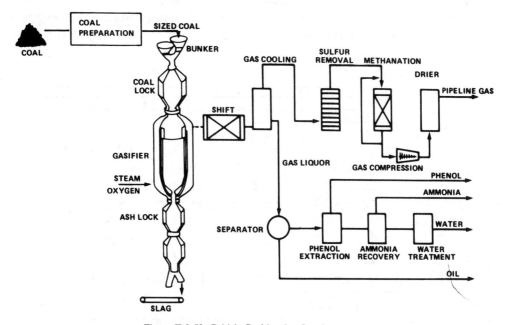

Figure II-3-58. British Gas/slagging Lurgi process.

riods on their normal 300 T/D prototype semiworks scale slagger at Westfield, Scotland, first, under private funding, and in 1978 under DOE funding. Successful operation was achieved on caking U.S. coal (Pittsburgh 8) and implied on another (Ohio 9) under the DOE program.

COGAS Process. Dried crushed coal is treated in three (and in some cases four) fluidized-bed stages at successively higher temperatures until a major fraction of the volatile matter of the coal is evolved. Heat for this pyrolysis is obtained from the gasifier by burning a portion of the char with air. Hot gases from the gasifier then flow countercurrently to the coal and constitute the fluidizing gas and heat supply for the third and second stages in order. Hot char from the last stage may be recycled to supplement the heat from the gases. The first-stage fluidizing medium is supplied as flue gas from the gasifier where a portion of the char is burned with air. Gas and oil are recovered by cooling and condensing the volatiles from the pyrolysis.

The char product of pyrolysis is fed to the gasifier while the remaining product, the raw oil, may be upgraded by hydrogenation to a high-grade synthetic crude oil, or by using less hydrogen in this step, to a low-sulfur No. 4 or No. 5 fuel oil. The hydrogen for this oil hydrotreating is supplied by reforming a portion of the product gas. From the gasifier, the synthesis gas is raised to a minimum pressure and cleaned to reduce particulates and sulfur compounds to a level acceptable for methanation. The resultant product gas is then methanated, dried, and compressed for use as a pipeline gas. (The process is illustrated in Figure II 3-59.)

Successful pilot plant operation has been achieved for both the COED and COGAS parts of this process. A pilot plant for the pyrolysis of 36 T/D of coal operated between 1970 and 1975 in Princeton, New Jersey. A 50 T/D char gasifier pilot plant at Leatherhead, England, completed a test program in 1978, validating the technical concepts of the gasifier section.

HYGAS process conceptual design project. This project, contracted to Procon, Inc., will provide the conceptual design of both commercial-scale and demonstration-scale HYGAS plants. It was completed in April, 1979. The primary advantage of the HYGAS process is the formation of up to 60 percent of the methane in the gasifier itself, which reduces the downstream equipment and oxygen requirement.

IGT, the process developer, is operating the pilot plant to confirm design parameters and

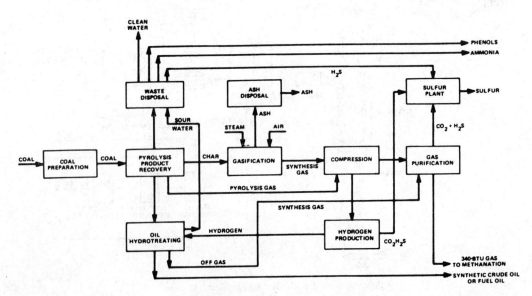

Figure II-3-59. COGAS process.

determine the inherent operability of the process technology. This work may be continued beyond the contract end date, June 1979.

The HYGAS process can be illustrated by the HYGAS pilot plant configuration shown in Figure II 3-60.

Coal is crushed, screened for oversize, weighed, and fed to an agitated tank where it is slurried in light oil. A positive displacement pump feeds at high pressure into the gasifier. This reactor operates at 500 to 1200 psi. It has four internally connected, fluidized-bed in which the reaction of gases with the coal is staged and in reverse flow. The upper bed dries the coal slurry. From there, the coal flow by gravity into a dilute phase riser stage which is the first step of hydrogasification. Here, coal particulates are heated to 1100°F by hot gases, which react with about 20 percent of the coal to produce methane. The partially reacted coal (now char) flows to the second gasification step, where it is heated in a fluidized bed to about 1700°F and is further gasified by the steam and hydrogen-rich gas rising from the steam-oxygen gasification stage below. The third gasification stage, at the bottom of the reactor, receives the feed gases made up of steam and oxygen. The temperature here of 1800°F results from the balance of the heat in the feeds, the exotherm from the oxygen reaction and the endotherm from the steam-carbon reaction. This ascending stream of hot gas provides heat and hydrogen to the test of the reactor. Fluidization is sufficiently vigorous in this stage to prevent slagging of the ash.

The high-ash spent char from this lower bed of the reaction system vessel is discharged through a solids control valve and carried away by stream. A circulating pump mixes it with water and maintains a slurry of even consistency with up to 30 weight percent solids. The coal ash slurry is let down in pressure using a special tungsten-carbide coated valve. The slurry is filtered at low pressure, and the filtrate is recycled to the quench vessel.

Low/Medium Btu Demonstration Plant, Industrial. This program includes the following demonstration plants:

- Fuel gas, industrial (A&B Conceptual Designs)
- Fuel gas, utility
- Fuel gas, small industrial

Fuel gas technologies have widespread potential for substitution for natural gas, especially for industrial and electric utility users. In February 1976, DOE issues an RFP to invite consideration of plants to demonstrate coal-to-fuel conversion for: (1) industrial use (i.e., a large gasifier supplying an industrial community, synthesis gas for chemical production); (2) small industrial/institutional use (a single industrial, a hospital, a commercial/residential complex, etc.); and (3) utility use (to substitute for recaptured natural gas for

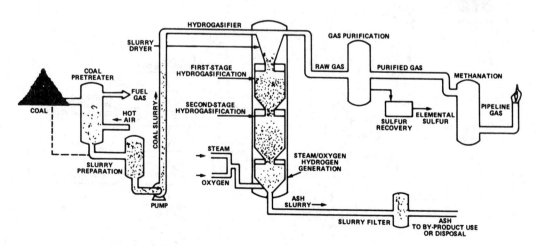

Figure II-3-60. HYGAS pilot plant configuration.

direct boiler firing, or as a source of energy for combined cycle (steam/gas turbine) generation).

Contract negotiations for the large industrial-user-dedicated plant have been completed with Grace/EBASCO and with the city of Memphis (Memphis Light, Water, and Gas) for a design competition which will lead to construction and operation of a demonstration plant. The two competing projects, Fuel Gas Industrial A and B, are described below.

Fuel Gas Industrial—Conceptual Design A. The Memphis Light, Gas and Water Division (a department of the city of Memphis), in association with Delta Refining Company, Foster Wheeler Energy Corporation, and the Institute of Gas Technology (IGT), has a fuel gas demonstration program based upon the U–GAS process to produce and deliver to Memphis industrial consumers, via a pipeline distribution system, 175 million CFPD (50 billion Btu/day) of industrial fuel gas having a nominal gross heating value of 284 Btu/scf. This is equivalent to 50 million CFPD of natural gas. The IGT U–GAS process is a fluidized-bed, oxygen-steam gasification process operating under conditions which promote the formation of ash agglomerates in the lower part of the bed.

The conceptual design of the gasifier for this project is an extention of the U–GAS reactor concept under development by IGT. The principal new development is operation of the gasifier at 90 psig and scale-up of a 3-feet diameter reactor to one in the range of 13 to 15 feet in diameter.

Crushed coal is fed to a pretreater (Figure II 3-61) and then contacted in the gasifier with an oxygen-steam mixture in a fluidized bed. Operating temperatures are in the range of 1870° to 2050°F. Fines entrained from the fluidized bed with the raw fuel gas are returned to the gasifier through external cyclones. The ash agglomerates produced in the bottom of the gasifier fall out through the bottom injection throat or venturi. The agglomerates are quenched with water and the resulting slurry is transported to ash separation and water reclamation. The hot, raw fuel gas is cooled in a waste heat recovery exchanger and fed to a venturi scrubber for the removal of ammonia, hydrogen sulfide, and coal dust. The fuel gas is then compressed to 195 psig and treated in a Seloxol acid-gas absorption process to remove essentially all of the hydrogen sulfide and organic sulfur compounds and part of the carbon dioxide. The purified gas from the Seloxol unit is ready for distribution.

Acid gas containing hydrogen sulfide is fed to a conventional Claus process. In the Claus unit the H_2S is partially oxidized to form elemental sulfur and water. The off-gas from

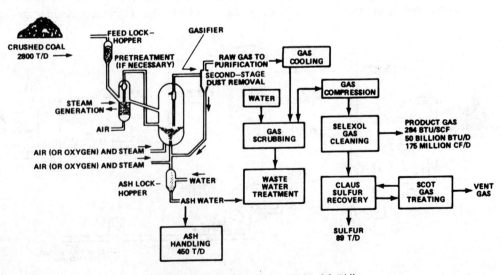

Figure II-3-61. Low-Btu fuel gas, industrial "A".

the Claus unit, is processed to a SCOT tail-gas unit, to recover remaining H_2S and recycle it back to the Claus plant. The off-gas from the SCOT process is environmentally acceptable to vent.

Fuel Gas Industrial—Conceptual Design B. W. R. Grace & Company proposed a facility to produce synthesis gas (a mixture of hydrogen and nitrogen) for ammonia production. The demonstration plant would be located in western Kentucky. Because ammonia formation is favored at high pressure, the synthesis step requires pressure exceeding 3,500 psig. Accordingly, a Texaco gasifier capable of operating at pressures up to 2,500 psig is the process that has been chosen for this project. Studies have been conducted that indicate that 1,200 psig is the preferred operating pressure.

Coal feedstock is prepared by grinding the coal and slurrying it in water, with the coal content of the slurry ranging from 50 to 60 percent by weight. (See Figure II 3-63.) The slurry is blended with oxygen for high-pressure injection into the gasifier. The resultant gas containing carbon and fly-ash flow to a scrubber where carbon is recovered and recycled to the gas generator. The synthesis gas from the gas generator, rich in carbon monoxide, is converted to hydrogen by the water gas shift reaction. Hydrogen sulfide and carbon dioxide removal is accomplished by physical absorption in refrigerated methanol. Liquid nitrogen scrubbing is utilized for final removal of residual unconverted carbon monoxide, residual methane and argon gases. The project gap is blended with nitrogen and is the feedstock for the ammonia synthesis step.

A contract has been awarded for the large industrial-user-dedicated plant to Memphis Light, Gas, and Water Division, City of Memphis, for a design that could lead to construction and operation of a 175 million CFD low-Btu Fuel Gas demonstration plant using 2,800 T/D of Kentucky No. 9 coal. Major milestones are given in Figure II 3-62.

A contract has been awarded to W. R. Grace and Company for conceptual design of a demonstration plant using 1,700 tons/day of Kentucky No. 9 coal feed in an entrained-bed oxygen blown pressurized gasifier, to produce 1,200 T/D of ammonia at an adjacent facility.

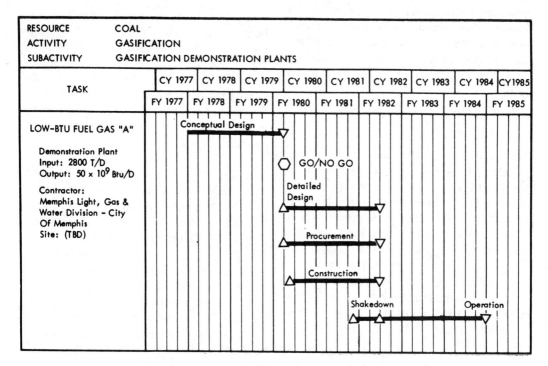

Figure II-3-62. Major milestones for low-Btu fuel gas "A".

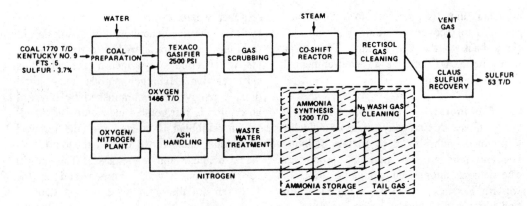

Figure II-3-63. Low-Btu fuel gas, industrial "B".

A single contract award for construction of a demonstration plant was to be based on evaluation of the commercial potential of the competing W. R. Grace and City of Memphis plant designs. However, the Administration now believes that the construction of the demonstration plant is not warranted given the market outlook for low-medium Btu gas. A rescission of the available construction funds was prepared in FY 1979. Phase I conceptual design efforts will be completed to provide definitive cost estimates, economic reassessment and environmental analyses to be used should future circumstances warrant a decision to construct a low-medium Btu coal gasification demonstration plant. No additional funds are requested for this project in FY 1980.

Major milestones are presented in Figure II 3-64.

Fuel gas utility plant. The objective of this project is to produce a fuel gas suitable for use in utility applications which is economically and environmentally acceptable. The contractor responsible for this project will be required to use second generation coal gasification technology (which is defined as technology in the pilot plant stage that has not been scaled up to commercial-size).

Potential markers for utility industry applications of coal gasification technology are as follows:

- Existing boilers currently capable of burning coal but that are not doing so because of the cost of environmental controls
- Existing boilers currently burning oil and gas where fuel gas produced from coal may be less expensive, or where an assured supply is imperative
- Future coal-fired boilers where fuel gas produced from coal may be less expensive than stack gas cleaning or the transporting of low sulfur coal
- Coal gasification combined-cycle or intermediate base-load generation

The fuel gas utility project proposals have been evaluated and Foster-Wheeler Development Corporation and Combustion Engineering Corporation have been selected for negotiations, which are scheduled for conclusion during the second quarter of FY 1980. At that time, reasonable schedules will be established. However, no funding for this project has been requested in FY 1980.

Low Btu Fuel Gas Demonstration Plant, Small Industrial. In this project, the contractor, Erie Mining Co., will use commercially available gasification systems; however, the objective is to expand the process to insure that the project gas is environmentally and economically acceptable.

Erie Mining has selected a two-stage, fixed-bed air blown gasifier (Figure II 3-65) to convert 500 T/D of coal to 7.4 billion Btu/day of 170 Btu-cubic foot fuel gas. The fuel gas will be used in the heat treatment of Taconite pellets.

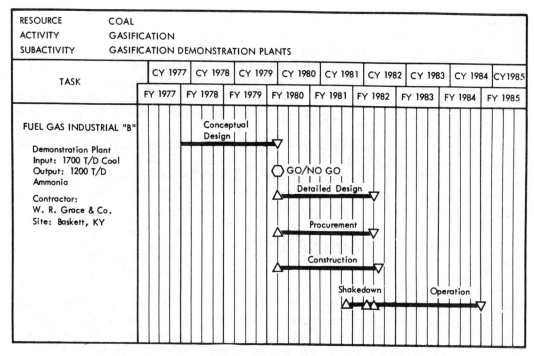

Figure II-3-64. Major milestones for fuel gas industrial "B".

Coal is brought to the coal handling area by railroad cars and is either transferred via feeders and belt conveyors to a line storage pile or to a dead storage pile. The coal is crushed in a roll crusher to ⅜-inch by 1-¼ inch size. The sized coal is fed to a two-stage gasifier through a water cooled revolving grate which also removes ash. A portion of the gas flows from the top of the gasifier at about 250°F and contains volatilized tar and oils and water vapor.

The tar is pumped from the precipitators to

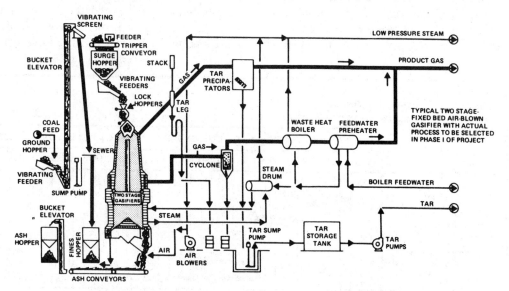

Figure II-3-65. Fuel gas, small-scale industrial.

a storage tank. This tar can be used as a standby fuel in the event of gasifier shutdown. The remaining portion of the produced gas flows from the gasification zone, existing at 1200°F. The hot gas is passed through cyclones to remove particulates and then through a waste heat recovery system to produce steam.

Ash is discharged from the lock hoppers at the base of the gasifier and collected on belt conveyors. A bucket elevator lifts the ash to the top of the ash storage silo which is sized to hold a day's ash production, and ultimately ash will be disposed of in a landfill.

The two product gas streams are mixed and the 225°F gas from the gasification system is cooled to about 100°F and compressed. The compressed gas is again cooled down to 100°F and fed to a Stretford unit. In the Stretford process, hydrogen sulfide is removed from the gas and converted to elemental sulfur. About 8 T/D of sulfur is produced in the process, which could be sold or used as landfill. The clean gas from the Stretford unit is piped to the Taconite pellet plant for direct use in the furnaces.

Erie Mining was selected for contract negotiation in May 1977, and a contract was awarded in October 1977. The work will be accomplished in three phases. Phase I is conceptual design; Phase II is detailed design and construction; and Phase III is operation and evaluation. The government will provide all funds for Phase I, while Phases II and III will be cost-shared on a 50/50 basis by the government, and Erie Mining. At the completion of Phase III, Erie Mining is expected to negotiate for the government's share of the gasification plant. (See Figure II 3-66.)

Erie Mining will use the fuel gas to produce taconite iron ore pellets. They currently produce 13 percent of the U.S. output. This technology application has potential use with other taconite producers, who currently account for 67 percent of the total U.S. iron ore production.

IN SITU COAL GASIFICATION

More than 90 percent of our nation's vast coal resource is not economically recoverable by conventional underground or strip mining. Underground coal conversion processes could recover a portion of the estimated 1,800 billions tons of unmineable coal by converting it to clean-burning gaseous and liquid fuels (Figure II 3-67). As a complement to mining plus surface-based coal conversion processes, the development of underground conversion processes could significantly expand the economic extraction of energy from coal in environmentally acceptable ways.

Underground Coal Gasification (UCG) can produce low-Btu gas for local use in electric power generation and as an industrial fuel. Medium-Btu gas can be produced for use as a chemical feedstock and can be economically transported to nearby markets. High-Btu syn-

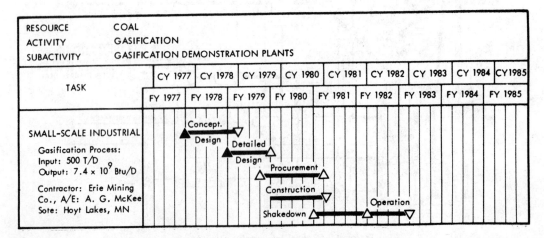

Figure II-3-66. Major milestones for small-scale industrial fuel gas.

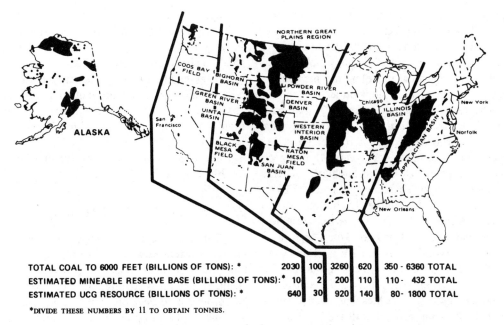

TOTAL COAL TO 6000 FEET (BILLIONS OF TONS): *	2030	100	3260	620	350 - 6360 TOTAL
ESTIMATED MINEABLE RESERVE BASE (BILLIONS OF TONS):*	10	2	200	110	110 - 432 TOTAL
ESTIMATED UCG RESOURCE (BILLIONS OF TONS): *	640	30	920	140	80 - 1800 TOTAL

*DIVIDE THESE NUMBERS BY 11 TO OBTAIN TONNES.

Figure II-3-67. Coal fields of the United States.

thetic natural gas (SNG) can be produced by upgrading the medium-Btu gas in surface facilities and can be pipelined directly into existing natural gas transmission systems. One recent study of SRI* has shown that the UCG-derived power can capture up to 1 quad/year of low-Btu gas and 0.5 quad/year of SNG in the 19-state region around the Rockies.

The objective of the DOE In Situ Coal Gasification Program is to develop commercially viable underground conversion processes for extracting energy from coal. Through Government and industry joint participation, the technology will be developed, proved on a large-scale, and transferred to the private sector. Data will also be provided to predict the economics of a commercial operation. The near-term goal is to develop at least one commercial underground conversion process by 1985 to 1987 and ensure technology transfer to the industrial sector. The mid-term goal is to develop advanced concepts; i.e., increased resource recovery, reduced water usage, and reduced dependence on underground characteristics.

The program goals are to develop technology that will lead to commercial processes that could produce either low- or medium-Btu gases from a wide variety of Eastern and Western coals to address area markets such as electrical generation, synthesis gas, conversion to liquid fuels, or conversion to high-Btu gaseous fuels.

UCG offers a number of significant potential advantages. Some of these advantages have not yet been proved, but FE's program is designed to show that, in addition to tripling our recoverable coal reserves, UCG minimizes health and safety problems associated with conventional coal extraction techniques since no mining is required; produces less surface disruption than strip mining and brings less solid waste to the surface; consumes less water and generates less atmospheric pollution; reduces the socioeconomic impact since fewer operators are required; and reduces both capital investment and gas costs by at least 25 percent.

The primary disadvantages of UCG are possible contamination of surface water and underground aquifers and subsidence of the surface. In order to minimize the impact of

---

* A.J. Molland and D.L. Olsen, "Preliminary Evaluation of Western Market for UCG-Derived Fuel Gas," SRI International, Proceedings of the Fourth Annual Underground Coal Conversion Symposium, Steamboat Springs, Colorado, July 17–20, 1978.

these potential problems, extensive environmental monitoring before, during, and after each field test along with the development of environmental control technology, lab simulations, and analytical modeling are included as integral parts of each field project.

The technical feasibility of UCG has been proven by the British and, most notably, the Soviets, who have had commercial UCG plants in operation for 20 years. Building on the available UCG technology, DOE's Laramie Energy Technology Center (LETC) and Lawrence Livermore Laboratory (LLL) have conducted highly successful subpilot-scale field tests at Hanna and Hoe Creek, near Gillette, Wyoming, respectively. Low- and medium-Btu gases have been produced for sustained periods and successful control of important process variables has been demonstrated. Proving economic feasibility is the key to commercialization of UCG in this country and will require demonstration of efficient use of the coal resource, controllability of the underground process (including environmental effects), reliable long-term operation at high production rates, and verification of the process design at several sites.

These requirements are being fulfilled through an integrated program of field tests, instrumentation, and process model development, and laboratory-supporting research.

The three major UCG process options being developed as part of DOE's In Situ Gasification Program address the wide range of chemical and physical properties of the coal seams and surrounding geology found in this country. The Western Low-Btu Gas project is concentrating on gasifying Western sub-bituminous coal with air injection and has potential for commercialization in 1985. The Western Medium-Btu Gas project is concentrating on gasifying Western subbituminous coal with oxygen and steam injection, instead of air, using a directionally drilled hole. The Steeply Dipping Beds (SDB) project is concentrating on coal seams which dip more than 35° and are not exploitable with conventional mining technology. The Environmental Support and the Supporting Research projects contribute to all three major projects by bringing new technology into the program. One example is the new university research study being conducted to determine the feasibility of underground coal liquefaction. The Eastern Coal Technology project is concentrating on gasifying highly swelling Eastern bituminous coals, which are more difficult to gasify but are closer to large consumer markets. This project is being funded out of the Supporting Research category.

Table II 3-6 summarizes the funding levels for FY 1978 to FY 1980 period.

Western Low-Btu Gasification

The Laramie Energy Technology is conducting the Western Low-Btu Gas Project in a 30-foot thick subbituminous coal seam in the Hanna Field in Wyoming. The objective of this project is to develop a process for gasifying coal in place to produce a low-Btu gas for utility power generation or industrial use. Linked Vertical Wells (LVW) with air injection is one of the processes to execute. It requires minimal seam preparation and has a good probability of success in coal seams that have adequate natural permeability.

The LVW process is applied by drilling vertical wells into the coal seam to provide for injection of air and collection of product gases. Reverse combustion is used to link the wells through the coal seam, followed by forward gasification of coal between the linked wells.

In Figure II 3-68, the center and right wells are linked by reverse combustion. High pressure air is injected in the right well and flows to the center production well through naturally occurring cracks and paths in the coal. The combustion zone advances from center well to right well (injection well) against the gas flow, creating a hot char channel between the wells. Once two wells are linked, the gasification fron reverses direction moving with the gas flow. This second stage, called forward gasification proceeds along the linked passage, which widens and caves in to expose more coal.

As illustrated, one pair of the wells (one module) can be linked while the coal between a second pair is being gasified. When the gasification zone passes the center well, that

Table II-3-6. Funding for in situ gasification for FY 1978 - 1980.

IN SITU COAL GASIFICATION	BUDGET AUTHORITY (DOLLARS IN THOUSANDS)		
SUBACTIVITIES	ACTUAL FY 1978	APPROPRIATION FY 1979	ESTIMATE FY 1980
Western Low—Btu Gasification (Linked Vertical Wells)	$ 4,200	$ 4,300	$2,700
Western Medium—Btu Gasification (Packed Bed)	2,200	2,450	3,200
Eastern Coal Technology	1,200	0	0
Steeply Dipping Bed	1,100	4,550	3,000
Environmental Support	1,200	900	600
Supporting Research	1,100	2,100	400
Capital Equipment	1,600	700	100
TOTAL	$ 12,600	$ 15,000	$ 10,000

well is switched to injection and the right well is switched to production. A link is then started to a new well, yet further to the right. Thus, the gasification proceeds in a line although it expands appreciably to either side of the line. Lines of wells are drilled close enough so the gasification cavities overlap, using as much of the coal as possible. Success of the process depends on controlled movement of the reaction zone between vertical wells and efficient use of the coal resource.

Development of the LVW process started in 1972. In Hanna 1, the first field tests, gas with an average heating value of 126 Btu/scf was produced for about six months, ending in March 1974. The test was quite successful since energy production was four times energy use. Hanna 2, completed in the summer of 1976, produced an average of 8.5 million scf/day of 172 Btu/scf gas. This is equivalent to the electrical needs for a town with a population of 6,000 producing 6 to 8 megawatts of electrical power. This test demonstrated a very high areal sweep efficiency by gasifying a total of 6,700 tons of coal from a four-well square pattern that enclosed 4,600 tons of coal.

The Hanna 3 environmental test was completed in July 1977. This was a two-well experiment designed primarily to measure the effects of the process on underground water flows and compositions and to further refine the LVW operating and control procedures. Analysis of the results will continue for an additional one year.

Beginning with Hanna 2, a major instrumentation effort was fielded by Sandia Laboratories, making these the most instrumented UCG tests ever conducted. Development of a detailed math model of the process was initiated, using a combination of field and laboratory data plus theoretical considerations. Refinement of this model will provide design information for applying LVW to other sites.

The Hanna 4 test is designed to determine the relationship between well spacing and sweep width over long distances and to gasify coal in a relay 5-well pattern with gasification followed by linking in a straight line. This test began in January 1978 and ran into difficulties due to breakage in casing, resulting in an override situation thereby bypassing a substantial amount of coal. The test was scheduled to restart as soon as corrective actions were completed and ran until March 1978.

The next test would be a combined air/ steam O_2 test at Hoe Creek and would be started in July 1979 with steam/O_2 followed by air in the early part of 1980. This test will be based on a mechanically produced link using directional drilling to get the link at the

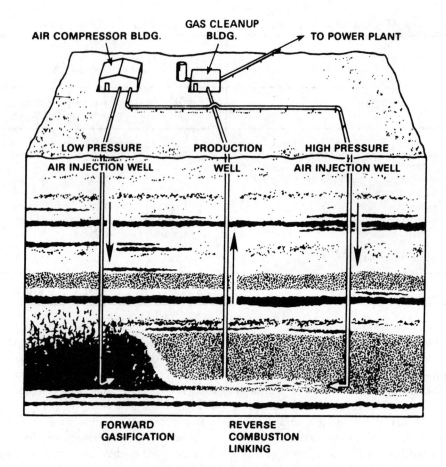

Figure II-3-68. Linked vertical wells process.

bottom of the seam and jet drilling to connect the vertical wells with the bottom link (see Figure II 3-69). First part of the test was performed by LLL and the second part by LETC. The next test will be prepilot test, the site selection for which will start in 1980. It will begin with a two-hole link test for site suitability and will follow with a large-scale test jointly by LETC, LLL, and Sandia. It will be designed to produce 30 to 60 million scf/day of low-Btu gas (equivalent to 800 to 1,500 bbls of oil/day), and will run for 1 year. Industry participation will be sought to facilitate technology transfer. Successful completion of this test will lead to an active industrial partner for pilot-scale test. This will be a 18-month multi-well test producing 90–120 million scf/day of low-Btu gas or 70 million scf/day of medium-Btu gas (equivalent to 2,500 to 3,300 bbls of oil/day) which will be used to generate 50 to 60 megawatts of electric power. The successful completion of the pilot-scale test should make this process ready for commercialization by 1987.

Sandia Laboratories' program for instrumentation and process control development addresses the problem of obtaining a thorough technical understanding of the chemical and physical mechanisms of the in situ reaction

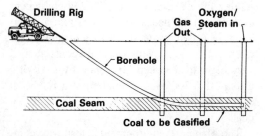

Figure II-3-69. Hoe Creek #3 Test using directional drilling.

and the problem of process control. Preliminary instrumentation efforts on the Hanna 1 experiment during 1973 to 1974 demonstrated that process related effects could be detected by electrical, acoustic, and thermal techniques. Based upon these results, a comprehensive instrumentation effort was designed as an integral part of the Hanna 2 experiment conducted in 1975 to 1976. Currently they are actively involved in the instrumentation effort for the Hanna 4 test. The planned milestones for the project are shown in Figure II 3-70.

Western Medium-Btu Gasification

Lawrence Livermore Laboratory is conducting the Western Medium-Btu Gas project in a 25-foot thick subbituminous coal seam at Hoe Creek, Wyoming. The objective of this project is to develop a commercial process using oxygen and steam injection to produce medium-Btu gas for use as a chemical feedstock or upgrading to SNG. Well-linking methods, other than reverse combustion, will be developed for application of this technology to a wide variety of resource/process combinations (especially very deep, thick coal seams).

LLL has been running the Western Medium-Btu Gas (WMBG) project since its inception in FY 1975. The suitability of a site at Hoe Creek in the Powder River Basin near Gillete, Wyoming, was verified. In FY 1977, a preliminary test of a two-shot fracture and forward gasification with air was run in a shallow coal seam at the Hoe Creek site. Although the desired permeability distribution was not achieved, the reverse linking step was successfully omitted. Data reduction, groundwater monitoring, and post-burn coring activities are in progress.

One site applicability test for LETC's LVW process was completed in the fall of 1977. This test (Hoe Creek 2) used coal that was more uniform, considerably wetter, and more reactive than the coal at Hanna. During this

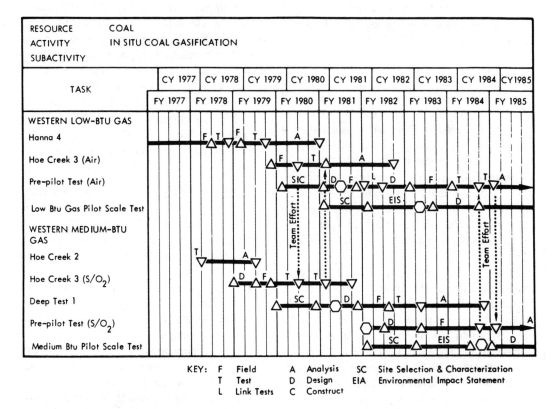

Figure II-3-70. Major milestones for western low- and medium-Btu gas.

test, oxygen and steam were injected for 3 days. Medium-Btu gas was successfully produced with a heating value that averaged 260 Btu/scf and ranged up to 300 Btu/scf over a period of 2 days.

Currently, a process with steam-oxygen injection is being developed using a mechanically derived hole (Hoe Creek 3). This has been previously described in the Western Low-Btu gas section. Alternative techniques are also being investigated. Hoe Creek 2 was completed in December 1977, and used the reverse combustion method of linking to demonstrate medium-Btu gas production with oxygen and steam injection. The Deep Test 1 will be conducted in a deep site, preferably with an industrial partner. The test will start in 1982 and will run for a six-month period. The prepilot test for this project will be combined with that of Western Low-Btu Gas project, also described in the previous section. At this point the project will be moved to Technology Development and Transfer section leading to a medium-Btu gas pilot-scale project assisting industry towards larger scale tests.

Major milestones are presented in Figure II 3-70.

Steeply Dipping Beds

The steeply dipping beds process is designed for seams that dip greater than 35° relative to the earth's surface. This coal is generally not recoverable by existing mining methods. The use of slant drill holes will require the use of sophisticated directional drilling techniques. Subsidence problems are less for the SDB process than for the other in situ gasification processes. The SDB project is cost-shared with industry. It is planned to produce a low-Btu gas.

Over 100 billion tons of SDB coal are estimated to exist in the United States. A large percentage of the Pacific coast coal is steeply dipping, so commercialization of an SDB process could contribute to the future energy supply of the populous West Coast. There are also substantial SDB deposits in the Rocky Mountain area and lesser amounts in the Appalachian and interior coal regions.

In order to develop the necessary cost data for the SDB pilot-scale field test, this project incorporates the execution of several test "burns". The first burn (Rawlins #1 Experiment shown in Figure II 3-71) will serve the purpose of developing the necessary skills and procedures for demonstrating gasification using the simplest module configuration at the most convenient location. The second burn (WS 1) will take place in the State of Washington, where the markets and the infrastructure exist. The third burn (WS 2) would demonstrate the expansion of these operations to a coal unit which contains more than one linked module. At this point it is expected that industry will pick up this technology and move it toward pilot-scale test with assistance from DOE.

The design phase of Rawlins #1 test and fielding is currently underway. The test is scheduled in the first quarter of FY 1980, about six months ahead of schedule. As shown in Figure II 3-72, the completed items are (1) selection of a gasification site, (2) evaluation of any environmental and permit acquisition problems which might be associated with gasification at the site, (3) development of a test plan for execution of this burn, and (4) definition of the facilities and instrumentation required to support the burns evaluation of any special problems (such as extinguishing the fire) which might be associated with project activities. Foreign tests and full-scale operations to gasify SDB's underground have been run with good success.

The project is divided into four phases as follows:

- Phase I: Feasibility Study and Program Plan Preparation (October 1977–March 1978) included site selection, environmental and parametric cost analyses of the design concept, and detailed project and cost plan preparation
- Phase II: Site Characterization, Environmental Assessment (April 1978–September 1979) included detailed site characterization, baseline environmental monitoring, definition of critical process parameters, scaling factors, field test systems requirements, and Environmental Impact Assessement preparation
- Phase III: Field Test (October 1979–March 1982) includes field site in-

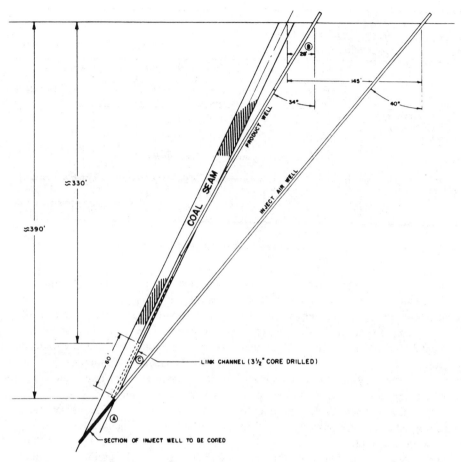

A The air inject well will be drilled first spudding in at 40° and at the point where the bit enters the coal seam, drilling will stop and 6⅝" casing set, at this point the seam will be cored in order to establish seam location at this depth.

B In order to avoid thermal/leak well problems, product well will be spudded in at ~ 34° and deflected towards the coal seam foot wall 2½" per 100 feet, once inside the seam, the bore hole will be straightened so that the product well will be located parallel to the seam foot wall and within the lower third.

C The link zone between the inject well and product well will be cored with a small size bit to a point near the inject air well, at least within 10-15'.

Figure II-3-71. Steeply dipping beds, Rawlins #1 Experiment.

stallation, Field tests 1, 2, and 3; and technical, economic, and environmental analysis of test results
- Phase IV: Pilot Plant Design Estimate (April–September 1982) includes preparation of a preliminary design and cost for an SDB pilot plant

Environmental Support

Environmental problems associated with in situ coal conversion may be less serious than those with mining and surface gasification. The biggest problem could be the disruption and contamination of local aquifers. This area requires further research; however, overall water usage should be lower for in situ than for surface gasification. Air emissions should be no worse than those from surface gasification and will be controllable by similar methods. Surface disruption due to subsidence should be negligible and easily repaired in comparison to strip mining.

Environmental support work determines the

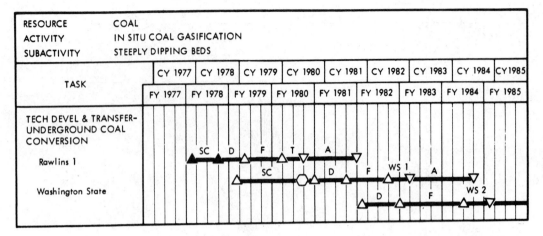

Figure II-3-72. Major milestones for steeply dipping beds.

true impact of environmental problems associated with in situ coal gasification and methods for controlling them. Since environmental problems are common to all in situ concepts, environmental support cross-cuts all three major projects.

The UCG process, by its nature, has potential environmental advantages. It may be possible to minimize surface disruption due to subsidence by adjusting the process design. In any case, surface disruption will be much less than that for strip mining. Most of the solid wastes remain underground. Since there is less equipment in a UCG plant versus and SG plant, there is lower cooling water demand, an important factor in the arid West. In addition, the seam water will enter the underground reactor and be produced. Part of this will be recovered and can be used as process water. Since coal mining is avoided or at least reduced, mined safety problems will also be reduced.

On the other hand, uncontrolled subsidence can leak gas to the atmosphere through cracks to the surface. Subsidence can also break overhead aquifers and allow their waters to enter the reaction zone during and after gasification when pollutants can be picked up and carried away by the subsurface flows. And the location of the best UCG resources in a part of the United States that has limited portable water and population could lead to socioeconomic problems. The Western deposits in regions with almost pristine air quality that are protected by strict laws will require stringent control technology.

An Environmental Development Plan has been issued. It will provide for the implementation of a comprehensive environmental plan for in situ coal gasification projects to ensure that all environmental issues are fully considered in decision-making.

Measurement of gaseous emissions was made during the Hanna 2 test of LVW. Emissions were equal to or less than those from surface gasifiers. Base-line water quality and post-burn charges have been measured in both the Hanna and Hoe Creek tests. Detailed hydrologic and geologic studies have been made in the Pricetown site area to provide base-line environmental data. LLL has run laboratory tests which show that surrounding coal can cleanse subsurface water of harmful pollutants. The tar produced in the Hanna 2 test revealed potentially carcinogenic compounds. Evidently these components were trapped and reacted underground.

The primary objective of Hanna 3 is measuring underground water quality and determining the effects of in situ coal gasification on it. Longer and more detailed air-emission measurements will be made on Hanna 4 and extensive subsidence measurements on prepilot. A small test of gas cleanup will be tried on the prepilot test.

LLL has a cooperative agreement with EPA to conduct a more extensive laboratory program to determine the amount of water cleanup required to remove organic and inorganic contaminants.

Supporting Research

The program's approach to determining the technical and economic feasibility of UCG in coal seams includes laboratory experimentation, process modeling, and the investigation of alternative linking techniques and gasification concepts in well-instrumented field tests. The objective of the laboratory research and modeling efforts is to support the field program by investigating process parameters, alternative linking methods, ignition techniques, and developing instrumentation and software for field use. This work is being performed as follows:

- Morgantown Energy Technology Center's laboratory and modeling program has centered on determining the mechanical and physical properties (especially natural permeability and fracturing characteristics) of coal under overburden stress. A series of experiments that simulate the field tests using the Longwall Generator concept was completed. The supporting research program also includes theoretical modeling and developing of methods to control the gasification process
- The ANL project provides kinetic data on the important chemical reactions that occur during in situ gasification and to determine the important process parameters for the control of these reactions
- the ORNL project is collecting experimental data on the thermal conductivity of the coal samples of interest
- The LASL project is studying the technical and economic feasibility of a two-stage CO_2–O_2 underground coal utilization scheme with minimal water requirements
- The University of Pittsburgh project is measuring and rationalizing the gas-liquid relative permeabilities of coals from lignite samples
- An energy contractor is providing planning support for the costs of a UCG pilot plant facility

- Stanford Research Institute has completed a market study for low- and medium-Btu gas produced from UCG in the Green River, Fort Union, Powder River, and San Juan River region.

Eastern Coal Technology: Morgantown Energy Technology Center (METC). Another supporting research effort is the Eastern Coal Technology Project. Its objective is to develop a method of gasification for Eastern coals that cannot be easily mined. Such reserves amount to approximately 220 billion tons. Thin, low-permeability, high-swelling eastern coal seams are most difficult to gasify by in situ methods. Rugged terrain and densely populated areas in the East make drilling closely spaced (100 feet or less) vertical wells difficult. Directional drilling and the Longwall Generator (LG) processes may solve this problem. Deviated holes provide the best means of developing the underground manifold if they can be drilled economically.

Field experimentation has concentrated on site, resource, and environmental impact assessment (EIA). A site near Pricetown, West Virginia, was selected for the initial test in the bituminous Pittsburgh seam. An LVW minitest (Pricetown I) was prepared for mid-FY 1979 (Figure II 3-73). Several techniques will be applied to initiate communications between the wells if reverse combustion linkage does not succeed. This will consist of hydraulic fracturing and pneumatic techniques. It will also give data on the maximum well spacing that can be used in future tests using directionally drilled wells.

The program will terminate at the end of the Pricetown I test. Depending on program priorities, the program will be reinitiated in FY 1981 with a site selection on several possible options. One option can be recovery of coals beyond the economic limit by mining; a second option can be to work on nonswelling bituminous coals at a shallow site; a third option can be to work with industry at their coal sites, and so on.

In the long run, the project goals are advancing directional drilling technology to ensure a flow capacity within the coal seam sufficient to accommodate the reactant gases,

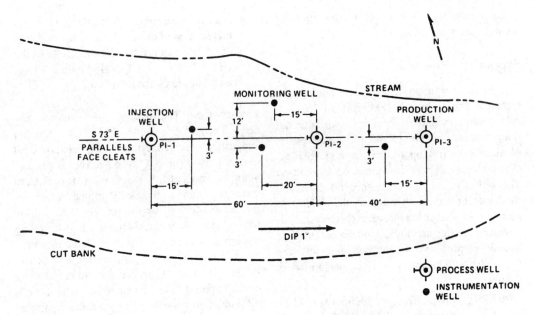

Figure II-3-73. Pricetown 1 Experiment.

testing of forward and reverse gasification modes, and scaleup and application of the process to various thin coal seams.

Major milestones are presented in Figure II 3-74.

ADVANCED RESEARCH AND TECHNOLOGY DEVELOPMENT

The Advanced Research and Technology Development (AR&TD) Activity covers a diverse group of activities ranging from applied research to R&D supporting activities. These are performed by the various technoloygy program divisions, the Division of Systems Engineering, and the Division of Program Control and Support.

Applied research in the subactivities of Materials and Components, Conversion Processes, and Direct Utilization are the responsibility of the respective technology divisions. (See Figure II 3-75.) Specifically, materials is the responsibility of the director of systems engineering; components and conversion processes are the responsibility of the Division of Fossil Fuel Processing; direct utilization is the responsibility of the Division of Fossil Fuel Utilization. In the Program Development and Coordination Subactivity, the Division of Program Control and Support provides long-term strategy and planning, program review and analysis, financial planning, program integration support. In addition, the Division of Systems Engineering provides support in process engineering, economics, and all aspects of environmental planning and assessment covering all Fossil Energy facilities and programs.

Principal objectives of this budget activity are as follows:

- To provide for a central applied research focus for all program areas of Fossil Energy
- To provide a foundation for innovative technology leading to advanced processes through programs in the DOE Energy Technology Centers (ETC's), national laboratories, other Government agencies, private industry, and universities
- To facilitate reliable and efficient operation of synthetic fuel plants through materials and components research
- To accelerate direct utilization of coal or coal-derived synfuels through technology development for combustion systems, heat exchangers and control systems, including applications to heat engines and fuel cells
- To ensure an adequate supply of trained

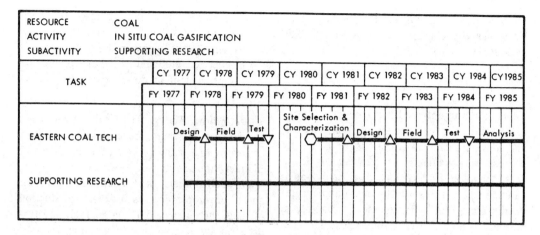

Figure II-3-74. Major milestones for supporting research.

technical personnel from the nation's university system
- To assess the viability of Fossil Energy processes under development in terms of national needs, economic, social and environmental constraints and benefits

The current research program explores the following areas discussed in subsequent sections:

- Advanced fossil fuel conversion processes, involving coal gasification, liquefaction, and refining of shale and coal-derived syncrudes to distillate fuels and chemical feedstocks

- Advanced, environmentally acceptable direct combustion utilization processes for both coal and synfuels
- Supporting research on materials and components, waste use and disposal, and fossil fuel science and engineering
- Systems studies contributing to development of a logical, integrated strategy and rationale for selection and development of Fossil Energy options

AR&TD projects are carried out approximately 50 percent by industry, 25 percent by ETC's, national laboratories, and other Government agencies, and 25 percent by universities.

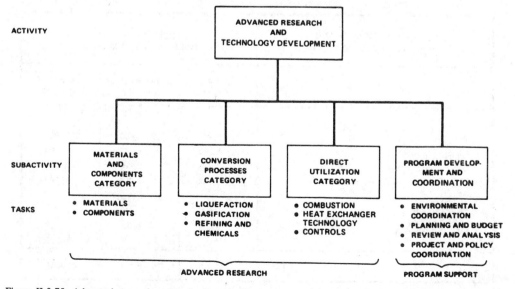

Figure II-3-75. Advanced research and technology development — activity, subactivity, and task support structure.

Table II 3-7 summarizes the funding levels by subactivity for the FY 1978 to FY 1980 period.

Processes

There is an urgent need for a better scientific and engineering base for the ongoing research and development programs as well as for process improvements; i.e., advanced, lower cost technology, which has fewer steps and is simpler and permits substantially higher production rates per unit volume of reactor, at reasonably high thermal efficiencies. To meet such needs, a comprehensive basic and exploratory process research program is being supported in the areas of coal gasification, coal liquefaction, and refining of coal-derived syncrudes and shale oil to distillate fuels, including gasoline.

The overall objective of this process research are as follows:

- To develop basic chemical and engineering knowledge in order to facilitate the successful development of current and/or modified coal gasification and liquefaction processes to provide leads to new processes
- To determine under steady-state conditions the technical feasibility of new process concepts
- To carry out supporting research in selected areas, such as catalyst development, to fill gaps in known processes under development, and/or to provide improvement in these processes

Liquefaction. The goals of this research are to develop, through bench-scale, advanced processes that show promise for the direct or indirect liquefaction of coal to low sulfur, liquid boiler fuels, and distillate syncrudes; and to develop processes for direct production of fuels such as gasoline, diesel fuel, and furnace oil by upgrading and refining coal-derived syncrudes.

The major research categories are as follows:

- Extraction
  — Hydroextraction / desulfurization process research
- Catalytic hydroliquefaction
  — Exploratory evaluation of catalysts
  — Slurry catalyst process
- Pyrolysis and indirect liquefaction
  — Flash hydrocracking / dilute phase hydrogenation
  — Indirect liquefaction from syngas
- Refining and chemicals
  — Exploratory refining process
  — Refining of coal derived syncrudes

Table II-3-7. Funding for AR&TD for FY 1978 - 1980.

ADVANCED RESEARCH AND TECHNOLOGY DEVELOPMENT	BUDGET AUTHORITY (DOLLARS IN THOUSANDS)		
SUBACTIVITIES	ACTUAL FY 1978	APPROPRIATION FY 1979	ESTIMATE FY 1980
Processes	$ 5,201	$ 7,850	$ 14,800
Direct Utilization	5,972	9,450	7,100
Materials and Components	7,564	9,290	8,900
Program Development and Coordination	8,363	13,056	7,500
Capital Equipment	300	375	400
Construction	9,600	6,350	12,050
TOTAL	$ 37,000	$ 46,371	$ 50,950

- Supporting Research
  — Basic chemical and engineering studies; e.g., structure of coal, preasphaltenes, mechanism of coal hydroliquefaction, coal / solvent /hydrogen mixing studies, catalytic reactor modeling / design studies, pre-heater studies

Specific needs include development of a better understanding of the basic chemical and engineering aspects of the preheater stage in direct liquefaction; development of a lower cost reliable process for de-ashing nondistillate fuels; definition of the stages in direct liquefaction and relative rates; definition of the applicability of catalysis in direct liquefaction and of effective catalysts; development of a more cost-effective Lewis acid catalyst than zinc chloride for effecting the hydrocracking of coal to low sulfur, low nitrogen distillate fuels; and determination of the behavior of coal on rapid heat-up in the presence of hydrogen or synthesis gas relevant to the production of distillate fuels.

Major accomplishments in FY 1978 and FY 1979 included:

- Demonstrated technical feasibility of converting methanol to high-octane gasoline in nearly quantitative yield in a fixed-bed, bench-scale unit
- Demonstrated, on a small scale, a novel catalyst system capable of converting syngas to high octane gasoline in one step
- Uncovered new information indicating that coal undergoes irreversible reactions with certain solvents such as pyridine
- Provided experimental evidence that donor solvents such as tetralin, to a certain degree, become bound to coal in addition to transferring hydrogen
- Developed a detailed economic comparison between gasoline via Sasol Fischer-Tropsch technology and via methanol

Figure II 3-76 shows the key projects. These efforts will be closely coupled to the technology development programs and, will, in part, focus on key issues and knowledge requirements identified in those programs. Major thrusts in FY 1980 will include: (1) innovative process concepts involving advanced catalysts for liquefying and gasifying coal, advanced synthesis gas conversion, upgrading of coal liquids and flash hydropyrolysis; (2) fundamental studies to develop new coal chemistry and better understanding of coal conversion processes, the structure of coal, and the role of mineral matter in coal on conversion processes; and (3) essential studies on mechanisms, kinetics, and the effects of reaction constituents on the catalyst as a function of time and temperature. Such research could lead to the development of combination catalysts for steam-carbon gasification, water-gas shift, and methanation which would enable these reactions to take place simultaneously in the same reactor at reduced overall cost.

Gasification. Second-generation coal gasification processes are highly capital intensive; end product costs are therefore very sensitive to conversion efficiency and operating conditions. The current state of the art offers room for improvement, through better understanding of process fundamentals, as well as a prudent data base for uses in both ongoing programs and identification and education of advanced processes. A comprehensive basic and exploratory process research program and an early development program are in progress.

The objectives of this project include:

- To provide an early assessment of the technical and economic potential of new, novel, and advanced gasification processes through small bench-scale and early process development studies
- To develop basic chemical and engineering knowledge to facilitate the successful development of current and future generation coal gasification processes and to suggest new approaches to coal gasification
- To carry out supporting research in selected areas such as catalyst development and gas cleanup to provide major improvements

Major accomplishments in FY 1978 and FY 1979 included:

- Expanded the data on the effectiveness of lime and lime/caustic pretreatments of a caking bituminous coal for fluid-bed gasification with steam or hydrogen at high specific gasification rates and high methane yields
- Completed the technical feasibility of elements of an advanced, catalytic coal gasification process through bench-scale experiments
- Completed initial evaluation of the technology for rapid rate hydrogasification concepts which produce synthetic natural gas (SNG) or SNG-light hydrocarbon liquid (BTX) as coproducts. The evaluation established process product yield structures as a function of reactor conditions and feed stream composition and characteristics
- Initiated programs to evaluate advanced gas cleanup processes for small particulate removal, tar removal, H_2S removal, and alkali metal controls
- Expanded the thermodynamic data base for inorganic compounds related to coal conversion processes

The Advanced Gasification Process Research program contains projects which are directed at conversion of coal to SNG, intermediate-Btu gas, synthesis and hydrogen. The main program elements are:

- Advanced gasification processes for high-Btu gas

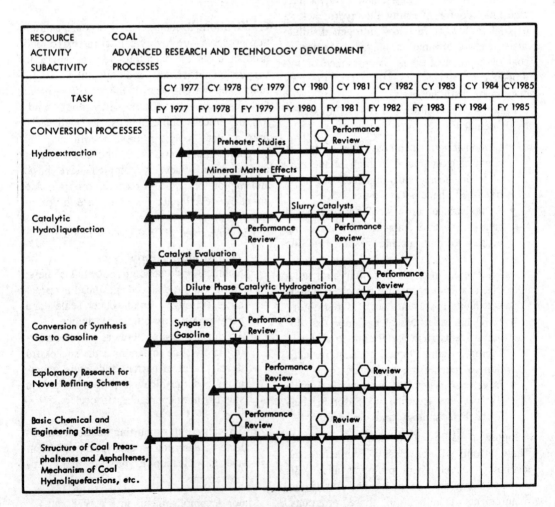

Figure II-3-76. Major milestones for conversion processes.

- Catalytic gasification
- Catalytic methanation
- Rapid rate hydrogasification
• Advanced gasification processes for low- and intermediate-Btu gas and hydrogen
  - Short residence time (fast pyrolytic) gasification
  - Catalysts gasification for hydrogen
• Supporting research
  - Gas cleanup and separation
  - Reaction kinetics and thermodynamics
  - Preliminary engineering and economic assessments

The overall program structure is provided in Figure II 3-77.

Specific current research needs include development of a better understanding of the fundamental controlling factors associated with coal/steam, coal/hydrogen gasification processes; development of an advanced catalytic gasification process; development of an advanced, short residence time hydrogasification process; development of cost effective, efficient hot-gas cleanup systems for removal of sulfur species, alkali metal species and particulates, and a further expansion of the data base related to process chemistry, kinetics, and thermodynamics.

FY 1980 goals include:

• Provide preliminary engineering assessments of advanced gas cleanup processes for both high- and low-Btu gasification
• Develop an inproved understanding and data base related to the kinetics and thermodynamics of gasification processes
• Initiate exploratory studies to identify advanced processes for dewatering peat
• Identify and evaluate improved gasification catalysts

A milestone chart for key projects is shown in Figure II 3-78.

Direct Utilization

Timely substitution of coal and coal-derived fuels (synfuels) by industry requires the solution of a variety of problems related to the end-use system. Techniques to upgrade the raw fuel characteristics, the development of new or modified equipment for more effective and environmentally acceptable combustion, and technologies for more cost-effective treatment of system effluents offer significant potential to improve the competitive position of these fuels and thus accelerate their market penetration. The program structure is illustrated in Figure II 3-79.

In FY 1976 and 1977, improved sorbent capabilities for SO_x removal and potential chemical comminution of coal have been demonstrated, quantitative ash-fouling tests have been utilized toward conservative boiler design, and the effectiveness of combustion modifications to lower the NO emissions from pulverized-coal combustors has been demonstrated. In FY 1978 and FY 1979, the emphasis was on laboratory and bench-scale development of advanced coal beneficiation processes, alternative preparation methods of coal/oil mixtures, and advanced hot-gas cleanup concepts. The environmental impact of these technologies is an important aspect of all these studies.

Program emphasis in FY 1980 will be changed to reflect the fact that work in the areas of fluid bed sorbents, coal beneficiation, coal/oil mixtures, and hot-gas cleanup have reached a point of development which allows responsibility to be transferred to the appropriate FE mainline program activity. Major thrusts in FY 1980 will focus on: advanced research to relate solid and liquid synfuel composition to combustion characteristics, including fundamental studies on gaseous pollutant formation and particulate agglomeration; development of combustion control technologies, including scaling studies for application to industrial-size equipment; and generic heat exchanger technology to identify and assess promising candidate techniques for improving cost and performance effectiveness in current and advanced systems.

Pulverized Coal and Synfuel Combustion. The contribution of this effort to accelerated use of coal-fueled systems is the development of improved combustion components capable of optimizing system performance for different industrial and utility applications.

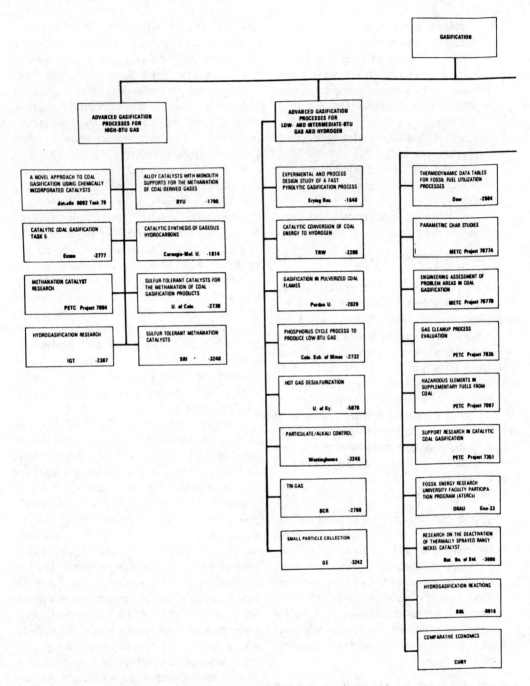

Figure II-3-77. Program structure for gasification detail of the conversion processes.

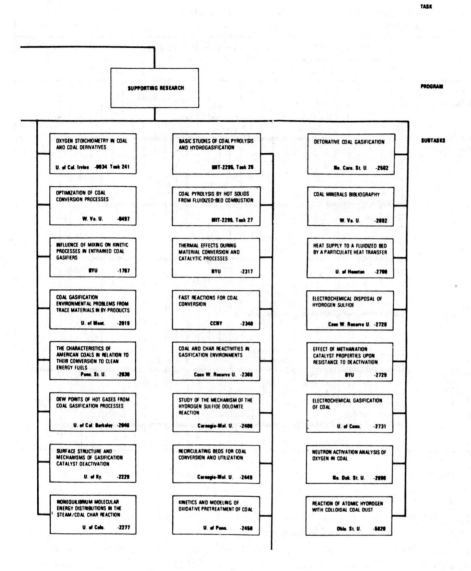

Figure II-3-77. (Continued).

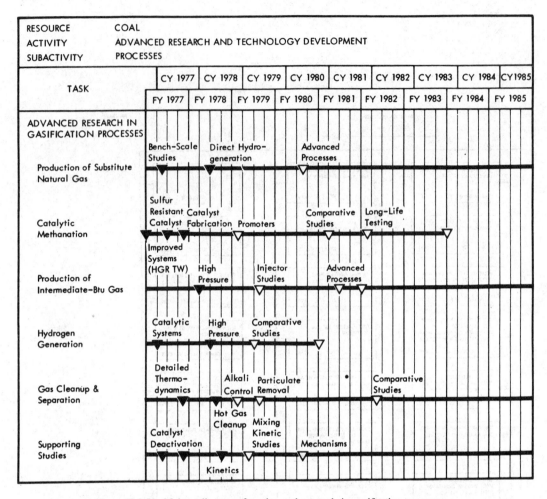

Figure II-3-78. Major milestones for advanced research in gasification processes.

This is a generic technology project whose goal is to provide the combustion technology base required for optimum substitution of coal into the industrial sector. Significant penetration of this sector presents the equipment designer with the problem of producing reliable combustion equipment for a larger variety of coals and coal chars in a broad range of equipment sizes, and in the face of severe economic and environmental constraints on system performance. The present technology base, which is focused primarily on a narrow range of coal types and equipment sizes, is inadequate to support the timely development of the required equipment. During FY 1979, efforts under this project focused on elucidating fundamental mechanisms for pollutant formation during pulverized coal combustion.

The FY 1980 effort will be a coordinated project involving participants from industry, universities, and national laboratories as well as Energy Technology Centers, whose ultimate objective is application of advanced diagnostics, and computational modeling, to apply the results of bench-scale studies to full-size pulverized fuel-burning systems.

Major milestones are given in Figure II 3-80.

The synfuel technology project will ensure the existence of the technology base required for optimum economic utilization of coal derived fuels, focusing on coal liquids. The goals of current FE liquefaction programs are to produce fuels which approach traditional fuel specifications sufficiently to allow use in existing combustion equipment designed for petroleum fuels. Cost penalties associated with upgrading raw coal liquids to petroleum fuel specifications have been estimated to be as high as 30 percent, a figure which has an obvious adverse effect on their comptetive position. The development of combustion equipment designed specifically for use of minimally processed coal liquids offers significant potential to reduce refinery/combustion system costs. This development is severely limited by the fact that the current state-of-the-art is incapable of determining those combustion characteristics required by the equipment designer from the elementary fuel composition data available from the various fuel processing investigations. In FY 1979, the project focused on problems associated with utilizing coal liquids in industrial and utility boilers.

During FY 1980, the scope of the project will be broadened to provide an effective technology base supporting coal-derived fuel utilization in the full spectrum of combustion devices in which they much be burned, including engine applications. (See Figure II 3-81.)

Heat Exchanger Technology. This generic technology development project has the goal of improving heat exchanger performance/cost characteristics for application in heat recovery and advanced energy conversion systems.

During FY 1979, some heat exchanger work was performed in the Heat Engines and Heat Recovery Activity. In FY 1980, it is included as part of the AR&TD program. The FY 1980 effort will focus on improved understanding of fundamental heat exchanger phenomena; e.g., flow-induced vibration (FIV), fouling, and corrosion, which affect design and cost and evaluation of new concepts that offer the potential of extended temperature capability and higher performance/cost ratios. Specifically, four main program areas will be pursued: (1) development and fluidized-bed heat exchanger (FBHX) design correlations and evaluation of FBHX concepts for heat recovery applications; (2) evaluation of low-cost tubing concepts including plastic heat exchangers; (3) development of ceramic heat pipe and tubular ceramic recuperator concepts; and (4) development of improved joining techniques and NDE/NDT techniques for ceramic heat exchangers. (See Figure II-3-82).

The objective of this technology project is to identify and develop new control system concepts required by advanced coal utilization systems now under development. Acceptance of complex and tightly integrated (e.g., combined cycle) systems will require control systems which equal or exceed those systems currently available for gas- or oil-fired systems in terms of safety and reliability.

The example variations in grade, rank, and physical preparation of the solid coal can cause perturbations in component performance and power output in a gasified coal-fueled gas turbine combined cycle which do not occur in conventional systems. Variable properties of the gasified coals can cause problems during startup, shutdown, and/or load following in addition to the recognized variations in gas stream contaiminants.

The project for FY 1980 will address three fundamental areas: studies to identify coal utilization systems which have nonconventional control system requirements; studies to identify promising new control system concepts based on emerging technologies such as laser fiber optics; and studies to develop performance criteria for control systems to meet new coal utilization system requirements. (See Figure II 3-83.)

Materials and Components

The success of coal conversion as a source for clean alternate fuels will hinge on both the efficiency and the continuity of operation of the conversion plant. Regardless of the process selected, continuity will be totally dependent on the reliability of peripheral and support equipment and on the ability to maintain a steady state of operation.

To assure a smooth transition from shakedown to productive plant operations, special emphasis is directed to programs in the following areas:

- Materials research and selection
- Development of specialized components and equipment subsystems
- Instrumentation and control

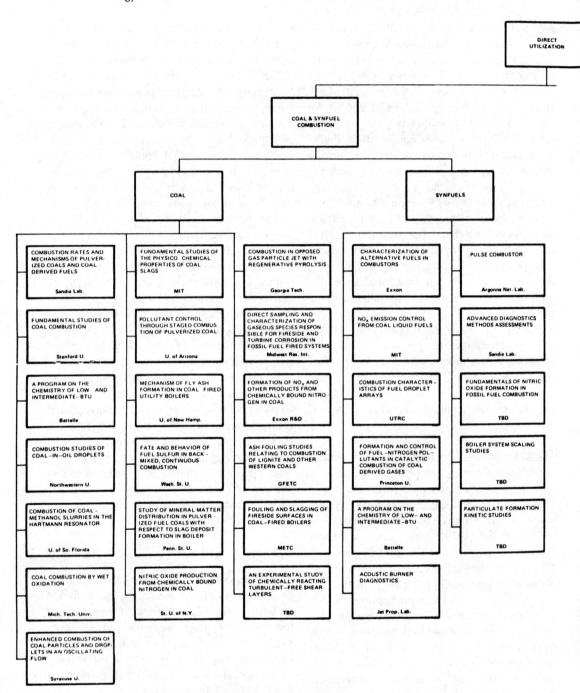

Figure II-3-79. Program structure for direct utilization subactivity and related details.

Activity Description—Coal 183

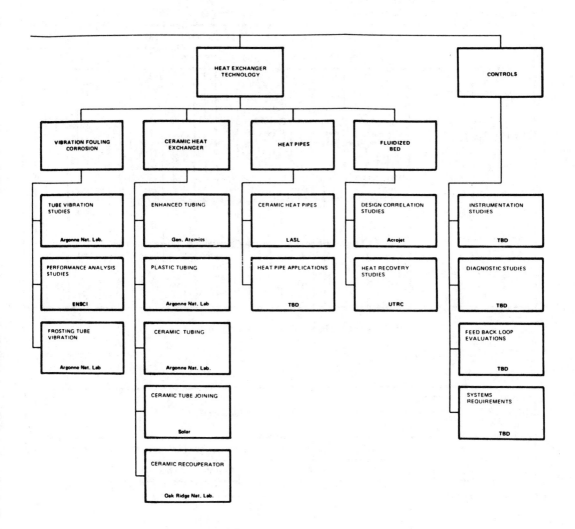

Figure II-3-79. (Continued).

184 Fossil Energy

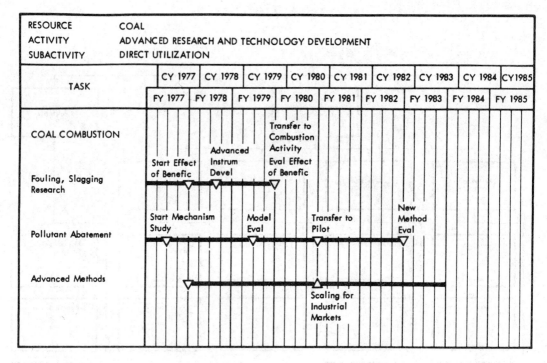

Figure II-3-80. Major milestones for coal combustion.

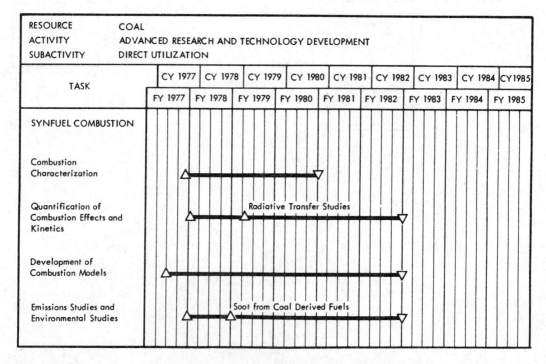

Figure II-3-81. Major milestones for synfuel combustion.

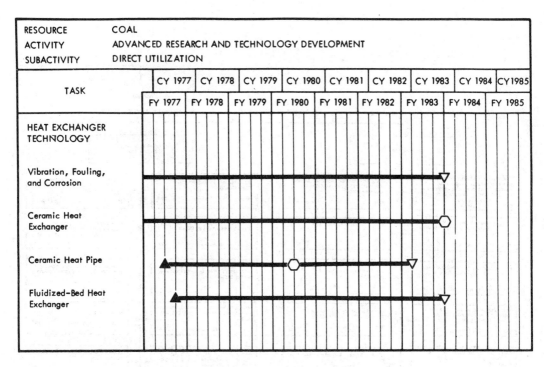

Figure II-3-82. Major milestones for the heat exchanger technology.

- Development of operating, maintenance and training manuals
- Development of personnel staffing guides specifically for coal process plants

Since no large or commercial coal plants have been constructed in the United States, extra emphasis is being directed to the environmental impacts of coal conversion and how these will influence the health and safety aspects of coal plant operations.

The overall program structure is provided in Figure II 3-84.

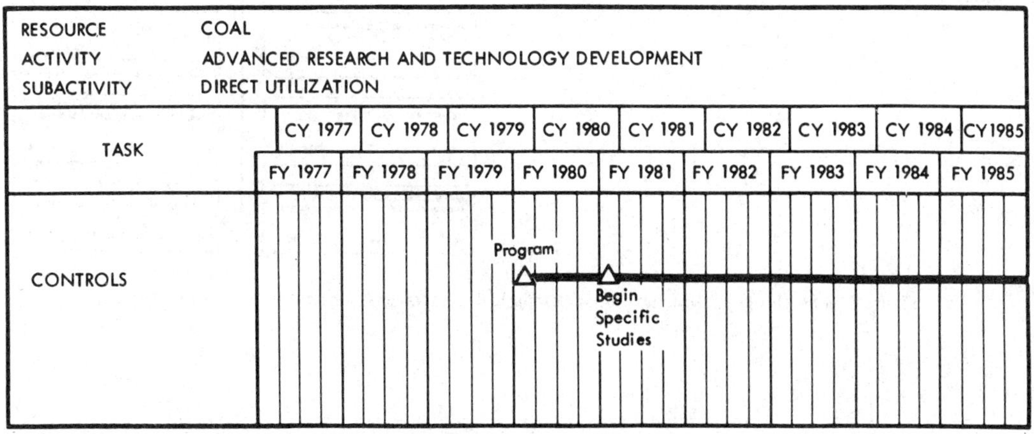

Figure II-3-83. Major milestones for controls.

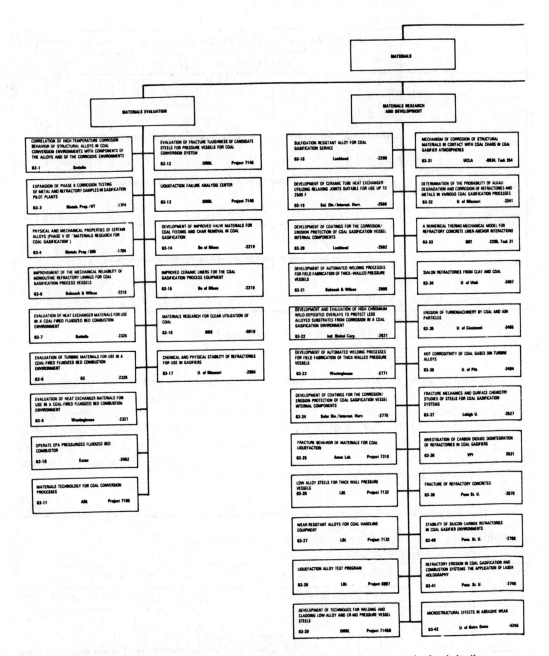

Figure II-3-84. Program structure for materials and components subactivity and related details.

Activity Description—Coal 187

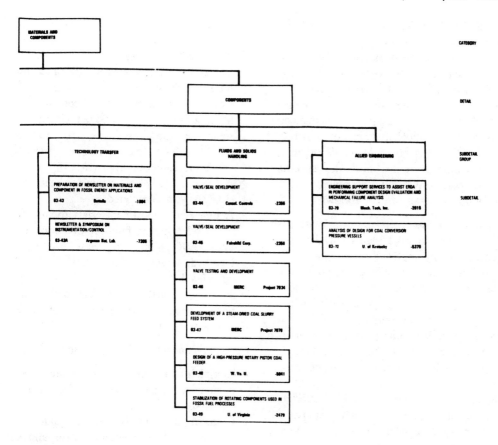

Figure II-3-84. (Continued).

Materials for Coal Conversion. Historically, every developing technology has been keyed to the prior development of critical materials. Efficiency and economy of coal conversion are enhanced at elevated pressures and temperatures; however, erosive and corrosive effects are greater. Successful commercialization requires compatible materials of construction. Accordingly, a broad materials program has been undertaken to meet the specific needs of coal-conversion technology. Elements of this program are as follows:

- Conduct failure analysis of materials now in use in pilot plants, PDU's, and coal-handling applications
- Conduct research and development on the mechanisms of erosion, corrosion, abrasion, fatigue, and other forms of deterioration as encountered under coal plant operating conditions and environments. Included will be metals, special-purpose alloys, ceramics, and refractories
- Select and develop compatible materials for economy of cost and endurance against wear or deterioration; steels, special-purpose alloys, ceramics, and refractory linings will be included
- Operate one or more materials test facilities as required to obtain realistic material-test data
- Investigate and identify extra-strong and weldable steels for thick-walled pressure vessels
- Provide for the timely distribution of materials information by seminars and newsletters

A significant portion of these tasks are being investigated at the national laboratories, including Ames, Argonne, Oak Ridge, Idaho National Energy Laboratory (INEL), and Lawrence Berkeley.

Significant accomplishments reported in FY 1978 include:

- Sulfidation has been shown as the most frequent cause of metal failure in gasification reactors.
- High-temperature alloys of FE–Ni–Cr are not resistant to erosion/corrosion in gasifiers over long periods of exposure.
- Alloys of Fe–Cr–Al show marked promise as corrosion-resistant claddings for internal use in gasifiers.
- Refractory concretes of 95 percent alumina are degraded in gasifiers containing steam at high pressures; 50 percent Al_2–O_3 refractories increase in strength under the same environment.
- A design study of ceramics heat exchanges was completed.
- A first evaluation of materials for use in heat exchangers in fluidized-bed combustions (FBC's) and in pressurized fluidized-bed combustions (PFBC's) was completed.
- A study of heat flow through multicomponent refractory lined gasifier pressure vessel walls was completed.

Accomplishments in FY 1979 included:

- Complete the selection and/or development of compatible coatings for the internal surfaces of gasifiers
- Complete a feasibility study to guide development of sulfidation-resistant alloys and metals for use in high-temperature gasifiers
- Complete a design study of ceramic heat exchangers
- Initiate materials program for third-generation coal conversion processes

Anticipated accomplishments in FY 1980 are as follows:

- Initiate materials test program at IEA's PFBC facility at Grimethorpe, United Kingdom
- Complete materials testing for second generation gasifier internals
- Operate ceramic heat exchanger module
- Operate pilot plant test unit for refractories for slagging gasifiers
- Issue data book on materials for gasifiers
- Determine corrosive species in coal liquefaction processes

Major milestones are given in Figure II 3-85.

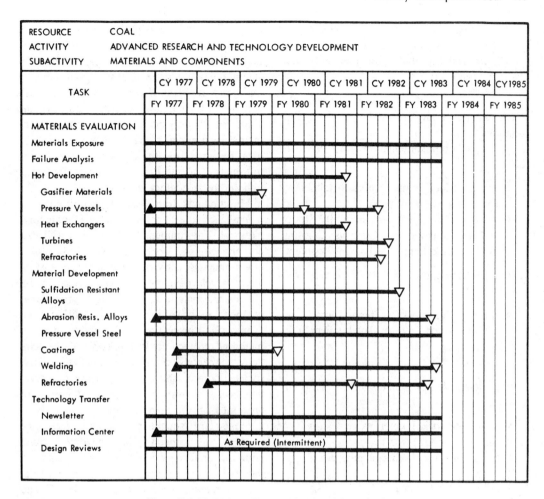

Figure II-3-85. Major milestones for materials evaluation.

Equipment Subsystems for Coal Conversion. All processes to convert coal to synthetic fuels are essentially similar in the basic chemistry. The difference among processes lie in operating pressures and temperatures, residence times, physical contact between reactants, and composition of in-plant streams. Improvements in conversion processes and composition and efficiency will result from plant scale-up, higher pressures, higher temperatures, faster reaction rates, and better contact between reactants.

The coal plant can be no more reliable, however, than its support equipment and peripheral subsystems. Many current operating problems and shutdowns in coal pilot plants are directly attributable to the use of commercial components never intended for the higher temperature and harsh environments of coal plants and in-plant streams. To provide equipment which will endure these environments over extended periods, a vigorous program has been undertaken to adapt, modify, and develop special components, heavy equipment, subsystem and control instrumentation compatible with and scaled up to conversion plant needs with proper emphasis on critical plant geometries.

The following subsystems and components are under active investigation and development:

- Advanced techniques for the continuous feeding of coal to pressurized reactors

- Development and testing of lock hopper valves from materials which will perform reliably under adverse and harsh environments

Major portions of the above program elements are being implemented by national laboratories.

Major accomplishments in FY 1978 included:

- Installed lock hopper valve in laboratory for certification of valve
- Awarded contracts to Fairchild-Stratos and to Consolidated Controls Corporation on lock hopper valve development

Accomplishments in FY 1979 were as follows:

- Initiate high-pressure, high-volume centrifugal slurry pump development program
- Develop a hot char discharge system

Expected accomplishments in FY 1980 include:

- Complete lock hopper valve testing
- Start O_2 compressor test
- Start hot char discharge program

Major milestones are given in Figure II 3-86.

Program Development and Integration

The four major supporting functions performed under this subactivity include:

- Environmental coordination
- Planning and budget
- Review and analysis
- Project and policy coordination

Environmental Coordination. The application of advanced research and the development of new technologies are important measures if coal is to play its part in achieving the goals set forth in the National Energy Plan. These efforts to find ways to use coal more efficiently must, however, consider any potential impact upon the environment.

Planning activities are carried out at both the programmatic and project levels to ensure that environmental problems are addressed at the earliest possible point. Programmatic considerations are addressed in the Environmental Development Plans (EDP's) prepared for each activity. FE planning activities address the ability of FE projects to meet current environmental requirements, the development of technical resource, and the economic performance information about FE technologies for input into future regulatory decisions.

FE conducts four basic types of environmental activities: policy analysis, planning, National Environmental Policy Act (NEPA) compliance, and operations. Generally, these environmental activities are directed toward areas which develop new knowledge in FE chemistry, materials, technology, processing, engineering, analysis, sampling, worker management and protection, and property management and protection. In this context, the term "environmental" refers to health and safety, socioeconomic, and energy process conservation issues as well as the physical impact of processes on the environment.

Inputs are sought from other interested groups and agencies and considered along with the environmental study data generated internally. Implementing the environmental strategy requires coordination between FE and its industrial partner. To facilitate this coordination, selected responsibilities are assigned to each of the co-sponsors. FE establishes the criteria necessary to carry out these activities, reviews and approves the environmental protection plan prepared by the industrial partners, monitors the effort and undertakes supporting and complementary research to assure the completeness and validity of the information being gathered.

The results of the environmental policy analysis activities serve as input to the technology goals. FE sponsors studies which consider how current and emerging laws, standards, and regulations impact its programs and technologies and the potential for commercialization. The technical performance, resource requirement, and economic information resulting from such studies is used to review new regulations and is exchanged with regulatory agencies.

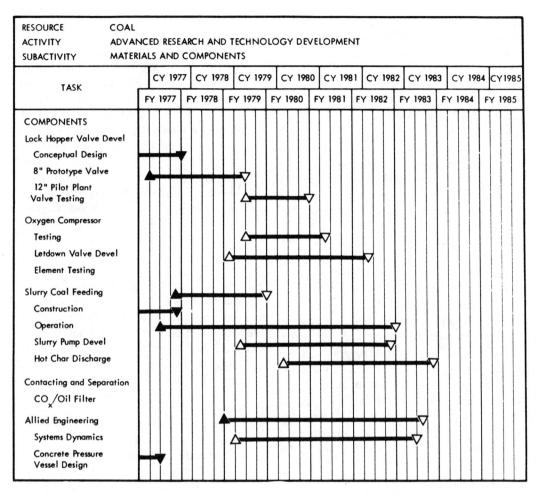

Figure II-3-86. Major milestones for materials and components.

Planning and Budget. The program planning support area includes many cross-interdepartmental coordination activities and provides the major focal point for interaction with the DOE staffs. Major activities include: participation in the development of DOE policy; interrelation of policy into programmatic terms; coordination between Headquarters, field and ET staffs in developing multiyear program plans; development of cross-cutting fossil energy strategies that relate the FE program to general policy guidance; and market analysis for new technologies. In addition, several critical studies are underway to provide supporting information to the decision process. The studies include policy and mission analysis and program strategy development.

Policy and mission analysis involves evaluating new and existing energy studies and forecasts that impact FE programs. This includes the energy information agency annual report to Congress as well as more speculative work done in support of National Energy Plan development. The products of these studies are analysis that interpret this work into FE programmatic terms.

The current focus for program strategy development is the development of detailed cross-cutting strategies for liquid, solid and gaseous fuels. These strategies will be developed in conjunction with Headquarters program management and field product management. The activity will use inputs from Policy and Evaluation in developing the strategies and will include sufficient analytical support to ensure adequate documentation of programmatic decisions. The effort will identify major goals and their priorities.

Major documents produced by planning, in addition to specific analysis, include the Multi-Year Program Plans for each major FE division and the strategic portions of the spring planning documents. Because of its strong analytical capability, the group is fully called upon to provide special analysis on a quick reaction basis. Finally, the group's broader overview of policy and planning makes it the ideal candidate for departmental task forces.

Planning and support documents are prepared by the program planning area including coordinating the preparation of the annual Fossil Energy Research Program (Gold) Book, Management Review and Control, and spring planning documents and assisting in the preparation of other technical documents.

The budget development support area includes formulation and execution of the Fossil Energy budget, development of financial systems, issuance of financial plans, development of integration networks for Fossil Energy program plans and budgets. In addition, the budget support area oversees responsibility for the following:

- Developing and planning FE resource requirements to carry out R&D programs. This includes coordinating all FE budget estimates, preparing calls and providing guidance and technical assistance to all program divisions and field offices; coordinating analysis and review of the budget; coordinating the preparation of testimony and supporting justification or information; coordinating reprogramming, amendment, and supplemental budget actions; and coordinating responses to internal Office Management and Budget (OMB) and congressional inquiries
- Coordinating resource requirements of the Energy Research Centers and national laboratories. This includes providing the field operations with planning and budget guidance, consolidating and analyzing their resource requests, and providing them with approved operating Financial Plans
- Ensuring effective utilization of available resources. This includes providing management with up-to-date financial reports, coordinating all contract procurement actions; certifying availability of funds in conjunction with approved procurement and financial plans; maintaining an on-line Data Management System to provide timely data to serve management in resource decision-making; and coordinating Energy Research Centers/National Laboratories budget matters with area Operation Offices.

Review and Analysis. The Systems Engineering and Analysis efforts provide program control options to the Fossil Energy Program Director based on technical analyses and provide technical support to the Fossil Energy operating divisions and their field activities. Specific efforts include:

- Analysis of large-scale systems to ensure proper functioning of the total system. Using a top down approach, the total input-output requirements of the system are determined in terms of functions required to be performed. The total system is further divided into successively smaller subsystems with input-output requirements and the method(s) of interconnection defined. From these analyses, an evaluation of the capability and effectivity of the system to accomplish the overall objectives is made.
- Provision of primary process design studies to provide definitive comparisons in the areas of liquefaction, low-medium Btu gasification, atmospheric fluidized-bed combustion, pressurized fluidized-bed combustion, flue gas desulfurization, and coal preparation.
- Incorporation in the system data books of environmental and social impacts on a regional basis and characterization of coal mining, preparation, and support programs.
- Preparation, test, and use of a new gen-

eration design and cost estimation simulation model. The simulation model will handle coal and shale solids and liquids unit operations.
- Development of consistent engineering and economic standards for performing and reporting design and cost engineering work.

The Fossil Energy Performance Assurance System (PAS). In recognition of its responsibilities under P.L. 95–91, in which DOE is charged to "increase the efficiency and reliability in the use of energy," and an effort to attain National Energy Goal Number IV to "increase the efficiency and reliability of the processes used in energy conversion and delivery systems," DOE–FE has established and implemented a formalized Performance Assurance System (PAS).

PAS is essentially a management technique to help achieve the early, reliable, and cost-effective performance of fossil projects. It consists of systematic applications of design considerations, data collection, analysis, and communicative networks and procedures.

The disciplines incorporated into PAS are reliability, maintainability, quality assurance, availability, operational service life, life cycle costs, safety engineering, and standardization. PAS uniqueness lies in the manner in which these techniques are applied to fossil programs.

PAS is designed to be applicable to all project maturity levels, from small material and component developments to the large demonstration plant. The Performance Assurance Program will be tailored to each project using basis criteria such as maturity level, dollar value, technical feasibility, and commercialization potential in determining the exact scope and size of the program.

The goals of the system are accomplished through:

- The PAS Office, an element of the Systems Engineering Division, FE, acquires, generates, maintains, and disseminates information, and makes available procedures, tools, and methodology to assist in assuring that systems availability, life-cycle cost, and the other performance assurance elements are adequately treated in the development, construction, and operation of fossil energy RD&D systems. The PAS Office provides technical support to project managers and contractors on an as-needed basis. PAS will have the capability to perform independent RAM assessments, life-cycle cost estimates, and special studies as required.
- A Fossil Energy Equipment Data System (FEEDS) is established for the collection, evaluation, and storage of data on the failure, operability, and maintenance of fossil energy components and materials, and for the analysis and dissemination of this information to assist architect-engineers, technicians, and managers in the selection, assessment, operation, and maintenance of fossil energy components and systems.

Conceptually, PAS has been developed as a support service. Project managers continue to exercise full management control, supervision, and responsibility over all aspects of their individual PA programs.

If properly applied, PAS will ensure or provide for the following:

- The identification of all reliability critical, maintenance critical, and cost critical items in a plant/project and the recording/documentation of the performance of these critical items throughout the design, construction, and operational phases
- Failure modes of these critical items have been properly identified and documented
- Actions to enhance plant operability have been taken (i.e., redundancy, improved maintenance, improved quality)
- Life-cycle cost analysis has been accomplished
- Service life estimations of equipment/components made
- Standardization through use of common items/materials stressed

- Plant configurations properly documented and updated as required
- Operational/maintenance procedures developed and documented
- Materials suitably monitored and documented
- Complete plant/project operability data collected

The conceptual design of the PAS was completed during 1978. This was accomplished through close coordination with industrial contractors to assure that the program approach selected was meaningful and realistic. These industrial contractors have confirmed the need for such a program. Other governmental and private agencies and organizations have expressed enthusiasm and delight that DOE–FE was developing such a program. No other program of this kind and character exists, nor is in development within or outside of DOE.

Future development, refinement, and initial application of the PAS program has been accomplished in 1979.

FE demonstration plant projects are *required* to address PAS. Application of PAS to remaining FE projects will be on a voluntary basis, as determined by each project manager.

ADVANCED ENVIRONMENTAL CONTROL TECHNOLOGY

The ultimate goal of the coal strategy is to decrease the nation's dependence on imported oil through a range of cost-effective uses of coal. Because close to 90 percent of the coal consumed in this country is, and will continue to be, burned directly, a primary objective in achieving this goal is to assure that stationary facilities now burning coal can continue to do so while meeting applicable environmental standards and coal combustion can become an increasingly viable option for new facilities in the utility and industrial sectors during the next decade. Environmental standards pose the greatest constraint on increasing direct coal use. Unless new technologies are available for meeting these standards, many facilities that might use coal will turn to other fuels.

Coal in inherently dirty. Sulfur, nitrogen, alkali and halogen compounds, and volatile trace metal species are found in abundance in coal and are released through gasification and combustion processes. These substances can degrade the performance of the energy-producing systems and degrade the quality of air and water. The environmental goals set out in the Clean Air Act, the Federal Water Pollution Control Act (FWPCA), and the National Environmental Policy Act are also high national priorities. It is important to achieve both the national energy goal of coal utilization and the environmental goals without endangering the public health or degrading the environment, and with minimum economic impact. The Advanced Environmental Control Technology (AECT) program is a key element in meeting these goals.

The environmental performance required for both utility and industrial coal-burning facilities depends largely on the age and location of the facility. Most existing facilities must meet the emission limitations in the Clean Air Act's State Implementation Plans (SIP's). New facilities which commence construction before proposal of EPA's revised New Source Performance Standards (NSPS) must meet the current NSPS. However, future coal use will be increasingly affected by the stricter standards in upcoming NSPS revisions.

There are three methods of removing polluting substances from coal: coal cleanup, gas stream cleanup, and flue gas cleanup. Coal cleanup research and development is described in the Fossil Energy Mining Research and Development Program. Flue gas and gas stream cleanup are the key areas addressed in its program. Figure II 3-87 shows the various coal utilization process streams.

The goals of the Advanced Environmental Control Technology Program are to identify, research, develop, refine, and demonstrate a range of engineering approaches capable of the following:

- Removing flue gas pollutants for compliance with emission standards
- Removing undesirable components from process streams produced during gas-

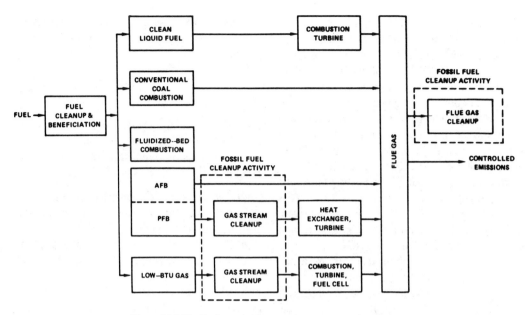

Figure II-3-87. Environmental process stream control options.

ification and/or combustion, thus protecting utilization equipment such as turbines, fuel cells, and heat exchangers

The technology program is organized into three subactivities:

- Flue Gas Cleanup
- Gas Stream Cleanup
- Technology Support.

Flue Gas Cleanup addresses the removal of pollution-casusing contaminants from the stack gases of conventional combustion units to meet environmental standards. Efforts will be focused on improving and demonstrating the reliability of conventional lime/limestone scrubbers, developing second generation Flue Gas Desulfurization (FGD) technologies that avoid wet sludge disposal, and the initiation of advanced technologies for removal of NO_x, particulates, and heavy metals. Gas Stream Cleanup includes the technology for removal of contaminants during the combustion process or from the process stream prior to its utilization. Both hardware and environmental protection are the key concers. Primary emphasis will be to develop technologies to clean gas streams produced by coal gasifiers or fluidized bed combustors that are used in gas turbines, fuel cells, and heat exchangers.

The objectives are the removal of sulfur compounds, particulates, and alkali metals. Technology Support will develop the "cross-cutting" technologies in the areas of waste management, instrumentation and process controls, innovative concepts, and systems and economic comparisons.

The role of the Advanced Environmental Control Technology Program is to develop cleanup systems which can be optimally integrated into process and power systems. The final goal is to demonstrate that the technology works and is economically, technically, and environmentally acceptable. The mechanism for achieving commercialization includes advancement of the technology through process development units (PDU's), use in pilot plants, and evaluation in full-scale existing plants. As the technology matures, private industry will be asked to cost-share development with the government. Information on progress will be disseminated via reports, symposia, and workshops.

The Advanced Environmental Control Technology Program is a new effort which initiated in FY 1979. Certain related efforts, however, have been conducted in previous years within the DOE Fossil Energy Combustion Systems and Advanced Research and Technology Development Activities. In FY

1980, these efforts will be integrated into this program, and are included as part of this budget request. As a new program, primary emphasis in FY 1979 was directed at developing detailed program and project plans to enable a smooth and organized transition into FY 1980, at which point, the program will have sufficient resources necessary to make significant strides in meeting its goals. These detailed plans will utilize the input and will require the review of the many organizations deeply involved and necessary to the success of this program (EPA, EPRI, utilities, private industry, TVA, government laboratories and technology centers, universities, etc.).

A wide range of existing facilities are available to support the experimental programs. These include 10-MW scrubber complexes, pressurized fluid-bed (PFB) facilities, atmospheric fluid-bed (AFB) facilities, and low-Btu gas-producing facilities (AFB facilities can be used to simulate PFB effluents). (See Table II 3-8.) During FY 1979, a plan for use and commitment of these facilities was initiated. Substantial use of larger facilities is also anticipated.

Table II 3-9 summarizes the funding levels by subactivity for the FY 1978 to FY 1980 period.

Flue Gas Cleanup

Flue Gas Cleanup (FGC) addresses the removal of pollution-causing contaminants from the stack gases of combustion units. The justification for the effort is as follows:

- Conventional combustion is the only available utility-scale approach in the near-term through 1990.
- FGC is required in most combustion applications (new construction) to meet EPA regulations.
- NSPS are already in force for utilities (stricter standards to be promulgated this year) and in preparation for industry.
- The Resource Conservation and Recovery Act (RCRA) may preclude the use of current lime/limestone technology due to sludge disposal problems.
- Environmental standards are a moving target which must be matched through

Table II-3-8. Facilities for experimental programs.

SCRUBBERS
- 10 MW TVA, SHAWNEE (3)
- 10 MW GFETC

PRESSURIZED FLUID-BED COMBUSTORS (PFBC)
- CURTISS-WRIGHT
- GRIMETHORPE (IEA)
- AMERICAN ELECTRIC POWER
- CURL, LEATHERHEAD

ATMOSPHERIC FLUID BEDS
- METC CTIU
- RIVESVILLE

LOW-BTU GAS
- CHEMICALLY ACTIVE FLUIDIZED BED (CAFB)
- METC PRODUCER GASIFIER

Legend

IEA: International Energy Agency
GFETC: Grand Forks Energy Technology Center
METC: Morgantown Energy Technology Center
CTIU: Component Test and Integration Unit

Table II-3-9. Funding for advanced environmental control technology for FY 1978 - 1980.

ADVANCED ENVIRONMENTAL CONTROL TECHNOLOGY	BUDGET AUTHORITY (DOLLARS IN THOUSANDS)		
SUBACTIVITIES	ACTUAL FY 1978	APPROPRIATION FY 1979	ESTIMATE FY 1980
Flue Gas Cleanup	$ 0	$ 2,700	$ 25,050
Gas Stream Cleanup	0	2,400	10,400
Technology Support	0	1,900	7,000
Capital Equipment	0	0	800
TOTAL	$ 0	$ 7,000	$ 43,250

improved capability of control technologies.
- Poor scrubber performance, sludge disposal problems, or high cost are court-accepted reasons for exemptions to burn oil or exemptions from oil to coal conversions.
- Poor scrubber availability results in plant shutdowns with subsequent increased usage of oil and gas in turbines and possible power shortages.

The technical objectives are stated in terms of meeting the following performance standards:

- 85 percent sulfur dioxide removal, 24-hour average; 0.2 lb/MMBtu limit
- 99 percent particulate removal; 0.03 lb/MMBtu limit
- 65 percent NO_x reduction; 0.6 lb/MMBtu limit
- Continuous control; no by-pass
- State control of sludge handling and storage
- EPA regulation of hazardous waste handling, storage, and disposal

The magnitude of the task becomes evident, considering that lime/limestone units are the principal field operational units (a few regenerable and Na_2CO_3 throwaway units are operating). Small-size units (less than 50-MW) do operate satisfactorily on low-sulfur coal. Larger plants, however, require multiple trains of 100- to 150-MW each; these are unsatisfactory. Operational availability (on-time) is 40 to 80 percent but is typically 50 percent. They are characterized by chemical plant rather than power plant operation, requiring large equipment and maintenance labor (up to 50-men crews), redundant units, etc. The units typically do not adapt to changes in coal supplies, sorbent grades, or load variations; are damaged on shutdown; and are difficult to "chemically" control.

Resultant sludge disposal problems are severe. Fixative agents are required to make the sludge harden, it cannot be allowed to seep into ground or runoff water, it has excessive space requirements for sludge ponds, and it has problems peculiar to humid areas.

Evolving particulates are excessively troublesome, and in many cases require bag collection systems and/or precipitators. Yet the lime/limestone system is perceived as the principal near-term solution (through 1985).

The Flue Gas Cleanup Subactivity is divided into three task areas:

- Lime/Limestone Scrubber Reliability
- Advanced Flue Gas Desulfurization
- Advanced Flue Gas Cleanup

Lime/Limestone Scrubber Reliability. The improvement to the functional characteristics and reliability of lime/limestone systems is the major thrust for the very near-term. The objective is to demonstrate reliable systems in full-scale operation in the near-term. Experimental facilities will be made available on a priority basis in support of this subtask.

The fundamental problems are perceived as a misunderstanding of the basic principles, inadequate design scale-up data, and the inability to measure and control key process parameters.

The strategy is to first collect, correlate, and evaluate detailed experience and then conduct parallel experimental programs using Eastern and Western coals. After technology development, final demonstrations will be conducted at a minimum size of 100-MW in facilities which already have controlled emissions. Thus, temporary inconvenience will not adversely affect the plant output. The results will be increasingly reliable "base case" modes, with off-base conditions explored and limits identified. A key to success will be utility industry participation.

The project will address economic issues and will provide utilities with technical support through the development of design, operational, and maintenance manuals. The project will supplement the current EPRI lime data book project.

Specific subtasks will address the following:

- *Materials erosion and corrosion:* Information for these key malfunction modes will be cataloged in conjunction with EPRI and EPA. Required data will be generated through experimental programs.
- *Process chemical paramters:* The grades of either the lime or limestone and the use of additives affect the SO_2 removal and reliability. The results are dependent upon the coal source and pretreatment. This effort will generate economic and purchase specifications for use with specific (pretreated) coals.
- *Sludge improvements:* Waste problems will be addressed using forced oxidation of sulfite to form gypsum and/or to reduce the liquid throughout due to conventional open-loop operation without causing solid-induced hardware damage. A number of new techniques will be considered.
- *Maintenance manual:* Most FGD plant downtine is due to regular maintenance to remove scale, unplug nozzles, and replace lines. The requirements for these are established by experience. However, these and unscheduled maintenance could be specified by monitoring process parameters which identify the occurrence, extent, and significance of degradation. This approach is analogous to the real time "onboard checkout" system of modern aircraft and modern telephone switching networks. Studies will be formulated which will result in operational guidelines for surveillance maintenance predictions. These will entail instrumentation studies and documentation of procedural guidelines for implementation.
- *Performance improvement teams:* A dedicated team of FGD specialists will be established to investigate performance inadequacies at existing FGD installations. The team will diagnose problems and provide recommendations to FGD operators.
- *Design manual for FGD applications at new power plants:* DOE will assist an ongoing EPRI program to develop a design manual for lime FGD at new power plants. An initial version of this manual will be prepared with available information supplemented by data generated in the previously described subtasks.

Major milestones for this task are presented in Figure II 3-88.

Advanced Flue Gas Desulfurization. Lime/limestone systems essentially trade air pollution for solid waste pollution problems. Smaller, cheaper, less energy-consuming systems that minimize solid waste problems will be developed under this subtask. A large number of processes have been proposed, many of which are sufficiently developed to

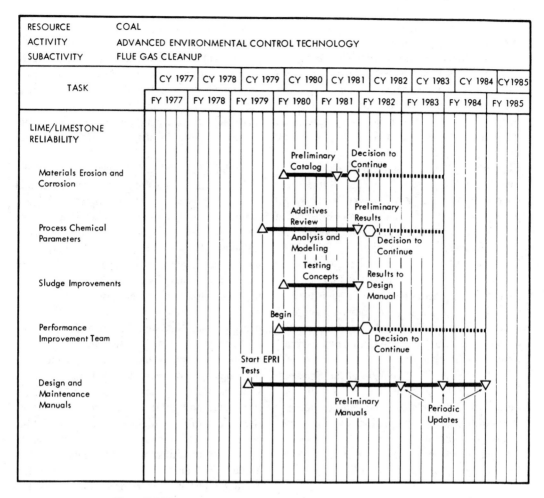

Figure II-3-88. Major milestones for the lime/limestones reliability task.

permit a narrowing of choices for application in the early 1982 to 1985 period. The processes are grouped into three categories: dry processes (nonregenerable), double alkali processes (nonregenerable), and regenerative processes.

Dry processes offer potential simplicity and reliability, producing dry solid wastes with no sludge disposal problems. Candidate sorbents include limestone, nacholite, trona, and high-alkali fly ash (Western coals). Addition of sulfur sorbents to pulverized coal has not yielded desirable improvement in the quality of effluent streams. It is not clear whether the problem is one of basic combustion interactions or a result of hardware malfunctions. A reassessment of this technology is needed.

Double alkali processes use clear liquids for SO_2 removal (i.e., sodium-based solutes), minimizing plugging and fouling problems. The liquid is regenerated with lime or limestone, producing gypsum, a relatively benign waste. The technology is mature and is commercial in industrial sizes. However, the economics and reliability are unproven, and operation with limestone needs to be demonstrated.

Regnerative processes minimize solid waste pollution, conserve natural resources, and generate merchantable products. Both wet and dry processes are in various stages of development. Wet processes are more mature and will be emphasized. They absorb the SO_2 in a scrubbing solution containing alkali, ammonia, magnesia, or citrate. The sulfur compounds are regenerated and the concentrated SO_2 is converted into elemental sulfur, sulfuric acid, or liquefied SO_2. These processes can operate in a closed-loop manner. In many processes, natural gas is used as a purge reductant. Reduction using coal gases or through the EPRI-funded RESOX process are primary approaches for alternate reductants. EPA initiated an evaluation of the Atomics International Aqueous Carbonate System now underway at Niagara Mohawk Power Company. This wet regenerative process is being cost-shared with the Empire State Electric Energy Research Corporation (ESEERCO), EPRI, and the New York State ERDA.

Concurrent with the advanced FGD experimental studies, economic analysis will be performed on candidate and other conceptual systems. These will include solid waste disposal costs and all environmental effects which can be quantified.

The three 10-MW scrubbers at the Shawnee facility will be used in support of the advanced FGD projects on a noninterference basis with, or after completion of, the lime/limestone reliability project.

Major milestones for the Advanced FGD Task are given in Figure II 3-89.

Advanced Flue Gas Cleanup. The presence of SO_x may cause nondesirable synergistic effects on NO_x, particulate, and trace metal removal from coal combustion. This task extends coal flue gas technology beyond sulfur removal to NO_x, particulate, and trace metals in anticipation of possible impacts caused by the increasingly extensive use of coal. The target is a more complete control of the total spectrum of pollutants with technology to become available in the mid-term (late 1980) period.

Technologies identified address dry and wet NO_x processes and dry and wet combined NO_x/SO_x processes. (The Catalytic/IFP/Chisso ammonia-based SO_x/NO_x system is an example of a wet combined process.) No technologies for combined SO_x, NO_x and particulate removal are presently viable, but new concepts may be in the offing. Combined processes must have the potential of being more cost effective and require less energy than separate NO_x and SO_x control systems.

Particulate control research is directed toward assessing and extending the capabilities of conventional particulate control systems (electrostatic precipitators (ESP's), scrubbers, and fabric filters) to abate emissions. The effort will include both evaluation of the devices for use at higher efficiency levels and modification of the devices where needed in order to make them suitable for operation under more rigorous conditions for operation on unproven but essential applications.

A continuing activity in advanced concepts

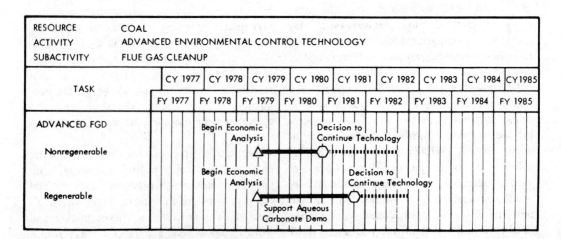

Figure II-3-89. Major milestones for the advanced FGD task.

will be maintained to ensure that promising technology and ideas can be evaluated. Several concepts under study propose use of irradiation techniques using electron beams, lasers, or plasma-jets.

In FY 1980, a program will be initiated to develop advanced centrifuge, filter, and electrostatic precipitation methods for fine particulate control. These methods will be tested at the bench scale or in a slip stream at a pilot plant. Detailed design, construction, and test plans will be prepared for a combined NO_x/SO_x cleanup process PDU, selected in FY 1979. A program will begin to construct and test on-line sensors for fine particle organics and NO_x cleanup process control. Similar devices for sorbent and burn-up cell control will be designed and tested in fluidized-bed combustion slip streams. Modification of electrostatic precipitator techniques for use when burning low-sulfur, high-ash coals will be studied. Performance and cost modeling of alternative flue gas cleanup techniques will continue. Efforts will be made to determine control parameters and assess control techniques for NO_x formation and decomposition in combustion zones and flue gas streams. Advanced techniques for trace elements, organics, and NO_x control will be investigated.

This task will also include studies of engine (turbines, diesels, etc.) exhaust controls when fueled by high-sulfur residual oils, coal liquids, or solid coal. The market for such engines will be surveyed. The results will evaluate the environmental impact from such sources, if uncontrolled, and define the need for future work in this area.

Major milestones are shown in Figure II 3-90.

Flue Gas Cleanup Status. As a new program starting in FY 1979, only a few of the critical activities were initiated during the first year. FY 1979 efforts primarily led to an expanded effort in FY 1980 during which the program will be fully underway. In FY 1979, detailed program plans were formulated to ensure that goals and objectives will be achieved. These detailed plans will use the input and require the review by the many organizations involved and necessary to the success of this

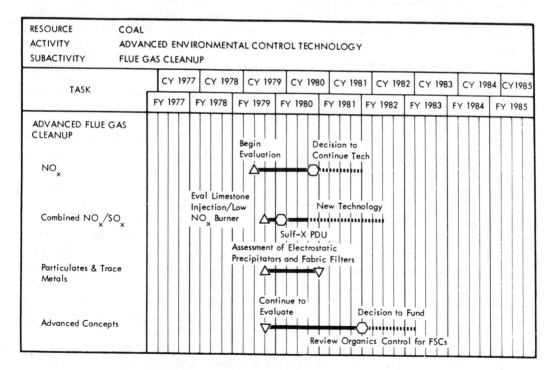

Figure II-3-90. Major milestones for the advanced flue gas cleanup task.

program (EPA, EPRI, utilities, private industry, TVA, etc.).

In the Lime/Limestone Reliability Task, DOE participated in the EPRI-sponsored FGD Characterization Tests starting in the second quarter of FY 1979. Data from these tests will provide a basis for the maintenance guide and design manual efforts which will be pursued in parallel with the characterization program. Preliminary manuals will be prepared in late FY 1981 and will be updated periodically as information becomes available through FY 1984.

The Process Chemical Parameter effort also started in FY 1979 and will perform limited laboratory-scale analyses or process chemistry modeling to determine rate-limiting steps and efficiency-limiting phenomena in lime/limestone scrubbers. A review of the effects of selected additives on scrubber process chemistry will also be conducted. Efforts to be conducted in FY 1980 include the following: laboratory-scale analyses and process chemistry modeling to determine factors which can be controlled and manipulated to improve process efficiency and reliability. Pilot-scale testing of both organic and inorganic additives for improved SO_2 capture will be initiated. Included will be investigations of methods for fixation and disposal of conventional lime/limestone sludge in an environmentally acceptable manner. Selected commercial FGD systems will be monitored in cooperation with industry to obtain operability and maintenance data required for pinpointing problem areas for concentrated applied effort. An important effort will be demonstration of forced oxidation as a means of converting calcium sulfite to the sulfate form in scrubber sludge, thus greatly reducing dewatering problems, waste volume, and disposal problems. Based on the results of the above work, a program of scrubber hardware improvement will begin to design alternative components and to replace those found to cause reliability and efficiency problems, including mist eliminators and reheaters.

In the Advanced Flue Gas Desulfurization Task, two efforts were initiated in FY 1979. An economic analysis of regenerable flue gas desulfurization techniques reviewed performance data and perform cost analyses of selected regenerable or dry SO_2 processes including waste disposal costs. Secondly, the Atomics International Aqueous Carbonate System Demonstration was supported on a cost-shared basis with the Empire State Electric Energy Research Corporation, EPA, EPRI, and the New York State ERDA. Both efforts will continue into FY 1980.

Specifically, in FY 1980, the performance and cost assessments of alternative dry, regenerable, and minimal waste FGD systems initiated in FY 1979 will be completed. Based on these studies, both a regenerable dry sorption FGD and a modified non-regenerable lime/limestone FGD system will be selected for scale-up tests and PDU design. The use of nacholite as a dry SO_2 absorber will be investigated. There will be a cost-shared demonstration of nacholite with a Western utility.

In the Advanced Flue Gas Cleanup Task for FY 1979, an evaluation was conducted to review the performance data, identify R&D requirements, and perform process cost analysis for selected NO_x control techniques. A program to evaluate the injection of limestone with coal in a low NO_x burner to simultaneously reduce SO_x and NO_x emissions was also initiated. This effort will be extended in FY 1980 to include additional SO_2 removal by use of fabric filters and spray drying the combustion gases using lime as a reactant. An assessment of the capabilities and costs of fabric filtration and electrostatic precipitation to control various levels of respirable and fine particulates will be conducted. Assuming successful laboratory and bench-scale testing and economic analysis of the combined NO_x/SO_x (SULF–S) process, a PDU was initiated late in FY 1979.

In FY 1980, a program will be initiated to develop advanced centrifuge, filter, and electrostatic precipitation methods for fine particulate control. These methods will be tested at the bench scale or in a slip stream at a pilot plant. Detailed design, construction, and test plans will be prepared for a combined NO_x/SO_x cleanup process PDU, selected by FY 1979. A program will begin to construct and test on-line sensors for fine particle organics, and NO_x cleanup process control. Similar devices for sorbent and burn-up cell control will be designed and tested in fluidized-bed combustion slip streams. Modification of electrostatic precipitator techniques for use when burning low-sulfur, high-ash coals will be studied. Performance and cost modeling of alternative flue gas cleanup techniques will continue. Efforts will be made to determine control parameters and assess control techniques for NO_x formation and decomposition in combustion zones and flue gas streams. Advanced techniques for trace elements, organics, and NO_x control will be investigated.

Gas Stream Cleanup

Combined-cycle generating plants powered by high-temperature, low-Btu gases from direct coal gasification or high-temperature gases from pressurized fluid-bed combustion may prove to be an extremely attractive approach for large-scale coal utilization.

Here the opportunity exists to provide gases at temperatures and pressures to match the operating conditions of the energy conversion or process equipment providing high overall system thermal efficiencies without degrading the hardware performance. Of particular interest will be applications to advanced turbines and fuel cells as well as heat exchangers used to thermally optimize an energy conversion cycle. The technology derived may also be applicable to the chemical process industries, for example, for producing better streams for catalytic hydrogen production. (These aspects are not an objective of this task.)

A range of inlet gas conditions for advanced turbomachinery and molten carbonate fuel cells (MCFC) is listed below. (Heat exchangers are passive devices and require operation at the same conditions.)

	Temperature (°F)	Pressure (Psia)
Turbine	1200–1900	200–500
MCFC	1000–1600	75–150

Outputs from both coal gasifiers (CG) and pressurized fluid-bed combustors (PFBC) can presently be tailored to meet these requirements of temperature and pressure. The concern is the type and amounts of impurities in the gas which can be accepted without undue corrosion, erosion, and/or contamination of system components. There are no definitive guidelines or specifications regarding gas cleanup requirements. In fact, when established, these may be more severe than EPA emission standards, especially in the area of particulates.

Indications are that addressing the gas cleanup problems at this stage rather than at the flue gas stage may provide large, total system economies. For example, NO_x problems caused by the high fuel-bound nitrogen content of coal or coal-derived liquids could be minimized. This subactivity, to be successful, will be closely coordinated with the Division of Fossil Fuel Utilization Combustion Systems, Heat Engines, and Fuel Cell Activities.

These activities are developing systems that require minimum hot gas cleanup. This subactivity will provide technology for "clean" gas streams. Commercialization will probably require mating of hardware and cleanup technology to provide cost-effective reliable systems. Materials compatibility is a key issue. The DOE Energy Materials Advisory Committee (EMAC) will be consulted to ensure that results of research from other projects (i.e., corrosion compatibility of materials in high-temperature SO_2/SO_3 environments) are utilized and not duplicated.

The strategy for this subactivity is to address the technology needs separately for turbines and MCFC and address other needs as a generic "process modification" technology.

Turbine Cleanup Systems. Hot gas inlet "cleanup" requirement specifications for turbines from a hardware standpoint are presently unavailable. The objective of this task is to quantify the limits of particulate and gaseous corrodants/erodants for acceptable gas turbine service life, to define the limits of operation of economically practical gas cleanup systems, and to develop the technology to provide reliable systems. Both hot low-Btu gasifier/gas turbine and PFBC/gas turbine systems will be studied.

Coal gasification produces primarily carbon monoxide and hydrogen and smaller quantities of hydrocarbons, carbon dioxide, and water. Existing coal-bound sulfur, nitrogen, alkalis, halogens, and volatile trace metals produce compounds which are toxins, erodants, corrodants, and catalyst poisons. For example, trace quantities of sodium and/or potassium sulfates can corrode conventional turbine blades at temperatures of 1250° to 145°F, and alkali metals, independently, can cause corrosion above 1500°F.

Particulates entrained in the gas and condensed metal vapors can cause hardware erosion and cause visible emissions problems.

The PFBC products, unlike a gasifier, are completely oxidized. Sulfur generally is removed but the process increases the particulates and still produces some sulfur and alkali contaminant particles. Otherwise, PFB systems have the same problems as low-Btu gasifiers.

Initial efforts will be made to define system requirements of the turbine, define potential outputs of the gasifiers and PFB combustors, and plan an experimental program to solve the derived interface problems.

The major effort in FY 1980 will be a continuation of the development contracts awarded in the FY 1979 hot gas cleanup PRDA. The PRDA will seek new and innovative ideas for removal of particulates and contaminants from low-Btu gasifier and PFB gas streams to protect the gas turbines used in these advanced combined cycle power systems. This effort will continue through FY 1983 and, for the more advanced or mature concepts, will include cost-sharing. Additional efforts will be initiated to identify and investigate cleanup techniques for the protection of heat recuperators in combined-cycle systems, to analyze the controlling parameters in low-Btu gas wet scrubber performance and secondary aerosol generation, to analyze the material entrainment performance of wet chemical low-Btu gas desulfurization units, and to assess techniques for the protection and removal of deposition in gas turbines.

Figure II 3-91 shows major milestones for this task.

Molten Carbonate Fuel Cell Cleanup Systems. Hot gas inlet "cleanup" requirements specifications for Molten Carbonate Fuel Cell

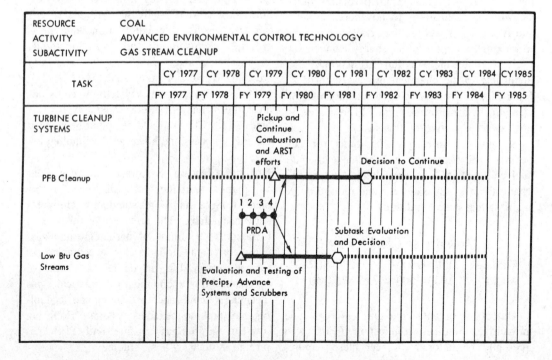

Figure II-3-91. Major milestones for the turbine cleanup systems task.

Systems Cleanup (MCFC) are presently unavailable. The objective of this task is to quantify the limits of particulates and gaseous pollutants aceptable for operation of a MCFC, define the limits of operation of economically practical gas cleanup systems, and develop the technology to provide reliable systems.

The fuel supply is derived from hot, low-Btu gasifiers. The MCFC problems using this source of fuel differ from turbine systems. The aerodynamics/materials problems linked to the high-speed turbomachinery do not exist with the fuel cell. However, the MCFC is susceptible to catalyst poisoning, electrode damage, and electrolyte contamination.

Initial efforts will be to define system requirements of the MCFC, define potential outputs of the gasifiers, and plan an experimental program to solve the derived interface problems.

Efforts to be initiated in FY 1980 include the bench-scale testing of cleanup systems for the removal of low levels of sulfur from low-Btu gas, the identification and initial testing of techniques for the removal of chlorine from the gas stream and entrained tars and oils from the reformer, and the selection and design of a cleanup train test system for multiple 1-kW molten carbonate fuel cell tests using a coal gasifier slip stream.

Major milestones are presented in Figure II 3-92.

Process Modification Technology. This task groups together systems to control process streams through innovative combustion techniques. Excluded are fluidized-bed techniques, which are covered in the Combustion Systems Activity. Included are studies of low-Btu gas-fueled combustion engines.

A data base for this effort is provided from past studies of burner design alterations, modifications of firing practices, and additions of sorbents. Subtasks will address the following:

- Ash/slag retention: Various designs have been proposed to retain mineral matter inside the combustor. The designs are often efficient for reducing the generation of fly ash, but at the cost of unacceptable increases in NO_x emissions. Evaluation of new combustor-modification concepts is needed.
- Combustion modification: Several techniques for combustion modifications for NO_x abatement are being explored by EPA: low excess air firing, staged combustion, multiple-register burners, etc. These technologies may provide a quick path to solving emission problems in existing installations. Development of modification techniques will be continued in this DOE activity.
- Combustor design: Existing conventional combustor design will be updated and improved by development of improved air/fuel ratio controllers and improvement in major components to further the state-of-the-art.
- Low-Btu gas/combustion engines: Direct

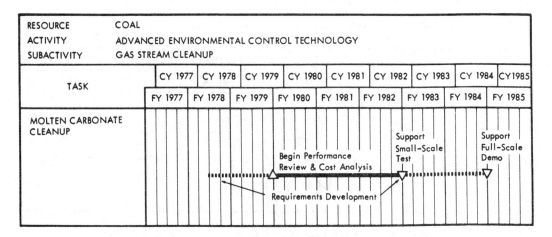

Figure II-3-92. Major milestones for the molten carbonate cleanup task.

combustion of low-Btu gases in engines other than turbines (Stirling engines) will be studied to define the gasifier-engine and gas cleanup requirements for such applications.

Major milestones are shown in Figure II 3-93.

Gas Stream Cleanup Status. Although this was a new subactivity in FY 1979, several efforts in support of hot gas cleanup for PFB combustors have been conducted in both the Combustion Systems and Advanced Research and Technology Development activities. All of these efforts will be integrated within this subactivity in FY 1980.

In the Turbine Cleanup Systems Task, FY 1979 efforts included the analysis, detailed design, and bench-scale testing of high-temperature, high-pressure electrostatic precipitators; bench-scale testing of vortex and impeller-type advanced centrifuges for removal of fine particulates; and performance analyses for humidification/condensation steps and fine aerosol generation in particulate, tar, and alkali scrubbers for high-pressure, low-Btu gas streams. In addition, a Program Research and Development Announcement (PRDA) was issued in early FY 1979. The PRDA will seek to identify new and innovative hot gas cleanup approaches for turbine applications (part of Combustion Systems Activity in FY 1979).

The major effort in FY 1980 will be a continuation of the development contracts awarded in the FY 1979 and, for the more advanced or mature concepts, will include cost-sharing. Additional efforts will be initiated to identify and investigate cleanup techniques for the protection of heat recuperators in combined cycle systems, to analyze the controlling parameters in low-Btu gas wet scrubber performance and secondary aerosol generation, to analyze the material entrainment performance of wet chemical low-Btu gas desulfurization units, and to assess techniques for the protection and removal of deposition in gas turbines.

Efforts in the Molten Carbonate Fuel Cell Cleanup Task will be initiated in FY 1980 and will include the bench-scale testing of cleanup systems for the removal of low levels of sulfur from low-Btu gas, the identification and initial testing and techniques for the removal of chlorine from the gas stream and entrained tars and oils from the reformer, and the selection and design of a cleanup train test system for multiple 1-kW molten carbonate fuel cell tests using a coal gasifier slip stream.

During FY 1979, a facility usage plan was initiated. The candidate facilities are identified in Table II 3-8. Major use of these facilities in support of this effort is expected to start in FY 1981; however, support for the Chemically Active Fluid Bed facility was continued in FY 1979 as part of the Process Modification Technology Task.

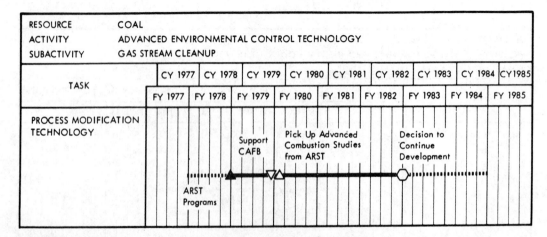

Figure II-3-93. Major milestones for the process modification technology task.

For FY 1980, efforts will be initiated to develop and test staged burner and dry burner additives for NO_x decomposition; to review and assess the use of catalytically and chemically active gasification beds for sulfur, alkali, nitrogen, halogen, organics, and trace metal release control; and to analyze, test, and assess combustor design modifications for oxidant temperature and residence time control of NO_x and organics formation and release.

Technology Support

The Technology Support Subactivity will develop the "cross-cutting" technologies in areas such as waste management, instrumentation and process controls, and systems and economic comparisons which are common to both flue gas and gas stream technology requirements. In addition, innovative concepts will be pursued within this area.

Waste Mangement. Methods to improve handling, reuse, and disposal of combustion waste materials will be developed: fly ash waste, FGD waste, and FBC waste.

The composition of the large volume of coal ash being produced presents opportunities for the economic recovery of mineral by-products. DOE will identify and develop those recovery processes which offer economic potential. It is noted that fly ash contains appreciable quantities of alumina and ferric oxide. In addition, fly ash has good potential as an additive in cement manufacture materials.

FGD sludge disposal costs represent a major part (up to 20 percent) of the capital and operating costs of an FGD system. These costs can be drastically reduced by improved sorbent (e.g., limestone) use controlled solids quality, and by improved sludge dewatering equipment. Currently under study are techniques to develop sulfite solids and forced oxidation processes to produce gypsum. Ocean disposal of FGD sludge also is being assessed because many plants in the Northeast may lack disposal sites.

The FBC process produces a spent-bed residue which is recovered in dry condition. DOE has initiated a program to evaluate whether this spent residue can be used commercially. Plans for field demonstrations include the use of the alkaline FBC residue as an SO_2 sorbent, a stabilizing medium for scrubber sludge, a neutralizer of acid-mine drainage discharge, and a component of a stabilized road-base pavement system in a highway construction project. Agricultural uses of FBC wastes are possible since they contain large amounts of nutrients necessary for plant growth. The major concern, however, is the possibility of contamination with biotoxic elements, such as lead, cadmium, and chromium. DOE and USDA are now investigating this aspect in a comprehensive program which includes animal feeding studies.

Major milestones are presented in Figure II 3-94.

Instrumentation. Instrumentation improvements are vital for reliable process diagnostics and automatic control. Equipment for measuring flow rates, pH, chemical species, pressure, and temperature in the hostile combustion, exhaust, and flue gas environments will be developed. An improved continuous SO_2 monitoring system is a high-priority need. As new systems become available, they will be evaluated in suitable experimental facilities. Figure II 3-95 shows the major milestones.

Applied Research. A continuing effort in materials will be pursued as a follow-on to the accelerated effort on materials for lime/limestone systems. Long-term materials testing will be conducted on a variety of materials in the presence of the large variety of chemical compounds found in flue gas cleanup systems. Accelerated aging tests will be developed and used.

Applied research on the basic chemical effects (reactions, solubility, crystal formation, etc.) and physical effects (mixing, droplet entrainment, demister action) will be carried out. The impact of various coal characteristics and coal preparation on FGD operations will also be studied.

Major milestones for the Applied Research Task are shown in Figure II 3-96.

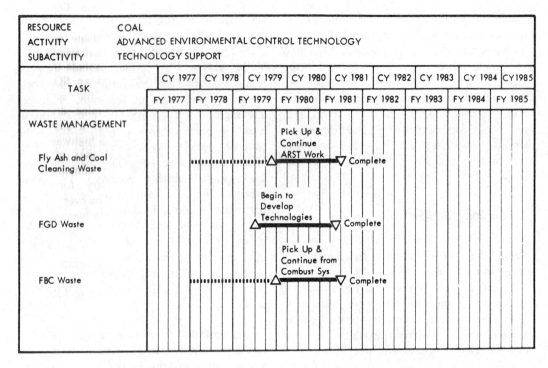

Figure II-3-94. Major milestones for the waste management task.

Systems and Economic Comparisons. A comprehensive analysis effort will be conducted to determine the trade-offs among various control technology options and between control technologies and power systems components. In the case of high-sulfur coal combustion/steam turbine system, for example, environmental standards could be met by using chemically cleaned coal, fluidized-bed combustion (FBC) (high limestone requirement) with particulate control, conventional coal combustion with high-efficiency FGD, or physically cleaned coal in a conventional combustor with FGD. An analysis of the economics of the options and comparisons with other systems (PFB, Closed-Cycle Gas Turbine, fuel cell, etc.) containing multiple options will be conducted.

Overall system efficiency and cost minimization require integration of the cleanup system into the overall power plant. For example, scrubber water flows can be integrated with other system water flows. This subtask will consider these factors and their impact on maximizing total system reliability.

Gas stream cleanup technologies will fill the gap between the next generation of energy conversion equipment and the next generation of fossil fuel utilization equipment. As such, the requirements for these technologies are highly dependent on the requirements for conversion and utilization equipment. Requirements analyses of the following systems are essential:

- *Low-Btu gasifier/gas turbine:* The objective of this effort is to quantify the limits of particulate and gaseous corodants/erodants for acceptable gas turbine service life, to define economically practical gas cleanup systems requirements, and then define requirements for gasifier products.
- *Low-Btu gasifier/molten carbonate fuel cell:* This effort will address the trade-off between fuel cell design, gasifier design, and gas cleanup requirements. Again, fuel cell tolerance will be determined experimentally and the outputs will be gasifier gas stream quality requirements, fuel cell design, goals, and cleanup technology requirements.

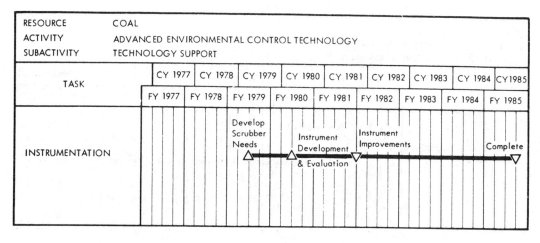

Figure II-3-95. Major milestones for the instrumentation task.

- *Low-Btu gas/combustion engines:* Another possible application for coal gasifier products is direct use in combustion engines other than turbines; i.e., Stirling engines. This effort will address the definition of engine, gasifier, and cleanup requirements for such applications.
- *Pressurized fluidized-bed combustion/gas turbine:* This effort will serve to define the gas cleanup requirements for PFBC/gas turbine systems. Efforts for both DOE and EPA already have begun to characterize the products from PFB combustors, and this work will continue using EPA, DOE, and British facilities.

The requirements of existing and anticipated environmental regulations and their impacts on fossil fuel energy conversion systems will continue to be incorporated with the outputs of the trade-off analyses and systems integration efforts in the ongoing planning of the AECT activity. DOE policy recommendations will depend heavily on the outcome of these systems and analyses.

Major milestones for this task are presented in Figure II 3-97

Technology Support Status. Waste management efforts began in FY 1978 in both the ARST and Combustion Activities and included studies in coal waste utilization, regenerable fluidized-bed sorbents, aluminium from fly ash, chemical properties of coal

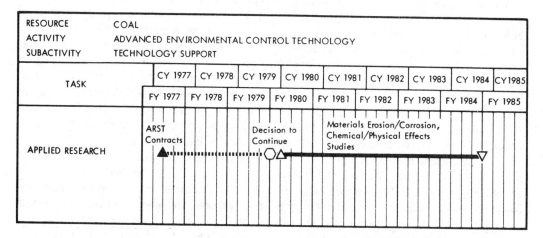

Figure II-3-96. Major milestones for the applied research task.

210 Fossil Energy

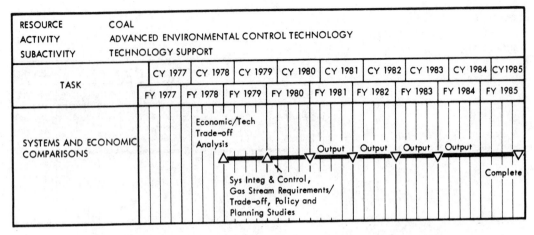

Figure II-3-97. Major milestones for the systems and economic comparisons task.

slags, and utilization of FBC spent-bed materials. These activities continued in FY 1979 and will be incorporated into this activity in FY 1980. A new effort was initiated in FY 1979 to characterize the properties and develop control technologies for the disposal of fly ash and sludge wastes in compliance with the RCRA. Specific FY 1980 efforts in waste management will include the continuation of control technology development for disposal of waste materials from lime/limestone scrubbers to satisfy RCRA requirements, continuation of an effort with the Department of Agriculture to evaluate the use of fluidized-bed spent bed material as a lime replacement for Eastern United States crops, the evaluation of spent bed material and ash as a construction material, the development and testing of methods for regeneration of FBC sulfur sorbents to increase overall sorbent utilization and reduce waste outputs, and the initiation of studies for treatment utilization and disposal options for wastes associated with advanced flue gas cleanup processes.

Efforts in the instrumentation area which were initiated in FY 1979 include the identification of critical lime/limestone scrubber process parameters for which on-line instrumentation is needed to control scrubber performance and reliability, determine the instrumentation needs for advanced flue gas cleanup and gas stream cleanup, and identify new and modified instrumentation techniques to meet on-line monitoring needs. Instrumentation development and evaluation will begin in FY 1980, with instrument improvements expected in FY 1981. The task will be completed by FY 1985.

Three Applied Research projects were started in FY 1977 and FY 1978 in the ARST activity. They are the Exxon Research efforts in determining the sources of NO_x from coal, the United Technology Research Center (UTRC) effort in coal volatilization, and he Aerodyne, Inc., effort in combustion of pulverized coal. Specific efforts in FY 1980 will include the determination of corrodant formation control paramters in coal gasification and combustion, development of hydrodynamic and reaction models of contaminant composition in gasifiers and pressurized fluidized bed combustor gas streams, measurement of thermodynamic parameters and kinetic data for contaminant reaction with gasification and combustion species, analysis of rates of liquid and solid particle formation and conglomeration under process stream conditions, and the measurement of thermodynamics and kinetics parameters for advanced sorbents for organic or alkali removal techniques.

Systems and economic comparison efforts were begun during the last quarter of FY 1978. Modification of the computer model of coal process technology to include control technology was begun. An analysis of the results of Phase I of the SULF–X process flue gas cleanup experiments was also started. These were completed near the end of FY 1979. Continuing trade-off analysis and systems integration and control efforts were initiated in late FY 1979 to determine the viability of various control technology and power systems options to best meet coal combustion and environmental requirements in the future. Specific efforts in FY 1980 will include completion of a modified computer model of coal process technology which will incorporate control technology, initiation of gas stream requirements/trade-off studies for low-Btu gasifier/gas turbine, low-Btu gasifier/molten carbonate fuel cell, low Btu gasifier/combustion engine, and PFB/gas turbine systems; an efficiency analysis and economic comparison of advanced cleanup devices in complete utilization systems including identification of potential significant process improvement; and the initiation of an evaluation of the potential for desulfurization scrubbers and particulate removal devices for meeting expected NSPS regulations for industrial coal-burning boilers.

Capital equipment funds will be required at the Energy Technology Centers and national laboratories to provide on-stream analyzers to measure flue-gas composition and scrubber performance; to provide necessary instrumentation for on-line monitoring of particle size distribution and organics in flue gas streams; and to procure necessary laboratory equipment for on-line alkali measurement of gas streams from gasifiers and PFB combustors.

HEAT ENGINES AND HEAT RECOVERY

The goal of the Heat Engines and Heat Recovery Activity is to develop technology for maximizing energy savings in the utility, industrial, and residential/commercial market sectors. There are three major objectives for achieving the goal:

- Increase the direct utilization of coal in the utilities by developing advanced, highly efficient integrated coal conversion/utilization systems that can efficiently displace critical oil and gas fuels.
- Improve fuel utilization in the industrial and residential/commercial sectors by developing advanced cogeneration systems that can supply both electric and thermal needs for the equivalent fuel energy now being used to generate just their electrical requirements.
- Develop advanced heat recovery component technology for producing useful work from waste heat rejected by the utilities, industrial and residential/commercial sectors.

The interdependent fuel usage and energy consumption patterns, along with the interrelated technology needs in the stationary power markets of the utilities, industrial, and residential/commercial sectors, were analyzed under the Energy Research and Development Administration (ERDA). It was determined that a focused program addressing both heat engines and heat recovery technology was needed to best achieve the energy savings potential in these markets. Consequently, the Department of Energy (DOE) reorganization centralized these previously separate research and development (R&D) activities within the DOE Fossil Energy (FE) Program in October 1977, under the direction of the Division of Fossil Fuel Utilization (FFU).

There are three subactivities included under the Heat Engines and Heat Recovery Activity:

- Integrated Coal Conversion and Utilization Systems
- Advanced Cogeneration Systems
- Heat Recovery Component Technology

Technology development pursued in the Heat Engines and Heat Recovery Activity is targeted for commercialization in the stationary power markets of the utility, industrial, and residential/commercial sectors. Opportunities for improved technology in this tri-market sector are shown in the following figure, which illustrates both the interrelation between the quads of fossil fuel consumption (Q_f) in each market and the associated quads of heat rejection (Q_r).

As shown in Figure II 3-98, the utilities currently consume over 18 quads of fuel energy to produce a little over 6 quads of electrical energy (Q_e). This electrical energy is

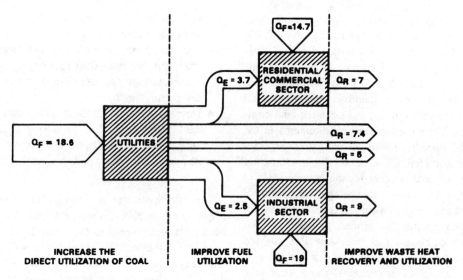

Figure II-3-98. Opportunities for heat engine and heat recovery technology.

distributed to consumers in the residential/commercial and industrial sectors, while two-thirds of the fuel consumed by the utilities is rejected as waste heat. Meanwhile, the industrial sector consumes more fuel than the utilities in order to produce heat for various process needs. The residential/commercial sector consumes a signficant quantity of fuel energy ($Q_f = 14.7$) for space heating, hot water, etc. A substantial portion of that energy consumption is also rejected without extracting any useful work. The opportunities for improving this wasteful tri-market energy pattern are the focus of the subactivities within the Heat Engine and Heat Recovery Activity.

Integrated coal conversion and utilization systems show the most promise for efficiently use of coal in an economic and environmentally acceptable manner. Successful development of these advanced technology systems would permit the utilities to increase the utilization of our abundant coal resources, thereby displacing critical oil and gas for use in other domestic markets.

Advanced cogeneration systems have been shown to be one of the most efficient means of improving fuel utilization by the residential/commercial and industrial sectors. The concept of simultaneously producing both electricity and process heat on the industrial or residential/commercial site is known as cogeneration. Examples of residential/commercial cogeneration are small dispersed total energy systems for multifamily buildings, or larger integrated energy systems for communities that can efficiently supply both the electrical and heating needs of residential/commercial complexes. The waste heat recovered from the electrical generation process in these total energy systems can supply space heating hot water and even cooling by combining absorption air conditioning operating directly from low-grade waste heat. Cogeneration in industrial plants can also supply their electrical needs and heat for the various processes. Industrial processes generally require higher temperature heat than residential/commercial applications, so the equipment for the two applications differ in their overall size as well as in the temperature of the heat produced.

Heat recovery component technology can provide a means to effectively recover waste heat and convert it to useful purposes thus offsetting fuel now burned to produce that equivalent work. Considering the current total 52 quads of fuel usage by the tri-market sector, it is estimated that approximately 28 quads are rejected to the atmosphere and rivers in the form of waste heat. Effective recovery and utilization of even a fraction of this reject heat could reduce dependence on foreign energy imports, alleviate the associated balance of payments problem, and help reduce the cost of energy to the American public. The technology related to improved waste heat recovery and utilization would have important applications in all three market sectors.

Table II 3-10 summarizes the funding levels by subactivity for the FY 1978 to FY 1980 period.

Integrated Coal Conversion and Utilization Systems

The major applications of technology under development in Integrated Coal Conversion and Utilization Systems is for base-load electrical generation by central station utilities. Coal and coal-derived fuels are projected to be the predominant fossil energy resource for electric utilities in the future. The effort is divided into three tasks: Open-Cycle Gas Turbine; Closed-Cycle Power Systems; and Support and Additional Programs, Engineering Analysis, Evaluation and Assessment. They are aimed toward providing technology for increasing the utilization of our abundant coal fuel resources in order to offset the use of critical oil and gas in the utility market.

Integrated coal conversion and utilization systems have been shown to be a most promising option for efficiently using coal in an economic and environmentally acceptable manner. Specifically, the integrated high-temperature gas turbine/low-Btu coal gasification combined-cycle holds the most promise for achieving the high efficiency needed to offset energy losses in gasifying and cleaning the coal; therefore, electricity can be produced at a cost comparable to those of oil-fired steam and nuclear base-load systems.

Closed-cycle power systems integrated with a primary heater or a fluidized bed combustor represent another environmentally promising approach for utilizing coal directly in utility base-load service. Though with current technology this option is potentially less efficient than the advanced combined-cycle gasification plant, closed-cycle systems using advanced technology primary heaters can be considerably more efficient than conventional coal-fired boiler systems. Closed-cycle systems have found widespread use in Europe, particularly in municipal central station cogeneration applications where waste heat is recovered and hot water distributed to residential/commercial customers, a concept known as district heating. If the cycle temperature could be increased, advanced closed-cycle power systems could have the potential for competing on the basis of efficiency and cost of electricity with other advanced base-load options.

Table II-3-10. Funding for the heat engines and heat recovery activity for FY 1978 - 1980.

HEAT ENGINES AND HEAT RECOVERY	BUDGET AUTHORITY (DOLLARS IN THOUSANDS)		
SUBACTIVITIES	ACTUAL FY 1978	APPROPRIATION FY 1979	ESTIMATE FY 1980
Integrated Coal Conversion Utilization Systems	$25,500	$26,700	$24,800
Advanced Cogeneration Systems	8,625	14,920	13,000
Heat Recovery Component Technology	7,975	16,180	8,000
Capital Equipment	235	200	200
TOTAL	$42,335	$58,000	$46,000

The following major technical problems are related to integrated coal conversion and utilization systems:

- Materials compatibility
- High-temperature turbine cooling for open-cycle systems
- Heat exchanger components for closed-cycle systems operating at high pressure and temperature
- Bearings, seals, and auxiliaries of adequate durability for these systems
- Efficient and economic low- and/or high-temperature gas cleanup for open-cycle systems
- Efficient economic low-Btu gas source in open-cycle systems.

Projects initiated in FY 1976 using open-cycle gas turbines as the topping portion of combined-cycles were continued in FY 1977. These projects started with system definition studies to determine both technical and economic factors for promising gas turbine designs. Efforts in blade and vane cooling, ceramic component evaluation, and materials corrosion and reliability studies, which were inititated in FY 1976 and FY 1977, continued during FY 1978. Projects started to evaluate improved blade durability, cleanup of particulate matter from turbine inlet gas, and coal-derived fuel cleanup methods continued during FY 1978. Extended testing in these areas (12,000 hours plus) is planned for FY 1980. Testing commenced in FY 1978 (blade cascades, materials, metallics and ceramics, testing rig, shakedown, etc.) and continued through FY 1979, including at least 8,000 hours of component and materials testing.

In the closed-cycle power systems area, work has proceeded on the high-temperature heater, on materials evaluation, and on system studies to evaluate overall power generating costs.

The overall environmental impacts of the advanced power systems technologies should be much less than those for conventional boilers because of the high efficiencies associated with them, and the fact that the turbines do not exhaust directly into the environment.

There are, nevertheless, two major environmental concerns associated with the Integrated Coal Conversion and Utilization Systems project: (1) possible NO_x emissions greater than conventional coal-fired boilers if high-nitrogen content synthetic fuels are used; and (2) control of particulate matter under the high-temperature conditions associated with these technologies. The latter concern is both an environmental problem and a technology need to reduce particulate levels prior to the hot gas entrance into the turbines. The environmental problem is being studied within DOE and by EPA and is common to many Fossil Energy technologies.

Open-cycle Gas Turbine Systems. The concept of this advanced system is to burn coal or coal-derived gaseous or liquid fuels in air to produce a high-temperature gas (combustion products at 2600°F or higher) before expanding it through the turbine to produce electricity. After expansion, remaining hot gases can be used to generate steam in conventional steam bottoming plants to produce additional electricity. Such a supplemental power-producing stage is called a bottoming cycle. The Open-Cycle Gas Turbine (OCGT)/steam system (Figure II 3-99) forms a combined-cycle power plant with potentially greater efficiency than today's standard steam plant. The approach is designing a gas turbine to burn coal having been identified as most promising from the standpoints of efficiency, cost of electricity, and emissions involves a combined-cycle system integrated with a low/Btu coal gasifier.

With this system, crushed coal, water, and compressed air are fed into the gasifier. Hot gases are passed through a cleanup system for removal of particulates and chemical contaminants, and are burned in a combustor prior to driving the gas turbine electric generator. From the turbine, the expanded gases are routed through a steam generator, relinquish their remaining heat, and are exhausted to the atmosphere. Steam thus generated is expanded through a steam turbine driving an electric power generaor. Exhausted steam is condensed, and a pump arrangement feeds the fluid back through the steam generator.

The OCGT base is quite advanced with clean fuels such as oil and natural gas. Valua-

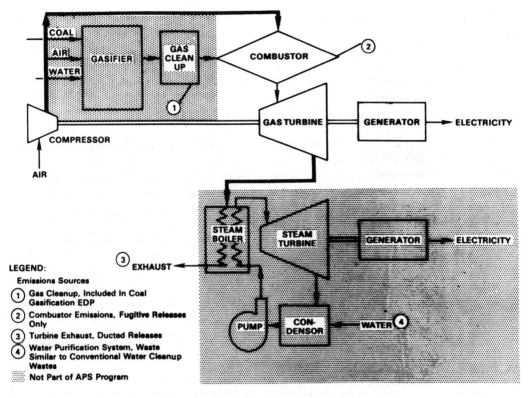

Figure II-3-99. Open-cycle gas turbine/steam turbine system.

ble information has been developed through military and commercial R&D, including maritime applications. Industrial units with clean fuels are presently in wide use for utility peaking service, as are clean fuel-fired gas turbine/steam turbine combined-cycles for intermediate-duty service.

Adaptation of new gas turbine technology for use with coal and coal-derived fuels will require development of three key capabilities:

- Compatibility with coal and coal-derived fuels
- Improved endurance of materials for turbine blades, vanes, and combustors to meet long-term base-load service
- Practical turbine cooling methods to withstand higher turbine inlet temperature, which will in turn produce higher efficiencies.

Turbine component durability is dependent on a combination of factors such as stress, temperature, and hot corrosion/erosion. The strategy for the open-cycle gas turbine project is to achieve technology readiness by developing key components for incorporation into a prototype advanced high-temperature gas turbine.

The open-cycle turbine project will be implemented by advancing the technology of a 2600°F multistage turbine subsystem to technology ready status within a 10-year period. The project is divided into three phases:

- Phase I: Project and System Definition
- Phase II: Technology Test and Test Support Studies
- Phase III: Technology Readiness Verification Tests.

Successful completion of these phases will provide technology readiness, and commercial demonstration is expected to follow.

The key area of this project concerns development of advanced cooling concepts, such as air film and transpiration cooling and water/steam cooling of turbine blades and nozzles, along with development of material alternatives (superalloy metals and ceramics) for combustor liners, transition sections, and appropriate hot section parts. An assessment of ceramics for key high-temperature, open-cycle turbine components is included as a project element, since impurities that may be present in coal-derived fuels may seriously degrade the metals and coatings conventionally used in gas turbines.

An additional project element is the development of advanced low-Btu gas combustor designs to ensure that the gas temperature profile is carefully controlled to prevent uneven temperature distribution or "hot spots" on turbine components and to ensure that the exhaust meets emission standards. Design data for combustors using coal-derived gases must be developed. Two specific approaches aimed at the "temperature profile" goals are to improve combustion control by premixing of fuel and air in a separate upstream section of the burner, and to use catalytic combustion. Approximately 7,500 hours of testing were conducted in FY 1979, and 1,100 hours of testing are planned in FY 1980.

Major milestones for the OCGT Task are given in Figure II 3-100.

Closed-Cycle Power Systems. Two cycles have been investigated in the Closed-Cycle Power Systems task. These are the closed-cycle gas turbine power cycle and the alkali metal vapor topping cycle. Both can be used in conjunction with a steam turbine bottoming system to use coal fuel more efficiently in central station utility power generation. The closed-cycle gas turbine system (primary heater) is continuing into Phase II.

As in the open-cycle gas turbine system,

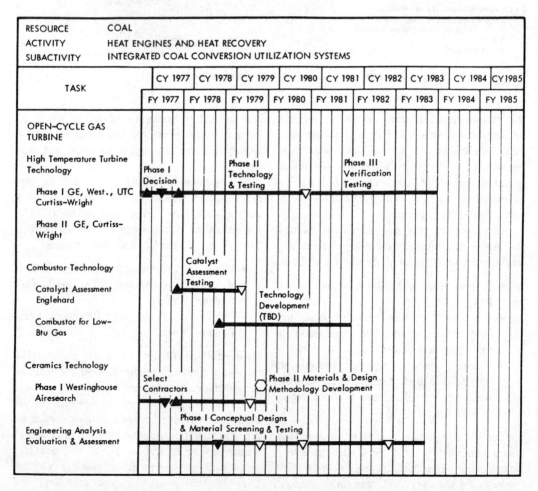

Figure II-3-100. Major milestones for the OCGT task.

the closed-cycle system can burn coal or coal-derived fuels. Because the working fluid is contained within a sealed system and is recirculated with negligible loss, working fluids (such as helium, nitrogen, carbon dioxide, and air) can be used.

Since this gas turbine can be driven by relatively noncorrosive or inert working fluids such as helium, it can operate at higher temperatures (1550°F and higher) than a standard steam system. After expansion through the turbine, remaining heat can be used in a conventional steam bottoming cycle or in a lower temperature bottoming cycle with an organic working fluid.

The gaseous working fluid is compressed and then heated by the primary heat exchanger. It is then expanded through the gas turbine driving an electric power generator. The expanded gas rejects heat to the steam generator and heat rejection heat exchanger. The fluid then is recycled through the system by being recompressed. Steam thus generated is expanded through a steam turbine driving an electric power generator, the exhausted steam is condensed, and a pump arrangement feeds the fluid back through the steam generator. (See Figure II 3-101.)

Approximately 500 hours of testing were conducted in FY 1979 and 1,000 hours of testing planned in FY 1980. Major milestones are presented in Figure II 3-102.

Support and Additional Programs. Supporting technology for the Integrated Coal Conversion and Utilization Systems is required in many areas. Within DOE there must be close-coordination in this technology development to avoid duplication. In addition, strong coordination is required and will be maintained with other Federal agencies, NASA, EPRI, and industry.

Studies will be conducted to determine the overall economics of various closed-power system approaches. Coal-derived liquid fuel/open-cycle gas turbine interface studies are

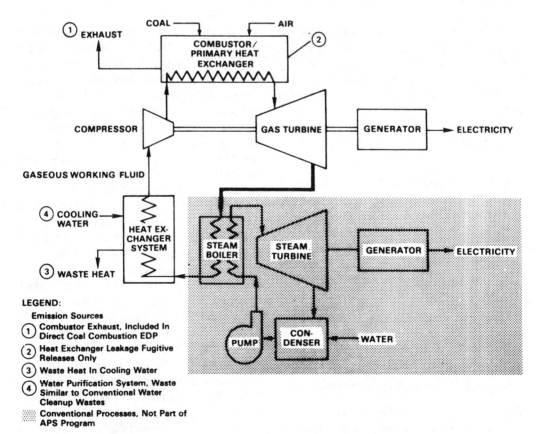

Figure II-3-101. Combined closed-cycles turbine/steam turbine system.

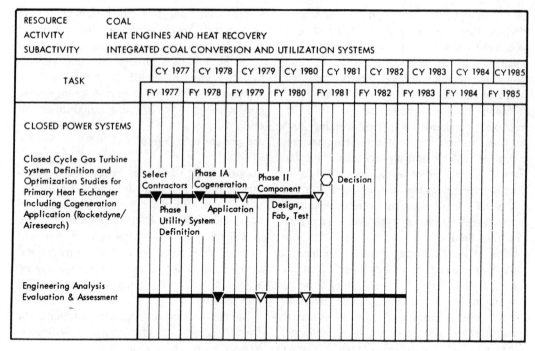

Figure II-3-102. Major milestones for the closed-cycle power systems task.

also planned. These studies will assess the economical technological trade-off for coal fuel conversion and gas turbine technology development (e.g., optimum level of coal liquid fuel cleanliness considering the projected turbine technology).

Efforts will continue on analysis assessment and engineering evaluations relating to the following areas:

- Key component studies and design of turbine bearings, seals, rotors, heat exchangers, controls, and instrumentation
- Power system technology testing and application of component design and development
- Coal-based fuel-fired advanced combustor design

These engineering analysis, evaluation, and assessment efforts are of a continuing nature and are efforts under the two major programs.

Advanced Cogeneration Systems

The technology for the efficient use of coal-derived liquid fuels rather than petroleum in stationary power plants is under development of Advanced (Coal) Cogeneration Systems (ACS) Subactivity. The sectors being served are the industrial and commercial/residential sectors which are the customers for electricity and fuels for process heat, and could benefit from their simultaneous generation (cogeneration). Because cogeneration systems can convert 70 to 85 percent of their fuel energy into useful electricity and process heat as compared to 15 to 40 percent for electric power generation only, cogeneration systems could use substantially less fuel and also produce substantially less emissions. Advanced cogeneration systems which use coal rather than petroleum are viewed as a means of reducing both emissions and life cycle plant fuel costs. Such systems are viewed as a very strong contender for marketing the minimally processed coal derived fuels beginning in the mid to late 1980s. The primary thrusts of work in the ACS Subactivity are to define the optimum cogeneration systems using these fuels, and to develop the engine technology for clean, reliable, and efficient operation on minimally processed coal-derived fuels.

Opportunities for Coal Generation in the new market are given in Figure II 3-103.

Between 1980 and 2000, the nation's electrical power demand could grow from the current level of 7 quads to a projected level of 14 quads. If this growth is met by burning fossil fuels, principally coal, it will require a new annual fuel demand of 25 quads/year (1,200 × 10^6 tons/year). During this same period, however, the fuel demand for heat in the commercial/residential and industrial sectors will also grow by about 16 quads per year by 2000. If the electrical growth requirements could be met by new cogeneration facilities at the users' sites, approximately half of the increased fuel demand for heating could be met by recovered waste engine heat as a by-product of the electric generation. While national energy growth would only be somewhat reduced, the important feature to note is that generating the electricity from coal and then using waste heat to displace some boilers and furnaces permits coal to displace petroleum, without cost or environmental penalty, in both power generation and in process heat production. The engine developments in ACS are intended to provide the fuels flexibility needed for this form of cogeneration so that it can proceed with coal-derived fuels in advanced cycled such as the diesel, gas turbine, and Stirling, which are more amenable to compact and smaller power generation at dispersed sites than are large utility coal-fired steam turbines.

In keeping with this strategy, the ACS effort consists of Prototype Systems Assessments to define optimum coal-fired cogeneration systems and develop critical nonengine components for cogeneration, and two types of coal-fired power plants (i.e., directly fired heat cycles and externally fired heat cycles).

Directly Fired Heat Cycles. The objective of the directly fired heat cycles task is to provide the engine technology for promoting a smooth transition from today's clean light distillate fuels to future minimally processed coal-derived liquid fuels.

Major near-term emphasis is placed on technology improvements for directly fired diesel and small- to medium-sized gas turbines to facilitate retrofit of these present

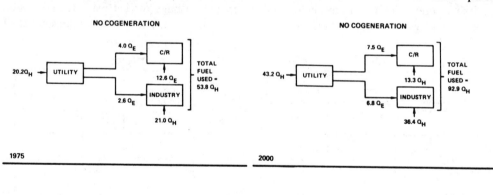

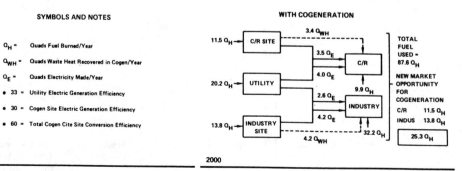

Figure II-3-103. Opportunities for coal generation in the new market — year 2000.

long-life engines that now require distillate or natural gas fuels. The R&D is directed toward permitting durable and low emission operation on minimally processed, more economically priced coal-derived liquids. Those are expected to be more like today's residual oils with respect to corrosive agents and emission sources than the more highly processed and expensive coal liquids that have these items removed by extensive hydrogenation. In addition to the economies of being able to operate on the residual type oils from coal, there will probably be more of these type fuels from a barrel of coal oil than there are from a barrel of crude oil. A comparison of the fuel characteristics of concern between light distillates, residual oils, and future coal-derived minimally processed liquids for gas turbines and diesels is shown in Table II 3-11.

A 10-MW directly fired heat cycles is illustrated in Figure II 3-104.

The reliable advanced liquid fuel turbine project is planned as a cooperative DOE/EPRI R&D effort. It is envisioned as a six-year joint program leading to commercialization of engines compatible with minimally processed coal-derived liquid fuels. The technology advances in coatings, coolings, combustors, and engine components are planned to be incorporated into a reference engine design. The results of these cofunded efforts are being directed toward a joint DOE/EPRI decision-point target in late FY 1981. At that time the need for a new engine development program will be determined.

An assessment of alternate means of incorporating this technology will also be thoroughly investigated by DOE/EPRI and engine manufacturers at the decision point milestone. It may be proven equally as effective to pursue "dash" engines, or modified versions of today's engines which incorporate new technology for improved efficiency and fuel flexibility, in lieu of a new engine development. Either the new engine development program or a development program involving modified "dash" engines would be structured to best support the needs of both industrial and utility gas turbine users, where liquid fuels are predominant.

Gas turbine technology now undergoing component development for improved performance and the capability to operate reliably and cleanly on coal-derived fuels include (1) ceramic coatings for hot part protection from alkali metals; (2) low NO_x combustor devel-

Table II-3-11. Future fuels and engine problems.

CHARACTERISTICS OF CONCERN	TODAY'S FUEL CLEAN NO. 2 OIL	NEAR-TERM RESIDUAL OIL	FUTURE COAL-DERIVED LIQUIDS	PROBLEM IN TURBINES & DIESELS	R&D COMMON TO RESIDUAL AND COAL-DERIVED LIQUIDS
Fuel Bound Nitrogen	Low	High	Higher	Environmental	Combustor Design, Fuel Cleaning
Hydrogen/Carbon Ratio	High	Low	Lower	Luminous Flame	Combustor Cooling, Coatings and Material
Alkali Metal Content	Low	High	High	Corrosion	Component Cooling, Coatings and Material
Viscosity	Low	High	High	Pumping & Atomization	Fuel Handling, Cleaning, Additives, Fuel Injection and Combustor Design
Ash	Low	High	High	Erosion/ Deposition	Fuel Cleaning, Additives, Coatings, and Materials

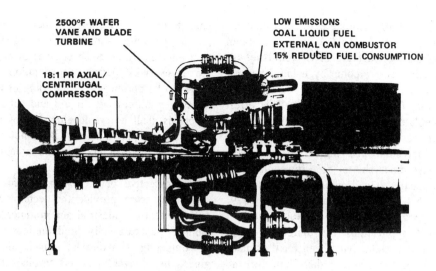

Figure II-3-104. Directly fired heat cycles — 10-MW industrial.

opment for burning fuels high in fuel-bound nitrogen without exceeding NO_x emission standards; (3) advanced soil surface convectively cooled turbine airfoils for enhanced cooling and no clogging with ashbearing, coal-derived fuels, and (4) advanced compressor technology for higher efficiency, lower wheel speeds, and up to 50 percent fewer parts for greater reliability and reduced manufacturing costs. During FY 1979, this compressor concept was used in the EPRI new centerline design.

An experimental evaluation of a promising convectively cooled airfoil design was completed in FY 1978 where the highest known cooling effectiveness for a convectively cooled airfoil was demonstrated in heat transfer cascade tests. This airfoil concept is now undergoing coal-derived fuel tests for EPRI. A major effort to screen promising concepts in ceramic coatings and low NO_x combustors through design, fabrication, and concept hot tests was initiated in late FY 1978 and will complete the Phase 1 concept screening effort in early FY 1980. The next phase for the coatings and the combustor program will be ready to start as a major procurement in late FY 1980 with the objective of making and starting engine testing with full-scale hardware by FY 1983.

The stationary diesel program is comprised of two major elements. The first major effort was initiated in FY 1978 and has the objective of defining the capability of current diesel operation, with minimum modifications, on fuels from coal—principally coal-derived liquids. This effort continued in FY 1979 along with selected efforts in low heat materials for high ash-bearing fuels. The engine test program competitively awarded in FY 1979 emphasized large-, medium-, and small-sized diesel engine capabilities to ignite on low-cetane number fuel, wear and corrosion due to ash and alkali metal content and emissions of NO_x. It is anticipated that this coal-derived fuel evaluation in diesels will last at least through FY 1980.

The objective of the second major element of the diesel program is the major redesign of diesels for coal-derived fuels, and it will be initiated only if the first evaluation phase of the program proves that this is necessary. Some progress was made during FY 1978 to indicate that slow speed diesels may already have a major amount of the desired capability to operate satisfactorily on coal-dervived liquids. Specifically, during FY 1978 a large, two-cycle, slow speed, single cylinder diesel was run on clean de-ashed COED fuel as well as on SCR-II fuel. Both fuels ran with less than 1 percent gain in heat rate (Btu/kWh). The COED test ran with less than 35 percent conversion of fuel-bound nitrogen to NO_x, which amounted to less than a 10 percent gain in NO_x over light distillate petroleum fuel. By the end of FY 1978, the only engine problem encountered was that the SRC would not ignite, which was overcome by using 5 percent pilot injection of light distillate before introducing the 95 percent SRC fuel. This effort was completed in early FY 1979. Whether or not such successes can also be obtained on smaller, medium, and higher speed diesels which have much shorter time for combustion and are usually valved engines, remains to be proven; this effort was started in FY 1979. Higher speed diesels are of interest because to date most applications for diesels in U.S. cogeneration systems have been for the small- and medium-size range rather than for the very large machines.

Major milestones are shown in Figure II 3-105.

Externally Fired Heat Cycles. The development of a prototype externally fired engine capable of virtually complete fuel flexibility is the longer range thrust of the cogeneration heat engines R&D. Such engines in appropriate sizes will permit cogeneration power plants to be optimized for a variety of domestic coal resources both liquid and solid. Here emphasis shall be on development of a common engine with a family of combustor options to provide a maximum degree of fuel flexibility on coal.

The objective of the externally fired heat cycle task is to provide the technology and development of industrial and commercial/residential size externally fired engines capable of competing with diesels, spark ignition engines, and conventional gas turbines for stationary power applications while operating on coal fuel. Within the objective are two primary thrusts: (1) the development of large Stirling engines as prime movers for industrial and residential/commercial total energy systems; and (2) the development of an exter-

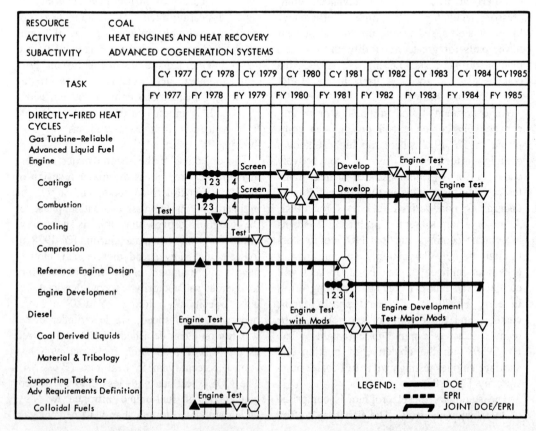

Figure II-3-105. Major milestones for the directly-fired heat cycles task.

nally fired Brayton engine for industrial cogeneration applications. Both of these engines are capable of using a variety of fuels, including liquid, solid, and gaseous fuels, to provide the necessary thermal input.

Figure II 3-106 illustrates a total energy (cogeneration) system with a Stirling engine as a prime mover. The Stirling engine extracts heat from the external combustor and uses it to drive a piston and crankshaft to generate electricity for use at the site which may be a hospital, apartment complex, or other similar application requiring electricity, heating, and cooling. The engine reject heat and the unused heat from the combustor is passed through a heat exchanger to transfer energy to a water circulator loop that can provide either space heating or cooling via absorption air conditioning. The working fluid in the Stirling never leaves the engine and all heat transfer is done by heat exchangers.

Only Stirling units directed toward automotive applications have been developed, although the concept has existed for many years. The major development effort will be to progress from the light-weight, compact automotive-type engine to the stationary power generation engines in two major size ranges with quite similar technologies—20 hp and 1000 hp where efficiency and long life are important. The low-cost small engine, in addition to being useful in itself for residential/commercial applications is also five years ahead in its development compared to the larger machine and will provide concept verification information for both size ranges.

Figure II 3-107 shows the externally fired Brayton cycle in an industrial cogeneration application. Heat is supplied to the working fluid (air) in the Brayton cycle through a heat exchanger. The clean, dry hot air exhaust from the power turbine is then sent directly to an industrial process requiring large quantities of clean, hot air (paper drying, foods, etc.).

Few externally fired Brayton cycle engines are in use because of the limit in turbine inlet temperature as a result of metal temperature limits in the heat exchanger, which results in

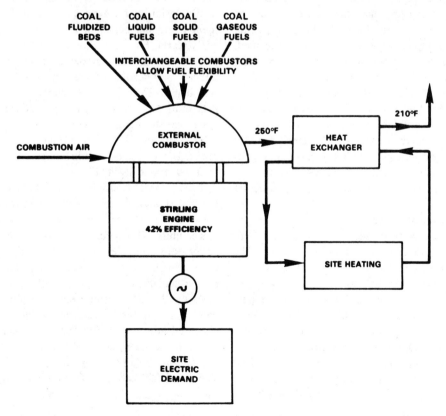

Figure II-3-106. Stirling engine total energy system.

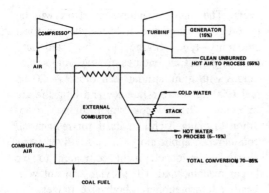

Figure II-3-107. Externally fired Brayton cycle.

relatively low efficiency where the only output is electric power. Using the exhaust air directly in a cogeneration installation however, boosts the overall fuel utilization from 15 percent up to over 80 percent and makes this cycle particularly viable as a means of substituting coal in industrial processes normally burning light distillate or natural gas.

The large Stirling engine development program is envisioned as a five-year competitive effort, going through key phases of design and critical technology development, after which a project evaluation will be performed. Subsequently, a prototype construction phase and an engine testing phase are planned to establish engine performance and capability with coal fuel sources. The present work involves defining the practicality of scaling up Stirling engine technology to sizes of 1,000 hp or greater, as well as determining economic viability of the stationary Stirling. The first contracts for conceptual design were let in mid-FY 1979.

An evaluation has been completed on the theoretical performance of Stirling cycle engines in various total energy systems. The best designs for residential and hospital sites indicated that Stirling systems offer very attractive rates of return (greater than 25 percent) when purchased electricity is greater than $0.04/kWh. In FY 1978, a preliminary engine evaluation was completed to determine the critical engine design problems in scaling up a Stirling engine to the greater than 1,000 hp level. A second evaluation will have defined the technical problems that must be solved to develop commercially competitive Stirling engines in the 500 to 3,000 hp range. Compared to the diesel and gas turbine engines, Stirling engines offer high efficiency, high reliability, quiet operation, and multifuel capability. Work has also been completed on the Stirling's applications, which include an adjunct to fluidized-bed combustors, small municipal utility power generation, total and integrated energy systems, and the utilization of coal fuels. Another evaluation was completed for the investigation of the feasibility of Stirling engines for industrial applications that utilize a vapor working fluid in lieu of monatomic gases that are now commonly used. By using a condensing vapor as a working fluid, many of the problems of gas Stirling engines could be solved. These include uniform heat transfer to heater tubes, engine control, unswept volume caused by heat exchangers for cooling the working fluid, and the criticality of loss of gaseous working fluids. The information gathered from these various sources was used by selected contractors in the design and critical component development work that began in FY 1979.

By early FY 1979, fabrication was completed on two different design small Stirling engines and testing of the full-scale engines will begin. The test results will be used as a data source for the industrial team selected for the large Stirling program.

The project to develop efficient and cost-effective small steam turbines was under contract by the end of FY 1978. This project addresses the low efficiency and high cost of today's state-of-the-art small steam turbines now available for dispersed cogeneration power units. The specific objective is to raise current efficiency levels of approximately 30 to 50 percent up to 75 percent or greater and reduce specific costs to that small steam turbines can be efficient and economically competitive as a prime mover component in small dispersed power sysytesm. The range of horsepower to be addressed is from 500 to 6,000 hp. During FY 1979, the design was completed and fabrication of feasiblity verification hardware begun. Fabrication and testing will be completed in FY 1981. Further development will depend upon the desirability of small steam turbine cogeneration systems.

One of the key cycles under investigation for advanced coal-fired cogeneration systems in the Cogeneration Technology Alternatives Study (CTAS), discussed under the Prototype Systems Assessment task, is the externally fired Brayton engine. While the applications and sizes for this cycle are pending in the CTAS program, certain critical technologies have been identified for early hardware investigation, which began in FY 1978. A small contract has been let to modify the combustor of an existing small turboshaft gas turbine to make it externally fired. This effort will investigate power modulation controls, startup and shutdown, and will use steam injection as a means of varying the ratio of thermal to electrical power, an essential parameter when considering the thermal and electrical power requirements of an industrial plant. The power plant was tested in FY 1979 (further investigation in the detailed systems design for industrial cogeneration applications is described in the prototype Systems Assessment Subactivity.)

Major milestones for this task are shown in Figure II 3-108.

Prototype Systems Assessment. The emission reductions and reduced costs to burn coal-derived fuels can range as high as 25 to 75 percent in cogeneration systems. (See Fig-

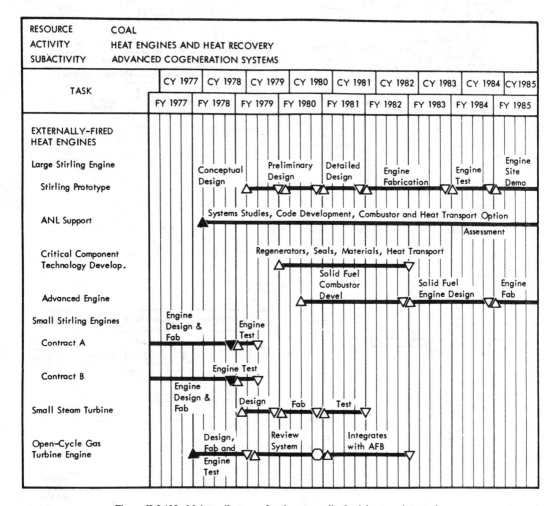

Figure II-3-108. Major milestones for the externally-fired heat engines task.

ure II 3-109.) However, whether or not such benefits can be considered feasible depends heavily on which type of engine is matched up with a particular industrial or commercial/residential process. Factors in this consideration include broad variations in engine power and waste heat characteristics, process electrical and heat requirements, engine size and cost characteristics, return on investment, and projected fuel prices. To assess this feasibility, the objective of the Prototype Systems Assessment task is to perform the systems analysis necessary by FY 1980 and to integrate all components and subsystems into advanced coal-burning prototype cogeneration systems after FY 1980.

Two stationary sectors are being examined, the industrial and the commercial/residential sectors. Two major studies began in FY 1977–1978 to perform this analysis. The in-

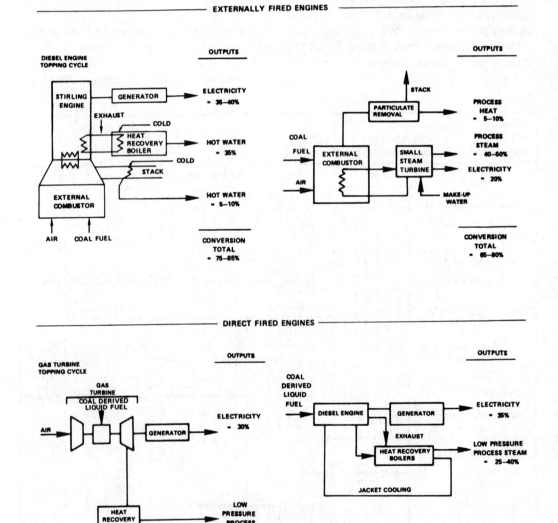

Figure II-3-109. Examples of cogeneration concepts.

dustrial effort is called the Cogeneration Technology Activities Study (CTAS), and the commercial/residential effort is called the Total Energy Technology Assessment Study (TETAS).

Typical cogeneration schemes consider two types of coal-fueled power plants—direct and externally fired heat engine cycles. Note that the overall efficiency of fuel utilization in cogeneration is several times higher in most cases than that of the isolated power plant. Of interest to the utilization of coal-derived fuels is the fact that emissions to provide heat and electricity and life-cycle plant costs for fuel are correspondingly reduced because of the increased output on the coal-derived fuel.

After the first generation of studies to define the optimum matches between engine types and process heat and electrical requirements for the 1985–2000 time frame are completed in FY 1980, the next step will be to examine the best systems in more detail to define all the critical system requirements besides engines operating on coal-derived fuels.

The Total Energy Technology Assessment Study (TETAS) was initiated in late FY 1977 to define the most promising advanced total energy systems for commercialization in the residential/commercial market sector. During FY 1978, system energy demands were defined, along with the evaluation of fuel-flexible heat engines (gas turbines, Stirling engines, diesels, and steam turbines) as engines for these total energy systems. Work in FY 1979 will cover the evaluation of heat pumps to enhance overall efficiency and definition of optimum systems. A market penetration analysis of the TETAS systems was completed in FY 1979, which will determine market acceptance of these systems fired by coal-derived fuels with particular emphasis upon emission reductions, life cycle cost savings, and scarce fuel displacement with coal. TETAS will be completed in FY 1980 and final reports published and distributed.

In the industrial sector, the focus is on defining advanced cogeneration opportunities for the six major industries that currently consume 70 percent of the nation's industrial process heat. This effort, the CTAS study, was started with two prime contractors and their industry teams in mid-FY 1978 and was completed in mid-FY 1979. During FY 1978, very specific industrial process requirements were defined and computer-modeled on an energy balance basis. The energy balance characteristics of current technology and advanced energy conversion systems (gas turbines, diesels, Stirlings, steam, organic Rankine cycles, fuel cells, heat pumps, and energy storage) were also defined, computer modeled, and matched against the industrial process requirements to define the technical impact of advanced engines in cogeneration. Efforts in FY 1979 included completion of the evaluation in cogeneration. Efforts in FY 1979 included completion of the evaluation by identifying optimum cogeneration systems with emphasis upon coal burned in lieu of scarce fuels, industry specific economics, and R&D requirements and emissions reductions.

Final reports were published and a public briefing was given on the outcome of the study in the spring of 1979. After the completion of the CTAS contracts, to meet the electrical and process heat requirements of several major Federal facilities (USN bases) and will also examine the applicability of cogeneration on coal fuels to boost the process efficiency of coal liquefaction plants. Work in the last quarter of FY 1979 and most of FY 1980 will assess the impact of successful cogeneration using coal fuels on the utility and residential/commerical sectors for the more successful systems in the evaluation. An RFP will be prepared for future issuance to examine the best systems in more detail in order to identify additional critical components and systems requirements needing R&D.

Both TETAS and CTAS focus on new technologies that can be commercialized in the 1985–2000 time period. To evaluate a promising advanced system with near-term potential for industrial cogeneration bottoming cycle applications, the design and analysis of a prototype 2-MW organic Rankine cycle PDU was started in late FY 1978 and will be completed in late FY 1979. Pending successful completion of the system studies, it is anticipated that any further pursuit of non-coal-fired systems such as this and other bottoming systems will come from activities outside of fossil energy, probably from conservation and solar applications.

Major milestones for this task are shown in Figure II 3-110.

Heat Recovery Component Technology

The goal of the Heat Recovery Components Technology Subactivity is to develop the required technology base and to demonstrate the technological feasibility of cost-effective components and systems for the recovery and utilization of waste heat. Widespread use of waste heat as an equivalent fuel source is at present limited by the availability of cost-effective recovery and utilization systems and the lack of sufficient technological options. Technology development is organized in three tasks: (1) Low Grade Heat Recovery, (2) High Grade Heat Recovery, and (3) Heat Exchanger Technology. The dividing line between low- and high-grade heat has been arbitrarily defined as 200°F. A breakout into these two broad areas was made because the types of utilization systems and the associated technological problems are quite different for each. Heat Exchanger Technology has been singled out for emphasis because of its widespread use in all waste heat recovery and utilization systems. In many cases, it is the most expensive component of the system. Technology developments will often be applicable to both the low-grade and high-grade applications. It is therefore highly desirable to maintain a coherent, directed effort in heat exchanger technology with input from this task to the other two tasks.

Primary emphasis in the low-grade heat task is to study the recovery and offsite utilization of energy rejected by Federal facilities using available technology. In addition, technology development has been initiated to expand the number of available utilization options as well as to improve the efficiency of the components in heat recovery systems so that sizes and costs can be reduced or performance can be increased to make the systems economically viable. Low-cost, high-effectiveness heat exchangers for low-temperature heat recovery are being developed under the heat exchanger task in support of this goal. Other major areas of R&D are novel

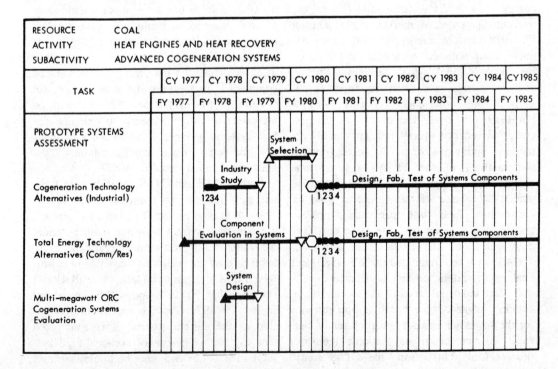

Figure II-3-110. Major milestones for the prototype systems assessment task.

heat engine concepts for power generation that offer the potential of reduced costs per kWh output, as well as an effort to develop the capability of providing space cooling from lower temperature waste heat sources. The area of low-temperature waste heat utilization also includes a focused R&D effort to develop industrial heat pumps. These heat pumps would take the energy from the low-temperature stream and, using either additional energy from the stream or an external energy source, pump it up to a temperature suitable for industrial process use.

With regard to higher grade heat, the major technology need is to provide capabilities that do not exist. At the present time, we are limited by heat exchange materials technology and economics to using fuels in the generation of steam at about 1000°F maximum temperature. Since the energy available in the combustion process is generally in the range of 3000°F, the thermodynamic availability, or the capability to convert thermal to mechanical energy, is severely inhibited by degrading it to 1000°F before extracting work. Technology development in the area of thermionics is being conducted for application to high-temperature topping cycles. In the area below 1000°F, representative of the exhaust profiles of stationary diesels and gas turbines as well as many industrial processes, the efficiency of power recovery cycles using steam as a working fluid drops off sharply. Cycles using organic working fluids are being developed for application in this regime. Also, many industrial processes today discharge high-quality waste heat from furnaces in the 2000°F to 3000°F temperature range. Presently, there are no durable heat exchangers available to effectively recover this energy for either power generation or preheating incoming combustion air. The application of high-temperature materials to heat exchangers is the major technology effort to develop this capability under the Heat Exchanger Technology Task.

Low-grade Heat Recovery. Low-grade heat recovery technology is being developed for application to heat sources with temperatures generally at 200°F or less. This effort is directed at using the very abundant, replenishable energy from low-temperature sources such as oceanic, geologic, unfocused solar, and waste heat from industrial processes and power production.

The major objectives of this effort which are directly related to DOE national goals are elimination of avoidable energy losses to reduce energy consumption; improvements in the efficiency of existing and future energy converters; and development of alternate energy sources to reduce both consumption rates of domestic fuel sources and importation rates of foreign fuels.

The Low-Grade Heat Recovery Task is divided into two major efforts: use of energy rejected from Federal facilities, and low-temperature technology development. In the former effort, feasibility evaluations to determine the technical, economic, and institutional viability of utilizating energy currently being rejected by DOE-owned facilities have been completed. The efforts are directed to the three uranium enrichment gaseous diffusion plants located in Oak Ridge, Tennessee; Paducah, Kentucky; and Portsmouth, Ohio; and the Savannah River Facility in South Carolina. These facilities reject a combined total of 12,000-MW of low-grade thermal energy. It is expected that the solution of using these large quantities of virtually steady state energy will not be trivial. Because of institutional barriers predicted for possible applications, direct use applications such as agriculture and aquaculture will not comprise the total solution. Figure II 3-111 represents a possible complete approach to utilizing this source. It is apparent that complex and innovative definition and full-scale optimization must be successfully accomplished to ultimately exploit this energy source.

The 200°F range of waste heat is a relatively uncharted area in terms of productive heat utilization. The Low-Grade Heat Recovery Task is designed to determine the feasibility of developing heat recovery systems which can economically convert the thermal energy available at these temperatures to productive use. Because of the paucity of such systems, a major project effort must, of necessity, be devoted to determining the requirements that have to be met to utilize low-temperature energy and to identify opportunities for applying the low-temperature heat utilization systems that are the object of the development program.

In recent years, a few sporadic attempts have been made to develop low-temperature heat utilization systems. For the most part, however, sufficient information has not been available to adequately evaluate the technological problems involved in scaling up these prototype systems to sizes suitable for commercialization.

In FY 1978, a Low-Temperature Technology Development Program was initiated to address the evaluation and conceptual design aspects of candidate heat engines and recovery devices. In addition to low-temperature heat exchanger enhancement and effectiveness improvement, novel heat engine concepts such as nitinol heat engines and engines employing the concept of pressure-retarded osmosis are being evaluated. Other areas being investigated are heat pumps and absorption cooling.

The feasibility evaluation phase of the Federal facilities program was completed in FY 1979. A "best mix of uses" system was defined for each of the four sites. A DOE-wide task force committee was appointed to integrate the findings of the feasibility evaluation phase and select one or more sites for subsequent implementation.

With the shift in emphasis from utilization to supply, the Low-Temperature Technology Development Program is currently under consideration for transfer to the Conservation and Solar Applications Office.

Major milestones for this task are presented in Figure II 3-112.

High-grade Heat Recovery. High-grade heat recovery technology is being developed for application to heat sources available at temperatures generally well above 200°F. This effort is directed at recovering and using energy ejected by prime movers such as diesel engines and gas turbines and energy expended in large central station power plants and industrial process heat streams.

In FY 1976, a six-month technology assessment, the Industrial Applications Study, was performed to develop a profile of energy consumption and waste heat availability for all Standard Industrial Code (SIC) two-digit industries. Because of the success of this effort, a follow-on study, to extend the data bank to SIC four-digit industries, was initiated late in FY 1977. This is a major undertaking which,

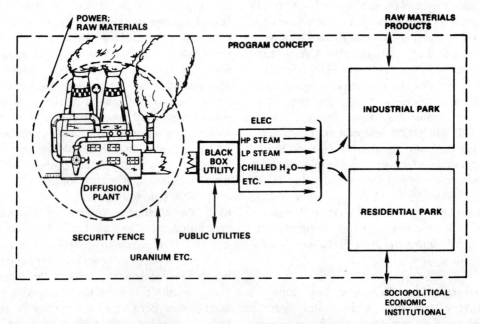

Figure II-3-111. Low-grade heat recovery-schematic concept for use of energy rejected by Federal facilities.

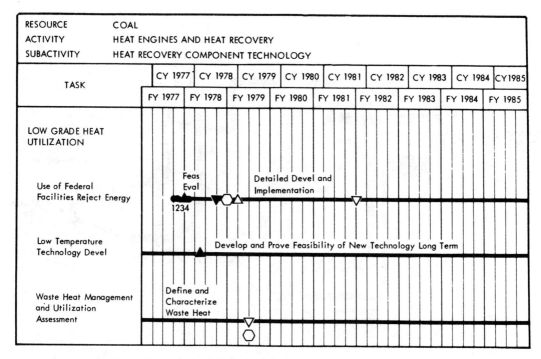

Figure II-3-112. Major milestones for the low-grade heat utilization task.

when completed, will provide the most complete and detailed data bank of waste heat resources to date for the planning of future energy conservation efforts. The effort is expected to take approximately 1.5 years. It is planned to use the results as a basis for the determination of additional high payoff areas of the higher grade waste heat utilization.

The Bottoming Cycle Systems Program covers technology suitable for application to waste heat streams with temperatures between 200°F and 1000°F. This temperature range includes most existing industrial and commercial waste heat sources. The program is directed towards the near-term utilization of these sources through the development of waste heat recovery systems. Efforts have been initiated to design, develop, and field test three competing Rankine-cycle-based systems suitable for operation with public and municipal utility and industrial diesel power plants. A typical bottoming system that employs an organic working fluid is shown in Figure II 3-113. The waste heat stream is passed through a heat exchanger (vaporizer) to vaporize the organic fluid; this vapor is expanded through a turbine to produce a shaft work output; the vapor at a lower temperature and pressure is then condensed back to a liquid state, compressed, and then revaporized. In general, higher availability and recoverability is provided by using organic working fluids than for steam in this temperature regime.

The one major effort presently underway in the Topping Cycle Systems area is the Thermionic Energy Conversion Development project. Thermionic energy conversion has several distinct advantages in addition to its high heat rejection temperature that make it attractive as a topping cycle. Thermionic converters have no moving parts, a characteristic important for high reliability at high temperatures. Since thermionic converters operate with very low internal gas pressure, material stress is minimized, also an important characteristic at high temperatures where material strengths are lower. In addition, thermionic converters are well suited to fabrication as self-contained modules, an approach which provides both higher overall system reliability and significantly lower development costs.

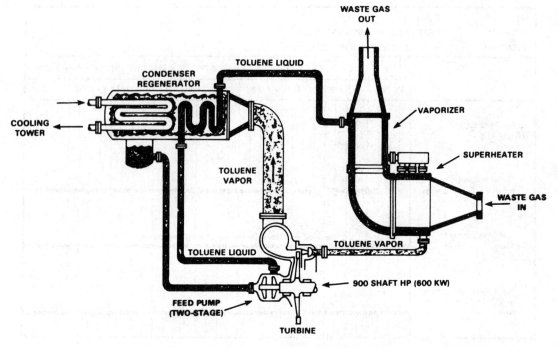

Figure II-3-113. Organic Rankine Bottoming cycle.

Thermionic conversion is a method for converting heat directly into electrical work through the use of thermionic emission. The thermionic device consists of, as a minimum, one electrode connected to a heat source and a second electrode connected to a heat sink separated from the first by an intervening space; there are also leads connecting the electrodes to an electrical load and an enclosure. The enclosure is either highly evacuated or filled with a suitable rarefied vapor. The concept has had widespread use in space propulsion technology. The thrust of this effort, however, will be to develop thermionics as a viable power-producing device for ground-based energy needs where high-termperature heat is utilized as the primary energy source.

The overall objective of the thermionic energy conversion development effort is to define, develop, and demonstrate thermionic modules to the degree necessary to encourage commercial participation and acceptance. Accomplishment of this objective could provide the United States with a new and attractive energy conversion alternative that could potentially save a substantial proportion of our energy resources. The task scope includes all aspects of thermionic energy conversion: research on fundamental processes affecting performance, development of suitable converter materials and assemblies, demonstration of component and module performance and reliability, and studies of the use of thermionic modules in promising applications. The task is closely coordinated with programs supported by other Government agencies, such as the current NASA program.

Studies performed by DOE has defined many potential benefits available from thermionic energy conversion. Strong interest and encouragement have been provided by major power plant component manufacturers, A&E firms, and private sector research organizations such as EPRI. However, the technology is still too young and the risks too high for industry to make a major commitment in this field. Rapid development of the thermionic technology will require Federal assistance to demonstrate its basic feasibility in ground-based power applications and to ensure its maintenance as a potentially significant energy conversion alternative for the mid-term.

The Rankine Bottoming Cycle Program has

been underway since FY 1976. The major thrust has been the development of systems in the 500–600 kW output range for application to waste heat utilization from stationary power generating diesels. In FY 1979, six field test demonstration systems were delivered, installed, and brought on stream. The fuel savings attributable to these field demonstrations units alone is in excess of 40,000 barrels of oil per year. Because of a programmatic shift in emphasis from utilization to supply, existing contracts and cooperative agreements with industry to monitor and evaluate the results from the field test program and to develop systems in different size ranges for application to other types of power generation equipment as well as to industrial processes are under consideration for transfer to other DOE offices.

In FY 1979, a major breakthrough was achieved in the thermionics program. A thermionic converter driven by a simulated heat source delivered a current of 6000 amperes. This value is over four times greater than that produced by any other thermionic device to date. Efforts over the next three years will focus on the development and demonstration of a first generation, prototype thermionic topping cycle module to establish the feasibility of this concept.

The FY 1980 effort is to develop technology for the utilization of waste heat in the temperature range of 200°F to 1000°F in the industrial and utility sectors. The bottoming cycle projects initiated in FY 1976 will be continued. Specifically, tests and evaluations were performed at the six demonstration sites completed in FY 1979, and development of advanced systems for future applications will be continued. Thermionics is also included in this project.

Major milestones for this task are presented in Figure II 3-114.

Heat Exchanger Technology. The recent energy crisis has sparked renewed interest in heat exchangers for both waste heat recovery and as components in advanced energy systems since they are often the highest cost components in the systems. A number of key areas have been identified to improve performance capability or reduce costs.

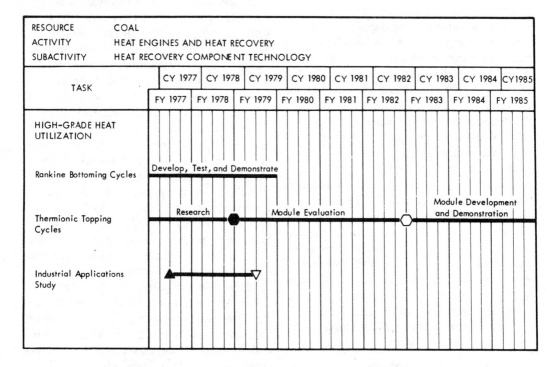

Figure II-3-114. Major milestones for the high-grade heat utilization task.

Flow-induced tube vibration is a key factor in limiting heat exchanger reliability and lifetime. Numerous failures have been recorded in industrial operations. Current practice is to over-design for size, therefore driving up the price. Since heat exchangers are the major cost items in heat recovery/utilization systems, the cost penalty incurred by vibration will limit system market penetration potential and hence energy savings.

A growing concern with efforts to efficiently recover waste heat in diesel exhaust streams is the lack of reliable information with which to accurately predict acid dewpoints, corrosion rates, and the extent of fouling for various heat exchanger configurations and materials.

One of the driving forces in this task is the need to reduce costs. By increasing heat transfer effectiveness, the size and cost of an exchanger for a given duty can be reduced. Laboratory tests indicate that local heat transfer coefficients can be improved by one or two orders of magnitude through the application of fluidized-bed technology. This is as yet a relatively unexploited area of great potential.

Another heat exchanger deficiency uncovered in the desire to rapidly implement advanced energy systems is the lack of a high-temperature capability. This capability is required in order to implement closed Brayton topping cycles and industrial recuperators for utilization of high-grade waste heat. Potential developments in ceramic materials technology offer promise in extending the upper temperature limit of heat exchangers.

One more area of concern is the inefficiency of low-temperature heat exchangers. Development of more effective heat exchangers would represent a major step toward the recovery of energy from one of the nation's largest waste heat sources: the cooling water from the DOE gaseous diffusion plants. Promising concepts for efficiency enhancement involving advanced heat transfer surface design configuration will be evaluated.

The objectives of the Heat Exchanger Technology task are as follows:

- Increase upper temperature limit capability of heat exchangers to greater than 2000°F
- Increase heat exchanger reliability and lifetime by up to 50 percent. Demonstrated reliability and lifetime capability are prerequisites to the commercialization of heat exchangers per se, or as system components
- Reduce heat exchanger costs by as much as 50 percent
- Increase heat transfer effectiveness by up to 50 percent. Gains in effectiveness can also be traded off to reduce costs.

Milestones for the projects comprising the Heat Exchange Technology Task are provided in Figure II 3-115. The initial project shown was conducted as part of the International Energy Agency Heat Transfer and Heat exchanger project under the Energy Conservation Working Group.

This subtask is testing variable geometry, segmentally baffled, shell-and-tube heat exchangers in a typical industrial site for the purpose of obtaining tube vibration data under controlled conditions. The data are required to evaluate the validity of presently used heat exchanger predictive methods and design criteria. The purpose of the evaluation is to assure that reliability and lifetime requirements are met with minimum design safety margins and associated cost penalties. Concurrently, a data bank of field operating experience is being developed to evaluate the validity of existing and new empirical correlations.

The presence and extent of fouling and acid corrosion of candidate heat exchanger configurations in stationary diesel exhaust streams, along with the experimental determination of sulfuric acid dewpoint, was determined. Methods were devised for predicting and alleviating the resultant adverse effects on exhaust side thermal resistance and pressure drop. The compilation of a reliable waste heat boiler corrosion and fouling data base for diesel exhaust heat recovery was completed.

The approach being taken to achieve the high-temperature capability (2000°F) required to recover heat from high-temperature industrial furnaces is the exploitation of ceramic materials, primarily silicon carbide. Other applications for this technology are in coal-fired, closed Brayton, and indirect-fired, open-Brayton cycles where this temperature capability would make these concepts very attractive for power generation from a thermodynamic efficiency standpoint. The high-temperature project is conducted in close coordination with EPRI. Active project elements include the following:

- Evaluation of thermal, mechanical, and physical properties of tubing manufactured by a number of processes
- Development of stress-relieving joining techniques for ceramic tubing
- Assessment of nondestructive test and evaluation techniques for application to ceramic heat exchangers and components
- Development of a ceramic heat pipe using a liquid metal working fluid. This is an attractive alternative to a conventional heat exchanger configuration from the standpoint of minimizing thermal stress. In FY 1979, a ceramic heat pipe using sodium as the working fluid was successfully tested at 950°C.

In support of the low-grade heat recovery task, a concerted effort was made to develop and evaluate heat transfer augmentation techniques such as enhanced surfaces to improve heat transfer effectiveness and reduce costs. Surveys to identify promising approaches are underway. Heat transfer measurements of new concepts are being made in the laboratory. Another approach evaluated to reduce costs is the application of plastics to heat exchanger fabrication.

Laboratory studies of heat transfer in fluidized beds have demonstrated heat transfer coefficients of two orders of magnitude greater than for conventional heat exchanger configurations. Thus, it is an attractive technological option for the improvement of heat transfer effectiveness which can then be traded off, if desired, against cost. An experimental evaluation of a unit commercially marketed for waste heat recovery has been completed. Based on these results, areas of potential performance improvements were identified and the evaluation of a second generation design is underway. Another application under evaluation is to the evaporator and condenser of

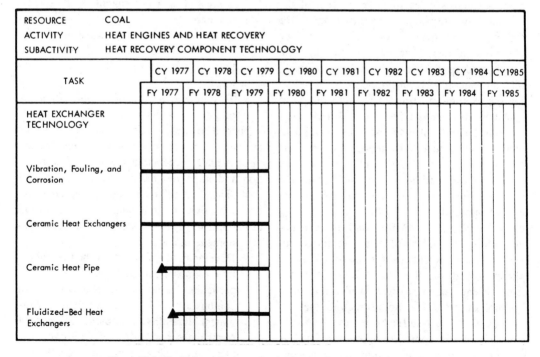

Figure II-3-115. Major milestones for the heat exchanger technology task.

residential heat pumps. By improving the airside transfer coefficient, the temperature drops across the heat exchangers can be increased. This will reduce the cycle irreversibility and increase the coefficient of performance. Further increases in performance will be attained by removal of frost from the evaporator in winter. This will eliminate the need for a defrost cycle with its consequent energy penalty.

Because this is a generic technology area with potential applications much broader than Heat Engines and Heat Recovery, it will be funded in FY 1980 under the Advanced Research and Technology Development budget line item.

Major milestones for this task are presented in Figure II 3-115.

COMBUSTION SYSTEMS

Combustion is involved in over 90 percent of fuel utilization processes in this country and, even under the most optimistic predictions for alternatives, will continue to dominate our energy usage. Because of the former low price of oil and gas, our most abundant resource, coal, accounts for less than 20 percent of our current fuel utilization. The tasks under the Combustion Systems Activity are selected on the basis of their potential for burning coal clearly and more efficiently and for reducing oil and natural gas consumption by developing the technology required for substitution of coal and coal-derived fuels for oil and gas.

Even though coal is the nation's most abundant domestic fuel, coal represents less than 55 percent of the fossil fuel consumed by steam-electric utilities. National emission standards for stationary coal-fired steam generators presently limit sulfur dioxide (SO_2) emissions to 1.2 lb/MMBtu, and nitrogen oxides (NO_x) to 0.7 lb/MMBtu. Proposed revised emission standards for coal-fired utility plants, based on the 1977 Clean Air Act Amendments requiring Best Available Control Technology (BACT), would require that 85 percent of the fuel sulfur be removed prior to emission of the combustion products. Sulfur dioxide is present only in relatively low concentrations, and the high cost of removing it from large volumes of stack gases has restricted the use of high-sulfur coals. Indications are that future standards will require greater cleanup of combustion products. Because fluidized-bed combustion (FBC) boilers, atmospheric and pressurized, absorb the majority of SO_2 in the combustion process and produce considerably less NO_x than conventional coal-fired boilers, the technology should enable increased coal utilization in an environmentally acceptable manner (i.e., reduce SO_2, NO_x and particulate emissions to acceptable levels). The particulates emitted from the combustion zone of FBC's can be controlled to acceptable levels with existing technology. The major environmental concern with FBC's is, however, disposal of the large quantities of solid wastes produced from the fly ash and the spent bed itself. Many programs are presently underway to study the characteristics and demonstrate possible uses for the materials.

The objectives of the Combustion Systems Activity are as follows:
- To develop fluidized-bed combustion systems capable of directly burning high-sulfur and other coals of all ranks and quality in an environmentally acceptable manner
- To develop technology to substitute coal for a substantial portion of oil in combustors capable of firing coal-oil mixtures
- To improve the reliability and efficiency of present boilers and furnaces

To realize these objectives, research efforts are being directed toward development of environmentally acceptable methods of combustion and improved technology for the use of coal. The total effort has been integrated into two subactivities: technology and component development. Specific areas of research and development include:
- Development of fluidized-bed combustion (atmospheric and pressurized) with particular attention to achieving high combustion efficiency, acceptable component durability, minimum emission of particulates and sulfur and nitrogen oxides, and reliable operation of combined-cycle systems

- Determination of combustion and heat transfer characteristics of chars, coal-oil slurries, solvent-refined coal, and coal-derived liquid fuels when burned in conventional equipment and the application of such data to improved combustion component design
- Determination of causes of adherent slag and ash deposits, and development of methods for minimizing these reliability- and efficiency-degrading problems
- Identification and control of toxic elements evolved during the direct combustion of coal

Results due to the application of these objectives are described below.

The 30-MW$_e$ atmospheric fluidized-bed boiler project, located at the Monongahela Power Company plant to Riversville, West Virginia, had an improved coal feed system installed in early 1978, and was in steady operation delivering superheated steam under utility controlled conditions of temperature and pressure on several occasions, including runs of 50, 25, and 24 hours. Preliminary analysis of the data obtained during 1,054 hours of operation in 1978 shows the SO_2 absorption to be in agreement with the design basis of the unit.

The AFB Component Test and Integration Unit (CTIU) at Morgantown, West Virginia, has proceeded to the stage where the building is partially complete and the research boiler is in the detailed design phase. It has been decided that the PFB CTIU will not be built in light of the progress on the International Energy Agency (IEA) PFB facility and the successful operation of Curtiss-Wright's Small Gas Turbine (SGT) PFB test rig. The PFB CTIU functions have been split up among the IEA, SGT and Leatherhead PFB facilities. The 100,000 lb/hr saturated steam AFB boiler for Georgetown University was nearing completion for startup in early 1980. The other four AFB industrial applications boilers and heaters have progressed thru sub-scale testing and design and are beginning construction. Three contractors have begun design work on AFB boilers to burn anthracite culm/anthracite.

The Commercialization Task Force report on AFB industrial boilers concluded that this product was ready for commercialization and recommended such a program.

Accomplishments in the pressurized fluidized-bed program include the operation of the Small Gas Turbine rig to test the PFB pilot plant combustor design and to obtain data on PFB combustor and gas turbine blade material life and corrosive attack mechanisms, and progress towards the completion of the design of the PFB pilot plant such that, subject to confirmative data on materials behavior, construction was started in 1979. The IEA PFB Test Facility was completed in 1979 and will commence combustion testing with a (stationary) turbine blade cascade. The IEA PFB facility scope of work was expanded to add a stationary gas turbine blade cascade with tertiary particulate removal and a design study to determine the merits of adding a (rotating) gas turbine to the combustor exhaust gas path.

The coal-oil mixture (COM) program progressed with the operation of the 700 hp COM test facility and COM preparation laboratory at the Pittsburgh Energy Technology Center (PETC). Construction of the demonstration projects has started. An AFB boiler materials test apparatus was contracted for to have a facility in which to perform long-term materials testing. Programs to demonstrate utilization of AFB spent bed materials as highway roadbase aggregate, agronomic soil nutrient, and as a source of lime for neutralizing acid soils and as a scrubber reagent produced initial encouraging results.

In addition to these projects, laboratory studies, engineering studies, and economic analyses are underway in support of the development programs.

Table II 3-12 summarizes the funding levels by subactivity for the FY 1978 to FY 1980 period.

Atmospheric Fluidized Bed (AFB)

Systems currently being investigated and tested include:

- Fluidized-bed boiler, 30-MW$_e$
- Industrial applications
- Atmospheric Fluidized-Bed Component Test and Integration Unit (AFB/CTIU)

Table II-3-12. Funding for the combustion systems activity for FY 1978 - 1980.

COMBUSTION SYSTEMS SUBACTIVITIES	BUDGET AUTHORITY (DOLLARS IN THOUSANDS)		
	ACTUAL FY 1978	APPROPRIATION FY 1979	ESTIMATE FY 1980
Atmospheric Fluidized Beds	$24,500	$23,600	$25,900
Pressurized Fluidized Beds (PFB)	15,229	14,234	20,000
Advanced Combustion Technology	13,036	7,342	6,500
Alternative Fuel Utilization	1,915	9,400	2,500
Combustion Systems Demonstration Plants	11,000	3,600	2,500
Capital Equipment	465	725	0
TOTAL	$66,145	$58,901	$57,400

- Atmospheric Fluidized-Bed Coal Combustion Closed-Cycle Gas Turbine Technology Test Unit (TTU) [formerly Modular Integrated Utility Systems (MIUS)]
- Anthracite applications

Coal-fired, fluidized-bed boilers are more attractive for commercial use than conventional coal-fired boilers with stack gas scrubbers. They show potential for improving net power generation efficiency and heat transfer rates while lowering emissions of sulfur oxides, nitrogen oxides and trade metals, and eliminating problems associated with ash slagging and fouling. The capital cost of such systems is expected to be less than that of conventional systems with stack gas scrubbers.

Emission control of SO_2 and NO_x for fluidized-bed boilers is centered in the combustion zone. Sized coal is burned in a fluidized-bed of inert ash and limestone or dolomite. The limestone or dolomite calcines and then reacts with SO_2 in the presence of oxygen to form a solid sulfate material that can be regenerated or disposed of with the ash as an inert granular landfill. Regeneration will produce a reuseable sorbent and either a stable sulfur compound or a marketable sulfurous material. Fluidized-bed boilers operate at combustion temperatures appreciable lower than conventional boilers, thus inhibiting formation of nitrogen oxides. An atmospheric pressure, coal-fired, fluidized-bed boiler having a capability of 5,000 lb/hr of sream has been operated successfully at furnace temperatures of approximately 1600°F since 1967. It has been demonstrated that all types of coal, char, coal wastes, and other low-grade combustible materials can be burned in an environmentally acceptable manner.

An investigation is currently being conducted concerning utilization of FBC solid waste material for use in agriculture, for neutralization of acid mine spoils, and for subsoil stabilization and aggregate for highway construction use. Building blocks have been produced using only FBC waste and conventional fly ash as ingredients. Wastes from currently operating fluidized-bed combustors are being disposed of as landfill in an environmentally safe manner.

AFB Boiler, 30-MW$_e$. The objective of the current DOE Fossil Energy 30-MW$_e$ AFB boiler task is to design, construct, test, and evaluate a 300,000 lb/hr capacity multicell fluidized-bed boiler (MFB) as a pollution-free method of burning high-sulfur content or slagging coals in a practical manner under actual electrical utility conditions. Steam pressure, temperature, and circuitry of the boiler, therefore, were designed to meet requirements of the selected Monongahela Power Company (Allegheny Power System), Riversville, West Virginia, power station.

Successful development of the atmospheric fluidized-bed boiler is seen as a necessary first step toward the better use of high-sulfur and slagging coals. Based on the successful performance of this first 30-MW_e unit and a subsequent demonstration plant, it is expected that several electric utilities will commit to the installation of atmospheric fluidized-bed boilers in the 200- to 800-MW_e capacity range.

As shown in Figure II 3-116, the MFB consists of four adjacent combustion cells. This unit of cells is 21-ft high × 39-ft long × 12-ft wide. The three primary cells (A, B, and C) operate at 1550°F while the carbon burnup cell (CBC) operates at up to 2000°F. Boiler tubes are arranged in waterwalls and horizontal bundles to control bed temperature, generate and superheat steam. Each of the four cells has economizer tubes in the upper convection region, which is heated by exhaust gases. The three primary cells have boiler tubes immediately above the fluidized-bed section, and cell C has boiler tubes deep within the bed. Cells A and B both have superheater tubes immersed deeply into the bed region. This fluidized-bed boiler has produced over 300,000 lb/hr of steam at 930°F and 1,280 psi; the temperature and pressure are determined by the requirements of the boiler installation site.

Crushed coal (¼ in top-size) and limestone (⅛ in top-size) are supplied to the system by weigh hoppers and blown into the combustor unit through multiple feeders with streams of air. Coal and limestone are fed to cells A, B, and C at a 4:1 weight ratio. Unburned carbon captured by the exhaust cyclone separator above the primary cells is fed to the carbon burnup cell. The total coal feed rate is about 15 tons/hr. Air for bed fluidization is supplied to the burner cells by a common plenum. Moderate levels of excess oxygen is maintained in the cells. The two-stage exhaust gas cleanup system contains cyclone separators followed by an electrostatic precipitator which operates at 700°F.

There are several advantages to the fluidized-bed boiler:

- Sulfur dioxide and nitrogen oxide emissions are well within EPA standards using high-sulfur coal; expensive, difficult to operate and maintain, and energy-

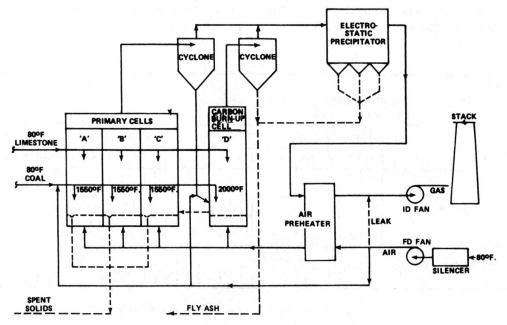

Figure II-3-116. Fluidized bed boiler, 30-MWe.

consuming flue gas cleaning equipment is not required
- NO_x emission levels are lower than those from conventional equipment
- Low-quality coal can be burned without danger of slagging and fouling because of low combustion temperatures
- The heat release and heat transfer coefficients are high, reducing boiler size, weight, and cost. Analyses of conceptual designs indicate that the utility-size, fluidized-bed boiler will weigh less and will cost proportionately less than conventional boilers
- The overall operating efficiency of fluidized-bed boiler power plant is projected to be 36 percent, compared to approximately 34 percent for a conventional coal-fired plant with stack-gas cleanup operating with the same coal and at the same steam pressure and temperature.

Since 1965, Pope, Evans and Robbins, Inc., has been investigating AFB combustors in its Alexandria, Virginia, laboratory facility. A 0.5-MW_e AFB PDU has been operated in Alexandria to support the pilot plant design, construction, and operation. Construction of a 30-MW_e MFB pilot plant was completed in 1976 at the Monongahela Power Company's palnt at Riversville, West Virginia, by Pope, Evans and Robbins, Inc. Initial light-off of the boiler with coal was successfully accomplished on December 7, 1976. Operation and testing of the pilot plant to demonstrate the applicability of the MFB for steam-electric power generation has been carried out since 1977 and is continuing. On September 30, 1977, the unit achieved its full nameplate rating. In the fall of 1977, it produced steam for use in commercial power generation on several occasions. Modifications of the coal feed system were made in 1978 to improve the system reliability. Instrumentation was also added.

Between April 12, 1978, and August 9, 1978, 1,054 hours of test operation were obtained. This included five periods of commercial power generation operation, including periods of 50, 25, and 24 hours continuous operation supplying superheated steam within the narrow control bank required for utility operation. Preliminary data show that the SO_2 removal verifies the design basis of the unit. On August 9, 1978, a fire developed in the air preheater. It is believed that the fire was due to the high carbon content of the fly ash during the lightoff procedure. The air preheater, induced draft fan, and associated duct work was damaged beyond repair and, together with water damaged equipment, are being replaced. Due to the intermittent nature of the developmental testing, the electrostatic precipitator is not able to control the particu-

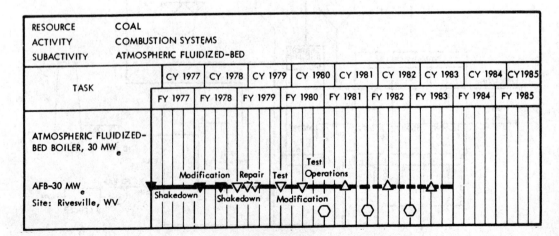

Figure II-3-117. Major milestones for the 30-MWe AFB.

late emission during off-design operation and a baghouse was added in the fall of 1979. The unit was back in service in early 1979.

The Riversville facility is the first major step toward commercial size central station power generation. The combustor for Riversville represents an initial design to be used as the basis for 200-MW_e and eventually up to 800-MW_e plants.

AFB milestones are shown in Figure II 3-117.

AFB Industrial Applications. These projects consist of development and demonstration AFB boilers and heaters for industrial applications, using the atmospheric fluidized bed concept, and operation of prototypes on industrial sites for such purposes as steam raising, process heat, and cogeneration. Objectives are as follows:

- To design, build, operate, test, and demonstrate prototype boilers and heaters to establish the practicality of burning high-sulfur coals and other fuels for industrial applications
- To obtain sufficient data from prototype operations to enable the industry to design, construct, and operate full-sized units in industrial plants

Although industrial boiler and process heater needs are quite diversified, a relatively small number of equipment configurations define the majority of applications. Thus, key components will be constructed on an industrial scale to determine applicability of fluidized-bed combustors to industrial units.

Three areas of industrial applications are being investigated:

- *Saturated Steam Boiler:* capable of producing steam of the quality and in the quantity suitable for process heating and space heating requirements of industrial or institutional facilities. The individual boiler capacity is predicated on the intended application
- *Superheated Steam Generator:* capable of producing steam of the quality and quantity for both cogeneration via back-pressure turbines and industrial process and heating requirements
- *Indirect-Fired Process Heater:* using tubes or other heat-exchange devices to heat process liquids or gases.

As listed in Table II 3-13, five contractors (or groups) were selected to design, construct, and test industrial boilers or industrial heaters using AFB combustors:

Table II-3-13. Contractors selected to design, construct, and test industrial boilers or heaters.

CONTRACTOR	STEAM RATE	COAL RATE
Georgetown University/Fluidized Combustion Co., Washington, D.C.	1000,000 lb/hr saturated steam	5 tons/hr
Combustion Engineering, Great Lakes Naval Training Center	50,000 lb/hr superheated steam	2.5 tons/hr
Battelle/Fluidized Combustion Co., Columbus, Ohio	25,000 lb/hr saturated steam	1 ton/hr
Fluidyne	Heated Air Metal Working, 780°F, 28 MMBtu/hr	1.5 tons/hr
Exxon Research & Engineering Co., Linden, New Jersey	Tube Still, Petroleum Crude Heater, 15 MMBtu/hr	0.75 ton/hr

Arthur G. McKee & Co. is providing technical assistance and services to DOE in developing the industrial applications of AFBC.

Figure II 3-118 presents major milestones for this effort.

AFB Closed-cycle Turbine Test Unit (TTU)
The development of an Atmospheric Fluidized-Bed Coal Combustion for Cogeneration Hot-Air Gas Turbine Technology Test Unit (TTU) was initiated by Oak Ridge National Laboratory (ORNL) under technical direction from ERDA/FE (the DOI/OCR) as a coal-fired application to the existing Housing and Urban Development (HUD)–MIUS program.

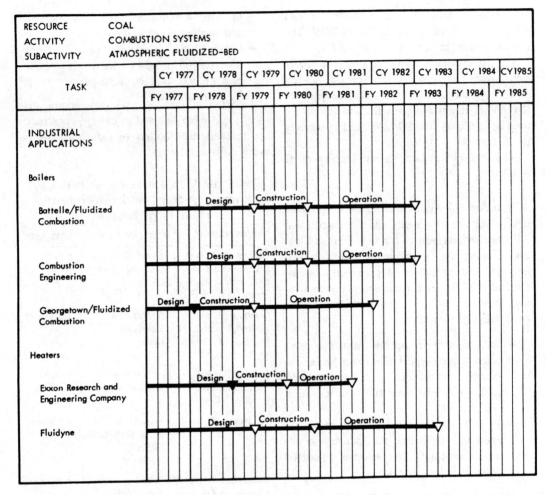

Figure II-3-118. Major milestones for industrial applications.

(See Figure II 3-119.) The TTU effort complements the overall DOE fluidized-bed combustion effort by providing investigation of a potentially viable technology option that is not being addressed elsewhere in the existing program.

The MIUS concept was originally based on efficient use of mechanical and thermal energies derived by burning a variety of fuels ranging from natural gas to inferior coals and municipal and process wastes in an environmentally acceptable manner. Such a total energy utilization system concept can result in a 60 to 85 percent overall thermal efficiency with a large portion (one-third) of the energy as high-grade electricity. Prior to the oil boycott, coal was not a serious economic contender for application to MIUS concepts. Post-boycott conditions have put AFB systems in the forefront for many diverse applications of coal utilization.

Based on the original MIUS objectives and goals, ORNL has completed two of the total four phases. Phase I, which included several technology studies and preliminary engineering and economic analyses, concluded that coal-fueled atmospheric FBC unit coupled to a closed-cycle, externally fired, hot-air turbine is the most promising subsystem for MIUS application.

Phase II was undertaken to further investigate the systems and configurations chosen in Phase I and to complete the conceptual design for the most promising configuration. As a result of this study, the work to date has yielded a conceptual design for the TTU.

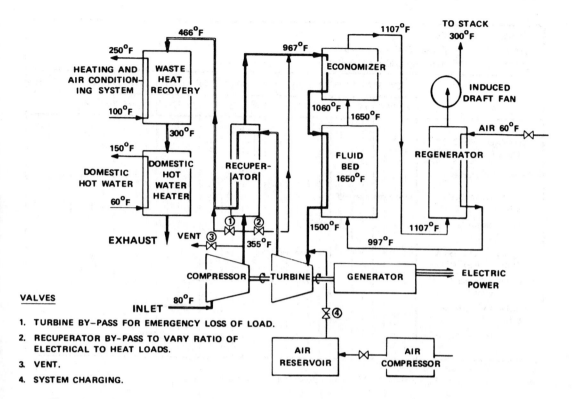

Figure II-3-119. Atmospheric fluidized-bed coal combustion externally-fired gas turbine technology test unit.

The TTU will consist of a coal-burning atmospheric FBC system coupled to an externally fired gas turbine and capable of ultimately being coupled to a closed-cycle, hot-air turbine generator. This test unit will have basic electrical capabillity from 300 to 500 kW plus approximately 2.5 × 10 Btu/hr of recoverable waste heat. This size test unit will provide credible scale-up data to the 5- to 50-MW range for commercial application in the industrial/institutional sector for cogeneration applications.

In 1978, the TTU Program was reoriented with the goal of earlier commercialization of the technology. The AFB combustor and gas turbine integration is being contracted for with an industrial team rather than having the combustor build to Government design. Upon successful completion of the program, therefore, an industrial team will be in a position to supply such combustors for cogeneration commercially because they can use their own design and fabrication method.

Major milestones for the externally-fired gas turbine are shown in Figure II 3-120.

AFB Component Test and Integration Unit (CTIU). The AFB/CTIU is a flexible, instrumented research and development facility at the smallest scale at which data apply to full-size systems. As shown in Figure II 2-121, it consists of a stacked, three-cell fluidized-bed combustor with its heat recovery system. Ash removal from flue gases will be in cyclone separators with reinjection for carbon burnup in a separate carbon burnup cell or within either coal-burning cell; final gas cleaning will be effected by bag filter collectors. Provision is made for coal, limestone, and additive receiving, preparation, and feeding into the combustor cells. The spent bed material is removed, purified by size classification and magnetic separation, and stored hot for reinjection into any of the combustor cells. Air for fluidization and injection is providing by primary and auxiliary forced-draft fans. An induced-draft fan moves the flue gases through the dust removal system and vents to the stack. The complete system is designed with sophisticated analytical instrumentation, computerized data acquisition, and electronic

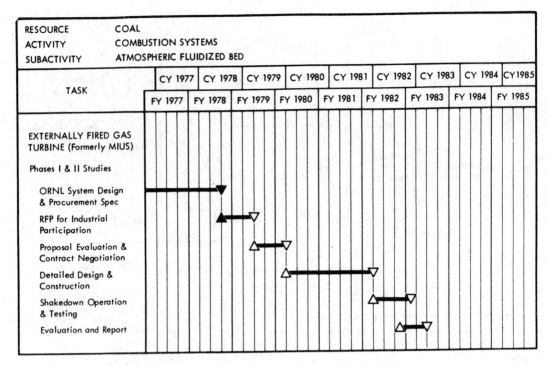

Figure II-3-120. Major milestones for the externally-fired gas turbine.

control systems so that the maximum amount of useful data can be obtained in the operation of the unit.

This CTIU is located in Morgantown, West Virginia, to obtain data on bed fluidization, combustion, emissions, sorbent performance and to test new atmospheric fluidized-bed combustion systems and components at an intermediate-size scale and to act as an investigative facility for problems developed in the 30-MW$_e$ boiler test project. Design of the CTIU started in FY 1976. Fabrication and field erection started in FY 1978, and construction will be completed in FY 1982. The CTIU will test improved solids-handling systems, develop technology for vertical stacking of multiple beds, and serve as a boiler development laboratory to test-tube bundle geometries and materials in operating fluidized-bed combustors. During 1978, the boiler contract was let, boiler detailed design begun, and the building shell closed in.

CTIU milestones are given in Figure II 3-122.

Anthracite Applications. This effort is designed to investigate the combustion of anthracite culm in fluidized beds and other direct culm-burning processes in an environmentally acceptable manner in industrial/utility boilers and heaters. The waste material resulting from the mining and cleaning of anthracite coal (principally in northeastern Pennsylvania) contains a substantial amount of fuel value, which can be recovered in fluidized-bed or other direct combustion boilers/combustors in an environmentally acceptable manner.

In the anthracite region, this refuse material is piled in a densely populated, geographically small area of about 480 square miles. It is estimated by the U.S. Bureau of Mines (BOM) that 800 banks containing 900 million cubic yards of refuse can be found within the anthracite mining fields. These spoil banks contribute to acid drainage and other environmental problems. The recovery of the fuel value of coal remaining in them while converting the remaining minerals to inert material provides a needed environmental service while obtaining useful energy from waste material.

DOE has begun a program of extending the combustion characteristics of these refuse ma-

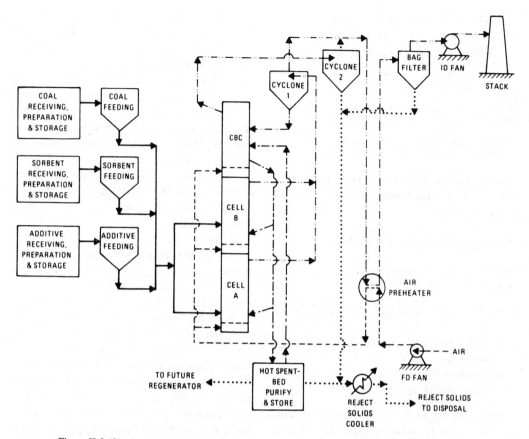

Figure II-3-121. Atmospheric fluidized-bed combustion test and integration unit (AFB/CTIU).

terials by fluidized-bed combustion. The value of this refuse as a fuel is important in the anthracite region because of the decline in mining in the area, which has led to a growing dependence on oil and gas.

A Program Opportunity Notice was released in the second quarter of FY 1977. The major activity in FY 1977 was the evaluation of proposals. The effort during FY 1978 was the selection of contractors and contracting negotiations, which will proceed to prototype equipment development. The selected contrac-

RESOURCE	COAL								
ACTIVITY	COMBUSTION SYSTEMS								
SUBACTIVITY	ATMOSPHERIC FLUIDIZED-BED								
TASK	CY 1977	CY 1978	CY 1979	CY 1980	CY 1981	CY 1982	CY 1983	CY 1984	CY 1985
	FY 1977	FY 1978	FY 1979	FY 1980	FY 1981	FY 1982	FY 1983	FY 1984	FY 1985
ATMOSPHERIC FLUIDIZED-BED TEST & INTEGRATION UNIT Morgantown Energy Research Center	Design	Building Construction		Equipment Installation			Shakedown Operation		

Figure II-3-122. Major milestones for the AFB/CTIU.

Table II-3-14. Contractors and equipment characteristics.

CONTRACTOR	STEAM RATE	COAL RATE
Wilkes-Barre, Pa/Fluidized Combustion Co.	100,000 lb/hr	5 tons/hr
Shamokin, Pa./Curtiss-Wright/ Dorr-Oliver	20,000 lb/hr	1 ton/hr
Towanda, Pa./Fluidyne	20,000 lb/hr	1 ton/hr

tors and equipment characteristics are noted in Table II 3-14. Major milestones are shown in Figure II 3-123.

Pressurized Fluidized Bed (PFB)

Projects currently underway in this area include:

- PFB Combined-Cycle Pilot Plant, 13-MW$_e$
- PFB Combustor and component development
- Granular-Bed Filter/Hot-Gas Cleanup
- International Energy Agency

Pressurization of the fluidized-bed combustor is a means of reducing emissions of SO_2, NO_x, CO, and hydrocarbons, while simultaneously effecting an increase in power-generation efficiency with the possibility of reducing the cost of power generation.

The basic technology requires that the coal and sorbent be fed into a pressurized combustion chamber where it burns in a fluidized bed with the off gas going to a gas turbine. The plant is operated as a combined-cycle, gas turbine, steam turbine system. The pressurized system further reduces the size of the combustor and thus the capital cost. The use of combined-cycle systems increases overall plant efficiency.

Objectives for the pressurized program are similar to the atmospheric program, except that developmental problems are greater and more complex, and the first commercial units are expected to be operational at an appreciably later date. Limited tests of American coal have been performed at five atmospheres in a test unit at Leatherhead, England, and have shown that this approach is technically feasible and that NO_x emissions are reduced below the level of AFB systems and that greater SO_2 removal is possible than with AFB systems. Additional work is required and is being per-

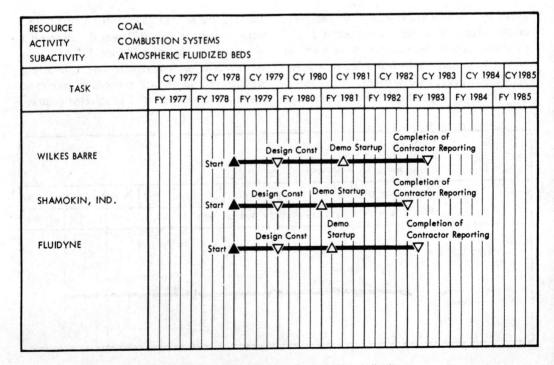

Figure II-3-123. Major milestones for anthracite applications.

formed to accomplish cleanup of the hot gases produced from pressurized FBC. This work is necessary to produce the high turbine reliability and long life required for practical power-generation application and control of emissions to the atmosphere. The use of gas turbines at low turbine blade temperatures (1200°F to 1300°F) to avoid the hot corrosion problems associated with the higher temperatures. The use of protective coatings is also being investigated.

In the same respect as atmospheric fluidized-bed boilers, the utilization of solid waste material is being considered in agriculture and highway programs.

PFB Combined-cycle Pilot Plant. The combination of PFB combustor technology with the gas turbine, steam turbine, combined-cycle power system offers the unique opportunity for the production of clean cost-competitive electric power from the combustion of high-sulfur coal.

The Curtiss-Wright Corporation is designing, constructing, operating, and evaluating a coal-fired gas turbine pilot electric power plant. The pilot plant, which will produce an equivalent of 13-MW$_e$, will address a number of key technical issues associated with PFB development. Environmental considerations, maintenance, reliability, and durability of the air-cooled PFB combustor and turbine will be evaluated. The program will provide design data to verify scale-up of the gas turbine and PFBC units to commercial plant size for the utilities industry.

The project objective is to conduct a pilot plant test program to obtain data on gas-turbine operation on coal combustion products and to evaluate the validity of the commercial plant design concept. The test program will identify and seek solutions for design or component deficiencies, establish operating characteristics under normal and off-design conditions, and provide a firm engineering base for full-scale plant development decisions.

The pilot plant program is a five-year, multiphase cost-shared effort. Phase I is a 14-month program to provide the following:

- Conceptual commercial design of a coal-fired 300- to 500-MW$_e$ combined-cycle generating station using the PFB combustion technique
- Pilot plant design that is representative of the commercial concept
- Site and environmental assessment study
- Technology support program that resolve technical issues and provide data on the performance of the selected PFB design.

Phase II is an eight-month effort to complete the detailed design drawings and specifications of the pilot plant, prepare a construction bid package, and obtain and evaluate bids.

Phase III is a twenty-month construction period for the manufacturing, buildup, and installation of the pilot plant.

Phase IV is a twenty-four-month effort to operate and evaluate the pilot plant operating parameters and durability on several coal types. At the conclusion of this phase, a reassessment of the commercial plant conceptual design will be completed.

As shown in Figure II 3-124, air is compressed to approximately 8 atm and fed into the pressurized fluidized-bed combustor (PFBC) where combustion takes place at 1650°F. The gases are passed through a two-stage cyclone separator and a hot-gas cleanup system prior to mixing with clean, heated air and subsequent expansion through a gas turbine, which drives an electric power generator. From the gas turbine, the gases pass through a heat recovery boiler, generating steam which in turn drives a steam turbine electric generator. The pilot plant does not include the steam turbine cycle, which is available technology. Its gas turbine electric generator will produce 7.15-MW$_e$. Six-MW of waste heat is also recovered in the form of 49,000 PPH 175 psig steam for inplant use, as a total energy system.

This project includes design, construction, and operation of a 13-MW coal-fired, PFB combined-cycle pilot plant to be located at the Curtiss-Wright Corporation, Woodridge, New Jersey, facility. This prototype module is to be designed as a basic element of a central station power plant so that operating, engineering, and economic data can be fully evaluated. An existing oil-fired combined-cycle electrical power-generating station is

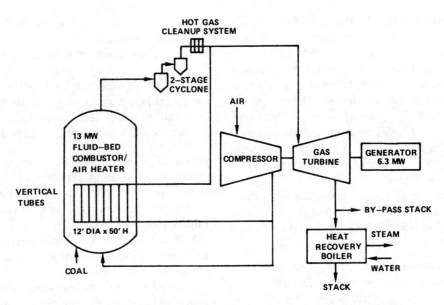

Figure II-3-124. Pressurized fluidized-bed combined-cycle pilot plant, 30-MW.

planned for conversion to a PFBC system. The pilot plant facilities will include a 12-ft diameter by 50 ft-high pressurized fluidized bed combustor, the CW 6515 gas turbine, and a waste heat boiler capable of generating 175 psig/377°F steam. The basic module will be about one-fifth the scale of commercial plant components. Phase I is a study of comparative commercial systems and a preliminary design of the pilot plant. Phase I also includes the operation of a small gas turbine test rig, which is a 3-ft diameter section of the full combustor, together with a turbine materials test program. Data are being obtained on the corrosion and erosion of turbine materials and the protection offered by transpired air. Phase I testing with the 3-ft diameter rig continued through 1978. Phase II is the detailed design of the pilot plant and is essentially complete. Phase III, construction of the pilot plant began in FY 1979, and shakedown testing will occur in FY 1981. Phase IV, operational testing and evaluation, will be performed in FY 1982 and FY 1983.

Major milestones are shown in Figure II 3-125.

PFB Component Test and Integration Unit. The PFB Program is based on the marriage of fluidized-bed combustion, at pressures of 6 to 16 atm, and gas turbine combined-cycle plant operation. So that commercial plants will be rapidly forthcoming, the development of improved components and their proof in integrated operation is needed. In May 1975, ANL completed a conceptual design of a 3-MW_e Component Test and Integration Unit to enable DOE to perform development of improved components and instrumentation and to provide a flexible facility to investigate alternative PFBC systems, including air-cooled and steam-cooled combustors.

In 1978, DOE reassessed the PFB CTIU project in light of the mission, available funding, capabilities of other PFB facilities and project needs. It was decided that, rather than dilute the program by constructing another PFBC research unit, it would be advantageous to apportion the component development mission of the PFB CTIU among the various existing PFBC facilities, namely the Curtiss-Wright SGT rig, the CURL Facility at Leatherhead, England, and the IEA Facility in construction at Grimethorpe, England, and to allocate the systems integration mission of the PFB CTIU to the PFBC pilot plant and the Engineering Development System (EDS) plant. The EDS is the first complete integration of a PFB combustor/boiler and a gas turbine. With such a reallocation of mission, greater emphasis can be placed on developing and improving hot gas cleanup equipment, solids handling (feed, removal and bed-level control) controls, gas turbine materials evalua-

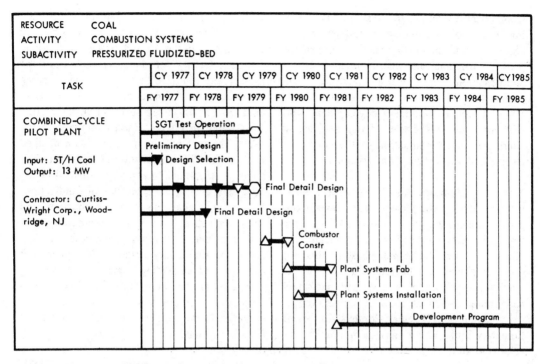

Figure II-3-125. Major milestones for the combined-cycle pilot plant.

tion and further analytic and experimental evaluation of the pressurized boiler-power recovery turbine PFB system configuration. If DOE proceeded with the PFB CTIU, the work on component development would have had to have been slowed and the whole program would have suffered delays. The PFB CTIU mission has been apportioned as follows:

- Small gas turbine rig
  — Instrumentation test bed
  — Hot gas cleanup test bed
  — Gas turbine blade cascade testing
  — Transpiration air film blade protection testing
  — PFB combustor materials testing
- CURL
  — PFB boiler configuration development
  — Hot gas cleanup test bed
  — PFB combustor/boiler configuration development
  — Gas turbine cascade testing
- PFBC pilot plant
  — Testing of the PFBC plant system with uncoupled gas turbine and steam turbine
  — Testing of plant control
  — Testing of gas turbine on cleaned and diluted combustion products
  — Testing of air-cooled vertical tube PFBC combustor configuration
- IEA facility
  — Combustion characteristics and profiles in a steam-cooled bed of substantial size (2m × 2m)
  — Testing of commercial-scale solids handling systems
  — Testing of gas turbine on undiluted PFB combustion products
  — Obtaining data on gas turbine materials over a range of blade temperatures
- Engineering Development System
  — Integration of PFB boiler/combustor with coupled steam turbine and coupled power recovery gas turbine
- PFBC gas turbine materials long-term test apparatus
  — Dedicated materials test rig

DOE has terminated further work on the PFB CTIU with the Title I design being completed. The project is otherwise ready to begin construction.

During 1979, the PFB Combustor and

Component Development Task involved testing of improved combustors and gas turbine blade cascades in Leatherhead, testing of improved particulate and alkali measurement instrumentation in several facilities, and performance of materials evaluation. During FY 1980, such component development work will continue with testing of hot gas cleanup alternative and continuation of the combustor testing and instrumentation development.

Hot Gas Cleanup. In FY 1977, a contract was awarded to the Combustion Power Company to determine the scientific and engineering principles upon which granular bed filtration operates. The first phase of this program developed the theoretical analysis of the moving-bed filtration process and performed verification testing at ambient temperature.

In January 1977, the Combustion Power Company initiated a 12-month program to develop the following:

- Theoretical analysis, mathematical model development, computer simulation and test data correlation
- Design and construction of a cold-flow, moving-bed granular-bed filter device (GBF)
- Design and implementation of a GBF test program
- Determination of filter screen deposition mechanisms, test and evaluation of various mechanical deposit removal devices

Following successful completion of the cold-flow model and the tests of front face cleaning for GBF, the program proceeded with the design and construction of a high-pressure, high-temperature granular-bed filter, to correlate the mathematical model with experimental data and to evaluate GBF performance under long-duration testing. The GBF was built in FY 1979 for testing at one of the PFB combustion facilities.

This project has been transferred to the Advanced Environmental Control Technology Program (ECTP), where it will be included in the Gas Stream Cleanup Subactivity.

International Energy Agency. As part of the activities of the International Energy Agency (IEA), consisting of a joint effort by the United Kingdom, the Federal Republic of Germany, and the United States, a combustion test facility for the development of the pressurized fluidized-bed combustor is being constructed in Grimethorpe, Yorkshire, England. As shown in Figure II 3-126, the facility was started in FY 1977 and completed in FY 1979. The unit will be used to obtain detailed data on the PFB combustion process in equipment large enough to apply to utility systems. Such data will include profiles of fluidization, bed-mixing, coal injection, combustion and sorbent kinetics, particle and alkali elutriation, and associated information necessary to engineer reliable utility systems. The plant will be expanded by adding a gas turbine test rig with a tertiary cleanup that will allow data to be produced on gas turbine performance for a range of blade temperatures and a variety of blade materials.

Engineering Development System (EDS). The Engineering Development System (EDS) will allow integration of the boiler-steam turbine plant components with the PFB combustor-gas turbine PFB system elements to proceed. Such a plant will have a PFB combustor/boiler supplying steam to a utility steam turbine and will be operated with compressed air supplied by the power recovery gas turbine. The unit will be designed for start-up and operation without any auxiliary boiler and will be capable of stand-alone operation.

In FY 1980, the EDS will be designed in cooperation with industry and the utilities.

Major milestones are shown in Figure II 3-127.

Advanced Combustion Technology

Tasks under this subactivity are as follows:

- Engine combustion technology
- Improved oil and gas burners
- Coal combustion support

The present generation of production engines evolved during a period of relatively inexpensive petroleum-derived fuels when performance and cost rather than efficiency or

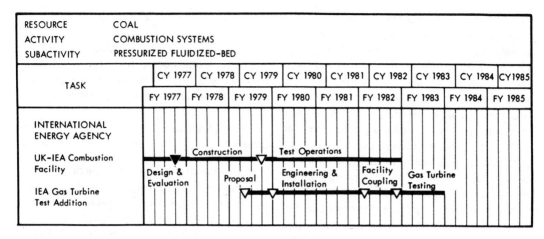

Figure II-3-126. Major milestones for IEA projects.

emission constraints governed their development. As a result neither the engines themselves nor the design techniques by which they have been incrementally improved are readily adapted to emerging requirements in terms of efficiency, emissions, and broadened fuel tolerance. Although there have been advances in the past few years with the introduction of stratified charge, and lean-burn engines, only a fraction of the 20 to 30 percent efficiency improvements inherent in these concepts has been obtained. In cooperation with industry, investigations are currently being conducted on prototype design concepts for a generation of engines specifically tailored for realizing these potentials in different engine classes.

Present industrial design practices for fur-

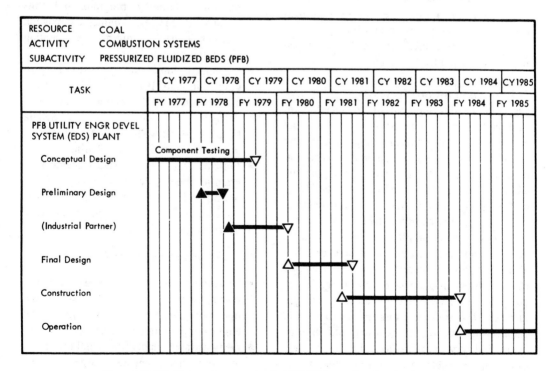

Figure II-3-127. Major milestones for the PFB Utility engineering development system.

naces and boilers are based on experience with specific business situations which traditionally stressed low cost rather than high efficiency. This R&D effort is concerned with the evaluation and the development of combustion technology and equipment for boilers and furnaces in residential, commercial, and industrial use. Such technology development should improve the efficiencies, controls, and heat transfer and fuel utilization properties of boilers and furnaces. This activity was started in the Increased Conversion Efficiency (ICE) Program in ERDA and was transferred to the Fossil Energy Program in DOE. The improved boiler and furnace program will be transferred to an appropriate group in DOE to handle the improvement of oil- and gas-burning equipment.

Laboratory and design studies now in progress support the direct coal combustion development projects. These studies are designed to promote optimum coal utilization by supplying data essential for application of direct combustion units to public utility and industrial systems. Key components of the total system, including processing, handling, and transport equipment, will be considered in the context of operability and design optimization. Component corrosion and fouling studies will be made and methods will be developed for minimizing these problems.

Engine Combustion Technology. This task supports the development of advanced combustion technology necessary to increase fuel efficiency, control emissions, and broaden the range of usable fuels for automotive and stationary engines. The approach is to apply the advanced research capabilities of the DOE multipurpose laboratories, the universities, and selected private R&D laboratories to the solution of critical problems in engine systems under advanced development in industry. The effort is directed primarily toward four engine classes: direct injected stratified charge, lean-burn homogeneous charge, diesel, and gas turbines. The subtasks complement traditional engineering research in industry and involve major industry participation. Relevance and technology transfer to the manufacturers' development laboratories are accomplished by their direct involvement in program definition, execution, and review. Research results from this task have immediate impact on industry design commitments for their near-term developments and will accelerate the energy savings to be achieved with these engine design concepts.

Accomplishments in FY 1977 and FY 1978 included:

- Conducted cooperative research program in direct injected stratified charge (DISC) engines involving Sandia, Lawrence Livermore Laboratory (LLL), Los Alamos Scientific Laboratory (LASL), Princeton University, and General Motors Research Laboratories. Generated noncombusting model of DISC engine aerodynamics
- Demonstrated stratified-charge diagnostic systems using combustion bomb facility loaned to Sandia by Volkswagen Research as part of international cooperative program
- Received cooperative financial support from the Motor Vehicles Manufacturing Association (MVMA) for engine combustion technology program and thus formulated MVMA engine combustion research review committee representing all of the auto industry
- Organized joint engine research program, through IEA, with U.S., U.K., and Sweden participation. Staff exchange has occurred between Sandia (U.S.) and Harwell (U.K.) laboratories to accelerate technology sharing
- Initiated major cooperative research program in direct combustion aimed at identifying and controlling particulate formation, noise, and carcinogenic aromatic hydrocarbon emissions

FY 1979 accomplishments included:

- Developed a 2-dimensional, noncombusting working model of DISC engine operation capable of providing simple engineering design tradeoff calculations
- Demonstrated enhanced combustion in lean-burn homogeneous charge engines
- Assessed improved injection for diesels

- Completed catalytic combustion characterization for continuous combustion systems

Major milestones are given in Figure II 3-128.

Improved Oil and Gas Burners. The overall objective of this project is to improve the utilization of gas and oil by modification (including retrofit) of the combustion equipment and processes. Of particular importance is the ability to provide means for switching to more readily available fuels (e.g., low-Btu gases for natural gas), and to improve the efficiency of boilers and furnaces burning low-grade residual oils so that they will be capable of burning the synthetic fuels of the future cleanly and efficiently.

This project was started in the second quarter (see Figure II-3-129) of FY 1977. The following efforts were completed in FY 1979:

- Pulse-combustor designs, offering great promise for residential and industrial heating (but currently unacceptable for that purpose), were improved
- Scaling criteria for oil and gas furnaces, leading to more confident transfer of information from small scale to boiler design, were evaluated
- The design of boiler components (e.g., atomizers) was evaluated with the goal of achieving minimum excess-air operation consistent with clean exhaust requirements

Coal Combustion Support and Engineering Evaluations. Laboratory and design studies now in progress which support the combus-

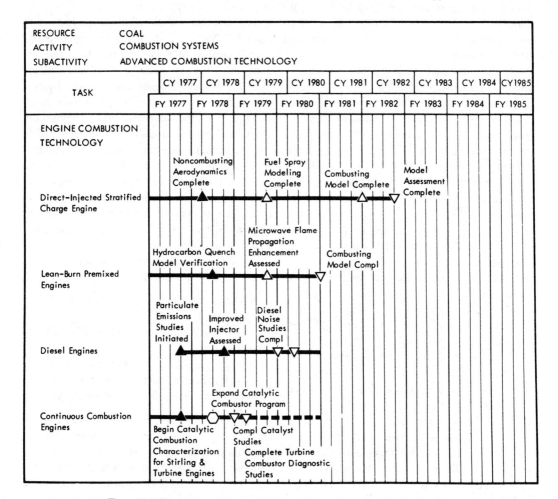

Figure II-3-128. Major milestones for the engine combustion technology task.

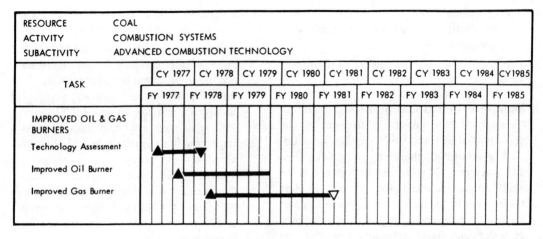

Figure II-3-129. Major milestones for the improved oil and gas burners task.

tion systems development projects are designed to promote optimum coal utilization by supplying data essential for application of direct combustion units to public utility and industrial systems.

Key components of the total system, including processing, handling, and transport equipment will be considered in the context of operability and design optimization. Component corrosion and fouling studies are being made, and methods will be developed for minimizing these problems. Optimum design of combustors capable of efficiently burning synthetic fuels will be promoted by characterizing fuels such as chars, coal wastes, low-Btu gas, solvent refined coal, and synthetic liquid fuel in terms of heat transfer and combustion properties.

Studies currently underway in the support studies and engineering evaluations area include engineering support studies; nonfluidized-bed combustion; environmental studies and supporting projects; and regeneration and utilization of fluidized-bed sorbents.

Laboratory studies, feasibility studies, and system analysis support development projects for atmospheric and pressurized fluidized-bed coal combustion are being conducted. Problems of regeneration, waste disposal, and emission control are also being addressed. An analytical model of the elemental processes of such combustors will be developed and experimentally verified. This model will assist design optimization of fluidized-bed combustors and advanced conversion systems.

The objective of the nonfluidized-bed combustion effort is to demonstrate the commercial feasibility of using direct substitution of coal, coal-derived chars, and solvent refined coal for oil and gas in existing industrial processes. This is expected to involve minor modification of the burners and combustion chambers and the addition of necessary fly-ash collection equipment. Federal participation is required due to the high costs and risks associated with the development of improved technologies and to ensure wide dissemination of the demonstrated technology. The program will be cost-shared with industry in the construction and operation phases.

As envisioned to date, the program is divided into the following major areas:

- Direct substitution of coal for short supply fuels
- Increased utility boiler reliability and efficiency
- American Boiler Manufacturers Association industrial stoker program
- In-house, solid fuel combustion research

ANL will continue FBC regeneration process development studies. For combustion-regeneration studies, ANL has a 6-inch diameter combustor and 4-inch diameter regeneration process development units as well as laboratory-scale equipment for supporting studies. The combustor can be operated at bed temperatures up to 950°C and pressures up to 10 atmospheres. The regenerator can be operated at 1,100°C and up to 10 atmospheres.

Environmental studies provide support to generic studies applicable to many or all direct combustion projects and provide support to new project starts. Such studies will consist of, but are not limited to, environmental assessments, process emissions, and effluent characteristics.

Special projects include process economic evaluations; interprogram trade-off analysis; new technology development (coal preparation, pretreatments, gas cleanup, coal feeder, ash disposal); process design and analysis for overall process optimization; and process reliability and maintainability.

Other support includes efforts to assist with detailed technical support of project planning, monitoring, and evaluation, and the development and integration of R&D technical programs. The support services include assistance in developing and supporting technically sound and economically defensible programs.

Some specific projects within the area of support studies include:

- *Particulate Analysis Instrumentation:* Leeds and Northrup Co. and Spectron Development Laboratory are developing an on-line instrument for measuring particulate loadings and particle size distribution in high-temperature pressurized gas streams for application to PFBC.
- *Stoker-Fired Boiler Testing Program:* A testing program to obtain basic data on emissions and to update equipment specifications and design criteria for stoker-fired boilers has been undertaken by the American Boiler Manufacturers Association. This contract is jointly funded by DOE and EPA.
- *Development of Pulverized Coal-Fired Glass Furnace:* This study is being performed by the Burns and Roe Industrial Services Corporation to develop the necessary engineering data and equipment design criteria to enable the U.S. glass industry to convert from petroleum-based fuels to the direct-firing of pulverized coal in contemporary glass furnaces and melters and to define the environmental constraints imposed by converting to coal-firing and assess the capital and operating impact of these constraints.
- *Agriculture Utilization Studies: Fluidized-Bed Combustion Waste Materials:* Working through the Agricultural Research Service (ARS), U.S. Department of Agriculture, DOE will study the utilization of waste material from the fluidized-bed combustion process. ARS will perform greenhouse and growth chamber testing at selected agricultural experiment stations. Where indicated, field tests of FBC waste material will be conducted to determine its effects on land reclamation, forage, plant growth response, and soil conditioning.
- *Nonagricultural Utilization Studies: Fluidized-bed Combustion Waste Materials:* A comprehensive program involving both laboratory and field demonstration activities has been initiated to evaluate nonagricultural utilization of fluidized-bed combustion waste materials. Applications include use of the material in flue gas desulfurization, water and sewage treatment, highway subsoil stabilization, acid mine drainage neutralization, and cementitious mixtures.
- *Atmospheric Fluidized-Bed Combustion: Bench-scale Studies (Morgantown Energy Technology Center):* The METC bench-scale atmospheric pressure FBC facilities will be operated to (1) define combination characteristics of low-grade fossil fuels such as shales and mining and coal cleaning wastes in FBC in sufficient detail to design a fluidized-bed boiler for commercial demonstration; (2) examine the effects of input coal composition and operating parameters on emission of sulfur oxides, nitrogen oxides, carbon monoxide, hydrocarbons, and trace metals as they relate to environmental and process considerations; (3) study heat transfer, bubble formation, sulfur sorbents, etc., in support of DOE-sponsored modeling and development programs; and (4) study advancnced heat

transfer concepts for power cycles utilizing the FBC as the heat source.

The program tasks will be carried out in two 18-inch combustors. One unit will be used to continue the testing of low-quality fuels for combusting and fouling behavior. The second combustor will be used to study fluidized-bed chemistry and fluid dynamics as they affect combustion efficiency, heat transfer, and sulfur sorbent utilization.

- *Western Coals Utilization (Grand Forks Energy Technology Center):* The GFETC support studies are aimed at developing improved technology for direct-firing Western coals in an environmentally acceptable manner. Ongoing goals include (1) to develop improved methods for using alkaline coal ash as a reagent for wet scrubbing and for disposal or use of the waste produced; (2) to determine the extent of sulfur retention on coal ash during fluidized-bed combustion; and (3) to develop design information for electrostatic precipitators for removing high-resistivity Western fly-ash from stack gases.

  The 120 scfm pilot scrubber at GFETC will continue to be used to evaluate a wide variety of fly-ashes. Alkali availability, SO_2 removal efficiencies, and scaling rates will continue to be investigated.

  Installation of an 18-inch fluidized-bed combustor will be completed and will be used with the existing 6-inch fluidized-bed combustor to provide data on heat transfer, corrosion, erosion, deposition, and optimum operating conditions for the application of AFBC to low-rank Western coals. Additional tests will be performed on the 7 square-foot Combustion Power Company fluidized-bed combustor to provide data on the effect of calcium to sulfur removal, and on the effects of firing the high-alkali Western coals on the corrosion and erosion of heat transfer surfaces within the bed and freeboard. Plans will be laid for the scale-up of FBC for Western coals.

- *Pressurized Fluidized-Bed Combustion: Bench-scale Studies (Argonne National Laboratory):* These ongoing studies support the development program for atmospheric and presurized fluidized-bed coal combustion. Laboratory and bench-scale studies aimed at providing needed information on combustion optimization, regeneration process development, solid waste disposal, synthetic SO_2-sorbent studies, emission control, and other tasks are included. Characterization of a variety of limestones and dolomites from various parts of the country for suitability in fluidized-bed combustors is also included. Reduction in solid waste volumes to reduce the environmental impact of the waste-sulfated limestone is one of the major goals of this program. These studies are designed to supply data essential for the application of fluidized-bed combustion units to public utility and industrial systems.

Alternative Fuel Utilization

Tasks currently underway in the Alternative Fuel Utilization Subactivity involve coal-oil mixtures and alternative fuel combustion.

The combustion of coal-oil mixtures in combustors traditionally fired with oil offer a near-term opportunity for extending fuel oil resources. The goal of the Alternative Fuel Combustion Task is to provide the fuels technology to assist the timely switching from premium fuel (petroleum and natural gas) to alternatives. This will be accomplished by developing the combustion technology for fuel specification and combustion equipment design and operation. Two groups of fuels are being addressed: (1) agricultural and forecast residues and industrial wastes (RW), and (2) principal alternative fuels (PAF) derived from coal and shale. (Municipal and urban wastes utilization are excluded since they are covered by other DOE divisions.)

Coal-Oil Mixtures. The objectives of this task are to modify or retrofit, operate, and test existing boilers, heaters and furnaces to demonstrate combustion technology and practicability of burning coal-oil mixtures.

Combustion of coal-oil mixtures in existing oil-fired combustors will be investigated to determine the extent to which their retrofit technology can be implemented practically. (See Figure II 3-130.) The goal is to substitute coal for an appreciable fraction of oil in appropriate industrial and utility combustors within the near-term.

Three contractors were selected to apply coal-oil mixture (COM) combustion technology to modified/retrofit existing equipment beginning in early FY 1977. The selected contractors and their project completion dates are:

New England Power Co.
Salem, Massachusetts
   Coal-Oil Mixture Combustion, Utility Steam Generator, 80-MW, Originally Designed for Coal and Converted to Oil—Early FY 1980 Completion

Acurex Aerotherm
Mountainview, California
(Site: Danville, Virginia, plant of Lorillard Tobacco Co.).
   Coal-Oil Mixture Combustion, Process Industrial Steam Generator, 80,000 lb/hr Steam, Originally Designed for Oil/Gas—Early FY 1980 Completion

Interlake, Inc.
Chicago, Illinois
   Coal-Oil Mixture Combustion, Blast Furnace—Early FY 1980 Completion

A 700-hp combustion test facility using coal-oil mixture for combustion has been

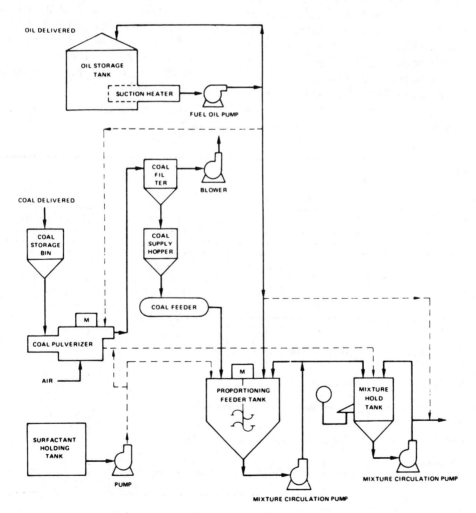

Figure II-3-130. Coal-oil mixture combustion.

completed at Pittsburgh Energy Technology Center (PETC). Preliminary COM information has been and is presently being obtained from an existing 100-hp firetube test unit at PETC. The objectives of the PETC projects are to develop in-house technical capability in coal-oil mixture technology and to supply direct technical support to the total coal-oil mixture combustion program.

Major milestones are presented in Figure II 3-131.

Alternative Fuel Combustion. The objective of this task is the timely substitution of alternative fuels (other than direct coal combustion) for premium fuels. The task is divided into two subtasks: Residue Fuels (RW), including agricultural and forest residues and industrial wastes, and Principal Alternate Fuels (PAF), including coal-derived liquids and more plentiful petroleum fractions. In FY 1980, the appropriate parts of the RW subtask will be transferred to ET Division of Distributed Solar Energy Technology and to Conservation and Solar as being more in line with their respective missions in areas of renewable energy. Figure II 3-132 shows alternate fuel combustion of a wood-fired spreader-stoker burner.

The potential fuels addressed by the Residue Fuels Subtask represent a renewable energy resource of well over four quads/year with less than one quad/year currently used. Fuel transportation economics and site-specific energy requirements indicate attractive possibilities for small-scale utilization for residential and industrial heating, off-grid power generation, and limited automotive use. The subject area builds on a data base of current European efforts and, in this country, technology developed mainly prior to 1940. The approach taken under this subtask is an assessment of present utilization opportunities and options, identification of critical technological barriers to expand utilization, and prosecution of the appropriate RD&D. Included under this subtask are direct combustion, gasifier/conversion systems, and use of pyrolysis-derived products. Major subtask elements are as follows:

- Regional resource and economical impact assessments

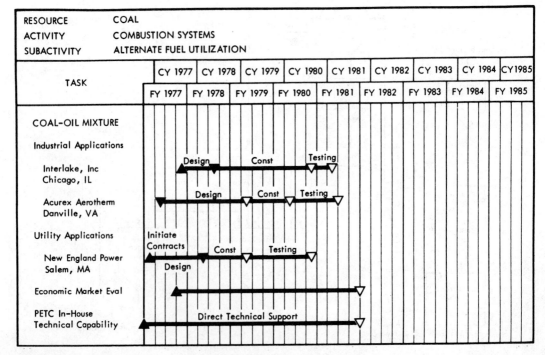

Figure II-3-131. Major milestones for the coal-oil mixture task.

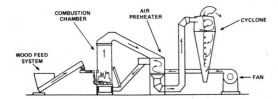

Figure II-3-132. Alternate fuel combustion schematic of wood-fired spreader-stroker burner.

- Fuel wood (wood-burning home heaters)
- Farm and forest residue fuels
- Processed fuel utilization
- Industrial waste fuels

Program plans were formulated and partially implemented. In preparation for the transfer of the RW program to other DOE divisions in FY 1980, all FY 1979 contract actions were scoped to provide definitive endpoints with the FY 1979 funds provided.

- A comprehensive R&D plan, which will include market resource assessments, is being completed by Brookhaven National Laboratory.
- The Maine Audubon Society is completing a fuel wood utilizing study to define approaches to economically increase fuel wood availability from the small woodlot supply sector.
- The fuel wood project was implemented. Contracts are underway at Auburn University and the Center for Fire Research on the National Bureau of Standards which will provide much needed recommendations in the areas of wood heater product certification and updated fire and building codes. A contract will be completed at the University of Maine addressed to space heating-hot water systems and the Georgia Institute of Technology will provide a modern compilation of fireplace performance data.
- At Oregon State University, techniques for improving combustion efficiency and reducing emissions of wood residue fuel spreader stoker boilers have been demonstrated and are being incorporated by commercial installations.
- The Babcock and Wilcox Co. has characterized the capability of co-firing shelled corn with high sulfur Eastern coals in pulverized coal fuel utility boilers. The conclusion was that 15 percent corn, on a Btu basis, could be co-fired with a minimum of risk.
- Science Applications Inc. (SAI) assessed the state-of-the-art and technology gaps for small-scale power system using the RW fuel. The BNL National Center for Analysis of Energy Systems has conducted "market potential" studies for use of the RW fuels. The results of both clearly identify the potential of the approach and the need for a DOE role in its implementation.

Projects in on-site industrial waste combustion (Peabody Gordon-Piatt Inc.) and turbine combustion of pyrolytically produced liquid-solid-two phase fuel (Teledyne CAE) were initiated.

Major milestones for this task are given in Figure II 3-133.

The expanded use of minimally processed, heavier petroleum fractions offers an energy-efficient strategy of significantly extent oil reserves while coal-derived liquids are leading candidates to furnish the bulk of our liquid fuel requirements after the mid-1990s. The resulting fuel stocks will have notably different characteristics than their petroleum based analogs and require significant energy consumption in processing prior to use in existing combustion equipment. The goal of this subtask is to provide the fuel combustion technology base required for the design of equipment capable of burning these fuels with a minimum of energy intensive processing, and thus allow an energy optimization of the fuel/combustion device system. The overall approach is to determine the relationship between the fundamental fuel composition and those combustion characteristics which are the necessary data for equipment design. The subtask is grouped into the following four elements:

- Combustion characterization
- Quantification of combustion effects and kinetics
- Development of combustion models
- Emissions and environmental studies

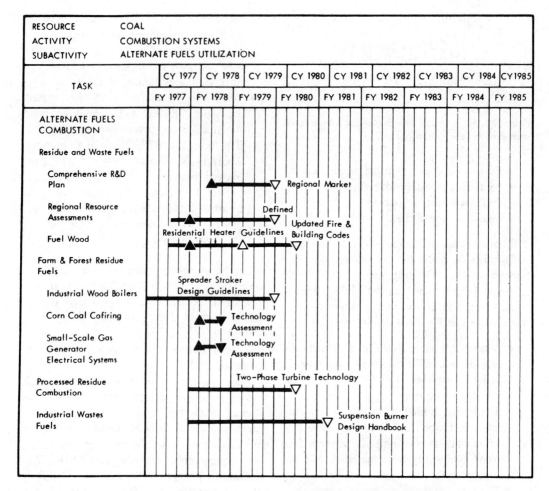

Figure II-3-133. Major milestones for the alternate fuels combustion task.

This subtask was inititated in late FY 1977. Seven projects are currently underway, six in the area of combustion characterization and one in NO_x emissions control techniques. In FY 1980, these activities will be conducted under the direct utilization portion of the AR&TD effort.

Major milestones are presented in Figure II 3-134.

Coal-Bed Methane Utilization. The coal-bed methane utilization program, which appeared in the Fossil Fuel Utilization Program in 1979, has been transferred to the Enhanced Gas Recovery Program of the Division of Fossil Fuel Extraction.

Combustion System Demonstration Plants

Only one demonstration plant in fluidized-bed technology has been authorized to date. The demonstration of a coal-fired fluidized-bed combustion boiler at a near-commercial size will be required to implement commercial adoption by the utilities. Congress authorized a direct combustion demonstration plant in FY 1976. The initial plant is expected to be an atmospheric pressure unit. The unit is to be a nominal 200-MW size and will provide high-pressure steam for utility power generation.

The planning level powerplant conceptual design contracts for complete AFB

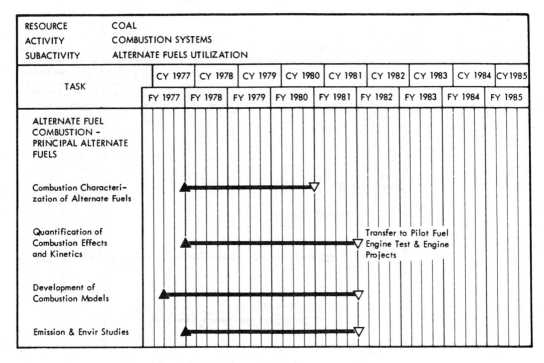

Figure II-3-134. Major milestones for the PAF subtask.

powerplants that were awarded to Stone and Webster and Burns and Roe was completed in FY 1978. Plant designs use preliminary boiler designs funded by DOE. This work was accomplished by three major utility boiler manufacturers and was completed in FY 1978. Sensitivity analyses were conducted to determine the effects of the EPA's revisions of the New Source Performance Standards based on the Best Available Control Technology (BACT) section of the 1977 Clean Air Act Amendments, which in preliminary form require 85 percent sulfur removed.

Preprocurement activity for the AFBC Utility Demonstration Plant has been suspended until the technology base is more fully developed.

Technology Support objectives in FY 1980 are to obtain engineering data on boiler materials testing and fuel facilities to assist in making decisions, technical and economic, on utility fluidized-bed demonstration plants.

An activity schedule for the plant is shown in Figure II 3-135.

FUEL CELLS

A fuel cell is a device that electrochemically converts potential energy of a fuel into electricity in a highly efficient and environmentally acceptable manner. Because of their unique features, fuel cell powerplants can be factory built and located close to the point of electricity demanded, reducing energy losses associated with transmission and distribution and permitting effective use of waste heat. The modular construction of fuel cell generators provides a size flexibility from kilowatt to megawatt capacities. The essentially non-polluting, practically noiseless operation permits location in large load centers and even on-site in urban, rural, and residential areas. Measured emissions from an experimental fuel cells powerplant were much lower than the EPA standards for modern conventional fossil-fueled central station generators. For the fuel cell system, nitrogen oxides were less than one-tenth of the EPA standard, and sulfur oxides and particulates were even less.

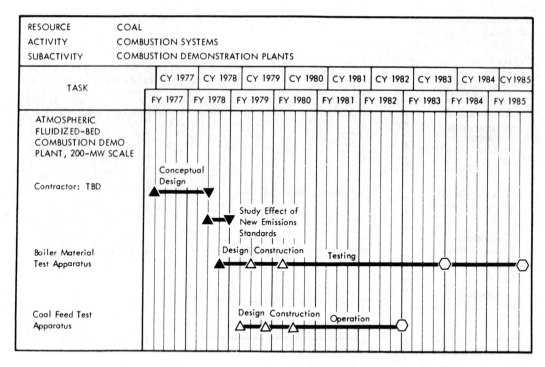

Figure II-3-135. Major milestones for the AFB combustion demonstration plant.

A fuel cell powerplant is comprised of three sections: the fuel processor, the power section, and the inverter. (See Figure II 3-136.) The fuel processor converts a hydrocarbon fuel into a gaseous mixture of hydrogen and carbon dioxide and/or carbon monoxide. If the product gas is to be used in first generation phosphoric acid fuel cells, the carbon monoxide is reacted with steam recovered from the power section to generate hydrogen and carbon dioxide. If the product gas (such as gasified coal) is to be used in second generation molten carbonate fuel cells, the carbon monoxide is used directly as fuel by the power section.

The fuel cell power section consists of many individual cells in which the hydrogen or hydrogen/carbon monoxide gas mixture and oxygen from the air are electrochemically reacted to produce dc electricity. Because the conversion of fuel to electricity is an electrochemical process, the conversion efficiency is very high and is achievable for any plant size and at part load. The currently available feedstocks which can be processed to generate hydrogen or hydrogen and carbon monoxide include methane, liquid petroleum gases, alcohols, petroleum distillates, and hydrogen or carbon-monoxide rich waste gas streams. With minor modifications, fuel processors designed to handle these existing fuel supplies could be made to handle coal-derived gases (high-, medium-, and low-Btu) and certain coal-derived liquids. A long-term objective of the fuel cell program is to commercialize fuel cell powerplants operating on coal-derived fuels. Fuel cells will allow for utilization of coal and coal-derived fuels in several applications in a more efficient and environmentally acceptable manner than present technology will allow. The last subsystem of the fuel cell powerplant is a static inverter that converts the dc output to ac. The inverter produces a waveform compatible with conventional electrical supplies generated by rotating machinery.

Fuel cell systems evolving within the next 5 to 10 years are potentially viable in the following applications:

- Industrial cogeneration where advantage is taken of the high electrical and ther-

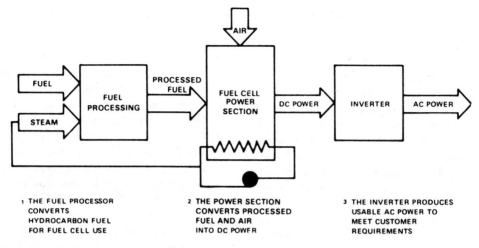

Figure II-3-136. Major sections of a fuel cell powerplant.

mal efficiencies with operation on coal-derived fuels
- Integrated energy systems for residential and commercial buildings and district heating, including operation from waste and coal-derived fuels
- Electric generation, including peaking and intermediate cycling loads; spinning reserve; small utilities; and replacement of old, inefficient units.

In addition to the above energy-saving applications, the foreign market potential for phosphoric acid fuel cells is quite large. It fits into four markets:

- Developed countries where environmental considerations are important
- Developing nations because of the small unit size and ability to run without operating personnel
- Oil- and gas-rich countries where the low capital cost, efficiency, and ability to operate without trained personnel make fuel cells attractive
- Countries that are shifting to an alcohol fuel-based economy

The objective of the DOE Fossil Energy Cell Activity is to establish the commercial feasibility of fuel cell powerplants for electric utility applications, industrial cogeneration, and for building total energy systems in the near-term and to develop advanced, higher efficiency, economically competitive, fuel cell technologies for all end-use applications in the longer term. As a result of the initiatives taken by this fuel cell activity, the nation can experience, over the next 3 to 6 years, the development of a fuel cell industry with its significant associated national benefits. Specifically, the fuel cell activity has accomplished the following:

- Consolidated Edison of New York City has been selected as the host utility for the demonstration of a 4.8-MW fuel cell powerplant on a utility grid. The powerplant, manufactured as a result of a three-way, cost-shared contract, was delivered in the fall of 1979.
- A market analysis and commercialization strategy for phosphoric acid fuel cells in residential and commercial building on-site integrated energy systems applications was completed.
- A cost-shared development of a 40-kW fuel cell powerplant for on-site integrated energy systems was completed on schedule and within projected cost.
- Two cost-shared programs with potential competitive manufacturers of phosphoric acid fuel cell systems were initiated.
- An analysis of the economics and energy savings of utilizing waste-derived fuels in dispersed fuel cell powerplants was completed.
- Integrated phosphoric acid fuel cell stacks were sucessfully tested. This is

considered significant because the integrated stack concept has the potential of significantly reducing cell manufacturing cost because of its compatibility with mass production techniques.
- A potentially economical and effected solution to wet seal corrosion in molten carbonate fuel cells, one of the previous life-limiting problems, was developed.
- Molten carbonate cells made with sheet metal hardware were successfully operated as an essential step toward economical production of these cells.
- Molten carbonate electrolyte tiles (the size-limiting component) were selected from 1 square foot to 2-¾ square feet.
- A five-cell solid oxide stack was fabricated and tested for over 700 hours through several thermal cycles meeting performance goals at the end of the test. This is considered a significant accomplishment because failure of this cell interconnection material had prevented such cycling operation in all past efforts.
- A feasibility analysis of the use of fuel cell powerplants in electric hybrid vehicles was completed.

In order to achieve the overall DOE Fossil Energy Fuel Cell objective, the activity is divided into four subactivities. Each of these areas has a specific objective as follows:

- *4.8-MW Electric Utility Powerplant Demonstration:* Test, on a utility grid, the operational feasibility of electric utility fuel cells which are nearing readiness for commercialization.
- *Phosphoric Acid Systems Development:* Develop fuel cell systems for on-site integrated energy system applications and provide technology to lower cost and increase reliability of phosphoric acid fuel cell systems.
- *Molten Carbonate Systems Development:* Advance the state-of-the-art of molten carbonate fuel cells in order to achieve the earliest possible commercialization of coal-fueled powerplants in electric utility baseload and industrial cogeneration applications.
- *Fuel Cell Applied Research:* Support emerging systems with a sufficient technology base, examine advanced fuel cell systems, and broaden the spectrum of acceptable fuels.

Table II 3-15 summarizes the funding levels for the Fuel Cell Activity for the FY 1978 to FY 1980 period.

Table II-3-15. Funding for the fuel cell activity for FY 1978-1980.

FUEL CELLS	BUDGET AUTHORITY (DOLLARS IN THOUSANDS)		
SUBACTIVITIES	ACTUAL FY 1978	APPROPRIATION FY 1979	ESTIMATE FY 1980
4.8 MW Electric Utility Power Plant Development	$16,500	$ 5,500	$ 0
Phosphoric Acid Systems Development	13,000	14,000	5,000
Molten Carbonate Systems Development	3,000	17,000	14,500
Fuel Cell Applied Research	2,700	4,000	0
Capital Equipment	200	500	500
TOTAL	$35,400	$41,000	$20,000

4.8-MW Electric Utility Powerplant Development

This subactivity consists of design and fabrication of a 4.8-MW fuel cell powerplant with subsequent integration of the hardware into a utility grid for testing. (See Figure II 3-137.) The system is being constructed at United Technologies Corporation (UTC). The project is cost-shared by DOE, the Electric Power Research Institute (EPRI), and UTC. Consolidated Edison has been selected as the host utility, and the powerplant is to be installed and tested in New York City. The overall objective of the project is to establish the operational feasibility of a fuel cell system in the technically and economically competitive market that faces all candidate equipment for the generation of electricity.

The 4.8-MW fuel cell powerplant is a modularly constructed, truck transportable, energy conversion device consisting of fuel processor, power section, power conditioner, and ancillary equipment to maintain appropriate process balance. Because of its modularity, as well as its virtual absence of pollutant generation, it has unlimited application for the production of electric power. There is unlimited siting flexibility because of its operating characteristics. It can serve as a component of central station generation, with particular application for intermediate and peaking loads, and as a dispersed generator for utilities, industries, buildings, and community systems. In the latter mode, the conservation of electricity and dollars is maximized since cogeneration becomes an available option. The operating temperature of the system, 375°F, results in the availability of process steam and low-grade heat for environmental conditioning. Thus, the installation of boilers and the consumption of additional energy for these purposes would be eliminated.

The technological feasibility of the phosphoric acid fuel cell for producing electric power has been demonstrated in systems up to 1-MW in size. The 4.8-MW module is the basic building block for a powerplant that can satisfy the minimum capacity requirements of a utility.

Accomplishments to date include execution of contracts for construction of the 4.8-MW

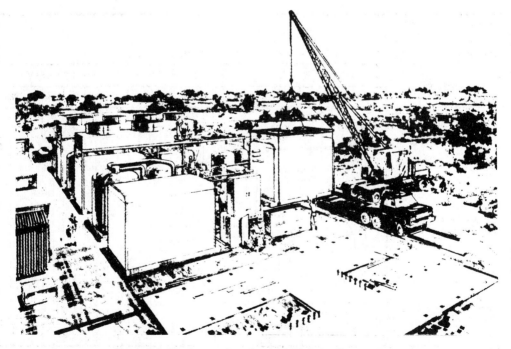

Figure II-3-137. Construction of a 4.8-MW fuel cell powerplant.

fuel cell powerplant and the spare parts required for validation and supplemental testing in a host utility. The design and design configuration phases on the program have been completed, and projected performance, meeting specification goals, has been substantiated through subcomponent testing and modeling. Another contractual action, selection of a host for integration of the powerplant in a utility grid, has been completed. The powerplant will be sited in lower Manhattan, and site approval by the New York City Board of Standards and Appeals has been received. Principally due to heat exchanger and cell component vendor delivery schedule slips, the powerplant delivery was slipped to late FY 1979.

Major milestones are shown in Figure II 3-138.

Phosphoric Acid Systems Development

Phosphoric acid fuel cells are in the transition phase between development in laboratories and the commercial marketplace. These systems require improved performance, longer life, and an extension of their range of application. Furthermore, additional industrial suppliers of phosphoric acid fuel cells are required in order to provide a competitive environment. Technology demonstrations (such as the 4.8-MW powerplant previously described and the 40-kW powerplant described below) will be useful, but significant improvements must come in the areas of cost reductions, endurance extensions and reliability, and performance improvements.

This subactivity emphasizes the development of kilowatt size powerplants in the on-site/integrated energy systems (OS/IES) area, with the objectives of developing fuel cell systems for specific applications as the last step toward prototype demonstration and providing technology to lower cost and increase reliability of these systems. Potential applications include fuel cell powered total energy systems for residential, commercial and industrial applications, including the utilization of waste stream fuel supplies; and OS/IES powerplants integrated with urban, industrial, and agricultural waste conversion systems. In order to meet these objectives, the

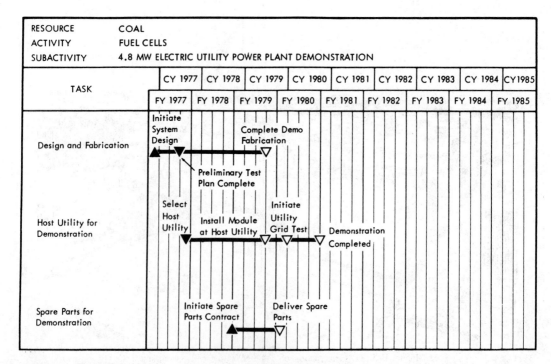

Figure II-3-138. Major milestones for the 4.8-MW electric utility powerplant demonstration.

subactivity is divided into two areas: OS/IES powerplant development and phosphoric acid technology development. A description of each of these areas and the strategy involved follows.

On-site/integrated energy systems, in which the prime mover meets both the electrical and thermal energy demands of the user, have the potential of saving over 20 percent of the present fuel requirements of conventional systems. An integrated energy system using a fuel cell as the prime mover could meet all of the thermal and electrical demands of a residential building complex using no more fuel than is presently used for heating alone. (See Figure II 3-139.) This is made possible because of the availability of waste heat at the application site and the reduction in distribution and transmission losses. The fuel cell power system is a favorable prime mover for the integrated energy system application because of its modularity, flexibility, and minimum impact to the environment. The integrated energy systems have application in residential and commercial buildings and industrial plants. The powerplant performance specifications are different for those applications; however, the improved fuel utilization efficiency is equally applicable. The primary limitations associated with conventional powered total energy systems in the past have been reliability and siting flexibility. Fuel cell systems provide an opportunity for overcoming these limitations. Fuel cells powered total energy systems would replace natural gas and oil burning systems which provide space and process heat. They would also replace conventional grid-supplied electrical power. These fuel cell systems could be transitioned to new fuel supplies such as synfuels or coal gasification products. In this area, the powerplant capacity requirements are small relative to a utility application and, therefore, potential suppliers are not faced with the prospect of long and costly developed and scale-up programs, and users would not be subjected to lengthy and costly field construction. Most of the emphasis to date has been on utility applications, thereby leaving the field of on-site applications open to competition. This effort has been structured to provide this competition.

The achievement of significant cost reduction via technology is also receiving major emphasis. Specific efforts involve: (1) improving cell endurance; (2) developing an integral cooler for lower cost and greater durability; (3) developing a design which eliminates the

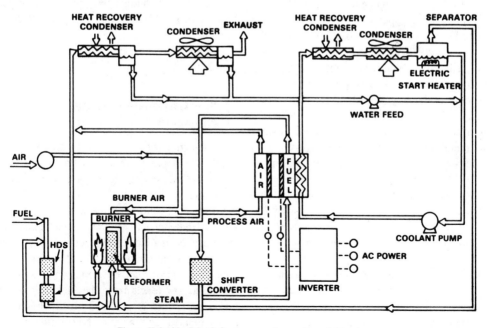

Figure II-3-139. PC 18 System — air supply and inverter.

need for nitrogen inert pressurization; (4) improving performance via higher operating temperature and pressure; and (5) developing a new low-cost integral cell stack concept. It is necessary to continually verify that changes which are proposed for purposes of increased performance or reduced cost do not also reduce stack endurance.

The complex process of determining system and technology requirements for OS/IES applications is vital to the success of the effort. Requirements involving electrical efficiency, heat quality, size, operational characteristics, cost, and fuel availability and compatibility are some of the items being addressed. The phosphoric acid fuel cell technologies being directed toward potentially competitive OS/IES applications are in the early stage of development. These system studies provide a focus for the technology effort and assess the range of applications that the technology serves.

In the OS/IES area, the 40-kW modification and development effort is in the most advanced state (engineering and development). The 40-kW pilot powerplant (PC 18), pictured schematically above, was originally developed under gas utility and United Technologies Corporation sponsorship and was the starting point for this development effort. This pilot powerplant has operated for over 15,000 hours. The current effort, cost-shared with the Gas Research Institute, is directed toward cost, reliability, endurance, and performance improvements of the pilot powerplant design in order to develop a viable, properly functioning 40-kW field test powerplant. Effort in this area also involves the work of several manufacturers to foster a competitive economic situation for continued development of the OS/IES concept.

Two major cost reduction tasks were initiated in FY 1978 and continued through FY 1979. The first, platinum catalyst activity enhancement, represents an alternative to higher temperature and pressure for increasing performance. This effort will focus primarily on the cathode. The success of this task will have an important impact upon phosphoric acid fuel cell systems being developed for OS/IES applications. The second effort involves the development of a low-cost integral cell stack concept to be incorporated into the 40-kW powerplants under development for field testing.

Several systems studies were completed during FY 1979 identifying market and regulatory restraints, capital requirements, quantified benefits, and associated costs. In addition, a study of the feasibility of a fuel cell integrated with an urban, industrial or agricultural waste conversion system was completed during FY 1978. This study compared the economics of several fuel cell/waste conversion system configurations with a conventional system and led to a developmental plan. An industrial cogeneration request for proposals (RFP) utilizing fuel cells has been let. This RFP will result in a number of site-specific studies from which an initial field demonstration will be selected.

In FY 1978, a study to assess the market potential for vehicles powered by phosphoric acid fuel cells was completed. The purpose of the study was to identify the advantages, problem areas, and prime market of fuel cell powered vehicles. The most promising markets for fuel cell vehicle applications was found to be forklift trucks and delivery vans. No further funding was available in FY 1979.

The FY 1980 budget only provides for continuation of the utility-size integral cell development.

Major milestones are presented in Figure II 3-140.

Molten Carbonate Systems Development

The objective of the Molten Carbonate Systems Development Subactivity is to bring about the earliest feasible commercial use of molten carbonate fuel cell systems, so that the environmental and resource conservation benefits of these clean, efficient, coal-fueled systems can be realized. Replacement of gas and oil with the minimum amount of coal is facilitated by high system efficiency. Sulfur is removed to low levels from the fuel stream, and because the fuel is oxidized electrochemically rather than by combustion, NO_x emissions are very low. The environmental impact of this

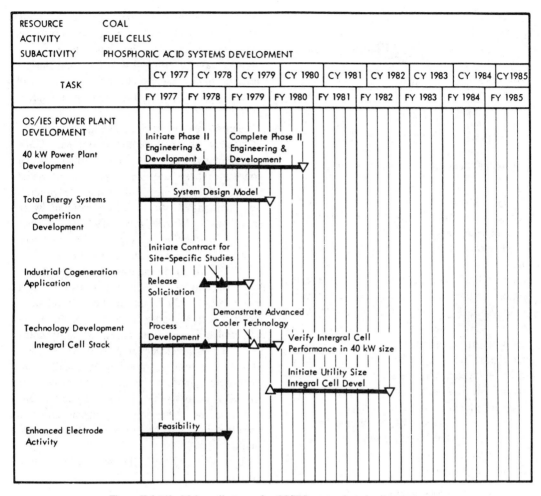

Figure II-3-140. Major milestones for OS/IES powerplant development.

type of coal-fueled generator is therefore relatively small.

The primary application of molten carbonate fuel cell generators is for baseload electrical generation and industrial cogeneration. The program leading to this product is well defined. With appropriate liquid fuels, such as petroleum or coal-derived hydrocarbons, it would be possible to meet intermediate and peaking demand as well with similar benefits. The fuel cell technology development required for this application is essentially the same as for baseload powerplants.

In a molten carbonate fuel cell, which would be thermally integrated with a coal gasifier, coal is gasified with steam and air or oxygen to form a fuel gas stream rich in CO and H_2. (See Figure II 3-141.) The gas stream is purified of particulate and sulfur in the cleanup section, then fed to the fuel cell where it is combined electrochemically with air to form CO_2 and H_2O. This process produces dc electric power, which is transformed to ac in the inverter and stepped up to transmission line voltage levels. Sensible heat from the gasifier and the fuel cell outlet gas stream is recovered by raising steam which is used in a steam turbine generator. About two-thirds of the powerplant electric output comes from the fuel cell and one-third from the steam turbine. Where plant sizes smaller than about 500-MW are needed, a gas turbine can be used instead of the steam system with about 5 percent less plant efficiency. The sensible heat in this system is also at a high enough temperature (600°C) to be used di-

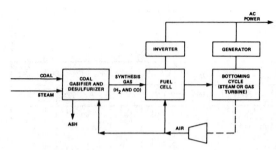

Figure II-3-141. Second generation fuel cell powerplant.

rectly in many industrial processes to replace gas or oil furnaces. A schematic of this system appears below.

The strategy for implementation is the same for all applications. The approach is to conduct parallel interactive efforts in system definition component development. The system defintion task provides guidance and direction to the development work. Similarly, the results of the component development work are fed back into the definition effort. For the next few years, the paramount effort will be in cell development since the cells are the least developed part of the system.

In order to meet the requirements of an electric utility base load system, the following goals were established for the molten carbonate fuel cell stacks:

- Life of 40,000 hours to meet low maintenance requirements
- Current density of 150A/ft^2 at 0.85 V
- First cost (stack only) of $60/kW at the current density and endurance design points

Successful achievement of these goals will allow for a system design of 50 percent overall efficiency at a competitive cost.

Cell performance is rapidly approaching the goal specification, and may well exceed that level before other goals are met. The present current density at 0.85 V is 100 A/ft^2. This specification must be met under system operating conditions of temperature (923°K), pressure (10 atm), CO_2 recycle (currently by burner), and in-cell fuel utilization (currently 85 percent).

The best cell life to date is 16,000 hours. Obviously, the cell was built with technology that is a couple of years old. Since then, satisfactory solutions to the wet seal corrosion problem (which was the apparent cause of failure) have been devised. The next hurdle appears to be electrolyte management, and several approaches to this problem are under study.

Capital cost is of fundamental concern in all aspects of component development. Three manufactured cost-related areas are of special concern. These are scale-up, use of sheet metal cell hardware, and electrolyte production methods. Critical cell components were scaled up to 2.75 ft^2 in FY 1978 and further scale-up to 4 ft^2 is in progress. The first true sheet metal cells were successfully operated in FY 1978. An anode material of improved surface area having no discernible loss of area for at least 3,000 hours has been developed, together with a synthesis process for the material, which allows wide control of alloy composition and electrode area, and which does not involve toxid carbonyl compounds. The electrolyte manufacturing advances initiated in FY 1977 were further developed, and it appears that both electrolyte synthesis and tile manufacturing will be amenable to mass production.

Figure II 3-142 gives the milestones for this subactivity.

Fuel Cell Applied Research

The Fuel Cell Applied Research Subactivity encompasses two efforts: (1) technology advancement, and (2) advanced concept investigation. Technology advancement efforts perform phenomenological research in areas where fuel cell development has shown that there are basic problems or technological limitations which adversely affect the cost, efficiency, reliability, and/or lifetime of fuel cells currently under development. The improved know-how may also accelerate demonstration and commercialization programs by pointing the way to improvements that can be incorporated without major engineering changes. Technology advancement deals with fundamental physics and chemistry in depth far beyond that of engineering development. For this reason, the most advanced sources of sci-

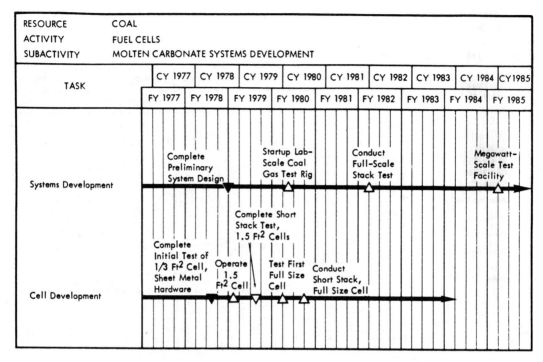

Figure II-3-142. Major milestones for molten carbonate systems development.

entific expertise are used in universities and specialized research organizations. Advanced concept investigation efforts consist of exploratory research and development to provide sufficient information to evaluate the development costs, risks, and benefits of new fuel cell concepts. Emphasis is on concepts with the potential for high national benefits but which cannot be evaluated on a sound basis without obtaining additional laboratory, experimental, and/or engineering data.

Advanced concepts explore as many promising concepts as can be accommodated within the budget. After eliminating the less desirable ones, additional research is carried out on the candidate technologies with the greatest likelihood of being attractive to industry. It provides the information necessary to evaluate in greater depth how promising the technology is for commercialization. The same technological data provides a springboard for industrial product development, or in some cases, for government development work outside of the applied research subactivity. The advanced concepts now subject to applied research are shown in Figure II 3-143.

Government support of these applied research efforts is appropriate not only because the results will be widely used in the national interest but also because it is exceedingly difficult for a single company to earn a satisfactory and timely rate of return on investment in applied research.

Technology development efforts have been initiated in FY 1978 and FY 1979 to investigate the areas of electrocatalysis, cell materials behavior, accelerated testing of fuel cells, technology for scale-up of fuel processing reactors and coal cleanup technology. Advanced concepts currently being evaluated include solid-oxide fuel cells and fuel processing concepts such as autothermal reforming and advanced steam reforming for processing of a broad range of coal-derived liquids. The applied research efforts in solid-oxide fuel cell development and advanced fuel processing have been transferred to the Molten Carbonate Subactivity in FY 1980. The other applied research activity will be completed in FY 1980.

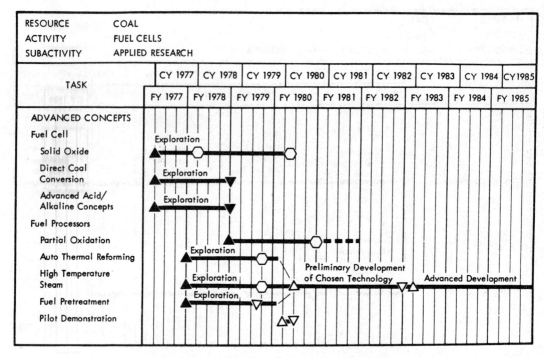

Figure II-3-143. Major milestones for fuel cell applied research.

MAGNETOHYDRODYNAMICS (MHD)

The objective of the MHD Program is to develop MHD technology in order to determine the commercial viability of MHD/steam power generation systems. This is being pursued because of the potential of MHD as a future source of low-cost, central station, electric power. Assessment of the market situation clearly shows the position occupied by MHD, provided MHD technology can be successfully developed. It will be the only advanced system available to an increasingly coal-dependent economy that burns coal directly to produce electricity; it can be coupled to a steam electric generator to provide an overall superior baseload plant or it can be coupled to an industrial thermal load to provide unique cogenerating capabilities. Environmentally, MHD offers an advantage in that sulfur is removed as an inherent part of the process, thereby reducing the need for extensive subsidiary clean-up equipment. Recent experiments support theoretical predictions of low NO_x emission. Another environmental advantage is that because MHD may produce at least 50 percent more power from a given quantity of heat than other fossil fuels or uranium-based power-generators, less heat is thrown away and less cooling water required. Studies indicate that a mature, coal-burning, utility-sized, MHD/steam powerplant could have the potential to do the following:

- Achieve overall cost-to-busbar power conversion efficiencies in the 50 percent range
- Meet all existing or proposed Federal standards for SO_2, NO_x, and particulate emissions with reduced thermal pollution
- Achieve a cost of electricity lower than potential alternative power systems

Table II 3-16 summarizes the funding levels by subactivity for FY 1978 through FY 1980.

MHD milestones are presented in Figure II 3-144.

Open-Cycle Plasma Systems

The MHD Process. MHD power generation is based on the direct conversion of heat to electricity by passing a high-temperature,

Table II-3-16. Funding for MHD for FY 1978 - 1980.

MAGNETOHYDRODYNAMICS	BUDGET AUTHORITY (DOLLARS IN THOUSANDS)		
SUBACTIVITIES	ACTUAL FY 1978	APPROPRIATION FY 1979	ESTIMATE FY 1980
Open Cycle Plasma Systems	$67,800	$78,500	$70,000
Closed Cycle Systems	3,200	1,000	2,000
Capital Equipment	500	500	0
TOTAL	$71,500	$80,000	$72,000

high-velocity, electrically conducting fluid through a magnetic field. (See Figure II 3-145.) The principle is similar to a conventional turbine-generator system, but differs in that the rotating conductors of a turbogenerator are replaced in open-cycle MHD by a partially ionized combustion gas which is accelerated to interact with the magnetic field. The interaction of the accelerated conducting fluid with the intense transverse magnetic field induces an electric field within the fluid. If electrodes are present to collect the current, then electric power can be supplied through an inverter to a utility power grid.

Potassium seed is injected into the combustor in the form of potassium carbonate to achieve adequate levels of electrical conductivity in the combustion products at the temperature involved. This requirement provides a unique build-in capability for removing sulfur pollutants produced during the combustion of sulfur-bearing coals. The potassium seed reacts preferentially with the sulfur, and the products are then removed from the system in the form of potassium sulfate, which is then regenerated to potassium carbonate to recover the seed and separate the sulfur.

The combustion products leaving the MHD generator are no longer appreciably conductive but still have a temperature in the range of 3000°F. The remaining large amounts of usable thermal energy downstream of the MHD generator are used to generate additional power by the utilitization of a bottoming steam-turbine powerplant. A schematic of a typical MHD/steam powerplant is shown in Figure II 3-145.

The concentration of NO_x products in the exhaust gas are controlled by a combination of two-stage combustion and gas stream cooling. High-temperature combustion can be initially "fuel rich" to minimize NO_x formation. Additionally, by controlling the cooling rate of the gas stream, the NO_x products can be allowed to decompose to the equilibrium values.

The Open-cycle Program. The current open-cycle baseline program plan assumes a three-step approach to commercialization. The first step, Phase I, lays the engineering groundwork for the design of an MHD pilot plant, termed the Engineering Test Facility or ETF. Commercial feasibility is then demonstrated, in Phase II, by construction and operation of the ETF. The ETF (baseline Phase II) will be a fully integrated, combined-cycle MHD/steam system operating at a power level of 250-MW_t. This power level represents the minimum scale with can demonstrate and verify the engineering design and operational characteristics of an open-cycle MHD powerplant without incurring unacceptable financial risk or performance penalties. This is

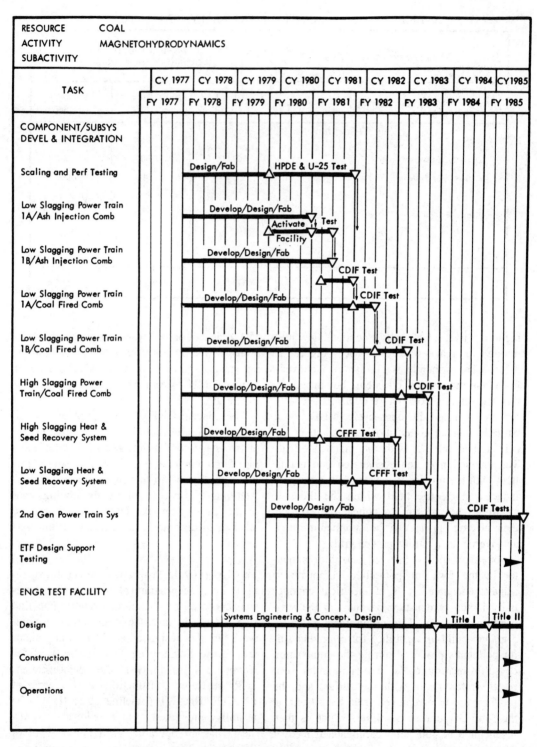

Figure II-3-144. Major milestones for MHD.

followed, in Phase III, by design, construction, and operation of a full-scale commercial plant.

Phase I focuses mainly on development of engineering components and subsystems, generally at up to a 20-MW$_t$ size, by industrial

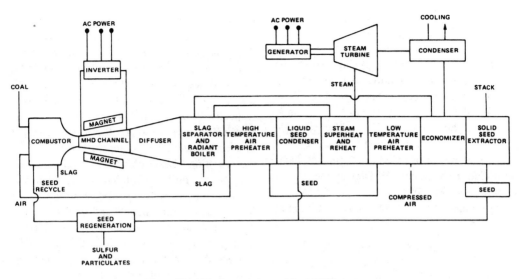

Figure II-3-145. Open-cycle coal-fired MHD system.

contractors. Engineering development progress in Phase I is monitored through a succession of critical 50-MW_t tests in the Component Development and Integration Facility (CDIF). Successively improved component and subsystems designs are selected from the 20-MW_t development work, scaled-up, and integrated in these 50-MW_t CDIF tests. This progession culminates in the determination of an ETF design, leading to Phase II. At this stage, construction and operation of the ETF is undertaken by an industrial contractor team at approximately 250-MW_t. Commercial demonstration follows in Phase III with the design, construction, and operation of a utility plant. Under this plan, commercialization is approached conservatively; viz., development progress at 20-MW_t is validated at 50-MW_t (CDIF) and when appropriate, translated to the ETF at 250-MW_t before commerical demonstration is attempted. This is a sound baseline approach to development of any new power system.

Since 1976, when MHD development emphasis was shifted more toward experimental engineering, progress has exceeded expectations. Impressive gains have been achieved in demonstrating generator efficiency, channel durability, and superconducting magnet performance. In addition, substantial progress has been made in proving the feasibility of coal combustors, air heaters, and heat and seed recovery subsystems. Major accomplishments in the MHD program include:

- Generator—successful completion of 20-MW_t 500 hours endurance tests under electrode loading conditions simulating commercial service and achieving order-of-magnitude improvements in electrode corrosion/erosion resistance
- Generator—successful subsonic generator operation at a high-magnetic field (5-tesla) representing the largest (30-MW_t) high field test ever run
- Combustor—slag rejection in excess of 95 percent with carbon conversion of 97 percent achieved in first-stage prototype of a two-stage combustor for MHD application
- Environmental—achievement of 95 percent sulfur removal from high-sulfur coal as well as NO_x reduction to 20 ppm
- High Temperature Air Preheaters—achieved 1000-hour cumulative testing at 2700°F on a direct-fired regenerative air preheater element under heat transfer, fluidynamic, and seed/slag conditions simulating MHD service; 100-hour continous run on a complete subscale heater at 3000°F using simulated coal firing with ash carryover into heater matrix
- Superconducting Magnet—design principles confirmed for large superconducting

magnets by operation of a 40-ton, 5-tesla magnet at design conditions

Major anticipated accomplishments for FY 1980 include: (1) completion of facility activation at the CDIF; (2) testing operation at the Coal-Fired Flow Facility (CFFF); (3) channel testing at the Soviet U-25 facility with the U.S. channel to be delivered in late FY 1979; and (4) testing in the High Performance Demonstration Experiment (HPDE) at the Arnold Engineering Development Center (AEDC).

Two Phase II ETF alternatives are being considered:

- Modernization of an existing fossil plant of up to 100–500-MW_t by the addition of a 250-MW_t topping generator
- Design and construction of a new, modern 500–750-MW_e coal-fired plant incorporating a 250-MW_t MHD topping generator

Under either of these options the plant, with its MHD topping generator, would be operating within a utility system producing revenues for that utility. Design and economic studies are being initiated to assess the potential in these concepts. In both alternatives the cost of the basic steam plant might be borne by the utility while the cost of the modifications, required by the addition of the MHD topping generator, could be borne by the government. This potential for cost-sharing along with direct utility participation in Phase II could result in a less extensive Federal role in Phase III and result in considerable savings of Federal money over the life of the program. Phase I development where the program now resides is common to these two alternatives as well as the baseline plan.

The potential offered by open-cycle MHD is supported by data obtained from the Energy Conversion Alternatives Study (ECAS). This study, jointly sponsored by ERDA, NASA, and the National Science Foundation, showed that MHD promised the highest efficiency of all the baseload power systems considered. ECAS predicted that a combined-cycle, MHD steam powerplant could achieve a coal-to-busbar efficiency approaching 50 percent at a cost of approximately 32 mils/kWh. This compares with an efficiency of aproximately 35 percent and a 1975 cost of approximately 37 mils/kWh for a conventional coal-fired powerplant. Furthermore, sulfur and NO_x emission levels below 0.55 pounds per million Btu and 0.3 pounds per million Btu, respectively, were predicted without requiring significant preprocessing of coal or stack gas cleanup. These levels compare favorably with the existing EPA emission levels of 1.2 pounds per million Btu and 0.7 pounds per million Btu, respectively, for sulfur and NO_x. More recent experimental and analytical evidence indicates the promise of virtually eliminating NO_x emissions. Indeed, the MHD process is capable of meeting the percentage reduction requirements and limits of 1.2 pounds per million Btu and 0.5 pounds per million Btu that have been recently proposed by the EPA for the emissions of sulfur and NO_x, respectively.

The ECAS study, performed by industry for the Federal Government, was directed mainly to baseload, power system concepts. Neither the potential benefits of cogeneration in the energy-intensive industries, retrofit of existing plants, nor application to peak loading were considered. Each of these benefits is postulated to be very significant in special circumstances.

Two additional studies, Operational Analysis of MHD and Comparative Study and Evaluation of Advanced Cycle Systems, have recently been performed for EPRI by major industrial contractors. The results of these studies further support the technical conclusions of ECAS and also indicate strong potential market penetration for MHD if the technology can be fully developed.

MHD open-cycle development will maintain a strong technology base while moving toward integrated engineering development aimed toward the demonstration of commercial feasibility. The main concerns in the current, initial phase are to establish basic design data, to apply these data to the preliminary engineering development of components, and, finally, through progressively stringent test screening, to select specific designs for integrated subsystem tests. Guided by environmental and economic requirements, various design choices and operating modes will be tested and refined to provide the technical specifications necessary to meet commercial cost and performance criteria. These experimental activities will be backed up by comprehensive and continuing systems analyses to establish realistic technical requirements and cost data. Tradeoff studies are required to direct the overall system concept towards the most advantageous designs. Such realistic system studies are essential to focus the experimental work on commercially valid goals.

MHD Program Implementation. The program uses numerous facilities at various locations in 15 states. Two major Government test facilities are also being constructed to accommodate key development and integrated engineering tests at larger scales than are available elsewhere in the United States. The first of these is termed the Coal-Fired Flow Facility (CFFF) and is located at the University of Tennessee Space Institute near Tullahoma, Tennessee. It will be capable of continuous operation at an input of 20-MW$_t$ and short duration testing at 50-MW$_t$. The mission for the CFFF is development testing of the components simulating the complete MHD/steam system, in particular, under conditions of 100 percent slag carry-over from the combustor. In addition, it is intended to be used for the development testing of the heat recovery and seed recovery system now being competitively procured by DOE. User testing can also be accommodated to provide a test capability for contractors who may not have either access to their own facility or a facility of 20-MW$_t$ input. Should circumstances require, checkout of test hardware destined for CDIF can be accomplished at the 50-MW$_t$ level prior to shipment to the CDIF. The CFFF will be equipped with both superconducting and conventional iron core magnets. The CFFF construction is currently over 80 percent complete.

The second major Government test facility, the CDIF, is located in Butte, Montana. This facility will be the principal facility within the MHD program for continuous testing of engineering-scale power trains of 50-MW$_t$, an intermediate level between existing facilities and the 250-MW$_t$ ETF. The CDIF will be the largest coal-fired MHD facility in the United States capable of continuous operation. It will be a highly flexible test facility for MHD component and integrated power train testing. The purposes of the CDIF are as follows;

- To evaluate the performance of MHD components and systems and compare the performance of different designs
- To provide a basis to scale or extrapolate MHD system designs to larger sizes or more complicated designs, in particular, the ETF

It is expected that a major portion of the data for the ETF design will be validated through CDIF testing. The facility is to be equipped with both a superconducting and conventional iron core magnet. The coal processing and feed system is designed to handle both Eastern and Western coals. Additionally, the facility can accommodate multistage combustors that may have a slag rejection of up to 90 percent. The CDIF is now over 65 percent complete. Major test components are now arriving at the site.

Two other key projects are directly related to the orderly development of MHD. One of these, at the Arnold Engineering and Development Center (AEDC) of the U.S. Air Force, is directed toward the quantification of scaling effects. Data generated through these tests will provide a firm basis for translating successful CDIF subsystems design to the 250-MW$_t$ scale of the ETF. In addition, operational data which are being made available through the joint U.S./U.S.S.R. Cooperative Program are expected to strengthen engineering design confidence in scaling to the ETF size.

With regard to the ETF, activities will be focused on system studies to provide tradeoffs, definition and requirements for this pilot plant facility.

Closed-Cycle Systems

Closed-cycle is differentiated from open-cycle MHD in that for closed-cycle the working fluid operates in a closed loop rather than being exhausted to the atmosphere as in open-cycle. The closed-cycle plasma system is being pursued in the United States. In this approach, a noble gas (argon) is seeded with cesium to enhance its electrical conductivity, is cycled through the MHD flow-train and steam system and is then recompressed and recirculated through the system. Because of "nonequilibrium ionization," the cesium seeded argon plasma attains useful electrical conductivity, the argon flow must be kept free from contamination by combustion products. The contaminate carry-over into the argon occurs in the heat exchanger process of heating the compresed argon to 3000°F prior to its passage through the closed-cycle MHD generator. Typically, a ceramic regenerative heat exchanger is used to transfer the coal-generated heat energy to the argon gas. Successful design and development of this heat exchanger has been identified as the pacing technology for closed-cycle MHD.

Current work in closed-cycle includes study of the compatibility of a coal-fired combustor with a regenerative heat exchanger. Furthermore, parallel systems engineering studies initiated in FY 1979 will provide an assessment of the performance potential of closed-cycle MHD powerplants and associated technology requirements. Future work addresses continued study of the noble gas approach, efforts aimed at a closed-cycle blowdown experiment, and further analytical studies of closed-cycle channel requirements and technology.

4.
Activity Description—Petroleum

Petroleum will continue to be one of the nation's major energy sources for many years. Recent estimates indicate that production of domestic petroleum will drop rapidly in the late 1970s and 1980s without new sources or the stimulating techniques of enhanced oil recovery.

Present extractive technology is grossly inefficient. The potential target for Enhanced Oil Recovery (EOR) is the estimated 334 billion barrels of oil in place which are not economically recoverable with conventional technology. This represents about two-thirds of the original oil in place. EOR technology under development has the potential to recover 5 to 10 percent of this currently unrecoverable resource. The remainder of this resource is the target of opportunity for new EOR techniques. The production by present enhanced oil recovery methods is about 240,000 barrels per day. DOE-sponsored programs are designed to encourage and accelerate domestic production and are parallel and supplemental rather than competitive to oil industry efforts.

Three processes show promise for recovering more oil from existing reservoirs: micellar-polymer, carbon dioxide, and thermal (steam and in situ combustion). A total of 22 major field tests will be conducted in FY 1979, composed of 12 micellar-polymer/improved water flood projects, 4 CO_2 projects, and 6 thermal projects. In addition, 3 CO_2 mini-tests, 2 thermal pilots, and a solar steam generator demonstration are planned to start in FY 1979.

A logical progression from the laboratory to develop alternate technologies is to go first to field mini-tests involving a few wells (up to five), to a pilot test of 25 to 50 acres, and then to a commercial demonstration test involving 200 or more acres. Pilot tests take 4 to 5 years, generally, and commercial demonstration takes 10 or more years. The most significant data, however, are usually developed in the first 3 to 5 years. All of these technologies require liquid or gas injection through porous media to release oil and/or move it to production wells. A definition of required chemicals, equipment, and manpower is part of this job. The current DOE program consists of applied research mini-tests and pilot tests. An active technology transfer program is conducted to encourage industry to begin and/or expand commercial demonstrations.

An integral part of the EOR Activity is to establish baseline data on all major U.S. reservoirs and to select the best candidates for the different EOR processes. A major emphasis was placed on this effort in 1978 and 1979. This baseline reservoir data then becomes part of a broader task of gathering, collating, and analyzing data coming from field tests of all varieties and finally performing inter-field analyses. This is part of the program that leads to technology transfer from DOE to the oil industry. A particular need for this kind of technical assistance is for the small- and medium-sized oil producers who do not have ready access to this kind of technology.

The nation's oil shale resources constitute the most promising alternate source of petroleum fuels to crude oil. Although coal, tar sands, and other resources are additional alternate resources for petroleum-based fuels, cost estimates consistently indicate that shale oil is the next higher cost alternate to natural petroleum.

Two basic recovery and processing techniques for oil shale are conventional mining plus surface retorting and in situ recovery. Surface retorting technology for the higher grade oil shales is generally considered to be ready for demonstration and not to need Government-sponsored research and development support.

In situ technologies, however, are still developmental and receive the major amount of Government-sponsored research and development support. This support involves true and modified in situ recovery, where the latter requires that an initial fraction be removed by mining prior to in situ recovery. Government-sponsored R&D projects include true and modified in situ technologies for the recovery of shale oil and true in situ techniques for the generation of low- and medium-Btu gases.

The potential advantages of in situ versus surface retorting technologies are lower production costs and environmental impacts. Lower production costs eventually may make it possible to economically process lower grade shales.

The most promising areas for major discoveries and production of conventional petroleum and natural gas lie in the offshore areas and in the deeper onshore zones. Among other nontechnical factors inhibiting activities in these areas is the need for improved and more efficient drilling technologies. The Government-sponsored R&D activities described here are involved in the development of these advanced drilling technologies, increasing the productivity of conventional drilling technologies, and in mitigating the environmental impacts of drilling.

Advanced process technology is oriented toward exploratory research in advanced process technologies, refining processes for shale oil and petroleum product characterization and utilization.

Advanced exploratory research is conducted in techniques to increase the permeability of rock matrices, advanced in situ recovery investigations, and the improvement of chemical and microorganism processes for enhanced oil recovery.

A goal of shale oil refining research is to determine the relationship between the nitrogen content of shale oil, and fuel stability, nitrogen emissions resulting from combustion and refining costs.

Product characterization and utilization research is oriented toward improvement of data on crude oils and products, establishing the chemistry of full stability, and developing asphalt materials with increased road life.

Table II 4-1 summarizes the funding levels by activity for the FY 1978 to FY 1980 period.

ENHANCED OIL RECOVERY

The objectives of the EOR Activity are to encourage and support industry participation in developing and demonstrating technologies to enhance the production rate and recovery of original oil-in-place by:

- Developing or improving EOR techniques and chemicals
- Testing these improved processes through mini-tests and pilots and transferring the technology to industry to ensure early commercialization.

One of the most promising opportunities for increasing domestic energy output in the near-term is EOR. After conventional (primary and secondary) production is depleted, substantial quantities of oil remain in oil fields. This unrecovered oil is estimated to be up to 70 percent of the original oil-in-place. (See Figure II 4-1.) Because of the long lead-time until current DOE EOR projects begin to yield incremental oil, the economic benefits of EOR have not yet been fully demonstrated. Due to the uncertainty of benefits to be gained, the present oil price structure, and government regulations and taxes, the individual members of the industry are reluctant to provide the necessary funds for extensive EOR research and development. A Federal

Table II-4-1. Funding for petroleum resources for FY 1978-1980.

PETROLEUM RESOURCES	BUDGET AUTHORITY (DOLLARS IN THOUSANDS)		
ACTIVITIES	ACTUAL FY 1978	APPROPRIATION FY 1979	ESTIMATE FY 1980
Enhanced Oil Recovery	$ 46,240	$ 54,072	$ 21,400
Oil Shale	28,782	48,600	28,200
Drilling and Offshore Technology	1,600	2,600	3,000
Advanced Process Technology	1,400	2,200	4,000
Program Direction	595	679	780
TOTAL	$ 78,617	$108,151	$ 57,386

RD&D program targeted at mitigating the economic and technological uncertainties and reducing the lag between project initiation and commercial production is thus needed. This is an area of R&D in which DOE is active.

In the enhanced recovery area, technology needs, priority efforts, and the DOE role have been identified. The technical, economic, and environmental feasibilities of the most promising tertiary oil recovery techniques are being evaluated through cost-shared field tests. Reservoir characteristics vary widely and require a variety of enhanced recovery techniques. Consequently, the DOE program incorporates a sufficient number of tests to develop and prove technologies that are applicable to principal reservoir types.

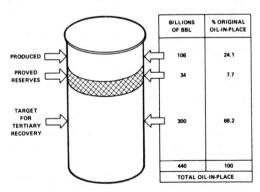

Figure II-4-1. Production, reserves, and residual oil-in-place (total United States, including offshore and Alaska).

Currently, 22 cost-sharing contracts are being conducted to test the feasibility of EOR technologies in various types of domestic reservoirs. Locations of these projects are shown in Figure II 4-2.

The EOR Activity includes university and industry research, both laboratory and field, process development, and demonstration. The objectives of the following three subactivities are to support the development of various processes, assure technology transfer to industry, and assist the private sector in selecting commercially acceptable processes. The milestone charts depict the timetables for the field tests currently in progress.

The Enhanced Oil Recovery Activity is an R&D program targeted at producing at least 5–10 percent more of the residual oil that remains after depletion of the primary and secondary production. This task will be accomplished by accelerating the development of an improved scientific, engineering, and economic knowledge base and by accelerating the transfer of DOE-supported technology to the public.

The Enhanced Oil Recovery RD&D Program is built on the following major components: target orientation, basic definition of the resource base, continuing research, and mini-tests and technical pilot tests. The ultimate measure of the EOR Program's effec-

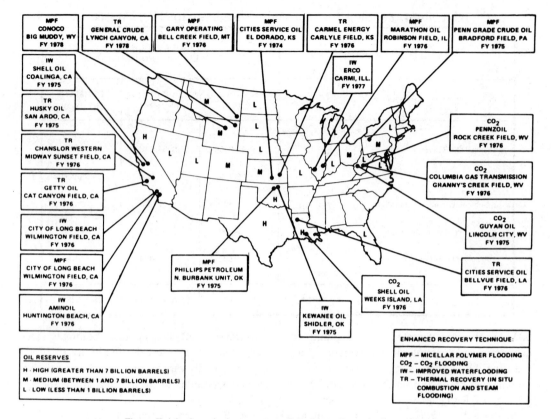

Figure II-4-2. Cost-sharing contractor locations enhanced oil recovery.

tiveness will be the incremental oil produced by private industry using advanced EOR technique.

The processes currently being supported by the activity are classified according to the fluid injected into the reservoir to drive out incremental oil, as discussed in the following subsections.

The EOR Program is being significantly redirected in FY 1980. The new emphasis will be on longer range, higher risk, advanced technology development. Past programs that were basically commercialization oriented were phased out in FY 1979. Activities of this nature will be covered in the future by new ERA pricing regulations. The affected programs are (1) California emissions problems; (2) alternate fuel boilers; (3) California heavy oils alternate uses; (4) steam drive in light oil; (5) CO_2 supply studies.

The bulk of the FY 1979 program involved continuing the activities began in FY 1978.

Solar steam generation for thermal EOR was a new area of investigation in FY 1979. A feasibility study will be conducted to analyze the problems associated with this application. A demonstration of this process was procured during FY 1979 as part of the Industrial Process Heat Program under the Assistant Secretary for Conservation and Solar Applications. This program will receive technical and financial (depending on feasibility study results) support from the EOR Program.

Two new areas of activity are planned to begin in FY 1980: a demonstration of solar steam generation for thermal EOR and research into microbial decomposition as a potential EOR method.

Work was also initiated to evaluate the possibilities for EOR in Outer Continental Shelf reservoirs. This work includes identification of optimum target reservoirs and evaluation of specific environmental and structural problems.

Table II 4-2 summarizes the funding levels by subactivity for the FY 1978 to FY 1980 period.

Micellar-polymer/Improved Waterflooding Process

Micellar-polymer flooding consists of injecting an appropriate surfactant dissolved in water to displace the residual oil from pores by preferentially wetting the solid matrix. (See Figure II 4-3.) This is followed by a polymer solution that forms a mobility-control bank or buffer in front of the final drive water; the buffer is needed to prevent dilution of the surfactant. Seven micellar-polymer flooding projects are underway in the United States, each in a different state. This process may be simplified, with some decrease in effectiveness, by adding chemicals such as only polymers or caustic soda to the water of a normal waterflood. The oil recovered may be less than from the micellar-polymer process, but may cost less per barrel. Five improved waterflooding projects are underway: three in California, one in Oklahoma and one in Illinois.

An extensive micellar-polymer applied research program is being conducted to evaluate such areas as chemical stability, displacement efficiency, injectivity, and overall performance of alternative systems. The reservoirs of interest include sandstones in California, the Rockies, the mid-continent, and the Gulf Coast. The main goal is to determine the technical validity of the micellar system under a variety of reservoir conditions.

The research program which also includes such problems as polymer adsorption onto rock, caustic precipitation, emulsion studies, and product improvement, began in FY 1978. A reservoir identification for California and mid-continent sandstone reservoirs was initiated, as well as a supply and demand study for process chemicals. A contract was signed with CONOCO to conduct a large pilot in Wyoming, and the 11 ongoing micellar-polymer and caustic flooding pilots continued.

The micellar-polymer flooding program is being redirected beginning in FY 1980. Emphasis in the past has been on mid-continental and California sandstones. Beginning in FY 1980, the primary objective will be to develop a viable micellar-polymer process for Gulf Coast and Outer Continental Shelf (OCS) sandstones. These reservoirs typically have higher temperatures and salinity levels which current micellar-polymer flooding chemicals cannot tolerate. In FY 1980, this program will be composed of applied research programs

Table II-4-2. Funding for EOR for FY 1978-1980.

ENHANCED OIL RECOVERY	BUDGET AUTHORITY (DOLLARS IN THOUSANDS)		
SUBACTIVITIES	ACTUAL FY 1978	APPROPRIATION FY 1979	ESTIMATE FY 1980
Micellar–Polymer Flooding	$12,500	$10,360	$ 5,200
Carbon Dioxide Flooding	6,600	6,720	3,200
Thermal Recovery	12,000	19,430	4,900
Improved Waterflooding	3,500	4,680	0
Environmental Studies and Support	11,340	12,172	7,600
Capital Equipment	300	700	500
TOTAL	$46,240	$54,072	$21,400

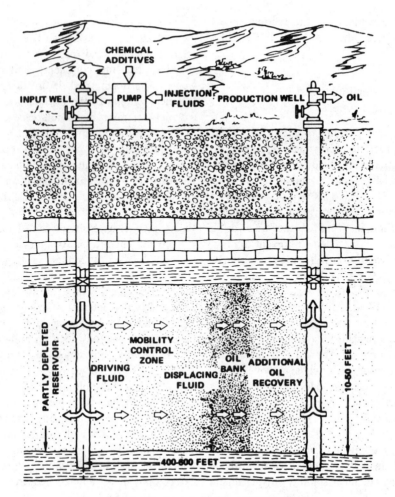

Figure II-4-3. Micellar-polymer/improved waterflooding.

aimed at developing polymers and surfactants capable of withstanding these severe conditions.

Because there is still considerable room for improvement in current micellar-polymer processes, an active applied research program will be maintained for further development. Caustic flooding studies will continue to determine precipitation effects and corrective measures. Analytical techniques will be developed to reliably determine chemical adsorption in reservoirs which will allow the design of optimum slug size and concentrations. Solutions to emulsion problems in target reservoirs will continue to be developed. Investigations to determine optimum preflush requirements and the actual mechanism of micellar-polymer flooding will continue.

These funds will be spent through the Bartlesville Energy Technology Center.

The closely related improved waterflooding projects are considered a variation of the micellar-polymer process and were originated in FY 1975 and 1976. Emphasis since FY 1977 has been on the two-step micellar-polymer process.

The polymer-augmented waterflooding process consists of using a chemical additive (usually a high molecular weight polymer) in waterflooding to enhance oil recovery. The polymer increases the viscosity of the water resulting in a thickened material that is less likely to disperse into non-oil bearing zones, and improves mobility control by attempting to achieve a piston-like displacement between injection and production wells. The amount of oil displaced from the reservoir and recovered is thereby increased.

These projects are economically submarginal at today's oil prices, due to the high costs of injection fluids, facilities, and additional drilling. DOE's financial participation with industry is necessary to improve and test both the improved waterflooding process and the micellar-polymer process.

The planned FY 1980 program will emphasize process development research, particularly in Gulf Coast (and OCS) and midcontinent sandstone reservoirs.

The objective of these micellar-polymer and improved waterflooding field tests is to demonstrate the efficiency and economics of recovering tertiary oil from reservoirs already depleted by waterflooding. Laboratory tests conducted by the Bartlesville Energy Technology Center (BETC) and DOE-sponsored universities indicated that, under optimum conditions, a high percentage of residual reservoir oil can be recovered by a micellar flood. To a lesser extent, improvement can be had by adding chemicals. Data were collected on the recovery from cores using various micellar and polymer fluids, and injectivity tests were carried out with positive results. This and similar work in industry laboratories led to contracts to evaluate the technique on a larger scale (in parallel with similar private field testing by oil companies).

The objective of the micellar-polymer field project test by Gary Operating Company at Bell Creek Field, Montana (Micellar-Polymer), is to demonstrate the technical and economic feasibility of recovering tertiary oil from a successfully depleted waterflood using a micellar-polymer flooding process. The drill sites have been approved and the plan for drilling completed. Process research and simulation are continuing, and a one-week injectivity test has been completed with successful results. The project is on schedule. Activities planned for FY 1980 include completion of the polymer injection phase and initiation of water injection.

The objective of the test being performed by Cities Service Oil Company, El Dorado Field, Kansas, is to demonstrate the efficiency of recovering tertiary oil from a successfully depleted waterflooded reservoir by using a micellar fluid followed by a mobility-controlled buffer zone to be displaced by water. Preflush and micellar injection into the formation is completed, and polymer injection is in progress in two side-by-side field patterns using two competing technologies. The project is running slightly behind the original schedule. Injection will continue throughout FY 1980.

The objective of the micellar-polymer test being performed by Phillips Petroleum Company, Burbank Field, Oklahoma, is similar to that of the Cities Service project discussed above. The entire surfactant and polymer slugs have now been injected, and the injection of fresh water is continuing. The project is on schedule, and encouraging stimulation of oil is appearing. The project was completed late in FY 1979.

The objective of the test being performed by Penn Grade Crude Oil Association, Bradford Field, Pennsylvania, is similar to those above, except that this test will demonstrate the effectiveness of the process in a low-permeability reservoir. Injectivity studies made during Phase I have established the overall feasibility of the project. Seventeen new wells have been drilled and completed so that the tertiary flood on the 25-acre Lawry tract can be accomplished. The micellar solution injection is complete, and polymer injection is continuing. The project is running slightly behind schedule, with completion planned for FY 1981.

The objective of the field demonstration being performed by the City of Long Beach, Wilmington Field, California, is to design a micellar-polymer system that will displace economic quantities of viscous oil otherwise not recoverable from unconsolidated sands. All wells have been drilled and lab tests of the process are continuing. Contracts are being negotiated with suppliers of the sulfonate, co-surfactant, and polymer.

The objective of the demonstration being performed by Marathon Oil Company, Robinson Field, Illinois, is to conduct a commercial-scale test of the micellar-polymer flooding process for improved oil recovery. It is 402 acres in size. The project has progressed smoothly and on schedule. All facilities including injection and production wells have been completed, and the micellar slug is being injected into both the 2.5- and the 5-acre patterns. Micellar slug injection was completed in November 1978, with polymer injection beginning at that time.

The objective of the micellar-polymer field test being performed by Continental Oil Company, Big Muddy Field, Wyoming, is to demonstrate the commercial feasibility of the micellar-polymer process in a typical Rocky Mountain reservoir with low matrix permeability, low salinity, and which fractures at bottom hole injection pressures near hydrostatic. Drilling and design of the chemical injection plant are complete. Injection of the preflush began in FY 1979, and chemical injection will begin during FY 1980.

The objective of the improved waterflood field test of Kewanee Oil Company, Stanley Stringer Field, Oklahoma, is to demonstrate the efficiency and economics of recovering tertiary oil from a highly heterogeneous reservoir after waterflooding by using a polymer slug of tapered concentration to improve the sweep efficiency of the reservoir. This project is on schedule and is showing promising increased production rates. Polymer injection is complete and brine injection will continue through FY 1982.

The objective of the improved waterflood field project of Shell Oil Company, East Coalings Field, California, is to demonstrate the relative merits of polymer flooding and waterflooding in a medium-viscosity oil reservoir which has an unfavorable water-displacement-mobility ratio. Polymer injection was initiated in May 1978. This project is moderately behind schedule but has required additional engineering and control facilities because of the multiple oil sands.

The objective of the improved waterflood test by the City of Long Beach, Wilmington Field, California, is to demonstrate the effect of injecting caustic solution and salt solution to sweep the residual oil from a formation that has been waterflooded. Laboratory core floods and modeling are continuing. The laboratory work has taken more time than expected and the project is behind schedule. Construction of the preflush injection facilities is complete, and installation of the caustic tankage was complete by the end of CY 1978.

Aminoil's improved water flood test in the Huntington Beach field in California was started in FY 1976 and joined by DOE in FY 1977. It is a caustic soda flood in a reservoir moderately different from that of the City of Long Beach with respect to reservoir characteristics and oil properties. Injection of the salt water preflush began in May 1978 and is continuing; caustic injection was initiated in late FY 1979.

Energy Resources Company, Inc. is performing a field test in the Storms Pool Unit at Carmi, Illinois, which will use polymer and water to produce light gravity oil from a depleted waterflood. Energy Resources Company (ERCO) is a high-technology group that has joined an independent oil producer. The project was funded in FY 1976, but delayed until FY 1977 by the need for Small Business Administration funding by the contractor. The final contract between the landowner and ERCO has been signed. Due to the financing and land problems, this contract is significantly behind schedule. Planned FY 1980 activities include the initiation of polymer injection.

Major milestones for each of these products are given in Figure II 4-4.

Carbon Dioxide Flooding

There are seven mechanisms in the CO_2 displacement process believed to contribute to improved oil recovery, with oil viscosity reduction and oil swelling (up to 40 percent) being the most important. It is possible to introduce CO_2 into the reservoir by several means: pure CO_2, carbonated water, acid gas (H_2S and CO_2), separator gas (CO_2 and hydrocarbon gas), flue gas, or CO_2 enriched with intermediate hydrocarbons CO_2–C_5). The process shown in Figure II 4-5 can be either miscible or immiscible, and ancillary materials may be added to improve mobility ratio displacement and sweep efficiencies.

The main DOE CO_2 project is to develop and improve the CO_2 process in carbonate reservoirs having a high residual oil saturation. Some of the important goals are to establish the range of physical conditions under which CO_2 is miscible with the crude oils in target reservoirs; the nature of additives that might promote miscibility; and, where miscibility is not achieved on first contact, the point in the reservoir at which miscibility can be expected to be achieved. It is possible that miscibility need not be achieved at any point to be effective. The DOE program in carbonates began in FY 1978; no field tests have begun yet. Three of the ongoing field tests are in West Virginia sandstones, and one (Shell) is in a Louisiana sandstone.

The Shell pilot is unique in that this reservoir is very similar to many offshore reservoirs in the Gulf. As such, the results of this test can be extrapolated to predict the performance of CO_2 flooding in offshore reservations.

Applied research into phase behavior and mobility control of carbon dioxide was initiated in FY 1978. A series of interfield analyses began as part of a reservoir identification effort to determine the optimum target reservoirs for both DOE pilots and eventual commercialization of the process. Three mini-tests in Southwest and Montana carbonates were initiated to evaluate the potential of the carbon dioxide process in high residual oil carbonate reservoirs. Four ongoing carbon dioxide pilots in West Virginia and one in Louisiana were continued. The Louisiana pilot is particularly significant in that its results can be extended to offshore fields due to similarity of reservoir characteristics.

This program in phase behavior of CO_2-Brine-Crude systems will be near completion in FY 1980 and will continue at a minimal level as required for necessary additions to the data base. The primary emphasis will be on the development of mobility control agents to improve the displacement and sweep efficiencies of the process, which are the major process problems. Two field mini-tests will begin in FY 1980 to test mobility control methods that have been developed. These tests will determine the effectiveness of the new processes and identify areas where additional research and development are necessary.

These funds will be spent through the Morgantown Energy Technology Center.

The objective of the Pennzoil Company (Rock Creek Field, West Virginia) project is to determine the efficiency and economics of recovering oil from a shallow, low-temperature reservoir using carbon dioxide and water to displace oil. The injection wells have been completed in the producing formation, and field pressure is being raised by water injection. CO_2 was injected in mid-1978; water injection began in FY 1979 and will continue through FY 1981.

The objective of the Guyan Oil Company (Griffiths Field, West Virginia) project is to demonstrate the efficiency and economics of recovering oil from a shallow, low-temperature reservoir using carbon dioxide and water to displace the oil. The new wells are drilled and the CO_2 storage facility is operable. Significant delays have resulted from the necessity to perform sizable environmental protective steps such as prevention of siting and erosion and plugging of old wells. Water injection to build up pressure is continuing. The project is running roughly 18 months behind its original schedule.

The objective of the project by Columbia Gas Transmission (Granny's Creek Field, West Virginia) is to demonstrate the efficiency and economics of recovering oil from a shallow, low-temperature, watered-out reservoir using carbon dioxide and water to displace the oil for tertiary recovery. Carbon dioxide injection is complete and chaser water injection is continuing. There has been some increase in oil production as a result of this injection. Results are not conclusive, however, and it is necessary for further geological and computer data to be evaluated to improve production efficiency. The project was completed during FY 1979.

The project by the Shell Oil Company (Weeks Island Field, Louisiana) was funded late in FY 1976 and is the only EOR project in the deep, high-temperature and high-pressure sands of the Gulf Coast. This project is to use the new technique of employing CO_2 (mixed with methane to make it gravity-

288 Fossil Energy

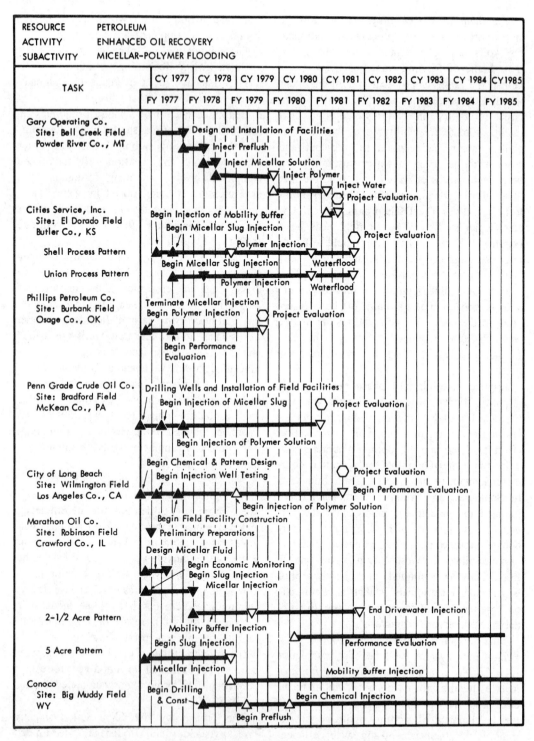

Figure II-4-4. Major milestones for the micellar-polymer flooding subactivity.

stable) as a miscible agent to displace residual oil in the "S" Sand. The objective is to determine the feasibility of the CO_2 miscible displacement process in the steeply dipping, loosely consolidated high-pressure and high-temperature oil zones of the Gulf Coast. The

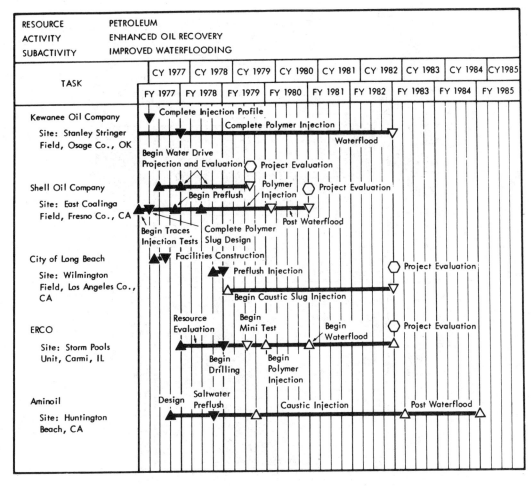

Figure II-4-4. Major milestones for the micellar-polymer flooding subactivity. (Continued).

field work was completed, and CO_2 injection began during the last quarter of FY 1978. The project was delayed roughly seven months by an audit and other contractual details.

Major milestones for these projects are presented in Figure II 4-6.

Thermal Recovery

Thermal processes are effective in the recovery of heavy oil and tar sands by using heat to reduce oil viscosity and cause it to flow in the reservoir. Two basic thermal-recovery methods are steam flooding and in situ combustion.

In steam displacement, steam is injected continuously into a heavy oil-bearing formation. As steam passes through the formation, it gives up heat, thereby reducing viscosity of the oil to make it more mobile and displaceable. Steam is effective in relatively shallow, heavy oil reservoirs with high oil content but very low primary recovery. Steam-displacement tests began in FY 1976 and continued throughout FY 1979.

In situ combustion (forward or reverse, dry or wet) is a process involving high-temperature oxidation in the reservoir itself by continuous injection of air or air and water. (See Figure II 4-7.) During the process, part of the oil is converted to coke which is then consumed to produce more heat. The water in the formation is turned to steam that, in turn, distills some of the oil ahead, making it more mobile and causing it to move more readily to the producing wells. Six thermal recovery projects are currently underway.

Pilot tests began in FY 1979 to test the

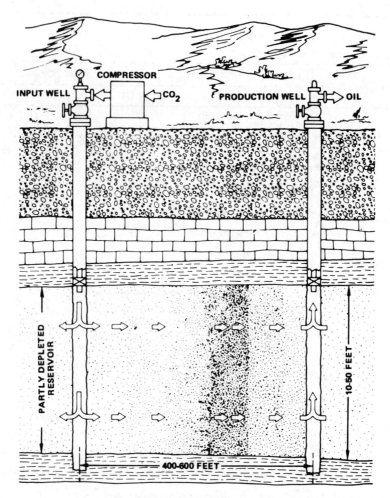

Figure II-4-5. CO_2 injection process.

steam drive process in a light oil reservoir and in a heavy oil reservoir using additives to improve sweep efficiency. (See Figure II 4-8). FY 1980 program emphasis will be on the development and testing of additives to improve sweep efficiency, on the development and testing of downhole steam generators for deep reservoirs, and on the steam drive process in tar sands.

In FY 1978, studies were initiated to assess the impact of air quality regulations and West Coast market constraints on thermal EOR. Three demonstrations of alternative fuel boilers for EOR use were also initiated. Applied and supporting research in developing steam additives, studying steam distillation in light oil, and testing the steam drive process in tar sands were initiated or continued. Procurement packages were initiated to conduct pilot tests of the steam drive process in a light oil reservoir and in heavy oil reservoirs using additives. An in situ combustion pilot in a heavy oil reservoir also was started. Development of downhole steam generators and better well completion techniques to extend steam drive beyond the current 2,500-foot depth limit also began. Four ongoing steam drive pilots and two ongoing in situ combustion pilots were continued.

The DOE thermal recovery program has been greatly scaled down in FY 1980. The prior year programs involving steam generator emission problems, alternate fuel boilers, steam drive in light oil reservoirs, and market constraints on California heavy crudes will be discontinued. The steam drive in light oils

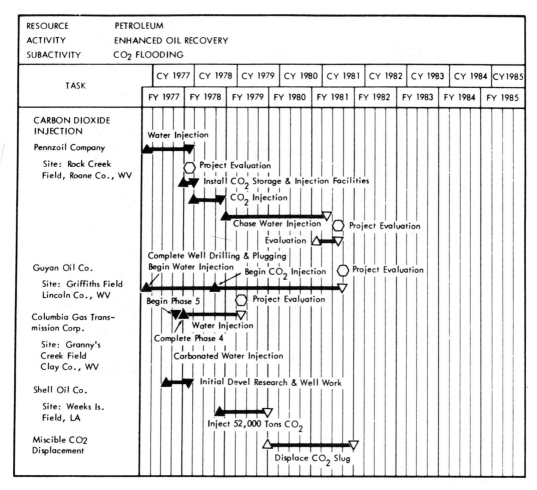

Figure II-4-6. Major milestones for the CO_2 flooding subactivity.

project is felt to be restricted by economics, not technology problems; therefore no DOE role is necessary. The other projects are no longer thought to be research and development problem areas.

Three areas in thermal recovery remain where an active DOE role is necessary because of little or no industrial activity. These three areas are the development of additives to improve the sweep efficiency of steam drive, extension of the depth limits on steam drive (deep steam project) and development of a viable thermal process for tar sands. Applied research will be continued to develop additives to improve the efficiency of steam drive. This will include continuing the ongoing research at Stanford and the University of Southern California as well as any new concepts that may develop. The Stanford research is expected to be ready for a field mini-test in FY 1980. This will be conducted on a cost-sharing basis with an interested oil company; several companies have already indicated interest. In addition, two mini-tests of industry-developed processes are scheduled for procurement in FY 1979. The ongoing project to extend the depth limit of steam drive will continue with laboratory tests of alternate downhole steam generator concepts and well-bore insulation techniques with selection of the best alternative(s) for each planned at the end of FY 1980. Applied research on thermal techniques for recovering oil from tar sands will continue, as will a steam drive mini-test planned for procurement in FY 1979.

The steam additives funds will be spent

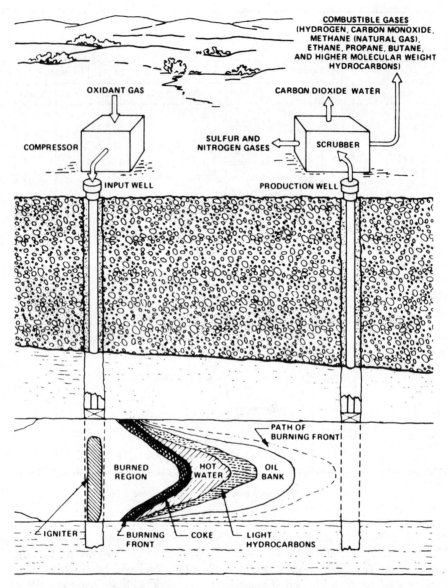

Figure II-4-7. Oil process in situ.

through the San Francisco Operations Office. The deepsteam funds will be spent through Sandia Laboratories in Albuquerque and the tar sands funds will be spent through the Laramie Energy Technology Center.

The Husky Oil Company is field testing the combination thermal drive method of enhanced oil recovery and has completed preparation for full field testing at the Paris Valley Field Project, Monterey County, California. Surface facilities, including the air compressor hook-up and a new steam generator/SO_2 emission scrubber, were completed in early FY 1977. The formation has been ignited, and air is being injected to advance the burn front. Due to several compressor problems and failures and the necessity to have the local utility install a new electrical substation (unplanned), this project is approximately two years behind the original schedule.

The Cities Service Oil Company is field testing improved oil recovery by in situ combustion in the Bellevue Field, Louisiana. All facilities were completed and were in produc-

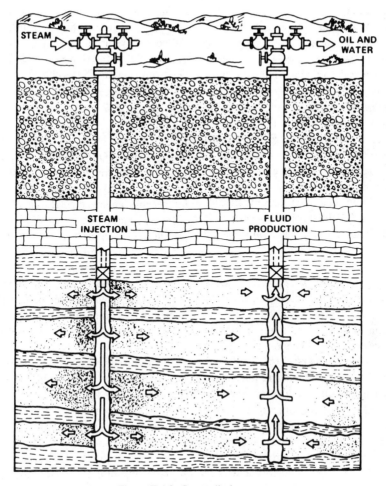

Figure II-4-8. Steam displacement.

tion on schedule. In situ combustion has been initiated in several air injection wells, and promising increases in production have been obtained. A combination of air and water are being injected. The project is on schedule.

The General Crude Oil Company, Lynch Canyon Field, California, is conducting a demonstration of the combination thermal drive process in a reservoir containing a very viscous crude oil. Due to the oil properties, it will be necessary to stimulate the producing wells with cyclic steam treatments prior to the arrival of the heat front. The process to be used is similar to the DOE/Husky pilot test mentioned above. However, the greater depth of the General Crude project will permit technical and economic evaluation at a more typical depth for fireflood operation. The project is also making a concerted effort to use directional permeability to enhance the process efficiency. General Crude has also developed an extensive data accumulation system which will allow burning front velocities and directional movements to be monitored. The contract was signed in September 1978, and formation ignition is planned for FY 1980.

Getty Oil Company has started a steam drive project in Cat Canyon Field, California. Four inverted five-spot patterns are involved that encompass 20 acres. All producing wells are being cyclically steam stimulated to improve the oil mobility until steam from the injectors breaks through. The estimated recovery from the project is 850,000 bbl of oil, and the project is well underway, although delayed a few months by the approval of an environmental permit for the steam generators. Chanslor-Western Oil Development Company has begun a steam drive project in Midway Sunset Field, California, to demonstrate the operational and economic aspects of steam-flooding a typical, low-pressure, flat, heavy oil reservoir that had unfavorable response to cyclic steam stimulation. Four inverted seven-spot drive patterns are under test with expansion to 14 patterns planned if justified by the results. Encouraging incremental oil production increases are now appearing, but they are about one year later than predicted. As a result, the project expansion is being delayed until more production data are obtained.

Carmel Energy, Inc., is demonstrating the efficiency and economics of recovering heavy oil by using the Vapor-Therm process. The process utilizes a mixture of combustion gases and water vapor to stimulate the production of low-gravity viscous oil. Construction of the commercially sized Vapor-Therm Unit was completed, and the injection of the hot vapors was applied successfully in Kansas. The results were significantly encouraging to stimulate private expansion in surrounding areas (265 additional wells are planned). To demonstrate the applicability of this process to other heavy oil reservoirs, Carmel Energy and DOE are conducting a second test of the Vapor Therm process in Vernon County, Missouri. The wells for this test are currently being drilled and cores evaluated. Injection began in early FY 1979, with project completion planned for FY 1980.

Major milestones for in situ combustion and steam flooding are given in Figure II 4-9.

Improved Waterflooding

This process is now considered to be a commercially viable technique, and no further DOE involvement is deemed necessary in view of the changed emphasis to a longer range, higher risk, advanced technology program. There are currently five projects in the improved waterflood category. All were started under the somewhat different objectives of prior years; three will be completed within several months, the remaining two will be allowed to run their course and no new improved waterflooding projects will be started. All projects are considered successful, particularly the polymer flood by Kewanee Oil Company in Oklahoma.

Environmental Studies and Other Support

The Enhanced Oil Recovery Activity includes controls to ensure that every field project will be conducted in accordance with existing and foreseeable environmental, health, and safety (EH&S) requirements. An EDP is planned to (1) identify and define EH&S requirements, problems, and milestones; and (2) to plan how these requirements will be fulfilled, how problems will be solved, and how milestones will be completed. The plan will be conducted in accordance with Federal, state, and local statutes and regulations for EH&S protection.

Enhanced oil recovery projects are generally undertaken in oil fields that have gone through years of primary and secondary production. Consequently, significant baseline environmental information is available for enhanced oil recovery projects to ensure that adverse environmental effects are not overlooked. Planning for resolving potential environmental problems will be part of this work, and projects to solve specific environmental problems will be initiated as required.

The primary objective of supporting research is to provide the appropriate contract and project support to assist the Enhanced Oil Recovery Activity. The strategy is to sponsor basic and applied research at various universities, national laboratories, Energy Technology Centers (ETC's), and industrial organizations. It includes research support and contract monitoring by the ETC's and development and analysis of EOR programs. Other supporting research includes residual oil determination studies and comparisons and an aggressive technology transfer program.

Environmental studies and other support activities, including residual oil studies and technology transfer, will continue at a level commensurate with the reduced program level.

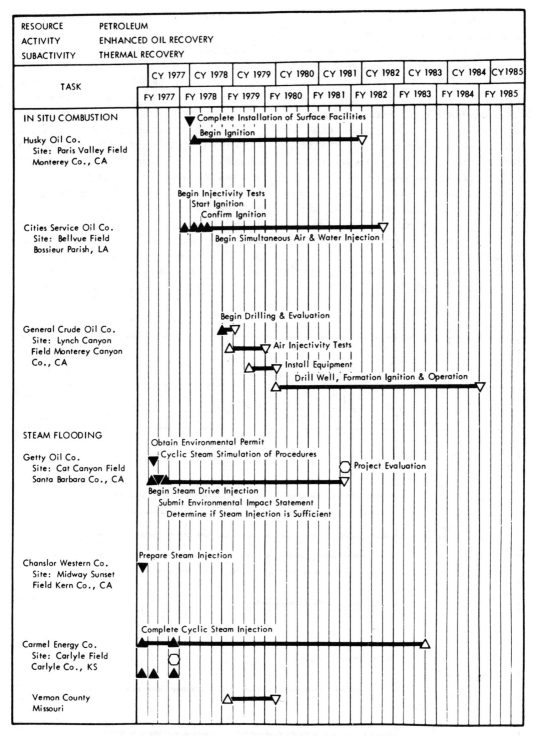

Figure II-4-9. Major milestones for in situ combustion and steam flooding.

Two new areas of investigation are included. Investigations will begin into microbial decomposition as a potential enhanced oil recovery technique. The initial activity will be a state-of-the-art survey followed by an applied research program directed toward the needs identified by the survey. Funds are also included for solar energy applications for enhanced oil recovery. The main activity will be for support of a solar steam generation project for thermal enhanced oil recovery. This project was procured in FY 1979 as part of the Industrial Heat Program under the Division of Conservation and Solar Applications with both technical and financial support from the EOR activity. Preliminary investigations will also be made of the potential of certain novel approaches such as inert gas injection, the application of microwaves and electrical enhanced oil recovery systems. Laboratory work would be subject to promising results from the preliminary findings.

Major milestones for this subactivity are shown in Figure II 4-10.

OIL SHALE

The nation's abundant oil shale resources are a major potential source of energy. The technology for conventional mining and surface retorting of oil shale has advanced to a level believed to be capable of early commercial application to the richer grades of oil shale, averaging over 25 gallons per ton, although commercial-scale equipment and operations have not been demonstrated. Present DOE activity in this area is focused mainly on R&D to develop in situ extraction processes that are capable of exploiting the leaner and deeper shales as well as offering an alternative to the surface technologies. In addition, DOE is conducting laboratory-scale supporting research on problems relevant to general oil

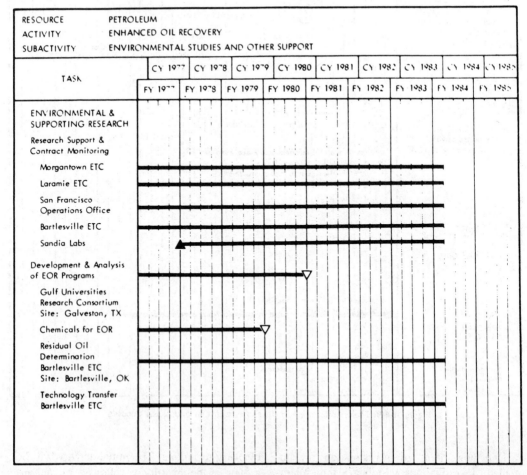

Figure II-4-10. Major milestones for environmental studies and other support.

shale processing and environmental concerns, both surface and in situ; contracting for development of advanced processing concepts at the bench-scale; providing general consultation to industry and technical information transfer via research publications; resource characterization; and making the Anvil Points facility available for industry-sponsored projects.

The near-term objective of the Oil Shale Activity is to provide the technology base for initial commercial implementation by 1985 of first generation process concepts that have progressed to pilot stages of development. This objective will be promoted by focusing research and development on critical technology problems associated with present industry process development efforts, in active cooperation with industry.

The longer-term objective (1985–2000) is to provide substantial improvements in the economics and environmental acceptability of shale oil and gas recovery through research on advanced surface and in situ process concepts that show promise of significant breakthrough in energy efficiency, resource recovery, and/or reduced environmental emissions and residuals.

The present Oil Shale Activity includes oil production research involving true in situ methods that require no mining; modified in situ methods that enhance the void volume and permeability, by mining a portion (15–25 percent by volume) of the mineral material from the oil shale formation and creating a rubblized volume; advanced processing technology with potential for reducing environmental impacts and improved energy efficiency of second-generation commercial plants; production research targeted to both Eastern and Western oil shales; environmental studies related to the in situ processes being developed; and supporting research for oil shale processing that is not tied to any specific technology but rather to oil shale development problems in general.

Advantages of in situ methods are relatively lower water, manpower, and shale disposal requirements. The potential application to lower grade shales is important in that two-thirds of the total estimated resource of 1.8 trillion barrels of shale oil in the western Green River formation is in low-grade deposits that may never be recovered by techniques that involve conventional mining. Laboratory-scale research on new process technology includes applied research on methods potentially offering similar advantages for above-ground processing.

Although in situ conversion of oil shale to liquid products offers several potential advantages over conventional mining and surface processing, no single version of in situ technology is applicable to the full range of oil shale deposit types. True in situ techniques rely upon an interconnected system of cracks artificially produced by hydraulic and/or explosive fracturing to provide the required permeability for passage of gases and liquids. No mining or other methods of enhancing void volume by mineral removal are required. Several techniques for creating the required fracture system are under active investigation in both Eastern and Western oil shale field projects. Although true in situ methods are generally considered limited to oil shales grading less than 30 gallons per ton, in thickness up to approximately 250 feet, and at depths less than 1,000 feet, these parameters describe about half of the Western oil shale resource and are characteristic of much of the eastern organic shales of Devonian and Mississippian age that range from Pennsylvania to Texas in a broad but poorly defined belt. True in situ techniques offer lower environmental impacts than most modified in situ methods, but will probably be the most difficult in situ technology to develop successfully.

Vertical modified in situ techniques are believed to offer the most promise for early development, and are especially promising for deeper and very thick (more than 300 feet) deposits. Certain modified methods also are potential alternatives for the shales to which true in situ methods would apply. Four major variations of modified in situ are of interest, the choice of the particular variation being partially governed by actual thickness of oil shale to be processed. Two of these variations are conceptually similar in that an initially mined void volume is subsequently redistributed throughout a much larger mass of rubblized shale. Physical dimensions of the completed retort modules have led to the general description of these processes as "vertical" or "horizontal." Two versions of the vertical process are under active development, with one having reached the stage of testing modules (Occidental contract) of approximately the size required for commercial operations. Both would utilize a downward moving combustion zone to process the shale. Two processing concepts have been proposed for "horizontal" retorts, one using a downward moving combustion zone and the other using a laterally moving combustion zone.

Another modified horizontal technique (Geokinetics contract) applicable only to very shallow depths (up to 100 feet of overburden) does not use conventional mining but instead uses proprietary blasting patterns to expand a bed of rubblized shale into space originally occupied by overburden. The raised overburden is sealed to prevent gas leakage before processing commences. This method applies to deposits typical of large areas of Utah oil shale. Field tests to date in PDU-size modules have been very promising.

A semivertical modified in situ process (Equity Oil Co. contract) is under development for processing the Leached Zone in Colorado that utilizes the void space created by dissolution of water-soluble salts originally distributed throughout this approximately 500-foot thick section of oil shale. That zone contains an in-place resource equivalent to over 1 MMbbl/acre of shale oil. Because this zone is a salt-water containing aquifer, direct-combustion methods are not considered applicable. The concept being tested involves passage of superheated steam through the full thickness of the Leached Zone. Preliminary evaluation tests and calculations of energy balances and economics indicate probably economic success if at least 50 percent of the in-place kerogen is recovered as shale oil.

The direct products expected from in situ retorting operations are as follows:

- Shale oil syncrude that can be refined into the various products (refinery fuelgas, jet fuels, LPG, and coke) needed in the commercial market
- Low- to medium-Btu gas that can be used on-site in a gas turbine generator to provide electricity for process operation and for sale
- Ammonia and sulfur products that are recovered from a normal shale oil refining scheme

The mechanism for technology transfer derives from DOE participation in all phases of the technology development. The final phases of development are the pilot field demonstrations, which DOE cost shares with industry. This provides maximum flexibility and utilization. The in-house supporting research program is designed to be all-inclusive with respect to developing fundamental data, laboratory and small pilot testing, and technical analysis of results.

Table II 4-3 summarizes the funding levels by subactivity for the FY 1978 to FY 1980 period.

Shale Oil

Major Field Tests. In the production of oil from oil shale, major accomplishments for FY 1979 and anticipated progress during FY 1980 are as follows:

Four field tests are underway via four major in situ Government/industry cost-sharing contracts.

Phase I of the contract to Occidental Oil Shale, Inc. (Debeque, Colorado) will be completed with the completion of retort No. 6. Phase II, a two-module demonstration originally planned for Colorado lease tract C-6, has been postponed due to mine shaft development delays. Alternative Phase IA proposal is under consideration for FY 1980.

Phase IV of the contract to Geokinetics Oil Shale Group (Uintah County, Utah) was completed during FY 1979. Phase V, consisting of four larger retorts in a group, will be completed during FY 1980. Oil yields as high as 70 percent Fischer assay have been achieved.

Injection of superheated steam into the leached zone began in mid-FY 1979 by Equity Oil Company (Rio Blanco County, Colorado). Oil production is expected to begin during FY

Table II-4-3. Funding for oil shale for FY 1978-1980.

OIL SHALE SUBACTIVITIES	BUDGET AUTHORITY (DOLLARS IN THOUSANDS)		
	ACTUAL FY 1978	APPROPRIATION FY 1979	ESTIMATE FY 1980
Shale Oil Production	$24,032	$43,350	$23,500
Shale Gas Production	3,850	4,350	3,500
Capital Equipment	900	900	1,200
TOTAL	$28,782	$48,600	$28,200

1980 and should continue throughout FY 1980-81.

Phase I of the contract to Talley Energy Systems (Rock Springs, Wyoming) was completed during FY 1979 with the fracturing and evaluation of the site. Phase II retorting contingent on evaluation results was completed in mid-FY 1979.

The locations of DOE and industry energy projects in the oil shale region are pictured in Figure II 4-11.

Laboratory Field Tests. During FY 1979, a number of research projects were conducted to support the national program. These include projects in modified in situ, true in situ, shale gasification, and environmental aspects. Laboratory and pilot-scale retorting studies are being conducted in above-ground equipment by the Laramie Energy Technology Center and the Lawrence Livermore Laboratories (LLL). The primary objective of these studies is developing and testing mathematical models that will be sufficiently general for application to large-scale in situ retorts.

A continuing program of field studies of in situ oil shale processing is being conducted by the LETC. Fracturing the oil shale in place without mining or removing material by men working underground is a major problem to be solved before true in situ retorting can become technically feasible. The fracturing problem is under investigation with much of the explosive fracturing design and evaluation work being conducted by Sandia Laboratories (SL) and Los Alamos Scientific Laboratories (LASL). SL also participated in instrumental evaluation of the Geokinetics, Talley, and Dow projects in FY 1979.

In another effort to speed the development of shale technology, working relationships have been established with several major Western universities to conduct research on problems related to oil shale development. Training university studies in energy technology on major in situ oil shale projects is an important part of technology transfer.

Information on oil shale technology developed by DOE is readily available through reports and papers published in the open literature and through reports submitted to the Program Administrator for Fossil Energy. To ensure efficient technology transfer from the industrial concerns operating the major in situ oil shale projects described in more detail on the following pages, Government employees will be assigned to the projects for monitoring and evaluating the results of the contract research. This approach will provide the Program Administrator for Fossil Energy with the maximum amount of information and ensure that DOE maintains sufficient in-house exper-

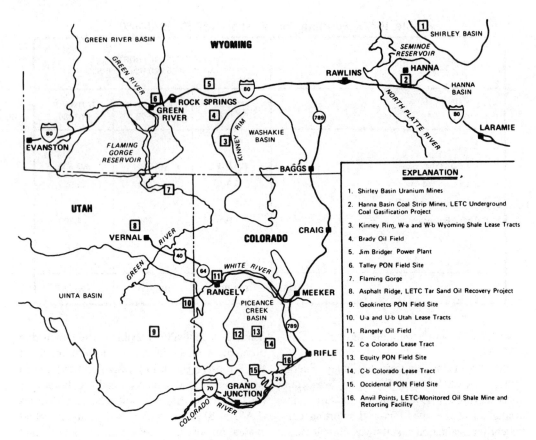

Figure II-4-11. DOE and industry energy projects in oil shale region.

tise to conduct an efficient energy development program.

The true in situ shale oil recovery process shown in Figure II 4-12 is characterized by fracturing techniques that require no mining or removal of major amounts of oil shale. Past experiments conducted by the Government used a combination of hydraulic and explosive fracturing to produce permeability in the oil shale bed as well as a particle-size distribution capable of sustaining combustion reactions. In these experiments, sand-propped hydraulic fractures were loaded with liquid or slurried explosives. The explosives were then detonated to further fracture the oil shale. Government and industrial experiments have also used explosive charges emplaced in drilled holes and detonated to fracture the oil shale. Other fracturing techniques have been used with somewhat less success; new and better methods may develop from the current research.

The fractured oil shale bed can be retorted by two general methods. The shale can be ignited at the bottom of the injection well and combustion sustained by air injection, in which case hot combustion gases retort the shale. In some cases, it is advantageous to supplement the air supply by injecting propane, recycled gas, or some other fuel to enhance combustion. In the second method, energy for retorting the shale can be supplied by injecting heated gases. The gases considered for use in this process are steam, natural gas, nitrogen, and others.

In either method, products of retorting are recovered from the production well. Liquid products collected in the bottom of the well can be pumped to the surface. Liquid entrained in the exit gas stream can be separated and collected on the surface. Depending on the heating value of the gas stream, it can be used as recycle gas, burned as a source of fuel on the surface, or discarded through a flare to prevent pollution.

True in situ retorting research is considered

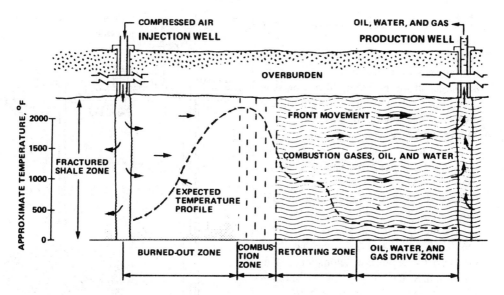

Figure II-4-12. True in situ retorting.

a high-risk research effort appropriate for Government support. If some of the problems can be solved satisfactorily, the benefits of true in situ processing may be large in terms of resource recovery and environmental acceptability.

The objective of this activity is to demonstrate the technical and economic feasibility of true in situ retorting techniques. These techniques promise to produce shale oil with minimal environmental impacts.

True in situ processing of oil shale is most likely to be applied to shales deposited in thin beds, possibly interspersed with barren rock. LETC has an experimental in situ project underway on such a shale deposit near Rock Springs, Wyoming. The retorting phase of one experiment (Site 9) was completed early in FY 1977, and efforts to recover additional oil were completed by the end of FY 1977. A second retorting experiment by LETC was planned for by FY 1978 at the Site 12 fracturing experiment completed in early FY 1978. Site 12 was retorted during FY 1979. These retorting experiments provide realistic field data for input to process control models and are not designed to produce extensive amounts of shale oil.

Additional projects are underway to continue the investigation of true in situ technology. As shown on the accompanying milestone chart, both fracturing and retorting experiments are in progress. Coordinated research by LETC, SL, and Los Alamos Scientific Laboratory (LASL) is addressing the problem of oil shale fracturing. In addition, projects to develop a model that interrelates fractures with rubble size in oil shale and to determine the effects of oil shale particle-size distribution on pressure drop are being conducted by LLL and LETC.

The difference between the modified in situ shale oil production processes and true in situ retorting methods is that between 15 and 25 percent of the oil shale or other minerals are mined or otherwise removed from within the retort to provide the void space for enhanced permeability when the remaining shale is rubblized, as previously discussed. If mined shale is oil rich, it will be sent to surface retorting; but if it is low-grade shale, it will be discarded. This can greatly influence mine designs and detailed development plans. Wells are drilled and prepared prior to fracturing the shale. After the oil shale is fractured through explosive techniques, a porous medium remains and retorting is begun.

Four general concepts of modified in situ techniques can be identified:

- Vertical modified in situ with partial mining in which the relative dimensions of the retort are larger in the vertical direction than in the horizontal, such as

a column (Occidental contract). (See Figure II 4-13.)
- Horizontal modified in situ with partial mining in which the relative dimensions are larger in the horizontal direction than in the vertical, such as a bed (in-house project). (See Figure II 4-14.)
- Modified in situ retorting of a zone in which minerals contained in the shale have been removed by naturally occurring groundwater (leached zone in Colorado) or by solution mining (Equity Oil Co. contract). (See Figure II 4-15.)
- Horizontal modified in situ retorting of a rubblized oil shale bed that has been prepared by explosive detonation resulting in noticeable surface uplift (Geokinetics contract). (See Figure II 4-16.)

Modified techniques that require partial mining followed by massive rubblization are believed to offer the most promise for deeper and very thick shale deposits. In these deposits the vertical configuration is most useful. A horizontal technique will be more useful in somewhat thinner deposits or as a secondary recovery method in a previously worked mine.

The objective of modified in situ retorting activities is to develop and test these underground retorting techniques to determine their technical, economic, and environmental feasibilities.

Major areas of research and development of modified in situ retorting include the following parameter investigations:

- Optimum setting of control parameters, such as flow and recycle, to ensure maximum energy recovery
- Mining and explosive blasting techniques to ensure desired void volume and optimum distribution of rubble permeability to maximize resource recovery
- Optimum underground support design to minimize surface subsidence
- Range of shale grade that best utilized the technology
- Retort configuration to maximize resource recovery with minimum pillar dimensions and spacing separating underground retorts

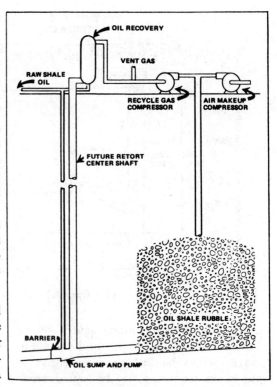

Figure II-4-13. Occidental oil shale process retort operation.

- Control of vertical or horizontal sweep of the retorting gases
- Environmental effects and control technology
- Economic feasibility
- Technology for sensible heat and additional hydrocarbon recovery from spent retort rooms
- Recovery of oil from the pillars.

Major problems still to be solved in the modified processes include scale-up to optimum size, some additional fracturing work, and determination of best levels of control variables. Three of the Program Opportunity Notice (PON) No. 2* contracts and a contract project in Michigan shale, include research in these areas.

A similar process for underground pyrolysis of oil shale has been the object of research at

---
*PON was issued in early 1976 regarding a Government/industry cost-shared project for the development of several in situ technologies. Four contracts were negotiated and awarded by October 1977.

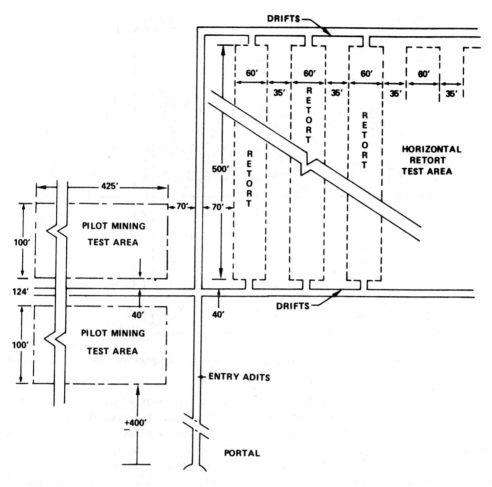

Figure II-4-14. Conceptual mining plan for modified horizontal in situ.

LLL. Laboratory work to investigate the chemical reactions involved in oil shale retorting as it relates to various-size shale particles, the change in permeability of deep oil shale beds at various temperatures, and how these and other variables contribute to an overall mathematical model of the modified in situ retorting process are currently under development. Results are being incorporated in designs of the Federal lease tract C-a development.

A preliminary research project by LETC on a horizontal configuration retort are being continued. Site selection investigations in Utah are continuing and additional engineering studies are pending further investigation and other contract results.

Major milestones for shale oil production are given in Figure II 4-17.

Environmental Aspects. There are several environmental concerns relative to oil shale development for both above-ground and in situ operations. Major concerns include significant land disturbance, spent shale disposal (including possible surface and ground-water contamination), water availability, water usage, and socioeconomic effects. Liquid effluents from oil shale processing such as retort waters and runoff from retorted shale disposal areas will be analyzed. Gaseous and particulate emissions from mining, drilling, surface preparation equipment, fines from shale processing, and off-gas from retorting operation will also be of concern. Underground water contamination from drilling, in situ burns, and underground migration of associated products will be a major environmental attention area.

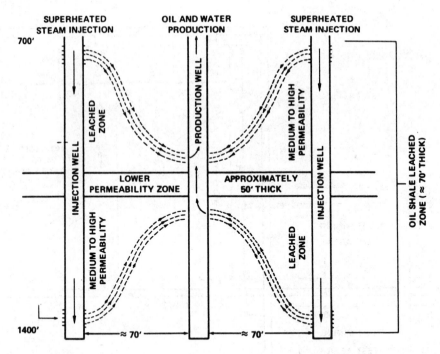

Figure II-4-15. Equity in situ project, idealized injection/production (current project is backed-up 5-spot pattern).

The environmental research being conducted at LETC and at selected universities is designed to address environmental aspects that are peculiar to in situ processing of oil shale. These include disposal of in situ water, migration of fluids during and after in situ processing, surface thermal changes, identification of the materials that can be extracted from in situ processed spent shale, and other environmental research that applies to site-specific problems of in situ processing. Site-specific environmental research includes geohydrological evaluation, socioeconomic studies, health effects determinations, occupational health studies, ecological effects studies, meteorological and air quality monitoring, compliance monitoring and support research, control system research, and total assessment of the environmental issues associated with the process development.

Environmental activities that address these issues and problems are an integral part of all field work conducted either in-house by DOE or by industry contractors. Preoperational assessments and baseline studies are used to analyze the environmental characteristics for in situ and modified in situ oil shale research sites, including DOE's Green River and White Mountain sites. Also, in-house studies of the effects of field operations are being conducted at Rock Springs Sites 9 and 12. Environmental research is a significant aspect of each of

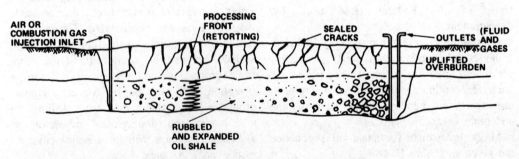

Figure II-4-16. Modified horizontal in situ process-GSG version.

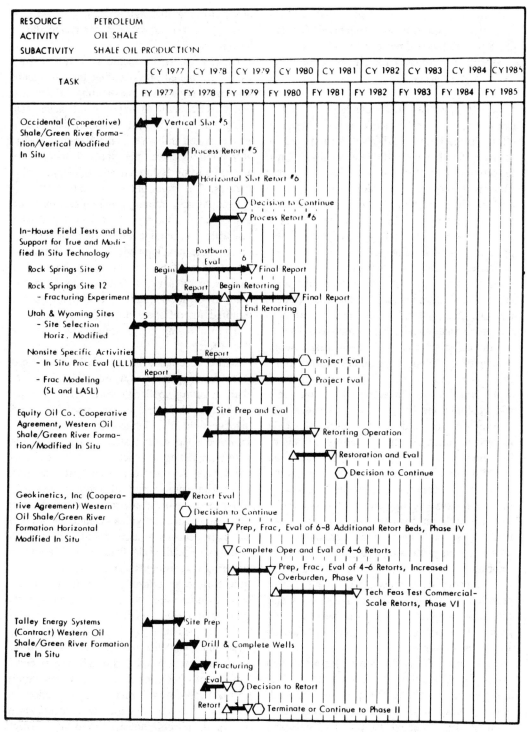

Figure II-4-17. Major milestones for shale oil production field tests.

the PON contracts. Each contractor has prepared an environmental research plan.

An Environmental Development Plan (EDP) has been formulated for the Oil Shale Activity. This plan provides for a comprehensive oil shale environmental strategy and an implementation plan to ensure that all environmental concerns are addressed for the Oil Shale Activity. Environmental Impact Assessments are prepared for all major DOE oil shale field projects.

Supporting Research. Research work on the characterization of oil shale is a continuing effort. In the future, LETC will be required to increase the characterization of oil shale cores to provide data for ongoing field projects. This will require that additional wells be drilled. The laboratory investigations will continue to provide for the development of new analytic process technology. A part of the project is to determine the characteristics of oil shales and oil shale deposits to support the energy development efforts. Under this project, a detailed knowledge of the composition and chemical, physical, and thermal properties of oil shale has been accumulated. This is a cumulative process with each year's work adding to the funds of available information. Each year, 20,000 to 25,000 individual oil shale samples are characterized for oil yield and geologic and mineralogic properties.

Another task is devoted to new oil shale process technologies. It is currently investigating innovative techniques for increasing the solubility of oil shale kerogen and developing the basis for economic and technologic assessments of oxidative upgrading of fossil fuels. The research on kerogen solubilization may lead to second generation technology for above-ground processing that does not require retorting. Oxidative upgrading of shale oil may improve economics by reducing hydrogenation requirements for removing sulfur and nitrogen compounds. Vacuum retorting and hydrogen retorting projects were initiated in FY 1978 and 1979.

Early in FY 1977, the Navy contracted with the industrial group at Anvil Points to produce a large quantity (100,000 barrels) of shale oil. This quantity of oil was to be used for a refining run designed to produce sufficient quantity of distillate products for specification tests. DOE personnel are continuing to observe this operation. DOE and Navy are planning on a cooperative shale oil refining program pending approval from Congress. Oil production required nine months into 1978. A refinery run on the oil was conducted during November 1978.

Broader activity includes major efforts which are aimed at expanding the knowledge base with regard to cross-cutting information which is required to improve oil shale technology as it is known today. This work is accomplished in-house at the Laramie Energy Technology Center, the national laboratories and supporting industrial and university laboratories. The work includes applied research in geophysics, mining operations, chemical engineering, environmental sciences and other scientific and engineering disciplines related to oil shale development. A small portion of the effort is expended on nontechnical constraints to oil shale development such as the analysis of local, State and Federal permits required for oil shale developments.

The following major centers are used in this effort.

LETC conducts a broad-based research program in support of several methods of oil shale retorting applicable to different portions of the total shale reserves and provides technical oversight for contracts with industry. The in-house supporting research includes fundamental modeling of reaction kinetics and rock mechanics; development of analytical methods for on-line monitoring of product properties and process events; environmental monitoring and compliance activities; characterization of oil shale deposits to assist in site selection for field tests, small-scale field fracturing and recovering experiments, and operation of engineering-scale retorts that simulate underground processing to provide data for the process modeling activities.

LLL's objective is to provide a technical base and complementary support for modified in situ retorting projects through further development of a retorting model, including experimental laboratory work and operation of laboratory retorts; and limited field efforts to obtain data from industrial retorting projects and provide advisory assistance to industry regarding application of government-developed measurement and retorting technology.

Objectives of Lawrence Berkeley Laboratory's (LBL) efforts are to (1) determine the distribution of major and minor elements between shale oil, process water, off gas, and spent shale as a function of retort operating conditions; (2) identify and quantify the organo-metallic species that are released to retort waters and off gases during in situ retorting; and (3) develop and apply an in-place monitoring technique and sampling system for gaseous emissions from an in situ retort. Data from these investigations will be used to help assess the impact of field in situ operation on the environment.

LASL's objectives are to provide dynamic material property data and explosive characterization data within the context of a cooperative project with Sandia Laboratories, to develop the ability to control and predict the result of explosive blasting in oil shale. The end result of this cooperative project is expected to be improved capability of preparing in situ retorts with desired properties of rubble size and permeability distribution. A related effort under this project is the study of techniques for altering in situ stress.

Sandia Laboratories' objectives for this project are (1) to develop oil shale material response models and numerical computational techniques to aid in design and analysis of modified and true in situ fracturing; (2) to develop advanced instrumentation systems and measurement techniques to monitor, control, and evaluate both fracturing and retorting processes; and (3) to apply these systems and techniques in support of both in-house field projects conducted by LETC, and DOE-funded industrial field projects where such support materially enhances quality and understanding of experimental data.

In addition to the supporting research, advanced concepts such as hydrogen retorting and microwave shale oil production are being studied.

Advanced surface and in situ retorting processes show promise of substantial economic improvements and/or reduced environmental impacts or problems relative to present progress that have undergone large-scale development testing. Such advanced processes have typically undergone bench-scale development or are presently in an active laboratory or small PDU-development phase that has provided promising data. Certain concepts require laboratory research to provide the first level of confirmation of potential benefits.

Figure II 4-18 presents major milestones for this effort.

Shale Gas

The major goals of the In Situ Shale Gas Production Subactivity are as follows:

- To study means of producing large quantities of low-Btu (100 to 300 Btu/scf) gas from oil shale by an economically feasible in situ process
- To study the feasibility of producing medium-Btu (300 to 500 Btu/scf) gas from oil shale for upgrading to pipeline gas and of developing production methods
- To define the operating parameters necessary to ensure maximum utilization of the resource
- To determine the yield and composition of liquid and gaseous products from oil shale retorting and their relationship to process variables.

An underground operation to produce gas from oil shale can be identical to one designed to produce shale oil. Changes in the levels of operating variables, resulting in higher shale bed temperatures, tend to favor gas production over liquid production. The heating value of the gas can be changed by altering the composition of the injected gas.

Past research at LETC has demonstrated that low- to medium-Btu gas may be obtained from oil shale at low pressure by proper selection of control parameters (e.g., gas flow rates and oxygen concentration). Work on gasification of Eastern and Western oil shale varies in nature because of the differences in the potential oil content of the shale and the desired products. Low-grade Eastern oil shale exists in great abundance in close proximity to highly populated areas. For this reason, the preferred products are low-Btu gas suitable for gas-turbine generators and/or medium-Btu gas for upgrading for home heating purposes. Some current work at LETC has demonstrated that proper control of process variables allows usable gas to be produced concurrently with

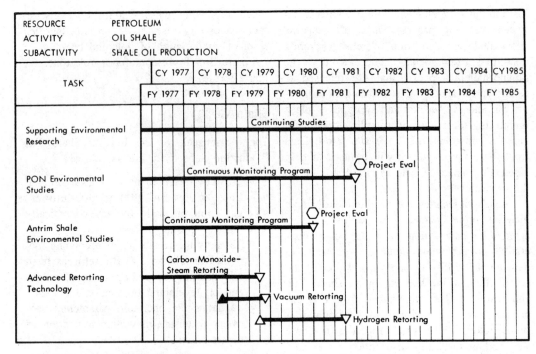

Figure II-4-18. Major milestones for shale oil production.

high quality shale oil. Bench-scale retort work to develop statistical models to relate operating parameters to products has been underway for more than two years at LETC. An intermediate-scale (0.5-ton batch) gasification retort was in operation during FY 1979 to verify the models. Products expected from the gasification of Western oil shale, concurrent with shale oil production, are:

- Low-Btu gas as off-gas from in situ retorting plants. A single 50,000 bbl/d shale oil plant should generate over 1 billion scf/d of adequate calorific value to produce over 500-MW when put through a combined-cycle, gas-turbine powered generating system
- Medium- to high-Btu gas suitable for upgrading and use as pipeline gas to fill any usual natural gas need.

A side benefit is greater utilization of the energy in the resource (up to 20 percent) through oil and gas production.

Problems facing the commercial gasification of oil shale in situ include those common to retorting; that is, the creation of sufficiently permeable surface area through the shale to allow efficient retorting and gasification to proceed. In addition, the proper level of field controllable parameters must be established and a full technical and economic feasibility study completed to determine the best methods of utilization of the resource.

Work is currently underway at LLL using small- to medium-size retorts equipped with adiabatic shields to determine the yield and composition of products and the optimum level of operating variables for the production of either gas or oil from shale. Supporting work in five smaller nonadiabatic retorts is also underway.

Work on Western oil shales will be continued by LETC. The Eastern shales are being investigated under a contract by Dow Chemical Company and members of the Michigan Energy Resources Research Association. LETC and Sandia Laboratories are also involved with Dow in the work on Eastern shales.

DOW completed a series of three fracturing tests of their two in situ gasification sites (Antrim shale—Michigan) during FY 1979.

After evaluation, one was selected for retorting during FY 1980. By the end of FY 1980, final evaluation and site restoration should be completed. It is entirely funded by DOE, excluding the value of equipment and land provided by Dow, and excluding consideration of the prior $7,500,000 expended by Dow on Antrim shale research over a 20-year period.

The project consists mainly of evaluating four potential in situ fracturing techniques, including one fracturing test conducted under prior Dow funding. A combustion test of this prior fractured site and a combustion test of the best of three new fracture concept tests are included in this project. Combustion retorting of Antrim shale preferentially produces synthetic gas because of a relatively low hydrogen-to-carbon ratio compared to Western shales. Low-Btu gas from this process would likely be used for industrial fuel. Economics relative to Western shale would be enhanced by nearness to markets, available infrastructure and labor, and availability of plentiful water.

The project presently represents long-term advanced research concepts and does not include development and demonstration.

Major milestones for shale gas production are given in Figure II 4-19.

DRILLING AND OFFSHORE TECHNOLOGY

Estimates of potential petroleum and natural gas resources in the 17 U.S. Outer Continental Shelf (OCS) areas range from 10 to more

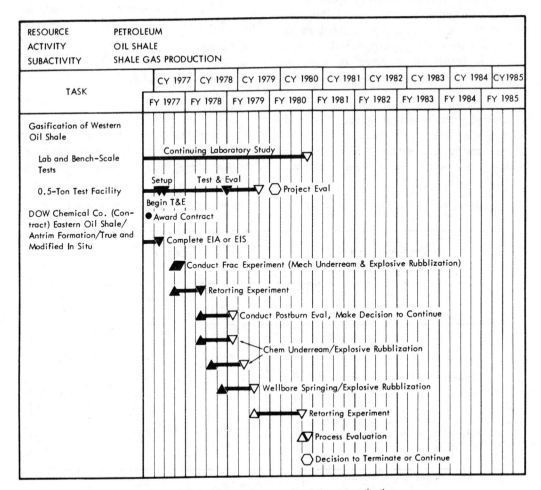

Figure II-4-19. Major milestones for shale gas production.

than 45 billion barrels of oil, 35 Tcf of natural gas, and 40 to 180 Tcf of undiscovered natural gas resources. DOE drilling and offshore technology activities in the near- and mid-terms will be focused on the following objectives:

- Increasing efficiency of drilling operations with a goal of speeding overall rate of penetration by 30 percent by 1985
- Developing technologies for frontier drilling on a commercial basis by 1985. Frontier areas defined as:
  — Drilling in petroliferous horizons below 30,000 feet
  — Drilling in deep offshore waters
  — Drilling in the Arctic (Land Arctic and Ocean Arctic)
- Reducing the cost of drilling and production (in the 5000-foot and deeper range)
- Providing support to other DOE programs requiring drilling technology applications; e.g., programmatic support for geothermal resource development, coalbed methane drainage, in situ coal gasification, etc.

This program is heavily impacted by technical, nontechnical, environmental, socioeconomic, and legal constraints which increase as industry moves into deeper water and more hostile frontier areas. The pace at which offshore energy resources are developed is constrained by insufficient data, inadequate technology, lack of long-term investment capital, long lead-time requirements, and wariness by industry of cooperative efforts resulting from the need to protect proprietary interests and maintain a competitive business environment among participants.

Studies currently being conducted under drilling and offshore technology include drilling, offshore technology, and environment and support.

Table II 4-4 summarizes the funding levels by subactivity for the FY 1978 to FY 1980 period.

Drilling

Five projects are currently being conducted under this program area:

- Downhole Telemetry
- Electrodril
- Drilling Technology
- Horizontal Drilling
- Project Monitoring

The objective of the Downhole Telemetry Project is to accelerate the development and commercial availability of wellbore telemetry.

Table II-4-4. Funding for drilling and offshore production technology for FY 1978 - 1980.

DRILLING AND OFFSHORE PRODUCTION TECHNOLOGY	BUDGET AUTHORITY (DOLLARS IN THOUSANDS)		
SUBACTIVITIES	ACTUAL FY 1978	APPROPRIATION FY 1979	ESTIMATE FY 1980
Drilling	$ 1,000	$ 1,509	$ 1,500
Offshore Technology	0	585	1,000
Environmental and Supporting Research	600	506	500
Capital Equipment	0	0	0
TOTAL	$ 1,600	$ 2,600	$ 3,000

This project addresses the need of the oil and gas industries to develop a downhole measuring and telemetering system for parameters, such as inclination, bit-wear, and weight-on-bit, while drilling is being conducted. The mud-pulse telemetering of these data contributes to more efficient and economic drilling as well as to a real-time characterization of the drilling activity as it is being conducted.

This project is being conducted in three distinct phases: Phase I, System Development; Phase II, System Manufacture; and Phase III, Field Demonstration.

The design for the downhole telemetry system has been selected and manufactured. Field demonstrations of the complete system were conducted in FY 1978. The system became commercial in September 1978.

This is a cost-shared project between DOE and Teleco with significant participation by six oil companies.

The Electrodril is designed to drill with a downhole motor that uses a retrievable power cable and employs a telemetry system that makes and transmits to the surface various drilling and safety measurements.

The objectives of the Electrodril two-phase project are as follows:

- Phase I—to test the Electrodril Directional Drilling System under actual drilling conditions
- Phase II—a three task Deep Drilling (Straight Hole) program.
  - Task A—began in July 1977, encompassed the engineering design and preprocurement documentation activity required for long lead time equipment.
  - Task B—last half of FY 1978, included final design review of key system elements and the completion of all engineering documentation.
  - Task C—is planned to follow the completion of the system verification test and will be a series of system demonstrations under operational conditions at three drilling sites.

During FY 1977, the directional drilling system operated successfully with tapered connectors used to drill through an aggregate at Brown Oil Tools. Initial tests were conducted with a directional drilling system to determine the system's ability to drill directionally to a predetermined target in a 600-foot deep hole. This testing was discontinued as the result of connector failure. Failure data analyses indicated that water seepage was the major problem. Testing was later resumed during the end of the first quarter of FY 1977, and all elements of the cable/connector, downhole-sensor, and telemetry system operated successfully to the 142-foot depth with some indication of mud leakage.

In addition to the cable/connector problem, motor-system mud/oil seal interface problems have also been identified. This project was scheduled to be completed at the end of the first quarter of FY 1978, but a no-cost extension was granted through the second quarter of FY 1978 to conduct further analyses of the failure data and resolve the problems.

Phase II Tasks A and B were completed in FY 1978, Phase II Task C, has not been funded; when funded, Task C will be to assemble and field test the General Electric Deep Drilling System. This is a cost-shared project between DOE and General Electric with significant participation by industry.

The objective of the drilling technology project is research to enhance the development of drilling technology. Sandia Laboratories in Albuquerque, New Mexico, is currently conducting research in the areas of (1) high-performance bit developments, (2) high-temperature mud-test equipment, and (3) improved HT/HP materials. In the high-performance bit development area, two activities (Stratapax, a synthetic diamond, bonding technique, and bit design) are currently being conducted. In the Stratapax bonding study, the results of two tests of two failed stud assemblies were examined to determine point wear and fatigue. Both assemblies were intentionally caused to fail during very heavy cuts in granite. Multiple fractures were observed in both the diamond and the stud. The failures of both assemblies were similar except that all of the diamond was broken from the stud of one assembly. Bonding experiments were done in the second quarter of FY 1977. A metalization technique for experimental test was transferred to industry in the first quarter of FY 1978. FY 1979 activities included development of a high-strength bonding technique to attach Stratapax to the bit. This technology transfer should occur during FY 1980.

In the bit design activity, several new studs were made to replace standard-type studs. Four gas-pressure diffusion-bonded Stratapax/ stud assemblies were evaluated and the data reduction scheduled for completion in February 1977. Hybrid drill bits were fabricated, assembled, and tested in the third and fourth quarter of FY 1978. FY 1979 activities included development of special performance bits for such tasks as downhole motors, coring, and ultradeep holes, which have limited market due to specialized users. Laboratory and field testing should occur in FY 1980.

In the high-temperature mud-test equipment research area, Sandia is using a rotating viscometer as an alternative to a telescoping bob viscometer. Tests were conducted during FY 1977 and further investigations are scheduled for the remainder of the experiment. Supply and tooling problems were encountered, but testing was completed in the third quarter of FY 1978. The device/technology is now ready to be transferred to the industry.

In the HT/HP materials project, various elastomer seals were obtained from industry and coated with plasma polymerized Teflon and returned for testing in FY 1978. The companies tested them under simulated drilling conditions in late FY 1978. In FY 1979, coated packers, seals, and other elastometric items field tested by industry were evaluated and recommendations were made. Mechanical strengthening (to help to eliminate degradation during drilling) of elastomers will be investigated. This technology transfer should be completed during FY 1980.

Technology assessments in jet-erosion drilling and drilling automation began in mid-FY 1979 to determine the feasibility of these advanced drilling systems. During FY 1980, these and other advanced drilling assessments/ feasibility studies will either be discontinued or expanded dependent upon FY 1979 recommendations.

A two-phase technology assessment and recommendation program relative to horizontal drilling will start in mid-FY 1981 at the Morgantown Energy Technology Center, Morgantown, West Virginia. Phase I is an effort to appraise existing technology and equipment. Phase II, based upon Phase I recommendations, will assist the division to identify near- and long-term research needs for horizontal drilling technology. If FY 1979 recommendations are positive, FY 1980 investigations will continue to determine the feasibility of directional drilling capability for some advanced applications (horizontal drilling in methane recovery for unminable coal seams) and transfer of recommended technologies.

Presently, the Bartlesville Energy Technology Center, Bartlesville, Alabama, is monitoring projects which are sponsored in full or cost-shared by DOE. These projects now encompass the following research areas: downhole electric drilling motor and mudpulse telemetry.

Major milestones for the Drilling Subactivity are given in Figure II 4-20.

Offshore Technology

Projects currently being conducted under this program area include:

- Sea Floor Geotechnical Instrumentation
- Data Acquisition and Dissemination.

Major milestones for this subactivity are given in Figure II 4-21.

Sea Floor Geotechnical Instrumentation. The objective of this program is to develop and demonstrate an acoustic data system for undersea measurements of geophysical phenomena. This instrumentation system, Seafloor Earthquake Measurement System (SEMS), must be capable of measuring strong motions of the sea-bottom sediments caused by earthquakes to provide inputs to the structural design of offshore platforms.

The project was designed to incorporate previous experience with marine sediment penetrators (MSP) and seismic instruments implanted in the ocean bottom with a high data-rate telemetry link. The initial project conducted in FY 1976 consisted of assembling

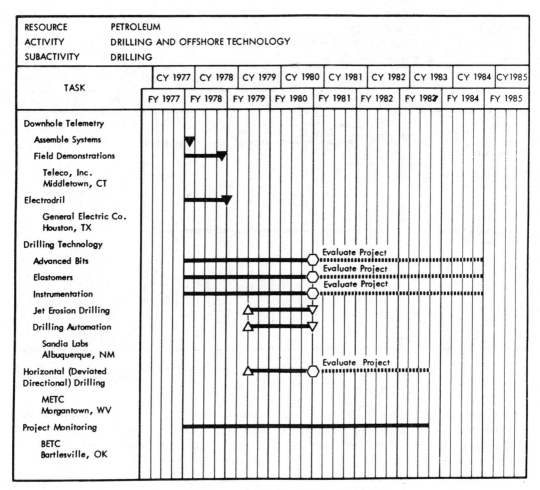

Figure II-4-20. Major milestones for the drilling subactivity.

and testing the equipment to demonstrate whether the concept was feasible. Following an evaluation of the results of the initial testing, additional acoustical equipment was ordered and installed in preparation for the next phase of testing. The equipment and test procedure will allow maximum flexibility in varying the test conditions and signals propagated, direct monitoring of the transmitted wave, and direct comparison of the received signal to the transmitted signal.

Sandia Laboratories, Albuquerque, New Mexico, successfully tested an acoustic transmission system for transferring data at up to 2,400 bits per second from bottom sediments to surface recording equipment in the Gulf of Mexico (in up to 600 feet of water) at the end of FY 1977 and in Icy Bay, Alaska (in up to 300 feet of water), in late FY 1978. In FY 1979, the SEMS technology was transferred and the Geotechnically Instrumented Seafloor Probe (GISP) project, which will develop and test an instrumentation system to measure and transmit seafloor engineering data (stresses on piles/structures, sediment shear strength), was begun. Laboratory and field test, evaluation, and recommendation for GISP is scheduled for FY 1980.

Data Acquisition, Analysis, and Dissemination. The objectives of this ongoing project are to (1) identify users and their needs for information relevant to offshore oil and gas resource development (exploration, development, production, transportation, and onshore impacts); (2) identify data, data sources, and

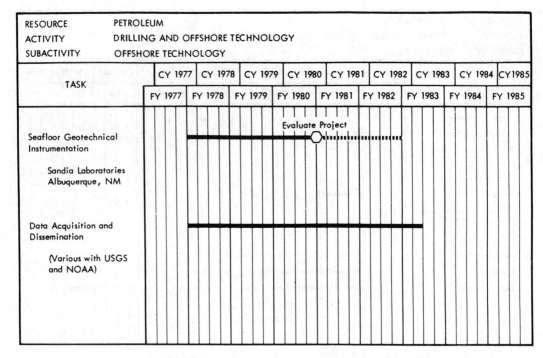

Figure II-4-21. Major milestones for the offshore technology subactivity.

acquisition methods; and (3) assess the completeness of data for satisfying requirements and developing a plan for DOE involvement in assisting users in acquiring timely data in a useful format. DOE, USGS, and NOAA cost-share these projects, which are performed by various contractors.

Environment and Support

Projects currently active in this area are technology assessment and program management plan.

Various ongoing technology assessments done by various contractors to meet the Drilling and Offshore Technology Program propose to (1) increase the efficiency and lower the costs of drilling and (2) continue development of technologies for unconventional drilling onshore and offshore will continue. Development of a comprehensive drilling plan will identify and prioritize these assessments.

The House Interior Appropriations Committee requested "a comprehensive plan similar to those of the enhanced oil recovery and enhanced gas recovery be developed for the research requirements of both onshore and offshore drilling" by September 30, 1979.

Major milestones for these two projects are presented in Figure II 4-22.

ADVANCED PROCESS TECHNOLOGY

The Advanced Processes Technology (APT) Program has been organized into three subactivities: Advanced Exploratory Research, Shale Oil Refining and Utilization, and Product Characterization and Utilization. Since these elements are different in technology and approach, each is described separately.

Table II 4-5 summarizes the funding levels by subactivity for the FY 1978 to FY 1980 period.

Advanced Exploratory Research (AER)

Although domestic petroleum reserves are declining, petroleum still contributes a large fraction of U.S. energy consumption, and large resources of coal and unconventional energy sources exist. The AER element is directed at developing advanced and innova-

Activity Description—Petroleum 315

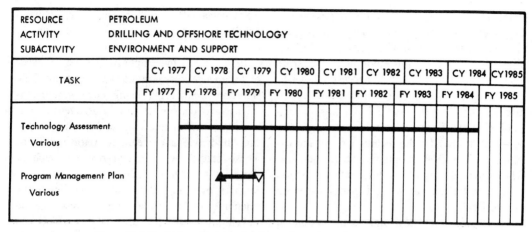

Figure II-4-22. Major milestones for the environment and support subactivity.

tive technology which could make significant improvements in extraction of resources. Specific areas for which these technologies will be sought are enhanced oil and gas recovery (EOR and EGR) in situ conversion of coal and oil shale, recovery of heavy petroleum deposits, and shale retorting.

The objectives of AER are to develop a better understanding of methods to produce increased permeability in rock matrices, which is a key factor in all in situ processes; initiate efforts in advanced technology for in situ processes with emphasis on use of synthesis gas and application of microorganisms to petroleum and oil shale recovery; initiate efforts to develop improved chemical techniques for enhanced oil recovery; and initiate studies in kinetics, mechanisms, and process dynamics of in situ processes.

The AER Subactivity will pursue two major thrusts: Basic Technology Development and Process Development. Basic Technology Development includes developments in rock mechanics and fragmentation and hydrology. Rock mechanics and fragmentation (the development of enhanced permeability in the rock matrix) is a controlling factor in virtually all in situ processes and is thus applicable to oil

Table II-4-5. Funding for the advanced processes technology program for FY 1978 - 1980.

ADVANCED PROCESS TECHNOLOGY SUBACTIVITIES	BUDGET AUTHORITY (DOLLARS IN THOUSANDS)		
	ACTUAL FY 1978	APPROPRIATION FY 1979	ESTIMATE FY 1980
Advanced Exploratory Research	$ 0	$ 0	$ 1,700
Shale Oil Refining and Utilization	0	1,000	500
Product Characterization and Utilization	1,400	1,200	1,300
Capital Equipment	0	0	500
TOTAL	$ 1,400	$ 2,200	$ 4,000

shale retorting, enhanced oil and gas recovery, and in situ coal gasification. The properties and characteristics of the resource-bearing formations in the in situ environment must be understood to be assured of success of in situ processes. DOE, through facilities and expertise at LLL, LASL, and LETC, has the capability to perform a key role in this generic area and to transfer the techniques and knowledge to other DOE programs and to industry. The potential payoff of this activity is too long-term for industry to do the required work on its own.

Hydrology (the flow and composition of formation water) also has an impact on the success or failure of in situ processes involving EOR by micellar and surfactant flooding, EGR, oil shale retorting, and in situ coal gasification. This work is performed by environmental scientists, hydrologists, and earth scientists at LETC, LLL, and LBL. Process development will stress advanced oil shale retorting and improved recovery by use of microorganisms.

Process Development includes new approaches to oil shale retorting and improved recovery of conventional hydrocarbons. Advanced oil shale retorting will develop techniques using synthesis gas and steam mixtures and is based on existing knowledge at LLL and LETC. The use of synthesis gas and steam should provide higher yields and better quality products than are currently obtained by air-based systems. The option of producing hydrogen or high-Btu gas from the product gas also exists.

Improved recovery by the use of microorganisms is in the experimental stage. It is well known that certain bacteria can break down the structure of kerogen and heavy oil into more valuable forms. Application of this phenomenon, in situ, may result in high resource recovery at low cost and high energy efficiency. This will be a long-term effort in which industry is not active. Thus, the formation of a DOE program is required to develop this novel approach.

These advanced technology projects will undergo regular review directed toward deletion of unsuccessful effort and concentration of resources on processes that exhibit promise. Special effort will be made to maintain communications with other groups involved in development of less-advanced technology.

In FY 1980, the emphasis will be to accelerate present research on advanced oil-shale retorting technology, using such systems as microwave conversion, radio frequency conversion, vacuum retorting, and the use of carbon monoxide and steam as retorting fluids. These studies will include research such as the heat transfer characteristics of oil shale, kinetics of shale oil reactions and fracturing phenomena of oil shale. In-house research and development will be performed at LETC and LLL.

Also, work will be performed to increase the knowledge base for the development of improved chemical flooding technology, using more efficient and less costly chemicals and basic research on the use of microorganisms for enhancing oil recovery. The Bartlesville (Oklahoma) Energy Technology Center (BETC) will be the focus for this effort.

Major milestones for AER are shown in Figure II 4-23.

Shale Oil Refining and Utilization (SORU)

In the area of Shale Oil Refining and Utilization, a number of uncertainties exist which cause major impediments to shale oil commercialization. These uncertainties result from the fact that all crude shale oils have a high nitrogen content, and the known methods for nitrogen removal are costly. Also, crude shale oil fractions are unstable and therefore, not currently suitable for direct use as fuels. In addition, there are significant differences in the physical properties of the crude shale oils produced by different recovery processes. Finally, the relationship between product specification and end-use performance for conventional petroleum fuels is essentially empirical and may not be applicable to shale oil-derived fuels, and most crude shale oils have pour points and viscosities that make conventional pipelining costly or unfeasible.

Preliminary indications are that conventional refinery procedures, using high-pressure, high-severity hydrotreating, can be

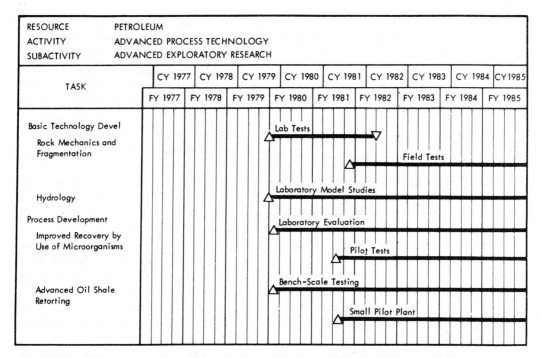

Figure II-4-23. Major milestones for the advanced exploratory research subactivity.

applied to shale oil refining but that the cost of these techniques would be much higher than that of refining conventional crude. In addition, since all crude shale oil fractions are unstable and not suitable for direct use as fuels, optimum levels of stabilization and nitrogen removal for combustion and transportation uses must be determined to establish appropriate trade-offs between refining severity and combustion procedures. Since petroleum contains very little nitrogen, no nitrogen specifications exist for conventional refined transportation fuels. To optimize shale oil refining, it is essential to establish definite relationships between nitrogen content and fuel stability and NO_x emissions for combustion.

The objectives of the SORU program are to continue the state-of-the-art shale oil refining studies, determine the nature of stability of shale oil fuels, and develop technology to remove nitrogen with minimum H_K consumption.

The strategy of the SORU is to continue studies that will enhance the ultimate utilization of liquid products derived from shale oil. The shale oil refining studies include defining capabilities of existing technology, developing a lower pressure hydrotreating, hydrocracking catalyst, and evaluating hydrotreating-physical separation processing schemes with the use of the extract for hydrogen manufacture.

Since the chemical nature of fuel stability is not well understood, studies will be initiated to characterize the heteroatomic compounds in hydrotreated shale oils and to correlate fuel performance with the content of these heteroatomic compounds.

Work planned for FY 1980 includes continuation of research contracts with Mobil Oil Company on the development of a more active and stable hydrotreating catalyst to permit the production of high yields of middle-distillate transportation fuels at lower cost; and with Suntech for the development of a refining process using hydrotreating combined with physical separation through extraction or adsorption to decrease the severity of hydrotreating. This could result in lower hydrogen consumption and the production of stable transportation fuels at lower cost. The Pittsburgh Energy Technology Center (PETC) will manage and monitor the contract research. The work also includes evaluation of projects on catalyst-life performance during refining runs on shale oils produced by different processes. This work was conducted and the funding included under the Oil Shale Activity in FY 1979.

Major milestones for the SORU Subactivity are given in Figure II 4-24.

Product Characterization and Utilization (PC&U)

The PC&U element focuses on the analysis and characterization of crude oil, heavy oil, and tar sands, including analysis of asphalt as a function of road life. For many years, research in these areas has been conducted at BETC and LETC.

The objectives of the PC&U subactivity are to maintain and improve the existing data bank on crude oils and products, establish the chemistry of fuel stability, and develop asphalt materials with increased road life.

Expected results are increased utilization of heavier petroleum fractions in producing quality fuels, matching of enhanced oil recovery chemicals with oil type resulting in more efficient resource recovery, design of improved processes for heavy oil and tar sand recovery, data on the chemistry of fuel stability, and efficient utilization of asphalt through improved road life. Characterization of the materials described above will provide important data on metals and heteroatom content. These are important environmental parameters impacting on utilization. Maintaining data on the quality of crude oils will provide a unique and valuable resource of data on domestic and foreign crudes for use by Government and industry. Correlation of asphalt wear with changes may yield new asphalt compositions with increased road life.

The PC&U strategy involves (a) the continuation of the analyses and tabulation of data on crude oils and products, (b) stability studies on blends of synthetic and conventional crudes and products, and (c) analytical studies of asphalt road life and correlation of road life with chemical breakdown of asphalt structure. Item (a) is carried out with relatively conventional analytical techniques but aims at developing characteristic data useful to the production of low-quality petroleum deposits. Also, more sophisticated techniques

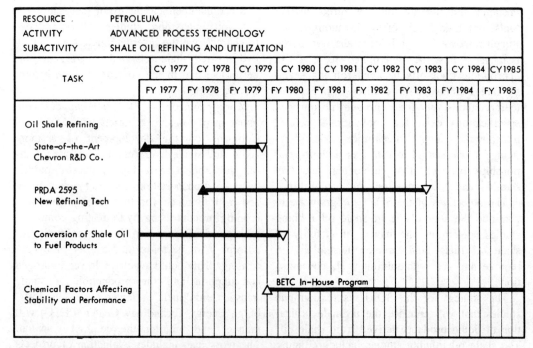

Figure II-4-24. Major milestones for the shale oil refining and utilization subactivity.

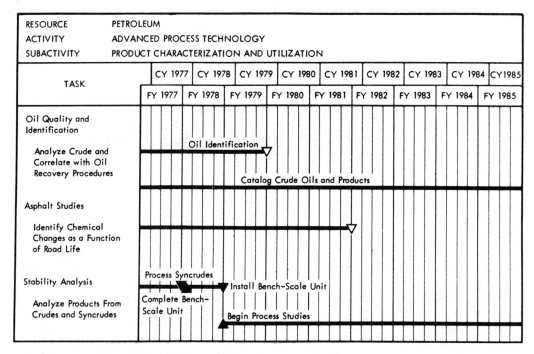

Figure II-4-25. Major milestones for the product characterization and utilization subactivity.

have been developed which can now be used on a routine basis to allow differentiation between old and new oil. Item (b) applies a vast background of expertise on fuel stability to resultant stability problems from blending petroleum crudes and products with synthetic crudes and products. Item (c) provides composition and chemical reactivity data on asphalt with reference to properties and useful road life. A number of compositional factors affecting road life have been identified. Additional information to be determined will include aging mechanisms, component compatibility, and the potential for recycling used asphalt.

The data collection aspect of this technology is carried out in close cooperation with the petroleum industry to ensure the transfer of data to industry. To ensure that the generated data are the type needed for improving the efficiency of EOR techniques, close contact with the EOR program is maintained. The transfer of technical data to industry is ensured through publications and participation in scientific and professional meetings. FY 1980 funding will provide increases for on-going in-house projects at the BETC and two projects at LETC. Projects at BETC include oil-spill identification, characterizing the heavy ends of petroleum, determining the quality of crude oils and products, storage stability of products, and refining process technology. Projects at LETC are correlation of asphalt characteristics with road stability and characterization of heavy liquids. All of the research and development is conducted in-house. These research projects are basic in nature and provide a base of fundamental data for advancing petroleum-research efforts.

Major PC&U milestones are presented in Figure II 4-25.

5.
Activity Description—Gas

Natural gas currently supplies approximately 25 percent of the total energy consumed in the United States. Proven producible reserves of gas (209 trillion cubic feet) only amount to a 10-year supply at the present consumption rate.

The bulk of these proven reserves are found in conventional reservoirs (porous, high-permeability sandstone and carbonate rock formations). Existing technical and economic situations permit production from these reserves at rates which yield acceptable profit incentives for the gas-producing companies.

Two additional sources of natural gas are available to augment dwindling supplies from the conventional reservoirs—imports (LNG and Canadian or Mexican pipeline gas) and unconventional resources. The obvious problems with imported gas are the negative balance of payments situation and reliability of supply, as was so dramatically demonstrated by the oil embargo in the winter of 1973–74 and the subsequent three-fold increase in Canadian gas prices. Unconventional gas resources appear to be the more viable option for the long-term as they are located within the borders of the United States and are responsive to developing extraction technologies. Vulnerability to the adverse effects associated with importing a large percentage of our energy will be lessened by a total commitment to develop technologies to allow full-scale commercial gas production from domestic resources.

A report issued by the National Research Council, Board of Mineral Resources Commission (1976) cites four unconventional geologic sources of natural gas:

1. The gas-bearing shale formations of the eastern United States
2. The low-permeability (tight) sandstone reservoirs of the Rocky Mountain region
3. The free methane present in coal beds
4. The high-pressure, methane-saturated aquifers of the Gulf Coast region (geopressured aquifers)

The DOE Gas Recovery Program is a multiyear effort aimed at developing the full potential of these unconventional gas resources. The major strategic objectives of this program call for:

- Shifting the price-supply curve through development of more efficient technology, thus providing increased production at any given gas price
- Accelerating the timing of R&D investments and subsequent application of improved technology
- Undertaking high-risk/potentially high-return R&D in areas currently deferred by industry

Successful conclusion of this program will provide a significant contribution to the domestic energy base and help alleviate anticipated near-term and long-term shortages of natural gas—an efficient, environmentally benign fuel source.

Three of the four unconventional gas resources (geopressured aquifers are a target of the DOE Geothermal Program) are addressed by individual projects within the overall Enhanced Gas Recovery (EGR) Program. Each project is a cohesive RD&D effort directed at a single resource target or geologically distinct subdivision of that resource.

Ongoing projects and their respective target resources include:

- The Eastern Gas Shales Project (EGSP) —the Devonian age, gas-bearing shales of the Appalachian, Illinois, and Michigan Basins
- The Western Gas Sands Project (WGSP) —the low-permeability (tight) sandstone reservoirs of the Rocky Mountain region
- Methane from Coal Project—the methane gas associated in coal seams and adjacent rock formations.

Estimates of the recoverable gas present in each resource vary considerably, as shown below.

RESOURCE	POTENTIALLY RECOVERABLE GAS
Eastern Gas Shales	10–520 trillion SCF
Tight Gas Sands	50–320 trillion SCF
Methane from Coalbeds	16–500 trillion SCF
Geopressured Aquifers	150–2000 trillion SCF

The considerable range in each estimate strongly implies that accurate geological and engineering data are not presently available. The true magnitude of the recoverable reserves and a comprehensive assessment of those physical and chemical properties which govern reservoir behavior (and the effectiveness of various stimulation techniques) need to be determined.

The EGR Program is designed to provide answers to key technical problems impeding development of these resources. Successful implementation of the program will stimulate a vigorous response from private industry, leading to large-scale, commercial production from these unconventional gas resources.

Table II 5-1 summarizes the funding levels for the FY 1978 to FY 1980 period.

ENHANCED GAS RECOVERY

The unconventional gas resources can make significant contributions to the national gas supply in both the near- and long-terms. The potential of these resources can be realized at a lower cost than most other supplemental gas sources and can be produced most effectively through a combination of technology development, resource characterization, and economic incentives.

The goals of the EGR are as follows:

- To accelerate the development of an im-

Table II-5-1. Funding for the gas resource activity for FY 1978-1980.

GAS RESOURCE	BUDGET AUTHORITY (DOLLARS IN THOUSANDS)		
ACTIVITIES	ACTUAL FY 1978	APPROPRIATION FY 1979	ESTIMATE FY 1980
Enhanced Gas Recovery	$27,085	$33,602	$27,310
Program Direction	316	261	283
TOTAL	$27,401	$33,863	$27,593

proved scientific, engineering, and economic basis for prompt, orderly development of enhanced recovery technologies to accelerate production from the nation's natural gas resources
- To transfer DOE-developed technology in EGR to those sectors of private industry that need such information to establish reserves and production goals and/or significantly reduce investment risks

The target resources for the current EGR Program include:

- The gas-bearing, Devonian shales of the eastern United States (Eastern Gas Shales Project)
- The low-permeability (tight) sandstones of the United States (Western Gas Sands, Northern Great Plains, and Southern Tight Basins Projects)
- The free methane (natural gas) present in coalbeds (Methane Recovery from Coalbeds Project).

The success of Enhanced Gas Recovery is predicated upon developing accurate resource characterization data and perfecting various recovery technologies (advanced hydraulic fracturing, chemical explosive fracturing, and directional drilling techniques). Achieving these technical objectives will significantly impact upon industry decisions leading to full-scale commercial development of the unconventional gas resources.

Table II 5-2 below summarizes the funding levels by subactivity (project) for the FY 1978 to FY 1980 period.

Eastern Gas Shales Project

The Eastern Gas Shales Project is a multi-disciplinary effort directed toward increasing natural gas production from Devonian shales of the Appalachian, Illinois, and Michigan Basins of the eastern United States.

Specific goals to be achieved by the project are:

- Developing a reliable scientific data base leading to an accurate methodology for estimating gas reserves and reservoir-producing mechanisms in the shale formations
- Developing a technique(s) for locating fractured reservoirs within the shale sequence
- Develop efficient extraction techniques for producing gas from the shales
- Demonstrating the commercial feasibility of evolving production technology as applied to Devonian shales

Specific task activities will include determining the true magnitude of the potentially

Table II-5-2. Funding for EGR for FY 1978-1980.

ENHANCED GAS RECOVERY	BUDGET AUTHORITY (DOLLARS IN THOUSANDS)		
SUBACTIVITIES	ACTUAL FY 1978	APPROPRIATION FY 1979	ESTIMATE FY 1980
Eastern Gas Shale Project	$14,000	$18,000	$ 9,000
Western Gas Sands Project	4,300	7,500	8,800
Methane from Coal	2,000	4,000	5,000
Environment and Support	4,985	3,902	4,110
Geopressure Aquifers	1,500	0	0
Capital Equipment	300	200	400
TOTAL	$27,085	$33,602	$27,310

recoverable reserves within the shale, thoroughly characterizing the physical/chemical make-up of the shale, and improving upon state-of-the-art exploration and extraction technologies. Results will be continually evaluated to determine areas in which new or additional studies may be needed and the desirability of planning and conducting a more extensive RD&D program. The project will concurrently monitor and protect the environment from possible damage resulting from natural gas development activities.

Major milestones are given in Figure II 5-1.

Resource. The resource evaluation portion of the project includes surface and subsurface mapping, structural studies, geochemical studies, clay mineralogy studies, borehole-gravity surveys, and data acquisition and processing for an appraisal of potential gas reservoirs in Eastern shales. Much of the needed data and analyses will be obtained by universities and state geological surveys in the study area. Field work will consist of coring and logging wells to provide data curves for laboratory work. Laboratory work will consist of chemical, physical, elemental, and mineralogical studies; fracture-orientation determination from cores; statistical analyses of fracturing data; and cost-effectiveness studies of fracturing techniques. Technology development and demonstration will be accomplished by field projects proposed and cost-shared by private industry. Some projects will be initiated by issuing requests for proposals for testing techniques such as massive hydraulic fracturing, chemical explosive fracturing, foam hydraulic fracturing, gas hydraulic fracturing, and drilling of directionally controlled wells.

Certain highly specialized analytical laboratory and field techniques have been contracted with both national and nonprofit research laboratories. Contracts were underway (FY 1979) which included activities with the universities, DOE national laboratories, and state geological surveys.

The U.S. Geological Survey (USGS), operating under an emergency agreement with DOE, is providing technical expertise in the areas of stratigraphic, structural, geochemical, and remote-sensing analyses for the Eastern Gas Shales Project. The USGS will also serve

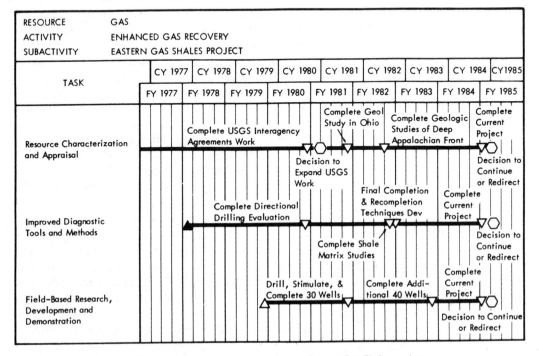

Figure II-5-1. Major milestones for the Eastern Gas Shales project.

as the technical coordinator for the numerous support contractors engaged in resource characterization work.

Technology. Various extraction technologies (drilling, stimulation, and production) will be evaluated through a series of carefully designed field experiments. Viable techniques will then be tested and verified by a series of multiple well tests conducted within limited geologic locales.

These Technology Testing and Verification experiments will be conducted in two phases:

- Phase I entails drilling sets of three or more "screening" wells at 20 locations. These screening locations will be used to collect adequate data to characterize the resource, evaluate geologic control, and test evolving stimulation technology.
- Phase II will involve selection of prime candidate sites at which full-scale commercial demonstrations (10 to 15 additional wells) will be conducted.

Major field contracts expected to be in force during FY 1980 are described below.

Mitchell Energy Corporation (Gallia County, Ohio) will drill, stimulate, and test up to 11 Devonian shale wells in the Ohio Shale formation of southern Ohio. A principal goal of this experiment is to demonstrate a new, unconventional technique for selectively locating and economically recovering gas from fractured portions of the shale formations. Additional research activities in the fields of remote sensing, geophysical logging, advanced hydraulic fracturing, and production testing will be conducted as an integral part of this contract. Initiated in FY 1979, experimental results will be available in early FY 1981.

The five wells in Phase I of this program have been drilled and are in some stage of pre-frac or post-frac evaluation by September 1979. At this time, a management decision will be made either to expand to Phase II (additional six wells) or terminate the work with completion of Phase I activities.

Several field demonstration contracts will be awarded to investigate the technical and economic feasibility of dual completions (Devonian shale and another reservoir) in individual wells. Contract awards were made in mid-1979.

Additional RFP's to evaluate chemical explosive fracturing and directional drilling techniques will be issued in FY 1980.

Western Gas Sands Project

WGSP is directed toward commercial development of the low permeability ("tight") sandstone reservoirs of the Piceance, Uintah, and Green River Basins of the Rocky Mountain region.

Specific goals set for this project include:

- Accurately define the resource base
- Identify the relationships between the physical and chemical properties of the sandstone reservoirs
- Develop effective stimulation and production technologies
- Assess potential recoverable gas reserves and demonstrate economic productivity leading to full-scale commercial development

Achieving these objectives will require a highly coordinated effort consisting of resource assessment studies, development of resource and stimulation testing techniques, field demonstrations of various stimulation options, and effective transfer of all technical and economic data to private industry.

Resource. The major portion of the resource assessment work is being performed by the USGS via an interagency agreement with DOE. Unlike the eastern United States, where much of the geological expertise resides in state geological surveys and universities, the bulk of the geologic data on the Rocky Mountain region resides in the USGS.

Resource assessment includes geological and geophysical studies to better understand the target resource base. One of the tasks will be to continue the gathering and synthesis of available well data taken from both an existing computerized data bank and other technical information libraries. Continued effort in general and detailed mapping in particular are needed to improve the understanding of the

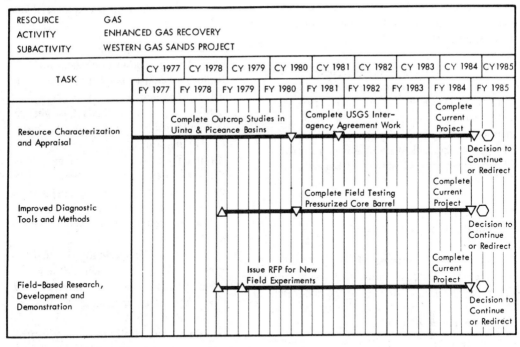

Figure II-5-2. Major milestones for the Western Gas Sands project.

gas-bearing formations and trapping mechanisms. This work will lead to selection of sites where subsurface information will be collected via cores, geophysical logs, and gas-production tests. Some of these sites will likely become the locations for field tests; when sufficient information is accumulated, resources and reserve estimates will be possible. The appraisals will delineate and characterize the reservoirs of each area that are the most promising for economic development.

Separate contracts with universities, research laboratories, and private consulting firms will be negotiated where specialized areas of research beyond the USGS capabilities are required.

Technology. Laboratory R&D activities are directed toward new logging and coring tools, rock mechanics investigations, mathematical reservoir simulation, core analysis, and fracturing technology advances. These activities have primarily been performed by the Energy Technology Centers, the national laboratories, and the USGS.

The field test and demonstration activities will evaluate advanced gas well stimulation techniques and determine the criteria for successful, cost-effective field applications to specific geologic targets. Some of the critical objectives of this research are as follows:

- Determine under what conditions the fracture will traverse a significant portion of the sandstone reservoir, which is actually penetrated by the wellbore
- Determine under what conditions a fracture will effectively intersect a sandstone lens which is not initially contacted by the wellbore
- Control the height of the fracture, such that the induced energy serves to propagate fracture length rather than ineffectual fracture height
- Successfully make use of numerous massive hydraulic fractures from the wellbore
- Create and maintain effective fracture conductivity, particularly in the deeper formations, where confining pressures tend to close the fracture and crush the propping agents

The ultimate goal of the field test and demonstration activities is to provide a statistically soundproof leading to full-scale commercial development of these tight sandstone reservoirs.

Major contracts in force during FY 1980 are predicated upon awarding work to respondents to several RFP's issued in FY 1979. No RFP has been issued and the potential contractors are unknown. Up to four major field tests are anticipated to be ongoing in the Uintah, Piceance, and Green River Basins during FY 1980.

Methane from Coal Project

During the natural process of coal formation, methane (the principal constituent of natural gas) is generated and trapped in the coal seam as well as the adjacent rock strata. All coal deposits contain methane although the concentrations vary from seam to seam and within the seam. The total magnitude of the U.S. coal-associated resource has been estimated at approximately 800 trillion cubic feet (800 quads). Given current and conservatively projected economic and technological factors, the recovery of some 300 trillion cubic feet of this resource appears readily feasible. This is equal to a 10- to 12-year supply based on present consumption.

The shortage of natural gas during the winter of 1976–77 focused attention on the urgent requirement to use this energy resource. Safety considerations in active mines have led to much development work on techniques for methane removal by the U.S. Bureau of Mines.

These techniques are now practiced by some mine operators with approximately 250 MMcfd of methane being emitted to the atmosphere. This gas is now being irretrievably wasted. The utilization of this valuable resource, including the methane available from "unminable" coal beds, is the objective of this task. RD&D necessary for economical utilization is provided.

Due to the recent natural gas shortage, economics are becoming more favorable for commercial exploitation of this resource. There are, however, still many barriers to extensive recovery and utilization on a commercial basis:

- Although a variety of technologies and techniques give promise of profitable utilization of coal-associated methane, the technical, operational, and economic viability of these methods has not yet been sufficiently demonstrated to attract private investment.
- The quality of coal-associated gas varies from essentially pure methane for pre-drained gas, to variable combinations of methane and air for gob gas, to extremely diluted methane-air mixtures in ventilation air.
- Gas sources are generally located remotely with respect to demand and individual wells have relatively low production rates compared to this resource.
- Gas prices historically have been insufficient to attract interest to this resource.
- Although coal operators have a legal right to release methane in the course of mining, they are wary of the legal implications of gas recovery, since generally, natural gas rights are held by others.
- Since the market value of a ton of coal is on the order of 100 times the value of the methane contained therein, coal mining companies have scant interest in gas-derived revenues relative to their primary objective of coal production. (Recent interest in gas recovery by a few coal companies is due primarily to the prospect of increased coal production. The methane-derived revenues are of secondary importance.)
- Low gas prices, low productivity, and the lack of a requirement for additional gas sources in the past have resulted in little or no interest in recovering methane from unminable coal.

The ranges of quantitative availability, gas quality, and geographic location of coal bed methane sources make it apparent that no single solution is appropriate for all cases. Studies conducted for ERDA (now DOE) and the Bureau of Mines indicate a high probability of economic gas recovery/utilization for several approaches, including direct pipeline injection, LNG production, on-site power generation, and petrochemical production. A variety of specific techniques and technologies have been considered under each of the preceding major headings. In some cases, off-the-shelf technology which could be modified for this application is available (e.g., gas turbines, LNG production units, and ammonia production units). However, further investigation, research, development, and demonstration is necessary.

The Methane from Coal Project (MCP) will develop methods and systems that establish technically and economically viable means for the recovery, conservation, and use of the methane gas associated with both minable and unminable (deep and/or thinly bedded) coal beds.

Major project objectives include:

- Location and characterization of methane resources
- Development of improved, cost-effective methane recovery and utilization technology
- Development of methane conservation techniques and systems
- Development of methane recovery prediction/projection techniques (models for well productivity)
- Development of field tests for pilot systems
- Investigation of legal and institutional constraints
- Transfer of applicable technologies to private industry

These project objectives will be realized by successful completion of each of the following project elements:

- Resource Characterization—identification and definition of the coal bed methane resource so the most attractive targets may be selected
- Research and Development—development of improved, more cost-effective methods and subsystems for conservation, recovery, and utilization of methane
- Pilot System Applications—demonstration of the technical and economic viability of a number of different system coal bed combinations to accommodate the variety of specific site conditions that will be encountered in large-scale commercialization.

Major MCP milestones are given in Figure II 5-3.

Resource. The characterization of the methane content of the U.S. coal beds has been done on a very limited basis, mostly in conjunction with active mining. Previous work includes only a small percentage of the coal resources and does not provide the knowledge needed to locate recovery and utilization projects in coal beds with the greatest potential for methane production.

The resource characterization effort will involve:

- Development of an overall methane resource delineation plan
- Development of core/sample drilling plans
- Acquisition of resource characterization from existing sources
- Analysis and evaluation of existing data
- Identification of potential recovery system sites
- Compilation of an all-inclusive data base for transfer to private industry

The existing sources of public information on methane in coal beds include the USGS, U.S. Bureau of Mines, several state geological surveys, various universities, and private research foundations.

Some of the major contractors engaged in this research as of FY 1979 include:

- Colorado Geological Survey
- Intercomp, Inc.
- TRW Energy Systems
- Utah Geologic and Mineral Survey
- Iowa State University

Technology. Two related areas of research are addressed in the technology development project element: recovery technologies and utilization systems.

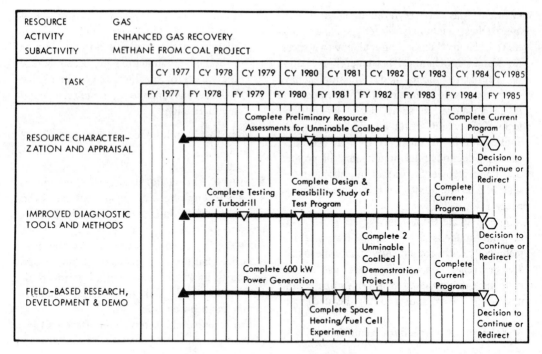

Figure II-5-3. Major milestones for the methane from coal project.

A number of techniques are available for producing methane from coal beds:

- Drilling of vertical wells with the option to stimulate methane production using hydraulic fracturing techniques
- Drilling of horizontal holes from the bottom of vent shafts or from the headings within the mine
- Drilling of directionally controlled wells from the surface so as to intercept the major continuous fracture system (face cleats) within the coal bed

None of these techniques has been developed to the degree that they are efficient and cost effective.

Utilization options for methane from coal beds include space heating, power generation, and use as a chemical feedstock. Selection for pilot systems will consider operational, institutional, and economic factors such as volume and characterization of gas production, environmental aspects, and cost/benefit of the particular method. Typical utilization system concepts include:

- Direct injection into commercial gas transmission pipelines
- Conversion to LNG
- Heating
- Electric power generation
- Conversion to ammonia

Major technology research and demonstration contracts expected to be ongoing in FY 1980 are described in the following paragraphs.

Westinghouse Electric Corporation (Pittsburgh, Pennsylvania) was awarded a contract (FY 1977) to test the feasibility of operating a 600-kW turbine/generator on predrainage (in advance of mining) and "gob" (impure methane produced after mining operations) gas. The 1-1/2 year field testing period will be concluded in September 1980. Successful results of this experiment could have significant impact upon on-site power generating systems for supplying operating coal mines with electricity. The turbine will have been on-line for approximately six months by the start of FY 1980.

Intercomp, Inc./Taiga Energy (Rio Blanco,

Colorado) was awarded a contract (FY 1979) to demonstrate recovery and utilization subsystems relating to methane from a number of unminable coalbeds in western Colorado. The anticipated three-year contract will involve drilling a series of vertical wells, hydraulically fracturing the wells, monitoring reservoir performance, developing completion techniques, and perfecting predictive reservoir modeling of methane production. Any commercial quantities of gas would be compressed and injected into existing gas transmission lines. The design phase will be completed and drilling of the wells will be underway by the start of FY 1980.

The contract let to Mountain Fuel Resources, Inc. (Book Cliffs, Utah) in FY 1979 involves a similar research effort to that of the Intercomp/Taiga contract. A series of thin coalbeds will be drilled, stimulated, and production-tested for methane. This three-year experiment will yield valuable information relative to the energy potential of methane from coalbeds in the Uintah Basin. The design phase will be completed and drilling underway by the start of FY 1980.

The Westinghouse contract (FY 1978) involves testing the recovery technology of using vertically drilled and hydraulically fractured wells in advance of mining (predrainage of the methane). The produced methane will be used for space heating and as a feedstock for a fuel cell. All wells will be drilled, completed, and delivering into the system by early FY 1980.

TRW Energy Systems (Morgantown, West Virginia) is an integrating contractor for selection, design, and coordination of the resource characterization effort and the recovery/utilization systems technology contracts. In addition to normal project planning and coordinating functions, monitoring and technical/economic evaluation must take place between the several pilot systems and demonstrations in order to provide effective project management.

Several as yet unreleased RFP's will solicit additional contracts to test recovery/utilization systems for both minable and unminable coal beds. Contract awards are anticipated in late FY 1979 and FY 1980.

Resource Characterization

Resource characterization work involves an evaluation of the magnitude and distribution of the resources plus characterization studies to determine the physical and chemical properties affecting reservoir behavior. These support programs involve comprehensive surface and subsurface geological mapping, structural and rock mechanics studies, geochemical studies, mineralogy studies, geophysical logging and interpretation techniques, reservoir engineering studies, and computer data analyses.

Due to present uneconomic aspects of the gas resources under investigation, the natural gas industry is not putting any effort into defining either the size of the resources or their producing characteristics. DOE must provide the initiative by developing a reliable data base of technical information. The development of this data base will provide incentives for future field tests of the various stimulation technologies.

Environment and Support

This EGR Subactivity includes controls to ensure that all field projects will be conducted in accordance with existing and foreseeable environmental, health, and safety (EH&S) requirements (See Figure II 5-4). A thorough understanding of the environmental implications of the new drilling and fracturing techniques is important to the success of the EGR. Environmental assessments will be made of all planned field activities to ensure proper consideration of all potential problems and to minimize adverse environmental impacts. Data compiled and analyzed during the program should provide a sound basis for predicting the environmental impact of commercial operations utilizing similar technologies.

Additional support work, such as economic analyses to assess cost effectiveness of various techniques, will provide the basis for defining and refining the overall EGR activities. Management support from professional systems analysts and engineering consultants will be used extensively to develop a coordinated, coherent management plan for EGR. (See Figure II 5-5 and II 5-6)

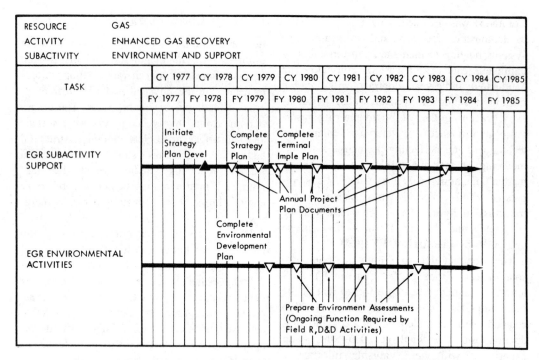

Figure II-5-4. Major milestones for the environment and support subactivity.

In addition to conducting field experiments, DOE, the national laboratories, and several independent research laboratories are providing technology support for various advanced hydraulic fracturing (AHF) and chemical explosive fracturing (CEF) experiments.

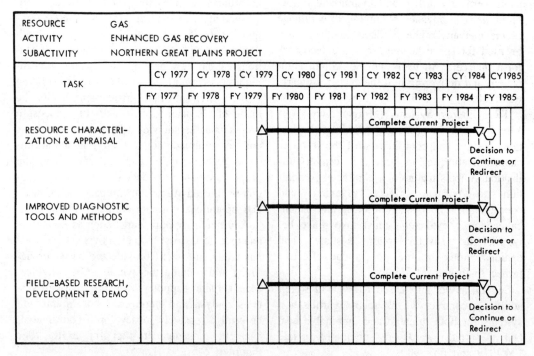

Figure II-5-5. Major milestones for the Northern Great Plains project.

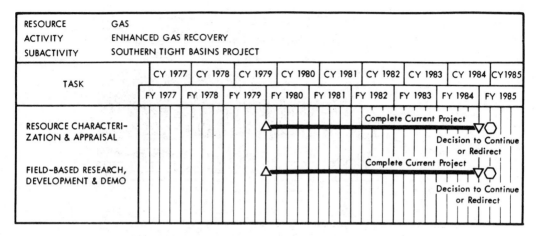

Figure II-5-6. Major milestones for the Southern Tight Basins project.

SL, Albuquerque, New Mexico, is developing a technique for mapping the length and direction of AHF-induced fractures using an electropotential resistivity technique. SL is also conducting field experiments aimed at relating actual AHF- and CEF-induced fracture geometry to theoretical rock mechanics predictions. The fractures are delineated by physically mining back into the fractured rock and mapping the fractures. These tests are being carried out at the Nevada Test Site. The ultimate goal of these experiments is to enable engineers to correctly predict the direction and extent of hydraulically induced fractures prior to actually performing the AHF job. Sandia is also investigating various log techniques to determine formation water saturation in the presence of shales.

LLL has developed one- and two-dimension computer codes to model fracture behavior and is in the process of developing three dimensional codes. LLL is also experimenting with geophysical logging techniques to aid in detecting natural fractures and predicting resultant reservoir performance.

LASL will experiment with shaped-charge explosives as a technique for fracturing Devonian shales as a means of increasing permeability or providing conduits for emplacing liquid explosives for fracturing. LASL is also conducting research into dynamic mechanical characterization and calculational (computer) design methods for optimizing chemical explosives fracturing techniques.

The Energy Technology Centers in Bartlesville, Oklahoma, and Morgantown, West Virginia, provide both field management and oversight responsibility and in-house supporting research to the DOE EGR Program. Bartlesville provides direct support to the Western Gas Sands Project. Morgantown supports the Eastern Gas Shale and Methane from Coal Projects.

Support Activities

Processes and efforts currently being developed and tested under the EGR Activity to recover the unconventional gas resources previously outlined include:

- Advanced Hydraulic Fracturing (AHF) Process
- Chemical Explosive Fracturing (CEF) Process
- Deviated Wells and Earth Fracture Systems Process
- Resources Characterization
- Environmental and Supporting Research

Advanced Hydraulic Fracturing Process. Hydraulic fracturing is a mechanical process for creating an extensive fracture in a reservoir formation. In the single-well technique illustrated in Figure II 5-7, fluid is injected through the wellbore to overcome natural stresses within the reservoir rock, thus causing a fracture to develop. Coarse sand is car-

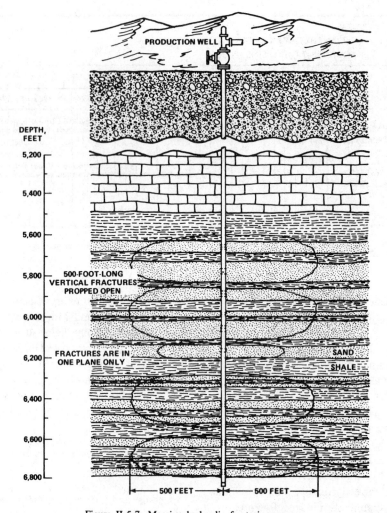

Figure II-5-7. Massive hydraulic fracturing process.

ried along with the fluid and acts as a propping material to hold the fracture open after fluid injection is stopped and the attendant pressure is released. These fractures increase formation permeability by permitting faster migration of gas into the wellbore, thus increasing gas production.

The first commercial hydraulic fracturing job was performed in 1947. This first small-scale project involving the injection of 2,000 to 3,000 gallons of fluid has gradually evolved to the point where fracturing jobs requiring 500,000 gallons of fluid, 1 million pounds of sand, and injection pressures in excess of 15,000 psi are not exceptional. Primary advances in AHF technology have been in equipment development (large-volume, high-pressure portable pumping units) and refinements in the science of fracture mechanics and reservoir engineering.

Although many of the AHF techniques being investigated by DOE are not totally unique to the gas industry, the techniques still need further refinement and have not been extensively employed to develop the gas resources under consideration. DOE is taking an active part in developing reliable AHF stimulation technology with the ultimate goal of transferring successful techniques to the oil and gas industry.

DOE is funding contracts in AHF to test new techniques and fracturing concepts in several different geologic formations. The projects were selected on the basis of their demonstrated technical merit. In addition, all the contractors are actively engaged in drilling and exploration within the geographic areas in which DOE's program activities are concentrated (the Devonian shales of the eastern United States and the tight gas sands of the Rocky Mountain region).

Chemical Explosive Fracturing Process. There are two basic approaches to the chemical explosive fracturing of gas wells. Borehole shooting, which consists of fracturing a portion of the open hole with nitroglycerin or a shaped-charge explosive, has been employed in some areas for many years. A newer concept in CEF, shown in Figure II 5-8, involves pumping a chemical explosive slurry into a formation and then detonating the mixture. In some cases the injection pressure is sufficient for the explosive slurry to actually induce hydraulic fracturing of the reservoir. The explosive slurry can also be pumped into existing natural fractures or fractures resulting from hydraulic fracturing. Current fracture mechanics theory predicts that the explosive shock wave resulting from detonation extends the existing fractures, thus increasing permeability and gas production.

Borehole shots using nitroglycerin have been used routinely for more than 60 years. Development of more sophisticated explosive compounds and shaped-charge propellants began in the mid-1940s. The results of crude borehole shooting with nitroglycerin are generally unpredictable and not particularly cost-effective, in terms of incremental gas production, over the long-term well life. Many of the newer techniques have evolved from R&D

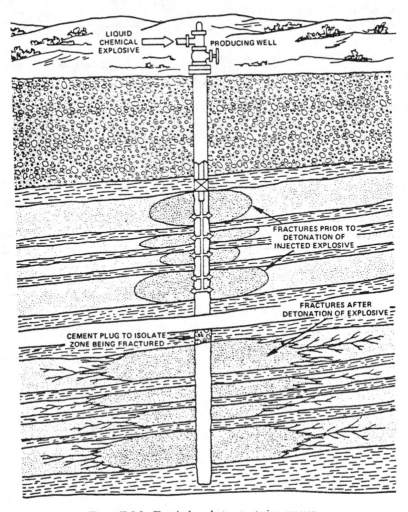

Figure II-5-8. Chemical explosive fracturing process.

activities in missile and military explosives technology. The DOE program in CEF provides the stimulus for development of more sophisticated explosives (rocket propellants and other tuned explosives) and design concepts based upon modern rock mechanics studies. The goal of the DOE-sponsored research is to achieve better (in terms of production), more cost-effective, and predictable final results. The DOE program fills a technological void that industry is not actively pursuing.

The DOE projects funded for CEF research and development were selected because the proposed techniques are unique when compared with current commercial CEF techniques.

Directionally Controlled Wells and Earth Fracture Systems Process. Directionally drilled (deviated) wells attempt to use existing natural fracture systems to increase gas production. The natural fracture system is thoroughly studied and mapped prior to drilling. Once the orientation of local fracture systems is determined, an optimum drill site is selected and a well is directionally drilled to maximize the probability of intersecting the fracture system. Each fracture cut by the wellbore represents a path of higher permeability for gas migration; the greater the number of fractures, the larger the gas flow. In addition, the fracture system will provide conduits for hydraulic or explosive fracturing fluids for subsequent MHF or CEF stimulations. The ultimate effect is an extension of existing fractures, an increase in reservoir permeability, and improved production of gas above the production rate expected by drilling conventional vertical wells. The Devonian shales of the eastern United States should be excellent candidates for directional drilling techniques, since the shales are intensely fractured in several areas. (See Figure II 5-9.)

Directional drilling of multiple wells from a

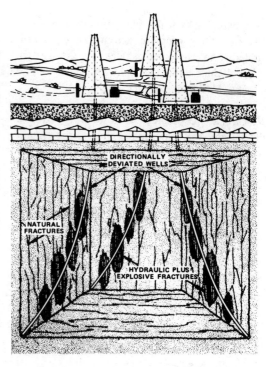

Figure II-5-9. Directionally controlled wells and earth fracture systems process.

single platform is a standard technique for producing oil and gas reserves in the offshore areas. Directional drilling of wells in the gas fields located onshore is rarely done except as a means of avoiding some surface obstacle or as a safety relief well for a blowout or a well on fire. The gas-producing formations onshore generally are also much harder than poorly consolidated sediments in many offshore areas. Generally speaking, there have been few incentives and, subsequently, little expertise developed by the oil and gas industry with regard to directional drilling in the Appalachian or Rocky Mountain regions where DOE is conducting its program activities. DOE believes it should actively sponsor R&D efforts in this area, especially as the technique can be applied to development of the Devonian shales.

6.
Field Activities

The significant technology center and national laboratory support of DOE fossil energy activity elements are discussed in this chapter. Table II 6-1 summarizes these activities.

ENERGY TECHNOLOGY CENTERS

The ETC's play an important role in achieving DOE goals in Fossil Energy technology. They have demonstrated the ability to achieve solutions to specific technology-base problems affecting the timetable of major Fossil Energy systems development. In addition to the in-house technology capability efforts at each center, ETC technical personnel have functioned in supervisory and liaison capacities in the management of contract research work with universities and laboratories in every major area of the Fossil Energy development program. The mission of the ETC's is being expanded greatly to encompass the role of manager for Fossil Energy projects. Lead centers have been identified for major Fossil Energy technologies. Although relatively small, the ETC's are staffed by highly skilled scientists, engineers, and technicians with a long history of accomplishments in R&D solutions to technology-base problems and in the production and utilization of fossil fuels. They have produced new ideas, new data, and new processes for better utilization of Fossil Energy resources.

Although the ETC's are a part of Fossil Energy, a few projects are conducted for other DOE efforts, primarily in the Conservation, Environment and Safety, and Physical Research program areas.

Table II 6-2 presents resource levels devoted to each program at the ETC's.

Locations of the ETC's are provided in Figure II 6-1.

Bartlesville ETC (BETC)

Research at BETC is focused primarily on project management of EOR and internal combustion engines (spark/compression/ignition activities). Work also centers on the production of petroleum and natural gas with in-house efforts on processing and utilization of petroleum. A lesser effort is devoted to the characterization and use of liquid fuels manufactured from nonpetroleum sources. The center plays a major role in providing in-house support and research and technical cognizance for DOE's field demonstration projects in EOR. An increasing emphasis is being placed on the enhanced recovery of petroleum and natural gas from presently known U.S. reserves, especially those with near-term potential.

Research and development activities in the area of oil and gas extraction in projects cost-shared with industry are centered on developing improved secondary and tertiary methods for the recovery of oil and gas. Field operations are being conducted in micellar-polymer flooding, caustic flooding, and thermal recovery. The effectiveness of massive hydraulic fracturing and the use of chemical explosives

Table II-6-1. Fossil energy development energy technology centers/mining technology center/national laboratories support activities.

TECHNOLOGY CENTER/NATIONAL LABORATORY/ MINING TECHNOLOGY CENTER	LEAD LABORATORY	LIQUEFACTION	GASIFICATION		
			HIGH-BTU	LOW-BTU	IN SITU
Ames Laboratory					
Argonne National Laboratory		Support studies for synthoil process.	Instrumentation and process control for demonstration plants.		Gasification of Chars.
Bartlesville Energy Technology Center	Enhanced Oil Recovery Internal Combustion Engines (spare/compression ignition)				
Carbondale Mining Technology Center	Mining Research and Development				
Brookhaven National Laboratory					
Grand Forks Energy Technology Center	"Applications Center" for low-rank coals	Development of CO-Steam process.		Fixed-bed gasification; slagging ash method.	
Idaho National Energy Laboratory					
Laramie Energy Technology Center	Oil Shale Non-oil Shale In Situ Processes				In Situ coal gasification technology. Linked vert well process.
Lawrence Berkeley Laboratory					

Table II-6-1. (Continued).

	COAL				
HEAT ENGINES AND HEAT RECOVERY	COMBUSTION SYSTEMS	FUEL CELLS	ADVANCED ENVIRONMENTAL CONTROL TECHNOLOGY	ADVANCED RESEARCH AND DEVELOPMENT TECHNOLOGY	MHD
				Coal beneficiation alloy evaluation.	
Shape memory alloy studies Total Energy Technology Assessment study; stationary Stirling engine support; ceramic components for heat exchanges.	Pressurized Fluidized-bed combustion. CTIF. Support sudies for fluidized bed combustion; pulse burner design principal alternate fuel combustion.	Fuels utilization; molten carbonate systems support.		Materials technology for coal conversion processes.	CDIF design & component development support, U-25 magnet test support and super conducting magnet system.
				Heat content of coal chars, characterization of syncrudes. Automative testing of synthetic fuels.	
	Secondary alternate fuel combustion.	Transportation applications for fuel cells; solid oxide & electrocatalysis.		Flash hydropyrolysis of lignite. Desulfurization of fuel gases.	
	Sulfur oxide emission control. Electrostatic precipitation of fly ash.			Ash fouling studies. Peat utilization.	
				High temperature materials and welding.	
	Continuous combustion engine studies.			High temperature erosion corrosion resistance materials development. NMR determination of coal composition and structure.	

Table II-6-1. (Continued).

TECHNOLOGY CENTER / NATIONAL LABORATORY / MINING TECHNOLOGY CENTER	LEAD LABORATORY	LIQUEFACTION	GASIFICATION		
			HIGH-BTU	LOW-BTU	IN SITU
Lawrence Livermore Laboratory					Development of packed-bed process.
Los Alamos Scientific Laboratory					Water requirements in coal gasification.
Morgantown Energy Technology Center	Unconventional Gas Recovery Fluidized—Bed Combustion Gas Stream Cleanup (includes Flue Gas) Coal Gasification, Feeding and Testing		Hydrane process scaleup.	Fixed-bed gasification. Hot gas cleanup absorbent method.	Underground gasification of eastern coals.
Mound Laboratory				Technical support for low-Btu Fuel Gas Demonstration Plant.	
Oak Ridge National Laboratory	Materials Technology	Hydrocarbonization in fluidized-bed reactors process modeling.	Process Modeling.	Process Modeling.	Pyrolysis of large blocks of coals.
Pacific Northwest Laboratory					
Pittsburgh Energy Technology Center	Coal Liquefaction Synthetic Liquids Characterization Coal Mixtures, Direct Combustion, MHD Combustion.	Development of synthoil process hydrogenation of coal. Solvent Refined Coal.	Dilute phase hydrogasification to produce synthetic natural gas. Synthane Process Dev.		
Sandia Laboratory		Support studies for synthoil process.			Instrumentation and process control development.
San Francisco Office of Program Coordination and Management					

Table II-6-1. (Continued).

HEAT ENGINES AND HEAT RECOVERY	COAL COMBUSTION SYSTEMS	FUEL CELLS	ADVANCED ENVIRONMENTAL CONTROL TECHNOLOGY	ADVANCED RESEARCH AND DEVELOPMENT TECHNOLOGY	MHD
	Direct-injected stratified-charge engine combustion.				
Ceramic heat pipes.		Fuel cells for transportation.			
	Atmospheric Fluidized bed CTIU* Fouling of surfaces Coalbed Methane Utilization.	Coal gasification/ fuel cell testing.		Coal liquefaction Catalytic Hydrogenation.	
				Valve and component test facility.	
Thermodynamic working fluids.	Alkali metal boiler with fluidized-bed. Fluidized-bed combustion. AFB test unit.	Molten carbonate testing.		Separations (liquefaction) technology; desulfurization; inspection techniques for coatings.	
Adiabatic diesel coal combustion	Solid fuel combustion research facility. Combustion of coal-oil slurries.		Removal of Sulfur from coal. Environmental Guidelines. Environmental Impact and control technology for CCU.	Comprehensive basic research program in coal fossil fuel development.	Combustor development for coal-fired MHD.
	Direct-injected stratified-charge engine combustion.			Protective sulfidation of alloys.	

*Component Test and Integration Unit

Table II-6-1. (Continued).

TECHNOLOGY CENTER/ NATIONAL LABORATORY	PETROLEUM				GAS	
	ENHANCED OIL RECOVERY	OIL FROM OIL SHALE	DRILLING AND OFFSHORE TECHNOLOGY	CHARACTERIZATION AND UTILIZATION	ENHANCED GAS RECOVERY	GAS FROM OIL SHALE
Ames Laboratory						
Argonne National Laboratory						
Bartlesville Energy Technology Center	Enhanced Oil Recovery; micell. or properties reservoir engineering; Project management.		Improved drilling and coring methods; telemetry; offshore technology.	Characterization; engine efficiency.	Production by massive hydraulic fracturing; chemical explosives.	
Brookhaven National Laboratory						
Grand Forks Energy Technology Center						
Idaho National Energy Laboratory						
Laramie Energy Technology Center	Recovery from tar sands; thermal recovery support.	In situ and modified in situ retorting; evaluation of oil shales and shale oils. Environmental studies. Project management.		Characterization of heavy liquids; asphalt studies.		In situ shale retorting and shale gasification R&D.
Lawrence Berkeley Laboratory	Mobility and surface tension control agent R&D.	Water requirements in situ retorting R&D.				
Lawrence Livermore Laboratory		Modified in situ oil shale retorting technology.			Massive hydraulic fracturing support.	

Table II-6-1. (Continued).

TECHNOLOGY CENTER/ NATIONAL LABORATORY	PETROLEUM				GAS	
	ENHANCED OIL RECOVERY	OIL FROM OIL SHALE	DRILLING AND OFFSHORE TECHNOLOGY	CHARACTERIZATION AND UTILIZATION	ENHANCED GAS RECOVERY	GAS FROM OIL SHALE
Los Alamos Scientific Laboratory		Explosive fracturing R&D.			Characterization of Devonian shale and explosive stimulation.	
Morgantown Energy Technology Center	Carbon dioxide injection technology.				Resource characterization Devonian shale; Program management.	
Mound Laboratory	Chemicals for EOR; Multiple tracers and ion exchange effects in EOR.				Physical and chemical characterization of Devonian shale.	
Oak Ridge National Laboratory						
Pacific Northwest Laboratory						
Pittsburgh Energy Technology Center						
Sandia Laboratory		In situ bed preparation. Advanced instrumentation and rock mechanics support.	Offshore drilling technology; subsurface properties.		Massive hydraulic fracturing R&D. Instrumentation and ADP.	
San Francisco Office of Program Coordination and Management	Project monitoring, Stanford Petroleum Recovery Institute.					

to increase recovery of natural gas from tight formations are being investigated. Research studies will continue on the characterization of crude oil and its products and on improved process methods.

Characterization studies of the properties of syncrudes from coal were initiated in FY 1976 to provide the bases for processes to upgrade syncrudes to refinery feedstocks and finished fuels. In FY 1979, research continued on the

Table II-6-2. Energy technology center and mining technology center funding levels.

	ESTIMATE FY 1978	ESTIMATE FY 1979	ESTIMATE FY 1980
BARTLESVILLE ENERGY TECHNOLOGY CENTER			
Operating Expenses:			
Coal	831	460	850
Petroleum	26,055	16,510	8,800
Gas	1,057	5,500	4,900
Conservation			
Total Program Operating Expenses, BETC	27,943	22,470	14,550
GRAND FORKS ENERGY TECHNOLOGY CENTER			
Operating Expenses:			
Coal	5,987	6,814	4,575
Petroleum	0	0	0
Gas	0	0	0
Conservation			
Total Program Operating Expenses, GFETC	5,987	6,814	4,575
LARAMIE ENERGY TECHNOLOGY CENTER			
Operating Expenses:			
Coal	4,013	3,425	2,000
Petroleum	20,867	21,053	7,200
Gas	534	529	0
Conservation			
Total Program Operating Expenses, LETC	25,414	25,007	9,200
MORGANTOWN ENERGY TECHNOLOGY CENTER			
Operating Expenses:			
Coal	17,030	18,413	24,060
Petroleum	1,720	5,370	3,000
Gas	15,181	16,000	9,800
Conservation			
Total Program Operating Expenses, METC	33,931	39,783	36,860

Table II-6-2. (Continued).

	ESTIMATE FY 1978	ESTIMATE FY 1979	ESTIMATE FY 1980
PITTSBURGH ENERGY TECHNOLOGY CENTER			
Operating Expenses:			
Coal	19,599	20,691	11,180
Petroleum	0	0	0
Gas	0	0	0
Conservation			
Total Program Operating Expenses, PETC	19,599	20,691	11,180
ENERGY TECHNOLOGY CENTERS			
Operating Expenses:			
Coal	47,460	49,803	42,665
Petroleum	48,642	42,933	19,000
Gas	16,772	22,029	14,700
Conservation			
Total Program Operating Expenses, ETCs	112,874	114,765	76,365
CARBONDALE MINING TECHNOLOGY CENTER			
Operating Expenses:			
Total Program Operating Expenses, CMOD	225	2,560	5,010
PITTSBURGH MINING TECHNOLOGY CENTER			
Operating Expenses:	260	2,850	7,014

properties, combustion, and engine performance characteristics of liquid fuels derived from coal.

Grand Forks ETC (GFETC)

GFETC is concerned primarily with near- and mid-term technologies for the utilization of low-rank coals, including lignite and subbituminous coals, in an environmentally sound manner to alleviate urgent energy supply problems in the western and midwestern United States. GFETC has been established as an "Application Center" for this technology-base activity.

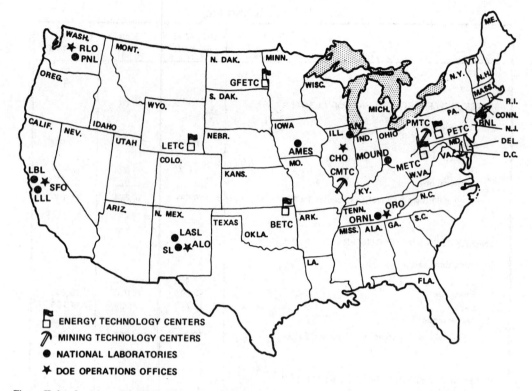

Figure II-6-1. Location of DOE fossil energy technology centers, mining technology centers, and national laboratories.

Research and development activities are continuing in the special problem areas of coal liquefaction, gasification, and combustion and in the control of pollution arising from utilization of the relatively abundant lower Btu-value coals. The effect of mineral matter in lignite on liquefaction will be studied in the CO-steam project. Since the fixed-bed slagging gasifier is a major project, detailed investigation of liquid effluents is being initiated. Development of these data affects the design of scaleup systems and is important for environmental considerations.

Laramic ETC (LETC)

LETC serves as the lead center for technology-base activities on the production and utilization of vast oil shale resources. Additionally, it is responsible for all work on non-oil shale work in situ processes.

In oil shale, current emphasis is being placed on the development of in situ retorting processes for recovering oil from oil shale deposits. Process support and field studies focus on developing fracturing techniques and methods for recovering oil from the fractured shale and on refining the shale oil. Modified in situ processes involving industry contracts and monitoring of above-ground retorting studies also are receiving major emphasis. A major effort is underway to develop underground coal gasification processes. A field test site is located near Hanna, Wyoming, in a seam of subbituminous coal 30 feet thick at a depth of about 400 feet. In FY 1979, the main program thrust was development and demonstration of the modular concept for expanding underground coal gasification technology to a commercial scale. Environment and safety research activities will also be part of the in situ coal gasification and oil shale efforts in FY 1980. Projects jointly managed with Fossil Energy activities are designed to identify environmental control needs for effluents from in situ operations. Data will be obtained from an in situ environmental burn at the Hanna site and other field studies, and an analysis of environmental control needs will be made.

In FY 1979, principal petroleum and gas technology-base activities were aimed at the development of processes for recovering oil from tar sand deposits, characterization of heavy liquids derived from fossil fuels, and methods for improving the performance and utilization of asphalts.

Morgantown ETC (METC)

Research, development, and technology-base activities on unconventional gas recovery fluidized-bed combustion gas stream cleanup, coal gasification, and coal-feeding systems are major activities at METC. A diverse program is carried out to develop technology for direct combustion and low-Btu gasification of coal. Enhanced recovery of oil and gas is also being investigated, as is in situ gasification of coal.

Current studies involve operation of a fixed-bed coal gasifier and an accelerated hot gas cleanup effort at the pilot plant scale. Construction of the atmospheric fluidized-bed component test and integration unit, which was begun in FY 1977, will continue. Operation is scheduled to begin in FY 1980. In the direct combustion program, fouling and corrosion of fireside surfaces are being studied.

Characterization and resource inventory of U.S. natural gas reserves will continue, with areas in Illinois, Michigan, and the Appalachian region being emphasized. Results of FY 1978 studies will be analyzed for major impacts on needed recovery technology. Additionally, the center provides technical oversight for DOE field demonstration projects on enhanced oil and gas recovery. Environmental considerations will be emphasized, and efforts in the utilization of methane recovered from coal beds will continue.

Program objectives are to demonstrate new technology, including directionally deviated wells that can be combined to degasify, devolatilize, carbonize, and gasify relatively thin, deep Eastern coal seams. In FY 1979, laboratory studies continued, and longwall generator flow testing and techniques for reverse burn and forward burn were completed.

Pittsburgh (ETC (PETC)

PETC is the largest Federal laboratory devoted solely to maintaining a technology-base R&D program on coal. The center carries out a diversified program on coal, including applied research to obtain new knowledge of the chemical and physical properties and reactions of coal and other technologies; these include coal liquefaction, synthetic liquids characterization, coal mixture, direct combustion, and MHD combustion.

Major efforts are underway in the development of processes to liquefy and gasify coal. Back-up research and technical direction are provided for the 75 tons/day Synthane coal gasification pilot plant. A 10 tons/day Synthoil process development unit is being operated to study the conversion of coal to low-sulfur fuel oil. PETC will demonstrate continued process reliability, develop improved components, and obtain process design data for the scaleup of the Synthoil process. In the coal hydroliquefaction process using disposable catalysts, improved catalyst blends will be developed and tested. Additional research is being conducted on combustion of coal-oil mixtures, coal combustion in adiabatic diesel engines, development of a coal-fired two-stage gasifier-combustor to supply a high-temperature plasma for an MHD power plant, and on chemical coal desulfurization. Research is being done on analytical techniques to characterize coal and coal products. Other major products will be in the R&D of technology in the Synthane process and in other advanced research and supporting technology. Environmental control technology, conservation, and safety of coal conversion processes will be studied intensively.

MINING TECHNOLOGY CENTERS (MTC's)

MTC's formerly under the Bureau of Mines Mining RD&D program, became an addition to Fossil Energy in accordance with the Department of Energy Organization Act of 1977, P.L. 95–91.

The two mining technology centers are located at Pittsburgh, Pennsylvania (PMTC), and Carbondale, Illinois (CMTC). The joint mission of these two centers is to conduct the research required to more efficiently mine and prepare coal (for direct combustion and for conversion to metallurgical coke and liquid and gaseous fuels) in ways that will:

- Conserve our nonrenewable coal reserves through high coal bed recovery and efficient coal preparation
- Minimize adverse environmental impacts through good mine plant design, proper choice of mining method, and proper operation of preparation plants
- Minimize the hazards to the mine worker and guarantee a healthy work environment in the mine and in the preparation plant
- Minimize overall production costs
- Maximize production of salable coal
- Be in the best public interest

Locations of the MTC's are shown in Figure II 6-1.

NATIONAL LABORATORIES

In addition to the significant contributions made by the Energy Technology Centers and the Mining Technology Centers to the overall Fossil Energy technology-base program, 11 national laboratories provide supportive contributions. These DOE laboratories possess unique facilities and in-house expertise which can contribute significantly in providing specialized task support to projects managed by Fossil Energy. These laboratories, like the ETC's, submit proposals for R&D work relating to Fossil Energy development in their respective areas of expertise. Many of these proposals are accepted and provide significant benefits to the various Fossil Energy technology-base programs.

A brief summary of planned activity in support of Fossil Energy programs is provided in the following 11 subsections. Locations of the national laboratories are given in Figure II 6-1.

Ames Laboratory (AL)

Ames Laboratory support concentrates on two basic research tasks: coal beneficiation and alloy evaluation for fossil fuel plants.

Argonne National Laboratory (ANL)

Argonne's work focuses on the coal resource, including a major effort on construction of the pressurized fluidized-bed (PFB) component test and integration unit (CTIU) as well as other studies in fluidized-bed combustion in support of Combustion Systems Activity technology, and work on high temperature/pressure samplers (high-Btu gasification). Argonne provides engineering support to the demonstration plant efforts, and has a second major effort in the MHD Component Development and Integration Facility as well as to other areas of open- and closed-cycle MHD technology and is involved in the U.S.–U.S.S.R. effort in MHD. Materials research for coal conversion processes is also included in ANL's task for Fossil Energy. Argonne also provides R&D support for the Heat Engines and Heat Recovery and the Fuel Cell Activities.

Brookhaven National Laboratory (BNL)

BNL provides limited basic research support in the coal R&D area, specifically relating to flash hydropyrolysis of coal and regenerative processes for desulfurization of high-temperature combustion and fuel bases. Additional efforts in principal alternate fuel combustion and fuel cells are being conducted.

Idaho National Energy Laboratory (INEL)

The Idaho National Energy Laboratory provides research support in the area of high-temperature materials, and it manages the construction of the MHD-CDIF.

Lawrence Berkeley Laboratory (LBL)

LBL provides support for two task areas: research in high-temperature erosion/corrosion resistance materials development and tertiary recovery with mobility control and surface active agents for enhanced oil recovery.

Lawrence Livermore Laboratory (LLL)

LLL is involved with supporting tasks in the massive hydraulic fracturing process for stimulating natural gas recovery. Supporting work in modified in situ oil shale recovery techniques and development of the packed-bed process for the In Situ Coal Gasification Subactivity are also being accomplished at LLL. In addition, efforts in direct injected stratified charge engine combustion are being conducted.

Los Alamos Scientific Laboratory (LASL)

LASL has two tasks in the Oil Shale and In Situ Technology Subactivities. Specifically, these efforts are explosively produced fracture of oil shale and in situ coal gasification of southwest coals with minimal water requirements. LASL also is investigating the explosive stimulation of natural gas and laser pyrolysis for characterization of Devonian shale, ceramic heat pipes, and fuel cell transportation applications.

Mound Laboratory (ML)

Mound Laboratory is providing technical support for the fuel gas demonstration plant project and has a project for characterization of Devonian shale.

Oak Ridge National Laboratory (ORNL)

Oak Ridge is the principal support national laboratory for coal technology and is performing many support tasks in the areas of demonstration plants by providing technical assistance. ORNL also has been selected as the lead laboratory for materials technology. Several research support tasks primarily in materials and component development are also planned, along with coal liquefaction research support. In the Gas and the Petroleum Resources Activities, ORNL is providing support in the enhanced oil recovery and processing and utilization areas. ORNL is supporting In Situ Coal Gasification technology by pyrolysis of large coal blocks, the Heat Engines and Heat Recovery Activity, and molten carbonate fuel cell efforts.

Pacific Northwest Laboratory (PNL)

PNL had two tasks planned for FY 1979 involving hot gas cleanup with molten salt and an electrode/insulation program for MHD technology development. In addition, PNL currently has a major role in support of the mining research program.

Sandia Laboratory (SL)

Sandia provides major support work for the Gas and Petroleum Resources Activities and In Situ Gasification Subactivity. Other minor task support is being provided in the MHD and advanced power systems technology areas. Major efforts include work in the massive hydraulic fracturing technology for enhanced gas recovery and instrumentation; diagnostic, bed preparation work for in situ oil shale recovery and coal gasification; and offshore drilling technology development. In addition, studies on direct injected stratified charge engine combustion are being conducted.

7. Commercialization

Throughout the development phase of a project, process or technology commercialization aspects must be considered. Incorporated within the general term "commercialization" are the procedures taken during the development phase, the efforts necessary to determine requirements, and the actions needed to introduce the technology into the marketplace. Commercialization will tell what the payoff is—the "bottom line"—of the time, effort, and funds that are being devoted to a technology.

Issues

The following major issues must be considered to determine if a technology is ready for commercialization:

- Is it technically ready? One must consider whether the technology risks are within industry's normal bounds: Are demonstrations needed to show technical feasibility or to show operations, maintenance, reliability?
- Is the market ready? One must determine if the private market is ready, if it needs stimulus for accelerating timing and/or increasing size of market.
- Is it environmentally acceptable? The technology under consideration must meet environmental requirements before it is ready for commercialization.
- Is it cost effective from both a Government view and ultimately from a private view?

In recognition of the importance of commercialization, DOE has established the Commercialization Committee comprised of Assistant Secretaries and other selected members. The committee, which is chaired by the Deputy Undersecretary for Commercialization, reviews commercialization issues and reports to the Undersecretary.

Task forces are convened to examine energy technologies in detail. Each task force focuses on one technology. The membership of each task force is made up of individuals selected for expertise in appropriate disciplines. Members are drawn from Energy Technology, Resource Applications, Conservation and Solar, Energy Research, and other offices within DOE that have individuals with desired backgrounds. Industry, university, and Government experts outside DOE are consulted on an as-needed basis. The task force gathers information on all aspects pertaining to commercialization of the technology under review, prepares an analysis, and presents findings and recommendations to the Commercialization Committee, the Deputy Undersecretary for Commercialization, and the Undersecretary.

Technical factors include the following:

- Decide what the private sector will do on its own and determine at what point technology will be ready
- Determine if the Government could and should help with demonstrations, laboratory developments, or research
- Define the worst problems or unknowns,

and decide if they are more or less than the industry normally solves itself
- Determine the schedule of technical projects, and what and how close the competing technologies are

Market and economic factors are as follows:

- Determine the composition of industries that will be users and suppliers
- Determine the quality of existing information, what needs to be generated in market analysis and venture analysis, what suppliers and users have done to date, in what areas DOE should stimulate industry and obtain desired information, and what efforts DOE should do in-house
- Define the size of the market and the rate of its development
- Identify critical factors involved, including DOE policies, complexity of process, status of technology, economics, environmental concerns, interest of industry, regulatory issues, demonstrations required, and the Government's role in providing incentives

The market assessment should include, as much as possible, information from the industry that would be investing in and using the product, and from the equipment suppliers who will develop, sell, and warrant the product.

In assessing the environmental suitability and determining the problems inherent to commercializing a technology, the following general procedures are required:

- Determine the process characteristics in terms of temperature, pressure, fuel feed, fuel quality and quantity, other ingredients, etc., that may affect resultant pollutant levels
- Identify appropriate pollution controls to reduce air, water, and waste related emissions
- Define environmental data available on all aspects of the technology as it exists and as forecast in commercial applications. The confidence levels associated with current data and forecasts should be stated
- Determine environmental impacts associated with location in nonattainment regions, expected cumulative environmental effects of colocation of facilities and other possible siting restrictions
- Review laws and regulations and other environmental requirements to determine applicability to the technology under consideration

Of particular concern are the uncertainties posed by future environmental legislation and regulations that may be changed in four-year cycles. Development time for industrial processes and technologies may in some cases take two or three times longer and be unable to accommodate the changes that could be imposed. Also of significant concern is the ability of law suits to stop implementation of a project until legal settlement, thus causing delay or termination.

Current Federal environmental laws include the Clean Air Act and Amendments, the Federal Water Pollution Control Act and Amendments, the Occupational Safety and Health Act (OSHA), the Resource Conservation and Recovery Act, the Toxic Substances Control Act, and the Safe Drinking Water Act.

Procedural requirements include NEPA Assessments (Environmental Impact Statements) and Section 13 Water Assessments.

Institutional factors include:

- For the private sector, determine what the technology will replace, possibility of resistance to change, and what needs to be done to make change attractive
- For the Government, determine what institutional barriers exist, and what actions need to be considered at local, state, and Federal levels to assist introduction of technology.

Is commercialization economically attractive to private industry? If the Government helps, is it worth it? What is the Government's cost to participate versus the cost saved by reducing imported oil? Analysis should be conducted to determine if the Government should actively support commercialization in some form, such as demonstration projects, capital grants, investment tax credits, fuel subsidies, fuel taxes, environmental regulations, and product indemnity guarantees.

Decision Procedures

Each task force is required to include in its report conclusions on the issues discussed, a rating matrix of barriers to commercialization, and a recommended strategy for commercialization activities to be undertaken. The Commercialization Committee may decide to proceed with commercialization, delay it, or shelve it altogether based on the findings of its task force.

Where applicable, resource managers are appointed to provide a DOE-wide point of focus for the integration of all activities required to achieve the earliest possible date for commercialization of specific technologies.

In general, when the R&D for a technology is nearing completion, responsibility for

Table II-7-1. Technologies and demonstrations ready for commercialization.

TYPE OF TECHNOLOGY	

Gaseous Fuels
Low Btu
Medium Btu
High Btu
  First Generation
  Advanced Technology
Enhanced Gas Recovery
 	
Liquid Fuels
Coal Liquefaction
Oil Shale - in situ
Surface Retort
Enhanced Oil Recovery
 	
Direct End Use Applications
Urban Waste
Cogeneration
Industrial Atmospheric Fluidized
  Bed Combustion
Conservation Products Marketing
  Oil Burners
  High Efficiency Motors
  Air Fuel Ratio
Solar
  Passive
  Hot Water
Electric and Hybrid Vehicles
  First and Second Generation
  Third Generation (Hot Battery)
	Electric Market
	Utility Atmospheric Fluidized Bed Combustion
	Combined Cycle With Integrated Gasifier for Utility Application
	Fuel Cell Power Plant
	Hydrothermal Geothermal
	Low-Head Hydro
	Large Wind
	Small Wind

"marketing" activity is transferred from Energy Technology (ET) to the Assistant Secretary responsible for commercialization [either Resource Applications (RA) or Conservation and Solar (CS)]. Where technology is somewhat mature, but additional engineering support is required, the Resource Manager will report to RA or CS and will call on ET for technology support in a matrix management system. Where large technology demonstrations are still needed, or the technology is not ready for commercialization because of cost or other barriers, the technology will not be transferred. In these cases, market analysis, cost analysis, and marketing research will be conducted by RA and CS as appropriate to ensure ET's efforts are guided toward a logical end market. No resource manager will be appointed.

Commercialization Activities

The Commercialization Committee, during the summer and fall of 1978, examined over 20 technologies, for readiness for commercialization. Results of the extensive review are shown in Table II 7-1.

8.
Environmental and Socioeconomic Implications

Energy Technology seeks to advance the availability of environmentally and socially acceptable energy resources. The strategy to achieve this goal incorporates independent activities conducted within each program and close cooperation among its industrial partners; the Office of the Assistant Secretary of Environment; Federal, state, and local regulatory agencies; and the public.

The Fossil Energy strategy recognizes the need to develop a broad range of environmentally sound technology options for utilizing domestic fossil energy resources. This means that the technology should have acceptable impacts on the physical and biological environment, protect the health and safety of workers and the public, enhance socioeconomic values, especially for the communities most affected, and contribute to the high efficiency of energy production and use. Consequently, environmental, health and safety, and socioeconomic issues associated with the operation of fossil energy processes are identified as early as possible, and research projects are developed to resolve these issues.

In the Fossil Energy area, coal can be used as an energy source for many diverse applications. It can be combusted directly to produce steam for power generation; it can be gasified to yield a fuel gas or a pipeline-quality gas; it can be liquefied to produce a fuel or alternate petrochemical industry base; and it can be used to fuel advanced power systems that are thermally efficient.

Environmental Implications

Most coal technologies can be associated with common impacts:

- Requirements and techniques for solid waste disposal
- Consumption of water
- Accidental discharge of toxic or hazardous products or by-products
- Carbon dioxide emissions and unknown cumulative effects
- Air pollutant emissions
- Surface water pollution
- Land and water disturbances from mining
- Occupational and health and safety effects

Major common issues are as follows:

- Application of air and water control technologies
- Potential limitations upon commercialization posed by environmental protection regulations

The wide variety of projects in FE results in many instances of unique and site-specific implications that are best handled on a case-by-case basis. For example, because synthetic coal conversion plants have not been constructed in the United States, there are apprehensions about possible environmental, safety, and health impacts associated with such plants.

The specific issues for a given plant include:

- The ability of a site to accommodate the plant physically (e.g., water and coal availability)
- The ability of a community to absorb the socioeconomic impacts of a plant (e.g., adequate housing and schools)
- The ability of a plant to meet existing standards (e.g., air and water quality)
- The ability of a plant to avoid introducing new pollutants in the ambient environment or the worker environment (e.g., trace metals and carcinogens)
- The ability of a plant to meet future standards

Operations other than those based on coal are also under investigation in Fossil Energy. These include enhanced oil and gas recovery and the extraction of liquid products from oil shales. Many of the same types of environmental, safety, and health implications apply. Because such unknowns exist, DOE is aware of possible problems and has initiated procedures and programs designed to define possible hazards and to implement or develop mitigative measures.

FE conducts several basic types of environmental activities; i.e., those related to policy analysis, planning, National Environmental Policy Act (NEPA) compliance, development of suitable control technology, and contract implementation. Generally, these environmental activities are directed toward areas that develop new knowledge in FE chemistry, materials, technology, processing, engineering, analysis, sampling, worker management and protection, and property management and protection. In this context, the term "environmental" refers to health and safety, socioeconomic, and energy process conservation issues as well as to the physical impact of processes on the environment.

Policy Analysis. The results of the environmental policy analysis activities serve as input to the technology goals. FE sponsors studies that consider how current and emerging laws, standards, and regulations affect its programs and technologies and their potential for commercialization. The technical performance, resource requirement, and economic information resulting from such studies is used to review new regulations and is exchanged with regulatory agencies.

Planning. Planning activities are carried out at both the programmatic and project levels to assure that environmental problems are addressed at the earliest possible point. Programmatic considerations are addressed in the Environmental Development Plans (EDP's) prepared for each activity. FE planning activities address the ability of FE projects to meet current environmental requirements, the development of technical resources, and economic performance information about FE technologies for input into future regulatory decisions. FE works closely with the office of the Assistant Secretary for Environment in the preparation of Environmental Development Plans, Environmental Readiness Documents, and Energy Systems Acquisition Program Plans.

NEPA Compliance. NEPA requires that an environmental impact statement (EIS) be prepared for each major Federal action that could have an adverse effect on the human environment. FE prepares environmental assessments for all field projects; EIS's are prepared for those projects that could have major environmental impact or are considered controversial. In addition, FE must consider the cumulative impacts of the programs and the ultimate commercialization of the technologies; these topics are addressed in programmatic environmental impact statements. Finally, FE reviews NEPA documents prepared by other DOE offices and other agencies.

Control Technology Development. Fossil Energy has a positive program underway with regard to control technology development. Substances such as sulfur; nitrogen, alkali and halogen compounds; and volatile trace metals are all abundant in coal, and are released in various forms under conversion and combustion conditions. Fossil fuel cleanup technology is currently available for compliance with environmental standards controlling emissions from coal combustion. However, many of the technologies are costly, mechanically unreliable, or highly energy consumptive. Fuel cleanup, gas stream cleanup, and flue gas cleanup are therefore all being pursued for the development of viable and improved hardware systems to remove contaminants, control emissions, and protect equipment.

The following priorities have been established for control technology development:

- Lime/limestone flue-gas desulfurization (FGD)
- Advanced flue-gas/cleanup hardware
- Gas stream cleanup hardware
- Particulate control and NO_x control
- Waste disposal techniques

An indicator of the increasing emphasis in this area is the increase in authorized budget (from $14 million for FY 1979 to $50 million for FY 1980). The FY 1980 budget includes:

- Evaluation of gas cleanup systems
- Development of lime/limestone FGD prototype
- Test and evaluation of alternate FGD systems
- Hot gas cleanup systems development
- Waste management of FGD sludge and waste

Project Development. Since environmental performance is an integral criterion of technology performance, pertinent environmental considerations are included in each step of the development of a project.

The preprocurement step involves the preparation of the RFP, PON, or other type of procurement package. An Evaluation Board (EB) is established with rules for the evaluation and selection of the best proposals. The FE Procurement Procedures Manual provides for consideration of NEPA requirements at the preprocurement stage by requiring that the responsible program official (Division Director) complete an Environmental Provisions Form. For projects involving construction activity, emissions and associated activities, the program official must indicate what steps will be taken for conducting environmental reviews, assessments, or impact statements. Once the relative environmental significance has been established for the proposed project, environmental goals, criteria, and requirements are established and incorporated in the document (RFP, PON, etc.). Instructions are also included for preparing environmental data as part of the proposal and proposal evaluation criteria. Thus, the preprocurement step is critical in establishing the quantity and quality of the information submitted in proposals and in establishing the ground rules for the evaluation of the environmental aspects of a project.

The procurement step involves evaluating and grading the proposals. For large projects, environmental review committees are established to conduct environmental reviews with relative weight of environmental factors established by the SEB. Site evaluations and oral presentations are also conducted to further assist in the evaluation process. The site evaluation process includes preparation of a preliminary site evaluation report for each proposal in order to determine, at a minimum, the air, water, and solid waste permitting requirements for each location and potential environmental impacts of the proposed facility. In the later contract phases, more intensive site studies are conducted. Local officials are usually contacted as well to determine the existence of local support or opposition to the project. Oral presentations are frequently scheduled to request further information regarding technical, economic, and environmental aspects of each proposal.

The contract implementation stage involves establishment of preliminary design, detailed design, construction, and operation of the facility. NEPA compliance activities are normally scheduled during the design stage, which extends for 12 to 18 months for large projects and is concluded prior to the start of construction. Large projects are incrementally funded, with decision points within and at the conclusion of each phase. EIS preparation during the design stage enables tradeoffs to be made regarding the need for major pollution control hardware. The emphasis in major projects has been to complete the EIS prior to the detailed design phase, resulting in obvious advantages with regard to reducing the environmental and financial risks to the project.

Some variation from these procedures occurs in small-to-medium-sized projects.

During the construction and operation phases, environmental activities continue. Most of the field activities, actual data gathering, and control technology development activities performed by FE are done in conjunction with or in support of these phases. Typical and specific activities are described in appropriate sections of this document under the individual programs. Examples of such activities follow. Monitoring is initiated prior to construction and continues through the operational phase to verify the true impacts on the environment. Sampling and monitoring activities are conducted to obtain compliance-oriented data and to determine whether new pollutants are introduced into the environment. Projects are carried out under representative environmental conditions to allow adequate assessment of their consequences. Processing conditions are varied to obtain a range of environmental, safety and health, and resource efficiency data. Environmental and safety control technology is developed when adequate technology does not exist.

Implementing the environmental strategy requires coordination between FE and its industrial partner. To facilitate this coordination, selected responsibilities are assigned to each of the co-sponsors. FE establishes the criteria necessary to carry out these activities, reviews and approves the environmental protection plan prepared by the industrial partners, monitors the effort, and undertakes supporting and complementary research to assure the completeness and validity of the information being gathered.

Socioeconomic Aspects

The implementation of new energy technologies will have various impacts on society. The social impacts can be beneficial or adverse. Impacts of the deployment of a given technology will differ as a function of the size of the industry, its location, and the rate of implementation. That is, the social impact of building 50 coal liquefaction plants is different from that of building one; if all the plants are built in one area, the impacts are greatly different than if the plants are dispersed over the entire country. Similarly, the social impact of building 20 coal gasification plants in 10 years is quite different than if 20 plants were built in 20 years. Socioeconomic impacts during the construction phase are also usually different from those associated with operation.

Socioeconomic impacts will also change as a function of other technology deployment activities which are concurrent, as well as other economic activity in general. For example, obtaining financing for five oil shale surface retorts would be much easier if no other construction projects were being undertaken simultaneously than if financing were being sought at the same time for a number of coal conversion plants.

The objectives of social impact analysis are (1) to provide information on significant social impacts to decision-makers so that these factors can be considered in planning the Fossil Energy Program, and (2) to ensure that social impacts of Fossil Energy projects (pilot and demonstration plants, etc.) are adequately considered in designing the project. The kinds of questions that social impact analysis address include the following: How much will energy cost from different technologies and what will be their relative impact on inflation? What will be the impact on regional economies? What infrastructure changes will be associated with the deployment of new technology? What changes in lifestyle will be associated with the deployment of new technologies?

In addition to the analysis of socioeconomic impacts conducted in the preparation of NEPA documents, the Fossil Energy Program includes several projects related to social impact analysis, a few of which are identified below:

- Yale (The Mapping Project on Energy and the Social Services) Engineering Test Facility (ETF) cooperatively funded this project with EV (in FY 1979) and with CSA (in FY 1978 and FY 1979). The activities of the project include the preparation of an academic research agenda on social science/energy issues;

preparation of a framework for social impact analysis of energy technologies; consultation with DOE personnel on energy/social science questions; and operation of a clearinghouse for energy/social science literature.
- The Dartmouth Systems Dynamics Group will be building a data base and an associated simulation model that will be used to estimate the impact of Government programs and policies on the commercialization potential of fossil energy technologies.
- The purpose of the Future's Group (Strategic Backdrop Analysis) work is to develop a strategic analysis capability. The method being developed begins by identifying three views of the future (in terms of social and economic characteristics) and then examines alternative pathways to these target futures.

Role of the Office of Environmental Activities/Division of Systems Engineering

Within FE, the technology divisions and the Office of Environmental Activities (EA) conduct complementary projects. The technology programs are primarily responsible for site-specific activities, while the Office of Environmental Activities serves as advisor to supplement and coordinate the activities of the technology programs. The environmental component of each Energy Technology Center provides support to the Office of Environmental Activities and performs similar functions in the environmental area for those technologies for which the center has lead responsibilities.

The primary goal of the Office of Environmental Activities is to help ensure the environmental acceptability of FE technologies by working to attain these specific objectives:

- Incorporation of environmental activities, including health, safety and socioeconomics, into FE funding proposals
- Full consideration of environmental values when projects are selected from among competing alternatives for FE funding
- Compliance by FE program projects with applicable Federal and state environmental laws
- Collection and interpretation of environmental data to provide a basis for future environmental regulatory decisions.

Increased efforts to analyze major environmental policies are underway in order to ensure that FE is successful in projects with early pay-off potentials. The desire to rapidly commercialize such technologies frequently must be balanced with proper concern for the environment. In particular, increased utilization of coal, as proposed in the National Energy Plan, requires increased attention to the environmental and health implications. Also, the impact of new legislation and regulations upon FE technologies must be considered. For example, work to understand the implications of new standards of performance and offset policies, as prescribed by the Clean Air Act Amendments of 1977, is already well underway. Preliminary analyses of provisions of the Toxic Substances Control Act, the Resource Conservation and Recovery Act, the Water Pollution Control Act, the Occupational Health and Safety Act, and other legislation are also underway, and these efforts will be intensified as FE processes become better understood and new regulations are formulated.

Preparing the EDP's and assisting the technology programs in planning the environmental portions of their budgets will continue to be high priority projects for EA in FY 1980. EDP's are the principal planning mechanism for coordination of planning activities at the activity level. EDP's examine the environmental impacts of the energy system, define the uncertainties and assess the current knowledge about those impacts, and identify areas for further research. The information developed in the EDP process is used to structure projects for addressing the environmental problems associated with a given process and to assist the project managers in selecting methods of resolving these problems.

Activities in support of the FE Technology Divisions include participation on project teams for FE demonstration plants at all stages of project development and review of all environmental, health, and safety activities. In this manner, potential sites for fossil demonstration plants are analyzed to evaluate both environmental and socioeconomic impacts.

Programs to develop an integrated approach to potential occupational health problems associated with FE development have been started. The upgrading of worker protection and worker health data collection programs at key sites is a high priority item for FY 1980. Several projects are underway to strengthen environmental guidelines and review criteria for site evaluations of pilot, demonstration, and commercial-scale projects. As program development and project support activities proceed, additional issues for study are identified. Examples of such studies include the development of environmental control technology; the definition of resource requirements of FE projects; identification of the need for additional monitoring activities and characterization studies; and determination of the impact of FE technologies on health, the environment, and the socioeconomic structure of the communities involved.

9.
International Programs

A mission of the Department of Energy is the continuing development of international cooperation to effectively increase world supplies of energy through greater production of energy resources and improved efficiency of resource utilization. In support of this mission, DOE/FE seeks to establish formal international agreements for R&D cooperation in fossil energy related technologies. Such agreements enhance domestic and foreign efforts by:

- Increasing the technology base
- Increasing applied technical manpower and expertise
- Reducing financial burdens on all participants
- Accelerating technical progress
- Reducing wasteful redundancy

Cooperation may be bilateral, multilateral, or government/industry and may take the form of information or personnel exchange, equipment loan or exchange, sharing of work at existing experimental facilities, joint construction and operation of facilities or experiments, activities involving a coordinated program but conducted by national teams in their own country, or a combination of the above. Although most of the agreements made so far by DOE/FE are primarily concerned with exchange of information, several of the larger agreements involve technology development and demonstration projects. These are significant in that they show substantial participation by private industries from each country.

Increased efforts are underway in coordinating the planning of broad areas of various national R&D programs to ensure that they are complementary without sacrificing any special strengths or the pertinence of any nation's efforts to its own needs.

DOE/FE international activities include continuing assessment of worldwide fossil energy technology developments in order to identify new areas where international cooperation would be beneficial to U.S. energy programs, and then to negotiate suitable cooperative agreements. Following ratification of agreements by participating countries, detailed implementation plans are developed and executed by designated experts on each side. The most recent agreements were signed in 1978—one, with Australia, is for information exchange in coal technology; the other is with the Federal Republic of Germany and involves major industrial participation from each country in Solvent Refined Coal (SRC-II) technology demonstration.

In addition to cooperative agreements with technically advanced countries, DOE/FE is exploring the possibilities of cooperation with the Less Developed Countries (LDC) in order to transfer to them energy technologies that are appropriate to their developmental needs.

To complement its major emphasis to date on coal technology, DOE/FE is also seeking to expand cooperative activities in oil, gas, and oil shale technologies. As any early step, cosponsorship of a forthcoming United Nations symposium on heavy oil is being arranged.

A complete listing of ongoing DOE/FE formal multilateral and bilateral activities and a summary of cooperative programs is given in Table II 9-1.

UNITED STATES—INTERNATIONAL ENERGY AGENCY (IEA)

The International Energy Agency, established at U.S. initiative in 1974 to promote the development of more secure energy supplies, has undertaken an extensive program in the development of technologies for the utilization of a wide range of energy resources. DOE/FE has been participating in development and implementation of cooperative IEA programs in the area of coal technology.

In 1975, five multilateral agreements in this area were signed by interested IEA member countries. As the U.S. representative, DOE/FE participates in the following four:

- Pressurized Fluidized-Bed Combustion (PFBC)
- Economic Assessment Service for Coal (EAS)
- Technical Information Service (TIS)
- Mining Technology Clearing House (MTCH)

Two additional multilateral agreements for combustion and heat transfer were signed in 1977.

In 1977, two bilateral commitments were concluded under IEA auspices between the United States and the Federal Republic of Germany:

- Memorandum of Understanding (MOU) for National Planning Coordination in the Field of Coal Hydrogenation Technology
- Letter of Intent (LI) for Cooperation in Coal Technology.

The responsibility for carrying out the first four programs has been delegated to a specially created organization, the U.K. National Coal Board (NCB), IEA Services, Ltd. The policy and program of work for each project are controlled by an Executive Committee (EC) consisting of one member from each participating country. The EC meets biannually to approve work programs and budgets and to review progress. Technical committees, with members appointed by each EC member, provide support to the EC.

The United Kingdom is the lead country for the PFBC, EAS, TIS, MTCH, and Heat Transfer Projects, and the United States is the lead country for the Combustion Project. No lead country has been designated for MOU and LI, which are still in developmental stages.

Responsibility for the development of coal technology cooperative programs has been delegated to the IEA Working Party on Coal Technology (WPCT) by the IEA Committee on Energy Research and Development. The WPCT is continuing to develop additional cooperative programs, and DOE/FE is actively participating.

Fluidized-bed Combustion

Countries participating in fluidized-bed combustion are the United States, United Kingdom, and Federal Republic of Germany.

The major objective of this program is to design, build, and operate a flexible experimental facility (at Grimethorpe, Yorkshire, U.K.) for investigating combustion, heat transfer, corrosion, and energy recovery in pressurized fluidized-bed systems, which require more advanced technologies than systems operating at atmospheric pressure.

The PFB combustor/boiler (80-MW_t) is expected to be the first of its kind to operate in the world and to be the first available source of data on such systems. The system will produce steam sufficient for a 20-MW_e turbine and pressurized combustion gases for operating a turbine up to 3.5-MW_e. The facility is expected to add a 3.5-MW_e gas turbine test unit for operation in 1981. Results from these experiments will be used in the design of commercial-scale pressurized fluidized-bed combustion plants.

The duration of the project is eight years (1975–83), with start of plant construction in late 1977 and operation by early 1980; construction and operating costs will be shared equally by the United States (DOE/FE), the United Kingdom, and the Federal Republic of Germany. The experimental program will make use of coals from the United States, the Federal Republic of Germany, and the United Kingdom in a test program now being decided upon by the participating nations. A full participant of the IEA project, DOE/FE has complete access to the plant's design and performance data through drawings, reports, and operating logs, as well as through direct participation in all aspects of the program. The Argonne National Laboratory participates

Table II-9-1. US/DOE/FE international program.

RESOURCE/SYSTEM	ACTIVITY	PARTICIPANTS IEA	PARTICIPANTS NON-IEA	PERIOD
Coal	Pressurized Fluidized-Bed Combustion	FRG UK		1975–1983
	Mining Technology Clearing House	Belgium Canada FRG Italy New Zealand Spain UK		1975–1982
	Economic Assessment Service	Canada FRG Italy Netherlands Spain Sweden UK		1975–1982
	Technical Information Service	Austria Belgium Canada FRG Italy Japan Netherlands New Zealand Spain Sweden UK		1975–1982
	National Planning Coordination in Coal Hydrogenation Technology	FRG		1977–1978
	Coal Technology	FRG		1977–1978
	Solvent Refined Coal (II)		FRG	1978–1979
	Coal Mining and Utilization		National Coal Board (UK)	1974–1980
	Coal Mining		USSR	1974–1979
	Coal Mining and Utilization		Poland	1974–1982
	Coal Mining and Utilization		Australia	1978–1983
	MHD		USSR	1972–1979
	MHD		Poland	1974–1979

Table II-9-1. (Continued).

RESOURCE/ SYSTEM	ACTIVITY	PARTICIPANTS IEA	NON-IEA	PERIOD
Oil	Enhanced Oil Recovery		USSR	1974–1979
Natural Gas	Enhanced Gas Recovery		USSR	1974–1979
Shale Oil	None			
Combustion	Energy Conservation in Combustion	Italy Sweden UK		1977–1980
Heat Transfer	Heat Transfer Systems	Sweden Switzerland UK		1977–1980

in the design of probes, instrumentation, and control systems.

Accomplishments as of the end of 1978 were as follows:

- Plant design completed
- Long lead major components ordered
- Construction 85 percent complete
- Initial coals to be tested have been chosen
- Details of the experimental program developed.

Commissioning tests began in the summer of 1979. The facility will soon be ready for experiments.

Economic Assessment Service for Coal

Participating countries in the Economic Assessment Service (EAS) for Coal include the United States, Canada, Federal Republic of Germany, Italy, Netherlands, Spain, Sweden, and United Kingdom.

The aim of the EAS is to assess the likely economics of current and projected worldwide coal extraction, processing, and utilization technologies in order to provide adequate background for the formulation of R&D projects by member countries. To make the results of such assessments applicable to the particular economic and environmental circumstances of each member country, the EAS analyses are designed to be "context free." Thus, an important feature of the service is that its work takes into account the economic standards and conventions in different countries. The service is currently focusing on the following areas: coal conversion and pollution control; coal supply and trade; and the transport of coal and coal-based energy.

By March 1979, a series of reports were issued to member countries and, as a result, EAS has done the following:

- Reviewed the status and performance of coal conversion technology and pointed out the strengths, doubts, and areas of concern in process operation. (Reports: Parts II–IV of Study A already issued.)
- Reviewed economic and financing conventions in different countries and indicated a central view for comparison of different processes. (Report: Part I of Study A already issued.)
- Calculated for the circumstances of different countries (for example, different economic conventions, plant costs, and coal costs) relative costs of different forms of secondary energy from coal. (Study E: in progress.)
- Reviewed the likely worldwide availability and broad costs of coal up to the year 2000, taking into account not only the declared intentions of countries, but also what we perceive to be the likely constraints on these intentions being realized. (Study C: in progress. First report already issued.)
- Studied future trends in transport costs and the possible constraints of transport costs on coal trade. (One report issued; others in progress.)

- Presented the view on the demand for coal in various uses, taking account of the likely costs of coal-based energy in relation to other primary energy. (Studies G and H in progress.)
- Presented a view of some of the pollution problems associated with coal use, in particular sulfur oxide and effluents from gasification plants. This will include the pollution to be anticipated without control, the legal situation in different countries, control technology, and the costs of control. (Studies in progress.)

The expenditure incurred in the operation of the Economic Assessment Service is borne by the contracting IEA countries in the following proportions: Canada, 4.7 percent; Federal Republic of Germany, 17.2 percent; Italy, 3.2 percent; Netherlands, 2.8 percent; Spain, 3.9 percent; Sweden, 2.6 percent; United Kingdom, 16.6 percent; and United States, 49.0 percent. The duration of the first agreement was three years (1975–1978). It has been renewed for three more years.

Technical Information Service

Countries participating in the Technical Information Service (TIS) are the United States, Austria, Belgium, Canada, Federal Republic of Germany, Italy, Japan, Netherlands, Spain, Sweden, and United Kingdom.

The objective of TIS is to report on worldwide developments related to all aspects of coal utilization and to facilitate exchange of information between participating countries.

The service is collecting and disseminating information relevant to coal RD&D in the areas of geology; prospecting and mining; coal utilization, including coal preparation and waste disposal; environmental and health problems; coal transport; legislative measures; and Government policy statement. It is not the intent to replace existing information services in member countries which serve their own coal industry and are tailored to their particular needs. The service aims to supplement them by making available a complete collection of technical information on all aspects of coal production and utilization and by ensuring that all sources of useful information are being tapped. Input is received from member countries in the form of abstracts of articles, technical periodicals, conference papers and reports; contracts are made for scanning literature of nonmembers such as Eastern European countries.

The essential basis for the TIS is a central information storage and retrieval system. Specific activities in support of its objectives include:

- Constructing and maintaining a computerized data base containing abstracts of current and recent past documents concerned with coal R&D
- Publishing the monthly document "Coal Abstracts" from the data base
- Publishing reviews of recent literature on subjects of great current interest, as requested by the TIS Executive and Technical Committees and by others with committee approval
- Responding to inquiries from individual members, using the abstracts in the data base and microfiche copies of original source documents, as available
- Providing Selective Dissemination of Information (SDI) service from the data base, using computer systems
- Providing "on-line access" to the data base
- Providing knowledge of other information services relating to coal science and technology
- Providing coal technology information from all sources worldwide, not just member countries
- Developing a worldwide directory of ongoing coal research as information is provided by member countries
- Publishing a regular news bulletin
- Cooperating with other IEA energy R&D projects, to support their information needs, and minimize overlaps/conflicts

Expenditures incurred in the operation of TIS are borne by the contracting countries in the following proportions: Austria, 1.9 percent; Belgium, 2.7 percent; Canada, 3.4 percent; Federal Republic of Germany, 13.5

percent; Italy, 2.2 percent; Japan, 6.0 percent; Netherlands, 1.9 percent; New Zealand, 1.9 percent; Spain, 2.8 percent; Sweden, 1.7 percent; United Kingdom, 13.0 percent; and United States, 49.0 percent. The duration of the first agreement was three years (1975–1978). It has been renewed for three more years.

Mining Technology Clearing House

Countries participating in the Mining Technology Clearing House (MTCH) include the United States, Belgium, Canada, Federal Republic of Germany, Italy, New Zealand, Spain, and the United Kingdom.

Under the 1977 Energy Reorganization Act, this cooperative activity was transferred to DOE/FE from the Bureau of Mines. The basic aim of the MTCH Service is to collect, collate, and disseminate information on research and development in all aspects of coal mining (both surface and underground) and coal preparation technologies, including the improvement of operational efficiency and safety and health. As it develops, the service will cover all relevant R&D, arrange contacts between experts in particular disciplines, and assist in the formation of new cooperative projects.

Initially, the MTCH Program is aimed at establishing a series of Project Registers of ongoing work and at investigating the situation of certain technical subjects in member countries.

The registers currently being compiled cover four technical fields:

- Mechanized drivage of roads, slopes, and shafts
- Coal seam exploration techniques
- Monitoring and control of machines and mining systems
- Reliability of equipment

Reports of technical investigations have been prepared for the following subjects:

- Underground transport of solids
- Optimization of cutting techniques
- Trackless transport of men and materials underground
- Mining methods to improve reserve recovery
- Certification of equipment for underground use

The expenditure incurred in the operation of the Mining Technology Clearing House is borne by the contracting IEA countries in the following proportions: Belgium, 3.6 percent; Canada, 5.0 percent; Federal Republic of Germany, 16.3 percent; Italy, 3.2 percent; New Zealand, 2.7 percent; Spain, 4.0 percent; United Kingdom, 16.1 percent; United States, 49 percent. The duration of the first agreement was three years (1975–78). It has been renewed for three more years.

National Planning Coordination in the Field of Coal Hydrogenation Technology

The United States and the Federal Republic of Germany are the two participants in the National Planning Coordination effort.

The objective is to eliminate duplication of R&D efforts in the field of coal hydrogenation technology. Exchange of planning information began as scheduled and arrangements for continuing exchange and agreed upon workshops have been completed.

Cooperation in Coal Technology

The United States and the Federal Republic of Germany are participating in the Cooperation in Coal Technology Agreement.

The objective of the agreement is to share in developments of each country's coal processing technologies through joint participation in design, construction, and operation of some of each country's planned experimental facilities.

Combustion

Participating countries in this effort include the United States, United Kingdom, Sweden, and Italy.

In 1976, energy conservation in combustion was established as a separate project area with the objective of improving the energy efficiency of combustion technologies and developing the fuel-switching capability of combustion equipment. Cooperative research and development efforts and information exchange among participating parties are currently underway.

In March 1977, an implementing agreement and an annex entitled Energy Conservation in Combustion were signed. The contracting parties will investigate methods for improving the information, instrumentation, and calculating procedures used in the design, manufacture, and operation of combustion equipment. Within the agreement, three specific areas for investigation have been identified: (1) combustion system modeling; (2) instrumentation and studies of fundamental processes in combustion; and (3) resource exchange.

Research in combustion-system modeling is directed toward developing combustion technology for engines and furnaces. In the work related to fundamental processes in combustion, participants will develop instruments and experimental techniques for measuring fundamental parameters and properties of combustion systems, and investigate the basic physical phenomena relevant to the combustion process. The purpose of the third area, resource exchange, is to provide for information exchange among the contracting parties on such subjects as the objectives and results of past programs, numerical analysis methods, laboratory analysis, and experimental test facilities.

The Executive Committee is preparing a new annex to become effective in January 1980. The new annex will provide for continuation of cooperative R&D in fundamental combustion processes, engine combustion, and furnace combustion. The annex also will provide for jointly funded projects, and a number of proposals for such projects are being considered by its committee.

Heat Transfer

Participating countries include the United States, United Kingdom, Sweden, and Switzerland.

A three-year implementing agreement was signed in June 1977 incorporating three annexes: heat exchange network synthesis and optimization, extended surface heat transfer, and flow-induced tube vibration. The United States participates in the latter two annexes.

The participants in the heat transfer project are seeking, through cooperative R&D and information exchange, to improve the design and operation of heat-transfer systems, particularly heat exchangers, as a means of conserving fuel. The specific objectives of the project are to (1) increase the efficiency of thermal-energy conservation through the development of more effective heat-transfer systems; (2) reduce free energy losses in processes by facilitating the operation of heat exchange systems at reduced temperature differences without unacceptable capital-cost requirements; and (3) investigate mechanical and other design constraints to the achievement of higher thermal efficiencies.

UNITED STATES—AUSTRALIA

In the summer of 1978, a broad Memorandum of Understanding for exchange of information concerning coal conversion, extraction, and processing was signed by representatives of the United States and Australia. Program definition is underway; implementation will follow.

UNITED STATES—UNITED KINGDOM

On January 24, 1974, an International Agreement covering both coal mining and utilization technologies was executed by the Department of Interior with the National Coal Board, England. The agreement was written for a three-year period, with an automatic extension for future periods of two years each and a provision for discretionary termination by either country.

Although pertinent areas of the agreement have been transferred to DOE/FE under the 1974 and 1977 Energy Reorganization Acts, the NCB has been actively seeking a redraft of appropriate programs (of the 1974 Agreement) as a separate document between DOE and NCB.

Coal Mining and Utilization

The mining technology program activity includes the exchange of information in the field of coal mining. During 1976, seven detailed joint projects were formalized in this area. Exchanges of specialists occur approximately on a six-month basis. The exchange to date has produced one joint project involving fabrication of mining equipment, one project involving exchange of prototype equipment developed by the respective countries, and five projects involving exchange of instrumentation.

The coal utilization program activity includes exchange of personnel, equipment, research material, and basic information on all aspects of utilization of coal from resource identification to end-use. Currently the United States is studying several new areas of joint activities proposed by the National Coal Board, including supercritical solvent extraction and production of methane from coal mines.

UNITED STATES—FEDERAL REPUBLIC OF GERMANY

Solvent Refined Coal-II

In the fall of 1978, the Department of Energy and the Federal Ministry for Research and Technology of the Federal Republic of Germany entered into a memorandum of understanding for a cooperative and cost-shared effort in the Solvent Refined Coal-II (SRC-II) demonstration plant program. Other parties in the proposed SRC-II program include industrial firms.

Cooperation between the two countries envisages a multiphase project. The memorandum of understanding provides guidelines for subsequent phases of the project to be entered into, subject to approval by U.S. congressional and executive branch authorities.

The result of this proposed project would be a single, 6,000 ton/day module located near Morgantown, West Virginia. Presently, the United States is in the process of carrying out phase zero, to be completed on or about April 30, 1979.

This memorandum of understanding currently serves as the initiating step toward the project agreement between the United States and the Federal Republic of Germany, subject to satisfactory negotiations with industrial partners.

UNITED STATES—UNION OF SOVIET SOCIALIST REPUBLICS

On June 28, 1974, a broad Agreement on Cooperation in the Field of Energy was signed by representatives of the United States and the Soviet Union. Cooperative activities include the development and implementation of programs that would accelerate R&D efforts in both existing and alternative sources of energy. To implement the agreement, a joint U.S./U.S.S.R. Committee on Cooperation in the Field of Energy was established. The committee's functions include approval of specific joint projects and establishment of working groups to perform specific assignments. The cooperative MHD program was officially transferred to this agreement in June 1974; programs in the areas of coal, oil, and gas were officially established in December 1976. The agreement was signed for a five-year period (1974–1979); an automatic extension for an additional five-year period was included.

Magnetohydrodynamics

The international cooperative program on MHD power generation is based on the complementary nature of MHD development in the two countries. In the United States, emphasis in MHD has been in gaining a scientific understanding of the process and through this, learning how to predict the behavior of the key components, especially the MHD generator channel itself. In the Soviet Union, emphasis has been on the modeling and construction of plants of increasing size and complexity. A further balancing factor is the Soviet Union's concentration on natural gas as a fuel for MHD, while U.S. emphasis has moved increasingly to the utilization of coal as the primary fuel. The objectives of the cooperative program are to obtain jointly as much engineering experience as possible, utilizing both U.S. and Soviet facilities, where practical, for testing components and materials, and to share the techniques developed for analysis of key MHD components. Two significant installations developed in the Soviet Union are the U-02 systems test facility and the U-25 MHD pilot plant, both located in Moscow. In addition, there is a small plant located in Kiev in the Ukrainian Republic.

The cooperative effort has focused on joint work contributing to the development and operation of the MHD channel, the major component of a commercial MHD power plant. By exchanging experience and information in this way, MHD development can be pursued more quickly, more economically, and with less risk of technical failure than would be possible if the two countries pursued the work separately. A determined effort has been made to avoid the program becoming merely an exchange of information, and work is being focused on a few specific tasks to assist in the technical development of MHD in both countries. During the past year, work progressed smoothly in all areas. The first three series of tests at the joint U-25B facility incorporating the U.S.-built superconducting magnet have been completed and the results analyzed, yielding valuable engineering data. The fourth series of U-25B tests was scheduled for December 1978. The Joint Status Report on MHD Power Generation was published in 1978. The fabrication of a large-scale MHD channel for the testing in the U-25 facility was completed during the first half of 1979. Phase III of joint U-02 electrode/insulator tests was completed in May 1978 and concluded the planned electrode systems development program. A joint combustor diagnostic test, using a U.S. combustor and U.S.S.R. diagnostic instrumentation was conducted at the University of Tennessee Space Institute in January–February 1979 in further implementation of the plasma diagnostics program.

The U.S.S.R. side has been extremely cooperative in providing information on its current work and future plans in developing commercial MHD power systems. A number of U.S.S.R. reports have been translated and published by DOE. These contain valuable information on U.S.S.R. MHD development. The Fourth U.S.–U.S.S.R. Colloquium on MHD Electrical Power Generation was held in Washington, D.C., in October 1978 and provided for extensive exchange of engineering experience and technical information in the field of open cycle MHD. The Soviets are continuing with the preliminary design of a 500- to 600-MW_e commercial demonstration plant and indicated a willingness to share the results of this work with their U.S. colleagues.

Oil

In December 1976, the U.S.–U.S.S.R. Joint Committee on Energy established a Joint Working Group of Oil Experts and charged it with the responsibility of identifying specific R&D activities of interest to both countries in the field of exploration and production of oil. The first meeting of the group was held in the United States in the fall of 1977. The group agreed on cooperative activities in the following five areas: geochemical exploration, fundamental properties of petroleum, pressure maintenance, enhanced oil recovery methods, and heavy oil production. These were approved by the Joint Committee in December 1977. Project groups have been established and exchange of information has been taking place as scheduled.

Gas

Although a Joint Working Group of Gas Experts was formally established by the U.S.–U.S.S.R. Joint Committee on Energy in December 1976, a number of discussions on subjects of mutual interest took place earlier in both countries. These preliminary meetings resulted in identification of a number of topics of mutual interest, which were spelled out in the February 1976 protocol. Since no protocol has been concluded by the formal Joint Working Group of Gas Experts, the 1976 document currently serves as the guideline document for renewed efforts in this area.

On December 7, 1977, the Joint Committee affirmed mutual interest in the development of cooperation in a number of suggested areas (e.g., increasing gas and gas condensate recovery from productive formations) and made a number of recommendations to the Working

Group. It is hoped that a formal working agreement will be reached during the next meeting of the Joint Working Group.

Coal

Although preliminary discussions have taken place between the representatives of both countries, the specifics of cooperation in the area of coal mining and utilization have not been agreed upon. A working group of coal experts was officially established by the Joint Committee in 1978. The group proceeded with the identification of specified R&D activities of interest to both countries in the areas of coal mining and preparation. This program of work, now awaiting approval by the Joint Committee, is expected to begin in FY 1979.

UNITED STATES—POLAND

In accordance with the agreement between the governments of the People's Republic of Poland and the United States, signed on October 8, 1974, the Marie Sklodowska-Curie Fund was established to finance joint scientific and technological research projects, including joint activities in the area of coal mining and utilization technologies and MHD systems. Joint activities utilize P.L. 480 special foreign currency funds. Specific projects agreed upon by the end of 1978 are described below.

Magnetohydrodynamics

The main thrust of the Polish MHD program is to use the exhaust gases from an MHD generator for coal gasification. This is an idea that originated at AVCO Everett Research Labs in Massachusetts but was never fully explored. Cooperation with Poland will allow the United States to investigate this concept without having to initiate and fund such a research effort in the United States. The agreement for mutual cooperation in these areas was signed in June 1977. Two or three joint experiments per year are expected to be performed in Poland.

The MHD research under this agreement will include (1) design and testing of a 5-megawatt thermal input coal-fired combustor with preheated air at temperatures up to 1200°C alkali seed injection; (2) evaluation of construction materials, slag retention on combustor walls, seed loss in slag and vitiated air injection; (3) testing of two different U.S. coals in the combustor; and (4) evaluation of the scaleup criteria and the designing of a 20-megawatt thermal coal-fired combustor for possible fabrication and testing.

Coal Liquefaction

An agreement for mutual cooperation in the following six coal liquefaction projects was signed in June 1977:

- Catalysts of Hydrogenation Processes
- Coal Extraction and Ash Removal from Coal Liquefaction Processes
- Carbonization of Solid Residues from Coal Liquefaction Processes
- Effects of Hydrogen Donor Solvent in Extraction Process
- Noncatalytic Coal Liquefaction in the Presence of Hydrogen
- Test on the Suitability of Pumps for High-Pressure Operation.

Combustion of Synthetic Fuels for Power Generation

This project was included in the June 1977 document (above). The first meeting to plan joint experiments was held in the United States in January 1978.

Coal Mining and Preparation

Programs for projects in the area of coal mining and preparation technologies were developed and finalized by the U.S. Department of Interior and Poland in October 1976. Four of these projects are being monitored by DOE/FE:

- Removal and recovery of fine-size coal from coal preparation plant waters
- Classification of roofs in U.S. mines for the selection of suitable mechanized support for longwalls
- Modern underground mining technologies for pitching coal bed extraction
- Subsidence prediction and control

10.
University Activities

Fossil Energy's University Activities Program has two principal objectives:

- To ensure a foundation for innovative technology through the use of the capabilities and talents in our academic institutions
- To provide an effective, two-way channel of communication between the Department of Energy and the academic community to ensure trained technical manpower is developed to carry out basic and applied research in support of DOE's mission.

The nation's current energy situation requires that the academic community play a critical role in both energy-related research and in training our young people for future manpower demands. Universities and colleges, both large and small, have been called into play through the following:

- Sponsorship of meritorious projects based on unsolicited proposals, which provides engineering, scientific, and baseline data that can be useful in explaining and developing processes related to Fossil Energy's goals and objectives
- Sponsorship of faculty and student participation programs throughout the country, which provides summer salary support for college professors, as well as

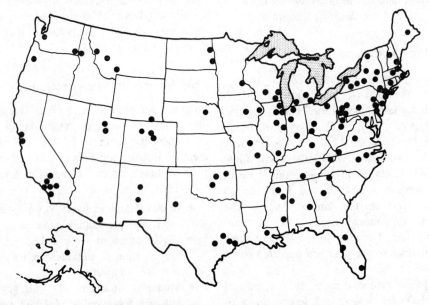

Figure II-10-1. Location of University activities.

students, to perform research of mutual interest at an Energy Technology Center
- Grants programs, which provide investigations related to longer range objectives. They are used to explore potentially useful, but higher risk, technical processes
- Teach workshops and summer institutes, which provide high school and college educators with up-to-date information on the nation's energy problems and helps them develop educational materials for classroom use.

There has been a constant interchange of information. New areas of research have been identified. New and improved technologies have been developed, all as a direct result of the close association developed between the academic community and the Federal Government. This interchange with college and universities has:

- Generated a variety of models for underground coal gasification aimed at predicting product gas composition, sweep efficiencies, and the nature of roof collapse (University of Wyoming, West Virginia University, and University of Texas)
- Obtained relative reactivities of sulfur and nitrogen compounds in coal liquids under practical operating conditions of processes for upgrading of such liquids (University of Delaware)
- Found catalysts with improved activity for conversion of carbon monoxide and hydrogen to methane after severe sulfiding treatment (University of Kentucky)
- Completed exploratory work on a new high-temperature chlorination process for recovering metals from power-plant fly ash (Iowa State University)
- Demonstrated that treatment (prior to recycling) of coal-derived mineral residue can drastically reduce the amount of hydrogen required for SRC processing (Auburn University)
- Prepared a state-of-the-art report, based on detailed reviews with A–E firms, that identifies advantages, disadvantages, and technical needs for low-cost, field-fabricated, high-pressure vessels (University of Kentucky)
- Developed a nickel-membrane probe for in situ measurement of hydrogen concentration under process conditions in coal liquefaction process streams (University of Pittsburgh)
- Found that titanium hydride, when added to coal prior to flash hydrogenation, markedly increases the yield of hydrocarbon liquids (City University of New York)

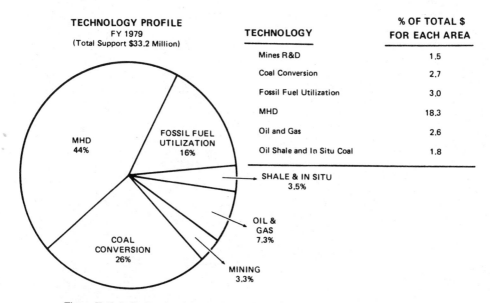

Figure II-10-2. Budget breakdown for the fossil energy University activities program.

- Demonstrated that gas-phase hydrogen generated by chemical reaction of carbon monoxide and steam is as effective as pure hydrogen in promoting the rate of coal liquefaction (Colorado School of Mines).

The relative geographical location of university work supporting the Fossil Energy activities is provided in Figure II 10-1. The budget breakdown for this program is given in Figure II 10-2.

Part III

Solar, Geothermal, Electric and Storage Systems

1.
Introduction

Part III of this handbook presents a comprehensive overview of the Solar, Geothermal, Electric and Storage Systems research and development (R&D) activities that will be performed. The four major solar technology programs have been established with the goal of making substantive contributions to the Nation's future supply of energy.

On April 29, 1977, the Administration submitted to Congress the National Energy Plan and accompanying legislative proposals designed to establish a coherent energy policy structure for the United States. After lengthy deliberations, Congress approved, with major revisions, the National Energy Act (NEA). The passage of the NEA and the establishment of the Department of Energy represent specific steps by the Administration and Congress to reorganize, redirect, and clarify the role of the Federal Government in the formulation and execution of national energy policy and programs. These energy technology R&D programs are an important part of the Federal Government's effort to provide the combination and amounts of energy resources needed to ensure national security and continued economic growth.

CURRENT U.S. ENERGY POSTURE

The United States is at a pivotal point in its energy/economic history. In the past, economic growth has been fueled by abundant supplies of inexpensive, readily available domestic fossil energy; currently, the United States competes in an international marketplace to purchase the supplies of oil and gas on which the domestic economy has become dependent. In the past, economic planners could assume an uninterrupted supply of energy from a variety of sources; currently, the United States must search for new supplies of energy needed for economic expansion. In short, the cost and availability of once inexpensive and plentiful energy have become major obstacles to the orderly growth of the U.S. economy.

In 1977, the United States consumed over 75 quads of energy in the residential/commercial, industrial, and transportation end-use sectors. Almost 50 percent (37 quads) of 1977 energy demand was supplied by petroleum, while natural gas supplied 26 percent (19 quads), coal supplied 19 percent (14 quads), and nuclear/hydropower supplied over 6 percent (5 quads).

At current world oil prices, domestic oil and natural gas resources are not capable of supplying the growing U.S. demand for oil, gas, and their products. In fact, U.S. imports of foreign oil and gas that are needed to augment domestic energy supplies have been increasing steadily since World War II. For example, in 1960 imported oil supplied 18 percent of U.S. oil demand, and imported gas supplied 1 percent of U.S. gas demand. By 1977, these percentages had increased to 45 percent and 5 percent, respectively.

Because the United States currently imports over one-fifth of its annual energy supply, serious questions arise concerning national security, the domestic economy, and political stability. Furthermore, the situation is not expected to improve in the near future. By 2000, U.S. energy demand is projected to exceed 100 quads, and although oil and gas will supply a proportionally smaller share of that demand, the United States may still depend on foreign sources for approximately one-half of its oil.

CURRENT U.S. ENERGY POLICY

Given the potential problems associated with the current and projected U.S. energy situation, the Federal Government has begun to develop an energy policy that will gradually reduce current dependence on imported oil and, at the same time, ensure enough energy supply to support continued orderly economic growth. As an important first step in the development of a coherent national energy strategy, the Department of Energy (DOE) was formed on October 1, 1977, to consolidate Federal energy-related actions and to serve as the focal point for the implementation of evolving U.S. energy policy. One of the responsibilities assigned to DOE in the Department of Energy Organization Act was to carry out the planning, coordination, support, and management of a balanced and comprehensive energy research, development, and demonstration (RD&D) program. The DOE Assistant Secretary for Conservation and Solar Energy (CSE) has been assigned a major responsibility for fulfilling this mandated responsibility, and within CSE the Offices for Solar, Geothermal, Electric and Storage Systems have the responsibility for researching and developing technologies related to four energy sources:

- Solar energy and biomass
- Geothermal energy
- Electricity transmission and distribution
- Energy storage

As a result of inevitable time lags in technology development, capital stock turnover, etc., DOE energy policy has the flexibility to implement different options in different time frames. Specifically, one option may produce near-term benefits but become less effective over the long-term; another may promise long-term benefits but not affect near-term problems. Nevertheless, energy program and policy options affecting the near-term (1978–1985), mid-term (1985–2000), and long-term (2000–2020) all are directed toward four interrelated goals:

- Reduced overall energy demand
- Reduced demand for imported energy
- Increased supplies of conventional energy resources
- Increased supplies of nonconventional energy resources

CSE recognizes these policy considerations and goals in formulating and managing its R&D programs. Consequently, CSE programs will result in increased savings of nonrenewable energy, especially imported oil, and increased supplies of both conventional and nonconventional energy through the near-, mid-, and long-terms.

CURRENT ENERGY LEGISLATION

The most significant energy legislation is the National Energy Act (NEA) of 1978, passed by the Congress on October 15, 1978. The NEA comprises five pieces of legislation:

- The National Energy Conservation Policy Act of 1978
- The Powerplant and Industrial Fuel Use Act of 1978
- The Public Utility Regulatory Policy Act of 1978
- The Natural Gas Policy Act of 1978
- The Energy Tax Act of 1978

The accumulated energy savings resulting from these five laws may amount to 2.4 to 3.0 million barrels of oil a day by 1985.

One of the major roles of CSE is to carry out NEA mandates related to areas of CSE responsibility. For example, Section 209 of the Public Utility Regulatory Policy Act mandates the Secretary of Energy, together with the Federal Energy Regulatory Commission (FERC), to study the reliability of the national electricity transmission and distribution system. Consequently, the ETS Division of Electric Energy Systems (EES) will work with FERC to complete the reliability study.

Energy legislation other than the NEA will also provide direction for solar programs. For instance, Federal market buys of photoelectric systems, called for by the Solar Photoelectric Energy Research, Development and Demonstration Act of 1978, will be supported by CSA. Signed into law by the President on November 4, 1978, the act recommends an accelerated program of RD&D of solar photovoltaic energy technologies that will lead to early, competitive commercial applications.

2.
Solar, Geothermal, Electric and Storage Systems Overview

The Offices of Solar, Geothermal, Electric and Storage Systems, organized under the Assistant Secretary for Conservation and Solar Energy (CSE) serves an essential function in overall Federal Government energy strategy. Programs carried out by the CSE office will lead both to increases in energy supply due to research and development (R&D) of solar and geothermal technologies, and also to reductions in energy demand resulting from more efficient electricity distribution and energy storage systems.

This section presents an overview of the Solar program. First, this overview provides a summary description of the Program, including a breakdown of the activity structure, budget data, and the basic technological, environmental, and socioeconomic requirements and considerations which must be resolved. Second, the Federal Government strategy for achieving goals is outlined together with the energy savings that the strategy is intended to achieve. Finally, the DOE management approach, including DOE headquarters and field responsibilities, is described to define the method by which CSE will implement its strategy to achieve its goals. The internal DOE process by which an energy technology is researched and developed to a state of technical readiness within CSE and then transferred to other DOE program offices for commercialization is also included in the discussion of DOE management.

PROGRAM SUMMARY

Activities and Structure

Activities are carried out in four main programs: Solar Technology, Geothermal Energy, Electric Energy Systems, and Energy Storage Systems. The overall objective of the Solar Technology and Geothermal Energy Programs is to increase the supply of energy available from renewable energy sources. The overall objectives of the Electric Energy Systems and Energy Storage Systems Programs are to reduce energy demand by increasing the utilization efficiency of a wide range of energy sources and to provide supporting technology for solar and geothermal energy systems.

Solar Technology. The Solar Technology Program consists of two programs: 1) Solar Electric Applications, and 2) Biomass Energy Systems. In each of these programs, CSE is developing solar technologies which will have a wide range of both centralized (energy generation at a distance from users) and dispersed (energy generation near or at the point of use) applications.

The Solar Electric Applications Program includes all new solar technologies that have as their sole or primary objective the production of electricity, including, in some cases, hybrid systems that also produce mechanical or thermal energy or both. These technologies and their applications are the following:

- *Solar Thermal Power:* The sun's heat is concentrated and used to heat water or some other fluid to drive a turbogenerator or to provide steam for industrial and agricultural processes. Since these systems can be sized to suit specific needs, they can be used in either centralized or dispersed applications.
- *Photovoltaic Energy Conversion System:* Sunlight is converted into electricity by the use of solar cells. Since photovoltaic cells can be combined in arrays of any size to suit specific needs, they can be used in either centralized or dispersed applications.
- *Wind Energy Conversion Systems:* These systems employ some form of wind machine to drive an electric generator. Again, since wind machines of different sizes may be employed to suit specific needs, they may be considered either a centralized or a dispersed technology.
- *Ocean Energy Systems:* Various kinds of "potential" energy stored in the oceans can be used by centralized facilities. Ocean Thermal Energy Conversion (OTEC), which can produce electricity from the difference in temperature between surface and deep-sea water, is currently the most advanced concept. Other ocean energy sources currently being researched and developed are salinity gradients, ocean currents, and wave energy.

The Biomass Energy Systems Program involves the use of biological materials (requiring sunlight for photosynthesis) to produce heat for generating electricity, or to be converted into fuels such as methane, wood alcohol, or chemicals by fermentation, chemical extraction, pyrolysis, etc. Biomass systems also can be scaled to suit specific needs. Wood, for instance, can be used for producing steam to generate electricity in a centralized facility; it also can be burned to provide direct residential space heating or steam for industrial processes.

Geothermal Energy. Geothermal energy is the natural heat stored beneath the Earth's crust. It occurs in many forms and has potential for both direct thermal use and electricity generation. (Typical geothermal powerplants are shown in Figure III 2-1.) Geothermal energy resources have been discovered in 33 states in the United States and are also believed to exist in others, where geothermal exploration has not yet been undertaken.

The Geothermal Energy (GE) Program is divided into three subprograms: 1) Hydrothermal Resources, 2) Geopressured Resources, and 3) Geothermal Technology Development. Through the R&D activities carried out in each of these areas, three principal types of geothermal energy resources may be developed through 2000. In increasing order of estimated magnitude, they are:

- Hydrothermal (vapor- and liquid-dominated)
- Geopressured (including dissolved natural gas)
- Hot dry rock

The Hydrothermal Resources Program provides R&D support to the Hydrothermal Geothermal Commercialization Subprogram that has been transferred to the DOE Assistant Secretary for Resource Applications. The support provided includes:

- Regional Planning—accelerate hydrothermal commercialization by supporting site-specific participatory planning
- Resource Definition—quantitatively assess the hydrothermal resource potential on a regional and national basis
- Facilities—design, construct, and operate facilities to test and evaluate new hydrothermal equipment, materials, and techniques
- Engineering Applications—demonstrate economic and technical viability of hydrothermal systems by field experiments and demonstration projects
- Environmental Control—address environmental and health and safety issues related to hydrothermal development

The Geopressured Resources Program is composed of four main subprograms related

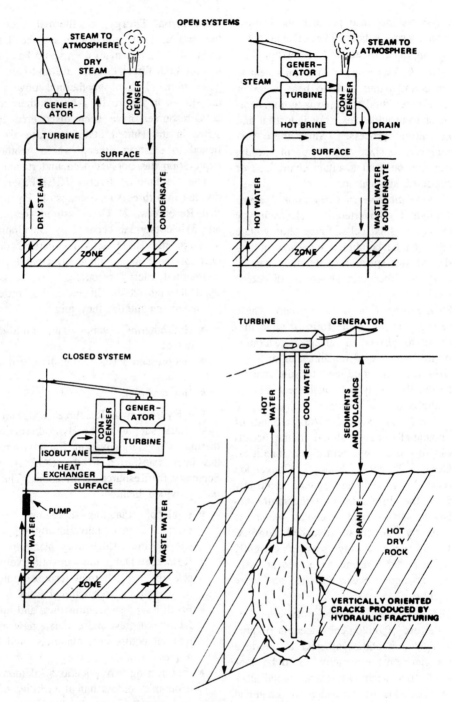

Figure III-2-1. Typical geothermal power plants.

to the development of geopressured geothermal resources:

- Resource Definition—determine the location, size, temperature, and pressure of geopressured sandstone aquifers on the Gulf Coast
- Existing Well Tests—collect geopres-

sured aquifer production data from wells previously drilled to search for oil and gas
- New Drilled Wells—collect data related to long-term fluid production rates to determine capacities and longevities of large geopressured reservoirs
- Reservoir Engineering—determine the behavior of geopressured aquifers under production conditions, establish the recoverability of methane, and develop a predictive capability for reservoir performance

The Geothermal Technology Development Program is composed of six subprograms which correspond to the delivery and resource exploitation process related to geothermal technologies. These subprograms are:

- Drilling and Completion Technology—reduce drilling costs by developing improved drill bits, down-hole motors, and drilling fluids
- Energy Conversion Systems and Stimulation—reduce geothermal electric generating costs, particularly for moderate temperature geothermal resources
- Geochemical Engineering and Materials—solve problems related to geothermal water, scale, and gas control, and system materials
- Geosciences—develop programs related to exploration technology, reservoir engineering, logging instrumentation, and log integration
- Hot Dry Rock—determine technical and economic feasibility of exploiting the hot dry rock resource
- Interagency Coordination and Planning—effect adoption of regulatory and policy measures needed to enhance attainment of national power-on-line objectives

Electric Energy Systems.

The current regional electric systems, which evolved to their present status without benefit of a systematic planning effort, allow a uniform energy supply to flow from a variety of primary sources (coal, oil, gas, nuclear, hydroelectric) to a wide range of end-users. Although regional electric energy systems already exist, future difficulties may arise as loads increase under greater constraints resulting from fuel availability, difficulty in siting generating plants and transmission lines, and national emergencies. The electric system will have to accommodate emerging generating and storage systems, many of which will be dispersed at consumer locations and need to be integrated into the grid. The EES program will support the integration of future generating and storage technologies into the existing electric energy transmission and distribution system and, at the same time, provide electric power from traditional energy sources more efficiently and reliably.

The EES Division is organized into two primary programs: (1) Power Supply Integration and (2) Power Delivery. The FY 1980–81 Power Supply Integration Program is composed of two subprograms that are designed to achieve increased fuel substitution for oil and gas, improved efficiency of electric energy transfer, and increased electricity production by decentralized sources. These FY 1980–81 subprograms are:

- Load Management
- Solar Generic/Integration of New Source Technologies

The Power Delivery Program addresses problems expected to arise as U.S. electric energy supply grows through the addition of both large central generating plants, sited at a distance from users, and dispersed generating and storage facilities, located at or near users. The two FY 1980–81 subprograms in the Power Delivery Program have specific objectives related to the delivery and control of large blocks of power over long distances, as well as smaller amounts over relatively short distances. These FY 1980 subprograms are:

- Electric Field Effects
- Underground Transmission and Compact Stations

The Power Delivery Program is being

closely examined in view of the recently passed National Energy Act and other considerations. This examination may result in a reorientation of program priorities if it is determined that such steps are needed.

Energy Storage Systems. Energy storage systems may be included in a broad spectrum of applications. To the extent that they are able to reduce or eliminate the need for backup systems, energy storage technologies will expedite the commercialization of solar energy. Energy storage will also enable electric utilities to use their facilities more efficiently. Potential applications of energy storage technologies also include transportation, building heating and cooling, and industrial processes. Furthermore, fuel substitution resulting from the use of energy storage will reduce the need to import expensive oil and can help to reduce environmental pollution.

The STOR Program is divided into two major programs:

- Battery Storage
- Thermal and Mechanical Storage

The Battery Storage Program accelerates the development of commercial technology for low-cost, long-life, high-performance batteries for use in transportation, solar, and electric network systems, and provides related electrochemical technologies for energy and resource savings. Subprograms included in the Battery Storage Program are Near-term Batteries, Advanced Battery Development, Solar Applications, Electrochemical Systems Research, Supporting Research, Electrolytic Technology, and Dispersed Battery Applications.

The Thermal and Mechanical Storage Program seeks to provide reliable, low-cost, high-performance energy storage technologies which allow various energy delivery systems to meet demand loads at time of need. The six subprograms in this STOR program include Thermal Storage, Chemical/Hydrogen Storage, Mechanical Storage, Magnetic Storage, Utility Applications, and Technical and Economic Analysis.

The broad range of energy storage technologies and their end uses are shown in Figure III 2-2. The types of energy to be stored include heat, electricity, and mechanical energy (energy of motion). Energy sources, shown in the left portion of the diagram, include solar energy, hydroelectric power, and waste heat from powerplants and industrial processes. Storage technologies appropriate to each energy source are shown in the center of the figure. For example, electricity produced by hydroelectric, nuclear, or fossil fuel powerplants may be stored using electrochemical storage, mechanical storage, or thermal storage for various end uses.

The DOE end-use organizations that will benefit from the development of thermal and mechanical storage technologies are the organizations that have prime responsibility for developing related end-use technologies. Within Energy Technology these are: Central Solar Technology, Distributed Solar Technology, Nuclear Power Development, Fossil Fuel Utilization, Magnetic Fusion Energy, and Electric Energy Systems. Within the Conservation and Solar Applications organization these are: Transportation Energy Conservation, Solar Applications, Buildings and Community Systems, and Industrial Energy Conservation. Within Resource Applications, the power marketing functions, also benefit.

The staff of the Energy Storage Systems Division also hold memberships on several DOE committees for program coordination. Jointly funded projects are used to coordinate energy storage activities with other Federal agencies. Coordination is also being achieved through interagency agreements with other Federal agencies including the National Aeronautics and Space Administration (NASA), Department of Defense (DOD), and the Department of the Interior (Bureau of Reclamation and the U.S. Geological Survey).

Budget Estimates

Estimated funding for the four Solar technology programs and their subprograms are presented in Table 2-1.

The FY 1980 funding requested for Solar Technology is 437 million, or about 65 percent of the total. Geothermal Energy represents about 21 percent of the total (or 139

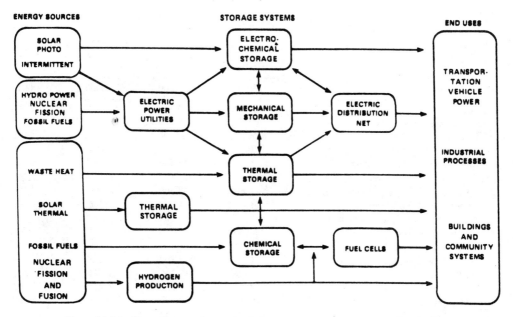

Figure III-2-2. Flow diagram of energy sources, energy storage systems, and end uses.

million), Electric Energy Systems represent 4 percent (or 26 million), and Energy Storage Systems represent 10 percent (or 65 million). Figure III 2-3 presents the requested FY 1980 funding levels broken down by the percentage of the total budget that each major program represents.

Significant Solar Projects

As an important part of its responsibility to research and develop emerging energy technologies, CSE funds, either by itself or in cooperation with other Federal Government offices or private industry/utilities, significant large-scale projects. These major projects often are used to demonstrate the technical and economic feasibility of commercial-scale systems, or to generate the data necessary to scale up systems for commercial applications. Table III 2-2 lists significant CSE projects, FY 1978–1980 funding levels, and the Total Estimated Cost (TEC) of each significant project.

Major RD&D Requirements and Considerations

The integration of technologies being developed by ETS into the U.S. energy supply system will be dependent on the resolution of a number of concerns related to those technologies. These concerns can be grouped into four general areas:

- Scientific and technological requirements

BUDGET AUTHORITY
(DOLLARS IN MILLIONS) *

	ESTIMATED FY 79	ESTIMATED FY 80	INCREASE (DECREASE)
SOLAR TECHNOLOGY	354.9	437.0	82.1
GEOTHERMAL ENERGY	156.2	139.0	(17.2)
ELECTRIC ENERGY SYSTEMS	40.0	26.0	(14.0)
ENERGY STORAGE SYSTEMS	58.0	65.0	7.0
TOTAL	609.1	667.0	57.9

* Totals may not add exactly due to rounding.

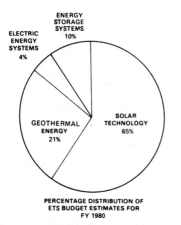

Figure III-2-3. Estimated FY 1980 funding relationships.

Table III-2-1. Estimated budget authority.

(DOLLARS IN THOUSANDS)

		FY 1978	FY 1979 ESTIMATE	FY 1980 ESTIMATE	INCREASE (DECREASE)
*SOLAR TECHNOLOGY	TOTAL	230,400	354,900	437,000	82.1
Solar Electric Applications	Total	189,700	312,500	380,000	67,500
Solar Thermal Power		64,600	100,100	121,000	20,900
Large Power Systems		25,800	54,100	65,000	10,900
Small Power Systems		29,100	32,500	33,000	500
Advanced Research and Development		9,700	13,500	23,000	9,500
Photovoltaic Energy Conversion		60,300	103,800	130,000	26,200
Advanced Research and Development		7,600	39,300	47,000	7,700
Technology Development		36,000	38,000	56,000	18,000
Systems Engineering and Standards		5,400	13,300	14,500	1,200
Test and Applications		11,300	13,200	12,500	(700)
Wind Energy Conversion Systems		35,400	60,700	67,000	6,300
Project Development and Technology		6,900	14,400	16,200	1,800
Small Scale Systems		7,600	19,700	20,500	800
Intermediate Scale Systems		4,600	7,250	6,150	(1,100)
Large Scale Systems		16,300	19,350	24,150	4,800
Ocean Systems		29,400	38,200	35,000	(3,200)
Project Management		2,100	4,700	5,100	400
Definition Planning		1,700	2,100	1,000	(1,100)
Technology Development		10,900	12,000	7,400	(4,600)
Engineering Test and Evaluation		8,100	15,700	15,400	(300)
Advanced Research and Development		6,600	3,700	6,100	2,400
Technology Support and Utilization		22,000	6,700	—	(6,700)
Solar Energy Research Institute Facility		—	3,000	27,000	24,000
Biomass Energy Systems	Total	18,700	42,000	57,000	14,600
Technology Support		2,600	16,200	16,000	(200)
Production Systems		4,800	4,300	5,700	1,400
Conversion Technology		6,700	15,900	25,000	9,100
Thermochemical			10,200	17,000	6,800
Biochemical			5,700	8,000	2,300
Research and Exploratory Development		2,400	5,200	8,800	3,600
Support and Other		2,200	800	1,500	700
GEOTHERMAL	TOTAL	105,800	156,200	139,000	(17,200)
Hydrothermal Resources	Total	55,000	70,900	59,100	(11,800)
Regional Planning		3,900	5,845	6,100	255
Resource Definition		14,400	25,470	9,000	(16,470)
Engineering Applications		7,800	10,500	9,831	(669)
Environmental Control		1,600	516	1,300	784
Facilities		26,100	27,169	32,069	4,900
Capital Equipment		1,200	1,400	800	(600)
Geopressured Resources	Total	16,400	27,700	36,000	8,300
Resource Definition		1,900	1,900	1,350	(550)
Existing Well Tests		3,900	3,600	6,950	3,350
New Drilled Wells		8,200	18,500	22,900	4,400
Reservoir Engineering		900	1,000	1,000	0
Other		1,500	2,600	3,500	900
Capital Equipment		1,500	100	300	200
Geothermal Technology Development	Total	34,400	57,600	43,900	(13,700)
Drilling and Completion Technology		2,300	6,000	7,000	1,000
Energy Conversion and Stimulation		11,100	13,100	10,000	(3,100)
Geochemical Engineering and Materials		3,600	6,000	3,700	(2,300)

Table III-2-1. (Continued).

		FY 1978	FY 1979 ESTIMATE	FY 1980 ESTIMATE	INCREASE (DECREASE)
Geosciences		7,100	11,700	4,200	(7,500)
Hot Dry Rock		5,900	15,000	14,000	(1,000)
Interagency Coordination and Planning		3,100	4,300	2,900	1,400
Capital Equipment		1,300	1,500	2,100	600
ELECTRIC ENERGY SYSTEMS	TOTAL	40,300	40,000	26,000	(14,000)
Power Supply Integration	Total	11,800	15,500	17,000	1,500
Load Management		4,000	4,600	6,900	2,300
Solar Generic		300	6,800	10,000	3,200
System Control and Development		7,500	4,000	0	(4,000)
			100	100	
Power Delivery	Total	28,500	24,500	9,000	(15,500)
Electric Field Effects		2,300	2,200	2,600	400
Underground Cables and Compact Stations		19,200	16,200	6,000	(10,200)
High Voltage Direct Current		3,200	3,000	0	(3,000)
High Efficiency Equipment and Systems		3,800	1,200	0	(1,200)
Capital Equipment			1,900	400	(1,500)
*ENERGY STORAGE SYSTEMS	TOTAL	45,631	58,000	65,000	7,000
Battery Storage	Total	20,676	25,400	33,100	7,700
Near Term Batteries		1,230	1,040	2,100	1,060
Supporting Research		1,206	2,140	3,000	860
Advanced Batteries		13,839	11,700	14,400	2,700
Electrochemical Systems Research		2,235	2,500	2,600	100
Solar Applications		442	800	5,000	4,200
Electrolytic Technology		591	1,320	1,500	180
Dispersed Battery Applications		1,133	5,900	4,500	(1,400)
Thermal and Mechanical Storage	Total	24,955	32,600	31,900	(700)
Thermal Storage		5,036	8,225	20,300	12,075
Chemical/Hydrogen Storage		3,883	7,400	2,500	(4,900)
Mechanical Storage		5,904	6,975	4,700	(2,275)
Magnetic Storage		1,750	2,300	700	(1,600)
Utility Applications		3,823	4,100	2,000	(12,100)
Technical & Economic Analysis		4,559	3,600	1,700	(1,900)
ETS PROGRAMS	TOTAL	422,131	609,100	667,000	57,900

*Capital equipment included in program, subprogram, etc. totals

- Commercialization considerations
- Environmental impacts
- Socioeconomic concerns

The primary objective of the R&D activities carried out by solar is to resolve the scientific and technological requirements related to each of its program areas. However, solar also has the responsibility to identify and address the specific environmental concerns and socioeconomic considerations associated with solar energy technology R&D.

The commercialization of hydrothermal energy resources is an example of Government efforts to encourage the development of energy technologies. Technology development and engineering development have been completed successfully, and the technology has been placed under the DOE Assistant Secretary for Resource Applications. Progress is being made on the preparation of a Cooperative Agreement for the construction and operation of a demonstration plant, the cost of which will be shared with private industry.

The process of developing a technology and commercializing it is discussed in greater de-

384 Solar, Geothermal, Electric and Storage Systems

Table III-2-2. Significant solar projects.

ETS PROGRAM	SIGNIFICANT ETS PROJECTS	ESTIMATED FUNDING (THOUSANDS OF DOLLARS)			TOTAL ESTIMATED COST
		FY 78 BA/BO	FY 79 BA/BO	FY 80 BA/BO	
SOLAR TECHNOLOGY	SERI Permanent Facility (Golden, Colo.)	......../........	3,000/2,000	27,000/22,000	30,000
SOLAR THERMAL	10-MW Power Plant (Barstow, Calif.)	41,000/525	28,000/25,000	36,500/50,000	108,000
	Shenandoah Large Scale Experiment (Shenandoah, Georgia)	2,755/2,755	3,975/3,975	5,000/5,000	13,497
	Small Community Systems Applications (TBD)	2,160/1,285	2,325/2,645	5,280/5,685	24,600
	Fixed Mirror Project (Crosbyton, Texas)	1,637/1,637	1,000/1,000	100/100	3,887
WIND	MOD-2 Wind Turbine Project (TBD)	9,800/7,800	11,500/12,500	16,200/16,100	27,300
	WEC Test Center Building (Rocky Flats, Colo.)	......../........	......../........	2,140/2,140	2,140
OTEC	OTEC-1 Ocean Test Facility (Keahole Point, Hawaii)	8,800/........	13,500/19,000	10,700/10,700	33,000
BIOMASS	Thermochemical Gasification Experimental Facility (TBD)	......../........	500/300	3,100/2,800	12,000
	Regional Silviculture Farm (TBD)	......../........	400/200	1,000/400	1,400
GOETHERMAL	50-MW Demonstration Plant (Valles Caldera, N. Mex.)	12,000/7,457	7,500/6,500	20,500/19,000	54,500
	Raft River Pilot Plant (Malta, Idaho)	10,026/6,979	9,561/12,577	4,727/6,845	23,672
	Geothermal Wellhead Generator (Puna District, Hawaii)	1,450/84	2,439/3,123	3,642/4,224	6,725
ELECTRIC ENERGY	Battery Energy Storage Test Facility (Hillsborough, N. J.)	0/950	1,000/2,100	1,200/2,442	....
ENERGY STORAGE	Storage Battery Electric Energy Demonstration (TBD)	0/0	900/200	1,200/600	....

(TBD = To Be Designated)

tail in Part III, Chapter 8. The major DOE decisions associated with the progress of a technology through the development cycle are enumerated. Commercialization considerations are presented for technologies currently being developed within CSE.

The specific requirements and considerations relevant to each of the CSE programs are summarized in Table III 2-3.

PROGRAM OBJECTIVES AND STRATEGY

Background

The energy supply system that has fueled U.S. economic growth for the last 40 years is becoming increasingly strained. Domestic supplies of oil and natural gas, the main source of U.S. energy since World War II,

Table III-2-3. Solar program areas versus requirements and considerations.

ETS PROGRAM	MAJOR RD&D REQUIREMENTS AND CONSIDERATIONS			
	SCIENTIFIC AND TECHNOLOGICAL	COMMERCIALIZATION	POTENTIAL ENVIRONMENTAL IMPACTS	POTENTIAL SOCIOECONOMIC CONCERNS
SOLAR THERMAL ELECTRIC SYSTEMS	ADVANCED COLLECTOR SYSTEMS ADVANCED STORAGE SUBSYSTEMS	COMPETITIVE ECONOMICS HELIOSTAT COST REDUCTIONS	LAND USE LOCAL ECOSYSTEMS AFFECTED BY SHADING OF LAND	UTILITY RATES AND BACKUP/BUYBACK RATES SUNRIGHTS
PHOTOVOLTAIC ENERGY SYSTEMS	INCREASED CELL EFFICIENCY TO REDUCE REQUIRED CELL AREAS ADVANCED STORAGE SUBSYSTEMS	COST COMPETITIVE WITH CONVENTIONAL SYSTEMS FABRICATION TECHNIQUES THAT PERMIT MASS PRODUCTION	LAND USE FOR CENTRAL APPLICATIONS LOCAL ECOSYSTEMS EFFECTS CONTAMINATION WITH PV MATERIALS IN MANUFACTURE	UTILITY RATES AND BACKUP/BUYBACK RATES SUNRIGHTS
WIND ENERGY CONVERSION SYSTEMS	IMPROVED BLADE PERFORMANCE ADVANCED CONCEPTS	COMPETITIVE ECONOMICS MANUFACTURING ECONOMICS	INTERFERENCE WITH NEARBY RADIO AND TV RECEIVERS	UTILITY RATES AND BACKUP/BUYBACK RATES LAND USE AND SITING INSURANCE COVERAGE AND SAFETY STANDARDS
OCEAN SYSTEMS	ADVANCED HEAT EXCHANGERS ADVANCED CYCLES ADVANCED TRANSMISSION CABLES	COMPETITIVE ECONOMICS RELIABLE OPERATION	ACCIDENTAL LEAKS OF WORKING FLUIDS POTENTIAL CLIMATIC EFFECTS OF LARGE SCALE DEPLOYMENT IMPACTS ON BIOTA	ABILITY TO MOVE LARGE BLOCKS OF POWER OR ENERGY INTENSIVE PRODUCT TO PLACE NEEDED
BIOMASS ENERGY SYSTEMS	UNDERSTANDING OF MECHANISMS OF CONVERSION PROCESSES ADVANCED TECHNOLOGIES FOR PRODUCTION AND CONVERSION TO LIQUID AND GASEOUS FUELS	OPTIMAL DESIGN FOR LOW-BTU GASIFICATION IMPROVE EFFICIENCY OF FERMENTATION AND DEVELOPMENT OF TECHNOLOGY FOR USING CELLULOSIC FEEDSTOCKS EXTEND ANAEROBIC DIGESTION TECHNOLOGY TO MIXED FEEDSTOCKS	LOCAL GROUND AND WATER CONTAMINATION SOIL EROSION EMISSIONS FROM LARGE-SCALE COMBUSTION	LAND USE TRADEOFFS INFRASTRUCTURE REQUIRED TO MOVE PRODUCTS TO MARKETS DISPERSED BIOMASS SYSTEMS OWNED BY FARMERS, COOPS, ETC.
GEOTHERMAL ENERGY SYSTEMS	ADVANCED DRILLING TECHNOLOGIES FOR HOT, DRY ROCK FORMATIONS ADVANCED TECHNOLOGIES FOR GEOTHERMAL EXPLORATION, FLUID PRODUCTION AND ENERGY CONVERSION	COMPETITIVE ECONOMICS AND EXTENT OF HYDROTHERMAL RESOURCE COMPETITIVE ECONOMICS AND EXTENT OF GEOPRESSURED RESOURCE	TREMORS AND LAND SUBSIDENCE RESULTING FROM FLUID EXTRACTION RELEASE OF TOXIC VAPORS GROUNDWATER CONTAMINATION FLUID DISPOSAL	INTRUSION ON SCENIC AREAS IMPACT ON PRESENT ENERGY SUPPLY INFRASTRUCTURE
ELECTRIC ENERGY SYSTEMS	INTEGRATING DISTRIBUTED TECHNOLOGIES ADVANCED CONTROL SYSTEMS TO REDUCE BLACKOUTS AND INCORPORATE DISTRIBUTED GENERATION HIGH-VOLTAGE, HIGH-CAPACITY UNDERGROUND LINES FIELD EFFECTS OF OVERHEAD HIGH VOLTAGE LINES COMPACT SUBSTATIONS FOR UNDERGROUND LINES AND OFF-SHORE STATIONS	COMPETITIVE ECONOMICS OF UNDERGROUND TRANSMISSION OPTIONS	ECOLOGICAL AND BIOLOGICAL EFFECTS FROM HIGH VOLTAGE LINES ACOUSTICAL AND ELECTRO-MAGNETIC NOISES	REDUCED BLACKOUT POTENTIAL LAND USE AND RIGHTS-OF-WAY ESTHETIC AND LEGAL ACCEPTANCE
ENERGY STORAGE SYSTEMS	IMPROVED ENERGY DENSITY OF BATTERIES ADVANCED PHASE-CHANGE STORAGE AND HEAT PUMP MEDIA IMPROVED FLY-WHEEL SYSTEMS—COMPOSITE ROTORS, ADVANCED BEARINGS, AND MOTOR GENERATORS MECHANICAL STORAGE FOR PHOTOVOLTAIC/WIND SYSTEMS	COMPETITIVE ECONOMICS IN ALL APPLICATIONS ACHIEVEMENT OF MANUFACTURING ECONOMICS FOR ADVANCED BATTERIES	GROUNDWATER CONTAMINATION FROM AQUIFER STORAGE WATER CONTAMINATION FROM ADVANCED BATTERY PRODUCTION OPERATOR HEALTH AND SAFETY	UTILITY RATES AND STRUCTURES IMPACT ON UTILITY PEAKS

are being depleted faster than new reserves are being discovered. The cost of imported oil, which in 1977 constituted almost one-half of total U.S. oil consumption and almost one-fourth of total U.S. energy supply, has quadrupled since 1973 and in all probability will continue to increase in the future. Besides contributing to balance-of-payments deficits and domestic inflation, continued large-scale imports of oil will exacerbate U.S. energy dependence at a time when increasing world oil demand is threatening to outrun world oil productive capacity. World oil demand pressures, coupled with international political and socioeconomic uncertainties, dictate that the United States reduce its reliance on foreign oil as soon as possible.

One method of reducing U.S. demand for oil and gas, especially imported oil, is to increase the availability and utilization of coal and nuclear energy. However, environmental problems (e.g., disposal of wastes) and infrastructure impediments (e.g., licensing procedures) may hinder the growth rates of technologies that use coal and nuclear energy resources. Even if these problems were resolved, there would still be a need for imported energy.

Consequently, as a part of U.S. strategy for long-term energy supply, inexhaustible resources are projected to serve an increasingly important function. As alternative technologies are developed and proven to be reliable and economically competitive, they will capture that share of the energy market otherwise supplied by nonrenewable, and often imported, fuels. This displacement of imported energy by domestic renewable resources, together with coal and nuclear energy, not only will improve the U.S. international economic posture, but will also aid the domestic economy by providing jobs for a domestic construction, operation, and maintenance work force serving centralized and dispersed renewable energy technologies.

However, if such expectations are to be realized, several problems inherent in the nature of renewable energy must be resolved. For instance, the widespread utilization of solar energy in decentralized applications (energy generated near or at the point of use) currently is impeded by the added costs associated with the intermittent solar energy source. Because solar systems supply energy only when the sun shines, either conventional back-up systems or unique energy storage systems must be integrated with the solar collectors. They will provide energy when and where needed, regardless of daily and seasonal fluctuations in the amount of sunshine.

Furthermore, because of the intermittent nature of the energy source, solar electric generating facilities cannot be integrated easily into existing distribution grids. If centralized solar applications (energy generated at a distance from point of use) are to be used to displace base load capacity, unique load management techniques and energy storage systems, together with system controls and control strategy, must be developed.

The same energy storage and electric management techniques needed to foster the widespread application of centralized and dispersed solar energy systems provide further benefits independent of their value to intermittent energy sources. For instance, improved load management techniques will allow more efficient utilization of traditional fuels (oil, gas, coal, nuclear) used to generate electricity. Thus, efficient electric energy distribution and energy storage systems have the potential to reduce demand pressure for exhaustible, and often imported, fossil fuels. At the same time, they allow solar energy to satisfy a wider range of centralized and dispersed energy needs.

Strategy Overview

The ETS Program has been established with the goal of contributing substantially to the long-term energy supply of the United States. Because programs carried out under ETS will help to increase the efficiency of traditional fuel utilization and also to research and develop technologies which use alternative, renewable energy resources, the four interrelated programs that constitute the overall ETS Program will serve a primary function in the U.S. energy strategy. The combined organization of these different programs under CSE is especially important, because technologies that result in increased energy efficiency (e.g., energy storage) often are indispensable to the widespread application of renewable, but intermittent, energy sources.

Consequently, CSE Program objectives are designed to produce a coordinated strategy for integrating fuel efficiency and renewable energy technologies in order to improve both the near- and long-term U.S. energy posture. The specific objectives of the CSE Program are:

- To research and develop technologies that use diverse solar and geothermal energy resources to displace imported and exhaustible fuels
- To develop electric transmission and distribution and energy storage systems necessary to integrate intermittent energy sources into the national grid and to maximize the efficient use of traditional and alternative energy sources
- To test and demonstrate technologies that are energy efficient or that use renewable energy, so that they are capable of competing in the commercial marketplace beginning as early as the mid-1980s and maturing as major energy sources within this century
- To begin to obtain from these sources a significant contribution to energy supply annually by the late-1980s

The following sections detail the specific strategies for each of the four solar Programs: Solar Technology, Geothermal Energy, Electric Energy Systems, and Energy Storage Systems. First, the goals and objectives set for each of the solar technology programs are described, and then current program priorities are reviewed in the context of longer term program strategy to demonstrate the long-range improvement in the U.S. energy posture, including energy savings, which the strategy will serve to attain.

Solar Technology Program

Solar technologies hold the promise of reducing U.S. dependence on imported, exhaustible fossil fuels by increasing the utilization of inexhaustible, relatively pollution-free solar energy. Solar technologies seem especially attractive because they may be used in both centralized and dispersed applications, producing electricity, industrial and agricultural process heat, residential and commercial space conditioning and hot water, and liquid and gaseous fuels.

The Domestic Policy Review (DPR) of Solar Energy, initiated by the President on May 3, 1978, produced a status report (August 25, 1978) that represents an initial effort to develop a national strategy emphasizing the use of renewable resources, such as solar and geothermal energy. The DPR also provides a range of policy options from which a national strategy can be derived.

As part of its analysis, the DPR projected that in a base case scenario (based on a world oil price of $25 per barrel), solar technologies (including process heat, solar thermal, photovoltaic, and wind technologies) and biomass technologies could displace as much as 4.9 quads of the more than 100 quads of energy needed annually by 2000 (Table III 2-4). In a "Maximum Practical" scenario, in which the Federal Government aggressively pursues RD&D programs for renewable energy technologies, solar technologies (including process heat, solar thermal, photovoltaic, and wind technologies) and biomass technologies could displace as much as 11.3 quads of energy by 2000. Although such projections are based on assumptions about future economic development and are meant to be modal values representing a wide range of potential rather than precise measurements, they nevertheless indicate the level of importance being attached to renewable energy resources in long-term national energy strategy.

Program Goals and Objectives. The overall objective of the Solar Technology Program is to prepare for the commercial market a range of technologies that produce clean fuels or

Table III-2-4. Estimated fuels displaced by solar technologies by 2000.

SOLAR TECHNOLOGY	BASE CASE ($25/BARREL)	MAXIMUM PRACTICAL (QUADS)
Biomass	3.1	5.5
Solar Thermal	0.1	0.4
Wind	0.6	1.7
Photovoltaics	0.1	1.0
OTEC	–	0.1
Process Heat*	1.0	2.6
TOTAL	4.9	11.3

*DONE IN SUPPORT OF C S.

generate electricity and heat using energy from biomass, the sun, wind, and ocean. Because all such systems ultimately derive their energy from the inexhaustible solar supply, solar systems are an attractive alternative to systems that depend on exhaustible energy resources.

Each of the programs contained within Solar Technology has been assigned specific objectives, as have the individual R&D projects within each program level and activity. The accomplishment of these specific objectives will result in the achievement of the four primary goals of Solar Technology:

- To develop and demonstrate solar systems producing industrial process heat, mechanical and electrical energy, and other products at commercially competitive costs starting by the mid-1980s
- To establish the environmental acceptability and operational safety of solar energy systems
- To stimulate early resolution and clarification of legal and institutional issues
- To aid in the creation of the necessary institutional infrastructure

The objective of the Solar Electric Applications Program is to stimulate the development of economically feasible, environmentally acceptable, and operationally safe solar electric systems which will contribute significantly to national energy supplies. More specific objectives have been established for each program within the Solar Electric Applications Program:

- *Solar Thermal Power:* To establish the technical readiness of cost-competitive solar thermal power systems for both small (under 10 MWe or 100 $MBtu_{th}$/hr) and large (over 10 MWe or 100 $MBtu_{th}$/hr) energy production applications, including nonelectric applications such as industrial process heat
- *Photovoltaic Energy Conversion:* To develop photovoltaic energy systems to the point where they will be commercially competitive in centralized and dispersed applications
- *Wind Energy Conversion Systems:* To develop wind energy systems so that they can provide economically competitive energy for both centralized and dispersed applications
- *Ocean Systems:* To develop environmentally acceptable, technically sound, and economically competitive technologies for converting ocean energy into a usable resource that can contribute significantly to national energy needs

The objectives of the Biomass Energy Systems Program are to stimulate commercially feasible production of biomass and to convert it directly into clean fuels or energy-intensive products. This may have potentially significant long-term impacts on projected national energy needs.

Program Strategy and Impacts. The overall Federal strategy for solar technologies is designed to create a viable solar power industry. Specifically, the strategy is: 1) to define market opportunities for displacing critical exhaustible resources, such as oil and natural gas, and provide an early market environment favorable to the adoption of solar systems; 2) to establish through R&D market-ready solar systems for these opportunities; 3) to reduce costs at least to a level where economic incentives can provide parity with conventional systems; 4) to accelerate market development through procurement steps that establish manufacturing, distribution, and servicing capabilities; 5) to perform studies on national policy for inexhaustible energy resources and systems; and 6) to promote rapid commercialization through technology transfer activities.

The Assistant Secretary for Policy and Evaluation (PE) has the lead responsibility for implementing the policy aspects of this strategy. The Assistant Secretary for Conservation and Solar Energy (CSE) has the lead responsibility for technology development, and for the commercialization of near-term technologies.

Several traditional Federal policy options are available to implement this overall strategy. They may be classified into six general categories:

- Economic incentives (e.g., tax credits, loan guarantees, etc.)

- Government funding for RD&D (including cost-sharing)
- Product support
- Regulatory policies
- Interaction with State and local governments
- Information dissemination

The different incentives have been addressed in the Domestic Policy Review of Solar Energy (August 25, 1978).

The selection of policy options for a particular solar technology will depend on the relative state of technological development, the end-use sector to be affected, and the time frame when the technology under consideration is expected to be economically competitive. Because the Federal Government currently is involved in the RD&D of a range of solar technologies, a strategy has been developed to ensure that the most appropriate policy options are applied to each technology alternative. More specifically, CSE intends to:

- Aggressively support those solar technologies that will be economically competitive in the near term
- Continue the R&D of those solar technologies with significant market potential but which cannot yet be commercialized because of technical or economic uncertainties
- Identify and evaluate long-term solar technologies whose potential costs and impacts are uncertain

The most promising near-term (1978–1985) technologies being developed by the Solar Technology Program are in its Biomass Energy Systems Program. For instance, because the direct burning of wood in both centralized and dispersed applications is already economically competitive under certain circumstances, forest and agricultural biomass are being cultivated extensively for near-term impacts. Small-scale units to produce process steam and electricity from biomass are already in operation.

In order to increase the mid-term contribution of solar energy, the Solar Technology Program is furthering the technological development and attempting to reduce the costs of several technologies with significant potential. A procurement program is planned for small (<40-kW) wind machines because they can be used in various dispersed applications. Larger wind systems (>1-MW) also are being developed for mid-term applications. Furthermore, R&D is also being carried out to increase the mid-term production of liquid and gaseous fuels from biomass.

Ocean thermal conversion, the most advanced ocean energy systems concept, is being tested in a small 1-MW unit; eventually larger scale tests (10-MW) will be required to determine scale-up costs and potential. A limited demonstration of central receiver electric generation is being carried out. Conceptual studies will be done for advanced technologies and new market applications.

In addition, the R&D of more cost-effective photovoltaic cells is being emphasized. The Solar Photovoltaic RD&D Act of 1978 and the National Energy Act are being supported by R&D programs conducted by CSE.

The Solar Technology Program is also evaluating the long-range potential of new solar technologies with uncertain technological and economic characteristics. For example, feasibility studies on advanced photochemical conversion processes and ocean salinity gradients and waves will be carried out to determine their future energy potential.

Finally, long-term technologies, for which adequate information is not available, are being identified and evaluated. For instance, feasibility studies on advanced photochemical conversion processes and ocean salinity gradients and waves will be carried out to determine their future energy potential.

This strategy will ensure that an increasing number of solar technologies are available for commercialization as the twenty-first century approaches and traditional supplies of energy begin to decline. However, Federal strategy also recognizes that extensive introduction of solar energy systems depends on the growth of a service industry and the maintenance of R&D incentives until private industry is able to assume the full burden of future R&D. Because extensive use of solar energy products also depends on the development of energy storage technologies and technical and management developments in electric energy distribution systems, the Solar Technology Program is coordinated with the Electric Energy Systems and Energy Storage Systems Programs within CSE.

Also supporting the solar program is the Solar Energy Research Institute (SERI) at Golden, Colo. SERI, which was established by the Solar Energy Research, Development and Demonstration Act of 1974, supports DOE's national program of RD&D and commercialization of solar energy technologies. The institute also carries out activities that will help create a solar energy industrial base to foster the widespread use of solar technologies.

To accomplish this mission, current plans call for SERI to perform a variety of functions. These include: 1) analysis and planning support for DOE; 2) research directed at both the near- and long-term potential of solar energy, with emphasis on high-risk research that the private sector is unwilling or unable to undertake; 3) programs in solar energy information dissemination, education, and training and technology commercialization; 4) assumption of a major role in the coordination of international collaborative programs in solar energy RD&D; 5) assistance to DOE in the technical and administrative management of selected elements of the national solar energy effort; 6) interaction with universities in areas related to solar energy.

In summary, the preferred solar technology strategy is indicated by the dark line in Figure III 2-4.

The rate at which the solar market and industry will develop is strongly influenced by the relative economics of alternative sources of energy, the technological advances in solar systems, and Government (Federal, State, and local) policies toward energy use. In general, this rate depends on regional and national decisions related to the pricing of conventional energy sources, the support provided for solar systems development, and the perceived societal impacts of all forms of energy use.

It must be remembered that solar resources are widely distributed and diverse. The mix of solar resources varies from region to region, and devices which convert or utilize them are located at or near the resource in most instances. Thus, a complex spectrum of local, regional, State, and Federal influences come to bear on the process of applying solar technologies to specific market needs.

The Domestic Policy Review of Solar En-

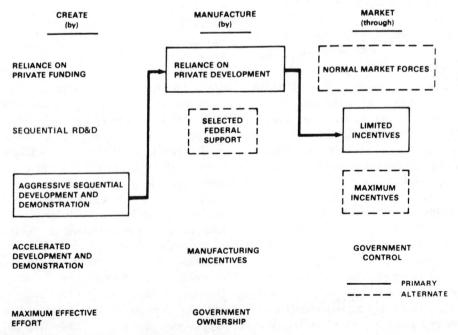

Figure III-2-4. Government role options.

ergy has projected that solar technologies being researched and developed within CSE may supply as much as 11.3 quads of energy by 2000 (assuming a world oil price of $25/barrel and maximum practical Federal Government programs). If such expectations are to be fulfilled, the R&D of solar technologies must be an important element of current DOE programs and long-range strategic planning.

Geothermal Energy Program

Although geothermal energy has been used for electricity generation in Europe since 1904 (Larderello, Italy) and in the United States since 1960 (The Geysers, California), only a very small fraction of this potentially extensive energy resource has been tapped to date. Nevertheless, geothermal resources can be applied to a range of energy end uses (e.g., electricity generation, production of hot water and steam for nonelectric applications, such as residential and commercial space heating, industrial process heat, etc.); furthermore, the resource occurs in widely dispersed locations (for instance, in at least 33 states in the United States). Consequently, the widespread use of geothermal energy may be an important element of the long-term U.S. energy posture. Strategic planning is necessary to develop and demonstrate geothermal energy technologies so that they may begin to be used extensively as U.S. energy demand grows and supplies of traditional, exhaustible energy resources decline.

Program Goals and Objectives. The overall purpose of the Geothermal Energy Program is to stimulate commercial development, by private industry and local public power authorities, of geothermal resources in the United States. With this goal in view, the program has the following general objectives:

Time Impact	Objective
(1978–1985)	• Confirm the magnitude and longevity of hydrothermal resources
	• Determine the extent and recoverability of geopressured resources
	• Establish the technical and economic feasibility of using geothermal heat for electric power generation andd direct heat applications with available technology
(1985–2000)	• Develop advanced technology for the discovery, confirmation, production, and application of geothermal energy
	• Define the hot dry rock resource base
	• Develop extraction technology for hot dry rock resources
(2000–2020)	• Develop an economical technology for exploiting normal gradient geothermal heat

More specifically, the Geothermal Energy Program is divided into three programs: 1) Hdydrothermal Resources, 2) Geopressured Resources, and 3) Geothermal Technology Development. The specific goals of the efforts in these three programs are as follows:

- Hydrothermal Resources: To support the commercial development of hydrothermal energy in a range of direct-heat and electric power generation applications.
- Geopressured Resources: To achieve commercial development, by the private sector, of geopressured-geothermal resources as an economical, reliable, safe, and environmentally acceptable energy source
- Geothermal Technology Development: To reduce geothermal development costs, improve resource utilization efficiency, define the hot dry rock geothermal resource, and develop technologies appropriate for its exploitation

Program Strategy and Impacts. The Division of Geothermal Energy (DGE) provides support for the strategy that leads to the near-term commercial development of the most advanced goethermal resource technologies (hydrothermal); continued R&D of geothermal technologies and resource types with significant potential, but with technological uncertainties that need to be resolved (geopressured); and preliminary resource and technological analysis of geothermal technologies and resources whose technical and economic characteristics are not well defined (hot dry rock). Accordingly, DGE will assist in developing a growing geothermal energy industry that will add significant quantities of process heat, steam, and geothermal electricity to U.S. energy supplies by the turn of the century.

More specifically, the geothermal development strategy involves:

- Reduction of uncertainty about the location, magnitude, and character of U.S. geothermal resources on regional as well as site-specific bases
- Provision of technology and resource definition support for the commercial development of hydrothermal resources expected to begin by the mid-1980s and to grow rapidly thereafter
- Reduction of the resource and technical uncertainties associated with geopressured resources by the mid-1980s
- Provision of technology and resource information required for initial commercial development of the hot dry rock resource in the 1990 time-frame and rapid development by the turn of the century

The Federal program for geothermal development is a comprehensive effort, dispersed among several Federal agencies. It includes RD&D, loan guaranty, leasing and management of Federal geothermal land, and development planning with State and local entities. Thus, a strategy for commercialization of geothermal resources will be implemented by a set of Federal actions, which will provide impetus to both the Federal and private sectors. The Interagency Geothermal Coordinating Council (IGCC) provides direction and coordination for the Federal program.

For the near- to mid-term commercialization of hydrothermal resources, DGE provides support to the commercialization activities of the hydrothermal geothermal Resource Manager in the Office of the Assistant Secretary for Resource Applications. This support includes resource definition, technology development, and demonstration projects required to bring about commercial development of hydrothermal resources.

The geopressured resource is not understood adequately enough to be a candidate for immediate commercialization. The task for the near term will be to complete the definition of the resource base and to establish the technical and economic feasibility of exploiting it, thus removing one of the primary impediments to non-Federal development.

Technologies that produce steam from hot dry rock are even further from commercial development. The magnitude and characteristics of the resource base have yet to be examined, and creation of an energy extraction technology has only recently begun. The program will move toward resolving these uncertainties by the early 1990s.

The overall strategy of the Geothermal Program is to bring each of these resource types and technologies to a point of technical readiness for commercial application. At that point, they will be transferred to the Assistant Secretary for Resource Applications, who is responsible for commercialization activities related to geothermal technologies.

The DGE strategy, if implemented successfully and supported by State and local governments and the private sector, will result in substantial energy savings. Table III 2-5 presents estimates of the impact of the full range of geothermal technologies or near-, mid-, and long-term U.S. energy consumption.

Electric Energy Systems Program

The electric power system is an important element of national energy supply, converting nuclear energy, coal, oil and gas, hydropower, and other primary energy sources into electricity and delivering it to all sectors of the economy. In 1977, approximately 7 quads of electricity, produced from roughly 22 quads of primary energy, were delivered to customers. The system has relied on large generating plants, bulk power transmission, distribution to individual customers, and interconnections with neighboring systems. Most utility systems cooperate with neighboring systems to reduce reserve requirements and provide emergency backups; these cooperating systems are organized into power pools. As a

Table III-2-5. National geothermal utilization estimates.

	QUADS/YEAR NEAR TERM 1985	MID-TERM 2000	LONG TERM 2020
Electrical Applications	0.2–0.3	1.5–3.0	5.0–10.0
Direct Thermal Applications	0.1–0.2	0.5–2.0	6.0–8.0
Geopressured Methane	0 – .02	2.0–4.0	5.0–10.0
Total	0.3–0.5	4.0–9.0	16.0–28.0

result of the Northeast blackout of 1965, regional reliability councils, organized under the National Electric Reliability Council, have been organized to oversee the reliability and adequacy of the Nation's bulk power system.

The Nation's power pools, though individual systems, have considerable transmission capability between pools. There are three intraconnected systems in the U.S.—one east of the Rocky Mountains, one west of the Rocky Mountains, and a third within the state of Texas. For all practical purposes, there are no ties between them.

Recent emergencies, such as the severe winter of 1977, the coal strike of 1978, and the western droughts of 1975 through 1977, have demonstrated the desirability of emergency capability to transfer power from one region to another. The historic trend to large, central generating plants may require modification as advanced energy systems—including solar, geothermal, energy storage, and load management—are introduced.

Program Goals and Objectives. The objective of the EES Program is to provide technological options and solutions to significant national problems of the existing and evolving electric network. Key problems include:

- Integration of new technologies into the electric grid
- Economically and environmentally acceptable transfer and control of large blocks of power

Solar electric generating systems (e.g., photovoltaic and wind systems, etc.) have system operating characteristics significantly different from conventional generating systems. For one thing, their output is intermittent. Furthermore, some solar electric generating systems produce direct current (DC), while end users generally require alternating current (AC). Integration of these systems into the distribution grid requires new system protection strategies, new control strategies, and transmission and distribution systems capable of incorporating a multitude of small generators.

Present interregional electricity transfer capability is limited. Moreover, there is increasing concern over the environmental effects of high-voltage transmission lines and also a general resistance to siting additional lines. Therefore, it is necessary to increase the power-carrying capacities of existing transmission corridors. The potential environmental effects of these systems are uncertain.

The FY 1980 EES Power Supply Integration Program is composed of:

- Solar Generic/Integration of New Source Technologies
  — Definition and testing of the technical characteristics of new generating technologies; development of the necessary integration procedures
- Load Management
  — Definition of technical options to modify the apparent use of electricity as a function of time, including storage

The Power Delivery Program addresses the problem of transmitting power from generating stations to loads during normal and emergency conditions. The two FY 1980 subprograms are as follows:

- Electric Field Effects
  — Development of guidelines for designing environmentally safe high-voltage overhead lines

- Underground Transmission and Compact Stations
  — Development of superconducting and gas-insulated solid-dielectric underground cables
  — Development of safer, more environmentally acceptable substations

Program Strategy and Impacts. The EES Program strategy is to assess the research needs of evolving electric power systems and to conduct research on key components and problems. This strategy emphasizes developing the framework for effective integration of new technologies, particularly solar. The approach is to assess the integration requirements for each technology, to define appropriate technology/power system tests, and to develop control strategies and system planning information to permit integration on a cost-effective basis. The component research effort is focused on expediting the development of higher risk technologies with long-term benefits. EES is also actively supporting the Economic Regulatory Administration in activities, required by the National Energy Act (Public Utilities Regulatory Policies Act), related to wheeling, reliability, and interconnection. EES also is participating in the National Grid Study.

Energy Storage Systems Programs

Growing dependence on imported oil and gas, escalating energy prices, and the environmental and socioeconomic costs associated with exhaustible energy all indicate that U.S. energy strategy must include the more efficient use of current energy sources, along with the increased use of inexhaustible energy. However, the realization of this combined strategy poses two different problems: 1) exhaustible energy often is available when or where it is not needed (e.g., electric generating capacity); and 2) inexhaustible energy often is needed when or where it is not available (e.g., solar technologies such as photovoltaic or wind systems provide power only when the sun shines or the wind blows).

Energy storage systems can assist in solving both problems. Efficient storage systems would allow utilities to use their facilities more efficiently (e.g., to store excess energy from more efficient base load generation for use during peak demand periods, thus reducing the need for peak generating capacity). Efficient storage systems also would allow inexhaustible but intermittent energy sources to be used when needed (e.g., by storing excess photovoltaic energy generated during periods of sunshine). Consequently, strategic planning is necessary if various forms of electrochemical, thermal, chemical, mechanical, and magnetic storage are to be used to conserve exhaustible resources and to increase the applicability of inexhaustible resources.

Program Goals and Objectives. The long-range goal of the STOR Program is to develop and demonstrate, in cooperation with industry, reliable, cost-effective, safe, and environmentally acceptable energy storage systems that will provide one or more of the following benefits:

- Enable solar, wind, and other intermittent but renewable energy sources to service continuous energy loads, with reduced need for conventionally fueled backup systems
- Increase the substitution of coal and nuclear energy for petroleum and natural gas in transportation and utility systems
- Allow the use of electricity, generated from coal or nuclear resources, in mobile applications, especially vehicles
- Enable electric utilities to use their facilities more efficiently, thereby retarding current rates of growth in the cost of electricity
- Allow the utilization of energy that otherwise would be lost as waste heat
- Promote conservation by recovering waste heat from utilities and industries and by improving process efficiency

To achieve this goal, the STOR Program is developing energy storage technologies which can be applied to transportation, building heating and cooling, industrial processes, and utilities, as well as individual homeowners. Because the different applications require storing varying amounts and forms of energy for various duty cycles, the following individual program objectives are application-specific:

- For vehicles:
  — Batteries having an energy density of 140 Wh/kg or more
  — Composite flywheels having an energy density of 88 Wh/kg or more
- For heating and cooling of buildings:
  — Seasonal storage of hot and cold water in aquifers
  — Customer-owned thermal storage of solar or off-peak electrical energy
- For utility load leveling, peak power generation, and solar electrical systems:
  — Small (up to 1,000 kWh) flywheel and compressed-air energy storage devices
  — Long-life (10 years) batteries for solar missions (including individual homeowner applications) and utility load leveling
  — Thermal, direct compressed-air energy storage, and underground pumped hydro storage systems not requiring oil
  — A 10,000-MWh superconducting magnetic system
- For industrial uses:
  — Energy-saving, industrial electrochemical processes
  — Retrofitted thermal storage for industrial heat recovery
- For the transport of energy:
  — Small, quick-response magnetic storage units for stabilization
  — Combined heat-storage and thermochemical pipeline systems
  — Processes for producing hydrogen from water

Program Strategy and Impacts. Energy storage systems may be economically advantageous in numerous applications (solar and conventional heating and cooling, transportation, industrial heating and feedstock substitutes, and loadshifting for coal, nuclear, and solar thermal electric powerplants). Thus, the development of storage systems supports the end-use missions of several DOE organizations. For example, Solar Technology and Electric Energy Systems Programs within ETS, Solar Applications (SA), Transportation Energy Conservation (TEC), and Buildings and Community Systems (BCS) within Conservation and Solar Applications (CS), and the power marketing functions within Resource Applications (RA) all depend to some degree on the development of reliable, cost-effective energy storage systems. Consequently, a basic element of STOR long-range strategy is close cooperation and interaction with appropriate organizational elements within DOE.

The STOR Program strategy consists of five parts:

- Identify energy storage technology and application options having the greatest potential markets
- Conduct analytical and experimental programs to select the best options for further development
- Perform proof-of-concept tests to validate laboratory test results on a pilot-plant scale
- Where appropriate, transfer successful technologies to the responsible DOE end-use organization
- Encourage and stimulate rapid commercial implementation of storage technologies through cost sharing and other incentives to industry

In most cases, technology transfer from the STOR Program to the end-use DOE organization occurs when the first full-scale or subscale commercial prototype is successfully operated. To clearly delineate the schedules and responsibilities for both technology development and technology implementation, written agreements are being prepared between STOR and the appropriate DOE organizations.

The STOR strategy for achieving its long-range goals is based on the assumption that the Federal role is to support high-risk R&D in areas where private firms are unwilling or unable to invest and in which national benefit is anticipated. The initiatives created by Federal involvement will increase private sector participation in R&D until commercialization of the storage technologies is achieved.

Figure III 2-5 shows the strategy and timetable for energy storage technology commercialization.

The STOR strategy and delivery timetable will result in significant fuel savings (including fuel shifting) in the near term, and increased utilization of intermittent, renewable energy resources in the long term. The STOR Program office has estimated that technology developments in the Battery Storage Program may save as much as 0.5 quads (measured in direct fuel savings and fuel shifts) in 1985 and 3.1 quads in 2000. The Thermal and Mechanical Storage Program may save 0.2 quads (measured in direct fuel savings and fuel shifts) in 1985, and 6.5 quads in 2000.

MANAGEMENT STRUCTURE

When the Department of Energy Organization Act was signed into law on August 1, 1977, DOE was assigned the responsibility to plan, coordinate, support, and manage a balanced and comprehensive energy RD&D program. In order to fulfill this responsibility, DOE has implemented a management structure which clearly establishes responsibility for overall program management in headquarters, and project implementation in field organizations. The following sections describe the interrelated functions of program and project managers.

CSE Headquarters Functions

Within DOE, the Assistant Secretary for Conservation and Solar Energy (CSE) carries out responsibilities in four solar programs:

- Solar Technology
- Geothermal Energy
- Electric Energy Systems
- Energy Storage Systems

The CSE organization, shown in Figure III 2-6, consists of five line divisions, two staff divisions, and a small senior special staff group.

The special staff assists in expediting decisions and actions within the Office of the Program Director, and advises the Program Director on scientific, environmental, and international matters.

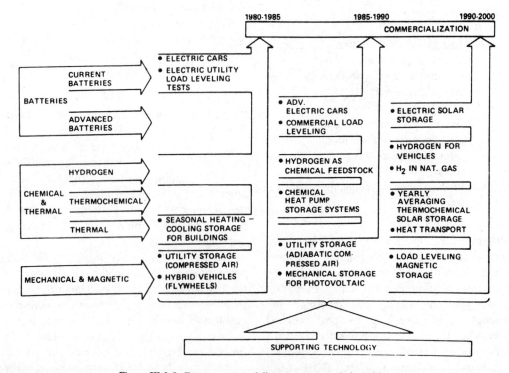

Figure III-2-5. Energy storage delivery strategy and timetable.

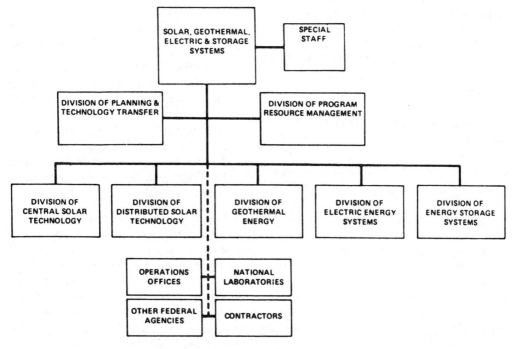

Figure III-2-6. CSE organization.

The Division of Planning and Technology Transfer performs cross-program analyses, relates program strategies to broad national interests, and recommends program priorities. It analyzes characteristics of technologies and market applications, performs program issue analyses, and manages the preparation of program plans and assists in the development of commercialization plans for solar technologies. It uses these special studies, plans, and policy guidance to support the budget formulation process within CSE.

The Division of Program Resource Management provides centralized support for formulating, executing, reviewing, and analyzing the budget. It also provides other program support, including developing management information, coordinating financial arrangements with organizations external to CSE, and providing general support functions related to personnel, administration, mail and records.

The five line divisions direct their attention toward making their programs available for commercial (public or private) application as early as possible. Their predominant problems are technical in character. Throughout CSE, increasing emphasis is being placed on moving more technical management responsibilities to laboratory and other centers-of-excellence in the field (e.g., decentralization of project management).

The Divisions of Central Solar Technology (CST) and Distributed Solar Technology (DST) are responsible for the development of the market-ready products of solar electric power systems and biomass conversion systems; consequently, they direct the efforts of both industrial contractors and national laboratories. The lead management responsibility is centered in CST and DST. In addition, CST and DST coordinate the efforts of other Federal agencies such as NASA, DOD, the Department of Agriculture (USDA), and the General Services Administration (GSA) in support of the Solar Technology Program. They rely on several DOE Operations Offices for procurement and management support.

The Division of Geothermal Energy (DGE) is organized to provide an effective structure for supporting the goals of commercial development of geothermal energy sources. The organization emphasizes cooperative planning efforts involving other DOE and Federal Agency entities, working together and with the non-Federal sector to develop and update a national plan for the achievement of geothermal goals. Much of the program itself is managed by staff personnel at DOE field offices, while overall program planning and management is directed by DGE/HQ.

The Division of Electric Energy Systems (EES) occupies a unique position in solar. Since EES activities contribute to the improved operation of the existing electrical power system, all electricity generating technologies being developed by DOE will be affected by the EES Program. Consequently, EES is the only DOE-systems-oriented group whose responsibilities cross-cut both centralized and dispersed technologies.

The Division of Energy Storage Systems (STOR) also is responsible for researching and developing storage technologies which will have both centralized and dispersed applications. STOR investigates promising energy storage technologies and develops and cooperates in the demonstration of energy storage systems for all end-use sectors. Because energy storage is a family of several technologies, the management structure is organized by technology areas. The Division Director exercises program responsibilities through two Program Managers. Progress reports for major contracts are made monthly to Program Managers and quarterly to the Director. Formal reviews of major programs are held semiannually.

In order to assist in the smooth transition of energy technologies from RD&D through commercialization, a Resource Manager is established within RA or CSE for each technology seen as a candidate for commercialization. The Resource Manager provides a DOE-wide point of focus for the coordination of all activities required to achieve the earliest possible penetration of the technology into the energy supply marketplace.

The Resource Manager's scope of responsibility will include, where appropriate, environmental and regulatory coordination, the planning and budgeting of commercialization programs, industry/Government liaison coordination, marketing research and marketing, cost/pricing and market penetration estimates, and development of means to overcome implementation barriers. These responsibilities are fulfilled, in order to achieve the major objectives of the Resource Managers Office:

- To provide more effective program integration, bringing together in one group the management, technical, and other skills necessary for total commercialization planning to assure the efficient achievement of DOE goals
- To provide a single point of focus within DOE for each technology approaching commercialization

Resource managers have been named for three technologies currently being researched and developed within CSE: 1) goethermal/hydrothermal; 2) wood (biomass); and 3) small wind machines. The resource manager for geothermal/hydrothermal technologies will report to the Assistant Secretary for Resource Applications, while the resource managers for wood and small wind technologies will report to the Assistant Secretary for Conservation and Solar Energy.

Field Activities

In accordance with DOE policy to decentralize project management, the solar Programs are transferring responsibility for projects to the field. The DOE field offices are responsible for the technical management of all programs, including the control of management plans for all fabrication and construction projects.

Although overall direction is provided from Headquarters with supporting field administrative management, responsibility for program implementation is decentralized to the maximum extent possible in a program that remains dominated by R&D considerations. At present, each of the line divisions has formal management plans with field organizations.

In many instances national laboratories have been given these management responsibilities. For the Divisions of CST and DST, the Solar Energy Research Institute at Golden, Colo., has been assigned responsibilities in the areas of research, analysis and assessment, information, education, international programs, and technology commercialization.

SERI is managed and operated by the Midwest Research Institute under a contract to DOE. Participants in Operations Offices include San Francisco, Albuquerque, Richland, Chicago, and Oak Ridge Offices. DOE laboratories perform significant R&D programs in support of both CST and DST. These laboratories include Oak Ridge, Los Alamos Scientific, Lawrence Berkeley, Lawrence Livermore, Sandia, Argonne, Brookhaven, and Pacific Northwest.

The Division of Geothermal Energy relies heavily on several DOE Operations Offices for contract administration and support, especially San Francisco, Idaho Falls, Albuquerque, Nevada, and Oak Ridge. Lawrence Berkeley, Idaho National Engineering, Los Alamos Scientific, Pacific Northwest, and Lawrence Livermore are the major laboratories that support GE's program execution.

For the Division of Electric Energy Systems, national laboratories have been given the responsibility for managing numerous efforts, and additional management responsibility is being transferred to the field in 1979. Project management responsibility for AC superconducting power transmission has been assigned to Brookhaven National Laboratory (BNL). NASA's Jet Propulsion Laboratory, Pasadena, California, was selected to perform the technical management and analysis work associated with distribution automation. Oak Ridge National Laboratory has been assigned the responsibility for managing load management demonstrations of thermal storage systems on the customer's side of the electric meter. Utility systems also manage several projects. These include the Battery Energy Storage Test Facility, operated by the Public Service Electric and Gas Company, and the 4.8-MW Fuel Cell Power Plant, operated by the Consolidated Edison Company. The Aerospace Corporation acts as managing field agent for all environmental electric field work.

Other important but relatively less expensive projects, such as field tests of alternative communications systems and automatic generation control, are jointly managed by various utility systems. In conversion and storage, EES has defined the requirements and jointly developed the storage application and fuel cell programs with the Office of Fossil Fuel Utilization and the Energy Storage Systems Division. The latter divisions conduct feasibility assessments, including laboratory investigations. Cooperative agreements are under way with the Division of Solar Technology to address the electrical systems requirements of ST plans and projects.

The Division of Energy Storage Systems relies heavily on several DOE operations offices, especially Chicago, San Francisco, Albuquerque, and Oak Ridge, for contract administration and support. Technical direction for selected major projects has been assigned to field organizations. National laboratories perform a technical management function and do considerable subcontracting to industry and universities. The laboratories organize and direct industrial groups in specific storage activities. Lead laboratories have been designated for each energy storage technology, except magnetic storage; these include Argonne, Brookhaven, Oak Ridge, Sandia Livermore, Sandia Albuquerque, Pacific Northwest, and Lawrence Livermore Laboratories, and the NASA Jet Propulsion Laboratory and Lewis Research Center.

3.
Solar Technology

The Solar Technology Program, encompasses the development of energy supply systems (other than solar heating and cooling and low-temperature agricultural and industrial process heat), which derive energy directly or indirectly from the sun. The Solar Technology Program consists of two major programs: 1) Solar Electric Applications and 2) Biomass Energy Systems. The Solar Electric Applications Program in turn is composed of four programs: 1) Solar Thermal Power, 2) Ocean Systems, 3) Photovoltaic Energy Conversion, and 4) Wind Energy Conversion Systems.

The management responsibility for these programs rests with two divisions: 1) the Division of Central Solar Technology (CST), and 2) the Division of Distributed Solar Technology (DST). (An organizational chart is shown in Chapter 2, Figure III 2-6.) The program responsibilities for these two divisions are:

- Central Solar Technology
  — Solar Thermal Power
  — Ocean Systems
- Distributed Solar Technology
  — Photovoltaic Energy Conversion
  — Wind Energy Conversion Systems
  — Biomass Energy Systems

Briefly the five major programs encompass the following:

- *Solar Thermal Power Systems:* The sun's heat is concentrated and used to heat water or some other fluid to provide industrial process heat or to drive a turbogenerator. The primary objective is to provide an alternative to fossil fuels for industrial and utility applications. Applications which provide both heat and electricity or "total energy systems" are also included in this area. Solar Thermal Power Systems are described in detail in this chapter.
- *Ocean Systems:* Ocean-stored solar energy has potential as an inexhaustible and renewable source of substantial quantities of base-load electricity and energy-intensive products. Systems currently being studied and developed include Ocean Thermal Energy Conversion (OTEC), Salinity Gradients, Ocean Currents, and Wave Energy. These concepts, currently at various stages of development, use resources that are available in different geographic areas. These processes are described further in this chapter.
- *Photovoltaic Energy Conversion:* Sunlight is converted into electricity by the use of solar cells. A number of different semiconductor materials can be used, and intensive research is underway to create improved, high-efficiency, lower cost photovoltaic devices. In general, there are two generic types—flat-plate arrays that operate on direct sunlight at normal intensity, and "concentrators" that increase the intensity of sun-light as much as 2,000 times. The distinguishing characteristics of these photovoltaic devices are described in this chapter.

- *Wind Energy Conversion Systems:* systems employ a wind-driven ma[chine] to turn an electric generator. (Wind [de]rives its energy from the sun's heating [of] the atmosphere and the earth's surface.) New applications are discussed in this chapter.
- *Biomass Energy Systems:* Biomass is organic material such as terrestrial or aquatic vegetation, manure, and agricultural and forestry residues. This material contains stored chemical energy produced by plants from solar energy. Biomass can be collected and burned directly or converted to liquid fuels (fuel oils, alcohols) or gaseous fuels (medium-Btu gas, synthetic natural gas) or other energy-intensive products (hydrogen, ammonia, petrochemical substitutes) that can supplement or replace similar products made from conventional fossil fuels. The primary objective of the Biomass Energy Systems Program is to develop the capability for converting these organic materials into usable fuels or energy-intensive products. The biomass Program that is part of the Solar Technology Program is detailed in this chapter.

The solar technology strategy to implement these programs is threefold:

- Foster product development and improvement by supporting research and development (R&D)
- Enlarge manufacturing capacity and improve production techniques through direct product support or other incentives.
- Stimulate the formation of market channels and institutions through a program of selected tests and applications coupled, as appropriate, with other incentives.

The following subsections describe in detail the two Solar Technology Programs: 1) Solar Electric Applications, and 2) Biomass Energy Systems. The descriptions of these programs are structured according to the FY 1980 budget; i.e., each is divided into its subprograms, which are briefly described in the form of two-page modules, except when a particular subprogram is unusually complex or relatively simple. Each module includes a general [des]cription of relevant technical aspects, an [illust]ration (where appropriate), a summary of pr[ogram] status, and a milestone chart which [depicts] 1984 major decision points through FY 198[4, fu]nding levels for FY 1978 through FY 19[80].

Because [the] Solar Technology Program is the largest [and] most complex program within CSE, most of [the] modules describe activities within subprogra[ms]. For example, each of the four Solar Electric [Ap]plications subprograms is described by a gr[oup] of modules, one for each of the main activ[iti]es within each subprogram. Other program a[sp]ects are described in order to provide an over[vi]ew of program activities which are not a part of any specific subprogram. For example, because solar energy may be used in decentralized applications, program emphasis is being placed on regional, dispersed, and small-scale systems. Consequently, the regional activities related to the various solar programs are discussed separately.

SOLAR ELECTRIC APPLICATIONS

The Solar Electric Applications Program attempts to stimulate the development of economically feasible, environmentally acceptable, and operationally safe solar electric systems that will contribute significantly to national energy supplies. (For the purposes of this document, a single description of both electric and nonelectric applications of Solar Thermal Power Systems is included in this chapter.)

The objectives of this program are described by the following five established goals:

- Develop and demonstrate solar systems producing mechanical and electrical energy and other products at commercially competitive costs by the mid-1980s
- Ensure operational compatibility of solar energy systems with existing systems and utility operations
- Establish the environmental acceptability

402 Solar, Geothermal, Electric and Storage Systems

- and operational safety of solar energy-based systems
- Stimulate early resolution and clarification of legal and institutional issues
- Aid the creation of the necessary institutional infrastructure

The four major programs within Solar Electric Applications are as follows:

(1) Solar Thermal Power Systems*
(2) Photovoltaic Energy Conversion
(3) Wind Energy Conversion Systems
(4) Ocean Systems

Each of these programs is detailed in subsequent sections of this chapter.

Table III 3-1 summarizes the funding levels by programs for FY 1978 through FY 1980.

Solar Thermal Power

The objective of the Solar Thermal Power Program is to develop solar thermal systems as a technically, economically, and environmentally acceptable technology for use in a variety of applications. The principal thrust of the program is directed toward the early industrial process heat and utility market sectors. The potential for both early and ultimate market penetration has been enhanced by focusing development of specific types of mid- and high-temperature concentrator systems on those sectors of the national energy market to which they are best adapted. The program is structured to take advantage of the versatility of solar thermal technology, including its ability to respond to a broad range of temperature and load requirements. This versatility permits consideration of a variety of market sectors including industrial process heat, electric power generation, total energy applications, mechanical power production and, in the long run, fuel and chemical applications. This objective is accomplished through research, development and demonstration (RD&D) in three program areas:

- Large Power Systems
- Small Power Systems
- Advanced Technology Research and Development

The Large Power Systems Subprogram focuses on facilities with capacities of more than 10-MWe or 100 MBtu$_{th}$/hr. Both new plants and retrofits of old plants now fueled with oil or natural gas are included. Presently, primary areas of emphasis are:

- Central Industrial and/or Utility Applications
- Central Receiver Facility and Component Development

The Small Power Systems Subprogram (systems with capacities under 10-MWe or 100 MBtu$_{th}$/hr) concentrates on applications in which the energy supply system can be integrated at the point of use. This can include a large variety of system applications including industrial process heat, total energy generation, small community electric generation, and remote site uses.

An Advanced Technology Research and Development Subprogram supports both of the other subprograms by defining the needs for likely future applications. Specifically, this involves identifying cost-effective candidate systems and assisting in the development of key advanced subsystems. In addition, the current state of solar technology, along with developing trends are assessed. These functions are performed by the development or characterization of:

- Innovative components and subsystems
- New materials to support the various applications
- Analyses and evaluation techniques

Table III 3-2 summarizes the funding levels by subprogram for FY 1978 through FY 1980.

Large Power Systems. Large power systems will be used either as powerplants in electric utility networks or will supply industrial process heat. The plants will generally range in size from 10 to 1000 MBtu$_{th}$/hr (30- to 300-MWe), although some applications may be as small as 10-MW. Development activities include storage-coupled plants, solar conversion

---

*Solar Thermal Power Systems appear potentially attractive for nonelectric applications involving the generation of industrial process heat. Both nonelectric and electric uses are described in this chapter.

Table III-3-1. Funding levels for solar electric applications.

SOLAR ELECTRIC APPLICATIONS	BUDGET AUTHORITY (DOLLARS IN THOUSANDS)			
ACTIVITIES	ACTUAL FY 1978	ESTIMATE FY 1979	ESTIMATE FY 1980	INCREASE (DECREASE)
Solar Thermal Power	64,600	100,100	121,000	20,900
Photovoltaic Energy Conversion	60,300	103,800	130,000	26,200
Wind Energy Conversion Systems	35,400	60,700	67,000	6,300
Ocean Systems	29,400	38,200	35,000	(3,200)
TOTAL	189,700	302,800	353,000	50,200

of existing fossil-fueled plants and solar/hybrid plants.

In a solar thermal system, the sun's radiant heat is collected, concentrated, and used to heat a working fluid such as water or steam. In a storage-coupled system, the heated fluid is used either directly; e.g., to provide heat for industrial processes or to drive a turbine generator and produce mechanical energy or electricity, or it is used indirectly to charge the thermal storage subsystem. The storage subsystem then maintains output during periods of insufficient solar radiation.

In one promising approach known as solar retrofitting/repowering, an existing fossil-fueled industrial plant or utility is provided with a solar-fired system. The plant can then operate on solar-supplied thermal energy, fossil-supplied thermal energy, or a combination, because the ability to operate on fossil fuel is retained, and full backup capability is ensured.

Another alternate approach, the solar/hybrid, is essentially the same as solar retrofitting/repowering, except that rather than using already existing equipment, an entire new plant is developed. Although both approaches may incorporate thermal storage, the basic storage device is the fuel.

The baseline system chosen for large power applications is the central receiver concept. In this system, the sun's energy is reflected from an array of sun-tracking mirrors (heliostats) to a central receiver located on top of a tower, where the working fluid is heated. Systems studies have identified this concept as being more cost-effective than others for large-scale systems.

Table III-3-2. Subprogram funding levels, FY 1978 through FY 1980.

SOLAR THERMAL POWER	BUDGET AUTHORITY (DOLLARS IN THOUSANDS)			
PROJECTS	ACTUAL FY 1978	ESTIMATE FY 1979	ESTIMATE FY 1980	INCREASE (DECREASE)
LARGE POWER SYSTEMS	25,800	54,100	65,000	10,900
SMALL POWER SYSTEMS	29,100	32,500	33,000	500
ADVANCED RESEARCH AND DEVELOPMENT	9,700	13,500	23,000	9,500
TOTAL	64,600	100,100	121,000	20,900

An alternate concept uses a high-temperature line-focusing collector system. In this type of system, the working fluid is heated at each collector and then transported by pipe to the turbine-generator.

Figure III 3-1 depicts a central-receiver solar thermal power system.

Federal involvement in Large Power Systems began in FY 1973, with initial system studies; it was accelerated in FY 1975 when three parallel contracts were funded to provide preliminary designs and subsystem experiments for a 10-MW system experiment. A fourth contract was awarded solely for heliostat development. These contracts were completed in mid-1977. One concept was chosen for followup, and the justification for proceeding with the project was reviewed and affirmed. The site near Barstow, Calif., was offered along with substantial cost-sharing by the Southern California utilities. Construction is expected to be completed in 1981.

Other system experiments include the International Energy Agency (IEA) project and the EPRI/DOE Brayton project. A cooperative agreement has been signed by several countries, including the United States, under the auspices of the IEA to jointly fund a 500-kWe sodium-cooled central receiver system to be constructed in Spain. Construction is scheduled for completion at the end of FY 1981.

EPRI has been working toward a 2.5-MWe hybrid pilot plant using a Brayton cycle. A joint program will be established in FY 1980 between EPRI and DOE for the development and demonstration of this system. A candidate air-cooled receiver has been tested.

A central receiver solar thermal test facility (CR–STTF) for evaluating system components has been built at Sandia Laboratories in Albuquerque. This facility will provide concentrated solar radiation for tests of central receiver hardware, materials evaluation, and general research activities. A 1-MW receiver was demonstrated in May 1977. The full 5-MWth capacity was achieved in 1978 and testing was initiated in late 1978. Design and construction of the 10-MWe system experiment is in progress, and low-cost heliostat designs, together with development of efficient processes for their manufacture, are being pursued.

The conceptual design studies for utility repowering/retrofit applications initiated in FY 1979 will be completed by the end of FY 1980 and, along with the strategy study by the Solar Energy Research Institute (SERI) in 1979, will provide an important input for a decision on whether or not to proceed further with the detailed design of a repowering experiment. Studies of solar/hybrid markets and concepts will also be made during 1979. A

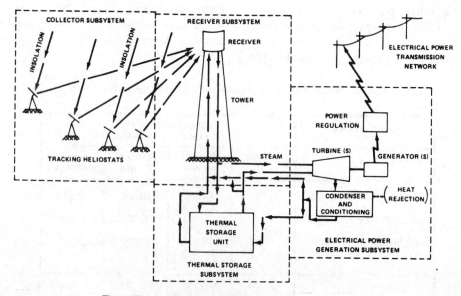

Figure III-3-1. Central receiver solar thermal power system.

decision on whether to proceed with the design and construction of a hybrid system demonstration will be made in early 1982, with possible completion in 1986.

Activities have been initiated which are expected to lead to significant cost reductions. These include low-cost heliostat designs, development of efficient process for the manufacture of heliostats, advanced receiver designs, and distributed collector system designs.

Milestones and funding for Large Power Systems are presented in Figure III 3-2.

Small Power Systems. This subprogram focuses on the development of solar thermal technology for applications in which the energy supply system can be integrated at the point of use. Such onsite systems are typically much smaller than those required for utility powerplant operation. Comparable conventional systems in current use rely heavily upon fuels such as natural gas, propane, and oil. The use of solar thermal power systems for these applications offers the potential for economically competitive energy production, reduced environmental intrusions, and reduced consumption of critical fossil fuels.

Small power systems differ from large power systems in size and in proximity to users. Small power systems applications are smaller than 100-MWth and are usually located within 3 miles of users. Small and large applications, however, may not differ in the form of energy supplied. The thermal energy generated by the solar system can be used for process heat, space heating and cooling, converted to direct mechanical energy or electricity, or can be used in any combination of the above.

Efforts in the Small Power Systems Subprogram include:

- Small- and intermediate-scale industrial process heat applications
- Total energy systems to supply both heat and electricity for industry, commercial building complexes, agriculture, and small communities
- Electric power systems for small communities or isolated loads
- Remote systems for specialized applications in locations far from, or inconvenient to, an electrical grid

These applications complement the developmental effort pursued under the Large Power Systems Subprogram. Small power systems applications can use either central-receiver or

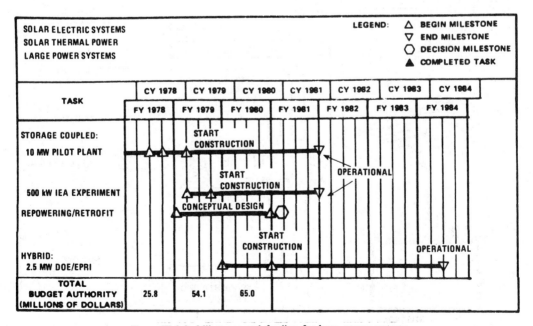

Figure III-3-2. Milestones and funding for large power systems.

distributed-collector systems, but the tendency in small systems is toward the latter. For electrical generation, power conversion subsystems under consideration include centrally located Rankine cycles (both steam and organic) and small collector-mounted heat engines (either Brayton or Stirling cycles). Either thermal-storage or fossil-fuel backup can be incorporated in small systems. Many current designs call for both thermal storage and an auxiliary fossil-fueled boiler.

Figure III 3-3 shows a typical Total Energy Application.

Applications analyses and definition studies, performed in-depth for large central stations, are currently being extended to examine small solar power systems applications, and an R&D program has been started to produce the component and modular subsystem technology required for this power range. Paralleling these efforts, early system experiments have been designed and are being constructed using currently available techniques. The purpose is to gain much needed knowledge of the operation and maintenance of small solar thermal power systems and to provide practical experience with the institutional interfaces. Combined data from large-scale applications, total energy systems, and early system experiments will be used to select promising baseline systems for further development work.

Power conversion subsystems being considered include centrally located Rankine cycles (both steam and organic) and small collector-mounted heat engines (either Brayton or Stirling cycles). The collector-mounted heat engine concept is attractive because of its modularity, with electricity produced locally at each collector.

Total energy system experiments are presently being carried out at the Solar Total Energy Systems Facility in Albuquerque. Substantial progress has also been made in the design of two total energy system experiments: one at Shenandoah, Ga., and a second at Fort Hood, Tex. The preliminary designs of the Shenandoah and Fort Hood projects have been completed. After a review of the Ft. Hood design, a decision was made not to continue this project. The Shenandoah project is currently being reviewed prior to making a decision to proceed to detailed design and construction. Each total energy system experiment encompasses a separate application: one industrial and the other military. Additional system experiments may be initiated.

Two developmental solar irrigation experiments have been initiated as part of the Re-

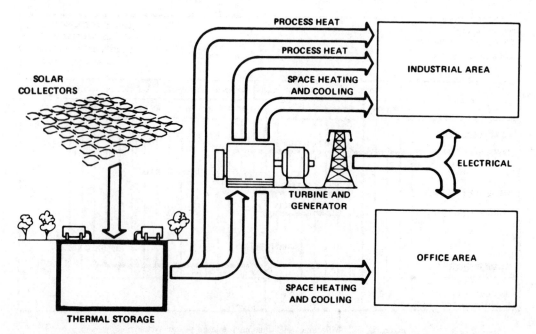

Figure III-3-3. Total energy application.

mote System Applications Program. Both experiments are intended to provide realistic performance and cost information on solar irrigation pumping applications. One of these, a 50-hp shallow-well irrigation system became operational during FY 1977 through 1978; the other, a 200-hp deepwell experiment is scheduled to be operational in 1979. In FY 1979, an advanced prototype module will be designed. Two system experiments for small community systems will be designed, one in FY 1979 and the other in FY 1980. It is expected that these efforts will be followed by the development of additional experimental systems, as more cost-effective components become available.

Initial studies of an experiment to use solar thermal power for enhanced oil recovery will be undertaken in FY 1979. This is being developed in conjunction with CSE and RA.

Milestones and funding for Small Power Systems are shown in Figure III 3-4.

Advanced Technology Research and Development. This subprogram focuses on identifying new, economically attractive systems to meet the long-term requirements of the Large and Small Power Systems Subprograms. Research experiments are performed to determine the technical feasibility of the key subsystems and components, and to conduct the generic research common to both the large and small power systems. The major activities of the Advanced Research and Development Subprogram are: Materials Supporting Technology, New Component Demonstrations, Advanced Systems Development, and Technology Assessment and Support.

Innovative component and process concepts will be explored under the New Components

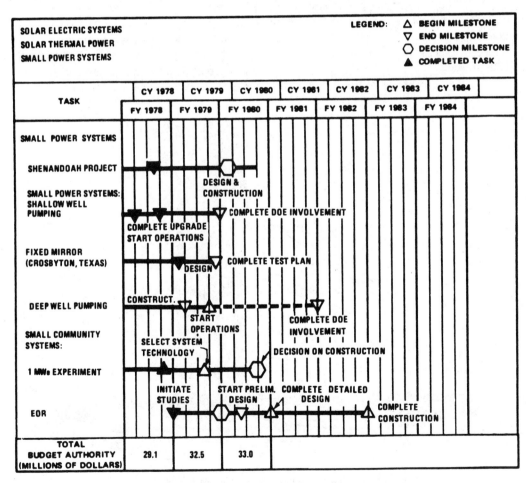

Figure III-3-4. Milestones and funding for small power systems.

Demonstration Activity. Emphasis will be placed on high-temperature receivers and new or improved power conversion subsystems, such as Brayton cycles and Stirling heat engines. Advanced central receivers, high heat-flux heat pipes, and high-temperature pumped liquid-metal and pumped molten-salt receivers will also be explored.

The Materials Supporting Technology Activity will provide improved absorber coatings, reflective surfaces, and construction of materials required for future thermal power systems. The Advanced Systems Development Activity will undertake component development for selected advanced solar thermal power systems. The Technology Assessment and Support Activity is likewise responsible for support and coordination of nonprogrammatic research experiments at the Solar Thermal Test Facility (STTF) at Albuquerque and the 400kW Advanced Component Test Facility (ACTF) at the Georgia Institute of Technology in Atlanta. Figure III 3-5 shows an advanced Sander's Receiver mounted on the ACTF tower.

The ACTF, capable of testing small materials up to about 400 kWth at a flux density of about 90-W_{th}/cm² has been made operational. The ACTF has already been employed to test a ceramic receiver, and the receiver achieved an operating temperature of 1,950° F prior to completion of the tests.

Other ongoing projects include development of selective absorber coatings, improved reflective materials and substrates, development technology for high-temperature receivers, and design studies for a dish-Stirling engine advanced small power system. A program for fuels and chemicals production, using the clean, high-temperature heat from solar thermal systems is being formulated under this project. Milestones and funding for this subprogram are presented in Figure III 3-6.

Photovoltaic Energy Conversion

Photovoltaic energy systems provide a clean, simple method for direct conversion of sunlight to electrical energy. Photovoltaic systems do not require complex machinery or moving parts, and they operate silently. Because these systems are also modular, a wide

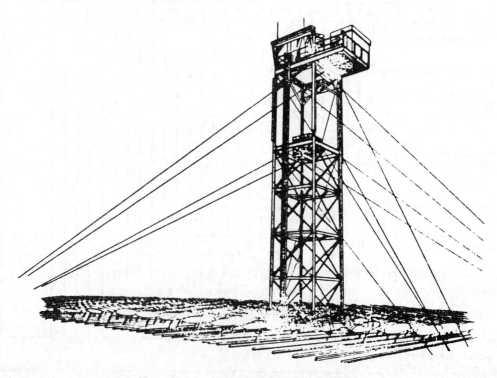

Figure III-3-5. Advanced components test facility.

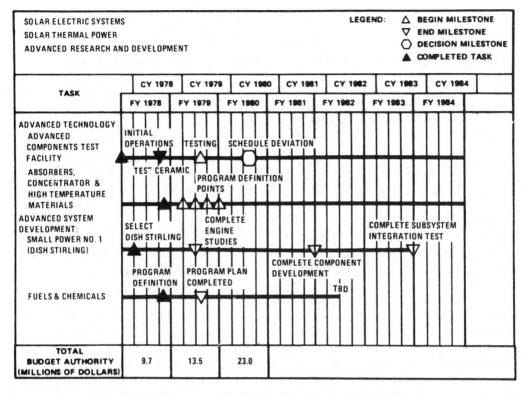

Figure III-3-6. Milestones and funding for advanced research and development.

range or application sizes and types can be designed to fit almost any need.

Photovoltaic devices absorb sunlight and convert it directly into electricity. When light energy or photons strike certain materials called semiconductors, internal voltages are created. In devices specially designed for the purpose, wires may be attached and electricity may be drawn off. The basic scientific principles are well known, and the technology for specialized uses is relatively advanced.

The major component in all photovoltaic systems is the "array" that collects the sunlight and converts it into electricity. The array is composed of a number of electrically interconnected sealed panels, each panel containing many solar cells. The F.O.B. array cost now accounts for about one-half of the total system cost, the amount being dependent on the particular system. Other system cost components—called the balance of system (BOS)—are for power conditioners, storage elements (batteries), controls, and structural members, as well as installation. Figure III 3-7 illustrates the basic operation of a solar cell.

Arrays of photovoltaic cells can be sized to suit many different applications. For example, they can provide electricity for individual residences which may be connected to the electric utility grid for backup power. They also can provide electricity for intermediate-size, onsite applications such as industrial facilities and residential communities. Finally, large arrays of photovoltaic cells can be connected in a central power station.

A major impediment to widespread commercial energy applications of photovoltaic systems is the high cost of cells and arrays. Today's costs are at least ten times more expensive than required. Progress in array cost reduction has been very significant over the past several years and detailed engineering studies show that over the next several years cost reduction progress will continue at significant rates. As array cost reduction continues, attention is being turned to reducing the cost of BOS. The objectives of the Photovoltaic

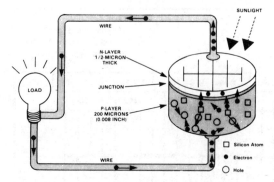

Figure III-3-7. The solar cell.

Program, then, are to achieve major cost reductions in installed photovoltaic systems to stimulate market penetration, and to foster establishment of an industrial infrastructure for the design, manufacture, distribution and installation of these systems. Each of these aspects will be promoted to the point where a significant portion of the Nation's electric energy requirements may be supplied by photovoltaic energy systems.

The Solar Photovoltaic Energy Research, Development and Demonstration Act of 1978 (Public Law 95-590) was signed by the President on November 4, 1978. The Act provides for an accelerated program of research, development, and demonstration of solar photovoltaic energy technologies. This is intended to lead to early commercial viability of such technologies. A 10-year Federal Government expenditure of $1.5 billion is contemplated in implementing terms of the program specified in the Act.

A specific target is toward reduction of the average cost of installed photovoltaic systems to $1 per peak watt by 1988.* The Act is intended to stimulate market growth in the private sector, so that 90 percent of purchases will be by this sector by 1988. The Federal Government is authorized to furnish manufacturers and consumers of photovoltaic systems with monetary incentives for pursuing these goals. These incentives could amount to as much as 75 percent of the costs of purchase and installation of such equipment.

---

*All power and energy costs in this section are expressed in 1975 dollars.

DOE is simultaneously charged with the responsibility of conducting a testing program to assist in the development and demonstration of prototype photovoltaic systems. Procedures are to be established to prescribe performance criteria and to enumerate techniques for product testing and certification. To implement these objectives, coordination of efforts with appropriate representation by scientific and professional societies, and with industry will be sought. The Act also calls for the formation of an Advisory Committee to study and advise the Secretary of Energy on the scope, pace, and course of the numerous projects being conducted. Technological, economic, and environmental implications of the use of photovoltaic systems are to be assessed in this context.

The Secretary is directed by the President and the Congress, within one year to make recommendations relating to barriers to the early, widespread acceptance of photovoltaic systems. Within this same time frame, DOE is to submit to the House and the Senate a plan for demonstrating and promoting widespread use of photovoltaic energy systems in other nations.

There have been numerous accomplishments within the Photovoltaics Program in recent years. Owing to the increased impetus received by this program, the cost of solar arrays has declined significantly—from $22 per peak watt in 1976 to $7 per peak watt in 1978. Several projects have been successfully conducted in diverse photovoltaic applications. Noteworthy among these are:

- A pumping system for an agricultural project in Mead, Nebr.
- Two forest lookout stations in the Lassen and Plumas National Forests in California
- A highway warning sign in Casa Grande in southern Arizona
- A remote radar station operated by the Naval Weapons Center at China Lake, Calif.
- A refrigerator experiment in Mount Royale National Park, Mich.
- Electrification of the Schuchuli Indian

Village on the Papago Indian Reservation, Ariz.
- A water purification system at Fort Belvoir, Va.
- Electric vehicle recharging demonstrations at the American Folk Life Festival in Washington, D.C.

Also planned or under construction are:

- The Mississippi County Community College Project (Total Energy System) in Blytheville, Ark.
- The Mt. Laguna Air Force Station project (60 kW) in California
- The Natural Bridges National Monument project (100 kW), at the Canyon Lands National Park, Utah
- Some of the recent 29 concentrator and flat-panel designs for applications of various sizes at diverse locations

The Photovoltaics Program will achieve the major cost reductions necessary to obtain competitive lifecycle costs by aggressively pursuing advanced research and technology development, and conducting real-world testing. Through this strategy, technologies and infrastructures will be developed that yield technically, economically, and environmentally viable energy-producing photovoltaic systems suitable for a wide range of applications.

In order to implement this strategy, CSE has developed a four-part Photovoltaic Program that will contribute to achieving the overall objectives. These strategy elements and their specific goals are as follow:

- Advanced Research and Development (AR&D)
  — Creates the technical feasibility for new or improved photovoltaic arrays that can provide the basis for further cost reduction
- Technology Development
  — Use the feasibility established in the AR&D element and develop manufacturing technology that will permit U.S. industry to make the components at a reduced cost
- Systems Engineering and Standards
  — Integrate the cost-reduced photovoltaic components into energy systems to serve specific applications, and to identify those applications that can have the earliest and greatest impact on the U.S. energy economy.
- Test and Applications
  — The first experiments selected for these applications are performed as part of this element.

Table III 3-3 presents the FY 1978 through FY 1980 funding levels for each program element. Subsequent sections describe these program elements in detail.

Advanced Research and Development. The Advanced Research and Development Subprogram will support some of the advanced photovoltaic materials and cell concepts—identified from theoretical and laboratory experiments—that have not yet been thoroughly investigated or tested. Because some of these advanced materials offer superior performance characteristics and have the potential to lower production costs, additional work is needed to confirm their promises to achieve higher efficiencies or lower costs or both. When research and analysis confirm that the technology has the potential, when mass-produced, to meet program goals, technology feasibility has been demonstrated. The technology will then be transferred to the Technology Development Subprogram.

Major areas of the Advanced Research and Development Subprogram include:

- Advanced materials/cell research
- High-risk research
- Research support and fundamental studies

Numerous materials such as cadmium sulfide, gallium arsenide, amorphous silicon, and polycrystalline silicon are being studied under the advanced materials/cell research area. The purpose of the studies is to determine which of these candidates may be carried through to the developmental phase on the basis of proven technical feasibility.

Table III-3-3. Funding levels for photovoltaic program elements.

PHOTOVOLTAIC ENERGY CONVERSION	BUDGET AUTHORITY (DOLLARS IN THOUSANDS)			
PROJECTS	ACTUAL FY 1978	ESTIMATE FY 1979	ESTIMATE FY 1980	INCREASE (DECREASE)
Advanced Research & Development	7,600	39,300	47,000	7,700
Technology Development	36,000	38,000	56,000	18,000
Systems Engineering & Standards	5,400	13,300	14,500	1,200
Test & Applications	11,300	13,200	12,500	(700)
TOTAL	60,300	103,800	130,000	26,200

The high-risk research efforts are directed toward those materials and concepts that are perceived to have a high risk for achieving the goals of the program, but which nevertheless demonstrate very high efficiencies, very low cost, or a combination of both. Amorphous materials, advanced multijunction concentrator cells, and electrochemical photovoltaic cells constitute a few of the concepts being studied in this category.

The support and fundamental studies effort seeks to furnish research tools and to anticipate potential problems in the advanced materials/cell research and high-risk research areas. Basic mechanism studies, tests and measurements, and materials screening constitute this area of activity.

Figure III 3-8 illustrates a technique to obtain high conversion efficiencies. This is done by using two solar cells of different materials, each of which is responsive to particular frequencies of the solar spectrum. A filter-mirror spectrum-splitter together with a concentrating lens are used to accomplish this.

Efforts are being pursued actively, in the Advanced Materials/Cell Research area, consisting of studies of thin-film concepts in materials such as cadmium sulfide, gallium arsenide, polycrystalline silicon and amorphous silicon having high conversion efficiencies.

Emerging materials, amorphous materials, advanced concentrator concepts and electrochemical photovoltaic cells, and innovative concepts constitute areas of investigation in High-Risk Research. Additional investigations concern materials that appear to have the potential for development into thin-film solar cells. These are expected to achieve conversion efficiencies of at least 10 percent. Also being investigated are nonolithic multijunction concentrator cells that may be able to offer efficiencies of more than 30 percent.

Research into electrochemical photovoltaic cells will be initiated during FY 1979, and objectives call for achieving 12 percent conversion efficiencies by FY 1985. Innovative concepts, including exotic materials, devices, and processes that constitute high-risk areas, will also be studies; those which offer promise of efficient, low-cost photovoltaic conversion will receive added emphasis in subsequent efforts. These ideas will be solicited through RFP's at regular intervals.

Basic mechanism research is aimed at identifying and reducing the mechanisms which limit conversion efficiencies of solar cells. It

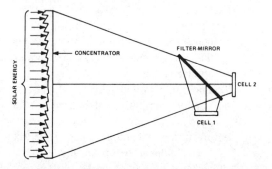

Figure III-3-8. Two-cell system using filter-mirror spectrum splitter and concentrating lens.

should be recognized that higher conversion efficiency systems are important due to their effect on the tradeoffs with balance of system costs. This area is performed under the research support and fundamental studies element. Test and measurement activities to evaluate cell performances and development of a sophisticated laboratory facility for analyzing and testing thin-film and other photovoltaic materials/cells are also planned within this area.

Other areas of research address problems related to reliability, availability, and stability of materials, and to environmental effects.

Major milestones and funding for Advanced Research and Development are presented in Figure III 3-9.

Technology Development. The Technology Development Subprogram focuses on techniques to improve photovoltaic system performance while reducing production costs and maintaining product quality. A number of technology options which continue to show promise of meeting program goals will be pursued simultaneously in order to increase the probability of success.

Both flat-plate silicon array and concentrating array technologies have currently demonstrated feasibility, and they appear to hold considerable promise of meeting program goals.

Flat-plate silicon arrays are being developed under the Low-Cost Solar Array (LSA) Project, which addresses all steps in the array production process. These steps are: production of raw polysilicon crystals of adequate purity, creation of an individual photovoltaic cell, encapsulation, cell fabrication, and high-volume automated array assembly. There are plans to build cost-shared experiments to evaluate different approaches to producing low-cost polysilicon. These experimental units are required to validate the feasibility of meeting the cost goals for silicon arrays. Emphasis is placed on improving quality while reducing cost during each phase.

Developmental work in concentrating sys-

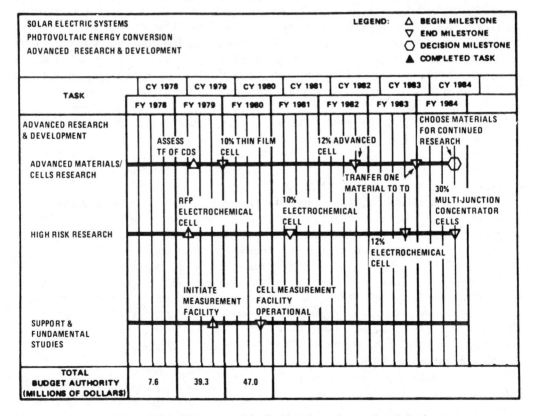

Figure III-3-9. Milestones and funding for advanced research and development.

tems is twofold. The first efforts are aimed at developing more economical concentrators, the second at producing solar cells capable of operating at higher efficiencies in concentrated sunlight.

Photovoltaic total energy technology, another area currently under development, is expected to achieve early cost-effectiveness. In this area, both electrical and thermal energy are extracted from the same system. The concept applies to both flat-plate and concentrator arrays and finds applications in residential, agricultural, industrial, and commercial areas.

Additionally, significant advanced technology concepts which emerge from R&D and which also have demonstrated technical feasibility will be carried through the developmental phase. One such candidate that is being actively pursued is the thin-film cadmium sulfide cell, which holds suitable potential for development.

The Technology Development Subprogram is also responsible for development and coordination of the BOS components. These components include the array structures and installation, power conditioners, and storage subsystems. Greater effort will be directed toward providing innovations and improving production techniques, with subsequent cost reduction of these items so that overall photovoltaic system goals can be met.

Figure III 3-10 depicts the interrelation of major components of a typical photovoltaic system.

The feasibility of achieving low-cost silicon flat-plate arrays has been established through detailed production analysis, and concerted efforts are being made to bring $2,000/kW array technology to commercial readiness by FY 1982. Efforts are also being made toward realizing a $500/kW advanced array by FY 1986. Concentrator array development is being pursued simultaneously and is also expected to achieve $2,000/kW commercial viability by FY 1982. Progress toward these goals may be accelerated by unexpected, but possible, advancements in solar cell research.

Research into thin-film cadmium-sulfide cells has indicated that it is possible to achieve commercial readiness at $500/kW by FY 1986. Plans are underway for the demonstration of $250/kW module technology readiness by FY 1987. Design improvements, such as the use of multijunction cells and concentrating photovoltaics in advanced concentrator arrays, are expected to enhance performance and thus allow $250/kW advanced concentrator technology readiness to be achieved by the mid- and late-1980s.

Initial studies of photovoltaic total energy systems in distributed community or district-sited systems have indicated the possibilities of lowering the "effective" array cost by a factor of two. This potential for lower cost results from savings in collector materials, support structures, land use, installation, and maintenance costs. Systems studies for regional residential designs and photovoltaic Total Energy Systems are to be completed in FY 1979, with a residential community load-center study to be completed in FY 1980.

A number of projects are underway for development of low-cost BOS components, including inverter/controller subsystems for operation in the 5 kW to 50 kW range, and storage subsystems which include batteries, flywheels, and similar components. Commercial readiness for residential BOS technology is expected to be achieved by FY 1985, to correspond to array commercial readiness.

Tests to demonstrate suitable applications of photovoltaic systems will be performed on a limited scale within technology development.

Major milestones and funding for this subprogram are presented in Figure III 3-11.

Systems Engineering and Standards. The Systems Engineering and Standards Subprogram aims at the identification and study of applications requirements for photovoltaic systems and for ensuring that other elements of the program and industry focus their efforts toward appropriate system and subsystem technologies. The scope of work under this area of activity is described by the following three program elements:

- System definition
- System development
- Performance criteria and test standards

System definition involves defining the per-

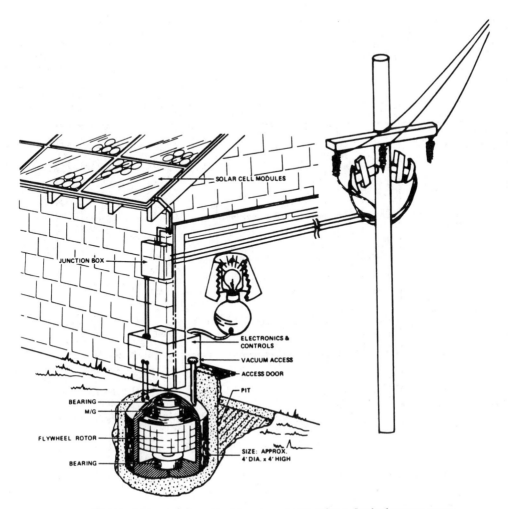

Figure III-3-10. A photovoltaic residential system concept using a flywheel storage system.

formance and interface requirements of promising photovoltaic applications. These include user preferences, application interface and hardware performance requirements, institutional and legal constraints, and the characteristics of the competing technologies, which comprise residential, onsite load centers and central power applications. This element will also furnish design guidelines, standard practices, and design algorithms and will provide subsystem—array and BOS—technical performance requirements and cost goals.

System development includes system prototype development, and testing and evaluation of BOS engineering approaches such as installation and maintenance techniques. Advanced system design concepts showing a high degree of promise will be evaluated carefully. Proven system and subsystem designs will be made available for expanded evaluation in application experiments and commercial systems.

Performance criteria and Test standards ensure that the necessary criteria and standards are developed and made available for programmatic use and for establishing nationwide uniformity in the specification of products and processes pertaining to the photovoltaic industry. Working groups, created with representation from major professional, trade, and consumer organizations, provide an interface to the consensus process. Other activities include recommendations for accreditation of test laboratories and product certification.

Figure III 3-12 displays the elements of a

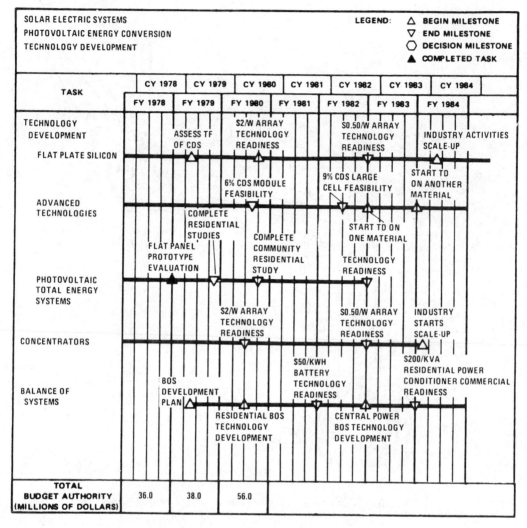

Figure III-3-11. Milestones and funding for technology development.

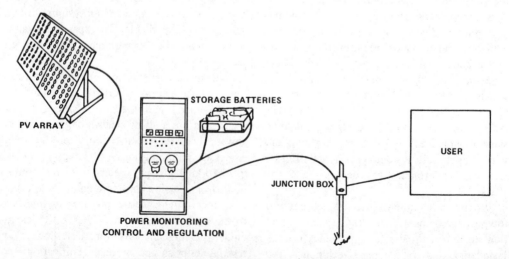

Figure III-3-12. Major components of a photovoltaic system.

photovoltaic system, including BOS components, which may be used in a residential application. The energy storage system depicted is a flywheel, which is a possible alternative to a battery or other storage device.

Extensive studies are being planned through FY 1979-80 to identify various application areas and to prescribe the requirements for three groups of experiments to be conducted during the next several years. Specific efforts are toward evaluation of design and system concepts relating to the following areas of application of photovoltaic systems:

- Residential
- Community and load center
- Central station

Efforts will also be directed toward ensuring that the hardware required for these experiments will be tested adequately before commencement of the experiments. Field tests of residential and community system prototype arrays will occur from FY 1979 to FY 1983. The balance of system (BOS) hardware tests will be initiated by mid-FY 1979.

The performance criteria and test standards responsibilities will be integrated with these other activities, so that standardized analyses and system designs will be available for use in the aforementioned experiments. A Performance Criteria and Test Standards Plan will be issued in FY 1979. An interim Performance Criteria document will be issued in FY 1980 and both will be updated periodically.

Major milestones for this subprogram are depicted in Figure III 3-13.

Test and Applications. The purpose of the Test and Applications Subprogram is to obtain practical experience with complete photovoltaic systems in a range of applications. These experiments are designed to:

- Permit evaluation of the performance and reliability of photovoltaic components and systems under field conditions
- Permit evaluation of BOS engineering approaches; e.g., site preparation and installation techniques
- Provide data on actual system costs, including BOS costs
- Provide a means of verifying the models and system design/evaluation techniques used by other photovoltaic tasks
- Establish the environmental acceptability and operational safety of photovoltaic systems
- Provide information regarding legal and institutional problems and interfaces
- Aid in the creation of the industrial infrastructure that will be needed for commercialization of photovoltaic systems

In order to demonstrate commercial readiness, three groups of experiments will be conducted in applications representing potentially significant markets. The first group will concentrate on individual utility interactive residences. The second group will stress distributed residential community and other on-site load centers (inclusive of remote applications). Based on the results of the earlier experiments, improved system designs will be incorporated in these. A further set of experiments will be designed to demonstrate the applicability of photovoltaic systems in utility central power stations.

Figure III 3-14 depicts an experimental project for powering a crop irrigation system at a farm near Mead, Nebr.

The problems of assembly and operation of large-scale systems are also being presently explored. A project is being funded for a photovoltaic system to provide some of the electrical and thermal requirements of a 50,000 square foot classroom building at the Mississippi County Community College (MCCC) in Blytheville, Arkansas. Another showcase project being funded, is the Paired Experiment at the Northwest Mississippi Junior College in Senatobia, Miss.

Each of the aforementioned groups of experiments are currently being planned in three stages, each of which demonstrates particular states of readiness. These stages are described by the following:

- Initial system evaluation experiments (ISEE)
    — to identify systems which would con-

418 Solar, Geothermal, Electric and Storage Systems

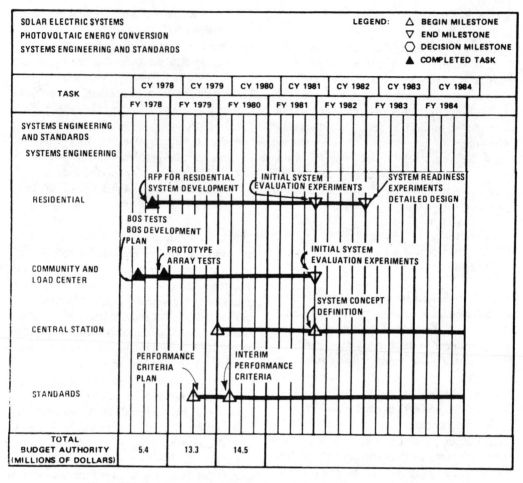

Figure III-3-13. Milestones and funding for systems engineering and standards.

tribute the greatest impact in the smallest timespan, and to establish technology readiness for each of these
- System readiness experiments (SRE)
  — to establish the efficient functioning of all of the units of an entire system
- Commercial readiness demonstration projects (CRDP)
  — to establish attainment of projected cost goals in particular applications.

Initial emphasis is being placed on utility-interactive residential and load-center applications. For residential, and for community and load-center applications ISEE are planned until FY 1982 and SRE from FY 1982 to the end of FY 1984. In the case of central station applications, ISEE's are planned for initiation at the beginning of FY 1982 and to be continued through FY 1984.

Among the large-scale experiments, the Mississippi County Community College project is scheduled for completion in the summer of FY 1980, and the Paired Experiment is to be completed in FY 1980. Plans are also being made to support photovoltaic aspects of international projects, such as joint U.S.–Italy and U.S.–Saudi Arabia activities.

Figure III 3-15 illustrates major milestones for this subprogram.

Wind Energy Conversion Systems

One promising way to tap the sun's energy is by using wind systems. A small portion of

Figure III-3-14. Solar photovoltaic power system for agriculture in Mead, Nebr.

solar energy received by the Earth is converted into surface winds. This energy source is continually renewable, and power systems driven by wind will have no major adverse effects on the environment.

Wind energy systems have been in use for many centuries, traditionally for irrigation and milling operations. Since the turn of this century, wind energy systems have also been used for generating electric power. Generally, these systems have been small, ranging in output from a few watts to a few kilowatts. The commercial availability of these machines has declined since the advent of inexpensive and reliable power from the Rural Electrification Administration (REA) in the 1930s.

The objective of the Wind Energy Conversion Systems (WECS) Program is to accelerate the development of reliable and economically competitive wind energy systems to enable the earliest possible commercialization of wind power. To achieve this objective, it will be necessary to advance the technology, develop a sound competitive industrial base, expand user awareness and acceptance, and address those nontechnical issues that could delay the introduction of new wind machines. Thus, the program's primary challenge is the development of economical wind systems capable of providing up to 30 years of reliable, safe, relatively maintenance-free service.

Wind systems of three relatively discrete sizes are being explored: small machines of less than 100-kW for use with light loads such as farms and rural residences; an intermediate size with a nominal 100- to 200-kW rating for a variety of intermediate-load applications; and megawatt-scale (1- to 3-MW) units for large loads, including central power grids.

Because of the relatively advanced state of aerodynamics technology, basic research is not stressed in the WECS program. Instead, the program centers on mission and applications analyses, the development of more effective machines, and the accumulation and analysis of site-specific wind data. The basic technology features horizontal-axis, propeller-type systems, although a variety of innovative concepts are being investigated, and several

420 Solar, Geothermal, Electric and Storage Systems

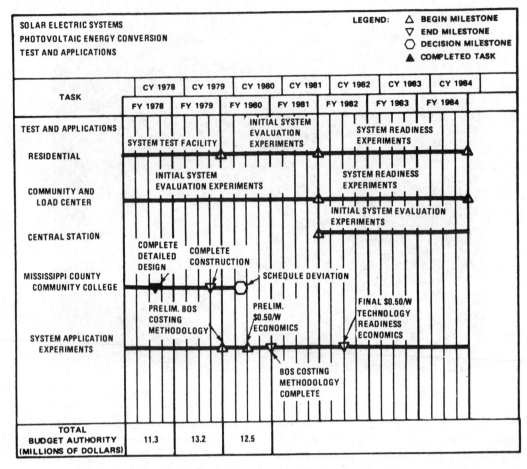

Figure III-3-15. Milestones and funding for test and applications.

of these concepts have successfully entered the competitive small systems development cycle.

The WECS Program consists of the following subprograms: Project Development and Technology, Small-scale Systems, Intermediate-scale Systems, and Large-scale Systems. Activities formerly accounted for under the title Multiunit Systems are now accounted for in other program areas.

The budget request for FY 1980 is $67 million. Table III 3-4 summarizes the funding levels by subprograms for FY 1978 through FY 1980. These funds will enable the initiation of a limited number of small systems for the third cycle of development, tests of small systems (both commercially available models and models from the second development cycle), continued tests of first-generation intermediate- and large-scale systems, development of second-generation systems of intermediate and large scales, and the continuation of research and supporting studies.

Project Development and Technology. The Project Development and Technology Subprogram is directed at enhancing performance and reducing the costs of WECS, as well as improving economic and site analysis capabilities and providing necessary information for manufacturers and users to support the commercialization of wind systems. Present technological emphasis is on rotor development and rotor-load prediction. Research and laboratory-scale testing will continue on Darrieus rotor, innovative vortex, diffusor, and other novel concepts.

Studies and research in conjunction with utility companies and other users continue in increasing depth on economic and operational

Table III-3-4. Budget for wind energy conversion systems.

WIND ENERGY CONVERSION SYSTEMS	BUDGET AUTHORITY (DOLLARS IN THOUSANDS)			
SUBPROGRAMS	ACTUAL FY 1978	ESTIMATE FY 1979	ESTIMATE FY 1980	INCREASE (DECREASE)
Project Development and Technology	6,900	14,400	16,200	1,800
Small Scale Systems	7,600	19,700	20,500	800
Intermediate Scale Systems	4,600	7,250	6,150	(1,100)
Large Scale Systems	16,300	19,350	24,150	4,800
TOTAL	35,400	60,700	67,000	6,300

issues, environmental effects, institutional and market issues, and meteorological and wind characteristics.

The objectives, activities, and key products of the components in this subprogram—mission analysis, applications/system studies, legal/environmental/social issues, wind characteristics program, technology development, and advanced innovative concepts—are presented in Table III 3-5.

A study of remote and isolated area markets was made during FY 1978. A marketing study for the high-potential wind systems applications, identified in earlier mission analysis efforts, was initiated in mid-FY 1978. These studies provide the first realistic estimates of the location and size of wind systems markets for use by wind systems manufacturers and hardware distributors, and thus they will provide potential marketing strategies. A study of potential economic incentives for wind systems commercialization will aid Federal planners by providing (1) scenarios for systems implementation and (2) estimates of the cost and impact of various Government incentives that could be used to stimulate commercialization.

With the completion of first-generation studies of the potential of integrating wind systems with regional utilities in New England, Hawaii, Michigan, Minnesota, western Texas, and California, this subprogram entered a new phase of increased specialization. Studies of the application of wind systems to REA cooperatives, local utilities, and combined wind-hydroelectric systems were initiated during FY 1978. These studies, in addition to providing information relevant to a specific utility, provide information for all utilities of a particular class. The studies address the complex operational and economic questions associated with integrating wind power systems into larger power-generating and delivery systems.

The vertical-axis Darrieus system is moving from the research to the development stage. At the small scale (30- to 50-foot diameter), systems capable of being used commercially are under development. A major technical review of the state-of-the-art of the Darrieus and its potential for competing with the propeller-type system was undertaken in late 1978 with favorable results. Planning options are being developed for a division point in mid-FY 1979 to develop larger scale vertical-axis systems.

The Wind Characteristics Program will focus in the future on the most promising of the current modeling methods for continued developments to enable cost-effective selection of single-system and multiunit sites. A comprehensive methodology will be developed for utilities and other users, so that optimal wind energy sites under a full range of terrain and climate conditions for machines of various types and sizes may be selected. Close cooperation between the Pacific Northwest Laboratories (PNL) and designers and users will be an important aspect of the efforts to provide wind characteristics data for

Table III-3-5. Objectives, activities, and key products of the project development and technology task.

TASK COMPONENT	OBJECTIVE	ACTIVITIES	KEY PRODUCT
Mission Analysis	Assess Economics of Wind Energy Assess Applications Assess Barriers to Acceptance Assess Incentives Study Market Needs	Marketing Assessments Ventures Analyses Incentives Studies Economic Analyses	Information for Decision Makers Estimates of Market Size vs. Cost
Application/System Studies	Investigate operational and economic questions associated with use of wind systems	Analysis of Specific Applications Economic Model of Utility Interface	Planning Workbooks for Users, Utilities
Legal/Environmental/Social Issues	Assess and monitor impacts of legal, environmental and social issues	Addressing Impacts of Widespread Use of Wind Systems Monitoring of Field Tests	Information for Decision Makers
Wind Characteristics Program	Improve capability of evaluating good wind sites Provide design requirements for wind systems	Collecting Wind Characteristics Data	Large-area Wind Survey Techniques National Wind Resource Assessments Siting Handbooks Experimental Design Data
Technology Development	Improve Performance Reduce System Costs Reduce Component Costs	Systems Designs Utility Interface Research Rotor Development Research	Higher performance, more reliable, lower cost systems and components
Advanced Innovative Concepts	Determine Potential of Alternative Wind Systems	R&D, Darrieus, Gyromill/cycloturbine, Self-twisting blade	Concepts with potential for further development

use in design, performance, and system operation. Milestones and funding data for Project Development and Technology are displayed in Figure III 3-16.

Small-scale Systems. Wind machines in the power range up to 100-kW have numerous potential applications. These applications include stand-alone systems in areas not reached by the utility grid as well as systems that supplement utility power. The early markets for small wind systems are for applications where utility interconnection charges are prohibitive or where the need for self-sufficiency dominates the cost factor. Wind systems that provide power where storage requirements are minimized (such as irrigation) or where storage can be made part of the energy system (such as hot water heating) are most suitable for early markets. The range of small systems is illustrated in Figure III 3-17. Initial market analyses have shown that farm, rural, residential, and agribusiness applications where utility backup energy is available constitute the largest market.

The Small-scale Systems Subprogram has two major components: small wind systems development, testing, and evaluation managed by DOE's Rocky Flats Wind Energy Test Center, and farm and rural applications testing and evaluation of wind system requirements by the USDA's Agricultural Research Service. The development program is designed to involve private industry in the study, design,

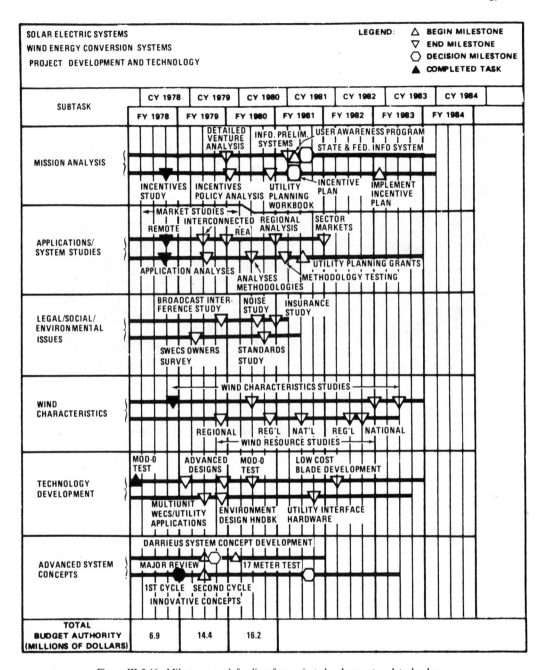

Figure III-3-16. Milestones and funding for project development and technology.

and fabrication of systems, creating a competitive industry and allowing development of an industry infrastructure. Potential users are also involved to increase user awareness.

The DOE Rocky Flats Wind Energy Test Center, located near Denver, Colo., has the responsibility for managing the development, testing, and evaluation of new small-scale systems. The center is currently performing controlled testing of commercially available small wind machines. Additionally, nine contractors have been selected to design and test parallel approaches to the development of high-reliability, economical wind turbine sys-

Figure III-3-17. Some commercially available small wind machines.

tems for farm and rural scale applications. The first development cycle of 1-, 8-, and 40-kW systems for farm and rural applications is underway and includes units once classified as "advanced concepts." Artists' conceptions of the 8-kW systems are shown in Figure III 3-18. Prototypes of these systems are scheduled to be tested in FY 1979, as is the initiation of a second cycle of new systems development.

The USDA Agricultural Research Service, which is responsible for managing projects to identify requirements for farm and agricultural wind systems, continues to test and evaluate small machines in farm and rural applications. Disseminating information to the farm and agricultural community is a major function of this subprogram.

Major milestones for Small-scale Systems are presented in Figure III 3-19.

Intermediate-scale Systems. The intermediate-scale wind systems (nominally 100- to 200-kW) are intended for applications such as irrigation, agribusiness, islands, remote communities, and small-scale industrial users. They also provide valuable information that supports the development of larger machines. The initial 100-kW wind machine (designated "MOD-O") was first operated in October 1975 at NASA's Plum Brook Station in Ohio and has provided extensive test data. This system operates as a test bed for new and advanced components. To support technology development efforts, MOD-O has been used to test a "soft tower" configuration and "pole spar" blades, with tip control/teetered hub and induction generator tests scheduled for FY 1979. This system is depicted in Figure III 3-20.

Development of intermediate-size systems (100- 150-foot diameters) is directed toward reducing costs and improving performance and reliability. The number of systems to be developed, the parallel contracts, and their timing have been planned to promote a competitive industry with continuity of design experience.

Systems, designated "MOD-OA," using first-generation technology are currently being

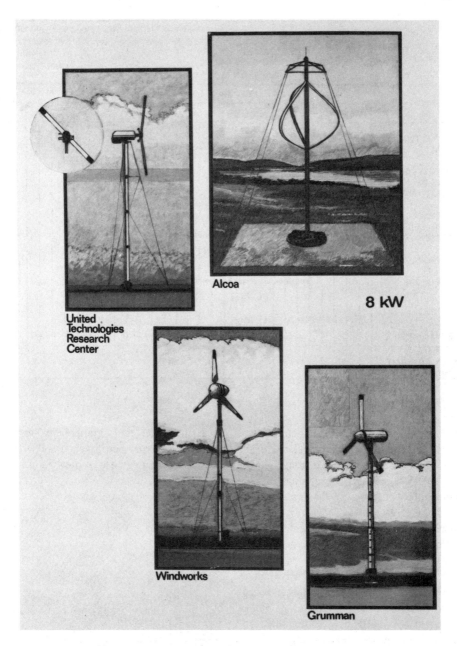

Figure III-3-18. Conceptual drawings of 8-kW wind systems.

tested. One MOD-OA system (an upgraded 200-kW version of MOD-O) was dedicated in January 1978 for experimental field testing in utility operation at Clayton, N. Mex. and is operating, unattended, with the municipal electric utility. This system was the first user-operated field test unit. It has demonstrated the technical feasibility of fully automatic synchronized operation. Remaining issues for investigation include fatigue, durability, electrical control and stability, and operating and maintenance costs. Testing will continue through 1980. An additional unit was installed at Culebra, P. R. in 1978. A third MOD-OA system will be installed at Block Island, R. I. during FY 1979. A fourth system is being fabricated for installation in Oahu, Hawaii in FY 1980.

426 Solar, Geothermal, Electric and Storage Systems

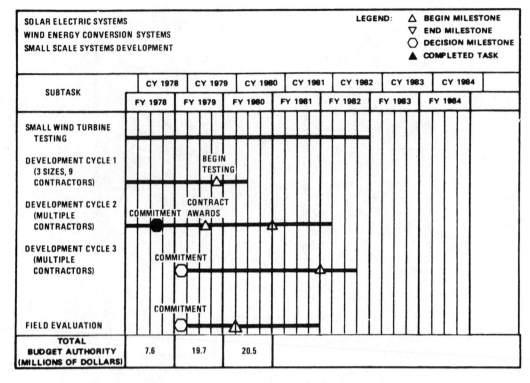

Figure III-3-19. Milestones and funding for small-scale systems.

In FY 1978, an industry competition was held for a second-generation system at the 125-foot diameter/200-kW size range to be called "MOD-4." While representing a considerable improvement over the initial MOD-OA (and the large system, MOD-1), the proposed machines would not represent a significant improvement in expected energy cost over 300-foot diameter/2.5-MW large-scale MOD-2. The rapid pace and aggressive nature of the program, including overlapping schedules, did not allow for the development or incorporation of sufficient technology advances. Based on the limited improvement expected in system economics and the probable cost, no contracts were awarded as a result of this competition. This project is now being replanned with an approach permitting significant technological and performance improvements to be developed and incorporated. The intent is to ensure adequate time to assimilate the latest technological advances and experience from other development efforts, expanded conceptual and preliminary design efforts to permit greater optimization, and increased blade and component testing to achieve improved performance and greater reduction in energy cost. A new industry com-

Figure III-3-20. MOD-O wind turbine.

petition is planned as soon as possible using the funds designated for MOD-4.

Major milestones for this subprogram are presented in Figure III 3-21.

Large-scale Systems. Large-scale wind machines that have a power rating in the 1- to 3-MW range offer significant potential as an alternative method of generating electricity.

The objective of the Large-scale Systems Subprogram is to develop durable and economical systems for eventual use by large power producers. Preliminary conceptual studies of large wind systems have examined alternative configurations and designs and have predicted that energy will be produced at the lowest unit cost by large wind machines having rotor diameters ranging from 200 to 300 feet. This prediction will be verified by developing and testing several large wind turbines.

Technology development of wind systems will permit industry to gain experience in system design and fabrication. Fabrication experience will be provided by the production of one or more units of each design. Finally, field testing of the units by utilities will determine actual performance and costs, identify and solve technical and operational problems, and alert potential users to economic and other advantages. Multiple-unit systems as well as single units will be tested by utilities to determine the increase in capacity credit provided by machine arrays through the statistical averaging of wind fluctuations.

General Electric was selected to perform the design and fabrication of the 200-foot diameter MOD-1 (2.0-MW) wind turbine to be installed at Boone, N.C. (Figure III 3-22) in FY 1979. Boeing Engineering and Construction Co. was selected as the industrial contractor to design and fabricate the MOD-2 (approximately 300-foot diameter, 2.5-MW) wind turbine with contract options for two or three additional machines. Design studies of

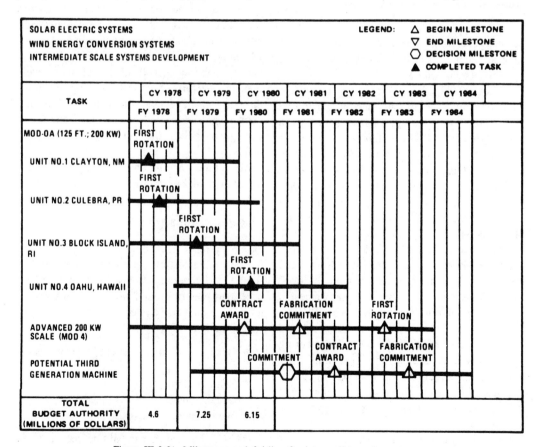

Figure III-3-21. Milestones and funding for intermediate-scale systems.

428 Solar, Geothermal, Electric and Storage Systems

Figure III-3-22. Largest U.S. wind energy conversion system installed at Boone, North Carolina.

second-generation machines in these size ranges, seeking to develop more cost-effective systems, began in FY 1978.

Although the first system of this size, MOD-2 actually represents second-generation technology using the experience gained from MOD-OA and MOD-1. A three-unit system is planned for operation in FY 1980 at a site to be selected in FY 1979. A full preliminary design review of the MOD-2 has been completed, and the MOD-2 is entering the final design and fabrication phase. This system includes advanced concepts such as: the blade tips rather than the entire blade are controlled, the rotor blades "teeter" to reduce loads, and the overall structure is lighter and "softer" than prior systems as a result of advances in vibration reduction techniques. The MOD-2 should produce about twice the energy from a system no heavier than MOD-1. Engineering projections indicate that the first units after development should produce energy at about 7 cents/kWh, while the mature product may be expected to meet an energy cost goal of 4 cents/kWh.

Operation of these machines in a user environment is expected to provide valuable information about wind turbine design features and electrical control features of the wind system interface with utilities. Economic data and information about the institutional issues and requirements of utility-based wind systems will be important products of these experimental projects.

An industry competition was held in FY 1978 to develop MOD-3, a system with 200-foot diameter and 2-MW power rating. Because only marginal increase in capability was likely to be achieved for a very large cost under the original plan and schedule, no contracts were awarded. Revised plans are being developed to allow for the development and incorporation of more advanced technology so as to make a more significant advance toward the energy-cost goal of 2 to 3 cents/kWh for large wind systems. An evaluation is being made of the benefits of a new industry competition to select contractors to design and fabricate an advanced megawatt-scale wind system using the funds previously designated for MOD-3. As with the intermediate-scale system, the intent of the new plan is to take advantage of the recent technological advances, experience with other machine development efforts, improved optimization through expanded conceptual and preliminary design, and increased blade and component testing to realize the desired cost improvements.

Milestones and funding for this subprogram are shown in Figure III 3-23.

Ocean Systems

Ocean systems have the potential for providing an inexhaustible and renewable energy source for generating substantial amounts of base-load electricity and producing energy-intensive products. Systems currently under study, include:

- Ocean thermal energy conversion (OTEC)
- Salinity gradients
- Ocean currents
- Wave energy

OTEC, the largest part of the program, is the most advanced of these technologies and

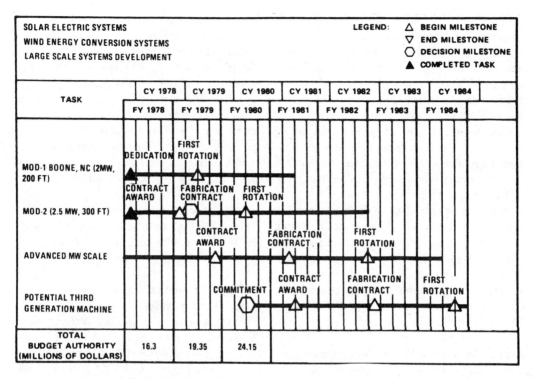

Figure III-3-23. Milestones and funding for large-scale systems.

appears to have the greatest potential. However, work on the three additional systems concepts is increasing. These alternate systems could serve regions and markets in addition to those planned for OTEC.

The OTEC concept uses the temperature difference between warm surface waters and cold water from the depths to operate a heat cycle and to generate electricity.

Two power cycles using this concept are:

- Closed cycle—a fluid (such as ammonia) with a low boiling point is used as the working medium (Figure III 3-24)
- Open cycle—seawater is used as the working medium

The most promising approaches in this mode of energy conversion involve use of the following:

- Moored platforms that supply electricity by cable to a grid on shore
- Floating platforms that make use of optimum temperature differences, by "grazing" for the warmest surface waters and generating energy for onboard process conversion of energy-intensive products

Power may be extracted from salinity gradients by using the energy potential that exists

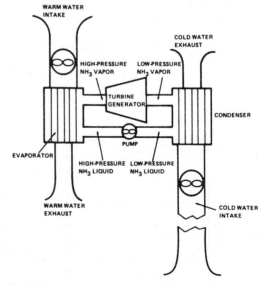

Figure III-3-24. Ocean thermal energy conversion (OTEC).

across a selective membrane between two solutions of differing salinity. Methods include the use of sea water for the saline solution and fresh water for the less saline solution. Alternatively, brine from salt domes may be used for the saline solution and sea water may be used as the "fresh" water source.

Ocean currents furnish another source of energy that may be exploited. In this application, large-diameter hydroturbines are held below the ocean surface by tension moorings and driven by the current.

Wave energy systems extract mechanical energy from ocean waves and convert it into electricity. Recent system concepts have been proposed which focus the wave energy and appear to offer cost-effective advantages.

Table III 3-6 summarizes the funding levels, by subprogram, for the FY 1978 through FY 1980.

Project Management. The Project Management Subprogram provides essential services in planning, programming, budgeting, and implementing activities within the Ocean Systems Program. In addition, this subprogram includes cost, schedule, and technical analyses of estimates and reports received from other Federal agencies, laboratories, and various contractor sources and also includes participation in preliminary design reviews, critical design audits, physical configuration audits, and formal qualification reviews to ensure that the contractor design will meet task requirements. It includes reviews, field-reported problems, proposed engineering changes, as well as contractors make-or-buy plans and subcontract management plans. It also includes establishing and maintaining a performance measurement system and a program control system for cost, schedule, and technical control. When the design is sufficiently stable, it includes the establishment and support of a configuration management system.

The agencies responsible for the various tasks within this organization, together with the individual duties of each, is shown in Figure III 3-25.

Definition Planning. The Definition Planning Subprogram provides essential services for mission analysis, mission-need determination, systematic and progressive identification of alternative concepts (through system engineering and analysis), sensitivity analysis, risk analysis, tradeoff analysis, conceptual and technical planning, acquisition strategy planning, legal, political, institutional, environmental and social impact analysis, and business and economic analysis. In addition, this subprogram includes modeling simulation required to estimate cost, schedule, and technical performance, to develop preliminary optimization of the OTEC system for minimum acquisition cost during the early development

Table III-3-6. Subprogram funding levels FY 1978 through FY 1980.

OCEAN SYSTEMS SUBPROGRAMS	BUDGET AUTHORITY (DOLLARS IN THOUSANDS)			
	ACTUAL FY 1978	ESTIMATE FY 1979	ESTIMATE FY 1980	INCREASE (DECREASE)
Project Management	2,100	4,700	5,100	400
Definition Planning	1,700	2,100	1,000	(1,100)
Technology Development	10,900	12,000	7,400	(4,600)
Engineering Test and Evaluation	8,100	15,700	15,400	(300)
Advanced Research and Development	6,600	3,700	6,100	2,400
TOTAL	29,400	38,200	35,000	(3,200)

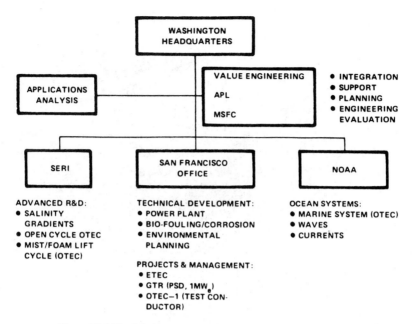

Figure III-3-25. Ocean systems management plan and organization.

phase, and to develop a minimum lifecycle cost for maximum return on investment during the commercial development phase. This modeling and simulation effort includes, but is not limited to, optimization of electrical generation and its transmission to shore, and electrical power use for onsite manufacture of energy-intensive products. The mission analysis effort includes considering product mix, site-specific economics, definition of appropriate commercial powerplant sizes, cost-benefit, market analysis and penetration, and net energy consideration.

Figure III 3-26 is a conceptual illustration of an offshore system which uses ocean thermal energy to derive energy-intensive products.

Technology Development. Technology Development activities include engineering design and development of all elements of a system, including the components and subsystems. Major developmental efforts are being devoted to the following:

- Power systems, which comprise turbines, heat-exchangers, pumps, etc.
- Biofouling, corrosion control, and cleaning systems
- Ocean systems, which include the ocean platform; anchoring, mooring, and dynamic positioning devices; and cold water pipe (CWP)
- Electrical systems in which the principal element is the transmission cable

Ocean engineering studies are underway to select appropriate hull configurations, a major cost component of OTEC systems. Figure III 3-27 shows suggested configurations.

Power system development of candidate shell-and-tube heat-exchanger configurations were initiated in FY 1977. A second effort for candidate plate heat exchangers started in FY 1978. An objective of each developmental activity is to design the ultimate module configuration and derive overall economics for the powerplant system.

The ammonia turbine research includes selection of materials for blades, selection of designs and noncorrosive materials for seals, and demonstration of bearing life. Turbine procurement to support a module test will be initiated in FY 1980.

A major technology effort is being implemented to define the CWP design and methods for deployment. Concrete is considered a leading candidate for the CWP. Indications are that flexible joints, lightweight concrete aggregates, and low-bending moments will be

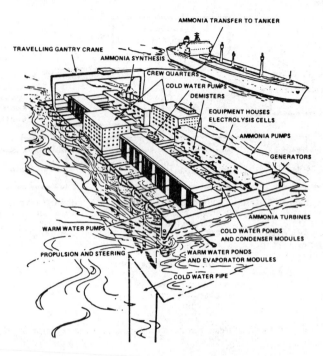

Figure III-3-26. General configuration of a sea-based fuels or chemicals OTEC plant.

required. Other leading candidate materials include steel and more compliant structures such as fiberglass and rubber. These CWP options will be subjected to a comparative evaluation that commenced in FY 1978. Test programs are being developed to define the strength, modulus, and porosity of lightweight concrete. Dynamic simulation programs are being developed, and model-basin tests are to be conducted. At-sea tests of a large-scale pipe (5 feet in diameter and 1,000 feet long) will be conducted in 1979.

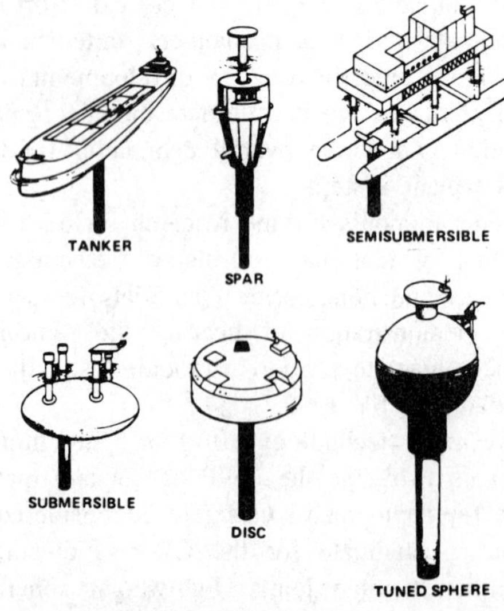

Several CWP deployment approaches are being studied: vertical assembly, onsite manufacture, and float and flip.

Present assessments of the OTEC electrical offshore application indicate that the OTEC submarine cable (except for island locations) will be high-voltage DC, with conversion subsystems at each terminal. Single or multiple submarine cables must be capable of transmitting in the 500-MWe range. Submarine cable development OTEC was initiated in 1977 through competitive procurements, and the contractors are now in the early stages of their studies of the riser and bottom cables. The most promising cable systems from among several alternatives for various plant sites will be selected for further work.

Phase II of submarine cable development will entail detailed design, fabrication, and testing of the selected cable systems. Phase III will involve the preparation of final specifications for use with OTEC plants at various sites.

Major milestones for Technology Development are presented in Figure III 3-28.

Figure III-3-27. Basic OTEC hull alternatives.

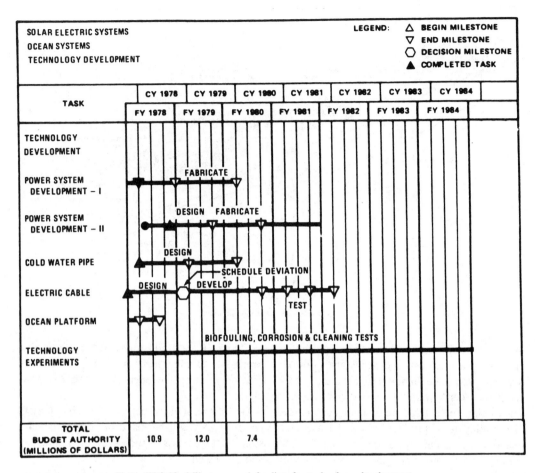

Figure III-3-28. Milestones and funding for technology development.

Engineering Test and Evaluation. The Engineering Test and Evaluation Subprogram provides for the construction of an engineering test facility, OTEC-1, by converting a suitable existing vessel. OTEC-1, depicted in Figure III 3-29, is strategically planned to furnish the earliest ocean-based performance of an OTEC power system of 1-MW; it also includes operation, component testing, and reporting results. Test data on cold water pipe, biofouling, corrosion, and cleaning will be collected in early 1980. OTEC-1 will satisfy the need for early testing of heat exchangers at large flow rates of warm and cold water in an ocean environment. It will be used to assess multiple-heat-exchanger configurations, materials, cleaning, and production approaches. The facility will provide a mechanism for obtaining long-term operational data in an actual sea environment where the required flow rates can be obtained. Information on cold water pipe loads and effluent dispersion patterns will also be obtained to test OTEC subsystems.

The OTEC-1 engineering test facility will be used for significant scaling experiments in an ocean environment. Five heat-exchanger configurations, including two shell-and-tube and three flat-plate types will be tested in this facility. By December 1979, the 1-MW shell and tube heat exchanger will be installed in OTEC-1. Tests will commence in April 1980. The vessel will incorporate the other four advanced heat exchangers from the power system development efforts starting in October 1979. Several materials and cleaning approaches will be explored. Three 4-foot diameter cold water pipes and the warm water system will supply enough water for an integrated 1-MW system—i.e., the equivalent of 6,000 tubes.

The OTEC-1 facility will begin operation

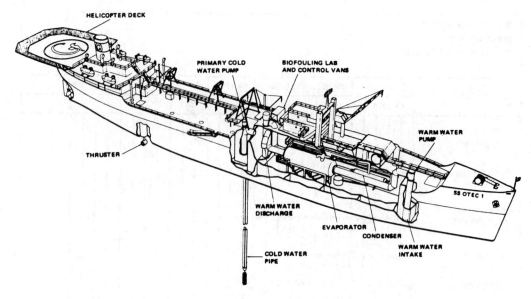

Figure III-3-29. OTEC-1 facility.

by early 1980 and will provide substantial operational experience and necessary information to determine long-term cleaning characteristics, corrosion effects, and performance degradation for a large number of candidates. The results from this facility will be used for selection of power cycle subsystems for the ensuing larger modular experiments and will complement those obtained from the parallel Advanced Research and Technology efforts.

Figure III 3-30 presents milestones and funding for the Engineering Test and Evaluation Subprogram.

Advanced Research and Development. Advanced Research and Development activities concentrate on studies relating to advanced heat exchangers and advanced open cycle concepts. Results of these studies will be used in the design and development of power systems for OTEC projects of the future. Alternative ocean energy concepts such as salinity gradients, wave energy, ocean currents and mist/foam OTEC are also being investigated to determine their technical feasibility and potential for significant production of baseload electricity or other applications. Included is research in the areas of membrane technology, solar ponds, osmotic power plants and electrodialytic batteries.

Figure III 3-31 illustrates a schematic of a process to extract electrical energy via salinity gradients, using the open-cycle osmotic principle.

Under this area of activity, specific project commitments will be made only upon conclusion of exploratory studies being conducted and reviewed.

A contract has been awarded for construction of a 100-watt solar pond, which will be used to extract electrical energy via salinity gradients. A design study has also been awarded for hydroelectric analysis and experiments on ocean currents. Turbogenerator design has been completed for use in an international experiment for the exploitation of wave energy.

Milestones and funding for Advanced Research and Development are presented in Figure III 3-32.

BIOMASS ENERGY SYSTEMS

Biomass is organic material (except coal, oil, etc.) used for energy and includes terrestrial and aquatic vegetation, agricultural and forestry residues, and animal wastes. It can be viewed as solar energy which has been collected by photosynthesis and is stored as chemical energy of the organic material. Bio-

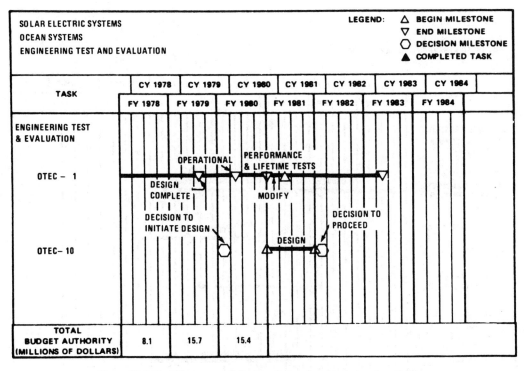

Figure III-3-30. Milestones and funding for engineering test and evaluation.

mass can either be collected and burned directly or converted to other usable energy forms, such as gaseous or liquid fuels, or energy-intensive chemical feedstocks. Principal applications for biomass products include industrial process heat generation, electrical generation, production of heat and power for onsite residential, industrial, or agricultural applications, and the production of liquid fuels for the transportation sector. Figure III 3-33 summarizes the scope of the sources and uses of energy from biomass available for national energy needs.

The primary objective of the Biomass Energy Systems Program is to develop the capability for economically converting renewable biomass resource into direct energy, fuels, or energy-intensive products, thereby displacing the use of fuels and products derived from petroleum and natural gas. In order to achieve this displacement, the program will provide technical support for early and broad adoption of biomass technologies. As technologies approach cost levels competitive with conventional energy sources, commercialization activities will be conducted through the Office of Conservation and Solar Energy (CSE) or the Office of Resource Applications (RA) with technical support from the Biomass Energy Systems Program.

The program strategy is based on the

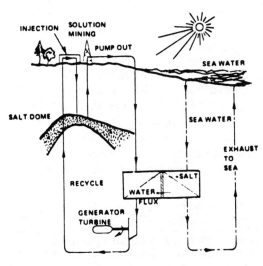

Figure III-3-31. Open cycle osmosis

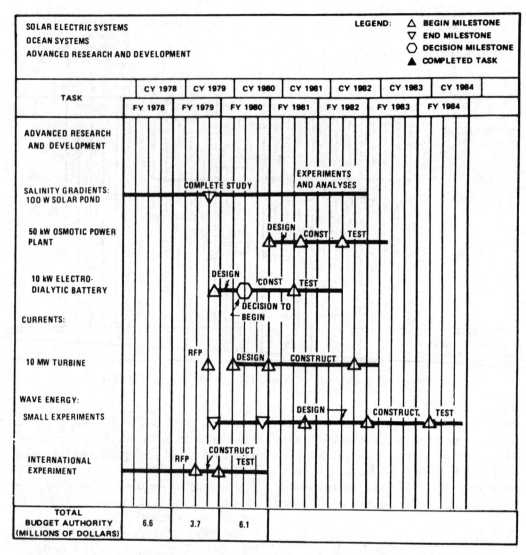

Figure III-3-32. Milestones and funding for advanced research and development.

grouping and phased development of biomass technologies and their applications according to the time frame in which they are expected to enter the commercial market. For technologies that are expected to enter the market in the near term (by 1985), efforts are directed toward process improvements and, where necessary, demonstrations to exhibit the technology in commercial-scale applications. Projects supporting technologies expected to enter the market in the midterm (1985–2000) are now primarily directed toward laboratory-scale investigations, process economics studies, and engineering design models using process development units (PDU's) and, if successful, large experimental facilities (LEF's).* Technologies expected to have long-term commercial impact (after the year 2000) are now being developed through applied research and exploratory development. The Biomass En-

---

*PDU's are small working models, larger than laboratory-scaled models, that are built to establish the physical parameters of operations. LEF's are engineering models, larger than PDU's and smaller than commercial units, that are built to provide further development of the technology.

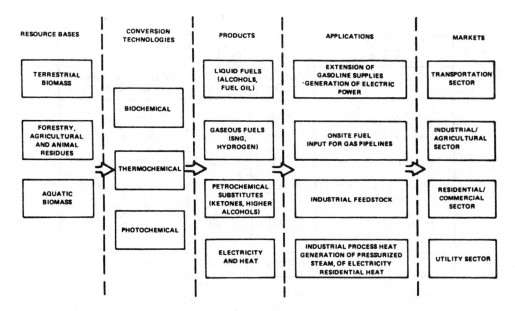

Figure III-3-33. Scope of markets and applications for products from biomass resources.

ergy Systems Program cooperates with USDA and other Government agencies pursuing related research.

Technologies expected to enter the commercial market in the near term include: direct combustion of wood and wood residues as a source of industrial process heat, electricity, and residential heat; fermentation of starches and sugars to grain alcohol that can be mixed with gasoline to extend gasoline supplies; and anaerobic digestion of manures to produce a gaseous fuel, and direct combustion of agricultural residues for use by the agricultural sector.

Direct combustion of wood is expected to have the largest quad impact on the energy supply of the three solar technologies that are ready now for commercialization. Biomass supply and conversion technologies being developed for midterm entry into the commercial market include energy farms to provide reliable resource supplies and the production of medium-Btu gas (MBG) and fuel oil by thermochemical conversion. Technologies expected to become commercial in the long term include aquatic energy farms, development of exotic species, and biophotolysis and other advanced conversion technologies. Requested FY 1980 funding for biomass systems is $57 million, up $15 million from the current year. Table III 3-7 summarizes the funding levels by subprograms for the FY 1978 through FY 1980 period.

Technology Support

The Technology Support Program element will provide technical support, systems demonstrations, and environmental research in support of activities nearing or already involved in commercialization activities. Technologies expected to have commercial impact by 1985 include: direct combustion and low-Btu gasification of wood residues; on-farm direct combustion of agricultural residues and anaerobic digestion of manures; and fermentation of grains and sugar crops to produce ethanol. Specific projects pursued will emphasize improvements in process performance, economic competitiveness, and reliability; assurance of acceptable process operating characteristics; reduction of environmental impacts; assessment of resource availability and demand; and investigation of international development opportunities.

The most significant near-term conversion process is the direct combustion of wood or other low moisture biomass for production of

Table III-3-7. Budget for biomass energy systems.

BIOMASS ENERGY SYSTEMS SUBPROGRAMS	BUDGET AUTHORITY (DOLLARS IN THOUSANDS)			
	ACTUAL FY 1978	ESTIMATE FY 1979	ESTIMATE FY 1980	INCREASE (DECREASE)
Technology Support	2,600	16,200	16,000	(200)
Production Systems	4,800	4,300	5,700	1,400
Conversion Technology	6,700	15,900	25,000	9,100
Research and Exploratory Development	2,400	5,200	8,800	3,600
Support and Other	2,200	800	1,500	700
TOTAL	18,700	42,400	57,000	14,600

steam for driving turbines and industrial process heat. As fuel prices have risen over the last decade, the forest product industry has been switching to biomass fuels such as wood and wood process residues. A portion of this program element is aimed at ensuring economic availability of terrestrial residues such as noncommercial standing forests, logging and mill residues at conversion sites. The distribution of U.S. wood resources shown in Figure III 3-34. Direct combustion technology for using wood residues to produce process steam or electricity for use on or near the site is relatively well established.

Anaerobic digestion of manures to produce a medium Btu gas is a technology sufficiently well developed to be marginally competitive for onsite power generation, hence qualifying as a near-term technology. The process takes place at essentially ambient conditions (to 65°C) and requires a moisture content of 80 percent or greater. Technology development studies are being directed toward increasing yields and modifying reactions to decrease costs, and increasing the potential for using byproducts as cattle feed. Efforts will be directed toward the development of simple and relatively inexpensive facilities for use on cattle feedlots and individual dairy farms. The primary resource is cattle waste, which is readily collected from feedlots.

The near-term focus in fermentation is R&D for improving the net energy balance and process economics of producing grain alcohol from sugar and starch crops. Figure III 3-35 shows the process and inputs.

In direct combustion of wood and wood residues, commercialization efforts are already underway. In the first quarter of FY 1979, a Resource Manager in CS was identified and the decision to shift demonstration responsibilities to CS will be made in the last quarter of FY 1979. This shift, if decided upon, would take place in FY 1980 with ET continuing with supporting research.

During FY 1980, low-Btu gasification units will be evaluated for process heat requirements. A number of small pilot projects in residue harvesting, preparation, and combustion will be supported to prove the technical and economic viability of specific systems.

Several PDU's (handling 1/2 ton/day to 3 tons/day of oven-dried feedstock) for digesting animal manure are in operation now, and data will be collected and analyzed by FY 1980. In FY 1980, a decision on commercialization readiness will be made. Depending on this decision, a shift of responsibility for lead development of anaerobic digestion to CS would be made. ET would continue to supply supporting research. Current projects include anerobic digestion of animal manures to produce methane at a Bartow, Florida feedlot/anaerobic digestion facility designed to proc-

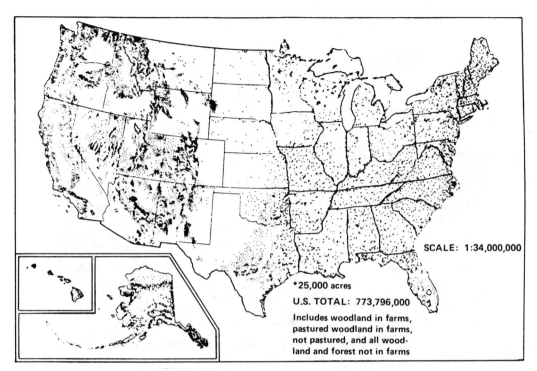

Figure III-3-34. U.S. forest and woodland distribution.

ess 25 oven-dried tons of feedstock per day and at a dairy farm digestion facility at Cornell University.

In fermentation, equipment design, testing and demonstrations are part of the near-term program. Feedstocks development and process improvement studies have been ongoing since FY 1977 with increasingly larger scale cultivation of sweet sorghum as a feedstock scheduled for FY 1979 and completion of process improvement testing in FY 1982. In anticipation of the near-term commercial impact of alcohol as a gasoline extender, DOE is developing a policy for alcohol fuels in FY 1979.

Milestones and funding for this program element are presented in Figure III 3-36.

Production Systems

The Production Systems Program element encompasses expansion of the resource base for biomass energy production by developing specific species and by developing intensive management techniques. Biomass resources include energy crops grown expressly for their energy content, as well as agricultural or forest residues suitable for conversion processes. These resources may be classified as terrestrial or aquatic. Terrestrial biomass resources include silvicultural resources (wood plants grown for use as biomass), residues (field crop residues, animal manures, and forest residues), and herbaceous resources (grasses, grains and other nonwoody plants). Aquatic biomass resources include algae and other aquatic plants grown in either fresh water, coastal brackish water, or marine water.

For the midterm, the development of herbaceous and silvicultural energy farms will supplement existing supplies of biomass for conversion to fuels and industrial feedstocks. The objectives for these energy farms include selection of the most promising species and development of site preparation, planting, and management techniques to obtain economical energy production. In the longterm, aquatic energy farms may provide additional biomass resource.

Research on species and production of biomass on herbaceous (soft-stemmed crops) and silvicultural farms was initiated in FY 1978.

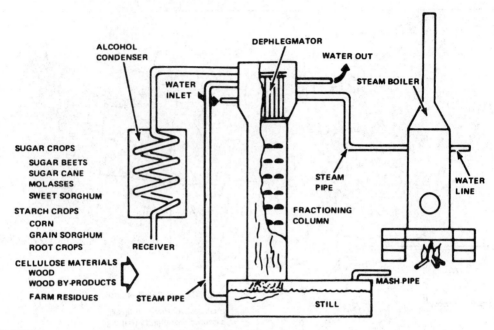

Figure III-3-35. Ethanol production from farm products.

Three regional silvicultural farms, including one located at the DOE Savannah River, S.C. site are scheduled. Under a recently signed memorandum of understanding (MOU) with the United States Department of Agriculture (USDA), the USDA will support the Biomass Energy Systems Program in the development of farm and forest crops for high energy value and the development of working energy crop management practices and systems. Completed research includes a prototype machine developed by the USDA Forest Service, Southern Forest Experiment Station, for harvesting forest residues for fuel. Other completed studies relate to economical planting patterns for sugar cane, evaluation of a new process for separating solid components of sugar cane stalks before extracting sugar juices, and investigation of the agronomic feasibility of mechanized, year-round production of solar-dried sugar cane and napier grass.

Major milestones and funding for this program element are presented in Figure III 3-37.

Conversion Technology

Biomass feed material can be converted to liquid fuels (fuel oils, alcohols) or gaseous fuels (medium Btu gas, synthetic natural gas) or other energy-intensive products (hydrogen, ammonia, petrochemical substitutes) that can supplement or replace similar products made from conventional petrochemicals. Technologies for converting biomass into liquid and gaseous fuels fall into the general categories of thermochemical and biochemical conversion.

The goals for further developing conversion technologies are to improve system efficiencies, to decrease process costs and eventually to increase the size of biomass converters. The goals will be met by conducting market assessments to identify optimal product strategies and necessary cost goals, by performing laboratory-scale experiments, by developing PDU's for promising technologies, and by scaling up to LEF's where necessary.

Thermochemical Conversion. In thermochemical conversion processes, biomass is decomposed by the use of various combinations of elevated temperatures and pressures, and chemical reactants. Midterm thermochemical conversion technologies include gasification and liquefaction.

Gasification and liquefaction produce gas-

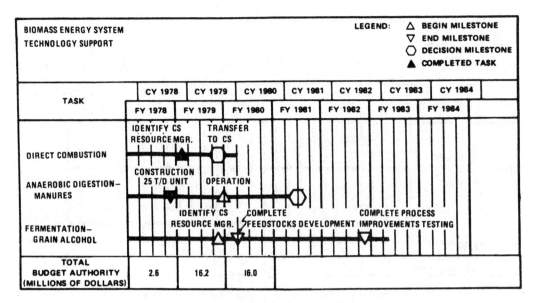

Figure III-3-36. Milestones and funding for technology support.

eous and liquid fuels from organic matter. A number of processes may be used to produce a wide range of products. For example, gasification produces medium Btu gas (MBG) to be used directly or further upgraded into a wide range of derived fuels, such as synthetic natural gas (SNG), methanol, ammonia, and petrochemical substitutes. The program seeks the development of processes which yield a large volume of products in a shorter time, which can compete with products of the larger scale facilities that can be built for other feedstocks such as oil, natural gas, and coal.

A major midterm objective is to improve conversion efficiency and productivity and to reduce production costs of the gasification process to produce MBG. MBG, predominately carbon monoxide and hydrogen, is expected to be economically attractive for use near the production facility either as gas or an intermediate to liquid fuels or feedstocks. Another major area of thermochemical biomass conversion for the midterm is liquefaction. In the liquefaction process, biomass is converted directly to fuel oils by reaction with carbon monoxide and hydrogen under high temperature and pressure in the presence of catalysts. Figure III 3-38 shows a process development unit for the conversion of wood to oil.

A group of seven of the most promising gasification processes has been chosen as the basis for constructing seven PDU's. Three units were in operation in the second quarter of FY 1979, with the remaining four anticipated to be operating by early FY 1980. After an 18-month to 2-year period of the PDU's operation, a decision whether to proceed to commercialization or pilot plant construction will be made for the most promising. In a parallel program designed to accelerate the time of commercialization of this technology, a gasification LEF will be initiated in FY 1979 as a cost-shared project with industry, and construction is scheduled to start in early FY 1981. This project can be described as an engineering prototype and is expected to provide information on system dynamics and scale-up of auxiliaries. The selected gasification process may differ from those for the PDU program.

An experimental liquefaction facility for production of fuel oil from biomass has been operated since 1977 in Albany, Ore., (Figure III 3-38). Based on a Bureau of Mines process, developed at the Pittsburgh Energy Research Center, this plant operates at a scale that permits analysis of process economics. It is equipped for pretreatment of biomass and for reaction with carbon monoxide and hydrogen at elevated temperatures and pressures.

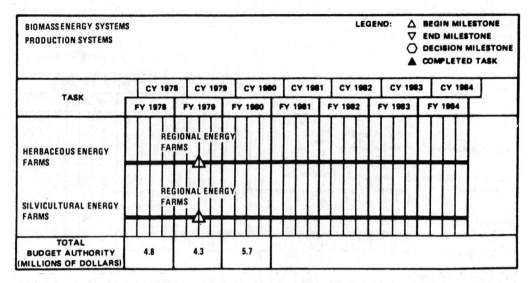

Figure III-3-37. Milestones and funding for production systems.

The unit sufficiently flexible in design to permit technoeconomic evaluation of a variety of biomass-to-oil processes with minimum equipment modification. This facility will be operated until the last quarter of FY 1980, when a decision will be made whether to proceed with pilot plant construction.

Milestones and funding for Thermochemical Conversion are presented in the Figure III 3-39.

Biochemical Conversion. Anaerobic digestion and fermentation are the two conversion methods included in this subactivity.

Anaerobic digestion, as discussed under near-term development, involves controlled decomposition of organic matter in the absence of air to produce methane and carbon dioxide. This medium-Btu gas can be burned directly or upgraded to pipeline-quality gas, such as SNG, or converted to other materials. In addition, the chemical industry uses methane as a starting material for many other chemicals. Anaerobic digestion has great potential because it yields a wide range of products and byproducts which can be put to many uses. Figure III 3-40 shows an anaerobic digestion unit.

For the midterm, the technology is being expanded beyond the use of manures toward anaerobic digestion of more abundant and widespread feedstocks such as agricultural and forestry residues. A major midterm objective is to develop pretreatment technology that will enable efficient anaerobic digestion of lignocellulosic (woody) feedstocks. Other midterm goals involve engineering improvements and cost reductions in anaerobic conversion systems and their components, and demonstration of equipment.

Approximately 10 percent of all ethanol used today is produced by the fermentation of grains and molasses. Ethanol is a versatile product because it can be blended with gasoline to extend supplies of fuel or used directly in modified internal combustion engines. The use of fermentation technology to provide the 10-percent alcohol content of gasohol may expand rapidly with the introduction of the Federal highway excise tax exemption of $0.04/gallon for gasohol. Other possible fermentation products include butanol, acetone, and acetic acid and other organic chemicals that can be used in place of compounds now obtained from petroleum.

A major goal of fermentation conversion technology research has been to broaden the feedstock base beyond those available for near-term fermentation systems. The main opportunity for cost reduction in the ethanol fermentation process involves development of cheap sources of sugars to ferment. For mid-

Figure III-3-38. Wood to oil process development unit, Albany, Ore.

term market impact, FY 1979 fermentation efforts have concentrated primarily on the use of lignocellulosic agricultural and forestry residues as low-cost feedstocks that, when processed, yield sugars, which can be fermented to alcohols by conventional technologies. To achieve this objective and reduce fermentation costs, current research activities are aimed at the use of acid or enzymatic hydrolysis of biomass feedstocks to produce fermentable sugars.

Anaerobic digestion of biomass residues other than manures to fuels is expected to reach the commercial market after 1985. In

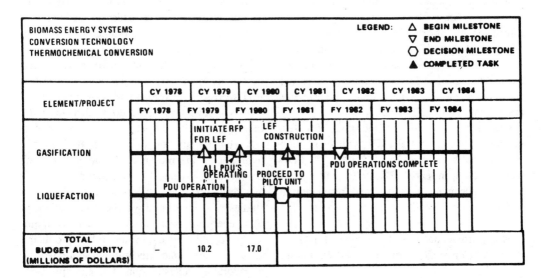

Figure III-3-39. Milestones and funding for thermochemical conversion.

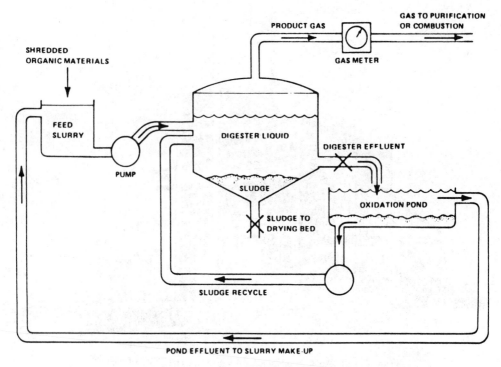

Figure III-3-40. Continuous unit for production of methane by anaerobic digestion.

FY 1977, bench-scale experiments were initiated. In FY 1979, anaerobic digestion research efforts were directed toward use of agricultural and forestry residues as digestion feedstocks. The goal was to broaden the resource base for anaerobic digestion beyond manures as a source of synthetic natural gas. A contract for a PDU to scale-up laboratory efforts on digestion of non-manure feedstocks was awarded in FY 1979, and expansion of this effort in FY 1980 is expected to accelerate technology development and enable earlier commercialization. Operational experiments are scheduled for FY 1982.

Research on cellulosic feedstock for fermentation began in FY 1977 with evaluation of progress scheduled for FY 1983. A PDU was initiated in early FY 1979, and operational experiments for advanced fermentation techniques are scheduled to start in late FY 1981.

Major milestones and funding for Biochemical Conversion are presented in Figure III 3-41.

Research and Development

This program element seeks to identify and investigate those technologies for biomass production and conversion that are expected to contribute to the energy supply after the year 2000. Research on innovative conversion systems and advanced feedstock development and studies of aquatic systems will be supported.

Research will fall into two broad areas: new conversion technologies and new terrestrial feedstock development. One example of a new conversion technology that is under investigation is biophotolysis, the process whereby certain organic materials generate hydrogen in the presence of sunlight. Though this process has been demonstrated, the yields are very low. The objective of biophotolysis studies is to determine whether this method of solar energy conversion can produce economically feasible yields. Some other promising advanced conversion technologies are worthy of laboratory investigation. For example, cer-

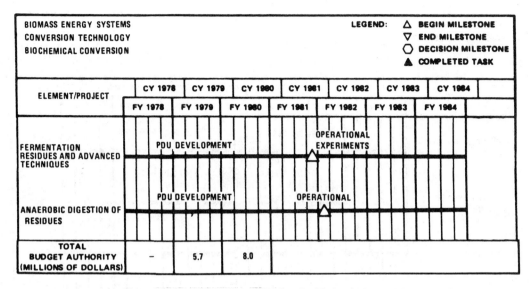

Figure III-3-41. Milestones and funding for biochemical conversion.

tain micro-organisms can feed on biomass suspended in water and produce large quantities of complex fuels or petrochemical substitutes, such as ethylene glycol. Results from all of this research can lead to the full exploitation of the many alternative ways of obtaining energy from biomass. In the area of advanced feedstock development, basic research is underway for developing new varieties of plants that are well suited for energy yield is based on genetic research patterns already developed to improve food and fiber yields.

Aquatic systems studies include fresh water and marine production, harvesting and conversion research, as well as some large-scale marine biomass farming technology.

Projects in new conversion technologies include advanced biochemical conversion research conducted primarily by universities. An example is the development of single-step fermentation of lignocellulosic material at Massachusetts Institute of Technology. Another example is biophotolysis research underway at SERI. In FY 1979, a plan will be developed for applied research issues to be addressed and promising conversion techniques to be studied.

Projects supported in advanced feedstock development include the breeding and evaluation of "exotic" crops (e.g., Lawrence Berkeley Laboratory work on latex-bearing plants), as well as basic research on plant improvement.

Completed aquatic production studies include a detailed physiological study of the growth of giant kelp, an investigation of the growth of filamentous algae in ponds, an engineering design analysis of growing and harvesting algae biomass in a 10-square-mile system, and an investigation of the cultivation of macroscopic marine algae and freshwater aquatic weeds. Aquatic systems projects include some large-scale marine biomass farming technology as well as fresh water and marine production, harvesting, and conversion technologies.

Figure III 3-42 reflects the milestones and funding currently associated with this program element. Additional milestones will be specified after the scope and nature of research plans have been defined.

Support and Other

This program element involves the performance of various systems studies and environmental assessments, as well as general program support. Mission analysis studies have been performed for several biomass

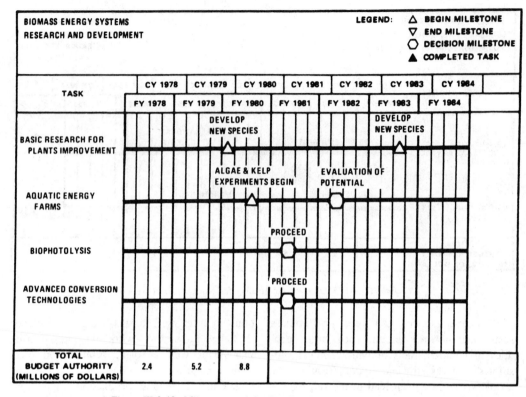

Figure III-3-42. Milestones and funding for research and development.

technology development programs. Market analyses have indicated that 19 of 50 missions could penetrate the market from 1985 through 2020. Future mission analysis will be used to evaluate the technical paths available to the biomass program and to assess their potential commercial impact.

Environmental and resource assessments have been the other major portion of this subactivity. Projects dealing with the issue of resource availability at both the national and regional levels have been implemented. Environmental assessments will be scheduled in close conjunction with technology development to assure that environmental costs and effects will be considered before widespread deployment of near-term technologies.

Additional activities being funded under this program element include program management support and consultation for assistance in program planning, preparation and evaluation of program solicitations, and participation in overall program technical evaluations, as well as coordination monitoring.

OTHER PROGRAM ASPECTS

The Solar Technology Program under the direction of DOE, contains certain other aspects not discussed previously. These aspects, addressed in detail in subsequent chapters of this document, are as follows:

- Regional activities
- Commercialization
- Environmental implications
- Socioeconomic aspects

Numerous Federal and private laboratories and other institutions have been delegated responsibilities toward achieving specific objectives of the Solar Program. The Solar Energy Research Institute (SERI) has coordinated a considerable amount of field efforts.

Commercialization considerations include successful development and demonstration of solar systems for technical feasibility, reliability, and economic viability with conventional power generation techniques.

Each phase of solar technology development will continue to be coordinated with environmental aspects—air, water quality, land use, ecology, health and safety, and esthetics, among others.

The rate at which the solar market and industry will develop is strongly influenced by the relative economics of alternative sources of energy. In general, this rate will depend on decisions related to the prices of conventional energy sources, the support provided for solar systems development, and the impacts of all forms of energy use on society.

Activities

Regional activities are an integral part of the solar program. Over the past few years, Congress and the Administration have become increasingly interested in decentralized applications of solar energy technologies as an alternative to the present common use of solar energy in centralized energy systems. Emphasis, therefore, is being placed on regional, dispersed, and small-scale systems. For the most part, this decentralized emphasis is being carried out through the various field activities.

The Solar Energy Research Institute (SERI) at Golden, Colo., is supporting RD&D and commercialization of DOE's solar energy technologies.

The Photovoltaic Program Office (PVPO) within SERI is providing the impetus for activities outside of SERI to develop options for the single-crystal silicon solar cell. The PVPO will provide management of the DOE Photovoltaic Advanced Materials Research Program.

The SERI Clean Fuels from Biomass Program is considering the problems of biomass production and conversion of biomass into gaseous and liquid fuels. The program scope enables SERI to develop the expertise and coverage necessary to support the DOE program, aid in the technical evaluation of contractor programs, and identify emerging technologies.

The SERI Wind Energy Systems Program includes activities related to advanced and innovative wind energy systems. It encompasses engineering analysis of the costs and benefits of various advanced and innovative wind energy systems, management of the DOE Innovative Systems Program, and commercialization and support activities.

Commercialization of Solar Technologies

Solar thermal power systems are classified by size. Small power systems produce up to 30-MW of electricity, and large power systems are in the 30-MW to 300-MW category. From a commercialization standpoint, engineering feasibility has been proven, and the demonstration phase of the development cycle has been reached for small systems. Information obtained from the demonstration projects can be used to determine whether incentives should be offered to manufacturers and users. Large systems are undergoing engineering development aimed at improved efficiency and reducing costs.

The technical feasibility of photovoltaic electricity generation has been established. Commercialization depends on whether such systems can be manufactured and installed so as to be competitive with alternative energy sources. Major efforts are directed toward cost reduction of photovoltaic systems. Research and development is being undertaken on advanced materials and manufacturing techniques that offer the possibility of greater cost reductions in the long term. An important issue in the commercialization of photovoltaic systems is the utility interface.

Wind energy systems may be divided into three categories: small (less than 100-kW), intermediate (100-kW–200-kW), and large (more than 1-MW). Small wind systems are commercially available, but at present they are cost-effective only for some remote and specialized applications. Current efforts are aimed at developing systems with improved performance and lower costs, which will have a larger market. Intermediate and large systems are in the engineering development phase. Performance improvement and cost reduction are prerequisites to the commercialization of these systems. The participation of manufacturers and users is being emphasized in the development of wind systems.

Systems that use the energy available from the ocean are in an early stage of development. Ocean thermal energy conversion (OTEC) systems are in the technology development stage, and other systems that use ocean currents, wave energy, and salinity gradients are in the conceptual stage. The recognition and resolution of specific commercialization issues depends on the emergence of approaches that are technically feasible.

Biomass systems encompass an array of resources, technologies, and products. Near-term (before 1985) commercialization efforts are aimed at general adoption of direct combustion of wood and wood residues, fermentation of agricultural crops to produce alcohol, and anaerobic digestion of manure. The technical feasibility of these processes has been established. It is important to convince users that the economics are favorable, and manufacturers that there will be no major technical problems in scaling up the processes. The exemption of gasohol from the Federal highway excise tax under the National Energy Act will provide an incentive for commercialization; so will the provision of additional investment tax credits for solar capital investment under the same Act. Technologies to be commercialized in the mid-term (1985–2000) include improved methods for producing gaseous fuels, fermenting woody plant matter to produce alcohol, and growing herbaceous crops. These technologies are in the technology development and engineering development phases. Technologies to be commercialized in the long term (after 2000) include energy and fuel production from exotic plants, applications using aquatic biomass, and advanced fermentation. All these are currently in the applied research stage.

A brief description of DOE's approach to commercialization and further details on the commercialization of solar technologies are presented in Chapter 8 of this part.

Environmental Aspects of Solar Technologies

Issues. Most of the direct environmental impacts of solar thermal, solar photovoltaic, wind energy conversion systems (WECS), ocean thermal energy conversion (OTEC) and fuels from biomass are of minimal concern. Solar thermal and solar photovoltaic systems may release cleaning materials, optical coatings, and thermal components to the atmosphere, and the release of detergent solutions, optical coatings, and vegetation suppressants results in leaching of the soil, leading to surface and ground water contamination. The impacts of photovoltaic systems are usually associated with the production of cells. WECS can create electromagnetic wave interference, obstruct flow, and create moderate to extremely loud noise. The environmental impacts of OTEC include the release of working fluids, chlorine, lubricants, and corrosion products to the water. One impact from biomass is the release of stack gases from biomass conversion and combustion operations. Other impacts include waste water discharges and runoff, which can cause surface and ground water problems, the disposal of large amounts of waste products, and the potential for widespread ecosystem destruction, with extensive removal of vegetation.

Environmental Activities. Activities in solar technologies consist of two major tasks: environmental and resource assessments. These tasks reflect the environmental issues discussed previously. Under the Environmental Assessments Task, program assessments of environmental factors for individual solar technologies have been compiled. These include ongoing studies of ocean thermal systems, fuels from biomass technologies, and health and safety assessments of materials/components of photovoltaic cells. Other programmatic environmental assessments will also be completed during the current fiscal year. In addition, the Solar Environmental Development Plans governing environmental aspects of solar technologies will also be completed in the current fiscal year.

Specifically, the activities will be centered in the following areas:

- Acquisition, analysis, and dissemination of insolation and spectral data; wind direction and speed; ocean thermal; and design and siting requirements of solar thermal applications

- Preparation of programmatic impact assessments of solar technologies and technology assessments of unintended or unanticipated societal effects that could occur from widespread application of solar systems

Socioeconomic Aspects of Solar Technology

Because energy is basic to the economy, new energy systems must be widely produced and marketed to achieve a significant impact on overall energy consumption. The efforts to introduce inexhaustible energy sources, including RD&D, constitute essential elements of the National Energy Act. As stated in the National Energy Plan, "Some technologies, such as solar hot-water and space heating, can make contributions now, . . . others, such as solar electric technologies . . . have great promise for the future."

Significant opportunities exist for accelerating the deployment of solar energy options. The rate at which the solar market and industry will develop is strongly influenced by the relative economics of alternative sources of energy; these are, in turn, affected by the technological advances in solar systems and Government (Federal, State, and local) policies toward energy use. In general, this rate depends on regional and national decisions related to pricing of conventional energy sources, the support provided for solar systems development, and the perceived impacts on society of all forms of energy use.

The best role for the Federal Government is the promotion of sound market growth. Clearly, the development and marketing of new energy systems is a function that can best be performed by the private sector. Proponents of economic efficiency and individual freedom of choice argue that government pricing policies toward crude oil and natural gas have served merely to inhibit the development of solar technologies. As demand for these fuels increases, effecting an increase in consumer costs, the economic process through which solar technologies become competitive will be accelerated. Under these circumstances, the value that society places on a given fuel, as reflected in its price, fully covers society's cost of producing that fuel. This economic arrangement is "equitable" and "fair." It minimizes the cost of producing all forms of energy, and it raises productivity, economic growth, and employment through the more efficient use of energy resources. Finally, it helps in the fight against inflation through the conservation of energy resources and also by reducing the size of the Federal budget deficit.

4.
Geothermal

Geothermal energy is the heat stored in the Earth that can be recovered by using current or near-current technology. Geothermal energy can be used to produce electric power, as well as for residential and commercial space heating and cooling, industrial process heat, and agricultural applications.

Three principal types of geothermal resources can be exploited through 2000. In order of estimated magnitude, these resources are:

- Hydrothermal
- Geopressured (including dissolved natural gas)
- Hot dry rock

Hydrothermal resources, systems of hot water or steam or both, heated by contact with relatively shallow hot masses of rock, are trapped in fractured rocks or porous sediments overlaid by impermeable surface layers. When such systems are tapped by drilling, the geothermal fluids become usable.

Geopressured resources, located primarily in sedimentary basins along the Gulf Coast, consist of water and dissolved methane at high pressure and moderately high temperature. These resources have the potential for recoverable methane, thermal energy, and mechanical energy.

Hot dry rock resources are geologic formations at accessible depths that have abnormally high heat content but contain little or no water. Thus, to extract the usable energy, heat transfer fluid, such as water, must be circulated through fractures made in the rock.

Table III 4-1 provides the funding levels by programs for FY 1978 through FY 1980.

The following sections describe in detail each of the three geothermal programs: (1) Hydrothermal Resources, (2) Geopressured Resources, and (3) Geothermal Technology Development. The descriptions of these programs are structured according to the FY 1980 budget; i.e., the subprograms that make up each program are described briefly in individual modules. Each module includes a general description of relevant technical aspects, an illustration (where appropriate), a summary of project status, and a milestone chart which presents major decision points through FY 1984 and funding levels for FY 1978 through FY 1980.

Finally, after each geothermal program is described in terms of its subprograms, other program aspects are described in order to provide an overview of program activities which are not part of any specific subprogram. For instance, because the commercialization of hydrothermal energy resources has been transferred to the Assistant Secretary for Resource Applications, that information has been included in the discussion of geothermal commercialization. Other program aspects also include regional activities, environmental aspects, interagency coordination/planning, technology transfer, and the international geothermal energy program.

HYDROTHERMAL RESOURCES

This program provides RD&D support to the hydrothermal commercialization activities

Table III-4-1. Program funding levels FY 1978 through FY 1980.

GEOTHERMAL ENERGY PROGRAMS	BUDGET AUTHORITY (DOLLARS IN THOUSANDS)			
	ACTUAL FY 1978	ESTIMATE FY 1979	ESTIMATE FY 1980	INCREASE (DECREASE)
Hydrothermal Resources	55,000	70,900	59,100	(11,800)
Geopressured Resources	16,400	27,700	36,000	8,300
Geothermal Technology Development	34,400	57,600	43,900	(13,700)
TOTAL	105,800	156,200	139,000	(17,200)

transferred to the DOE Assistant Secretary for Resource Applications. Basically, this support includes:

- Confirmation of geothermal reservoir prospects and definition of their characteristics in cooperation with industry
- Field experiments for geothermal direct heat applications to demonstrate the practical and economic aspects of the use of moderate temperature geothermal resources by non-Federal entities; the participants are selected by competitive solicitation that requires cost-sharing
- Experimental facilities, maintained to perfect equipment, materials, and techniques designed to resolve technical and/or economic impediments to the exploitation of hydrothermal resources, particularly for generating electric power
- A demonstration plant that will produce 50-MW of electric power from high-temperature, moderately saline geothermal fluids. This plant will establish the technical and economic viability of the technology at full commercial scale, built and operated by industry; the cost-shared project was awarded on the basis of responses to a competitive solicitation

Table III 4-2 provides the funding levels by subprograms for FY 1978 through FY 1980.

Regional Planning

The Regional Planning Subprogram is designed to accelerate commercialization of hydrothermal resources for direct heat applications and electric power generation. This will meet national power-on-line goals by:

- Supporting site-specific participatory planning
- Providing incentives
- Removing institutional barriers
- Establishing economic feasibility
- Defining market penetration of geothermal energy as a viable alternative to imported fossil fuels

This subprogram has now been placed under the Assistant Secretary for Resource Applications. Thus, further discussion of this subprogram in this document is not appropriate. The funding for Regional Planning is shown in Table III 4-2.

Resource Definition

A quantitative assessment of the regional and national hydrothermal resource potential and the confirmation of hydrothermal prospects through a program of selective drilling are crucial to attaining near-term goals for hydrothermal energy. The objectives of the Resource Definition Subprogram are as follows:

- Evaluate the hydrothermal resource potential within the United States
- Determine the geographical distribution of the resource
- Confirm the existence and commercial potential of high-temperature reservoirs suitable for electric power generation
- Assess and confirm low- and moderate-

Table III-4-2. Funding levels by subprograms FY 1978 through FY 1980.

HYDROTHERMAL RESOURCES SUBPROGRAMS	BUDGET AUTHORITY (DOLLARS IN THOUSANDS)			
	ACTUAL FY 1978	ESTIMATE FY 1979	ESTIMATE FY 1980	INCREASE (DECREASE)
*Regional Planning	3,900	5,845	6,100	255
Resource Definition	14,400	25,470	9,000	(16,470)
**Engineering Applications	7,800	10,500	9,831	(669)
Environmental Control	1,600	516	1,300	784
Facilities	26,100	27,169	32,069	4,900
Capital Equipment	1,200	1,400	800	(600)
TOTAL	55,000	70,900	59,100	(11,800)

*The Hydrothermal Commercialization effort is transfered to the Assistant Secretary for Resource Applications
**Includes $4 million for second demonstration plant. DOE is considering whether to request Congressional authorization for this plant in FY 1979, or to request authority for the reprogramming or recision of these funds.

temperature prospects that show potential for direct heat application

The strategy selected for achieving these objectives is to perform regional and national assessments of the hydrothermal resource in cooperation with the U.S. Geological Survey to confirm high-temperature reservoirs with near-term commercial potential in cooperation with resource developers and to confirm low- and moderate-temperature reservoirs in cooperation with individual states. In addition, a "pre-commercial" program has been initiated to finance exploratory drilling where areas of highly speculative prospects precede commercial interest.

The status of the Resource Definition Subprogram is as follows:

- National Geothermal Resource Assessment Update—With DOE support, the U.S. Geological Survey has completed a major update of its assessment of U.S. geothermal resources. The result will be published in January 1979 as U.S.G.S. Circular 790. The study identifies very large energy contents for hydrothermal geopressured and hot dry rock resource bases.

- Industry-coupled Case Study—In FY 1978, the Industry-coupled Case Study was begun to accelerate confirmation of geothermal reservoirs with apparent commercial potential for producing electricity. Exploratory drilling costs are shared with industry in exchange for publication of reservoir data. Nine companies are participating. This activity will be concluded in FY 1980, at which time adequate reservoir capacity to support the 1985 goals for geothermal electrical power will have been confirmed. The National Energy Act of 1978 contains incentives to accelerate development of hydrothermal resources through tax credits, expensing of intangible drilling costs, and depletion allowances. These incentives are expected to stimulate further industrial exploration and confirmation of hydrothermal reservoirs.

- State-coupled Program—Presently, low- and moderate-temperature resources for direct heat applications are being defined in cooperation with 28 of the 37 States that have been identified as having resource potential. This effort consists of two phases. In Phase I, existing geological and geophysical data are analyzed to establish the probability and distribution of hydrothermal resources. As promising resources are identified, Phase II is initiated to provide a more detailed assessment of target areas. This phase may include the drilling of deep holes to confirm the existence and nature of the resources. An example is the program being conducted in the Atlantic Coastal Plain. Scientists from the Virginia Polytechnic Institute, eight participating States, and the U.S. Geological Survey have delineated a number of probable reservoir targets as part of the Phase I activities. Beginning in FY 1979 and extending into FY 1980, deep test wells will be drilled to provide a basis for estimating resource potential along the East Coast.

- Pre-commercial Program—This cooperative effort with the U.S. Geological Survey confirms the existence of suspected hydrothermal reservoirs. Included are drilling at Mt. Hood, Ore. and the Snake River Plain, Idaho. The program will be discontinued in FY 1980 because new incentives in the NEA are expected to encourage industry to pursue the confirmation of hydrothermal resources.

Milestones and funding for Resource Definition are shown in Figure III 4-1.

Engineering Applications

To date, the private sector in the United States has made little use of geothermal energy for nonelectric purposes. Nevertheless, there is a large potential market in industrial sectors such as processing, agribusiness, and space and water heating in commercial and residential buildings. Demonstration projects (field experiments) are needed: (1) to provide visible evidence of the profitability of various nonelectric applications in a number of geographical regions, and (2) to obtain reliable, objective, and definitive technical and economic data under field operating conditions.

To demonstrate the economic and technical viability of direct utilization of geothermal energy resources by actual field experiments, the first solicitation for direct-use field experiments was issued during summer 1977, and 22 proposals were received. Eight of these proposals were selected for subsequent con-

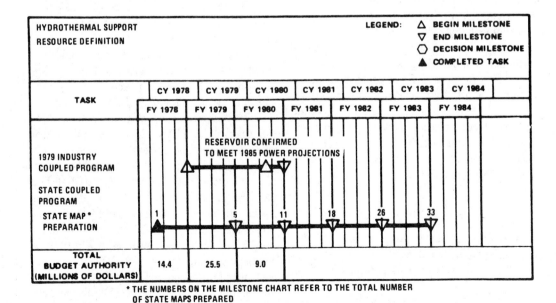

Figure III-4-1. Milestones and funding for resource definition.

tracts, with the Government's share of the cost varying from 46 percent to 80 percent. A total of $2.1 million was obligated in FY 1978 for the initial phases of these projects, resulting in commitments of $2.9 million in FY 1979 and $650 thousand in FY 1980 to complete these projects.

A second solicitation, issued in April 1978, resulted in 40 proposals, 15 of which were selected for initial FY 1979 funding of approximately $4.9 million. Consequently, approximately $1.0 million will be required in FY 1980, and approximately $5 million in 1981, to continue projects started in FY 1979. Government cost-sharing generally amounts to approximately 60 percent of total project cost.

Of the 23 contracts expected to be underway in FY 1979, the majority are for space heating, while a few are directed at agriculture and aquaculture, and three involve industrial processing. Although most of the projects to date are in the Western States, greater emphasis will be placed on locating future demonstration sites in the East, as suitable geothermal resources are defined there.

Milestones and funding for Engineering Applications are shown in Figure III 4-2.

Environmental Control

The Environmental Development Plan (EDP) for Geothermal Energy Systems* identified environmental, health, and safety issues affecting geothermal development. Release of hydrogen sulfide (H_2S) is a major air quality problem which causes corrosion and odors. Withdrawal of fluids from geothermal reservoirs may cause another serious problem, namely, land surface subsidence. In addition, potential seismic disturbances may result from geothermal fluid extraction and injection processes. These and other issues are discussed below.

Air

- Hydrothermal fluiids containing dissolved, noncondensible gases, such as hydrogen sulfide (H_2S), ammonia (NH_3),

---
*U.S. Dept. of Energy "Environmental Development Plan (EDP) Geothermal Energy Systems 1977," DOE/EDP-0014, March 1978.

boric acid (H_3BO_3), and radon (Rn), could produce toxic effects. Hydrogen sulfide could also produce a disagreeable odor.

Water

- Disposal of hydrothermal and geopressured fluids, which may contain toxic substances and large quantities of dissolved solids, is a major waste management problem.
- Geothermal brines may affect ground water quality due to accidental surface spills and leaks in underground pipes.

Land

- Land subsidence may occur when large quantities of fluids have been extracted from the ground.
- Earthquakes may be induced by the injection of water at high pressures.

Health and Safety

- Well blowout may occur during geothermal development activities.
- Noise could result from the drilling of wells during the exploratory and development phases and from venting during the testing phase.

DGE has examined each of the major issues affecting geothermal development, and environmental control research programs have been established as needed. Issue definition studies are being conducted in liquid waste disposal, noise generation, and well blowout. The first two studies were begun in cooperation with the Division of Environmental Control Engineering (EV). Environmental control research programs currently deal with H_2S emissions. One such system, a scrubber using copper sulfate, was successfully tested at The Geysers geothermal field in FY 1977. Removal efficiencies for H_2S exceeded 99 percent with this scrubber; it also extracted NH_3 and H_3BO_3 in large quantities. In FY 1979, a commercial-scale scrubber will be tested under operating conditions at The Geysers. Should this test prove successful, the system will be ready for full-scale commercialization.

In addition to environmental control stud-

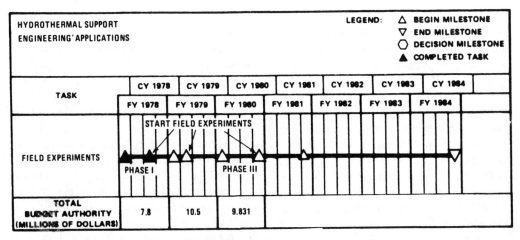

Figure III-4-2. Milestones and funding for engineering applications.

ies, DGE sponsors environmental monitoring for each of its major field projects. The monitoring is done in part to support preparation of Environmental Impact Assessment/Statements (EIA/EIS) as required by NEPA regulations. To assist with EIA/EIS preparation, DGE has developed guidelines for environmental reports by contractors. Finally, as a means to promote geothermal development, DGE encourages the adoption of consensus standards and environmental performance criteria. The work in standards definition is done in cooperation with the Division of Operational and Environmental Compliance.

The funding for Environmental Control is shown in Table III 4-2.

Facilities

Raft River Pilot Plant. This project is a 5-MW binary cycle plant that uses a Rankine cycle to convert energy from a moderate-temperature hydrothermal resource (300°F) to electric power. The results from this plant will supply valuable information on the geothermal reservoir and plant equipment, operations, and economics for future commercialization of moderate-temperature resources.

Plant design has been completed, and construction began in FY 1978. Major long-lead equipment will be delivered by February 1980. The supply and injection wells and surfacce piping have been completed. Testing of the wells has begun; operation of the plant is scheduled to begin early in FY 1981.

Hawaii Geothermal Wellhead Generator. The objective of this project is to establish the feasibility of using wellhead generators to produce base-load electrical power. The technology would be applicable in remote areas of the Western States and in Hawaii.

The 3-MWe test generator will be supplied by an existing geothermal well in the Puna district of Hawaii. Major plant components are portable.

Plant design has been initiated, and construction began in the third quarter of FY 1979. Operation will start in FY 1980 and continue for about two years.

Geothermal Loop Experimental Facility (GLEF). This facility will establish the feasibility of flash-steam/binary systems in the production of electric power from high temperature/high salinity resources. The project is cost-shared on an equal basis with the San Diego Gas and Electric Co.

The first GLEF was constructed in 1975. The facility operated 1,000 hours before it was shut down for removal of accumulated scale. Present scale removal techniques take approximately two weeks. Faster techniques are being evaluated. New plant designs incorporating redundant flash trains can increase the plant's production capacity from well below 75 percent to over 85 percent and reduce energy production costs to below 38 mills/kWh in the 50-MWe size.

Recent studies have produced effective pre-injection treatments to eliminate injection clogging problems. The new system will be installed in FY 1979. At the same time the power cycle will be modified to incorporate a direct-contact heat exchanger. The system will be tested in FY 1980.

Geothermal Component Test Facility (GCTF). This facility provides high-temperature, moderate-to-low salinity geothermal fluid and supporting services to experimenters for R&D testing of equipment and components to be used in advanced geothermal systems.

The GCTF is currently being used by several industrial firms; it will be operational until demand diminishes.

Demonstration Plant. The major purpose of this activity is to design, construct, and operate a commercial-size, geothermal electrical powerplant in the United States using a liquid-dominated hydrothermal reservoir. This is intended to provide maximum stimulus to non-Federal development in the United States of the widest spectrum of liquid-dominted hydrothermal resources that can be used for generating electricity.

The plant will demonstrate that producing electric power from geothermal resources in the United States can be economical, environmentally sound, and socially acceptable. The project will address the following objectives:

- Demonstrate reservoir performance characteristics of a specific liquid-dominated hydrothermal reservoir
- Demonstrate the validity of reservoir engineering estimates of reservoir productivity (capability and longevity)
- Demonstrate commercial-scale energy conversion system technology
- Provide Federal assistance to initiate commercial development at a resource site with great potential
- Act as a "pathfinder" for the regulatory process and other legal and institutional aspects of geothermal development
- Provide the financial community with a basis for estimating the risks and benefits associated with geothermal investments

The plant will generate statistically reliable engineering and cost data on the performance of a reservoir and on plant construction, operation, and maintenance for a 5-year period. This in turn will demonstrate predictable technical, economic, and environmental performance with acceptable risk at a commercial scale.

If the information derived from the project is to be useful, Federal involvement must not distort the data from normal business practices. Therefore, DOE will delegate management of the demonstration project to industrial participants. DOE will also ensure that the project provides a realistic basis for the private sector to assess commercial feasibility.

Under a cooperative agreement, both DOE and the competitively selected industrial participant will finance the design, construction, and operation of a 50-MWe commercial-scale geothermal demonstration plant. DOE plans and budget, and legislative communications have been based on an assumption of equal cost-sharing. According to plans, equal cost-sharing will not be extended to overruns; the award instrument will contain maximum costs on plant and well field capital and operating costs for which the Government will be obligated. It is anticipated that the title to facilities and equipment will remain with the industrial participant, who will be given the right of first refusal to acquire any Federal equity if the Government stops participating in the project. The arrangement also involves paying back to the Government its portion of the investment.

On September 30, 1977, ERDA issued Geothermal Demonstration Powerplant Program Opportunity Notice (PON) EG-77-N-03-1717. Proposals for the plant were submitted by the closing date of January 31, 1978. On July 6, 1978, the Under Secretary determined, following presentation of the Source Evaluation Board findings, that negotiations for the construction and operation of the 50-MWe commercial-scale geothermal demonstration plant should be initiated with Union Oil Company of California and the Public Service Co. of New Mexico, under a cost-shared Cooperative Agreement. The proposed site is the Baca Ranch in the Valles Caldera, N. Mex.

A cooperative agreement was executed between DOE, Union Oil, and Public Service Co. of New Mexico on September 29, 1978. The final Cooperative Agreement is scheduled to be completed on January 31, 1979, and the objective is to operate the plant by the second quarter of FY 1982.

Milestones and funding for Hydrothermal Support Facilities are shown in Figure III 4-3.

GEOPRESSURED RESOURCES

Geopressured resources, hot water aquifers contained at high pressures, may contain appreciable amounts of dissolved methane gas. These resources offer the potential for significant methane recovery, as well as for associated thermal and kinetic energy. Figure III 4-4 illustrates a possible way of recovering all three energy forms in the geopressured extraction process.

Dissolved and associated methane is the major target in the recovery process. Thermal and kinetic energy from the fluids produced will provide additional energy. The principal uncertainties to be resolved are: (1) the number, location, permeability, and size of individual aquifers; (2) their ability to sustain large flows of water over long periods of time; (3) the amount of recoverable methane; (4) the means and cost of disposing of large quantities of saline water; and (5) a means for preventing land surface subsidence.

Reservoir modeling and mapping of the Gulf States has delineated much of the resource and identified numerous candidate sites for exploration. Resource studies of this type are continuing, using existing well logs and seismic survey data to determine geopressured zone boundaries, temperatures, pressures, and gas content. Extensive data from thousands of deep wells, drilled originally for oil and gas, prove the existence of these high-pressure, high-temperature zones and identify the geo-

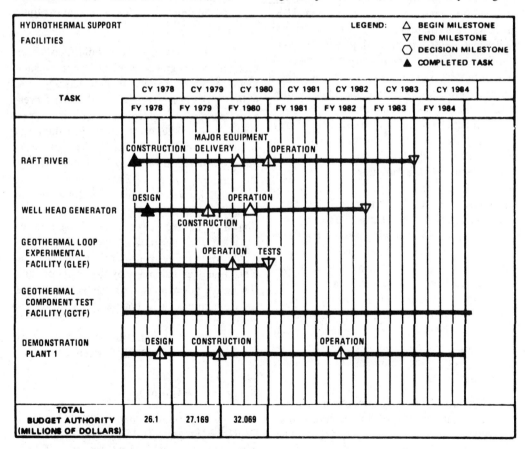

Figure III-4-3. Milestones and funding for facilities.

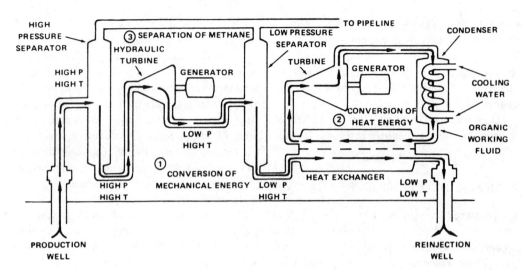

Figure III-4-4. Geopressured energy recovery.

logic conditions under which they occur. These studies indicate the possibility that thousands of trillions of cubic feet of methane are dissolved in the brine contained within geopressured sandstone aquifers. Current U.S. methane consumption is about 20 trillion cubic feet per year; thus, this resource base constitutes a major energy target. The actual volumes and permeabilities of the aquifers, however, have not been demonstrated by prolonged flow tests. Consequently, the economic practicality of recovering their energy is not yet known, and the quantity that might be recovered is highly uncertain.

The focus of the program is to perform flow tests of wells drilled into geopressured reservoirs. The results will reveal the basic drive mechanisms that cause fluid and gas production, thereby allowing reservoir longevity to be estimated. Because the characteristics of geopressured aquifers vary widely, and because reservoir performance under production conditions is a complex phenomenon, a substantial number of tests will be required before the potential of the total resource can be estimated reliably. The program will provide answers to the following key technical and economic questions:

- How much fluid and how much methane will be produced over the economic lifetime of a well drilled into a particular geopressured aquifer?
- What is the smallest reservoir that can be produced economically?
- What fraction of the resource base is contained within reservoirs that exceed the minimum size for economic exploitation?

After the geopressured fluid has been brought to the ground surface, the associated methane can be recovered with existing technology. The conversion of the heat and pressure energy to electricity (or the direct application of the heat) apparently does not present significant technical barriers. In Texas and Louisiana the recovery of thermal energy is required by law. Some improvements, however, in well completion technology and in methods for managing the reservoir production may be required. Another issue to be considered is disposal of the brine, which would be produced in large volumes. The most practical method may be injection into subsurface formations, but the economics of such operations have not yet been firmly established.

Table III 4-3 provides the funding levels by subprograms for FY 1978 through FY 1980.

Resource Definition

The purpose of the geopressured Resource Definition Program is to determine the location, size, temperature and pressure of geo-

Table III-4-3. Funding levels for subprograms FY 1978 through FY 1980.

GEOPRESSURED RESOURCES ACTIVITIES	BUDGET AUTHORITY (DOLLARS IN THOUSANDS)			
	ACTUAL FY 1978	ESTIMATE FY 1979	ESTIMATE FY 1980	INCREASE (DECREASE)
Resource Definition	1,900	1,900	1,350	(550)
Existing Well Tests	3,900	3,600	6,950	3,350
New Drilled Wells	8,200	18,500	22,900	4,400
Reservoir Engineering	900	1,000	1,000	
Other*	1,500	2,600	3,500	900
Capital Equipment		100	300	200
TOTAL	16,400	27,700	36,000	8,300

*Includes planning, engineering applications and environmental conformance.

pressured sandstone aquifers on the U.S. Gulf Coast. This complex analysis is based primarily on geophysical log data from tens of thousands of wells drilled into geopressured formations in the course of past exploration for gas and oil. Seismic exploration data and core data from the wells are also used. After the raw data have been interpreted by geoscientists, maps and cross sections are produced showing individual reservoirs, distributions of temperature and pressure, and permeability information.

The mapping of onshore geopressured formations in Texas is almost complete; the mapping of onshore and parts of offshore Louisiana formations to depths of 19,000 feet has been completed. Also, over 70 large geopressured reservoirs have been identified. The continuity of these reservoirs, however, has not been established; many may be divided into smaller units by offsetting subsurface growth faults. The critical question of size distribution of geopressured aquifers is therefore not yet fully addressed by the Resource Definition Subprogram. Size remains an important consideration because the smaller aquifers may not constitute economically attractive targets.

Although these maps allow a reasonably accurate estimate of the volume of fluid contained within the sandstone aquifers, they do not reveal the amount of natural gas content and do not shed much light on the questions of recovering methane and possible sinking or subsidence of land surface.

Mapping efforts, including the assembly of a computerized data base, will continue in FY 1979 and FY 1980. Emphasis will be placed on higher resolution of the larger identified reservoirs, mapping deeper formations, and extending the work to previously unmapped areas.

Also the sites proposed for tests of existing wells and for new drilled wells will be evaluated within this subprogram.

Milestones and funding for Resource Definition are presented in Figure III 4-5.

Existing Well Tests

A substantial number of data points are needed to define the resource base because the area of the Gulf Coast geopressured resource is up to 800 miles long and 300 miles wide and because the characteristics of the resource vary across the region. The quickest and least expensive way to obtain geopressured aquifer production data is to use existing wells that have been drilled to search for oil and gas. Many such wells penetrate Gulf Coast geopressured formations. Both "dry" wildcat wells and production wells that have depleted their conventional gas pockets are candidates. Some candidate wells have been shut in and

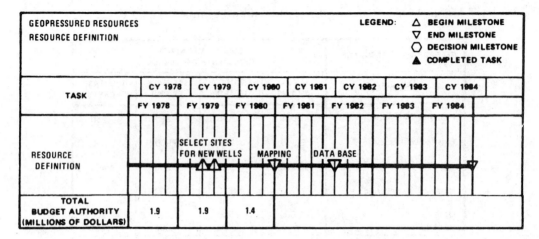

Figure III-4-5. Milestones and funding for resource definition.

plugged; others (wildcats) are in the planning or drilling stages and could be converted for geopressured aquifer tests when presently scheduled operations are completed.

The manner in which these wells are drilled and completed makes them unsuitable for the high-volume fluid production believed necessary for the economic production of geopressured brines containing dissolved methane. However, these wells can provide useful information at lower production rates in tests lasting a few weeks. The re-entry of an old well is economically risky because the deterioration and corrosion resulting from years of service may require expensive repairs. Consequently, the final cost of such a job is always uncertain until the well has been re-entered.

The first publicly documented production test of a geopressured aquifer, conducted under Division of Geothermal Energy (DGE) sponsorship in the Edna Delcambre No. 1 well in Vermillion Parish, La., produced encouraging results. The cost of reentering, refurbishing, and testing this well was more than $3 million. For approximately two weeks, production tests were conducted in two separate geopressured aquifers penetrated by the well. Flow rates of up to 12,000 barrels per day were achieved. Gas recovery varied up to about 80 scf/bbl (standard cubic feet per barrel of fluid). Because laboratory tests showed that only about 22 scf/bbl could be contained in solution at reservoir conditions, the test raised questions about the source of the additional gas. Possible explanations include:

- The presence of a small gas cap
- Leakage outside the well casing from a gas bearing formation
- Preferential production of solution gas

There is inadequate information to distinguish between these hypotheses. Only additional tests of wells in other areas will reveal whether the Delcambre well performance was typical.

DGE plans to test above five existing wells in FY 1979 and five per year thereafter. Attempts to renovate two existing wells in Louisiana failed early in FY 1979, and the wells were abandoned.

Figure III 4-6 presents milestones and funding for Existing Well Tests.

New Drilled Wells

Long-term (2 years) fluid production tests at high flow rates are required to properly define the productive capacities and longevities of large geopressured reservoirs. When proven mathematical models are available, it will be possible to reliably predict productivity and longevity from short-term initial well test data, but long-term tests are essential to the development of the models. Such tests are also needed to determine the technical and economic aspects of the disposal of large amounts of produced brine and to define the

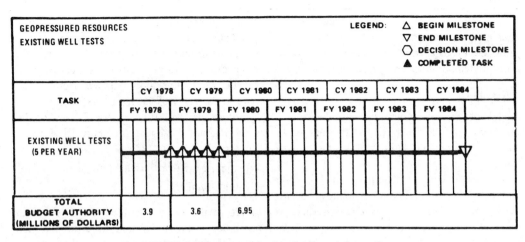

Figure III-4-6. Milestones and funding for the existing well tests.

probability of land surface subsidence triggered by the compaction of the geopressured aquifers.

Conventional oil and gas wells, designed for different production conditions, are not suitable for long-term tests of fluid production at high flow rates. Their basic designs, however, with larger diameters to accommodate higher flow rates, are expected to be suitable for the long-term tests. No new technological advances are believed necessary for drilling and testing such wells.

The first geopressured aquifer test well is being drilled at Pleasant Bayou in Brazoria County, Tex. Prospective sites for several other drilled wells are currently under evaluation in Louisiana and Texas. DGE plans to initiate drilling at two of these sites in FY 1979 and subsequently to drill about five wells each year.

Milestones and funding for New Drilled Wells are presented in Figure III 4-7.

Reservoir Engineering

The Reservoir Engineering Subprogram is aimed at: (1) determining the behavior of geopressured aquifers under production conditions, (2) establishing the recoverability of methane, and (3) developing a predictive capability for reservoir performance and longevity. Two basic elements involved in the establishment of a predictive capability are rock mechanics and numerical simulation.

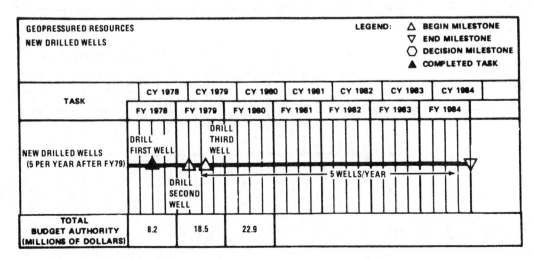

Figure III-4-7. Milestones and funding for new drilled wells.

The rock mechanics effort defines the mechanical properties of the reservoir rock in the presence of reservoir fluid under the conditions of temperature and pressure that prevail in the reservoir.

The numerical simulation is a computerized model of the reservoir that calculates flow rate, temperature, pressure, and the amount of gas in the produced fluid, all as functions of time.

A nearly completed experimental project will define the solubility of methane in brines as a function of salinity, temperature, and pressure. This will allow a more precise estimate of the methane contained in solution in geopressured aquifers.

A rock mechanics laboratory has been set up to measure the mechanical properties of cores from geopressured formations under conditions of elevated temperature and pressure, in the presence of gas-saturated brines. The data obtained will be relevant to the questions of reservoir mechanics and of possible formation compaction, a precursor to land surface subsidence.

The rock property data will be used in computer codes that predict the lifetime production from a geopressured well based on initial short-term production results and on core and log data. Similar computer codes are used by the petroleum companies to develop conventional gas reservoirs. The computerized theoretical models, however, must be refined by comparing their predictions to actual well production data. The models for geopressured reservoir production will have to undergo the same process before industry accepts them as reliable productive tools that will significantly reduce the risks to investigators.

Milestones and funding for Reservoir Engineering are presented in Figure III 4-8.

GEOTHERMAL TECHNOLOGY DEVELOPMENT

Geothermal energy can be exploited with technology similar to that used for oil and gas. Oil field and water well equipment can be used safely and economically for some low-temperature geothermal applications, but moderate- and high-temperature geothermal resources have conditions that exceed the capabilities of existing equipment.

The objective of the Geothermal Technology Development Program is to remove the technical impediments to geothermal exploitation by encouraging industry to develop and market equipment and technology for all geothermal environments. The program consists of categories that follow the discovery and exploitation activities of a geothermal resource. Drilling and completion technology improvements could reduce the cost of geothermal wells 25 percent by 1983 and 50 percent by 1986. This would affect the projected 8,000 wells that must be drilled in order to

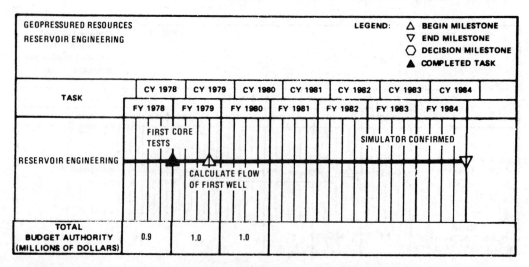

Figure III-4-8. Milestones and funding for reservoir engineering.

bring 20,000-MWe of geothermal power online. The Energy Extraction, Conversion, and Stimulation Technology Program is developing pumps, heat exchangers, and systems to use moderate-temperature geothermal fluid for economical production of electricity. Stimulation is a way to increase production from an individual well, thereby reducing the number of wells required to exploit a reservoir. The Geochemical Engineering and Materials Subprogram addresses the special character of geothermal fluids and their interaction with other materials. Program areas include fluid handling to control scale formation and injection well plugging, high-temperature seals, new instrumentation sensors for high-temperature corrosive environments, and development of improved materials for geothermal service, such as well cement. The Geosciences Subprogram focuses on improving the technologies for exploration, reservoir engineering, logging instrumentation, and log interpretation. The Hot Dry Rock (HDR) Subprogram assesses the potential of the HDR resource and supports development of new technical approaches for extracting HDR energy content. Presently, this subprogram consists of a successful experimental loop at Fenton Hill, N. Mex., and an assessment of the HDR potential on a national basis.

Table III 4-4 presents the funding levels of subprograms for FY 1978 through FY 1980.

Drilling and Completion Technology

This subprogram will reduce the high cost of geothermal drilling in two stages. An early impact is anticipated as a result of improving traditional rotary drilling technology for a 25 percent reduction in the cost of geothermal drilling by 1982. This reduction would come primarily from improved drill bits, down-hole motors, and improved drilling fluids. A 50-percent reduction in drilling costs will require the development of a new drilling system. An example of an improved drill bit is the Stratapax bit, which incorporates manmade diamonds. One stud from such a bit is shown in Figure III 4-9.

The status of major efforts aimed at meeting the well cost-reduction goals is as follows:

- A diamond chain-drill bit capable of high penetration rates in hard igneous rocks, typical of many geothermal areas, has been successfully designed and labo-

Table III-4-4. Funding levels by subprograms FY 1978 through FY 1980.

GEOTHERMAL TECHNOLOGY DEVELOPMENT	BUDGET AUTHORITY (DOLLARS IN THOUSANDS)			
ACTIVITIES	ACTUAL FY 1978	ESTIMATE FY 1979	ESTIMATE FY 1980	INCREASE (DECREASE)
Drilling And Completion Technology	2,300	6,000	7,000	1,000
Energy Conversion Systems And Stimulation	11,100	13,100	10,000	(3,100)
Geochemical Engineering And Materials	3,600	6,000	3,700	(2,300)
Geosciences	7,100	11,700	4,200	(7,500)
Hot Dry Rock	5,900	15,000	14,000	(1,000)
*Interagency Coordination And Planning	3,100	4,300	2,900	(1,400)
Capital Equipment	1,300	1,500	2,100	600
TOTAL	34,400	57,600	43,900	(13,700)

*Interagency Coordination and Planning Activities are being transferred to the DOE Assistant Secretary for Resource Applications

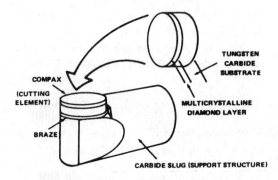

Figure III-4-9. Stratapax/Slug system.

ratory-tested. The bit hydraulically advances an indexed chain to bring new cutting surfaces into place down-hole without the costly removal of the entire drill string. The concept, if completely successful, could reduce total drilling costs by 10 percent or more. The first field applications of the diamond chain-drill bit were undertaken in FY 1979.

- Experimental roller cone bits which incorporate materials with high strength at elevated temperatures have been successfully field tested. Preliminary analysis indicates significantly improved performance.
- Several drag-type diamond bits incorporating manmade diamonds have been tested in the laboratory. Field tests were conducted in FY 1979.
- The geothermal application of down-hole drill motors requires the development of improved bearings and seals to operate under the severe conditions. This development has been underway since FY 1976, and field testing of the first improved seal/bearing assembly began in September 1979.
- Additional developments are underway in the areas of geothermal drilling fluids and well completion techniques.
- The development of an advanced drilling system will be initiated in a January 1979 workshop.

Milestones and funding for Drilling and Completion Technology are shown in Figure III 4-10.

Energy Conversion Systems and Stimulation

The objective of this subprogram is to reduce geothermal-electric generating costs, particularly for moderate-temperature geothermal

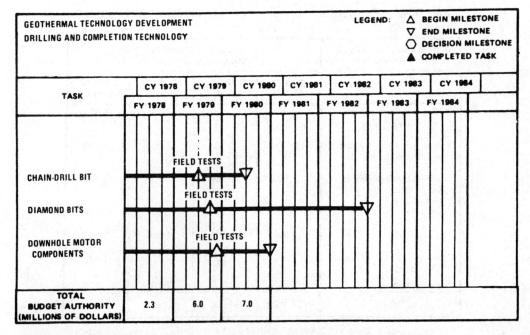

Figure III-4-10. Milestones and funding for drilling and completion technology.

resources. Energy conversion technology for moderate-temperature resources is being emphasized because they constitute a much larger resource base than high-temperature resources and are also inherently more expensive to produce. The greatest potential benefits are identified with binary conversion systems, toward which the programs are strongly directed. Heat exchangers account for 50 to 70 percent of binary plant costs and also strongly influence the efficiency of binary systems. Several current programs will improve performance and reduce costs of binary heat exchangers. Other efforts under this subprogram include testing a prime mover such as the helical screw expander, which is illustrated in Figure III 4-11. Wells account for up to 50 percent of geothermal power costs. Well-stimulation technology to increase productivity will have a highly beneficial effect by reducing the numbers of wells required. The high temperatures and geologic conditions found in geothermal reservoirs pose a very different set of requirements for stimulation technology than that which currently exists for oil and gas. Thus, new equipment and techniques to improve formation permeability are required, as are techniques for identifying geothermal reservoirs suitable for stimulation.

A 500-kWe skid-mounted binary power system employing direct contact heat exchangers was designed during FY 1978 and is presently being constructed. This system will be tested in FY 1979 at the GCTF.

A 100-kWe direct contact power system for low-temperature (210°F) geothermal applications is under construction and will undergo a field test by Arkansas Power & Light (AP&L) during FY 1979. Following successful completion of the 100-kWe test program, a 3-MW commercial system coupled to the AP&L grid will be designed in FY 1980.

A 60-kWe binary pilot plant has been constructed at Raft River and is presently operational. Performance testing will continue through FY 1979.

The first phase of a program to develop a 3- to 5-MWe binary well-head generator system, the "thermal pump binary," was initiated in FY 1978 and will continue through FY 1979. Additional phases including drilling of a large-diameter well and final design, construction, and testing will occur in FY 1980 and FY 1981.

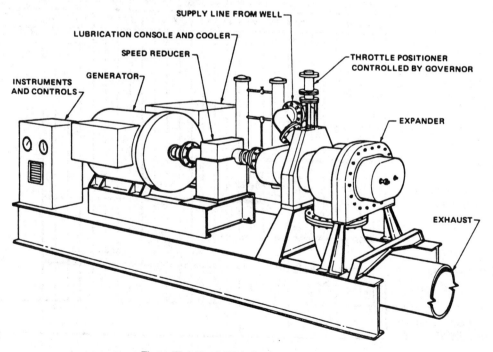

Figure III-4-11. 1-MW helical screw expander.

The initial test and evaluation of a 1.2-MWe helical screw expander well-head generator was completed at Roosevelt Hot Springs, Utah, in FY 1978. System modifications will be completed in early FY 1979 and the machine will be sent to Mexico, Italy, and New Zealand for additional testing and evaluation, in cooperation with the IEA (International Energy Agency).

Two major efforts in stimulation were initiated in FY 1979:

- A $4.5 million contract to plan, manage, and implement stimulation research and field testing which will support laboratory and field studies of formation acidizing and fracturing
- Explosives are being used to stimulate a Union Oil geothermal well at The Geysers to increase the steam flow rate from 80,000 to 150,000 lb/hr. This effort is cost-shared by DOE and several developers at The Geysers.

Milestones and funding for Energy Conversion Systems and Stimulation are presented in Figure III 4-12.

Geochemical Engineering and Materials

This subprogram seeks technical solutions for the problems of geothermal water, scale and gas control, as well as for material-related problems. Materials and chemistry considerations are closely related to individual resource geology and fluid characteristics. Plant and reservoir chemistry conditions must be carefully controlled to achieve overall economy in materials of construction, operation and maintenance.

Figure III 4-13 identifies technical management responsibility and technology transfer mechanisms for a typical materials development project.

During FY 1978, electric materials and sampling and analysis handbooks were completed. Industrial oversight of projects to im-

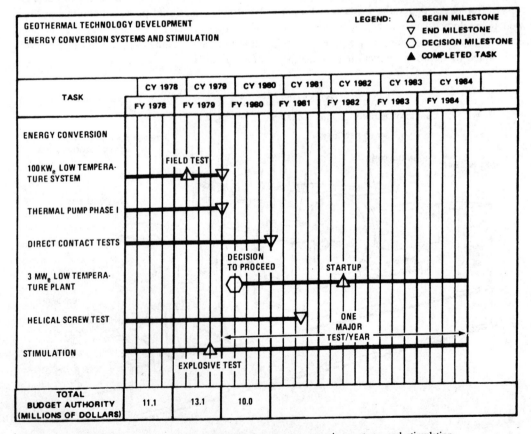

Figure III-4-12. Milestones and funding for energy conversion systems and stimulation.

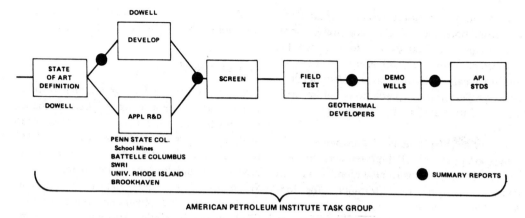

Figure III-4-13. Management responsibilities and typical technology transfer mechanisms.

prove materials in high-temperature cements and seals were initiated with the cooperation of the American Petroleum Institute (API) and the American Society for Testing and Materials.

Corrosion resistant polymer concrete and 260°C elastomer seal materials were successfully tested. Future plans include developing fluid monitoring and control instrumentation for fluid and gas handling and disposal, and developing technology necessary to establish fluid handling and systems materials standards.

Milestones and funding for Geochemical Engineering and Materials are presented in Figure III 4-14.

Geosciences

This subprogram is aimed at removing technical and economic barriers to geothermal resource exploration and assessment. The two principal areas of effort are: (1) exploration technology, to develop efficient methods for finding and assessing resources, and (2) logging technology, to develop reliable and accu-

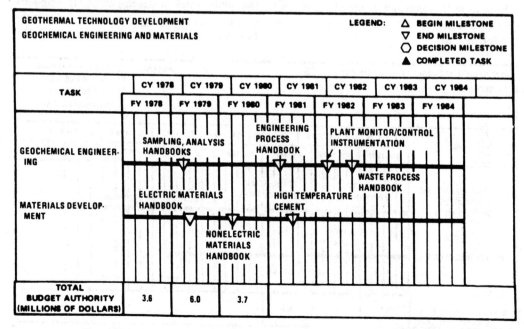

Figure III-4-14. Milestones and funding for geochemical engineering and materials.

rate methods to measure reservoir characteristics from boreholes. The exploration technology program objectives are to improve the cost-effectiveness and reliability of methods used to locate well sites, to establish accurate reservoir performance predictions before drilling, and to transfer the technology to the industry.

The principal elements of the logging technology program are: (1) high-temperature hostile environment instrumentation, and (2) advanced interpretation techniques for logging. The objectives of this program are to upgrade logging tools, cables, and seals to permit operation in temperatures up to 350°C, and to interpret the acquired data. At present, most instruments are rated to only 180°C, and interpretation technology is based on oil and gas exploration requirements that differ significantly from geothermal requirements.

The capability to build logging tools for 275°C operation has been accomplished through development of high-temperature electronic circuitry. These circuits and upgraded tools to measure temperature, pressure, and fluid flow were successfully demonstrated in a 275°C well at Valles Caldera, N. Mex. The program for log interpretation will provide calibration of test wells for industrial use, comparative commercial logging of test wells, and depositories for cores and log records.

Milestones and funding for Exploration Technology are presented in Figure III 4-15.

Hot Dry Rock

The hot dry rock (HDR) geothermal resource is qualitatively defined as the heat contained in rocks that do not contain sufficient water to qualify as hydrothermal systems. The HDR resource base is extremely large, and its exploitation could be significant to the Nation. Program objectives are: (1) to confirm technical feasibility and obtain economic data for HDR energy extraction systems in the commercial-size range by 1985, (2) to complete resource definition by 1982, and (3) to support commercial development, beginning in the early 1990s.

The technical key to the extraction of energy from one type of HDR resource is the establishment of a circulating fluid loop through the hot rock by fracturing the rock to provide a large heat exchange surface. This is a very difficult technical problem because circulating water may have to continuously produce fractures to expose fresh heat transfer surfaces. Figure III 4-16 shows the existing loop at the Fenton Hill site.

In FY 1978 the HDR Subprogram followed two major lines of development: (1) long-term operation of the energy extraction loop at the Los Alamos Scientific Laboratory Fenton Hill

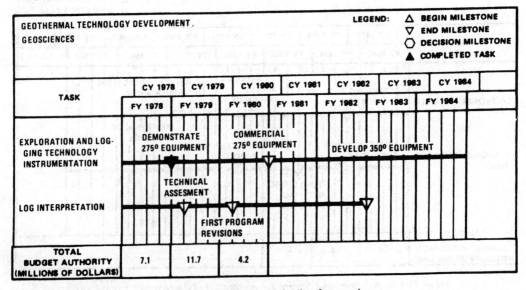

Figure III-4-15. Milestones and funding for geosciences.

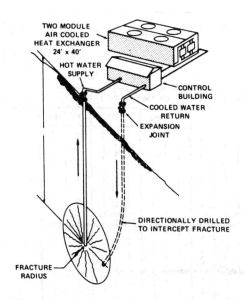

Figure III-4-16. Hot dry rock extraction experiment.

site and development of supporting technology, and (2) initiation of program planning, economic study, and resource assessment activities directed toward a national HDR program.

The thermal loop at Los Alamos Scientific Laboratory (LASL) operated for 3 months to assess temperature drawdown characteristics, acquire system performance data, evaluate fluid loss, and determine energy input requirements. The test showed that a thermal loop could be operated successfully.

In the context of a broader HDR Subprogram, regional and national activities were initiated to quantify and define the HDR resource. A draft plan for a national HDR program was developed, and technical and economic research on alternate HDR systems was initiated.

FY 1979 funding supports efforts leading to a determination by 1985 of the technical and economic feasibility of HDR as a significant new energy source. During this period, at least one new site will be developed to establish feasibility in a different geologic setting. In FY 1979, the HDR program will participate with the Hydrothermal Resources Program in the drilling of a deep hole on the Atlantic Coastal Plain. The existing resource definition activity will be expanded to meet HDR needs. The development of exploration, drilling, and logging technology specifically for HDR will be supported as required to meet program goals.

Figure III 4-17 presents milestones and funding for Hot Dry Rock.

Interagency Coordination and Planning

Although the Department of Energy has a lead Federal role in geothermal energy resource development, numerous other agencies have significant responsibilities. The geothermal interests of all these entities are coordinated and managed by the Interagency Geothermal Coordinating Council (IGCC) chartered in 1977 in response to Public Law 93-410, the Geothermal Steam Act of 1974.

The IGCC, which is chaired by the DOE Assistant Secretary for Energy Technology, includes representatives of more than 20 Federal agencies. The IGCC organizational structure is depicted in Figure III 4-18.

The Staff Committee, chaired by the Division of Geothermal Energy, formulates overall Federal geothermal program plans, and directs the activities of the Budget and Planning Working Group (BPWG) and the three panels.

The BPWG formulates long-range geothermal energy utilization goals, together with comprehensive Federal program plans to meet those goals; provides interagency coordination of geothermal programs and budgets; prepares the IGCC Annual Report; and identifies relevant program and policy issues related to Federal program status and non-Federal progress toward National geothermal goals.

The Resource Panel oversees the assessment and Federal leasing of U.S. geothermal resources.

The Research and Technology Panel ensures effective exchange of technical geothermal information within the U.S. geothermal energy community.

The Institutional Barriers Panel identifies legal, institutional, and policy issues that impede geothermal energy development and recommends initiatives to overcome those barriers.

The funding for Interagency Coordination and Planning is shown in Table III 4-4.

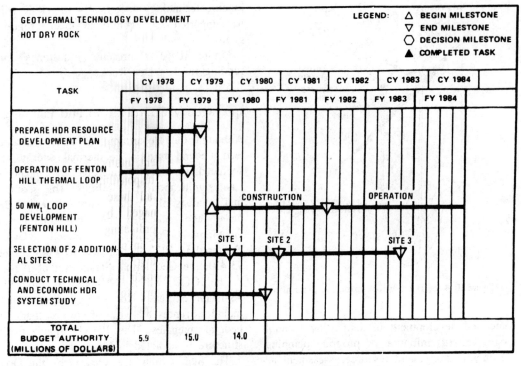

Figure III-4-17. Milestones and funding for hot dry rock.

OTHER PROGRAM ASPECTS

The technical and hardware aspects of the geothermal program have been described in the previous sections of this chapter. Other equally important, but often less visible, program activities are essential to a coherent program. These include the program's relationship with other DOE units, other Federal entities, interested foreign nations, and the domestic non-Federal geothermal community, as well as the overall program management machinery. In addition, compliance with certain aspects of Federal law requires special emphasis on environmental concerns and participation by small and disadvantaged businesses in the program. The funding for these efforts is generally included in allocations for the Hydrothermal Resources, Geopressured Resources, and Geothermal Technology Development Programs.

Certain program activities related to the commercialization of hydrothermal resources have been identified for transfer to the Assistant Secretary for Resource Applications and thus are not discussed in this section.

Regional Activities

The characteristics of potentially usable geothermal resources vary regionally. This tends to place increasing importance on regional activities which will result in the successful development of the geothermal program. The geographic suitability contributes to the geothermal program activities in a number of regions. Examples include Hawaii, which contains volcanic hydrothermal resources and the Gulf Coast, which contains geopressured resources with large quantities of methane. The Western States are generously endowed with geothermal resources, and in the Rocky Mountain area, a large number of moderate-temperature hydrothermal resources, as well as a considerable amount of hot dry rock potential, have been delineated. Recent investigations in the Eastern United States have targeted several probable moderate-temperature resouces in the Atlantic Coastal Plain of the Mid-Atlantic States.

In general, however, it is the commercialization aspects of geothermal development that will involve close coordination between

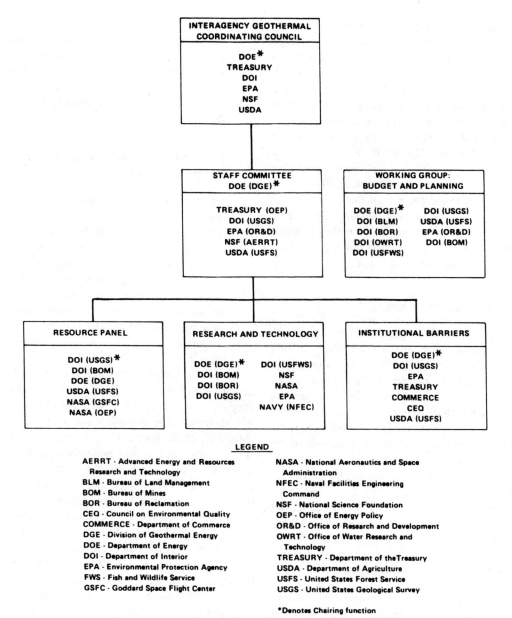

Figure III-4-18. IGCC organizational structure.

the Federal program, and state, local and regional entities. These activities will be the responsibility of the Assistant Secretary for Resource Applications.

Regional and field activities are also discussed in Chapter 7.

Commercialization

A cooperative agreement for the cost-shared construction and operation of a 50-MWe demonstration plant is being negotiated between DOE and a competitively selected industrial firm (see above). Beyond this, the commercialization of systems using hydrothermal energy resources has been placed under the Assistant Secretary for Resource Applications. Because these efforts are beyond the scope of the program described in this document, further discussion here is not appropriate.

There are no major technical barriers to the

utilization of geopressured heat and mechanical energy. Some of the equipment developed for the hydrothermal program is expected to be adaptable to geopressured resources. As the character of the geopressured resources and their potential for methane production become established, field experiments will be conducted for the recovery of heat and mechanical energy. These efforts are intended to lead to the utilization of geopressured resources, beginning in the mid-1980s.

Efforts to extract energy from hot dry rock are in the technology development phase. Current work is intended to establish technical feasibility and provide data for a preliminary evaluation of process economics.

Technology Transfers

Much of the DGE program is aimed at developing information that will ultimately modify the risks associated with the resources, technology, and economics that are perceived by potential geothermal investors. An integral part of the DGE program strategy is to ensure the transfer of knowledge and technology by involving end-users in the development of the information. Conventional information dissemination techniques, such as report distribution, workshops and conferences, new releases, etc., are used fully in the program, but essentially they supplement rather than constitute the actual technology transfer.

Several examples that illustrate the application of this policy are as follows:

- Industrial firms with commercial oil and gas operating experience were solicited on a competitive basis to manage the geopressured "wells of opportunity" project and the Atlantic Coastal Plain drilling project; a less desirable alternative would have been to assign these tasks to Government-owned contractors or universities.
- Field experiments to test and demonstrate the technical and economic feasibility of geothermal direct heat applications are conducted cooperatively with non-Federal end-users selected on a competitive basis.
- A criterion for authorizing most projects to develop hardware, materials, and processes for geothermal applications is the involvement of one or more commercial entities that are likely to ensure the future commercial availability of the perfected products.

International Geothermal Energy Program

The United States participates in international geothermal energy programs through multilateral and bilateral agreements with other nations. Generally, the agreements call for the exchange of information, exchange visits of scientists and examination of appropriate areas of cooperative activities. In some cases these agreements have led to more active cooperative research and development programs.

The only current multilateral agreement activity involving the United States is the Program of Research and Development of Manmade Geothermal Energy Systems (MAGES) of the International Energy Agency (IEA). This agency was established in 1974 under the sponsorship of the Organization for Economic Cooperation and Development (OECD). Five countries (the United States, the Federal Republic of Germany, Sweden, Switzerland, and the United Kingdom) of the nineteen IEA participants are active in the MAGES program; the Federal Republic of Germany is the lead country. The MAGES program will explore hot dry rock system in rock structures different from those involved in the LASL program; it will also investigate alternative methods of recovering energy from hot dry rock formations.

A recently completed cooperative international geothermal program, the Geothermal Pilot Study, was carried out under the auspices of the North Atlantic Treaty Organization's Committee on the Challenges of Modern Society (CCMS). Participants were Canada, the Federal Republic of Germany, France, Italy, the Netherlands, New Zealand, Sweden, Switzerland, the United Kingdom, and the United States. This program resulted in recommendations for: (1) continued international exchange of information; (2) international cooperation in studies of reservoir engineering, reinjection, and subsidence; (3) international cooperation in small-scale geothermal powerplant development; and (4) establishment of a cooperative international research and a development program in hot dry rock technology under the IEA. A conference is planned for 1980 to review the status of international geothermal development and to consider the need for additional NATO-CCMS activity.

Significant cooperative international programs are underway under bilateral agreements with Italy, Mexico, and the Federal Republic of Germany. There are also two separate agreements with Italy. The first involves reservoir stimulation and computer modeling, environmental control technology data exchange, and seismic studies. The second provides for continued data exchange through computerized information systems.

The United States-Mexico bilateral agreement resulted in a cooperative program to collect and analyze field date for reservoir engineering studies. Mexico shared its field work results concerning the structure, hydrodynamics, and chemistry of the Cerro Prieto, Mexico, geothermal field with the United States in the First Symposium on the Cerro Prieto Field held in San Diego, Calif., in September 1978. Work under this agreement, which continued in 1979, was directed toward the development of a dynamic model of the Cerro Prieto field. This model should aid in understanding the field interactions with regional aquifers, particularly as affected by reinjection of geothermal fluids.

The bilateral agreement with the Federal Republic of Germany (FRG) provides for full-time participation by FRG scientists in the LASL hot dry rock program. It has been proposed that FRG will provide about 25 percent of the funding for this program. An implementation agreement for this program is expected in the near future.

Other bilateral agreements provide for information exchange, resource assessment support, cooperative economic analyses, and exchange visits of scientists between the United States and El Salvador, Taiwan, New Zealand, Egypt, and Peru. Continuing participation by the United States in the IEA activities, in possible future NATO-CCMS cooperative studies, and in other international meetings is expected to lead to additional bilateral agreements.

5.
Electric Energy Systems

The National Energy Act (NEA) directs the research and development (R&D) initiative of the United States in attaining the goals of reduced dependence on foreign oil, increased utilization of solar generation at the household level, and conservation of limited resources. The electric energy network, composed of 260,000 miles of transmission and almost 4,000,000 miles of distribution lines, will contribute to accomplishing the NEA goals.

As new technologies are developed to produce electric power, the electrical transmission and distribution system must be capable of accommodating the energy in an efficient, flexible, and environmentally acceptable manner. The Division of Electric Energy Systems (EES) has program responsibilities for emerging technologies that are relevant and applicable to the systems aspects of U.S. electric networks. EES is the only program division within DOW that specifically assumes a systems approach to the electric energy networks.

The EES research development, and demonstration (RD&D) effort is performed in conjunction with the electric utilities, the Electric Power Research Institute (EPRI), equipment manufacturers, and other Federal agencies (e.g., Environmental Protection Agency, Department of Defense, Nuclear Regulatory Commission, and the Department of the Interior). EES expedites technologies with short-range energy and capital paybacks and sponsors development of higher risk, long-term, high-payback technologies, especially those that have a regional or national impact on NEA objectives.

The EES Program consists of two programs: (1) Power Supply Integration and (2) Power Delivery. Together, these programs are designed to achieve four major goals:

- Support implementation of National Energy Policy, in particular the Public Utility Regulatory Policies Act
- Ensure that all elements of future electric energy systems (oil, coal, nuclear, geothermal, solar, storage, fuel cells, conservation, etc.) are integrated as required to achieve an overall energy-efficient system with particular emphasis on the new source generation technologies
- Ensure the continued capability of the system to delivery electric energy, as needed, from the source to the user
- Develop the system design and control techniques required to maintain reliability as the system increases in size and complexity

The main elements of the existing two-part EES Program including the major 1980 subprograms, are shown in Figure III 5-1.

EES has undergone a shift in program emphasis for FY 1980 towards integration of new technologies and the resolution of environmental issues associated with high voltage power transmission. Consequently, several areas are being de-emphasized:

- DC underground cables and compact sta-

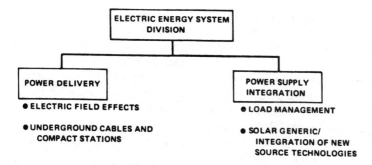

Figure III-5-1. EES program elements.

tions phased out; work on AC is continuing
- High voltage direct current technology (converters)
- High efficiency equipment & systems
- System control and development

Program priorities are under continuing study, and may, at a later date, result in further shifts in emphasis. In particular, the Federal role and the implications of the NEA for power systems research are being closely examined.

Program funding levels for FY 1978 through FY 1980 are presented in Table III 5-1.

The following subsections describe in more detail each of the two EES programs: (1) Power Supply Integration, and (2) Power Delivery. The descriptions of these programs are structured according to the FY 1980 budget; i.e., the subprograms which make up each program are described briefly. Each includes a general description of relevant technical aspects, an illustration (where appropriate), a summary of project status, and a milestone chart which presents major decision points through FY 1984 and funding levels for FY 1978 through FY 1980.

After each electric energy systems program is described in terms of its subprograms, other program aspects are described in order to provide an overview of program activities which are not a part of any specific subprogram. These program aspects also include regional activities, commercialization, environmental aspects, and socioeconomic aspects.

As noted previously, the elements of the EES program are being reevaluated in view of the NEA and other considerations.

POWER SUPPLY INTEGRATION

The Power Supply Integration Program consists of three subprograms: Load Management, Solar Generic/Integration of New Source Technologies, and System Control and

Table III-5-1. Program funding levels FY 1978 through 1980.

ELECTRIC ENERGY SYSTEMS	BUDGET AUTHORITY (DOLLARS IN THOUSANDS)			
PROGRAMS	ACTUAL FY 1978	ESTIMATE FY 1979	ESTIMATE FY 1980	INCREASE (DECREASE)
Power Supply Integration	4,300	15,500	17,000	1,500
Power Delivery	28,500	24,500	9,000	(15,500)
TOTAL	40,300*	40,000	26,000	(14,000)

*Included $7,500 for Systems Management

Development. The R&D activities carried out in these subprograms will assist in the continued improvement of the U.S. electric energy system.

Load management refers to the systems concept of altering the pattern of electricity use to optimize generation, transmission, and distribution capabilities of the electric power system. It is expected to produce near-term benefits, since much EES input is either being field-tested or actively promoted by utility companies. Moreover, Load management through automated distribution and control will enable dispersed solar technologies to be effectively integrated and controlled within a system originally designed to operated with central generation.

Solar generic/integration of new source technologies helps to remove current power system constraints so that the transition from technology development and demonstration to commercialization can be made when large numbers of new dispersed electricity sources are connected to the power system. Because energy savings realized at the source level must accrue to the consumer through proper integration into the transmission and distribution system, EES will continue to refine and model the expected characteristics of the technology/power system relationship and to test for model verification.

System control and development provides the design procedures that allow effective and reliable control of electric power systems in the face of increasingly complex operating conditions and constraints. As illustrated in Figure III 5-2, when new technologies start to become a significant part of the electric power system, complexity increases considerably, thereby requiring new controls and control strategies. The effort to design new procedures will be systemwide and long-term because it will be directed at commercialization levels of market penetration that have significant system impacts.

Load Management

Elements of the NEA, especially the Public Utility Regulatory Policies Act of 1978, promote more reliable, efficient production and delivery of electric energy. A near-term approach toward meeting this goal is load management, which calls for electric utilities and their customers (the latter for the first time) to share responsibility for improving the energy and capital efficiency of the electric supply/demand system.

The principal objective of load management is to optimize the system's performance by:

- Increasing the efficiency of system operations, particularly fuel use
- Reducing the need for capital investment
- Improving the ability to use fuels other than oil or gas
- Increasing the reliability of service to essential loads

Historically, utilities have met their customer's instantaneous demand by building and operating the most economical generation and transmission systems. Under load management, the utilities attempt to modify customer demand for electricity by direct control of loads, customer or local storage, production of incentives to customers to change their consumption patterns (e.g., use measurement), and modification of the apparent customer demand through system interconnections or bulk power system energy storage (e.g., supply management) or both. Time-of-use electric rates provide an incentive for users either to adopt load management practices or to modify behavior patterns.

Two new technologies are associated with load management. They are: (1) energy storage and (2) communications systems that connect utility operations directly to the equipment to be controlled. Both result in more efficient utilization of existing generating equipment, thus reducing the demand for primary energy (coal, oil, and gas).

Energy storage can be installed at the generation level, at distribution substations, or at the point of use. Storage does not reduce total energy use, but rather shifts the time of power demand to off-peak periods so that base-load generators (fueled primarily by coal or uranium) can supply more of the total energy. Consequently, the requirements for peak-load generators, which usually burn oil or gas, can be reduced. As shown in Figure III 5-3, load

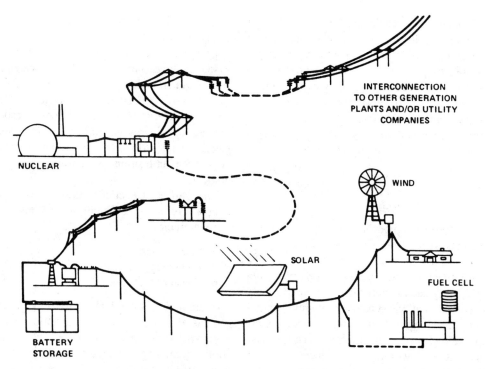

Figure III-5-2. Electric power supply integration.

management storage could transfer a portion of peak-energy demand in area "A" to the off-peak time presented in area "B."

Communications systems that link the utility to the equipment to be controlled allow the utility to switch certain loads on and off and to provide numerous new operating benefits such as remote metering. This can increase the use of available generating capacity and improve system reliability. Furthermore, the monitoring of loads and system functions allow automatic reconfiguration of the distribution system. This, too, promises increased system efficiency.

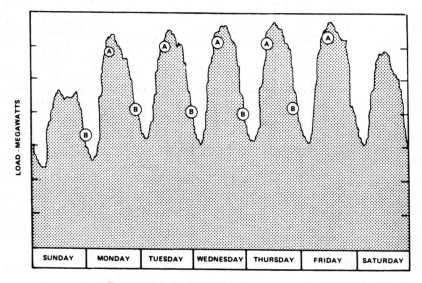

Figure III-5-3. Weekly load management curve.

Benefits expected in the near term through wide application of load management are: a noticeable shift of fuel demand away from gas and oil, improved reliability, increased productivity, reduced reserve margins, and effective integration of dispersed generation into the existing energy system.

Milestones and funding for Load Management are presented in Figure III 5-4.

Solar Generic/Integration of New Source Technologies

There are many models that simulate the existing U.S. electric network. All of these, however, are inadequate for representing the future electric grid because they were designed for electric networks composed solely of central station generating plants. Large-scale adoption of new source technologies, especially dispersed intermittent technologies, will require new models and techniques for evaluating electric systems. EES efforts under solar generic/integration of new source technologies have the objective of defining the expected characteristics of the future electric systems, modeling the possible technology/power system relationships, and testing these models for verification. As a result, system-wide characteristics as well as generic aspects relevant to systems integration can be investigated, thus contributing to more cost-effective large-scale utilization of new source technologies.

The Solar Generic/Integration of New Source Technologies Subprogram consists of three activities:

- Solar generic
- Technology specific
- Nonsolar/fuel cell

The entire FY 1980 budget for this subprogram is allocated to the solar generic activity. Although EES manages these subprogram efforts, technology specific funds are provided from outside the division, and Nonsolar/fuel

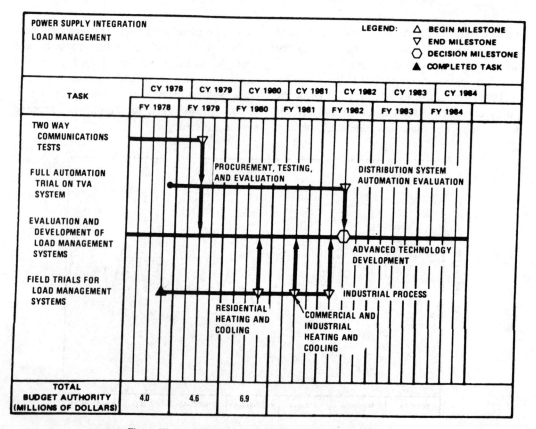

Figure III-5-4. Milestones and funding for load management.

cell funds come from the Fossil Energy Program.

Present nationwide electric energy systems are designed around central electric power generation stations. With the emergence of dispersed electric energy generation and storage devices, electric energy systems must undergo substantial design change to meet the required safety and reliability standards. Models and simulation techniques presently used for network analysis and synthesis cannot represent the future, more complex systems. In order to ensure safe, reliable, and efficient electric power delivery, new methodology must be developed to include inherent characteristics of such complex systems. EES along with the technology divisions share responsibility for successful development and demonstration of new source technologies for the electric energy system. Analysis from the standpoint of an individual source technology is limited in scope inasmuch as it tends to consider the salient features of that particular technology. EES, on the other hand, will undertake a broader approach by investigating systemwide characteristics as well as generic aspects relevant to systems integration. This will contribute to more cost-effective large-scale utilization of new source technologies. To this end, EES will structure programs to permit control of complex systems and develop communication apparatus and technologies for nonconventional source control, develop power conditioning and systems support devices, and undertake complementary systems for integrating new source technologies.

The technology specific activity focuses on four new solar technologies that are expected to attain near-term benefits in generating electricity: (1) storage, (2) wind, (3) photovoltaics, and (4) OTEC. Projects may also be undertaken to assess the integration of geothermal power and its transfer to load centers. Storage applications will be used in load leveling; recent studies have shown that unless storage capacity is needed to provide backup to an isolated, intermittent, renewable source (e.g., photovoltaics or wind), maximum economic benefit accrues when the storage device operates as a part of the power system. The Central Storage Applications Program is evaluating compressed air energy storage and underground pumped hydro storage preliminary designs and site studies for demonstration/commercial facilities to aid potential users in evaluating total system needs. Wind, photovoltaics, and OTEC systems have similar R&D requirements for system interface and are organized within the Technology Specific Activity to facilitate information exchange. Analysis of EES system integration is coordinated with expertise from the specific programs in the following principal R&D projects:

- Wind—development of needed interface hardware, dynamics of multiunit wind systems, analysis of power system operation with wind generators, wind farm electric stability study, wind turbine/power system simulation, and small wind turbine evolution
- Photovoltaics—development of power conditioning subsystems and interface equipment (to be continued beyond hardware and system tests)
- OTEC—dynamic stability studies, utility planning and operating assessments, and development of cable technology

The third activity within the Solar Generic/Integration of New Source Technologies Subprogram is the nonsolar/fuel cell. EES has the responsibility for application of the 4.8-MW fuel cell power plant on the utility network. This plant will use fuel cells in a commercially operating environment. Fuel cells provide numerous additional benefits because they are environmentally benign and thus can be located almost anywhere. If successfully demonstrated, fuel-cell generators could be the first application of dispersed sources. Although this project has been subject to extensive budget changes, technical tests are expected to continue through FY 1980.

The milestones and funding for Solar Generic/Integration of New Source Technologies are presented in Figure III 5-5.

System Control and Development

No funds are being requested in the FY 1980 budget for this program element. As men-

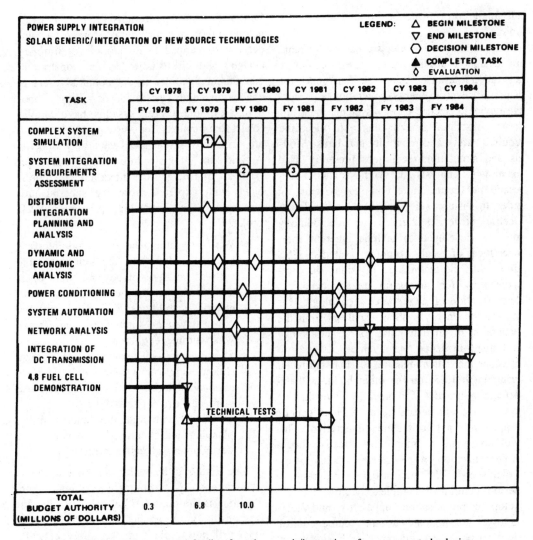

Figure III-5-5. Milestones and funding for solar generic/integration of new source technologies.

tioned earlier, this is one of the areas which is currently being closely reexamined in the context of the NEA and other considerations. Thus, the future direction of this activity is uncertain until this reexamination has been completed.

A brief description of this program element follows.

The objective is to develop operational and design procedures that allow reliable control and economical operation of increasingly complex energy systems which must operate under increasingly restrictive constraints and in an uncertain future environment. Such problems will become more complex with the introduction of new energy production and storage technologies. The importance of the objectives is underscored by two recent events:

- National Energy Act requirements (Sec. 209) pertaining to studies of ways to improve reliability of service to electricity customers
- The FERC's Congressional testimony which states that there are no procedures now for dealing with this type of issue

Methods are needed for assessing the viability and operational reliability of the system during emergency situations. Major disruptions of electrical service, which are frequently related to weather conditions, have occurred due to failure of large items of equipment. Furthermore, solar and wind energy sources are expected to intensify the weather-related stresses normally encountered by electric energy systems. Operators need assistance to evaluate quickly the effects of these failures and to assess the vulnerability of the remainder of the system. Power system emergencies sometimes lead to widespread blackouts. Operators currently have little information or control available to them which can be used during the period between the onset of an emergency and the point when loss of service occurs.

POWER DELIVERY

The EES Power Delivery Program is concerned with enhancing the transmission capability of the Nation's electric energy transmission networks. Whereas the EES Power Supply Integration Program, previously discussed, is primarily concerned with developing the "software" to analyze systems problems and to facilitate the integration of new source technologies into the electrical grid, power delivery is primarily an advanced "hardware" development program.

The FY 1980 budget request reflects increased priority in EES for the Power Supply Integration Program, stressing the integration of new source technologies into the grid. The current budget request also emphasizes work on determining problems associated with electric field effects of high voltage lines on humans, plants, and animals. Work on underground AC cables is to be continued, but at a lower level of effort than in FY 1979. The FY 1980 request phases out the work in high voltage direct current (HVDC) technologies and high efficiency equipment and systems.

Analysis of the power delivery technologies is continuing. This may result in a reconsideration of the priorities in this program at a later date, as was previously discussed.

Electric Field Effects

Many energy technologies being developed by DOE require a high-voltage electrical transmission system to deliver energy to the consumer. Recently, there has been concern about the appearance, land use, and safety of overhead transmission lines. The most serious of these concerns is the belief that strong electric fields may affect biological processes. Although 70 years of overhead transmission history have not produced any noticeable biological effects, recent studies conducted in Russia and in the United States have claimed to uncover subtle biological effects in animals and cell cultures exposed to high electric fields. Unfavorable disturbances have also been reported in switchyard workers occupationally exposed to high-voltage gradients for long periods of time. To date, most of these studies have not been widely accepted by the scientific community. Many case studies have been judged to be inconclusive because of poor or nonexistent controls and unhealthy and improperly fed test animals. However, additional reports recently released through the U.S. technical exchange program appear to substantiate the earlier Russian reports.

The Electric Field Effects Subprogram involves an intensive broad-based research program in all electric and magnetic field and corona effects from overhead transmission lines. Efforts will be made to quantify these effects through direct laboratory experimentation and by measurement under transmission lines. These measurements will provide data, in a format similar to that shown in Figure III 5-6, that can be used to develop guidelines to protect biological and environmental conditions. This program is conducted in conjunction with the Division of Biological and Environmental Research, the Division of Environmental Control Technologies, and the electric utility industry, including EPRI.

The purpose of the EES Electric Field Effects Subprogram is to determine guidelines by 1982 for preventing detrimental environmental effects from overhead transmission lines. These guidelines would then be used to specify parameters for prototype powerlines. These guidelines will also aid the nuclear, solar, and fossil programs with environmental impact statements that they can use for siting large electric generation facilities.

Areas which have been or are being satisfactorily addressed by the electric utility industry will not be included in the DOE

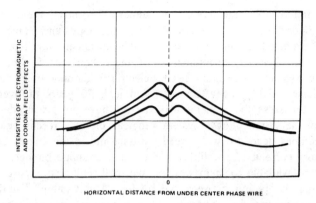

Figure III-5-6. Measurement of electric field effects.

program. Such areas are radio and television interference and right-of-way ecological management. During foul weather, most extra-high voltage transmission lines (345-kV and higher) produce corona, which is a local breakdown of the air at the surface of the conductor. This is the source of radio and television interference and audible noise from transmission lines. Minute amounts of ozone are also produced by corona, but in recent studies these have been shown to produce insignificant effects on the environment.

The electric utility industry has also done considerable work in the area of audible noise from transmission lines. However, acceptable levels of audible noise have not been determined. Therefore, EES has undertaken a comprehensive program to develop greater understanding of the psychoacoustic effects of audible noise from transmission lines. This program will lead ultimately to new methodologies for making measurements that directly correlate with human perception and thus to criteria for evaluating the effects of noise from transmission lines.

The Electric Field Effects milestones and funding are presented in Figure III 5-7. As shown, DOE is placing increased emphasis on the field effects work in the FY 1980 budget request to help resolve major environmental issues about the effects of high voltage transmission.

Underground Cables and Compact Stations

The FY 1980 budget requests reduced emphasis on the HVDC portion of this area while continuing work on advanced AC options in order to focus limited resources on the problems of integrating the new source technologies and resolving problems associated with electric field effects, as previously discussed.

Two competitive advanced underground options that are being studied are AC superconducting cables and the compressed gas insulated cable. These will be continued.

Work on compact substations is being reduced. The underground DC option is primarily to facilitate transfer of large blocks of power to specific locations within a complex network as may be eventually required to create an effective "national grid"; the option also may facilitate movements of large amounts of power (e.g., "coal by wire") from remote generating stations to load centers. State-of-the-art capabilities in this area are primarily at lower voltage cables for underwater applications in Europe. An analysis of the implications of funding requests in this area are continuing and may result in further reorientation of priorities at a later date.

The two advanced cable options—superconducting and gas insulated cables—will for the first time move large blocks of power across long distances underground. DOE is, thus, continuing to develop advanced underground power transmission options. The advantage of superconducting AC cable systems is increased transmission voltages. These systems have a metallic conductor that is cooled by liquid helium, which is supplied by refrigeration units installed about every 10 miles along the route. A schematic of a superconducting cable system is shown in Figure III 5-8. The

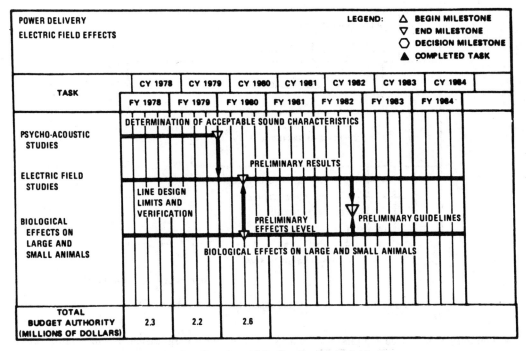

Figure III-5-7. Milestones and funding for electric field effects.

increased power carrying capacities and energy transmission efficiency of superconducting cables is compared to other single-circuit transmission cables in Figure III 5-9.

An advantage of the compressed gas insulated cable system is its ability to match high voltage (up to 1200-kV) systems. As the transmission system increases in voltage, changes in the insulation of substation equipment from air to compressed gas will be required. The size of the equipment, support insulation, and electric buswork have forced utility companies to realize that air insulation is not the practical medium. Compressed-gas insulated substations, however, would require totally new equipment designed specifically for use in a gas environment.

The milestones and funding for Underground Cables and pact Stations are shown in Figure III 5-10. Schedules after FY 1980 depend on special program review presently underway and on resulting budget levels.

High Voltage Direct Current Technology

The FY 1980 budget request eliminates ETS support in this area on the basis of its low priority relative to other areas in the EES budget. A preliminary assessment indicates that the current state-of-the-art in DC overhead system and terminal equipment may be adequate for the near future (through 1990) and that private industry work already underway will produce needed cable technology in the appropriate time frame (before 1990). We

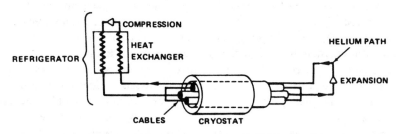

Figure III-5-8. Superconducting cable system.

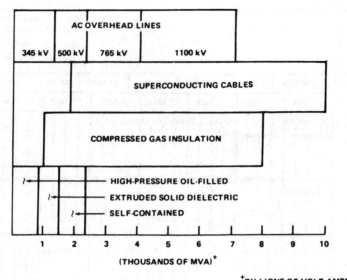

Figure III-5-9. Power-carrying capacities of single-circuit cables.

are continuing to evaluate this preliminary judgment.

The power transfer capability and apparent environmental advantages of DC transmission lines have been demonstrated; the principal barriers to more widespread use are the high cost of required conversion equipment and the lack of a DC breaker to protect a network system. This cost also figures prominently in underground transmission, where DC cables have some significant advantages over AC cables, such as a four-to-one capacity advantage for comparable conventional cables. Transformers, filters, valves, and controls unique to DC transmission also need improvement.

Presently, 12 large-scale direct current transmission lines are operating throughout the world, some using mercury arc valves and others using thyristor (solid-state) valves. About 12 more lines, all using thyristor valves, are planned for operation by 1980. The most ambitious HVDC transmission system in the United States is the 846-mile Pacific Northwest-Southwest Intertie, energized in 1970. Paralleling and supplementing alternating current transmission lines, it operates at ± 400 kV and can carry 1,440 MW of power—enough to accommodate the average needs of one million people.

ETS recognizes the unique ability and potential economy of DC transmission, especially over long distances, to control more directly power flows between converter terminals. There may also be increased applications of DC for point-to-point use. However, a DC network will most likely be slow to develop and there may be no compelling need to proceed with DC breaker development at this time. Of course, the results of the National Grid Study and ETS's ongoing analysis might suggest a revision in priorities. Analyses of the implications of the requested funding in this area are continuing and may result in a further reorientation of priorities at a later date.

High Efficiency Equipment and Systems

The FY 1980 budget request eliminates ETS support for work in this area because the technologies will require considerable time to develop and the benefits may only be marginal. For example, successful development of a superconducting generator would result in an improvement in efficiency (reduction in generator losses) of from 98.5 percent today to 99.3 percent. Reduction of transformer losses would be the same order of magnitude, but would require a long time to develop.

These technologies offer the potential for size reduction of transformers and generating equipment; they also present the potential for

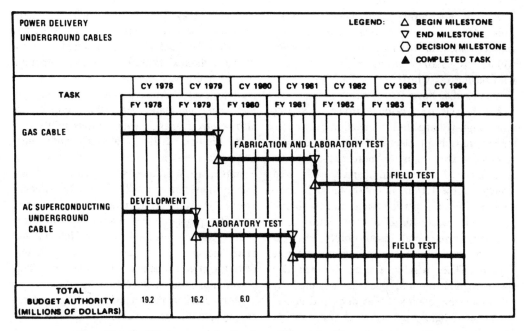

Figure III-5-10. Milestones and funding for underground cables and compact stations.

improving system stability and reliability. On balance, the priority accorded to these areas has been reduced. An R&D program has produced a conceptual design for a 5- to 10-MVA superconducting generator. Other work has produced advanced concepts for transformers, including superconductivity and low loss amorphous magnetic metals. These new concepts and approaches may not be carried forward to further development and testing at this time; analysis of the long-range potential of high-efficiency equipment and systems is continuing and may result in reconsidering their priority at a later date.

OTHER PROGRAM ASPECTS

EES is concerned with long-run, high-risk energy system projects that show promise of helping to attain NEA objectives. Recent policy directives give greater responsibility to EES for administering these RD&D projects, so that proper attention is given to related issues that may evolve as each program matures. For example, successful projects will be adopted by the commercial sector and introduced into the competitive market. Issues which affect the timing of commercialization of electric system technologies are examined by EES. Because the implications of siting a single component of the overall electric energy system may involve significant social or cultural change, socioeconomic and environmental issues are investigated separately for each project. The discussion of regional activities provides a brief description of responsibilities that have been assigned to national laboratories and of other interfaces in the program.

Most of the electric system RD&D efforts are being performed by national laboratories, universities, and utility companies. Brookhaven National Laboratory, NASA's Jet Propulsion Laboratory, and the Oak Ridge National Laboratory all have multiproject responsibilities. EES meets regularly with representatives of EPRI and the universities to ensure proper communication of new developments and access to work currently underway. Many utility companies are eager to demonstrate innovative technologies in their operating systems; Public Service Electric and Gas Co. (New Jersey) and Consolidated Edison Co. (New York) are two utilities with large demonstration projects, which they are cofunding with DOE.

Commercialization efforts focus on accelerating development and demonstration of projects with near-term energy benefits. Typically, the projects with longer term payback characteristics are funded wholly by EES, while cost-sharing arrangements are made with businesses which accept liabilities on investments that have less risk.

As the electric system grows, requiring more high-voltage interconnections, environmental impacts of electric energy system development are becoming more prevalent. An environmental plan accompanies each construction project from initiation through completion. Because participating utilities build, own, and operate all EES construction projects, they have overall responsibility for documenting environmental and socioeconomic impacts. There will be a need for expanded environmental reports, each documented by the responsible contractor, as dispersed generating technologies, underground transmission lines, larger central station, and other new technologies are used in the electric energy system.

Human displacement issues are inherent in electric system decision-making. Transmission lines are the topic of many community meetings; proximity to central gathering stations entail the assurance of public health and safety; utility load management affects people's behavior and work style. These kinds of socioeconomic impacts are evaluated to maintain Federal awareness of potential abridgements on citizens' rights.

Regional Activities

National laboratories have been given the responsibility for managing several efforts and additional management responsibility was transferred to the field in FY 1978. Project management responsibility for AC superconducting power transmission was assigned to Brookhaven National Laboratory. NASA's Jet Propulsion Laboratory was selected to perform the technical management and analysis work for the Power Distribution Systems' Lighting Project and Distribution Automation. Oak Ridge National Laboratory will administer load-management projects and the demonstration of thermal storage systems.

Utility systems are also engaged in managing a number of projects. These include the Battery Energy Storage Test Facility operated by Public Service Electric and Gas Co., and the 4.8-MW Fuel Cell Power Plant operated by Consolidated Edison Co. Other important but relatively less expensive projects, such as field tests of alternative communications systems and automatic generation control, are also jointly managed with various utilities.

In addition, EES is working with other DOE divisions in a number of areas. For example, EES is interfacing with the Solar Energy Division on OTEC and the Energy Storage Division on activities of mutual interest. The program also interfaces with other Federal agencies as well as other National and State agencies. Also, there is United States-Soviet Union cooperation in superconducting power transmission development.

More detailed information on field and regional activities is contained in Chapter 7.

Commercialization

The natural customers for the technologies developed by EES are the electric utilities. As industrial customers of significant size, technical sophistication, and financial resources, utilities are favorably disposed to new technologies designed to improve the efficiency and economics of their operations. Thus, issues related to consumer awareness and adoption of devices, retail channels, installation and repair services, and the like are not particularly important for electric energy systems.

Utility awareness of technological advances being sought and achieved in EES is enhanced by the participation of EPRI. Much of the work is done in collaboration with university groups for theoretical projects and with utilities for applied projects. Meetings in which universities, professional societies, and utilities participate are held to ensure communication of new developments.

Utilities are eager to demonstrate innovative technologies in their operating systems. By securing utility participation in engineering development and demonstration, EES plans to make progress toward commercialization in two ways:

- Make full use of the technical knowl-

edge of the utility regarding existing equipment and operating requirements
- Through successful demonstrations, the utility becomes immediately aware of the technical and economic advantages and has already acquired the practical experience needed to implement the new technology more extensively

Environmental Aspects

Issues. A principal concern of EES activities is to determine the biological effects from electric and magnetic fields associated with high voltage transmission facilities. The electric fields from high voltage overhead lines have an unknown impact on human health, as well as animal and plant life. In addition, land-use requirements may be substantial. Issues relating to overhead power line corridors are more fully discussed in the socioeconomics section of this chapter.

Milestones and Goals. The objectives of the Electric Field Effects Subprogram are to conduct an intensive broad-based research program into all electric and magnetic field and corona effects from overhead transmission lines, and to conduct this program in conjunction with the Division of Biological and Environmental Research, the Division of Environmental Control Technologies, and the private electric utility industry, principally through EPRI.

The purpose of the program is to determine reasonable guidelines for obviating detrimental environmental effects from overhead transmission lines by 1982. These guidelines would then be used to specify parameters for some of the prototype lines that will be coming out of the compact line research programs of EES and to aid the nuclear, solar, and fossil programs with environmental impact statements that they can use for siting large electric generation facilities.

Those areas which have or are being satisfactorily addressed by the electric utility industry will not be included in the DOE program. Such areas are radio and television interference and right-of-way ecological management. During foul weather, most extra-high voltage transmission lines (345-kV) produce corona, which is a local breakdown of the air at the surface of the conductor. This is the source of radio interference, television interference, and audible noise from transmission lines. Also, minute amounts of ozone are produced by corona, but in recent studies this has been shown to be of no environmental consequence. The industry has also done significant work in the area of audible noise from transmission lines. However, there is still much disagreement in determining acceptable levels of audible noise. Therefore, EES has undertaken a rather comprehensive program to develop a fuller understanding of the psychoacoustic effects of audible noise from transmission lines. This program will ultimately lead to new methodologies for making measurements that directly correlate with human perception and thus led to criteria for evaluating the effects of noise from a transmission lines.

Milestones (presented in Figure III 5-9) for the environmental activities in EES are the same as those for Electric Field Effects Subprogram activities.

Socioeconomic

A large number of the provisions of the Public Utility Regulatory Policies Act of 1978 and the Power Plant and Industrial Fuel Use Act of 1978 provide policy direction for electric energy systems RD&D. These provisions are aimed at improving the functioning of U.S. electric energy networks. The role of EES is to tie together, improve, and enhance the linkage among the producers, delivery systems, and users of electricity.

From an economic standpoint, a number of the provisions, if pursued vigorously, could make a significant contribution to reducing the cost of producing electricity and could contribute to improving system reliability. First, the act recommends the prohibition of declining block rates and encourages the implementation of time-of-use rates. Second, the act encourages the interconnection or wheeling of electric power between utilities. This requirement would help, through the efficiencies of large-scale operation, to improve system reliability and would help to reduce unit costs of production.

6.
Energy Storage Systems

The goal of the Energy Storage Systems Program is to develop and demonstrate in cooperation with industry, reliable, cost-effective, safe, and environmentally acceptable energy storage systems that will provide one or more of the following benefits:

- Increase the substitution of coal, nuclear, and solar energy for petroleum and natural gas
- Enable solar, wind, and other intermittent energy systems to provide continuous service
- Promote conservation by recovering waste heat from utilities and industries and by improving process efficiency

These goals are being achieved by developing energy storage systems and components for use in transportation, building heating and cooling, industrial processes, solar systems, and utilities. Since different applications require different amounts and types of energy to be stored for various duty cycles, several energy storage technologies are needed. The major thrusts for development in specific applications are as follows:

- Electric and hybrid vehicles—improved propulsion batteries and flywheel regenerative braking systems to upgrade vehicle performance to the level required for significant market penetration
- Heating and cooling buildings—aquifer storage of hot and cold water for seasonal applications, phase change thermal storage for daily cycling applications, and phase change and chemical heat pump storage for solar applications
- Solar and conventional electric utility applications—batteries, thermal, mechanical, and underground (compressed air, pumped hydro) storage for utility load leveling to substitute for natural gas and oil-fired peaking turbines, and for use in solar electric systems
- Industrial and multipurpose uses—energy saving electrochemical processes, retrofit thermal storage for industrial heat recovery, and economical processes for producing hydrogen

As shown in Figure III 6-1, energy storage activities support the efforts of a number of other DOE organizations. Jointly funded projects are being used to coordinate energy storage activities with other DOE organizations. In addition, the staff of the Energy Storage Systems Division has been active on several DOE committees for program coordination. Examples include chairmanship of the Hydrogen Energy Coordinating Committee in FY 1976, FY 1977, and FY 1978, and the ERDA Materials Coordinating Committee in FY 1977.

Coordination is also being achieved through interagency agreements with other Federal agencies such as NASA, DOD (the U.S. Navy), and the Department of Interior (the Bureau of Reclamation and the U.S. Geological Survey).

The program of the Division of Energy Storage Systems (STOR) is organized into

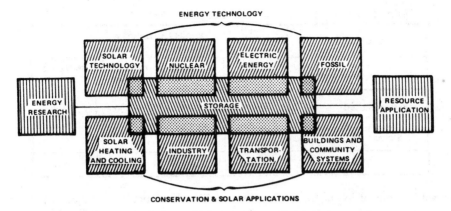

Figure III-6-1. Energy storage systems interfaces with other DOE organizations.

two major program areas: Battery Storage and Thermal and Mechanical Storage. These are organized into the programs shown below:

- Battery Storage Systems
  — Near-term Batteries
  — Advanced Battery Development
  — Solar Applications
  — Electrochemical Systems Research
  — Supporting Research
  — Electrolytic Technology
  — Dispersed Battery Applications
- Thermal and Mechanical Storage
  — Thermal Storage
  — Chemical/Hydrogen Storage
  — Mechanical Storage
  — Magnetic Storage
  — Utility Applications
  — Technical and Economic Analysis

Distribution of the FY 1980 budget between the two major programs and its relation to the FY 1979 budget is shown in Figure III 6-2.

The funding levels by programs for FY 1978 through FY 1980 are summarized in Table III 6-1.

The following subsections describe in detail each of the two energy storage programs: (1) Battery Storage, and (2) Thermal and Mechanical Storage. The descriptions of these programs are structured according to the FY 1980 budget; i.e., the subprograms which make up each program are described briefly in individual modules. Each module includes a general description of relevant technical aspects, an illustration (where appropriate), a summary of project status, and a milestone chart which presents major decision points through Fy 1984 and funding levels for FY 1978 through FY 1980.

Finally, after each energy storage program

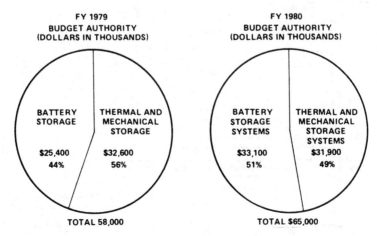

Figure III-6-2. Funding distribution by programs.

Table III-6-1. Program funding levels FY 1978 through FY 1980.

ENERGY STORAGE PROGRAMS	BUDGET AUTHORITY (DOLLARS IN THOUSANDS)			
	ACTUAL FY 1978	ESTIMATE FY 1979	ESTIMATE FY 1980	INCREASE (DECREASE)
Battery Storage	20,676	25,400	33,100	7,700
Thermal and Mechanical Storage	24,955	32,600	31,900	(700)
TOTAL	45,631	58,000	65,000	7,000

is described in terms of its subprograms, other program aspects are described in order to provide an overview of program activities which are not a part of any specific subprogram. For instance, since the commercialization of energy storage systems is closely linked to the commercialization of technologies that use these storage systems, the implications of the energy generation/storage relationship is included in the discussion of energy storage systems commercialization. Other program aspects also include regional activities, environmental aspects, and socioeconomic aspects.

BATTERY STORAGE

The Battery Storage Program includes both near- and long-term projects. In the near-term, lead/acid, nickel/zinc, and nickel/iron batteries are being developed to support the Electric and Hybrid Vehicle Demonstration Act as candidates for use in electric vehicles in the early 1980s.

The development of longer term advanced battery systems for electric vehicle and utility applications is concentrated on lithium/metal sulfide, sodium/sulfur, and zinc/chlorine batteries to be commercialized in the late 1980s. In addition, electrochemical processes are being developed to conserve both resources and energy in industrial processes. The relative emphasis of the Advanced Battery Subprogram and supporting subprograms is shown in Figure III 6-3; the Near-term Batteries Subprogram and related efforts, funded with passthrough funds from the Transportation Energy Conservation Division, and the Dispersed Battery Applications Subprogram are not depicted.

The principal objectives of the Battery Storage Program are as follows:

- Support near-term demonstration programs by improving the performance of selected existing batteries
- Reduce battery costs from the present range of $70–90/kWh for lead/acid bat-

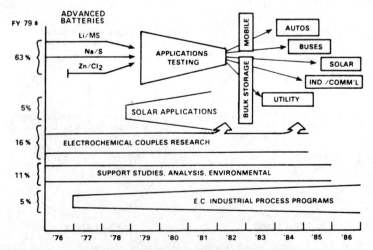

Figure III-6-3. Advanced battery development and supporting subprograms thrust and emphasis.

teries to $30/kWh for advanced batteries for utility applications
- Increase energy density in batteries from the present 26 Wh/kg for lead/acid batteries to 140 Wh/kg, and power density from 20 W/kg to 70 W/kg sustained and 200 W/kg peak for advanced batteries for vehicle applications
- Develop more efficient, safe, and flexible industrial electrolytic processes for resource and energy savings

During FY 1979, the Battery Storage Program expects to achieve the following goals:

- Construct a full-size lithium/metal sulfide battery suitable for testing in a battery driven van
- Test a utility submodule-size 50-kWh zinc/chlorine battery prototype
- Construct and test the first multicell, multi-kWh batteries using full-size sodium/sulfur cells
- Test improved near-term batteries to support vehicle purchases legislated by the Electric and Hybrid Vehicle Demonstration Act

Table III 6-2 summarizes funding levels by subprogram for FY 1978 through FY 1980.

Near-term Batteries

The Near-term Battery Subprogram is aimed at improving current battery technology for use in electric vehicles. Lead/acid, nickel/zinc, and nickel/iron types have been selected for improvements because facilities can be equipped to build them in the time frame and quantity required for near-term applications. Near-term battery projects are presented in Figure III 6-4. The Near-term Battery Subprogram is divided into two phases: Phase 1 (1977–81) and Phase 2 (1982–84). Phase 1 emphasizes the following:

- Substantially improving near-term battery technologies that will extend and improve the performance of current vehicles
- Developing pilot-line manufacture of a limited number of full-size (20–30-kWh) batteries for technological evaluations in Transportation Energy Conservation Division test and demonstration vehicles in 1980–84

During Phase 2, the three activities are as follows:

- Reducing the number of different types of battery systems being developed
- Upgrading the pilot line technologies of the chosen systems
- Beginning to develop battery design variations for use in hybrid vehicles

Efforts in this subprogram will be directed toward meeting the goals of the Electric and Hybrid Vehicle (EHV) Research, Develop-

Table III-6-2. Battery storage funding levels.

BATTERY STORAGE SYSTEMS SUBPROGRAMS	BUDGET AUTHORITY (DOLLARS IN THOUSANDS)			
	ACTUAL FY 1978	ESTIMATE FY 1979	ESTIMATE FY 1980	INCREASE (DECREASE)
Near-Term Batteries*	1,230	1,040	2,100	1,060
Advanced Batteries	13,839	11,700	14,400	2,700
Solar Applications	442	800	5,000	4,200
Electrochemical Systems Research	2,235	2,500	2,600	100
Supporting Research	1,206	2,140	3,000	860
Electrolytic Technology	591	1,320	1,500	180
Dispersed Battery Applications	1,133	5,900	4,500	(1,400)
TOTAL	20,676	25,400	33,100	7,700

*Does not include Near-Term Electric Vehicle Battery Applications funds budgeted by Division of Transportation Energy Conservation. ($5.2 million in FY 1979, $8.0 million in FY 1980).

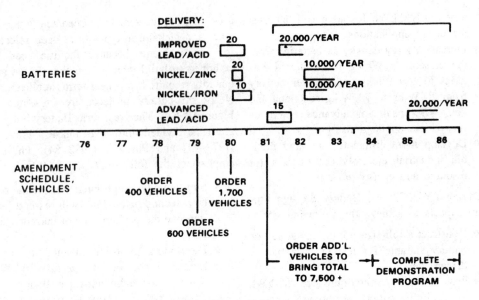

Figure III-6-4. Near-term battery projects.

ment and Demonstration Act of 1976, Public Law 94-413 (as amended). Tests of improved lead/acid, nickel/zinc, and nickel/iron cells, and multicell batteries were conducted in FY 1979. Subsequently, batteries will be delivered for testing in the National Battery Test Laboratory, and battery packs will be installed in electric vehicles for the initial engineering demonstration.

Milestones and funding data for Near-term Batteries are presented in Figure III 6-5.

Advanced Battery Development

Three key advanced batteries (lithium/metal sulfide, sodium/sulfur, and zinc/chlorine) are being developed. Vehicle field tests are planned for 1979 through 1985. Evaluation of utility load leveling by the Battery Energy Storage Test (BEST) facility is scheduled for 1982 through 1986.

The lithium/metal sulfide battery, which operates at about 400° to 425°C, consists of a series of cells containing lithium in the negative electrode and a metal sulfide in the positive electrode. The electrolyte is a molten salt. Present lithium/metal sulfide cells having capacities of 150- to 300-Wh achieve a storage density of about 60 to 90 Wh/kg. Experimental versions have demonstrated much higher energy densities of up to 150 Wh/kg.

Prototype engineering cells are presently achieving a 400 to 500 charge-discharge cycle life (about 2 years), and some cells have been cycled more than 1,000 times.

The sodium/sulfur battery shows promise for meeting the requirements of electric vehicles, utility and solar storage missions. Operating at a temperature of 300° to 350°C, it consists of molten sodium in the negative electrode and a mixture of sulfur and sodium polysulfide in the positive electrode, with a solid ceramic electrolyte. Laboratory cells with capacities of 200-Wh have demonstrated energy densities of more than 100 Wh/kg.

The zinc/chlorine battery operating at near-ambient temperatures has an aqueous electrolyte. The reactants are zinc and chlorine gas. The chlorine gas is chilled in the presence of water to form an ice known as chlorine hydrate (Figure III 6-6). This unique form of storage provides added safety in case of accidents.

In Fy 1979, the life of newer, higher energy density lithium/metal sulfide, sodium/sulfur and zinc/chlorine batteries will show over 500 cycles with tests continuing. This achievement will bring the cycle of life of these advanced batteries within 50 percent of the program goal of electric vehicles and within 30–40 percent of the program goal for utility applications.

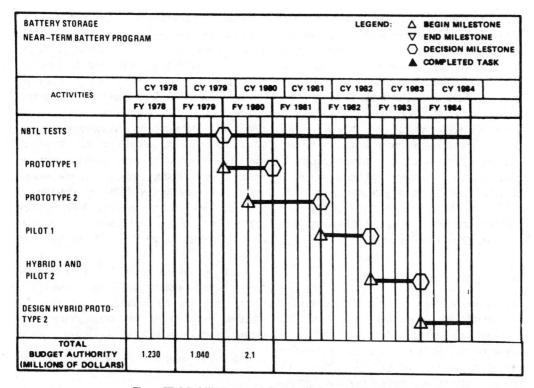

Figure III-6-5. Milestones and funding for near-term batteries.

Engineers plan to test a 40-kWh experimental lithium/metal sulfide battery in an electric van in 1979 and the Mark II lithium/metal sulfide battery in 1983. The load-leveling version of this battery will be tested at BEST in 1986. The BEST facility will provide the capability for testing battery installations of up to 10,000-kWh in an actual utility environment. Under joint sponsirship of DOE and EPRI, the BEST facility is being built and will be operated by the Public Electric and Gas Co. of New Jersey.

During FY 1979, the first multi-kWh sodium/sulfur battery will be constructed and tested. Particularly gratifying is the progress in fabricating high-quality beta-alumina tubes, a vital battery component. Reproducibility is now established, justifying the FY 1978 investment in a tube pilot line. Work is also in progress on hollow-fiber sodium/sulfur bat-

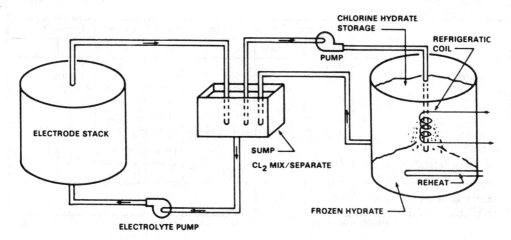

Figure III-6-6. Zinc/chloride battery system schematic.

teries. This effort focuses on a unique design approach for the sodium/sulfur system involving thousands of hollow glass fibers in place of the traditional single beta-alumina tube. The cheaper glass results in a potentially lower cost sodium/sulfur battery.

Qualification testing of a 100-kWh prototype zinc/chlorine module began in FY 1979. If progress continues according to current plans, the zinc/chlorine battery could become the first advanced type to be tested in the BEST facility. It will be delivered to the facility in 1981 and tested in 1982.

Milestones and funding for Advanced Batteries are shown in Figure III 6-7.

Solar Applications

The availability of low-cost energy storage can aid in developing the full potential of solar and wind energy to produce electricity. Advanced storage batteries offer a promising way to assist photovoltaic (PV) (Figure III 6-8) or wind energy technologies to harness these sources. Excess energy produced during a sunny or windy day would be stored for use when it is either dark or cloudy or when the air is calm, smoothing the output of these intermittent energy sources.

Four potential application areas have been identified for battery storage in PV energy systems. These are: remote, residential, commercial, and utility/industrial. In remote systems having no backup power (such as microwave repeaters and weather stations), the PV array and battery storage must be sized to meet the load during cloud cycles. Various solar PV system configurations are under consideration for residential applications. These include all-electric systems, thermal hybrid systems, and systems with or without onsite battery storage. Recent analyses of the economic viability of batteries in residential PV systems indicate that, depending on the specific application, break-even costs generally range from $0 to $80 per kWh.

The major efforts in FY 1979 are PV-application battery assessment studies, which are being integrated by Sandia Laboratories. In addition to planning and management, the major efforts include:

- Mission analysis of dedicated storage economics
- Laboratory evaluation
- Development of test standards and initiation of the testing program

During FY 1980 there will be increased efforts in the area of mission and application analysis. Milestones and funding for this subprogram are shown in Figure III 6-9.

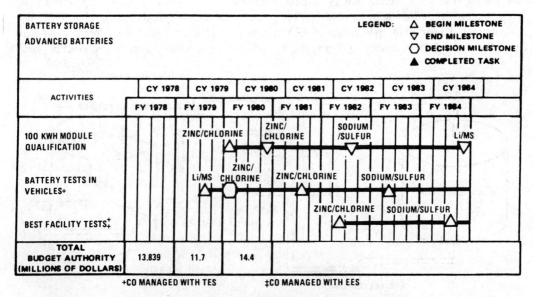

Figure III-6-7. Milestones and funding for advanced batteries.

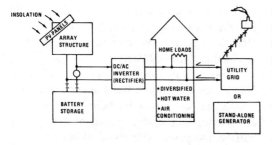

Figure III-6-8. Residential photovoltaic system schematic.

Electrochemical Systems Research

During FY 1980, about 8 percent of the Battery Storage Program budget is devoted to exploratory battery research. These new systems, if proven energy-efficient and economically viable at the prototype level, would then be considered for full-scale engineering development for bulk or mobile storage applications.

Iron/iron, zinc/bromine, and iron/chromium are reduction/oxidation (redox) cells using ambient-temperature aqueous acid electrolytes. The power and storage capacities of these battery systems could make them especially applicable to weekly storage required by certain solar electric or dispersed electric missions. Research in the iron/iron redox system, composed of low-cost materials, is currently supported by passthrough funds from the Division of Distributed Solar Technology. The zinc/bromine system, which is similar to the zinc/chlorine battery described in advanced battery development, may be safer because bromine can be stored more easily than chlorine.

At present, metal/air batteries are the only long-range candidates for electric vehicle batteries (outside of zinc/chlorine) that do not require an elevated temperature for operation. The theoretical energy density of metal/air batteries is much greater than that of any other near-term or advanced batteries.

Calcium is being investigated as a substitute for lithium in the lithium/metal sulfide battery because of its greater abundance in the earth's crust, which will lower costs when these batteries are used widely. The calcium battery will have applications in the electric vehicles, solar electric, and dispersed electric load-leveling missions.

The sodium/sulfur system, uses a sodium ion-conducting solid electrolyte (beta-alumi-

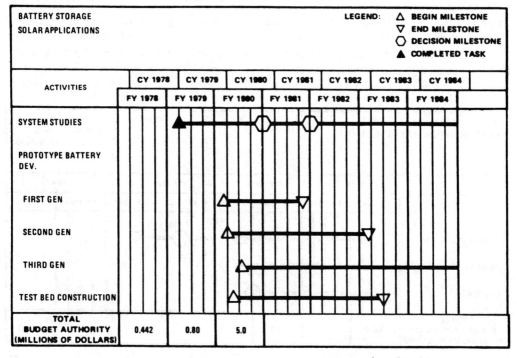

Figure III-6-9. Milestones and funding for solar applications.

na). NASICON, a new solid electrolyte developed by the Supporting Research Subprogram, has been manufactured in small sizes and tested in small cells. Because this electrolyte is insensitive to moisture and can be sintered at lower temperatures, it may be less expensive than beta-alumina.

The redox flow system uses two fully soluble couples separated by a conducting membrane and storage of the solutions external to the cell stack. Improvements in electrode kinetics and in the conductivity and selectivity of the membrane have increased power density tenfold in FY 1978, and interim goals of >75 percent energy efficiency and >50 watts/ft² maximum power output have been exceeded.

Definitive tests are being run to compare lithium/air and aluminum/air batteries. Lithium has a higher energy density; on the other hand, an aluminum industry is in operation, while a lilthium industry to reprocess the reaction products of the battery would have to be established. During FY 1979, the results of these tests, along with an economic analysis of both types, will enable STOR and the contractors to decide which system should be developed further.

Major advances in identifying cell components of calcium/metal sulfide batteries were made in early FY 1979. The negative electrode chosen was Ca_2Si; positive electrodes, Ni_2S or $Fe_{0.9}Co_{0.1}S_2$; electrolyte, a mixture of LiCl, NaCl, $CaCl_2$, and $BaCl_2$. Engineering-scale cells will be fabricated in FY 1979 and FY 1980 with the goals of high performance (160 Wh/kg) and low materials cost in mass production ($10–15/kWh).

A project to fabricate and test full-scale NASICON electrolytes has been initiated. Sodium/sulfur cells will be built with the new electrolyte, and cost and life of these cells will be compared with those of current beta-alumina cells. At the engineering development stage, these systems will be transferred to the Advanced Batteries Development Subprogram.

Milestones and funding for this subprogram are shown in Figure III 6-10.

Supporting Research

The Supporting Research Subprogram supports the Near-term Battery, Advanced Battery Development, and Solar Applications Subprograms. It also supports research in electrochemistry and electrochemical engineering.

Five new battery and battery system components are being developed to enable more efficient performance of various batteries. These components are as follows:

- New gauges for determining the state of charge are required for those battery systems in which the specific gravity is constant. Phenomena such as electrolyte or electrode conductivity are being investi-

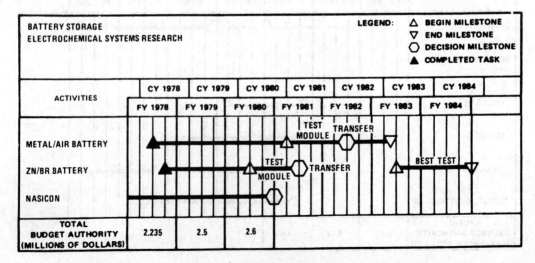

Figure III-6-10. Milestones and funding for electrochemical systems research.

gated to provide state-of-charge gauges for these batteries.
- Improved chargers are mandatory for lithium/metal sulfide and sodium/sulfur batteries to ensure that they are never overcharged. A lighter-weight charger for electric vehicle applications is also being developed.
- Control circuitry is also being developed to ensure balancing of cells during charge and discharge. Otherwise, small cell differences will become more severe as the battery is cycled, and eventually the cells will be forced into reversal, causing battery failure.
- Variations in temperature in large batteries, resulting from less than 100 percent efficiency, can shorten battery life and cause sudden failure. Effects of temperature increase in specific batteries are being measured, and methods of disposing of heat and maintaining normal operating temperatures are being sought.
- The high-temperature batteries require a hot environment in order for the system to function. Improved insulating jackets are being designed to control heat loss and to maintain constant temperatures. This will be especially important in the electric vehicle mission in which space limitations preclude the use of conventional bulky insulation.

A number of studies are being performed to support this subprogram:

- A field survey is underway to determine the amount of lithium ore that is available in the United States for commercialization of lithium/metal sulfide and lithium/air batteries. Cofunded with the Office of Fusion Energy, this survey includes test drilling and analysis
- Uniform testing, costing, and data reporting methods are being developed to allow comparison of alternative battery types.
- Environmental and hazard assessments are being made as required by the Environmental Development Plan (EDP). Presently, the zinc/chlorine battery and hydrogen hazards resulting from off-gassing by certain batteries when charging are being studied.
- The energy required to produce batteries and the energy input-output totals over the lifecycle of a specific mission are now being analyzed. These findings will be compared to alternate methods of achieving the same goal.
- The increase in battery manufacturing capability and the maintenance infrastructure required to support the introduction of electric vehicles will be analyzed.
- Management tools will be developed to aid in selecting alternative batteries for the electric vehicle, solar electric, and dispersed electric load-leveling missions. The parameters to be considered are: cost, life, energy density, power density, environmental effects, safety, and land area.

Milestones and funding for Supporting Research are presented in Figure III 6-11.

Electrolytic Technology

Electrolytic processes used by the industrial sector consume approximately 6 percent of the Nation's electrical energy. The goal of this subprogram is a 10-percent reduction in the specific energy consumption of these industrial processes by 1985 and 30 percent reduction by 2000. Emphasis is being placed on the aluminum and chlorine industries, which use 65 percent and 25 percent, respectively, of the electrical energy consumed by this sector. Tasks are also proposed in other areas. Overall energy savings (expressed in electrical energy equivalent) could be between 11 and 18 billion kWh per year by 1985 and between 33 and 110 billion kWh per year by 2000.

The Electrolytic Technology Subprogram supports the energy and resources conservation mission and currently is jointly funded by Industrial Energy Conservation and STOR. Some of the advances in electrochemical engineering made in this subprogram (e.g., in membranes, zinc electrodes, and fluid dynamics of cells) are applicable to both the Near-term Battery and Advanced Battery Development Subprograms.

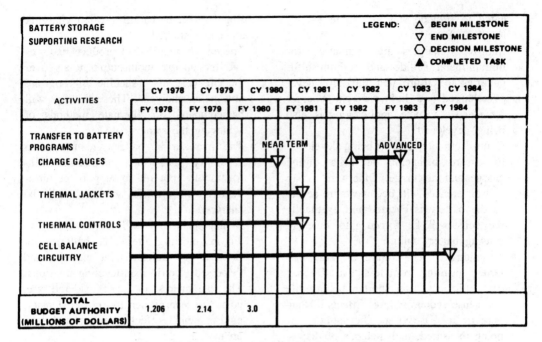

Figure III-6-11. Milestones and funding for supporting research.

RD&D efforts are directed toward developing a fundamental understanding of the electrolytic processes involved, collecting the required data base, improving the energy efficiency of the present processes, developing new and more efficient processes, and widely disseminating the developed technology.

Near-term technology development is directed primarily toward the improvement of existing processes in the aluminum and chlorine industries, leading to specifications and recommendations for improving existing technology in FY 1980.

Long-term technology development is directed toward developing advanced technology, such as dimensionally stable anodes in the aluminum process, and practical oxygen electrodes for the chlorine process and identifying those electrochemical technologies that could replace energy-intensive chemical processes. The end product of the long-term effort will be recommendation and specification of advanced process technology for commercial development.

Survey studies include the processes for aluminum production, chlorine/caustic production, metal winning, metal recycling, and electro-organic and inorganic synthesis.

Milestones and funding for Electrolytic Technology are shown in Figure III 6-12.

Dispersed Battery Applications

This subprogram is funded by the STOR budget, but its day-to-day management is provided by the Electric Energy System (EES) Division.

The objective of the Dispersed Battery Applications Subprogram is to incorporate developed battery storage technologies into electric utility load-leveling systems. There are substantial short-term (hourly) and long-term (seasonal) fluctuations in the demand for electrical energy. Power companies must have generating equipment available for periods of peak electrical consumption, such as daylight hours during weekdays. This generally means operating "peaking" turbines to provide supplemental generating capacity during times of high demand. Such turbines are expensive to operate and normally are fueled by either natural gas or highly refined oil.

The use of batteries for utility load leveling is illustrated in Figure III 6-13. Battery storage systems may be applied at utility central station, substation, and dispersed (customer)

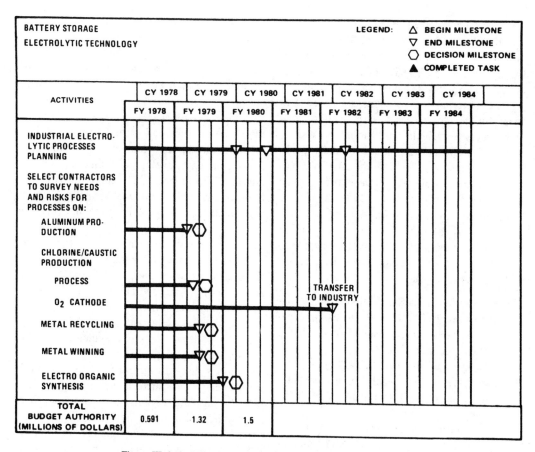

Figure III-6-12. Milestones and funding for electrolytic technology.

levels to achieve optimum results from each of the various technologies available.

This subprogram is heavily dependent on the success of the Advanced Battery Development Subprogram and contributes significantly to the Solar Electric Applications Program. The Advanced Battery Development Subprogram develops batteries to the point of qualifi-

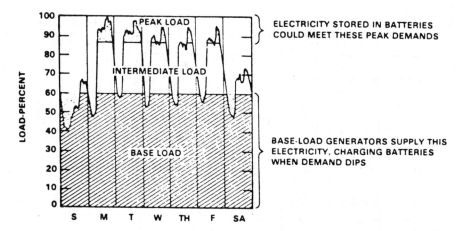

Figure III-6-13. Load leveling with batteries.

cation testing, after which they are installed in the BEST facility. The pilot lines used to produce the commercial prototype modules for examination in BEST, having been built at contractor expense as "items of contribution," are then available to produce advanced batteries for the Solar Electric Applications Program, and for vehicle applications as well.

Batteries appear best suited for daily peaking operations. By 1985, 5- to 10-MWh batteries, with a projected 10-year life and an anticipated cost of less than $35/kWh, will be demonstrated for this application.

Much work has already been accomplished in the development of battery systems applicable to utility load leveling. Specific qualities of these battery systems are as follows:

- Lithium/metal sulfide cells now consistently reach 400 to 600 cycles, and selected cells have reached over 1,000 cycles (the goal is 2000 cycles).
- Prototype utility-size (0.2-kWh) sodium/sulfur cells now reach more than 300 cycles at over 80-percent energy efficiency.
- Zinc/chlorine batteries (50-kWh) are being laboratory-tested. BEST facility testing could begin in early FY 1982.

Milestones and funding for this subprogram are shown in Figure III 6-14.

THERMAL AND MECHANICAL STORAGE

The Thermal and Mechanical Energy Storage Program of STOR develops reliable, efficient, and inexpensive energy storage technologies to support other DOE end-use divisions in their energy substitution and energy savings missions. In addition to thermal and mechanical technology development, this program includes a Technical and Economic Analysis Subprogram which supports both the Thermal and Mechanical Storage Program and the Battery Storage Program.

Thermal and mechanical storage technologies are most advantageous in applications such as solar and conventional heating and cooling, transportation vehicle shaft power, industrial process efficiency, chemical fuel production, solar and wind shaft power, and load-shifting for coal, nuclear, or solar thermal electric plants. The DOE end-use organizations that will benefit from development of thermal and mechanical storage technologies are the organizations having prime responsibility for satisfying the above applications. These are: Central Solar Technology (CST), Distributed Solar Technology (DST), Solar Applications (SA), Transportation Energy Conservation (TEC), Buildings and Community Systems (BCS), Industrial Energy

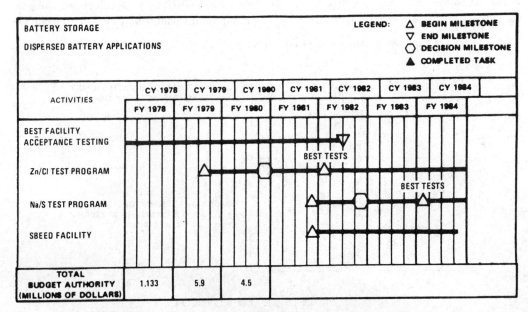

Figure III-6-14. Milestones and funding for dispersed battery applications.

Conservation (INDUS), and Resource Applications (RA—power marketing functions). Technology development for some of these applications also helps to satisfy missions within the divisions of Nuclear Power Development (NPD), Fossil Fuel Processing (FFP), Magnetic Fusion Energy (MFE), and Electric Energy Systems (EES).

Major accomplishments within the Thermal and Mechanical Storage Program include the following:

- A model of long-term aquifer storage of industrial waste heat with 70-percent recovery has been verified experimentally in field testss. Design studies for a demonstration site are underway.
- The first multicell solid polymer electrolyte (SPE) electrolyzer assembly to permit production of hydrogen for use as a chemical feedstock (in place of natural gas) has been tested.
- Special conductor cables for a 10,000-MWh superconducting magnet have been developed.
- Conceptual designs for electric utility peaking demonstration facilities using underground pumped hydro and compressed air storage have been completed.
- A 5-Wh/kg flywheel for regenerative in a Postal Service vehicle has been installed.

Table III 6-3 presents the recent and proposed budgets for the Thermal and Mechanical Storage Program.

Thermal Energy Storage

The goal of the Thermal Energy Storage Subprogram is to develop inexpensive and reliable technologies for storing or transporting energy as sensible heat, latent heat, and reversible chemical reaction heat. Sources of heat include the sun; fossil, biomass, and nuclear fuels; off-peak electricity; industrial processes; and utility cogeneration. The primary sources of cold are winter-chilled air and vapor compression chillers powered by off-peak electricity. Near-term applications of thermal energy storage include:

- Heating and cooling buildings
- Solar thermal power generation
- Industrial heat recovery

Applications in the longer term may include:

- Long-distance heat transport
- Base-load powerplant load following

Potential annual energy savings offered by these technologies through fuel substitution and energy recovery are 30 million barrels of oil equivalent per day (BOE) (about 0.2 quads) in 1985 and 1.1 billion BOE (about 6 quads) in 2000.

Thermal energy storage (TES) systems have varied applications. Some applications systems and components are commercially available and ready for use, but in other cases only theoretical studies concerned with the feasibility of using a particular system have been

Table III-6-3. Recent and proposed budgets for thermal and mechanical storage program.

THERMAL & MECHANICAL STORAGE SYSTEMS SUBPROGRAMS	BUDGET AUTHORITY (DOLLARS IN THOUSANDS)			
	ACTUAL FY 1978	ESTIMATE FY 1979	ESTIMATE FY 1980	INCREASE (DECREASE)
Thermal Storage	5,036	8,225	20,300	12,075
Chemical/Hydrogen Storage	3,883	7,400	2,500	(4,900)
Mechanical Storage	5,904	6,975	4,700	(2,275)
Magnetic Storage	1,750	2,300	700	(1,600)
Utility Applications	3,823	4,100	2,000	(2,100)
Technical and Economic Analysis	4,559	3,600	1,700	(1,900)
TOTAL	24,955	32,600	31,900	(700)

done to date. All commercially available TES systems use sensible heat storage. Whether these systems are used in a particular application depends on economic rather than technical factors. All phase-change material systems are presently undergoing laboratory experiments in which technical, instead of economic, problems are the primary concern. Large-scale underground storage of thermal energy has progressed to the stage of engineering design of field tests.

TES technologies are expected to have near- or mid-term impact on the electric utilities' peaking problem. This can be accomplished by load leveling in buildings and by operating the base-load central stations in a load-following mode. There is great potential for mid-term applications in the industrial sector. By 1990, these applications will make a significant contribution toward achievement of national energy goals.

Technology transfer is of central importance to the commercialization of TES technologies. Technology transfer is implemented by involving industrial and utility contractors and co-funders (such as EPRI) at very early stages in the development of each technology. The private sector cost-shares a small but increasing fraction of the Thermal Storage Subprogram.

Approximately 30 percent of the funds for thermal storage projects managed by STOR are entirely or jointly provided by other DOE divisions. The major thermal technologies being developed for systems that are the responsibility of other DOE divisions are:

- Aquifers for seasonal storage—Buildings and Community Systems and Solar Applications
- Heat storage for industrial heat recovery—Industrial Energy Conservation and Buildings and Community Systems
- Off-peak electric and solar daily heating and cooling storage—Resource Applications through Electric Energy Systems, Solar Applications, and Building and Community Systems
- Heat storage for power generation—Central Solar Technology and Electric Energy Systems

Most projects will be in the technology development phase in FY 1980. By 1984, work in operational systems development will be underway.

Table III 6-4 presents the recent and proposed budgets for the Thermal Energy Storage Subprogram.

Seasonal Storage in Aquifers. Seasonal storage refers to the storage of heat or cold for long periods of time to heat or cool buildings. Heated water (from utility or industrial reject heat sources or solar heat) can be stored in an underground aquifer during one period of the year and recovered and used to heat buildings at another time. Similarly, winter-chilled water can be stored for air conditioning in the summer.

Such a storage system has been studied for the JFK International Airport in New York; Figure III 6-15 illustrates the operation of this aquifer storage system. In winter, water would be withdrawn from the aquifer through a well, chilled to 34°F, and returned to the aquifer through a second well. The existing cooling towers (which are presently idle during the winter) would be used to chill the well water. The stored chilled water would, in turn, be used during the summertime to cool the circulating fluid in the existing air-conditioning system. Thus, the same water would be used year after year by simply reversing seasonally the direction of flow between cold and warm wells.

Because of the high potential for near-term energy savings by using aquifers for seasonal heating and cooling, transfer of this technology to the private sector will be expedited by initiating parallel field tests in several regions of the country. In FY 1980, multiple site-specific studies for regional field tests of summer cooling with aquifer-stored winter-chilled water will be initiated concurrently. This will provide a broad base of technical, economic, and environmental data. Several regional field tests will also be conducted for winter heating using either stored summer plus winter solar heat or industrial or utility waste heat. The effects of climatic and geologic variations on the economic viabililty of aquifer storage must be studied in diverse regions. These types of studies will expedite technology transfer to the private sector and will also help to achieve commercialization.

Table III-6-4. Recent and proposed budgets for thermal energy storage.

THERMAL ENERGY STORAGE ACTIVITIES	BUDGET AUTHORITY (DOLLARS IN THOUSANDS)			
	ACTUAL FY 1978	ESTIMATE FY 1979	ESTIMATE FY 1980	INCREASE (DECREASE)
Seasonal Storage in Aquifers	1,736	2,500	9,100	6,600
Industrial Waste Heat Storage	1,250	2,025	4,800	2,775
Solar Heating and Cooling	800	1,300	3,200	1,900
Customer Storage for Load Leveling Electric	400	700	1,200	500
Storage for Solar Thermal Power Generation	850	1,700	2,000*	300
TOTAL	5,036	8,225	20,300	12,075

*Does not include $5.5 million passthrough from central solar technology

The capital cost goal for the seasonal storage system is $0.01/kWt for heating and $0.10/kWt for cooling by 1981. With successful commercialization, the projected energy savings resulting from seasonal storage of winter-chilled water, solar heat, industrial waste heat, and utility reject heat are expected to be about 1 million BOE annually per unit. The potential annual savings are 10 million BOE (about 0.06 quad) by 1985 and 350 million BOE (about 2 quads) by 2000.

Milestones and funding data for seasonal storage in aquifers are shown in Figure III 6-16.

Industrial Waste Heat Storage. Industrial rejected process heat can be captured and used at another time, either for space or process heat or for generating electricity. One potential system for the steel industry, a cogeneration system to reduce peak electrical load requirements, is illustrated in Figure III 6-17. The system would use the exhaust gas steam from electric arc furnaces, store the heat in a packaged bed of slag, and then use this as the heat source to drive a steam turbine. Such a system could be cycled once or twice a day.

The objective of this activity is to identify, develop, and validate TES systems for near-

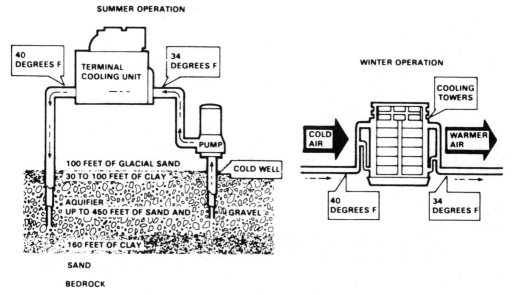

Figure III-6-15. Proposed natural cooling system for JFK Airport.

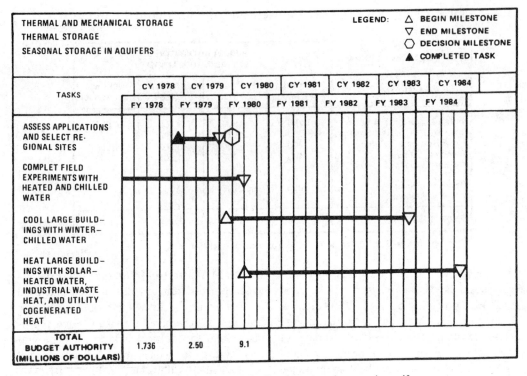

Figure III-6-16. Milestones and funding for seasonal storage in aquifers.

term application. Studies completed in FY 1978 identified and indicated technical and economic feasibility of TES in the food processing, aluminum, iron and steel, paper and pulp, and cement industries. These industries account for more than 25 percent of the total national industrial energy usage.

Studies of waste-heat storage systems in two of the energy-intensive industries (food processing and paper) could result in retrofit systems in selected plants, with subsequent technology transfer to INDUS by FY 1983. A third project, involving the use of the waste heat from an aluminum plant for district heating, is expected to be transferred to BCS in FY 1980. Projects in at least two other industries are planned. The cost goal is an initial investment payoff within 3 years. The annual energy savings from using storage in these industries are expected to average 250,000 BOE per plant. Assuming 20 operational plant retrofits by 1985, the annual energy savings would be 5 million BOE per year (0.03 quads). With expansion of this task to other industries (iron and steel, and cement) that have been shown to have high energy-savings potential, plus extension to smaller plants, the energy savings could be 105 million BOE annually (about 0.6 quads) by 2000.

Milestones and funding for this task are shown in Figure III 6-18.

Storage for Solar Heating and Cooling. The objective of this activity is to develop phase-change and chemical heat pump storage technologies for application to solar heating and cooling systems. The storage subsystem capital cost goals range from $5 to $10/kWh thermal by FY 1982, depending on the specific application and the effects of storage on overall system costs.

Phase-change materials can be used either in a separate storage module or integrated directly into building materials. Modular phase-change storage system concepts and materials are being developed for active solar heating and cooling. Building materials with phase-change storage capability are intended primarily to be used with passive solar heating and cooling systems.

A chemical heat pump (Figure III 6-19) is a device that operates as a result of the spon-

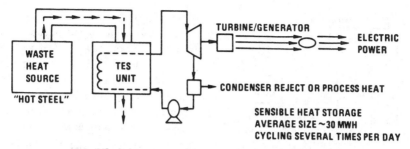

Figure III-6-17. Storage for heat recovery in the steel industry.

taneous movement of vapors between two different vapor pressure-temperature relationships. In solar heating and cooling systems, chemical heat pumps perform the functions of thermally driven heat pumps, chillers, and storage systems. Chemical heat pumps make possible lower overall cost and greater efficiency than separate solar heating, cooling, and storage units.

Methods of adding phase-change materials to concrete are being developed, and their economics evaluated. Prototypes will be developed for selected bulk storage concepts.

Laboratory testing is underway on several chemical heat pump storage concepts. One or two of these concepts will be selected for prototype development, performance testing, and economic evaluation. Chemical heat pump units could save about 33 BOE per residence annually. Assuming 150,000 units by 1985, the estimated energy savings are 5 million BOE annually (about 0.03 quads). Assuming 3,500,000 operational units by 2000, the estimated potential energy savings are 115 million BOE annually (about 0.65 quads). Energy savings from these advanced storage technologies will accrue mainly through their acceleration of the market penetration of solar heating and cooling.

Milestones and funding data for this activity are shown in Figure III 6-20.

Customer Storage of Off-peak Electricity. Off-peak electricity can be stored as thermal energy and released during peak-load hours to heat or cool buildings. Conservation results because off-peak electricity is generated by efficient base-load powerplants. Also, because

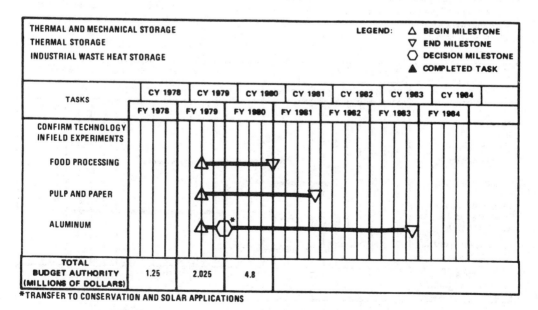

*TRANSFER TO CONSERVATION AND SOLAR APPLICATIONS

Figure III-6-18. Milestones and funding for industrial waste heat storage.

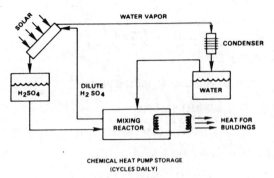

Figure III-6-19. Chemical heat pump storage system.

base-load plants generally use coal or nuclear fuel, storage of off-peak electricity will result in fuel switching from scarce liquid fossil fuels generally used by peak-load plants. Storage systems can be designed for both electric resistance and heat pump applications. The off-peak heating concept is depicted in Figure III 6-21.

The objective of this activity is to develop and demonstrate technologies that store off-peak electricity for heating and cooling at a later time. Solid sensible storage of electric heat is already commercialized in some areas of Europe. Investigations center largely on improving European systems and on developing systems that use phase-change materials.

This effort is being managed by Oak Ridge National Laboratory (ORNL). Until now, much of the work has been directed toward the technical aspects of the problem, particularly finding suitable phase-change storage materials. Economic analysis, which has been done to provide only an indication of potential competitiveness, was substantially expanded in FY 1979 to show the benefit/cost impact of such systems.

Customer-side-of-the-meter resistance heat storage systems for electric load leveling are being experimentally evaluated in residential environments in cooperation with the Technical and Economic Analysis Subprogram. By FY 1981, in agreement with EES, the commercial feasibility of customer-owned components for storing off-peak electric energy will be demonstrated for the 10- to 100-kWh range at a capital cost of $8/kWh. Experiments with off-peak storage units for heat pumps and air-conditioning are also planned. The projected annual energy shifting by displacing oil and natural gas is expected to be about 33 BOE per residence. Assuming 300,000 units in operation by 1985, the potential annual energy shifting is 10 million BOE (about 0.06 quad). With additional market penetration, about 115

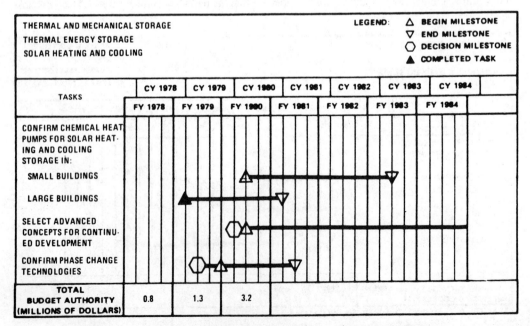

Figure III-6-20. Milestones and funding for storage for solar heating and cooling.

million BOE (about 0.65 quad) could be shifted by 2000.

Milestones and funding for this task are shown in Figure III 6-22.

Storage for Solar Thermal Power Generation. The objective of developing daily storage for solar thermal power generation is to support the implementation of solar electric power and to allow its penetration into a larger portion of the utility market.

Investigations are underway for both solar total energy system applications and for alternative solar central power station applications, such as proposed concepts using high-temperature Brayton and Rankine cycles. A closed-cycle Brayton system, an example of an alternative solar central power system, is displayed in Figure III 6-23. Preliminary studies of latent heat storage units suggest that these systems may provide a compact, cost-effective means of storing energy for solar total energy system applications.

The development program is closely coordinated with the Division of Central Solar Technology, in consonance with a plan jointly generated with that Division.

Major project areas with applications in solar thermal power generation (for transfer to CST) include hourly (buffering) and daily

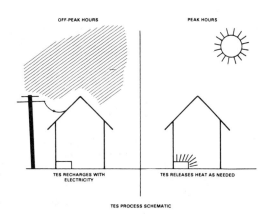

Figure III-6-21. Daily load leveling with thermal energy storage.

storage using sensible heat and latent heat. Smaller efforts are also planned to evaluate peak-load-following sensible storage for conventional base-load powerplants.

In the development of latent heat storage systems prior to FY 1979, contracts were awarded for laboratory studies and development of specific latent heat storage technologies selected on the basis of desirable physical properties, costs, performance, and material availability. Particular emphasis is now being placed on the development of active heat exchange concepts for phase-change storage, which will be used with advanced solar receivers.

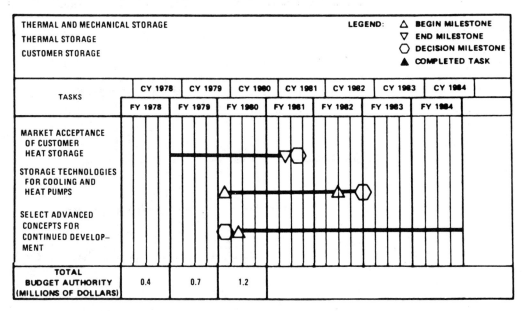

Figure III-6-22. Milestones and funding for customer storage of off-peak electricity.

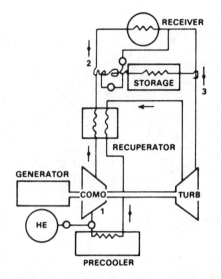

Figure III-6-23. Closed-cycle Brayton system.

ergy savings would result from the increased market penetration of solar thermal power made possible by the storage technologies.

Milestones and funding for this task are shown in Figure III 6-24.

Chemical/Hydrogen Energy Storage

The goal of the Chemical/Hydrogen Energy Storage Subprogram is to develop a technology that will economically substitute hydrogen produced from water for hydrogen produced from oil and natural gas. As a result of an increased demand for hydrogen as a chemical feedstock in recent years, more than 1 quad per year of natural gas, or about 5 percent of the total U.S. natural-gas consumption, is currently used to produce hydrogen. Successful commercialization of hydrogen energy systems technology can save a significant fraction of the natural gas that would otherwise be consumed to produce hydrogen.

The use of hydrogen as a fuel is also expected to increase rapidly after 2000. Hydrogen can be derived from water using nonfossil

By FY 1982, in agreement with CST, thermal storage components for solar and conventional power generation will be developed. These components will have a capital cost of $20/kWh and a service life of 30 years. With successful commercialization, significant en-

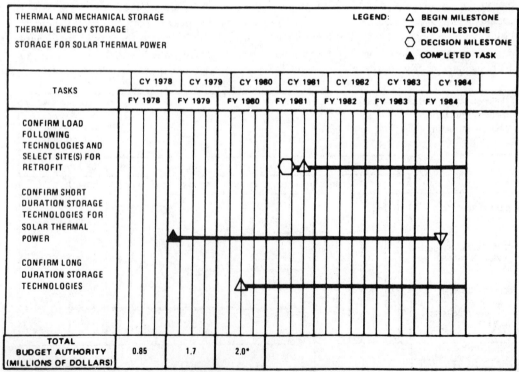

Figure III-6-24. Milestones and funding for daily storage for solar thermal power.

primary energy resources such as nuclear, geothermal, hydro, solar, or readily available fossil resources such as coal and oil shale. Hydrogen can also be used as a supplement or replacement for natural gas. Several studies indicate that this technology could result in natural-gas savings of about 50 million BOE (about 0.3 quad) per year by 2000 and 800 million BOE (about 4.5 quads) per year by 2025. This estimate includes the advanced electrolysis technology required to produce hydrogen at a cost of $6/million Btu, with process efficiency of 90 percent and capital cost of $100/kW.

Several technologies for production of hydrogen from water are being explored, including water electrolysis, closed-loop thermochemical cycles, and direct production of hydrogen using sunlight (in the applied research stage).

The hydrogen production activity satisfies legal commitments with nine other countries under an IEA agreement. This agreement provides that all technologies developed under its aegis are available for licensing by all members.

The electrolytic method of hydrogen production is a relatively simple process in which the hydrogen and oxygen products can be easily separated from the water feedstock, and no byproducts or pollutants are emitted at the production site. The main impediment to early large-scale commercialization of this technology is the availability and increasing cost of the input electric power. Both liquid alkaline and solid polymer (Figure III 6-25) electrolytic systems are being developed for low-cost, high-efficiency operation.

Thermochemical water splitting is another method of providing a nonfossil renewable source of hydrogen. The general method of thermochemical water splitting is to drive a series of chemical reactions by supplying energy, usually heat, and thereby accomplishing hydrogen and oxygen release in separate chemical steps. Objectives of this process are to avoid high temperatures (above 2,500°C) of direct thermal decomposition of water and to reduce the electricity requirements for the electrolysis of water. Development and testing of the sulfur iodine and hybrid sulfur cycles,

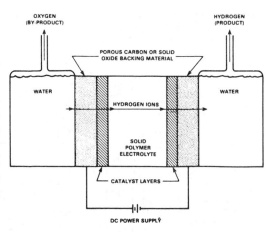

Figure III-6-25. Solid polymer electrolytic water-splitting process.

the two most promising approaches, will continue in the early 1980s.

A 500-kW solid polymer electrolyzer was designed and constructed during FY 1979 and will continue through FY 1981. Design and construction for a 5-MW electrolysis plant will begin in FY 1980, with field testing to be initiated in FY 1984. In FY 1982, a decision will be made whether to field-test a unit for producing hydrogen for chemical feedstock. Bench-scale testing of the hybrid sulfur cycle and of the sulfur iodine cycle, the most promising thermochemical cycles, will be completed in FY 1981 and FY 1982, respectively. In FY 1982, one cycle will be selected for scale-up to 50-kW size.

Milestones and funding for this subprogram are shown in Figure III 6-26.

Mechanical Energy Storage

The near-term objective of the Mechanical Energy Storage Subprogram is to provide safe and economical flywheel systems for regenerative braking in electric/hybrid vehicles (Figure III 6-27) by FY 1984. Such flywheels will improve vehicle acceleration, save energy, and increase the life of batteries in vehicles.

The energy storage capability of a flywheel depends primarily on the mechanical properties of the materials from which it is fabricated. For applications in which storage-system weight is a limitation, a high strength-

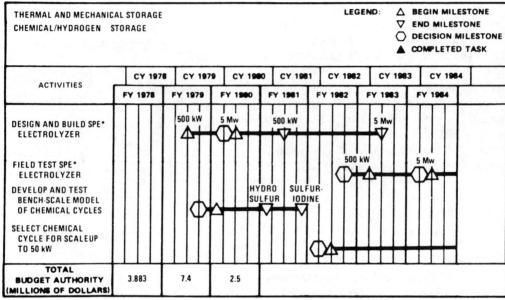

Figure III-6-26. Milestones and funding for chemical/hydrogen energy storage.

to-density ratio is an important material property. Because of their high strength-to-density ratio, composite materials are particularly attractive for flywheel fabrication.

The energy storage capability of a flywheel is also influenced by the stress distribution in the rotor, which depends on the rotor cross-section (the moment of inertia depends on geometry) and on the way in which it is assembled. Possible flywheel shapes include the flat disk/cyclinder, concentric rings, thin rim, and tapered disk.

A flywheel regenerative braking system currently being developed under joint funding with the U.S. Postal Service is expected to increase from 350 to 500 the number of stops that Postal Service jeeps can make on one battery charge. In FY 1978, characterization of high-strength composite materials was completed for flywheel regenerative braking systems. By FY 1984, flywheel storage density could increase from the current level of 14 to 40 Wh/lb. These systems can yield up to a 30-percent energy saving when incorporated into electric vehicles for city driving. Laboratory and on-the-road testing of automotive flywheel systems are planned during FY 1981 to FY 1984, with the objective of achieving initial large-scale technology demonstration by FY 1984.

By Fy 1985, net energy savings of approximately 0.05 million BOE are projected for flywheel systems. This estimate is based on a savings of 2.5 BOE per vehicle annually and on the assumption that 20,000 electric/hybrid vehicles will be operational. By FY 2000, assuming 1.8 million operational vehicles, savings could be approximately 20 million BOE (about 0.1 quads). The fuel-substitution potential of flywheels is approximately ten times the amount of net energy savings.

Milestones and funding for this subprogram are shown in Figure III 6-28.

Superconducting Magnetic Energy Storage

The goal of the Superconducting Magnetic Energy Storage (SMES) Subprogram is to

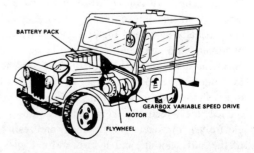

Figure III-6-27. Battery-flywheel hybrid electric car.

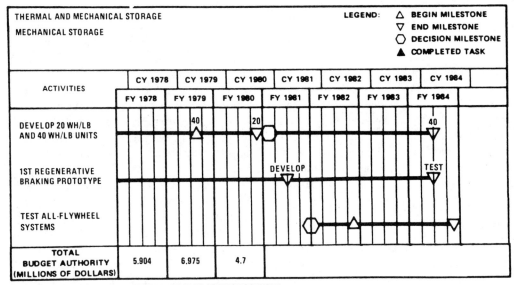

Figure III-6-28. Milestones and funding for mechanical energy storage.

produce the technology for two types of energy storage devices, one for utility load leveling and the other for transmission line stabilization.

Energy can be stored in an electromagnet, which consists of a cylinder or torus wound with conducting wires that carry electric current. At very low temperatures (near absolute zero), certain conductors lose all electrical resistance to direct current. A direct current at such temperatures will persist indefinitely and maintain the associated magnetic field for future energy withdrawal. The efficiency with which stored energy can be recovered in such a system (Figure III 6-29) is very high (85 to 90 percent) and is one of the main advantages of this concept for utility load leveling. The magnets would be capable of storing 1,000- to 10,000-MWh of electricity and would have short response times. The ability to store off-peak energy in efficient, cost-effective, large magnetic units could result in annual fuel shifts of about 3 million BOE per year for each 500-MW unit. Assuming 12 plants in operation by 2000, total fuel shifting of approximately 35 million BOE per year (equivalent to 0.2 quads) would be achieved.

The second objective of this subprogram is a much smaller state-of-the art magnet. It will be used by electric utilities to dampen system oscillations caused by short-term mismatches between customer demands and generating equipment. The addition of a small, 10-kWh magnetic energy unit to a system is expected to improve the transmission capability of a potential user by approximately 25 percent.

Currently, for central station energy storage, less expensive conductors and less expensive means of transmitting electromagnetic forces to bedrock are being developed within this subprogram. High-purity aluminum could substitute for copper as the metal stabilizer around the superconductor, resulting in savings of millions of dollars per unit. Polyester-glass could substitute for epoxy-fiberglass as the low thermal-conductivity structural member and transfer the electromagnetic loads

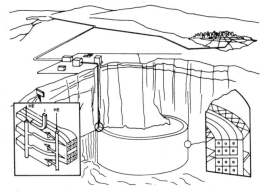

Figure III-6-29. Superconducting magnetic energy storage facility.

from the conductor to the bedrock, yielding savings of over $100 million per plant. If there are successful results from the conductor and support studies by FY 1984, a decision would be made to proceed with a prototype subscale experiment.

The design phase of the smaller magnetic storage system for transmission line stabilization is planned for completion in FY 1979, and fabrication of components will be initiated in FY 1980. Milestones and funding for this subprogram are shown in Figure III 6-30.

Utility Applications

The goal of this subprogram is to accelerate commercialization of underground pumped hydro storage (UPH) and compressed air energy storage (CAES) in electric utility systems through design and cost studies and large-scale field tests at utility sites.

UPH is being considered by DOE and the electric utilities as an alternative to the gas turbine for load-leveling and peak-shaving applications. Energy is stored during off-peak hours by pumping water from a lower (underground) reservoir to an upper (surface) reservoir (Figure III 6-31). When energy is needed to help satisfy a utility's peak power-generation demand, the water is released to the lower reservoir, turning a water turbine and electrical generator as it falls. The round-trip efficiency of such a plant is 70 to 75 percent. Off-peak energy for pumping is generated by base-load coal or nuclear powerplants.

The upper reservoir of a UPH station may be an existing body of water or an artificial lake formed by dikes and dams. The lower reservoir may be created by excavating, in hard rock, a cavern at the desired distance below the upper reservoir. The cavern would be a room and pillar design with passage dimensions approximately 50 feet wide by 75 feet high. It would be cut into a hard rock formation; softer rocks would be unacceptable because of the erosive nature of turbulent water flow.

The power station would be in a separate cavern housing both the pumping/generating equipment and associated control and service facilities. Reversible pump/turbines would be employed to perform the dual functions of generating and pumping in a pumped-storage project. The balance of plant, consisting primarily of the switchyard, would be located above ground.

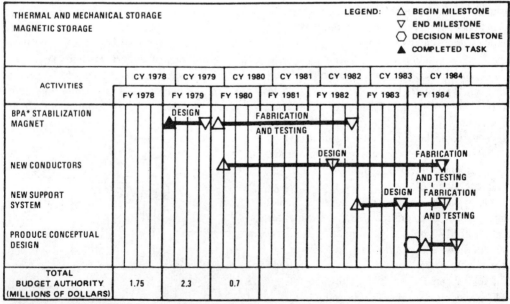

Figure III-6-30. Milestones and funding for superconducting magnetic energy.

Energy Storage Systems 513

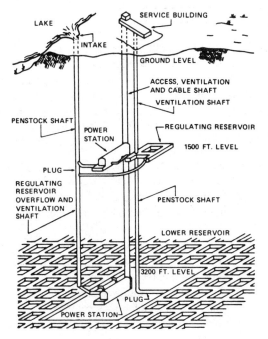

Figure III-6-31. Underground pumped hydroelectric storage.

CAES systems have near-term applications for utility load leveling. Air is compressed using off-peak electrical power; the compressed air is then stored in large, naturally occurring underground reservoirs, aquifers, rock or salt dome caverns, or depleted oil and gas fields (Figure III 6-32). During peak-demand periods, the compressed air is recovered and used to operate gas-turbine generators. The result is an estimated 60-percent oil saving over the simple gas-turbine systems that are currently used.

Two system designs being evaluated are: (1) an isothermal and (2) a no-oil system. In the isothermal design, any heat associated with compression is lost to the atmosphere and must be replaced during expansion by combustion of fossil fuel. In the more advanced no-oil design, the heat is removed from the compressed air and stored for later reinjection into the compressed air stream prior to expansion through a turbine.

Compressed-air storage systems have two major components: underground air storage chambers and above-ground generating systems. Aquifers, mined caverns, and salt domes appear to be the most promising storage chamber options because of their stability and the widespread occurrence of suitable geologic strata. The above-ground facilities

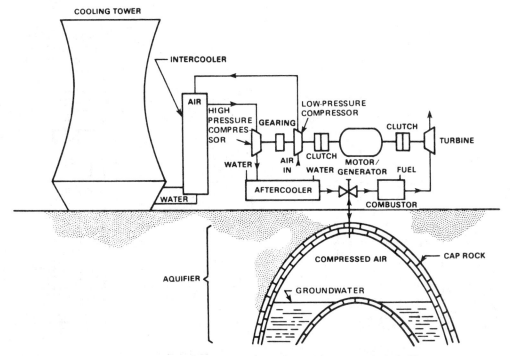

Figure III-6-32. Hydraulically compensated underground compressed air facility.

will include equipment similar to that commonly used by electric companies.

Requested funds are expected to allow completion in FY 1981 of CAES work initiated in FY 1977 by three utility companies as prime contractors for three preliminary designs, including cost analysis and specific site evaluations for compressed air storage in an aquifer, a salt dome, and a hard rock cavern.

The technology readiness of UPH precedes CAES, but the storage system is more restricted by site availability, since only mined rock caverns are suitable. The preliminary design work, initiated in FY 1978, is planned for completion in FY 1980.

In FY 1981 the UPH and CAES projects will be considered for transfer through EES to RA and the private sector for demonstration facility construction and testing. The estimated energy shifting for petroleum and natural gas using these technologies is 1 million BOE per plant per year. The potential annual energy shifting is 2 million BOE (about 0.01 quad) by 1985 and 175 million BOE (about 1.0 quad) by 2000.

The milestones and funding for this subprogram are shown in Figure III 6-33.

Technical and Economic Analysis

The Technical and Economic Analysis Subprogram provides STOR with the information needed to establish priorities and to evaluate RD&D objectives. This function includes project evaluations at several stages in the energy storage technology RD&D cycle, such as:

- Identifying RD&D needs for energy storage technologies and systems
- Evaluating competing energy storage options for contribution to national energy goals
- Assessing potential environmental concerns and benefits for new, emerging energy storage technologies
- Coordinating commercialization strategies for technology development and systems integration with energy storage customers

Two major activity areas have been created to develop information and criteria for energy storage research and development. These are:

- Applications analysis
- Technology information systems

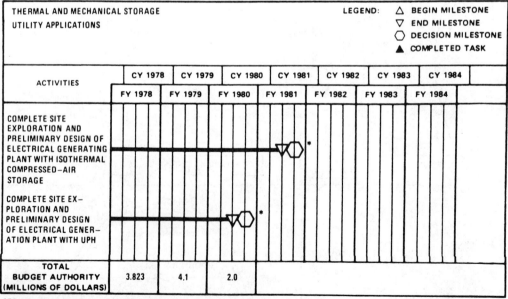

Figure III-6-33. Milestones and funding data for utility applications.

Applications analysis provides information related to management responsibilities in each of the principal energy storage applications areas. Applications analysis addresses the feasibility of technical proposals, the environmental impact of specific technologies, the need for new research in each application area, and the contribution of energy storage systems to the achievement of national energy goals.

The Technology Information Systems Activity provides decisionmakers with the information to meet program objectives. The interaction of these activities is illustrated in Figure III 6-34.

To date, a number of RD&D evaluations and applications analyses have been completed. Several of these have resulted in the identified technologies that show promise for development in specific application areas. Others have contributed to decisions to defer support for technologies having apparently low technical and economic feasibility.

Major accomplishments and impacts in FY 1978 and FY 1979 include:

- Analysis of specific energy storage systems for vehicle propulsion

- Completion of study and field evaluation plans for customer-owned thermal energy storage

Resource support for the activities of the Technical and Economic Analysis Activity has been included as a separate line item of the STOR FY 1980 budget. The recent and proposed budgets for the Technical and Economic Analysis Subprogram are shown in Table III 6-5.

Applications Analysis. Applications analyses quantitatively determine the suitability of energy storage technologies for specific end-use applications. To identify customers and commercial opportunities for energy storage technologies, information is required on the technical risk, expected benefits, and RD&D costs of developing a new energy storage technology. The Technical and Economic Analysis Subprogram develops this information by assessing the potential impacts of storage technologies on:

- Energy savings and fuels substitution
- Costs of energy services
- Health, safety, and the environment
- Implied changes in lifestyles

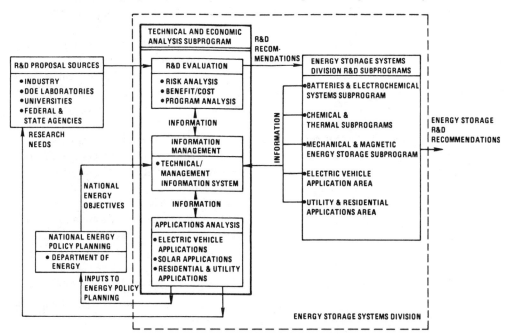

Figure III-6-34. Technical and economic analysis program flow chart.

Table III-6-5. Technical and economic analysis budgets.

TECHNICAL AND ECONOMIC ANALYSIS ACTIVITIES	BUDGET AUTHORITY (DOLLARS IN THOUSANDS)			
	ACTUAL FY 1978	ESTIMATE FY 1979	ESTIMATE FY 1980	INCREASE (DECREASE)
Applications Analysis	4,059	3,200	1,400	(1,160)
Technology Information Systems	500	400	300	(100)
TOTAL	4,559	3,600	1,700	(1,900)

In FY 1977, a multilaboratory study was initiated to evaluate a wide range of energy storage devices (batteries, mechanical, hydrogen and thermal). Using decision analysis techniques, the study identified storage devices and power system configurations potentially capable of providing viable alternatives to current petroleum-fueled vehicles.

In FY 1979 and FY 1980, the Technical and Economic Analysis Subprogram coordinated, with other DOE Divisions and SERI, a multilaboratory study to assess the performance characteristics and requirements of energy storage for a wide spectrum of solar energy applications. During FY 1979, a study plan was developed, and several high-priority solar applications were selected for detailed analysis in FY 1980. The Division of Energy Storage Systems plans to use the information from these analyses to establish priorities for new programs to develop solar energy storage systems.

Analysis efforts in FY 1977 and FY 1978 identified technically feasible energy storage technologies that will be commercially available in the near, intermediate, and long-terms. Assessments to be conducted in FY 1979 and FY 1980 will determine optimum size criteria for storage units, energy storage RD&D needs, and necessary electric rate structures to minimize total costs for both the customer and the utility. Assessments of European experience indicate that the joint introduction of time-of-day pricing and customer-side-of-the-meter space heating and cooling storage systems in the United States could save 0.042 and 1.14 quads of oil annually by 1985 and 2000, respectively, by shifting the peak load to base-load plants. Moreover, it is estimated that the introduction of customer-owned storage for space heating and cooling systems could provide utilities with annual capital savings of $120 million by 1985, and $3.4 billion by 2000.

In a related project, energy storage devices are being installed and will be tested and evaluated in regional project sites during FY 1979 and FY 1980. This study, using previously installed metering equipment and data reduction and analysis facilities from ERA rate experiments, is designed to test distributed and hybrid energy storage systems and to reduce RD&D costs significantly. Also, technology characterization data, to be developed for comparing distributed energy storage technologies for residential applications, will be widely disseminated through the energy storage integrated information system.

Milestones and funding for Applications Analysis are shown in Figure III 6-35.

Technology Information Systems. The key to effective RD&D investment decision-making is the availability of accurate data and information. To provide complete and reliable data for analysis, a computerized technical/management information system is being developed. The system will contain standardized data bases for the following energy storage technology areas:

- Cost data
- Technical performance and design data
- Material properties

In addition, the data elements for an environmental, health, and safety data base are being

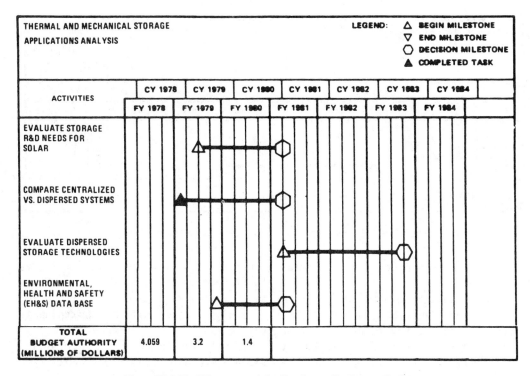

Figure III-6-35. Milestones and funding for applications analysis.

defined with a view toward establishing an information support system for analyses of all data base components.

When completed, this data base will contribute to making sound technical, economic, and environmental decisions regarding energy storage RD&D projects at each stage of the RD&D cycle.

A data base for physical properties of materials for thermal energy storage and flywheel applications was completed in June 1978. In FY 1979, new data bases are being designed for energy storage technology characterizations in solar and dispersed applications. Elements in the data base include: technical performance; cost and design specifications; environmental, health, and safety impacts; and energy savings and fuel switching benefits of energy storage technologies. The integrated energy storage information system supports RD&D activities in both the Battery Storage and Thermal and Mechanical Storage Programs by providing accurate, timely information for RD&D decisionmaking.

Presently, a computerized Storage Program Evaluation and Review System (SPERS) is being developed. This system will provide a risk-analysis model and a data management and control system to provide state-of-the-art evaluations of new RD&D proposals.

Milestones and funding for Technology Information Systems are shown in Figure III 6-36.

OTHER PROGRAM ASPECTS

Four other aspects of the Energy Storage Systems Program are discussed in other chapters of this document. These aspects are: field activities, commercialization, environmental implications, and socioeconomic aspects.

Field activities are widely dispersed because application of energy storage systems is region-dependent. For example, diurnal heat storage is employed mainly in winter-peaking utilities located in northern climates.

Commercialization of the energy storage program presents numerous problems. In some cases, such as daily heat storage, CAES, and near-term batteries, existing hardware experience can be applied to the problem. Many of the other technologies will require special

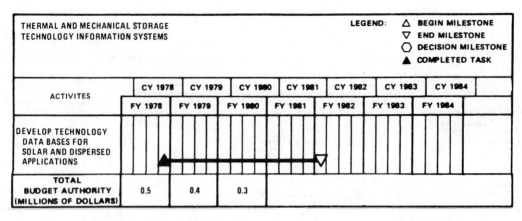

Figure III-6-36. Milestones and funding for technology information systems.

commercialization programs to achieve widespread utilization.

The environmental implications of energy storage systems are generally not severe. Potential hazards exist in the manufacture of advanced batteries and in the utilization of thermal, chemical, and flywheel energy storage systems. As with most large systems, adverse ecosystem effects are possible with large energy storage systems on adjacent properties.

Some of the storage technologies may introduce socioeconomic problems. One question involves the ownership of customer-side-of-the-meter storage devices that benefit the utility system. Other areas of uncertainty include the impact of subsurface energy storage systems on adjacent properties.

Regional Activities

The Energy Storage Division relies heavily on several DOE operations—especially Chicago, San Francisco, Albuquerque, and Oak Ridge—for contract administration and support. Technical direction for many selected major projects has been assigned to field organizations. In addition to pursuing RD&D, the National Laboratories perform a technical management function while subcontracting to industry and universities. The laboratories organize and direct industrial groups in specific storage activities. Lead laboratories, including Argonne, Brookhaven, Oak Ridge, Sandia Livermore, Sandia Albuquerque, Pacific Northwest, Lawrence Livermore Laboratory, NASA Jet Propulsion Laboratory, and Lewis Research Center, have been designated for each energy storage technology (except magnetic storage).

More detailed information concerning regional activities is presented in Chapter 7.

Commercialization

Commercialization of energy storage technologies is closely linked to the systems that use these technologies.

Electric storage batteries are a major cost element of electric vehicles. Economies of scale resulting from expanded battery production can be secured only with the increasing market acceptability and market penetration of electric vehicles. A number of incentives to promote market penetration are being studied. These incentives include tax credits, production subsidies for the construction of battery plants, financing plans, and battery lease/exchange/recycle plans.

Commercialization of thermal storage technologies is implemented through the participation of industrial and utility contractors in technology development, beginning at an early stage. Private companies are working with DOE on a cost-sharing basis for a small but increasing portion of the thermal storage program.

Technology transfer activities are also planned for the Chemical/Hydrogen Storage Program in FY 1983 and FY 1984 to assist eventual commercialization. Nuclear Power Development (NPD), Central Solar Technology (CST), and Resource Application (RA) will be involved. Funds to convert high-cost water electrolysis technology, developed for space applications, into a large-volume, low-cost means for hydrogen production are provided by STOR and by General Electric (the contractor), with additional funds from Niagara-Mohawk Power Co., Empire State Electric Energy Research Company, and the New York State Energy Research and Development Authority.

Private companies have been involved in the commercialization of mechanical energy storage systems, primarily in defining the needs and requirements for technology development. They are participating in the R&D program through Government contracts or joint-funding agreements.

Environmental Aspects

Potential adverse impacts associated with energy storage systems (STOR) are related primarily to health and safety aspects of production and operation. Toxic or hazardous substances may be accidentally released by batteries, posing a threat to exposed workers and to the public. Wastes generated during the manufacture of components and at the end of the useful life of batteries present a solid waste disposal problem. Finally, there is potential for impairing water quality during operation of CAES and UPHS systems. STOR efforts are coordinated with other DOE divisions and Federal organizations to minimize duplication of efforts and to ensure progress toward resolving energy storage environmental concerns.

Activities in the STOR area reflect the environmental issues previously discussed.

Socioeconomic Aspects

As consumers and industry seek to conserve energy, energy storage systems become increasingly important. Potential market applications include photovoltaics, wind power, electric vehicles, building heating and cooling, and utility load leveling for small- and large-scale uses. Optimum use and wide deployment of these systems depend, to a significant degree, on the share of energy expenditures in the budget of an individual firm. The industrial sector, which has demonstrated a much greater propensity for conserving energy than other sectors, is likely to adopt the use of such systems more readily than other sectors.

7.
Regional Activities

INTRODUCTION

Regional and field activities are integral parts of DOE policy and program planning. Consequently, the Federal Solar, Geothermal, Electric and Storage Systems Program development efforts include using research talent and facilities throughout the Nation. National laboratories, universities, and numerous private companies contribute to the design, development, and construction of experimental devices and also conduct basic research in related areas.

RESEARCH BUDGET ALLOCATION

The geographic distribution of federally supported solar, geothermal, electric, and storage systems research is determined by several intrinsic technological factors. These include:

- Existence of high-technology experimental facilities and technical personnel in universities, national laboratories, and private firms
- Selection of sites for construction of research devices, leading to the necessity of concentrating Federal expenditures for hardware, operation, and follow-on research at particular locations

The design, construction, and operation of new experimental machines has been accompanied by extensive work in applied research and development of new technologies for further research. These activities take place at DOE laboratories, several educational institutions, and numerous private firms. These institutions often possess specialized research facilities and personnel, and research may therefore be effectively distributed over a wide geographic range of contractors.

The research budget for ETS design research or equipment fabrication is distributed among DOE laboratories, universities, and private firms. Table III 7-1 shows the distribution of the estimated FY 1979 research budget by program areas and organization recipients. Much of the funds allocated for major projects are dispersed to architect-engineers, instrument manufacturers, engineering designers, hardware fabricators, and private-sector researchers at each research location. In addition, funds that are allocated to other Federal agencies, national laboratories, industry, and universities are subcontracted, in part, to small businesses, disadvantaged businesses, and other companies.

DECENTRALIZATION POLICY

In October 1977, Energy Technology began the difficult task of formulating requirements for decentralization of project management. The knowledge gained by developing initial decentralization plans and information has contributed to the promulgation of the Under Secretary's May 31, 1978, memorandum on an overall program and project management

Table III-7-1. Distribution of FY 1979 (estimated) research budget (dollars in millions).

ETS DIVISION / ORGANIZATION	DISTRIBUTED SOLAR	CENTRAL SOLAR	GEOTHERMAL	ELECTRIC ENERGY	STORAGE	TOTALS (DOLLARS IN MILLIONS)
OTHER FEDERAL AGENCIES	53.9	18.5	1.3	1.2	3.6	79.5
NATIONAL LABORATORIES	64.1	35.4	40.2	9.9	24.5	178.1
INDUSTRY	15.1	27.4	99.6	23.0	28.3	127.6
UNIVERSITIES	24.0	4.6	12.1	3.7	1.6	46.0
TOTALS (DOLLARS IN MILLIONS)	157.1	85.9	153.2	38.0	58.0	431.2

system for DOE outlay programs. It also provided the basis for "Decentralization of Project Management for all DOE Outlay Programs."

The responsibilities of the DOE Headquarters personnel are related to the development of overall plans, establishment of priorities, and completion of analyses to ensure coherent, goal-oriented programs that are related to the NEA, NEP, related plans, and legislation. In addition, efforts are directed toward the development of the overall budget, presentation of DOE programs to Congress, and then assessment of the adequacy and progress of approved efforts to meet the stated goals and objectives. This essentially is "program management."

Detailed management of projects, which includes managing resources to accomplish a given objective within prescribed funding, performance, and schedule constraints, is the responsibility of the "field." This essentially is "project management." A major element of this division of effort is the execution of a Project Charter, approved by headquarters, which sets forth the responsibilities, authority, and accountability for the work to be performed.

Efforts have been underway to evaluate, plan, and implement the decentralization of programs. Approximately 54 percent of the total expenditure was managed in the field in FY 1978. In the next 2 years, it is estimated that 70 percent and 77 percent, respectively, will be under field management. At that time it is expected that, with the exception of a few projects which remain at headquarters, programs will have been completely decentralized.

Programs are being decentralized at three levels: programs, major segments of programs, and projects. Operations offices and technology centers are being restructured to execute programs, lead laboratories are being restructured to execute major segments of programs, and all DOE field establishments are being reshaped to effectively conduct project management.

Decentralization is well underway, but this is a difficult and time-consuming activity, particularly because program degradation is not affordable. This plan will be updated quarterly, but monitored on a continuing basis to ensure that problems are identified and resolved quickly, and that the plan is executed as scheduled.

GEOGRAPHIC PROGRAM DISTRIBUTION

Activities in the Solar, Geothermal, Electric and Storage Systems Program have ramifications beyond the development of viable technologies. The development of new technologies will have an impact on the various regions of the country, depending on factors such as climate, existing patterns of energy use, projected energy demand, and regional resources.

Activities in the solar and geothermal programs have physical factors that determine their regional impacts. In solar technologies, these factors include insolation, wind availability, access to suitable ocean thermal sites with adequate temperature differential, and suitability for growing biomass. Activities in the geothermal program depend on water temperature, salinity, and other factors characterizing the geothermal resources of an area. Certain energy storage technologies, such as compressed air and underground pumped hydro, require specific geologic formations. The following sections of this chapter provide a summary of some of these physical factors that determine regional emphasis.

Other regional factors include the types and amount of equipment installed for power generation and transmission, present and projected load characteristics, and existing patterns of fuel supply. These considerations determine which energy sources can be integrated most rapidly with the existing system and where that energy can be used.

The Department of Energy's administrative and research operations as well as various levels of program and project management are dispersed regionally. Various levels of administrative, program, and project management support are received from field offices in San Francisco, Calif., Albuquerque, N. Mex., Richland, Wash., Chicago, Ill., and Oak Ridge, Tenn. DOE research laboratories include Oak Ridge, Los Alamos, Lawrence Berkeley, Lawrence Livermore, Sandia, Argonne, Brookhaven, and Pacific Northwest. A notable development in this regard is the formation of the Solar Energy Research Institute at Golden, Colo., where a broad program of technical support is being conducted.

Solar Technology Program

Generally, solar thermal technology and photovoltaic concentrating systems are best suited to regions of high total insolation. The annual variation of sunshine in the United States indicates that solar thermal and concentrating photovoltaic systems will function most efficiently in the Southwest, with reasonable expectations across the South and much of the West.

Photovoltaic flat systems, which make more effective use of diffuse sunlight, are more broadly applicable across the southwestern, south central, and southeastern regions of the Nation.

Wind Energy Conversion Systems (WECS) will be most applicable in regions that have the best combinations of high wind velocity and maximum number of days of availability. In general, wind power is greatest in the coastal Northwest, the Northeast, and the high Central Plains.

Ocean thermal technology will be applicable generally along the Gulf Coast, where ample sites are available within the required distance of land. In addition, Puerto Rico and Hawaii are well endowed for the exploitation of such systems.

Fuel produced from biomass will be widely distributed throughout the country. In terms of production, biomass cultivation and processing will be most effective in areas that are best suited for growing wood, namely regions with long growing seasons and adequate rainfall. Also, the land used must not be a prime agricultural resource. The south central and southeastern regions are best endowed for supporting this technology. Throughout the Nation, residues from the lumber industry and biomass from substandard and diseased forests will be used, with maximum effect in the near term.

Figure III 7-1 provides a geographic plot of all projects, valued at more than $200,000 value, funded by the FY 1978 solar program budget.

Geothermal Program

The natural distribution of geothermal resources lends itself to geothermal program activities with a strong regional nature. Descriptions of the geographic suitability of various regions follow.

- Hawaii, geographically isolated from the coterminous United States, has near-surface hot water as the result of the islands' volcanic origin. Because Hawaii has few alternative indigenous energy sources, geothermal energy may supply much of the islands' electricity in future years. The wellhead generator in Hawaii, sponsored by the Division of

Figure III-7-1. Solar projects valued at more than $200,000, FY 1978 solar technology program budget.

Geothermal Energy, is expected to spur further development in both Hawaii and in its sister islands.

- The Far Western States, because of their relative geologic youth, are also generously endowed with geothermal resources. California has hot brines in the Imperial Valley and dry steam at The Geysers. The Cascade Range in Oregon and Washington, with its volcanic activity, promises hot-water resources associated with subterranean chambers. The eastern slope of the Sierra Nevada Range, from southern California northward through Nevada, has significant geothermal potential. Program planning and resource definition activities in these areas are conducted on a regional basis.
- The Rocky Mountain area, from Canada to the Mexican Border, has a large number of geothermal prospect areas, some with hot water, many more with moderate temperatures, and quite a few with hot dry rock potential. The initial resource definition work is carried out under the state-coupled programs.
- The geopressured resources of the Gulf Coast lend themselves naturally to regional program activities. The State universities, State geologists, and State governments in Louisiana and Texas have been working together and with DOE since the inception of the geopressured program.
- Because of its relatively advanced tectonic age, the eastern United States was, until quite recently, believed to contain few geothermal resources with near-term development potential. In 1978, however, geologic investigations revealed several probable moderate temperature resources in the Atlantic Coastal Plains in the Mid-Atlantic States. The prospects for similar resources in the coastal regions of the Southeastern States seem promising.

Figure III 7-2 plots the location of all projects valued at more than $200,000, funded by the FY 1978 geothermal program budget.

Electric Energy Systems Program

Several activities in Electric Energy Systems (EES) have regional implications. These activities include RD&D aimed at preparing for future sources of electricity generation. It is likely that electricity will be generated from both large and small generating sources. Large central-station generating facilities will probably be located in rural areas remote from urban centers. There will be a large demand for energy in urban areas, and it will be necessary to transmit large blocks of energy effi-

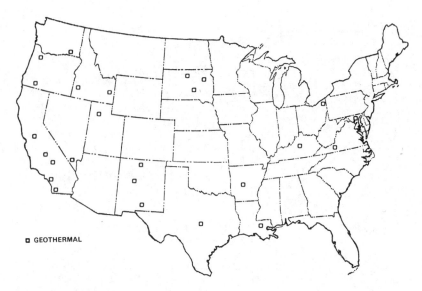

Figure III-7-2. Geothermal projects valued at more than $200,000, FY 1978 geothermal program budget.

ciently over long distances. Activities are therefore supporting the development of rural energy sources, without restricting the pattern of population distribution.

Demographic phenomena which result in regional population concentrations underlie the central station/dispersed source initiatives. Geologic phenomena form the basis of the Compressed Air Energy Storage (CAES) and Underground Pumped Hydro (UPH) activities.

The aim of these activities is to develop the ability to store energy in off-peak periods for use in peak periods (load leveling). At present, the uncertainties involved in these technologies make private development unlikely, even though their potential contribution to load leveling is considerable. The north-central, northeast, and the southeast regions show the greatest promise for this technology, and projects are planned for each of these areas.

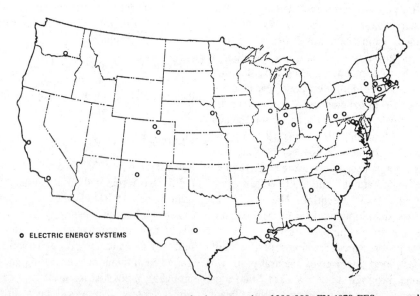

Figure III-7-3. Electric energy system projects, valued at more than $200,000, FY 1978 EES program budget.

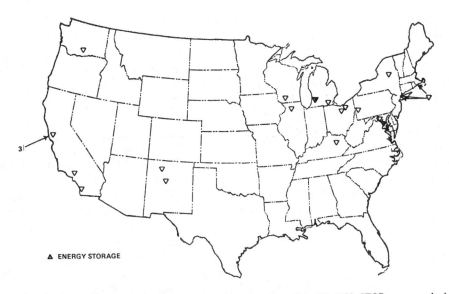

Figure III-7-4. Energy storage projects valued at more than $200,000, FY 1978 STOR program budget.

All projects, valued at more than $200,000, funded by the FY 1978 EES budget, are plotted in Figure III 7-3.

Energy Storage Systems Program

The major regional impacts to be derived from energy storage will result from the use of CAES systems and aquifer energy storage systems. CAES requires the underground storage of compressed air at high pressure. Certain geologic formations, such as salt domes, are especially suitable for this purpose. Other formations, such as porous rock, would allow the compressed air to leak out. Therefore, the regions that are suitable for the implementation of CAES technology are limited to those with suitable geological formations. Fortunately, there are many of these throughout the United States.

The load patterns of many utilities vary considerably among regions. The utilities are

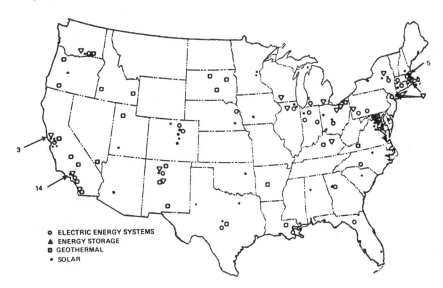

Figure III-7-5. Distribution of solar research activities by state.

either summer or winter peaking, and the design of an appropriate energy storage system must take this into account. Therefore, some energy storage technologies will be more appropriate for the southern regions of the country, while others are more appropriate for northern locations.

Energy storage will be an important consideration in the utilization of alternate energy sources such as solar and wind energy. These energy sources vary significantly among regions, and this variation will influence the choice of an appropriate energy storage system.

Although the impact of any one energy storage technology may not appear to vary regionally, when the technology is compared to other energy storage systems, a strong pattern of regional variation will emerge. For example, an underground pumped hydro or superconducting magnetic energy storage facility may be more cost effective in areas where CAES systems are only marginally feasible. As an ongoing activity, the Energy Storage Systems Division examines each energy storage technology within the overall context of the technology.

All projects valued at more than $200,000, funded by the FY 1978 energy storage budget, are plotted in Figure III 7-4.

State Distribution of Research Activity

Figure III 7-5 provides an organizational breakdown by State of the significant projects (over $200,000) within ETS. Also indicated in the table are the technology area in which the research is being performed, a description of the work, and the amount of money being spent. This information provides an overview of the distribution and diversification of ETS research programs throughout the country. This distribution is determined not only by several factors discussed previously, but is also dependent on the existence of high-technology experimental facilities and technical personnel in DOE laboratories, universities, and industry.

8. Commercialization

Commercialization is the final phase in the development cycle for technologies being developed by the Solar, Geothermal, Electric and Storage Systems Program. It is the adoption and introduction of a technology into the marketplace by the private sector.

Early phases of the development cycle, are influenced by the consideration that the technology is eventually intended to enter the marketplace. For this reason, and for completeness, the DOE view of the entire development cycle and the process of commercialization are presented first. Commercialization of specific solar technologies follows. This chapter concludes with a discussion of the commercialization of geothermal, electric and storage systems.

THE PROCESS OF COMMERCIALIZATION

The Technology Development Cycle

For the purposes of DOE programs, the technology development cycle is treated as consisting of seven phases:

1. Basic research—the performance of fundamental studies to advance scientific knowledge in subjects related to energy.
2. Applied research—using the knowledge gained through basic research, the systematic investigation of problems related to specific applications. These problems may be in the physical, biological, or social sciences.
3. Exploratory development—efforts comparable in content to applied research but directed toward a particular energy technology.
4. Technology development—efforts directed toward establishing the technical feasibility of a particular energy technology. This includes formulating alternative approaches and developing laboratory-scale models.
5. Engineering development—efforts directed toward the design and testing of pilot plants. Emphasis is placed on engineering and systems analysis. Major facilities may be developed for testing and improving pilot plant components.
6. Demonstration—construction and operation of a first-of-a-kind facility to exhibit the acceptability of the technology in an operating environment.
7. Commercialization—the adoption and introduction of a technology into the marketplace by the private sector. This involves the participation of private industry, but allows the Government to provide a combination of incentives and information dissemination services that are considered most appropriate for a given technology. When the technology gains market acceptance, it is expected to merge into the mainstream of the general economy without further Government action.

Requirements for Commercialization

Commercialization is not merely something that occurs at the end of the development cycle; requirements for successful commercialization influence efforts in earlier phases. Commercial introduction of a technology requires the following:

- All technical and engineering problems associated with the technology must have been solved.
- The economics of the technology must be such that manufacturers can obtain a fair return on investment at prices that offer users an attractive alternative to conventional technologies.
- Users must accept the technology as effective and reliable.
- The public must be satisfied that the socioeconomic and environmental effects of the new technology have been properly analyzed and that actions are being taken to eliminate adverse effects. (Environmental effects are discussed in Chapter 9).

When these conditions are satisfied and commercialization proceeds, an institutional infrastructure (producers, distributors, suppliers of raw materials, and industry associations) will evolve. The development of this infrastructure can be aided by certain Government actions, such as the early establishment of industry standards for system performance, subsystems, and components.

Major Decisions in the Development Cycle

Technology commercialization, within DOE, is currently viewed as requiring four major decisions. They are discussed in this section to complete the overview of the technology development cycle, and because development efforts within CSE are aimed at providing information to support these decisions.

The major decisions concern the readiness and suitability of a candidate technology to enter the phases of: (1) technology development, (2) engineering development, (3) demonstration, and (4) commercialization. The progress of a technology from research through engineering development is shown in Figure III 8-1, and from engineering development through commercialization in Figure III 8-2. The four major decisions in this process are emphasized in Figures III 8-1 and 8-2. These figures also depict situations for rejecting candidate technologies and situations where technologies could be placed in reserve. A decision to resume work on a technology placed in reserve could be made at a later stage, possibly on the basis of progress (or the lack of it) in other technologies.

The Energy Systems Acquisition Advisory Board (ESAAB) will make these major decisions regarding the progress of all technologies being developed in DOE. This arrangement will allow high-level DOE policymakers to make comparisons across technologies and classes of technologies and to select the most suitable options for further development.

COMMERCIALIZATION CONSIDERATIONS FOR SOLAR TECHNOLOGIES IN THE OFFICE OF ENERGY TECHNOLOGY

Solar Thermal Power

Solar thermal power systems use the sun's energy to heat a working fluid that is used to drive a turbine that generates electricity. It is useful to discuss the commercialization of such systems with reference to their size.

Small power systems produce up to 30-MW of electricity. They are useful for localized applications such as small community systems and agricultural systems in remote areas, particularly for shallow-, and deep-well pumping. Community systems are frequently designed as total energy systems, in which the "waste heat" of the working fluid is used for space and water heating.

Small power systems are in the demonstration stage of the development cycle. The engineering feasibility of these systems has been proven and will be well established for regular operating conditions in the course of the demonstrations. Before these systems can

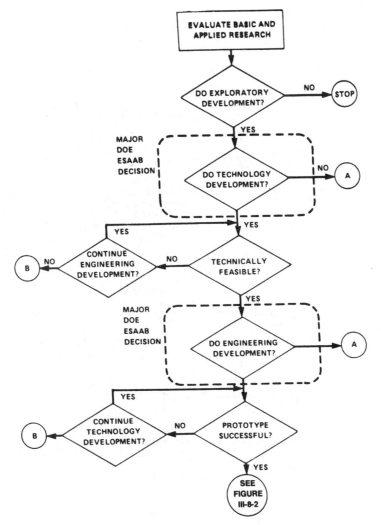

Figure III-8-1. Progress of a technology from research through engineering development.

achieve acceptance in the marketplace, their long-range performance characteristics must be determined. A system life of 20 years is considered acceptable from the standpoint of users. Accelerated lifecycle testing is planned for the demonstration system under construction at Shenandoah, Ga. Data obtained from demonstration projects can be used to determine whether certain incentives, either to manufacturers or to users, will be desirable in the commercialization phase.

Large power systems producing 30- to 300-MW are intended to meet peaking requirements in utility grids. These systems are undergoing engineering developing; although the physical principles by which they work are well established, additional work is necessary to improve efficiency and to decrease costs. A

530 Solar, Geothermal, Electric and Storage Systems

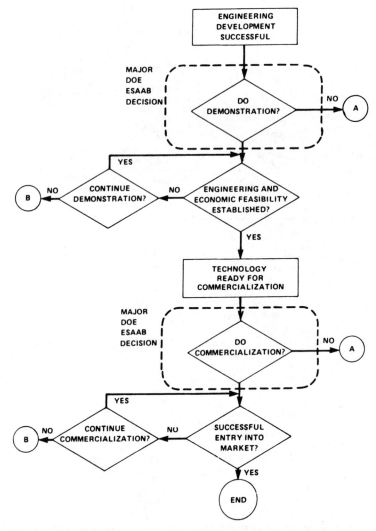

Figure III-8-2. Progress of a technology from engineering development through commercialization.

pilot plant, now under construction, will allow economic performance to be gauged when the system is operational. Work is also being performed on advanced designs for certain system components which, depending on when they become available, will be incorporated in pilot or demonstration plants. Also, as the engineering development phase proceeds, cost improvements and other commercialization requirements can be identified more specifically.

Photovoltaic Energy Conversion

Photovoltaic (PV) systems are already being manufactured commercially on a limited scale in the United States. Systems with a total power output of approximately 0.9-MW peak were produced in 1977. In 1978, production increased to approximately 1.4-MW. There are approximately 15 companies currently in the market (for example, Arco Solar Inc., Sensor Technology Inc., Solarex Corp., and Solar Power Corp.). Their market consists primarily of specialized applications, many of which are in remote locations. These include cathodic protection of bridges, power for small electronic devices, submersible pumps, and a number of defense applications. This market, although sufficient to allow a number of companies to operate successfully, is not large enough to bring about price decreases resulting from economies of scale in production. Furthermore, the market is too small to produce any significant savings in fossil fuels by substituting photovoltaic energy as a power source. Economies of scale in production and significant savings in fossil fuels become attainable at array prices of approximately $2.00 per peak watt, at which new and much larger market segments (in the range of hundreds of megawatts per year) are expected to open up. At array prices of approximately $0.50 per peak watt, the market is expected to be several thousands of megawatts per year.

The entry market segment for these larger markets consists of distributed residential and commercial applications in the southwestern region of the United States, using utility grid connections for backup. Such systems would displace a significant part of the centrally generated electricity used in single-family residences, multiple residences such as apartment buildings, and commercial establishments, such as shopping centers and small industrial parks. Acceptance of photovoltaic systems in the Southwest, where high insolation (abundant sunshine) makes the economics of such systems more favorable, is expected to lead to increased consumer interest and expansion of the industry infrastructure. Acceptance in the southwest region is also expected to lead to expanded production and consequent cost reductions, which would improve the cost-effectiveness of such systems in other regions of the United States.

The technical feasibility of photovoltaic electricity generation has been well established. The major question concerning residential and commercial systems is not whether they will work, but whether they can be manufactured and installed for a price that is cost-effective to users. Initial penetration into this market is expected to occur after 1982, as array prices drop below $2.00 per peak watt and as users take advantage of incentives provided by legislation such as the National Energy Act.

Major efforts are underway to develop manufacturing methods that will decrease the cost of photovoltaic systems. Research and development are also being undertaken on alternative materials and techniques that offer the possibility of still greater cost reductions in the long run.

An important issue in the commercialization of distributed photovoltaic systems is the utility interface. Utility charges for backup connection and power supplied, utility buyback of excess power, and the effect on utility generating capacity requirements are among the matters to be addressed. The National Energy Act provides for regulations to ensure that small power producers will be able to sell or buy power from utilities at just and reasonable rates. Even before the National Energy Act was passed, electric utilities, such as Southern California Edison, had started work on policies and rate changes that would encourage the installation of dispersed systems.

Wind Energy Conversion

Wind energy conversion systems can be grouped in three categories according to size: small systems of less than 100-kW rated power, intermediate-scale systems of 100- to 200-kW rated power, and large-scale systems of 1- to 3-MW rated power.

Small systems are well adapted to dispersed power markets, such as rural residences, agricultural applications, and applications in remote locations. Systems of this type were used from 1850 to 1930, but they were displaced by inexpensive, centrally generated electric power. The Federal program seeks to encourage the development of economical small wind energy systems and the emergence of a wind system industry. Currently (in 1979), there is a market for 50 to 100 units per year, ranging from 1- to 15-kW rated power.

Small wind systems have the greatest economic advantage in remote locations, where the alternative of electric power from a utility requires the construction of expensive powerlines to the nearest point on the utility grid. For such applications, which constitute a very limited market, small wind systems are already more economical than conventional electric power. Other applications are expected to become cost-effective as system performance improves, system prices decrease, and issues associated with utility grid backup and utility buy-back of excess power are resolved. It is possible to use onsite storage rather than grid backup to compensate for the intermittent nature of wind energy, but current costs for storage are so high that this option is, in most cases, uneconomical. Efforts to improve storage technologies are discussed in Chapter 6 of this document.

The first development cycle of small wind systems under the Federal program includes units of 1-kW, 8-kW, and 40-kW rated power. Prototypes of these units were scheduled to be tested in FY 1979. Work is also scheduled to begin on a second development cycle of small wind systems with improved performance and lower costs. In FY 1980, second-cycle systems are to be tested, and work is to begin on a limited number of third-cycle systems.

Intermediate-scale systems are suitable for use in irrigation and agriculture. They can also supply electricity to remote communities and small-scale industries. Unlike small systems, intermediate-scale systems are not currently produced and sold commercially even on a limited basis. The Government program is aimed at improving the performance and cost-effectiveness of these systems and encouraging their adoption. As an added benefit, this development effort provides valuable information regarding the development of large-scale systems, which are discussed below.

First-generation intermediate-scale systems are currently being tested. Units have been installed at Clayton, N. Mex., and Culebra, P. R., in 1978. Plans call for the installation of a unit at Block Island, R. I., in FY 1979 and another at Oahu, Hawaii, in FY 1980. Efforts to develop second-generation systems are influenced by the consideration that such systems should take advantage of the knowledge and experience gained, not only from first-generation intermediate-scale systems, but also from early large-scale systems. Preparations are being made for an industry competition for second-generation intermediate-scale systems.

Large-scale systems offer the capability to supply electricity directly to existing power grids. Their primary application is in the electric utility industry. Units may be used singly or in groups. Wind energy is variable over time; the variations follow seasonal and daily patterns, but cannot currently be predicted with complete accuracy. The electricity generated by wind energy systems is subject to corresponding fluctuations. The demand for electricity has also been observed to follow daily and seasonal patterns without being entirely predictable using current knowledge. There may be a mismatch between the supply of wind energy and the demand for electrical energy. The use of statistical analysis and storage technologies, however, may permit large-scale wind systems to be integrated into a utility power system in such a way that fuel is saved and the utility is able to reduce the amount of new generating capacity required. Large-scale wind systems show great promise for utility applications in regions that have favorable winds and now depend on oil-fired generation. These systems can also be used in conjunction with hydroelectric power generation.

A system of 2.0-MW rated power is to be installed at Boone, N. C., in FY 1979. A more advanced system of 2.5-MW, based on the experience gained in developing first-generation intermediate-scale wind systems, is entering the final design and fabrication phase. A site is to be selected in FY 1979, and installation is to take place in FY 1980. Testing and evaluation will follow. This machine is expected to be the first commercial prototype. Efforts are underway to plan an industry competition for the development of the next generation of large-scale wind systems, which will be held if such systems are shown to represent significant advances in technology and cost.

Ocean Systems

Systems for using the energy available from the ocean are in the early stages of the development cycle.

Ocean thermal energy conversion (OTEC) systems are in the technology development stage. Laboratory-scale results on closed-cycle OTEC systems are currently being scaled up in OTEC-1. Work on open-cycle and advanced systems is still in the applied research stage. Because current work on OTEC is still directed primarily at establishing technical feasibility, efforts directed at specific commercialization issues would be premature. The possible environmental effects of OTEC, as well as the economic and institutional aspects, will have to be addressed if the technology continues to show promise as development work proceeds.

Other systems for obtaining energy from the oceans, such as those based on tides and salinity gradients, are at the conceptual stage. Alternative approaches are being pursued in laboratory-scale experiments. Of these, the more promising approaches will be selected for technology development. The enumeration of specific commercialization issues must await the emergence of approaches that are technically feasible.

Biomass Energy Systems

There is a wide array of resources, technologies, and products within the scope of Biomass Energy Systems. Thus, a number of different energy needs can be served with resources that are renewable.

Commercialization of Biomass Energy Systems can be described according to the time frame in which various end-products are expected to be introduced in the marketplace: near-term (before 1985), mid-term (1985-2000), and long-term (after 2000).

Three technologies are ready for near-term commercialization:

- Direct combustion of wood and wood residues
- Fermentation of agricultural crops to produce alcohol
- Anaerobic digestion of manure to produce methane gas

The technical feasibility of these processes has been well established, and the products are cost-competitive with conventional alternatives for certain applications. It is now important to convince potential manufacturers that the economics are favorable and to show that no special problems will be encountered in scaling up to commercial-scale facilities.

Within DOE, a Resource Manager has been appointed for the direct combustion of wood. This manager will lead efforts to extend the combustion of wood and wood residues from the forest products industry, where the practice is already widespread, to small utilities and to other industries. The National Energy Act provides incentives for the production of alcohol by fermentation for use as a transportation fuel. The alcohol is blended with gasoline to roduce gasohol. The act exempts gasohol from the Federal highway excise tax of 4 cents per gallon and also provides general incentives such as an additional investment tax credit for all solar capital investments.

In the technology development and engineering development phases of the development cycle, techniques that have proved successful in the laboratory are scaled up, and engineering and economic performance are evaluated. Improved methods for producing gaseous fuels, and for fermenting woody plant matter (not currently usable) to produce alcohol, are among the technologies being developed. Also under development are tree farms and herbaceous crops. Commercial introduction is anticipated in the mid-term.

Work on technologies that are still in the applied research stage includes exploratory work on exotic species of plants suitable for energy or fuel production, on aquatic biomass, and on advanced fermentation technology. Technologies that are developed as a result of these efforts are expected to be commercialized in the long-term.

COMMERCIALIZATION CONSIDERATIONS FOR GEOTHERMAL, ELECTRIC AND STORAGE SYSTEMS IN THE OFFICE OF ENERGY TECHNOLOGY

Geothermal

The commercialization of systems using hydrothermal energy resources has been placed under the DOE Assistant Secretary for Resource Applications. Progress is being made in the preparation of a Cooperative Agreement for the construction and operation of a demonstration plant, the cost of which will be shared with private industry. These efforts are not described in this chapter because they are beyond the scope of the program described in this document.

There are no major technical barriers to the utilization of geopressured heat and mechanical energy. It is expected that some of the equipment developed for the hydrothermal program will be adaptable to geopressured resources. As the character of the geopressured resources and their potential for methane production are established, field experiments will be conducted for the recovery of heat and mechanical energy. These efforts are intended to lead to the utilization of geopressured resources, beginning in the mid-1980s.

Efforts to extract energy from hot dry rock are in the technology development phase. Current work is intended to establish technical feasibility and provide data for a preliminary evaluation of process economics.

Electric Energy Systems

Efforts in Electric Energy Systems (EES) are directed toward the more efficient use of existing equipment for generating, transmitting, and distributing electricity and toward the development of advanced techniques and components. The natural customers for the resulting products are the electric utilities. As industrial customers of significant size, technical sophistication, and financial resources, utilities are favorably disposed to new technologies designed to improve the efficiency and economics of their operation. Thus, issues related to such matters as consumer awareness and adoption of devices, retail channels, and installation and repair services, are not particularly important for EES.

Utility awareness of technological advances being sought and achieved in EES is enhanced by the participation of the Electric Power Research Institute (EPRI). Much of the work is done in collaboration with university groups for theoretical projects and with utilities for applied projects. Meetings in which universities, professional societies, and utilities participate are held to ensure that new developments are publicized.

By securing utility participation in engineering development and demonstration, EES plans to make progress toward commercialization in two ways:

- Make full use of the technical knowledge of the utility regarding existing equipment and operating requirements
- Use demonstrations, so that participating utilities will be immediately aware of the technical and economic advantages and will have already acquired the practical experience needed to install the new technology on a broader scale

Energy Storage Systems

Commercialization of Energy Storage Systems is closely linked to the commercialization of technologies that use these storage systems.

Electric storage batteries are the major cost element of electric vehicles. The economies of scale resulting from expanded battery production can be secured only with the increasing market acceptability and market penetration of electric vehicles. A number of incentives to promote market penetration are being studied. These incentives include tax credits, production subsidies for the construction of battery plants, financing plans, and battery lease/exchange/recycle plans.

Technology transfer is critical to the commercialization of thermal storage technologies. Beginning at an early stage, technology transfer is implemented through the participation of industrial and utility contractors in technology development. Private companies are working with DOE on a cost-sharing basis on a small but increasing fraction of the thermal storage program.

Technology transfer activities are also planned for the Chemical/Hydrogen Storage Subprogram to assist eventual commercialization by Nuclear Power Development (NPD) and Central Solar Technology (CST). Funds to convert high-cost water electrolysis technology, developed for space applications, into a large volume, low-cost means for hydrogen production are provided by STOR and by General Electric (the contractor), with additional funds from Niagara-Mohawk Power Co., Empire State Electric Energy Research Co., and the New York State Energy Research and Development Authority.

Private companies have been involved in the commercialization of mechanical energy storage systems, primarily in defining the needs and requirements for technology development. They are participating in the R&D program under either Government contract or joint-funding agreements.

The superconducting magnetic energy storage program is being redirected in FY 1979 because no budget authority is planned for FY 1980. Two types of companies will be involved in the commercialization of magnetic storage systems: (1) small, high-technology companies that design and manufacture superconductors, and (2) large-system vendors. The small companies will be involved early in the project; the system vendors will compete for contracts for the stabilizing magnet and the large-scale experiment.

9.
Environmental Aspects

INTRODUCTION

This section provides a summary of the environmental aspects of four energy technology programs: Solar Technology, Geothermal, Electric Energy Systems, and Energy Storage Systems. Much of the information has been gathered from Environmental Development Plans (EDP's). Prepared annually as basic planning documents, EDP's briefly describe the status, program goals, potential applications, and potential areas of environmental concern. Environmental concerns involve both the media—air, water, and land—and the groups—occupational workforce, public-at-large, and biological communities—that are affected. Environmental concerns differ in order of magnitude, that is, in their degree of impact. This chapter discusses only the major environmental aspects.

ENVIRONMENTAL ASPECTS OF SOLAR TECHNOLOGY PROGRAMS

The environmental issues associated with solar- and biomass-based technologies can be examined within each of the following programs:

- Solar Thermal Power
- Photovoltaic Energy Conversion
- Wind Energy Conversion Systems
- Ocean Systems
- Biomass Energy Systems

The objectives of the environmental efforts are: (1) to identify social and environmental problems early in the decision-making process so that adverse impacts can be avoided or mitigated, and (2) to collect, standardize, certify, process, and disseminate information for each technology.

Issues

Solar Thermal Electric Conversion. The FY 1980 Environmental Development Plan (EDP) will include detailed assessment of all relevant environmental programs, RD&D needs, and strategies to resolve major environmental problems. The most serious environmental issue is the amount of land required for the collector subsystem. Other effects include health and safety risks to workers around heliostat fabrication plants, possible contamination of drinking water supplies, and disruption of flora and fauna patterns. For each of the media, the environmental issues and potential problems are identified as follows:

Water
- Handling and disposal of system fluids and wastes could contaminate drinking water supplies.
- Construction and operation of heliostat fields could impact local hydrology patterns.

Land
- Land utilization by solar dispersed power facilities can restrict alternative land uses and significantly modify community land-use patterns.

Health and Safety
- During all phases of heliostat fabrication

and use, fires and resultant burns and eye injury may occur.
- Working fluid additives may contaminate the potable water supply, in particular with chromates and nitrates.

Biological Communities
- Construction and operation of the heliostat field could affect local flora and fauna patterns.
- Alteration of the microclimate, resulting from solar central power systems operations, may adversely affect native wildlife and vegetation.

Photovoltaic Energy Conversion. Photovoltaic systems produce impacts similar to those of solar thermal conversion systems: large amounts of land required for the collector subsystem, local effects on flora and fauna, and possible contamination of water supplies. The principal concerns, however, involve worker exposure to hazardous substances during preparation of semiconductor materials and photovoltaic cells. Because some of the semiconductor materials used to make photovoltaic cells are toxic, major occupational health and safety issues are created during their production. These issues are as follows:

Health and Safety
- Toxic products of material decomposition generated by off-gassing during overheating and fire can result in asphyxiation.
- Fires involving gallium arsenide cells can cause poisoning and skin irritation; cadmium sulfide cells can cause respiratory problems, kidney damage, and cancer.
- Workers may be exposed to silicon dust that can cause respiratory disease, cadmium compounds that could lead to lethal poisoning, and arsenic compounds that could result in lung cancer.

Wind Energy Conversion Systems. Potential physical impacts of Wind Energy Conversion Systems (WECS) are limited. WECS produce no major air or water pollutants or solid waste products. Principal environmental concerns are structural safety and electromagnetic radiation interference. Although extensive land tracts are needed for WECS, it is possible to use the land simultanously for other purposes. A listing of the issues for the Wind Energy Conversion Systems Program follows.

Health and Safety
- Safety hazards may be associated with structural failure, electric transmission equipment, or construction and maintenance of WECS towers.
- Operational noise from WECS systems may include sounds that are both audible and inaudible to humans and thus might annoy people or interfere with their activities.

Biological Communities
- Bird collisions with towers and moving blades can be a hazard to local or migratory bird populations.

Ocean Systems. The major environmental issues associated with implementing Ocean Thermal Energy Conversion are: chemical releases, metallic discharges, potential hazards to marine species, use of biocides, working fluid leaks, and climatological impacts. An elaboration of each issue follows.

Health and Safety
- Accidental release of chemicals could expose workers to health hazards.

Biological Communities
- Metallic elements resulting from the erosion and corrosion of heat exchangers may be toxic to indigenous marine species.
- Marine species may be trapped against the screens covering the cold and warm water intakes. They may also be swept along through the system and subjected to rapid pressure and temperature changes.
- Possible use of chlorine as a biocide and ammonia as a working fluid may be toxic to marine species or may have adverse effects on marine ecosystems.
- Ocean mixing may alter the air-to-surface water-temperature ratio, thus affecting the microclimate by influencing winds and currents, and potentially in-

creasing the concentration of CO_2 in the atmosphere.

Biomass Energy Systems. The Environmental Development Plan that has been prepared includes detailed assessment of the many potential environmental problem areas associated with DOE Biomass Energy Systems. The key environmental issues of DOE projects involving biological and thermochemical conversion, and biomass production and collection for each of the media are as follows:

Air
- Fugitive dust emissions, aggravated by extensive removal of crop residues on biomass plantations, can adversely affect air quality.
- Direct combustion of biomass can emit nitrogen dioxide, carbon monoxide, and particulates.
- Thermochemical biomass conversion processes produce small amounts of hydrogen sulfide and phenols that affect air quality.

Water
- Water requirements of terrestrial biomass plantations may restrict the development of competing uses of water in some regions.
- Sediment loads to waterways and fugitive dust emissions from biomass plantations may affect water quality.
- Anaerobic digestion produces a sludge which, if not disposed of properly, may cause pollution of surface and groundwaters.

Land
- Large land requirements* of terrestrial biomass plantations may restrict the development of competing uses of land in some regions.
- Removal of agricultural and silvicultural residues and total harvesting or clear-cutting schemes reduce the amount of organic matter that decaying residues contribute to the soil; this may limit future crop or forest growth.

Health and Safety
- Thermochemical conversions produce tar and oil products that superficially resemble coal tar, a known carcinogen.

Environmental Activities

The two major activities in Solar Technology are: Resource and Environmental Assessment. These activities reflect the environmental issues discussed previously. Under the Resource Assessment Activity in the current fiscal year, both renewable and exhaustible resources are being assessed. Also, the Department of Energy/National Oceanic and Atmospheric Administration solar radiation network is currently being expanded from 35 to 38 sites to include Guam, Hawaii, and Puerto Rico. A study of radiation from available meteorological statistics for 170 additional locations is being performed in the United States. These data will be disseminated during FY 1980.

Under the Environmental Assessment Activity, environmental factors for individual solar technologies have been compiled. These include ongoing studies of ocean thermal systems, biomass technologies, and health and safety assessments of materials/components of photovoltaic cells. Other programmatic environmental assessments will be completed during the current fiscal year. In addition, the Solar Environmental Development Plans covering environmental aspects of solar technologies will be updated during the current fiscal year.

During FY 1979, a wide range of geophysical data for design criteria and performance evaluation—essential to the successful implementation of various solar systems—will continue to be collected. Other Federal agencies, as well as non-Governmental sources, will also assist in performing these evaluations.

Specifically, efforts will be centered in the following areas:

- Acquisition, analysis, and dissemination of the following: insolation and spectral data, wind direction and speed, ocean thermal gradients, and design and siting requirements of solar thermal applications

---

*The total acreage needed to supply 1 percent of the Nation's present energy requirements ranges from 1.5 to 4.5 million acres, or 32 to 96 square miles of land per quad of energy grown as biomass (1 quad = 10^{15} Btus).

- Preparation of programmatic impact assessments of solar technologies, and technology assessments of unintended or unanticipated societal effects that could occur as a result of widespread application of solar systems

ENVIRONMENTAL ASPECTS OF GEOTHERMAL RESEARCH, DEVELOPMENT AND DEMONSTRATION

Issues

The Environmental Development Plan (EDP) for Geothermal Energy Systems* identifies environmental, health, and safety issues affecting geothermal development. Release of hydrogen sulfide (H_2S), resulting in corrosion and odors, is a major air quality problem. Withdrawal of fluids from the geothermal reservoir may cause another serious problem—sinking or subsidence of land surface. In addition, potential seismic disturbances may result from geothermal fluid extraction and injection processes. A discussion of these and other issues follows:

Air
- Hydrothermal fluids containing dissolved, noncondensible gases such as hydrogen sulfide (H_2S), ammonia (NH_3), boric acid (H_3BO_3), and radon (Rn) could produce toxic effects. Hydrogen sulfide could also produce a disagreeable odor.

Water
- Disposal of hydrothermal and geopressured fluids, which may contain toxic substances and large quantities of dissolved solids, is a major waste management problem.
- Geothermal brines may affect groundwater quality as a result of accidental surface spills and leaks in underground pipes.

Land
- Land subsidence may occur when large quantities of fluids have been extracted from the ground.
- Earthquakes may be induced by the injection of water at high pressures.

Health and Safety
- Well blowouts may occur during geothermal development activities.
- Noise could result from the drilling of wells during the exploratory and development phases and from venting during the testing phase.

Environmental Activities

The Division of Geothermal Energy (DGE) has examined each of the major issues affecting geothermal development, and environmental control research programs have been established as needed. Issue definition studies are being conducted in liquid waste disposal, noise generation, and well blowout. The first two studies were begun in cooperation with the Divion of Environmental Control Engineering (EV). Environmental control research programs currently deal with H_2S emissions. One such system, a scrubber using copper sulfate, was successfully tested at the geothermal field at The Geysers in FY 1977. Removal efficiencies for H_2S exceeded 99 percent with this scrubber, which also extracted NH_3 and H_3BO_3 in large quantities. In FY 1979, a commercial-scale scrubber was tested under operating conditions at The Geysers. This test proved to be successful, and the system is ready for full-scale commercialization.

In addition to environmental control studies, DGE sponsors environmental monitoring for each of its major field projects. The monitoring is done, in part, to support preparation of Environmental Impact Assessments/Statements (EIA/EIS) as required by NEPA (National Environmental Policy Act) regulations. To assist with EIA/EIS preparation, DGE has developed guidelines for environmental reports by contractors. Finally, to promote geothermal development, DGE encourages the adoption of consensus standards and environmental performance criteria. The work in standards definition is performed in cooperation with the Division of Operational and Environmental Compliance (EV).

---

*U.S. Dept. of Energy "Environmental Development Plan (EDP) Geothermal Energy Systems 1977," DOE/EDP-0014, March 1978.

ENVIRONMENTAL ASPECTS ASSOCIATED WITH ELECTRIC ENERGY SYSTEMS

An EDP that has been prepared details the environmental and safety issues associated with the generation, transmission, distribution, and storage of electricity. Although a large number of potential problem areas exist, only a few are expected to be significant. Environmental research is necessary, however, to prevent the identified problem areas from becoming constraints.

Electric Energy Systems (EES) technologies for which environmental RD&D will be reported include:

- Overhead AC/DC transmission
- Underground transmission
- Generation technology

Issues

The aforementioned EDP includes a detailed assessment of relevant environmental problems, R&D needs, and strategies to resolve environmental problem areas. These problem areas, for the most part, are of a low order of magnitude.

Overhead AC/DC Transmission. The primary environmental concerns for overhead transmission technology relate to the physical, ecological, health, and safety impacts of high voltage transmission lines and equipment under normal and fault conditions. In addition, land-use requirements are substantial. The issues related to locating generating facilities long distances from anticipated loads contribute to the opposition to new powerlines on legal, aesthetic, and health and safety grounds. These issues are also discussed in Chapter 10.

Land
- Large-scale land utilization by new overhead lines can restrict alternative land uses and can also modify community land-use patterns.

Human Health Effects
- High voltage transmission lines produce corona-generated acoustical noise and field-generated electromagnetic noise.
- Sulphur hexafluride*, freon, and the products of their thermal decomposition produce highly toxic fumes of fluorides and SO_x.
- Possible effects of high voltage transmission lines on pipeline corrosion, communication systems, gasoline explosions, brushfires, and electric shocks increase risks to public safety.
- During normal and fault conditions, workers are exposed to noxious and hazardous materials, high voltage parts, etc.
- Breakup of the DC circuit breaker increases risks to workers and to the public.

Ecosystem Effects
- Animal and plant life are adversely affected by 1200-kV lines.

Underground Transmission. In underground transmission technology, the low temperature associated with the cryogenic-resistive and superconducting cables leads to worker safety concerns, especially under fault or malfunction conditions. Therefore, the primary environmental concern for underground transmission technology is as follows:

Human Health Effects
- Toxic emissions from cryogenic materials, especially during fault and accident conditions, could increase risks to workers and to the public.

Generation Technology. The key generation technology projects are the 4.8-MW fuel cell demonstration facility and the superconducting generator. The superconducting generator has no primary identifiable concern. Worker safety is the primary environmental concern for the fuel cell demonstration facility.

Human Health Effects
- Noxious gases, fluids, and materials (e.g., naphtha, phosphoric acid) utilized in the fuel cell facility could adversely affect worker and public safety.

---

*Sulfur hexafluoride and freon are under consideration as insulating gases to be used in 1200-kV gas-insulated equipment.

Environmental Activities

The objective of the Environmental Field Effects Activity is to conduct intensive broad-based research into all electric and magnetic field and corona effects from overhead transmission lines, in conjunction with the Division of Biological and Environmental Research, the Division of Environmental Control Technologies, and the private electric utility industry, principally through EPRI. Increased emphasis is being placed on this activity in FY 1980.

The purpose of the research is to determine, by 1982, reasonable guidelines for preventing detrimental environmental effects from overhead transmission lines. These guidelines would then be used to specify parameters or some of the prototype lines that will be developed through the compact lines research efforts of Electric Energy Systems (EES). These guidelines would also aid DOE nuclear, solar, and fossil programs with environmental impact statements that they can use for siting their large electrical generation facilities.

Those areas which have been or are being satisfactorily addressed by the electric utility industry will not be included in the DOE program. For example, these areas include radio and television interference and right-of-way ecological management. During foul weather, most extra-high voltage transmission lines (above 345-kV) produce corona, which is a local breakdown of the air at the surface of the conductor. This is the source of radio and television interference and audible noise from transmission lines. Also, minute amounts of ozone are produced by corona, but in recent studies this has not had any environmental consequence. The industry has also done significant work in the area of audible noise from transmission lines. However, much disagreement still exists in determining acceptable levels of audible noise. Therefore, EES has undertaken a rather comprehensive program to develop a greater understanding of the psychoacoustic effects of audible noise from transmission lines. This program will ultimately lead to new methodologies for making measurements that directly correlate with human perception and, thus, to criteria for evaluating the effects of noise from a transmission line.

The specific objectives of these broad-based efforts include research on: pathology, growth effects, blood chemistry, cardiac characteristics, the central nervous system, genetics, and perception and avoidance of behavioral effects. Research projects in these areas will be rigorously scientific to ensure investigation of all reasonable indicators of biological effects from electric and magnetic fields, and to ensure that those effects discovered can be converted to design guidelines for overhead transmission facilities. In the planning and implementation of the aforementioned DOE-funded studies, multidisciplinary teams consisting of engineers, physicists, biologists, and behaviorists will be involved.

The following sections describe the work that will be continued or initiated to meet the objective of the EES research efforts.

In the area of corona phenomena, the following efforts are planned:

- Develop an acceptable criterion for measurements on operating lines
- Collect precise long-term measurements on operating lines
- Complete the psychoacoustic analysis of transmission line noise, and present design suggestions for lessening the impact of audible noise from transmission lines

In the Electric Field Effects Activity, the following work is planned:

- Continue intensive investigation of possible exposure level or effects of the duration of exposure with small animals
- Continue joint management of the biologically based programs with the Division of Biomedical and Environmental Research (DBER)
- Complete the design and cost-benefit evaluation of the Non-human Primate Research Program, and initiate the program if desirable
- Initiate research on possible low-level field effects on circadian rhythm and brain membrane functions
- Continue a carefully engineered test to determine central nervous system effects
- Develop detailed profiles and characteris-

tics of electric fields in the vicinity of DC transmission lines operating at voltages up to ± 1200-kV
- Develop techniques for calibrating biological test cages used in both AC and DC electric field research

Planned magnetic field effect activities include:

- Continue to jointly manage programs initiated by DBER on the effect of high magnetic fields
- Initiate exploratory methodologies to study possible effects of weak AC and DC magnetic fields

ENVIRONMENTAL IMPLICATIONS ASSOCIATED WITH ENERGY STORAGE SYSTEMS

Issues

An EDP has been prepared and is currently being updated. This document details the environmental and safety issues associated with the following energy storage (STOR) programs:

- Battery Storage
- Thermal and Mechanical Storage

The major areas of concern relate to the human safety and environmental hazards created by working fluids during accidental spills, exposure to fire, and abnormal working conditions. The occupational hazards during manufacture of energy storage systems are not clearly understood, and workable handling, storage, and containment procedures need to be developed. A summary of environmental concerns and the program involved follows.

Water
- Contamination of surface and groundwaters may result from compressed-air energy storage (CAES) and underground pumped hydro storage (UPH). (Thermal and Mechanical Storage)
- The alternate heating and cooling of aquifer strata may result in aquifer instability and may cause the leaching of salts and trace elements, thus contaminating local ground water supplies. (Thermal and Mechanical Storage)
- Pressure and temperature fluctuations in underground caverns may cause structural deterioration of cavern walls, stimulate seismic activity, and alter groundwater flow patterns in aquifers. (Thermal and Mechanical Storage)

Land
- Disposal of hard-rock mining wastes and saline mine water during the mining and excavation phases may be one impact of CAES and UPHS. (Thermal and Mechanical Storage)
- Seismic activity and land subsidence may result from UPHS and superconducting magnets. (Thermal and Mechanical Storage)
- The manufacture of sodium sulfur, zinc chloride, and lithium metal-sulfide batteries may require the disposal of wastes which may adversely impact water systems, landfills, and the ecosystem. (Battery Storage)

Human Health Effects
- Accidental spillage of hot fluids or chemicals may result in thermal or caustic burns to workers. (Thermal and Mechanical Storage)
- Continued exposure of workers to hazardous substances from latent heat storage and thermochemical storage (Thermal) and the production of zinc bromine, zinc chlorine, sodium sulfur, lithium metal-sulfide, and nickel zinc may cause adverse health effects. (Battery Storage)
- Toxic gas releases from fiber-epoxy materials at elevated temperatures and flying objects and debris may result from the uncontrolled releases of stored energy in flywheels. (Thermal and Mechanical Storage)
- Hydrogen use may involve the hazards of fire and explosion. (Thermal and Mechanical Storage)
- The charging and operation of lead acid batteries results in the emissions of toxic gases such as arsine and stibine, both adversely affecting human health. (Battery Storage)

Ecosystems
- Leakage of chemicals from thermochemical or latent heat systems can adversely affect local aquatic ecosystems through pH changes. (Thermal and Mechanical Storage)

Environmental Activities

Potential adverse impacts associated with energy storage systems are related primarily to health and safety aspects of production and operation. Toxic or hazardous substances may be accidentally released, posing a threat to exposed workers and to the general public. Wastes generated during the manufacture of components andd at the end of the useful life of batteries present a solid waste disposal problem. Finally, the operation and manufacture of energy storage systems have the potential for impairing air and water quality.

Efforts in the STOR area reflect the environmental issues discussed previously. These efforts are coordinated with other DOE divisions and Federal organizations to minimize duplication of efforts and to progress toward resolving energy storage environmental concerns.

10.
Socioeconomic Aspects

The manner in which Americans use and pay for energy is a fundamental cultural characteristic which is being challenged by the current energy crisis. As a result of this crisis, changes in the mix of energy supply resources and technologies that Americans use are inevitable. These changes will have social and economic consequences that cannot be predicted precisely. DOE recognizes the importance of identifying the socioeconomic aspects of its activities and of taking steps toward resolving these aspects within the framework of national policy.

CSE realizes that the socioeconomic aspects of the energy technologies for which it is responsible must be considered when alternative future energy supply strategies are compared. Consequently, in accordance with DOE policy, CSE addresses the socioeconomic issues that may arise due to the commercialization or application of these technologies. Some of the major socioeconomic aspects of these technologies are outlined in this chapter.

SOLAR TECHNOLOGY

In its several forms, solar energy represents an environmentally sound, renewable, and essentially inexhaustible resource. Although it is not a conventional energy resource, solar energy is expected to play an increasingly important role in the future. Currently, such technologies as solar hot-water and space heating are making growing contributions. CSE, therefore, will address both the technical and socioeconomic issues that may inhibit the commercialization or effective application of the solar technologies being investigated.

Solar Thermal Power and Photovoltaic Energy Conversion

The expanding use of solar thermal collectors and photovoltaic cell will tend to blur the existing distinction between energy producers and energy consumers. Individuals will be able to reduce their reliance on utilities or oil companies for energy by producing it themselves, thereby altering the character of the demand for conventional energy. This, in turn, will raise a number of new socioeconomic issues, which are discussed in the following sections.

Utility Rates. The increased use of solar energy will affect the business practices of suppliers of conventional forms of energy. Until more effective methods of storing solar energy are devised, there will be increased demand for conventional sources during cloudy weather and at night. Consequently, trends in the demand for power from electric and gas utilities as well as the demand for petroleum fuel may be altered.

This increased demand for power to backup domestic solar systems may require electric utilities to increase their reserve peak-power capacity. To pay for this increased capacity, utility rate structures may have to be adjusted. This in turn may affect the economic competitiveness of solar energy and retard its competitiveness of solar energy and retard its commercial growth, especially if solar energy users are asked to pay a premium price for electricity. In addition, if utilities invest in large central solar power stations, "time-of-day" rates may have to be adjusted to encourage peak consumption during daylight hours.

Land Use and Zoning. A serious socioeconomic consideration related to the development of centralized solar electric facilities is the amount of land needed for the solar collector subsystem. For instance, as much as one square mile of land may be needed for a 100-MW solar thermal electric facility. Such extensive land-use requirements for centralized solar thermal electric and photovoltaic facilities will not only restrict siting options, but will also raise socioeconomic issues at the locations of the facilities.

Land required for the collector subsystems of centralized solar thermal electric and photovoltaic generating stations will be precluded from alternative uses. Specifically, the land will not be available for farming, housing, or commercial/industrial development. Consequently, the construction of such facilities could increase land prices and also conflict with community land-use plans.

Local zoning restrictions also may inhibit the use of solar energy in homes or in the generation of electricity for communities. Some people view solar collecting devices as unattractive or inconsistent with conventional architecture, and local laws or covenants in property deeds sometimes prohibit the construction necessary for solar power. Such conflicts need to be addressed if the potential of solar energy is to be realized.

Rights-of-access. As the use of solar energy becomes more widespread, a number of unique legal questions will have to be addressed. For instance, new laws may be required to determine rights-of-access to sunlight. The most important of these questions is the relationship between property rights and rights-of-access to sunlight, and the ability of such rights-of-access to block any construction that might shade a present or future solar collector site.

Regulation. The installation and maintenance of solar systems will provide additional demand for unskilled or semiskilled labor, because installing and maintaining solar systems are labor-intensive processes. Because solar collectors are relatively simple to install and maintain, many small businesses may enter the industry. On the other hand, as many companies enter the growing field, industry-wide quality control may become difficult. Therefore, to protect both the consumers and the integrity of the expanding industry, legal standards may be needed for both the equipment and its installation.

Wind Energy Conversion Systems

Like solar thermal electric and photovoltaic systems, wind energy conversion systems raise socioeconomic issues related to the intermittent nature of the energy source and to use of wind systems in either centralized or dispersed applications. The following paragraphs discuss the socioeconomic considerations related to wind systems.

Legal/Regulatory Problems. The legal problems associated with the use of the wind as an energy source will need to be addressed. For example, it must be decided whether property ownership includes the rights-of-access to the winds that cross a property, and whether an individual may file a suit stemming from the blockage of winds across an area.

Jurisdictional problems may occur if wind machines are located in offshore areas. Legislation may be required to empower either appropriate State governments or the Federal Government with the authority to regulate offshore windmills.

Utility Rates. The expanded use of wind energy will affect the business practices of those supplying conventional forms of energy. Until more effective methods of storing wind energy are devised, there may be an increased demand for conventional forms of energy, mainly electricity, as a backup energy source for wind systems. For the electric utilities, this will require extra peak-power generating capacity for use during periods of low wind velocity. Consequently, utility rates may require adjustment to finance extra generating capacity and to permit the use of electricity during such periods.

Public Safety. Although wind energy conversion systems will be designed and constructed for safe operation, unusual stress may cause fractures in wind machine blades. If this happens, pieces of broken blade sections may fly several hundred yards in any direction, posing a threat to public safety. Fabrication and safe operation standards will be needed, along with requirements for adequate insurance coverage to protect the public from the risk of injury from flying blade sections.

Ocean Systems

Inasmuch as Ocean Thermal Energy Conversion (OTEC) systems are most likely to be used as utility base-load plants, they are expected to have minimal socioeconomic impacts. Some economic growth may be expected to occur at the port or shore sites where OTEC systems are constructed and serviced, but because such sites will probably be chosen on the basis of previously demonstrated construction capabilities (e.g., deepwater drilling-rig construction sites), socioeconomic impacts may be negligible.

Biomass Energy Systems

The production of energy from biomass could create a new type of farmer—one who grows crops for fuel rather than food. Such farming, by giving farmers a new source of income, could stem the current decline in the farm population.

Extensive energy farming, however, may reduce the amount of land available for recreation or for the production of other crops. Previously untilled land may be cultivated, and land left fallow under "set-aside" programs may be used for energy production. On the other hand, a profitable energy farming industry may reduce production of food crops and farm exports, thereby raising grocery prices. The expansion of the agricultural sector may also create an additional demand for irrigation, a problem which could be especially acute in areas of chronic water shortage.

GEOTHERMAL

A number of socioeconomic impacts can be associated with deriving useful energy from the heat within the earth. Although its use is limited at present to several areas in the western States, geothermal energy may prove to be an environmentally sound, renewable, and essentially inexhaustible resource that will become increasingly important in the future.

Local Economy

The construction and operation of any large-scale, centralized geothermal facility in previously undeveloped areas may produce "boom-town" syndrome effects. The impact of a particular geothermal power station will depend on the size of both the facility and the community in which it is located. Although the local economy will expand during the station's drilling and construction phases, when construction activity declines, the local economy could suffer.

"Boom-town" effects also have a number of indirect consequences. For example, rapid economic growth is often attended by increases in property values and greater demand for public services and facilities. The decline in the local economy that occurs after construction is often accompanied by decreases in employment, business property values, and population. Careful planning is required to stem the negative consequences of both the "boom" and "bust" phases of this syndrome.

Water Use

In many areas of the western United States where geothermal resources are available, water resources are scarce. In these areas, the need for water to extract the geothermal energy from hot dry rock might conflict with the needs of others. Farmers, for example, may believe that using water for irrigation is more important than using water to supply power. Thus, before a geothermal plant can be built in a particular community, the problem of water allocation among these kinds of competing demands must be resolved.

Seismic Activity

Reinjection of process water may cause earthquakes if geothermal resources are located in potentially seismic areas. Clearly, the construction of geothermal power stations should not be permitted in earthquake-prone areas, but local building codes may have to be revised to allow for geothermal-related earth tremors or, if necessary, land subsidence.

Aesthetics

A geothermal power station, particularly the actual structure and its appearance, may be regarded as an unattractive intrusion on a scenic area. Also, the noise and sulfuric odor which it may cause could be considered offensive. These social concerns must be addressed before a geothermal power station is constructed.

ELECTRIC ENERGY SYSTEMS

The U.S. electric energy transmission and distribution grid has evolved to its present state without the benefit of a systematic planning effort. However, as the system becomes more complex and larger amounts of power are transmitted between different areas, a systems approach must be taken to ensure that the expanding electric network serves consumers efficiently and reliably.

Rights-of-Way

The need for efficient and reliable transmission and distribution of electricity is complicated because conventional large central generating facilities are being located farther from load centers. For example, as a safety precaution, nuclear generating facilities are often established in remote areas, while large coal-fired generating facilities are situated near the fuel source; e.g., in remote western areas. The remote siting of centralized electricity-generating facilities has led to a growing need for long-distance, high-voltage transmission lines to link electricity producers to consumers.

However, as the U.S. population expands and the demand for land becomes more intense, powerline rights-of-way will become more difficult to obtain. Presently, for example, legal processes are often necessary to resolve conflicting interests among utilities, farmers, and electricity consumers. Furthermore, even when utilities allow the land under powerlines to be used for farming, grazing, etc., the uncertain biological and ecological effects of transmitting high voltage electricity complicate the land-use problem. Finally, transmission line rights-of-way are often contested on aesthetic grounds, especially when lines are planned for scenic or recreational areas.

Utility Interface

Decentralized solar technologies will also raise socioeconomic concerns related to the electric energy system. Not only will rate structures be affected by the need for additional generating capacity to backup residential solar systems, but dispersed solar electricity generating systems (e.g., photovoltaic arrays) will have the capability of producing extra power during periods of high solar insolation. This power could be added to the grid system, if the technical and rate structure problems can be resolved.

Finally, an electricity transmission and distribution system must be prepared to operate reliably during bad weather, generating plant failures, or spot fuel shortages. The system, therefore, must be able to transmit large amounts of power from one area of the country to the other if the socioeconomic consequences resulting from blackouts are to be avoided.

ENERGY STORAGE SYSTEMS

Energy storage systems will not affect socioeconomic concerns directly. However, because these systems will be used in conjunction with energy-producing technologies to level loads and to reduce the need for backup systems, they are believed to have positive socioeconomic effects. For instance, energy storage systems may lower the capital requirements of utilities and lessen their need to use oil and gas to produce peak power. This will reduce the need for imported energy, thereby strengthening the dollar in foreign exchange markets and increasing the viability of nonconventional energy resources.

Improved energy storage also will allow continuous use of intermittent energy sources. At present, solar power can be used only when the sun shines or when the winds are of sufficient velocity. Consequently, a backup system must supply power when the intermitten sources are unavailable. To the extent that storage systems encourage the use of solar or wind power, they could considerably increase the effectiveness of these resources.

11. International

BACKGROUND

During the past 100 years, the industrialized world has come to depend very heavily on fossil fuel systems as basic power resources. Historically, this occurred because the industrialized countries discovered that a sequence of fossil fuels could be used abundantly and conveniently; many countries also discovered large indigenous fossil fuel resources that seemed adequate for centuries to come.

These countries believed that short supplies in their own resources could be increased through purchases at low prices on the world market. This in turn spurred large capital investments for developing, commercializing, and installing a wide range of power systems based on fossil fuels; the industrial nations also incurred large research and development (R&D) investments to improve the performance and efficiency of conventional fossil fuel systems.

Developing countries, attempting to optimize their own economic development efforts, found it expedient to spend their limited capital on developed fossil-fuel power systems. This appeared to be an obvious approach because fossil fuels were thought to be widely available at a relatively low cost, even though most developing countries had minimal indigenous fossil fuel resources.

Presently the economic development of developing countries, as well as that of the industrialized countries, is being threatened by the recent crisis in availability of one fossil resource, namely, oil. Moreover, their economic development is being jeopardized by the negative patterns in the discovery and production of the most desirable fossil fuels relative to world energy needs and the resulting increases in the prices of all fossil fuels. Because of a lack of resources and embedded economics, the developing countries are less able to make the necessary adjustments to continue their programs for social and industrial development.

Industrial countries in their search for alternative energy resources, are now making substantial investments, especially in solar energy. The United States has taken an overall lead position in developing alternative energy systems. In particular, the Nation has a foremost commitment to develop solar energy technologies, including direct low- and high-temperature thermal energy conversion systems for heating and cooling buildings, agricultural and industrial process heat, and electric power generation.

The United States also has established a lead position in photovoltaic systems, for direct conversion of sunlight to electricity, and in conversion systems, which provide electrical, thermal, or mechanical power from wind, ocean, and biomass-based systems (indirect forms of renewable solar energy).

There is potential for substantial benefits through cooperative R&D agreements with other industrialized countries that are developing solar energy technologies. Such cooperation can reduce costs of projects, fill gaps in national programs, and provide a larger pool of new ideas for improving systems. Agreements can include information exchange, visit and staff exchanges, coordinated research performed in national laboratories, joint R&D projects performed in one of the participating countries, and joint industrial activities.

In general, developing countries around the world have substantial native energy resources, including direct and indirect solar, geothermal, waves, and tidal power. Their solar resources are sufficient for much of their foreseeable development needs. Figure III 11-1 shows the widespread availability of solar energy over most of the heavily populated areas of the Earth.

The United States is one of the few industrialized countries with substantial solar resources. While developing solar technology alternatives for domestic needs, the U.S. can cooperate with other industrialized countries interested in solar technology and can provide assistance to developing countries so that they, too, can use abundant solar resources. This can be accomplished through multilateral and bilateral agreements. The Department of Energy, directs the implementation and review of international cooperative participation. The following are recent international commitments to global energy research, development and demonstration (RD&D) that fall either wholly or partly under the CSE charter:

- Title V, Nuclear Non-Proliferation Act of 1978 (Public Law 95-242)
- Section 119, International Development and Food Assistance Act of 1977 (Public Law 95-88)
- Department of Energy Reorganization Act (Public Law 95-91)
- Geothermal Energy Research Development and Demonstration Act of 1974 (Public Law 93-410)
- Bonn Summit commitment to renewable energy assistance to developing countries
- President Carter's Export Promotion Program
- International Energy Development Program
- Foundation for International Technological Cooperation

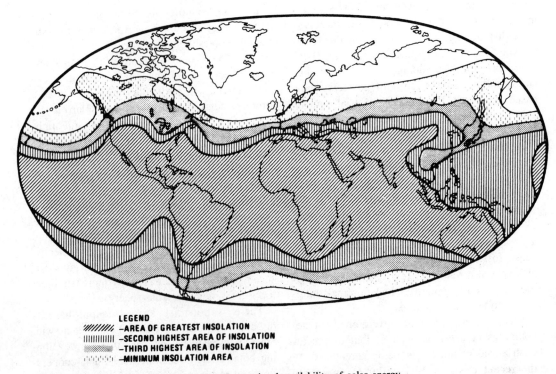

LEGEND
/////// –AREA OF GREATEST INSOLATION
|||||||||| –SECOND HIGHEST AREA OF INSOLATION
▓▓▓▓▓ –THIRD HIGHEST AREA OF INSOLATION
∷∷∷∷∷ –MINIMUM INSOLATION AREA

Figure III-11-1. International availability of solar energy.

- Solar Energy Domestic Policy Review, International Panel

The U.S. is involved in only one substantial multilateral energy agreement. This agreement is administered through the International Energy Agency (IEA), an autonomous body within the Organization for Economic Cooperation and Development (OECD). Headquartered in Paris, IEA was founded as part of the 1974 Agreement between principal industrialized countries for an International Energy Program.* Its objectives are to provide international RD&D to ensure the development and application of new energy technologies which will provide significant contributions to near-term energy needs and will also reduce long-term dependence on oil. As R&D project areas are developed under IEA sponsorship, each member country can elect to support and participate in a given project. Participants in each topic area provide all of the project financial support; thus only those activities considered mutually beneficial are undertaken. A listing of major activity areas which IEA administers is shown in Table III 11-1.

GOALS AND OBJECTIVES.

CSE has established an overall management office and an international multiyear plan to coordinate U.S. participation in international RD&D agreements involving solar, geothermal, electric and storage systems. In this way, the various CSE international activities are integrated as a unified effort focused on clearly stated and ranked mission goals. The principal goals addressed in the CSE management plan in order of scope and the resulting benefits are the following:

(1) Advance the objectives of the U.S. domestic program through cooperative agreement activities and projects
   - Decreased RD&D costs through shared information and project cost
   - Increased breadth of information and experience
   - Reduced time required for some phases of domestic projects through the use of existing facilities overseas
(2) Promote the balance of U.S. trade by increasing exports of U.S. goods and services reducing unnecessary imports
   - Reallocation of resources from consumption to investment
   - Increased opportunities for small and large U.S. businesses
   - Expanded U.S. manufacturing capacity that parallels or prepares for the emerging U.S. market
   - Increased domestic jobs through export expansion
(3) Increase conservation and use of renewable energy resources among allies and trading partners
   - Promote increased independence of oil imports
   - Provide conservation of nonrenewable resources among the greatest energy consumers
(4) Provide development assistance through promotion of non-fossil based industrial systems* and decentralized rural energy systems
   - Independence of oil imports
   - Resource conservation
   - Attainment of development assistance objectives through application of longer-range policy tools
(5) Promote bilateral good will and decreased international tensions through cooperation and assistance
(6) Protect the global environment

These goals will be accomplished through CSE support of those projects which show the greatest promise for international commercial

---

*As of July 1, 1977, IEA countries are Austria, Belgium, Canada, Denmark, Federal Republic of Germany, Greece, Ireland, Italy, Japan, Luxembourg, Netherlands, New Zealand, Norway, Spain, Sweden, Switzerland, Turkey, United Kingdom, and United States.

*While development plans of less developed countries are based on the replication of the essentially fossil-based systems in use throughout the industrial world, it is widely believed that this path will never provide the per capita consumption goals ultimately sought. However, these countries may be in a position to turn a liability into an asset. Since they have not yet committed themselves to a conventional energy base, they can pursue an entirely different path that promises to make their goals attainable, and provide as well, longer term reliability, self-sufficiency, and improved environmental quality.

Table III-11-1. IEA activity areas.

ENERGY CONSERVATION
- Urban planning
- Building energy loads
- Heat pumps
- Combustion
- Cascading
- Heat transfer/exchangers

COAL TECHNOLOGY
- Technical information service
- Economic assessment service
- World coal resources data bank
- Mining technology clearing house
- Fluidized bed combustion
- Low Btu coal gasification
- Coal pyrolysis
- Treatment of coal effluents

NUCLEAR POWER
- Reactor safety experiments

FUSION ENERGY
- Superconducting magnets
- Plasma/wall interactions
- Intense neutron source

GEOTHERMAL ENERGY
- Manmade energy systems

SOLAR ENERGY
- Heating system performance
- Cooling system performance
- Development of heating and cooling components
- Solar collector testing
- Instrumentation package
- Meteorological data
- Small solar demonstration

BIOMASS CONVERSION
- Technical information service

OCEAN ENERGY
- Wave power

WIND ENERGY
- Large scale wind demonstration
- Technology assessment

HYDROGEN
- Hydrogen production from water

application. CSE is currently designing guidelines for project evaluation. The results of evaluations will assist CSE in determining the extent of U.S. participation in cooperative agreements as well as further project funding. Evaluation guidelines are to include qualitative assessments of each agreement's impact on domestic programs, expected cost savings, balance of payments, conservation, environmental quality, and possible impacts on international relations. The assessments will use a numeric weighting scale for each agreement.

CURRENT INTERNATIONAL AGREEMENTS

CSE is currently supervising 30 RD&D international agreements. Each activity is either ongoing, under negotiation, or inactive. A summary listing for CSE divisions by technology activity (there are more than 30 line items because of the multiactivity nature of some agreements) is presented in Table III 11-2.

The following tables present brief descriptions of the energy RD&D agreements concluded at the departmental level or above and for which lead responsibility rests within or is shared by CSE. Multilateral U.S. agreements administered through IEA are presented in Table III 11-3; a summary listing appears in Table III 11-4. U.S. bilateral agreements are described briefly in Table III 11-5; Table III 11-6 contains a summary listing.

COMMERCIALIZATION POTENTIAL

The multilateral and bilateral technical exchange agreements in which CSE participates can assist U.S. marketing efforts in the participating countries as well as nonparticipating developing countries. By expediting promising research and development, CSE indirectly influences international commercialization. This influence is an inherent characteristic of the links which join basic research, applied research, development, demonstration, and commercialization. However, the effects are still relatively small because there have been few large cooperative projects involving construction and operation of experimental or commercial facilities in CSE technology areas. There is great potential, however, for CSE International Program RD&D funding to aid U.S. export interests. The impact on each technology area is discussed in the following sections.

Table III-11-2. RD&D agreements by technology.

ETS Division / Technology	Sponsoring Organization or Cooperating Country	Status
1. Solar Electric Applications		
Small Solar Power Systems	IEA	on-going
Solar Thermal Conversion System	France	on-going
Solar Thermal	USSR	on-going
Solar Thermal Central	Spain	on-going
Solar Thermal	Saudi Arabia	on-going
Solar Thermal	Italy	under-negotiation
Photovoltaics	Saudi Arabia	on-going
Photovoltaics	Spain	on-going
Photovoltaics	USSR	on-going
Photovoltaics	Italy	under-negotiation
Wind Energy Conversion	IEA	on-going
Large-Scale Wind Energy Conversion	IEA	on-going
Wind Energy	Spain	on-going
Wind Energy	Saudi Arabia	on-going
Large Wind -- 200 kw	IEA	under-negotiation
2. Fuels from Biomass		
Forestry Energy	IEA	on-going
Biomass	IEA	on-going
Biomass Conversion	Saudi Arabia	on-going
Fermentation	Brazil	under-negotiation
3. Geothermal Energy Development		
Geothermal Information	Italy	on-going
Geothermal Technology	Italy	on-going
Geothermal Survey	Mexico	on-going
Man-Made Geothermal Energy System	IEA	on-going
Geothermal	Iceland	inactive
Geothermal	New Zealand	inactive
Geothermal	USSR	inactive
Geothermal	Japan	under-negotiation
Man-Made Geothermal	Federal Republic of Germany	under-negotiation
4. Energy Storage Systems		
Thermochemical H_2 Production	IEA	on-going
High Temperature Thermochemical H_2	IEA	on-going
Hydrogen Market Assessment	IEA	on-going
Electrolytic H_2 Production (using alkaline or solid polymer electrolytes)	IEA	on-going
Electrolytic H_2 Production (using solid oxide electrolytes)	IEA	on-going
Thermal Storage	IEA	on-going
Hydrogen Storage-Vehicle Use	U.S., Federal Republic of Germany	under-negotiation
Battery Storage	U.S., Japan	under-negotiation
5. Electric Energy Systems		
AC/DC UHV Transmission	USSR	active
Superconducting Transmission	USSR	active

Table III-11-3. U.S. multilateral agreements administered through IEA.

SUBJECT AND BASIC TERM*	SIGNATORIES	DESCRIPTION
1. SMALL SOLAR POWER SYSTEMS OCT. 1977–OCT. 1983	AUSTRIA, BELGIUM, FED. REP. OF GERMANY, GREECE, ITALY, SPAIN, SWEDEN, SWITZERLAND, UNITED KINGDOM, UNITED STATES	Jointly funded design study (Phase I), leading to construction and operation (Phase II) of a 500-kW distributed collector solar power plant and a 500-kW central receiver solar plant in Almeria, Spain
2. LARGE-SCALE WIND ENERGY CONVERSION OCT. 1977–OCT. 1979	DENMARK, FED. REP. OF GERMANY, SWEDEN, UNITED STATES	Exchange of information and personnel, including periodic meetings of program directors, to coordinate execution of national projects to design, construct, and operate machines with a rated power of 1 MWe
3. WIND ENERGY CONVERSION SYSTEMS OCT. 1977–OCT. 1980	AUSTRIA, CANADA, DENMARK, FED. REP. OF GERMANY, IRELAND, JAPAN, NETHERLANDS, NEW ZEALAND, SWEDEN, UNITED STATES	Exchange of information, joint funding of studies, joint development of models and workshops on (I) environmental and meteorological aspects, (II) evaluation siting models, (III) integration into electricity supply systems, and (IV) rotor-stressing and operation of large-scale systems
4. WAVE POWER APR. 1978–OCT. 1980	CANADA, JAPAN, UNITED KINGDOM, UNITED STATES	Jointly funded testing and demonstration of wave-powered air-turbine generating systems on Japanese wave-breaking buoy, the "Kaimei"
5. FORESTRY ENERGY APR. 1978–APR. 1981	BELGIUM, CANADA, IRELAND, SWEDEN, UNITED STATES	Exchange of information for national RD&D activities, for planning national programs, and developing proposals for future RD&D projects in the following areas of forestry energy: system feasibility studies, growth and production, harvesting, on-site processing, transportation, and conversion
6. BIOMASS TECHNICAL INFORMATION SERVICE MAY 1978–MAY 1981	IRELAND, SWEDEN, UNITED STATES	Establishment of jointly funded central information service to collect and disseminate scientific and technical data in all areas of biomass energy technology. The Service will provide regular abstract bulletins, bibliographies, and literature reviews

* IEA Agreements typically cover a specified period with the provision for automatic extension.

(CONT.)

Table III-11-3. (Continued)

SUBJECT AND BASIC TERM*	SIGNATORIES	DESCRIPTION
7. MAN-MADE GEOTHERMAL ENERGY SYSTEMS OCT. 1977–OCT. 1979	FED. REP. OF GERMANY, SWEDEN, SWITZERLAND, UNITED KINGDOM, UNITED STATES	Jointly funded study to identify possible technical systems and recommended possible future laboratory studies, hardware development, and field testing for extracting thermal energy from the earth's crust
8. GEOTHERMAL WORKING PARTY	AUSTRIA, CANADA, FED. REP. OF GERMANY, ITALY, JAPAN, NETHERLANDS, SWEDEN, SWITZERLAND, TURKEY, UNITED KINGDOM, UNITED STATES	Formation of four panels which relate to different areas of geothermal R&D, including manmade geothermal energy systems (listed above), low-enthalpy geothermal sources, high-enthalpy geothermal sources, and geophysical exploration techniques
9. HYDROGEN PRODUCTION FROM WATER OCT. 1977–OCT. 1980	BELGIUM, CANADA, EUROPEAN ECONOMIC COMMUNITY, FED. REP. OF GERMANY, ITALY, JAPAN, NETHERLANDS, SWEDEN, SWITZERLAND, UNITED STATES	Exchange of information and personnel, including workshops on specific process steps for thermochemical production of hydrogen
10. HIGH TEMPERATURE THERMOCHEMICAL H_2 PRODUCTION OCT. 1977–OCT. 1980	FED. REP. OF GERMANY, UNITED STATES	Exchange of information and personnel, including workshops on specific process steps for interfacing thermochemical hydrogen production with high temperature nuclear reactors
11. H_2 MARKET ASSESSMENT OCT. 1977–OCT. 1980	BELGIUM, CANADA, EUROPEAN ECONOMIC COMMUNITY, FED. REP. OF GERMANY, ITALY, JAPAN, NETHERLANDS, SWEDEN, SWITZERLAND, UNITED STATES	Exchange of information and personnel, including workshops on the market assessment by country, of hydrogen
12. ELECTROLYTIC H_2 PRODUCTION (ALKA. AND S.P.) OCT. 1977–OCT. 1980	BELGIUM, CANADA, EUROPEAN ECONOMIC COMMUNITY, ITALY, JAPAN, NETHERLANDS, SWITZERLAND, UNITED KINGDOM, UNITED STATES	Exchange of information and personnel, including workshops, on electrolytic hydrogen production using alkaline or solid polymer electrolytes
13. ELECTROLYTIC H_2 PRODUCTION (SOLID OXIDE) OCT. 1977–OCT. 1980	CANADA, EUROPEAN ECONOMIC COMMUNITY, FED. REP. OF GERMANY, ITALY, SWEDEN, UNITED KINGDOM, UNITED STATES	Exchange of information and personnel, including workshops on electrolytic production of hydrogen using solid oxide electrolytes

*IEA Agreements typically cover a specified period with the provision for automatic extension.

Table III-11-4. International energy agency agreements in areas of CSE responsibility.

TECHNOLOGIES / PARTICIPANTS	AUSTRIA	BELGIUM	CANADA	DENMARK	E.E.C.	GERMANY	GREECE	IRELAND	ITALY	JAPAN	NETHERLANDS	NEW ZEALAND	SPAIN	SWEDEN	SWITZERLAND	TURKEY	UNITED KINGDOM	UNITED STATES
SOLAR ELECTRIC																		
Small Solar Power Systems	*	*				*	*		*				*	*	*		*	*
Wind Energy Conversion Systems (WECS)	*		*	*		*		*			*	*		*				*
Large-scale WEC System				*		*								*				*
Wave Power			*						*								*	*
FUELS FROM BIOMASS																		
Forestry Energy	*	*							*					*				*
Biomass Technical Information Service									*					*				*
GEOTHERMAL																		
Man-Made Geothermal Systems						*								*	*		*	*
Working Party	*		*			*				*	*	*		*	*	*	*	*
ENERGY STORAGE																		
Thermochemical Hydrogen Production	*	*			*	*				*	*	*		*	*			*
High-Temperature Thermochemical H_2						*												*
H_2 Market Assessment	*	*			*	*			*	*	*			*	*			*
Electrolytic H_2 Production (Alka. and S.P.)	*	*			*				*	*	*				*		*	*
Electrolytic H_2 Production (Solid Oxide)				*		*			*						*		*	*

Solar Technology

The most immediate overseas market appears to be that for relatively small onsite solar technology. The consensus of U.S. industry and international organizations that decentralized solar technologies (mainly photovoltaics) can now competitively meet energy needs for water pumping, remote (oil platform, navigational, railroad, highway, etc.) signals and communications, corrosion control, educational TV, and, shortly, for off-grid village power. Somewhat later, a market should develop for exports of specialty hardware in the non-photovoltaic fields, sale of systems design engineering and advanced technology under licensing, and royalty or joint venture agreements.

In developing countries, many areas have no access to power grids and, where they receive electricity, kilowatt costs are often very high. As a result, large-scale commercialization of solar electric technologies in developing countries should be possible much sooner than in the United States. The specific need of developing countries is for low-cost, easily maintainable systems that do not require the construction of expensive transmission networks. Pilot testing of U.S. decentralized, renewable systems in these countries could help penetrate this market. An unresolved issue is the extent to which the United States should provide such projects as aid and those for which the recipients should share costs.

Demonstration of a wide range of U.S. solar energy technologies is planned in Saudi Arabia, with whom the United States has a $100 million agreement (costs are shared equally). Hopefully, demonstrations and application of solar energy systems in Saudi Arabia can help develop that market as well as improve U.S. marketing prospects in other OPEC countries.

Solar Thermal Power. Competitive opportunities for small-scale solar thermal plants are emerging in remote areas where fuel and electric costs are high and energy supply is insufficient. In such areas in the United States and overseas, the best commercialization approach may be the use of dispersed solar systems in combination with fossil fuel plants or as part of a total energy system.

Additional CSE support of RD&D for small-scale thermal plants will raise the United States' current low profile in this near-term market. This may be accomplished by placing increased emphasis on the IEA small-scale solar thermal demonstration project. This agreement could give U.S. firms in the field some international market exposure with possibilities of establishing joint commercial ventures with foreign firms. Another area of emphasis for which additional CSE support could generate export earnings is the agriculturally based dispersed solar technologies. These small-scale facilities, one of which is shown in Figure III 11-2, have immediate promise in developing countries because they can provide most of the thermal energy requirements for a farm not connected to a conventional electric grid.

Commercialization of large-scale thermal plants is expected to take considerably longer than that for small-scale plants. A large-scale thermal demonstration plant is not expected until at least 1986. Another main goal is to reduce heliostat costs to $14/square foot by the early 1980s and eventually, to $7/square foot for reflector surfaces. At the latter price, thermal electric plants are expected to be competitive with other technologies in areas of high insolation, such as the southwestern United States and most tropical areas.

Photovoltaic Energy Conversion. There is considerable near-term market potential for a variety of decentralized photovoltaic applications, with the bulk of the market in developing countries. On December 16, 1978, the United States dedicated a 3.5-kW photovoltaic system in Arizona (Figure III 11-3). This photovoltaic system will provide base power to pump water and operate lights and appliances in an isolated Indian village, which is not connected to the electric grid. A project of this type has great potential applicability in less developed countries with similar isolated villages, which are either not serviced by traditional forms of electric generation or are dependent on small-scale diesel engines to produce power.

A 350-kW peak photovoltaic system is included among projects to be performed under

Figure III-11-2. Agricultural application of solar thermal technology.

Table III-11-5. U.S. bilateral agreements.

COUNTRY	SCOPE, TERM AND DESCRIPTION
1. BRAZIL	Fermentation: under negotiation. The agreement will add to U.S. knowledge in fermentation aspects of biomass conversion. Brazil has extensive experience in fermentation and conversion techniques.
2. FEDERAL REPUBLIC OF GERMANY	Hydrogen storage for vehicle use: under negotiation. The agreement will provide exchange of information on improved hydride hydrogen storage and vehicular demonstration. The improved storage techniques will result in smaller physical volumes for storage and less complex hydrogen regenerative techniques.
	Energy Storage Systems (Heat Pipes), Sept. 1977–Sept. 1980. Development in the United States and evaluation in West Germany of chemical heat pipe components (aspect of agreement also concerning high-temperature reactors).
3. FRANCE	Solar Thermal Conversion Systems, May 1976–May 1978. The agreement provides an exchange of information, joint tests, and performance evaluations of a U.S.-designed and fabricated cavity boiler tested jointly at the French solar furnace at Odeillo.
	Solar Power Towers, Sept. 1977–Sept. 1979. The agreement provides an exchange of information, joint tests, and performance evaluations to assess the risks associated with heliostat fields and towers. Tests are being conducted to measure heliostat flux intensity and light characteristics reflected into the airspace above the fields.
	Geothermal: under negotiation.
4. ICELAND	Geothermal: inactive (Nov. 1973, amended 1974). The agreement provides for cooperative projects and technical information exchange on Iceland's extensive experience in geothermal space heating.
5. ITALY	Solar Thermal Conversion Systems: under negotiation.
	Photovoltaics: under negotiation.
	Geothermal Data Base, May 1976–May 1981. The agreement provides for the exchange of information in computer format dealing with location, size, and characteristics of geothermal wells and fields, bibliographic information, and heat transmission data.
	Hydrothermal and Hot Dry Rock, June 1975–June 1980. The agreement covers RD&D of applications in such project areas as stimulation of hot dry rock and hydrothermal reservoirs, reservoir physics and engineering, and environmental control technology.
	Hot Brines Reservoir, June 1975–June 1980. The agreement provides for cooperative efforts in RD&D in the utilization of hot brine resources.
6. JAPAN	Geothermal: under negotiation. The agreement will provide an exchange of information on a hot dry rock energy system.
7. MEXICO	Geothermal Survey, July 1977–July 1982. The agreement provides for a cooperative survey of the Cerro Preito geothermal fields to determine the most economical and productive methods in further developing the reservoir. The Mexicans have several years experience operating a 75-MWe generator at the Cerro Preito site, adjacent to Imperial Valley. The field similarities are such that the United States can benefit from their operational experience, while the Mexicans feel U.S. instrumentation can improve their operation.

(CONT.)

Table III-11-5. (Continued)

COUNTRY	SCOPE, TERM AND DESCRIPTION
8. NEW ZEALAND	Geothermal: inactive (1974–1979). The agreement provided for the technical information exchange concerning the development and application of geothermal energy. Progress has been restricted as a result of lack of funding and reduced interest in New Zealand caused by large discoveries of natural gas.
9. SPAIN	Solar Thermal Central Receiver, Sept. 1976–Sept. 1981. The agreement provides, as a supplement to the U.S.-Spain Treaty of Friendship and Cooperation, solar research, development, and testing of a 1-MWe solar central receiver system.
	Photovoltaics, Sept. 1976–Sept. 1981. Under the context of the U.S.-Spain Treaty of Friendship and Cooperation, R&D information exchange will be provided under photovoltaic technology and applications.
	Wind Energy Conversion Systems, Sept. 1976–Sept. 1981. Under the context of the U.S.-Spain Treaty of Friendship and Cooperation, information exchange will be provided for wind energy conversion and meteorology technologies.
10. SAUDI ARABIA	General Solar Energy Agreement, Oct. 1977–Oct. 1982. This Solar Energy Agreement involves a 5-year commitment of $100 million with each country providing one-half of the funding. This agreement provides for solar technology, RD&D, and facilitation of technology transfer in all types of solar systems and technologies, including centralized, dispersed, solar thermal, photovoltaics, biomass conversion, wind and ocean energy. The Preliminary Technical Program Plan stresses solar R&D in six major areas: solar energy availability in Saudi Arabia, thermal processes, storage, fuel production, solar electricity generation, and other alternative solar-related sources.
11. U.S.S.R.	Solar Thermal Conversion Systems, June 1974–June 1979. The agreement provides for cooperative solar thermal research and development and facilitation of technology transfer; it is part of a larger program to develop various non-conventional fuel sources.
	Photovoltaics, June 1974–June 1979. The agreement provides cooperative R&D and facilitation of technology transfer in photovoltaic technologies and is part of a larger program to develop nonconventional fuel sources.
	Geothermal: inactive. The agreement provided for joint development of geothermal technology including drilling methods, reservoir modeling, reservoir production, utilization technology, and environmental protection. If this agreement had been actively pursued, the United States could have benefited from Soviet experience in operating a geothermal energy plant. However, U.S. representatives were denied access to the only operational geothermal plant in the U.S.S.R.
	AC/DC UHV Transmission: active. The agreement is for information and personnel exchange on design and development of AC/DC Ultra-High-Voltage Transmission.
	Superconducting Transmission: active. The agreement is for information and personnel exchange on design and development of commercial superconducting transmission lines.

Table III-11-6. U.S. bilateral agreements in areas of CSE responsibility.

TECHNOLOGIES	BRAZIL	FEDERAL REPUBLIC OF GERMANY	FRANCE	ICELAND	ITALY	JAPAN	MEXICO	NEW ZEALAND	SPAIN	SAUDI ARABIA	U.S.S.R.
SOLAR ELECTRIC											
Solar Thermal Conversion Systems		*	◊							*	*
Solar Power Towers		*									
Solar Thermal Central Receiver									*		
Photovoltaics			◊						*	*	*
Wind Energy Conversion Systems									*	*	
Ocean Energy										*	
FUELS FROM BIOMASS											
Fermentation	◊										
Biomass Conversion										*	
GEOTHERMAL											
Geothermal Data Base					*						
Hydrothermal & Hot Dry Rock					*						
Hot Brines Reservoir					*						
Geothermal Survey							*				
Geothermal		*	□		◊		□				□
ENERGY STORAGE											
Hydrogen Storage For Vehicle Use		◊									
ELECTRIC SYSTEMS											
AC/DC UHV Transmission											*
Superconducting Transmission											*

Note: * Indicates Active Bilateral Agreements.
□ Indicates Inactive Bilateral Agreements.
◊ Indicates Bilateral Agreements Currently Under Negotiation.

the United States/Saudi Arabia Solar Energy Agreement. It is expected that the first 50-kW ystem will be in operation in 1980, the larger system by 1981.

The national photovoltaic goals seem modest, given the estimate that a photovoltaic market of 6 gigawatts exists solely for water pumping in developing countries, with significant market penetration possible by 1981–1982. However, many potential users are unaware and unsure of the benefits and readiness of solar cell power for their applications. While photovoltaics are already competitive for many applications, experimental programs overseas by CSE may be necessary to satisfy potential users' concerns about geographic compatibility and economic acceptability of the systems in their specific setting.

Wind Energy Conversion Systems. There appear to be attractive commercial opportunities in developing countries for wind energy conversion systems which meet the need for dispersed electric uses and which can efficiently operate in the 6- to 10-mph winds common to tropical zones.

The two IEA Wind Energy Conversion System agreements focus on the availability of data concerning large wind systems appropriate for centralized power grid applications. This is a reflection of the domestic program emphasis on reducing the costs of intermediate- and large-size wind energy systems. The United States is participating in a project with Denmark to obtain data on operating efficiencies of the Gedser windmill, shown in Figure III 11-4. This windmill was built by Denmark

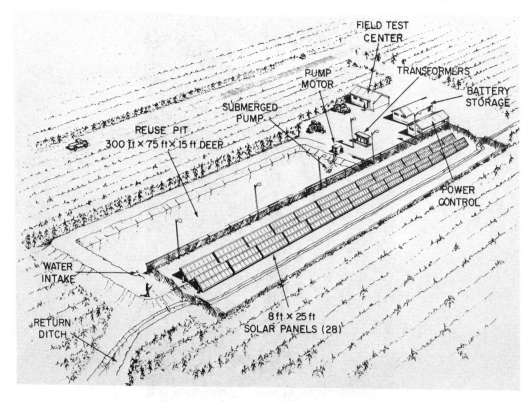

Figure III-11-3. Photovoltaic system for agriculture.

in 1961 to evaluate a wind-turbine design. The facility was not operated for more than a decade until the recent renewed interest in wind-turbine development emerged in both the U.S. and Denmark. The wind-turbine was renovated in 1977 and is now being operated to provide the United States and Denmark with performance data for an improved wind-turbine design.

For the United States to take advantage of international commercialization possibilities, CSE is directing more RD&D for smaller dispersed wind electric systems; attention must be given to designing systems capable of efficiently operating in 6- to 10-mph winds instead of designing systems only for the 10- to 15-mph winds common to temperate climates. Also, local demonstrations of how wind system applications may meet the needs of developing countries may assist the United States in penetrating this market.

Ocean Systems. International commercialization considerations should be particularly important for ocean systems projects as a result of the relative scarcity of viable operational sites off the U.S. coasts.

At present, the United States is participating in an IEA wave-power technical exchange agreement and also has an agreement to conduct ocean thermal energy R&D in cooperation with Saudi Arabia. The IEA wave agreement should give the United States an opportunity to become knowledgeable, at low costs, about recent wave technology advances such as Japan's Masuda buoy.

More CSE technical exchange programs with foreign countries that have advanced wave and tidal technology are necessary if the United States is to be involved in the eventual commercialization of these energy systems. Planned commercialization of ocean thermal energy conversion systems (OTEC) may be accelerated by CSE initiation of RD&D projects with countries that have promising test and operational sites and that are also potential customers.

The most technologically advanced CSE

Figure III-11-4. Gedser windmill in Denmark.

ocean energy system is OTEC. However, the most promising sites for such systems are located off tropical coastal areas, south of the U.S. land mass. In these warmer latitudes, the water gradient differentials tend to be larger and biofouling, a major obstacle to OTEC systems. Initiation of RD&D agreements with the appropriate tropical States could lead to increasing the projected OTEC market, reducing costs, and accelerating planned commercialization.

Biomass Energy Systems. Developing countries have a particular need for biomass conversion equipment that is well adapted to the forms of biomass collectible and that can also produce good quality fertilizer along with other derivatives. In contrast to industrial country needs, the developing countries' market is oriented toward decentralized biomass conversion systems that minimize capital costs. The CSE long-term objective is to operate large experimental facilities which, it is hoped, will make biomass fuel costs competitive.

CSE support for RD&D of decentralized biomass technologies, particularly those which produce high quality fertilizer, is necessary to take advantage of the large market opportunities in developing countries.

Geothermal

The Italian government, the leader in geothermal power production, views Latin America as a $10-million annual market currently and as a $100 million-annual market within a short time. Presently, the United States is on the fringe of this market. U.S. firms are concentrating primarily on exploration, consulting, and management rather than the production of hardware and products.

Lack of resource information presents an important obstacle to geothermal commercialization. The information gained by the United States through international agreements on reservoir behavior is particularly crucial for industrial development. However, the type of geothermal international exchange agreements in which the United States has participated are not ideal for facilitating foreign market penetration. One effect of United States' emphasis on exchange agreements with Italy and Japan (under negotiation) may be to strengthen their leading export roles in geothermal energy.

Overseas sales may be effectively encouraged by international arrangements that provide a package of technical assistance, information exchange, resource assessment, and concessional export financing. Hopefully, the United States' recent agreement with a limited number of IEA countries to test certain geothermal equipment cooperatively will result in some penetration of South American and East European markets to which IEA countries have access. Through the United States' cooperative survey effort in Mexico, that government has become aware of and is buying some U.S. geothermal instrumentation. In November 1977, the Department of Commerce sponsored a geothermal seminar/industrial mission to Central America, which resulted in the purchase of valves from an American geothermal supplier. Also, the Agency for International Development has been involved in geothermal development in El Salvador, Nicaragua, Indonesia, the Philippines, and Kenya.

Energy Storage Systems

The rising cost of fossil-fuel electricity makes energy storage economically attractive both overseas and in the domestic market. Modular battery systems seem particularly amenable to overseas marketing, with subsequent reduction of domestic costs.

The objective of the IEA energy storage agreement, in which CSE participates, is to reduce the production costs of hydrogen to below $6/million Btu, at which point it becomes commercially viable. To meet this objective, joint RD&D is being conducted for a variety of thermochemical and electrolytic hydrogen production methods, along with a joint market assessment of hydrogen for each country. As each energy storage system moves closer to commercialization, CSE activities will concentrate on smoothing the way for industry to introduce commercial energy storage systems to the marketplace. Where appropriate, Federal impetus should be given to international as well a domestic commercialization.

Electric Energy Systems

The United States' only international agreement in Electric Energy System (EES) R&D is with the Soviet Union, with whom there are bilateral agreements concerning information and personnel exchange on AC/DC ultrahigh voltage transmission technology and superconducting transmission technology.

The broad objective of EES is to ensure that the national electric system is capable of meeting future demands in the most reliable and efficient manner. As the EEs mission is directly related to the configuration of the American electric energy system, it has little direct interest in international commercialization. However, awareness of overseas market opportunities and availability of such information to domestic companies could result in some international commercialization.

Part IV
Fission Energy

1.
The National Energy Plan and Fission Energy Policy

This chapter describes the relationship of the Fission Energy Program to the National Energy Plan and the Program strategy that involves extending our fissile fuel resources used in once-through fuel cycles and developing technology for possible future breeder reactors. It also introduces the structure for implementing the program.

THE NATIONAL ENERGY PLAN

The National Energy Plan (NEP) of April 1977 sets forth the current energy policy of the United States and establishes three overriding energy objectives for the nation:

> ... "as an immediate objective that will become even more important in the future, to reduce dependence on foreign oil and vulnerability to supply interruptions; in the medium-term, to keep United States imports sufficiently low to weather the period when world oil production approaches its capacity limitations; and in the long-term, to have renewable and essentially inexhaustible sources of energy for sustained economic growth." (NEP, p. 6)

The NEP proposes to achieve these objectives through a combination of measures, including:

- Conservation—reduction of energy waste and improvement in fuel efficiency
- Rationalization of pricing of the scarcer fuels in the direction of reflecting cost of replacement, thus:
  — Further encouraging conservation by consumers
  — Providing appropriate incentives to develop new sources of the scarcer fuels and increase domestic production
  — Improving the economics of converting existing energy consumption and conversion facilities from the scarcer fuels to the more abundant coal where technologically feasible
- Substitution of facilities to use coal and nuclear fission energy resources rather than scarcer fuels
- Development of renewable and essentially inexhaustible energy resource technologies, with appropriate energy conversion and storage methods, to replace oil and gas on a commercial basis as they enter their final stages of depletion

Nuclear fission energy has potential roles in the last two of these major measures. In the near and intermediate terms, coal and uranium must share the burden of electricity generation. National energy policy stresses the key role of the present generation of fission systems—the light water reactor (LWR) operating in a once-through fuel-cycle mode. The NEP expresses a resolve to eliminate technical and institutional impediments to maturing of the LWR system and to continue the evolution of LWR technological advances. Department of Energy (DOE) policy particularly stresses improvements in LWR uranium utilization efficiency.

In the longer term, the fission breeder concept offers an essentially inexhaustible electrical energy source. NEP defers demonstration and commercial deployment of the breeder and other advanced fission systems requiring reprocessing and recycle of spent fuel. This deferral is based on two factors:

- A lessened sense of urgency for extending natural uranium ore supplies, in the light of the fall-off in LWR orders in the past few years
- An increased sensitivity to the issue of proliferation of nuclear weapons to non-weapons nations

At the same time, however, the fission energy policy calls for maintaining a breeder system option through a strong research and development program. In the long-term, inexhaustible energy resources must take over a major fraction of our energy supply. Remaining U.S. and world resources of oil and gas must be preserved for future generations to the greatest extent possible, for use not as power-generation fuels, but for transportation and as industrial feedstock for manufacturing essential materials obtainable in no other way. The four basic categories of inexhaustible resources known today are solar, geothermal, fusion, and the fission breeder. The breeder has been developed to the point where a realistic schedule for significant deployment in the United States can be contemplated. It follows from the long-term objective of the NEP that the development of the breeder must be continued in parallel with resolution of the proliferation issue, as a hedge against the possibility that an inexhaustible resource will be required before another technology is technically or economically ready for deployment.

PROGRAM STRATEGY

The pause in the breeder system demonstration process allows time for a reexamination of alternative advanced reactor system concepts* in terms of resource utilization, economics, practicality, and relative nonproliferation strategic value. The principal vehicle for this reexamination has been the Nonproliferation Alternative Systems Assessment Program (NASAP). A goal of NASAP is to reduce the large number of possible fission concepts to a manageable and affordable set. Although the NASAP effort is still incomplete, preliminary results indicate that further development of U.S. nuclear power should continue to emphasize concepts that employ either light water or sodium as coolants. Therefore, it has been proposed to redirect the Thermal Gas Cooled Reactor Program toward the direct-cycle High Temperature Gas Cooled Reactor (HTGR) and the Very High Temperature Reactor (VHTR) for process-heat applications—applications for which it is uniquely qualified. NASAP has shown the steam-cycle HTGR system to be of marginal benefit because of its limited deployment "window."** limited industrial interest, and the fact that an expensive major scale-up of a third heat transport system technology—for gas cooling—would be required.*** This redirection would, in effect, concentrate further government-sponsored fission development effort on the two coolant systems in which the United States already has a major investment—light water and sodium.

Extending Once-through Fission Resources

The U.S. decision to defer commercial fuel reprocessing and breeder reactor demonstration has engendered increased interest in alternative fission system concepts that would extend the use of the uranium resource base, as well as interest in a more accurate deter-

---

*Systems offering improved fuel utilization efficiency as compared to the LWR once-through system.

**The period beginning when a significant commercialization is initially achieved and ending when ore resource depletion prevents further deployment.

***It should be noted that the principal element of cost and risk in a reactor research, development, and demonstration (RD&D) program is the scale-up of heat transport technology through a sequence of development and demonstration plants.

mination of the magnitude of the base. In particular, the following areas of effort are addressing these concerns:

- Improved fuel utilization in LWR's operating on a once-through fuel cycle
- Advanced isotopic enrichment techniques
- National Uranium Resources Evaluation (NURE)

LWR Technology Improvement Program. LWR's are already commercially deployed. No further scale-up of heat transport equipment is necessary. Sixty-nine commercial LWR plants now constitute 9-10 percent of U.S. electrical generating capacity and are providing about 13 percent of the nation's electric power. Fuel performance and fuel management improvements can result in an extension of the uranium base without requiring any modifications to heat transport systems or other parts of the plants. The earlier the improvements occur, the greater the cumulative impact on conserving the uranium resource base.

Incentives other than resource conservation for LWR once-through improvements include the following:

- Economics—Development of higher burnup fuel would reduce real ore and fabrication costs per unit of electrical energy delivered in inverse proportion to burnup. If utilities use the increased burnup to increase reactor-plant availability by operating longer between refuelings, the need for replacement power would be reduced, thereby decreasing the cost of electricity to the consumer and also reducing the need for imported oil.
- Strategic flexibility—The long-term future of U.S. nuclear power development remains uncertain as new U.S. policy is evolving. It follows, therefore, that additional options for continued use and improvement of the commercial LWR must be pursued wherever practical to increase the benefits from present technology and to allow maximum flexibility in planning for future development of the breeder reactor and other advanced fission energy systems.
- Reduced requirements for spent-fuel storage—Improved fuel utilization in current and planned LWR plants will result in fewer spent fuel elements to be stored.

Finally, if reprocessing and the breeder are reinvoked at some point in the future, there is the prospect that improved LWR plant designs may offer the additional potential of operating on U-233 fuels produced in breeder blankets through a synergistic nuclear economy of both breeders and LWR's. U-233, which must be bred from thorium, is a better fuel in thermal-neutron-spectrum reactors than in fast-spectrum machines.

Consistent with this perspective, the evolution of LWR improvements is envisioned to occur through four phases over time:

- Phase 1, which involves development of near-term retrofittable improvements that could begin to produce benefits in existing LWR once-through plants within the next ten years.
- Phase 2, which includes development of mid- and long-term retrofittable improvements that could produce benefits over the longer term (beyond 1988).
- Phase 3, which involves development of modified LWR designs, including both retrofittable and non-retrofittable improvements, that might be constructed over the longer term to obtain maximum benefits from the once-through LWR uranium fuel cycle.
- If reprocessing and the liquid metal fast breeder reactor (LMFBR) are employed in the future, Phase 4 would involve the option of using U-233 in LWR's in a symbiotic configuration with thorium blanketed LMFBR's.

Current DOE efforts address all four phases. Demonstration of the technical feasibility of a 15 percent reduction in ore requirements for LWR (Phase 1) is targeted for 1988. Further backfittable improvements to achieve an additional 10-15 percent savings could be demonstrated beyond 1988 with the cooperation of the reactor manufacturers. A conceptual design study being conducted in FY 1979 and FY 1980 is aggregating and optimizing the most attractive of both backfittable and nonbackfittable features. These improvements would require a new LWR market line but would not require a demonstration plant. The continued operation of the Shippingport light water breeder reactor (LWBR) is expected to confirm the technical feasibility of certain features, such as movable fuel control, that are beneficial in improving fuel use. These features will be evaluated further in support of the development in Phases 2, 3, and 4.

Advanced Isotope Separation Technology. Another approach to extending uranium resources for use in LWR's is the development of advanced isotope separation (AIS) methods that could be applied to the existing large stockpile of tailings, as well as future tailings, from uranium enrichment plants. These tailings still contain 0.2–0.3 percent fissile U-235, whose further extraction and utilization could increase the efficiency of total uranium resource use by about 20 percent. The program calls for developing three known AIS techniques through the preprototype phase. This will be followed by thorough evaluation of scaleability, economic potential, proliferation resistance and environmental effects, and a possible decision to proceed with engineering development of the best of the three techniques if the evaluation so warrants.

National Uranium Resources Evaluation. The NURE program has been established to improve our estimates of the availability and accessibility of both uranium and thorium domestic supplies. It is under the cognizance of the Assistant Secretary for Resource Applications in DOE, not the Assistant Secretary for Energy Technology, and therefore is not described in this document.

LMFBR Research and Development (R&D)

The deferral of breeder demonstration allows the continued evaluation of fuel-cycle alternatives for use in the LMFBR, which uses the second principal coolant technology—sodium. All practical combinations of the following alternative aspects are investigated in detail from the standpoints of resource extension, economics, practicality, and proliferation resistance:

- Driver-fuel forms—oxide, advanced oxide, carbide, etc.
- Blanket fertile material forms and isotopics—above forms plus thorium/U-233 and U-238/plutonium cycles.
- External fuel-cycle configurations—spiking, coprocessing, Purex and Civex processes, etc.

Here again the evaluation will result in a winnowing of options to a manageable set for further development effort.

As stated earlier, the national fission energy policy calls for maintaining a breeder system option through a strong research and development program. Pursuant to this policy, continuing development of the LMFBR is focused on an in-depth Conceptual Design Study (CDS) that will result in a design for an R&D plant. This design study will be completed by March 1981, at which time it will be decided whether to proceed with the detailed design and construction of a reactor plant. A parallel base technology program complements the conceptual study and is capable of supporting the expeditious pursuit of detailed design and construction if the decision is made to proceed.

The decision to build a breeder R&D plant and subsequent prototype plant(s) is quite different from the marketplace decision to deploy the breeder commercially. The first, a government decision, and the second, an industry decision, are made at different times. The industry decision for commercialization will only be made if the following determinations are made:

- The utility sector perceives that a breeder plant will be to its advantage
- The technical and financial risks, shared by the utility sector and its supplier industry, are tolerable

Normally, risks associated with scaling-up a new technology are absorbed by gaining experience through a sequence of plant projects of increasing size. Without this experience, it is unlikely that a utility would commit the necessary funds for a commercial-size plant. Ideally, the scale-up process ends with successful operation of a prototype plant(s), whose principal components and systems are of commercial size.

Once the prototype phase has been completed, the decision to deploy can be made concurrently with, or subsequent to, the prototypical operation phase, depending on the demand and the comparative availability and economics of other sources. This strategy to maintain the breeder option through an R&D program provides for deployment options consistent with the uncertainty in the uranium resource base and cost factors influencing the timing and need for the breeder.

The number of plants in this scale-up process and their relative timing are governed by a balance of technical and financial risks and the degree of urgency to deploy the concept commercially. Figures IV 1-1a and IV 1-1b show two alternative long-range breeder programs—one with a strategy based on sequential plant construction, the other with a compressed sequence employing overlapping plants—that lead through separate major decision points from an R&D plant through the prototype phase to commercialization. The paths shown depict separate decisions by the government to design and construct an R&D plant, followed by a prototype(s), leading ultimately to the option of commercial deployment. The sequential strategy is generally consistent with the view that a large uranium resource base exists, the nuclear growth rate will be low, breeder economics are marginal, and the urgency to deploy is low. Such a strategy would have relatively low technological risk. On the other hand, the overlapping strategy reflects the view represented by a more limited uranium resource base, a high demand for nuclear power, favorable breeder economics, and the need for early deployment. Such a strategy would entail a relatively greater technical risk.

The Conceptual Design Study augmented by parallel technology development will define a developmental plant concept that meets the NASAP/International Nuclear Fuel Cycle Evaluation (INFCE) criteria and is within the technical state of the art. The decision to

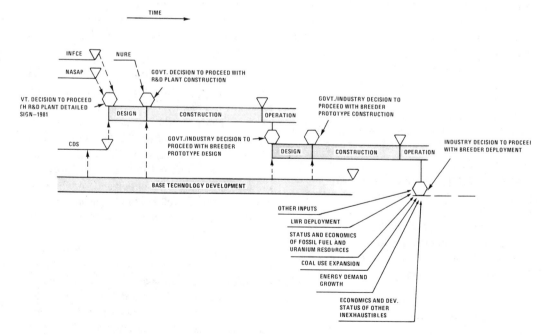

Figure IV-1-1a. Decisions leading to breeder commercialization via sequential plant construction.

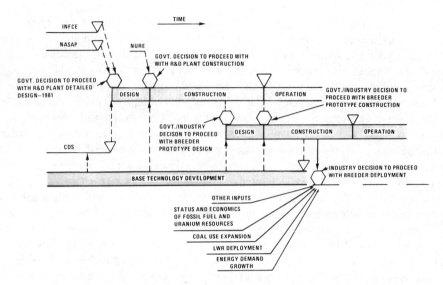

Figure IV-1-1b. Decisions leading to breeder commercialization via a compressed schedule with overlapping plants.

begin detailed design can then be made. Continuing parallel technology and detailed design efforts following this decision will result in a design that can be constructed and operated as a testbed to develop data required for the subsequent prototype and commercialization phases.

As detailed design efforts reach the point of a construction decision for each major type of plant, considerably more information will be available from the NURE program. These data, as well as additional information from the technology program, will support the decisions on R&D plant and follow-on plant construction.

Utility commercialization decisions can be made only after the scale-up process is complete. In addition to the economic and reliability factors for the LMFBR, the corresponding factors for all alternative sources of energy (i.e., fossil-fueled systems, LWR's, and advanced converters) and the status of other inexhaustible sources, as well as energy demand growth at the time, must be evaluated. Commercialization will proceed only if all these factors, in the aggregate, favor LMFBR deployment.

Implementation Structure

In summary, the Fission Energy Program strategy is structured to do the following:

- Enhance the potential for increased near-term commercial deployment of light water reactors by improving LWR technology and developing other means of extending the uranium resources base
- Assess advanced fission options for their nonproliferation, economic, technical, and institutional merits (during the time while a commitment to breeder and other technologies requiring reprocessing and recycle of spent fuel is deferred)
- Maintain the breeder option through R&D

To implement this strategy efficiently, the Fission Energy Program employs a work breakdown structure (WBS) approach. Details of this approach are discussed in Chapter 2, but Figure IV 1-2 shows the first three levels of the Fission Energy WBS. Level 1, the entire Fission Energy Program, includes all the activities under the cognizance of the program director. Level 2 defines the major subdivisions of the program, each of which has fission as a common element, but they differ from one another in basic application or uses. For example, converter reactor systems fulfill near-term and long-term energy needs, breeders are directed toward long-term needs, and defense applications cover military needs. Subdivisions at Level 3 are identified by a particular end product (e.g., LWR or LMFBR).

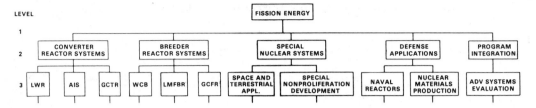

Figure IV-1-2. Fission energy program upper-level work breakdown structure.

Funding. The funding for the Fission Energy Program for FY 1979 and FY 1980 is shown in Table IV 1-1.

FISSION ENERGY SYSTEM FUNDAMENTALS

In all fission systems used or seriously contemplated today, the fuel contains both fissile material (in which energy is generated by neutron-induced fission) and fertile material (which is converted to fissile material by neutron absorption). In the LWR reference system, the fissile material is uranium-238, some of which is converted to fissile plutonium in the reactor.* In the reference once-through fuel cycle, fuel is loaded into the reactor, used until full-power operation is no longer practical and then discharged and disposed of without reprocessing to recover unused fissile material, either that remaining from the original charge or that generated from fertile material in the reactor.

The fuel loaded into the reactor is uranium oxide that has been enriched in its U-235 content from the naturally occurring 0.7 percent to about 3.2 percent. The enrichment is done in one of the government-owned gaseous diffusion enrichment facilities at Oak Ridge, Tennessee, Paducah, Kentucky, or Portsmouth, Ohio; they will be augmented in the next decade by a centrifuge-type plant. The process has the effect of removing almost 85 percent of the fertile material (U-238) originally mined, so that it never reaches the reactor. Instead, it is stored in tailings stockpiles (still containing 0.2 percent U-235) at the enrichment plant sites. The conversion ratio of a reactor is the ratio of the rate at which fissile material is generated from fertile material and the rate at which it is consumed in fission in that reactor. A reactor may be classified either as a converter, in which fissile material is consumed faster than it is generated (conversion ratio less than unity), or as a breeder, in which the reverse is true (conversion ratio greater than unity). Converters that are more efficient in fuel utilization than the LWR on a once-through fuel cycle are termed advanced converters.

Reactors may also be categorized by the predominant velocity or energy of the neutrons that induce fission. In the reactor, neutrons are produced at high velocities equivalent to several million electron volts. The neutrons collide with the atoms of the reactor's collant and structural materials, and these collisions reduce the neutron velocities. Light-weight atoms, especially hydrogen and carbon, are particularly effective in slowing neutrons. The slowing process can continue until the neutrons are in thermal equilibrium with surrounding materials (about 0.025 electron volts at room temperature). Reactors such as LWR's, that operate in this neutron energy range, are termed thermal reactors. Those reactors that operate with relatively little slowing down of neutrons are termed fast reactors, since the average neutron energy is very high. The inherent nuclear properties of materials make breeding readily achievable in fast reactors. However, because of the need to avoid neutron slowing, the choice of structural and other materials for fast reactors is more limited than for other reactors, and this complicates their design.

As a reactor operates, the additional fissile material generated from the fertile material is consumed along with the fissile material initially charged, thus extending fuel life in the

---

*A fuel cycle using thorium as the fertile material, which is converted to fissile uranium-233 in the reactor, is also being tested, but has not yet been demonstrated as commercially practical.

Table IV-1-1. Fission energy program funding for FY 1979–1980.

FISSION ENERGY PROGRAM	BUDGET AUTHORITY (DOLLARS IN THOUSANDS)		
PROGRAM	FY 1979	REQUEST FY 1980	INCREASE (DECREASE)
CONVERTER REACTOR SYSTEMS			
Thermal Reactor Technology			
Operating	$ 23,900	$ 36,500	$ 12,600
Equipment	0	500	500
Construction	0	0	0
Subtotal	23,900	37,000	13,100
Advanced Isotope Separation Technology			
Operating	46,200	46,200	0
Equipment	8,000	8,000	0
Construction	0	800	800
Subtotal	54,200	55,00	800
Gas Cooled Thermal Reactors			
Operating	39,500	0	(39,500)
Equipment	2,500	0	(2,500)
Construction	0	0	0
Subtotal	42,000	0	(42,000)
Advanced Reactor Systems			
Operating	0	7,200	7,200
Equipment	0	2,800	2,800
Construction	0	0	0
Subtotal	0	10,000	10,000
BREEDER REACTOR SYSTEMS			
Liquid Metal Fast Breeder Reactor			
Operating	460,001	348,900	(111,101)
Equipment	25,181	22,000	(3,181)
Construction	81,485	91,100	9,615
Subtotal	566,667	462,000	(104,667)
Water Cooled Breeder			
Operating	50,900	57,900	7,000
Equipment	3,100	2,100	(1,000)
Construction	9,000	0	(9,000)
Subtotal	63,000	60,000	(3,000)
Gas Cooled Breeder Reactor			
Operating	24,500	24,500	0
Equipment	1,500	1,500	0
Construction	0	0	0
Subtotal	26,000	26,000	0
Fuel Cycle Research and Development			
Operating	66,850	21,100	(45,750)
Equipment	2,000	2,500	500
Construction	6,750	6,400	(350)
Subtotal	75,600	30,000	(45,600)
ADVANCED NUCLEAR SYSTEMS			
Advanced Systems Evaluation			
Operating	10,300	2,600	(7,700)
Equipment	0	0	0
Construction	0	0	0
Subtotal	10,300	2,600	(7,700)
Space and Terrestrial Applications			
Operating	38,380	34,300	(4,080)

Table IV-1-1. (Continued)

FISSION ENERGY PROGRAM	BUDGET AUTHORITY (DOLLARS IN THOUSANDS)		
PROGRAM	FY 1979	REQUEST FY 1980	INCREASE (DECREASE)
Equipment	4,700	2,100	(2,600)
Construction	0	0	0
Subtotal	43,080	36,400	(6,680)
TOTAL	$904,747	$719,000	($185,747)

reactor. The degree of extension depends on the conversion ratio. However, fuel life is limited not only by the amount of fissile material available, but also by the buildup of waste products of the fission process, which tend to poison or retard the chain reaction. As the amount of fissile material decreases and the poisons build up, a point is reached beyond which the reaction can no longer be maintained at full power. At this point, even though the fuel still contains a significant amount of fissile material, it must be removed and replaced by new assemblies.

In the context of fission power systems, fuel utilization is defined as the amount of usable energy obtained per unit quantity of natural uranium (or thorium) mined. Although the relationship between conversion ratio and fuel utilization is complex, it may be said generally that the higher the conversion ratio, the higher the fuel utilization. However, if the higher conversion ratio is achieved at the expense of requiring a higher enrichment of fissile material in the fuel charge (and consequently a higher fraction of the fertile U-238 in the mined uranium discarded in the enrichment process), this may not necessarily be true.

The reference LWR once-through fuel cycle is shown schematically in Figure IV 1-3. Under this mode of operation, because of the fertile and some fissile material removed in the enrichment process, and the fissile and fertile materials remaining in the spent fuel, only about 1 percent of the energy content of the mined uranium is obtained.

There are several processes that can be invoked to significantly improve the fuel utilization in converter reactors. One of these is uranium recycle, illustrated in Figure IV 1-4. Here the spent fuel assemblies, after an initial cooling period, are dissolved and the residual uranium, plutonium, and fission products are chemically separated in a reprocessing plant. The recovered uranium, containing about 0.8 percent U-235, is then passed to a radioactive fuel refabrication plant where it is mixed with new enriched uranium to bring its U-235 content up to the power plant requirement of about 3.2 percent, refabricated into new assemblies, and returned to the same or another power plant. The refabrication processes are about the same as those for new (i.e., non-radioactive) fuel except that the operations are conducted using remote handling techniques because of the residual radioactivity in the

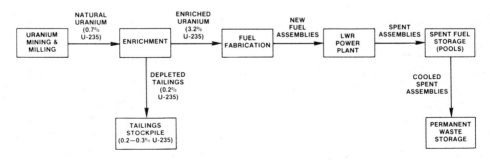

Figure IV-1-3. LWR once-through fuel cycle.

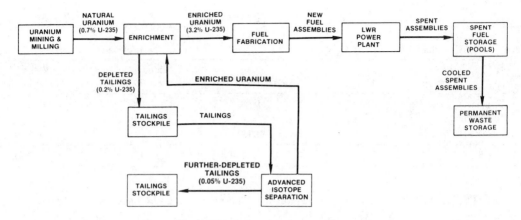

Figure IV-1-4. LWR fuel cycle using uranium recycle.

recovered uranium. Uranium recycle has the potential of improving fuel utilization by about 20 percent.

The plutonium recovered from the reprocessing step could also be recycled along with the uranium. However, because plutonium would be required for the initial fuel loadings in fast breeder reactors should their deployment be required in the future, and because the consumption of plutonium is more efficient in breeders, prudent planning would favor storing the plutonium for possible later breeder use rather than recycling it in converters.

It should also be noted that uranium recycle has been deferred indefinitely by the Administration's policy banning plutonium reprocessing, and therefore must be regarded only as an option for future consideration. However, it is important that, if a decision were made to permit recycle at some future date, all the spent fuel from past reactor operations could be reprocessed and recycled, provided the spent fuel had been stored in a retrievable storage facility.

A second method being considered for improving fuel utilization in converters is the use of advanced isotope separation (AIS) techniques to extract more U-235 from the stockpile of depleted uranium tailings from enrichment plants. In the current operating mode, these tailings contain about 0.2 percent U-235, and constitute a significant source of additional U-235 if it can be extracted economically. Processes using lasers and other advanced techniques are being developed to enrich these tailings to the natural uranium U-235 content of about 0.7 percent, leaving only about 0.05 percent U-235 in the tails. The product could then be used as feed in the more conventional separation plants to produce additional reactor fuel. This is illustrated in Figure IV 1-5. The AIS process, when and if deployed, can be applied to enrich the tailings produced subsequent to AIS deployment and to economically lower the tails assay of existing DOE enrichment plants. AIS has the potential of improving the long-term fuel utilization by about 30 percent, and can also be used in conjunction with uranium recycle if desired.

A thorium/U-233 fuel cycle, such as might be used in an HTGR—an advanced converter—is shown in Figure IV 1-6. Except for the addition of the thorium supply, it is generally similar to the LWR using uranium recycle, because reprocessing is required to recover and utilize the U-233 generated from the fertile thorium. In fuel fabrication, the recovered U-233 is supplemented with U-235 to obtain the proper fissile material fraction in the new fuel assemblies. This could be done by making some fuel assemblies with new U-235 and some with recovered U-233, or the U-233 and supplemental U-235 could be mixed. Both approaches are shown. Plutonium storage is indicated because some plutonium will always be formed from the U-238 fed to the reactor in conjunction with the U-235 makeup. This could also be returned to the reactor if separation is not desirable.

In a breeder reactor, the fuel discharged

The National Energy Plan and Fission Energy Policy 577

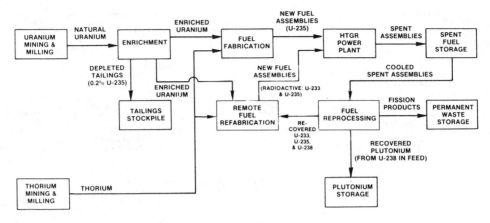

Figure IV-1-5. LWR fuel cycle using advanced isotope separation.

from the reactor contains more fissile material than the fuel initially charged. This fuel must be reprocessed to extract the fissile material for further use as reactor fuel. Fissile material in excess of that required to fuel the reactor in which it was produced can be withdrawn from the cycle and used in other reactors or as the initial fuel inventory for new breeders. The fuel life in the breeder reactor is limited by the desire to recover the newly produced fissile material and by fuel structural damage due to neutron bombardment.

The plutonium-uranium fuel cycle for a fast breeder is illustrated in Figure IV 1-7. In this case, since all fissile material requirements of the reactor are fulfilled by bred and recycled plutonium, no U-235 input is required. The fertile U-238 requirements can be met from the existing stockpiles of depleted uranium tailings from enrichment plants. The energy content of this tailings stockpile, already mined, refined and on the shelf, is comparable to that of all the world's recoverable petroleum (1,700 billion barrels) or all of the U.S. reserves of coal (400–500 billion tons). Hence no mining or milling of uranium would be required to supply breeders for hundreds of years. In comparison with the 1 percent energy recovery in the reference LWR once-through system, fast breeders can recover 60–70 percent of the energy content of the uranium originally mined.

Table IV 1-2 summarizes the characteristics of LWR's and other major types of power reactor of interest in the United States.

HISTORY OF U.S. FISSION ENERGY DEVELOPMENT

Policy Evolution

The potential benefits of civilian applications of fission energy were recognized in the 1940s. It was also perceived, in those early

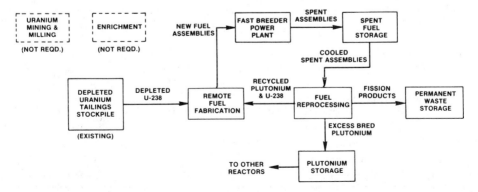

Figure IV-1-6. HTGR using Thorium/U-233 fuel cycle.

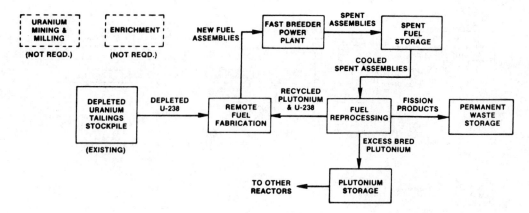

Figure IV-1-7. Fast breeder plutonium-uranium fuel cycle.

days, that, for fission power to have a lasting, practically inexhaustible role, breeder reactors would be necessary. Development of such technology began almost immediately. In fact, the world's first reactor plant to produce electricity (in December 1951) was a sodium-cooled fast reactor, the U.S. Experimental Breeder Reactor No. 1 (EBR-I). Soon thereafter, measurements confirmed that EBR-I was breeding—that is, producing more fuel than it was consuming.

The early official literature shows that the potential for diversion of fissionable materials to weapons purposes was recognized from the outset. The concern was expressed in the Smyth Report (November 1945), the Truman-Atlee-King Proposals (November 1945), the Anderson-Lilienthal Report (March 1946), the "Baruch Plan" (June 1946), etc. These pioneers identified the problems and proposed that procedural and institutional solutions be adopted in timely fashion.

Federal policies and programs in the 1950s promoted the application of nuclear energy to peaceful purposes. A sense of positive accomplishment, a problem-solving ethic, pervaded

Table IV-1-2. Characteristics of fission reactor systems.

					FUEL CYCLE		
					REACTOR LOADING		FISSILE MATERIAL GENERATED
REACTOR TYPE	NEUTRON ENERGY	STATE OF DEVELOPMENT	EXISTING U.S. REACTORS	COOLANT	FISSILE	FERTILE	
Light Water Reactor (LWR)							
Converter	Thermal	Commercial	Many commercial	Water	U-235	U-238	Pu
High Temperature Gas-Cooled Reactor (HTGR)							
Advanced converter	Thermal	Developmental	Fort St. Vrain	Helium	U-235 or U-233	Th	U-233
Canadian Deuterium-Uranium Reactor (CANDU)							
Advanced converter	Thermal	Commercial*	None	Heavy water	U-235	U-238	Pu
Light Water Breeder Reactor (LWBR)							
Breeder	Thermal	Developmental	Shippingport	Water	U-233	Th	U-233
Liquid Metal Fast Breeder Reactor (LMFBR)							
Breeder	Fast	Developmental	EBR-II & FFTF	Sodium	Pu	U-238	Pu
Gas-Cooled Fast Breeder Reactor (GCFR)							
Breeder	Fast	Developmental	None	Helium	Pu	U-238	Pu

*In Canada and by Canadian export only; not commercial in the United States.

the nuclear enterprise. The Atomic Energy Act of 1954 made it possible for the private sector to build and operate nuclear electrical plants. Commercialization of nuclear power was a national goal, and the Atomic Energy Commission launched a power reactor demonstration program in cooperation with industry, to explore various reactor concepts. The technological success of water-cooled reactors for naval propulsion and the existence of substantial fuel-enrichment capacity led to the commercial deployment of LWR's, beginning in the early 1960s.

As the commercial deployment of LWR's proceeded, increased attention was given to planning for the next generation of reactors. A major policy statement, "Civilian Nuclear Power—A Report to the President, 1962," urged increased emphasis on the development and ultimate demonstration of breeder reactors. As a contingency against possible delays in the breeder development program, continued work on a variety of advanced converter reactors was also proposed. Several years later, after exhaustive technical and economic assessments of the most promising advanced converter and breeder reactors, as well as strategic combinations of various reactor types, the fission development program was concentrated on the LMFBR.

The "classical" fission deployment strategy adopted at that time was based on the uranium/plutonium cycle and recognized the LWR-FBR symbiosis:

- The plutonium stockpiles accumulated from LWR operation would be used to start up a breeder economy.
- Enrichment plant tailings from the LWR fuel cycle would be used as fertile material for breeders.
- Eventually, depending on growth requirements, excess plutonium produced in breeders could be used to sustain LWR operation and substantially reduce the need for enriched uranium.

In this scenario, it was envisioned that breeder reactors would displace LWR's for new capacity additions as natural uranium resources were exhausted or became economically restrictive. Plutonium requirements for new breeder capacity would ultimately be supplied entirely by excess material generated by the breeders themselves. And since the breeders require no uranium isotopic separation, national enrichment plant capacity requirements would vanish. As enrichment tails become exhausted centuries in the future, the remaining low grade, high cost natural uranium (uneconomic in LWR's) could be exploited for breeder requirements, since the breeders would use only negligible amounts of ore. Although the classical strategy did not explicitly contemplate the use of thorium, it was recognized that the uranium resource base could be extended by using thorium in breeder blankets or in advanced converters.

The 1960s saw the startup and successful operation of EBR-II, which demonstrated both the delivery of reliable breeder-generated power to a utility grid and the in-plant recycle of fuel, and SEFOR, a special facility designed to test certain safety characteristics of the breeder. Fermi-1, a private venture under the AEC's power reactor demonstration program, was also completed in this period. Although operation of the plant was marred by a fuel assembly meltdown, the fact that the accident was sustained without undue public hazard and without permanent damage to the plant constituted an important demonstration of breeder safety.

The 1960s also saw the U.S. commitment to a high-performance fuel and materials test facility—the Fast Flux Test Facility (FFTF). Plans were also made for one or more plant demonstration projects—the first an intermediate-size plant in the 300–400-MWe range that ultimately became the Clinch River Breeder Reactor Plant (CRBRP) Project. Studies to determine the characteristics of future purely commercial 1000-MWe LMFBR plants were conducted. In 1971, the President assigned the highest priority to the breeder program, expressed support for a second demonstration project, and established a 1980 goal for demonstrating the LMFBR as a commercial power producer.

In the 1971–1974 period, a number of serious obstacles to early commercialization appeared, primarily stemming from increased public sensitivity to environmental and social impacts of industrial technology in general and of nuclear energy in particular, and increased competition from other energy technologies for RD&D resources. In 1972, the Atomic Energy Commission was directed by a federal court to develop a comprehensive LMFBR program environmental impact statement, assessing the overall effects of a mature, deployed LMFBR economy as opposed to a limited evaluation of the impacts of individual plants and facilities. The principal findings of the resulting exhaustive study were that a high-confidence decision on the suitability of the LMFBR for large-scale deployment could not and need not be made at that time, due to several residual uncertainties, but that the LMFBR program should proceed through the demonstration phase in order to obtain data for a decision at a later date. Based on these fundings, a highly focused and integrated plan was formulated and adopted in 1976. Its major features were as follows:

- Resolution of major environmental, technical, and economic issues by 1986:
  — Intermediate-scale demonstration of the complete LMFBR energy system, including the power plant and its supporting fuel cycle
  — A better estimate of the extent and quality of the uranium resource base
  — Resolution of remaining health, safety, and environmental issues
- Establishment of a state of preparedness, in case a decision were to be made in 1986 to deploy the breeder

To quantify and schedule this state of readiness, the plan was driven by uranium supply and LWR growth projections: a U.S. uranium resource base of 3.7 million short tons of U_3O_8 and the commercial deployment of 600,000-MWe of LWR capacity by the year 2000. On the basis of these assumption, it was clear that few LWR's could be committed for construction beyond the early 1990s, and significant commercial breeder deployment would have to commence at about that time.

The LWR-to-LMFBR transition could not be abrupt, however. To allow for an orderly LMFBR deployment, the first large commercial unit was assumed to come on line in 1993, with a buildup to 30,000-MWe in the year 2000. Fuel-cycle capacity to support this schedule was also assumed in the plan. The central path of the program was the scale-up of technology through the sequence of power-plant and fuel-cycle demonstration projects. All elements of the program were driven by requirements and schedules of the demonstration projects, and were focused on the following:

- Resolution of outstanding issues before deployment could occur
- Achievement of a state of readiness should deployment be required

Full-scale implementation of this program was in progress at the end of 1976. However, in 1977, the U.S. fission energy development policy was changed for the following reasons:

- An increased concern over the potential for proliferation of nuclear weapons through deployment in nonweapons states of fission energy systems involving the use of fuel recycle and plutonium
- A lessened sense of urgency for the breeder and other advanced fission systems because of Administration initiatives in other areas—increased reliance on coal, aggressive conservation measures, decreased projections of installed nuclear capacity in the 1985–2000 period (and consequently decreased consumption of uranium resources), and increased development attention to other inexhaustible energy sources such as solar, geothermal and fusion
- The perception of a need, prior to a commitment to a definitive demonstration and deployment schedule, for a reexamination of all promising advanced fission options, during which major planning uncertainties (electricity demand growth, uranium and thorium resource bases) could be resolved
- An emphasis on fission options offering near-term (pre-2000) benefits, and perception of a need to complete the technical, economic, and institutional validation of LWR's and their supporting fuel cycle

Consequently, in April 1977, the Administration announced that the commercial reprocessing and recycling of plutonium, as well as the commercial introduction of the plutonium breeder, would be deferred indefinitely. Further, the design of the CRBR would be completed and the testing of certain components already on order would proceed, but the construction of the plant would be cancelled, primarily because the facility would not result in a meaningful demonstration of commercial LMFBR technology.

Types of Reactors

The early commercialization of the LWR in the United States resulted from two major factors: (1) the availability of directly applicable technology spun off from the naval propulsion reactor program; and (2) the existence of uranium enrichment facilities built for the nuclear weapons program. However, even before LWR commercialization began, it was recognized that the breeder would eventually be necessary if nuclear fuel resources were to be fully exploited.

Two types of advanced converter reactors are also being considered: the gas-cooled thermal-neutron rector and the heavy-water-moderated natural-uranium-fueled reactor known as the CANDU (Canadian Deuterium Uranium) reactor.

Gas-cooled reactors have had a long history in the United States and abroad. Their initial development in the United Kingdom and France resulted in the construction of more than 30 such plants in Europe and Japan beginning in the late 1950s. They operated at low temperature and had low conversion ratios. Development to increase the coolant temperature, and consequently the plant efficiency, continued, and an advanced experimental plant called Dragon was built in the United Kingdom by an international consortium. In the United States, a small (40-MWe) HTGR power plant (Peach Bottom No. 1) was built in the late 1960s and operated more than seven years, and an intermediate-scale (330-MWe) plant (Fort St. Vrain) is now undergoing startup testing. There is hope that the economic performance and conversion ratios of future gas-cooled thermal power plants can be improved. Their potential for high temperatures (900–1000°C) should make them particularly useful for industrial process heat applications.

Heavy-water-moderated natural-uranium-fueled reactors also have a long history, both as research facilities and for producing weapons material. Canada adapted the concept to power production, and since 1968 23 CANDU-type plants in the 200–700-MWe size range have been built or are under construction in Canada, India, Pakistan, Argentina, and South Korea. Until recently, little consideration has been given to deploying this type of reactor in the United States, largely because the United States has enrichment facilities, so it is not limited to using only natural-uranium-fueled reactors. Questions about licensability and economics in the United States also remain.

Interest in the WCB concept, a thermal-neutron system, grew out of studies indicating that breeding might be achieved in existing and future LWR's by using the thorium/uranium-233 fuel cycle, in which the fissile material is uranium-233, generated from the fertile thorium. To test the projected breeding performance, an LWBR core was installed in the LWR at the Shippingport Atomic Power Station and put into operation in late 1977. After three to four years of operation the core will be removed for analyses to measure its breeding ratio; the analyses are expected to be completed after 1984.

Development of the LMFBR began in the United States in the mid-1940s and progressed through the construction and successful operation of six small-scale liquid-metal-cooled reactor facilities. One of these, EBR-I, produced the world's first electricity generated by nuclear energy, on December 20, 1951. Two facilities (EBR-II and Fermi-1) demonstrated the delivery of power to a utility grid; the EBR-II facility also demonstrated in-plant recycle of fuel. EBR-II continues to operate successfully as a testing facility. Construction of the next major liquid-metal-cooled reactor, the FFTF, which will be used for performance testing fuels and materials, is essentially complete. Twelve LMFBR plants have been built and operated abroad—in the USSR, United Kingdom, France, Japan, and West Germany—including three intermediate-scale (250–350-MWe) LMFBR demonstration plants in the USSR, France, and the United Kingdom. Ten additional LMFBR plants are being built or planned, including the 600-MWe BN-600 reactor in the USSR, the 40-MWe reactor being constructed in India, and the commercial-scale (1200-MWe) Super-Phenix under construction in France. LMFBR development in the United States now concentrates on the technology of large components and on the safety and economics of commercial-scale plants.

The GCFR concept has evolved from the long history of gas-cooled reactors, but its design would differ considerably from the gas-cooled thermal reactor system. Whereas the latter system uses graphite-based ceramic fuels in which the carbon slows the neutrons, the gas-cooled fast system would use metal-clad fuels operating at lower temperatures, and thus would require a considerably different thermodynamic cycle. Basic conceptual and safety studies are underway for this reactor type, as a backup to the LMFBR system.

It should be noted that this material on Nuclear Fission was finalized just before the historic events occurred at Three Mile Island. The nuclear accident at Three Mile Island will dramatically change all aspects of nuclear fission as presented in this section. Only time will tell the impact on nuclear power.

2.
Fission Energy Program Management

The scope of this document is restricted to the DOE programs managed by the Program Director for Nuclear Energy (ETN).

MANAGEMENT APPROACH

The management approach used in the Fission Energy Program is a product-oriented, management-by-objectives system. The principal management functions and their interrelationships for the Fission Energy development programs, as shown in Figure IV 2-1, are as follows:

- Establish program goals in accordance with national energy policy
- Establish a comprehensive plan and schedule to achieve these goals
- Determine and secure resources required to execute the plan
- Separate the total program scope into manageable work segments and assign responsibility and resources to contractors for execution of the work segments
- Monitor and measure progress of the program against resources expended, taking corrective action as necessary
- Revise the overall plan and redirect subordinate plans as necessary to accommodate changes imposed by external forces, (e.g., changes in national energy policy, projected need dates, and budget actions) and internal events (e.g., new technological findings in the program and unforeseen delays).

PROGRAM FUNCTIONAL ELEMENTS

Three generic functional elements in fission energy programs are demonstration projects, base technology programs, and supporting facilities.

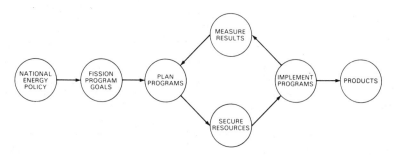

Figure IV-2-1. Fission energy program management functions and interrelationships.

584 Fission Energy

A complete development program contains all three elements. Development and demonstration projects bring all elements of the program together and provide the mechanism for transferring the technology to industry. The scope and timing of the projects drive the remainder of the program. Technology development (in such areas as nuclear fuels, safety, and components) supports specific project requirements and provides the basis for progress between successive projects. Test facilities are necessary to support both the base technology programs and to confirm performance of project systems. The relationships of projects, base technology programs, and facilities are illustrated in Figure IV 2-2.

The DOE Fission Energy Program elements include a varying range of balance among the three basic functional elements; i.e., some program elements include substantial work in all three, while others are limited to basic technology development with no plans for demonstration projects. However, the intent in most cases is to develop a supporting technology base through basic development activities, use test facilities as appropriate to contribute to the technology program, and, ultimately, have the option to demonstrate the technology on a scale that ranges from testing in existing commercial facilities to dedicated government-owned facilities.

WORK BREAKDOWN STRUCTURE (WBS)

The WBS is one of the most useful and important tools in the planning and control of Fission Energy programs and projects. By definition, a WBS is a product-oriented logic diagram depicting the hardware, services, and data that comprises a program. In such a diagram, the work packages relate to one another through an integrated hierarchy leading to the end product. Thus, the WBS diagram graphically represents the various efforts required, their interaction, and the sequence that is required for success.

The upper-level work breakdown structure for the entire Fission Energy Program, down through Level 3, was introduced in Figure IV 1-2. Levels 4 and below show projects and their supporting activities. A typical WBS for

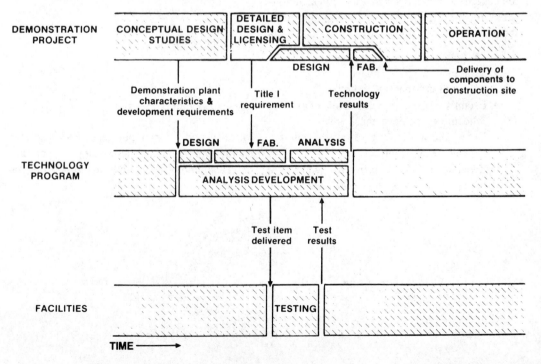

Figure IV-2-2. Relationships among projects, technology programs, and facilities.

Fission Energy Program Management 585

a major construction project, which would typically be represented at Level 4, is shown in Figure IV 2-3.

Starting with the broadest definition of a program (e.g., Fission Energy) or project, each of the work packages that is required for completion is successively subdivided into smaller and smaller units until, at the lowest level, each element in the WBS represents a discrete package of relatively short duration, with a definite beginning and end, and to which a specific budget can be assigned. Completion of all these packages will constitute completion of the program or project, and the summation of the costs of these subactivities will yield the total cost of the project.

Thus, a WBS is both a planning mechanism in which all activities required for the completion of a program are carefully planned in advance, and a control mechanism that enables close monitoring of progress in terms of objectives (milestones), schedule, and cost. The WBS approach has been adopted as an integral part of the technical and budgetary management of the Fission Energy Program and its subdivisions.

ORGANIZATIONAL STRUCTURE AND DELINEATION OF RESPONSIBILITIES

Within DOE, the Assistant Secretary for Energy Technology administers the research and development programs associated with solar, geothermal, fossil, nuclear, and magnetic fusion energies, and waste management. The Office of Nuclear Energy Programs implements the nuclear energy programs in accordance with policy established by ASET. Figure IV 2-4 is the organization chart showing the offices and divisions managing the nuclear energy program.

The relationship between the WBS and the ETN organizational structure is shown in Figure IV 2-5. The WBS program structure down to Level 3 is shown along the top, and the ETN organizational units (to the branch level) responsible for the work are shown along the left side. At the intersections of these, responsibilities are indicated by solid circles.

The relationship between the WBS and the budget and reporting structure is shown in Figure IV 2-6. The program structure, down to WBS Level 3, is shown along the top, and the FY 1980 budget structure titles are shown along the left side. At each intersection where specific work is funded, the total level of funding obligation for FY 1980 is indicated (in millions of dollars of budget authority).

It is DOE policy to decentralize program and project management activities. Overall program responsibilities will be retained at Headquarters; project management responsibilities will be moved to DOE field offices.

Headquarters Role

ETN Headquarters is responsible for the overall implementation of DOE fission energy

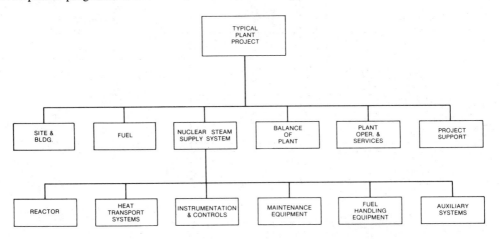

Figure IV-2-3. Work breakdown structure for a typical plant project.

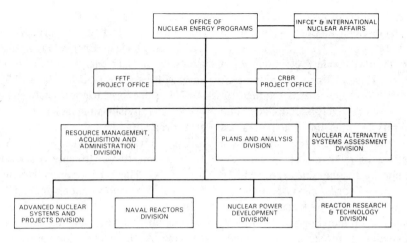

Figure IV-2-4. Organization of office of nuclear energy programs.

programs in accordance with ASET policies. Headquarters accomplishes this by doing the following:

- Developing program plans, budgets, and resource allocation requirements
- Working with the Office of Management and Budget, Congress, Federal and state agencies, industry, universities, the governments of other nations, and international organizations to coordinate fission energy activities
- Providing program guidance and resources to DOE field offices for implementation
- Monitoring program execution with DOE field offices to ensure that programs are being implemented as planned

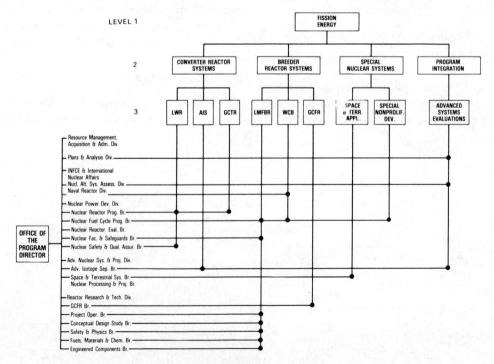

Figure IV-2-5. Relationship between ETN organizational structure and program WBS.

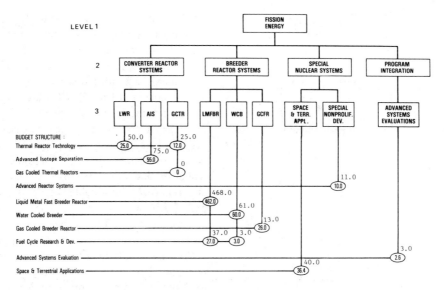

Figure IV-2-6. Relationship between budget and reporting structure and program WBS.

Field Office Role

The DOE field offices are responsible for implementing the programs in detail in conformance with the guidance and resources provided by ETN Headquarters. The field offices' activities are as follows:

- Develop detailed program plans, budgets, and resource allocations needed to execute the programs as planned

Contractor's Role

The contractors (national laboratories, universities, industry) are responsible for program execution. They are responsible for the following:

- Initiating proposals or responding to requests for proposals
- Assisting in the development of detailed program plans
- Carrying out the technical program in accord with the guidance provided by the field offices

PROGRAM PLANNING AND CONTROL

Fission energy program planning documents provide clear visibility as to the course to be followed and serve as contracts between organizational units—both vertically and laterally in the management structure. Plans and planning documentation are not static, however. They must accommodate changes resulting from events both external and internal to the program. Therefore, it is essential that program changes be made in an orderly and coherent manner, that they be documented and tracked, and that all programmatic and budgetary impacts be recognized. For these reasons, a formal management control system, with provision for a disciplined change approval process, has been established in the Fission Energy Program.

The documentation identified in Figure IV 2-7 constitutes a hierarchy of planning documents for managing the Fission Energy Program. Overall guidance is provided through national energy policy emerging from both the executive and legislative branches of the Federal Government which considers all energy requirements and potential resources and defines the overall role of each, including fission energy.

Detail and numbers of documents increase from top to bottom of the planning hierarchy. This architecture provides the structure through which policy, technical, and fiscal guidance is communicated downward and through which program progress, including

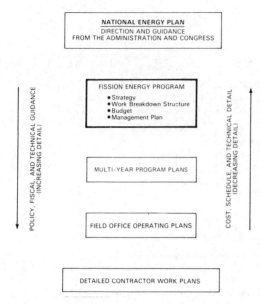

Figure IV-2-7. Hierarchy of management planning documents.

cost, schedule, and technical data, is reported upward. All major planning documents are updated no less than annually in accordance with the planning and budget cycle of the Federal Government.

The complexity and magnitude of the Fission Energy Program requires a formal Program Reporting and Change Control System (PRCCS). Application of the PRCCS to fission energy activities was begun in 1978, and all components of the PRCCS are now operational or being installed. The PRCCS provides the mechanism to carry out consistent integrated planning, execution, and management at all management levels. Milestones, against which the individual programs can be directed and reported, are established in the detailed multi-year program plans. These milestones are reviewed by ETN management and cannot be changed without the director's approval.

Progress in the attainment of the milestones is reported monthly to the Program Director for Nuclear Energy on milestone status sheets. A management-by-exception philosophy is adopted: unless informed otherwise, the program director can assume that program progress is in accordance with approved baseline plans. For those milestones that have slipped, an explanation of the cause of the slippage, the impact on the overall program, and recovery plans to minimize the impact are required.

3.
Converter Reactor Systems

This chapter describes four programs: Thermal Reactor Technology, Advanced Isotope Separation Technology, Gas Cooled Thermal Reactors, and Advanced Reactor Systems.

OVERVIEW

The Thermal Reactor Technology Program is developing and demonstrating advanced technology to meet the following objectives:

- Improve uranium use in once-through LWR's—involves in-plant irradiations of large quantities of test fuels, which are fabricated by accepted manufacturing techniques, licensed by normal procedures, and tested for several years in an operating environment typical of large commercial reactors.
- Reduce occupational radiation doses—involves
  — projects to develop and demonstrate chemical processes to decontaminate reactor systems and reduce the radiation fields,
  — studies to develop and demonstrate surveillance and diagnostic techniques capable of replacing manual inspection methods, and
  — tasks to identify other means of reducing radiation doses.
- Resolve critical technical problems that adversely affect plant productivity—presently limited to developing ways to alleviate the effect of "denting" in pressure water reactor (PWR) steam generators.
- Enhance the safety of LWR plants by improving systems and data, improving the man-machine interface, and using risk-analysis methods for design decisions.
- Support the international cooperative development of advanced high-temperature gas-cooled reactors for process-heat and direct-cycle applications; this work has been conducted under the Gas Cooled Thermal Reactor Program (see below) through FY 1979, but will be part of the Thermal Reactor Technology Program beginning in FY 1980.

The Advanced Isotope Separation Technology (AIST) Program, which is supporting the national policy to defer the reprocessing of spent commercial nuclear fuel, is developing other methods to increase availability of uranium-235. Specifically, it is investigating the economic recovery of U-235 from the "tails" material accumulated from the present enrichment plants. The goal is to raise the U-235 content of the depleted "tails" material accumulated from conventional enrichment plants to the 0.711 percent U-235 of natural uranium at a cost below natural uranium's current market price of $45/lb of U_3O_8; doing so would effectively increase the U.S. uranium resource base by about 20 percent. Three advanced separation technologies are being developed: two laser methods and a plasma technique. When the preprototype hase is completed at the end of FY 1981, an evaluation will determine which process should be developed further.

589

The Gas Cooled Thermal Reactor (GCTR) Program includes development of the High Temperature Gas Cooled Reactor steam cycle (HTGR-SC). Because of a lack of utility commitment to the system, the GCTR Program will be terminated in FY 1980. During FY 1979, ongoing conceptual-design tasks and a reference-plant cost estimate will be completed. All work will be documented so as to be readily retrievable is needed in the future.

The advanced Reactor Systems Program is developing systems that comply with U.S. nonproliferation objectives. The work involves development of the following:

- Low- or medium-enriched uranium fuels to replace the highly enriched fuels now used by many research reactors; this could prevent the highly enriched fuels from being diverted to make weapons
- Proliferation-resistant concepts for the design and operation of fuel-cycle facilities in support of NASAP-developed criteria

Funding

Table IV 3-1 shows the funding levels by program for the FY 1978 through FY 1980 period.

THERMAL REACTOR TECHNOLOGY

The Thermal Reactor Technology Program is developing and demonstrating advanced technology for use in commercial nuclear powerplants, particularly LWR's. These plants are already providing a significant share of U.S. electrical power needs for improving uranium use in once-through LWR's, for reducing radiation exposure to nuclear plant personnel, for resolving critical technical problems that adversely affect plant productivity, for enhancing the safety of nuclear plants, and for developing gas-cooled direct-cycle and process-heat reactor base technology for advanced gas-cooled reactor systems. The technologies developed in this program will

Table IV-3-1. Program funding levels, FY 1978–1980.

CONVERTER REACTOR SYSTEMS	BUDGET AUTHORITY (DOLLARS IN THOUSANDS)			
PROGRAM	FY 1978	FY 1979	REQUEST FY 1980	INCREASE (DECREASE)
Thermal Reactor Technology				
Operating	$ 18,800	$ 23,900	$ 36,500	$12,600
Equipment	0	0	500	500
Construction	0	0	0	0
Subtotal	18,800	23,900	37,000	13,100
Advanced Isotope Separation Technology				
Operation	43,617	46,200	46,200	0
Equipment	8,485	8,000	8,000	0
Construction	0	0	800	800
Subtotal	52,102	54,200	55,000	800
Gas Cooled Thermal Reactors				
Operating	32,200	39,500	0	(39,500)
Equipment	1,000	2,500	0	(2,500)
Construction	0	0	0	0
Subtotal	33,200	42,000	0	(42,000)
Advanced Reactor Systems				
Operating	0	0	7,200	7,200
Equipment	0	0	2,800	2,800
Construction	0	0	0	0
Subtotal	0	0	10,000	10,000
TOTAL	$104,201	$120,100	$102,000	($18,100)

be of significant value in ensuring that thermal reactors play the important role expected of them in helping to meet the nation's energy requirements.

Overview

During the late 1940s, the Atomic Energy Commission (AEC) began a program to build a number of experimental power reactors for operation in the 1950s. Early power reactors were built largely at AEC laboratories. The Atomic Energy Act of 1954—by permitting private ownership of reactors, private use of nuclear fuel under leasing arrangements, and the declassification of certain reactor technology—ushered in the development of nuclear power for industrial applications.

The year 1957 was significant because it marked the initial operation of a 60-MWe prototype PWR powerplant. That government-owned reactor facility, built at Shippingport, Pa., under a cooperative agreement with the Duquesne Light Co., was the nation's first civilian nuclear power reactor. After the AEC invited industry to join them in 1955 in developing power reactors by means of the Cooperative Power Demonstration Reactor Program, commercial nuclear power stations completely owned and operated by electric utilities followed. Some of these were built under the joint demonstration program, others were privately financed. These plants included:

- The first large boiling-water reactor (BWR), Commonwealth Edison's Dresden Unit One (200-MWe), completed in 1960
- Yankee Atomic Electric Company's Rowe PWR unit (175-MWe), completed in 1961
- Consolidated Edison's first Indian Point PWR unit (175-MWe), completed in 1962

Figure IV 3-1 shows a typical PWR nuclear steam supply system. Figure IV 3-2 shows a typical BWR nuclear steam system. Figure IV 3-3 shows the present commercial LWR nuclear fuel cycle in the United States with no reprocessing.

In the Cooperative Power Demonstration Reactor Program, the Government contributed technical and financial assistance, particularly for the fuel and the R&D programs. These

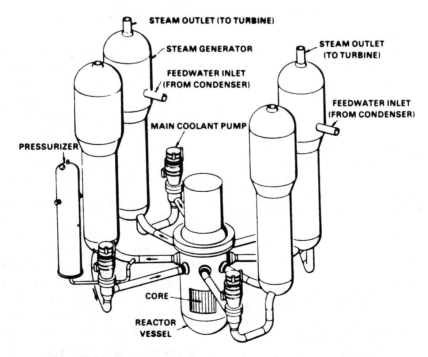

Figure IV-3-1. Typical pressurized-water reactor nuclear steam supply system.

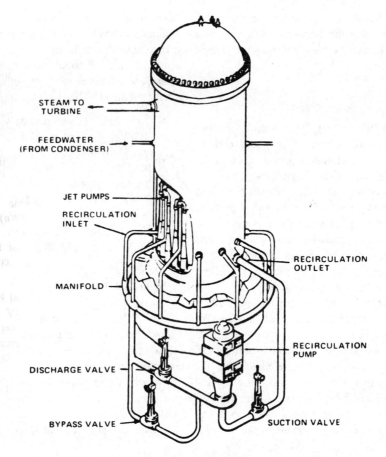

Figure IV-3-2. Typical boiling-water reactor nuclear steam supply system.

AEC initiatives, together with the larger R&D expenditures to develop PWR nuclear-propulsion systems for the U.S. Navy (which underlie much of the development of LWR systems for civilian use), constitute an investment of approximately $2 billion.

The nuclear industry began to develop and commercialize LWR's in the 1960s. As of February 1, 1979, 72 nuclear power reactors, of which 69 are commercial LWR's, are providing about 13 percent of the nation's electrical energy. By 1985, nuclear plants are expected to provide about 19 percent of the nation's electrical energy. By the year 2000, there will be an installed nuclear capacity of 255–395,000-MWe, consisting almost entirely of LWR's, compared to about 52,000-MWe at present. By 2000, nuclear power is projected to provide between 34 and 40 percent of the nation's electrical energy.

Despite the performance and growth record of nuclear power, the nuclear industry is not without problems. In the last four years, projections of nuclear capacity have declined markedly because of construction delays and utility deferrals and cancellations. The reasons for the lack of orders, illustrated in Figure IV 3-4, are multiple, but the causes mentioned most frequently include the following:

- Re-evaluations of electricity-use growth projections
- Utility financing and economic considerations
- Uncertainty in government policies regarding fuel reprocessing, enrichment capacity, and waste management
- Procedural delays in licensing, environmental approval, and public acceptance

In January 1975, the Energy Research and

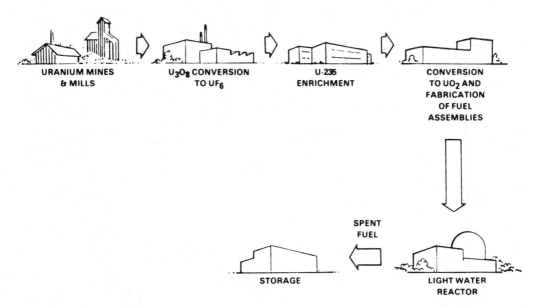

Figure IV-3-3. Commercial LWR fuel cycle with spent fuel storage.

Development Administration (ERDA—a successor agency to the AEC and the predecessor of DOE) established a task force to consider the need for government participation with industry in a cooperative LWR technology development program. As a result of that group's findings, ERDA initiated the LWR Technology Program in FY 1976.

Early in 1978, DOE reviewed all LWR Technology Program activities with respect to the validity of the governmental role, the short-term effect on nuclear powerplant productivity, and its relationship to the government's nonproliferation and enhanced safety objectives. As a result of that review, the program was substantially reoriented. It was decided that DOE would not initiate new projects directed solely toward obtaining nuclear-plant productivity improvements, and that industry would be encouraged to fund follow-on work in that area. In reorienting the program, priorities among competing possibilities were set based on the urgency of public need and the likelihood that the results would not be achieved without federal assistance. The future of productivity-improvement projects is being reassessed.

In mid-FY 1978, LWR safety development was initiated. Sandia Laboratories was designated as technical manager for this part of the program, and studies were undertaken to define the state of the technology and to identify potential areas for improvement. In FY 1979, safety tasks have been initiated in areas having the greatest potential for success.

The Direct Cycle HTGR (HTGR-DC) and VHTR are advanced gas-cooled fission converters that have evolved from the technology developed for the HTGR steam cycle in the United States and Europe over the past 20 years. These systems have the potential for the greatest exploitation of the unique high-temperature capability of gas-cooled thermal reactors.

Three power conversion modes are being considered in the HTGR Program:

- A closed direct gas turbine (Brayton) cycle—In this mode, heat is transferred from the reactor core to helium, which passes directly through a gas turbine to generate electricity. The high reject heat temperature of this cycle also permits its use with a bottoming cycle, which increases the efficiency to 40 or 45 percent, or for low temperature process heat applications.
- High-temperature process heat applica-

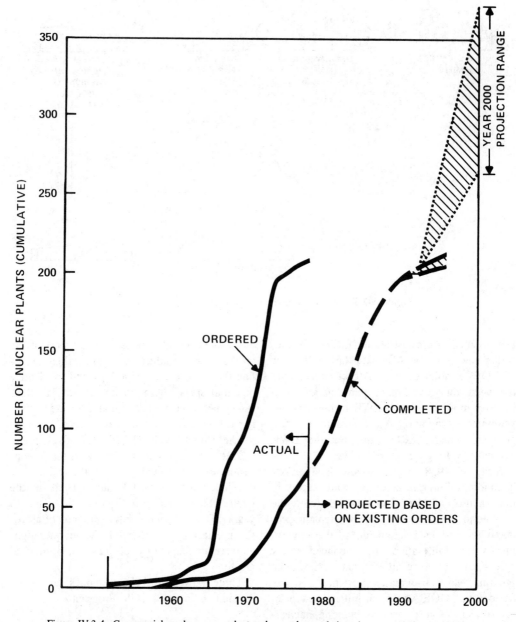

Figure IV-3-4. Commercial nuclear powerplant orders and completions between 1953 and 2000.

tion—For this application, heat is transferred from the reactor core to helium, which then passes through a heat exchanger where it is again transferred, either directly or through an intermediate loop, to a high-temperature industrial process such as coal conversion, hydrogen production, steelmaking, etc.

• A combined-cycle mode—In this mode, energy would be extracted through heat exchangers at high temperatures for process applications, and the lower temperature reject heat could then be used to generate electricity using either the conventional steam (Rankine) or direct gas-turbine (Brayton) cycle.

Gas-cooled thermal reactors have the potential of serving two distinct markets:

- For central station electric power applications, this reactor can produce high efficiency, economic, electric power generation with reduced uranium ore consumption, relative to LWR's.
- For advanced non-electric applications, this is the only fission system concept now under development with potential industrial high temperature (> 850°C) process heat applications.

In addition to the extensive HTGR development program conducted in the United States over the last 20 years, several other countries have also supported gas-cooled reactor development. Major programs have been conducted in European countries, including design, construction, and operation of several HTGR steam-cycle plants, namely the German 15-MWe AVR pebble-bed reactor at the KFA Laboratory in Julich, the German 300-MWe Thorium High Temperature Reactor (THTR) plant scheduled for operation in 1980–1981, and the 20-MWt Dragon reactor successfully operated from 1965 until 1976 in the United Kingdom as a joint project involving 12 members of the European Nuclear Energy Agency. France, Switzerland, and Japan are also conducting development work in support of the HTGR for both electric power and process heat applications. An important U.S. option currently being considered is to develop HTGR systems through a cooperative international program. In recognition of this, a quadrapartite "umbrella" agreement on gas-cooled reactors was signed in 1977 by the United States, Germany, France, and Switzerland to provide the framework for cooperation in this area.

The HTGR Program reflects the DOE recommendation to shift gas-cooled thermal reactor program emphasis from the HTGR steam cycle to the direct-cycle and VHTR (process heat) systems.

The objectives will be met by implementing subprograms for the following:

- Uranium utilization concentrating on developing and demonstrating technology for increasing the efficiency of uranium use of LWR's operating in a once-through mode
- Occupational radiation dose reduction emphasizing the development of methods to reduce radiation exposures of nuclear workers. Solutions can involve techniques either to reduce the sources of radiation or to decrease or eliminate the time workers spend in radiation fields
- Technology development to help resolve critical technical problems that adversely affect LWR plant productivity
- Development of and assistance in the implementation by 1990 of reactor systems and concepts that offer significant safety improvements. This subprogram will be coordinated with Nuclear Regulatory Commission's (NRC's) safety research program, as well as with work by U.S. industry and by other nations
- Development of low- and medium-enriched coated-particle fuel and advanced gas-cooled thermal reactor base technology while supporting an international cooperative development program involving the Federal Republic of Germany, Switzerland, France, and possibly other countries. Technology applicable from steam cycle activity such as fuels, materials, and fission-product studies will be modified in FY 1979 and 1980 to emphasize the high-temperature applications of the direct cycle and process heat reactors

In the pursuit of program objectives and in recognition of the commercial status of the LWR industry, DOE efforts for LWR uranium use are carefully formulated and implemented using the following basic ground rules:

- Industry participation to ensure commercial acceptance of new technology
- Use of utilities as prime contractors in carrying out projects, with equipment suppliers as subcontractors
- No displacement of private-sector initiatives or interference with competition

- Cost-sharing arrangements with project participants
- Projects designed to culminate in in-plant demonstrations

These ground rules provide assurance that results from this program will be accepted and used promptly in LWR plants operated by the electric utility industry.

The upper-level work breakdown structure for the entire Fission Energy Program has been shown in Figure IV 1-2. Further breakdowns of the Thermal Reactor Technology Program are shown in Figure IV 3-5 for its LWR parts and in Figure IV 3-6 for its HTGR parts, with matrices showing how the lower-level packages in the WBS relate to the arrangement of the program description in this section.

The Assistant Secretary for Energy Technology is responsible for the Thermal Reactor Technology Program. This program is assigned to the Director of the Division of Energy Programs who delegates management responsibility to the Director of the Division of Nuclear Power Development (NPD). In turn, NPD's Director delegates program responsibility to the Chief of the Nuclear Reactor Programs Branch for LWR and HTGR development and to the Chief of the Nuclear Safety and Quality Assurance Branch for LWR safety. The branch chiefs provide overall direction and program approval, and are responsible for successful implementation of the program within overall cost and schedules.

In LWR technology development, NPD provides the Chicago Operations Office (CH) with the overall policy and programmatic and budgetary guidance required to implement the program. CH is responsible for project negotiation and administration, including procurement, legal, budget, contractual, and contract liaison matters. The participating contractors are responsible for executing projects assigned to them. They develop proposed scopes of work, schedules, and budget recommendations for work proposals. They advise DOE of project problems and their proposed solutions, routinely report technical progress, cost and schedule status, and milestone achievements for their assigned tasks.

For LWR safety, NPD is responsible for the development of overall policy, objectives, approaches, and schedules. Albuquerque Operations Office (AL) is responsible for administration of the Sandia Laboratories (SL) contract provisions related to this activity. SL provides planning, implementation, subcontract administration and technical management of all research and development activities. A large portion of the specific activities within the subprogram are subcontracted by SL.

Table IV 3-2 shows the operating funding levels for the Thermal Reactor Technology Program by program elements for the FY 1978 to FY 1980 period.

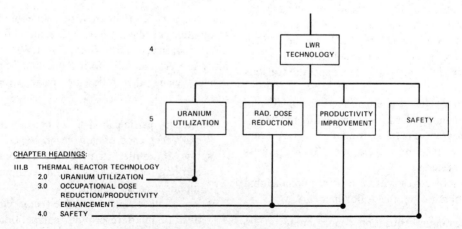

Figure IV-3-5. WBS for the LWR part of the thermal reactor technology program.

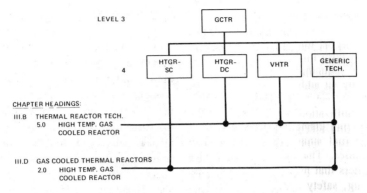

Figure IV-3-6. WBS for the HTGR part of the thermal reactor technology program.

Uranium Utilization

Current estimates of uranium reserves, plus reasonably assured and potential resources indicate a level that could support, throughout their anticipated 30 year lifetimes, all nuclear power reactors expected to be constructed in the United States through the rest of this century. Implementation of the President's nuclear policies requires that every effort be made to stretch the uranium resource base as far into the future as possible. Recent studies have shown that there is a significant potential for increasing the efficiency of uranium utilization of LWR's operating in a once-through mode. The technology to achieve a 15 percent savings in uranium may be possible within the next 10 years—and an additional 15 percent savings may be possible by the year 2000. It is expected that most of these improvements could be retrofitted to current-design LWR's.

The once-through improved LWR has a number of advantages. The proliferation resistance of its fuel cycle is acceptable if the spent fuel is safeguarded. There is no material in any part of the fuel cycle that is readily usable for weapons. The improved LWR discharges less plutonium than present LWR's. This reduces the incentives for, and potential consequences of, material diversion. Less spent fuel will be generated in the improved LWR, resulting in reduced storage requirements and lower storage costs. The LWR is the most widely used reactor in the United States and abroad, and it has proven itself to be a safe, reliable, and economic source of power.

The objectives of the uranium utilization program element are to develop and demonstrate near-term improvements to LWR fuel and fuel management practices that have the potential of reducing uranium consumption in

Table IV-3-2. Funding (BA) for the thermal reactor technology program.

THERMAL REACTOR TECHNOLOGY	BUDGET AUTHORITY (DOLLARS IN THOUSANDS)				
PROGRAM ELEMENT	FY 1978	FY 1979	REQUEST FY 1980	INCREASE (DECREASE)	MAJOR CHANGE*
Uranium Utilization	$ 8,600	$14,400	$18,000	$ 3,600	A
Occupational Radiation Dose Reduction/Productivity Enhancement	5,800	5,500	3,000	(2,500)	S
Safety	300	4,000	3,500	(500)	
HTGR**	0	0	12,000	12,000	
Other Light Water Reactor	4,100				
TOTAL	$18,800	$23,900	$36,500	$12,600	

*Key: A = Increase in technical scope and selective acceleration; S = scale-down and reduction in scope.   **Funding prior to FY 1980 was provided under the Gas Cooled Thermal Reactors program.

existing LWR's by as much as 15 percent by 1988, and longer-term improvements that have the potential of reducing uranium consumption in LWR's by an additional 15 percent by 2000.

The uranium utilization projects are implemented by selecting electric utilities as prime contractors and fuel suppliers as subcontractors to the utilities. The utilities are teamed with fuel suppliers that normally provide their fuel, engineering, safety analysis, and licensing support. In this way, warranty agreements and other commercial arrangements affected by the demonstration fuel can be accommodated in the fuel-supply contracts between the utility and the supplier.

Acceptance of the improvements requires demonstrations in commercial LWR's. These involve in-plant irradiations of large quantities of demonstration fuel with fabrication by industrial manufacturing techniques, normal licensing procedures, and multi-year testing in an operating environment typical of large commercial reactors.

To meet program objectives, developmental projects are carried out cooperatively with as large a segment of the utility industry as possible. The most straightforward near-term means for substantially improving uranium utilization is the development and use of higher-burnup fuel. Complementary high-burnup PWR fuel projects are underway with Duke Power Co. and Arkansas Power and Light Co., both with Babcock & Wilcox Co. as subcontractor. Similar projects involving other utility/fuel-supplier teams are being initiated.

A major problem that must be resolved if high fuel burnups are to be attained is pellet-clad interaction (PCI)—the interaction between irradiated uranium fuel pellets and the Zircaloy cladding. PCI is thought to occur through a stress-corrosion mechanism—i.e., for PCI to occur, it is believed necessary to have stresses caused by mechanical interaction of fuel and cladding, and chemical attack caused by release of corrosive fission products from the fuel. Other problems that must be resolved in the development of high-burnup fuel are fission gas-release, corrosion and hydriding of fuel cladding, and fuel assembly dimensional and structural changes.

A number of methods other than higher-burnup also show promise for improving uranium utilization in LWR's—fuel lattice changes, spectrum changes (without use of heavy water), enrichment zoning/blankets, re-insertion of startup core fuel, improved fuel management and control designs, cycle stretchout, fuel reconstitution, and more frequent refuelings. Projects will be supported as appropriate on the development and demonstration of these improvements.

In the Duke Power project, five fuel assemblies that were due to be disposed of as spent fuel in September 1978 were examined and reloaded, to be burned an extra irradiation cycle in the Oconee-1 reactor. Four of the five assemblies are expected to reach burnups of 40,900 megawatt days per metric ton (MWD/T)—well beyond the target goal of 38,000 MWD/T. The five assemblies will be discharged in late 1979, after which their performance will be evaluated. (Nuclear fuel is currently discharged from a typical LWR after it has achieved an average burnup of 25,000 to 33,000 MWD/T.) Figure IV 3-8 depicts the major milestones.

In the Arkansas project, fuel capable of achieving a burnup of 45,000 MWD/T will be designed and developed, in part by using information obtained from the Duke program. It is planned to load "conservative" lead test assemblies (LTA's) into the Arkansas Nuclear One Unit One (ANO-1) reactor in late 1980, irradiate them until 1985, remove them, and then examine them in a hot cell. It is planned that "optimized" LTA's will be inserted in ANO-1 in mid-1982, irradiated until 1986, removed, and examined. The information from these projects will be used to demonstrate an entire reload batch of about 57 subassemblies designed for 45,000 MWD/T.

Two major projects are underway to alleviate the PCI problem. The first, with Consumers Power Co. as prime contractor and Exxon Nuclear and Battelle-Northwest as subcontractors, involves testing annular pellets in conjunction with graphite-coated cladding as the principal remedy, and packed-particle fuel as

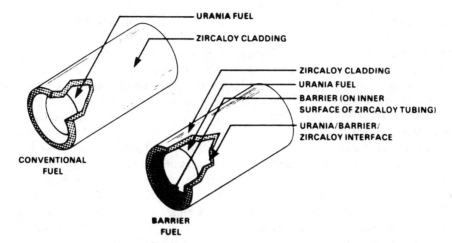

Figure IV-3-7. Barrier to minimize stress corrosion of cladding caused by interaction of pellet and cladding.

a backup. Tests of these fuels began in July 1978 in the Halden reactor in Norway, and are scheduled for completion in late 1983. Irradiations of about 360 test rods will be conducted in Consumers Power Company's Bic Rock Point reactor beginning in early 1979. These irradiations are scheduled for completion in 1985.

The second PCI project is being conducted with Commonwealth Edison Co. as the prime contractor and General Electric Co. as subcontractor to evaluate the effectiveness of either a copper barrier or a high-purity zirconium liner inside the cladding (see Figure IV 3-7). Ramp tests on this fuel have been conducted in the General Electric Test Reac-

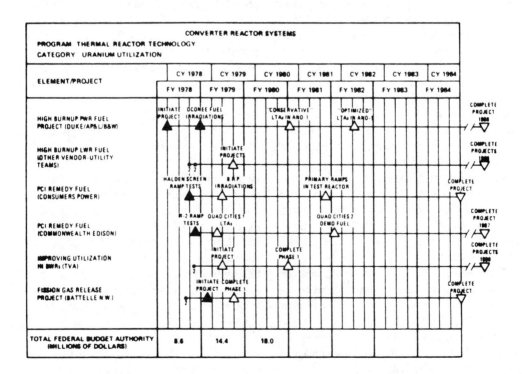

Figure IV-3-8. Major milestones and funding for uranium utilization.

tor at Vallecitos, California, and in the Studsvik reactor in Sweden. Lead test assemblies were to be inserted in the Quad Cities 1 reactor in 1979, and test irradiations are scheduled for completion in 1985. A full-scale demonstration in the Quad Cities 2 reactor is to be initiated in late 1981, with completion scheduled for 1987.

Several of the other proposed methods for improving utilization will be investigated in a Tennessee Valley Authority (TVA)/DOE interagency project. General Electric (GE) will be the subcontractor in this project. A cost-shared cooperative project is being conducted to investigate the fission-gas release characteristics of UO_2 fuel irradiated to high burnups, with Battelle-Northwest as contractor.

Occupational Radiation Dose Reduction/Productivity Enhancement

Radiation exposure to workers in LWR plants has increased since the inception of commercial nuclear power because of increasing levels of plant contamination, more severe regulatory requirements for in-service inspection, and an increasing number of installations. In general, the average exposure to individual workers has remained relatively constant since 1969 and is within current regulatory limits, but the yearly total exposure has grown markedly. LWR technology program efforts to reduce occupational radiation exposure support the government policy of minimizing such exposure.

Nuclear power's projected contribution to the nation's energy needs will not be fully achieved if technical problems are encountered that significantly impact their productivity. LWR technology program efforts are directed toward resolving such problems.

The objectives of this program element are to demonstrate the technology to reduce the occupational radiation exposure to LWR plant personnel and to develop solutions to critical technical problems which, if unresolved, could result in a severe short-term decrease in nuclear plant productivity.

The strategy being used for achieving these objective is similar to that described previously for the uranium utilization element of the LWR technology program.

Progress toward achieving the occupational radiation dose reduction objective is being made through projects to develop and demonstrate chemical processes to decontaminate reactor system and reduce the radiation fields; projects to develop and demonstrate surveillance and diagnostic techniques capable of replacing many currently required manual operations, thereby reducing the need for worker entry into radiation areas; and other means to be identified through studies for that purpose.

Work on resolving critical technical problems is presently limited to developing ways of alleviating the effect of "denting" in PWR steam generators, a problem expected to cause extended outages in several nuclear powerplants. The principal solution that has been explored to date is the development of a process to chemically clean the secondary side of PWR steam generators.

Several projects are underway to develop chemical decontamination techniques for removal of radioactive deposits from reactor primary systems of both PWR's and BWR's.

A feasibility study of processes for chemically decontaminating a PWR has been completed by the Consolidated Edison Co. of New York, As a result of the work done in this first phase, several decontamination methods have been selected for further development. An evaluation is underway to determine how to proceed with subsequent phases—including demonstration in a commercial PWR.

Two BWR decontamination projects are being carried out using two different approaches: a thorough removal of radioactive materials after removal of the fuel, and a rapid but partial cleanup without removal of fuel. Commonwealth Edison Co. is the prime contractor for both projects; Dow Chemical Co. and General Electric Co. are subcontractors for the thorough decontamination method; General Electric is the subcontractor for the rapid decontamination method. The first process was demonstrated in Commonwealth Edison's Dresden-1 reactor in mid-1979. The rapid process may culminate in a demonstration in 1981–1982, probably using Commonwealth Edison's Dresden-2 reactor.

The other major approach for reducing occupational radiation exposures is the development of surveillance and diagnostics technologies. Examination of occupational radiation-exposure records from operating plants shows that more than 70 percent of the total exposure to plant personnel comes from maintenance and in-service inspection activities during plant outages. Keeping plants operational is an effective way to reduce total exposure. Techniques are being developed and demonstrated for replacing, eliminating, or reducing the frequency or duration of current manual maintenance and safety inspections in high radiation areas.

An occupational radiation management study is being initiated in early 1979 to investigate the feasibility of government action on other methods for reducing radiation exposures to workers in nuclear plants.

Two surveillance and diagnostics projects are in progress. One involves the development of an acoustic-emission monitoring system for total pressure-boundary surveillance in PWR's. Consolidated Edison is the prime contractor for this project, and the Indian Point-1 reactor, which is no longer in nuclear operation, is being used as a testbed. The other project involves the development of an acoustic-emission monitoring system for surveillance of key components in BWR's. Philadelphia Electric Co. is the prime contractor for this project, with the Peach Bottom 3 BWR as the testbed. These demonstrations are scheduled to be completed in late 1980.

"Denting" has been found to occur in PWR steam generators because of the buildup of corrosion products in the annulus between the tube and tube-support plant (see Figure IV 3-9). After filling this clearance, the deposit grows due to corrosion of the carbon-steel support plate and exerts an inward force on the tube, causing it to dent. A project was initiated in April 1977 with Consolidated Edison as the prime contractor to develop a chemical cleaning process to remove the deposits causing the denting problem. Other participants in this project were United Nuclear Industries, Halliburton Services, and Wyle Laboratories.

The initial phase of the project involved the

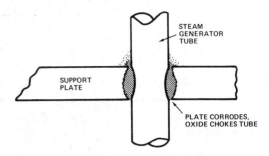

Figure IV-3-9. "Denting" in PWR steam generator.

development and testing of chemical reagents and application techniques, and the pilot testing of leading candidate chemical processes in steam generators at the nonoperating Indian Point-1 reactor. A full-scale demonstration had been planned for the Indian Point-2 reactor, but has been suspended because of uncertainties about the condition of the steam generators.

An evaluation is underway to determine the additional work on alleviating steam generator "denting" to be carried out in the Thermal Reactor Technology Program. A continuing effort is also being made to keep apprised of other LWR technical problems that adversely affect nuclear plant productivity.

Major milestones and funding are shown in Figure IV 3-10.

Safety

The primary objective is to improve the safety of LWR's by developing advanced safety systems, the use of computers in reactor operations, risk-based methods, and safety data for use in high-burnup fuels activities. More detailed objectives for these four areas of interest are as follows:

- Improved systems and fuel safety data to develop advanced concepts or improve existing designs and plant layouts so that plants can meet existing and new safety requirements more effectively, reliably, and economically; and to investigate the need for and to develop, where necessary, safety data to support the activities to increase the utilization of uranium fuels

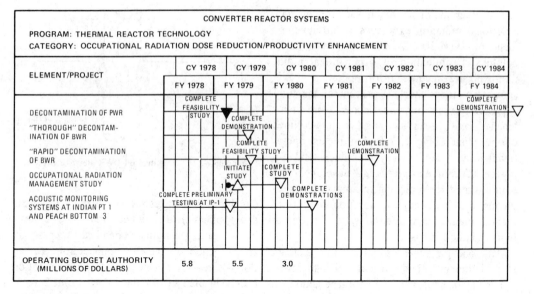

Figure IV-3-10. Major milestones and funding for occupational radiation dose reduction/productivity enhancement.

- Investigate the man-machine interface to improve safety in the control of normal operations, including fuel handling and maintenance, and operator response during emergency situations and during the early phases of accident sequences
- Risk-methods utilization to develop and use risk-based methods and studies to establish R&D priorities by focusing attention on sensitive systems, components, and processes, reduce risk through design and operations changes, and help influence licensing priorities

Concepts and systems to improve safety are being developed and will be implemented in LWR's. Potential areas have been defined through discussions with the NRC, the nuclear industry, national laboratories, and members of the technical community.

An integrated systems development approach is being used in which DOE has an active role from defining functional requirements through assisting a utility demonstration in an LWR. The requirement that the improved systems be accepted and used in LWR's has led to the additional constraints that they must reduce licensing uncertainty, be developed through cost sharing, or reduce the cost of safety systems. The safety work will be completed by 1990 so the output can be incorporated in reactors scheduled for operation about the year 2000.

The planning of this subprogram was initiated in FY 1978. Sandia Laboratories was selected as technical manager. During FY 1979, a detailed program plan was prepared and work was initiated in all four areas of activity. The initial requests for proposals for performing these tasks are being released. It is anticipated that many of the larger activities, particularly those directed toward systems development and testing, will be performed by industry. Scoping studies, analytical tasks, and methods development are likely to be performed by national laboratories, universities, or small engineering R&D organizations.

Specific safety-related LWR systems are to be improved to meet their functional requirements more effectively and reliably. Simplified plant layouts are being investigated to reduce the plant's sensitivity to common-cause accident initiators, such as incorrect maintenance activities, adverse environmental conditions, and fires.

For certain of the systems of interest, NRC is performing, in parallel with the above efforts, research to determine functional design requirements. Those NRC requirements will be drawn upon to formulate selected DOE systems development activities.

Work is planned in the areas of structural mechanics, valve improvement, advanced containment systems, and optimizing plant layout to meet a number of broad safety needs. These are principally scoping studies to identify: important features, relationships between features, available and needed data, available and needed analytical methods, and products and their benefits.

The results of these studies, which generally will be completed in FY 1980, will be used to define further development work.

Approaches are being developed for using computers to improve plant safety through improved operation. Their development will result in the following benefits: reduce demands on the plant-protection system, reduce plant-transient frequency, reduce spurious-shutdown frequency, and optimize maintenance activities. The work is directed toward developing automatic transient-control systems, on-line monitoring and diagnostics systems, and an automatic plant-protection system.

A study initiated in FY 1979 is identifying the required elements and their current status. The necessary follow-on tasks will be undertaken when resources permit. Also in FY 1979, planning and other tasks to improve maintenance were initiated. Idaho National Engineering laboratory (INEL) is beginning to apply modern control theory to the design of a total control system. This will provide better information on plant performance and the capability to identify, at an earlier time than now possible, off-normal trends of sensitive parameters. Close contact is being maintained with the Electric Power Research Institute (EPRI), which sponsors research that could contribute to the concept development. A decision on the potential of this control approach and the total task scope will be made when the planning task is concluded in FY 1980.

The "Reactor Safety Study" (WASH-1400, NUREG 77/104, October 1975) quantitatively analyzed the public risks due to postulated LWR accidents. The analysis used probabilistic analysis techniques extensively. The logical and orderly way of analyzing reactor accidents offered by these methods leads to the desire to capitalize on these characteristics to aid in making decisions during design and licensing. Thus, this complementary program is developing risk-analysis methods that utilities and plant designers can use as additional bases for safety decisions in design and licensing.

Work is being initiated in two specific areas: (a) the University of California at Los Angeles (UCLA) is comparing and evaluating three proposed models for choosing acceptable-risk criteria, and (b) a technique to incorporate safety and reliability into plant layout and design is being initiated at INEL.

Work in FY 1979 and FY 1980 includes data-base evaluation, statistical-methods development, and risk-methods utilization in the design and licensing processes.

A study is determining the safety implications of operating with high-burnup fuel. If a need for experimental safety data is identified, a test program will be planned in FY 1980.

Major milestones and funding are presented in Figure IV 3-11.

High Temperature Gas Cooled Reactor

The HTGR Program presently includes two basic concepts:

- Direct cycle HTGR
- Very High Temperature Reactor (VHTR) for process heat applications Figure IV 3-12 shows the flow diagram of the direct cycle gas-turbine HTGR (GT-HTGR). Several important features should be noted:
- Both the nuclear steam supply system (NSSS) and the power-conversion loop are contained within the prestressed concrete reactor vessel (PCRV)
- Direct helium turbine power conversion (no steam loop)
- Recuperator added to increase efficiency

As shown in Figure IV 3-13, the VHTR concept involves pumping helium through the reactor core by means of a circulator and then transferring the heat through a steam reformer or intermediate heat exchanger for use in a high-temperature process (e.g., hydrogen pro-

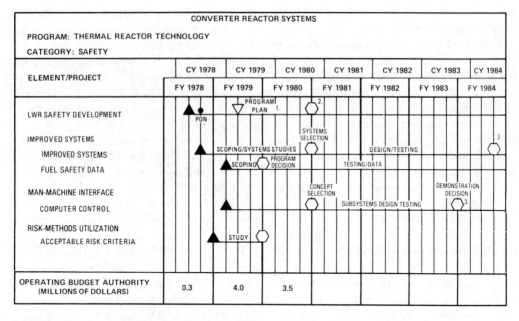

Figure IV-3-11. Major milestones and funding for the safety category.

duction), then through a boiler where steam is generated for either electric power production or low-temperature process heat application.

The specific objectives of the work in FY 1980 on the advanced high temperature reactors are as follows:

- To evaluate conceptual designs for both VHTR and GT-HTGR.
- To reorient the HTGR technology program to emphasize higher temperature applications in support of an international development program.
- To advance the technology of low- and medium-enriched HTGR graphite-coated particle fuels.

A key element in the Advanced High Temperature Reactor (HTR) strategy will be a reorientation of the HTGR technology programs in the areas of fuels, graphite, materials, safety and fission products to emphasize the higher temperature conditions required for the Advanced HTR.

The fuels program involves irradiation testing of low- and medium-enriched coated particle fuels to determine the effects of impurities and fission products on coating stability and fission product retention. The graphite program is aimed at defining the long-term characteristics of graphite under irradiation with particular emphasis on potential oxidation. The materials work involves a metals screening program to identify candidates for advanced HTR applications. Structural analysis and testing of the prestressed concrete reactor vessel and its associated thermal barrier are being performed as part of the structural components effort. The safety program involves seismic and containment response analysis along with probabilistic risk assessments. The reactor surveillance program involves data compilation during startup testing of the Fort St. Vrain Reactor.

Direct cycle and process heat design tasks will emphasize evaluations of the German-type pebble bed reactor for various applications. A key question requiring resolution in the near term is whether a pebble bed or a prismatic core is better for the advanced systems. Extensive analysis on this topic is anticipated over the next two to three years.

The following work is involved:

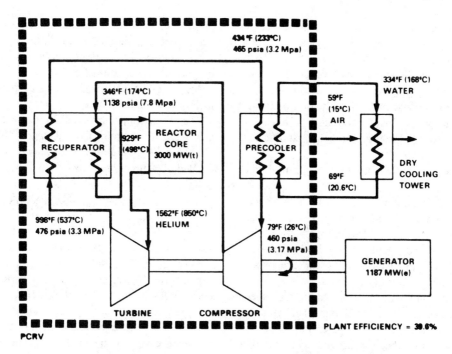

Figure IV-3-12. Typical HTGR with direct Brayton cycle.

- Evaluation of candidate metal alloys for application in the VHTR and direct cycle reactors, including studies on surface stability, structural stability, creep, stress rupture, and fatigue.
- Conceptual design and analysis of alternative direct cycle HTGR plant configurations, VHTR process heat reactor, and German pebble bed reactors.
- Cooperation with the Federal Republic of Germany, France, and Switzerland on fuels, graphite, materials, safety, systems, components, and HTGR-DC conceptual design.

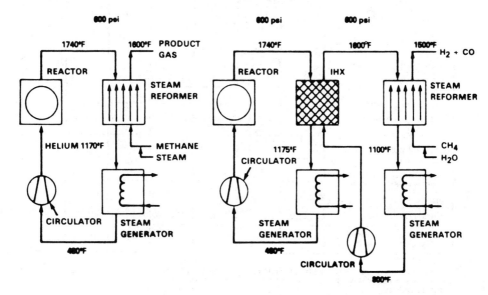

Figure IV-3-13. VHTR for producing process heat by direct coupling (left) or intermediate-loop coupling (right).

Conceptual design activities have been underway on both the direct cycle HTGR and VHTR for process heat applications for several years. Work on the VHTR has led to conceptual designs by GE, Westinghouse, and General Atomic and an evaluation of these by Oak Ridge National Laboratory (ORNL). The materials test program is being conducted by GE.

The gas-cooled reactor "umbrella" agreement between the United States and the Federal Republic of Germany was signed in Bonn, Germany, in February 1977, and established the basis for any cooperative development. France and Switzerland signed this agreement in October 1977 and are expected to become actively involved.

There are now approximately fifty specific pieces of technical work that have been identified as candidates for potential cooperation. Twenty have been approved and are underway; the remaining are in various stages of preparation and negotiation. This approach was adopted to initiate some limited areas of cooperation while awaiting major HTGR/HTR programmatic decisions.

Major milestones and funding are presented in Figure IV 3-14.

The contractors involved in this program are General Atomic Co., General Electric Co., Oak Ridge National Laboratory, and Gas Cooled Reactor Associates.

ADVANCED ISOTOPE SEPARATION TECHNOLOGY

The contribution of light water reactor technology to the nation's energy requirements depends on the availability of fuel enriched in uranium-235. One potential source of U-235 that requires no additional mining is the "tails" from the enrichment plants; the economic recovery of this fuel is a principal objective of the advanced isotope separation

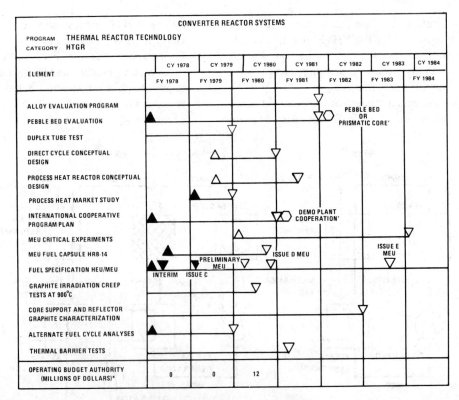

Figure IV-3-14. Major milestones and funding for the HTGR program.

program. Three advanced separation technologies—two laser methods and a plasma concept—are being developed.

Overview

This section summarizes the background, objectives, strategy, management, and funding for the Advanced Isotope Separation Technology (AIST) Program.

The fuel for the current generation of LWR's operating in the United States is uranium enriched to about 3 percent in the U-235 isotope. Currently, all U.S. uranium enrichment is accomplished by gaseous diffusion technology, a process developed in the Manhattan Project during World War II. Three gaseous diffusion plants were constructed to produce material for nuclear weapons, but now only about 5 percent of the capacity is used for all U.S. Government programs; the remainder is used to produce low-enriched uranium for nuclear power reactors in the United States and abroad. Two programs are underway to increase the capacity of the diffusion plants. The Cascade Improvement Program (CIP) and the Cascade Uprating Program (CUP) will increase the annual separative capacity of the three plants from 17.2-million separative work units (SWU) to about 27.3-million SWU in the early 1980s.

During the Manhattan Project, gas centrifuge technology was evaluated, but gaseous diffusion was judged to be more attractive. Advances in materials and machinery design during the 1950s increased the potential for centrifuge technology. Investigation of this method during the 1960s in both the United States and Europe led to long-term testing of advanced centrifuges in the early 1970s. The President announced on April 20, 1977, that the next increment of uranium enrichment capacity built by the U.S. Government would use gas centrifuge technology.

Selective laser-induced chemistry to separate isotopes was suggested in the United States in 1971 by the Los Alamos Scientific Laboratory (LASL). Since then, the U.S. program to develop isotope separation technology has expanded to include two laser methods and a plasma technique.

The major objectives of the AIST Program are as follows:

- Conserve uranium resources by developing technologies that will economically permit the lowering of the tails assay for current and future enrichment plants
- Evaluate the nonproliferation implications of new and alternative enrichment technologies
- Apply advanced isotope separation technologies to the solution of other energy related problems

The Advanced Isotope Separation Technology Program consists of two categories of effort: uranium enrichment, and proliferation assessment and applications of advanced technology.

The AIST Program strategy involves the development of three methods—two laser techniques and a plasma technique—through the preprototype devices, in conjunction with additional off-line experimentation will determine the components and scientific principles that would be used in a later engineering demonstration facility. During FY 1981, the operation of these preprototypes will generate a data base that will permit comprehensive evaluation of the scalability, economic potential, proliferation resistance and environmental consequences of each technology. A decision may be made at the end of FY 1981 to proceed to engineering development if the evaluation warrants such development.

The strategy of the nonproliferation and advanced applications effort involves the assessment of promising technologies and their implications with respect to nuclear weapons production.

The upper-level WBS for the entire Fission Energy Program was shown in Figure IV 1-2. The further breakdown of the AIST Program is shown in Figure IV 3-15 with a matrix showing how the lower-level packages in the WBS relate to the arrangement of the program description in this section.

The AIST Program is under the cognizance of the Assistant Secretary of Energy Technology who delegates authority to the division of Advanced Nuclear Systems and Projects. This division develops the overall program strat-

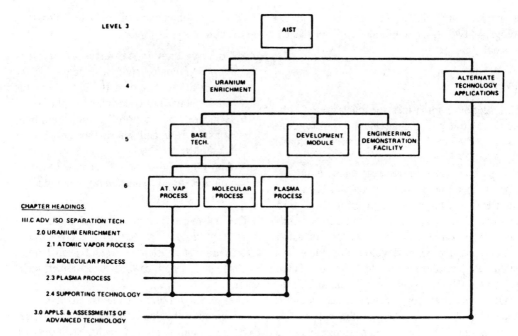

Figure IV-3-15. AIST program WBS.

egy, plans and budgets, and provides program management in consonance with policy guidance provided by ET and ETN.

The national laboratories and contractors provide the day-to-day management of the individual projects, develop project management plans, and recommend the resources required to meet the program objectives. They have the ultimate responsibility for performing the program and meeting the established milestones.

Table IV 3-3 shows the operating funding levels by category for FY 1978 through FY 1980.

Many of the accomplishments of the AIST Program are classified, so that detailed schedules and milestone charts have been omitted from this section. However, Figure IV 3-16 depicts the program schedule in terms of the DOE major acquisition phases.

Uranium Enrichment

Natural uranium is composed of several isotopes, the most important of which are U-235 (0.711 percent) and U-238 (over 99 percent). Before it can be used as fuel in current commercial nuclear powerplants (LWR's) in the United States, it has to be enriched; i.e., the concentration of the fissionable isotope, U-235, must be increased from its normal 0.711 percent to about 3 percent. The residue from the enrichment process, depleted tails, still contains on the order of 0.20 percent U-235. If the U-235 concentration in the tails can be economically reduced to 0.1 percent or

Table IV-3-3. AIST program funding for FY 1978–1980.

ADVANCED ISOTOPE SEPARATION TECHNOLOGY	BUDGET AUTHORITY (DOLLARS IN THOUSANDS)			
CATEGORY	FY 1978	FY 1979	REQUEST FY 1980	INCREASE (DECREASE)
Uranium Enrichment	$41,926	$44,775	$41,200	($3,575)
Applications and Assessments of Advanced Technologies	1,691	1,425	5,000	3,575
TOTAL	$43,617	$46,200	$46,200	0

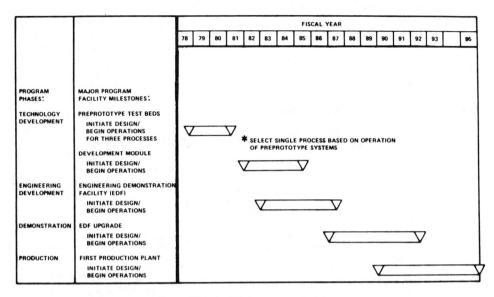

Figure IV-3-16. AIST program schedule.

0.05 percent, the effective uranium resource base would be increased by 16 to 19 percent.

The major objective of the AIST Program is to conserve uranium resources by developing technologies that will economically (i.e., cost 50 percent less than with existing technologies) permit lowering the tails assay for current and future enrichment plants.

Three processes are currently being studied—two laser techniques (atomic vapor and molecular) and a plasma technique. The primary responsibility for each technology and the work site are listed in Table IV 3-4.

In addition, the staffs at DOE's gaseous diffusion and Y-12 production plants at Oak Ridge, Tennessee, provide support in the areas of source and collector development, uranium metal handling, materials compatibility, physics and spectroscopy, process equipment development, conceptual engineering, and economic evaluations.

Three processes are being investigated in parallel through the applied research stage. At the start of FY 1978, laboratory-scale devices required to define the scientific issues of the three technologies were available. This work will continue until the end of FY 1981. Intensive development will lead to the fabrication and assembly of preprototype testbed systems by the end of FY 1980. During FY 1981, the operation of these large-scale testbeds will generate a data base that will permit extensive evaluation of the scalability, economic potential, proliferation resistance, and environmental consequences of each technology. A decision may be made at the end of FY 1981 to proceed to engineering development if warranted by the data.

The overall program strategy and phases are presented in Figure IV 3-17.

Conceptual engineering cost studies continue to indicate that these technologies have great promise for meeting the program objective. In general, the expected results of the R&D activities are as follows:

- FY 1979: component development work focused on lasers, magnets, collectors,

Table IV-3-4. Contractors for uranium enrichment studies.

PROCESS	CONTRACTOR, LOCATION	FORM OF ENERGY ABSORBED
Atomic Vapor	Lawrence Livermore Laboratory (LLL), Livermore, Calif.	Visible light
Molecular	Los Alamos Scientific Laboratory, Los Alamos, N.M.	Infrared and ultraviolet
Plasma	TRW, Inc., Redondo Beach, Calif.	Radio-frequency

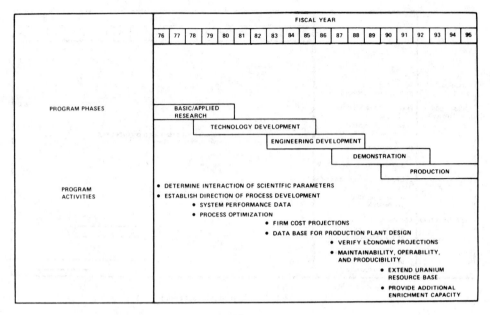

Figure IV-3-17. AIST program overall strategy and phases.

extractors, optics, and feed delivery systems installed in preprototype systems.
- FY 1980: system integration work will concentrate on integrating large-scale components into a preprototype system; major issues of engineering scaling for all parts of the process will be identified.
- FY 1981: systems operations work will concentrate on operating the preprototype facilities and solving operating parameter problems.

The operating funding levels by process for FY 1978 through FY 1980 are shown in Table IV 3-5.

Atomic Vapor Process. The atomic vapor process takes advantage of the small differences in the electronic spectra of uranium isotopes. The process requires the controlled vaporization of uranium atoms followed by selective excitation and ionization of U-235 using tunable lasers in the visible or ultraviolet regions of the spectrum. The resulting plasma of U-235 ions can be removed from the uranium vapor using electromagnetic methods. The large-scale generation of vapor under required process conditions, the handling of molten uranium, the demonstration of a scalable uranium ion extractor, and the development of full-scale lasers suitable for plant operation are under intensive investigation. Figure IV 3-18 illustrates the atomic vapor process.

The following major areas of development are required:

- Laser—increase the average power output per laser, and increase the number of

Table IV-3-5. Uranium enrichment funding for FY 1978−1980.

URANIUM ENRICHMENT	BUDGET AUTHORITY (DOLLARS IN THOUSANDS)			
PROCESS	FY 1978	FY 1979	REQUEST FY 1980	INCREASE (DECREASE)
Atomic Vapor Process	$13,739	$14,710	$13,400	($1,310)
Molecular Process	17,417	18,230	16,900	(1,330)
Plasma Process	9,913	10,260	8,930	(1,330)
Supporting Technology	857	1,575	1,970	395
TOTAL	$41,926	$44,775	$41,200	($3,575)

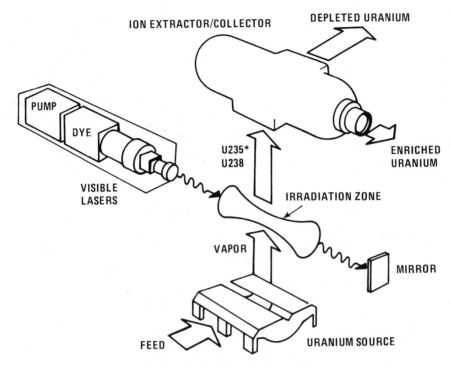

Figure IV-3-18. Atomic vapor laser isotope separation process.

times the laser flashes per minute (improved repetition rate)
- Material—improve the lifetime of material exposed to liquid uranium and uranium vapor
- Extractor—develop an ion extractor technique to operate at the densities required for a production module

An important component of the process development program is the evaluation of the uranium vapor generation source. Efforts during FY 1978 were highly successful in this area. The commitment to the preprototype systems requires that design and fabrication for the uranium handling system began in FY 1979. Toward the end of FY 1980, the uranium handling systems will be installed. The total system, including lasers and electro-optical components, will become operational during FY 1981.

Molecular Process. In the molecular process, UF_6 is expanded through a nozzle to reduce the temperature and simplify the spectrum of the gas. The UF_6 is then exposed to laser radiation in the infrared and ultraviolet yielding enriched uranium as solid UF_5 that can be separated from the gaseous UF_6. This process has intrinsically high throughputs and uses existing materials handling technologies. However, the high-power lasers at the specific wavelengths required are not yet available. Figure IV 3-19 illustrates the molecular process.

The following major areas of development are required for the molecular process:

- Laser—increase the average power output per laser, and increase the number of times the laser flashes per minute (improved repetition rate)
- Science—increase the basic understanding of the interaction between the laser power and the process gas

Major fundamental questions still exist on details of the molecular process involving the interaction of the laser energy and the gas. During FY 1979, experiments are defining these phenomena better.

High-power lasers for the preprototype system were ordered during FY 1978 from industrial vendors. These lasers are expected to be delivered during FY 1980 and installed on the

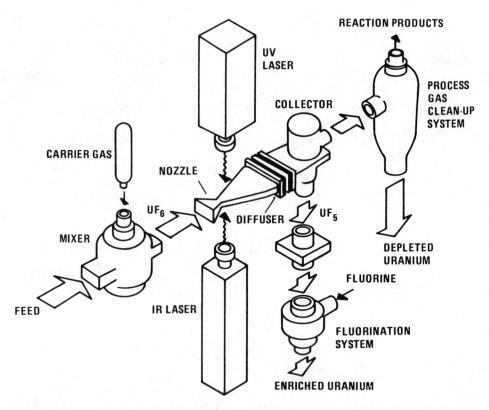

Figure IV-3-19. Molecular laser isotope separation process.

molecular process preprototype testbed, which will become fully operational during FY 1981.

Plasma Process. The plasma process is based on ion cyclotron resonance. The method involves generating a stable plasma in a magnetic field, thereby establishing different ion cyclotron frequencies for the individual isotopes. The frequency of a radio frequency field is then tuned to the ion cyclotron frequency of the U-235 where resonance is achieved and energy is absorbed by the U-235 ion. This results in an increase of the cyclotron orbit of this species; the difference in orbits is used to separate the nonresonant U+ from the resonant enriched U+. Figure IV 3-20 illustrates the concept of the plasma process.

The following major areas of development are required for the plasma process:

- Plasma density—increase the operating plasma density
- Material handling—increase the lifetime of material exposed to uranium
- Science—increase knowledge of the excitation and plasma process and determine its operational limits

Major science questions on the plasma process are still unresolved. Small magnet systems are being used as research testbeds. By mid-FY 1979, a larger magnet system began operation to aid in determining the process parameters.

During FY 1978, a contract was awarded to a magnet vendor to design and fabricate the preprototype magnet system. This will be built during FY 1979 and FY 1980, with delivery expected in late FY 1980; operation is scheduled in FY 1981.

Supporting Technology. This area of effort is conducted at the Oak Ridge Gaseous Diffusion Plant and is composed primarily of scoping conceptual engineering and cost analysis for the three separation technologies. For this study, detailed cost models for the atomic vapor and molecular processes were developed during FY 1977; during FY 1978, a

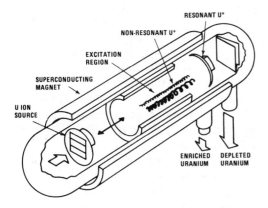

Figure IV-3-20. Plasma isotope separation process.

similar model was developed for the plasma process. In general, as more information is obtained for the three processes, the cost models are updated and sensitivity analyses are performed on selected variables. These analyses highlight the relative importance of the parameters and serve to direct the R&D programs by identifying the variables that have high leverage on process costs.

The major activities for FY 1979 and FY 1980 are to define better the costs of key systems requiring industrial expertise; examples are laser systems, electro-optical systems, and large superconducting magnets. Plant and equipment layouts will be updated to reflect the current reference design and to establish equipment and building costs that are more accurate. The computer models describing the process science will be updated to provide a more accurate analysis based on process experiments and updated theory.

Applications and Assessments of Advanced Technology

Advances in science and technology have opened new vistas for economical isotope separation using processes previously thought to be impractical, if not impossible. In addition, the development of laser isotope separations technology affords a new and powerful technique that can be used to solve problems in other areas of the nuclear fuel cycle. Research and development on these new concepts ensure that the United States maintains its leadership in isotope enrichment methods and exploits the use of these technologies to the solution of energy-related problems, such as the ultra-purification of materials that might be required for solar and biomedical applications and partitioning of nuclear waste material.

The objectives of the applications and assessments of AIST programs are as follows:

- Maintain DOE on the forefront of understanding uranium separation methods, evaluate new uranium enrichment technologies, and accurately assess their proliferation implications
- Apply advanced separation technologies to energy-related problems

Experiments are being performed to evaluate the theory of the uranium chemical exchange separation process, including the French technique, and to develop the scaling laws required to analyze the proliferation implications of these technologies. Efforts on the chemi-ionization were terminated at the end of FY 1979.

GAS COOLED THERMAL REACTORS

Overview

The Gas-Cooled Thermal Reactor (GCTR), referred to in the United States as the High Temperature Gas-Cooled Reactor (HTGR) and in Europe as the High Temperature Reactor (HTR), has evolved from more than 20 years of development.

The HTGR is an advanced fission converter; i.e., a reactor system that provides better uranium fuel utilization than the LWR but not as high conversion ratios as the breeder. It can operate on a variety of uranium or a uranium/thorium fuel cycles ranging in fissile enrichment from 10 percent to 93 percent. Cycles with enrichments greater than 10 percent employ thorium* as a fertile material; cycles with 10 percent or less enrichment use only uranium.

The HTGR steam cycle uses a conventional steam Rankine cycle, in which heat is trans-

---

*When thorium is charged into a reactor along with fissile material, it is converted to uranium-233, which is fissile and can be consumed to produce additional energy.

ferred from the reactor core to helium, which then passes through a heat exchanger (steam generator) where the heat is used to convert water to steam, which is then used in a conventional steam turbine-generator to produce electricity.

The GCTR is for use in central-station electric power applications where it can produce high efficiency, economic, electric power generation with reduced uranium ore consumption, relative to LWR's. The higher-temperature version of this cycle will be used to supply process heat for various chemical and metallurgical operations.

U.S. experience with the use of HTGR's for civilian power applications has included design, construction, and operation of two central-station electric plants. The 40-MWe Peach Bottom I HTGR was operated successfully by the Philadelphia Electric Co. from 1967 until 1974, when the plant was shut down for decommissioning after completing the planned demonstration program. The 300-MWe Fort St. Vrain HTGR is now undergoing startup testing by the Public Service of Colorado Co. and had operated at 68 percent full power to date. Although Fort St. Vrain is several years behind schedule at this time, the difficulties encountered with the project have been typical of large first-of-a-kind nuclear plants, rather than basic to HTGR technology per se.

In 1974, the General Atomic Co. (GA) had commitments, or near commitments, for 10 commercial-size (700–1200-MWe) HTGR steam-cycle plants. All were subsequently cancelled for a variety of business and economic reasons. Since GA terminated the last two of these commercial contracts in 1975 at a cost of approximately $250 million, a series of assessments has been sponsored by government and industry to determine the future role of GCTR's in the United States. These studies have consistently recognized the overall technical versatility of gas-cooled reactors for various applications, but have raised questions on the marginal economics of the HTGR steam cycle, and the business and institutional aspects of commercialization.

DOE proposes to close out the HTGR steam cycle program; any future work on this system is to be done in the Thermal Reactor Technology Program.

The program strategy is to complete ongoing conceptual design tasks and a reference-steam-cycle cost estimate in FY 1979 before closing out the program in FY 1980. All work will be documented in such a fashion that the program can be restarted expeditiously if needed in the future.

The upper-level WBS for the entire Fission Energy Program was shown in Figure IV 1-2. The further breakdown of the GCTR program is shown in Figure IV 3-21 with a matrix showing how the lower-level packages in the WBS relate to the arrangement of the program description in this section.

ASET is responsible for the GCTR Program. The program is assigned to ETN who delegates management responsibility to the Director of the Nuclear Power Development Division, who, in turn, delegates program responsibility to the Chief of the Nuclear Reactor Programs Branch. The branch chief provides overall direction and program approval, and is responsible for successful implementation of the program within overall cost and schedules.

Under the direction of the appropriate DOE field office, the national laboratories and contractors provide the day-to-day management of the individual projects, develop project management plans, and recommend the resources required to meet the program objectives. They have the direct responsibility for the performance of the program and meeting the milestones established.

Table IV 3-6 shows the operating funding levels by category for the FY 1978 to FY 1980 period.

High Temperature Gas-Cooled Reactor

The HTGR steam cycle (HTGR-SC) is the most extensively developed of the GCTR concepts. Figure IV 3-22 shows the flow diagram for a typical HTGR steam cycle plant.

Some important characteristics of the HTGR are as follows:

- Helium coolant
- Graphite moderator
- Ceramic fuel cladding

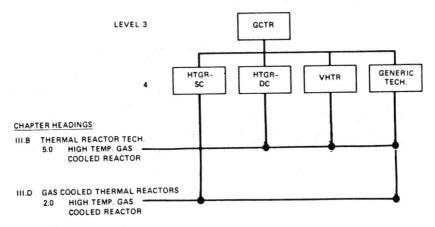

Figure IV-3-21. GCTR program WBS.

- Prestressed-concrete reactor vessel
- Fuel cycle flexibility, advanced converter (conversion ratio 0.6 to 0.9)
- Theoretically could be a thermal breeder

DOE is recommending closeout of the HTGR-SC Program due to the following:

- An insufficient near-term utility commitment to the HTGR-SC necessary to support commercialization without the government bearing a disproportionate share of the risk
- Marginal economic advantages of the once-through fuel cycle relative to an improved LWR
- Marginal potential for improved resource utilization relative to the LWR

The recommended program strategy is to complete ongoing conceptual-design tasks and a cost estimate for a reference HTGR-SC plant prior to closeout of the program in FY 1980.

During FY 1978, the HTGR plant design previously proposed by GA was assessed by the utility organization, Gas Cooled Reactor Associates. As a result of the assessment and changing government policy regarding fuel recycle and highly enriched fuels, several significant modifications to the GA design were proposed. These involved plant size reduction, configuration and systems modification, and the use of a low- or medium-enriched fuel cycle. Steam cycle design efforts in FY 1979 have therefore concentrated on incorporating these changes recommended by the utilities.

Major milestones and funding are depicted in Figure IV 3-23.

During FY 1979, three contractors participated in the HTGR steam cycle program: General Atomic Co., Oak Ridge National Laboratory, and Gas Cooled Reactor Associates.

ADVANCED REACTOR SYSTEMS

The purpose of the Advanced Reactor Systems Program is to develop technology for

Table IV-3-6. GCTR program funding for FY 1978−1980.

GAS COOLED THERMAL REACTORS	BUDGET AUTHORITY (DOLLARS IN THOUSANDS)			
CATEGORY	FY 1978	FY 1979	REQUEST FY 1980	INCREASE (DECREASE)
High Temperature Gas Cooled Reactor (HTGR)	$26,300	$34,000	0	($34,000)
Advanced High Temperature Reactor*	5,900	5,500	0	(5,500)
TOTAL	$32,200	$39,500	0	($39,500)

*Advanced High Temperature Reactors are the Direct Cycle and the VHTR.

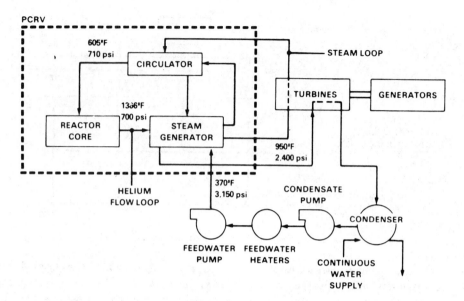

Figure IV-3-22. Flow diagram of typical HTGR with Rankine steam cycle.

both reactors and fuel-cycle facilities so as to reduce proliferation risk. This is a new initiative and is intended to translate the results of NASAP and INFCE studies into actual technology.

Overview

Several technology assessments have emerged from the NASAP and INFCE studies which, if coupled with political and institutional measures fostered by the Nuclear Nonproliferation Act, would result in reduced risk of nuclear weapons proliferation. These technologies, if they are adopted overseas, would have to be developed in a manner which preserves the essential qualities of nuclear energy that have made it attractive to the United States and foreign countries; i.e., the ability to produce economical safe, reliable, and environmentally acceptable electric power to meet increasing energy demands in the world.

To achieve these dual goals, the Advanced Reactor Systems Program embodies two short-term objectives:

- To develop and demonstrate the technology that will permit the world's research reactors to operate with fuel materials that are not usable for weapons, and still meet all of the operator's research needs
- To develop alternative and innovative systems designs and operating techniques for fuel-cycle facilities that would increase the proliferation resistance of vulnerable operations and key process equipment

The objectives are responsive to the Administration's nonproliferation policies as well as the Nuclear Nonproliferation Act of 1978 enacted by the 95th Congress.

The objectives will be met by implementing subprograms on the following:

- Reduced Enrichment Research and Test Reactors, concentrating on developing, fabricating, testing, and installing fuel of reduced enrichment in research and test reactors which now use highly enriched uranium (HEU) fuels
- Proliferation Resistant Concepts, evaluating generic work on advanced safeguards and proliferation resistant engineering concepts developed by other progress within DOE for application to fuel cycle facilities

The upper-level WBS for the entire Fission Energy Program was shown in Figure IV 1-2. The further breakdown of the Advanced Reactor Systems Program is shown in Figure IV 3-24 with a matrix showing how the lower-

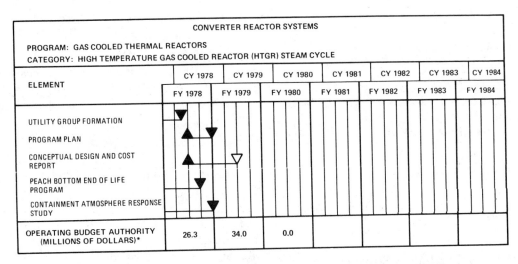

*THE FUNDING SHOWN HERE FOR FY 1978 AND FY 1979 ALSO COVERS WORK SHOWN AS PART OF THE HTGR CATEGORY (SEE SECTION III.B).

Figure IV-3-23. Major milestones and funding for the HTGR-SC.

level packages in the WBS relate to the arrangement of the program description in this section.

ASET is responsible for the Advanced Reactor Systems Program, which is assigned to the Director of Nuclear Energy Programs, who delegates management responsibility to the Director of the Nuclear Power Development Division, who, in turn, delegates program responsibility to the Chief of the Fuel Cycle R&D Branch. The branch chief provides overall direction and program approval, and is responsible for successful implementation of the program within overall cost and schedules.

Under the direction of the appropriate DOE field office, the national laboratories and contractors provide the day-to-day management of the individual projects, develop project management plans, and recommend the resources required to meet the program objectives. They have the direct responsibility for the performance of the program and meeting the milestones established.

The FY 1980 operating expense budget request for this program is $7.2 million, a $3.2 million increase over the FY 1979 effort included under the budget category of Fuel Cycle R&D. The increase reflects the transition of nonproliferation activities from a study mode to a hardware development mode. The funding breakdown is shown in Table IV 3-7,

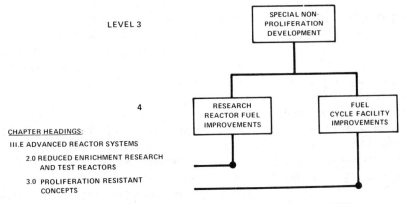

Figure IV-3-24. Special nonproliferation development program WBS.

Table IV-3-7. Advanced reactor systems program funding for FY 1978–1980.

ADVANCED REACTOR SYSTEMS	BUDGET AUTHORITY (DOLLARS IN THOUSANDS)			
CATEGORY	FY 1978	FY 1979	REQUEST FY 1980	INCREASE (DECREASE)
Reduced Enrichment Research and Test Reactors	0	0	$4,700	$ 700*
Proliferation Resistant Concepts	0	0	2,500	2,500
TOTAL	0	0	$7,200	$3,200

*The real increase is a net of $700 because this work was funded at $4,000 in FY 1979 under the Fuel Cycle R&D Program.

which summarizes the operating funding levels for the FY 1978 to FY 1980 period.

Reduced Enrichment Research and Test Reactors

Research reactors are used throughout the world for a variety of beneficial purposes related to energy development, medicine, industrial processes, and basic research. They range in size from very small (less than 1-MWt) to a few rather high-powered reactors (up to 30-MWt). Most research reactors are fueled with low uranium density (about 20 weight percent) uranium-aluminum alloys containing highly enriched (93 percent) U-235, a potential weapons material.

Materials Test Reactors (MTR's) have been operating for nearly three decades. Early missions for these MTR's required high-flux densities, stable and predictable neutron spectra, and reasonably long lifetimes for the replaceable fuel elements. The fuel fabrication technology developed for those early fuels was unique and expensive, so rather than undertake the expense of another development campaign, subsequent MTR-type reactors used the established fabrication technology and generic fuel design.

Many MTR-type reactors, having widely different missions and capabilities, use HEU contained in relatively long and thin envelopes (fuel plates) that are arranged lengthwise in a layered array that serves as a conduit for the water (usually) cooling medium that transfers the heat (released as the result of fissions) from the fuel assemblies (Figure IV 3-25). The amount of HEU in the fuel elements for a specific reactor was determined by the reactor's mission, core design, and planned operating schedule (e.g., high-power, high-flux reactors require fuel containing high concentrations of U-235 in fuel plates having relatively thin cross-sections so that heat transfer paths are short enough to control maximum temperatures of both fuel material and its cladding). Naturally, lower-power reactors with less demanding missions use lower concentrations of HEU and thicker fuel plates.

The TRIGA-type (test reactor and irradiation facility) reactors were developed as compact, safe, and effective research systems, using an altogether different fuel material than the MTR's, to serve a different range of missions. The TRIGA-type fuel contains a dense, stable rod of uranium-zirconium hydride enclosed within a stainless steel or incoloy tube (Figures IV 3-26 through IV 3-28). The amount of contained HEU and the fuel-rod diameter are determined by the fuel's mission (both neutron flux and heat removal requirements). The TRIGA fuel and reactor concepts were devised to facilitate pulsing the reactor between its rated steady-state power level and very high power levels (as much as 10,000-fold greater than the steady-state level) for short periods. (This pulsing capability is advantageous for specific missions but is not desired by all research reactor programs.) Naturally, the higher-power TRIGA-type reactor is fueled with an assembly containing a larger number of thinner-diameter fuel pins than a low-power TRIGA. Some of the very low-power TRIGA's are fueled for life by their initial fuel cores.

The technology exists (state-of-the-art) to permit most of the low-power research reac-

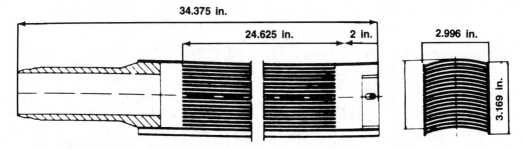

Figure IV-3-25. MTR-type reactor fuel element.

tors in the world to use higher density, low-enrichment (less than 20 percent U-235) fuel, and still meet virtually all their performance objectives. This same state-of-the-art technology can also be used to reduce the enrichment of many medium-power research reactors (e.g., 10- to 20-MWt) to about 45 percent U-235. This technology is based on the use of uranium aluminide (UAl_x–Al) fuel, which has been tested at uranium densities up to 45 percent in the Advanced Test reactor in Idaho, and on the use of uranium oxide dispersed in aluminum (U_3O_8–Al) that has been tested at uranium densities up to 35 percent in the High Flux Isotope Reactor at Oak Ridge. Confirmation is needed that these fuel systems can be used to reduce enrichment levels to 20 percent in low-power reactors, and 45 percent in medium-power reactors. This will be done by fabricating, installing, and testing two demonstration cores, one in a low-power research reactor and one in a medium-power system.

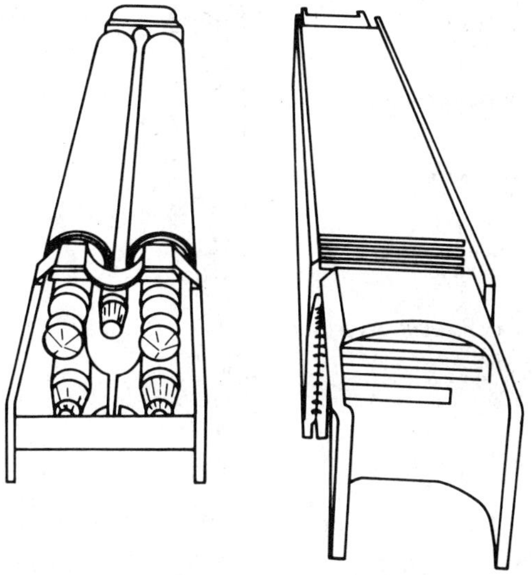

In the longer run, it is desirable to develop advanced fuels that will enable the U-235 enrichment to be lowered to 20 percent even in medium- to high-power reactors. The advanced fuel development is the second main thrust of the reduced enrichment research reactor program and involves consideration of several fuel forms including uranium-molybdenum and uranium silicide.

The principal program objective is to improve the proliferation resistance of nuclear fuels used in research and test reactors—in particular, to provide means for reducing the enrichment of uranium fuel used, in both foreign and domestic reactors, to substantially less than the current 90–93 percent enrichment, which will reduce the risk that enriched uranium might be diverted for non-peaceful purposes. More specifically, the program aims at the following:

- In the near-term, to apply current fuel fabrication technology to use reduced enrichment fuels. Enrichment can be reduced from 90–93 percent to 20–45 percent in many plate-type fuel designs without reducing the design margins, and to below 20 percent in some additional reactors (Table IV 3-8).
- In the long-term, to develop and apply high-uranium-density fuel technology that will permit the use of fuels enriched to less than 20 percent in all research and test reactors (except possibly only the highest power systems) and to pro-

Figure IV-3-26. TRIGA four-rod fuel cluster (left) and Standard MTR plate-type element (right).

620 Fission Energy

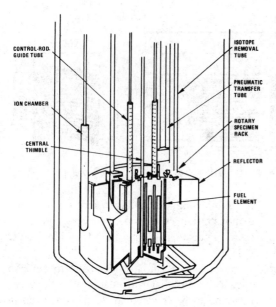

Figure IV-3-27. TRIGA Mark-I core and reflector assembly.

ment fuels in new and existing research and test reactors in the United States and abroad:

- Minimize any reactor performance reduction (neutron flux per unit power) relative to highly enriched fuel currently used in existing reactors or, for new reactors, relative to highly enriched fuel designs typically used for similar reactors of the same power level
- Minimize any fuel-cycle cost increases relative to operation with highly enriched uranium fuel
- Minimize the number of new safety and licensing issues raised by the conversion to reduced-enrichment fuel
- For existing reactors, minimize the pos-

vide the technical support necessary to make the high-uranium-density fuels commercially available.

A controlling objective is to demonstrate to the users and operators of research and test reactors throughout the world the operation of such devices with reduced uranium enrichment fuels meets all the required criteria of reliability, performance, safety, core on fuel elements fabricated by commercial companies, possibly from various countries participating in a cooperative enrichment-reduction effort.

Strategy and scope. The basic strategy of the Reduced Enrichment Research and Test Reactor (RERTR) Program is to develop, fabricate, test, and install fuel of reduced enrichment in research and test reactors now requiring HEU fuels. The primary technological means to be used to achieve the objective consists of increasing the uranium density in the fuel beyond that currently used in research and test reactor fuels, up to presently demonstrated maximum uranium loadings.

Because using reduced enrichment fuels might result in some performance and cost penalties, guidelines are being followed for the practical acceptability of reduced enrich-

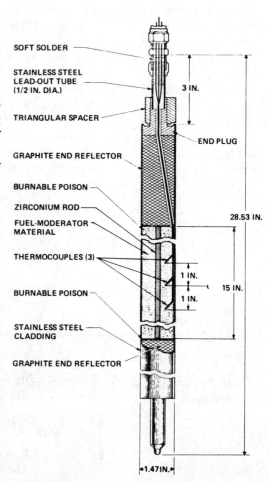

Figure IV-3-28. TRIGA instrumented fuel element.

Table IV-3-8. Estimated potential near-term reduction in number* of research reactors using highly enriched uranium.

REACTOR TYPE AND POWER	PRESENT ENRICHMENT 93%		NEAR-TERM POTENTIAL CONVERSIONS ENRICHMENT					
			<20%		45%		93%	
In the U.S.								
MTR-type								
Greater than 15-MWt	6	(645)*	—		1	(30)	5	(615)
5 to 15-MWt	7	(52)	—		5	(38)	2	(15)
1 to 5-MWt	8	(14)	7	(12)	1	(2)	—	
0.01 to 1-MWt	15	(1)	15	(1)	—		—	
TRIGA-type	5	(6)	5	(6)	—		—	
Other types	—		—		—		—	
Subtotal	41	(719)	27	(19)	7	(70)	7	(630)
In other countries								
MTR-type								
Greater than 15-MWt	8	(320)	2	(100)	6	(220)	—	
5 to 15-MWt	21	(148)	3	(29)	18	(119)	—	
1 to 5-MWt	12	(22)	12	(22)	—		—	
0.01 to 1-MWt	43	(3)	43	(3)	—		—	
TRIGA-type	4	(22)	4	(22)	—		—	
Other types	12	(480)	—		8	(160)	4	(320)
Subtotal	100	(995)	64	(176)	32	(499)	4	(320)
TOTAL	141	(1714)	91	(195)	39	(569)	11	(950)

*With thermal megawatts (MWt) shown in parentheses.

sible needs for reactor and facility modifications

The Research Reactor Fuel Program consists of three subprograms:

- State-of-the-art fuel fabrication, and demonstration is intended to establish that the operation of research reactors with reduced-enrichment uranium fuels will meet all existing criteria of reliability, performance, safety, core lifetime, and economics. The planned fuel demonstration activities include: (a) a near-term accelerated irradiation test in a high-flux facility of some elements of each relevant fuel type beyond its normal life or burnup limit; (b) a near-term whole-core demonstration of low-enrichment uranium fuel in a reactor in which detailed physics measurements can be made to assess and calibrate the neutronics and safety characteristics of the core; (c) long-term irradiation tests of the new fuel elements, and (d) a long-term whole-core demonstration of intermediate enrichment uranium fuel in a reactor appropriate for the adequate study of the physics and safety characteristics of the core throughout the entire fuel burnup cycle.
- Advanced fuel development, test, and demonstration (about a 7-year effort) is intended to yield fabrication techniques for research and test reactor fuels of significantly higher uranium density than is achievable today.
- Support activities, including generic reactor analyses and design, consultation on specific research reactor fuel export requests, interaction with individual research reactor operators, and planning of fuel commercialization.

The study of each export request will consider the reactor's potential for conversion to reduced enrichment fuel and will provide the Executive Branch with a series of options taking into account the pertinent trade-offs—experiment performance (or mission capability), core lifetime, fuel-cycle economics, and licensing issues.

Application of reduced enrichment replacement fuel to specific foreign and domestic reactors will be expedited by providing technical support and close cooperation with the reactor operating organization and fuel manufacturer. Thus, the work involves interaction with reactor operations, the engineering design of fuel elements, fuel procurement specifications, component modifications, and safety analysis revisions.

The intent of the commercialization planning effort is to ensure that the reduced enrichment fuels become commercially available on a worldwide basis without the need for significant continuing government financial support. The work will involve the identification of potential fuel vendors and support for technology transfer. Achievement of full international fuel commercialization will require political, economic, and diplomatic cooperation from other government entities such as the State Department and ACDA.

The FY 1979 fuel-fabrication development work involves screening to determine the limits of fabricability. Test samples of plate-type fuels will be fabricated to span the region between current and maximum fabricable loadings. The samples will be subjected to metallographic studies, temperature stability experiments, blister tests, U-ZrH hydrogen pressure measurements (where applicable), and corrosion tests. High-power and transient screening of rod-type elements will also be initiated.

In FY 1980 and 1981, extensive irradiation testing will cover the expected fuel-burnup conditions and exceed them in some cases, up to conditions of clearly measurable effects. Post-irradiation examinations will be completed.

During FY 1982 and 1983, the fuel development activity will be concluded with the completion of the U_3O_8–Al dispersion and the very high-density fuel mini-plate irradiations, post-irradiation examinations, and selection of candidates for the very high uranium density fuel demonstration during the following two years. International participation and cooperation in this program will be sought.

In the fuel demonstration program, the near-term work includes three parallel efforts. Two concern the confirmation of the capabilities of reduced-enrichment, high-uranium loadings for plate-type UAl_x–Al and U_3O_8–Al fuel elements. The third effort concerns the feasibility of reduced-enrichment, high-uranium loadings for rod-type U-ZrH fuel elements. These three demonstrations involve simple extensions of fuels that are currently used in research and test reactors; i.e., fuel elements that provide increased uranium loadings but without significant changes in the outer dimensions of current fuel-element designs.

The long-term work, directed toward advanced fuels, consists of four parallel efforts. The first three will extend the UAl_x–Al, U_3O_8–Al, and U-ZrH fuels technology considered in the near-term objectives. The fourth effort will develop very high density fuels, such as uranium silicide (U_3Si), U-Mo alloy, and, possibly, UO_2. Considerable fabrication and testing experience exists for these fuels, which were developed for other reactor applications, and much of this experience can be applied directly here. The Oak Ridge Research Reactor will be used for most of the accelerated burnup testing of the prototype fuel materials and assemblies; other reactor facilities will be selected depending on the tasks involved. The first full core of low-enrichment uranium plate-type fuel will be installed in the Ford Nuclear Reactor at the University of Michigan, and its performance documented. A foreign site for potential demonstration of an intermediate-enrichment uranium (about 45 percent U-235) core was determined in FY 1979. Reactors will be selected, when the fuel development warrants, for future demonstrations of promising advanced fuels.

Post-irradiation examination of reduced enrichment fuel elements will be conducted at appropriate facilities, either national laboratories or DOE contractor facilities.

DOE's NPD is the focal point with regard to the formulation and provision of overall programmatic guidance to the DOE Chicago Operations Office. NPD is also the principal division concerned with the establishment and development of intergovernmental interfaces. CH is the focal point for implementation of the program in accordance with broad guidance provided by NPD. CH is also responsible for the interfaces with all participating DOE field offices and CH prime contractors. Programmatic responsibility has been assigned to the Argonne National Laboratory (ANL). ANL interacts directly with participating DOE and private industrial contractors and university participants concerning definition, development, and progress of their assignments within the technical scope of their contracts.

The ANL Applied Physics Division has studied and documented the capabilities and limitations of current fuel fabrication technology. National laboratories and DOE contractors have been assigned to appropriate program activities.

Major milestones and funding are shown in Figure IV 3-29.

The participating subcontractors will be responsible for executing the portions of the program that are assigned to them. Subcontractors will work directly with ANL in the development of scopes, schedules, and budget recommendations for proposals. They will continuously advise ANL of technical problems and proposed solutions, and will routinely report directly to ANL and their respective field offices on the progress, cost, schedule status, and milestone achievements for their assigned tasks.

Fabrication technology development responsibilities have been assigned to EG&G Idaho, GA, and ANL's Materials Science Division for UAl_x–Al, U-ZrH, U_3O_8–Al and advanced fuels, respectively. Commercial production of prototype fuel elements and of full cores for higher power applications will be provided through competitive bids from commercial fabricators. As soon as adequate irradiation experience is accumulated to confirm fuel performance potential, commercial fuel vendors will be provided with access to, and assured transfer of, pertinent fuel fabrication technology to assure future availability of the resulting reduced enrichment fuels at costs acceptable to the reactor operators.

Proliferation Resistant Concepts

Nonproliferation criteria for key fuel cycle facilities are being defined in conjunction with

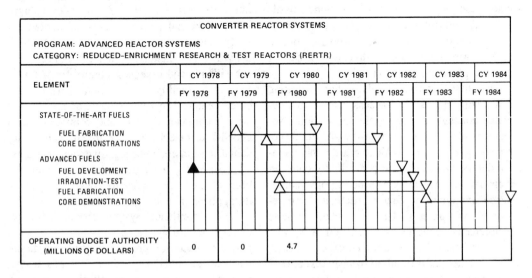

Figure IV-3-29. Major milestones and funding for RERTR.

NASAP. It is becoming increasingly apparent that the key determinants relate not so much to the fuel cycle concepts being evaluated, but rather to the specific equipment and systems designs for the fuel cycle facilities involved. This program provides for a new initiative to investigate the sensitivity of the proliferation resistance of facilities to the engineering of both critical equipment and process control.

The principal program objective is to develop alternative and innovative systems designs and operating techniques for fuel cycle facilities that would increase the proliferation resistance of vulnerable operations and key process equipment. This will allow refinement of proliferation resistance criteria to reflect realistic targets for preventing material diversion.

These activities draw on generic work on advanced safeguards and proliferation resistant engineering concepts developed in other programs within DOE and evaluate them in relation to their application to fuel cycle facilities. Such concepts include, for example, completely automated and computerized process control that can be remotely monitored.

The program consists of three principal activities that focus on the definition of inplant systems that could significantly increase the diversion resistance of fuel cycle operations by further reducing human access to process equipment and operations involving fissile material and by improving the capability to remotely monitor operations to obtain conclusive information regarding diversion threats or convert plant modification.

The first activity involves demonstrating and testing a near-zero-access remote maintenance concept for fuel cycle application. Advancements in the state-of-the-art of force-reflecting electromechanical master-slave manipulators now make it possible to consider their use in performing routine remote maintenance operations in fuel reprocessing and refabrication plants. The advanced remote maintenance concept is based on a special characteristic of the electromechanical master-slave manipulator: in addition to its exeptional dexterity, the slave arms are completely separable from the master arms and can be remotely located or mounted on a transportable carriage, which eliminates the need for routine human access to the process area and reduces to a minimum the penetrations of the process-canyon enclosure. The slave arms are self-powered and are controlled completely by electrical signals generated by motion of the master arms. The signals are communicated to the slave arms either by electrical cable connection or by radio or laser beam transmission.

The second activity involves development of a robotized process-stream sampler. In contrast with typical remotely operated process systems, in which stream samples for analysis are withdrawn through sample lines from the process vessels to shielded sampling stations outside the process enclosure (thus these lines are readily accessible through the sampling station and represent a finite fissile material diversion vulnerability), the advanced concept would have sample lines from process vessels terminated within the process enclosure. A robotized sampler that would withdraw a stream sample into a capsule and deliver it outside to an analytical facility. Such a system would virtually eliminate the potential for direct withdrawal of weapons-significant quantities of fissile material through sample lines.

The last activity involves extending distributive process-control systems to incorporate the surveillance of diversion-attempt information and passive use-denial actions. A study undertaken as a part of the NASAP activities has defined a system concept for surveillance and diversion control for use in a fuel cycle facility in conjunction with a secured center. The key elements of the concept are a command, control, and communications network to detect diversion threats and a use-denial system for aborting diversion attempts. The concept is of sufficient interest and has been defined adequately to consider seriously its application and demonstration.

The surveillance system provides for remotely monitoring: (1) process instrumentation normally used for process control and fissile material accountability; (2) safeguards instrumentation dedicated to the physical protection of the facilities; and (3) added in-plant instrumentation to indicate the state of the plant (valve positions, tank levels, pump condition, pipe flows, etc.) that are normally inactive during process operations. Visual and other forms of monitoring of sensitive process equipment and piping are also provided for. A computer would continuously analyze the data from the surveillance system to detect abnormal situations that might be interpreted as diversion threats. A data discrimination system will be devised to distinguish abnormal situations from signal noise, process transients, instrument error, system malfunction, and other spurious data.

A practical application of the surveillance system will be demonstrated in an existing process line through the extension of the in-plant distributive control system. A test of the completed system will be made during a planned operating campaign.

As a result of studies conducted at ORNL during FY 1979, a commercially available candidate manipulator has been tentatively identified for further development for the remote maintenance concept demonstration. In FY 1980, a detailed program plan will be developed for the complete demonstration and testing of the remote maintenance concept. A current-model manipulator will be purchased for endurance testing and design and fabrication will be started on new components.

Preliminary design studies have been made at ORNL on a robotized process-stream sampler. Design and fabrication of a development model will be initiated in FY 1980. A detailed program plan, to include mockup testing and demonstration of the full sampling system, will be prepared.

As a result of FY 1979 scoping studies conducted by ORNL, an interrogation logic has been defined for application to a typical solvent extraction operation in a reprocessing plant. Process monitor signals needed for a surveillance system have also been identified. In FY 1980, a test of the system will be

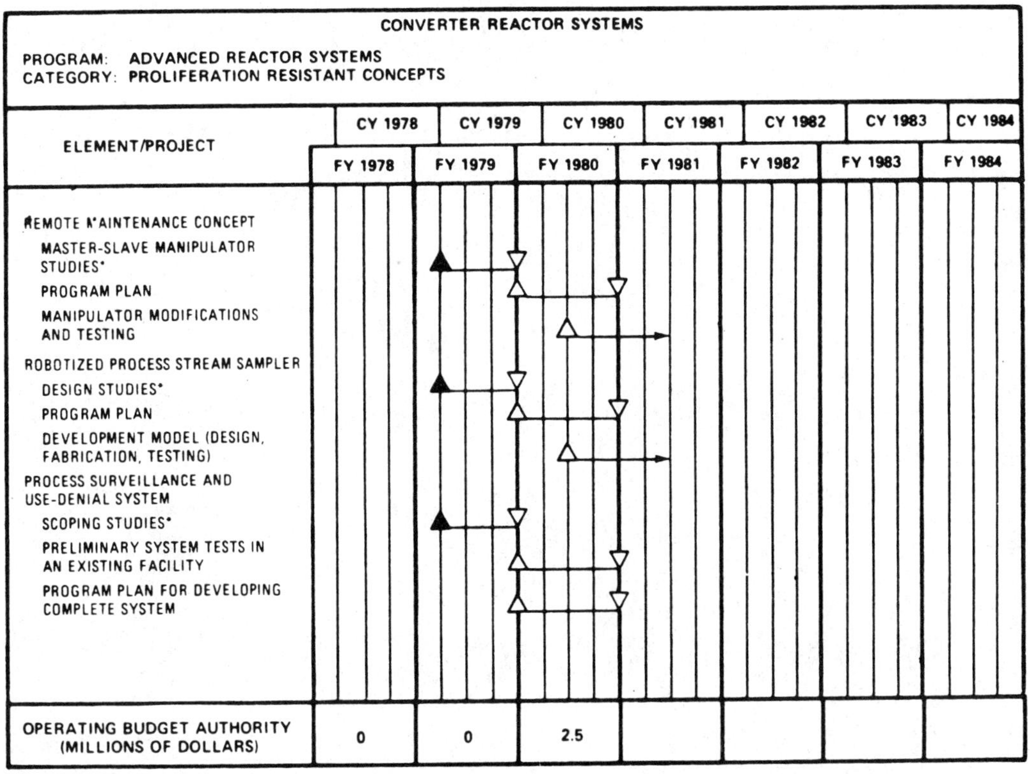

*These studies were conducted under the Nuclear Fuel Cycle Program.

Figure IV-3-30. Major milestones and funding for proliferation resistant concepts.

initiated using a planned operating campaign of the thorium/U-233 process line in Bldg. 3019 at the ORNL as a basis. The required additional instrumentation will be added and the existing computer complex will be used for data handling. A program plan for the development of a complete surveillance system and attendant use-denial system also will be prepared in FY 1980.

Major milestones and funding are presented in Figure IV 3-30.

These activities will be conducted by ORNL with the assistance of industrial subcontractors to be selected.

4.
Breeder Reactor Systems

The motivation for developing breeder reactors is the prospect of extending the uranium resource base to a practically inexhaustible source of electrical energy. Without the breeder, exhaustion of the uranium supply would ultimately truncate the nuclear fission option, which now relies on LWR's. Introduction of the breeder would permit virtually complete use of fertile uranium and thorium resources for the generation of electrical power.

The timing of the need for the breeder is governed principally by three factors:

- The magnitude of the uranium resource base
- The future nuclear-electric demand growth
- Relative economics

The uranium-resource constraint on the ability of LWR's to supply future nuclear-electric load growth establishes the latest date when a breeder would be needed. However, there could be commercial advantages in earlier breeder introduction.

This chapter described the programs for developing the Liquid Metal Fast Breeder Reactor (LMFBR), the Water Cooled Breeder (WCB), and the Gas Cooled Fast Reactor (GCFR), as well as their fuel cycles.

OVERVIEW

The U.S. interest in breeder reactors dates back to the Manhattan Engineering District in the 1940s when the possibility of breeding was first recognized by pioneers in the nuclear field. One of the earliest steps in this program was the construction of an experimental fast reactor called Clementine at LASL. It was used from 1946 to 1953 to demonstrate the feasibility of operating a reactor with fast neutrons, plutonium fuel, and a liquid-metal coolant. The Experimental Breeder Reactor No. I (EBR-I) in Idaho, which produced the first nuclear-generated electric power in 1951, was built and operated by ANL through December 1963 to prove the feasibility of breeding and to demonstrate the technology of liquid metal coolants.

These developments led to the construction of two fast reactors in the mid-1960s—the Experimental Breeder Reactor No. II (EBR-II) in Idaho and the Enrico Fermi Atomic Power Plant in Michigan. The EBR-II was built as a 20-MWe demonstration breeder facility and included the demonstration of onsite spent-fuel reprocessing and remote refabrication. EBR-II was later converted to a fuel test facility and is currently operating in that capacity. The Enrico Fermi plant was a utility-constructed commercial prototype reactor system. It experienced an accidental partial meltdown of its core in 1968, resumed operation in 1971, and was subsequently decommissioned for economic reasons.

In 1962, the Atomic Energy Commission reported the status of the Civilian Nuclear Power Program to the President. The report stated:

The future program should include the vigorous development and timely introduction of improved converters and especially of economic breeders; the latter are essential to long-range major use of nuclear energy.

A 1967 supplement to the 1962 report set forth the changes that had taken place in the interim, and reported that the work on the LMFBR had been given priority status while the effort on alternative-coolant fast breeder reactors was being maintained at a constant level. Thus, much of the development effort on the breeder reactor in this country, as in other countries, focused on the liquid-metal-cooled fast breeder. As an alternative, effort was initiated on the gas-cooled fast breeder reactor in 1963.

In April, 1977, the Administration announced a change in the Nation's nuclear energy policy. Concern about the potential for the proliferation of nuclear weapons capability emerged as a major issue in considering whether to proceed with the development, demonstratiion and eventual deployment of breeder-reactor energy systems. Plutonium recycle and the commercialization of the fast breeder were deferred indefinitely. This led to a reorientation of the nuclear fuel cycle program which was previously directed toward the commercialization of fuel reprocessing and plutonium recycle to the investigation of a full range of alternative fuel cycle technologies.

Two major system evaluation programs, the Nonproliferation Alternative Systems Assessment Program, which is domestic, and the International Nuclear Fuel Cycle Evaluation, which is international, are assessing the nonproliferation advantages and other characteristics of advanced reactor concepts and fuel cycles. These evaluations will allow a decision in 1981 on the future direction of the breeder program. In the interim, the technologies of three breeder-reactor concepts are being developed: LMFBR, WCB, and GCFR.

The interim goals of the Breeder Reactor Systems Program are as follows:

- To maintain and improve technical and engineering bases to enable selecting a breeder system consistent with U.S. nonproliferation objectives and anticipated national electric-energy requirements
- To provide technical support to the Administration's foreign policy initiatives on nuclear nonproliferation

Supplementary to these generic goals are the principal goals of each of the breeder programs:

- LMFBR—maintain the capability to commit to a breeder option through a strong R&D program, and investigate alternative fuels and fuel cycles that might offer nonproliferation advantages
- WCB—confirm that breeding can be achieved in existing and future LWR systems using the thorium/uranium-233 fuel system
- GCFR—provide a viable alternative to the LMFBR that will be consistent with the developing U.S. nonproliferation policy, and provide GCFR technology and other needed support on a schedule consistent with an internationally developed cooperative program.
- Fuel cycle—develop a reprocessing technology base emphasizing fast breeder systems but generic to all reactor types, and provide analytical support to the LWBR proof-of-breeding experiment.

Strategy and scope. The strategy of the Breeder Reactor Systems Program is to maintain the capability to commit to a breeder-reactor option in the future through strong and broad-based R&D programs on the technologies of three types of breeder reactors: LMFBR, WCB, and GCFR, as well as of their fuel cycles.

Strategies for the LMFBR are as follows:

- as a focal point, the initiation of a 30-month Conceptual Design Study (CDS) ending in March 1981
- continuation of the base technology program
- intensified search for technical alternatives to the conventional uranium/plutonium breeder fuel cycle through NASAP and INFCE

WCB strategies are as follows:

- continued operation and testing of the Shippingport Atomic Power Station
- development, manufacture, operation, examination, and evaluation of the Shippingport LWBR core
- development and dissemination of technical information on the LWBR concept

The following strategies have been developed for the GCFR:

- as a focal point, the initiation of the Program Definition and Licensing Phase for the supporting gas-cooled technology base activities
- support by private industry to the international cooperative program

Fuel cycle strategy involves development of processes applicable to the various fuel cycles (U/Pu, Th/U, LWR, LMFBR, etc.) to the maximum extent possible.

Funding levels for breeder reactor systems for FY 1978–1980 are listed in Table IV 4-1.

LIQUID METAL FAST BREEDER REACTOR

The LMFBR has had priority due to the excellent heat-transfer properties of sodium that enable the LMFBR to operate at high temperature without resort to pressurized systems. The practical result is a plant with high thermal efficiency—40 percent for an LMFBR as compared with 32 percent for LWR's.

Figure IV 4-1 shows the major heat-transport systems in an LMFBR power plant. The flowing sodium removes the heat produced in the fissioning fuel in the reactor core, transfers the heat to a second sodium piping loop in the heat exchanger, then returns to the reactor. The heat in the secondary sodium loop passes through a steam generator where it turns water into steam to rotate the turbine generator and produce electric power.

The LMFBR is a "breeder" because it produces more nuclear fuel than it consumes. This means that it not only produces enough fuel for continued operation, but also generates additional fuel for new reactors.

The LMFBR produces fuel by converting uranium-238 (a fertile material), which is not directly usable as fuel, into plutonium-239 (a fissile material), which is. This is especially significant since uranium-238 constitutes more than 99 percent of naturally occurring uranium. Current water-cooled nuclear power plants can only efficiently use uranium-235, which constitutes less than 1 percent of naturally occurring uranium. Therefore, the breeder makes available 60 to 79 percent of the energy potential or uranium—not just the 1 percent from uranium-235.

Of course, the LMFBR can operate on a variety of fuel cycles, determined largely by the choice of fertile material and fissile material in the fuel. The fuel is in the form of pellets that are sealed in tubes to form the fuel elements. The fuel elements are grouped into assemblies for leading into the reactor core.

Overview. The LMFBR program plan in effect in January 1977 prescribed a decision on LMFBR deployment in 1986; and pending an affirmative decision in 1986, was scoped technically and logistically to support deployment beginning in the early 1990s with a buildup to about 30-GWe by the year 2000.

Although a commitment to commercial deployment was not a part of the previous Administration's policy, planning included the contingency for such a decision in 1986. For planning the scope and pace of RD&D program, the following reference assumptions were made in 1976:

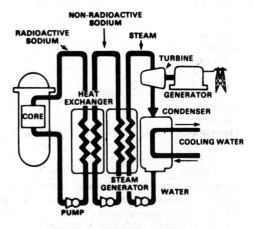

Figure IV-4-1. LMFBR heat-transport system.

- The total potential urnaium resource base is 3.7-million short tons U_3O_8.
- The number of LWR's in the United States would grow such as to reach a total nuclear capacity of 625-GWe in the year 2000.

If the fission option were to have a role as an essentially inexhaustible energy resource in the nation's energy economy, these two assumptions established the requirement to commence on-line deployment of breeder reactors in the early 1990s.

Since 1976, two factors were influential in the redirecting decision to slow the LMFBR Program:

- A reassessment of the timing of the need for the breeder
- Increased concern for the proliferation issue

The timing reassessment is based on substantial reductions in projected demand for nuclear generating capacity: the current projection for the year 2000 is 255–395-GWe. In 1971, when it was decided to build the Clinch River Breeder Reactor Plant, the year 2000 nuclear capacity was projected to be 900–1200-GWe. In addition, the estimated uranium resource base (reserves and probable resources) has been revised upward from 2.2 to 2.4 million short tons of $50 forward cost reserves and probable resources.

With more time available, the United States can pause for an examination of alternative institutional and technical non-proliferation strategies while maintaining the breeder option. Thus, the President, in April 1977, deferred indefinitely the commercialization the LMFBR Program and conventional fuel reprocessing.

Table IV-4-1. Funding for breeder reactor systems for FY 1978–1980.

BREEDER REACTOR SYSTEMS PROGRAM	BUDGET AUTHORITY (DOLLARS IN THOUSANDS)			
	FY 1978	FY 1979	REQUEST FY 1980	INCREASE (DECREASE)
LMFBR				
Operating	$412,805	$460,001	$348,900	($111,101)
Equipment	29,950	25,181	22,000	(3,181)
Construction	121,305	81,485	91,100	9,615
Subtotal	564,060	566,667	462,000*	(104,667)
WCB				
Operating	38,600	50,900	57,900	7,000
Equipment	3,100	3,100	2,100	(1,000)
Construction	0	9,000	0	(9,000)
Subtotal	41,700	63,000	60,000	(3,000)
GCBR				
Operating	14,670	24,500	24,500	0
Equipment	1,330	1,500	1,500	0
Construction	0	0	0	0
Subtotal	16,000	26,000	26,000	0
Fuel Cycle Research and Development				
Operating	91,200	66,850	21,100	(45,750)
Equipment	9,000	2,000	2,500	500
Construction	3,000	6,750	6,400)	(350)
Subtotal	103,200	75,600	30,000	(45,600)
TOTAL	$724,960	$731,267	$578,000	($153,267)

*LMFBR proposed composite budget authority is $504,000 in FY 1980 as follows:

LMFBR program as itemized above (includes operating, capital equipment, and construction)	$462,000
LMFBR fuel cycle research and development (includes operating, capital equipment, and construction, but excludes $3-million for WCB Proof of Breeding)	27,000
NRC fast reactor research (program submitted in the NRC budget)	15,000
	$504,000

The objectives of the LMFBR Program are as follows:

- To maintain the capability to commit to a breeder option through a strong R&D program
- To investigate alternative fuels and fuel cycles that might offer nonproliferation advantages

Prior to 1977, the central feature of the program for the development and demonstration of the LMFBR was the scale-up of LMFBR power-plant and fuel-cycle technologies to unit capacities and performance levels suitable for commercial application. This technology scale-up was to be achieved through a sequence of power-plant and fuel-cycle demonstration projects. Power-plant demonstration projects included the Fast Flux Test Facility (FFTF), Clinch River Breeder Reactor Plant (CRBRP), and Prototype Large Breeder Reactor (PLBR). Fuel-cycle demonstration projects included High Performance Fuel Laboratory (HPFL) and Hot Pilot Plant (HPP). Breeder technology development and facilities were tied explicitly to requirements of these projects; in that way, project objectives and schedules drove most of the LMFBR Program.

Since 1977, the LMFBR Program has been redirected to emphasize alternative fuels and fuel cycles that might offer strategy nonproliferation advantages. The Administration has proposed that CRBRP be discontinued except for design of those systems representing an advancement in the art and testing of key components. It is the Administration's objective that appropriate legislation for discontinuance be enacted as early as possible.

The redirected program intensifies the search for technical alternatives to the conventional uranium/plutonium breeder fuel cycle through NASAP. Also, to build an international consensus on the proliferation issues, INFCE has been initiated to study the form and management of the nuclear fuel cycle.

As a focal point for the LMFBR Program, it was proposed that a CDS be initiated to develop a conceptual design for the next developmental plant, if any. The objectives of this plant would be as follows:

- To maximize strategic nonproliferation value, that is, it should further U.S. nonproliferation objectives
- To build on CRBRP, FFTF and PLBR technology and incorporate changes to improve reliability, reduce capital costs, and improve the fuel performance and breeding while assuring safety
- To be of a capacity significantly closer to commercial size than CRBRP

The President has affirmed his policy of maintaining a breeder reactor program option through a strong base technology program. Thus, the program stresses the development of breeder technology in the areas of fast reactor components, fuels and core design, safety, physics, and materials and chemistry.

Another principal feature, then, of the LMFBR Program is a 30-month CDS ending with a conceptual design in March 1981. The CDS will benefit from, and reflect results of, the nonproliferation assessment studies. Thus, by identifying a set of plant design parameters, the CDS will provide a focus for technology development and will culminate in a report to Congress in 1981 treating estimated costs and schedules, specific design aspects including heat transport system concept, core and fuel design, fuel cycle and related proliferation and safeguards strategy.

The CDS, together with the nonproliferation studies (NASAP and INFCE), the uranium and thorium resource assessment program (NURE), and continued achievements in the base R&D program will provide the basis for a decision in 1981 as to the future fast breeder reactor role and development strategy, including the scope and pact of the LMFBR Program.

The adequacy and relevance of the base program is measured by relating base-technology outputs and schedules to the requirements of a hypothetical developmental plant whose commitment could possibly be an outcome of the 1981 decision point. The technology being developed in the redirected breeder program will ensure readiness if a decision is made as early as 1981 to initiate work on a developmental plant using an advanced high-gain driver fuel and an optional breeding blanket (U-238 or thorium).

It should be emphasized that a commitment now to a research and development plant is not part of the LMFBR Program redirection. Such a decision must await the outcome of the assessments, but this will be one of the options available to the President and Congress in 1981. If the decision is made to proceed, detailed engineering design (Title I and II) of a research and development plant could commence following the CDS as soon as funding is made available.

The LMFBR Program is composed of plant projects, breeder technology, and test facilities. Plant projects include the FFTF, CRBRP, and the CDS. R&D efforts in components, fuel and core design, safety, physics, and materials and chemistry compose the breeder technology element. The third element, test facilities, covers the operation and maintenance of the facilities required by the projects and breeder technology efforts.

The upper-level WBS for the entire Fission Energy Program was shown in Figure IV 1-2. The further breakdown of the LMFBR Program is shown in Figure IV 4-2 with a matrix showing the relationship between the lower-level packages in the WBS and the descriptive arrangement of the program material in this section.

The Assistant Secretary for Energy Technology is responsible for the LMFBR Program. This program is assigned to ETN, who delegates management responsibility to the Director of Reactor Research and Technology (RRT) Division. The RRT Director provides overall direction and program approval, and is responsible for successful implementation of the program within overall cost and schedules.

Under the direction of the appropriate DOE field office, the national laboratories and contractors provide the day-to-day management of the individual projects, develop project management plans, and recommend the resources required to meet the program objectives. They have the direct responsibility for the performance of the program and meeting the milestones established.

The major contractors in the LMFBR Program and their main areas of involvement are listed in Table IV 4-2.

Table IV 4-3 shows the operating funding levels for the LMFBR Program for the FY 1978 to FY 1980 period.

The overall LMFBR Program schedule is shown in Figure IV 4-3. This milestone chart indicates the close interaction among elements of the program and the support available from the technology effort for the R&D plant if the decision in 1981 is to proceed.

Plant Projects. The power plant developmental projects are FFTF, CRBRP, and CDS. These are major program projects that have clearly defined objectives and firmly established schedules. Plant projects are designed and constructed to provide the central focus for the research and development program and to provide technical information for the LMFBR concept.

The FFTF is a reactor being constructed in Hanford, Washington, at the Hanford En-

Table IV-4-2. Contractors and their areas of involvement in the LMFBR program.

	CONTRACTORS	PROJECTS
Reactor Manufacturers	General Electric	CRBRP, Components, CDS
	Westinghouse	CRBRP, Components, CDS
	Atomics International	CRBRP, Components, CDS, Technology
		Components, CDS
	Combustion Engineering	Components, CDS
	Babcock & Wilcox	
National Laboratories	ANL	Fuels, Safety, Components, Physics
	LASL	Fuels
	ORNL	Materials, Physics
Engineering Laboratories	INEL	Safety
	Hanford Engineering Development Laboratory	FFTF, Fuels, Components, Physics
	Energy Technology Engineering Center	Component Testing

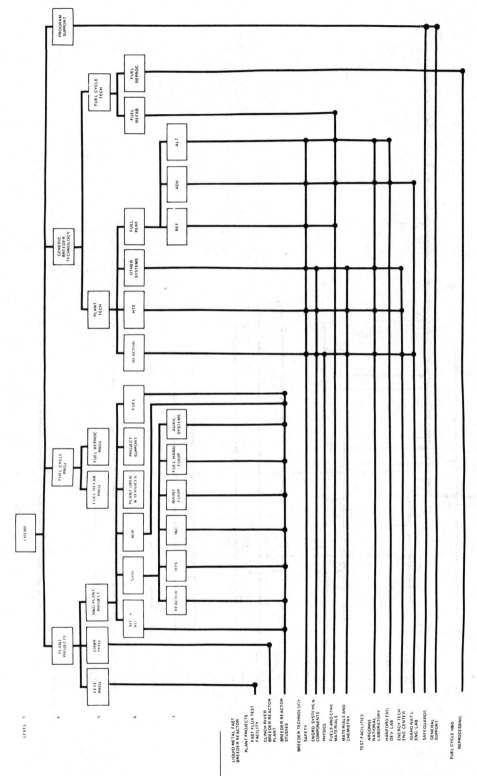

Figure IV-4-2. LMFBR program WBS.

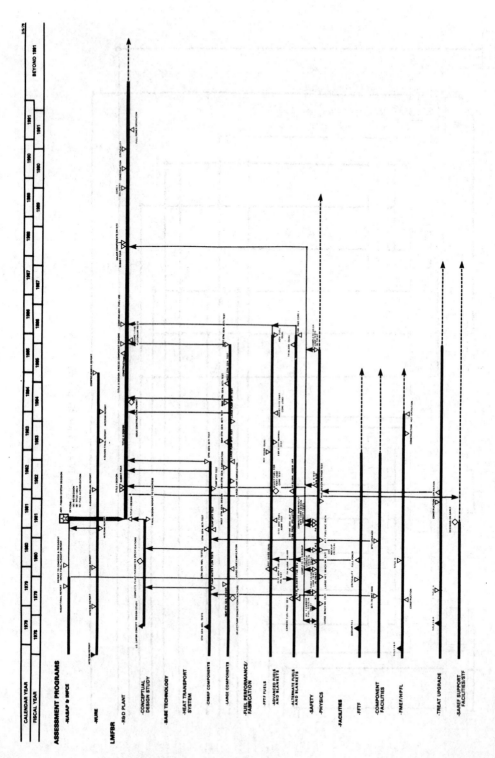

Figure IV-4-3. Summary of overall LMFBR program schedule.

Table IV-4-3. LMFBR program funding for FY 1978–1980.

LIQUID METAL FAST BREEDER REACTOR	BUDGET AUTHORITY (DOLLARS IN THOUSANDS)			
PROGRAM ELEMENT	FY 1978	FY 1979	REQUEST FY 1980	INCREASE (DECREASE)
Plant Projects	$159,254	$253,200	$131,200	($122,000)
Breeder Technology	178,223	136,801	142,400	5,599
Test Facilities	75,328	70,000	75,300	5,300
TOTAL	$412,805	$460,001	$348,900	($111,101)

gineering Development Laboratory (HEDL) as a fuels and materials test facility. The CRBRP is a joint government-industry effort on a 380 gross megawatt electrical demonstration breeder reactor power plant at Oak Ridge, Tennessee. The effort under the breeder reactor studies is a CDS of an improved LMFBR research and development plant.

Table IV 4-4 shows the operating funding levels by project for the FY 1978 to FY 1980 period.

FFTF Program. A high-priority part of the Fast Breeder Reactor Program is the completion of the FFTF and its subsequent operation. The FFTF is a 400-thermal-megawatt sodium-cooled fast-neutron-flux reactor designed specifically for irradiation testing of nuclear reactor fuels and materials for LMFRB's (see Figure IV 4-4). Construction is being completed at HEDL in Richland, Washington.

FFTF will serve as an advanced fuels and materials irradiation facility for Fast Breeder Reactor Program and as a baseline for scaling up systems and components. It has the capability of serving as an advanced fast-flux test bed for alternative fuels and materials experiments needed to continue DOE advanced reactor development programs.

Although FFTF, as a testing reactor, was not designed to breed or to produce electricity, it has provided and will continue to provide valuable information to follow-on LMFBR projects and base-technology programs in the areas of plant system and component design, component fabrication, prototype testing, and site construction.

In fulfilling its prime role of providing a test bed for demonstrating and evaluating the performance of future LMFBR-plant fuel assemblies and core designs at reference conditions, the FFTF will be used for the following:

- To test fuel elements up to and including failure under dynamic sodium-flow conditions to establish ultimate capability and failure modes; an understanding of failure modes is essential to establish LMFBR core safety, reliability, performance, and lifetime
- In tests to develop the advanced fuels (oxide, carbide, and nitride) and the advanced cladding and duct materials essential to attaining operational breeding ratios in the 1.25 to 1.45 range with fuel-doubling times of less than 15 years
- As a prototype irradiation test bed for

Table IV-4-4. Plant projects funding for FY 1978–1980.

PLANT PROJECTS	BUDGET AUTHORITY (DOLLARS IN THOUSANDS)			
PROGRAM	FY 1978	FY 1979	REQUEST FY 1980	INCREASE (DECREASE)
Fast Flux Test Facility Program	$ 79,254	$ 65,800	$ 76,200	$ 10,400
Clinch River Breeder Reactor Project	80,000	172,400	0	(172,400)
Breeder Reactor Studies	0	15,000	55,000	40,000
TOTAL	$159,254	$253,200	$131,200	($122,000)

Figure IV-4-4. Fast Flux test facility.

various fast reactor fuel and blanket materials considered in the nonproliferation fuel cycle studies
- To obtain experience in the operation of a LMFBR having coolant loops and components at temperatures and coolant flows typical of large LMFBR power plants

For controlled and instrumented fast-flux conditions for testing fuel specimens, rods, subassemblies, and cladding, as well as reactor structural materials, the design of the initial FFTF core has eight instrumented test positions (two closed loops and six open loops).

In closed test loops, test components that are inserted in the reactor core region, as well as their coolant, instrumentation, and heat-transfer systems, are completely separated from the main FFTF core, permitting the testing of fuels and materials over a wide range of temperatures in a controlled environment independent of the main reactor coolant system. In open-loop test positions, test components are integral with the reactor core and are cooled by the reactor primary-coolant system; these positions are intended for testing large quantities of candidate fuel pins and assemblies. Figure IV 4-5 shows the FFTF fuel system.

Table IV 4-5 lists the highlights of FFTF's history. Major construction activities have been completed and emphasis is on preoperational checkout and testing. The FFTF will play a key role in breeder-reactor technology development and testing. For FY 1980, an increase in the budget has been requested to provide for the following:

- Completion of the post nuclear acceptance testing
- Full-power demonstration
- Initiation of the fuels and materials irradiation program including hardware procurement, test article assembly and engineering and safety analysis
- Procurement of cores 3 and 4 and essential spare parts

The construction and operation of the FFTF reactor is on schedule with criticality planned for August 1979. All work assigned to the prime construction contractor (Bechtel) was

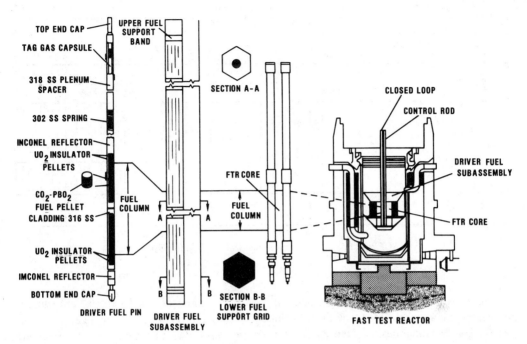

Figure IV-4-5. FFTF fuel system.

completed in September 1978. The Advisory Committee on Reactor Safety (ACRS) finished its review of the Nuclear Regulatory Commissions's (NRC's) August 1978 FFTF Safety Evaluation Report (SER) in November 1978; an addendum to the SER will be issued in April 1979 to complete NRC's review. Prenuclear acceptance testing and loading of the fuel will precede the achievement of criticality. Fuels and materials experiments are being prepared so that experiments can proceed when full-power operation begins in early 1980.

Major milestones and funding are shown in Figure IV 4-6.

DOE's Hanford Engineering Development

Table IV-4-5. FFTF schedule history.

YEAR	ACTIVITY
1963	AEC authorized Hanford Laboratoties to scope a Fast Test Reactor (FTR)
1966	Architect-engineering was authorized
1967	Full project was authorized
1969	Preliminary design of FFTF was initiated
1970	Westinghouse Electric Corp. was selected as AEC prime contractor to operate HEDL and manage the FFTF project
1972	The FFTF containment vessel was successfully pressure-tested; the reactor vessel was received on site; the AEC issued the Environmental Impact Statement on an LMFBR demonstration plant
1974	The 122-ton reactor guard vessel, to house the FFTF's reactor vessel and core, was installed in the containment
1976	The FFTF Final Safety Analysis Report was submitted to NRC; Title-II design was completed and delivered to NRC
1977	The reactor heat transport system was pressure-tested; prototype pump testing was completed; 90% of construction was completed
1978	The secondary loops (heat-dissipation system) were filled with sodium in July; NRC issued its FFTF SER in August; all work assigned to the prime construction contractor (Bechtel) was completed in September; the primary system (reactor vessel and associated loops) was filled in December

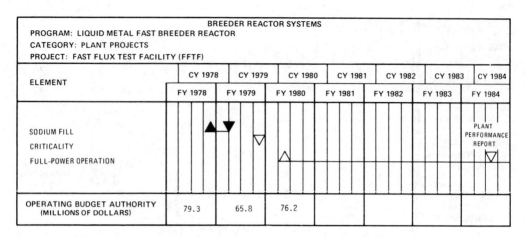

Figure IV-4-6. Major milestones and funding for FFTF.

Laboratory is overall project manager and will be the operator when the facility is completed. The Westinghouse Advanced Reactor division was the reactor-plant designer; the plant architect-engineer and construction manager was Bechtel Corporation.

CRBRP. The Administration has proposed that CRBR Project be discontinued except for selected design component and generic licensing activities of value to the overall LMFBR program. Implementation of the proposal is pending final resolution by the Administration and Congress.

The objectives of the CRBRP Project, pending final resolution of the Administration's proposal for discontinuation, are to design, license, build and test a 380-MWe liquid sodium cooled breeder reactor demonstration plant and to operate the plant as part of a utility system. Upon enactment of legislation authorizing the proposed revised objectives for CRBRP, appropriate implementing actions will be taken. Pending such legislation, project engineering, fabrication and support activities are continuing.

Project activities continue on a reasonable and prudent basis pending consideration of the project's future by the Administration and Congress. In summary, the status of project activities as of December 1978 was as follows:

- The engineering activities of research, development, and design are approximately 63 percent complete,
- The value of equipment on order was $424 million of which approximately $201 million has been costed,
- Approximately $631 million had been invested through December 31, 1978, including about $531 million of DOE funding and $100 million of funding by utilities and other participants,
- The Final Environmental Statement, the Site Suitability Report, and Design Criteria issuance had been obtained from the Nuclear Regulatory Commission, and
- Licensing proceedings necessary to secure the Limited Work Authorization were suspended on April 25, 1977.

The CRBRP Project is authorized under P.L. 91-273. Under that law, Congress authorized the AEC (later ERDA and now DOE) to enter into a definitive cooperative arrangement for an LMFBR Cooperative Demonstration Plant and provide financial assistance. That cooperative arrangement, called the Quadripartite Contract, has been in effect since 1973 among AEC/ERDA/DOE, Project Management Corp. (PMC), Tennessee Valley Authority (TVA), and Commonwealth Edison Co. (CE).

The federal role in the CRBRP Projects consists of both overall project management and major financial support. DOE financial support to the project is applied to all elements of the project funding in excess of the utilities' contributions of approximately $260 million and the reactor manufacturers' contributions of about $10 million.

The utilities are to provide both financial support and technical assistance to the CRBRP Project. More than 750 public, private, cooperative, and municipal electric systems pledged contributions totaling $260 million to the project. These funds are to be used with the government funds to meet any project expense. CE and TVA are the principal organizations that represent the utility interest through the PMC and the Breeder Reactor Corp. (BRC).

The CRBRP Project is managed by the CRBRP Project Office under the fuel and fuel form chosen, so the design goal is to: (a) accommodate both carbide fuels and oxide driver fuels, and (b) allow for optional blanket fuels (uranium-238 or thorium) and fuel management schemes—depending on relative production requirements for plutonium and uranium-233.

This plant design study, which will be carried out by several reactor manufacturers and architect-engineers, is to provide:

- A large plant conceptual design with necessary nonproliferation and safeguards characteristics to serve as a focus for structuring and pacing the fast breeder reactor base-technology research and development program, and
- An advance in breeder reactor technology by integrating design improvements that have evolved in the CRBRP, PLBR, the base program, and foreign technology.

The resulting plant conceptual design, including cost and schedule estimates, will be sufficient in detail to permit a subsequent decision on a project to design, manufacture, construct, and operate this type of plant.

The Conceptual Design Study will be accomplished in two phases. The first phase, of 18 months duration, will consist of trade-off studies of basic plant design alternatives (e.g., pool vs. loop, plant parameters), after which design concept decisions will be made as appropriate. The second phase, 12 months long, will develop an integrated plant conceptual design based on the design selections and requirements from the first phase. It will produce the design descriptions, drawings, requirements, and other engineering data DOE's Program Director for Nuclear Energy. The CRBRP Project Office is staffed by DOE and utility personnel and is located in Oak Ridge, Tennessee, near the planned plant site. Contract administration is performed by the Project Office with some administrative support from the DOE Oak Ridge Operations Office. The utility financial contributions to the project and the utility personnel assigned to the project are administered by PMC under the Quadripartite Contract.

The three prime contractors performing the project work are Westinghouse Electric Corp. as the lead reactor manufacturer, Burns and Roe as architect-engineer, and Stone & Webster as constructor. General Electric and Atomics International are major subcontractors to Westinghouse.

Breeder Reactor Studies (CDS). A key element in ensuring viability of the LMFBR option in the United States is the performance of a CDS. The CDS, a totally Government-funded and managed effort, will develop an integrated conceptual design based on evaluations and optimization of overall plant safety, environmental acceptability, proliferation resistance, economics, and reliability. This work will be accomplished in two phases over a 30-month period ending March 31, 1981, to provide an opportunity at that time for deciding whether to proceed with the detailed design and construction of such a reactor plant.

To accommodate various nonproliferation and resource constraints and other driving forces, fuel and fuel-cycle flexibility will be an explicit goal of the CDS activity. The design of the sodium heat-transport system and reactor internals can be largely decoupled from needed to define the conceptual design of a single plant.

By March 1981, technical, environmental, economic, nonproliferation, and other LMFBR specific information will be assembled and a comprehensive conceptual design will be complete. Supporting R&D will have been defined and initiated. The principal requirements for major plant and reactor components will be defined and the proposed project oganization, staffing, and plans for the detailed design will be complete. Consequently, a substantial basis will exist including cost and schedules estimates for an informed decision as to whether to proceed with plant design and construction.

The status as of December 1978 was as follows:

- The DOE management organization had been established.
- All major contractors (reactor manufacturers and architect engineers) had been selected and were working.
- A competitive process for selecting an objective technical integrator to serve as project technical coordinator was underway.
- Detailed study criteria, objectives, and guidelines had been prepared.
- Trade-off studies on key plant parameters were underway.

Continuation of trade-off studies and evaluations, and the selection of many of the key plant parameters were carried out in the remainder of FY 1979.

In FY 1980, trade-off studies will be completed, key plant parameters selected, and conceptual design of the plant initiated.

Key technical parameters to be determined by FY 1980 include the primary system configuration (pool/loop), steam cycle, plant size, fuel cycle, fuel handling scheme, and reactor containment/confinement design. The initial phase will also include the identification of LMFBR design improvements and simplifications, the development of safety and licensing criteria, the identification of R&D requirements and a compliation of principal plant requirements and criteria.

Figure IV 4-7 depicts the major milestones and funding for the CDS.

Reactor manufacturers and architect-engineering firms participating in the CDS program include Atomics International Division of Rockwell International, Babcock & Wilcox Co., Bechtel Power Corp., Burns & Roe, Inc., Combustion Engineering, Inc., General Electric Co., Westinghouse Advanced Reactors Division, and Stone & Webster.

Breeder Technology

This program, which now is function-oriented rather than system-oriented, consists of developing the technology of breeder reactor components, fuel and core design, safety, physics, and materials and chemistry. As the cross-cut of these elements and the LMFBR WBS in Figure IV 4-2 shows, they span the major reactor systems. These elements both support and derive from the CDS.

The breeder technology program, which provides engineering information and technology support for future LMFBR plants and

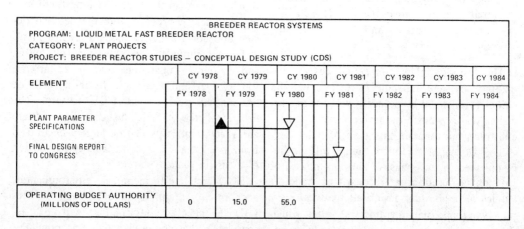

Figure IV-4-7. Major milestones and funding for CDS.

facilities, will increase the safety and reliability, reduce the cost, and facilitate licensing of future plants and facilities. In addition to the general development of breeder technology, the program will provide contingency alternatives to critical items.

During the 1979-1981 period, the breeder technology program is paced and scoped to permit a decision on the LMFBR in 1981, and allow flexibility in the choice of fuel and fuel cycle. Integration of the LPDS and base-technology program assumes, for planning purposes, a hypothetical plant that is to come on-line in 1991, with detailed design commencing in 1981. The scope includes technology in the areas of safety, engineered systems and components, physics, fuels and core materials, and materials and chemistry.

Table IV 4-6 shows the operating funding levels by technology area for the FY 1978 to FY 1980 period.

Safety. To attain and demonstrate the required degree of safety for fast breeder reactors without unacceptable economic penalities, a sound frame of technical reference is being developed for assessing the risk to public health and safety from postulated accidents with the full spectrum of breeder reactor designs and fuel cycles being considered. Analyses and experiments are providing the broad technology base with which designers can: (a) combine features to obtain requisite levels of safety while minimizing impacts on plant cost, performance, and schedule; and (b) demonstrate that the probability of a major accident is negligible and that minor accidents will not escalate into major accidents.

The safety program aims to demonstrate that even highly unlikely reactor accidents will pose no significant hazard to the public. Thus, a technology base is being developed that is fully responsive to safety considerations in the design, evaluation, licensing, public acceptance, and economic optimization of breeder reactor systems for electrical power generation.

A major reactor accident cannot take place unless a sequence of failures occurs within the reactor system. Therefore, the safety goal is being achieved by developing technology in four generic and integrated segments, each of which relates to a sequential barrier preventing accident progression, thus preventing release of any significant amount of radioactivity to the environment. Each technology segment is called a Line of Assurance (LOA). These are shown schematically in Figure IV 4-8 and described as follows:

- First LOA: Prevent accidents, so that fuel melting will not occur. System design can prevent the initiation of accidents. Plant features that prevent accidents include a design that is inherently stable under normal operation, has a high tolerance for abnormal operation and component malfunction, and has highly reliable protective devices to prevent core damage if plant systems and components fail. These features combine to produce a plant with an extremely low probability of accident initiation.

Table IV-4-6. CDS funding for FY 1978-1980.

BREEDER TECHNOLOGY	BUDGET AUTHORITY (DOLLARS IN THOUSANDS)			
TYPE	FY 1978	FY 1979	REQUEST FY 1980	INCREASE (DECREASE)
Safety	$ 52,562	$ 34,240	$ 36,500	$2,260
Engineered Systems and Components	42,261	41,561	42,000	439
Physics	9,953	8,000	9,000	1,000
Fuels and Core Materials	64,566	46,000	46,000	0
Materials and Chemistry	8,881	7,000	8,900	1,900
TOTAL	$178,223	$136,801	$142,400	$5,599

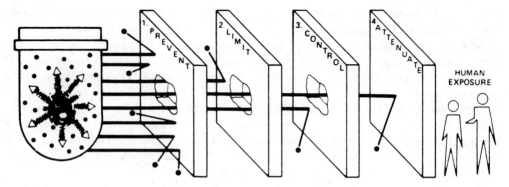

Figure IV-4-8. Safety lines of assurance.

- Second LOA: Limit core damage, so that whole-core disruption does not occur even in the unlikely event that a severe accident begins. The inherent response of the reactor system is being used to prevent extensive core melting. Thus, if accident progression terminates before the whole core is disrupted, significant amounts of radioactivity cannot be released.
- Third LOA: Control accident progression, so that the secondary containment does not fail even if whole-core disruption occurs. Significant amounts of radioactivity would not be released to the environment unless large amounts of accident products were to leak through the containment boundary. Such leakage could result if the containment were breached structurally by a strong pressure-generating event or by being melted through because of inadequately accommodated core debris. This program is demonstrating the inherent and engineered capability of the plant to accommodate a core meltdown.
- Fourth LOA: Demonstrate that the attenuating mechanisms (both inherent and engineered) for the radiological consequences of accidents can protect the public from significant radioactivity hazard even if LOA's 1 to 3 are breached.

The incremental consequences of failure of each barrier are known, and probability goals for each failure can be assigned. Thus, R&D objectives can be related to particular degrees of risk. A reference risk curve (i.e., a probability vs. consequences plot) can be constructed based on the goals established for the probability of reduction in those consequences. Figure IV 4-9 shows a risk curve constructed in this manner. The LOA approach provides a risk-assessment framework for making rational decisions for evaluation R&D choices based on cost-benefit analysis, and ensures optimal use of program resources by channeling R&D in directions that promise the greatest payoff for reducing overall risk.

Analyses and experiments are conducted to validate the probability goals. The degree of validation that can be attained, and thus, the level of confidence that can be projected in breeder reactor safety, are strongly influenced by the quantity and quality of safety experiments carried out in test facilities.

The experimental programs to attain the objectives of the four LOA's include the following:

- LOA-1—prevent accident:
  — Reliability
  — Local-fault prevention
  — Inherent stability
- LOA-2—limit core damage:
  — Transient overpower without scram accommodation
  — Transient undercooling without scram accommodation
  — Loss of shutdown heat removal accommodation
  — Loss of pipe integrity accommodation
  — Local-fault accommodation

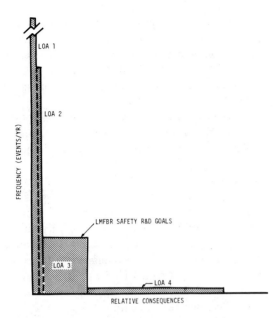

Figure IV-4-9. LMFBR accident reference risk curve.

- LOA-3—control accident progression:
  — Energetics accommodation
  — Core debris accommodation
- LOA-4—attenuate accident consequences:
  — Attenuation in failed containment
  — Attenuation in environment

Thus, the top-level objectives can be connected to specific experimental programs through a series of related sub-objectives.

With a reference risk curve and an understanding of the development that must be accomplished to attain its safety probability goals, the LMFBR safety R&D program can be managed by iteration on what is "acceptable," "do-able," and "cost effective." Acceptability of the LOA goals and plant design is obtained when the reference risk curve can be held below some recognized acceptance criteria, for example, "comparability to light water reactors." "Do-ability" is obtained when the development work breakdown structure can be defined and executed on a schedule. Cost effectiveness is obtained when the selected "acceptable" and "do-able" tasks represent the lowest cost path to the desired R&D goal.

The LOA approach has gained wide acceptance in the United States and internationally.

Intensive international cooperative programs are underway between DOE fast reactor safety and similar organizations in France, Germany, the United Kingdom, and Japan. These programs are aimed at developing an international consensus on safety issues required to be resolved in the licensing of LMFBR's. These activities are also aimed at developing joint programs to resolve the above safety issues and reduce individual program costs.

Breeder reactor safety technology is being developed by analyses and experiments at national laboratories, industrial contractors, and universities. The analysis development improves the basis and means for reliably analyzing the course of events in postulated accidents and the estimated consequences of such accidents. This improvement is being accomplished by the following:

- Developing the understanding of the basic phenomena needed for analytical modeling of postulated accidents
- Accumulating basic data on these phenomena so that the analytical models will have the appropriate realism or conservatism over the desired range of application
- Integrating these data and models into complete analytical descriptions of the accidents so that the quantities and parameters important to a review of safety can be accurately determined
- Establishing requirements for experiments to test the adequacy of the analytical models for predicting accident sequences and consequences
- Applying the analytical models to the analyses of generic questions

This data base will be used to demonstrate that the probability of a major accident is negligible, and that minor accidents do not escalate into major accidents.

The analytical-modeling and computer-code development work is being increasingly redirected to resolve the safety issues of alternative fuel technologies and advanced converters. The development of breeder reactor risk assessment methods will be continued; preliminary estimates of the environmental effects of using alternate fuels in breeder reactors have been completed.

Experimental work coupled to the analytical work contributes to the overall safety program by verifying the analytical tools and allowing the development of a fundamental understanding of the key phenomena controlling the progression of breeder reactor accidents. The experiments use the Sodium Loop Safety Facility (SLSF) installed in the Engineering Test Reactor (ETR), the Transient Reactor Test Facility (TREAT), and the Thermal Hydraulic Out-of Reactor Safety (THORS) Facility.

One activity for LOA-1 includes the demonstration of accident prevention with high confidence, reliability testing of full-size components and systems, including shutdown systems, to meet the reliability goals of LOA-1.

An important effort in demonstrating low probability of exceeding limited core damage (LOA-2) will be to test in the sodium environment core components and systems designed to be inherently safe. An articulated-control-rod concept is being tested at prototype sodium-temperature increase rates to demonstrate system response. Tests with a second inherently safe system, a floating absorber-ball concept, are also being conducted.

For low-probability occurrences (less than one chance in 100,000) leading to LOA-2 accident conditions, the inherent response of the core in limiting core damage is being further verified through additional in-reactor tests in TREAT and the SLSF. The experiments in TREAT include initial tests in the U.S.–U.K. safety exchange program. Preparations for FY 1980 loop tests in the U.S.–U.K. test program are also being completed, including acquisition of the Mark-III TREAT test vehicle for testing full-length, prototypically pre-irradiated fuel. These tests will be continued to verify the oxide fuel's favorable inherent response to accident transients.

Transient tests using carbide fuel are also required. The first tests were conducted early in 1979 to examine the potential for fuel-coolant interactions. To obtain pre-irradiated fuel for the other carbide tests to be conducted later, a vehicle capable of pre-irradiating carbide fuel in FFTF is being developed. Tests of carbide fuels, both out-of-reactor and in-reactor, will establish basic tie points to the data base available from the oxide-fuel tests so as to minimize the number of carbide tests that will be required. As part of this program, data on the high-temperature properties of carbide fuels are being obtained.

A test in the SLSF loop is expected to confirm that cooling of the reactor core can be maintained even if one of the cooling pipes loses its integrity.

These in-reactor tests will coupled with out-of-reactor experiments on fission-gas dispersal of molten fuel, large-scale fuel-sweepout tests, and boiling-stability tests to verify further the analytical accident codes used in safety analysis. Model development will provide an improved model for fuel and cladding behavior. Codes will be developed for use with a carbide fuel system.

Major milestones and funding are shown in Figure IV 4-10.

For LOA-3, two capsule tests are being run in TREAT to provide further evidence of limited accident energetics. Code and modeling work with supporting tests continue to show control of accident progression and attenuation of accident products. These test include a large-scale aerosol test to verify reduction in consequences due to particle agglomeration and fallout. Molten-fuel and concrete interaction tests are demonstrating the consequence-confining capability of concrete.

For LOA-4, large-scale, out of containment, sodium aerosol tests will provide experimental data for verification of currently available and future computer codes and analytic methods describing sodium aerosol attenuation both in and out of containment.

The major contractors participating in developing the technology of LMFBR safety and their areas of involvement are listed in Table IV 4-7.

Engineered Systems and Components. The Breeder Component Development Program, which is largely independent of fuel and fuel-cycle types, and thus of the nonproliferation issue, is consistent with the hypothetical plant schedule and includes CRBRP and alternative-component design, fabrication, and testing. The objective is to develop the steam genera-

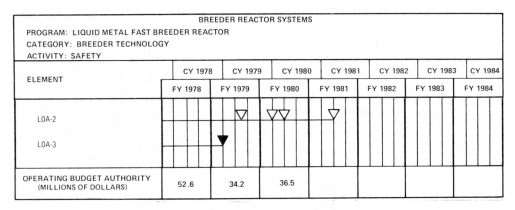

Figure IV-4-10. Major milestones and funding for safety.

tor, pumps and other components for the LMFBR system.

A 1974 review by reactor manufacturers, national laboratories, and utilities resulted in a consensus that the base technology program underway at that time needed better balance and, in particular, that it should be strengthened in areas that would reduce uncertainties in component performance and plant capital costs to levels acceptable to utilities and reactor manufacturers. As a result, the present vigorous component development and testing program was planned and launched in 1975. To provide the technology for reliable and economic equipment, models of critical components applicable to future breeder reactors are designed, fabricated, and tested, confirming the feasibility of the design for large plant application. Primary emphasis is on the development of steam generators and large pumps.

Table IV-4-7. Contractors and areas of participation for the safety activity.

	AREA OF PARTICIPATION			
	LOA-1	LOA-2	LOA-3	LOA-4
LASL			X	X
ANL	X	X	X	X
Westinghouse	X	X		
HEDL		X	X	X
ORNL		X	X	X
General Electric	X	X	X	X
Atomics International		X	X	X
EG&G			X	

The development work required to support these activities will be conducted.

Experience with breeder reactor plants in this country (EBR-II and Fermi) and abroad (PFR, Phenix, BN-350) has shown that the test-transport components have been the major source of plant operational deficiencies, so that development emphasis is on long-lead and unproven but essential components such as steam generators and main coolant pumps. Component vendors design and fabricate model components and perform the concept-dependent supporting development. Tests and generic-type development are conducted by the national and engineering laboratories and the reactor manufacturers.

Major milestones and funding are shown in Figure IV 4-11.

In the international LMFBR community, it is recognized that the sodium-to-water steam generator is the most complex and demanding component from engineering design and system reliability viewpoints. For this reason, the U.S. components program includes the design, fabrication, and testing of two different steam generator designs. A single-wall tube, helical-coil concept is being designed, with a 70-MWt model to be delivered in late 1982 to the Sodium Components Test Installation (SCTI) at the Energy Technology Engineering Center (ETEC) at Canoga Park, California, for installation and a 1-year test. A double-wall tube, straight-tube concept (Figure IV 4-12) is also being designed, with a 70-MWt model of that concept scheduled for delivery to the SCTI in late 1984 for installation and a

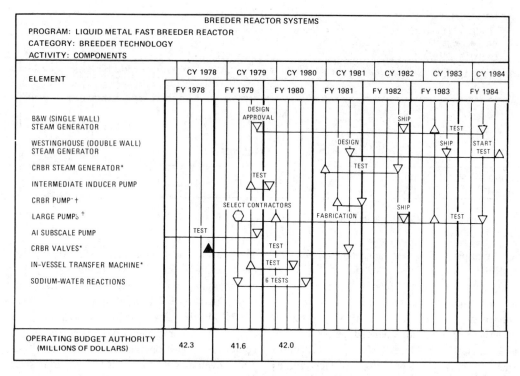

Figure IV-4-11. Major milestones and funding for component development.

1-year test. Supporting development, including feature and small-model testing and manufacturing development, are underway in support of these two concepts.

In support of the development of a CRBRP steam generator, a Few Tube Test Program is underway. It involves testing small models of the CRBRP evaporator and superheater (each containing seven full-length tubes) at General Electric's Steam Generator Test Rig facility. After completion of the full-scale CRBRP prototype steam generator (11-MWt) now being fabricated, steady-state and transient testing will be performed at the SCTI in 1981. The last of an initial series of six large sodium-water reaction tests was successfully completed in July 1978. These tests, managed by GE and conducted by ETEC in the Large Leak Test Rig (LLTR), simulate the effects (on a breeder reactor steam generator) of the sodium-water reaction resulting from an unlikely complete rupture of a steam/water tube. A second series of large leak tests using a full diameter (CRBRP) test vessel installed in the LLTR will be initiated in 1979.

Steam generator development has the highest priority under component development of FY 1980, and will include the following:

- For the single-wall steam generator being developed by Babcock & Wilcox Co. (Ohio), preliminary design of full-size and prototype detailed design and fabrication will be well underway, and much of the supporting development tasks will have been completed.
- For the double-wall steam generators being developed by the Westinghouse Tampa Division (Florida), preliminary design of prototype steam generators for two different steam conditions will be completed and all supporting development tasks will be near completion, including fabrication and testing of a few-tube model.
- The first series of large sodium-water

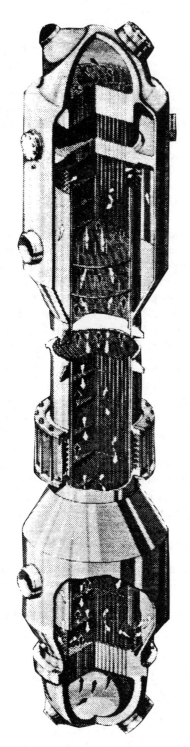

Figure IV-4-12. Double-wall-tube steam generator.

reaction tests utilizing a full-diameter test vessel will be completed in the LLTR, at ETEC.

Because of the difficult suction requirements for the primary-system pumps for large breeder plants, one concept being explored is an inducer-type pumping element in series with a conventional centrifugal pumping element. To evaluate the inducer's suitability for use with sodium, a small (3,000 gpm) inducer/centrifugal pumping element was installed on an existing pump and is being tested at ETEC in 1979. A modification of the 15,000 gpm FFTF prototype pump for incorporation of an inducer will be installed in the Sodium Pump Test Facility (SPTF) at ETEC and tested in 1979.

Fabrication of the full-scale (35,000 gpm) CRBRP pump will be completed and it will be tested in the SPTF at ETEC.

Proposals were solicited and received in 1978 from three sodium pump vendors for design, development, fabrication, and test of a primary- and an intermediate-system sodium pump (85,000 gpm) for large breeder reactors. Development and demonstration testing of primary pumps for large plants is necessary because the difficult hydraulic conditions and large capacity necessitate the use of pump concepts that are significant departures from those used for smaller plants. Also, development and testing of an intermediate-system pump for large plants is being pursued to take advantage of the major cost savings and simplicity of configuration possible for this component (as opposed to use of the same concept as the primary-system pump) since the hydraulic and other requirements are much less severe than for the primary pump.

Work is underway at the Borg Warner Corporation's Byron Jackson Pump Division, Carson, California (for design, fabrication, and test of the intermediate pump), Westinghouse Electric Corporation's Electro-Mechanical Division, Cheswick, Pennsylvania (to design, build, and test the primary pump), and Rockwell International's Energy Systems Group, Canoga Park, California (for the design and supporting development for an alternative primary pump). The reference primary and intermediate pumps will be tested in the SPTF following completion of the CRBRP pump test.

Technology is being developed, and key

features are being tested, for other components, including the following:

- Fuel handling equipment
- Intermediate heat exchangers
- Small valves

In addition, base technology is being developed in the following areas:

- Flow-induced vibration
- Structural-design methods
- Materials behavior
- Fluid mixing processes

This will provide needed input to the component development tasks.

Tests will be performed to develop and validate seismic and structural design methods and criteria; improved design rules and guidelines will be issued. Development of flexible pipe joints and Alloy 718 materials data will continue; prototype CRBR steam generator transition joints for SCTI test will be delivered; transition joint life test in the Thermal Transient Facility (TTF) at ETEC will be continued to failure; the Nuclear Systems Materials Handbook will be updated.

The major contractors in components development and their areas of involvement are shown in Table IV 4-8.

Physics. Critical experiments generic to large proliferation-resistant fast reactors are providing measurements of core integral performance and safety parameters to validate nuclear data and design methods available and being developed. Both homogeneous and heterogeneous cores are being studied using various core and blanket fuels of major interest to the CDS's.

The main aim is to develop design data and confirm computational methods with an accuracy and scope sufficient to enable specification of core loadings, shielding requirements, and control requirements for large LMFBR's with factors of conservatism consistent with low design costs, low plant costs, improved plant performance, and competitive power costs. Detailed nuclear properties of reactor materials are being measured; emphasis is on cross-sections for the thorium-cycle isotopes. The goals are to fill key gaps in nuclear-property data and to obtain accuracies comparable to those now achieved for the uranium-cycle isotopes.

For FY 1980, a slight increase over the FY 1979 budget has been requested. These funds will be used for the following studies:

- Complete large benchmark critical measurements for carbide fuels
- Measure remaining high priority cross-sections for thorium-cycle isotopes
- Establish a reference data file for alternative fuels
- Provide materials dosimetry to support the FFTF irradiation testing program
- Perform shielding assessments for CDS
- Complete advanced computational codes for accurate three-dimensional fuel depletion and transient analyses

The increase requested is required to provide timely critical experiments and methods

Table IV-4-8. Major contractors and areas of participation for component development.

	AREA OF PARTICIPATION		
	STEAM GENERATOR	PUMP	OTHER
ANL	X	X	X
Atomics International Div. of Rockwell Intl. Corp.	X	X	X
Babcock & Wilcox Co.	X		
Byron Jackson Pump Div. of Borg Warner Corp.		X	
Energy Technology Engineering Center	X	X	X
Foster Wheeler Co.			X
General Electric Co.	X		X
HEDL		X	X
ORNL			X
Westinghouse Electric Corp.	X	X	X

support to the conceptual design studies, to conduct fuel burnup and depletion measurements on samples from operating foreign reactors and FFTF, to complete an assessment of CDS shielding problems and to develop effective computational methods for detailed burnup and transient analyses.

The physics program encompasses integral measurements of core and shield properties and nuclear and dosimetry data, detailed nuclear data measurement via energy-dependent experiments, and development of computational methods for predicting core and shield design and safety parameters. These activities support the available options for fuel forms, reactor types and designs, and nuclear fuel cycles, as well as U.S. participation in international technical exchanges.

The critical and shielding experiments are planned to verify data files and computational methods for general design use and for direct experimental verification of first-of-a-kind U.S. facilities. Nuclear data are measured corresponding to detailed analytical models for a wide range of candidate reactor fuels and materials.

Integral experiments are mostly performed at ANL, where the sole laboratory capability for U.S. critical experiments generic to FBR's exist. However, planning of experiments and analyses of measured results is shared by most design and laboratory participants in the physics program. Similarly, most data (cross-section) measurements are made using other sophisticated ANL and ORNL facilities, but the review, selection, and formulation of measured valued is widely shared.

Computational methods and codes are being developed to enable more accurate calculations of fuel utilization, control, heat sources, core reactivity coefficients, blanket performance, transient behavior, and radiation shielding. Development of computer codes for nuclear design is carried out mainly within national laboratories with access to government-owned computers and is shared by ANL, LASL, and ORNL to take advantage of special expertise at each site and to maintain a broad overall perspective. Additionally, small computer implementation tasks are maintained generally to provide means for wide (designer, as well as developer) participation in planning developments, joint testing of codes, and site-to-site code conversions.

State-of-the-art experimental and analytical physics support is provided for accurate determination doses of fuels and materials irradiation tests in FFTF and EBR-II.

The Zero Power Plutonium Reactor (ZPPR), Figure IV 4-13 is being used to obtain benchmark measurements on a series of reference large core assemblies to determine the adequacy of neutronic design methods for applications to the CDS. Of current interest are heterogeneous core benchmarking, zone substitution of alternative fuels (uranium-233-thorium) for experiments to refine further the breeding potential of alternative systems under consideration in INFCE and NASAP, and experiments on large reference carbide thorium blankets to confirm the breeding characteristics of the uranium-233 producer/reactor concept. As the CDS progresses, the neutronic performance of detailed mockups of the proposed core configuration will be accurately determined.

Measurements of key cross section needs for U-233 and thorium began in FY 1979 and major parts of this work will be completed in FY 1980. Analyses of reactor physics data and heavy actinide burnup obtained through exchange agreements with other countries are performed as data become available.

Major milestones and funding for the Physics Activity are presented in Figure IV 4-14.

The major contractors participating in physics development and their areas of involvement are given in Table IV 4-9.

Fuels and Core Materials. This work identifies and develops breeder reactor fuel, blanket and absorber elements, and demonstrates their performance capabilities under steady-state and design-transient conditions. It includes developing fuel fabrication equipment and processes and analytical design and performance models, and verifying the results by various levels of irradiation testing.

In the development of FFTF fuels, absorbers, cladding, and duct materials, the major objectives are as follows:

- Obtain statistically reliable fuel and absorber pin steady-state and off-normal

Figure IV-4-13. Zero power plutonium reactor.

transient performance data to allow the reliable operation of the FFTF
- Determine the effects of irradiation on reference-fuel-pin-cladding and subassembly-duct materials
- Determine the mode and consequences of cladding breach and the advisability of continuing element operation beyond cladding breach
- Define the parameters limiting performance of fuel and absorber components as a basis for innovative improvements and design changes

In the development of advanced and alternative fuel, cladding, and duct materials, the major objectives are as follows:

- Determine the steady-state and transient irradiation performance, endurance capability, and cladding-breach modes of alternative breeder fuels and fuel pins with advanced cladding and duct materials so as to form a basis for the development of alternative fuel cycles
- Provide, by late FY 1979, an adequate data base (from EBR-II, TREAT, and fuel properties studies) for selecting a carbide or nitride fuel concept for detailed design and development

In the development of fuel fabrication processes, the major objectives are as follows:

- Develop fabrication technology for fuel manufacture in support of the domestic and international initiatives on the use of proliferation-resistant alternative breeder fuels

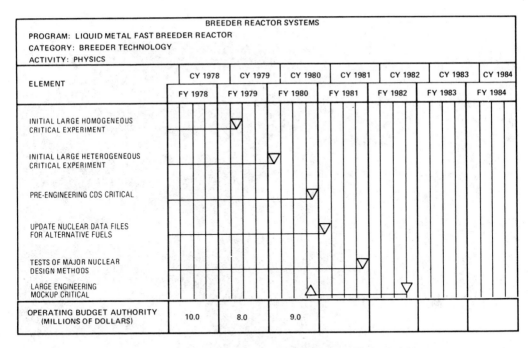

Figure IV-4-14. Major milestones and funding for the physics activity.

- Provide an assured source of fuel for FFTF

The overall strategy for the Fuels and Core Materials Program is to maintain two parallel activities: (1) the confirmation of previous development work on reference FFTF fuel-system materials (i.e., (U,Pu) oxide fuel and Type-316 stainless-steel cladding ducts) to support the safe and reliable operation of FFTF as a test reactor, and (2) the performance of sufficient EBR-II-, TREAT-, and FFTF-based investigations, as well as fabrication development activities, on advanced and alternative fuels and blanket concepts to proceed from the current situation in which a large number of concepts have been identified to the point where one fuel system can be selected and demonstrated for use in more advanced LMFBR's. Thus, this work aims to allow a choice among conventional fuels, high-performance fuels, and fuel-cycle alternatives (such as a thorium blanket) that might offer nonproliferation advantages over conventional fuels.

The development work is organized into three elements:

- Fuels, absorbers, and materials for FFTF
- Advanced and alternative fuels and materials
- Fuel fabrication processes

The program depends heavily on irradiation tests conducted in the EBR-II (see Figure IV 4-15) and in the FFTF.

Table IV-4-9. Major contractors and areas of participation for the physics activity.

	AREA OF PARTICIPATION			
	EXPERIMENTS	NUCLEAR DATA	NUCLEAR DESIGN METHODS	MATERIALS DOSIMETRY
ANL	X	X	X	
LASL		X	X	
BNL		X		
HEDL		X	X	X
ORNL	X	X	X	

Figure IV-4-15. Top of EBR-II reactor.

Significant aspects of the irradiations program in EBR-II include: completing advanced fuels tests to and beyond goal burnups to identify ultimate performance limits, continuing irradiations of advanced cladding and duct materials to high exposures, and performing Run Beyond Clad Branch (RBCB) tests on advanced fuels.

Key aspects of the irradiations to be conducted in FFTF include qualifying FFTF driver fuels, testing candidate advanced cladding and duct materials, and testing advanced fuels and thorium blanket pins. These experiments will lead to testing full-scale assemblies of prime-candidate advanced fuels and uranium and/or thorium blanket concepts; these will be followed by large-scale qualification runs with partial core loadings of advanced fuels and uranium and/or throium blanket assemblies.

Work on reference FFTF fuels and core materials in FY 1979 focuses on completing EBR-II and TREAT tests on FFTF-prototypic fuel elements and clad/duct materials, while preparing to start surveillance and testing of actual FFTF driver assemblies (see Figures IV 4-16 and IV 4-17) when FFTF begins operation. These activities include completing the fabrication of the first three FFTF-size test assemblies of candidate improved driver elements as well as starting the irradiation of two instrumented driver assemblies. The FY 1980 work scope will be focused on demonstration of early-in-life thermal performance of the driver fuel, primarily to show that results from the semiprototypic EBR-II tests also hold for FFTF.

The advanced alternative fuels and materials work is aimed at developing improved or alternative fuels that could be used in future breeder reactors. A parallel path approach will be followed in the development of advanced uranium-plutonium fuel. One path will be directed to development of higher breeding oxide fuels (relative to the current FFTF driver fuel), while the other will emphasize carbide

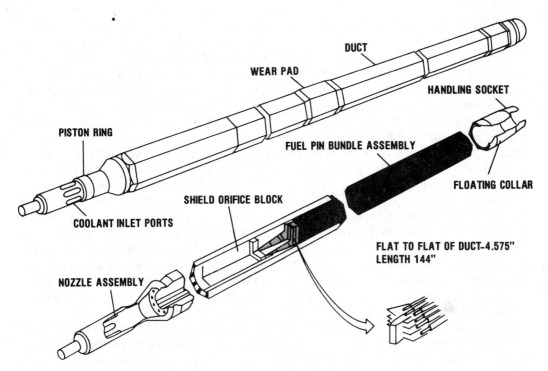

Figure IV-4-16. FFTF driver-fuel assembly.

fuel designs. The first advanced oxide and carbide tests will be inserted into FFTF in March 1980 when steady-state full-power operation begins.

The efforts (begun in FY 1978 and continued in FY 1979) to develop fuels providing alternatives to the uranium-plutonium fuels, including oxide and carbide fuels containing

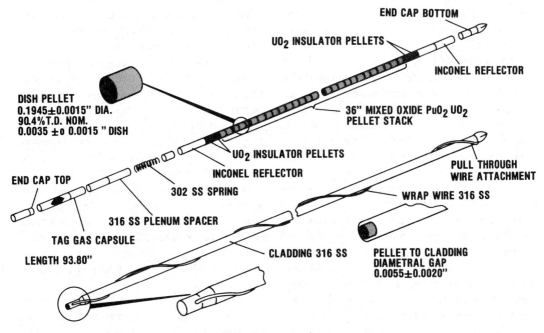

Figure IV-4-17. FFTF driver-fuel pin.

thorium and uranium-233, will continue in FY 1980.

Work on advanced and alternative fuels, blankets, and clad/duct materials in FY 1979 includes continuing EBR-II and TREAT tests of advanced uranium and plutonium fuel elements and clad/duct materials to and beyond goal burnups, while completing fabrication of initial FFTF carbide test elements in the program leading to a 1983–1984 partial loading. Also in FY 1979, fabrication of the final two EBR-II assemblies in a 4-assembly test series of alternative thorium/plutonium and uranium-238/uranium-233 fuels (using uranium-235 as a stand-in for uranium-233) is being completed and their irradiation will begin. Other work includes: (1) the design and fabrication of three additional FFTF-size blanket assemblies (containing ThC, UC, and ThO_2 test elements) to supplement the two previously fabricated UO_2 test assemblies, (2) expansion of the TREAT transient irradiation program for carbide fuel and blanket elements, and (3) acceleration of the development of the reconstitution capabilities required for testing improved clad/duct alloys in FFTF. Additionally, post-irradiation examination and data analysis activities on carbide fuels and improved cladding alloys now under irradiation in EBR-II will be expanded to support the selection of lead concepts for follow-on tests in FFTF.

Fabrication of UO_2 and ThO_2 radial-blanket rods of the CRBR design was completed in FY 1979. These rods were irradiated in FFTF coinciding with FFTF full power operation beginning in March 1980 and will provide data on uranium and thorium blanket performance that will be applicable to future breeder designs.

An assembly of fuel specimens for an irradiation test called the P-60 test were designed and fabricated by HEDL for insertion into EBR-II at the beginning of FY 1978. The fuel samples were clad with two commercially available advanced cladding alloys, I-706 and PE-16, which had been shown in cladding- and duct-materials irradiations to have improved in-reactor swelling and creep behavior relative to the reference cladding (20 percent cold-worked Type-316 stainless steel). The test aims to demonstrate the in-pile performance—specifically, greater burnup capability—of advanced-alloy cladding pins. The test assembly is now at about 50 MWD/kg burnup and reached goal burnup (80 MWD/kg) near the end of 1979; it will not be removed at goal burnup but will be irradiated to cladding breach.

Six advanced alloys were selected in March 1978 as the prime candidates for further development for cladding and duct applications in fast breeder reactors. The alloys were chosen from a final list of 10 alloys on the basis of improved in-reactor swelling and creep performance, evaluation of their potential in fuel element, duct, core restraint, and absorber assembly applications, and projections of performance including cost benefits in optimized core designs.

In September 1978, helium-bonded carbide pins and sodium-bonded carbide pins in assemblies attained a goal burnup of 12 atompercent in EBR-II. These results demonstrate that prototypic carbide pins designed for low-doubling-time fast reactors can achieve a high burnup without cladding failure.

The FY 1980 work scope in clad/duct development will build on the efforts begun in earlier fiscal years in EBR-II. Early tests in FFTF will include specimens of both reference and advanced cladding and duct alloys, which will be irradiated to determine their swelling behavior and mechanical properties during and after irradiation to high fluence. This effort is an important adjunct to the fuel performance activities inasmuch as it is believed that the use of advanced cladding and duct alloys will allow a step change improvement in the performance of breeder reactor fuels of all types.

The absorber materials development work scope in FY 1980 parallels the fuels development activities inasmuch as the reference FFTF absorber assemblies will also be under active surveillance and tests of an improved design will also be inserted into FFTF during this fiscal year.

Development work continues in two areas: the development of fuel fabrication equipment and processes, and the supply of driver fuel with which to operate FFTF.

Process development will focus on fabrication process and equipment development for both low gamma and high gamma alternate/advanced fuels, on both the pellet and particle fuel forms, including blanket fuels. The objectives are to ensure the availability of fabrication technology to support the breeder option, and specifically to support key decisions for a research and development plant. The program in FY 1980 will build on work initiated in FY 1979 and will encompass the following:

- Studies will continue to review, optimize, and improve carbide fabrication processes and equipment. Powder preparation, compaction, sintering, inspection, analysis, and loading studies will be continued. Orders for carbothermic and sintering furnaces will be placed. Required analytical standards and methods development will be carried out.
- Process automation of low gamma fuels will continue. Cold testing of powder operations will be completed and equipment interfacing/integration will start, leading to cold testing of pin line operations. Tests of the canister receiving, powder metering, pellet press, sintering operations and radiography will be completed.
- Development of equipment and processes for remote operation and maintenance for both the pellet and particulate fuel forms will continue. Modifications or redesigns of automated pellet fabrication equipment will be made where feasible. Where required, new concepts will be developed to achieve simplified operations and easier maintenance. Conversion to remote operation will continue for both ceramic and pin fabrication steps within the guidelines of the integrated remote system study completed in FY 1979. Studies on particulate fuels will focus on areas of highest technical fabrication uncertainty (effect of irradiation hydrolysis, stable Pu/U solutions, shipping and handling, pin loading and density) with initial screening completed by mid-FY 1980.
- Technical support programs including safeguards (real-time accountability), personnel exposure and effluent control, and close coupled chemistry will continue in support of the fuel fabrication development activities and the design and operation of the rescoped FMEF (i.e., the combined FMEF/HPFL). High gamma fuels introduce a new level of uncertainty with respect to fuel inspection and assay, chemical analysis, waste and effluent control, and facility design features and operating schemes to reduce personnel exposure. These activities are necessary to meet the major program milestones.

The above activities will provide the basis for merging the oxide and carbide fabrication technology in FY 1981 and accomplishing major activities necessary for selecting facility and equipment designs for high gamma (pellet or particulate) fuel fabrication. In addition, necessary process development activities for blanket fuels (thorium oxide, uranium or thorium carbides) will be examined.

Momentum will be maintained on development of equipment and processes for advanced and alternative proliferation-resistant fuel systems to support the breeder option. This work must provide the technical base for equipment and processes capable of being remotely operated and remotely maintained in order to handle highly radioactive fuels, and must incorporate safeguards improvements, low personnel exposure, and improved waste and scrap management. A rotary pellet press, designed and modified for automated and remote operations to facilitate maintenance, was delivered to HEDL. Tests on this press have been highly successful. An anvil press is also being tested. Ceramic and mechanical processes and equipment are being tested and evaluated to ensure their availability for the early 1984 startup of the rescoped FMEF, where tests will continue in the research and development of remote operations. Analytical and nondestructive inspection techniques for process control and materials accountability are also being tested and evaluated. The acquisition of fuel for FFTF operation involves outside procurement of fuel pins and subsequent manufacture of the pins into completed fuel assemblies by HEDL.

Major milestones and funding for the Fuels and Core Materials Activity are presented in Figure IV 4-18.

Participants. The major contractors in the development of fuels and core materials and their areas of involvement are presented in Table IV 4-10.

Materials and Chemistry. The materials and chemistry technology development task is primarily concerned with materials for use outside the reactor core, although the work also provides support for the development of alternative fuel technologies. The materials task includes establishing operational limits on current component materials and developing advanced structural alloys, hardfacing alloys, and cover-gas seal materials. The chemistry task develops methods for monitoring and controlling both radioactive and nonradioactive impurities in sodium and the reactor cover gas. Process development to support maintenance operations such as sodium removal, radioactivity decontamination, and subsequent component requalification are also included.

An adequate materials and chemistry base technology, which is required to achieve safe, economical, and reliable design and operation of FBR's, provides plant designers with the data needed for design and safety analysis. It also provides development information that can be used to select optimum materials and operational parameters for advanced designs. As FBR technology matures, this development work will provide improved materials and chemistry technology beyond the current state-of-the-art, thereby helping to improve the overall economics of the FBR.

The objectives of this task are as follows:

- Identify the operational limits of materials for near-term FBR applications
- Develop alternative structural alloys for more economic, safe, and reliable FBR's and determine the compatibility of advanced fuel-cladding alloys with sodium

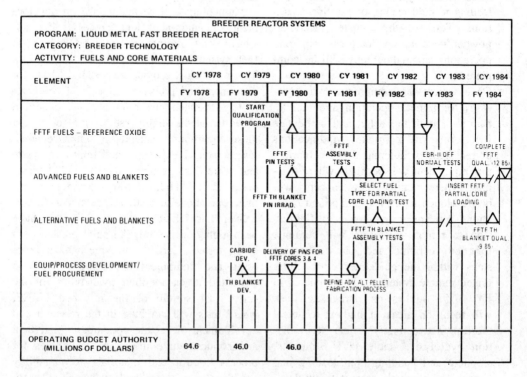

Figure IV-4-18. Major milestones and funding for the fuels and core materials activity.

Table IV-4-10. Major contractors and areas of participation in the fuels and core materials activity.

	AREA OF PARTICIPATION		
	FFTF FUELS & MATERIALS	ADVANCED & ALTERNATE FUELS	FUEL FABRICATION PROCESSES
LASL	X	X	X
ANL	X	X	
Battelle Columbus Laboratory		X	
Westinghouse Electric Corp.	X	X	
HEDL	X	X	X
Atomics International	X	X	
ORNL			X
General Electric Co.		X	
Combustion Engineering		X	

- Develop nondestructive in-service inspection methods
- Establish decontamination and requalification methods for FBR components
- Provide methods to monitor, reduce, and control coolant impurities, including radionuclides

The materials program emphasizes the development of technology required for the near-term FBR's, primarily by working directly with plant designers to define the immediate needs and schedule; this approach has proven highly profitable in that deficiencies in design have been identified and corrected before plant startup. Concurrently, the work provides the technology base needed for upcoming reactor experiments in FFTF and for understanding the experimental results and operational characteristics of EBR-II. The technology is also developed for materials to be used with advanced fuels and plant designs including alternative structural and fuel-cladding alloys, improved hardfacing materials, and high-temperature cover-gas seals.

The strategy for the chemistry program is similar in that it aims primarily at the near-term coolant-monitoring and impurity-control requirements of FFTF, followed by substantial emphasis on component decontamination and requalification needs of FFTF in the near future. Improvements in impurity monitoring and control instruments and components are also being developed.

This work is organized in six major elements, four of which concern primarily materials and two of which involve chemistry:

- Materials
  — Mechanical properties
  — Fabrication and inspection
  — Hardfacing and seal materials
  — Advanced clad compatibility and structural alloy development
- Chemistry
  — Radioactivity control
  — Sodium removal and decontamination

Basic test data are being obtained for such material properties as creep life, fatigue life, fracture toughness, cumulative damage relationships, and environmental (coolant and irradiation) effects on the properties of structural materials. The materials of interest are the present-day reference steels and a modified 9Cr-1Mo steel for future use.

Aging studies on stainless-steel Type-304 and -316 welds were completed in September 1978. The studies evaluated the elevated-temperature phase stability of stainless-steel welds made using 16-8-2 filler metal. Specimens aged up to 10,000 hours were examined to determine the extent of phase transformations in the weldments that could influence mechanical properties. Additional studies are required to separate the aging processes so as to understand the effect on mechanical properties.

Improved fabrication methods (e.g., tube-to-tubesheet and bimetallic welds) and inspection methods for use before and after reactor plant startup are being developed. Major emphasis is on the fabrication and inspection needs for reference plant materials; increasing emphasis will be on advancing the current state-of-the-art in these areas.

The operational limits of hardfaced components now in place at FFTF and in the design stage for other FBR's are established. Work is starting on improving hardfacing-alloy applications for optimizing component performance and reliability. Elastomer seals are being developed with lifetimes exceeding 5 years at temperatures above 105°C.

An improved structural alloy (modified 9Cr-1Mo steel) is being developed with greater economic, safety, and reliability characteristics. Alloy composition is being optimized. The stability of the alloy under the environmental and thermal conditions of FBR's is being studied. Near-term development will concentrate on optimization of composition, property evaluations, fabrication and codification. The work on advanced cladding alloys is limited to studies of sodium corrosion and mechanical properties in sodium.

On-line traps are being developed to reduce radionuclide levels in FBR coolant systems. In addition to radioactive-species control, this work includes improving sodium-system cleanup devices and on-line instruments.

In the area of sodium removal and decontamination, suitable cleanup methods are being established for component maintenance and failed-fuel shipping. Procedures are being developed for cleaning and chemically decontaminating components—removing deposited radionuclides from component surfaces with minimum deleterious effects on subsequent component performance. These developments are of immediate use for EBR-II, near-term use for FFTF, and future use in advanced designs.

Planned materials program accomplishments in FY 1980 include the following:

- Completion of 5000-hour mechanical property and corrosion tests on commercial ingot (20-ton) products of modified 9Cr-1Mo steel, as part of a program begun in 1976 to supply designers by 1983 with improved structural alloys for FBR application
- Completion of planned mechanical property data acquisition on reference FBR alloys (304 and 316 stainless steel, 2-1/4Cr-1MO and Alloy 718)
- Completion of evaluation of 6000-hour sodium corrosion tests of six advanced cladding alloys

The chemistry activity is directed toward the development of methods of monitoring and controlling radioactive and nonradioactive impurity levels in sodium and cover gas and to the development of processes needed to decontaminate FBR components of sodium and radioactive deposits in order to facilitate maintenance operations. Providing technical support to FFTF operation in these areas is an important activity.

Planned chemistry activity accomplishments in FY 1980 include the following:

- Completion of a report on the evaluation of a 6000-hour EBR-II test of a nickel radionuclide trap and evaluation of cesium trap performance
- Preparation of a process standard for radionuclide decontamination of components
- Demonstration of a technique for unloading plugged cold traps to improve system performance

Major milestones and funding for the Materials and Chemistry Activity are shown in Figure IV 4-19.

The major contractors in materials and chemistry development and their areas of development are presented in Table IV 4-11.

Test Facilities

Reactor test facilities provide support and are an essential part of the Fast Breeder Reactor Program. This category of activity includes the operation and maintenance of existing test facilities. It also includes certain facility management and operating activities for safeguards systems at ANL, HEDL, and INEL, and landlord support at HEDL, INEL, and ETEC.

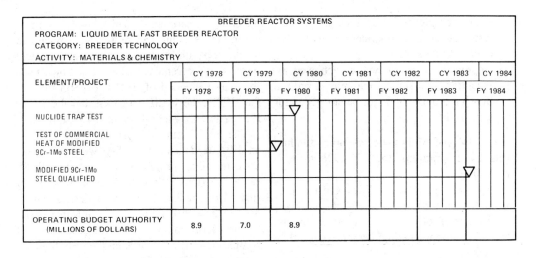

Figure IV-4-19. Major milestones and funding for the materials and chemistry activity.

A 1974 review by reactor manufacturers, national laboratories, and utilities concluded that an expanded program was needed to reduce uncertainties in component performance and resulted in greater demands on test facilities to proof-test components. The need for improvements in component reliability has also been reemphasized by the growing experience with light water reactors and fast reactors throughout the world showing that component failures have been the leading cause of plant downtime. With the decision to defer construction of developmental plants, the need for test facilities is intensified as a partial substitute for the component demonstrations that would have been obtained in the developmental plants. In particular, notwithstanding the proposed termination of the CRBRP project, many plant components will have been fabricated. Highly beneficial data can be obtained from these components by testing in non-nuclear test facilities.

Another factor that adds emphasis to the need for test facilities is the increase in size being considered for the next research and development plant. The size extrapolation from FFTF components to a factor of five or more will require adequate testing to identify and minimize risk.

The objective of this program element is to provide, operate, and maintain test facilities on schedules established by breeder technology efforts.

An improved testing capability is being achieved in several ways:

- Existing test facilities are being expanded in size to accommodate the required larger components
- Existing test facilities are being fitted

Table IV-4-11. Major contractors and areas of participation materials and chemistry activity.

	AREA OF PARTICIPATION			
CONTRACTOR	MECHANICAL PROPERTIES	HARDSURFACING AND SEALS	RADIOACTIVITY AND CONTROL	SODIUM REMOVAL AND DECONTAMINATION
ANL				X
Westinghouse Electric Corp.		X		
HEDL		X	X	X
ORNL	X			

with multiple test positions so several tests can be run simultaneously at a reduced cost per test
- Improved efficiency will be achieved by installing fuel-saving equipment on the breeder program's major steam generator test facility

Table IV 4-12 shows the operating funding levels by program for the FY 1978 through FY 1980 period.

The major operating facilities and the portion of the breeder program that they support are listed in Table IV 4-13, grouped by the laboratory or engineering center where they are located.

New facility projections planned or under construction are listed in Table IV 4-14.

ANL Facilities. ANL, operated by the University of Chicago and Argonne Universities Association, is located on a 1700-acre site near Chicago, Illinois. The major reactor facilities are located at INEL near Idaho Falls, Idaho.

The Argonne facilities include an operating breeder reactor (EBR-II), a unique device that provides reactor physics data on full-size liquid metal fast breeder reactor cores (ZPPR), and a large shielded facility used to examine and analyze highly radioactive fuels and materials (HFEF). The major ANL facilities are briefly described in the following paragraphs.

The EBR-II is a 62.5-MWt 20-MWe sodium-cooled pool-type fast reactor at ANL-West (ANL-W), INEL, Idaho. EBR-II is currently used primarily for safety, and fuels and materials experiments. This involves maintaining the existing capabilities with regard to reactor operation, engineering, analysis and material procurement. Major facilities used for total evaluation of EBR-II experiments are the HFEF, the LASL Hot Cell and the ANL-East (ANL-E) Alpha-Gamma Hot Cell.

The HFEF can be considered primarily as an EBR-II dedicated facility with secondary uses in other programs such as the SLSF safety program. With the present mission of EBR-II to irradiate fuels and materials experiments, the HFEF is used on a full-time basis.

The FHEF comprises two adjacent hot cell complexes—HFEF North and HFEF South—located at ANL-W. The south complex contains two heavily shielded hot cells, one with an argon inert atmosphere. Each complex includes unshielded repair areas, test laboratories, equipment rooms, storage areas, offices, and other facilities.

The HFEF provides: (1) operational support to EBR-II, (2) capability to examine and reconstitute experimental subassemblies irradiated in EBR-II for breeder reactor programs on fuels and materials, and (3) the capability to examine reactor safety transient experiments conducted in loops and/or capsules for the TREAT Program, the SLSF Program, and the SAREF Program.

The principal uses of HFEF are post-irradiations handling, examination, and analysis of fuels and reactor materials from breeder programs, including:

- Nondestructive (interim) examinations of experimental capsules and elements—

Table IV-4-12. Funding for test facilities for FY 1978–1980.

TEST FACILITIES CATEGORY	BUDGET AUTHORITY (DOLLARS IN THOUSANDS)			MAJOR CHANGE
	FY 1978	FY 1979	REQUEST FY 1980	INCREASE (DECREASE)
ANL Facilities	$27,095	$28,755	$30,088	$1,333
HEDL Facilities	9,609	4,836	5,750	914
ETEC Facilities	19,891	15,192	17,995	2,803
INEL Facilities	6,400	7,342	7,583	241
Safeguards	3,257	3,898	3,900	2
General Support	9,074	9,977	9,984	7
TOTAL	$75,328	$70,000	$75,300	$5,300

Table IV-4-13. Major operating facilities and supporting technology areas.

LABORATORY	FACILITY	TECHNOLOGY AREA
ANL	EBR-II	Fuels and Core Materials
	Hot Fuel Examination Facility (HFEF)	Fuels and Core Materials
	TREAT Facility	Safety
	ZPPR	Physics
	Zero Power Reactors -6 & -9 (ZPR-6 & -9)	Physics
HEDL	High Temperature Sodium Facility (HTSF)	Components
	Small Component Sodium Test Facility Complex	Components
	Hydromechanical Facilities	Fuels and Core Materials
	Fuels Development and Analysis Laboratory	Fuels and Core Materials
ETEC	TTF	Components
	SCTI	Components
	SPTF	Components
	LLTR	Components
	Small Components Test Loop (SCTL)	Components

subassembly dismantling and reassembly, visual and photographic examinations, gamma scanning, and weight dimensional measurements
- Examinations of fuel, cladding, and structural materials—cutting, fission-gas sampling, and preparation and examination of metallographic samples
- Handling, disassembly, examination, and assembly of TREAT and SLSF loops
- Preparation for off-site shipment of irradiated components for destructive examinations elsewhere
- EBR-II reactor support services—surveillance of driver-fuel and handling, examination, and disposition of spent driver-fuel, blanket and reflector subassemblies and of irradiated reactor components
- In-cell neutron radiography

Operation, maintenance, and upgrading of the Zero Power Reactor (ZPR) facilities include the necessary engineering, procurement, modifications, and construction activities for ensuring the safe and reliable performance of fast critical experiments supporting the ongoing and new requirements of the DOE breeder reactor physics/safety programs. Also, support is provided to other agencies (e.g., NRC) when necessary, and cooperative efforts are arranged.

ZPR-6 and ZPR-9 are located at ANL-E, while ZPPR is located at ANL-W. The ZPPR is the largest of the critical facilities with a matrix of 14 ft × 14 ft × 10 ft and is equipped with a matrix loading machine designed and fabricated by General Electric Company. The loading machine greatly reduces manual loading/unloading operations and thereby miminizes personnel exposure to radiation in accord with DOE policy.

The facilities provide information on reactor performance parameters from critical assemblies such as benchmark or engineering mockup experiments for both the sodium-cooled and gas-cooled breeder reactor programs. Critical experiments conducted at low power ("zero power") provide reference-design-related data such as fuel inventory and loading arrangements, control worths, power distributions, fuel conversion, and reactivity coefficients.

ZPR-6 will be needed to conduct Safety

Table IV-4-14. Facility projects.

LABORATORY	FACILITY	TECHNOLOGY AREA
ANL	Safety Research Experimental Facilities (SAREF)	Safety
HEDL	FMEF	Fuels and Core Materials
	Maintenance and Storage Facility (MASF)	Fuels and Core Materials
	Fuel Storage Facility (FSF)	Fuels and Core Materials

Test Facility (STF) experiments in support of the SAREF project because the STF criticals will require a matrix material (e.g., aluminum or other) different from the stainless steel presently in ZPR-9. Since ZPR-6 has been in standby for some time, no reactor shutdown is necessary and the extended time required for the special matrix preparation does not impede other programs. The SAREF Project is under review and consequently preparations for the use of ZPR-6 for STF criticals have been delayed.

TREAT at ANL-W is an air-cooled, thermal, heterogeneous reactor. The construction of TREAT was started in February 1958 and was completed in November 1958. The first criticality was achieved in February 1959 and transient testing was started in July 1959. Since that time, TREAT has been the primary transient test facility for the fuel of EBR-II, Hallam, SNAP, PBF, SEFOR, and FFTF. The first neutron radiography in TREAT was conducted in January 1967. Currently, TREAT provides the capability to irradiate reactor fuels and structural materials under conditions simulating various types of nuclear excursions and transient undercooling situations. It is also available for limited neutron radiography. (See Figure IV 4-20.)

TREAT Upgrade is part of the SAREF described below.

One of the breeder reactor support facilities under consideration for construction is SAREF. This project was initially planned to consist of four parts: a new reactor STF; upgrade and expansion of TREAT; required Support Facilities (SF); and the modification of EBR-II to a safety facility (EBR-II SRM). The breeder reactor safety program realignment has established the priority for construction of the SAREF Project so that it will be accomplished in phases and has eliminated the requirement for EBR-II SRM.

Phase I is the TREAT Upgrade (TU). It allows larger fuel pin bundles of fully prototypic preirradiated fuel for FBR's to be subjected to severe thermal transients to determine fuel, clad, and coolant behavior when subjected to low probability severe accident conditions.

As a result of the LMFBR program redirec-

Figure IV-4-20. TREAT reactor building.

tion, the need for Phase II (SF) and Phase III (STF) of SAREF will be considered on a continuing basis.

Major milestones and funding for ANL facilities are presented in Figure IV 4-21.

HEDL, at Richland in southeastern Washington, is operated for DOE by Westinghouse Hanford Company, a subsidiary of Westinghouse Electric Corporation. The laboratory is near the tri-city area of Richland, Kennewick, and Pasco (population 90,000), and is the engineering development laboratory for the LMFBR Program.

The primary facility at HEDL is the FFTF, which will be used for fuel development. Various other facilities at HEDL enable testing of reactor components, such as refueling equipment, control rod drives, core resistant devices, etc. HEDL facilities also have an important role in the design of reactor internals, heat transport system layout, and sodium chemistry, purification and cleaning. Other facilities are used to analyze nuclear fuel, examine irradiated fuel, and improve the processes involved in fuel fabrication.

Besides the FFTF, the major test facilities at HEDL are the HTSF, the Small Sodium System (SSS), and the Hydromechanical Facilities. The HTSF and SSS will be maintained in standby so that tests can be conducted, if necessary. Hydraulic and vibration tests will be conducted in the Hydromechanical Facilities to verify the design of LMFBR components.

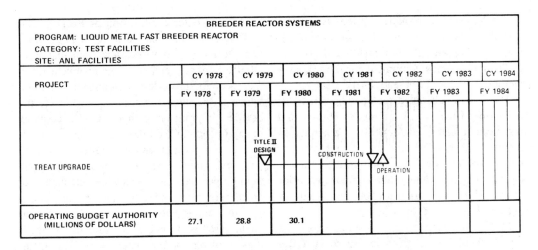

Figure IV-4-21. Major milestones and funding for ANL facilities.

The FMEF will be a major facility of the U.S. breeder reactor program for fuel development, fuel fabrication development, and post-irradiation examination of the full range of advanced and alternative reactor fuels.

This facility will provide essential support to the fuel performance and process development programs through its capability to fabricate test pins, support fabrication and process development, perform required analytical chemistry on fresh and irradiated fuel pins, disassemble and reconstitute fueled and non-fueled test assemblies, and examine test pins following irradiation. FMEF will also provide remotely operated and maintained equipment which can be used for testing proliferation-resistant fuel system processes, and also for evaluating real-time accountability and other safeguards features.

The FMEF and High Performance Fuel Laboratory (HPFL) projects were carried as separate line items in the FY 1978 budget. Subsequently, the emphasis on enhanced non-proliferation advanced/alternative fuels made it apparent that future fuel fabrication technology would have to handle highly radioactive feed material. A study was conducted to determine how best to accommodate these changing functional requirements, and a natural outgrowth was the combination of these two facilities. Except for the previously planned integrated prototype high throughput fabrication and related wet scrap reprocessing capabilities, the combined R&D facilities meet the essential functional requirements originally defined for the separate facilities. Added shielded work space has been included to house and test remotely operated and maintained equipment and processes for developing proliferation resistant fuel systems. Combining the two facilities also had the significant financial benefit of a 25 percent reduction in the total estimated cost (TEC) (i.e., $170 million vs. $228 million for separate facilities).

The FSF, when completed in FY 1981, will provide sufficient spent fuel storage capacity to permit five to eight years of FFTF operation (about 464 fuel positions). FFTF spent fuel is to be stored in FSF until the ultimate method of disposing of irradiated fuel is defined.

The FFTF has an in-containment storage capability for 112 irradiated fuel assemblies in the Interim Decay Storage Vessel. This capability will enable operations through mid-1981. Lack of or delay of the construction of FSF would thereafter interrupt FFTF operation.

The current cost estimate of $30 million is $5 million less than the previous cost estimate, realized through the use of simplified heat removal systems and less costly materials. The FSF will adjoin FFTF. Fuel handling within FSF will be accomplished, in part, using FFTF fuel handling equipment.

The MASF will provide, by the last quarter of 1982, a facility for cleaning, decontaminating, and repairing equipment and storing FFTF maintenance equipment for the 400-area facilities (FFTF, FSF, and FMEF).

The FFTF, scheduled for operation in 1979, will use MASF for handling sodium-wetted and contaminated components. Capability is required at the 400-acre site to remove sodium from nonfuel core components, maintenance equipment and other components prior to hands-on support for 400-area facilities does not otherwise exist.

Major milestones and funding for the HEDL facilities are given in Figure IV 4-22.

ETEC Facilities. ETEC is located at the Nuclear Development Field Laboratory in the Simi Hills of southern California. The 90-acre site is approximately 15 miles from the Los Angeles suburb of Canoga Park and about 40 miles northwest of the Los Angeles Civic Center. The ETEC is operated for DOE by the Energy Systems Group of the Rockwell International Corp. (Figure IV 4-23).

As has been stated in previous sections, reviews by reactor manufacturers, national laboratories and utilities, the growing experience with all types of power reactors, the decision to defer construction of R&D plants, and the size extrapolation from FFTF components to those of the next developmental plant all place greater emphasis on nonnuclear test facilities to proof-test components. ETEC, which is the component testing center for the breeder program, is composed of a complex of liquid sodium facilities for testing and evaluating components and instruments such as heat exchangers, steam generators, valves, piping, pumps, and mechanical elements for breeder reactors and other energy conversion systems employing high-temperature liquid-metal systems. Principal component test facilities available at the center are: SCTI, Components Handling and Cleaning Facility (CHCF), SCTL, and the Sodium Pump Test Facility (SPTF). ETEC was established as the prime facility for testing very large sodium components.

SPTF is a large nonnuclear sodium loop with auxiliary equipment that can test large sodium pumps, flowmeters, and other large sodium components. An associated CHCF has large cranes, sodium cleaning equipment, and maintenance facilities to support components tested in SPTF and elsewhere at ETEC.

SPTF is currently being expanded under the ETEC Facility Modification Project (Project

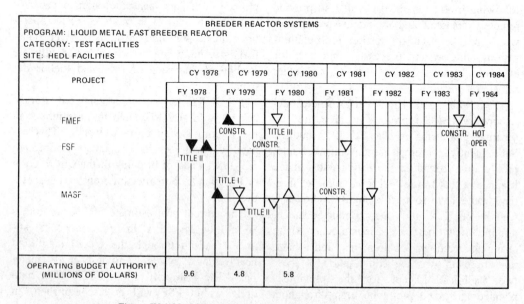

Figure IV-4-22. Major milestones and funding for HEDL facilities.

Figure IV-4-23. ETEC.

No. 78-6-e) to increase its capacity to over 85,000 gpm; this will make it the largest sodium pump test facility in the world. It can provide extensive flow mapping of the pumps and can impose thermal transients that would simulate the most severe conditions that would be encountered in an actual plant.

Testing has been completed on a prototype FFTF pump. This pump will soon be reinstalled with a new impeller design for tests of a new impeller concept. After completion of the expansion program, the CRBR pump will be tested, followed by testing of pumps in the 85,000 gpm range (Figure IV 4-24).

The SCTL is a 10-inch sodium closed circulation loop with a 3,500 gpm main pump and a 9,500 gallon expansion tank that can be used for immersion testing up to 1,200°F. There is also an 8-inch loop in which transient testing is conducted.

The SCTL has a wide range of capabilities and has tested a variety of components including a secondary isolation valve for FFTF, electromagnetic (EM) pumps, valves, and cold traps. At present, an experimental inducer-type pump is under test. At the conclusion of this test, the experimental pump will be replaced with a 3,500 gpm facility pump having a 1,200°F maximum temperature and 520-ft head rise. Plans are in progress to test CRBR components including small valves, EM pump, and an In-Vessel Transfer Machine (IVTM). (See Figure IV 4-25.)

The SCTI is a high-heat-throughput sodium-cooled heat exchanger test facility that is being expanded from 35-MWt to 70-MWt. In its expanded form, the SCTI will be the largest sodium heat exchanger test facility in the world. Previous tests include model steam generators, an intermediate heat exchanger, and the FFTF dump heat exchanger. Future scheduled tests include the CRBR steam generator, two advanced steam generators, and tests of other heat exchangers. Another addition to the facility is planned to facilitate testing of small components such as few-tube steam generator models and small auxiliary heat exchangers.

The LLTR is a sodium-water-reaction test facility comprised of a scaled-down LMFBR

Figure IV-4-24. Sodium pump test facility.

secondary-sodium heat transfer system, steam generator, and vent relief system. Capability is provided for realistic water/steam injection into the sodium side of a prototypical steam generator from an intentionally faulted water/steam tube. Test results are used to validate and/or modify analytical techniques suitable for predicting the consequences of major sodium-water reactions on the steam generator and other secondary system components (Figure IV 4-26).

The AI Modular Steam Generator was the first item to be tested in LLTR, with six tests completed in November of 1977. The facility has been modified to incorporate a large leak test vessel in which steam generator test internals, representing the reference design of the CRBRP steam generator, will be installed for testing. The first test of this series is scheduled for early 1979.

Major milestones and funding for the ETEC facilities are given in Figure IV 4-27.

INEL Facilities. INEL, located in southeastern Idaho between Idaho Falls and Arco, conducts advanced energy research in nuclear reactor and fuel cycle technology as well as environmental, geothermal and other non-nuclear energy research. The 893-square mile site includes eight multi-facility research areas operated by four major contractors: EG&G Idaho, Inc., Allied Chemical Corporation, ANL, and Westinghouse Electric Corporation. Research for the LMFBR Program is conducted in test facilities at two of the eight INEL research areas, the ANL-W Area operated by ANL and the ETR located at the Test Reactor Area and operated by EG&G.

The ETR is a 175-megawatt research test reactor that has been in operation since 1957. In 1972, ETR began preparations for a new role as a key test facility in support of DOE's breeder reactor safety programs. The objective is to use the ETR to conduct tests contributing significantly toward verification of the safety characteristics of reactor fuel and core designs for reactors using liquid metal as the primary coolant. The ETR was modified to accept a series of tests, performed by EG&G, in large packaged sodium in-reactor loops under the SLSF Program. Each SLSF loop provides the controlled environment in which reactor safety irradiation experiments can be conducted under conditions closely simulating the thermal-hydraulic conditions, both steady-state and transient, in fast breeder reactor systems.

ETR modifications started in 1973 included the addition of a new reactor top closure, a helium coolant system, and a sodium handling system. SLSF tests started in October 1975 and are scheduled through FY 1984.

Safeguards. An objective of the DOE safeguards and security program is to implement upgraded physical protection systems, techniques and procedures in conjunction with improved methods of materials control and accountability in order to achieve improved integrated safeguards systems for those DOE facilities that acquire, use, store, and dispose of special nuclear materials (SNM). It is intended that these upgrading activities will provide, to the maximum extent practicable, the protective measures necessary to detect and prevent loss or diversion of SNM from authorized use by insiders; prevent theft by outsiders; and protect nuclear materials and facilities against malevolent acts, attack, or intrusion.

Figure IV-4-25. SCTL and SCTI site.

The primary policies and guidelines for implementing DOE safeguards activities are set forth in DOE Orders which are currently being strengthened and updated to meet the more rigorous demands of today's environment. The overall scope of these guidelines encompasses a wide spectrum of operations, equipment, and construction activities which will ultimately be integrated and implemented in conformance with safeguards plans designed to meet the specific needs and characteristics of a given DOE facility or site.

Argonne National Laboratory-East and West (ANL-E/W), INEL, and HEDL represent typical DOE facilities/sites where intensive efforts are underway to upgrade safeguards in compliance with emerging new requirements in DOE Orders. A significant portion of these upgrading efforts consist of Line Item construction projects at the respective sites. Budgetary considerations have necessitated a phased approach such that major construction at ANL was approved for initiation in FY 1978, FY 1979 for initiation at HEDL, and it is planned to commence construction in FY 1980 at INEL.

ANL-E/W operating funds will support the following activities: continuation of SNM physical control, accountability and management; achievement of full use of a special materials information system (SMIS) program and nondestructive assay (NDA) laboratories; achievement of full use of upgraded security guards, arms and electronic devices and occupancy of new security facilities as they become available; maintenance of surveillance of the safeguards construction program to ensure conformity with the site specific safeguards plan.

HEDL safeguards operating funds will be used to operate systems to control, account for, report, measure, and inventory SNM, and to conduct studies and develop new systems to improve these capabilities.

INEL safeguards operating costs are liquidated through allocable overhead charges and are not a direct part of the funding indicated.

668 Fission Energy

Figure IV-4-26. Sodium-water-reaction test facility.

Major milestones and funding for the safeguards activity are presented in Figure IV 4-28.

General Support. The general support area consists of three subcategories: landlord activities, program assurance, and other support.

The landlord activities at ANL, HEDL, ETEC, and INEL including providing various general site services, providing general purpose equipment (GPE), and constructing general plant projects (GPP) and line item construction projects for general purpose facilities and certain other multi-program projects. The objectives of the landlord activities at the four sites are to achieve safe and efficient operation; to prevent further deterioration of site support facilities, systems, and equipment; and to provide and preserve facilities and equipment required to perform R&D for DOE assigned programs. Landlord activities are conducted by the major operating contractors at each of the four sites using subcontractors selected by competitive bidding for the larger construction projects.

Continuing operation of the Nuclear Quality Assurance Program Office at HEDL, the Nuclear Standards Office at ORNL, and Safety Assessment Offices at ANL and HEDL is supported under the program assurance subcategory.

Other support include facility fire inspections, INFCE coordination, and information transfer and assessment activities.

WATER COOLED BREEDER

This program is developing technology to improve significantly the use of nuclear fuel

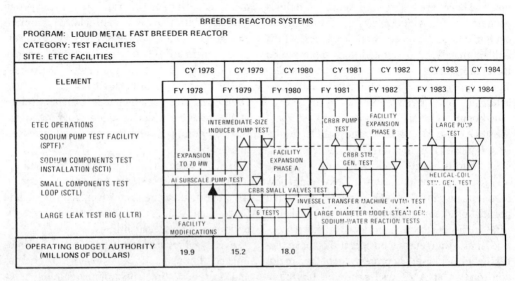

Figure IV-4-27. Major milestones and funding for ETEC facilities.

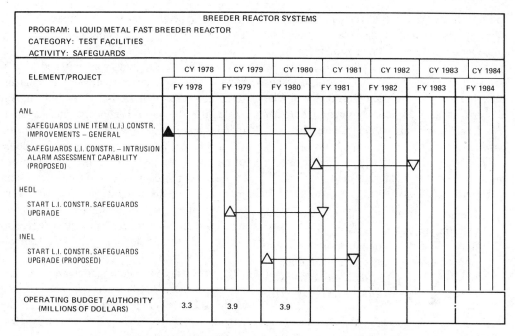

Figure IV-4-28. Major milestones and funding for the safeguards activity.

resources for generating electrical energy with water-cooled nuclear reactors.

Successful development of this breeder cycle would make available for power production about 50 percent of the energy potentially available in thorium reserves, a source of energy many times greater than known fossil-fuel reserves. This is the only known approach using light-water-cooled reactors that can make available for power production significantly more than 1-2 percent of the energy potentially available in nuclear fuel reserves.

Overview

In the early 1960s, the Knolls and Bettis Laboratories, under the direction of the Naval Reactors Division, showed that it might be possible to develop a practical self-sustaining breeder cooled with ordinary (light) water and fueled with uranium-233 and thorium. The principal development necessary was the breeder core itself. Such a breeder would have the advantage of using the well-established technology of LWR's, which is the basis of the present commercial nuclear power industry. This early work led to the establishment of the LWBR program in 1965.

The immediate objective of the Water Cooled Breeder (WCB) Program is to confirm that breeding can be achieved in existing and future LWR systems using the thorium/uranium-233 fuel system.

The LWBR concept has the potential of significantly increasing the amount of useful energy that can be produced from our nuclear fuel reserves using water-cooled reactors. One measure of this potential is the reduction in the amount of uranium that must be mined to produce a given amount of useful energy.

If it is assumed that a current commercial LWR with a plant electrical output of 1000-MWe requires mining about 130 tons of uranium ore concentrate (U_3O_8) per year, and that such a plant lasts for 40 years, then 5,200 tons of U_3O_8 would have to be mined (and enriched) to fuel the plant over its lifetime. An additional 5,200 tons of U_3O_8 would have to be mined to refuel a replacement plant over its 40-year lifetime, and so on.

Preliminary studies show that, using the LWBR concept, it might be necessary to mine only about 3,000 tons of U_3O_8 to make enough uranium-233 in prebreeder cores to fuel a 1,000-MWe breeder. Once enough uranium-233 is accumulated to put an LWR on the breeding cycle, the increment of electrical capacity represented by that reactor would become self-sustaining without further mining of uranium. The only makeup fuel required would be small amounts of thorium. Thus, each increment of capacity using the LWBR concept, instead of present types of commercial LWR's, would greatly reduce the long-term uranium mining requirements.

The water-cooled-breeder fuel cycle is based on the use of thorium and uranium-233 and consists of two phases:

- Prebreeder phase: U-235 or other fissile material would be used to fuel LWR's that would also contain thorium; operation of these reactors would produce fillile U-233 from the thorium while providing power for generating electricity.
- Breeder phase: U-233 produced in the prebreeder reactors would be used to fuel breeder reactors that are expected to be self-sustaining—that is, they would produce enough fissile fuel to replace the fuel used in generating electricity and require no mining or enrichment of uranium for fissile-fuel makeup.

The work to achieve the goal of the WCB Program involves the following:

- The Shippingport Atomic Power Station (PWR) Program—the continued operation and testing of the station
- The LWBR Program—the development, manufacture, operation, examination, and evaluation of the Shippingport LWBR core
- The Advanced Water Breeder Applications (AWBA) Program—the development and dissemination of technical information on the LWBR concept

An LWBR core has been developed, designed, fabricated, installed, and is now operating in the DOE-owned pressurized-water reactor plant of the Shippingport Atomic Power Station at Shippingport, Pennsylvania. It is expected that the core will be operated for three to four years, after which it will be removed and the spent fuel will be shipped to the Naval Reactors Expended Core facility in Idaho for detailed examination and determination of breeding performance. It is expected that the expended core will contain about 1 percent more fissile material than the initial loading.

The AWBA Program was initiated in FY 1976 to develop and disseminate technical information that will assist U.S. industry in evaluating the LWBR concept and its commercial application. Over the next five to ten years, the AWBA Program will explore some of the problems that will be faced by industry in adapting technology information and promptly disseminating it. In this way vendors, utilities, and other interested organizations will be able to determine the degree to which they desire to apply LWBR technology.

Reprocessing and refabrication work associated with the LWBR concept is not included in the WCB Program. Such work is under the cognizance of the DOE Division of Nuclear Power Development.

The Assistant Secretary for Energy Technology is responsible for the WCB Program. This program is assigned to the Director of Nuclear Energy Programs (ETN), who delegates management responsibility to the Director of the Naval Reactors Division (NR). The NR Director provides overall direction and program approval, and is responsible for successful implementation of the program within overall cost and schedules.

The WCB Program is carried out primarily at two government-owned laboratories: Bettis Atomic Power Laboratory (BAPL) in Pittsburgh, Pennsylvania, and Knolls Atomic Power Laboratory (KAPL) in Schenectady, New York. These laboratories are operated for DOE's Naval Reactors Division by Westinghouse Electric Corp., and General Electric Co., respectively. BAPL also manages the DOE's Expended Core Facility in Idaho, where most of the spent-core examination, including nondestructive proof of breeding tests, will be performed.

Table IV 4-15 shows the operating funding levels for the WCB Program by program elements for the FY 1978 to FY 1980 period.

Shippingport Atomic Power Station

The Shippingport Atomic Power Station, 25 miles northwest of Pittsburgh, Pennsylvania,

Table IV-4-15. Funding for the WCB program for FY 1978–1980.

WATER COOLED BREEDER PROGRAM ELEMENT	BUDGET AUTHORITY (DOLLARS IN THOUSANDS)			
	FY 1978	FY 1979	REQUEST FY 1980	INCREASE (DECREASE)
Shippingport Atomic Power Station	$11,100	$11,500	$12,600	$1,100
Light Water Breeder Reactor	15,800	25,300	30,900	5,600
Advanced Water Breeder Applications	11,700	14,100	14,400	300
TOTAL	$38,600	$50,900	$57,900	$7,000

was the first large-scale central-station nuclear powerplant in the United States and the first plant of such size (initial operation, 60-MWe net) in the world operated solely to produce electric power (Figure IV 4-29). This project was started in 1953 by the U.S. Atomic Energy Commission (now DOE) to confirm the practical application of nuclear power for commercial purposes. Initial criticality of the first Shippingport reactor core was achieved in December 2, 1957.

Operation of the Shippingport plant with PWR cores has provided much of the technology for designing and operating the commercial central-station nuclear powerplants now in use.

The objective of the Shippingport Atomic Power Station continues to be the advancement of PWR basic technology, including the development and testing of advanced reactor concepts. The operation of the LWBR core (Figure IV 4-30) in the Shippingport plant is an extension of the kind of development work that has been carried out at this facility since its initial operation in 1957.

The Shippingport Atomic Power Station reactor plant will operate for three to four years with the LWBR core (Figure IV 4-31).

Figure IV-4-29. Shippingport atomic power station.

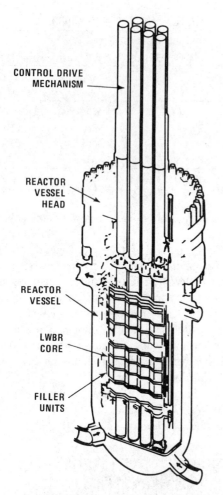

Figure IV-4-30. LWBR core in Shippingport reactor vessel.

The Shippingport Atomic Power Station reactor plant with the LWBR core installed was release for routine commercial operation to supply power to the Duquesne Light Co. distribution system on December 1, 1977 (Figure IV 4-32). During FY 1979, the plant continues to operate with the LWBR core installed. In addition, disposal of the fuel from the previously installed PWR Core-2 is being completed and examination of retained fuel assemblies continues. Preparations for disassembly and disposal of the PWR Core-2 upper core barrel are being completed, including procurement of the shipping cask.

Major milestones and funding for the Shippingport Atomic Power Station are given in Figure IV 4-33.

The Shippingport Atomic Power Station, which is operated under the WCB Program, is a joint project of DOE and the Duquesne Light Co. DOE owns the reactor-plant portion of the station; Dusquesne Light Co. owns the Shippingport site and the conventional turbine-generator portion of the station and operates the reactor plant under DOE contract. The Bettis Atomic Power Laboratory designed and developed the reactor portion of the Shippingport Station under the direction of the Division of Naval Reactors.

LWBR Program

The LWBR Program was established in 1965 to develop a breeder core using light water as the moderator. (See Figures IV 4-34 and IV 4-35.)

Operation of the LWBR core at Shippingport is being conducted primarily to accomplish three objectives:

- To prove that breeding can be achieved in a LWR nuclear powerplant using a thorium/uranium-233 fuel system
- To confirm a practical way to obtain energy from thorium
- To show that a LWBR core that uses thorium fuel can be installed in a pressurized-water nuclear powerplant using the same type of pressure vessel, pumps, heat exchangers, piping systems, etc., as are used in commercial PWR's.

The LWBR Program includes development of the supporting technologies required for the successful operation of a breeder reactor, including the following:

- Development of a thorough understanding of the nuclear and materials characteristics of the thorium/uranium-233 fuel system in water reactors
- Analytical and experimental development supporting improved performance of ZIrcaloyclad oxide fuel
- Engineering design of a practical movable-fuel control system that will accomplish all normal control functions without parasitic loss of neutrons
- Design of a reliable fuel-rod support system with minimum detrimental effect on neutron economy

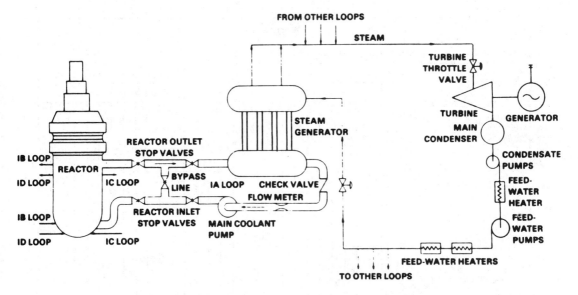

Figure IV-4-31. Shippingport plant operating cycle with LWBR core.

During FY 1979, operational inspections, tests, maintenance, and evaluation of the Shippingport plant LWBR core continue. The core thermal-hydraulic performance is being evaluated, documented, and periodically reviewed. The performance of the LWBR core-protection system is being checked. Periodic tests are being conducted in the areas of physics, thermal-hydraulics, and control drive mechanisms. Work to determine the actual useful life of the Shippingport LWBR core continues. Additional support and evaluation of fuel, cladding, and structural samples are being accomplished to determine the effect of irradiation on water-cooled reactors with the objective of providing data that could lead to improved fuel design parameters. Testing and operation of the plant provide valuable information on long-term performance of LWR's, systems, and components. The plant is being periodically operated in a swing-load mode as well as base-load mode, to prove out the fuel system under the most severe anticipated conditions.

Preparations for defueling and module disassembly of the spent LWBR core are in progress. Design of the fuel removal equipment, module disassembly equipment, fuel storage racks, and shipping equipment continues and procurement has been initiated.

Work on the shipping-container safety analysis reports for packaging continues. Plans for prototypical testing of equipment are being developed and preparation of defueling procedures is in progress.

After about three to four years of power operation, the LWBR core modules will be removed from the Shippingport reactor vessel and will be shipped to the Expended Core Facility (ECF) for detailed core examination

Figure IV-4-32. LWBR fuel installation.

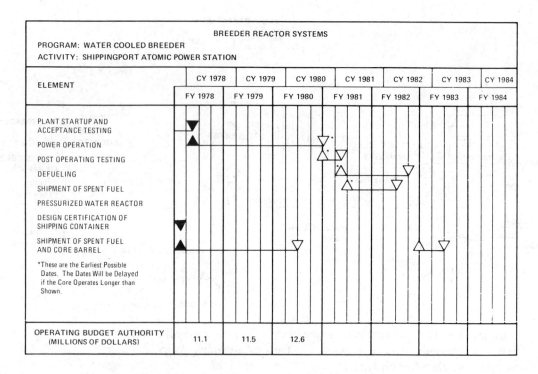

Figure IV-4-33. Major milestones and funding for the Shippingport atomic power station.

and determination of breeding performance. After the spent fuel is examined in detail, it will be turned over the the DOE Division of Advanced Nuclear Systems and Projects for storage and reprocessing. To support the core examination and determination of breeding performance, a construction project for extension of the ECF water pits, called the Breeding Nondestructive Assay Facility (BNAF) was approved as a part of the FY 1977 budget. Detailed design of the BNAF has been completed and construction was initiated during FY 1978. Evaluation of LWBR-type fuel and structural materials will continue to be carried out at the ECF.

After the LWBR core is removed, the Shippingport reactor plant will be turned over to the DOE Office of Waste Management for disposition.

Major milestones and funding for the LWBR are shown in Figure IV 4-36.

BAPL, operated by Westinghouse Electric Corp., provides technical support for and evaluation of tests performed at the Shippingport Atomic Power Station.

Advanced Water Breeder Applications

The Advanced Water Breeder Applications (AWBA) Program was initiated in FY 1976 as part of the WCB Program to develop and disseminate technical information that will assist U.S. industry in evaluating LWBR technology and applying it to their programs.

The program explores the technical problems that will be faced by industry in adapting technology confirmed in the LWBR Program. The focus of this program is on first developing technical information and then promptly disseminating it.

Information being developed in the AWBA Program includes concepts for commercial-scale prebreeder cores that will produce uranium-233 for light-water breeder cores while producing power for customers, improvements for breeder cores based on the technology developed for the fabrication and operation of the Shippingport LWBR core, and other information and technology to aid in the evaluation of commercial application of the LWBR technology.

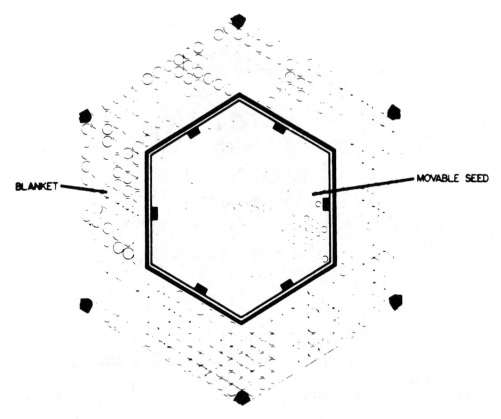

Figure IV-4-34. Cross-section of typical LWBR fuel module.

Enough uranium-233 must be available for the initial fuel in order to operate a breeder reactor on the U-233/thorium fuel cycle. The LWBR core for Shippingport was built using uranium-233 produced in DOE production reactors at Savannah River and Hanford. However, for commercial application, the U-233 would have to be produced using the prebreeder cores in commercial nuclear reactor plants.

The AWBA effort includes development of fuel-element concepts for producing uranium-233 in prebreeder cores that could be installed in existing commercial LWR's, as well as prebreeder cores for installation in new plants that are optimized for breeding. Selected concepts for advanced light- and heavy-water cooled breeders and prebreeders are also to be evaluated. Fuel-element support grids and advanced movable-fuel control systems that could reduce neutron losses and increase breeding potential for commercial applications are being evaluated (Figures IV 4-37 and IV 4-38). The direction of continuing efforts in later years will depend in large part on the results of current development work.

During FY 1979, selected concepts for advanced water-cooled breeders and prebreeders are being evaluated in conceptual design studies. Development of fuel technology and reactor component concepts is continuing to support these design studies. The results of new developments are being documented and disseminated to industry. Evaluations are also being completed for consideration by NASAP.

Fuel technology will be advanced through fabrication development to improve the understanding of thorium oxide (ThO_2) based fuels, by work on fuel systems that optimize the use of fissile materials, by evaluating possible advanced fuel elements to prove their value, and by work to qualify procedures for designing and analyzing fuel elements for both prebreeders and breeder reactors. Fuel-element

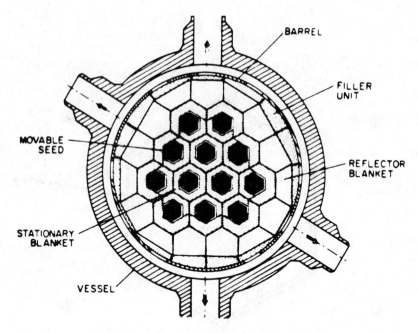

Figure IV-4-35. Cross-section of LWBR core.

support grids that could reduce neutron losses and increase breeding potential for commercial applications are being developed and evaluated. Basic corrosion and critical-heat-flux testing will be performed to better characterize heavy-water (D_2O) properties for

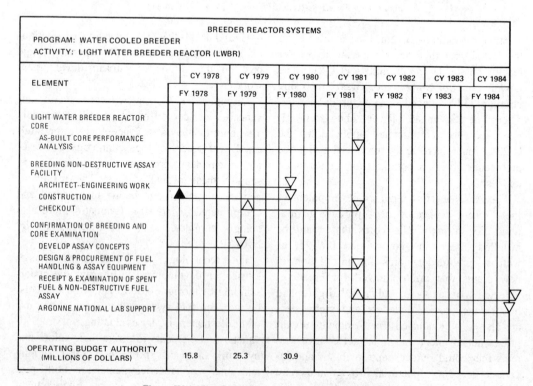

Figure IV-4-36. Major milestones and funding for LWBR.

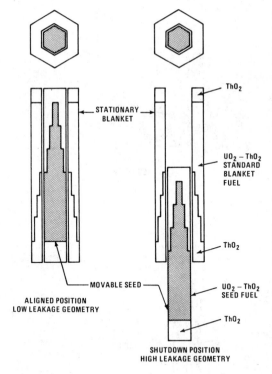

Figure IV-4-37. Variable-geometry nuclear control concept.

potential use in breeder reactors. Advanced moveable-fuel concepts that could reduce neutron losses, enhance breeding and reduce power peaking for commercial-size LWBR technology are also being developed and evaluated. Thermal-hydraulic, mechanical, and irradiation testing will be performed to evaluate the feasibility of these concepts.

Additional irradiation tests are being conducted in the Advanced Test Reactor at INEL and in the NRX reactor at Chalk River, Ontario, Canada. Future testing at these sites will be conducted on a bundle of long rods containing developmental dual-region pellets. Interim and final post-irradiation examinations in ECF and Bettis hot laboratories provide data on the effects of irradiation, which are analyzed to support performance predictions.

Major milestones and funding for AWBA are given in Figure IV 4-39.

BAPL, operated by Westinghouse Electric Corp., and KAPL, operated by General Electric Co., under the technical direction of the Division of Naval Reactors, explores the technical problems that will be faced by industry in adapting technology confirmed in the LWBR Program. Technical information developed is widely disseminated to U.S. industry. Review of industry programs and concerns and comments from industry are considered in establishing detailed program plans and priorities.

GAS COOLED BREEDER REACTOR

As in the case of the LMFBR, the Gas Cooled Fast Breeder Reactor (GCFR) has the potential for extending the uranium resource base to a practically inexhaustible source of electrical energy.

Historically, the GCFR concept has received lower priority than the LMFBR and, as a result, is at an earlier stage of development. However, much of the existing and developing breeder reactor technology and HTGR technology is applicable to the GCFR Program.

Attractive potential characteristics of the GCFR include a high breeding ratio, low capital costs, low maintenance costs, and the ability to incorporate alternative or advanced fuels.

Overview

For the GCFR Program, this section summarizes the background, objectives, strategy, scope, and funding.

U.S. development of the GCFR was initiated in 1963 under the AEC. Now under DOE, this program is one of three interrelated programs being carried out on the GCFR. The other two are: (1) a U.S.-industry program being supported by the Helium Breeder Associates (HBA), which represents utility companies that comprise about 30 percent of the electrical generating capacity in the United States, and the General Atomics Company, and (2) an international program being carried out primarily with the Federal Republic of Germany (FRG) and Switzerland under the U.S.–FRG–Switzerland–France Gas Cooled Reactor Umbrella Agreement. Cumulative program support from all sources (domestic

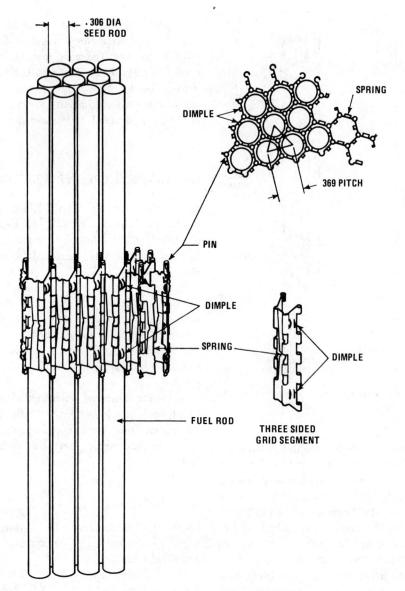

Figure IV-4-38. Typical LWBR blanket-rod support grid.

and foreign) has amounted to approximately $100 million.

Since the inception of the program in 1963, private sector and international interests have accounted for over half the funding support for the program. DOE has taken several actions to promote and encourage this support. An international cooperative research and development program on the GCFR has been established under the Gas Reactor Exchange Agreement with the United States, Germany, France, and Switzerland. The foreign GCFR effort of $5-6 million/year is about equally divided between FRG and Switzerland. It is expected that other countries may enter the agreement in the future. Existing major foreign contributions are in the design, fabrication, and testing of fuels in the Belgian BR-2 test reactor (Germany and Belgium), GCFR physics and performance of thermal-hydraulic experiments in the Agathe loop (Switzerland), safety analysis and design of nuclear steam supply system components (Germany). The FRG series of fuel irradiations in a high pressure helium loop in the reactor in Mol, Belgium, is a prime example of international

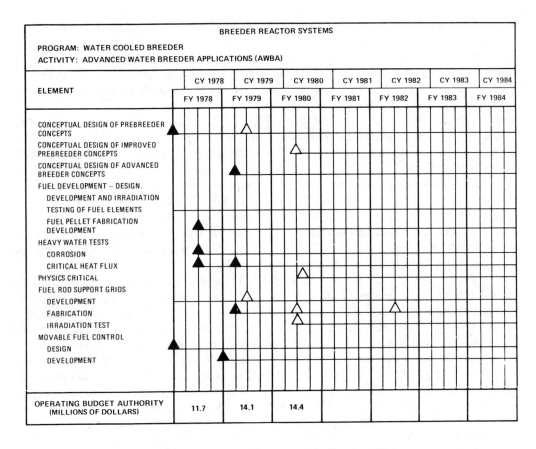

Figure IV-4-39. Major milestones and funding for AWBA.

cooperation on this program. No equivalent U.S. facility exists, and the cost of constructing a similar loop would exceed $50 million.

DOE is also a member of the Nuclear Energy Agency (NEA) GCFR Coordinating Committee, which is part of the Organization for Economic Cooperation and Development (OECD). This group shares the results of R&D performed in member countries and sponsors meetings among technical specialists.

The objectives of the GCFR Program are: (1) to develop the potential of the gas-cooled breeder system as a viable and long-term nuclear power option, and (2) to provide GCFR technology and other needed support on a schedule consistent with an internationally developed cooperative program led and financially assisted in by private industry. These are consistent with the objectives of the National Energy Plan to further evaluate alternative breeder reactor concepts.

In FY 1979, DOE contracted with HBA to provide overall technical management of the DOE GCFR Program and to integrate this work with that sponsored directly by HBA. The overall strategy of the program is to continue the GCFR Program Definition and Licensing Phase (PDLP)—an outgrowth of a utility proposal to ERDA in June 1976. The PDLP serves to guide and focus all research and development on systems, components, fuel and safety. Foreign participation under the U.S.–FGR–Switzerland–France Gas Cooled Reactor Umbrella Agreement is an integral part of the program and includes the development of a single international GCFR design.

For planning purposes, the key projects and developmental tasks are being stretched out in time and funding to correspond to about a 30-year commercial deployment schedule for the concept. This is consistent with a recently completed HBA study entitled, "Alternative GCFR Roles and Development Strategies."

Near-term milestone decision points scheduled for the PDLP period include the selection of a common U.S.–Federal Republic of Germany design, limited component testing in government and industry facilities for a selected few critical components, evaluation of the safety issues, determination of thermal-hydraulic performance of the core, and definition of the required development program.

Full advantage will be taken of ongoing and planned fuel development programs carried out by the Federal Republic of Germany at their Julich laboratory and in the MOL reactor in Belgium. Activities incremental to other ongoing fuels and materials efforts will be included for those tests uniquely needed by the GCFR Program. In many cases, developmental activities will be carried out by U.S. national laboratories to take advantage of the facilities and the expertise developed under other ongoing programs.

In general, due to the large base of gas-cooled technology that exists in the United States and abroad and the use of applicable breeder reactor technology developed for other applications, the GCFR requires a comparatively modest development program.

The GCFR Program is composed of three elements:

- GCFR development
- Core Flow Test Loop (CFTL)
- Gas Reactor In-Pile Safety Test (GRIST-2)

The upper-level WBS for the entire Fission Energy Program was shown in Figure IV 1-2. The further breakdown of the GCFR Program is shown in Figure IV 4-40, with a matrix showing the relationship between the lower-level packages in the WBS and the descriptive arrangement of the program material in this section.

ASET is responsible for the GCFR Program. This program is assigned to ETN, who delegates management responsibility to RRT. In turn, the RRT Director delegates program responsibility to the Chief of the Gas Cooled Fast Reactor Branch. The branch officer provides overall direction and program approval, and is responsible for successful implementation of the program within overall cost and schedules.

Under the direction of the appropriate DOE field office, the national laboratories and contractors provide the day-to-day management of the individual projects, develop project management plans, and recommend the resources required to meet the program objectives. They have the direct responsibility for the performance of the program and meeting the milestones established.

The operating expenses portion of the budget authority by element for the FY 1978 through FY 1980 period is listed in Table IV 4-16.

GCFR Development

The GCFR uses the inert, nonradioactive, and transparent gas helium under a pressure of about 1,300 pounds per square inch to remove heat from the reactor core and to produce

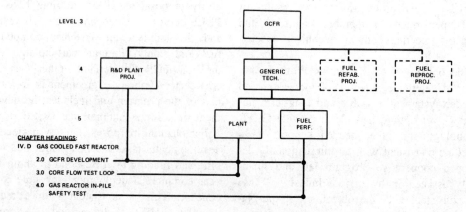

Figure IV-4-40. GCFR program WBS.

Table IV-4-16. Funding for the GCFR program for FY 1978–1980.

GAS COOLED BREEDER REACTOR

BUDGET AUTHORITY (DOLLARS IN THOUSANDS)

PROGRAM ELEMENT	FY 1978	FY 1979	REQUEST FY 1980	INCREASE (DECREASE)
GCFR Development	$10,256	$20,150	$19,188	($962)
CFTL	3,844	3,500	4,012	512
GRIST-2	570	850	1,300	450
TOTAL	$14,670	$24,500	$24,500	0

high-temperature steam for generating electrical power (Figure IV 4-41).

By building on HTGR technology, the GCFR Program can consider certain development work completed that otherwise would have to be addressed. The GCFR is at an earlier stage of development than the LMFBR and uses much of the existing and developing LMFBR technology. Attractive potential characteristics of the GCFR include a high breeding ratio, low capital costs, low maintenance costs, and the ability to incorporate alternative or advanced fuels.

In order to carry out the PDLP, the objective of the GCFR development program element are as follows:

- To advance the technology of the GCFR through reactor, fuel cycle, and systems studies for the initial and subsequent developmental plants
- To achieve an understanding of the GCFR system configuration, cost, and safety characteristics for the first developmental GCFR of a depth consistent with that required to produce a Preliminary Safety Analysis Report acceptable to regulatory bodies
- To integrate the foreign GCFR efforts and results into the U.S. GCFR Program
- To develop the GCFR Program information necessary to evaluate and decide upon the next phase of the GCFR development following completion of the PDLP in 1985

The GCFR Program incrementally builds on national and international reactor-component development work in the HTGR Program including: steam generators, circulators, and prestressed-concrete reactor vessels. Similarly, information obtained for ongoing LMFBR reactor fuels programs including stainless steel cladding tests provides a data base for input to the GCFR fuel element design.

The GCFR development program element is composed of five activities: fuels and materials, physics and shielding, components, safety, and alternative/advanced fuel cycles.

The fuels and materials activity provides the analytical, out-of-pile and in-pile experimental development program necessary to establish the basis for the design of the GCFR fuel (Figure IV 4-42), control and radial blanket assemblies, and pressure equalization system. Although the major operational parameters of the GCFR fuel rods (Figure IV 4-43) are similar to those used for the LMFBR fuel rods, the use of the pressure-equalized and vented fuel rods, ribbed cladding, fuel rod spacer grids, and a gas coolant requires a comprehensive development program to extend the LMFBR fuels and materials technology to the conditions and design features unique to the GCFR. FY 1980 activities include continuation of the construction of

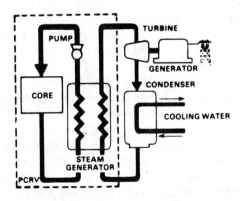

Figure IV-4-41. GCFR systems.

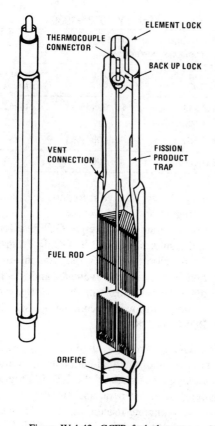

Figure IV-4-42. GCFR fuel element.

the core flow test loop to provide thermal hydraulic data on the performance of the GCFR fuel assemblies and continuation of the GCFR fuel irradiation experiments in EBR-II and the BR-2 reactor. Mechanical testing of the core elements for flow induced vibration will be initiated at General Atomic Co.

Fuel element development work in the Federal Republic of Germany and Belgium contributes significantly to ongoing U.S. core design efforts. An example of some subassembly hardware designed, built, and tested at Kraftwerk Union (FRG) is shown in Figure IV 4-44.

The purpose of the shielding work is to validate nuclear radiation shielding design methods and to reduce the uncertainties in design-related nuclear techniques. The purpose of the physics work is to confirm the core inventory, enrichment, reactivity, and breeding calculations. Both work areas will provide accurate specifications for licensing procedures and margins on core inventory,

and will permit more precise power, life, and breeding calculations for the core. The reactor shielding is used to ensure the integrity of vital internal structural components of the reactor, particularly for the grid plate that supports the core and blanket.

In FY 1980, shielding experiments will continue on a mockup of the GCFR radial shield configuration. The physics task will concentrate on planning for the critical experiments to be run in FY 1981 by ANL. Each experiment, of about six months' duration, will be followed by an analysis program of equal length. Thus, one experiment per year will be performed over the period FY 1981 through FY 1984.

NSSS component development activities have been funded at a relatively low level through FY 1979 due to the need to concentrate efforts on reaching a common design with the Federal Republic of Germany. Among the issues to be resolved in FY 1980 are coolant pressure and flow direction, steam vs. electric circulator drives, single cavity vs. multi-cavity PCRV and hot vs. cold vessel liner.

The GCFR will take full advantage of the component development activities conducted in the United States and Europe under the HTGR Program. Consequently, the component development program is limited to those components whose performance cannot be extrapolated to GCFR operating conditions. Specifically, because of the higher pressure (about twice that of the HTGR), the PCRV closures and helium circulators will require test programs. In FY 1980, scale models of the closures will be tested at ORNL. The conceptual design for a helium circulator test facility will be completed and bearing and seal tests will be initiated for the GCFR main circulator.

The GCFR safety program element will develop the safety analysis and experimental information required for evaluation of the GCFR concept. Specifically, information is being obtained to assess the probabilities and consequences of various postulated accidents in the GCFR, to evaluate post-accident fuel containment capability, to provide a technical basis for the integration of safety and re-

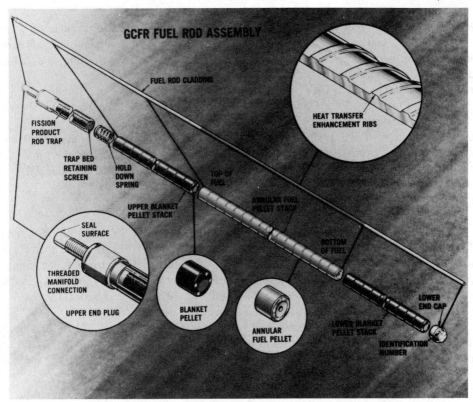

Figure IV-4-43. GCFR fuel-rod assembly.

liability targets into the GCFR engineering effort, and to verify the fuel and cladding behavior during accidents that may lead to core damage.

An out-of-pile, direct electrical heating test program is continuing at LASL to investigate the behavior of simulated GCFR fuel assemblies during loss of flow (LOF) and design-basis depressurization accident (DBDA) conditions. These tests will include full-size core assemblies.

Both analytical and experimental work will continue at ANL to investigate the consequences of various core disruptive accidents.

Figure IV-4-44. GCFR fuel fission-product vent manifold.

This will include transient reactivity effects, fuel compaction and/or dispersion mechanisms, and the thermal-chemical interactions among molten core materials and the various shielding, insulation, and PCRV materials. Design and construction of the Direct Electrical Heating (DEH) loop at ANL-E is scheduled for completion in FY 1980. Work will continue at General Atomic to integrate the results of safety work at LASL and ANL into the GCFR effort.

In support of the President's non-proliferation policy, efforts are continuing to develop the technology which could support a viable GCFR breeder option based on non-plutonium fuel cycles. These efforts will include specific reactor core designs, fuel irradiation experiments in available test facilities, shielding designs, an examination of compatibility with the present reactor systems components and a study of the unique aspects of uranium-thorium cores in GCFR's. Conceptual design activities at BNL will also continue on a once-through GCFR core design that appears to have attractive fuel utilization and non-proliferation characteristics.

Major milestones and funding for the GCFR Development Program are shown in Figure IV 4-45.

The major participants in the GCFR development program element and their main areas of involvement are listed in Table IV 4-17.

CFTL

The CFTL (Figure IV 4-46) is a non-nuclear pressurized helium filled, closed loop for testing of simulated fuel elements containing internal electric heaters. The CFTL is being built at ORNL and is scheduled for initial operation late in FY 1981.

The objective of the CFTL is to provide a test apparatus for verifying the GCFR core thermal hydraulic design analysis and safety

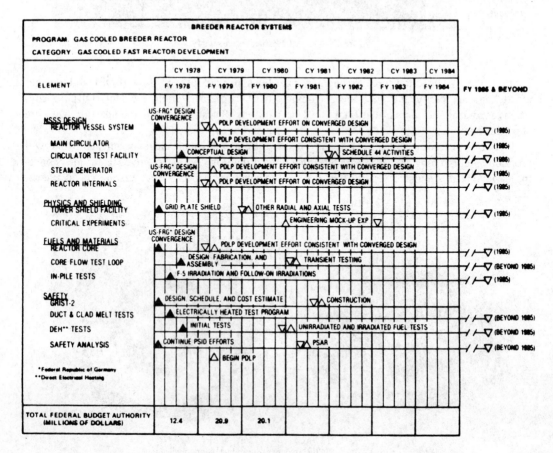

Figure IV-4-45. Major milestones and funding for the GCFR development program.

Table IV-4-17. GCFR development program participants and areas of participation.

PARTICIPANT	AREA OF PARTICIPATION
HBA	Represents utilities; responsible for overall program management and integration
ANL	Portions of safety, fuels and materials, and core physics
ORNL	Neutron shielding, prestressed concrete reactor vessel
LASL	Safety tests
BNL	Nonproliferation activities associated with conceptual design of Fast Mixed Spectrum Reactor
General Atomic Co.	All aspects of GCFR development
Westinghouse Electric Corp.	Early stages of determination
General Electric Co.	Early stages of determination

margins up to the design basis accident conditions. The CFTL will provide data through a series of out-of-pile simulation tests that will: (1) demonstrate the ability of the GCFR fuel, control and blanket assemblies to meet design goals; and (2) verify predictions of analytical models that describe design operation and accident behavior.

Accomplishments include the completion of the conceptual design, the issuance of the construction directive by Oak Ridge Operations Office, DOE, and the delivery of the pressure vessel for the prototype circulator test. R&D efforts included the development of a reference fuel rod simulator heater and the successful operation of the heater and thermocouples at temperatures prototypical of expected operating conditions. The type of power supply system (ac) and data acquisition system were specified. A building at ORNL was cleaned to make way for loop construction. The circulator test loop construction was completed.

In 1979, the assembly of the prototype helium circulator was completed and operation at design conditions in the Mechanical Tech-

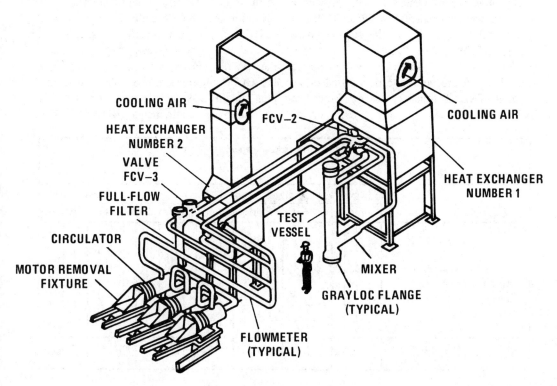

Figure IV-4-46. Core flow test loop.

nology, Inc. (MTI, Latham, N.Y.) test loop began. Modifications to the building housing the CFTL and addition of an overhead crane were completed. Title I design for the first test bundle were completed and procurement of the hardware for the fuel rod simulators was initiated.

Figure IV 4-47 gives the major milestones and funding for the CFTL.

ORNL has the responsibility for the design, construction, R&D, operation, analysis and quality assurance. General Atomic Company (GA) is responsible for criteria, requirements and evaluation. HBA, as technical manager of the GCFR Program, supervises schedules and plans. DOE HQ (Division of Reactor Research and Techology) and Oak Ridge Operations Office provide funding and program policy guidance.

GRIST-2

The GRIST-2 loop is a high pressure, circulating helium loop that will be inserted into the TREAT Upgrade reactor for transient testing of GCFR fuel.

The objective of GRIST-2 is to provide a test loop for experimental verification of the safety analysis used to predict fuel behavior during GCFR power transient conditions. GRIST-2 is being designed by EG&G in Idaho Falls, Idaho, and is to be inserted in the modified TREAT transient reactor facility at INEL.

Accomplishments include the completion of the conceptual design of the helium loop and in-pile test section by EG&G. ANL continues with the design of the test section and the design of the TREAT Upgrade reactor and facility modifications. A tradeoff study was conducted by EG&G to determine the cost of alternative loop designs and test train sizes.

In 1979, the Title I design was continued by EG&G. ANL continued the design of the test train. The system design description of the helium system was issued in the first half of FY 1979 by EG&G. ANL prepared a test plan and the system design descriptions on their portion of the design. HBA issued the final management plan, interface requirements and functional requirements documents. GA issued the fuel procurement plan. In 1980, preliminary design will be carried out.

GRIST-2 milestones and funding are shown in Figure IV 4-48.

EG&G has the responsibility for the design, construction, operation, and quality assurance of the helium loop and in-pile test section. ANL has the responsibility for the design and fabrication of the test train, handling equipment, post irradiation fuel examination and operation of the TREAT Upgrade reactor. ANL also is responsible for the analysis and evaluation of the test data. GA will procure the test fuel. HBA, as technical managers of the GCFR program, provide the management plan, quality assurance procedures, schedules and interface requirements. DOE HQ (Divi-

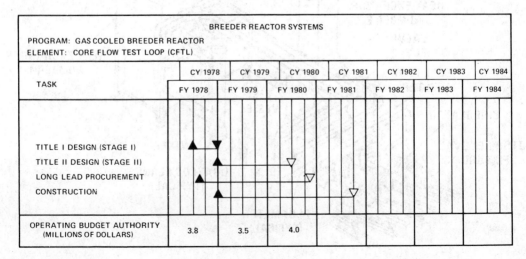

Figure IV-4-47. Major milestones and funding for CFTL.

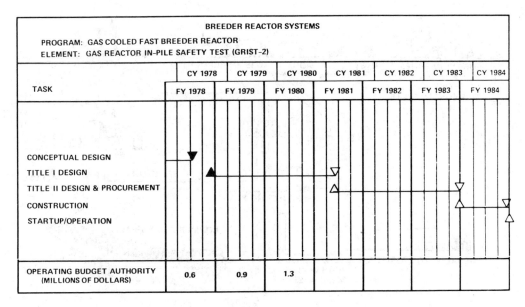

Figure IV-4-48. Major milestones and funding for GRIST-2.

sion of Reactor Research and Technology) and the Idaho Operations Office provide funding and overall program policy guidance.

FUEL CYCLE RESEARCH AND DEVELOPMENT

The Fuel Cycle Research and Development Program is concerned only with the reprocessing of spent reactor fuel and the refabrication of recovered fuel. The program is directed toward solving the engineering problems associated with reprocessing high-exposure commercial fuel and meeting new and proposed regulatory requirements for solid oxide product and effluent control.

Figure IV 4-49 shows the overall nuclear fuel cycle, which begins with the mining of uranium ore, includes power generation in nuclear reactors, and ends with final disposition of spent fuel, whether by reprocessing of disposal.

Overview

This section summarizes the background, objectives, strategy, scope, and funding of the Fuel Cycle R&D Program.

Nuclear fuel reprocessing was a necessary part of weapons material production during World War II. The basic technology for the Purex process evolved from early wartime plants as weapons-material production capability expanded in the early 1950s. That same Purex process if the basis for the technology adopted for reprocessing commercial fuel from nuclear power plants.

Several attempts have been made to commercialize solvent extraction reprocessing in the United States, but these have failed to lead to a viable commercial industry:

- The Nuclear Fuel Services (NFS) Plant at West Valley, N.Y., operated for about five years before its shutdown in 1972. The total fuel reprocessed there was about half the plant's nominal capacity, and most of that was low-exposure fuel from the government-owned N Reactor. The owners of the NFS Plant have abandoned all plans for restart.
- The General Electric Co. reprocessing plant at Morris, Ill., was never started due to engineering deficiencies identified during cold-testing of the plant.
- The Allied General Nuclear Services (AGNS) Plant at Barnwell, S.C., is essentially complete but stands idle pending further Government decisions on reprocessing. However, if a decision were made to proceed, major segments

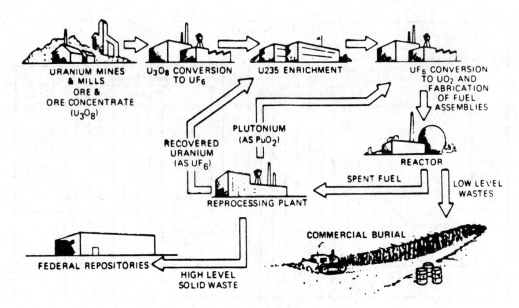

Figure IV-4-49. Nuclear fuel cycle.

would have to be added for waste solidification and plutonium conversion.

The major technical uncertainties that hindered these attempts to commercialize reprocessing and fostered the growth of the Fuel Cycle R&D Program do not relate to Purex flowsheet development, but rather to the engineering of the major equipment items required for handling commercial fuel. The Fuel Cycle R&D Program was initiated in FY 1976 to address technical uncertainties that precluded the commercialization. The elements of the program were drawn from existing programs for HTGR and LMFBR fuel recycle, which until then had been funded as part of those reactor development programs. A new element was added to address LWR fuel recycle exclusively.

On April 7, 1977, President Carter announced a new nuclear energy policy, which deferred indefinitely the reprocessing of commercial spent fuel so as to allow time to investigate more proliferation-resistant alternative fuel cycles. As the result of the new policy, the program was reoriented to investigate optional technologies involving a variety of combinations of fissile and fertile materials (uranium, plutonium, and thorium). The recovery and reuse of fissile materials from these fuels for various reactor configurations with minimum proliferation risk would involve the modification of familiar separations technologies (i.e., Purex) and/or the introduction of new technologies (i.e., pyrochemical and dry methods).

Budget reductions in FY 1979 caused the Fuel Cycle R&D Program elements to be consolidated so as to optimize the use of resources and to effect a more consistent method for setting program priorities. This consolidation combines:

- The reprocessing programs for various materials (U, Pu, and Th), reactor types (LWR, LMFBR, and HTGR), and technologies (aqueous, pyrochemical, and dry) into a single reprocessing program
- The refabrication programs for various materials (U, Pu, and Th), reactor types (LWR and HTGR), and technologies (particle and pellet) into a single refabrication program

In FY 1979, the program also included support of the Barnwell Nuclear Fuels Plant (BNFP) in South Carolina, and analytical activities to support the demonstration of proof of breeding (POB) for the LWBR.

For FY 1980, the program will take advantage of the consolidated management approach and will be oriented to support fuel-cycle concepts consistent with U.S. nonproliferation objectives. At this time, only breeder reactors appear to require reprocessing within this context. Thus, if fuel reprocessing is to be undertaken, it will be conditioned on eventual deployment of a breeder. For this reason the reprocessing activities within the Fuel Cycle R&D Program will focus on developing breeder fuel reprocessing technology. However, these activities will recognize the need for reprocessing fuel from other reactor concepts to (1) provide initial fuel loads for a breeder system as needed and (2) establish a synergistic fuel material balance in a proliferation-resistant "secured center" scenario.

Because the consolidated refabrication activities supported by this program apply only to converter reactors (work on refabrication of breeder fuels is conducted in the LMFBR programs), the above strategy results in the elimination of the refabrication program element from the program. Linking reprocessing to the breeder necessarily precludes near-term recycle for converter reactors.

The development of HTGR fuel recycle has been eliminated for FY 1980 because of termination in that year of HTGR steam-cycle work. As directed by legislation, development work supporting the BNFP is also eliminated for FY 1980. Analytical activities to support demonstration of POB for the LWBR will be continued in FY 1980.

The objectives of the Fuel Cycle R&D Program are to:

- Develop and demonstrate proliferation-resistant fuel cycles that enable nuclear energy to contribute to meeting the nation's energy needs while effectively using natural resources and protecting the environment
- Provide analytical support to the LWBR Proof of Breeding Program.

To meet the objectives of the reprocessing program, emphasis is placed on developing and demonstrating the reprocessing of spent fuel from breeder reactors. A variety of scenarios are considered, involving breeder reactors inside secured centers operating in conjunction with converter reactors outside secured centers on both uranium-based fuel cycles. Therefore, the program is flexible so that it can respond to the implementation of any of a number of reactor-fuel cycle systems, depending on NASAP and INFCE results and future breeder decisions. Fuel-cycle development activities will be conducted on an appropriate engineering scale such that uncertainties relating to the transition from laboratory to practice can be adequately addressed. Ultimate testing of the reprocessing techniques and equipment under actual operating conditions using irradiated fuels discharged from test and demonstration reactors will be required to establish suitability for practical application and to demonstrate the adequacy of applied proliferation-resistance measures; a Hot Experimental Facility (HEF) is planned for this.

To establish POB for the LWBR, a large number of fuel rods will be assayed using a rapid, though accurate, nondestructive assay method rather than destructive chemical assay methods. The needed accuracy for the assay gauge requires its calibration against chemical-based destructive measurements on significantly fewer irradiated LWBR fuel pins than needed for full chemical assay. The POB work within the Fuel Cycle R&D Program has been established to provide the chemical data necessary to effect this calibration.

The Assistant Secretary for Energy Technology is responsible for the Fuel Cycle Program. This program is assigned to ETN, who delegates management responsibility to the Director of the Nuclear Power Development Division (NPD). In turn, the NPD Director delegates program responsibility to the Chief of the Nuclear Fuel Cycle Programs Branch. The branch chief provides overall direction and program approval, and is responsible for successful implementation of the program within overall cost and schedules.

Under the direction of the appropriate DOE field office, the national laboratories and contractors provide the day-to-day management of the individual projects, develop project management plans, and recommend the resources required to meet the program objectives. They have the direct responsibility for the performance of the program and meeting the milestones established.

Table IV 4-18 shows the operating funding levels for the Fuel Cycle R&D Program for the FY 1978 to FY 1980 period.

Table IV-4-18. Funding for the fuel cycle R&D program for FY 1978–1980.

FUEL CYCLE R&D	BUDGET AUTHORITY (DOLLARS IN THOUSANDS)				
PROGRAM ELEMENT	FY 1978	FY 1979	REQUEST FY 1980	INCREASE (DECREASE)	MAJOR CHANGE*
Reprocessing	$57,623	$32,050	$18,100	($13,950)	R
Proof of Breeding	2,389	2,000	3,000	1,000	
Refabrication[+]	17,388	12,800	0	(12,800)	T
Barnwell	13,800	16,000	0	(16,000)	T
Reduced Enrichment Research and Test Reactors[++]	0	4,000	0	(4,000)	
TOTAL	$91,200	$66,850	$21,100	($45,750)	

*Key: R = program redirection; T = termination. [+]Includes NASAP and INFCE studies funding. [++]Funded in FY 1978 under Nuclear Energy Assessments program, and in FY 1980 under Advanced Reactor Systems.

Reprocessing

Fast reactor fuels of U.S. design are generically similar to LWR fuels and can be reprocessed using the same base technology; i.e., the shear-leach head-end followed by solvent extraction. There are, however, significant differences in such things as fuel cladding, fuel-geometry, fissile content, decay heat and radiation levels, which create special problems in directly applying the existing technology base. Although solutions to the technology needs that are unique to reprocessing fast reactor fuels will be obtained by building from the existing LWR technology base, there are many instances where new equipment concepts and additional process steps must be developed.

One breeder option being considered uses a plutonium fueled core, so the base technology being developed for core fuel recycle is being continued. In addition, several activities will be initiated to integrate thorium-based fuel concepts into the current technology.

The well-established wet chemical separation methods form the main basis for the evolving technology. However, phyrochemical methods are also being studied. These include high-temperature techniques that involve the use of molten metals and fused salts to effect mass-transfer separations. Safety, protection of the environment, and safeguarding weapons-usable materials are given maximum emphasis in all phases of the technology development.

Figure IV 4-50 schematically shows the steps involved in reprocessing spent breeder-reactor fuel by wet chemical methods.

The objectives of this program are as follows:

- To provide a sound reprocessing technical data base for making decisions on deployment of breeder reactor systems
- To facilitate the establishment of actual recycle capability when deployment is required

An additional goal is to evaluate the feasibility of using alternative methods of reprocessing breeder fuels that minimize the risk of diversion of weapons-usable fuel material.

To accomplish the primary goals, the development effort associated with each breeder fuel recycle mission is being carried through appropriate stages to produce a fully engineered reprocessing system.

The development of reprocessing technology requires a systems approach because all process steps must perform satisfactorily as a system to meet the basic requirement of separating and recovering or disposing of all fuel constituents. This has been the underlying principle for organization of the program element.

Breeder-fuel reprocessing technology builds on the aqueous-reprocessing technology developed for LWR fuels. Process and equipment concepts applicable to reprocessing a variety

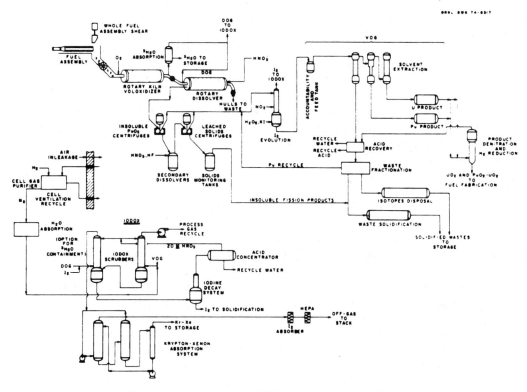

Figure IV-4-50. Schematic LMFBR fuel-processing flowsheet.

of breeder fuels are defined through laboratory, hot-cell, and bench-scale engineering tests. Then, these concepts are developed thoroughly through progressive phases of evaluation and testing, first as individual component systems and then as integrated process systems using engineering-scale equipment. It is essential that these systems be evaluated on an engineering scale before they can be accepted as a practical solution to the processing problems resulting from the unique breeder-fuel properties.

The scale of testing selected is such that uncertainties relating to the transition from laboratory to practice can be addressed properly, equipment scale-up laws adequately defined, and requirements for future development of full-scale prototype equipment identified.

Process and equipment concepts will be tested as an integrated system using unirradiated mockup fuels in the Integrated Equipment Test (IET) facility. Ultimately, it is planned to test the complete process line using actual breeder-reactor spent fuel in the HEF. As new fuel systems are defined for advanced recycle missions, the specific requirements for accommodating these fuels in the base process system will be addressed.

Spent fuel received at the reprocessing plant in shipping casks is unloaded, prepared for storage, stored for fission-product decay, and subsequently removed, cleaned if contaminated with sodium and prepared for reprocessing. System development addresses the problems associated with the storage and handling of sodium-contaminated fuel and the feasibility of the rapid fuel movement required to maintain a practical plant thoughput with the relatively small breeder subassemblies. The development effort, which now is in the evaluation and planning stage, encompasses evaluating fuel storage media and cleaning procedures, and the design of fuel-subassembly transport and handling equipment.

In the shearing system, whole cleaned subassemblies are cropped of end hardware, deshrouded, and sheared into 1-inch segments to expose the contained fuel for further processing. The system also provides capability for packaging scrap subassembly hardware produced in these and subsequent operations. Development of this system is aimed at adapting existing LWR fuel shearing technology to handle the unique breeder fuel properties. Of principal concern are the multiplicity of small-diameter fuel pins, the heavy-gauge shroud, the high radiation and decay-heat levels, the potential presence of sodium from leaking fuel, and irradiation-embrittled cladding. Development concentrates on designing, fabricating, and testing the key equipment components in the system—the fuel disassembly machine, fuel shear, scrap hardware compactor, and material transfer devices. Engineering-scale components are being studied. This entire system will be included in an integrated systems test using unirradiated fuel.

Figure IV 4-51 shows the fuel disassembly system that will be available for testing in FY 1980. Figure IV 4-52 shows the fuel shearing system whose features are being tested. The results of these tests will define the design requirements for the model to be used in the testing of subsequent systems scheduled for design and fabrication in FY 1979 and 1980.

The voloxidation system involves new technology for reacting the sheared fuel pieces with gaseous media at elevated temperatures to release tritium from the fuel and deactivate residual sodium in sodium-cooled fuel prior to exposure to aqueous media. It also provides the potential for converting carbide fuels to oxide form for reprocessing compatibility. The development work concerns defining the process and equipment design parameters. Effort is concentrated on the design, fabrication, and testing of the key equipment components in the system—a rotary-kiln-type voloxidizer, product cooler, and isolation valves.

Figure IV 4-53 shows the first engineering-scale model to be tested. In FY 1979 and 1980 results of test operations of this unit will be used to design and fabricate the unit to be used for testing. Small-scale process studies using irradiated fuel are also a significant part of the development effort to define process flowsheets.

The dissolution system uses nitric acid to leach exposed fuel from sheared pieces.

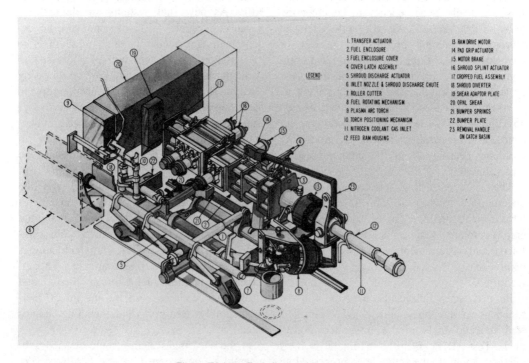

Figure IV-4-51. Fuel disassembly system.

Figure IV-4-52. Fuel shearing system.

Figure IV-4-53. Engineering-scale dissolver.

Breeder core mixed-oxide fuels are not completely dissolved in nitric acid and require a new approach to dissolution. The heart of the new system is a rotary dissolver that provides not only for continuously dissolving the bulk of the oxide fuel but also for separating a product stream (containing dissolved oxides, undissolved plutonium oxide, and fission-product residue) from a waste stream of solid cladding segments or hulls. The development work includes basic studies of the factors influencing fuel solubility, small-scale process studies using irradiated fuel, and the design, fabrication, and testing of the rotary dissolver, material transfer devices, and isolation valves.

Figure IV 4-53 shows the engineering-scale dissolver whose features are being tested. The first phase of testing is targeted for completion by mid-FY 1980. Only minor modifications are expected to be required to prepare this unit for systems testing. The adaptability of the dissolution system to thorium-base breeder blanket fuel will also be addressed.

Product solution from the dissolver system containing undissolved plutonium oxide and fission-product residues is processed in the feed preparation system to complete the dissolution of plutonium oxide and produce a clarified feed for solvent extraction. The solid residues are collected in a centrifuge of special design to accommodate the fission-product decay-heat load and maintain criticality safety. The development effort focuses on the design, fabrication, and testing of the centrifuge but also includes a significant laboratory investigation of the processes for the secondary recovery of the plutonium from the dissolution residue. This secondary recovery method most likely will require the introduction of hydrofluoric acid to effect complete dissolution. Design of the engineering-scale test unit will be undertaken in FY 1980.

The well-established Purex solvent-extraction system will be applied to separation and decontamination of breeder-core fuel. Flowsheet adjustments for the high plutonium and fission-product content of feed prepared from such fuels must be defined and tested as a part of the system development effort. Tests and evaluation of several models of centrifugal contactors specially designed for criticality safety will also be included in the integrated systems test. Testing is now being conducted in bench-scale equipment similar to that in Figure IV-4-54. In FY 1980, the design of an engineering-scale unit will be started.

Figure IV-4-54. Bench scale solvent extractor.

Figure IV-4-55. Microcomputerized control interface unit.

Studies are underway to define a reference Thorex solvent-extraction system applicable to thorium-base breeder blanket fuels; it will be tested in the bench-scale units in FY 1980.

Product solutions from solvent extraction are converted to solid form in this system preparatory to storage for shipment to refabrication facilities. The final product form for all fuel reprocessing is tentatively identified as solid oxide; the development now being planned consists primarily of studies to identify process and equipment parameters for several options for converting the nitrate-solution effluent from solvent extraction to the desired solid form. It is expected that in FY 1980 the desired form of products will have been determined from nonproliferation assessment activities and the development effort can focus on engineering-scale development of a single option.

The function of the off-gas control system is to maintain radioactivity levels in gaseous discharges from the hot pilot plant at levels "as low as reasonably achievable." Subsystems are required for separation and retention of the large quantities of iodine-131 present in breeder fuel, as well as xenon, krypton, tritium, iodine-129, ruthenium and particulates from the several plant off-gas streams. Development, which aims to reduce the discharge levels several orders below present regulatory requirements, involves testing and evaluating newly defined retention processes—tritium collection on solid absorbents or in cold traps, the Iodox process for iodine, and xenon, krypton, and carbon-14 adsorption in liquid refrigerant R-12.

The tritium collection and Iodox processes are being pilot tested on a bench-scale, which will continue in FY 1980. The liquid refrigerant adsorption process for xenon, krypton, and carbon-14 was pilot tested. This unit's operation was terminated in FY 1980 after completion of planned tests.

The radioactive waste management system concentrates, solidifies, and packages for transport the high-level radioactive wastes generated during reprocessing. It also provides for processing and disposing of other liquid and solid wastes containing lower levels of radioactive contamination. Generic processes and procedures for handling these wastes are to be developed in other DOE programs. Development in this program is required to ensure the compatibility of these

Figure IV-4-56. Model of remote operating and maintenance demonstration area.

generic processes with the wastes generated in breeder-fuel reprocessing, and to ensure adequate capability for in-plant waste handling. In FY 1980, the design of a test system will be initiated.

In general, the application of advanced instrumentation and control techniques to reprocessing of spent nuclear fuel has lagged that attained in the nonradioactive process industries. In the interest of ultimately improving plant availability, maintaining higher standards of safety and environmental protection, and improving safeguards against diversion of weapons-usable material, the application of advanced process sensors and computerized control is an objective of breeder reprocessing development. A number of process-operator microcomputerized interface units, such as that pictured in Figure IV 4-55, are being tested in conjunction with equipment development. The integration of these units into an overall distributive control system will be the principal focus of development in FY 1980.

Because of the higher concentrations of fissile material present in breeder fuels, special requirements for criticality control and materials management throughout all the process systems are recognized. Development being planned will concern primarily the identification and testing of hardware and procedures that make up the engineered safety and safeguards systems.

To attain the stringent radioactive effluent controls that will probably be imposed on reprocessing operations, ingress and egress of equipment in the process cells must be through controlled-access locks. This approach requires special maintenance techniques recognized to be new plant practice. Development now in the design study stage will involve fabricating and testing key equipment components involved in overall plant maintenance. In addition, several of the newly developed process-equipment components have features necessitating unique applications of established maintenance techniques that will require mockup testing as part of the

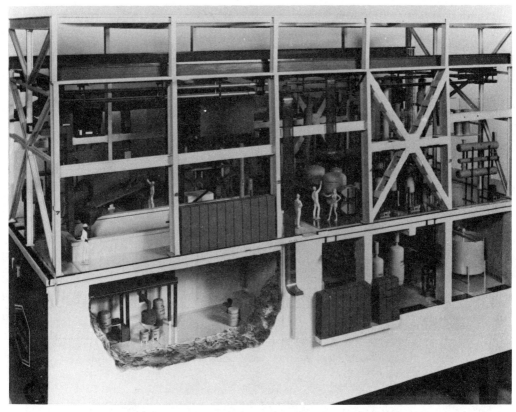

Figure IV-4-57. Model of integrated process demonstration area.

development. Demonstration of maintenance techniques will be a part of the integrated systems test. The proliferation-resistant engineering concepts place further constraints on the maintenance system that lead to additional development requirements.

Current activities aim to upgrade remote maintenance techniques—such as improvements in master-slave manipulator reliability, bridge-crane control, remote connector control, etc. Improved equipment models obtained from commercial manufacturers are being tested. In FY 1980, these tests will continue and design of an integrated maintenance system incorporating these techniques will be started.

The following major facility requirements are designed to demonstrate breeder fuel reprocessing technology on a scale amenable to extrapolation to a commercial plant.

The IET will provide the capability for testing the process equipment using unirradiated mockup fuel. The project will modify and add to existing facilities originally constructed for the Experimental Gas-Cooled Reacted (EGCR) complex and for equipment services to support the engineering development activities in the advanced recycle program. The EGCR complex had been idle since termination of the EGCR program in the mid-1960s. By using that facility's buildings, the reprocessing program has been consolidated at a single location and crowded conditions in other laboratory facilities at ORNL have been alleviated. Two areas are being provided: a Remote Operating and Maintenance Demonstration (ROMD) area (Figure IV 4-56), which will be used to develop and demonstrate remote operational and remote maintenance features of process equipment; and an Integrated Process Demonstration (IPD) area (Figure IV 4-57), which will be used to demonstrate operation of equipment components developed separately as an integrated process system using unirradiated feed. The development and testing is to be conducted on an

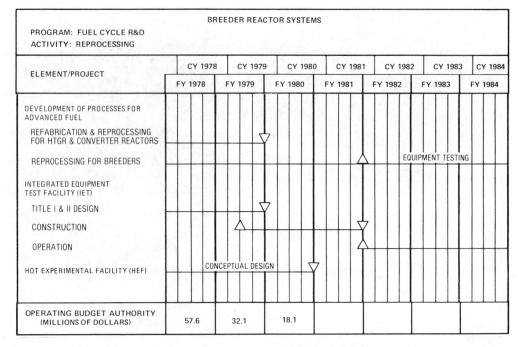

Figure IV-4-58. Major milestones and funding for the reprocessing activity.

Figure IV-4-59. Outside view of analytical test hot cell.

engineering scale, approximately one-tenth of projected full-scale requirements, the next logical step up from bench-scale.

The IET project total estimated cost is $16 million. Design was initiated in FY 1977 with $600,000 CP&D funding and continued in FY 1978 with $3 million authorized for architect-engineer design and advanced procurement. Authroization for $12.4 million to complete the project was requested in FY 1979. An appropriation of $6 million in FY 1979 was approved; an appropriation of $6.4 million is being requested in FY 1980 for project completion.

Ultimately, breeder fuel reprocessing technology will require engineering-scale testing using fully irradiated fuels discharged from test and demonstration breeder reactors. The HEF is planned to fulfill this need. The facility will comprise a complete irradiated-fuel reprocessing system and associated shielded containment structure. Provision will be made in the design of the facility for testing several alternative processes and for demonstrating unique safeguards measures. Support capability for fuel receiving and storage, radioactive waste management, equipment maintenance, and other operations-support services will be included to provide an essentially independent facility. The scale of operations is planned to be the same as for the engineering tests conducted in the IET program so that experience gained in the equipment and process tests with unirradiated fuel will be directly applicable to the design of the HEF. Facility conceptual design work now underway is planned for completion in FY 1980.

Pyrochemical methods of reprocessing various breeder-fuel types are also being investigated. The development is planned to provide several successive points at which the future course of the program is identifiable and subject to potential redirection. During the initial phase of the program in FY 1978, several classes of pyrochemical processes and materials developments were treated in the same way. A review of pertinent experience and literature to establish technical feasibility was

Figure IV-4-60. Prototype fuel feed and shear system within hot cell.

followed by the development of a process outline that included flowsheets, expected behavior of important elements, and engineering information as available.

In the second phase in FY 1979 and FY 1980, work on each of the process types under study will consist largely of experimental work aimed at providing process information. Experiments will define the behavior of specific important elements (e.g., uranium, thorium, plutonium), and broader experiments will establish physical and chemical properties of the processing methods, such as reaction rates and solubilities. In addition, attention will be given to scale-up of the more promising process candidates, including equipment, materials of construction, and unit operations as carried out remotely (e.g., mass transfer, separation of solids from liquids, or distillation of solvent metals). The number of candidate processes will be narrowed to two or three and these will be subjected to more intensive cold-laboratory testing to yield products that can be used for fuel fabrication and performance testing in other parts of the fuel-cycle program. Engineering and hot studies will be included in the process scale-up.

Studies underway in FY 1979 provided a focus for the ultimate application of nonaqueous technology. These studies established criteria against which the output of the development effort can be evaluated in FY 1980.

Figure IV-4-61. Prototype fuel dissolver within hot cell.

The evaluation will either identify a prime candidate system for more intensive development or provide the basis for a decision to terminate activities related to nonaqueous technology.

Major milestones and funding for the reprocessing activity are shown in Figure IV 4-58.

Under the direction of DOE's Oak Ridge

BREEDER REACTOR SYSTEMS							
PROGRAM: FUEL CYCLE R&D ACTIVITY: PROOF OF BREEDING (POB)							
ELEMENT/PROJECT	CY 1978	CY 1979	CY 1980	CY 1981	CY 1982	CY 1983	CY 1984
	FY 1978	FY 1979	FY 1980	FY 1981	FY 1982	FY 1983	FY 1984
PROOF-OF-BREEDING FACILITY DESIGN, CONST, TEST				▽			
ANALYTICAL OPERATIONS				△		▽	
CLEAN UP						△	▽
OPERATING BUDGET AUTHORITY (MILLIONS OF DOLLARS)	2.4	2.0	3.0				

Figure IV-4-62. Major milestones and funding for the POB activity.

Operations Office, ORNL functions as the lead laboratory. Other national laboratories, their operating contractors, and DOE's managing field offices provide program support. These include the Savannah River Laboratory (South Carolina), operated by the DuPont Corporation under the Savannah River Operations Office, and ANL (Illinois), directed by the Chicago Operations Office. Private supporting contractors include General Atomic CO. (San Diego, California), Exxon (Bellevue, Washington), General Electric Co. (San Jose, California), Westinghouse Electric Corp. (Pittsburgh, Pennsylvania), Bechtel National Corp. (San Francisco, California), and Gulf & Western Industries (Philadelphia, Pennsylvania).

Proof of Breeding

The Proof of Breeding Program supports the Naval Reactors Division's work to develop a WCB. In the WCB Program, and LWBR core is being tested in the reactor of the Shippingport Atomic Power Station. After that core is removed in 1981 and 1982, its fuel will be analyzed to determine whether the LWBR core has indeed "bred" fuel; i.e., converted thorium to fissile uranium-233 at a rate at least equal to fuel consumption. This LWBR Proof of Breeding (LWBR-POB) analytical program is performing precision wet chemical analyses that will aid in selecting and calibrating NDA devices that BAPL will use to measure whether breeding was achieved. Laboratory work on the project has proceeded through the equipment and methods development stage. Prototype equipment for fuel cutting and dissolution has been installed and is being tested in a hot cell at ANL. Figure IV 4-59 is an outside view of the test hot cell; Figure IV 4-60 is an in-cell view of the prototype fuel feed and shear system; Figure IV 4-61 is an in-cell view of the prototype fuel dissolver.

This work supplies the chemical analyses of irradiated fuel rods required for calibrating and verifying the performance of the Irradiated Fuel Assay Gauge (IFAG) being devised by BAPL to determine the breeding (fissile fuel content) of the Shippingport LWBR core at the end of its life.

The IFAG will be used by BAPL to nondestructively assay a large sample of the fuel rods that will be discharged at the end of life (EOL) of the Shippingport LWBR core. The results of this assay will indicate the extent of fuel breeding achieved.

To provide calibration data, ANL is shearing fuel-rod test sections and analyzing 7–12 specific segments of experimental LWBR-type fuel rods. In addition, ANL will assist in validating the IFAG data by chemically analyzing at least one experimental LWBR-type fuel rod or section after IFAG assay. At the EOL, ANL will remove and analyze up to 250 segments from 50 fuel rods and will completely dissolve and analyze five additional fuel rods. For this purpose, hot-cell equipment is being developed and installed for precisely cutting and crushing fuel-rod segments, dissolving the segments to provide samples for analysis, for analyzing the samples, disposing of scrap fuel material, and concentrating, solidifying, and disposing of wastes. BAPL prototype equipment will also be modified and installed. This LWBR-POB project covers the design, development, testing, installation, and use of the required equipment in existing hot-cell facilities at ANL to attain these objectives.

The first phase of the LWBR-POB project is essentially nearing completion. Shear and dissolver prototypes have been successfully tested with regard to equipment and methods with irradiated validation fuel rods. Design is progressing on full-scale equipment—a full-scale shear (FSS) and a multi-unit dissolver. These and supportive equipment items will be fabricated, installed in a hot-cell, and used to perform the necessary analyses.

The LWBR-POB project essentially involves three phases. Phase 1 is methods development and equipment design and testing. Phase 1 involves the completion of equipment fabrication and installation in preparation for the EOL campaign. Phase 3 is the EOL campaign in which a portion of the Shippingport reactor's spent fuel will be analyzed to calibrate the BAPL equipment. A final step of equipment and facility cleanout and waste disposal will follow.

Figure IV 4-62 shows the major milestones and funding for the POB activity.

Under direction of DOE's Chicago Operations Office, ANL functions as the lead laboratory.

5.
Advanced Nuclear Systems

OVERVIEW

This program area consists of two programs:

- Advanced systems evaluation—analyses to develop management tools for making decisions on:
  - The overall direction of nuclear programs
  - Priorities to be accorded to new concepts or improvements in existing concepts
  - Approaches and strategies for accomplishing the objectives
- Space and terrestrial applications—primarily the applications of nuclear isotopes to electrical power needs in space and to beneficial uses on earth.

Table IV 5-1 shows the funding levels for these two programs for FY 1978 through FY 1980 period.

ADVANCED SYSTEMS EVALUATION

Every energy production plant began with a concept that was investigated, developed, and ultimately brought to the market place. Nuclear technology, too, has nurtured many new ideas, some of which are improvements to existing technology, other which foresee completely new technologies. It is the function of the work described in this section to compile the data and to conduct the analyses required to evaluate these new concepts, thereby permitting program managers and policy makers to make informed decisions on nuclear program direction.

Overview

This section summarizes the background, objectives, strategy, scope, management, and funding of the Advanced Systems Evaluation program.

Table IV-5-1. Funding for the advanced nuclear systems for FY 1978–1980.

ADVANCED NUCLEAR SYSTEMS	BUDGET AUTHORITY (DOLLARS IN THOUSANDS)				
PROGRAM	FY 1978	FY 1979	REQUEST FY 1980	INCREASE (DECREASE)	MAJOR CHANGE*
Advanced Systems Evaluation					
Operating	$23,100	$10,300	$ 2,600	($ 7,700)	C
Equipment	0	0	0	0	
Construction	0	0	0	0	
Subtotal	23,100	10,300	2,600	(7,700)	
Space and Terrestrial Applications					
Operating	45,850	38,380	34,300	(4,080)	R
Equipment	4,680	4,700	2,100	(2,600)	
Construction	0	0	0	0	
Subtotal	50,530	43,080	36,400	(6,680)	
TOTAL	$73,630	$53,380	$39,000	($14,380)	

*Key: C = Phase completion; R = Program redirection.

From the earliest days of the development of nuclear energy, many different reactors and fuel cycles have been considered. During the 1950s and early 1960s, many of these concepts were converted into hardware and tested in an operating environment. These field tests eliminated the least promising ideas, focusing attention on a few technologies that continued to be developed and improved.

New ideas are continually brought forth. However, today's climate, particularly the economic climate, does not generally permit these ideas to be thoroughly tested in operating hardware prior to making a choice on full-scale development. Reliance must be placed primarily on "paper studies" that consider technical and commercial feasibility, economic and marketplace incentives, proliferation resistance, resource and environmental impacts, and other similar factors. These analyses are the primary basis for a decision to commit to development and possible demonstration.

The objective of this effort is to evaluate nuclear power systems, their applications and impacts, in support of management decision-making.

As new ideas and concepts are promulgated, they are evaluated initially as to their merits in comparison to existing concepts. Many are discarded at this point. The concepts that survive the initial screening are then analyzed in detail, considering all factors that bear on future acceptability. Upon completion of these analyses and if there is clear reason and need, the concept may enter the development and demonstration phase that requires the commitment of substantial money. In this phase, the concept is established in its own right as a major development program.

The three elements of this program are as follows:

- Nonproliferation studies were initiated in response to a particular need and perceived to require a total reevaluation of all nuclear reactor and fuel cycle concepts. The reevaluation centers primarily on the prevention of the proliferation of weapons-usable material.
- Technical and economic analysis is centered on maintaining a data base of technical and economic information on reactors and fuel cycles of interest to management. The data are valuable for monitoring the economic competitiveness and economic trends of nuclear power generation costs.
- Advanced technology systems assessments are directed primarily toward nonelectrical applications of nuclear power and toward improving the acceptability of nuclear plants in areas where siting considerations (water use, land use) may currently prevent their installation.

The upper-level WBS for the entire Fission Energy program was shown in Figure IV 1-2. The further breakdown of the Advanced Systems Evaluation Program is shown in Figure IV 5-1 with a matrix showing how the lower-level packages in the WBS relate to the arrangement of the program description in this section.

The Assistant Secretary for Energy Technology is responsible for the Advanced Systems Evaluation Program. This program is assigned to ETN, who delegates management responsibility for the Nonproliferation Studies to the Director of the Nuclear Alternative Assessment Division, for the Economics Evaluation to the Director of the Plans and Analysis Division, and for the Advanced Technology Systems Assessments to the Director of the Advanced Nuclear Systems and Projects Division, who, in turn, delegates program responsibility to the Chief of the Advanced Isotope Separation Branch. The branch chief provides overall direction and program approval, and is responsible for successful implementation of the program within overall cost and schedules.

Under the direction of the appropriate DOE field office, the national laboratories and contractors provide the day-to-day management of the individual projects, develop project management plans, and recommend the resources required to meet the program objectives. They have the direct responsibility for the performance of the program and meeting the milestones established.

The systems chosen for detailed evaluation

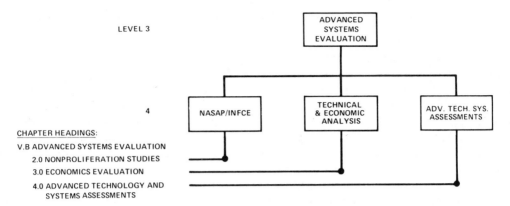

Figure IV-5-1. Advanced systems evaluation program WBS.

are selected by DOE Headquarters. The expertise available in the national laboratories is utilized to assist DOE in management and coordination of the programs. Joint programs are undertaken as feasible with the potential users of new applications. The ultimate decision on which systems to pursue beyond the study stage rests with DOE.

Operating funding levels for the elements of Advanced Systems Evaluation in the FY 1978 to FY 1980 period are given in Table IV 5-2.

Nonproliferation Studies

In April 1977, the President announced a new national policy regarding nuclear power and its relationship to nuclear weapons proliferation. A number of initiatives were established to help maintain the separation between civilian nuclear power programs and nuclear weapons programs. The Nonproliferation Policy Act of 1978, signed into law on March 10, 1978, provides legislative authority for implementation of certain elements of the policy directive, as well as establishing more stringent conditions for exort of U.S. nuclear materials and facilities.

The basic proliferation concern is that a non-nuclear weapons state possessing a civilian nuclear power program which employs any weapons-usable material, such as plutonium, could divert the material and build a nuclear weapon in a short time if it had made previous preparations to do so. Present international and U.S. supply agreements and safeguards need to be re-examined and strengthened to reduce the risk. Technical alternatives need to be examined which, in combination with strengthened institutional barriers, might make more difficult or time consuming the seizure of materials and fuel cycle facilities for use in making weapons.

To resolve this concern, two programs were set in motion:

- Nonproliferation Alternative Systems Assessment Program

Table IV-5-2. Funding for the advanced systems evaluation for FY 1978–1980.

ADVANCED SYSTEMS EVALUATION	BUDGET AUTHORITY (DOLLARS IN THOUSANDS)			
PROGRAM ELEMENT	FY 1978	FY 1979	REQUEST FY 1980	INCREASE (DECREASE)
Nonproliferation Studies	$16,037	$ 6,959	0	($6,959)
Economics Evaluation	1,489	650	600	(50)
Advanced Technology and Systems Assessments	5,574	2,691	2,000	(691)
TOTAL	$23,100	$10,300	$2,600	($7,600)

- International Nuclear Fuel Cycle Evaluation

NASAP was created to examine technical and institutional alternatives for reducing proliferation risk. NASAP is an analysis of nuclear fuel cycle alternatives, including new nuclear systems and modifications to existing systems, and is being conducted by DOE. Technical and other data are being developed to support U.S. international nonproliferation policy objectives, and to assist policymakers in framing future U.S. domestic nuclear developments. Technical and institutional approaches to reactor and fuel cycles with acceptable proliferation resistance will be recommended.

INFACE was proposed by President Carter in the April 1977 nuclear policy statement as a means of furthering international cooperation in reducing proliferation risks. Fifty-three nations and four international organizations are participating in scientific studies of alternative nuclear technologies to develop a common basis for evaluating systems and thereby contributing to reaching a consensus on the future regime for governing nuclear energy.

Table IV 5-3 shows the operating funding levels for nonproliferation systems analysis for the FY 1978 to FY 1980 period. The NASAP effort supports the INFCE work to some degree, but it is not possible to separately identify the costs for this work.

NASAP. The overall goal of NASAP is to recommend to energy policy makers such strategy options for implementing civilian nuclear power systems which, when deployed internationally, offer increased resistance to proliferation while maintaining the benefits of nuclear energy over the long-term.

In support of this goal, the principal objectives of NASAP are as follows:

- To provide technical support for U.S. participation in INFACE
- To identify commercially feasible and internationally acceptable technical systems options which provide improved proliferation resistance
- To identify institutional and nontechnical frameworks in which preferred system options can be deployed
- To develop strategies to implement system and institutional options

The strategy for achieving these objectives takes into consideration the following assumptions:

- A commitment to nuclear power with increasing nonproliferation constraints implies that a major change in the direction of presently planned nuclear power development programs may be required.
- For any nuclear option to be viable, it must be competitive with other options and available when needed. Therefore, when considering proliferation resistance, the tradeoffs among technical feasibility, economics, resource utilization, and commercialization must be analyzed.
- It is difficult to specify nonproliferation constraints in concrete operational terms and to assess objectively the absolute amount of proliferation resistance of a system. Recourse must ultimately be made to achieving a consensus of technical, economic, and political viewpoints

Table IV-5-3. Funding for nonproliferation studies for FY 1978–1980.

NONPROLIFERATION STUDIES			BUDGET AUTHORITY (DOLLARS IN THOUSANDS)		
CATEGORY	FY 1978	FY 1979	REQUEST FY 1980	INCREASE (DECREASE)	MAJOR CHANGE*
NASAP	$16,037	$6,959	0	($6,959)	C
INFCE	0	0	0		
TOTAL	$16,037	$6,959	0	($6,959)	

*Key: C = Phase completion.

of relevant quantitative and qualitative issues. A practical methodology for generating the information necessary for evaluating the proliferation resistance of alternative systems in the context of commercial feasibility and international acceptability must be formulated.

Based on these assumptions, a three-part program strategy has been developed:

- Individual skills of private contractors, DOE program divisions, and other public and private organizations are used to provide basic data needed to characterize fully the nuclear systems and to support evaluation of the technical, political, and economic factors.
- An integrated assessment of alternative systems is being conducted. All tradeoff analyses are being conducted using the frames of reference and data base created in part one of the strategy. Outside reviews, both peer and public, are being obtained to place NASAP in perspective with evolving nuclear policy. The output of the integrated assessment will be strategy options include: base technology program; research, development and demonstration projects; and domestic and international institutional initiatives and impacts.
- The necessary technical analyses and evaluations to develop and support U.S. participation in INFCE are being provided. An important input from NASAP to INFCE has been a proposed proliferation-resistance assessment technique. International input to NASAP is being provided through the INFCE communications channels established with other nations.

NASAP is being conducted according to the following progression of tasks:

- Select representative candidate systems for assessment. This process allows for a workable number of systems to be presented for characterization without peremptorily dismissing any alternatives, while ensuring that all potential classes are included. Results of the sequential evaluative process may lead to increased scrutiny of additional systems within a particular class based on significant advantages in proliferation resistance.
- Develop a methodology for evaluating the key criteria, which consist of proliferation resistance, technological feasibility, economics, resource utilization, environmental and safety considerations, commercial feasibility, and international acceptability. This task establishes the basic approach to evaluation and specifies the data needed to perform the subsequent assessments.
- Characterize candidate systems in standard form to permit comparative evaluation. This step defines the technical and institutional factors of candidate systems in sufficient detail in accordance with the data requirements in order to permit valid assessments.
- Evaluate and assess the preliminary candidate systems. Under this task, the relative merits of each individual system with respect to each key criterion are evaluated. Strong and weak points are identified and other factors for each system are summarized. The results of this task may require additional investigation of a selected system within a class, and will provide direct input to the integrated assessment.
- Perform an integrated assessment of all key factors of final candidate systems. This task represents the culminating effort of the program, providing the basis for the selection of the strategy option necessary to implement final recommendations.

The relationships of the above tasks and the interfaces with INFCE are detailed in Figure IV 5-2.

A methodology for assessing the relative proliferation resistance of nuclear systems has been developed by the NASAP. In an issue as complex as proliferation, it would, of course, be unrealistic to expect that no further refinements and modifications will be made. Fur-

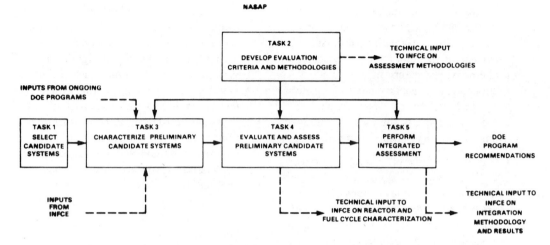

Figure IV-5-2. Relationship of NASAP tasks and interfaces with INFCE.

thermore, comparing the proliferation resistance of certain fuel cycles addresses only part of the issue. For example, the political influences affecting decisions to acquire nuclear weapons are not being addressed. Nonetheless, a significant advance has been made in our capacity to deal with and understand a very difficult subject. The main features of selected nuclear systems relevant to proliferation resistance have been identified and described, the principle activities for weapons pathways have been delineated, and some technical and institutional measures which could increase the proliferation resistance of systems have been identified. Consideration is also being given to the international response to proposed technologies and institutional procedures to improve proliferation resistance. A draft framework for assessing the proliferation resistance of alternative nuclear systems has been provided to the INFCE.

Systems being studies by NASAP are shown in Table IV 5-4. Preliminary analyses have been completed for the LWR on the once-through fuel cycle and with uranium-plutonium recycle for the HTGR with recycle, for the heavy water reactor (HWR), for the LMFBR, and for denatured uranium-233/thorium systems. Analyses of parts of the fuel cycle have also been completed for spent fuel storage, for reprocessing, and for existing and advanced isotopic separation technologies. There are residual proliferation vulnerabilities of the once-through fuel cycle that can and should be countered by improved safeguards and international institutional arrangements for spent fuel. Institutional and technical countermeasures can improve the proliferation resistance of breeder and recycle systems, although the cost and international deployment potential of these measures are open issues.

The technical characterizations of most reactor and fuel cycle systems have been completed. Studies of the denatured U-233/thorium fuel cycle have been substantially completed. Findings indicate that the denatured cycle, which would require reactors that burn plutonium to produce U-233, appears more attractive for the LWBR, HTGR, and HWR than for the LWR. However, use of the denatured U-233/Th cycle in thermal reactors would significantly reduce fuel utilization and would preclude the possibility of breeding in thermal reactors such as the LWBR.

Design studies have been completed for the molten salt breeder reactor, accelerator breeder reactor, gaseous core reactor, and fusion-fission hybrid reactor. Design studies for a proliferation resistant core for the LMFBR are being finalized.

The studies of advanced converter reactors (SSCR, HWR, and HTGR) found that, on a once-through fuel cycle, only marginal uranium savings compared to an improved LWR could be achieved by three systems and that it

Table IV-5-4. Systems of interest.

SYSTEM	ONCE-THROUGH CYCLE	RECYCLE		
		DENATURED U-233/TH	PU BURNERS AND U-233 CONVERTERS	U-PU RECYCLE
Systems characterized in detail				
LWR	X	X	X	X
SSCR	X	X	X	X
HWR	X	X	X	
HTGR	X	X	X	
LWBR/HWBR	X	X	X	
LMFBR		X	X	X
Systems characterized by scoping studies				
Accelerator breeder reactor				
Fast mixed-spectrum reactor				
Gaseous core reactor				
Fusion-fission hybrid reactor				
Molten salt breeder reactor				

would be expensive to introduce them. On the other hand, in a recycle the advanced converter reactors could have substantial impact in scenarios where the breeder reactor is delayed until the mid-twenty-first century. LWBR's could be advantageous in such scenarios.

Some of the most significant findings to date relate to potential improvements to the fuel use efficiency of the once-through LWR. Studies have shown that fuel efficiency improvements of up to 10 to 15 percent may be possible in about 10 years. The contemplated improvements would be retrofittable to existing reactors. Up to 10 percent in additional improvements may also be possible by the year 2000 and, if advanced isotopic separation is also used, total uranium savings of up to 45 percent might be possible. These improvements would be especially significant in view of the high proliferation resistance of the once-through fuel cycle. Contemplated improvements include:

- Increased fuel burnup—using fuel in the reactor for a longer time; discharge exposure increased from current 25–35,000 MWD/T to perhaps 45–55,000 MWD/T
- Improved fuel management and control designs—a variety of methods to reduce neutron absorption by control poisons, reduce residual poison at end of cycle, increase uniformity of discharge burnup, reduce neutron leakage, and use early fuel batches of startup cores
- Non-heavy water spectrum shift—changing the energy levels of neutrons by varying BWR void content, varying inlet temperature to the core, or selectivity removing fuel rods from assemblies during refueling
- Enrichment variation—varying the enrichment level of fuel in the axial or radial zones of the core, including blankets of natural or depleted uranium. This would permit a higher batch-average burnup for a given burnup limit and/or capture neutrons that normally escape the core
- Lattice changes—changing the design to optimize water-to-fuel ratio for once-through operation

The feasibility of these improvements must be proven through extensive R&D programs. In addition, the impact of the changes on systems safety, environmental effects, economics, and proliferation resistance are now being evaluated. Nevertheless, LWR improvements appear to be a very attractive near-term option for stretching available uranium supplies.

For research reactors, preliminary studies have found that high-density low-enriched fuel can be substituted for high-enriched fuel with little impact on fuel performance but with major impact on improving proliferation resistance. These results have been provided to INFCE.

The basis for a review by the Nuclear Regulatory Commission (NRC) of safety and environmental issues associated with alternative nuclear systems has been defined. Systems being reviewed by the NRC include the modified light water reactor on the once-through fuel cycle, the LWBR, the HWR, the HTGR, the GCFR, and the LMFBR on uranium-plutonium, thorium-uranium, and thorium-plutonium cycles. Interim safety and environmental assessments have been prepared and submitted to the NRC; final documentation was completed in January 1979.

In the area of commercialization, a U.S. market survey found that there is virtually no interest by utilities or vendors in introducing a new reactor system. A preliminary assessment of new reactors and fuel cycles found that only LWR-based systems could be expected to be available before the year 2000 on a commercial basis. Methodologies were developed for determining the introduction patterns and approach to mature-plant economics, and for assessing the commercial potential of new reactors and fuel cycles.

In the area of the INFCE support, papers were prepared on LWR once-through advanced converters and on economic and resource-use aspects of breeder reactors for Working Group 5. Also, a paper that contained a preliminary proliferation-resistance assessment of the LWR on the once-through cycle, of thermal recycle systems, and of fast breeder fuel cycles was submitted to the INFCE.

Major milestones and funding for NASAP are given in Figure IV 5-3. The NASAP schedule is constrained by both INFCE support tasks and the level of effort required. All the NASAP activities have been scheduled to achieve final recommendations by mid-1979. A draft public report was prepared by October 31, 1979, and a final report reflecting public comment was submitted to Congress by December 31, 1979.

Responsibility for program direction and control is assigned to the Division of Nuclear Alternative Systems Assessment, which is under the Assistant Secretary for Energy Technology. Other division within Energy Technology, Defense Programs, Environment, International Affairs, and the Office of Energy

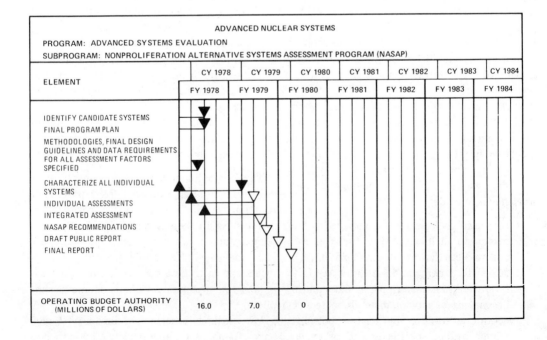

Figure IV-5-3. Major milestones and funding for NASAP.

Research are providing analytical input to NASAP.

Other Federal agencies with interests in nonproliferation matters are the Department of State and the Arms Control and Disarmament Agency (ACDA). The Department of State is responsible for directing U.S. efforts in support of INFCE. ACDA will review and approve individual evaluation criteria developed by NASAP.

Other participants in the NASAP study include the national laboratories, private contractors, the nuclear industry and its trade associations, universities, and the public. National laboratories, private contractors, and universities are performing specific technical analyses required by NASAP. Industry leaders, through trace associations, are consulted on their veiwpoints. Public inputs will be solicited as part of the review process of the draft NASAP report and public comments will be included in the final NASAP report to the Congress and the President.

INFCE

The INFCE Program was created to further the international cooperation needed to keep the nuclear fuel cycle from becoming an attractive proliferation path. The INFCE is a major forum for nations to reexamine assumptions and conclusions concerning the future direction of the nuclear fuel cycle. The focus of NASAP is the domestic policies for nuclear power development with due consideration to the nuclear weapons proliferation. INFCE is helping to build the foundation for an international consensus on the practices and institutions that are necessary to govern nuclear energy utilization and prevent misuse of the nuclear fuel cycle.

The stated objective of the INFCE, agreed upon by the participating states, is to "conduct technical and analytical studies of measures which can and should be taken at the national level and through international agreements to minimize the danger of the proliferation of nuclear weapons without jeopardizing energy supplies or the development of nuclear energy for peaceful purposes."

The implementing strategy to be pursued by the INFCE in meeting its objective consists of five points:

- All interested states and relevant international bodies may participate in the work and will have an equal opportunity to contribute.
- The INFCE represents a technical analytical study, not a negotiation. Participants will not be committed to the INFCE's results.
- Participation in the studies will not jeopardize participant's fuel-cycle policies or international cooperation, agreements, and contracts, provided that agreed-upon safeguard measures are applied.
- Special consideration will be given to the needs of developing countries.
- The evaluations will be conducted along lines set forth in the "INFCE Technical and Economic Scope and Methods of Work" and the "INFCE Proposed Terms of Reference."

The Department of State is the lead agency for U.S. participation in the INFCE. While no direct funding in DOE is provided for the INFCE, analytical studies conducted under the NASAP support the requirements of the INFCE. NASAP's contributions to the INFCE are described in the previous section.

INFCE milestones and funding are presented in Figure IV 5-4. In general, individual country analyses for input to the working groups have been essentially completed. First drafts of the individual working group reports were due in May 1979, and the overall INFCE report was scheduled to be published after approval by the final plenary conference, which was held in February 1980.

At the INFCE organizational meeting in October 1977, eight working groups were formed to deal with the areas of concern (Table IV 5-5). The technical tasks of each working group are coordinated by a Technical Coordinating Committee comprised of the co-chairmen of each working group.

Economics Evaluation

Nuclear power is one of the options a utility can select to satisfy its electrical generation

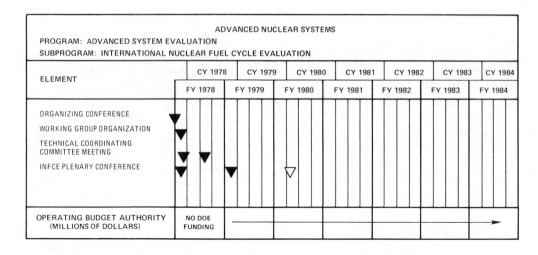

Figure IV-5-4. Major milestones and funding for INFCE.

needs. The choice between nuclear and nonnuclear alternatives is based on an evaluation of many factors, a most important one being economics. Therefore, the continued viability of nuclear power depends on maintaining its economic competitiveness with the alternatives. DOE maintains a data base of capital, fuel, and operating and maintenance costs for both fossil and nuclear plants. This data base serves as a means of comparing differences in costs, monitoring trends in costs, and identifying causes of the continuing cost escalation for both fossil and nuclear plants. Analysis of the data provides a basis for determining what actions, if any, can be taken to alleviate the upward pressure on nuclear electric generation costs.

Furthermore, DOE is developing alternative nuclear concepts, which, if they are to be commercially acceptable, must be economically competitive at the time they are introduced. Thus, the data base has recently been expanded to include the fast breeder, several advanced converter systems, and several alternative fuel cycle systems. The economic data on these concepts, which are not commercially viable at the present time, are not as detailed as for the LWR and coal-fired plants. However, sufficient design detail is available to provide a realistic comparison of costs and to identify plant design areas that, because of projected excessive costs, must be modified if the concept is to be economically competitive.

Overall, the data base is an essential management tool for decision-making both in the broad sense of selecting the concepts most likely to succeed economically and in the narrower sense of identifying specific areas of

Table IV-5-5. Working groups for INFCE.

WORKING GROUP	CO-CHAIRMEN
1. Fuel resources and heavy-water availability	Canada, Egypt, India
2. Enrichment availability	France, Federal Republic of Germany, Iran
3. Assurances of long-term supply of technology, fuel, heavy water, and supplies for all materials	Australia, Philippines, Switzerland
4. Reprocessing, plutonium handling, and recycle	Japan, United Kingdom
5. Fast Breeders	Belgium, Italy, USSR
6. Spent fuel and waste management	Argentina, Spain
7. Fuel conservation and fuel utilization in thermal reactors	Finland, Netherlands, Sweden
8. Advanced fuel cycle and reactor concepts	Republic of Korea, Romania, USA

individual concept designs that must be addressed from a cost reduction standpoint.

The objective of this program is to maintain and periodically update technical, capital costs, fuel cost, and operating and maintenance cost information of significance in planning U.S. civilian nuclear power programs.

The data produced during approximately 10 years of economic evaluations have been consolidated into a central data base. The data base is periodically updated as commodity prices, labor rates, regulatory requirements, and other factors change. Comparative analyses are conducted to determine targets that nuclear plants must meet to compete favorably with non-nuclear alternatives. The data base is expanded as required to increase the effectiveness in meeting the objective. Both ongoing tasks, which reflect the need for continual assessment of program direction and opportunities, and ad hoc tasks, which are initiated as specific needs dictate, are supported.

Most recently, the data base has been updated to reflect January 1, 1978 dollars and regulatory requirements. Table IV 5-6 is a summary of the latest capital cost estimates for the BWR, HTGR, PWR, pressurized heavy water reactor (PHWR), GCFR, and LMFBR. Also included for comparison are large- and intermediate-size high-sulfur coal plants (HS12 and HS8) with flue-gas desulfurization and large- and intermediate-size low-sulfur coal plants (LS12 and LS8). All estimates are on a consistent basis and normalized to a plant size of 3800-MWt. The estimates include direct and indirect costs only. Interest during construction, escalation, and items that are site or utility specific are not included. The cost ratio for each plant is given in terms of a comparison with the PWR.

Table IV 5-7 summarizes 30-year levelized fuel-cycle costs for 12 different fuel-cycle systems that are currently being discussed as candidate systems complementing the plants listed in Table IV 5-6. The estimates are given in January 1, 1978, dollars, are based on a year 2001 reactor startup date, and are levelized over an assumed 30-year reactor lifetime.

Major milestones and funding for the Economics Evaluation Subprogram are presented in Figure IV 5-5.

The work is under the technical management of United Engineers and Constructors, Inc., a large architect-engineering firm with extensive experience in designing and building power plants of all types. Oak Ridge National Laboratory provides consulting services on operating and maintenance costs and maintains the CONCEPT computer code, which is designed to provide summary level capital cost information from the data base and also to provide regional variations of capital costs.

Advanced Technology and Systems Assessments

A capability is maintained in the area of advanced technology systems assessments to ensure that management has comprehensive and timely data on which to make decisions as to possible programmatic emphasis on specific nuclear systems or applications of such systems. This assessments activity is structured to evaluate proposed new systems and systems applications and to provide a basis

Table IV-5-6. Normalized* capital cost summary by plant model.

	NUCLEAR PLANT MODELS						COAL PLANT MODELS			
COST	BWR	HTGR	PWR	PHWR	GCFR	LMFBR	HS12	HS8	LW12	LS8
Size MWe	1,264	1,546	1,268	1,162	1,440	1,390	1,419	1,368	1,371	1,322
Direct cost, $M	492	593	487	540	562	737	486	471	409	395
Indirect cost, $M	205	199	205	211	253	265	100	103	81	84
Base cost, $M	697	792	692	751	815	1,002	586	574	490	479
Unit cost, $/kW	551	512	546	646	566	721	413	420	357	362
Cost ratio	1.01	0.94	1.00	1.18	1.04	1.32	0.76	0.77	0.65	0.66

*Normalized to a plant size of 3800 MWt or its equivalent; costs are in January 1, 1978 dollars.

Table IV-5-7. Summary of 30-year levelized fuel-cycle costs.*

REACTOR/FUEL-CYCLE SYSTEM	$/MBTU
PWR/throwaway**	0.76
PWR/uranium & plutonium recycle**	0.67
HTGR/throwaway	0.76
HTGR/uranium-233 recycle	0.73
HWR/(natural uranium) throwaway	0.73
HWR/(low-enriched uranium) throwaway	0.40
HWR/(thorium) throwaway	1.04
HWR/(thorium) uranium (plutonium recycle)	0.62
LMFBR/oxide fuel, uranium blanket	0.39
LMFBR/oxide fuel, thorium blanket	0.48
GCFR/oxide fuel, uranium blanket	0.45
GCFR/oxide fuel, thorium blanket	0.43

*January 1, 2001, reactor startup/date; costs are in January 1, 1978 dollars. **Representative of either the BWR or PWR.

for selection of those with the potential of meeting both near-term and far-term energy needs in the United States. This is a continuing function in that new or advanced systems and their associated applications are assessed and work terminated or emphasized depending on the perceived potential for solving specific energy-related problems. The source of new approaches or ideas is not restricted, thus ensuring that promising concepts will not be eliminated or overlooked.

Studies are conducted on conventional, advanced, and specialized nuclear systems and their applications. These studies can be categorized as those directed toward improved thermal energy utilization and increased siting acceptability. The former category includes nuclear cogeneration (production of both electricity and thermal energy) and beneficial uses of waste heat. The latter category includes plant siting options, plant cooling methods to reduce water consumption, and meteorological effects of large heat and moisture releases from power plant cooling systems.

The assessment of advanced systems applications encompasses both short-term analyses of feasibility and long-term assessments that require complex studies and/or confirmatory research.

The overall objective of the advanced technology systems assessments program is to conduct studies that support management in decision-making on the development and deployment of conventional, advanced, or specialized nuclear systems for unique applications.

The basic strategy of the advanced technology systems assessments program consists of maintaining a strong capability for assessing all aspects of advanced nuclear systems and their applications so as to provide required data relative to systems potential for decision-making by management. In this regard, preliminary evaluation is emphasized to determine whether such systems or proposed systems applications warrant further, more detailed, investigation. Associated with such systems is the assessment of specialized system components (e.g., dry cooling towers) that may have a direct bearing on future system implementation either in terms of siting

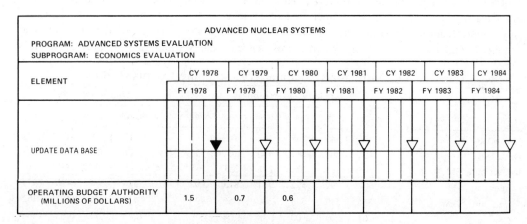

Figure IV-5-5. Major milestones and funding for economics evaluation.

acceptability or improved resource utilization. The maintenance of the assessments function permits evaluation of the impacts of nuclear energy development strategies and selection of the most favorable strategy for implementation. It should also be noted that the assessment function includes economic, environmental, and institutional aspects of systems applications in addition to technical evaluations.

Table IV 5-8 shows the operating funding levels for the two categories of advanced technology systems assessments in the FY 1980 period.

Increased Thermal Efficiency. As an outgrowth of past assessments, this activity is now focused on two major areas: nuclear cogeneration and the beneficial uses of power plant waste heat. Cogeneration is defined as the simultaneous production of electricity and thermal energy. Thermal energy may be contained in steam for industrial processes or desalting, or in hot water for district heating. Industrial process heat and district heat (space conditioning, including heating and cooling) account for almost 40 percent of total U.S. energy usage, or about 30 percent of oil use and 60 percent of natural gas use, or the energy equivalent to about 10 million barrels of oil per day. These needs could be largely served through the application of existing reactor technology, thus reducing dependence on fossil fuels. Therefore, highest priority is assigned to these efforts. A secondary effort is directed toward the assessment of effective use of nuclear power plant waste heat for aquaculture and agriculture.

The objective of this activity is to assess nuclear cogeneration and the beneficial use of waste heat so as to define those systems and systems applications that have the potential for major efficiency improvements in overall energy utilization.

A concerted effort is made to conduct the assessments in conjunction with federal, regional, state, and local organizations (including utilities and industry). This assures applicability of results to potential users and others who would benefit and permits the most rapid transfer of results into practical applications.

The scope of the activity includes:

- Cooperative studies with utilities and industry, primarily in highly industrialized areas such as the Gulf Coast, on the cost and feasibility of supplying industrial process heat (steam) by means of intermediate- and large-size nuclear power plants.
- Cooperative studies with other DOE divisions, Federal and State Government agencies, and private industry on assessments of large-scale regional district heating systems, such as ongoing efforts in Minneapolis-St. Paul.
- Cooperative studies with utilities, the Environmental Protection Agency (EPA), and others, of methods for using low-temperature power plant waste heat for beneficial purposes. This includes the study of specialized components (e.g., heat exchangers) for application to nuclear power plants to assure systems safety when heat is used for food production.

Studies have been completed of the applica-

Table IV-5-8. Funding for advanced technology systems assessments for FY 1978−1980.

ADVANCED TECHNOLOGY SYSTEMS ASSESSMENTS CATEGORY	FY 1978	FY 1979	BUDGET AUTHORITY (DOLLARS IN THOUSANDS) REQUEST FY 1980	INCREASE (DECREASE)
Increased Thermal Efficiency	$2,510	$1,471	$1,100	($371)
Increased Siting Acceptability	3,064	1,220	900	(320)
TOTAL	$5,574	$2,691	$2,000	($691)

tion of small- and intermediate-size nuclear power plants for supplying industrial steam to industry. These studies indicate that small (400-MWt) plants are not cost effective in the applications studied; however, intermediate-size (1200-MWt) plants appear to match process steam loads in a number of areas and may be economically attractive to steam users. Large plants are limited in their application because the large quantities of steam that can be produced exceed most industrial demand in a given region. However, when large plants are operated as cogeneration units, supplying electrical energy to the power grid while producing steam to meet nearby industrial demands, they appear to have economic advantages over electricity-only central station plants.

A major study of district heating in the Minneapolis-St. Paul area was initiated in April 1977 with the Minnesota Energy Agency, the Northern States Power Co., and several other local groups, and with participation by a number of DOE divisions. This study, which compares both nuclear and coal plants for this purpose, includes all aspects of a district heating system from energy generation to end use. Results of preliminary assessments were available in FY 1979.

Studies of beneficial uses of waste heat have been conducted by ORNL. These studies assess the benefit-to-cost ratios of various waste heat utilization schemes (e.g., soil warming, greenhouse heating, and fish growing). The results of this effort have indicated that aquaculture (fish growing) results in the highest return on investment as compared to other uses of waste heat. An additional study has been undertaken at the Vermont Yankee nuclear power plant on the possible use of waste heat for fish production in a very cold region.

The major efforts in desalting have been the biennial updates of dual-purpose (electricity and desalting) plant costs. In addition, data have been developed on coupling and control of such dual-purpose plants and on costs of large plants using advanced desalting technology.

The principal FY 1980 efforts will be devoted to: (1) site application studies, in cooperation with utilities and industrial companies, of large- and intermediate-size nuclear cogeneration plants and nuclear process steam plants designed as a integral part of an industrial complex; and (2) analysis and preliminary engineering design of a large regional cogeneration district heating system. In other process-heat related work, an update of dual-purpose plant economics will be conducted factoring in recent improvements in sea water desalting technology. Also, joint studies of beneficial uses of waste heat will be conducted with utilities (e.g., TVA) to establish optimum economic uses and component requirements unique to applying nuclear systems for this program.

Major milestones and funding for this activity are shown in Figure IV 5-6.

ORNL is the major participant in the studies related to increased thermal utilization (e.g., studies of waste heat utilization, desalting cost studies). In addition, they assist DOE Headquarters personnel in the management and coordination of the programs undertaken cooperatively with other entities. Other participants in these programs are discussed above.

Increased Siting Acceptability. One of the major concerns related to the use of nuclear power (in fact, all central station power) is plant siting. Although nuclear plants do not emit particulates, they do require large quantities of water for cooling and large areas of land for exclusion boundaries. Because their efficiency is somewhat less than that of fossil plants, they also reject large quantities of heat and moisture when evaporative cooling is used.

As a result of past studies done by the DOE and others, four major activities related to increased siting acceptability are being conducted. These are studies of nuclear energy centers, investigations of the meteorological effects of large thermal energy releases, studies of advanced power plant heat rejection systems, and maintenance of a cooling water availability/use data bank.

The objective of this activity is to assess improved methods of power plant siting and to evaluate possible system component modifications (e.g., cooling towers) so as to in-

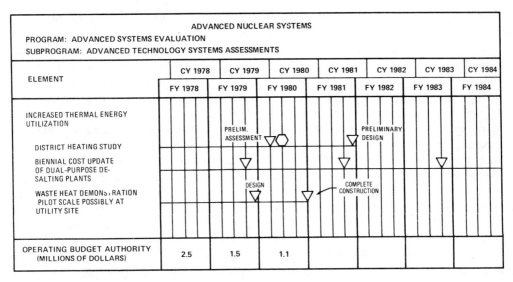

Figure IV-5-6. Milestones and funding for increased thermal energy utilization.

crease the flexibility of such siting. Associated with this objective is the delineation of the environmental impacts of siting large energy producing facilities and the development of means to mitigate such impacts.

The major strategy in the area of increased siting acceptability is to define those problems that restrict siting and to assess new approaches for their solution both in the near- and long-terms. The scope of the present activity is:

- Detailed analysis of the nuclear energy center concept in the south (large load growth region) and in the west (limited water resource region) to establish the feasibility and practicality of this siting option. These studies include technical, economic, institutional, and environmental factors related to centers having up to 12 units.
- Theoretical analysis and confirmatory research to establish the impact on the local meteorology of large heat and moisture releases from cooling towers serving a group of power plants. This program, jointly funded by the DOE Offices of Environment and Energy Technology, is of a long-range nature with validated atmospheric models planned for completion in FY 1982.
- Assessment of advanced power plant heat rejection systems. This includes dry and combined wet and dry cooling towers, high-performance heat transfer surfaces, unique heat transfer fluids, and nonconventional cooling tower components. All these methods may reduce water consumption. The design, construction and operation of an advanced systems test module, a major element of the program, is supported by the Electric Power Research Institute (EPRI). Supporting technology development is jointly funded by the DOE Offices of Environment and Energy Technology and by EPA.
- Assessment of water requirements for power plant cooling and the maintenance of a water availability/use data bank. These tasks provide data for the development of a predictive model for estimating where the siting of central station plants may be limited by water resource availability and where and when less-water-consumptive advanced cooling systems will be required. The U.S. Geological Survey (USGS) and the DOE are jointly supporting this program.

Under the energy center program, screening studies in the south and west were conducted in FY 1978. These resulted in the selection of specific sites in South Carolina and Utah. The detailed study Phase III of a site in South Carolina was initiated in late FY 1978 on essentially a 50/50 cost sharing basis between DOE and the state and local participants. The Utah Energy Office has conducted an initial study of the data base available for the Phase III study, and plans call for implementation of a detailed study in Utah in early CY 1979. As in the South Carolina study, the Utah study will be conducted on a cost-sharing basis. The regional energy center studies are planned for completion at the end of FY 1980.

Under the Meteorological Effects of Thermal Energy Release (METER) program, field experiments and physical and mathematical modeling studies are being conducted on the effects of heat and moisture releases on the atmosphere from large power plant cooling systems. The principal activity during FY 1980 will be the continuation of field studies to collect data to validate analytical models. This is a five-year program funded cooperatively with the DOE Office of Health and Environmental Research and is conducted jointly by ORNL, ANL, the Rand Corp., and Pennsylvania State University. The overall program is considered essential to developing siting criteria and concepts and other information on large concentrations of energy producing facilities irrespective of the fuel source. The siting criteria will be made available for use by utilities, states, and Federal agencies.

DOE has supported the assessment of advanced power plant heat rejection systems including economic analysis of various combinations of wet and dry cooled power plants, design studies of advanced systems, and bench-scale tests of improved surfaces. The effort reached the point in FY 1978 where PNL recommended that a demonstration cooling tower module (of about 15–20-MWt) be designed, constructed, and operated to evaluate the use of ammonia as a coolant, to investigate advanced low-cost heat transfer surfaces, and to evaluate the approach of water deluge of the cooling surfaces on hot days to maintain power plant performance. EPRI has assumed support for the design and construction phases. Associated technology development is supported by DOE and EPA.

The FY 1980 effort encompasses support of PNL in the development of ammonia cooling for power plants, the evaluation of new approaches to power plant cooling, and the economic studies of applying such approaches to water-short areas in the United States. The effort is directed toward assisting utilities and industry in developing and applying advanced heat rejection concepts, in order to reduce the demand on dwindling consumptive water resources and to greatly increase power plant siting flexibility. Present projections of consumptive water availability and restrictions on the use of such water indicate that such concepts will have to be employed in some regions of the United States prior to 1990.

The basic water availability/use data bank is in operation at HEDL, but is being updated to reflect desires by the USGS to increase the flexibility of this computerized system. In addition, the predictive model will be operational in the second quarter of FY 1979. This system will receive continued support in FY 1980 to improve the data base. It is so structured that a wide range of power plant and water data can be recalled in various formats depending on user need. Based on data requests received to date, the system will be an effective tool for utilities, water planners, and federal, state and local agencies in planning power plant siting.

Major milestones and funding are presented in Figure IV 5-7.

SPACE AND TERRESTRIAL APPLICATIONS

The unique characteristics of nuclear-powered electric generators—compact size, light weight, and long life—enable operation of the sensing, analytical, and communication systems of spacecraft, satellites, and other remotely located devices for long time periods without relying on external sources of energy. Thus, the space and terrestrial systems program, by its technical initiatives, supports the national security as well as the civilian scientific exploration and exploitation of space. By applying the technology developed for space applications, the program also fosters beneficial terrestrial uses of reactor byproducts and

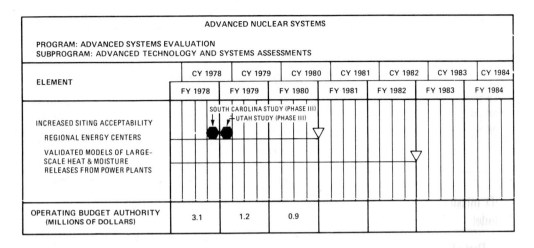

Figure IV-5-7. Major milestones and funding for increased siting acceptability.

of special nuclear materials recoverable from reactor wastes.

Overview

The development of nuclear power for aerospace applications was initiated in the early 1950s. Major efforts were concentrated on reactor propulsion systems for manned or unmanned aircraft and missiles for the Department of Defense (DOD). Ten years later, the corollary development of radioisotope-fueled Space Nuclear Auxiliary Power (SNAP) systems permitted the deployment of earth-orbiting satellites for DOD navigation systems. Also in the early 1960s, a series of strontium-90-fueled radioisotopes thermoelectric generators (RTG's) demonstrated their capability of safely and reliably producing electrical power in remote weather stations, buoys, lighthouses, and undersea beacon installations. Commercial production of the Sr-90-fueled systems followed in the 1970s. Tables IV 5-9 and IV 5-10 list the space and terrestrial missions that have used nuclear electric generators.

Based on these successes, development has proceeded on improved thermoelectric conversion, on dynamic systems for intermediate levels of power demand, and on reactor technology leading toward the anticipated higher power requirements of future NASA and

Table IV-5-9. Space missions that have used nuclear electric generators.

MISSIONS	LAUNCH	POWER SOURCE	POWER, WATTS ELECTRIC	MISSION PURPOSE
Transit 4A, 4B	1961	RTG*	2.7	Navigation satellite
Transit 5BN 1, 2	1963	RTG	25	Navigation satellite
Snapshot	1965	Reactor	500	Flight test
Nimbus III	1969	RTG	28	Weather satellite
Apollo 12	1969	RTG	74	Lunar science experiments
Apollo 14, 15	1971	RTG	74	Lunar science experiments
Apollo 16, 17	1972	RTG	74	Lunar science experiments
Pioneer 10	1972	RTG	80	Jupiter exploration
Transit	1972	RTG	36	Navigation satellite
Pioneer 11	1973	RTG	80	Jupiter/Saturn exploration
Viking I & II	1975	RTG	85	Mars lander exploration
LES 8/9[+]	1976	MHW[++] RTG	310	Communications satellite
Voyager I & II	1977	MHW RTG	475	Jupiter/Saturn/Uranus

*Radioisotope thermoelectric generator. [+]Lincoln Experimental Satellites. [++]Multi-hundred watt.

Table IV-5-10. Terrestrial missions that have used nuclear electric generators.

GENERATOR	DEPLOYMENT	POWER SOURCE	POWER, WATTS ELECTRIC	MISSION
Sentry	1961	RTG	5	Arctic weather station
SNAP-7A	1964	RTG	10	USCG light buoy
SNAP-7B	1964	RTG	60	USCG lighthouse
SNAP-7C	1962	RTG	10	Antarctic weather station
SNAP-7D	1964	RTG	60	Navy weather boat
SNAP-7E	1964	RTG	7.5	Undersea beacon
SNAP-7F	1965	RTG	60	Oil-rig navigation aid
SNAP-21	1974	RTG	10	Ocean current survey

DOD missions. Accomplishments in FY 1978 include:

- Periodic reports from Viking, Lincoln Experimental Satellites (LES) 8/9, and Voyager missions indicate RTG performance at or beyond design requirements
- Selenide 100-watt(e) ground demonstration converter evaluated vs. Galileo mission (orbiter and space probe of Jupiter) requirements
- Performance evaluation of Brayton and organic Rankine dynamic system ground demonstration units led to selection of the organic Rankine system for further development
- General-purpose heat source prototype module design available
- Initial design concepts for Solar Polar Mission developed and reviewed
- Technology transfer for multi-hundred-watt (MHW) fuel sphere accomplished; Savannah River Plutonium Fuel Facility enters production status
- Economic study completed showing commercial incentive for beneficial use of radioisotopes
- Sandia Irradiator for Dried Sewage Sludge completed and cesium-137 capsules delivered for installations
- Agricultural studies showed value of irradiated dried sewage sludge as fertilizer, soil conditioner, and ruminant animal feed supplement.

The principal objectives of the Space and Terrestrial Systems Program are as follows:

- To deliver qualified nuclear-electric systems to other Federal agencies for both outerplanetary space missions and earth-orbiting satellites
- To provide power units for scientific experiments, telemetry, navigation, and communication purposes
- To develop isotope-powered "nuclear batteries" for terrestrial deployment at remote or undersea sites
- To investigate various isotopes and their utilization using the skills developed for the handling and applications of plutonium-238

The objectives of the program are being met by implementing subprograms in the areas of space nuclear systems and terrestrial isotope systems.

By means of research and development, the program delivers environmentally acceptable, operationally safe, and technically qualified nuclear energy systems to Federal user agencies for earth-orbital and outerplanetary space missions, as well as for terrestrial applications. For national security missions, the program meets DOD requests for improved isotope and reactor power systems technology for future defense operations and communications for which a compact and reliable power supply is a principal factor in determining mission viability.

Improved power supplies are also essential for NASA's unmanned planetary explorations to obtain scientific data on the nature and origin of our solar system and universe. In 1976, nuclear electric generators provided the power for the first scientific exploration of the surface of Mars. The outerplanetary missions to the planet Saturn and beyond, launched in August and September 1977, are part of a program of scientific exploration of the remote regions of the solar system planned by NASA for the coming decades. These missions require higher-powered spacecraft with reduced weight and increased endurance and reliability at reduced cost.

The directions the program takes are determined both by user agency product requests and by initiatives indicated by user trends and technology developments. Figure IV 5-8 shows the general approach for developing systems for flight or terrestrial purposes. Before a major application effort starts, cost/benefit analyses are conducted when the development risks and potential benefits can be meaningfully identified in advance; these analyses include mission and commercial requirements, environmental, health, and safety requirements, and specific needs (such as reliability, longevity, and survivability), and consider non-nuclear as well as nuclear alternatives. Program work is undertaken in concert with appropriate interagency formal agreements. Liaison between DOE, NASA, and other user agencies is pursued at several management levels through the medium of coordinating boards with both technical and administrative membership. These interagency coordinating groups provide the information on which the direction of the advanced research and technology activities is focused.

As the technology and support programs advance and are integrated into a system technology, user hardware requirements are specified and the system development and fabrication activities begin. The appropriate environmental assessments and impact statements, preliminary and final safety analysis reports, safety testing, evaluations and fuel fabrication and assembly are done concurrently with the system development. Upon completion and acceptance testing, the end product is delivered to the user for operational use.

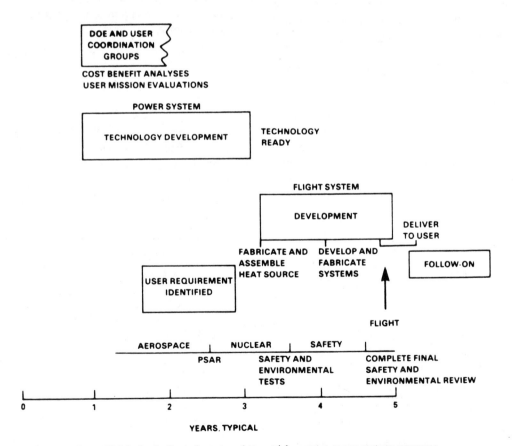

Figure IV-5-8. Logic flow of space and terrestrial compact power systems programs.

The upper-level WBS for the entire Fission Energy Program was shown in Figure IV 1-2. The further breakdown of this Space and Terrestrial Applications Program is shown in Figure IV 5-9, with a matrix showing how the lower-level packages in the WBS relate to the arrangement of the program description in this section.

The Assistant Secretary for Energy Technology is responsible for the Space and Terrestrial Applications Program. This program is assigned to ETN, who delegates management responsibility to the Director of the Advanced Nuclear Systems and Projects Division. In turn, he delegates program responsibility to the Chief of the Space and Terrestrial Systems branch. The branch chief provides overall direction and program approval, and is responsible for successful implementation of the program within overall cost and schedules. Contract administration is shared by Headquarters and the DOE field offices at Albuquerque, Chicago, Oak Ridge, San Francisco, and Savannah River. Plans being pursued now place more of this responsibility in the field offices.

For space nuclear systems, management policy permits funding of the first-of-a-kind nuclear systems, with the costs for follow-on flight systems being reimbursable to DOE. In certain cases, planning carries a developmental system through to a point of technology readiness only, the remaining flight development and demonstration costs are borne by the user agency.

Mission system development work is usually performed by industrial contractors. Fuel and safety research and development, special metals technology, capsule hardware fabrication, and heat-source assembly are done in government-owned contractor-operated laboratories and plants (such as LASL, ORNL, SRL, and the Mound Facility) because these operations depend on the nuclear facilities, equipment, and expertise maintained for nuclear fuel and weapons programs. Other activities performed by industrial or institutional contractors include:

- Quality assurance and reliability monitoring
- Isotope materials and capsule development
- Environmental impact and safety evaluation
- Advanced studies on thermoelectric materials, mission requirements projections, and technical consultation

Table IV 5-11 shows the operating funding levels for FY 1978 through FY 1980.

Space Nuclear Systems

The Space Nuclear Systems Subprogram supports other Federal agencies by providing progressively improved nuclear power sources

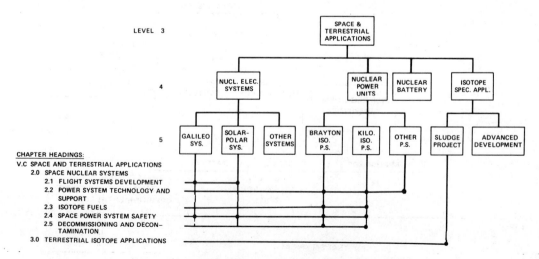

Figure IV-5-9. Space and terrestrial applications program WBS.

Table IV-5-11. Funding for space and terrestrial applications for FY 1978–1980.

SPACE AND TERRESTRIAL APPLICATIONS	BUDGET AUTHORITY (DOLLARS IN THOUSANDS)				
PROGRAM ELEMENT	FY 1978	FY 1979	REQUEST FY 1980	INCREASE (DECREASE)	MAJOR CHANGE[+]
Space Nuclear Systems	$31,008*	$34,380	$32,300	($2,080)	R
Terrestrial Isotope Applications	14,842	4,000	2,000	(2,000)	R
TOTAL	$45,850	$38,380	$34,300	($4,080)	

*Excludes inventories. [+]Key: R = Program redirection.

to meet mission objectives. Development work is undertaken to reduce weight, enhance power levels, improve reliability and mission lifetime, and reduce cost to the user.

The work in the Space Nuclear Systems Subprogram involves the following:

- Flight systems development consists of design, development, qualification, and fabrication of flight-qualified generators
- Power systems technology and support includes test and evaluation of materials and components for proposed flight systems, and also contains analysis, quality assurance, and reliability support functions.
- For isotope fuels, plutonium-238 forms and capsule hardware are prepared, fuel assemblies are delivered, and final assembly of the radioisotope heat source is performed.
- Power systems safety includes assessments to ensure safe and environmentally acceptable operations leading to flight approvals.
- Decommissioning and decontamination consists of the restoration of previously used fueling facilities to a condition suitable for other uses.

Table IV 5-12 shows the operating funding levels by category for the FY 1978 through FY 1980 period.

Flight System Development. Effort within the Flight System Development category includes designing, developing, fabricating, and qualifying nuclear-powered generators for approved missions of NASA and other Government agencies.

Within the last five years, significant progress has been made in the direction of increased thermal output from the radioisotope heat sources, and improved efficiency in the static thermoelectric converters. For example, the plutonia-molybdenum cermet heat sources aboard the Viking landers on Mars each delivered about 680 thermal watts to a modified lead telluride converter, yielding a power output of 40 watts, at a conversion efficiency of

Table IV-5-12. Funding for the space nuclear systems subprogram for FY 1978–1980.

SPACE NUCLEAR SYSTEMS	BUDGET AUTHORITY (DOLLARS IN THOUSANDS)			
CATEGORY	FY 1978	FY 1979	REQUEST FY 1980	INCREASE (DECREASE)
Flight System Development	$ 5,887	$11,983	$12,400	$ 417
Power System Technology and Support	9,140	6,340	7,400	1,060
Isotope Fuels	9,093	10,043	10,250	207
Space Power System Safety	3,598	3,014	2,250	(764)
Decommissioning and Decontamination	3,290	3,000	0	(3,000)
TOTAL	$32,008*	$34,380	$32,300	($2,080)

*Excludes inventories.

approximately 6 percent. In contrast, the MHW pressed plutonium oxide heat sources powering the LES 8 and 9 communications satellites for the USAF and the NASA Voyager spacecraft for outerplanetary exploration each delivered 2400 thermal watts to a silicon/germanium converter, yielding nearly 160 watts at an efficiency just under 7 percent. The flight systems development effort uses all available innovations in materials and design to accommodate the increasing specific-power requirements of near-term missions. In turn, this development activity stimulates and guides the related power systems technology and support activities.

The objectives of this development are as follows:

- To support the power-source requirements of the NASA outerplanetary exploration program or other planned missions
- To support the power-source requirements of the DOD in its space and/or terrestrial activities, including communication satellites and navigation aids

The strategy of the Flight System Development effort is to attain the objectives by the following:

- Providing safe, reliable, and long-lived nuclear-electric systems to space-mission agencies
- Fully using the accumulated experience in radioisotopes, thermoelectric elements, and specialized metals and alloys
- Developing, designing, analyzing, and testing materials and components to meet mission-performance requirements
- Maintaining rigorous standards of quality control and environmental safety
- Comparing actual performance with that predicted so as to guide improved design

The work includes:

- Reducing and analyzing the telemetry data on the performance of the Viking I and II, Lincoln Experimental Satellites, and the Voyager I and II RTG flight units so as to guide design improvements
- Designing, developing, fabricating, and testing selected components of the static Si/Ge isotope generator for the NASA Galileo mission (orbiter and surface probe of Jupiter) in December 1981
- Designing, developing, fabricating, and testing the components of the improved static thermoelectric generator, incorporating the modular general-purpose heat source, for the Solar Polar Mission for launch in February 1983.

The following activities are associated with current missions:

- Power monitoring and data reduction on Viking I and II were completed in FY 1978.
- Power monitoring and data reduction on the LES were completed in FY 1979. These data assist in qualifying the Multi-Hundred-Watt heat source (with possible design modifications) to support its use on the Galileo mission.
- The Voyager spacecraft attained their closest approach to Jupiter on March 5, 1979 (Voyager I) and July 9, 1979 (Voyager II). Current telemetry data indicate that all generators on board the two spacecraft are performing predictably to design requirements. Continuing data review will evaluate the stability and reliability of these generators in the outer planetary regions.

Preparations are underway for three future missions. For the first mission, the Galileo isotope generators are being designed to produce 120 watts or more of electrical power at 95 percent reliability at the end of 50,000 hours (5.7 years) after fueling. Three units are required for the Galileo to meet the spacecraft power requirements. These will be mounted on the spinning section of the spacecraft.

The Galileo RTG MHW unit is shown in Figure IV 5-10. Power/weight goals are 1.9 watts per pound at a nominal weight of 80 pounds and an anticipated converter efficiency of 6.7 percent.

Development of improved iridium alloys for thee fuel capsule and of advanced fibrous graphite composite aeroshell members is expected to augment the inherent safety capabilities of this generator system, under the

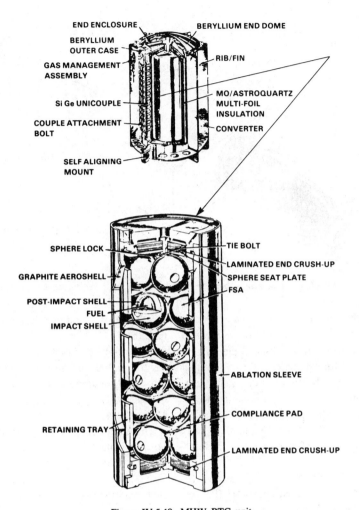

Figure IV-5-10. MHW RTG unit.

conditions of the NASA shuttle launch environment.

The Solar Polar Mission is jointly supported by NASA and the European Space Agency (ESA). Two spacecraft are planned to circumnavigate the sun, orbiting in opposite directions over the polar axis. A relatively short interval is available to design, qualify, fabricate, test, and deliver the generator systeems for this mission. In addition, a substantial configuration change is expected, because the modular General Purpose Heat Source, developed by LASL, will provide the thermal input for an improved selenide statis thermoelectric converter. The primary reference design is shown in Figure IV 5-11.

Performance requirements and specifications for these generators are:

Power Output:
  286 watts after 1000 hr at 30 volts DC
  275 watts after 17,5000 hr at 30 volts DC
  250 watts after 40,000 hr at 30 volts DC
Weight:
  110 pounds maximum at beginning of mission
Envelope:
  Height: 30 inches   Diameter: 60 inches
Operating conditions:
  Hot-junction temperature: 1035°C
  Cold-junctin temperature: 300°C
  Vacuum (pressure less than 10^{-6} torr)
Converter efficiency:
  6.7%, after projection to 1000 hr

In FY 1978, two competitive radioisotope-fueled dynamic isotope power systems for the

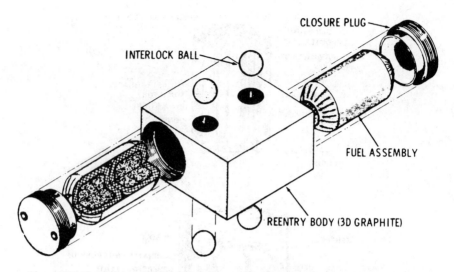

Figure IV-5-11. Modular general-purpose heat source for improved thermoelectric converter.

third mission were reviewed. These systems were the Brayton Isotope Power System (BIPS) and the organic Rankine-cycle system known as the Kilowatt Isotope Power System (KIPS). Both were being considered for ultimate application as a 1.3-kilowatt power source for demonstration on the Space Test Program for DOE. Mission launch was planned for the NASA space shuttle in January 1983.

Ground demonstration tests performed by the respective system contractors (BIPS by AiResearch Corp., KIPS by Sundstrand Energy Systems Corp.) were critically evaluted, which led to the selection of the KIPS for further development. Concurrently, DOD deferred the flight demonstration program. In lieu thereof, a technology verification program has been undertaken to assure technology readiness of the system components as of the end of FY 1980, thus permitting timely review of the nuclear option for power system requirements in 1–2 kilowatt range for missions that might be viable at that time. Salient features of KIPS are shown in Figure IV 5-12.

Major milestones and funding for Flight Systems Develoment are shown in Figure IV 5-13.

The pressed plutonium oxide spheres for the Multi-Hundred-Watt heat source are produced by the Savannah River Plant in its plutonium fuel facility. The research and development on these spheres was performed at the Los Alamos Scientific Laboratory (LASL).

The prime contractors for the RTG system for the Galileo mission remain to be selected. The heat source will be assembled at the Mound Facility, Monsanto Research Corp., Miamisburg, Ohio.

The systems contractor for the Solar Polar Mission has not yet been selected. The Mound Facility of the Monsanto Research Corp. will fuel and test the generator with other contractors participating.

Power System Technology and Support. Traditionally, work in this category provides the technological basis for meeting near-term needs for viable missions. In addition, it provides support functions that contribute to the fabrication and evaluation of specialized system components, to assessment and analysis, and to development of quality engineering and reliability approaches that ensure compliance with mission requirements and specifications.

Ground tests of the competitive dynamic isotope power systems (BIPS and KIPS) have been completed, as has the preliminary design of the selenide isotope generator, so current technology and support work contribute primarily to development of the static thermoelectric generator for the NASA Galileo and Solar Polar missions and to component technology for the advanced space reactor project.

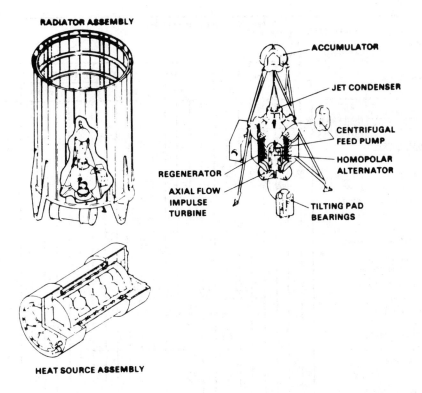
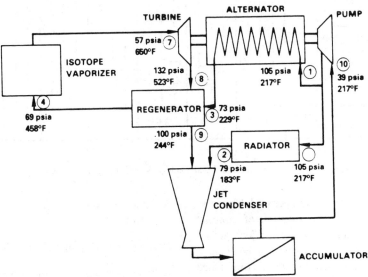

Figure IV-5-12. Kilowatt isotope power system.

The objectives are as follows:

- Developing technology for increased power, improved conversion efficiency and safety, and reduced cost
- Studying alternative power sources, such as reactors, to anticipate potential high-power demand applications
- Extending the range and potential applicability of nuclear power sources to meet future requirements

The objectives will be accomplished through the use of both industrial contractors and government-owned contractor-operated facilities; in addition, analytical studies will be

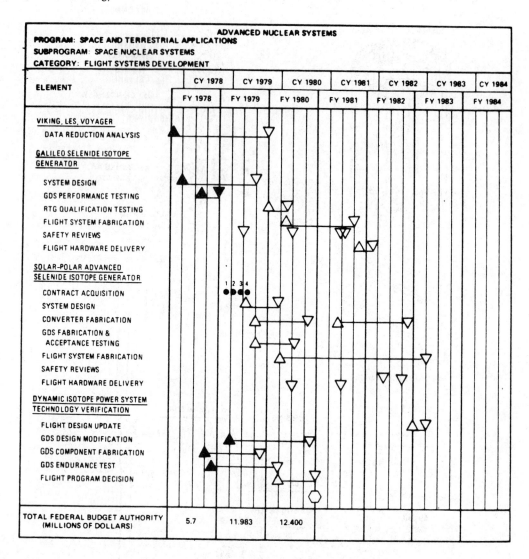

Figure IV-5-13. Major milestones and funding for flight system development.

pursued at a university engineering laboratory.

In view of the decentralization of contract and technical management of project activities, DOE site representatives will be located at industrial vendor sites to ensure compliance with engineering, quality, and reliability requirements for specific mission projects.

During FY 1979 and FY 1980, the work will include:

- Engineering analysis, conceptual design, and site-representative support
- Space-power analytical studies
- Quality assurance and reliability engineering, on-site inspectors, and related support
- Space-reactor technology studies and experimentation

Current activities include the following:

- Technical analysis, consultation, conceptual design engineering, assistance in program reviews, and on-site assessment of contractor progress in selected component areas. These activities relate to both the Galileo and Solar Polar Missions and embrace the flight system development effort.
- Development and use of analytical codes to assess the reentry capability of selected modular heat sources and to ex-

plore the impact of the NASA space shuttle on various missions.
- Quality assurance, inspection reviews, reliability assessment, assistance in design reviews, and site representatives at contractors for the Galileo and Solar Polar flight systems development.
- Identification and evaluation of applications for nuclear reactor electric power for missions up to the year 2000. Also, screening studies are to continue to determine the technologies necessary to support requirements in the 10–100-kwe range. Conceptual designs will be reviewed for feasibility. Properties of a selected 80 percent uranium oxide/20 percent molybdenum fuel form will be measured, the molybdenum heat pipe component will be studied further to elicit heat transport mechanisms, and the forming, purity, and compatibility of molybdenum and port mechanisms, and the forming, purity, and compatibility of molybdenum and molybdenum alloys will be investigated. Nickel-beryllium laminates will be explored as radiator materials. Thermoelectric elements of silicon/germanium will be baselined, but other couples will be evaluated; high-heat flux modules will be built and tested. Performance evaluation and long-term operation will be pursued as prototype units are available.

Major milestones and funding and given in Figure IV 5-14.

The Fairchild Space and Electronics Co., Germantown, Md., provides technical analysis and support. The Aerospace Engineering Department of the Pennsylvania State University is developing analytical codes for reentry studies. The Sandia Laboratories, Albuquerque, N.M., provides quality and reliability support. Los Alamos Scientific Laboratory coordinates and conducts studies on the space reactor system.

Isotopes Fuels. All static thermoelectric and dynamic power systems used or proposed for space applications during the last 15 years or more have been fueled with plutonium-238. This man-made radioisotope is primarily an alpha emitter and consequently requires minimal weights for shielding. Its half-life of 87.7 years enables design and operation of systems with reliable power outputs over a long lifetime. Fuel-form development now centers on improvements in high-density, high-power-

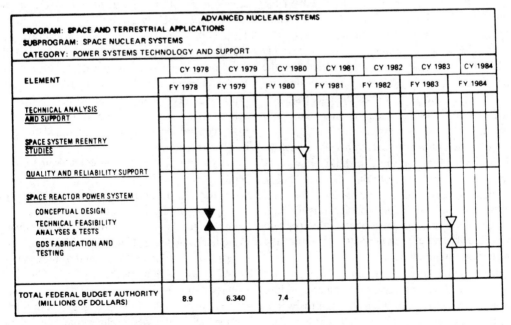

Figure IV-5-14. Major milestones and funding for power system technology and support.

output ceramic oxide shapes. Progress in encapsulation materials and procedures has resulted in continuing improvements in safety index under contingent reentry and/or abort impact conditions. MHW heat sources designed for the Lincoln Experimental Satellites, Voyager, and Galileo generators reflect this continuing learning process and product improvement. The modular General Purpose Heat Source incorporates additional refinements expected to result in improved performance and safety.

Objectives for this category are as follows:

- Develop, process, and produce isotope-fuel forms
- Develop capsule materials and metallurgical technology
- Encapsulate and assemble the hat sources

This work improves and produces fuel forms, capsules, and heat-source assembly.

Fuel-form process development aims to improve performance and safety in all heat-source applications. Research at the LASL plutonium-238 laboratories is translated by the Savannah River Laboratory (SRP) into a production-process flowsheet, which is tested further in a preproduction mode in the cell line of the Plutonium Fuel Fabrication (PuFF) Facility at SRP. Thus, studies with small pellets at LASL precede the manufacture of 252-gram 100-watt (thermal) MHW spheres at SRP for the heat sources destined for the static and dynamic systems. The resulting ceramic fuel form is more resistant to fragmentation and fines generation, and thus will enhance the safety margins.

All plutonium-238 fuel forms are encapsulated within one or more metallic containers to preclude release of the isotope during launch or orbital-reentry accidents. The hazards of explosion, high-velocity travel, high temperatures during atmospheric reentry, and possible impact are counteracted by protective materials, such as special tantalum or platinum-rhodium alloys, iridium, etc., together with Hastelloy or graphite outer-cladding payers. Process controls ensure quality wells, control of grain growth, adequate impact strength, and resistance to coorosion or oxidation in the event of a return to earth.

Development of improved platinum-rhodium or iridium alloys is pursued through the preproduction and production phases. Sheet and blanks prepared at ORNL are formed into hemispheres at the Mound Facility. When completed and fully characterized, matched sets of these hemispheres are sent to the PuFF Facility for encapsulation of the fuel spheres formed there; each sphere contains a pair of sintered iridium filters or vents to permit helium escape while retaining plutonium-238 fines. Each encapsulated sphere is returned to Mound for enclosure within a composite-graphite impact shell, which becomes the assembly unit for the MHW heat source.

The assembly of the heat source for a typical 2400-watt (thermal) MHW radioisotope heat source involves placing three groups of eight spheres each into special graphite retainer rings, each with a central bolt structure that holds the group of eight spheres together. The three "eight-packs" are loaded into a graphite aero-shell and the enclosure is locked into position. This heat-source assembly is sent in a sealed environmental-protection container to the systems contractors, who inserts the heat source into the thermoelectric or dynamic converter for a particular mission.

Fuel-form production and first-phase encapsulation activities have been transferred from the Mound Facility to PuFF so as to limit further the already very low probability of dispersion of plutonium-238 into an inhabited area, by moving such operations to areas with low population density. During FY 1978, the PuFF Facility began production of MHW spherical fuel forms for the qualification and flight generator heat sources scheduled for the Galileo mission.

Concurrently, blanks prepared from a recently developed improved iridium alloy have been supplied by ORNL to the Mound facility for hydroforming into hemispherical capsule members for later use by the PuFF Facility. Completed fueled sphere assemblies will be shipped appropriately to Mound for eventual assembly of the heat sources.

Early studies of improved MHW fuel spheres and applicable processes were performed at the LASL plutonium facilities. Technology transfer and further process development are carried out in the Alpha Materials

Laboratories and the Plutonium Experimental Facility (PEF), elements of the Savannah River Laboratory, operated by E. I. du Pont de Nemours, Inc. Fuel production and encapsulation for the MHW RTG's for the NASA Galileo mission qualification and flight system requirements are being pursued in the PuFF Facility, at the Savannah River Plant, also operated by du Pont.

Fuel-form development at LASL preceded the selection of a 62.5-thermal-watt hot-pressed PuO_2 cylindrical fuel form for the SPM heat-source module. Technology transfer for this process is proceeding in PEF and production is planned to meet the scheduled requirements as shown in the milestone chart. Again, ORNL provides capsule material as blanks for forming capsules at Mound, encapsulation in PuFF, and assembly at Mound. The current conceptual design of the RTG for the Solar Polar mission also necessitates fueling of the generators in the Mound Facility.

Ongoing process development and product improvement is expected as a continuing element of the Savannah River Laboratory/SRP operations to meet future mission performance and schedule requirements.

Figure IV 5-15 presents the major milestones and funding for Isotope Fuels.

Fuel and capsule material movement for heat-source fabrication occurs entirely within government-owned contractor-operated laboratories and plants. Assembly activities further utilize the special equipment and expertise available at a government facility. Transportation will be effected by the use of specialized systems and procedures. Safety and accountability are the watchwords in all such fuel-related events.

Space Power System Safety. A persistent awareness of the unique capabilities of nuclear power sources, a compelling regard for the safety of mission crews and the public, and prudent design engineering combine to establish a demand for a continuing safety effort of high caliber. In support of this demand, safety specifications have been developed and imposed on all flight system development programs and their terrestrial counterparts. Power systems technology and support are required to observe high levels of quality assurance and reliability and to meet safety criteria. Preparation and processing of

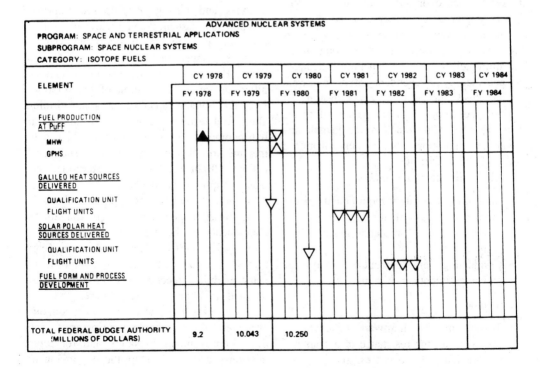

Figure IV-5-15. Major milestones and funding for isotopes fuels.

radioisotope fuels are performed in safeguarded facilities and under audited operating safety rules. Assembly of heat sources and qualification thereof are pursued to ensure safe handling and mission performance without undue risk to the public. Materials and design improvements are continuously caught and developed to maximize the safety of the mission even under contingent malfunction or abort conditions.

The objectives of this work include the following:

- Ensures that all fuel processing operations are conducted below the current radiological exposure limits
- Ensures delivery of safety-qualified RTG's for all missions, meeting performance requirements and national safety policy standards
- Provides safety analysis reports and evaluations to meet the requirements of the Interagency Nuclear Safety Review Panel.

Based on rigorous standards for safety in operation and under low-probability accident conditions, progressively more stringent goals are set for space or terrestrial missions. Improved engineering by system contractors, tighter controls in government fueling facilities, and incorporation of new components lead ultimately to a safer generator. Critiques and analyses by independent safety reviewers provide check and balance against design and materials development. Interagency safety review precedes official action for launch approval and final approvals are sought from the Office of the President.

Work in this category entails:

- Supporting the safety requirements of the flight and terrestrial applications
- Conducting safety research and development
- Conducting environmental, analytical, and experimental programs

It includes:

- Testing mission hardware to evaluate, for example, the resistance of a graphitic aeroshell to blast overpressure
- Evaluating the shelf-life aspects of mission hardware to determine, for example, the limitation of storage time to preclude excessive helium pressure in unvented capsules, or the long-term compatibility of an alloy when exposed to plutonium-238 fuel at operating temperature for extended times
- Reviewing the post-mortem analyses of fuel hardware to determine whether the fuel form and capsule hardware from the current manufacturing processes conform to the expected levels of impact resistance, fuel deformation, and fines development
- Participating in interagency reviews that support the work of the Interagency Nuclear Safety Review Panel
- Monitoring continuously contractor safety programs to ensure compliance with safety design specifications and all launch support activities

Present work involves the Galileo mission, the Solar Polar mission, and advanced safety reserch and development. For the Galileo mission, a standard regimen of program safety standards, design reviews and assessments, analysis and testing of proposed improved components, issuance and review of preliminary, updated, and final safety analysis reports is pursued before Interagency Nuclear Safety Review Panel activities. Of special significance are the evaluation of the improved fuel, the impact capability of a new iridium alloy, and testing and analysis of carbon-carbon composite aeroshell members planned for this mission. Upon completion of such evaluations, flight readiness is determined and launch approval is sought.

A similar procedure is applied to the Solar Polar mission generator design and to the components of the modular General Purpose Heat Source (GPHS). The introduction of new fuel shapes and carbon-carbon composite reentry module protection materials is expected to yield an improved safety index.

Advanced safety R&D to ensure ongoing awareness of the technology necessary to continue improvements in safety and for future missions, and to develop those components contributing to these improvements, is performed at LASL, the lead laboratory in this area.

Major milestones and funding for Space Power System Safety are shown in Figure IV 5-16.

LASL and Sandia Laboratories test mission hardware with analytical support from the Applied Physics Laboratory of John Hopkins University and from the Fairchild Space and Electronics Co. Interagency reviews include contributions to environmental impact studies performed by LASL, the Naval Sea System Command, the Naval Undersea Center, the National Oceanic and Atmospheric Administration, Nuclear Utilities Services Corp., and others. These efforts precede the final safety evaluation reports and solicitation of Presidential approval for launch of the mission system.

Decommissioning and Decontamination. Based on extensive planning and effective use of specialized techniques and equipment, the plutonium-238 fuel processing, fabrication, and recovery facilities at the Mound Facility are being decontaminated to a controlled-use level, thereby making available a substantial portion of the plutonium processing and related building areas for other purposes. A corresponding effort is planned for the hot-cells previously used to process and evaluate curium-244 fuel materials at the ORNL facilities.

Objectives for decommissioning and decontamination (D&D) are as follows:

- To complete the D&D at the Mound Facility and returning the building areas to a condition suitable for controlled nonradioactive uses
- To complete D&D activities at the ORNL curium facility and returning the hot-cells to a condition suitable for controlled radioactive uses

Based on approved planning and work phasing, the building areas of the Mound Facility previously involved with unencapsulated fuel material will be decontaminated, to ensure the maximum of safety and thoroughness of the

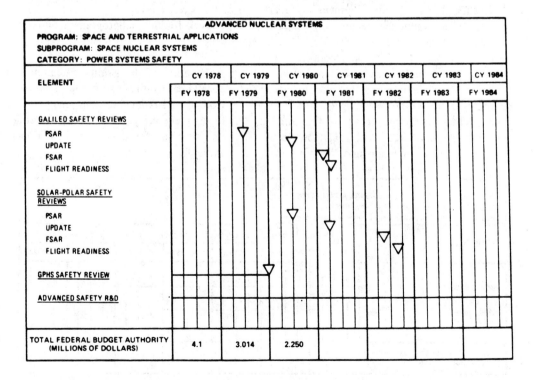

Figure IV-5-16. Major milestones and funding for space power system safety.

D&D operation. Health-physics monitoring is intensive and precautions are taken to ensure the safety of personnel and the return of the building to conditions suitable for nonradioactive uses. A similar regimen will be pursued with the ORNL curium-244 facility.

The Mound Facility is located in Miamisburg, Ohio, a suburb of the populous Dayton area. In support of increasingly stringent safety standards, all space nuclear systems operations involving unencapsulated plutonium-238 materials have been transferred from the Mound Facility. Fuel fabrication activities for space applications ceased in 1977. D&D efforts were underway during FY 1979 in those areas of the Mound complex previously occupied by space mission fueling operations.

Major milestones and funding for D&D are given in Figure IV 5-17.

Mound Facility and ORNL staff will be used, insofar as possible, at their respective facilities.

Terrestrial Isotope Applications. The current activities comprise the development of fuels, materials, and component technology for heat sources, and terrestrial isotope applications. Most of the work involves developing beneficial uses and applications for isotopes readily available from defense wastes, the residues remaining from the reprocessing of spent fuel from military reactors—cesium-137, strontium-90, and kryton-85 made available from separations plants in Idaho and Washington.

The cesium-137, strontium-90 and kryton-85 radioisotopes are derived from the reprocessing of military wastes by well known and developed processes. The cesium-137 isotope decays with a half-life of approximately 30 years and yields a 0.67-Mev gamma radiation output, suitable for disinfection processes. Cesium-137 is available as a chloride product; this material is readily soluble, and undergoes volume changes as it passes through its melting point.

The strontium-190 isotope is currently available as a fluoride; previous applications of this isotope in the SNAP series of radioisotope generators was made possible by using strontium titanate as the fuel material. The properties and compatibility of the fluoride alternative material are being explored to enable near-term licensability and use of the fluoride fuel capsule.

The krypton-85 isotope is a gaseous product recovered from fuel reprocessing. Applications are being evaluated; preliminary studies have been conducted to determine practicable means of enriching the gas to the 50 to 90 percent level.

The work for this subprogram includes the following:

- Application of the gamma radiation and decay-heat available from the cesium-137 isotope to irradiate sewage sludge so as to disinfect the product and demonstrate its utility as fertilizer, soil conditioner, cattle feed, etc.

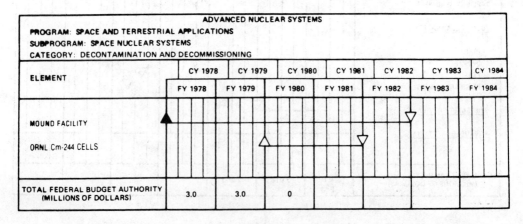

Figure IV-5-17. Major milestones and funding for decommissioning and decontaminating.

- General studies on the beneficial utilization of isotopes recoverable from reactor waste streams

The objectives are being attained by using the resources of the national laboratories and selected contractors where the necessary skills are available.

A cesium-137 sewage-sludge irradiation pilot plant dedicated early in FY 1979 will be activated; plans will proceed toward an EPA-supported facility and a program to demonstrate application of the irradiation principle and pilot-plant technology to composted sewage sludge. Experiments will continue to determine the product's applicability as agricultural fertilizer and animal feed.

Improved methods of forming and containing the cesium-137 source material are being explored. Additional applications of kryton-85 will be evaluated in nondestrictive testing methods.

In the area of the radioisotope, materials and components, lead laboratory and supporting technology functions are provided by the Sandia Laboratories, PNL, and ORNL.

Work on terrestrial isotope applications is limited to expanding the technical base and extending the applications of cesium-137, strontium-90, and krypton-85. The Sandia Laboratories at Albuquerque, N.M., provide the lead laboratory function for cesium-137 fuel material and encapsulation, with ORNL in a supporting role in the development of a stable cesium aluminum silicate ceramic fuel form. The Sandia Irradiator for Dried Sewage Solids, dedicated in October 1978, will be used to provide quantities of disinfected

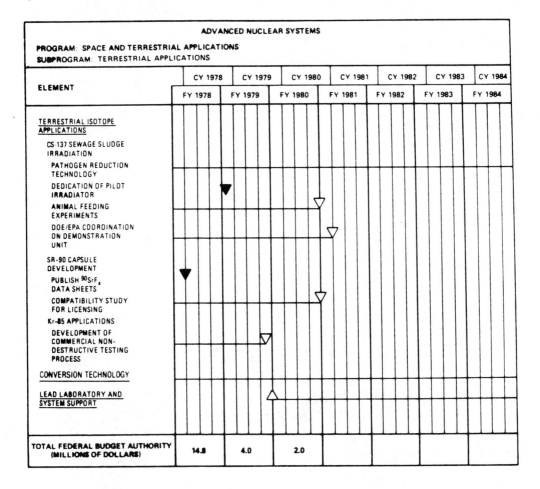

Figure IV-5-18. Major milestones and funding for terrestrial isotope applications.

sludge product for experiments and for guidance in the design and operation of a future demonstration plant. Continued studies on the pathogens in sewage sludge will be pursued at Sandia. The benefits of sludge application to farm plots will be evaluated further, and larger-scale experiments will be conducted to evaluate the effects of feeding irradiated sludge solids to ruminant range animals. These agronomic experiments are conducted at New Mexico State University. Assistance in the development of a demonstration irradiator will be provided by the Sandia staff.

The cesium-137 supply function of the PNL operation is expected to continue; the strontium-90 technology support is in the concluding stages.

Major milestones and funding for the Terrestrial Isotope Applications Subprogram are shown in Figure IV 5-18.

6.
Deployment Considerations

This chapter described factors affecting the deployment of civilian nuclear power: commercialization, environmental implications, socioeconomic aspects, and small and minority business utilization.

COMMERCIALIZATION

Commercialization, the last phase in the energy-system acquisition process, is reached only after the system proceeds through six previous phases: basic research, applied research, exploratory development, technology development, engineering development, and demonstration. Table IV 6-1 lists the present status of the major fission energy systems in terms of these development phases.

Defining commercialization so that the term can be applied uniformly to all major fission energy development is difficult. Possibly the most universal definition is that the first commercial plant is the first plant of its kind to be initiated by vendors and industry (for example, reactor vendors and utilities) and constructed, perhaps with government financial assistance. Successive commercial plants are assumed to follow this first plant rapidly, with some of these also probably receiving government assistance, but evolving into an industry that is solely privately financed. This definition implies that the preceding development program has fostered industrial capability and confidence to the point that private enterprise is willing to risk its capital on the commercial plant.

Breeder Reactors

The President's policy is to defer any U.S. commitment to advanced nuclear technologies that are based on the use of plutonium, so the commercialization of breeder reactors is deferred at present.

The LMFBR is already in the technology development stage; in keeping with the President's policy, commercialization is not being considered. Commercialization will depend on such considerations as: (1) future policy stud-

Table IV-6-1. Present status of U.S. fission energy systems.

SYSTEM	PHASE
Breeder Reactors	
Liquid Metal Cooled Fast Breeder	Technology development
Gas Cooled Fast Breeder	Technology development
Water Cooled Breeder	Technology development
Converter Reactors	
Present-generation Light Water Reactors	Commercialization
Improved Light Water Reactors	Technology development
Advanced Converters	Technology development
Fuel Cycle	
Gaseous-diffusion uranium enrichment	Production, operation
Centrifuge-separation uranium enrichment	Demonstration
Advanced isotope-separation uranium enrichment	Applied Research
Reprocessing	Technology development
Refabrication	Technology development

ies such as NASAP and INFCE; (2) economic considerations based on the demand for energy, uranium supply for current nuclear systems, and information obtained from R&D and demonstration breeder plants; and (3) the status of alternative inexhaustible energy supplies.

The WCB is in the technology development stage. If the current development efforts yield favorable results, additional demonstration activities to define the economics of this system will be required. Final commercialization will depend on these results and the compatibility of this system with the then-existing LWR's being deployed, because these two systems are highly synergistic. The WCB would operate with the thorium/U-233 fuel system, so its future use also depends on future thorium industry development.

It is too early in the program to consider seriously the commercialization aspects of the GCFR. In its role as backup to the LMFBR, its future depends on the future of the LMFBR.

Converter Reactors

LWR's operating on the once-through cycle are currently commercial; both improved LWR's and ACR's are in the technology development phase. As of January 1, 1979, 69 licensed LWR's (BRW's and PWR's) were supplying power to the nation's utility grid. It is expected that the present development program for LWR's will result in both lower costs and increased uranium utilization. If so, the Under Secretary will decide on demonstration and commercialization, which will take place as rapidly as possible thereafter. This will include retrofitting. The present generation of LWR's will be retrofitted and design changes will be incorporated into newly ordered reactors. There is a question as to whether ACR's offer any advantages over the improved LWR's from the standpoints of either uranium utilization or proliferation. If any of these concepts are carried through the demonstration stage, commercialization will depend primarily on the relative cost of power from these units as compared to the costs of power from other sources, particularly improved LWR's.

Fuel Cycle

There are no plans to commercialize uranium enrichment, reprocessing, or refabrication in the United States, although a number of firms mine, mill, and fabricate uranium.

Uranium Enrichment. In 1975, the Administration proposed enactment of the Nuclear Fuel Assurance Act, which would permit private financing, construction, ownership, and operation of all new uranium enrichment plants. In anticipation of enactment of this bill, a group of private companies, known as Uranium Enrichment Associates (Bechtel, Goodyear Tire and Rubber, and the Williams Companies), proposed to build a 9-million-SWU gaseous-diffusion plant. However, the bill was not approved by Congress. Thus, commercialization of uranium enrichment is not being actively considered. All uranium enrichment in the United States continues to be done by the DOE.

Reprocessing and Refabrication. In April 1966, the AEC issued a provisional license to Nuclear Fuel Services (NFS) Corp. and the New York Atomic Research and Development Authority (NYARDA) to operate a solvent extraction facility at West Valley, N.Y. From 1966 to 1972, about 625 metric tons of fuel were reprocessed. In 1969, NFS stock was acquired by Getty Oil Co. (majority) and Skelly Oil Co. In 1972 the plant was shut down to expand processing capabilities and to make modifications to reduce radioactive effluents and occupational exposure. The cost of such modifications to meet the new criteria and regulations was so great ($600 million) that in 1976 NFS announced its intention to withdraw from commercial reprocessing.

Other companies that have shown an interest in commercial reprocessing of LWR fuel have been General Electric Co., with its modified fluoride volatility plant at Morris, Ill., and Allied Chemical Co., with its Barnwell, S.C., solvent extraction plant. In addition, melt refining of metallic fuel has been successfully demonstrated by ANL at its EBR-II complex in Idaho.

However, the situation regarding reprocessing drastically changed on April 7, 1977, when President Carter announced that, because of proliferation concerns, the United States will defer indefinitely commercial reprocessing and recycling of plutonium. This policy resulted in a study to determine an acceptable fuel cycle consistent with this nonproliferation policy. Initial evaluations were done at the end of FY 1979, so it is too early to consider commercialization of any of these processes. Thus, there are no current plans for commercializing reprocessing or refabrication in the United States.

ENVIRONMENTAL IMPLICATIONS

Numerous studies have established that fission energy can be used with minimal adverse effects on the biosphere. Coal, which is considered to be the only viable alternative to nuclear fuel for the near-term future, has serious and well-defined environmental problems involving land use (strip mining), air pollution (sulphur and nitrogen oxides, particulates), water contamination (from mining operations), as well as some more poorly understood issues such as the long-term effects of CO_2 emission (the "greenhouse" effect). These problems are either absent or minimal in the case of fission energy.

The environmental effects of fission energy have been more thoroughly and exhaustively explored than those for any other source of power. The principal environmental questions involved in the use of the fission process for power production can be considered under the categories of radiation, air, water, and land use.

From an environmental viewpoint, the unique characteristic that distinguishes the use of fission energy from the traditional fossil-fuel methods of power generation is the production of radioactive species. This problem has been recognized from the beginning, and rigorous standards and criteria on radioactive emissions have been developed over the years. As a result of this, radiation as an environmental hazard of nuclear plants has been greatly decreased.

Under present regulations, which are enforced by NRC, the average dose to a person living at the boundary of a nuclear power plant site would be less than 1 millirem per year (Ref. 1), less than 1 percent of the dose from natural radiation—and considerably less than the exposure from a chest X-ray (50 millirem) or a transcontinental flight in a jet airplane (5 millirem, arising from increased exposure to cosmic rays at high altitudes). For perspective, it should be noted that the average exposure in the United States from background and man-made radiation is 130 millirem per year; an inhabitant of Denver receives 100 millirem more due to the high altitude. Table IV 6-2 shows the average annual doses from radiation sources in the United States. Recent studies have confirmed that the radioactivity emitted by a coal-fired power plant arising from the average levels of uranium (1 ppm) and thorium (2 ppm) naturally present in the coal exceeds that from a nuclear power plant, (Ref. 2).

Nuclear plants contribute only insignificantly to atmospheric pollution. For example,

Table IV-6-2. Average annual doses from radiation sources at sea level in the United States.

TYPE OF RADIATION	SOURCES	AVERAGE ANNUAL DOSE* (MREM/YR)
Alpha & gamma	Natural radioactivity (e.g., uranium) in solids, rocks, minerals	30+
Beta	Natural radioactivity (e.g., potassium-40) in soils, rocks, minerals	20+
	Television (an average of 1 hr/day)	0.5
	Natural radioactivity in the air (e.g., tritium)	2+
Gamma	Medical and dental X-rays	20
	Cosmic radiation at sea level	40+

*Dose to reproductive organs. +The sum of these four figures, 92 millirem, is the average background from natural radiation at sea level.

no solid particles (like the fly ash from coal plants) are emitted.

In the mining of uranium there is some release of hazardous radon gas that escapes into the air. Great care is taken to protect the uranium mine workers from this hazard, and the risk to the general public is negligible. For comparison, the release of toxic carbon monoxide in coal mining operations, and its resulting danger to the workers, is many orders of magnitude greater.

Nuclear plants use cooling water (generally from river, lake, or ocean sources) to condense the steam in their closed-cycle condenser cooling systems back into water. The discharges into the streams or other bodies of water are rigorously monitored and controlled, so that the incremental chemical species added by passage through the power plant are small.

A few years ago there was some concern over the fact that the discharge water from a nuclear power plant was warm (10 to 20°F hotter than the intake water) and that aquatic life might be adversely affected. In many cases, cooling towers and cooling ponds have been installed to minimize this problem and to meet water quality standards (Figure IV 6-1).

Nuclear power plants are inherently compact and, basically, require less diversion of land to industrial utilization than oil or coal plants do. They need no piles of coal in storage, and no extensive oil tank farms. For reasons of safety conservatism, regulations require that nuclear power plants have sizable "exclusion areas" around them; with these, the total land commitments for nuclear and fossil-fueled power plant sites are about the same.

The mining of uranium for such plants, like coal mining, is about equally divided between underground operation and surface mining. In both cases, after the mine site is exhausted, the land must be restored to beneficial use. But, because the tonnages of uranium needed are so small compared with the amount of coal required to produce the same amount of power, the land damage is much less for the nuclear plant.

Nuclear power plants require only a stable foundation free of active faults. Almost any kind of site meeting these criteria is suitable—desert, rock, etc. Remote locations away from population and transportation centers are more readily used for fission energy plants because it is not necessary to continuously supply them with large quantities of bulk fuel. Urban sites within and near large population centers are precluded by current exclusion-zone regulations. All power production installations, those that use nuclear as well as those that use fossil fuel, demand copious quantities of water for cooling, and this requirement is a serious restriction on the choice of plant sites.

The simplicity of form and lack of smokestacks of most modern nuclear power plants make them much less obtrusive and objectionable than most other industrial installations. Many reactor sites have been beautifully landscaped.

The environmental effects of specific types of nuclear power plants and installations are discussed in the following section.

Water- and Gas-colled Reactors

The bulk of U.S. nuclear power is generated with LWR's, either of the pressurized-water (PWR) or the boiling-water (BWR) type. The only other kind of commercial fission energy system in operation in the United States is the Fort St. Vrain gas-cooled reactor in Colorado. As of January 1, 1979, 52-GWe of nuclear power were installed in this country, constituting about 9.6 percent of our net national capacity (Ref. 4) and providing about 13 percent of our annual electricity production (Ref. 5).

Because of their favorable economics and the minimal effect on their surroundings, as outlined above, these plants have attained widespread utility acceptance.

Fast Breeder Reactors and Other Advanced Types

All studies of the future of nuclear power agree that, sooner or later, the supply of available uranium will be exhausted or this material will become so expensive that its use for power production will be marginal. A

Deployment Considerations

Figure IV-6-1. Cooling tower (background) at the 1130-MWe Trojan nuclear power plant (shown during construction in foreground) along the Columbia River near Prescott, Oregon, 42 miles northwest of Portland.

potential solution to this dilemma is the use of the FBR, which creates rather than uses up fuel. Because of the possibility of the eventual commercial deployment of breeder reactors, extensive studies have been made of their environmental effects.

The principal results of these studies have been summarized in the environmental statements on the LMFBR (Ref. 6) and the CRBRP (ref. 7). Those studies show environmental benefits (e.g., no mining) arising from a shift from an LWR to an FBR nuclear economy.

The principal environmental difference between the LWR and LMFBR is the greater inventory of plutonium in FBR fuel. There has been some concern regarding the hazards of plutonium, but the facts (Table IV 6-3) show that plutonium is less toxic than a number of poisons. While it must, of course, be handled with great care, there is no reason to believe that the environment is more likely to be endangered by this than by any other toxic material routinely handled in commerce.

Also, fast breeders use either molten sodium metal or helium as the heat-transfer medium instead of the water of the LWR. Although these materials require special handling and precautions to ensure their confinement, extensive experience here and abroad assures that no significant escape from the system, or subsequent damage to the biosphere, is likely to occure. Sodium-cooled plants (LMFBR's) operate at low pressure, which adds a significant element of safety and freedom from deleterious environmental effects.

Fuel Preparation

To prepare uranium for use in reactors, it must be mined, milled, enriched, and fabricated into fuel elements.

Underground mining, such as for uranium ore or coal, is one of the most hazardous occupations in American industry. However, because of the relatively small tonnage required as compared with coal for the same net electrical output [about 3 million tons of coal are required for the same amount of energy as can be obtained from one ton of uranium (Ref. 10)], many fewer fatalities can be expected from uranium mining than from mining coal of comparable energy content.

There is one special health hazard in uranium mining that does not normally exist for coal—the presence of radioactive radon gas. The dangers of this substance and the required protective devices are well known, and uranium mining operations are under the continuous scrutiny and regulation of NRC, EPA, the Occupational Safety and Health Agency (OSHA), the Mine Safety and Enforcement Administration (Department of Labor) and others. The perils are parallel with, but considerably less than, those in underground coal mining (per kilowatt of power produced), where carbon monoxide and methane are major sources of danger.

After raw ore has been minded, the milling operation extracts the uranium and concentrates it into relatively pure oxide U_3O_8 ("yellow cake"). The uranium content of the minerals extracted is only a few tenths of a percent or less, so milling leaves sizable residue, the mill "tailings." These are not only

Table IV-6-3. Comparison of reactor plutonium with highly toxic materials.

TOXIN OR POISON	LETHAL DOSE (MILLIGRAMS)	TIME TO DEATH
Ingested (swallowed)		
Anthrax spores	Under 0.0001	—
Botulism	Under 0.001	—
Lead arsenate	100	Hours to days
Potassium cyanide	700	Hours to days
Reactor plutonium	1,150	Over 15 years
Caffeine	14,000	Days
Injected		
North American coral snake venom	0.005	Hours to days
Indian king cobra	0.02	Hours to days
Reactor plutonium*	0.078	Over 15 years
Diamondback rattler	0.14	Hours to days
Inhaled		
Reactor plutonium*	0.26	Over 15 years
	0.7	3 years
	1.9	1 year
	12	60 days
Nerve gas (Sarin)	1.0	Few hours
Benzopyrene (1 pack/day of cigarettes for 30 yr)	16.0	Over 30 years

TOXIN OR POISON	LETHAL CONCENTRATION[+] (MILLIGRAMS/CUBIC METER)	TIME TO DEATH
Inhalation atmosphere		
Reactor plutonium (4-hr exposure)	0.026	Over 15 years
	1.3	60 days
Cadmium fumes	10.	Few hours
Mercury vapor	30.	Few hours
Phosgene	65.	Few hours

*Mixture of plutonium isotopes as plutonium oxide (more than five times more hazardous than plutonium-239).   [+]Exposure at different levels.

unsightly, like the steel residues and slag heaps around steel mills, but also contain some radioactive uranium and other radioisotopes. Although the level of radiation from these tailings is markedly less than that of the ore from which they were derived, they are on the surface of the earth rather than underground. Unfortunately, the dangers of these tailings have not always been realized in the past; there have been some cases of their use for land fill and similar purposes, resulting in exposure to their emissions. This problem is being studied and remedial actions are underway.

After the uranium in yellow cake has been extracted and concentrated, it must undergo several chemical and mechanical steps before it can be fabricated into fuel for use in a nuclear reactor. Perhaps the most important of these is isotope enrichment (it should be noted that uranium enrichment in the United States is, by law, as Federal Government monopoly, and no commercial facilities are permitted). As it occurs in nature, uranium is made up primarily to two isotopes—fissile U-235 at a concentration of 0.71 percent, and the rest mostly U-238. To use the uranium for power production in LWR's, the enrichment process increases the U-235 content to about 3 percent.

All processes for doing this first convert the oxide to the hexafluoride UF_6, a gas with a high vapor pressure. This conversion requires well-known chemical processing techniques, with no significant release of harmful species to the environment.

The next step, isotopic enrichment of the UF_6, can be carried out in a number of ways. The oldest process, first developed in the 1940s for the nuclear weapons program, utilizes the small but significant difference in the rate of passage of the lighter (U-235) and heavier (U-238) UF_6 molecules through a diffusion barrier. By cascading this process, moving the material through a large number of these barriers, the required enrichment can be achieved. Plants to carry out this separation are in operation at Oak Ridge, Tennessee; Portsmouth, Ohio; and Paducah, Kentucky. Because of the necessity for multiple stages, these diffusion plants require an enormous amount of space, and the plants are among the largest in industry. They also need very large amounts of electric power to pump the gas through the barriers and for other steps in the process. Diffusion separation in this way, while costly in terms of land and power use, has little or no harmful effect on the biosphere. These plants, for economic reasons, are among the tightest large systems ever constructed, and leakage to the environment is virtually nil.

Because of the disadvantages of diffusion enrichment, and for other reasons, the United States is now turning to another process for this step, the use of the gaseous centrifuge. This technique depends on a separation of the light and heavy UF_6 molecules in a high centrifugal field, the same principle that is used in a dairy cream separator. Plants for carrying this out are already operational in Europe, and similar installations are under construction here. Staging is necessary in centrifuge plants, as in diffusion installations, but much less space and electricity are required. Like diffusion, the impact of the centrifuge isotope separations plants on the external environment is negligible.

A number of other techniques for enrichment are in the research and development stage: laser separation, nozzle processes (West Germany), "stationary wall" centrifuge (South Africa) and chemical absorption method (France). Savings in land use and power consumption are the claimed advantages of these methods, and much smaller and more compact installations can be expected if these developments can be reduced to industrial practice. No serious environmental problems are foreseen with any of them.

After the required degree of enrichment (to about 3 percent U-235) has been achieved, the UF_6 is converted into uranium oxide (UO_2), the UO_2 is compacted and sintered to form ceramic pellets, the pellets are stacked in metal cladding tubes, and the tubes are sealed by welding end caps onto them. Then a number of these fuel rods are fastened in parallel in a matrix to form a subassembly that can be loaded into a reactor. These steps can be carried out safely and with minimal effect on the environment, and the rigid controls that are imposed (for economic as well as environmental reasons) by the fabricators and the government agencies involved leave little possibility of danger either to the environment or to people in these operations.

Fuel Handling After Use

Nuclear fuel, when first inserted into a reactor, emits little or no radioactivity. Once it has started producing power, however, it becomes strongly radioactive due to the formation of fission products and transuranium elements, plutonium for example. This high level of emission persists until the fuel is removed from the reactor (usually after about three years) and is replaced by fresh material, and for years beyond this time.

Until some of their radioactivity has decayed, the spend fuel elements are stored under water in pools dedicated to this purpose, either at the reactor site or elsewhere. These spent fuel elements, which had contained approximately 3 percent U-235 when they were first loaded into the reactor, end up with about 0.82 percent U-235 and about 0.68 percent plutonium (Ref. 11), both of which can be recycled and used for further power production. The procedure for removing the fission products so the U-235 and plutonium can be reused is known as "reprocessing".

The current Administration policy is to indefinitely postpone reprocessing, for reasons associated with the proliferation issue. But because the breeder reactor, which requires reprocessing, may eventually be deployed, and other nations are proceeding with reprocessing plans, it is appropriate to consider the impact of reprocessing plants on the environment.

There are no commercial reprocessing plants currently operating in the United States. When such installations are constructed, they will have to be licensed by NRC, and it is expected their regulation will be at least as rigorous as is applied to the operation of LWR's. If the standards are essentially the same, no significant environmental impact can be expected from either the spent fuel storage pools or the reprocessing installations.

For the duration of the reprocessing postponement noted above, spent fuel elements are considered as waste. But when fuel elements are reprocessed to recover their residual uranium and plutonium, there results a significant amount of waste that contains radioactive fission products (strontium, cesium, iodine, rare earths, etc.) and some uranium and transuranium elements (plutonium, americium, curium, etc.).

The problem is not new or unique to nuclear power. Radioactive wastes from the military weapons program have been accumulating since the early 1940s; these far outweigh the wastes produced by power plants to date and will continue to do so for some time into the future. "The solidified volume of the (high level) military wastes exceeds that expected from all U.S. nuclear power plants to the year 2000 by a factor of 10 (ref. 12)." Treatment techniques for power plant wastes have been developed. Solidifying the wastes into glass or rock-like ceramic, encasing them in an appropriate corrosion-resistant metal can, and burying them in a deep stable geologic repository should ensure that they will not get into the biosphere for the required thousands, and perhaps millions, of years.

Current public concern about waste disposal stems largely from two factors:

- For many years the military waste problem was virtually ignored. This has led to the erroneous public conclusion that nothing was done because no solution to the problem was known.
- At Hanford, carbon-steel tanks containing liquid wastes have leaked. "These leaks deplorable enough in themselves, are regularly cited as representative of the risks in current techniques. One is never told that the tanks in question are of early postwar design, and have long since have been superseded. It is as if the safety records of the railroads in 1845, when accidents were very common, had been used in 1925 to justify closing them down" (Ref. 12).

In the search for stable geologic formations in which to store radioactive wastes, subterranean salt formations have been prime candidates. Geologic evidence indicates that these have existed virtually unchanged for hundreds of thousands, sometimes millions of years; and this existence of a water-soluble material for so long is persuasive evidence that hydrologic conditions are unlikely to destroy these formations in the future.

Both salt beds and stable rock formations, basalt or granite, have been identified as suitable formations that could ensure complete isolation of the wastes from the environment for geologic eras.

The recent draft report of the Interagency Review Group (IRG) on waste disposal that carefully reviewed the present state of scientific and technical knowledge on this subject concluded that "Overall scientific and technological knowledge is adequate to proceed with region selection and site characterization, despite the limitations in our current knowledge and modeling capability. Successful isolation of radioactive wastes from the biosphere appears technically feasible for periods of thousands of years provided that the systems view is utilized rigorously" (Ref. 13). That report also concludes that there is no pressing urgency, from the point of view of the health and safety of the public, to construct such repositories in the immediate future. There are, however, political and institutional problems that remain to be solved.

Conclusion

Nuclear energy, like virtually all human activities, carries with it the possibilities of environmental damage. But its expected risks are slight and its potential benefits are enormous in comparison with the alternative methods of power production.

SOCIOECONOMIC ASPECTS

Nuclear power has many socioeconomic effect, both beneficial and adverse. The primary benefit is the supply of economical electricity for industrial, commercial, and residential users. The principal problem is probably the safe handling and disposal of radioactive wastes.

Nuclear plants contribute significantly to the electrical power and energy capability of the United States. As of January 1, 1979, 70 licensed nuclear units capable of producing about 52,000 megawatts (net) of electrical power were operable. This is approximately 9 percent of the total electrical generating capacity in the contiguous United States. In 1978, nuclear power plants generated nearly 25 billion kilowatt-hours (kWh) per month, or 13 percent of the total U.S. electricity production (Ref. 14; references are listed at the end of this section).

If nuclear power plants were not available, most of our electricity would have to be provided by a combination of fossil-fueled plants in the short term. To generate the current electrical output of nuclear power stations with coal-fired plants would require about 10 million tons of coal per month (Ref. 15). That amount of extra coal would raise U.S. electric utilities' present coal consumptiion by 25 percent, total U.S. coal consumption by 20 percent (Ref. 16). The available transportation facilities could ship this increased tonnage only with extreme difficulty (Ref. 17). If it is assumed that 80 percent of the deficit due to lack of nuclear power is obtained from oil-fired plants, the additional cost of this electricity would be $440 million per month. If this were added to the cost of supplying 20 percent of the deficit by coal, there would be an estimated total incremental cost of $475 million per month by not having nuclear power (Ref. 18).

There also would be significant economic impacts if the electrical energy provided by nuclear power plants were not available at all. The 25 billion kWh supplied per month by nuclear power is 20 percent of the total electrical power used by the commercial and industrial sectors. If these two sectors shared in absorbing the total shortfall of electricity, up to 20 percent of the affected industries might be required to close down, or all would have to reduce production by 20 percent. Considering the disruptions that occur when one industry closes down, as, for example, during a labor-management dispute, the loss of this 20 percent of our total commerce and industry would have a very significant impact (Ref. 18). This is particularly evident considering the historical parallelism between the gross national product and gross energy consumption (Ref. 19). A recent study concluded that the cost of electricity not provided during a 1977 blackout totalled $4.11/kWh (Ref. 20). That study also cited other estimates for electricity not provided that range from 33¢ to $91/-kWh. Applying the low end of this range for conservatism (33¢/kWh) to the 25 billion kWh per month of displaced energy results in a monthly cost penalty of about $8.3 billion/month. Even if only a part of the electrical needs went unmet, the cost could well be in the billions.

There would be additional direct and secondary impacts throughout the range of nuclear fuel cycle activities if these were to cease operation. In 1976, there were more than 130,000 workers employed in all phases of the nuclear fuel cycle (Ref. 21). The income derived from all phases of the nuclear fuel cycle was nearly $9 million in 1976, in terms of value added (Ref. 18).

Over the next 10 years, an additional 103 nuclear units totaling about 123,000-MWe are scheduled to be constructed. These additions constitute approximately 50 percent of the total projected net increase in U.S. electric capability, and by the end of 1987 are expected to bring nuclear's share of total U.S. electrical capability to about 20 percent and electrical output to 27 percent (Ref. 14). Furthermore, most of these scheduled nuclear additions are already in some stage of the licensing process; as of March 31, 1978, 69 of these proposed nuclear additions were already under construction (ref. 18).

Construction of nuclear power plants will add significantly to the total capital requirements of the nation. In addition, the economic and social impact of construction on selected regions and localities might be even greater because nuclear plants are usually built in rural locations that otherwise are not characterized by intensive capital investment or construction activity.

The accumulation of long-lived radioactive wastes associated with the operation of nuclear power plants raises unique questions of social responsibility to future generations, due to the very long time that must elapse before such wastes become innocuous to man. "Because of the serious consequences of improper handling of radioactive wastes and the time scale of concern, the Federal Government must play a strong role in the management and ultimate disposal aspects of nuclear wastes. In this connection, there is an obvious societal question of whether or not a possibly short-term use of fission energy by current and near-term future generations of man could impose an unprecedented caretaker burden on future generations for periods beyond man's experience" (Ref. 22). There are several options for permanent disposal of high-level wastes open to contemporary society, all of which need to be carefully evaluated on socioeconomic grounds, in view of the unprecedented period of future time involved. Societal questions are at least as important as technological considerations. Assuming, for example, that high-level wastes are to be sequestered in geologic formations, should the choice be one of foreclosing future accessibility? What options might future generations like to find left open to them? The technological testing of disposal options, as currently planned, will probably extend into the 1980s before commitments to a permanent disposal method are possible.

All energy systems, except perhaps highly decentralized ones, are possible concentrated targets for widespread social impact. Nuclear power systems add another dimension to this potentiality in that they create the means, in theory at least, of an entirely new form of antisocial behavior. A major difference between an electrical energy system based on nuclear fission and one based on other fuels arises from the presence, when the nuclear system is being used, of enriched uranium and plutonium in a number of secure fuel cycle installations or in transit between such facilities. The need to protect against diversion and misuse of fissile material required the implementation of effective safeguards systems.

Such systems would usually have little public impact because they would be in effect at relatively few locations in a highly specialized industry. Emergency measures following a hypothetical successful diversion would concentrate on rapid recovery of the stolen material. Such emergency situations, however, can be expected to be extremely infrequent and of short duration. Thus, it is unlikely that measures to safeguard nuclear materials would lead to significant socioeconomic impacts (e.g., invasion of privacy or suppression of certain civil liberties).

Overall, it is felt that the benefits of nuclear power greatly outweigh the potential adverse socioeconomic and environmental effects of its use. However, there is clearly a tradeoff involved in the production of electricity by nuclear power plants in that future generations will need to manage radioactive wastes being produced by the use of the nuclear fuel cycle by current generations.

REFERENCES

1. American Nuclear Society, *Nuclear Power and the Environment. Questions and Answers*, p. 50 (April 1976).
2. McBride J.P., et. al., Radiological Impact of Airborne Effluents of Coal and Nuclear Plants, *Science 202*, p. 1945 (December 8, 1978).
3. Rasmussen N., et. al., *Reactor Safety Study, An Assessment of Accident Risks in U.S. Commercial Nuclear Power Plants*, U.S. Nuclear Regulatory Commission, WASH-1400 (October 1975).
4. Atomic Industrial Forum, INFO, No. 125 (December 1978).
5. U.S. Department of Energy, *Energy Information Administration, Annual Report to Congress. Vol. III, 1977. Statistics and Trends of Energy Supply, Demand, and Prices*, DOE/EIA-0036/3 (May 1978).
6. U.S. Atomic Energy Commission, *Proposed Final Environmental Statement, Liquid Metal Fast Breeder Reactor Program*, Vols. I and II, WASH-1535 (December 1974).
7. Project Management Corp., Tennessee Valley Authority, and U.S. Energy Research and Development Administration, *Final Environmen-*

tal Statement Related to the Construction and Operation of Clinch River Breeder Reactor Plant, NUREG-0139, Docket No. 50-537 (February 1977).
8. Cohen B. L., The Hazards of Plutonium Disposal, Nuclear News 18(8), pp. 44-45 (June 1975); also The Hazards of Plutonium Disposal, University of Pittsburgh, Pittsburgh, Pa. (1975).
9. Minton S. A., Jr., Snakebits, Scientific American, pp. 114-119 (January 1957).
10. Lapp R. E., America's Energy, Reddy Communications, Inc., Greenwich, Conn. (1976).
11. INFCE Working Group 8. Improved Once-Through Fuel Cycles for Light Water Reactors, INFCE/WG.8/USE/DOC 12-A (November 30, 1978).
12. McCracken S., The War Against the Atom, Commentary (September 1977).
13. Interagency Review Group on Nuclear Waste Management, Report to the President, TID-28817 (October 1978 draft); see also Interagency Review Group on Nuclear Waste Management, Subgroup Report on Alternative Technology Strategies for the Isolation of Nuclear Wastes, TID-28818 (October 1978 draft).
14. 8th Annual Review of Overall Reliability and Adequacy of the North American Bulk Power Systems, National Reliability Council (August 1978).
15. Electric Power Supply and Demand, 1978-87 for the Contiguous United States, DOE/ERA-0018 (July 1978).
16. Annual Report to Congress, Energy Information Administration, U.S. Department of Energy, DOE/EIA-0036/2 (1978).
17. Transporting the Nation's Coal: A Preliminary Assessment, U.S. Department of Transportation (January 1978).
18. Proposed Staff Response to the Jeannine Honicker Petition for Emergency and Remedial Action, U.S. Nuclear Regulatory Commission, SECY-78-560 (October 1978).
19. Lapp R. E., America's Energy, Reddy Communications, Inc., Greenwich, Conn. (1976).
20. Impact Assessment of 1977 New York City Black-Out, U.S. Department of Energy, DOE HCP/T5103-01 (July 1978).
21. An Economic Definition of the Nuclear Industry, prepared for Office of Planning and Analysis, U.S. Nuclear Regulatory Commission (April 1978).
22. Proposed Final Environmental Statement, Liquid Metal Fast Breeder Reactor Program, WASH-1535 (December 1974).

Part V
Nuclear Waste Management

1.
U.S. Energy Policy

In recent years, the energy policy of the United States has been stated in greater detail. The most comprehensive statement of national energy measures is in the National Energy Plan, which was developed during consideration, revision, and passage of the National Energy Act. Radioactive waste management policy is designed to protect public health and safety and the environment from radioactive materials that result from national defense programs, energy R&D, and commercial activities. Chapter 1 discusses the general outline of national energy policy as it relates to Nuclear Waste Management.

THE NATIONAL ENERGY ACT AND NUCLEAR WASTE MANAGEMENT

One objective of the NEA and accompanying legislation is to enhance the potential for increased near-term commercial deployment of Light Water Reactors (LWR's). Because LWR's produce radioactive waste which must be managed by interim storage and ultimate disposal, the safe management of nuclear wastes is fundamental to meeting environmental protection goals associated with energy development goals. In addition, some recent State legislative actions have effectively precluded the nuclear option until safe nuclear waste management is demonstrated.

Under Federal statutes and regulations, long-term management facilities for high-level waste from both commercial and defense sources must be in Federally licensed, Federally-owned facilities are specifically exempted from such control by Federal law. (The applicable regulations for commercial waste are in 10CFR50, Appendix F, and for defense waste, the Energy Reorganization Act of 1974.) The major thrust of the commercial waste program is to construct several radioactive waste repositories in geologic formations to provide for disposal of high-level waste—HLW (in spent fuel) generated by civilian operations. In addition, these repositories could be available for the disposition of any high-level radioactive waste under Federal custody.

By law and in view of public concerns, the U.S. Department of Energy cannot unilaterally decide on and implement a course of action for the long-term waste disposal program. The public, the States, interest groups, and other Federal agencies are involved in the DOE decision-making process through the provisions of the 1969 National Environmental Policy Act (NEPA) and through regulatory and budget processes of Federal and State agencies. For example, the Energy Reorganization Act of 1974 requires that facilities for the long-term storage of high-level waste be licensed by the Nuclear Regulatory Commission (NRC). DOE's programs, therefore, involve decision-making phases followed by an implementation phase with provisions throughout for the involvement of the public and their representatives and appropriate Federal and State agencies.

Two policy decisions made by President

Carter in 1977 have major impact on the U.S. nuclear fuel cycle and waste programs:

- On April 7, the reprocessing of spent fuel from commercial reactors was indefinitely deferred.
- On October 18, the Federal Government offered to take title to spent reactor fuel.

These new policies emphasize fuel/resource conservation, more efficient separation of uranium-235 from natural uranium, the search for nuclear fuel cycles that can meet nonproliferation criteria, and long-term management of spent fuel. Accordingly, there is less need to develop commercial fuel reprocessing further and to provide for the disposal of wastes from such reprocessing.

DOE has management responsibility for all defense nuclear waste, commercial spent fuel, and high-level nuclear waste in Federal custody. This authority is derived through the Atomic Energy Act of 1954, the Energy Reorganization Act of 1974, the Department of Energy Organization Act of 1977, and Federal regulations. Thus, DOE is responsible for formulating strategies and implementing a program for the long-term management of highly radioactive wastes. The magnitude of this responsibility is shown when consideration is given to the current and projected inventories of defense and commercial waste volumes listed in Table V 1–1.

In addition, DOE is responsible for future wastes generated by defense, test, and research nuclear programs as well as the spent fuel and other wastes from the commercial nuclear power program.

The successful long-term management of nuclear wastes is dependent on satisfying institutional, political, environmental, and technical constraints. Over the last 30 years, management of defense wastes has provided a basis for the technology of nuclear waste disposal.

Although specific designs are incomplete and facilities are unavailable for permanent disposal of high-level radioactive waste, there is a consensus of scientific opinion that permanent disposal of high-level waste in geologic formations can be safely achieved and that the technology exists for the identification of specific sites for such disposal. A number of major institutional issues must still be resolved, however, before DOE can demonstrate the safe management of nuclear wastes. For planning purposes for permanent disposal, wastes are defined to include unreprocessed spent fuel and wastes that would result from

Table V-1-1. Volume of defense and commercial waste

		DEFENSE		COMMERCIAL	
		CURRENT[1]	PROJECTED TO YEAR 2000[2]	CURRENT[3]	PROJECTED TO YEAR 2000[3]
HLW 10^6 ft^3	SOLID	6.28	10.4[4]	0.2	4.0
	LIQUID	3.45	2.7[4]	.082	.082
TRU 10^6 ft^3		9.01	15 to 125[5]	.6[6]	.23[7]
LLW 10^6 ft^3		61.17	70 to 250[5]	17.91	100 to 280

Source: IRG Report, Appendix D except as footnoted below

[1] As of October 1, 1978.
[2] Defense HLW estimated only to 1985.
[3] Commercial solid HLW reflects spent fuel. Currently there are about 4,000 metric tons of reactor spent fuel being stored in reactor basins. There could be as much as 95,000 metric tons of spent fuel discharged by the year 2000.
 If the spent fuel were reduced and stored as HLW, the volume would be reduced by a factor of about 10.
[4] Maximum volume case assumed.
[5] Range depends on extent of D&D activities.
[6] Figure taken from Table 1 in "Sources, Production Rates, and Characteristics of ERDA LLW."
[7] Generated between 1977 and 2000.

fuel reprocessing if that option were exercised in the future.

DOE conducts research and development, testing, and design work to enable timely implementation of the most promising alternatives for commercial and defense wastes. This work is carried out with full public disclosure through public reports, information meetings, environmental impact statements, and regulatory proceedings, where required.

SPENT FUEL POLICY

On April 7, 1977, President Carter announced that the United States would defer indefinitely all civilian reprocessing of spent nuclear fuel. Therefore, the possibility that spent fuel will be discarded as waste must be seriously considered by waste management programs, and appropriate accommodations must be made. The fundamental reason for establishing this policy was a concern that the widespread separation of plutonium from other products of fission contained in spent nuclear fuel could lead to the proliferation of nuclear weapons. This significant policy change in the use of nuclear fuel stock was first identified in President Ford's October 26, 1976, policy statement which raised the possibility of a ban on reprocessing until a better institutional basis could be developed to avoid the risks of nuclear weapons proliferation. On April 7, 1977, President Carter announced indefinite deferral of all commercial reprocessing until both the institutional and technical aspects of proliferation could be considered.

This policy change stopped industry's plans for the development and use of reprocessing and ancillary processes. Currently available spent fuel storage space will be fully used within a few years, as reactors continue to discharge spent fuel. Although onsite commercial reactor spent fuel storage basin modifications (such as installing closely spaced storage racks) can provide additional near-term storage, by the mid-1980s some basins will still have inadequate capacity to handle the fuel being discharged. Additional storage could be provided by either an away-from-reactor (AFR) or a disposal facility

In order to support the nonproliferation policy and to meet the need for additional spent fuel storage, DOE, with the approval of the President, announced more detailed plans for spent fuel policy implementation on October 18, 1977. Pending payment of a one-time, usage-based storage fee, the Government offered to take title to and store spent nuclear fuel from private power reactors. This policy is intended to provide for interim spent fuel storage until a disposal facility is available or reprocessing is permitted. Consideration will also be given to the storage of spent fuel from foreign reactors.

Within DOE, spent fuel disposal activities and high-level waste disposal responsibilities are both assigned to the Office of Nuclear Waste Management. Accordingly, the National Waste Terminal Storage Program is directed to provide geologic disposal capability regardless of whether the material is in the form of spent fuel or solidified waste from a fuel reprocessing operation. Storage activities support the indefinite deferral on spent fuel reprocessing, and have separate budget and management controls for civilian and military nuclear waste programs.

It is important to note that the basic need for permanent disposal of nuclear waste in one form or another is independent of U.S. policy on reprocessing. Development of waste disposal capabilities is essential for a fuel cycle based on reprocessing as well as one without reprocessing.

NATIONAL ENVIRONMENTAL POLICY ACT AND DOE POLICY

The National Environmental Policy Act of 1969, implemented by Executive Order 11514 and dated March 5, 1970 (35 FR 4247), and the Guidelines of the Council on Environmental Quality (CEQ) of August 1, 1973 (40 CFR Part 1500, 38 FR 20500), require that all agencies of the Federal Government prepare detailed statements on the environmental implications of proposals for legislation and other major Federal actions significantly affecting the quality of the human environment. The objective of NEPA is to build into the Federal agency decision-making process, beginning at the earliest possible point, an appropriate and careful consideration of environmental aspects of proposed actions.

In its planning and decision-making processes, DOE considers the possible consequences of its proposed actions by the following:

- Evaluation of both the long-range and short-range implications of such actions to man, including his physical and social surroundings, and to nature
- Exploration, development, and analysis of alternative actions
- Identification of the potential environmental effects of such proposed actions to agencies and other decision-makers, including Congress and the President, and to the public.

Two types of NEPA documents identify the environmental implications of Federal programs: environmental assessments (EA's) and environmental impact statements (EIS's) (either generic or program specific). An EA is prepared to assess whether a proposed DOE action would be "major" and would "significantly affect" the quality of the environment. It serves as the basis for determining whether an EIS is required. One of two decisions can be made based upon an EA: a negative determination, which signifies that an EIS is not required for a proposed action, or an EIS is prepared in accordance with NEPA guidelines.

If a determination is made to prepare an EIS based upon the EA, guidelines developed by CEQ determine the basic structure of the EIS. Because Federal programs vary substantially among agencies, CEQ guidelines require that agencies develop specific procedures tailoring the EIS process to their particular activities. In the case of waste management, in place of or preceding a facility-specific EIS, a Generic Environmental Impact Statement (GEIS) may be prepared to discuss the impact of a program in which one or more facilities may eventually be involved and which has widespread implications that warrant a statement of overall impact. The decision to prepare either an EIS or GEIS to fulfill NEPA requirements is based upon internal DOE guidelines.

Frequent and continuous coordination is maintained with the DOE Office of Environment (EV) which has responsibility for implementing NEPA guidelines. Proposed revisions to CEQ guidelines for implementation of NEPA are under review by the DOE Waste Management Office for application as required.

2.
FY 1980 Nuclear Waste Management Overview

This Program Summary Document describes the program of the DOE Office of Nuclear Waste Management for fiscal year 1980. The budget and the program which this document supports are designed to address the technical, social/institutional, economic, environmental, and political issues in waste management which have been evolving in the period preceding FY 1980 budget preparation. The major recent influence on the program is the report to the President by the Interagency Review Group (IRG) on Nuclear Waste Management.

In his April 1977 presentation of the National Energy Plan to Congress, President Carter directed that a review of the entire waste management program be undertaken. DOE completed the first phase of this review with publication of the draft Report of the Task Force for Review of Nuclear Waste Management. The President subsequently formed the IRG to formulate recommendations for establishment of Administration policy. At this printing, the final Presidential review and approval of IRG recommendations cannot be fully anticipated. However, provisions have been made within the proposed FY 1980 budget to accommodate the technical recommendations contained in the IRG report.

The IRG report summarizes the state of knowledge of each of the primary disposal options, identifies and analyzes alternative policies for guiding radioactive waste management, and makes important recommendations on the nature of the R&D plans underpinning the waste management strategy. IRG recommendations suggest the next steps in the evolving understanding and appreciation of radioactive waste management problems.

Comments have been sought and encouraged on all issues addressed in the draft report. In certain instances, however, the IRG noted areas in which public comment is of special concern or interest. These areas include the following:

- The relationship between nuclear waste disposal and the future use of nuclear energy
- The extension of Nuclear Regulatory Commission (NRC) licensing authority to DOE facilities
- The acceleration of the schedule for development of Environmental Protection Agency (EPA) standards
- The role of intermediate-scale facilities
- The regional approach to the siting of waste disposal facilities
- The interim strategic planning basis for the first high-level waste repository

DOE/ETW has been a primary participant in the development of the IRG report. DOE has already begun implementation of some IRG recommendations where changes have been within the scope of ETW management activity and consistent with prevailing national policy. In areas where final decisions are still pending (such as the number of geologic media under investigation and the schedule for development), implementation awaits Presidential and congressional review. This budget and this program will evolve and will be amended as required as policy evolves. Such evolution is expected to result from continuing interaction among Federal agencies, Congress, State/local governments, public and private interest groups, and the general public.

NUCLEAR WASTE MANAGEMENT STRATEGY

The achievement of the Government's overall objective of providing safe management of nuclear wastes requires a broad-based strategy that includes careful attention both to the operation of existing waste storage and disposal facilities and to long-range plans for interim storage and disposal of wastes. In the first several decades of the Atomic Energy Commission's (AEC's) nuclear program, waste management strategy was dominated by concern for production of weapons materials, development and operation of research and test reactors, and development of naval propulsion systems. Nuclear wastes were handled in a safe manner, but only interim solutions were sought to the waste management problem.

Beginning in the late 1950s, waste management strategy turned from short-term solutions to an increasing emphasis on the development of waste disposal methods which would provide isolation from the biosphere for very long time periods with minimum requirements for maintenance and surveillance by man. At the same time, commercialization of nuclear technology began to take place with the use of radioisotopes in medicine and industry and with the construction of large nuclear power plants for generation of electricity. The non-Government sector began to produce nuclear waste in increasing quantities.

Increasing concerns with long-term environmental problems and the expanding role of nuclear technology in the non-Government sector have resulted in an evolving waste management strategy which has received new emphasis in recent years and become increasingly broad based. The Government's waste management strategy is continuing to evolve, and major new initiatives are under consideration. In particular, the recommendations of the DOE task force on waste management and the IRG draft report on nuclear waste management contain specific recommendations for changes in the current program.

The following discussion of waste management strategy considers:

- All Government-generated nuclear waste (usually referred to as "defense waste" although some of it results from non-defense Government activities)
- Civilian high-level waste
- Civilian transuranic (TRU) waste

In accordance with recent Presidential direction and congressional authorization, the following additional types of nuclear material have been incorporated into DOE's strategy considerations:

- Civilian spent fuel
- Civilian low-level waste (LLW)

Background

The Government has been faced with handling radioactive wastes since the days of the Manhattan Project. Three principal methods were used and are still in use today: shallow land burial of low-level waste, storage of high-level liquid wastes and sludges in shielded subsurface tanks, and storage of spent fuel and other materials in shielded water basins.

A limited amount of waste has also been stored in vaults and on surface pads, disposed of in deep ocean, held in settling ponds, and disposed of as grout in deep shale formations by shale fracturing.

High-Level Waste. HLW is a product of reactor fuel reprocessing and occurs in the form of an acid aqueous waste stream in the separation of plutonium and uranium from the spent fuel. Almost all of the present HLW has resulted from processing of military reactor fuels. At the Hanford Reservation and Savannah River sites, the neutralized liquid HLW is held as liquids and solids in large subsurface tanks surrounded by concrete shielding. At ODE's Idaho reservation, acidic liquid waste is converted to a solid by calcining and stored retrievably in stainless-steel bins inside concrete vaults.

In the past decade, considerable effort has been devoted to improving the safety of the high-level waste tanks. Some of the measures have included evaporation of excess water to produce a solid salt cake in the tank, provision of new double-walled tanks, improved instrumentation for leak detection, and removal of strontium (Sr) and cesium (Cs) to reduce heat production. Leaks have occurred in some of the older tanks releasing radioactivity to the surrounding earth. These leaks have not resulted in any measurable release of radioactivity to underground aquifers or in any injury to health, but they have caused concern about the use of the tanks for long-term storage.

In the late 1950s, work began on developing the technology for permanent disposal of HLW. The AEC focused its efforts on processes for solidification and immobilization of the liquid wastes (by calcining and vitrification) and on isolation of the solidified waste canisters in deep, stable geologic formations. Salt formations were identified as being most promising. In the mid-1960s, Oak Ridge National Laboratory (ORNL) carried out an experimental program in the Lyons, Kansas, area to develop the technology for disposal in bedded salt. The program was generally successful in accomplishing its technical objectives.

In 1971, AEC asked Congress to authorize construction of a repository for disposal of commercial HLW and defense TRU waste, with Lyons as a tentative site subject to the satisfactory completion of certain additional tests and studies. Congress authorized only additional studies. In view of political opposition and the recognition that two site-specific safety questions might not be resolved in a reasonable period of time, AEC abandoned work at Lyons in mid-1972 and began evaluation of possible replacement sites. In order to maintain the existing regulatory limit on generator custody of solidified HLW, AEC simultaneously began a program for Retrievable Surface Storage Facilities (RSSF's).

In 1974, a parallel program was initiated at Sandia to develop a Waste Isolation Pilot Plant (WIPP) at a bedded salt site in New Mexico. The WIPP facility is to receive TRU waste generated by the defense program and to provide experimental test capability for HLW. A proposal has also been made to emplace a limited amount of spent fuel in WIPP on a retrievable basis.

Shortly after the formation of the Energy Research and Development Administration (ERDA) and its assumption of responsibility for nuclear technology development in 1975, ERDA management reevaluated the waste management program. The ERDA analysis indicated that, due to delays in the buildup of commercial nuclear power and fuel reprocessing, geologic waste isolation could be available in time to satisfy the Nation's needs and an RSSF would probably not be required.

New emphasis was given to the geologic program with the formation of the Office of Waste Isolation (OWI) at Union Carbide Nuclear Division in Oak Ridge. OWI was assigned the tasks of performing extensive geologic exploration to identify potential repository sites in salt and to investigate igneous and argillaceous rock formations as possible disposal media. This included extensive geologic exploration at the Hanford Reservation to evaluate the potential of the area's basalt formations and at the Nevada Test Site (NTS) to assess the local geologic formations. The broad-based OWI effort included conceptual repository design studies, equipment development, waste migration studies, and detailed planning for licensing, construction, and operation of a repository. In 1977, the activities at Hanford and NTS were separated from OWI as specific site evaluation projects. In July 1978, the OWI program was transferred to the Office of Nuclear Waste Isolation at Battelle Memorial Institute, with essentially the same objectives.

When the Government announced in 1977 its decision to defer civilian fuel reprocessing indefinitely, the HLW disposal program was expanded to consider disposal of spent fuel in geologic repositories. Subsequently, the Government proposed a new spent fuel policy which extended its previous obligation to accept commercial HLW for disposal to spent fuel as well. Under the spent fuel policy, the Government will offer to accept and take title to spent fuel for a one-time charge which is to cover all the Government's costs for storage and disposal. Since it is expected that spent fuel receipts by the Government will begin before the availability of a geologic repository, a program has been initiated to provide temporary away-from-reactor (AFR) fuel storage in water basins.

Low-level Waste. Significant quantities of LLW began to be generated when the Government initiated reactor operations in support of the weapons program in the 1940s. At most of the major sites, land was set aside for disposal of this waste by shallow land burial. The techniques have changed little with time. Solid waste is placed in relatively shallow trenches which are filled to about one meter from the surface and backfilled with the excavated material.

Until the early 1960s, AEC accepted the relatively small amount of commercially generated LLW at two of its burial sites (Oak Ridge and Idaho). In 1962, the first commercially operated burial ground was opened (Beatty, Nevada). Subsequently, a total of six commercial sites went into operation around the country. (See Figure V 2-1.)

These commercial burial sites are State owned with the exception of the Richland site (Hanford Reservation), which is leased to the State of Washington by the Federal Government. Five of the burial ground operations are licensed by the States under the "agreement state" provisions of the Atomic Energy Act; the sixth, the Illinois burial ground, is under direct NRC regulation.

In the last several years, opposition to the commercial burial ground operations at a number of sites has emerged on both local and State levels. Due to restrictive actions by the States, only one (Barnwell) of the four Eastern sites remains in operation. The two Western sites continue to operate but are far from most of the reactors (which are predominantly in the East).

DOE operates burial grounds at 14 Government-owned sites for the Government's waste, five of which are considered major operations. These sites are not under NRC regulation but are under internal DOE regulations. A portion

Figure V-2-1. Low-level waste burial sites.

of the DOE wastes contains significant quantities of long-lived TRU isotopes. In 1970, the Government decided that these TRU wastes should no longer be disposed of in shallow land burial. The intended final disposal of TRU wastes is in a deep geologic repository. Until a repository is available, these wastes are being stored on the surface. A small quantity of comercial TRU waste continues to be accepted for burial at the Richland, Washington site.

Rationale for Current Program Priorities

Program planning for nuclear waste management has been segmented into commercial waste management, defense waste management, and spent fuel storage.

In many R&D areas, the Commercial and Defense Waste Management Activities deal with similar waste forms and waste handling technologies. Therefore, the two activities are planned and managed jointly to complement each other and to avoid duplication of effort. Basic R&D is funded with commercial or defense budgets depending upon where the first or most widespread application may occur. Applied work, such as adaptation of equipment and processes to existing wastes, is funded by the program that will use the technology.

Defense Waste Management Activity. Since 1943, radioactive wastes have been generated as a by-product of the national defense programs to produce and use nuclear materials in weapons and other national defense applications. The principal source of radioactive waste is the reprocessing of fuels irradiated in nuclear reactors to recover plutonium, uranium, and other products. Other major sources of wastes are the manufacturing operations to produce plutonium components for weapons, laboratory operations, and reactor operations.

The general objective of the Defense Waste Management Activity is to confine radioactive wastes so that safety, environmental, and public health standards are met and maintained. In the near-term, the specific objective is to provide confinement of radioactive waste in a way that is compatible with potential long-term management practices. The long-term objective is to provide the required confinement of radioactive wastes. This must be accomplished (1) with minimum reliance on future maintenance and surveillance by man, and (2) in a way that assures a high degree of isolation from man's environment over the period that waste is a potential radiation hazard.

The Defense Waste Management Activity consists of five primary subactivities:

- Interim Waste Operations
- Long-term Waste Management Technology
- Terminal Storage
- Transportation
- Decontamination and Decommissioning

DOE/ETW recognizes that information developed through the Defense Waste Management Activity may be applicable to achieving commercial waste management objectives. The commercial and defense programs are therefore managed jointly by the same organizational units.

Interim Waste Operations Subactivity. The strategy of the Interim Waste Operations Subactivity is to continue to provide safe handling and storage of DOE radioactive wastes pending implementation of the long-term waste management programs. A continuing improvement program for the confinement of existing and currently generated waste products will be maintained. Treatment and storage facilities will be compatible with alternatives or improvements under consideration for the long-term management of those wastes.

The design of improved facilities for HLW is based on the philosophy of multiple containment, as necessary, to minimize the consequences of possible accidents.

Low-level solid waste will continue to be disposed of by shallow land burial. Upgrading of shallow land burial practices for LLW will focus on implementation of improved technology, making operations subject to comprehensive criteria, and stabilization of sites that are no longer needed. Upgraded requirements and related criteria will be developed through assessment studies. Reductions in waste generation will be encouraged. Alternatives to shallow land burial will be developed to help manage potential increases in waste generation from decontamination and decommissioning and other activities. Intermediate-level liquid wastes generated at ORNL will continue to be disposed of by shale fracture.

TRU-contaminated solid wastes will continue to be stored retrievably, pending availability of any required treatment facilities and a final disposal mode (repository). The construction of a first treatment facility is expected to generate final waste form(s) by 1986 or 1987.

Long-term Waste Management Technology Subactivity. This subactivity covers R&D on the following categories of waste:

- High-level waste
- Contaminated solid waste
- Airborne waste

In recognition of individual site considerations and to minimize the economic impact and realize the benefits of experience, specific long-term management solutions for treatment and disposal of HLW will be implemented on a site-by-site basis with Savannah River as the first site, followed by Richland and Idaho. An early demonstration will be made of immobilization, packaging, and emplacement of defense HLW at a suitable repository. To prepare radioactive wastes for disposal, waste treatment and handling technology will be developed, and facilities constructed, based on health, safety, environmental, and economic considerations. Requirements will be based on applicable current and expected regulatory criteria standards and restrictions on releases.

Terminal Storage Subactivity. The strategy is to design, construct, and operate a disposal facility for TRU waste—the Waste Isolation Pilot Plant (WIPP). WIPP will be a geologic repository located in a deep-bedded salt formation; Sandia, in southwestern New Mexico, is a possible site. WIPP is proposed to be used for the permanent isolation of defense TRU waste and for conducting experiments with various forms of HLW in a geologic repository environment. Decisions are pending on licensing the WIPP and on providing for a demonstration of spent fuel storage using up to 1,000 spent fuel assemblies. As appropriate to meet technical and regulatory requirements, geologic studies and exploration, testing, design and construction, and technical, economic, and environmental evaluations will be conducted. This project has been designated a major systems acquisition to provide highest agency priority and management exposure to the project. An overview network for WIPP is presented in Figure V 2-2.

Transportation Subactivity. A systematic approach will be initiated to ensure the timely availability of commercial transport systems to meet operational requirements for DOE intersite transportation of radioactive wastes. Technology will be developed and implemented; facilities and equipment will be designed, constructed, and tested; and operations will be conducted. These activities will involve industry, DOE laboratories, and universities.

The objective of the Transportation Subactivity is to develop a national capability to cope with the growing complexities of transporting nuclear materials. Safe, reliable, publicly acceptable, and cost-effective systems must be made available on a schedule that is consistent with the transportation needs of WIPP and other national programs. A systems approach which addresses technology, institutional, and operational problems will be implemented to ensure that transportation hardware is available. Previous and existing transportation activities are being reviewed to preclude redundant activities. New activities (including establishment of a Transportation Technology and Information Center) will be initiated on a priority basis consistent with funding levels.

The Albuquerque Operations Office and Sandia Laboratories have been designated as the lead centers for DOE nuclear materials transportation activities. A program strategy plan has been prepared.

Decontamination and Decommissioning Subactivity. A systematic program of decontamination and decommissioning (D&D)

FY 1980 Nuclear Waste Management Overview 761

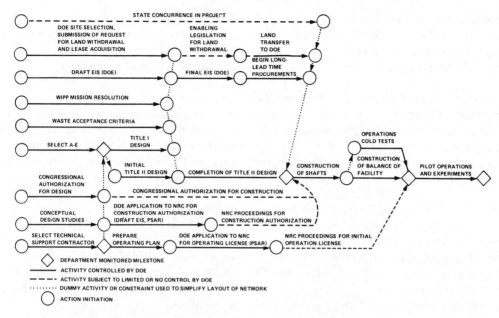

Figure V-2-2. Waste isolation pilot project overview network.

projects for ASET radioactive facilities will be developed and pursued as such facilities are declared excess. R&D will be conducted where required to support specific projects. The pace of the subactivity is established by the need for providing D&D services without unreasonable delays. Major systems acquisitions may be involved, in that management and budgeting techniques for multiyear, high-cost D&D projects will be planned and controlled in the same manner as major facility acquisitions.

Commercial Waste Management Activity. This activity encompasses management of all radioactive waste and spent reactor fuel resulting from civilian use of radioactive materials. The broad objectives of the civilian radioactive waste management activity are as follows:

- To develop and demonstrate technologies to provide safe forms and packaging of radioactive wastes from different fuel cycles for initial handling, interim storage, processing, and permanent isolation
- To demonstrate the geologic isolation of radioactive waste and to provide appropriate sites for waste isolation
- To provide for the increased spent-fuel storage requirements that have occurred as a result of our nonproliferation of nuclear power policy

The President's indefinite deferral of U.S. commercial fuel reprocessing has shifted the program emphasis toward the storage and disposal of spent reactor fuel, rather than only of wastes from the nuclear fuel cycle. Both the decline in projected nuclear electric generation and the increase in public demand for demonstrated safe waste disposal in a repository has led to a special focus on the first full-scale commercial repository. Current planning calls for this repository to be in operation as early as 1988.

The Commercial Waste Management Activity is composed of seven subactivities:

- Terminal Isolation
- Waste Treatment Technology
- Supporting Studies
- Solidification Process Demonstration
- Decontamination and Decommissioning
- Low-level Waste Management Operations
- West Valley Nuclear Activities

All commercial waste management activities will maintain close coordination with relevant defense waste activities to ensure timely and effective technology transfer between the two programs. Close coordination and frequent communication will be maintained with appropriate Federal, State, and local public agencies and officials to ensure public acceptance of the program.

Terminal Isolation Subactivity. Sites for the terminal storage of commercial high-level waste, spent fuel, and TRU waste will be sought in different regions of the country. Based on the technical status of the program, the site location for the first repository could be selected from candidate sites in salt. The location for the second repository, in a different region in either salt or crystalline rock, could be selected within 2½ years after the first has been selected and prior to NRC's approval of the construction authorization for the first repository.

Current development work on salt and basalt will continue, and work on alternate media will be accelerated. Evaluation of two existing DOE sites, the Nevada Test Site and Hanford, will be continued to determine if they are suitable for a repository.

The question of near-term and long-term safety will continue to receive attention including the analysis of sites and the complete evaluation of expected performance of each repository. Programmatic, environmental, and technical objectives will be met by conducting studies of geologic formations in numerous states, by conducting R&D projects relevant to each geologic media, and by planning and conducting the design and construction of facilities. The intent is to provide an adequate technology base for the NRC licensing of the site, the construction and operation of a repository, and closure of a filled repository.

Waste Treatment Technology Subactivity. The objective of the Waste Treatment Technology Subactivity is to develop and demonstrate waste handling and treatment technology that will provide waste forms for disposition which meet waste repository criteria. Preliminary design specifications will be developed for processes needed to immobilize each waste type and confirmed in radioactive pilot plant demonstrations when appropriate. The technology developed and demonstrated to handle defense-related wastes helps to establish methods for commercial waste handling and processing. Close coordination with the defense waste programs will ensure prompt and effective technology transfer.

Supporting Studies Subactivity. Information and data evaluations in the areas of health and safety, economics, risk assessments, quality assurance, and environmental impact statements will be developed and reviewed to support the waste management program. Evaluations will be prepared on alternative systems for treatment and disposal of radioactive waste from the conventional LWR fuel cycle and from other fuel cycles identified in International Nuclear Fuel Cycle Evaluation (INFCE) and Nuclear Alternative Systems Assessment Program (NASAP) studies as having nonproliferation potential. A program has been established to keep abreast of all international waste management programs. A committee of national experts to evaluate and report on all major waste management programs, the Committee on Radioactive Waste Management of the National Academy of Science, is being supported.

Solidification Process Demonstration Subactivity. This subactivity is intended to demonstrate immobilization of typical commercial high-level wastes by spray calcination and in-can melting to glass. The project was completed in FY 1979.

Decontamination and Decommissioning Subactivity. The objective of this portion of the commercial waste activity is to communicate to industry DOE's expertise in D&D through information centers, data dissemination, advice, and consultation.

Low-level Waste Management Operations Subactivity. DOE does not have operational responsibility for commercial burial grounds at this time. However, if Congress in the future modifies public laws, DOE may become responsible for the operation of such facilities. If such action occurs, the lead office for LLW (the Idaho Operations Office) would establish a program for extending Federal management appropriately.

West Valley Nuclear Center Activities Subactivity. DOE does not have any authorized responsibility in this area beyond the completion of a study on future options for the West Valley Nuclear Center, authorized in FY 1978. As a result of the study and congressional action, a future DOE role may evolve. The final report (TID 28905, Volumes 1 and 2) was submitted to Congress in February 1979.

Spent Fuel Storage Activity. The overall objectives of the Spent Fuel Storage Activity are as follows:

- To ensure that storage space is available for spent fuel so that operation of nuclear power plants within the United States can be maintained
- To support nonproliferation initiatives by offering a credible alternative to nuclear fuel reprocessing by foreign nations

Therefore, the Spent Fuel Storage Activity consists of the Domestic Spent Fuel Storage Subactivity and the International Spent Fuel Storage Subactivity.

The objectives of the Domestic Spent Fuel Storage Subactivity include the following:

- To plan and manage a program which provides interim spent fuel storage capacity until geologic storage or recovery is available
- To establish charge and contract requirements for DOE to provide storage and disposal services
- To identify and resolve safety and environmental issues on spent fuel storage
- To develop technology to increase the capacity of existing spent-fuel facilities and to develop advanced, improved high-capacity storage technologies

The objectives of the International Spent Fuel Storage Subactivity are as follows:

- To develop the capability to accept limited amounts of foreign spent fuel in order to meet nonproliferation objectives subject to necessary legislation and the approval of a required plan for foreign fuel storage in the United States
- To provide technical support to international bodies and limited direct assistance to individual countries to establish spent fuel storage as a credible alternative to reprocessing
- To establish and maintain a current assessment of foreign spent fuel storage needs, capabilities, and plans, and to develop and maintain a logistics plan for transferring fuel in critical or sensitive situations
- To augment, as appropriate, domestic spent fuel technology and demonstration activities to support the unique requirements of international or foreign transportation and storage

Many of the aspects of the Spent Fuel Storage Activity are applicable to both the domestic and international fuel disposal activities. These activities include the establishment of storage charges, enactment of necessary legislation, and the selection and acquisition of storage facilities. Since more requirements are currently expected for the use of spent-fuel storage facilities for domestic fuel, the majority of the funding and effort in FY 1980 is directed toward that part of the program. This is supplemented wherever necessary to support the unique aspects of the international spent fuel storage.

In order to meet its objectives, the Spent Fuel Storage Activity is divided into four subactivities:

- Program Management and Support
- Fuel Storage Technology Development
- Safety and Environmental Studies
- Facilities and Systems

The following items will be addressed in the strategy for meeting the program's objectives:

- Execution of a program of one or more options to provide required capacity by 1983, and subsequent years, until disposition capability for spent fuel is available
- Continuing assessment of fuel storage requirements and capabilities
- Evaluation of Government-owned AFR facilities

- Development of improved technology for spent fuel storage
- Continued evaluation of spent fuel behavior during extended storage

Strategic Planning and Alternative Scenarios for Nuclear Waste

Introduction. The basic strategies and present program priorities for nuclear waste management outlined above are in an evolutionary state. Within DOE and other interested agencies and at the interagency level, the existing strategies and plans are under intensive review. Specific conclusions and recommendations with significant impact on the present program are beginning to emerge. One major conclusion has been that the waste management program should be broadened in scope, that more emphasis should be placed on a systems approach to program planning and technical decision-making, and that supporting R&D programs should be broadened in scope. This would mean that management and technical decisions could rest on a broader technical base.

Interagency Review Group. The Interagency Review Group (IRG) on Nuclear Waste Management was formed at the President's direction in March 1978. In October 1978, a draft of the IRG's tentative conclusions and deliberations was released for public review and comment. The extensive documentation published by the IRG, along with the public's comments, will be a major element in DOE's strategic planning effort. The following excerpts from the IRG draft report delineate some of the major strategy and policy options that have resulted from this review:

INTERIM STRATEGIC PLANNING BASIS

It should be made clear that a number of different approaches to nuclear waste management are possible. There will be a continuing need for flexibility in planning as well as an opportunity to adjust the program to reflect the results of new developments in both the technical, social, and institutional areas of nuclear waste management. Not all decisions can or should be made now. However, a clear interim strategic planning basis must be set forth to develop near-term waste management programs, assign priorities, and plan R&D programs prior to completion of the NEPA (National Environmental Policy Act) process and selection of a strategy.

Such an approach is required to ensure that, taken as a whole, this country is moving along a course which, at its conclusion, will permit implementation of a nuclear waste management program meeting basic environmental and safety requirements in a manner which is socially acceptable, economically feasible, and consistent with general nuclear policies.

Implementation of an overall strategic planning basis will involve a series of major decision points based on environmental reviews, standards setting and licensing procedures, R&D findings, State and local decisions, and congressional actions. Successful implementation depends on satisfactorily passing such points, and failure to do so, whether because of technical limitations, policy constraints, or timing difficulties, would require modification of the strategy approach selected.

This IRG Report emphasizes the following topics:

- Proposed Objectives for Nuclear Waste Management
- Technical Findings and Conclusions
- Tentative Policy Recommendations (including issues not yet resolved by the IRG during its deliberations)
- Implementation Recommendations
- Legislative Requirements
- Work Plans

These are summarized, in turn, below.

PROPOSED OBJECTIVES

The primary objective of waste management planning is to provide assurance that:

- Existing and future nuclear waste from military and civilian activities (including discarded spent fuel from the once-through nuclear power cycle) can be isolated from the biosphere, and pose no significant threat to public health and safety

The national nuclear waste management policy, plan, and program must meet additional key subobjectives:

- The selected technical options must meet all of the relevant radiological protection criteria as well as any other applicable regulatory requirements; although zero release of radionuclides or zero risk from any such release cannot be assured, such risks should be within preestablished standards, and beyond that, be reduced to the lowest level practicable.
- The responsibility for establishing a waste management program shall not be deferred to future generations. Moreover, the system should not depend on the long-term stability or operation of social or government institutions for the security of waste isolation after disposal
- The capability to deal with a wide range of alternative situations in the future must exist. The basic elements of the program should be independent of the size of the nuclear industry and of the resolution of specific fuel-cycle or reactor-design issues of the nuclear power industry
- Appropriate cost of storage and disposal of any waste generated in the private sector should be paid for by the generator and borne by the beneficiary; budgetary and cost considerations, while important, should not dominate the design of the program or system

TECHNICAL FINDINGS AND CONCLUSIONS

Because of the need to isolate High-Level Wastes (HLW) and Transuranic (TRU) waste from the biosphere for relatively long periods of time, and because disposal in mined repositories is the nearest term option, the IRG carefully reviewed the present status of scientific and technological knowledge pertinent to mined repositories. The IRG review identified a number of important technical findings which it believes to represent the views of a majority of informed technical experts:

- A systems approach should be used to select the geologic environment, repository site, and waste form. A systems approach recognizes that, over thousands of years, the fate of radionuclides in a repository will be determined by the natural geologic environment, by the physical and chemical properties of the medium chosen for waste emplacement, by the waste form itself. These factors can and should provide multiple, and to some extent, independent, natural and engineered barriers to the release of radionuclides to the biosphere
- Overall scientific and technological knowledge is adequate to proceed with region selection and site characterization, despite the limitations in our current knowledge and modeling capability. Successful isolation of radioactive wastes from the biosphere appears technically feasible for periods of thousands of years provided that the systems view is utilized rigorously
- Detailed studies of specific, potential repository sites in different geologic environments should begin immediately. Generic studies of geologic media or risk assessment analyses of hypothetical sites, while useful for site selection, are not sufficient for some aspects of repository design or for site suitability determination
- The actinide activity in TRU wastes and HLW suggest that both waste types present problems of comparable magnitude for the very long term
- The degree of long-term isolation provided by a repository, viewed as a system, and the effects of changes in repository design, geology, climate, and human activities on the public health and safety can only be assessed through analytical modeling
- The effects of future human activity must be evaluated more carefully
- Reprocessing is not required to assure disposal of commercial spent fuel in appropriately chosen geologic environments. Moreover, current United States repository designs are and will continue

to be based on the ability to receive either solidified reprocessing waste or discarded spent fuel as a waste material. The question of whether commercial reprocessing will be initiated in the United States, while an important issue, is therefore not fundamentally related to the issue of safe waste disposal.

Because the necessary isolation periods for waste disposal are so long, no demonstration can prove the presumption of safety. Thus, a social consensus, based on scientific evidence, must be obtained through:

- Dissemination of fundamental scientific information
- The development, analysis, and near-term validation of long-term predictive models
- Extensive, independent, objective review of results by scientific experts, and of proposed facilities and operations through the licensing process
- Practical experience, including careful monitoring of the isolation systems
- A demonstrated capability to take any needed corrective or mitigating actions
- An ongoing R&D program to increase the state-of-the-art of knowledge

Only if such a social consensus is obtained, can disposal of HLW and TRU waste in geologic formations actually be implemented and the public be confident that nuclear waste can be safely isolated in this way over very long periods of time.

With respect to other waste types, technologies exist for the management and disposal of Low-Level Waste (LLW), uranium mill tailings, and for waste from Decontamination and Decommissioning (D&D). However, existing practice must be improved considerably to further reduce the potential for public hazard associated with these materials.

Most LLW, which is found in a wide variety of forms, remains radioactive for up to several hundred years. The current means of disposing of this waste is through shallow-land burial. Radionuclides in low concentrations have migrated from the burial trenches at a few LLW sites. Studies to date conclude that such migration does not pose any present significant threat to public health and safety. Monitoring of future radionuclide migration from these sites over periods of decades should be considered as a potential regulatory requirement. In the future, siting of LLW disposal facilities should give much greater attention to the hydrologic characteristics of proposed locations than has been the case in the past. NRC and DOE should take appropriate action to assure that this occurs.

Numerous technical approaches to LLW disposal have been proposed as alternatives to conventional shallow-land burial. Some of these require only a decision to use them and location of a suitable site to be implemented immediately. Others require additional technology development and perhaps demonstration before their feasibility and safety could be assured. R&D for improved methods of shallow-land burial of LLW and of alternative methods of disposal should be accelerated because shallow-land burial, as currently practiced, may not be an adequate disposal method for all LLW in the future.

Alternative Technical Strategies. An Interagency Review Group on Nuclear Waste Management has examined a number of technical strategies whose environmental impact can be described by the analyses contained in the IRG's report. The IRG has recommended the following key characteristics of a near-term interim strategic planning base for HLW disposal:

- The first disposal facilities for HLW will be mined repositories. Several geologic environments possessing a wide variety of emplacement media should be examined.
- Near-term R&D site characterization programs should be designed so that, at the earliest date feasible, sites selected for location of a repository can be chosen from among a set with a variety of potential host rock and geohydrologic characteristics. To accomplish this, R&D on several potential emplacement media and site characterization work on a variety of geologic environments should be promptly increased.

- With respect to R&D on technical options other than mined repositories, the nearer term approaches (i.e., deep ocean sediments and very deep holes) should be given funding support so that they may be adequately evaluated as potential competitors; funding for rock melting, space disposal, and transmutation would allow some feasibility and design work to proceed.
- A number of potential sites in a variety of geologic environments should be identified, and early action should be taken to reserve the option to use them if needed at an appropriate future time. Near-term actions should create the option to have at least two (and possibly three) repositories become operational within this century, ideally in different regions of the country.
- Initial emplacement of waste in at least the first repository should be planned to proceed on a technically conservative basis and permit retrievability of the wastes for some initial period of time.
- Work should proceed promptly to permit siting of one or more intermediate-scale facilities in different emplacement media and geologic environments.
- Waste disposal facilities should be sited on a regional basis insofar as technical considerations permit.

The IRG has defined four technical strategies for consideration in planning for disposal of HLW:

- *Strategy I*: Only mined repositories would be considered, and only geologic environments with salt as the emplacement media would be considered for the first several repositories.
- *Strategy II*: For the first few facilities, only mined repositories would be considered. A choice for the first repository site would be made from the types of environments that have been adequately characterized.
- *Strategy III*: For the first facility, only mined repositories would be considered. However, three-to-five geologic environments possessing a wide variety of emplacement media would be examined before the selection. When proven to be technologically sound and economically feasible, other technological options would be contenders.
- *Strategy IV*: The choice of technical option and, if appropriate, geological environment would be made only after information about a number of environments and other technical options has been obtained.

These strategies do not exhaust the possibilities for consideration but do provide a basis for strategic planning and for evaluation through the NEPA process.

Need for the Nuclear Waste Management R&D Program

DOE's waste management program has two fundamental responsibilities: the safe management of existing radioactive wastes, and the development of technology and facilities for long-term management of present and future wastes. The R&D activities described in this document are intended to satisfy the latter responsibility.

The IRG has stressed that the waste management program should analyze the entire system consisting of the waste form, the repository, and the environment of the repository. In particular, the IRG report states: "Successful isolation of radioactive wastes from the biosphere appears feasible for periods of thousands of years provided that the systems view is utilized rigorously to evaluate the suitability of sites and designs, to minimize the influence of future human activities and to select a waste form that is compatible with its host rock." The waste management R&D programs provide the supporting experimental data and analysis required to meet this objective.

The IRG has emphasized three major areas of consideration requiring extensive R&D effort: need for site-specific investigations, future human activities, and importance of modeling.

The assessment of the appropriateness of multiple natural and engineered barriers, of a host rock and its environment, of conservative engineering practices, and of any particular waste form or container requires detailed site-specific evaluations. Generic geologic studies and/or risk assessments of hypothetical sites, although useful for site selection and methodology development, do not constitute a sufficient basis for some aspects of repository design or for site suitability determinations. The natural variability of hydrogeologic, geochemical, and tectonic conditions, as well as the heterogeneity of rock mass properties, precludes the transfer of detailed earth science data from one region to another. Since site-specific investigations are needed to determine the suitability of any particular site, the importance of examining potential sites and their geologic environments cannot be overestimated. The need to gain access to potential sites is urgent to ensure the timely development of the first repository and, subsequently, a series of repositories.

Further attention will be devoted to the possible effects of future human activities on a repository or its hydrogeologic environment, and to the means available to the present generation for influencing such effects. Because it is impossible to predict or to restrict the activities of future generations, site selection guidelines, site suitability criteria, and repository design criteria must be developed to minimize potentially adverse effects on human activities.

The degree of long-term isolation provided by a repository, viewed as a system, and the effects of changes in repository design, geology, climate, and human activities are evaluated most effectively through mathematical modeling. Although additional work is needed to ensure that all potential release mechanisms are considered, to improve modeling of groundwater flow through fractured media, and to evaluate or remove other uncertainties, bounding calculations can usually be performed to place reasonable limits on the expected behavior of a repository.

Transportation

Although not a formal budget category, the transportation program cuts across and supports other programs.

The Nuclear Materials Transportation Activity was established to: (1) address the complex problems of transporting radioactive materials, (2) meet the transportation needs anticipated to satisfy the requirements of DOE facilities, and (3) support other DOE programs such as the Waste Isolation Pilot Plant (WIPP), the National Waste Terminal Storage (NWTS) Program, the Away-from-reactor (AFR) Storage Sites, the Breeder Reactor Program (BRP), international shipments, and research activities.

The Nuclear Materials Transportation Activity has the following objectives:

- Identify the needed transportation services, determine when they will be required, and take necessary steps to ensure meeting these requirements on schedule
- Integrate DOE transportation activities to deal more effectively with near-term and long-term problems
- Establish procedures for planning, reporting, reviewing and allocating resources
- Establish methods for evaluating transportation costs and benefits
- Establish a Transportation Technology Information Center
- Ensure incorporating public health and safety and environmental protection as key program objectives
- Develop and provide cost-effective operational hardware to meet DOE program schedules
- Ensure that the regulatory, societal, and other institutional factors are properly addressed so that the civilian sector can produce and operate the required transportation system.

The following management steps have been taken to ensure that the objectives will be met:

- The consolidation of DOE transportation activities has been initiated by designating the Sandia Laboratory (Albuquerque, New Mexico) as lead laboratory. As lead, Sandia holds technical management responsibility for the DOE Nuclear Materials Transportation Activity

- DOE management of this work has been decentralized to the Albuquerque Operations Office (ALOO) Transportation Program Manager who is responsible for carrying out DOE transportation program activities under DOE Headquarters policy guidance
- A Transportation Technology and Information Center has been established at Sandia Laboratories

The extent of DOE involvement in satisfying the needs of the civilian sector will depend on several factors. Firm national policies regarding the disposition of nuclear wastes and spent fuel must be established so that reliable estimates of transportation demand can be made. The subsequent response of the civilian sector to effecting the required transportation system to meet this demand will be a measure of the required involvement of DOE.

However, regardless of DOE's involvement in support of the civilian sector, DOE's responsibilities for nuclear materials transportation in the defense program, in support of DOE-funded R&D projects and research activities, and in support of possible future international commitments are unchanged. Therefore, management action to consolidate DOE transportation activity is essential.

PROGRAM STRUCTURE AND BUDGET

The programs of the Office of Waste Management (ETW) are divided into three activities:

- Defense Waste Management
- Commercial Waste Management
- Spent Fuel Storage

Figure V 2-3 shows the structure of ETW program elements within these program categories. Together, these programs are designed to meet the general objective of protecting public health and safety and the environment from the potential hazards of radioactivity, regardless of the source of such materials.

Figure V 2-4 shows ETW funding for FY 1978 through 1980.

A summary of waste management funding in the categories of operating expenses, capital equipment funds, and construction funds is presented in Figure V 2-5.

SIGNIFICANT ISSUES

The ETW program is organized to resolve the primary issues about radioactive waste management. These issues can be grouped into three general clusters:

- Technical
- Socioeconomic (including environmental)
- Institutional and political

Regardless of the type of waste, the significant issues can be grouped in these clusters. "Technical issues" are unresolved scientific questions of how natural or manmade materials or systems interact, and how to resolve such scientific questions. "Socioeconomic issues" are unresolved questions about cost (in monetary and social impact terms), who will pay, and the terms for payment or for handling the social impact. "Institutional issues" are questions about waste management organization, the political basis for programmatic and regulatory decisions, the relationships among political entities, and the authorities and responsibilities of program participants.

To date, the most succinct statement of the issues in waste mangement is found in the report of the Interagency Review Group. The previous discussion of program strategy in this chapter describes the organization and implementation of ETW programs to resolve the significant issues.

MANAGEMENT APROACH

This section is designed to cover briefly the main features of DOE/ETW staff organization and the primary management procedures which govern the ETW program.

ETW programs are conducted with an emphasis on decentralized implementation by field offices and contractors. Policy and overall budget priorities are established by ETW Headquarters, which gives appropriate direction to the field. Headquarters subsequently monitors the field office and contractor activities.

ETW coordinates its Headquarters and field activities with other Federal agencies, State and local governments, and other relevant organizations. Mechanisms are being established

DEFENSE WASTE MANAGEMENT

1.0 Interim Waste Operations

    1.1 Albuquerque Operations Office—LASL/Sandia
    1.2 Hanford Reservation—Richland, Washington
    1.3 Idaho National Engineering Laboratory (INEL)
    1.4 Nevada Test Site (NTS)
    1.5 Oak Ridge National Laboratory (ORNL)—Tennessee
    1.6 Savannah River Plant
    1.7 "Landlord" Responsibilities—Richland, Washington

2.0 Long-term Waste Management Technology

    2.1 High-level Waste (HLW)

        2.1.1 High-level Waste Technology
        2.1.2 Defense Waste Processing Facility—Savannah River

    2.2 Low-level Waste (LLW)
    2.3 TRU Waste

        2.3.1 TRU Waste Technology
        2.3.2 Waste Retrieval and Treatment Facility—Idaho

    2.4 Airborne Waste

3.0 Terminal Storage

    3.1 Waste Isolation Pilot Plant Research and Development
    3.2 Waste Isolation Pilot Plant Facility

4.0 Transportation
5.0 Decontamination and Decommissioning

COMMERCIAL WASTE MANAGEMENT

1.0 Terminal Isolation

    1.1 Office of Nuclear Waste Isolation (ONWI)
    1.2 Hanford Basalt Program
    1.3 Nevada Test Site (NTS)
    1.4 National Waste Terminal Storage (NWTS)

2.0 Waste Treatment Technology

    2.1 High-level Waste (HLW)
    2.2 Low-level Waste (LLW)
    2.3 TRU Waste
    2.4 Airborne Waste
    2.5 "Landlord" Responsibilities—Pacific Northwest Laboratory

3.0 Supporting Studies
4.0 Solidification Process Demonstration
5.0 Decontamination and Decommissioning
6.0 Low-level Waste Management Operations (Commercial Burial Sites)
7.0 West Valley Nuclear Center Activities

SPENT FUEL STORAGE

1.0 International Spent Fuel Storage
2.0 Domestic Spent Fuel Storage

    2.1 Program Management and Supporting Studies
    2.2 Away-from-reactor (AFR) Storage Facilities

Figure V-2-3. ETW program elements.

to receive input from Government organizations, industry, universities, and public interest groups in the development and implementation of waste management policy.

Business Approach

The Department of Energy has adopted a comprehensive approach toward energy technology development. Components include the following:

- Two high-level advisory groups (i.e., Energy Systems Acquisition Advisory Board (ESAAB) and the Research Council) to assist the DOE Under Secretary with major technology development decisions
- A Central Review Board established in May 1978 to help identify budget cuts so that new initiatives can be undertaken
- A concentrated DOE staff effort to assess project progress and to avoid unnecessary expenditures

Some of these components are described in more detail in the following sections.

The waste management projects favor a strong Government role in waste management rather than private industry commercialization of program projects.

Energy Systems Acquisition Advisory Board. The board serves as an advisory body to the Under Secretary and will provide supporting information and make recommendations when program or project decisions are necessary.

In addition to the Under Secretary, the board membership includes the Under Secre-

PROGRAM ELEMENT DETAIL	FY 1978 BA ACTUAL ($M)	FY 1979 BA ESTIMATE ($M)	FY 1980 BA ESTIMATE ($M)	FY 1980 BA INCREASE (DECREASE)
DEFENSE WASTE MANAGEMENT				
Interim Waste Operations				
Albuquerque Operations Office—LASL/Sandia	1.3	1.4	2.4	1.0
Hanford Reservation—Richland, Washington	87.2	60.1	62.8	2.7
Idaho National Engineering Laboratory	43.1	41.1	24.4	(16.7)
Nevada Test Site (NTS)	.4	.5	.7	.2
Oak Ridge National Laboratory—Tennessee	8.3	5.1	11.8	6.7
Savannah River Plant—South Carolina	73.2	53.4	66.2	12.8
"Landlord" Responsibilities—Richland	0	0	6.8	6.8
Subtotal	**213.5**	**161.6**	**175.1**	**13.5**
Long-term Waste Management Technology				
High-level Waste				
High-level Waste Technology	17.8	29.9	52.1	22.2
Defense Waste Processing Facility—SR	5.0	8.4	14.2	5.8
Low-level Waste	4.5	4.8	10.4	5.6
TRU Waste				
TRU Waste Technology	12.7	16.1	21.0	4.9
TRU Waste Retrieval and Treatment Facility—ID	0	.5	10.0	9.5
Airborne Waste	4.7	2.7	4.3	1.6
Subtotal	**44.7**	**62.4**	**112.0**	**49.6**
Terminal Storage				
Waste Isolation Pilot Plant (WIPP) R&D	12.0	18.3	21.2	2.9
Waste Isolation Pilot Plant Facility	22.0	10.0	55.0	45.0
Subtotal	**34.0**	**28.3**	**76.2**	**47.9**
Transportation	.4	2.6	5.0	2.4
Decontamination and Decommissioning	0	.6	2.0	1.4
Total	292.6	255.5	370.3	114.8
Personnel Resources*	1.4	1.5	1.7	.2
Grand Total	**294.0**	**257.0**	**372.0**	**115.0**
COMMERCIAL WASTE MANAGEMENT				
Terminal Isolation				
ONWI	88.3	81.3	109.9	28.7
Hanford Basalt Program	22.0	34.7	48.0	13.3
Nevada Test Site (NTS)	13.8	18.0	16.5	(1.5)
NWTS	10.0	25.0	0	(25.0)
Subtotal	**134.1**	**159.0**	**174.4**	**15.5**
Waste Treatment Technology				
High-level Waste	10.9	16.2	9.7	(6.5)
Low-level Waste	1.3	1.5	.6	(.9)
TRU Waste	6.8	3.3	3.5	.2
Airborne	2.6	2.7	2.0	(.7)
"Landlord" Responsibility—PNL	2.4	2.5	3.7	1.1
Plant Operations	0	0	2.0	2.0
Environment and Energy	0	0	0	0
Subtotal	**24.0**	**26.2**	**21.5**	**(4.8)**
Supporting Studies	4.4	4.1	1.6	(2.5)
Solidification Process Demonstration	20.4	0	0	0
Decontamination and Decommissioning	0	.1	.5	.4
Low-level Waste Mgt. Operation	0	0	0	0
West Valley Nuclear Center Activities	0	0	0	0
Total	182.90	189.5	198.0	8.5
Personnel Resources*	1.2	1.3	1.4	.1
Grand Total	**184.1**	**190.7**	**199.4**	**8.7**
SPENT FUEL STORAGE				
International Spent Fuel	5.0	3.0	4.2	1.2
Domestic Spent Fuel				
Program Mgt. and Supp. Studies	0	3.0	15.8	12.8
AFR Storage Facilities	0	5.0	0	(5.0)
Subtotal	**0**	**8.0**	**15.8**	**7.8**
Total	5.0	11.0	20.0	9.0
Personnel Resources*	.3	.4	.5	.1
Grand Total	**5.3**	**11.4**	**20.5**	**9.1**

*"Personnel Resources" covers Headquarters personnel expenses.

Figure V-2-4. ETW budget authority for FY 1979-1980.

FUNDING SUMMARY

EXPENDITURE CATEGORIES	(DOLLARS IN MILLIONS)													
	FY 1978		FY 1979		FY 1980		FY 1981		FY 1982		FY 1983		FY 1984	
	B/A	B/O	B/A	B/O	B/A	B/O	B/A	B/O	B/A	B/O	B/A	B/O	B/A	B/O
OPERATING EXPENSES*	291.5		328.5		450.6		480.3		505.0		518.3		515.8	
CAPITAL EQUIPMENT	19.8		19.1		25.8		28.2		30.7		31.5		32.0	
CONSTRUCTION	172.1		111.5		115.5		227.8		238.8		183.3		183.3	
TOTAL	483.4		459.1		591.9		736.3		774.5		733.1		731.1	

*Includes personnel resources, i.e., FY 78 = $2.9 M; FY 79 = $3.2 M; FY 80 = $3.6 M.

Figure V-2-5. ETW program funding summary (budget authority).

tary's two deputies; the Assistant Secretaries for Energy Technology, Resource Applications, Environment, Defense Programs, Policy & Evaluation, Conservation & Solar Applications; the General Counsel; the Controller; and the Director of the Office of Energy Research. Other Assistant Secretaries can participate in meetings if programs in their areas are to be discussed.

Research Council. The council, a standing committee chaired by the Director of DOE's Office of Energy Research,, provides the Under Secretary with information on all research-related areas by the following means:

- Identifying high-priority research areas
- Recommending allocation of DOE resources for research
- Recommending splits in the in-house vs. out-of-house research efforts
- Recommending the placement of new research facilities
- Assessing current research efforts
- Helping to shape the annual DOE research budget

Multiphase Contracts. Contracts are written in four general phases: project definition, design and development, fabrication and construction, and operation/production. These phases allow DOE to manage its contracts better and provide a mechanism to terminate ill-defined or uneconomic projects before too much money is expended.

State-of-the-art Assessment. DOE program managers are required to prepare periodic reports which detail how well selected programs have met their primary objectives. Problems, such as technology gaps and environmental difficulties, are also to be identified so that the Acquisitions Board can discuss any changes necessary for the program to proceed.

Business Strategy. At the earliest practical time in the development of a project, an analysis of the desired approach is prepared in order to commercialize the program, where commercialization is desirable. The analysis includes the following:

- Methods of obtaining and sustaining competition
- Selection of the type of contract or financial assistance best suited to each phase of the project
- Need for developing contractor incentives
- Identification of economic risks and their effects on the program

At present, the unique requirements for Federal Government operation of waste management make this strategy less applicable to ETW programs than to other ET programs.

Program Management—Headquarters

Under the direction of the Under Secretary, DOE is currently developing and testing a set of comprehensive program management guidelines. The guidelines, formally titled the "Program and Project Management System for DOE Programs," are described in detail in the Under Secretary's Memo to DOE management on May 31, 1978, and have been implemented by ASET through Management Directive ET-009. The system is composed of (1) objectives governing the system, (2) major concepts for program framework, (3) basic system elements, (4) system implementation schedules, and (5) definitions of key concepts.

Objectives of the management system include the following:

- Ensure that all major programs are developed with clear operational time-phased objectives (multiyear to relate to budget process)
- Ensure that all significant projects and activities have clear, relevant subobjectives
- Relate program/project and activity objectives to NEA, NEP, and other plans and legislation, where possible
- Provide for determining priorities among programs and projects, and relate these priorities to various levels of resource availability
- Avoid commitment of major resources prior to adequate project definition
- Provide key program and project managers with a clear view of related program and project objectives and plans
- Provide visibility on all key decisions and timely feedback for all levels of management
- Maintain accountability through all levels of the organization with a minimal amount of procedure and paperwork

Basic system elements include the following:

- Policy and fiscal guidance
- Program plans
- Energy systems acquisition project plans
- Energy Systems Acquisition Advisory Board
- DOE Research Council
- Financial management
- Procurement and contracts management
- Program/project review and management reporting
- Market Development Planning and Management System
- Laboratory Work Authorization System
- Institutional planning process for DOE laboratories
- Program and project management training
- Review of outlay programs—FY 1980 budget
- Criteria for selecting program/project approval and review
- Basic inventory of ongoing programs/projects, including construction
- Review of technology base

The organization of the Office of Nuclear Waste Management and the functions associated with each organizational element are shown in Figure V 2-6.

Following the ASET policy of decentralized management, all ETW projects are managed in the field. Formal management agreements are being established between ETW and specific field operations offices. These agreements set the arrangements for execution of ETW programs. They cover purpose, program direction, organization, responsibilities of key officials and participants, and the authorities of the participants. An example is the agreement with the Idaho Operations Office for management of the Airborne Waste Management Program.

More specifically, the roles and responsibilities at Headquarters, operations offices, and contractors are as follows:

Office of Waste Management (Headquarters ETW)

- Within the policy guidance provided by ASET, interprets policies and develops strategic and program plans encompassing scientific and technical activity content, goals, and objectives (including environmental and safety requirements)
- Within ASET, interacts, as appropriate, with DOE staff organizations and other Federal agencies

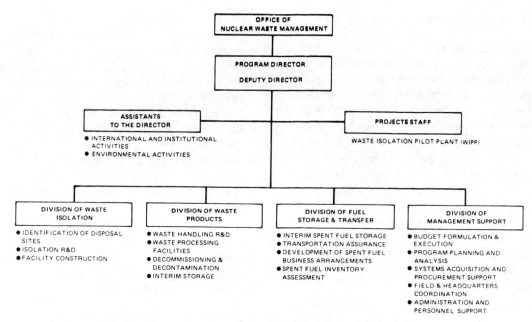

Figure V-2-6. Organization chart of the office of nuclear waste management.

- Develops and defines budget requirements and defends budget requests to the Office of Management and Budget (OMB) and congressional committees as appropriate
- Provides general policy guidance and technical direction to field offices and contractor organizations
- Monitors technical and programmatic progress, initiates programmatic reviews, evaluates results and modifies program direction, as appropriate (includes both research operations and major device fabrication funding activities), within the limits allowed by ASET

Operations Offices
- Manage projects with full authority in accord with management plan approved by Headquarters
- Provide day-to-day management on projects with periodic feedback to ETW
- Perform contract administration activities and other activities incidental to project management
- Direct contractors in financial and technical matters

Contractors (National Laboratories, Universities, and Industry)
- Initiate proposals, or respond to requests for proposals
- Assist in the development of strategic activity plans
- Execute the technical program within the guidance provided by DOE

Project Management—Field Activities

ASET seeks to decentralize its program activities. Field offices and their contractors are responsible for program and project execution. In waste management, the analyses, developmental work, operations, and project management are performed by national laboratories and other contractors who report to DOE field operations offices. In addition, decentralization involves assigning major program integration functions to DOE field organizations and their lead contractors. An example of such decentralization drawn from the Commercial Waste Management Activity is the program management of the HLW repository development effort. The reporting relationships are shown in Figure V 2-7.

More detail of the organization and operation of field activities is provided in Chapter 4.

PUBLIC PARTICIPATION

The Presidential directive to the Interagency Nuclear Waste Management Task Force, dated March 13, 1978, states that "the deliberations of the Task Force should include opportunity for appropriate participation by the interested public, industry, States, and Members of Congress."

Under the Atomic Energy Commission, the public participatory aspects of the Radioactive Waste Management Program were often handled at the general departmental level. Technical aspects of the program were the focus of waste management staff. More recently, as public concern over waste management has grown, public participation has become an integral part of the ETW program. DOE recognizes that the development of an acceptable waste repository, a Spent Fuel Storage Activity and a program to manage existing waste inventories requires direct and frequent interaction with the following parties:

- State and local governments
- Public interest groups (national, State, local community, environmental, consumer, and other representatives)
- Industrial representatives
- Nongovernment scientific experts
- Professional associations
- General public

Opportunities for interaction by these groups include the following:

- Congressional-DOE budget cycle hearings
- NEPA process public forums (e.g., EIS public hearings)
- NRC licensing hearings (for eventual specific waste disposal facilities)
- National Academy of Sciences/National Research Council—Committee on Radioactive Waste Management
- Advisory committees
- Technical meetings on waste management technology (e.g., the meetings sponsored by universities, professional societies, and governmental organizations
- Local public information meetings sponsored by DOE field offices

Additional mechanisms for public input are being developed with the assistance of regional DOE representatives.

Need to Improve Public Participation

The October 1978 draft summary of the IRG's report to the President states the following:

> The resolution of institutional issues required to permit the orderly development and effective implementation of a nuclear waste management program is equally important as the resolution of outstanding technical issues and problems. The solution of institutional problems involves difficult implementation issues and will require Federal action.

Because public participation will increase in the future, the current programs are likely to expand in the near future. This chapter describes current DOE/ETW programs as well as new activities suggested in recent reviews of the Nuclear Waste Management Program.

Current Mechanisms for Public Participation in the DOE Nuclear Waste Management Program

Several mechanisms for public review of the DOE Nuclear Waste Management Program

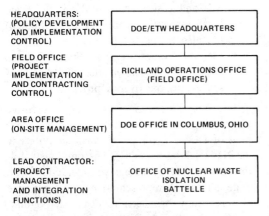

Figure V-2-7. Example of reporting relationships between headquarters, field offices, and lead contractors.

exist under current legislation and governmental operating practices.

Congressional Hearings. Congress annually reviews the DOE nuclear waste management budget and periodically holds oversight hearings on this program. Public testimony on this process can result in redirection and re-emphasis of DOE programs by Congress.

Federal and State Regulatory Processes. Many projects included in the Nuclear Waste Management Program are subject to review by other agencies. NRC will be required to license most waste disposal facilities. EPA will set the applicable general environmental standards required for these facilities. State and local agencies may also review certain aspects of facilities proposed within that State. All of these regulatory organizations may offer opportunities for participation on issues relevant to nuclear waste management.

NEPA Process. The NEPA process provides for direct public review of major projects of the sponsoring agency. This remains a principal mechanism for public input on major projects proposed in this program.

DOE-sponsored Public Hearings and Direct Communication with the DOE Program. In 1976, as part of the NWTS Program, ERDA originally initiated direct contact with 36 States in areas where potential waste disposal sites were being considered. Open cooperation on site selection criteria and procedures was offered to these States, and subsequently information developed by ERDA and later by DOE was made available to the States. Contractor-developed information on geological characteristics and surface characteristics of the study areas is shared with the States, so that the data base upon which decisions will be made is common knowledge. Project-specific public meetings have been held with regard to the proposed WIPP facility in New Mexicio, the future of the West Valley facility in New York, and site exploration activities in the Gulf Coast region.

Future Expansion for DOE Public Participation Programs

Executive Planning Council. The draft IRG report dated October 1978 makes the following statements:

The IRG recommends that the President establish, by Executive Order, an Executive Planning Council. The Executive Planning Council would consist of selected governors, selected Indian nation representatives, officials of national organizations of State and local government, and representatives of DOE and other Federal agencies.

The purpose of the Executive Planning Council is to:

- Identify joint Federal/State planning activities in nuclear waste management
- identify and agree on the appropriate existing or new mechanisms and timetables for carrying out such joint activities
- Develop mutually acceptable criteria for evaluating proposed nuclear waste management activities (from R&D and geological characterization through the possible siting of disposal facilities)
- Develop regional waste disposal facility siting plans and support State "consultation and concurrence" activities
- Support design, preparation, and evaluation of environmental impact statements covering waste management activities
- Develop a mechanism for planning which will effectively represent all interests and concerns not only at the State level (executive and legislative branch) but also at the local level
- Identify other Federal/State actions needed to maximize the likelihood of success of the overall program

The recommendation for the Executive Planning Council will greatly assist the planning of DOE's commercial waste activities in the NWTS Program. DOE has previously followed a policy of extensive interaction with State and local representatives. The purpose of this policy has been to establish communication between the Federal Government and State/local representatives on the issues and concerns raised about waste management activities and plans. Adoption of IRG proposals would accentuate this interaction and formalize the importance of the State and local governments in these activities.

Public Participation. The October 1978 draft report by the IRG states the following:

In addition to their interaction with State and local government entities (sic), the Federal agencies responsible for developing a nuclear waste management plan and program need to interact with the public. The IRG's own experience with public participation and the recommendations of many citizens appearing before the IRG indicate the urgent need for sustained, effective efforts to inform the public and to provide opportunities for discussion between the public and the government.

To generate a spirit of openness on the part of the government and of full participation on the part of the public, IRG recommends that, at a minimum, the President encourage, direct and/or coordinate a program to:

- Routinely update the status of scientific and technical knowledge on nuclear waste management and provide this information to the public-at-large in understandable terms.
- Increase interaction and discussions between Federal program managers and nationally or locally based institutions and organizations desiring such interaction and discussion.
- Support private sector efforts to generate a greater degree of social and technical understanding and agreement on nuclear waste management issues.

The IRG experience indicates that the DOE Nuclear Waste Management Program has to expand the opportunities for public participation in all of its activities.

3.
Subactivity Descriptions

This section describes each of the activity and subactivity budget items appearing in the Budget and Reporting Structure of the waste management portion of the DOE budget. Contents for each item include: name, brief description of the category, status of the program, a pictorial entry relevant to the item, and a chart identifying milestones and funding for the program.

DEFENSE WASTE MANAGEMENT ACTIVITY

Since 1943, radioactive wastes have been generated as a by-product of the national defense programs to produce and use nuclear materials in weapons and other national defense applications. The principal source of radioactive waste is the reprocessing of fuels irradiated in nuclear production reactors to recover plutonium, uranium, and other products. Figure V 3-1 identifies the role of fuel reprocessing and the primary products and wastes produced. Other major sources of wastes are the manufacturing operations to produce plutonium components for weapons, laboratory operations, and reactor operations.

In this program summary handbook radioactive wastes are classified as high-level waste (HLW), low-level waste (LLW), transuranic (TRU)—contaminated waste, and airborne waste. HLW is generated in the first phase of fuel reprocessing and contains practically all of the fission products, the small fraction of uranium and plutonium that is not recovered, and the bulk of the other transuranics present in the irradiated fuel. High-level waste is generated as a liquid waste in defense program fuel reprocessing operations which are carried out at Savannah River (SR), South Carolina; Hanford Reservation, Richland (RL), Washington*; and the Idaho National Engineering Laboratory (INEL).

All other nongaseous radioactive wastes are classified in this summary document as LLW or as TRU if significantly contaminated with TRU elements. LLW and TRU are solid wastes. Liquids are either converted to solids, decontaminated, and released under controlled conditions, or concentrated and managed as HLW. Intermediate-level liquid wastes, generated at Oak Ridge National Laboratory (ORNL), are disposed of by shale fracturing. Gases are treated by techniques, such as filtration and scrubbing, and then released under controlled conditions.

Typical low-level solid waste consists of discarded contaminated equipment, filters from the cleanup of gaseous wastes, ion-exchange resins from the cleanup of liquid wastes, liquid waste converted to solid form by incorporation in concrete, and miscellaneous contaminated trash such as paper, rags, glassware, and protective clothing. Table V 3-1 shows the present inventories of defense wastes and projections to 1985, when long-term disposition operations are expected to begin.

The general objective of the Defense Waste Management Activity is to confine radioactive wastes so that safety, environmental, and public health standards are met and maintained.

---
*The Richland reprocessing facilities are not currently being operated but are in a standby condition.

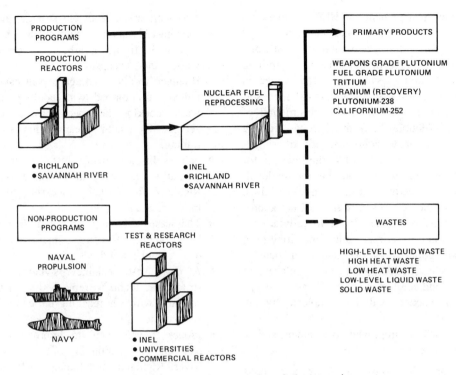

Figure V-3-1. Nuclear waste from fuel reprocessing.

In the near term, the specific objective is to provide the required confinement of radioactive waste in a way that is compatible with potential long-term management practices. The long-term objective is to continue to provide the required confinement of radioactive wastes (1) with minimum reliance on future maintenance and surveillance by man, and (2) in a way that ensures a high degree of isolation from man's environment while the waste is a potential radiation hazard.

The strategy of the Defense Waste Management Activity is structured to meet the needs for the safe management of the existing substantial inventories of radioactive defense wastes as well as wastes that may be generated in the future. The defense waste and commercial waste activities are closely coordinated by DOE/ET Headquarters.

Interim Waste Operations

The objective of this subactivity is to provide interim storage of waste in a way that is compatible with alternatives under consideration for long-term management and that minimizes the potential for release to the environment. Activities for waste management operations have been structured for the following waste categories:

- Interim management of HLW
- Management of low-level solid wastes
- Management of TRU-contaminated solid wastes
- Management of ORNL intermediate-level liquid wastes
- Management of airborne wastes

Table V-3-1. Defense Waste Inventory.

DEFENSE WASTE	VOLUME (10^6 FT^3)	
	CURRENT	PROJECTED TO 1985
HLW		
Hanford Reservation	6.1	6.2
Savannah River	2.9	2.6
Idaho	0.4	0.3
TRU Waste		
Buried	13.0	13.0
Retrievably Stored	1.7	3.76–35.95
LLW	51.0	75.06–102.25

Interim management of HLW requires (1) operation, surveillance, and maintenance of facilities for volume reduction of waste and for interim storage of the existing 73 million gallons of liquids, salt cakes, and sludge in underground tanks and granular calcined solids in underground bins; (2) construction of improved (double-shell) underground storage tanks to eliminate continued use of single-walled and other tanks of earlier design for storage of liquid waste; and (3) process development and facility construction to continue converting liquids to solids and making appropriate transfers to new tanks to provide improved interim storage until the long-term disposal option is selected and implemented. Facility construction for the tank replacement program and implementation of improved operational practices will be completed by the end of 1987.

The following improved containment activities are underway:

- Double-shell waste tanks are being constructed to replace old waste tanks and to provide additional storage capacity at Savannah River and Richland.
- Existing waste tanks are being evaluated to determine their acceptability for continued use at Savannah River and Richland.
- Improvements are being made to existing interim storage facilities at Savannah River, Richland, and Idaho.
- Stainless steel underground bins are being constructed for additional storage of calcined waste at Idaho.

Waste processing and solidification activities are as follows:

- Excess water is being evaporated from waste at Savannah River and Richland to reduce the volume requiring storage and to convert liquid to immobile salt cake to reduce the potential for releases to the environment; the waste backlog at Richland has been evaporated.
- New waste evaporation facilities are being constructed at Savannah River to replace existing facilities.
- Improved technology is being developed and demonstrated at Savannah River for removing waste from old storage tanks.
- Cesium-137 and strontium-90 are being separated from high-heat, generating waste and encapsulated separately at Richland to allow storage of salt cake in tanks not equipped with cooling coils; waste backlog has been processed to separate Cs and Sr; encapsulation is continuing.
- Salt wells are being installed in the crystalline salt cake in storage tanks at Richland for further processing to remove residual liquid.
- The waste calcining facility at Idaho is being operated to convert liquid waste to dry, granular solids.
- A new waste calciner is being constructed at Idaho to replace the existing calciner, which is reaching the end of its serviceable life.
- Process development to improve waste handling is continuing at Savannah River, Richland, and Idaho.

Low-level solid wastes are disposed of by shallow land burial. Continued reliance on disposal depends on development of an improved technology base, generation of acceptable and comprehensive operating criteria, and development of stabilization techniques for sites no longer needed to ensure that disposal areas will remain safe over the long term, with minimal reliance on continuing maintenance and surveillance. Upgrading and improvement actions are taken based on periodic site assessments in comparison with applicable criteria and standards. Stabilization of inactive disposal areas and compliance with criteria and standards will be completed in FY 1987. Alternate disposal technologies will be developed and assessed for potential increases in low-level waste quantities, particularly rubble and soil from decommissioning actions and other high-volume wastes containing very low concentrations of radioactivity.

The management of TRU-contaminated solid waste includes retrievable storage of currently generated waste (pending ultimate disposition) and development of techniques, construction of facilities, and operations to remove the wastes from storage and treat them, if necessary, to meet disposal acceptance criteria. Scheduling will be consistent with early shipment of wastes from Idaho to a repository (in FY 1986). Techniques and requirements for recovery and treatment of previously buried TRU wastes at Idaho are being developed pending a decision in FY 1981 on whether or not to recover these buried wastes.

Shale fracture is used for disposal of intermediate-level liquid waste at ORNL. A new facility and operational improvements are being implemented. In shale fracture the waste is mixed with grout and injected into a deep underground shale bed where it solidifies as thin grout sheets and fixes the wastes in place.

Regulations will require that improved handling and storage/disposal of "airborne" wastes (recovered from gaseous effluent systems) be implemented. The timing and degree of recovery depend on release limitations established by the EPA. (Not all of the limitations have been published.) DOE plans to provide engineered storage pending decisions on ultimate disposal techniques, with initial focus on krypton-85 and iodine-129.

Defense waste management operations are conducted on a site-by-site basis with overall program coordination provided by Headquarters.

Albuquerque Operations Office-LASL/Sandia. The Albuquerque Operations Office(ALOO) has management responsibility for nine weapons laboratories and facilities. Only two facilities—Los Alamos Scientific Laboratory (LASL) and Sandia Laboratories (SL)—have direct DOE/ETW funding for radioactive waste management. Both LASL and SL are involved in research and development; however, neither laboratory stores or handles HLW. TRU-contaminated wastes are either shipped off site or stored retrievably. TRU wastes disposed of prior to 1970 are buried in shallow landfills. LLW disposal is accomplished by burial following solidification of any liquid waste, encapsulation, dewatering, and ion-exchange processes.

The Control Waste Collection System at LASL is shown in Figure V 3-2.

Although neither LASL nor SL handles HLW, significant amounts of TRU and LLW

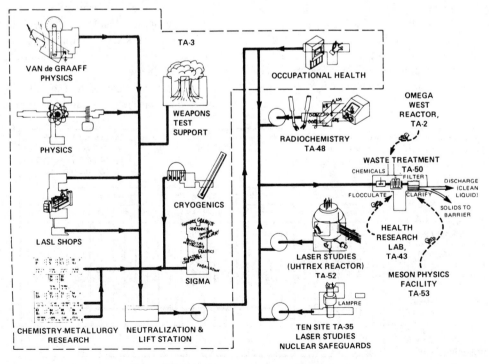

Figure V-3-2. LASL control waste collection system.

are handled at the sites. Current activities for waste management include the following:

- Improved volume-reduction techniques
- Upgrading waste treatment facilities to minimize releases
- Decontamination and decommissioning excess facilities
- Removal of abandoned, contaminated waste collection lines

LASL is directly involved in the TRU waste volume-reduction RD&D activities. A new TRU waste volume-reduction facility began test operations in FY 1979, and full-scale operations are scheduled to start in FY 1980.

Major milestones and funding are shown in Figure V 3-3.

Hanford Reservation—Richland, Washington. Since 1944, the Hanford Project, located near Richland, Washington, has been producing special nuclear materials (primarily plutonium) for defense and research. High-level radioactive waste has been and will continue to be accumulated as part of the process of producing plutonium.

HLW is initially stored in underground carbon steel-lined concrete tanks as alkaline liquids together with precipitated sludge layers. Most of the long-lived (~ 30-year half-life, high-heat-emitting isotopes, strontium-90 (^{90}Sr) and cesium-137 (^{137}Cs), are removed and doubly encapsulated (Figure V 3-4). After the heat generated by radioactive decay of the short-lived fission products (those with half-lives of approximately 2.5 years or less) has diminished appreciably, water is evaporated from the liquid waste to form damp salt cake to reduce volume and mobility. The existing storage facilities for HLW consist of large underground tanks for salt cake, sludge, and liquids, and stainless steel-lined concrete basins filled with water for the encapsulated ^{90}Sr and ^{137}Cs.

The liquid is being processed to less mobile solids, and residual liquids are placed in double-shell tanks (some are built, and others are under construction or planned) for safe storage for an interim period. Alternatives for long-term management of HLW are being evaluated. Interests of present and future generations must be considered. Many approaches to long-term waste management have been identified and studied.

Strontium and cesium capsules are pictured in Figure V 3-4.

The current waste management program is scheduled to be completed in the early 1980s, at which time the HLW will be in three forms and stored in the existing facilities. The damp salt cake and precipitated sludge will be in the single-shell tanks; the nonevaporable residual liquor will be in the double-shell tanks; and most of the high-heat-generating ^{90}Sr and ^{137}Cs will be sealed in double-wall metal capsules stored in the water basins.

The stabilization and isolation of the 149 old single-shell waste storage tanks is continu-

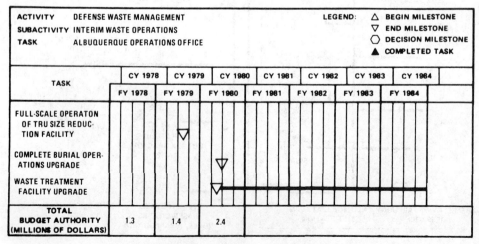

Figure V-3-3. Milestones and funding for ALOO.

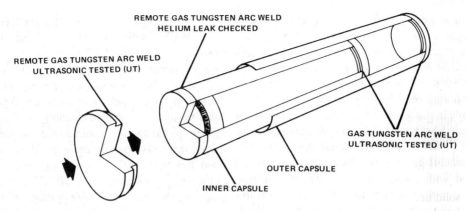

Figure V-3-4. Strontium and cesium capsules.

ing. Drainable interstitial liquid waste is being removed from the crystalline salt cakes and sludge which will remain in the single-shell tanks for the interim period. Fill-lines and other connections into the older tanks are being disconnected and sealed. These efforts are directed toward providing safe storage of the Hanford HLW until the long-term waste management program is implemented.

Major milestones and funding are shown in Figure V 3-5.

Idaho National Engineering Laboratory. The Idaho National Engineering Laboratory (INEL) was established in 1949 to construct, test, and operate nuclear reactors, and to support plants and equipment. Since 1949, more than 50 nuclear facilities, mostly reactors, have been built, or are under construction or design at INEL. These facilities include the Idaho Chemical Processing Plant (ICPP), which reprocesses irradiated spent enriched nuclear fuel.

Radioactive high-level liquid waste (HLLW) from the reprocessing of spent nuclear fuel elements has been converted to a dry granular solid by calcining and stored safely in underground stainless-steel bins at ICPP since 1963. The HLLW produced at the ICPP primarily originates from reprocessing fuel elements

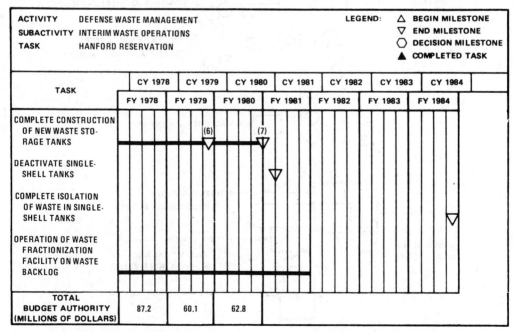

Figure V-3-5. Milestones and funding for Hanford Reservation.

from the national defense and reactor testing programs; however, some waste also is generated from reprocessing spent fuels from non-defense research reactors. LLW and TRU-contaminated wastes also are disposed of or stored in burial grounds at the site.

Waste streams requiring treatment or monitoring are HLLW from the extraction system, intermediate-level liquid waste (ILLW), contaminated organic wastes, dissolver and vessel off-gases, and solid wastes contaminated with radioactivity. All HLLW streams are solidified; ILLW is converted to HLLW and low-level liquid wastes, which are decontaminated prior to disposal. Contaminated organic liquids are steam-stripped. All dissolver and vessel off-gases are filtered to remove particulates; selected off-gases can be processed through the gas plant to recover krypton and xenon; the remainder is released to the atmosphere via the stack.

Figure V 3-6 shows a plot plan of the ICPP waste management facilities.

A New Waste Calcining Facility (NWCF) will replace the existing WCF, originally built as a demonstration plant, which is nearing the end of its production life. The NWCF, like the WCF, will convert high-level liquid (acidic) waste from ICPP spent fuel reprocessing operations to a dry granular solid form.

Specifically designed for remote maintenance to reduce personnel radiation exposure and to improve plant availability (on steam-time) by reducing maintenance turnaround time, the NWCF will process all HLW identified for the next 20 years of forecasted fuel-processing operations. The NWCF has a design net throughput of 3,000 gallons of liquid per day. The construction project completion date was October 1979.

Other activities at Idaho include the following:

- Continued solidification of HLLW to the calcine form
- Storage of TRU waste in a retrievable mode, and disposal of low-level solid waste by burial
- Development of improved waste handling processes and techniques

Major milestones and funding are shown in Figure V 3-7.

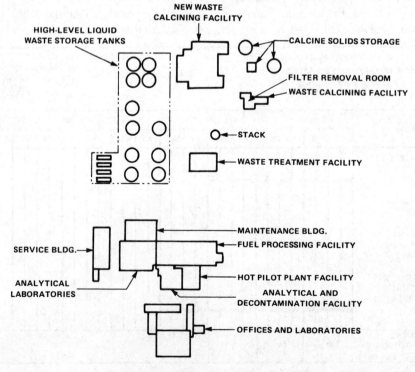

Figure V-3-6. ICPP waste management facilities.

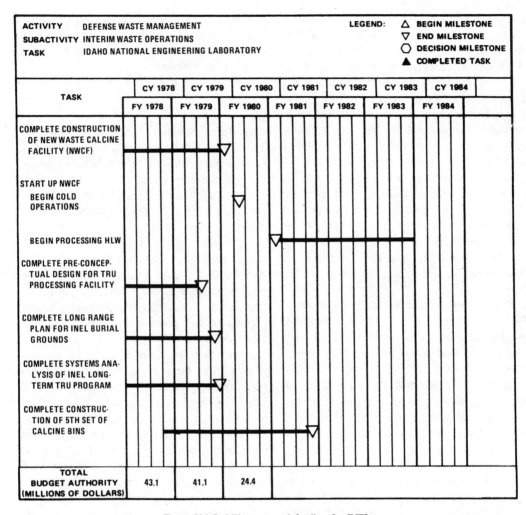

Figure V-3-7. Milestones and funding for INEL.

Nevada Test Site (NTS). The Nevada Test Site located near Mercury, Nevada, is a defense research and development facility. Although most of the radioactivity at the NTS remains at the locations deep underground where nuclear tests have been conducted, there are a variety of sources which contribute to radioactive waste on the surface. These include the following:

- Radioactivity and activated materials remaining from early atmospheric tests
- Waste generated from equipment decontamination
- Radioactive waste generated by and remaining from experiments and projects conducted at NTS not related to underground nuclear testing
- Radioactive materials in tailing piles and debris from post-shot tunnel reentry and rehabilitation activities
- Radioactive materials and wastes generated elsewhere by defense programs and related activities sent to the NTS for storage or disposal

Low-level liquid waste generated at the DOE Experimental Farm is collected in a holding tank and then transported to Area 8 (see Figure V 3-8) where it is pumped more than 1,000 feet down into the underground cavity created by an expended underground nuclear explosion.

A waste treatment facility is located in NTS Area 6. There, heavy construction equipment, critical tolerance instruments, and work clothing are processed to reduce the level of radioactive contamination. Runoff liquids from the laundry and the decontamination facilities drain into a settling pond. Contaminated waters from post-shot drill-back operations and from tunnel drainage are retained in other ponds in the operating areas. Because evaporation is continuous in this low-humidity, high-temperature environment, the residual radioactivity is concentrated in these ponds. Ultimately, the dried-up evaporation ponds are covered with the appropriate depth of clean dirt, posted, and protected as are other LLW management areas. Dry waste compactors are also located in the Area 6 laundry facility, at the Area 25 EMAD building, and within the Radioactive Materials Control Facility in Mercury.

Four active radioactive waste management sites (RWMS's) at NTS have been designated as disposal sites. In the Area 5 TWMS, deep (~20 feet) trenches and pits are used. Sealed drums and boxes containing soil, salt, tools, equipment, and assorted hardware are accumulated in these trenches and are ultimately covered with at least one meter of earth. In Area 3, the U3ax subsidence crater is used to accumulate bulky, unpackaged low-level radioactive debris. (See Figure V 3-8).

A new radioactive waste management plan is currently being implemented at NTS. The plan is based on the expansion of the waste management facility in Area 5 to a size of approimately 90 acres. The concept underlying the new radioactivity management plan is to collect contaminated debris from other locations on the NTS and consolidate this debris in the New Area 5 complex, the U3ax subsidence crater, and the lined drill hole at the U3fi site (see Figure V 3-8).

In the future, the storage and disposal of wastes generated from defense programs will continue to be managed at the Area 5 facility. The Nevada Operations Office has recently authorized construction of a 2,400-square-foot radioactive waste processing building at Area 5 RWMS. The facility will compact TRU-contaminated waste, permit consolidation and packaging of other LLW, and contain a record and documentation office. A decontamination capability will be available for both packages and people. The Area 5 facility will also permit evaporation and solidification of liquid waste and provide facilities and electric power for a waste incinerator.

Milestones and funding for NTS are shown in Figure V 3-9.

Oak Ridge National Laboratory (ORNL)—Tennessee. ORNL is primarily a research, development, and test facility. Routine operation of the test reactors and other nuclear facilities produces low-level waste, TRU-contaminated waste, and intermediate-level waste.

As defined by ORNL, "intermediate-level liquid waste" is aqueous waste produced in their R&D activities. Intermediate-level liquid wastes produced at other sites are either concentrated and managed as high-level waste or decontaminated and the residues managed as solid low-level waste.

Shale fracturing, with injection of a liquid waste/cement grout under high pressure, has been used to dispose of this waste in shale formations underlying the ORNL site. It is both the most economical and the lowest risk alternative for achieving the long-term waste management goals. This method of disposal will continue unless the program is redirected toward alternatives for long-term disposal.

One or two injections of intermediate-level waste/cement grout take place per year. This method of disposal has been evaluated for environmental consequences, and an EIS has been issued for management of ORNL intermediate-level waste. LLW is disposed of by shallow land burial. TRU-contaminated waste has been retrievably stored since 1970.

Figure V 3-10 shows the ORNL shale fracture facility.

The present shale fracture disposal facility is an old demonstration facility and is approaching the end of its useful life. Construction of a new shale fracture disposal facility will begin in FY 1979. The facility will be able to handle the waste generated for at least the next 15 years, as well as the sludge that has been accumulated in the interim storage tanks. A second waste evaporator, stainless steel storage tanks, and double containment transfer lines are being installed at ORNL at collect, reduce the volume, and store the intermediate-level wastes pending disposal by shale fracture. These items are replacements for existing handling systems that have exceeded their design lives. After the new shale fracture facility becomes available and liquid waste has been disposed of, the sludges will be disposed of and the old facility an storage tank will be decommissioned. Experiments are underway to determine the composition of the sludges in the waste tanks, methods to remove them, and the best means to inject them using the shale fracture process.

Major milestones and funding are given in Figure V 3-11.

Savannah River Plant (SRFP). The Savannah River Plant (Figure V 3-12) was established in 1950 by the United States Atomic Energy Commission to produce nuclear materials for the national defense. The production of these nuclear materials resulted in radioactive waste by-products. At present, SRP has stored approximately 16 million gallons of radioactive

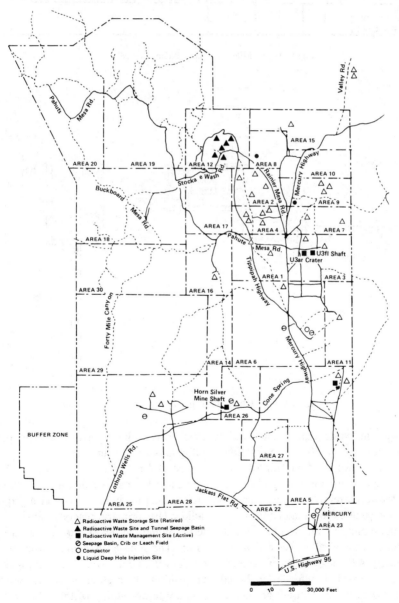

Figure V-3-8. NTS radioactive waste area locations.

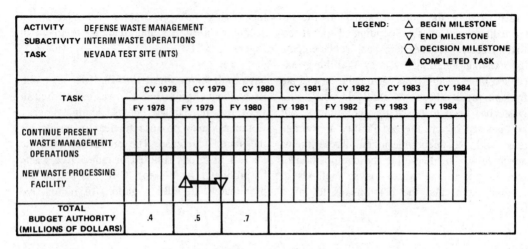

Figure V-3-9. Milestones and funding for NTS.

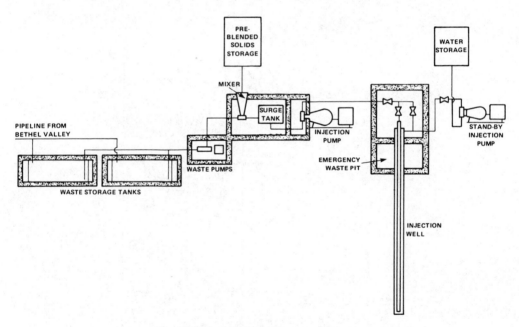

Figure V-3-10. ORNL shale fracture facility.

high-level waste in waste tanks. This waste consists of approximately 10 million gallons of high-level liquid waste, four million gallons of salt, and two million gallons of sludge. This waste currently contains approximately 660 megacuries of radioactivity. In addition to the liquid waste, a total of 270 thousand cubic meters of solid waste containing approximately 4.3 megacuries of radioactivity has been buried in the SRP burial ground. Also stored in the burial ground is 1,500 cubic meters of retrievable TRU low-level waste that contains approximately 52 kilograms of transuranics.

The Defense Waste Management Activty at Savannah River is subdivided into three major areas: interim liquid, long-term liquid, and solid. The primary responsibility of the interim liquid waste efforts is the waste tank farm. The long-term liquid waste program is concerned mainly with the solidification and disposition of high-level liquid waste. The primary responsibility of the solid waste activity is the burial ground.

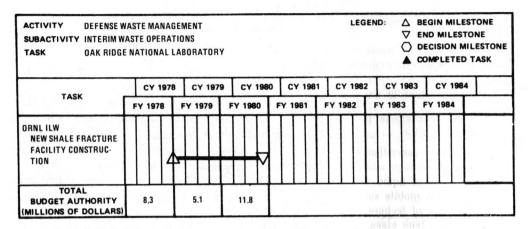

Figure V-3-11. Milestones and funding for ORNL.

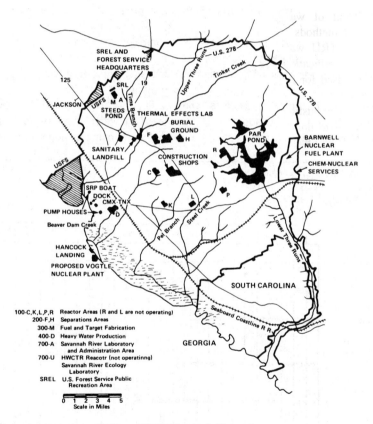

Figure V-3-12. Savannah River plant.

At Savannah River, HLW will be transferred from old obsolete tanks into new, improved double-shell waste tanks which have recently been built or are currently under construction. The objective of the program is to improve the HLW containment at SRP. In FY 1978, a demonstration to clean out one of the older tanks was initiated, and it will continue in FY 1979. In FY 1980, operations to transfer waste salt cake from old single-shell and double-shell tanks into new double-shell tanks will begin. These operations involve dissolving of the salt cake in the old tanks, transferring the liquid to waste evaporators, and recrystallizing of the waste into new double-shell tanks. The salt removal operation will continue through FY 1981 and will be followed by operations to transfer sludge waste to new tanks.

The current programs for waste management are as follows:

- Storage and surveillance of HLW in underground storage tanks
- Conversion of liquid wastes by evaporation to less mobile salt cake
- Refinement of technology for sludge removal and tank cleanout
- Replacement of 23 old tanks with 18 new tanks
- Improvement of waste monitoring and land burial methods
- Storage of TRU waste in a retrievable mode, and verification of the integrity of containers used for retrievable storage of TRU waste
- Storage of combustible liquid waste, and converting it to non-TRU solid waste by incineration or other means
- Completion of process and equipment assessments for retrieval, conversion, and long-term storage of TRU waste

Major milestones and funding are depicted in Figure V 3-13.

"Landlord" Responsibilities—Richland, Washington. The defense nuclear facility near Richland, Washington, is located on a 570-square-mile reservation. The operation and maintenance of this facility present a very large "landlord" responsibility which is not directly attributable to any particular waste facility or program at the site. Rather, landlord responsibilities provide direct support to the entire site operation and all waste facilities and programs at the site in general. Figure V 3-14 shows the Hanford Reservation and lists some of the major landlord responsibilities which are considered separate from waste management but directly support the overall program at the site.

Funding is provided by the headquarters program which supports the major portion of the work being conducted at the installation.

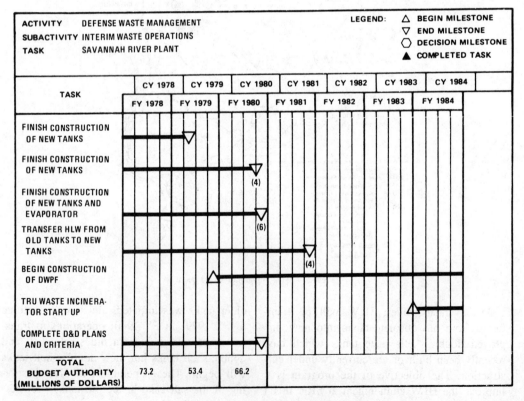

Figure V-3-13. Milestones and funding for SRP.

ETW has been assigned landlord funding responsibility for Rockwell-Hanford as part of the defense waste program.

Landlord responsibilities and activities will continue at Richland at a level necessary to support the DOE program activities at the site. An upgrading of equipment and general support facilities is necessary at Richland and is currently in progress. New heavy equipment is scheduled to be purchased to replace old unreliable equipment.

Major milestones and funding are shown in Figure V 3-15.

Long-Term Waste Management Technology

The program for long-term management of nuclear wastes requires technology development, engineering studies, conceptual design, tests, demonstrations, and evaluations as a basis for selecting site-specific alternatives consistent with the National Environmental Policy Act and for building associated facilities to implement them. The research and development program in progress is designed to fill the technology gaps for long-term waste management.

The long-term waste management R&D program is organized by waste category (HLW, TRU, LLW, airborne) and by site. Each site has been designated "lead" for the development of technology and techniques for a given waste category: HLW at Savannah River, TRU waste at Albuquerque (Rockwell International/Rocky Flats), LLW, and airborne waste at Idaho. Although R&D work on one particular waste category may be in progress

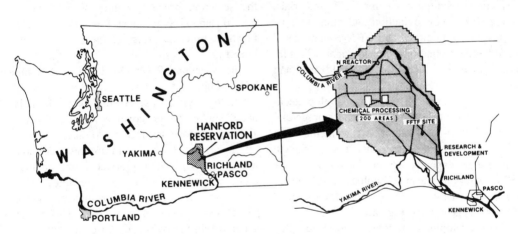

Figure V-3-14. Hanford Reservation and landlord responsibilities.

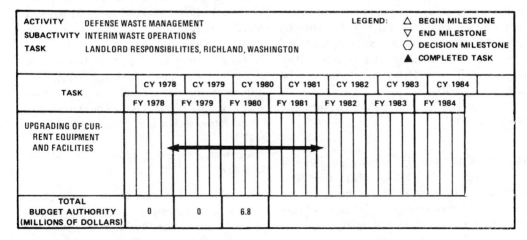

Figure V-3-15. Milestones and funding for landlord responsibilities.

at several sites, the lead site is responsible for coordination of the overall effort.

The HLW Task is developing immobilization processes such as the proposed reference process for the Savannah River Defense Waste Processing Facility (DWPF). This facility is planned to begin operation in FY 1989. Of the three DOE sites storing HLW, Savannah River will be the first to implement a long-term waste management option. Richland and Idaho will follow. The HLW program is consistent with DOE's stated plans to limit the use of the new waste storage tanks to approximately 20 years.

R&D on TRU-contaminated solids includes design and operation of facilities to develop technologies for safe and economic handling, treatment, and packaging for disposal. Incineration will be developed as the major volume reduction technique for combustible materials, along with decontamination and other volume reduction techniques. The TRU-contaminated solid waste technology development will support treatment facilities at Idaho and elsewhere.

Airborne waste R&D will develop techniques for capture, treatment, and storage/disposal of long-lived radioactive effluents from off-gas streams to meet the more stringent regulations expected to be imposed by the EPA. Airborne waste technology development must meet the 1983–85 schedule for expected EPA effluent control regulations.

The development of long-term waste management technologies has considered and will continue to consider future criteria and standards of the Nuclear Regulatory Commission and EPA. These criteria and standards and applicable NRC licensing requirements will be satisfied. All DOE HLW repositories will be licensed by NRC.

Major milestones and funding are shown in Figure V 3-16.

High-level Waste. The primary objective of DOE's long-term High-level Waste Task is to provide the required safety of man and his environment by long-term isolation of the HLW from the biosphere with minimum maintenance and surveillance.

HLW Research and Development. Three basic alternatives in the long-term HLW Task have been identified:

- Retrieve waste from interim facilities, process as required to produce a stable solid form, package in canisters, and dispose of the waste on site in surface or geologic facilities designed to minimize future maintenance and surveillance
- Retrieve, process, and package essentially as in the first alternative and dispose of waste permanently in an off-site geologic repository
- Continue ongoing practices indefinitely with evolutionary improvements

A comprehensive R&D program supporting each of the three alternatives is underway to develop the technical information needed in the selection process. Figure V 3-17 shows the conceptual flow sheet for a waste solidification process using glass as an immobilizer.

The Savannah River Operations Office has been designated as lead office for the HLW technology programs. Several very broadly based studies which cover several reasonable alternative processes leading to stable, immobile forms for HLW's are in progress. Possible waste immobilization product forms include glasses, artificial minerals, concretes, calcines, ceramics, and matrices such as calcine particles in a metal base material. Studies are underway in various laboratories to determine acceptable physical and chemical properties for these various forms and their epected radionuclide release rates in fabrication, shipment, and storage.

The technology base necessary for the evaluation of alternatives is being established. Technology is being developed and demonstrated at each of the three HLW storage sites for the retrieval of waste from interim storage facilities.

Alternative selection activities include the following:

- Technical alternative defense waste documents describing and assessing reasonable alternatives have been issued for each of the three sites.

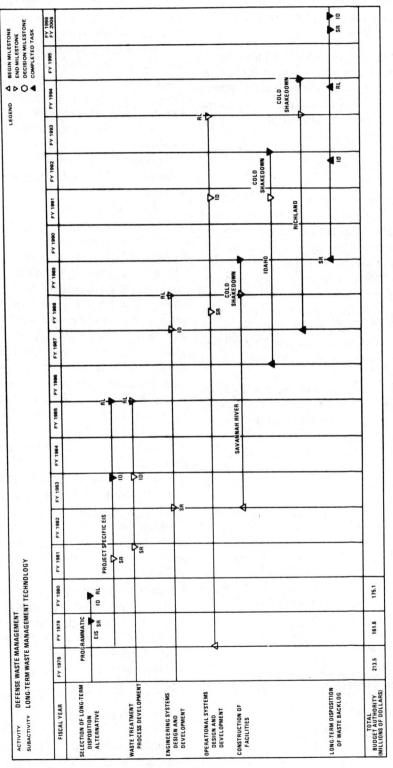

Figure V-3-16. Milestones and funding for long-term waste management technology.

- Programmatic Environmental Impact Statements covering reasonable alternatives are being prepared for each of the three HLW sites.

Conceptual design studies are underway on facilities needed at each of the three sites to implement reasonable alternatives. Engineering studies and tests of equipment for retrieval of waste are being conducted at each of the three HLW storage sites.

Major milestones and funding are shown in Figure V 3-18.

Defense Waste Processing Facility (DWPF). DOE's plan for the long-term management of defense HLWs is to put them into a form and location suitable for long-term isolation from the biosphere and to minimize dependence on institutional controls to effect such isolation. Because of the difference in waste properties and composition from site to site, DOE plans to handle each situation separately. Savannah River has been chosen as the first site to implement a long-term HLW management program.

The DWPF project will include facilities for retrieval, immobilization, and temporary storage of the Savannah River HLW. The reference DWPF process involves the following steps: (1) removal of waste from the tanks, (2) separation of the sludge from liquid by physical processes such as centrifugation and filtration, (3) removal of the cesium and some other radioisotopes from the liquid by ion exchange, (4) incorporation of the sludge and the cesium into a suitable waste form such as borosilicate glass, (5) evaporation of the residual solution to a salt cake (containing less than 0.1 percent of the total activity), (6) interim storage of containers of the HLW form in a natural convection air-cooled facility, (7) eventual transport to a DOE repository, and (8) storage of the residual salt on site.

Figure V 3-19 is a diagram of the reference process.

The technology development program now underway and proposed for the future is aimed toward timely conversion of the waste to glass and disposal in a geologic repository. The type of geologic formation and the specific site for the repository have not yet been chosen. The present technology development program does not foreclose any of the repository options being considered because the glass waste composition and container size would be compatible with any option. The outer container material may change depending upon the type of geologic formation, but there would be adequate time to implement any change, and the technology development program is not sensitive to the outer container material.

Research and development to date has included a waste tank sampling program to characterize the waste, preparation, and evaluation of promising waste forms in shielded cells using actual waste; evaluation of container materials using simulated waste; demonstration of cesium removal at small and intermediate scales; small-scale tests of centrifugal sludge-liquid separation using actual waste; and engineering and cost studies assuming concrete or glass at the product form for the high-activity fraction.

Major milestones and funding are shown in Figure V 3-20.

Low-level Waste (LLW). Low-level contaminated wastes do not contain significant amounts of transuranic nuclides and have relatively low levels of radioactivity. Such contaminated waste may come from a wide

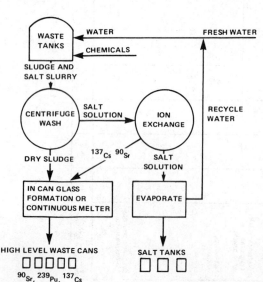

Figure V-3-17. Conceptual waste solidification process.

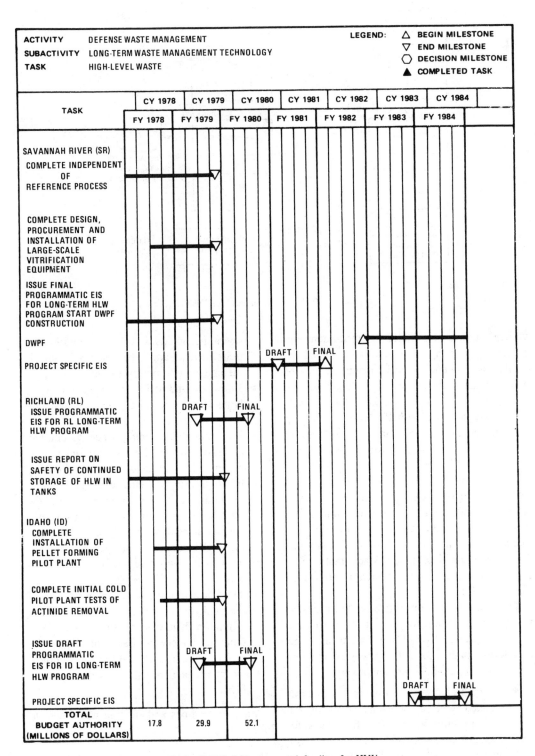

Figure V-3-18. Milestones and funding for HLW.

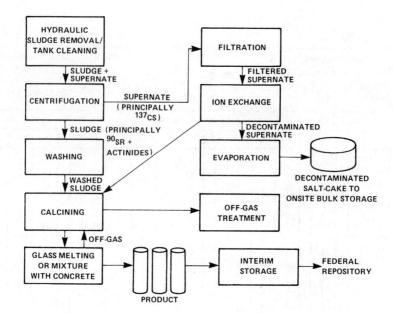

Figure V-3-19. Reference process for DWPF.

variety of sources, including manufacturing, processing, testing, research, and medial installations. Depending on its origin, it may be contaminated with almost any radioactive isotope in small amounts. The kinds of materials being handled are equally varied, ranging from paper towels and laboratory gloves to materials of construction from obsolete nuclear facilities.

The shallow land burial of LLW is an established practice for the disposal of such wastes. Stabilization of the waste burial site or other steps taken limit the potential for radioactive releases to the biosphere or migration of radionuclides to the water table.

The solid waste storage area is shown in Figure V 3-21.

The Idaho Operation Office has been designated as lead office for the LLW. In support of the objectives of this program and the long-term strategy for defense wastes, the following R&D activities are underway:

- A multilaboratory program was formalized with a published plan in FY 1977 to improve the scientific basis for disposal methods and develop disposal criteria by the end of FY 1980. Work is in progress on establishing a technical data file, site selection procedures, site operating practices, waste preparation requirements, and a systems analysis method to optimize procedures.
- Burial ground stabilization technology is being developed, and some field tests are being conducted.
- Preparations are being made to conduct at least two small-scale demonstrations of the shallow land burial technology being developed. One demonstration is to be in an arid environment, and the other in a relatively humid environment.
- An assessment of the adequacy of existing DOE shallow land burial sites has been made, and a report on the findings was issued in FY 1978.
- At each disposal site, characterization of the geohydrological environment and the migration behavior of the wastes contained is underway. Based on the results of these studies, the stabilization requirements of each burial site will be established.

Major milestones and funding are shown in Figure V 3-22.

TRU Waste. TRU-contaminated solid waste is a solid waste contaminated with long-lived transuranic radionuclides. Most transuranic

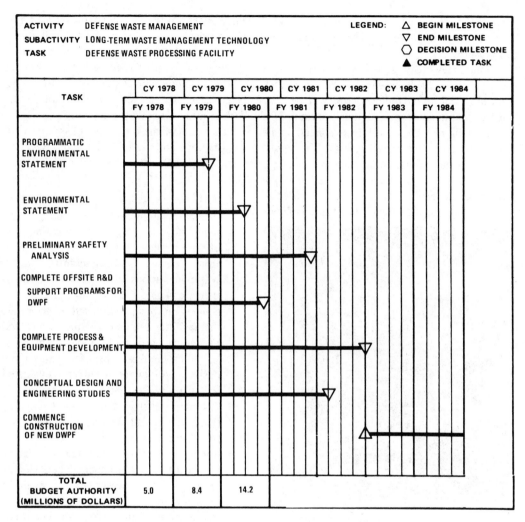

Figure V-3-20. Milestones and funding for DWPF.

nuclides show the unusual combination of physical properties of long half-life and high specific toxicity. The routine disposal of TRU solid waste at DOE sites by shallow land burial was stopped in 1970. Since then, the waste has been stored in a way that permits ready retrieval for a period of 20 years. This will allow access to the waste for further processing, if necessary, and ultimate permanent disposal.

TRU Waste Technology. Due to the special characteristics of TRU-contaminated solid waste, new technology must be developed for the safe handling and disposal of this waste. The R&D programs established for TRU waste management have the following specific objectives:

- To develop technology for treating TRU waste to meet the criteria being developed for disposal
- To provide facilities and equipment for treatment, packaging, and transport of TRU waste to a Federal geologic repository

The objectives pertain to the disposition of three categories of TRU waste: previously buried waste, retrievably stored waste, and future wastes. Figure V 3-23 shows one method of TRU waste incineration being investigated. This process is being developed to reduce substantially the volumes of TRU waste to facilitate packaging, transport, and disposal.

Some previously buried waste may require exhumation. Each burial location for TRU waste or TRU-contaminated soil will be evaluated. Exhumation of previously buried TRU waste will be very carefully evaluated in future environmental impact studies. The studies will analyze the environmental and occupational risks associated with exhumation versus the risks of leaving the TRU waste buried as is presently the situation. The Waste Isolation Pilot Plant (WIPP) is being developed in a southeastern New Mexico bedded salt formation to provide the capability for permanent geologic disposal of defense TRU waste.

The Albuquerque Operations Office has been designated as lead organization for the TRU waste technology program.

In support of the objectives of this program and the overall strategy for defense wastes, numerous R&D efforts are underway, specifically:

- TRU waste technology is being developed, including volume reduction systems, such as incineration, smelting, and electropolishing to remove contamination from metals, and immobilization of waste in various matrixes such as polymers, concrete, metals, and glass.
- Studies are being conducted to determine the facilities required for processing and packaging the retrievably-stored TRU waste for geologic disposal. Conceptual designs will be prepared and evaluated for specific sites and central facilities.
- A comprehensive plan is being prepared for retrieval and treatment as necessary for permanent disposal.
- Studies are being conducted to analyze the long-term environmental behavior of currently buried waste and to evaluate the alternatives for both exhumation and stabilization of the waste where buried.

Figure V-3-21. Solid waste storage area.

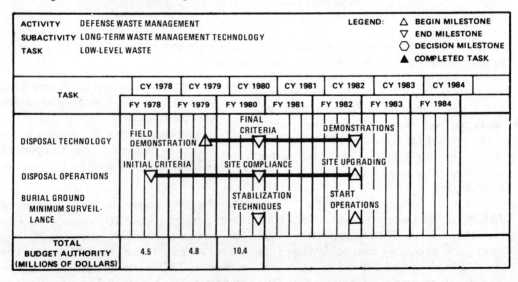

Figure V-3-22. Milestones and funding for LLW.

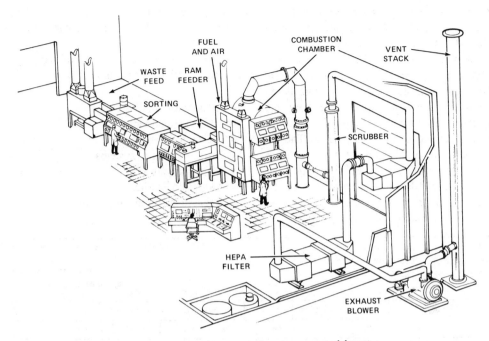

Figure V-3-23. TRU waste incinerator-conceptual layout.

Experimental exhumation projects are underway to determine methods for exhumation and the associated risk.
- A study is being conducted to evaluate container designs in terms of future transportation requirements, economics, and regulations. Other studies are underway to identify any major policy decisions needed to facilitate the transport of radioactive waste and to establish standardized methods for packaging and transporting TRU waste.

Major milestones and funding are shown in Figure V 3-24.

Waste Retrieval and Treatment Facility (WRTF). The Radioactive Waste Management Complex (RWMC) at Idaho has been receiving TRU waste since 1954. By 1988, the RWMC is estimated to contain 155,800 m^3 (5,500,000 ft^3) of TRU waste (see Table V 3-2). Between the years 1954 and 1970, waste was buried below ground with no consideration given to retrieval. In 1970, a policy of placing TRU waste in a 20-year retrievable storage mode was adopted. Since November 1970, waste has been placed on aboveground asphalt pads referred to as the Transuranic Storage Area (TSA).

Three alternate concepts for the long-term management of buried TRU waste are being studied:

- Alternative I—Leave the waste in place and retain existing waste management practices (complete the emplacement of a layer of densely compacted and highly impermeable lake-bottom clay over the waste disposal area)
- Alternative II—Leave in place as above, but provide engineered containment using entombment or encapsulation concepts
- Alternative III—Retrieve, process, and transport waste to a Federal repository

No large-scale DOE TRU waste processing is underway in the United States. The primary objective of the WRTF Project then is to immobilize chemically TRU waste in a form which will be acceptable at a Federal repository. Concurrently, the most widely accepted technical approach for processing is to incinerate the waste material and to immobilize the residue.

The base case for this long-term management plan has included provisions for buried and stored TRU waste to be retrieved, processed, and shipped to a Federal repository. The retrieval and processing of buried waste is the most complex alternative considered and thus provides an "upper limit" on the work effort and funding required for long-term management of the TRU waste. The retrieval and processing of stored waste is scheduled to commence in FY 1986. Following processing, the waste residue will be packaged in appropriate containers for transportation to a Federal repository. The waste will be processed at INEL, and the final form of the waste will be made acceptable to the repository operator.

In support of the WRTF, the following specific R&D activities are underway:

- Incinerator studies
- Off-gas treatment
- Waste handling equipment
- Instrumentation systems
- Slag analysis
- Migration studies
- Regulatory requirement study
- MOL Belgium R&D tests

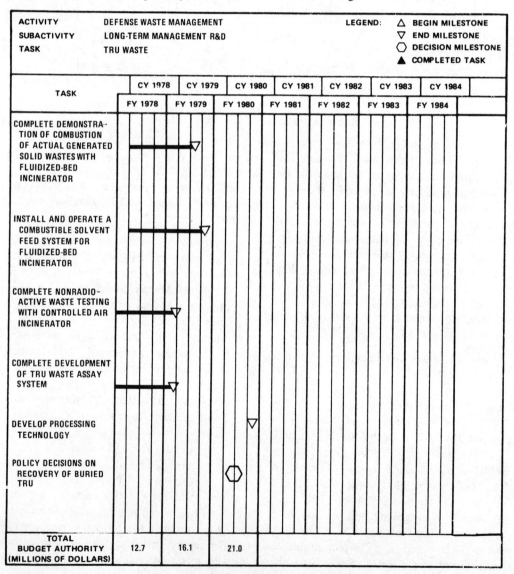

Figure V-3-24. Milestones and funding for TRU waste.

Table V-3-2. TRU Waste Volume Projections.

YEAR	VOLUME (FT³)	NO. OF CARTONS	NO. OF DRUMS	NO. OF BOXES	NO. OF BINS
1954-1970 (below ground)	2,300,000	13,000	198,000	6,200	0
1970-1977 (above ground)	1,200,000	0	82,000	5,000	172
1978-1988 (projected at 200,000 ft³/yr)	2,000,000	0	155,000	9,600	275
TOTAL BY 1988	5,500,000	(13,000) 0	(198,000) 237,000	(6,200) 14,600	(0) 447

- Full-scale demonstration plant test plan
- Remote operation

Major milestones and funding are given in Figure V 3-25.

Airborne Waste. Airborne waste originates from nuclear reactors, irradiated fuel storage, fuel reprocessing, weapon-related activities, and waste treatment processes such as waste incineration and vitrification.

Air flowing out of all process buildings must be filtered, since mists or small particles of dust in the air may contain radioisotopes. Particulate material in the gaseous waste streams is removed by scrubbing or filtration through sand, fiberglass, and/or high-efficiency media. Figure V 3-26 shows the filtration of the gaseous effluents from a spent fuel reprocessing facility. Each high efficiency filter stops approximately 99.97 percent of the particles. Some radioisotopes that exist as a gas require special treatment. Iodine, for example, is readily vaporized. It has two radioactive isotopes, iodine-129 and iodine-131, which require control to minimize release to the environment. Special absorbers are used to retain them in the plant. The very small quantities of radioactive material that escape to the atmosphere through high stacks are within governmental emission limits established by the Federal Government.

The objectives of this task are to increase the service life and reliability of filters; and to develop improved technology for additional

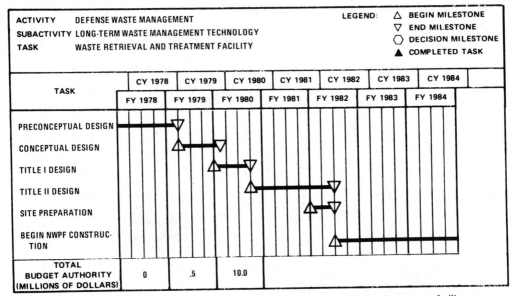

Figure V-3-25. Milestones and funding for the waste retrieval and treatment facility.

collection, fixation, and long-term management of gaseous radionuclides in anticipation of regulatory actions that would require further reductions in airborne radioactive releases from defense program facilities.

Available technology and practices are adequate to meet current health and safety standards. The Airborne Waste Task is aimed primarily at cost-effective improvements, quality assurance, and new capability in areas where more restrictive standards seem likely to apply in the future.

The Idaho Operations Office has the lead responsibility for the airborne waste technology program. In support of the objectives of this program and the long-term strategy for defense wastes, the following specific R&D activities have been started:

- To reduce cost and waste, iodine asorbents that have substantially longer service life and that can be regenerated are being developed.
- More durable filters are being developed for service in off-gas streams from radioactive processes that emit acid vapors and/or high temperature gases.
- Several prefilter concepts are being developed for capturing radioactive particulates near their source. These concepts will extend the service life of high-efficiency particulate air (HEPA) filters and thus reduce the costs for their replacement and disposal.
- Improved monitoring systems are being demonstrated for specific long-lived airborne radioactive constituents in the off-gas of waste processes, particularly iodine-129, carbon-14, and tritium.
- Technology is being developed to remove tritium from air and water, to immobilize it, and to store it safely.
- New systems for monitoring emissions of iodine-129, krypton-85, and carbon-14 are being developed for use at DOE sites.
- Improved filter test methods are being developed for DOE radioactive iodine and particle removal filtration systems. This work also supports the development of filtration standards, regulations, and test procedures.

Major milestones and funding are shown in Figure V 3-27.

Terminal Storage

As a consequence of the Government's production of plutonium for nuclear weapons over the last 35 years, major quantities of

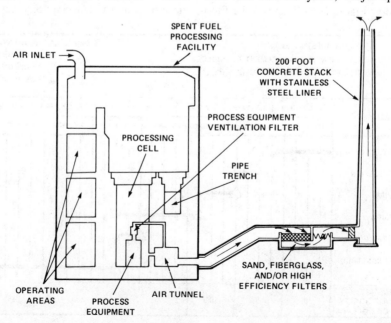

Figure V-3-26. Removal of airborne wastes.

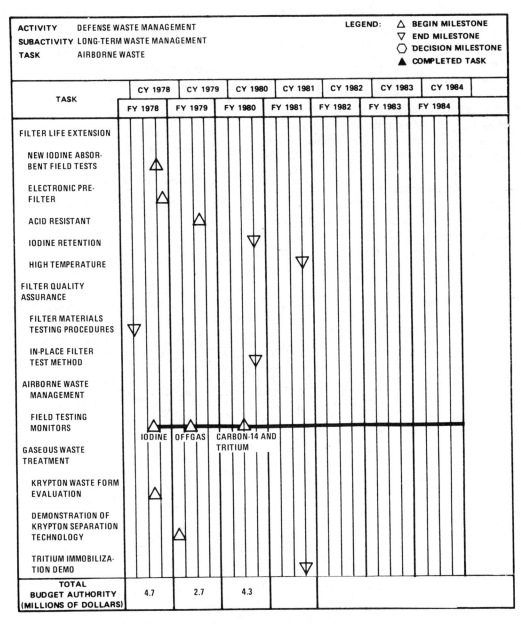

Figure V-3-27. Milestones and funding for airborne waste.

nuclear wastes have been produced. All of the wastes having low levels of radioactivity were disposed of in shallow land burial grounds until 1970. At this time it was decided that the portion of these wastes that were contaminated with transuranic elements (TRU waste) should be stored retrievably to facilitate further transfer to deep geologic disposal. The major function of the WIPP facility is the disposal of TRU wastes.

WIPP Research and Development. WIPP is the first facility specifically designed for the long-term management of defense TRU-contaminated wastes. The objective of the facility is to demonstrate that TRU wastes can be disposed of safely with minimum maintenance and surveillance for thousands of years. Because of the long periods involved, numerous R&D activities are being carried out to confirm expected results and ensure a thorough understanding of the long-term behavior of the waste, the disposal of the medium, the waste immobilizer, and the interactions of all three.

Various confirmatory R&D studies regarding the design of the proposed WIPP facility have been accomplished in the laboratories. However, laboratory tests and studies cannot totally simulate the actual conditions encountered at the WIPP facility. To ensure that no factors are overlooked, bench-scale R&D activities are underway and field/in-situ (i.e., full-scale) R&D activities are planned.

Figure V 3-28 shows an artist's rendering of the proposed WIPP facility.

In support of the WIPP Task, the following specific R&D studies and tests are underway or are scheduled:

- Transuranic-waste characterization
- HLW/container interaction
- Thermal/structural interaction
- Nuclide migration
- Permeability
- Brine migration
- Borehole plugging
- Operation and design
- Instrumentation development

The above constitutes only a small part of the over R&D program for WIPP. The R&D program is designed to support WIPP operation directly by FY 1986.

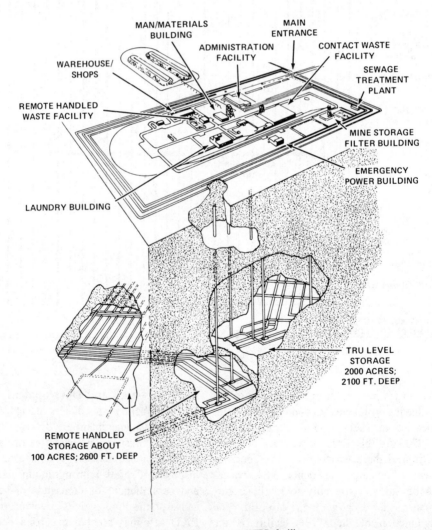

Figure V-3-28. Proposed WIPP facility.

Major milestones and funding are depicted in Figure V 3-29.

WIPP Facility. When the AEC withdrew its proposal to construct a nuclear repository in Lyons, Kansas, a nationwide search for a new site was begun. With the cooperation of the U.S. Geological Survey (USGS), the Los Medanos region in southern New Mexico was identified as a promising area. Field investigation were initiated in 1974, and exploratory holes were drilled. An initial site in this area was found to be unacceptable in July 1975. A new site, seven miles southwest of the first and 25 miles southeast of Carlsbad, was recommended by USGS and Sandia Laboratory, and investigation began in early 1976 (see Figure V 3-30).

The specific site under consideration would cover about 30 square miles (approximately

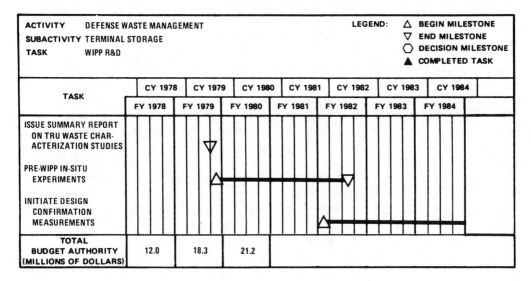

Figure V-3-29. Milestones and funding for WIPP R&D.

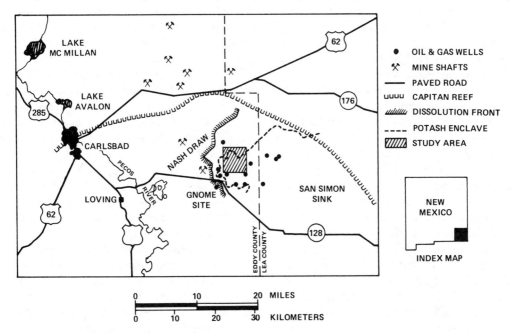

Figure V-3-30. Proposed WIPP facility site.

19,000 acres). The current ownership of the land and existing leases for oil, gas, or potash recovery have been identified. In cases where drilling operations by lease holders were imminent, the leases have been condemned by the Federal Government to preserve the site for potential use as a WIPP.

If the suitability of the site were confirmed by further investigation, the next step would be to withdraw the land from the public domain and reserve it for the Department of Energy's purposes. As the law currently stands, the federally-owned land could only be withdrawn from public use for a period of 20 years, which would be inadequate for waste disposal purposes. Consequently, legislation transferring the land permanently to DOE would have to be sought from Congress. The 1,760 acres owned by New Mexico would be purchased by the Federal Government via condemnation.

An artist's conception of WIPP is shown in Figure V 3-28. The facility on the surface would cover approximately 60 acres and be serviced by truck and rail transportation. Access to this 60-acre surface area would be restricted to authorized personnel by chain-link fencing and various security measures. Activities on the remaining surface area would also be subject to approval by DOE, but could be virtually unlimited. Any development of underground resources, however, would be under DOE's control to ensure that drilling or mining activities do not threaten the ability of the proposed repository to keep the waste isolated from man.

Waste disposal would be accomplished at two levels below ground in a bedded salt deposit. A majority of the wastes received at WIPP would be TRU wastes from the defense program, which do not give off significant heat. Twenty-one hundred feet below the surface, a maximum disposal area of 2,000 acres could be made available for those TRU wastes. The wastes can be handled directly (contact-handled wastes). The four million cubic feet of these wastes expected to have accumulated by 1985 would occupy about 94 acres. After 1985, these materials will be generated at the current rate, about 250,000 cubic feet per year. More of these wastes will arise from decommissioning of surplus DOE facilities.

At the 2,600-foot level, TRU wastes which must be shielded during handling (remote-handled wastes) would be deposited. By 1985, 100,000 cubic feet of remote-handled wastes occupying about 10 acres could be isolated in this area. If the current rate at which these materials are generated persists after 1985, an additional acre per year will be required. The 2,600-foot level at which these wastes are placed also could be expanded beyond 100 acres and made available for the contact-handled wastes.

Experiments with high-level defense waste and spent fuel. A detailed understanding of the behavior of deep geologic deposits, when subjected to conditions such as elevated temperatures, is necessary to continue the investigation of long-term HLW management options. Although various computer codes, design assumptions, and previously performed laboratory and bench-scale experiments have indicated the adequacy of geologic formations for HLW disposal, this work must be confirmed and validated by experiments in an actual geologic repository.

One purpose of the WIPP experiments would be to simulate the long-term disposal environment by subjecting the area to conditions, such as higher temperatures, caused by the high-level defense wastes and spent fuel. This would provide confirmatory information on the usefulness of the computer codes and the validity of design assumptions and would provide guidelines for operating large-scale, permanent repositories for these wastes at other locations. These wastes would also require remote handling.

A 20-acre area at the 2,600-foot level would be devoted to these R&D activities. All the material used in these R&D activities would be removed once the experimental program is complete.

Proposed demonstration of the disposal of spent nuclear fuel elements. A DOE task force studying the waste disposal program has proposed that the facility also be used to demonstrate the permanent disposal of up to 1,000 spent fuel assemblies discharged from commercial nuclear power plants.

One-to-three assemblies would be placed in a cannister. The spent fuel in each cannister would generate no more heat than about six 100-watt light bulbs at the time of emplacement.

Conducted at the 2,600-foot level, the demonstration of spent fuel disposal would require no more than 20 acres. The canisters would be placed inside protective sleeves. After the assemblies are in their protective sleeves, the demonstration area would be backfilled with salt. Nevertheless, these assemblies could be retrieved at any time during the 20 years, if desired. A decision on this proposal is pending implementation of the NEPA process and congressional approval.

Major milestones and funding through FY 1984 are shown in Figure V 3-31. Operation is scheduled for FY 1986.

Transportation

This program provides a balanced and integrated approach for management of DOE's defense waste transportation activities and requirements. The FY 1980 work will focus on continuing development of equipment and systems to support the WIPP facility, continuing base technology for transporting all DOE wastes, and a major effort to meet DOE responsibilities in dealing with the concerns of the public, citizen organizations, and State and local governments. Research will be conducted on cost-effective TRU waste transportation equipment, package integrity, system utility reliability, and vulnerability to malevolent attacks. Analysis and scale model testing will be conducted, and preparation for full-scale testing will be initiated.

Figure V 3-32 shows a rail car designed to transport radioactive waste.

The consolidation of DOE's ongoing and increasing activities in nuclear materials transportation initiated in FY 1979 must be strengthened to provide the following:

- An integrated systems approach to transportation operations, planning, and execution
- Technical management of transportation R&D in support of WIPP and other DOE programs
- A technology and information data bank
- Risk and environmental impact analyses
- Transportation factors input to waste acceptance criteria facility designs, handling hardware, and shipping systems
- Information to the public about waste transport, and achievement of societal acceptance
- Opportunities for public, industry, and State and local government discussion and participation

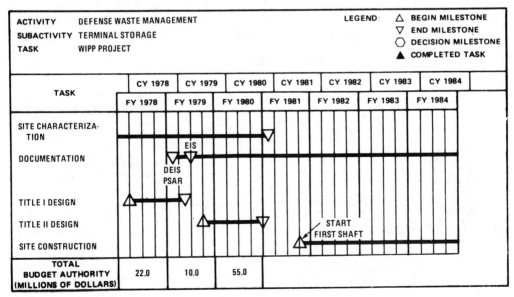

Figure V-3-31. Milestones and funding for the WIPP project.

The systems development effort specifically supports the availability of an operational defense waste transportation system by 1986 and, in general, is supportive of the transportation of all nuclear materials. The following activities were initiated in FY 1979:

- Develop a coordinated program for testing and evaluating the performance of current and future generation packaging systems during accident conditions
- Increase package integrity and system use and reliability, and extend scale-model tests to future container designs
- Initiate work toward full-scale testing
- Determine response characteristics of radioactive materials to accident environments
- Determine vulnerability of packages to violent attack and assess public hazard

Funding is also included for developing the Transportation Technology and Information Center into a reputable center of transportation expertise and information considered essential in meeting the recommendations of the Interagency Review Group on Nuclear Waste Management. The Albuquerque Operations Office and Sandia Laboratory are the lead organization for this activity, but work will be carried out by industry, universities, and other DOE organizations.

DOE has continuing responsibilities in the transportation of nuclear materials which have direct application to the commercial sector. These responsibilities were highlighted in the draft report to the President by the Interagency Review Group on Nuclear Waste Management dated October 1978. The findings and recommendations on transportation matters presented in this report state that DOE should do the following:

- Conduct a program to test and evaluate the performance of current and future generation packaging systems during accident conditions
- Assist in development of a data bank on shipment statistics and accident experience to be operational by 1982
- Ensure early State participation in barge, rail, and highway route planning
- By 1980, complete studies to determine the need for physical protection measures for nuclear waste transportation, and assist in rulemaking to be completed by 1981
- Expand Federal assistance in developing the capability to handle emergencies

The technology and systems developed to transport existing and future inventories of defense-related nuclear waste will also facilitate developing the capability for moving nuclear materials associated with other activities such as decontamination and decommissioning of facilities.

Figure V 3-33 shows the milestones and funding for the development of a transportation system supporting the WIPP project.

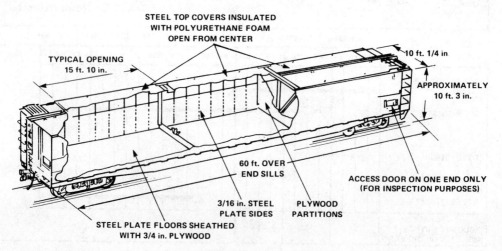

Figure V-3-32. General design of the ATMX-500 rail car.

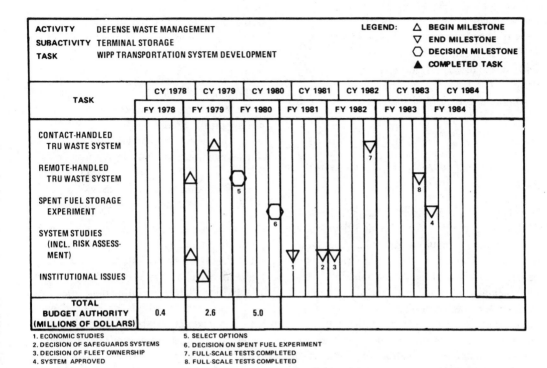

Figure V-3-33. Milestones and funding for WIPP transportation system development.

Decontamination and Decommissioning (D&D)

The defense nuclear program originated in the early 1940s. Since that time, numerous facilities constructed for test, research, and nuclear material production have outlived their useful lifetime and are now obsolete and in standby. The majority of these facilities are contaminated to varying degrees and can be regarded as defense wastes. The objectives of the D&D program is to decontaminate nuclear facilities declared excess to DOE nuclear programs for beneficial alternate uses or to decommission such facilities, where deemed appropriate, by such modes as mothballing, entombment, or dismantlement and removal.

Various studies and demonstration have been carried out for decontamination. However, no major facilities have been decontaminated and decommissioned, and, as such, this activity represents a complete RD&D program. The importance of this program is underscored when consideration is given to the fact that there are over 560 identified DOE nuclear facilities currently in excess or expected to become excess over the next few years. About 80 percent of these facilities are located at the Hanford site. The program to carry out D&D on facilities recently declared excess first identifies facilities with highest priority for D&D (i.e., facilities that are potentially hazardous and expensive to maintain in their current condition). First-round nuclear D&D projects, for which D&D will be carried out with FY 1980 operating-expense or line-item funds, are being identified, and planning action are being carried forward. As planning and engineering are completed and funds are allocated, first-round D&D projects will be implemented, followed by additional rounds of D&D projects. Table V 3-3 identifies the first three rounds of D&D projects as currently anticipated.

Current activities include establishing a lead field office (Richland, Washington) and lead contractor; planning and engineering for Round 1 and 2 projects; and conducting D&D for the Standard Pile, Dosimeter Applications Research Equipment, and Intermediate-level Waste Line (all at ORNL). The program will expand as major projects in Rounds 2 to 3 become active. At the present time, R&D activities are comprised of generic aspects of D&D. As specific technical issues arise, they will be subject to resolution by an R&D program.

Major milestones and funding are shown in Figure V 3-34.

COMMERCIAL WASTE MANAGEMENT ACTIVITY

The Department of Energy has the responsibility for providing facilities and developing the technology required for managing commercial radioactive wastes for which the Federal Government is responsible in an environmentally acceptable manner. In implementing this responsibility, DOE is currently faced with a major agency action—that of selecting an appropriate technical strategy for the safe, permanent isolation of commercial radioactive wastes. A permenent disposal strategy represents a programmatic approach by DOE to the development of a waste management program. The scope of the strategy consists of site investigations into candidate media, R&D into alternative treatment processes, and examination of alternative waste management technologies.

Table V-3-3. Candidate DD Facilities.

ROUND 1 (FY 1980 START)	LOCATION
Mound Laboratory Buildings PP, R	AL
Hanford Z Plant	RL
ORNL Curium Facility	OR
INEL Hot Waste Tank	ID
INEL Fuel Element Cutting Facility	ID
ROUND 2 (FY 1981 START)	
ANL 301 Hot Cell Facility	CH
ORNL Waste Holding Basin	OR
INEL Reprocessing Cells	ID
INEL Liquid Metal Furnace	ID
INEL Original Waste Calciner	ID
Shippingport Atomic Power Station	PNR
ROUND 3 (FY 1982 START)	
ORNL Gunite Storage Tanks	OR
ORNL Original Hydrofracture Facility	OR
INEL CPP Heating/Ventilating System	ID
INEL CPP Carbon Beds; Filters	ID
INEL Hot Waste Tanks	ID

Budgetary and programmatic classifications for the Commercial Waste Management Activity are structured under the following subactivities:

- Terminal Isolation
- Waste Treatment Technology
- Supporting Studies
- Solidification Process Demonstration
- Decontamination and Decommissioning
- Low-level Waste Management Operations (Commercial Burial Sites)
- West Valley Nuclear Center Activities

All commercial waste management subactivities will maintain close coordination with relevant defense waste efforts to ensure timely and effective technology transfer between the two activities. Close coordination and frequent communication will be maintained with appropriate Federal, State, and local public agencies and officials to ensure public acceptance of the activity.

Terminal Isolation

The Terminal Isolation Subactivity for disposal of radioactive waste is directed primarily at the development of facilities for the placement and storage of high-level and TRU wastes within a deep geologic formation to provide safe, long-term isolation of the waste from human activities and the environment. The following elements are included in the Terminal Isolation Subactivity:

- Geologic studies to identify potential sites in varied geographic regions and use of suitable geologic media
- Technology development to provide the capability to analyze behavior mechanisms for radioactive waste placed in candidate geologic structures
- Engineering design and development to derive specific designs for operating repositories and specialized equipment
- Development of packaging and storage capability for unreprocessed spent fuel elements
- Evaluation of the suitability of repositories located at the Hanford Reservation and Nevada Test Site

Based on the current technical status of the program, the site location for the first repository will probably be selected from sites with salt or basalt formations. The location for the second repository, in a different region, using either salt or crystalline rock formations, could be identified within 2½ years after the first has been selected and prior to the approval by NRC of the construction authorization for the first repository. Development

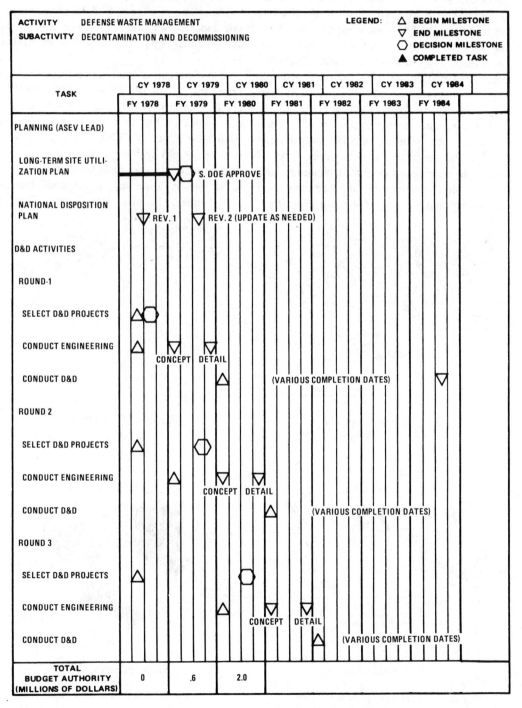

Figure V-3-34. Milestones and funding for D&D.

work on other nonsalt media is being accelerated.

The resolution of near-term operational safety considerations and long-term waste isolation mechanisms will continue to be stressed. Analysis of site performance characteristics and the complete evaluation of technical and safety topics in support of repository performance will be included. Programmatic, environmental, and technical objectives will be met by conducting geologic studies in numerous sites and in varied geologic formations; by conducting research, development, and demonstration projects relevant to each geologic media; and through design and construction of facilities. The intent is to provide an adequate technology base for the site, construction, and operation of a repository and closure of a filled repository. The technology base will include the development of the data in a form suitable for NRC licensing.

A program to ensure adequate transportation links vital to a successful repository operation is under development.

The maps in Figure V 3-35 show bedded salt deposits, granitic rock, shale formations, and potential repository basalts in the United States.

Salt formations that appear particularly promising, based on technical considerations, are the Permian Basin, Salina Basin, interior salt domes of the Gulf Coast, and Paradox Basin (see Figure V 3-35). Geologic exploration by geophysical methods has begun, and drilling to verify the characteristics of the salt formations is underway in the Gulf Interior Salt Domes, Permian Basin, and Paradox

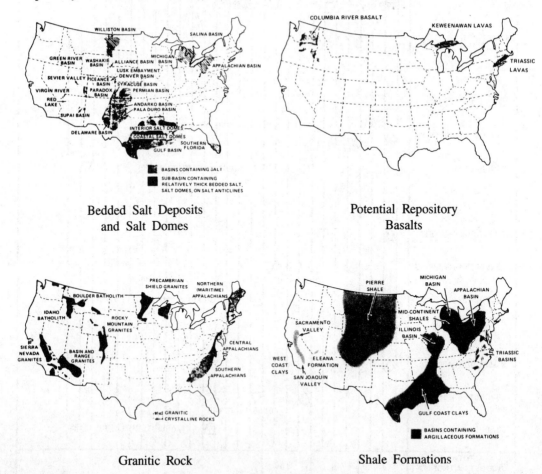

Figure V-3-35. Potentially suitable geologic formations for repository sites in the contiguous United States.

Basin. Figure V 3-35 also shows the location of other common geologic media that are being evaluated for repository siting.

A major new initiative is anticipated in FY 1980 to explore crystalline rock and other formations within the continental United States. This program will begin with a systematic evaluation of all candidate geologic systems in the United States prior to identification of specific host rock media.

Generic geologic, hydrologic, and engineering studies will continue in granite, basalt, shale, tuff, etc., to develop understanding of their suitability. The Hanford Reservation contains basalt. The Nevada Test Site contains formations of granite, shale and tuff and provides ideal locations to conduct generic evaluations of the media in an expeditious manner. Evaluation of these sites as a potential location of a repository has been accelerate. By the end of FY 1979, the existence of any major geologic or hydrologic impediments that would prohibit the use of the two DOE reservations will be determined.

In cooperation with the USGS, an earth science R&D plan is being prepared to refine the technology base for such disciplines as geochemistry, geophysics, hydrology, and the prediction and consequences of natural geologic events, as applicable to a repository.

Office of Nuclear Waste Isolation (ONWI). The activities under this task include the following:

- Site evaluation efforts related to qualification of the sites to comply with specific licensing criteria
- Technology development for identification, coordination, and direction of the development of the criteria and specific technologies required for the exploration, design, licensing, construction, operations, decommissioning, and monitoring of the nuclear waste repositories
- In-situ testing activities in support of design, construction, licensing, and operation of these repositories
- Facility and equipment design and development for the conceptual design and engineering for the nuclear waste repositories
- Spent fuel packaging and storage activities related to a packaging and storage system for placement of spent fuel in a geologic repository
- Assistance in providing information to the public regarding the program

Figure V 3-36 shows the ONWI work breakdown structure.

Management of the ONWI Task was assigned under DOE's Richland Operations Of-

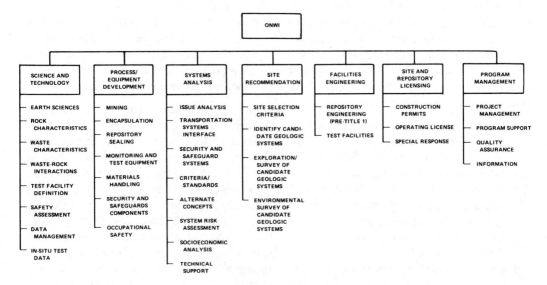

Figure V-3-36. ONWI work breakdown structure.

fice to Battelle Memorial Institute at Columbus, Ohio.

Site evaluation activities include geologic studies and supporting regulatory activities. The confirmation of a potential waste repository location requires a study of the geology, hydrology, and other technical characteristics of that location to establish, in the framework of licensing regulations, that no credible circumstances would be encountered which would result in releases of radionuclides from the emplaced waste to the biosphere in quantities that would constitute a hazard to the public.

Geologic disposal of radioactive wastes could be achieved in common geologic media such as salt domes and beds, granite, shale, and basalt found in geologic structures throughout the United States. These disposal media are meant to represent only a range of candidate host rocks. Other earth materials may also meet the requirements for media property and environmental distribution. Additional media have properties similar to those media identified. Associated disposal media are grouped as salt (anhydrite and gypsum), granite (general crystalline rock, granodiorite, periodotite, and syenite), shale (general argillaceous rock), and basalt (gabbro and some tuffs).

Since most investigations of geologic disposal to date have centered on salt formations, the primary emphasis of the program has been on promising salt domes and bedded salt formations. The studies include the following investigations of generic and specific factors required to identify and select locations in many candidate geologic regions:

- Specific geologic salt studies of the Gulf Coast Salt Domes and the Saline, Paradox, and Permian Bedded Salt Basins
- Geologic studies of other salt formations
- Geologic studies of sedimentary rock media as typified by shales in the Nevada Test Site, Illinois Basin, Pierre, and East Coast Triassic regions, Gulf Coast clays and rocks
- Geologic studies of crystalline rocks as typified by Columbia River Basalts, New England, Canadian Shield, and other igneous rock
- Geologic studies of other possible candidate geologic media
- Generic hydrogeologic studies for all regions of the continental United States

The motivation for choosing a geologic environment for a repository is the isolation provided by the geologic structure. Important geologic considerations are as follows:

- Depth of isolation to provide a barrier of geologic materials between the waste and biosphere and also to protect the repository from surface environmental processes
- Properties and characteristics of the host rock and its dimensions to provide barriers against the effects of radioactive waste; these include thermal properties as well as chemical and physiochemical interactions between rock and waste
- Tectonic stability to ensure against deformation of the host rock and possible loss of repository integrity
- Hydrologic properties of the host rock and surrounding geology to provide a barrier to waste transport
- Multiple natural barriers to waste transport to provide added confidence that the disposal media will isolate waste; geologic considerations for surrounding media are similar to those for the disposal media
- A low resource potential of the host rock to avoid the loss of a potentially valuable natural resource and possible future disturbance of the repository by prospecting activities

Site selection begins with tectonic and hydrologic considerations that apply on a broad national scale. The next stage enhances the data base for candidate regions identified in the initial stage. This includes extensive geologic mapping, evaluation of rock properties in available geologic structures, characterizations of regional hydrology, evaluation of climatic data, and accumulation of instrumental data such as that obtained from geodetic, geophysical, and microseismic networks. The next stage involves area investigations within the region. Finally, siting considerations are applied on a site-specific scale. Additional data base requirements for the final stage are detailed site exploration data obtained by drilling, geophysics, and opening of test tunnels. In-situ measurements of site-specific rock properties, state of stress, and hydrology will be conducted to the extent possible without compromising the future integrity of the repository.

The development of environmental investigations and evaluations parallels the geologic activity. Baseline data encompassing all aspects of the total environment (terrestrial ecology, aquatic ecology, meteorology, geography, hydrology, water resources and use, land resources and use, demography, societal, economic, aesthetic, institutional, etc.) will be gathered. Evaluations will identify potential environmental impacts and risks of radioactive waste repositories which might be located within the specific area.

Technology development includes science and technology data base development (e.g., generic characteristics of risk types), process and equipment development, and system analysis.

A systematic evaluation of the safety and reliability of deep geologic disposal of radioactive waste is required to ensure the viability of the design and to identify mechanisms and phenomena requiring more scientific information. This activity will establish and verify mathematical models, provide data accumulation for the models, and develop the appropriate techniques and instrumentation for data accumulation and verification in the following areas:

- Safety assessment models and concepts to assess the safety of a repository, including disruptive event analysis and prediction from natural, repository construction and operation, and man-induced events
- Thermal analysis models to predict time-temperature distributions resulting from the radioactive wastes. Outputs include mechanical stress due to thermal expansion of the host rocks resulting from predicted temperature distribution histories and thermal loading design criteria
- Waste/rock/water interaction studies to characterize the chemical, physical-chemical, geochemical, and radiochemical reactions and processes between emplaced radioactive waste, the surrounding rock, and the subterranean water
- Waste migration models to characterize the source term, model transport phenomena, and collect and evaluate transport data
- Borehole plugging studies to establish the required technology to plug the exploratory boreholes and to seal off shafts of the decommissioned repository; development of plugging materials and techniques to measure performance and verification by field emplacement and testing

System studies will assist in the guidance, coordination, and control of the terminal storage efforts. Topics addressed under this activity are as follows:

- Geologic systems analysis, including the host rock and associated geohydrology
- The waste management system, including waste characteristics, and treatment, interim storage, and transportation links
- Criteria development for waste acceptance, siting, operational and post-operational monitoring
- Alternatives to deep-mined geologic disposal, such as sea bed disposal, deep hole disposal, rock melting, transmutation, and space disposal
- Technology studies such as waste handling
- Spent fuel accommodation studies
- International repository study

In-situ testing will include the development of engineering information and data necessary to guide, direct, and support the conceptual design and engineering effort related to construction and operation of waste disposal facilities. This involves extensive field testing experiments to provide information on the in-situ behavior of the rock, as well as engineering studies to provide information for design of the facility and the hardware. In situ testing will provide geologic and engineering data on such topics as storage array spacing, mining techniques to minimize fractures, borehole plugging, backfilling, waste retrieval, shaft sealing, engineered barriers, storage liners and plugs, and isotopic dilution.

The program will guide design, specification/prototype development, and support work related to special handling and emplacement equipment, by establishing performance requirements. Also, it will provide information to study the nuclear material accountability needs at a waste isolation repository by relating these needs to existing nondestructive techniques and equipment and by performing verification testing.

Facility and equipment design and development includes all activities related to the conceptual design and engineering required for the construction and operation of the repositories, namely:

- Architect/engineer (A/E) conceptual design and support effort required to describe the physical plants that would meet national needs for both the nuclear fuel reprocessing and spent fuel disposal options
- Special A/E and preconceptual design studies to cover several technical and programmatic areas as conceptual design of facility progresses
- Development of specialized equipment
- Training and indoctrination of the operating contractor
- Facility site confirmation and certification
- Estimation of operating costs

Spent fuel packaging and storage includes activity related to the development of technology for spent fuel packaging and storage systems, in preparation for placement of spent fuel in geologic repositories.

Several options exist for encapsulating the spent fuel. The baseline option is placement of spent fuel in canisters with only an inert gas fill. Other more advanced methods are under study as technical alternatives. These include a metal matrix fill, sand fill, other glassy or ceramic materials, and multiple barrier encapsulation of the spent fuel and canister at the time the fuel is declared a waste for disposal.

One of the first steps involves an evaluation of the sensitivity of the waste form to the performance of the geology for containment in a repository. Technical information is being accumulated on the packaging and encapsulating options so that an assessment of the alternatives can be made by 1979–1980. The assessment is expected to provide a recommendation on whether any waste form other than the canistered spent fuel is desirable as a package for disposal of spent fuel as opposed to retrievable storage of spent fuel.

The program also includes the consideration of surface dry storage as it relates to interim storage requirements at the repository site. For example, spent fuel elements which have been packaged may need to be stored prior to placement in the repository.

Major milestones and funding for the ONWI Task are depicted in Figure V 3-37.

Hanford Basalt Program. DOE's Hanford Reservation, located in southeastern Washington, is being explored for potential terminal storage sites. The geologic system of interest in this region is layered formations of basalt (a dense fine-grained rock formed by the solidification of lava).

Activities within this task include the following:

- Site evaluations to determine preliminary site suitability and to identify specific candidate sites for a repository
- Technical support to determine waste-geologic system-mining interactions
- In situ testing in an underground facility of thermal-related and radiation-related effects to support the determination of the suitability of basalt as a repository medium
- Facility engineering in support of a full-scale basalt repository

The proposed Near Surface Test Facility (NSTF) is pictured in Figure V 3-38.

The lead management role of this program has been assigned to DOE's Richland Operations Office. Rockwell Hanford Operations (Rockwell International Corporation) at Hanford, Washington, provides technical assistance for management activities.

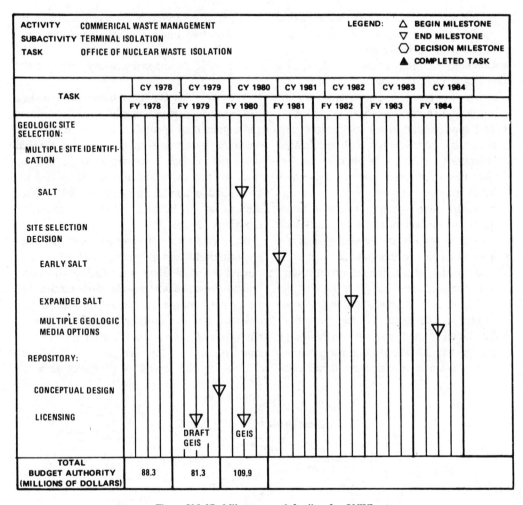

Figure V-3-37. Milestones and funding for ONWI.

Program management will include the following:

- Plan, organize staff, direct, and control combined efforts of the task and functional organizations
- Develop and administer program planning, budgeting, scheduling, and reporting policies, administrative systems, and practices
- Provide necessary quality assurance and control of all program activities, with emphasis on site investigation tasks
- Provide procurement support, including contract administration, procurement planning, and materials inspection
- Establish and maintain a data and documentation management system

Site evaluation will involve identification of three candidate sites by the end of FY 1979 and, possibly, selection of a suitable site by the end of FY 1981. These studies will include geologic investigations to include a comprehensive drilling and testing program to provide subsurface geologic, geophysical, and hydrologic data; surveying and mapping; regional and local surface and groundwater flow systems and analytical modeling; and seismic monitoring.

Technical support activities include the following:

- Waste-rock interaction studies to determine reactions through laboratory experiments and analytical modeling between spent fuel, canister, overpack material,

groundwater, and basalt; mathematical modeling of transport of radionuclide through the geologic media
- Borehole plugging studies to develop a borehole plugging system, including materials, machines, and techniques for the basalt/sedimentary geologic sequence
- Test support studies for rock testing and consulting services for rock mechanics; laboratory testing of basalt cores and in situ hydraulic fracturing to determine basalt rock properties (thermal and mechanical). Use mining consultants as required in support of in situ projects with respect to mining and modeling analyses
- System studies to integrate activities that lead to site identification, qualification, and licensing, including the current activities of site selection and risk assessment; in the area of site selection, development and use of methodology and criteria for identifying one potential site; in the area of risk assessment, studies to determine the consequences of radioactive nuclide release scenarios from a basalt repository; development of licensing documentation required by NRC

The NSTF on Gable Mountain will provide capability to perform in situ electrical heaters and commercial spent fuel tests in basalt. The electrical heater tests will define the thermal properties of basalt, determine the effects of fractures and discontinuities in the heated regime, and obtain in situ data to validate theoretical gross heating effects to assist in repository design.

Spent fuel tests will develop spent fuel storage criteria for basalt; develop technology and capability for handling spent fuel in an underground basalt environment; demonstrate the placement, storage, and retrieval of spent fuel in basalt; and demonstrate spent fuel monitoring capability. Overall, the testing will provide necessary information to assist in determining the suitability of basalt as a re-

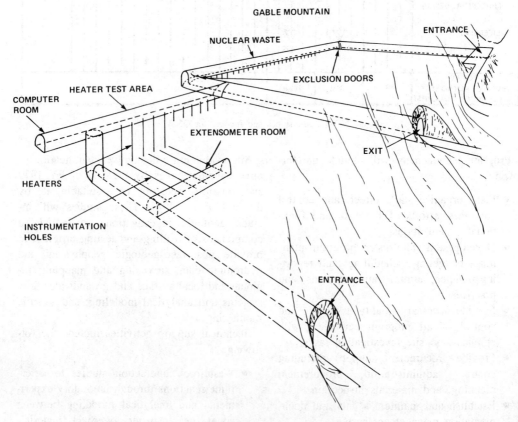

Figure V-3-38. Conceptual sketch for near surface test facility (in situ testing).

pository medium and in designing and constructing a full-scale basalt repository.

NSTF design was initiated in early FY 1978. Construction (tunnel excavation into Gable Mountain) began in July 1978. The schedule calls for heater testing to begin about December 1979 and spent fuel loading to begin in CY 1981. Spent fuel will be characterized, packaged, and shipped through support of the NTS EMAD facility. The tests will run for approximately three years each, followed by the removal of test hardware (including spent fuel).

Repository studies leading to the development of a full-scale facility include a facility project study and system studies. The facility project study will provide a conceptual design of a basalt repository, as well as support a decision concerning the engineering feasibility of a basalt repository. Current activities include several repository engineering studies (e.g., a parametric design trade-off study) and preparation of a preconceptual design. The current schedule calls for an architect/engineering firm to be selected and awarded a contract by the end of FY 1979 for performing conceptual design of a basalt repository. The effort will also include to-be-determined engineering design studies to support various aspects of the repository project, such as licensing and the remote handling system. It is anticipated that the conceptual design will be completed by September 1981.

Major milestones and funding are shown in Figure V 3-39.

Nevada Test Site (NTS). DOE's Nevada Test Site, located in southern Nevada, is being investigated for potential terminal storage sites and for compatibility with the predicted effects of continued weapons testing. Unlike other candidate sites, this site provides varied geologic systems, allowing investigations to be performed in granite, argillite (a compact clay rock cemented by silica), and tuff (a heat-fused volcanic ash).

Activities within this task are as follows:

- Site evaluations to determine if any major impediments exist to prohibit location of a repository at NTS and to identify specific sites and suitable geologic media for a repository
- Technical support to determine the characteristics of the available geologic media and weapons testing seismic effects
- In situ testing of thermal-related effects on an underground facility including the placement of encapsulated spent nuclear fuel

Core drilling for geologic samples is pictured in Figure V 3-40.

DOE's Nevada Operations Office was assigned the lead management role. Sandia Laboratories at Albuquerque, New Mexico, was chosen to provide technical assistance.

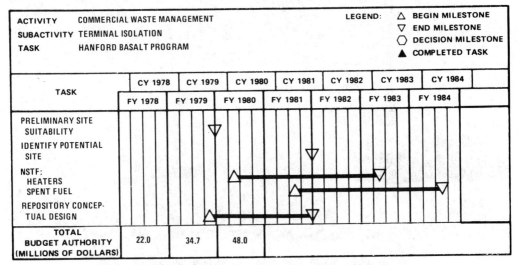

Figure V-3-39. Milestones and funding for the Hanford basalt program.

Program management will involve management, direction, initiation, and administration of all activities. For the NTS Task, this specifically includes the following:

- Program technical overview to provide for support by a technical management contractor (Sandia Laboratories) in preparation and maintenance of program and management plans, technical direction and management of technical program plans, review and evaluation of technical program, and technical status and cost reports
- Quality assurance to provide for development and establishment of the quality assurance program to be implemented on the investigative phases of the NTS Task

Site evaluations idenfied and resolved impediments in FY 1979 and, potential site locations will be determined in 1981. These studies include current geologic investigations to determine tectonics, stratigraphy, hydrology, natural seismicity, volcanism, and other characteristics. The Syncline Ridge was disqualified in mid-FY 1978 as a potential site due to geologic complexities caused by deep faulting; therefore, the program is examining the Yucca Mountain (tuffs) and Calico Hills (granite and shale) areas, as well as other promising formations off the NTS.

Technical support includes the following:

- Radionuclide transport studies (by Los Alamos Scientific Laboratory) to evaluate the effectiveness of shale, granite, and tuffs as a waste migration barrier
- Seismic investigations including data processing and analysis, seismic monitoring (surface and subsurface), and ground motion repository design criteria studies, to establish the potential effects of ground motion resulting from weapons testing at NTS

In situ testing will determine the response of the geologic media to thermal loading and the technology for the placement of spent fuel in an underground repository. These tests include the following:

- Electrical heater tests were initiated in early FY 1978 in the Piledriver facility 1,400 feet deep in Climax stock granite. These tests, completed in mid-CY 1978, provided in situ thermal-mechanical data and permeability data and verified the thermal analytical model. Further Climax tests with spent fuel will start in Fy 1980. Heater tests are underway in a shallow shale formation (Eleana) and are planned in the future in a tuffs formation.

Figure V-3-40. Core drilling for geologic samples.

- Commercial spent fuel is also being planned to be emplaced for a three-year test period into an extension of the Climax granite mine. Construction began in late FY 1978; spent fuel loading is expected to begin in mid-CY 1980. This demonstration test has objectives similar to the Hanford Basalt NSTF spent fuel test. Support for spent fuel acquisition, characterization, and packaging will be from the NTS EMAD facility.

Major milestones and funding are shown in Figure V 3-41.

National Waste Terminal Storage (NWTS). Based on the investigations performed and the results of evaluations made within the ONWI, Hanford Basalt, and NTS Tasks, a site will be selected. Appropriate criteria will be established for the design and construction of a repository, depending on the characteristics of the selected site.

A project organization with appropriate subcontractors will be established, and contractual arrangements will be made for accomplishing the work.

Figure V 3-42 depicts a conceptual waste repository.

Within the ONWI Task, conceptual designs of repositories in salt formations are being prepared to provide repository cost estimates and to identify the necessary elements of the engineering development program. Two separate studies have been completed: (1) design of a facility for spent fuel disposed in bedded salt, and (2) design of a facility for solidified HLW in domal salt. Both conceptual design efforts include below-ground and above-ground facilities to handle waste from either the reprocessing or the spent fuel cycle. The conceptual designs will provide engineering evidence of project feasibility, describe attainable facility performance levels, and include detailed cost estimates and schedules.

Other engineering development programs and studies which are included in the ONWI Task involve the following:

- Transporter vehicle for moving canistered wastes in the mine
- Drilling machines to drill waste emplacement holes
- Transportation requirement
- Alternate waste retrieval methods

Milestones and capital construction funding for the NWTS Task are shown in Figure V 3-43.

Transportation. The objective of DOE's Transportation Task is to ensure a national capability of coping with the growing complexity of nuclear material transport, including all types of radioactive waste.

The technology developed and demonstrated to transport existing and future inven-

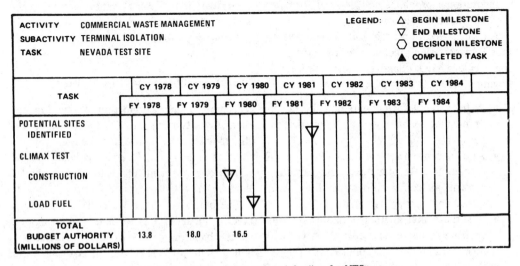

Figure V-3-41. Milestones and funding for NTS.

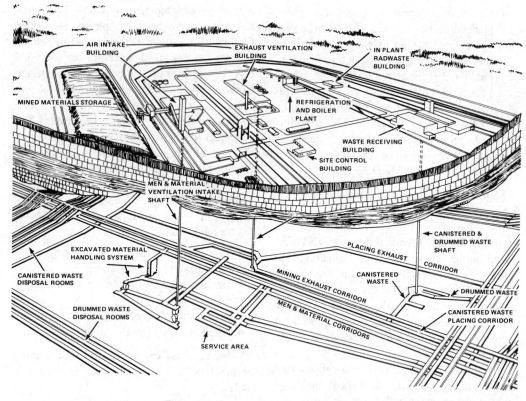

Figure V-3-42. A conceptual waste repository.

tories of defense-related wastes will also provide for establishing transportation methods and equipment for commercial wastes. Close coordination with the defense waste operation will ensure prompt and effective technology transfer.

Institutional problems are identical for commercial and defense wastes.

An integrated approach will be pursued by the Transportation Technology and Information Center (ALOO/Sandia) to provide the following:

- A focal point for encouraging the development of technology and industrial capability
- Adequate interface data and planning information data matching transportation systems to fixed facilities, handling hardware, and carrier parameters
- Assurance that all transportation systems are safe, reliable, and cost-effective, and meet with public acceptance
- Specific needs of commercially-oriented program as they are identified and satisfied

The aftermath of a shipping cask impact test within the DOE package testing program is shown in Figure V 3-44.

Because commercial transportation activities are currently not funded as a separate budget category, no separate milestone chart is shown here. Transportation activities will support the commercial waste management program milestones.

Waste Treatment Technology

This part of the Commercial Waste Management Activity deals with R&D associated with the treatment of radioactive wastes prior to their shipment to a Federal repository for terminal emplacement of disposal by shallow land burial of low-level waste.

The objective of the Waste Treatment Technology Subactivity is to develop and demonstrate handling and treatment technology which will provide waste disposition forms which meet appropriate waste disposal criteria. Preliminary design specifications will be developed in accordance with appropriate

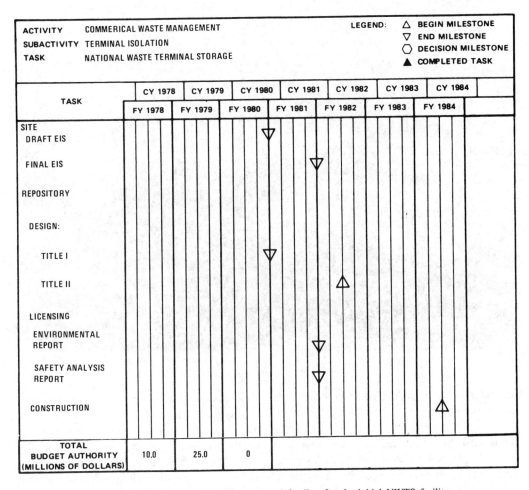

Figure V-3-43. Earliest possible milestones and funding for the initial NWTS facility.

processes needed to immobilize each waste type, based on radioactive pilot plant demonstrations. The technology to handle existing inventories of defense-related wastes will also provide a basis for establishing handling and processing methods for commercial wastes. Close coordination with the defense waste operation will ensure prompt and effective technology transfer. Support will be provided by the Transportation Technology and Information Center for related transportation issues.

Pilot-scale waste immobilization studies are planned to demonstrate that prespecified limits of waste containment can be attained using currently available processes. It is expected that this technology will satisfy all relevant storage handling, environmental, and repository criteria.

Figure V 3-45 shows borosilicate glass, which is one of the most promising forms for storing high-level radioactive wastes from reactors.

A high-level waste R&D program is designed to provide an immobilization process and a waste form for treating existing commercial wastes by 1981. It also supports the treatment of wastes which may arise from alternate fuel cycles in the late 1980s. Portions of the gaseous materials and contaminated solid waste activities will provide a technology base to help meet the new, more stringent EPA emission standards.

National program strategy documents will be established for each waste type and will be periodically updated. As soon as major treatment methods have been developed and demonstrated, and technology transfer documentation, including appropriate design information has been prepared, the methods and documentation will be made available to the commer-

Figure V-3-44. Package testing program.

cial industry. If it is decided to proceed with an alternative fuel cycle, including chemical processing of reactor fuel, appropriate programs will be established to build the first full-scale plants for converting waste to products which meet Federal repository acceptance criteria, NRC regulatory requirements, and all other applicable standards and regulations.

Techniques will be developed to handle existing and future wastes from fast flux reactors. Radioactive sodium coolant and sodium-contaminated components of fast flux reactors present unique waste handling problems. Treatment methods will be developed to convert these wastes into forms which can be routinely handled and stored.

Milestones and funding for the Waste Treatment Technology Subactivity are given in Figure V 3-46.

High-level Waste (HLW). As nuclear fuel is used in a reactor, the uranium fuel material is slowly converted by the nuclear reaction into solid fission products dispersed uniformly through the fuel material in the fuel rods or as gases entrapped within the fuel rod. High-level nuclear wastes are highly radioactive waste fission products which are stripped out of the fuel elements during the chemical extraction cycles used to recover the useful materials remaining in the spent nuclear fuel rods. The wastes are initially in a liquid form and must be converted into a dry, stable solid for final disposal to ensure that required safety, environmental, and public health standards are met and maintained.

This task is directed toward the development of immobilization technology for existing high-level radioactive wastes. Waste forms

Figure V-3-45. Vitrified waste samples (nonradioactive).

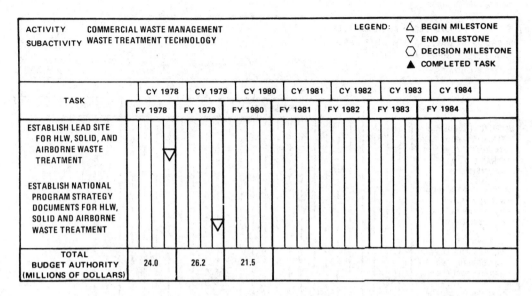

Figure V-3-46. Milestones and funding for waste treatment technology.

which increase the stability and minimize the possibility of dispersion are being developed. Borosilicate-type glasses continue to be the leading contender for high-level waste; in addition, synthetic materials, ceramics, concrete, clay, and metal matrix forms are also being evaluated. This activity is closely coordinated with defense waste immobilization activities. An HLW canister is depicted in Figure V 3-47.

DOE's Savannah River Operations Office has been assigned the lead management role. E. I. DuPont de Nemours & Co. at Savannah River Laboratory is providing technical assistance.

The R&D program for HLW includes the development of technology for a spray calciner/in-can melter process, fluid-bed calciner/continuous melter process, direct liquid-fed continuous melter process, fluid-bed calciner process, alternative waste form processes, and waste form quality verification. The program will confirm solidification and packaging technology by preparing full-scale canisters of vitrified waste. In the initial demonstration, pilot plant facilities at Richland will be used to vitrify and encapsulate the waste products. Processes being demonstrated are calcination, batch melting, continuous melting, in-can melting, liquid-fed melting, and various combinations.

Development of waste forms and associated processes are being expedited in FY 1979 as alternatives to glass; these include cement, special concrete, ceramics, and metal matrix materials.

Work to date has concentrated on nonradioactive demonstration of various processes and waste/glass formulations which can safely treat and contain the waste.

The current activities include the modification of existing facilities, construction of additional facilities needed, and installation of equipment. Plans were prepared for nonradioactive tests and radioactive waste operations began in FY 1979. Major milestones and funding are shown in Figure V 3-48.

Low-level Waste (LLW). Commercial LLW is a by-product of nuclear power reactor operation, medical and industrial use, and R&D activities. Typical examples of solid LLW are discarded equipment and metals, filters from the cleanup of gaseous wastes, ion-exchange resins from the cleanup of liquid wastes, liq-

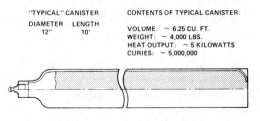

Figure V-3-47. Typical HLW canister.

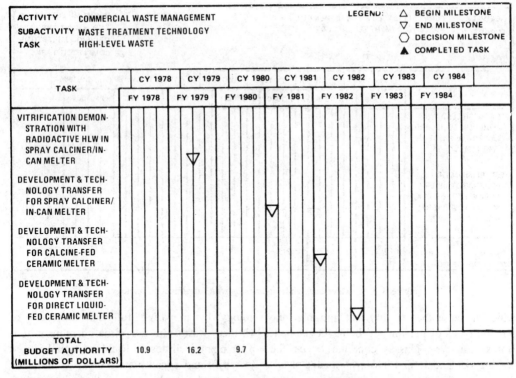

Figure V-3-48. Milestones and funding for HLW.

uid wastes that have been converted to solid form by techniques such as mixing with cement to make concrete, and miscellaneous trash such as paper, rags, glassware, and protective clothing. (Liquid LLW is decontaminated and released under controlled conditions.)

The objective of this task is to support commercial shallow land burial of low-level solid wastes.

Technology transfer will ensure that burial ground techniques developed under the defense LLW program will be applicable to the commercial program.

Figure V 3-49 shows a typical LLW burial ground.

The lead management role for this task was recently assigned to DOE's Idaho Operations Office, with associate lead to DOE's Oak Ridge Operations Office. Those groups chose Edgerton, Germeshausen, and Grier Corporation (EG&G) at INEL and Union Carbide at ORNL to provide technical assistance for their management activities.

In concert with the Defense Waste Management Activity, the R&D program for LLW encompasses all aspects of the technology. However, the funding provided by the Commercial Waste Management Activity is principally directed to the development of technology for concentration of waste radionuclides, immobilization of waste concentrates, and packaging and interim storage of LLW from effluents. Three low-level liquid waste purification processes are being evaluated for removal of radioactive contaminants from effluent streams (as alternates to evaporation). It is planned to demonstrate at least one process in pilot scale by 1984.

Current activities include investigation of ultrafiltration, selective absorbents, and biological concentration processes, as well as concreting as a method for solidifying resulting sludges.

Major milestones and funding are depicted in Figure V 3-50.

TRU Waste. TRU-contaminated solid waste is contaminated with transuranic nuclides above a threshold concentration measured in nanocuries per gram. This threshold was derived on the basis of certain natural uranium-

Figure V-3-49. Savannah River solid waste burial ground with surface stabilization test plots.

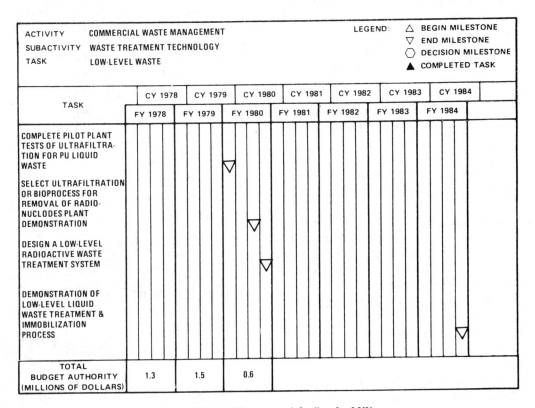

Figure V-3-50. Milestones and funding for LLW.

containing ores and is currently under review by DOE and NRC. Waste below this threshold level is acceptable for shallow land burial at controlled sites as LLW. Plutonium-239 is the nuclide of principal concern in TRU waste because of its long half-life of 24,000 years. Both in terms of volume and mass, the proportion of radioactive material to the contaminated material is minute.

This task involves development of technology for safely and economically treating TRU wastes to meet anticipated criteria for long-term isolation. This task also includes development of technology for spent fuel cladding hull storage and handling safety, hull decontamination, hull consolidation and immobilization, feasibility of hull material recovery, and criteria of acceptable storage forms. Such technology includes compaction, crushing, digestion or combustion, immobilization of residues, and decontamination of metallic surfaces to reduce waste quantities. Compacting is a single process that works reasonably well for low density materials. Crushing works for glass and large metal objects that can usually be cut into pieces for tighter packing. Better volume reduction for many types of wastes is obtained using some form of digestion or combustion performed in closed systems so that radioactive material is contained. Processes receiving attention are acid digestion and anaerobic bacterial digestion. Two of the processes under investigation are shown in Figure V 3-51.

Most sophisticated incineration processes result in an ash, a fine powder, or a sludge. Immobilization of resulting ashes, residues, and sludges may be necessary to meet the geologic repository waste acceptance criteria. Technology is being developed to produce a high-density solid that is rugged enough to be handled without damage, with individual pieces large enough that they cannot become airborne or scattered too widely for easy recovery and with physical and chemical properties that prevent release of the radioactive material to the surrounding geologic environment.

The management of this program was assigned to DOE's Albuquerque Operations Office. Rockwell International is providing technical assistance.

The R&D program for TRU-contaminated solid waste has the primary objective of developing and demonstrating solid waste treatment methods (incineration, immobilization, etc.) required to meet Federal repository acceptance criteria by FY 1981.

During FY 1978, testing of three prototype TRU-contaminated solid waste treatment systems (Controlled Air Incinerator, Fluidized Bed Incinerator, and Acid Digestion) was initiated, and radioactive tests begun if warranted. These tests were completed and documented in FY 1979.

Work is continuing on the demonstration of the incineration of combustible TRU waste. Studies on smelting as a method for reducing the spent fuel cladding hulls are continuing. Also, the use of electropolishing for decontaminating metal surfaces is being tested on a pilot scale.

Major milestones and funding are given in Figure V 3-52.

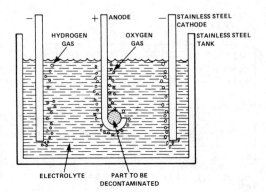

Electropolishing

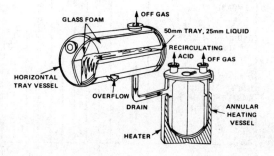

Acid Digestion

Figure V-3-51. Examples of TRU decontamination processes.

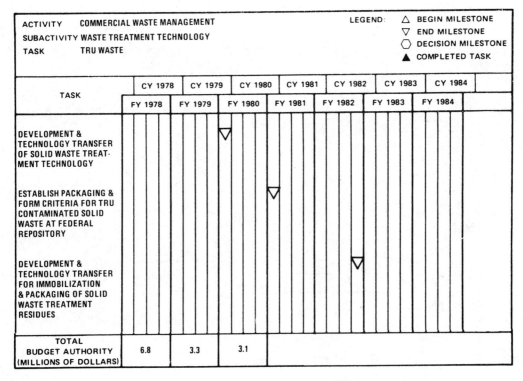

Figure V-3-52. Milestones and funding for TRU waste.

Airborne Waste. The Airborne Waste Task was developed to provide technology needed to capture and manage the radioactive materials found in the off-gas from nuclear facilities in a manner which protects the public health and safety, produces minimal environmental impact, and meets all applicable laws and regulations.

These wastes arise from nuclear reactors, irradiated fuel storage, reprocessing of spent fuels, and waste treatment processes such as waste incineration and vitrification.

This task includes development of technology for waste treatment criteria, noble gas immobilization and storage, carbon-14 immobilization, iodine immobilization and storage, and gaseous and other airborne waste measurements. One process under investigation is shown in Figure V 3-53.

DOE's Idaho Operations Office was recently given the lead management responsibility. Allied Chemical Corporation at the Idaho National Engineering Laboratory is providing technical assistance.

The DOE R&D work in this area encompasses all aspects needed to monitor, concentrate, collect, immobilize, package, transport, store, and prepare for disposal of these waste materials from commercial facilities. Specific activities currently being pursued are structured to accomplish the following objectives:

- To provide a national strategy plan for managing all airborne wastes by 1979
- By 1980, to develop storable waste forms for krypton-85, iodine-129, and carbon-14, and demonstrate suitable processes for producing these forms by 1984
- By 1982, to develop the capability to monitor in real-time airborne waste streams containing carbon-14, iodine-129, and krypton-85
- By 1983, to develop the conceptual design for a storage facility for krypton-85 and iodine-129
- To identify, treat, and manage airborne wastes which will be generated in alternative fuel cycles now under investigation

Major milestones and funding are shown in Figure V 3-54.

830 Nuclear Waste Management

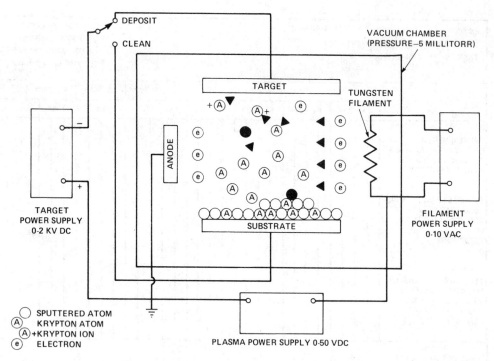

Figure V-3-53. Sputtering process for Kr-35 solidification.

ACTIVITY	COMMERCIAL WASTE MANAGEMENT							LEGEND:	△ BEGIN MILESTONE
SUBACTIVITY	WASTE TREATMENT TECHNOLOGY								▽ END MILESTONE
TASK	AIRBORNE WASTE								⬡ DECISION MILESTONE
									▲ COMPLETED TASK

TASK	CY 1978	CY 1979	CY 1980	CY 1981	CY 1982	CY 1983	CY 1984	
	FY 1978	FY 1979	FY 1980	FY 1981	FY 1982	FY 1983	FY 1984	
CONCEPTUAL DESIGN FOR KRYPTON LONG-TERM MANAGEMENT FACILITY				▽				
COMPLETE DEVELOPMENT OF WASTE IMMOBILIZATION TECHNOLOGY FOR Kr-85, I-29 & C-14						▽		
FIELD DEMONSTRATION OF REAL-TIME MONITOR FOR RADIOACTIVE GAS				▽				
TOTAL BUDGET AUTHORITY (MILLIONS OF DOLLARS)	2.6	2.7	2.0					

Figure V-3-54. Milestones and funding for airborne waste.

"Landlord" Responsibilities—Pacific Northwest Laboratory (PNL).

Landlord functions at the various DOE installations provide the general purpose equipment and facilities required in support of the overall operation at the site.

The funding is provided by the Headquarters program which supports the major portion of the work being conducted at the installation. ETW has assigned the landlord funding responsibility to Battelle Pacific Northwest Laboratory (Figure V 3-55) as part of the Commercial Waste Management Activity.

The FY 1980 program will provide for general plant projects related to the PNL site. Also included will be a Plant Operation Maintenance Facility. This project will provide a new centralized craft and operations support facility of approximately 16,000 square feet in support of the PNL research and development activities in the Hanford 300 Area. The proposed facility will consist of a one-story, pre-engineered metal building approximately 80 ft. x 200 ft. The building will contain various craft shops such as carpentry, electric, welding, painting, instrument, millwright, plastics, filter, a mechanical/electrical machine shop, plus a tool crib, change rooms, restrooms, a lunch room, offices, and building services areas. A detached, covered storage warehouse of approximately 2,000 square feet is also to be located adjacent to the main facility. In addition to the covered space provided by this project, outdoor assembly pads for large craft projects, loading docks and sufficient maneuvering space for delivery vehicles, and material storage racks will be provided.

Major milestones and funding are shown in Figure V 3-56.

Supporting Studies

Information and data evaluations of health and safety, economics, risk assessments, quality assurance, and environmental impact statements will be developed and reviewed to support the waste management program. Evaluations will be prepared on alternative systems for treatment and disposal of radioactive waste from the conventional LWR fuel cycle and from other fuel cycles identified in International Nuclear Fuel Cycle Evaluation (INFCE) and Nuclear Alternative System Assessment Program (NASAP) studies as having nonproliferation potential. A program has been established to keep abreast of all international programs dealing with waste management.

Battelle Pacific Northwest Laboratory (BPNL) is the lead site for preparation of the Generic Environmental Impact Statement

Figure V-3-55. Battelle Pacific Northwest Laboratory.

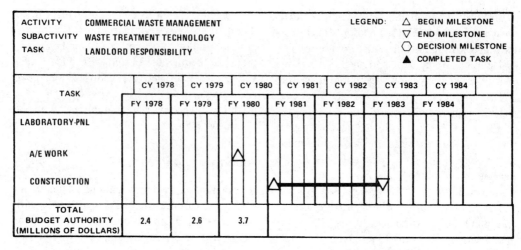

Figure V-3-56. Milestones and funding for landlord responsibilities.

(GEIS) on Management of Commercially Generated Radioactive Waste, coordination of certain aspects of international programs, performance of safety studies, and evaluations of alternate waste management technology including technology for changing the nature or reducing the amount of waste generated. This subactivity supports several staff members who serve as U.S. representatives to INFCE working groups 6 and 7, the Nuclear Energy Agency (NEA) Waste Management Committee, and subsidiary working groups. Cooperative research program agreements exist with the Commission of the European Communities (CEC), as well as multilateral and bilateral agreements with foreign countries.

A committee of national experts to evaluate and report on all major waste management programs and activities at all plant operating sites, the Committee on Radioactive Waste Management (CRWM) of the National Research Council (NAS/NAE/IOM), is being supported.

The nuclear fuel waste management cycle is illustrated in Figure V 3-57.

The principal objective is to prepare a comprehensive GEIS to support the Commercial Waste Management Activity. This statement will evaluate the environmental impacts of the various technologies in the activity. Cost estimates will be presented to respond to the frequent comment that high waste management costs will be a threat to the safe accomplishment of waste isolation.

The scope of the GEIS focuses on the post-fission radioactive wastes which are identified for Federal custody in present or proposed NRC regulations (high-level and TRU wastes). The activities include storage, treatment, transportation, and final disposition, regardless of whether they are carried out by industry or by DOE. The sites of the activities described will be typical or hypothetical. The GEIS will not eliminate the need for future environmental statements, either by NRC or DOE, to cover a specific waste repository or the installation of a specific waste treatment process at a specific commercial plant. Three possible fuel cycles are evaluated.

Major milestones and funding for the Supporting Studies Subactvity are depicted in Figure V 3-58.

Solidification Process Demonstration

This subactivity is intended to demonstrate the applicability of an existing process for the immobilization of typical high-heat generation radioactive wastes. It includes all activity for procuring spent fuel, preparation of hot-cell to process fuel, maintaining and upgrading vitrification equipment, converting plutonium nitrate to plutonium oxide, vitrifying the waste, reporting results, modifying existing

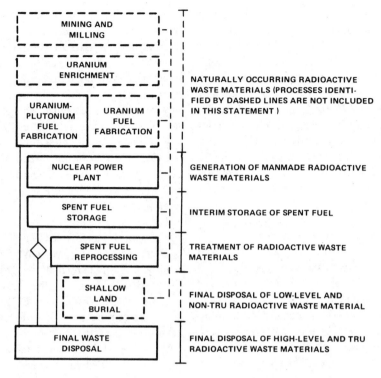

Figure V-3-57. Nuclear fuel cycle waste management.

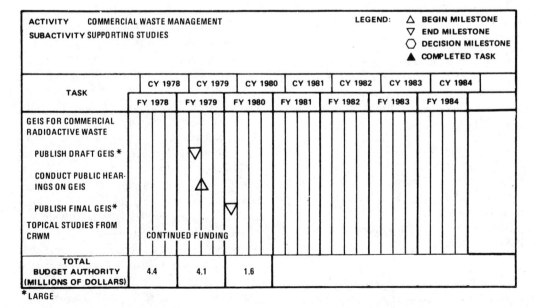

Figure V-3-58. Milestones and funding for supporting studies.

buildings, and relocating existing americium processing equipment. A process flow diagram for the project is shown in Figure V 3-59.

During FY 1978, the modification of all facilities to accommodate this demonstration was completed. Design manufacturing and construction was completed. Cold operation (utilization of nonradioactive feedstock) was begun. This project was completed in FY 1979 with the production of two canisters (8" in diameter by 8' high) of borosilicate glass.

Decontamination and Decommissioning (D&D)

This subactivity includes planning and developing methods for D&D of commercial nuclear reactor and fuel cycle facilities. DOE has a program to decontaminate nuclear facilities declared excess or to decommission such facilities where appropriate. Facilities for such action are being selected depending upon their maintenance expenses and their potential hazard. The candidate facilities (Figure V 3-60) provide a spectrum of representative facilities; therefore, the techniques developed to execute this activity are of value to decontaminating or decommissioning commercial facilities.

A program is being developed to address commercial D&D assistance that will provide the following:

- Data dissemination, consultation, advice, and research and development such as archival maintenance methodology
- A readiness capability to conduct direct D&D if required

The lead role for management of this subactivity was assigned to DOE's Richland Operations Office. United Nuclear Industries Inc., at Richland, Washington, is providing technical assistance.

Major milestones and funding are shown in Figure V 3-61.

Low-level Waste Management Operations (Commercial Burial Sites)

If public laws are modified to extend Federal management over all commercially-operated disposal facilities a program will be developed to establish a fair and timely change of responsibility. Activities to be performed would include a survey of existing facilities and equipment with appropriate upgrading recommendations, evaluation of waste and disposal environments to establish acceptance criteria and site operating procedures, evaluation of site capacity and throughput requirements, and fee schedule development. These Federal activities would be coordinated with private industry users to ensure an equitable and efficient disposal program.

Figure V 3-62 shows the low-level waste burial sites in the United States.

DOE does not have any direct responsibility for commercial burial grounds at this time. However, if in the future Congress modifies public laws, then DOE may be responsible for the operation of those facilities. If this occurs, the Idaho Operations Office supported by EG&G will establish a program for extending Federal management over commercial waste burial ground operation. Present programmatic effort are of a contingency nature.

West Valley Nuclear Center Activities

In accordance with the FY 1978 Authorization Act, DOE conducted a study to explore technical and institutional options for the possible decontamination and decommissioning of this site and possible future continued use.

The Western New York Nuclear Service Center (WNYNSC) is located in a rural area about 30 miles southeast of Buffalo. The small communities of West Valley, Riceville, and Ashford Hollow and Springville are located within 5 miles of the center. The center's facilities include the only commercial nuclear fuel reprocessing plant that ever operated in the United States, a fuel receiving and storage facility, burial areas for solid radioactive wastes, and tanks containing liquid high-level radioactive wastes. The commercial operator of the center reprocessed fuel there from 1966 to 1972. The plant is now being maintained in a shutdown condition. The facility is located on land leased from New York State. The terms of the contract between

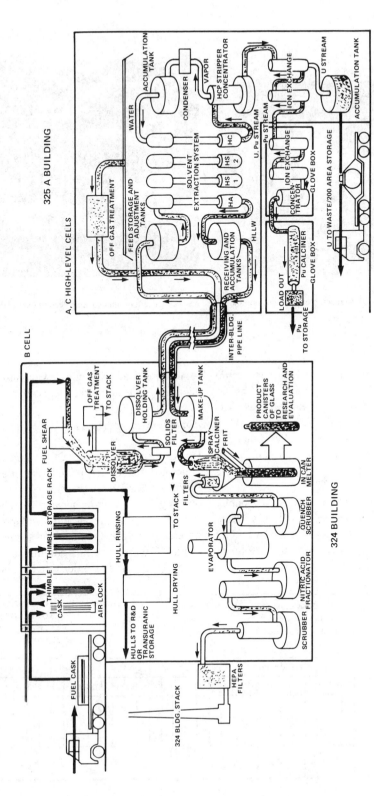

Figure V-3-59. Commercial nuclear waste vitrification project.

Figure V-3-60. A reactor experiment complex that has been decontaminated and decommissioned.

the State and the commercial operator require the State to assume responsibility for the facilities and the wastes (when the lease expires) at the end of 1980. The State of New York requested, in November 1976, that the Federal Government take over the site.

The reprocessing facility located on the site is shown in Figure V 3-63.

The study report was published for public comment in November 1978 and was transmitted to Congress (Document No. TID 28905, Volumes 1 and 2) with the results of the public review in February 1979.

Additional activity must await future policy decisions and supporting legislative action.

SPENT FUEL STORAGE ACTIVITY

On October 18, 1977, DOE announced its policy to offer to accept and take title to spent fuel from U.S. nuclear reactors and a limited quantity from reactors located in other countries. The spent fuel activity will provide the necessary storage facilities and supporting activities to carry out this commitment.

President Carter's order to defer indefinitely

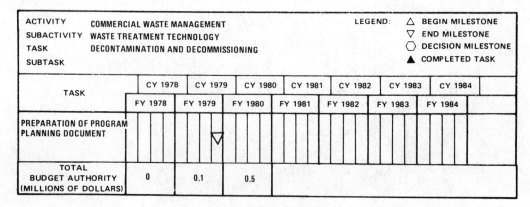

Figure V-3-61. Milestones and funding for D&D.

Figure V-3-62. LLW burial sites in the contiguous United States.

the reprocessing of spent fuel in the United States and his urging that other nations join in this deferral require retrievable storage of spent fuel for an indefinite period of time, pending a decision either to recover and reuse that fuel or to dispose of it permanently.

Extensive transportation requirements for spent fuel disposition stem directly from these policies. Separate investigation and planning is now being undertaken for domestic fuel and for foreign fuel in keeping with the overall program for moving all nuclear materials. DOE-owned spent fuel has been transported for many years, primarily under the Naval

Figure V-3-63. Western New York Nuclear Service Center.

Reactors Program. In addition, a few shipments of spent fuel from research reactors have also taken place. All of these shipments have proceeded safely and without adverse effect to the public or the environment.

Figure V 3-64 shows the flow diagram for the handling and storage of spent fuel.

The goal of the Spent Fuel Storage Activity is to provide for the safe storage of U.S. and some foreign spent fuel consistent with national nuclear energy goals and U.S. nonproliferation objectives.

The overall objectives of the program are as follows:

- To ensure that space is available for the storage of spent fuel so that operation of nuclear power plants within the United States can be maintained
- To support nonproliferation initiatives by offering a credible alternative to nuclear fuel reprocessing by foreign nations

The immediate objectives for domestic fuel storage activities include the following:

- To plan and manage a program that provides interim spent fuel storage capacity until geologic storage or recovery is available
- To identify and resolve safety and environmental issues and concerns over extended spent fuel storage
- To establish a charge and contract requirements for DOE to provide storage and disposal services
- To develop technology to increase the capacity of existing spent fuel facilities, and to develop advanced, improved, high-capacity storage technologies

The immediate objectives for international spent fuel storage are as follows:

- To accept limited amounts of foreign spent fuel in order to meet nonproliferation objectives subject to necessary legislation and the approval of a required plan for foreign fuel storage in the United States
- To provide technical support to international bodies and limited direct assistance to individual countries to establish spent fuel storage as a credible alternative to reprocessing
- To establish and maintain a current assessment of foreign spent fuel storage needs, capabilities, and plans, and to develop and maintain a logistics plan for transferring fuel in critical or sensitive situations
- To augment, as appropriate, domestic spent fuel technology and demonstration activities to support the unique requirements of international or foreign transportation and storage

The activity should reduce the uncertainty faced by utilities with respect to the disposi-

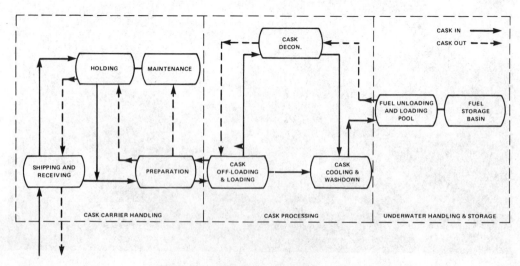

Figure V-3-64. Spent fuel handling and storage.

tion of spent fuel discharged from reactors, as well as foreign pressures to reprocess spent fuel because of full storage basins.

International Spent Fuel Storage

The capacity to accept and store foreign fuel will parallel the development of domestic spent fuel storage capacity and is included in planning for additional capacity demands. The storage of foreign spent fuel in DOE-approved facilities, the establishment of multinational and international fuel storage facilities, and technical assistance to international studies of fuel storage technology are potential efforts in support of U.S. non-proliferation initiatives.

Similar to those conditions associated with transportation of domestic spent fuel, the transportation of foreign fuel to storage sites, located either within the United States or without, will be subject to extensive regulation. In this case, both U.S. and foreign regulations and those of international regulatory bodies could apply. In addition, the transportation system configuration must be compatible with foreign, domestic, multinational, and international storage facilities (when and as they are established) and with foreign and domestic carriers.

Through interaction with other Government agencies, such as the Department of State, DOE is working to determine the estimated amounts of foreign fuel which the United States may receive. It is anticipated that the vast majority of foreign fuel will be from light water reactors (LWR's) and will be amenable to the same storage techniques as domestic fuel. Figure V 3-65 shows typical LWR fuel assemblies.

The International Spent Fuel Storage Subactivity includes R&D of new and existing processes, equipment, or facility layout where international spent fuel storage and/or its unique problems are involved.

In addition, R&D and other investigations are proceeding in the international transportation program area to resolve any unique problems and to ensure that consequent transportation systems will be supportive of U.S. nonproliferation goals.

Major milestones and funding are shown in Figure V 3-66.

Domestic Spent Fuel Storage

A one-time fee will be established and paid to the U.S. Government for storage and final disposal or "disposal only" services for spent fuel. The fee will be based on full-cost recovery to the Government. If the fuel is eventually retrieved and reprocessed, an appropriate adjustment to the fee or other compensation will be made after retrieval.

An essential element of this storage subac-

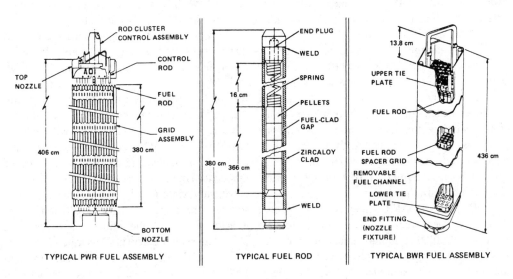

Figure V-3-65. Typical LWR fuel assemblies.

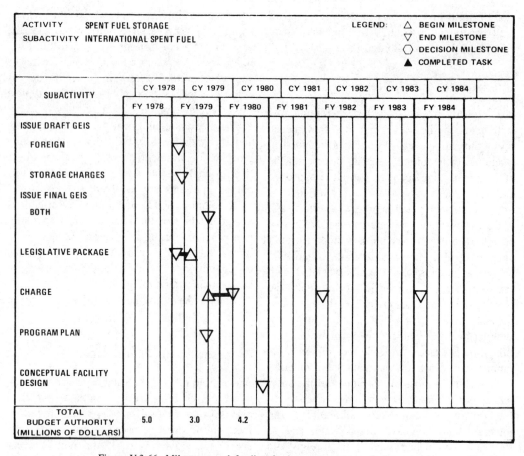

Figure V-3-66. Milestones and funding for international spent fuel storage.

tivity is the need to transfer spent fuel elements to the storage sites and, eventually, away from these sites when later disposition of the spent fuel is determined. The transportation systems and procedures which are to be employed—storage in privately owned or Government-owned AFR's—must be configured and operated not only in a way compatible with the storage facility itself, but also in a manner which complies with regulatory controls governing the movement of all nuclear materials. DOE is charged with assuring that such spent fuel transportation can be accomplished when required as part of its overall nuclear transportation program.

DOE-approved AFR water storage basins will store spent fuel in a retrievable mode. DOE's FY 1980 program plans to provide storage capacity by 1983 consistent with the spent fuel policy approved by the President and the recommendations of the IRG report.

Prior to 1983, DOE will consider accepting spent fuel in cases where reactors do not otherwise have sufficient storage capacity. Figure V 3-67 shows an AFR fuel storage pool.

Many utilities are facing a shortage of adequate spent fuel storage capacity. In some cases, this can be alleviated by more compact racking of their basins. In other cases, transfers of fuel from one basin to another within their own system are possible. While the increased storage basin capacities will provide relief for some utilities, others will face the prospect of inadequate storage. Estimates of the actual transfers the Government will receive vary significantly. Two significant data sources are a recent NRC GEIS on Spent Reactor Fuel and the utility responses to a DOE letter of inquiry in December 1977.

The NRC GEIS points out that, through 1990, the total combined storage capacity of all reactors basins, without expansion, exceeds cumulative reactor discharges. While this space can be used to some extent by trans-shipment of fuel between reactors and between utilities, there may be limitations. From a technical standpoint, some fuel cannot be stored in basins designed for and containing fuel from other reactors without some modifications. From a practical standpoint, it is likely that utilities would be reluctant to devote their limited capacities to the storage of spent fuel from other utilities. However, transfer of fuel between reactor basins owned by the same utility is possible and is being done to delay plant shutdowns.

In December 1977, DOE sent letters of inquiry to utilities with nuclear power plants operating or under construction to determine their interest in transferring fuel to the Government through 1990 under the terms of the Spent Fuel Storage Activity. Most utilities indicated a desire to transfer some fuel by 1990; some wanted to begin transfer as soon as the Government would accept it. Others indicated an interest in delaying transfer until their basin capacity was full or until the fuel could be disposed of directly into a repository. On the average, the amount of fuel which utilities appear to be interested in transferring lags the reactor discharges by about 7 years.

As part of its continuing effort to assess fuel storage requirements and capabilities, DOE has developed a system for assessing storage and transportation requirements for a variety of assumptions. This information is kept current with up-to-date utility plans. Based upon utility plans as of December 31, 1978, limited trans-shipments of spent fuel within the utilities and maintaining full core reserve, the following storage requirements exist:

YEAR	CUMULATIVE METRIC TONS
1979	70
1980	80
1981	200
1982	350
1983	560
1984	810
1985	1,180
1986	1,810
1987	2,770
1988	3,860

Figure V-3-67. Fuel storage pool.

Program Management and Supporting Studies. Many of the aspects of the Spent Fuel Storage Subactivity are applicable to both the domestic and international fuel disposal activities. These activities include the establishment of storage charges, legislation, or the design of a facility. Since greater requirements are currently expected for the utilization of spent fuel storage facilities for domestic fuel, most of the funding and effort in FY 1980 is directed toward that part of the activity. This is supplemented, wherever necessary, to support the unique aspects of the International Spent Fuel Storage Subactivity.

The Spent Fuel Storage Subactivity is structured into four work elements as shown in Figure V 3-68.

Program management and support includes studies and analyses that will be performed to provide the basis for the development of spent fuel disposal policy, legislation, and business and financial arrangements and to delineate general requirements. A significant fraction of this effort will be directed toward the development of GEIS's as required. This task includes the development of program plans and the establishment of a spent fuel storage fee.

Fuel storage technology involves provision of technical support for spent fuel storage technology development. This includes improvements in the efficiency and capacity of reactor storage basins and the development of concepts other than water basin storage.

Safety and environmental studies are being performed to improve calculational techniques and models used to assess the safety and environmental impacts of storage basins. Advanced storage technology will be performed and will include criticality and shielding investigations.

Facilities and systems provides for the acquisition of spent fuel storage capabilities by 1983, including the use of existing facilities and their expansion, technical support for the design of new industry basins and a Government basin. These activities will be carried out concurrently until the selection of the appropriate option or options to satisfy specific program needs. In particular, contingency storage at existing facilities to meet critical near-term requirements will be investigated.

The lead role for management of this task was assigned to the Savannah River Operations Office. E. I. DuPont de Nemours and Company at Savannah River Laboratory is providing technical assistance.

Major milestones and funding are shown in Figure V 3-69.

AFR Storage Facilities. Despite utility actions to increase storage capacity at the reactor sites, some utilities find it necessary to ship spent fuel off site. Under the Spent Fuel Policy, the Federal Government would accept this fuel and provide for its storage at AFR storage facilities.

Although a number of concepts have been studied for AFR storage, current interest has focused on storage of uncanistered fuel in an open water basin constructed of concrete with a stainless steel liner, a concept which has been used successfully for over 30 years in Government and commercial applications. All DOE AFR facilities are expected to be licensed.

Figure V 3-70 shows a typical plot plan for a conceptual AFR installation.

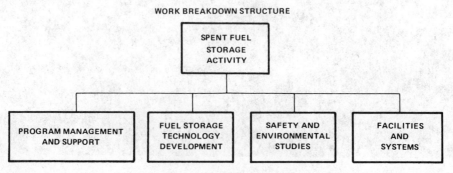

Figure V-3-68. Spent fuel storage major work elements.

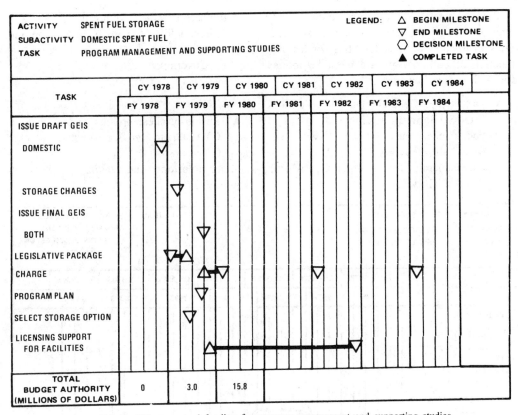

Figure V-3-69. Milestones and funding for program management and supporting studies.

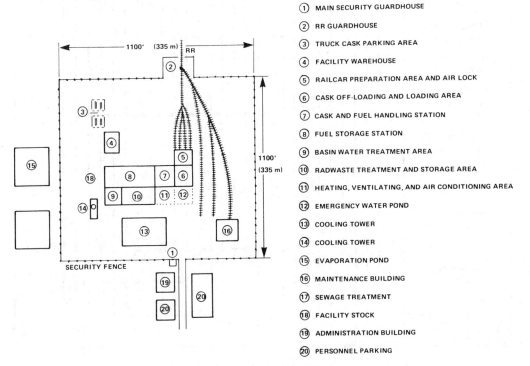

Figure V-3-70. Conceptual AFR plot plan.

The expected AFR storage requirements may be met by the use of existing facilities, the construction of new facilities, or a combination of both. Below are possible options for AFR storage capacity.

- The Allied-General Nuclear Services (AGNS) Barnwell Facility
- The Nuclear Fuel Services (NFS), Inc. West Valley Facility
- The General Electric (GE) Co. Morris Creek Facility
- Tennessee Valley Authority (TVA)
- "Greenfield" (A new facility at an unspecified site)
  — Government
  — Private

A decision on these options will be made in early CY 1981 to provide adequate storage capacity by 1985.

Major milestones and funding are shown in Figure V 3-71.

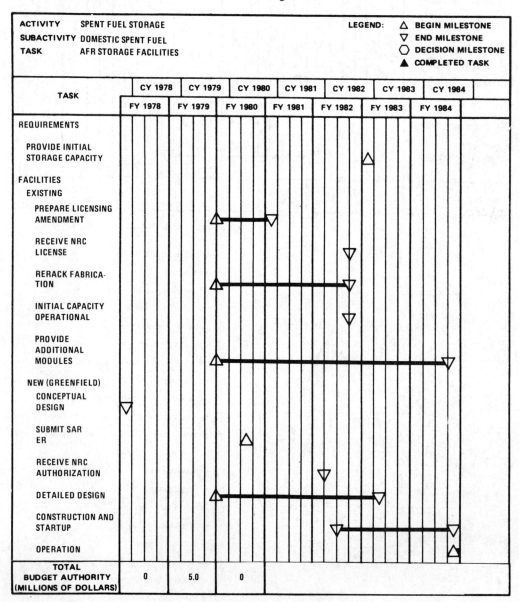

Figure V-3-71. Milestones and funding for the AFR storage facilities.

4.
Field Activities

The Government's Nuclear Waste Management Program has been a geographically decentralized program since its inception under the Manhattan Project. Each Government site has primary responsibility for management of the wastes generated at that site.

The responsibilities of DOE Headquarters personnel include development of overall plans, establishment of priorities, and completion of analyses to ensure that programs are coherent, goal-oriented, and related to the NEA, NEP, and other plans and legislation. In addition, effort are directed toward developing the overall budget, guiding DOE's programs through Congress, and then assessing the adequacy and progress of the approved efforts to meet the stated goals and objectives.

Under the DOE policy of decentralization, day-to-day project management is the responsibility of field offices. Detailed management of projects includes the management of resources to accomplish a given objective within prescribed dollar, performance, and schedule constraints. Specific field office authority and responsibility are detailed in a program agreement which is approved by Headquarters.

In addition, the management of waste technology development programs has been decentralized to lead field offices and contractors. As defined in management agreements, the lead offices provide overall technical program direction and management in specified program areas. Lead field offices and contractors are listed in Table V 4-1.

Figure V 4-1 illustrates the interface between Headquarters, field offices, and lead contractors. This interface provides the foundation for decentralization.

DOE LABORATORIES

The national and engineering laboratories traditionally have played important roles in achieving goals in the waste management program. They have provided solutions to R&D problems affecting the technical and economic

Table V-4-1. Lead Field Offices and Contractors.

TECHNOLOGY AREA	LEAD FIELD OFFICE	CONTRACTOR
HLW	SR	DUPONT
TRU	AL	ROCKWELL
LLW	ID	EG + G
AIRBORNE	ID	ALLIED CHEMICAL COMPANY
D + D	RL	UNI
WASTE ISOLATION	RL	BATTELLE COLUMBUS
SPENT FUEL	SR	DUPONT
TRANSPORTATION	AL	SANDIA

feasibility and the timetable of nuclear energy activities. In addition to the performance of in-house research and development at each installation, their personnel have functioned in liaison and supervisory capacities for the management of subcontractors such as universities, other laboratories, and industrial organizations. The laboratories are staffed with scientists, engineers, and technicians with long histories of accomplishments in research, development, and demonstration in nuclear energy. New ideas and processes evolve from these laboratories.

Argonne National Laboratory (ANL). ANL, located in Argonne, Illinois, is operated by the University of Chicago. Argonne participates in the waste management R&D program in the following areas: technology for converting high-level waste to forms suitable for permanent disposal, criteria for converting the cladding-hull waste from fuel reprocessing to a form suitable for permanent disposal, and studies of the potential for the migration of transuranic (TRU) elements through various soils and geologic media.

Brookhaven National Laboratory (BNL). Located in Upton, New York, BNL is operated by Associated Universities of New York. Brookhaven's contribution to the waste management program consists of R&D on the removal of tritium from effluents and its conversion to a solid form suitable for permanent disposal.

Hanford Engineering Development Laboratory (HEDL). HEDL is in Richland, Washington, and is operated by Westinghouse Electric Corporation. HEDL's role in the waste management program is R&D on processes for solidifying liquid wastes other than high-level waste and for treating TRU-contaminated solid waste by acid digestion to reduce its volume. HEDL also operates facilities for testing high-efficiency filters for removing particulates from gaseous effluents.

Idaho National Engineering Laboratory (INEL). INEL is located in Idaho Falls, Idaho, and is operated by Edgerton, Germeshausen, and Grier Corporation. INEL's activities include the development of techniques and the evaluation of the impact of recovering previously buried TRU-contaminated waste, the investigation of alternatives for the long-term management of the buried TRU waste, the development of technology for the conversion of Idaho high-level waste to a form suitable for permanent disposal, and the selection of

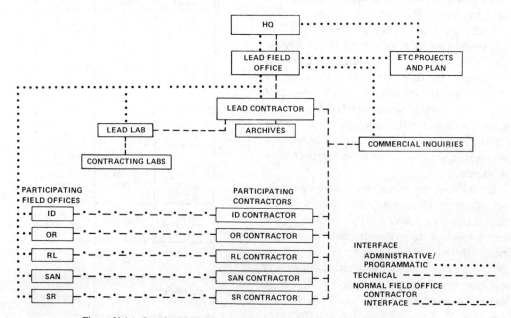

Figure V-4-1. Interface between headquarters, field offices, and lead contractors.

an alternative for implementing the long-term, high-level waste management program.

Lawrence Livermore Laboratory (LLL). LLL, located in Livermore, California, is operated by the University of California. LLL's activity consists of R&D to improve the performance and reduce the cost of particulate filtration systems.

Los Alamos Scientific Laboratory (LASL). Located in Los Alamos, New Mexico, LASL is operated by the University of California. LASL's role in the technical program for reducing the volume of TRU-contaminated waste consists of testing incinerators and other volume-reduction equipment under conditions representative of plant operations. LASL also is developing techniques for the long-term evaluation of TRU-contaminated waste burial grounds. In addition, LASL's waste management program includes R&D on methods for testing high-efficiency particulate filters and improved methods for the siting, operation, and long-term stabilization of shallow land burial grounds for the disposal of low-level waste.

Mound Laboratory. Mound Laboratory is located in Miamisburg, Ohio, and is operated by Monsanto Research Corporation. Mound Laboratory is developing ultrafiltration technology for application in decontaminating non-high-level liquid waste and is doing R&D on volume-reduction systems for TRU-contaminated waste.

Oak Ridge National Laboratory (ORNL). ORNL, located in Oak Ridge, Tennessee, is operated by Union Carbide Corporation. ORNL participates in the waste management program, specifically in: (1) studies for volume-reduction, packaging, and disposal of high-level, ORNL intermediate-level, TRU-contaminated, and tritium-contaminated wastes by various techniques; and (2) fuel-cycle analysis to determine quantities and characteristics of waste generated under various technology assumptions.

Pacific Northwest Laboratory (PNL). PNL is located in Richland, Washington, and operated by Battelle Memorial Institute. PNL has responsibilities for vitrification of high-level waste; development of waste management concepts and methods, monitoring risk assessment procedures, and preparation of GEIS's; and implementation of international program support plans.

Rockwell Hanford Operations (RHO). Located in Richland, Washington, RHO is operated by Rockwell International Corporation. RHO's waste management activities include the development of technology for conversion of Richland high-level waste into a form suitable for permanent disposal, the selection of an alternative for implementing the long-term high-level waste management program, and the investigation of alternatives for the long-term management of Hanford-contaminated soils and sediments.

Rocky Flats (RF). RF, located in Golden, Colorado, is operated by Rockwell International Corporation. RF's primary R&D activity in waste management consists of investigating the fluidized-bed incinerator concept for TRU waste treatment.

Sandia Laboratories (SL). SL is located in Albuquerque, New Mexico, and is operated by Western Electric Company. Sandia is the principal contractor in the Waste Isolation Pilot Plant program. Technical activity focuses on the selection of a specific site. Sandia's role includes exploratory drilling and the highly detailed studies necessary for site confirmation. In addition, Sandia participates in the gaseous-waste management program and is responsible for evaluating alternative long-term storage facility concepts for krypton-85. Sandia has a lead responsibility for management of transportation activities and will establish a Transportation Technology and Information Center.

Savannah River Laboratory (SRL). Located in Aiken, South Carolina, SRL is operated by E. I. DuPont de Nemours and Company. SRL's activities in the waste management program consist of developing technology for the conversion of Savannah River's high-level waste to a form suitable for permanent disposal and conducting studies related to minimizing potential radionuclide migration from shallow land burial grounds that are in a humid environment. SRL has a major role in the Spent Fuel Storage Subactivity in the area of technical management.

The laboratory is preparing the environmental impact statement for implementation of the spent-fuel storage policy. In addition, R&D is being conducted to increase spent-fuel storage capacity at existing basins. Alternative away-from-reactor storage concepts are being evaluated.

UNIVERSITIES

Universities are important to the waste management program as a source of ideas, people, and consulting services, and as contractors and subcontractors in the execution of the waste management technology programs. The universities participate in DOE's waste management program in several ways and at different programmatic levels. The Government-owned laboratories that participate in the waste management program and are operated by universities or university associations are shown in Table V 4-2.

Table V-4-2. Government-owned Laboratories operated by Universities or University Associations.

LABORATORY	UNIVERSITY OPERATOR
Argonne National Laboratory	University of Chicago
Brookhaven National Laboratory	Associated Universities of New York
Lawrence Berkeley Laboratory	University of California
Lawrence Livermore Laboratory	University of California
Los Alamos Scientific Laboratory	University of California

Other universities that participate in the DOE programs are listed below:

Montana Technical University
Colorado School of Mines
Idaho State University
University of Missouri
Oregon State University
University of Utah
University of Idaho
New Mexico State University
University of Arizona
Colorado State University
University of Texas at Austin
Indiana University
University of Tennessee
Alfred University
Washington State University
New Mexico Highlands University
University of Florida
University of Washington
University of New Mexico
North Carolina State University
Harvard University
University of California Davis
Pennsylvania State University
Virginia Polytechnical Institute
Oregon State University
Lousiana State University
Cornell University
Vanderbilt University
Stanford University
University of Miami
University of California-Irvine
Rensselaer Polytechnical Institute

INDUSTRY

Participation by industry includes operation of major Government facilities, major subcontracts for waste management activities, and procurements for design, construction, equipment, supplies, and services.

The major Government-owned facilities participating in the waste management program and operated by industry are listed in Table V 4-3.

Table V-4-3. Government-owned Facilities operated by Industry.

FACILITY	INDUSTRY OPERATOR
Hanford Engineering Development Laboratory	Westinghouse Electric Corp.
Idaho National Engineering Laboratory	Edgerton, Geremeshauser, and Grier Corp.
Mound Laboratory	Monsanto Research Corp.
Oak Ridge National Laboratory	Union Carbide Corp.
Office of Nuclear Waste Isolation	Battelle Memorial Inst. (non-profit)
Pacific Northwest Laboratory	Battelle Memorial Inst. (non-profit)
Rockwell Hanford Operations	Rockwell International Corp.
Rocky Flats	Rockwell International Corp.
Sandia Laboratories	Western Electric Co.
Savannah River Laboratory	E. I. DuPont de Nemours and Co.

5. Commercialization

In general, commercialization is the adoption of technology by the private sector and its introduction into the marketplace. For some technologies, including most of those related to the management of radioactive wastes, Government control is considered necessary for reasons of national policy. The final phase in the development of such controlled technologies is Government management of the operation of commercial-scale facilities. Commercialization in this context refers to the final phase of the technology development cycle rather than to the assumption of control by private industry.

The management of radioactive wastes is primarily an environmental program. This program is directed toward protecting people and the environment from exposure to toxic and radioactive wastes. The long-term nature of the hazard represented by some of these wastes makes it imperative that the public health and safety be fully protected for extended periods.

THE TECHNOLOGY DEVELOPMENT CYCLE

The technology development cycle for ASET and for the waste management program consists of seven phases:

1. Basic research—the performance of fundamental studies to advance scientific knowledge in subjects related to waste management
2. Applied research—the systematic investigation of problems that arise from the findings of basic research; these problems may be in the physical, biological, or social sciences
3. Exploratory development—efforts comparable to applied research in content, but directed toward a particular aspect of waste management
4. Technology development—efforts directed toward establishing the technical feasibility of a particular technology applicable to waste management, includes the formulation of alternative approaches and the development of laboratory-scale models
5. Engineering development—efforts directed toward the design and testing of pilot plants, with emphasis on engineering and systems analysis, may develop major facilities for testing and improving pilot-plant components
6. Demonstration—construction and operation of a first-of-a-kind facility to exhibit the acceptability of the technology in an operating environment
7. Government operation*—full-scale implementation of the technology in regular facilities controlled by the Government and operated by contractors on a routine basis under Government supervision

The progress of a technology through the development cycle requires major decisions

---

* For some classes of technologies, such as solar energy conversion, the final phase consists of commercial introduction in the marketplace by private companies rather than implementation on a commercial scale by the Government. Government operation, however, is appropriate for waste management technologies.

regarding the readiness and suitability of a candidate technology to enter the phases of technology development, engineering development, demonstration, and commercialization. The process by which these four major decisions are made depends on whether the technology requires Government control. For waste management, Government control is required by law.

FEDERAL CONTROL OF RADIOACTIVE WASTE MANAGEMENT

The Energy Reorganization Act of 1974 (P.L. 93-438) gives the U.S. Nuclear Regulatory Commission responsibility for controlling nuclear waste management through regulation. This law also gives DOE responsibility for R&D and for operational tasks in managing radioactive waste. Although current plans require control of waste management by the Government, they anticipate participation by private industry in parts of the waste management program. Areas that can be delegated to private industry without endangering the public interest are determined by DOE. Some areas that are likely to satisfy Federal law and public concern for adequate protection are as follows:

- Transportation of radioactive wastes. This is currently done almost exclusively by private companies that are required to conform to Federal regulations. Regulations will be modified as necessary to reflect technological advances
- Management of specific facilities and operations under contract to the Government.

6.
Environment Implications

OBJECTIVES OF THE WASTE MANAGEMENT SYSTEM

The draft "Generic Environmental Impact Statement (GEIS) on Commercially Generated Radioactive Waste" (DOE 1559, September 1978) states the following:

> The objective of any waste management system is the isolation of the radionuclides from mankind until they have decayed to a safe level. It is generally accepted that the first thousand years of the disposal period are the most critical. The radioactivity and thermal effects from the wastes continue to decay with time. Although our ability to predict future geologic events decreases with the time horizon of the prediction, the hazards of the wastes are likewise decreasing with time.

The objectives of the radioactive waste program are to ensure that radioactive waste materials from the nuclear fuel cycle and other DOE programs can be safely disposed of or stored pending disposition. This is a complex task because the waste materials have varying properties and, in some cases (i.e., spent fuel rods), residual value.

HLW is defined as the fission fractions and actinide by-products resulting from spent-fuel reprocessing. Spent fuel rods are also treated as HLW for disposal planning purposes. Spent fuel is currently being generated from commercial reactors and DOE programs. HLW's are the most difficult management problem because they are highly radioactive and have long decay half-lives. Isolation of these wastes, therefore, must be thorough and long-lasting. Short-term retrieval of the fuel rods is desirable to test the containment integrity of the disposal system.

Other parts of the reactor fuel cycle and other programs produce intermediate- and low-level wastes, with less severe disposal requirements. However, disposal methods must also be provided for these waste materials.

THE DOE WASTE MANAGEMENT PROGRAM AND TREATMENT OF ENVIRONMENTAL ISSUES

The waste management program is essentially an environmental program. Therefore, program objectives, milestones, assumptions, and strategy are directed at environmental problems. DOE's waste management program addresses environmental problems in each of its three activities: management of defense waste, management of commercial waste, and management of spent fuel. These programs address waste handling, disposal, and isolation to ensure that public health and safety and the environment will be protected. These three activities are described in detail in Chapters 2 and 3.

The three program activity categories are consistent with current budget categories. The management structure often cuts across these categories in order to group similar activities under the same management control. Thus, the management control and decision-making for environmental issues often cuts across these budget categories to ensure consistent and comprehensive treatment.

The Environmental Impact Statement (EIS) Process. The environmental issues involved in the Nuclear Waste Management Program are examined in the EIS process. The EIS process allows public participation in the review of major Federal projects. Major EIS activities in DOE/ETW which are already underway are discussed in the following sections.

Several alternate strategies have been proposed in the Report to the President by the Interagency Review Group (IRG) on Nuclear Waste Management. For interim guidance purposes, the IRG report has recommended a policy framework that will allow DOE to proceed with program planning and budgetary proposal concerning the implementation of one to several of these strategies in a manner that will not foreclose any of the disposal options.

For the commercial waste activity, a GEIS entitled "Generic Environmental Radioactive Waste" (DOE 1559) has been written to cover the potential technologies for nuclear waste disposal and their environmental impact.

The GEIS is focused on HLW's (in the form of discarded spent fuel elements or canisters of waste from the reprocessing of such elements) and TRU-contaminated wastes from all activities, including the decommissioning of commercial nuclear facilities. The current Federal plan is to store spent fuel elements retrievably, pending future decisions concerning the recovery of their residual energy values. An environmental statement analyzing the retrievable storage of these fuel elements has been issued in draft form. The scope of the GEIS also includes the environmental impacts associated with the potential diposal of spent fuel which may be shipped from other countries to the United States.

Historically, the GEIS is the successor to a draft EIS (WASH-1539), issued in September 1974 by the Atomic Energy Commission, on the program for developing interim and permanent repositories for high-level and TRU wastes. The statement was withdrawn in April 1975 by the Energy Research and Development Administration. The present document, which replaces WASH-1539, also addresses the public sector's comments received in response to the *Federal Register* notice on October 1, 1976, and at public meetings. The analysis in this draft GEIS provides the required consideration of environmental impacts to allow a deliberate and formal selection of a national strategy for the disposal of high-level and TRU wastes from the commercial fuel cycle. After comments have been received on this draft statement and a final GEIS is issued, approved recommendations will be included. Furthermore, the GEIS will be used as an input to future decisions which refer to construction of a disposal facility at a specific site.

Draft EIS's have been issued for the storage of domestic spent fuel, the storage of foreign spent fuel, and the charge for spent fuel. They will focus on the need for intermediate AFR storage until permanent disposal facilities are available.

A series of EIS documents are being prepared to cover the long-term management of defense wastes. These documents are site specific and deal with plans for disposal of high-level and TRU wastes at each major defense facility. Table V 6-1 summarizes the status of these EIS documents.

Table V-6-1. Status of EIS Documents.

TITLE	SITE	PUBLICATION DATE
TRU DEFENSE		
Long-Term Mgmt. of TRU Wastes	Sandia	Dec. 1978
Long-Term Mgmt. of TRU Wastes	INEL	Dec. 1979
Long-Term Mgmt. of TRU Wastes	SR	Jan. 1980
Long-Term Mgmt. of TRU Wastes	LASL	Oct. 1981
Long-Term Mgmt. of TRU Wastes	RL	Jan. 1982
HLW DEFENSE		
Long-Term Mgmt. of HLW	SR	Aug. 1978
Long-Term Mgmt. of HLW	ID	Feb. 1979
Long-Term Mgmt. of HLW	RL	Mar. 1979
AWAY-FROM-REACTOR		
Storage of U.S. Spent Power Reactor Fuel	SR	Aug. 1978
Storage of Foreign Spent Power Reactor Fuel	SR	Dec. 1978
Charge for Spent Fuel Storage	SR	Dec. 1978

METHODS OF WASTE MANAGEMENT

The foreword to the draft commercial GEIS states the following:

> All waste management systems rely on a multibarrier approach to achieve the required level of isolation. For each disposal concept considered there is a different set of barriers. Each set of barriers starts with the waste form.

The chemical and thermal characteristics of this waste form can be varied in many ways. Variations include aging (cooling) prior to production of the disposal form, dilution of the radionuclides in a matrix material, and partitioning and/or transmutation of certain isotopes. Design of the waste form and adjustment of the disposal canister dimensions provide the next barriers to radioactive material release. The available metals give various degrees of isolation depending on the particular disposal medium and exposure conditions. Absorptive overpack materials, such as zeolites and bentonites, are available as a further barrier. The disposal medium is a massive barrier to the migration of wastes into the biosphere. Choices range from stable, absorptive clays of the seabed, very deep hard rock formations, granite, shale, basalt, salt, and other rock formations to outer space. The final barrier is the institution(s) which control(s) the disposal system. (The integrity of the institutions determines, in part, the success of the disposal program.) Although no government or society is stable over long periods of time, works of early civilizations still exist today. Therefore, it appears that today's society can construct a repository which would survive. In the near-term, institutional controls would provide further assurances of the integrity of the disposal system.

With the variety of design variables available, it is reasonable to project that nuclear waste systems can be designed to ensure that this and future generations will be protected from radioactive wastes.

Federal regulations require that high-level wastes, if in liquid form, be converted to a solid form within 5 years and that they be shipped to a Federal repository within 10 years after separation from the spent fuel (if reprocessing is to take place). Disposal criteria for high-level and TRU-contaminated wastes are not fully defined, and a disposal facility to receive these wastes and/or spent fuel has not been demonstrated. Various methods are under consideration.

The IRG Subgroup on Strategies for Waste Isolation found that the technique of disposing of nuclear wastes in mined repositories is promising. The majority of scientists and engineers consulted by this subgroup agreed that successful isolation of radioactive wastes from the biosphere appears feasible for thousands of years provided that the systems view is used vigorously to evaluate the suitability of sites and designs, to minimize the influence of future human activities, and to select a waste form that is compatible with its host rock. A variety of geologic environments and emplacement media could be potential candidates for a repository. Although the engineering aspects of a repository in salt have been studied extensively, no particular environment or medium is currently preferred.

Interim storage facilities for fuel rods or other unreprocessed nuclear material must meet isolation criteria for the period the facilities are to be used. Methods for waste transport are also needed because disposal sites are close to the waste sources.

Final long-term waste disposal must also minimize the possibility of future accidental release of radioactive waste, either through natural forces or man's future activities. The initial approach by the U.S. Government is emplacement of the wastes in deep continental geologic formations (salt, crystalline rock, or argillaceous rocks).

Low-level waste materials are now disposed of by burial in shallow trenches. There are 6 commercial and 14 DOE burial sites. The total amount of commercial low-level nuclear waste is increasing at about 25 percent per year, mainly from the growth in commercial use of nuclear power reactors.* The provision

---

* U.S. EPA, *Consideration of Environmental Protection Criteria for Radioactive Wastes*, February 1978, p. 8.

of disposal capacity is a near-term issue for commercial sites.

RESOLVING ENVIRONMENTAL PROBLEMS OR ISSUES

General Approach

For each waste disposal scheme proposed, it is necessary to ensure that the environment is adequately protected. This means that demonstrations and analyses must be performed to show that potential waste migration from the disposal sites is acceptable in terms of concentration, probability, and current and future impacts on human health. The main purpose of the waste management program is to develop an environmentally acceptable waste management system. This involves determining the environmental criteria, as well as assessing storage system performance.

Radioactive effluents from each step of the nuclear fuel cycle can be controlled to reduce amounts which escape to the biosphere. The issue of the amount of reduction that is justified in terms of health impacts versus incurred costs, has been controversial. The EPA is responsible for setting the applicable standards to protect human health.

Analysis of dispersal of the waste materials placed in a disposal medium is also complex. The waste, the encapsulating material, and the geologic structure in which it may be embedded must be studied as they interact with each other. The potentials for chemical and physical transport of the waste material must be carefully analyzed. This is a complex and interactive problem because there are many waste forms (especially from different fuel cycles) and disposal schemes. The NRC licensing process provides a forum for public review and discussion of the safety analyses for sites and disposal techniques.

Proposed NEPA Decision-making Strategy for the National Waste Terminal Storage Program

A proposed NEPA strategy for the waste isolation program concerning repositories for high-level wastes and spent fuel developed by the Office of Waste Isolation is described in the following sections. This strategy is consistent with the findings of the IRG Subgroup on Strategies for Waste Isolation and with the discussion of program strategy in Chapter 2.

Key Decisions to be Made. 1. Should the first disposal facilities for HLW be geologic repositories using conventional mining techniques?

This is currently a recommendation of the IRG on an "interim planning basis." Formal adoption of this recommendation was supported by issuance of the draft GEIS on commercially generated wastes in January 1979 and by a final statement in late 1979.

2. Is any particular site (30 square miles) within any particular region or geologic system suitable or potentially suitable as a repository location?

This decision must be reached by DOE prior to consideration of whether to propose that a repository be built at any particular site. It would not be a major Federal action having major impact on the environment but would be a finding of technical fact based upon investigation.

Information necessary for DOE to make this finding falls into four specific areas, each of which will require well-documented work. The four areas are site characterization, safety, environmental impacts, and engineering. Site characterization might be compiled in a document similar to the site characterization report recently prepared for WIPP. Findings on safety aspects will be based on safety analyses, risk assessment studies, geologic and hydrologic data and the effectiveness of multiple barriers. The results of the safety analyses should be documented as closely as possible to NRC requirements for Safety Analysis Reports (SARs).

Environmental impacts and specific site data will be studied for each potential site and should be documented in a format to satisfy NRC requirements for environmental reports. Engineering information will derive from various design and tradeoff studies and should be available in the form of a completed conceptual design report and cost estimate.

Findings on the suitability of specific sites will proceed as rapidly as possible, as recommended by the IRG. The finding of a suitable site may not, however, lead to a simultaneous decision to proceed with development of a repository at that site. Present program progress indicates that sites may be identified as qualified in salt domes in early 1981, in bedded salt in late 1981, in basalt at the Hanford Reservation in 1981, and in other media or systems by 1984 or 1985.

3. Should DOE designate a preferred site and proceed with a request for licensing by NRC?

This decision and the resultant authorization of a specific repository project may be considered a major Federal action. Plans will be made to prepare a NEPA document to support this action. This EIS will focus on a comparison of possible sites on the basis of then-available data. A separate chapter in the NEPA document could be devoted to possible sites in each geologic basin that DOE/ETW has studied. Very detailed ecological characterization of sites would be in the Environmental Report (ER) and the document needed to support construction. The document, prepared by NRC, should be available in draft form at the time that full project authorization is requested from Congress.

If Strategy II of the IRG is adopted, work on this EIS should begin no later than January 1980. Data on potential alternate repository sites would be included even though no finding has yet been reached on their suitability. For example, information on a possible site at Hanford would be included even though a finding on its suitability may not be made until 1981.

If Strategy III of the IRG is adopted, this EIS would not be needed until 1984 or 1985.

4. Should a repository be constructed at a particular site?

This decision may be made jointly by at least two other parties in addition to DOE: Congress and NRC. The decision by Congress can be made through the authorization and appropriation process and could be supported by the EIS prepared by DOE contractors. The NRC decision will be made by their licensing action on DOE's application and will have to be supported by an EIS prepared by NRC. Information for use by NRC in this EIS will come from data collected in the various site investigations, which should be available in the prescribed SAR and ER formats.

7.
Socioeconomic Issues

Since nuclear power is providing a significant portion of the United States' electric energy, the management of nuclear waste will be essential. Further, existing waste inventories and national defense programs require a comprehensive waste management program. Regardless of the growth of the power reactor, handling radioactive waste will be increasingly important. These issues include the following:

- Waste facility siting
- Worker health and safety
- Use of other resources near waste sites
- Local economic impacts
- Assessment of storage costs
- Liability for accidents
- Long-range planning

Waste Facility Siting

There is sincere concern over having nuclear material stored or disposed of near one's home. A nuclear waste facility is often viewed as a community detriment, much as prisons or sewage treatment plants are perceived. The public must be convinced that the disposal method is safe, that the local communities will not be exposed to radiation or threatened with nuclear contamination, and that property values will not be significantly depressed by a waste facility. Communities located close to the repository may feel that they are bearing a disproportionate share of a national burden. They may not realize the benefits of nuclear power to their community, but they will realize that vast amounts of nuclear wastes are located nearby.

Public participation in deciding where to locate and how to operate repositories will be necessary. Cooperation between the Federal, State, and local governments must be supplemented by contact with community groups, for DOE to broaden public participation.

Worker Health and Safety

There may be accidents or injuries whenever heavy machinery is required to handle hazardous materials. Radiation will be maintained at low levels. Workers will be monitored to ensure that they do not receive injurious doses. Nevertheless, accidents may occur. DOE will continue to comply with all occupational health and safety rules.

It is difficult to quantify the risk to workers' health and safety because nuclear waste management is new and future risks cannot be determined accurately without direct working experience. Nevertheless, risks can be estimated from the experience of industries performing similar work and from other nuclear activities.

Use of Other Resources Near Waste Sites

Some land will be permanently committed to the management of nuclear waste. In geologic repositories, the radioactive life of the stored material will limit the use of the storage area for long periods. The land surface may again become usable after the repository is sealed. Recovery of underground resources near the repository site will be restricted for long periods to keep the repository intact.

Local Economic Impacts

Repositories are likely to be placed in remote areas. The local community can benefit economically from having a waste storage site nearby. The local economy is stimulated during construction and operation of a repository. Local workers may find jobs, and merchants will also benefit. It has been suggested that resources should be provided for public projects. Government projects do not generally result in local tax resources.

Both construction and operation will bring new people and money into the community. Of course, provision of additional local governmental services may require increased Government investment. Local governments may not be in a position to respond in a timely fashion to these needs. In some people's view, the community could become a less attractive place to live due to increased traffic and noise. The repository itself may be considered as a detriment to the area.

Assessment of Storage Costs

Determination of the proper fees (or charges) to a utility for spent fuel storage and disposal is affected by several issues. Because the technologies for storage of spent fuel have not been finally agreed upon, the storage system costs are not completely known. Maintenance and upkeep costs required to maintain the integrity of the storage system can only be estimated. Disposal charges would have to include provisions to cover the future costs of the storage repository.

Liability from Accidents

Liability inurance for the nuclear industry is largely governed by the Price-Anderson Act. Through a system of private insurance and Government indemnity, the Price-Anderson Act is designed to assure that the public would be protected in the event of a nuclear accident connected with a facility operated under a contract with or license issued by the Government. Such an accident is and has always been considered extremely unlikely. However, Congress wanted to ensure that there would be protection should an accident occur and the resulting loss or damage exceed that which could be expected by private insurance.

Under the Price-Anderson Act, the Department of Energy and the Nuclear Regulatory Commission are authorized to enter into indemnity agreements with contractors and licensees operating nuclear facilities. Through these indemnity agreements, financial protection is currently afforded up to a limit of $560 million per accident. If damages should exceed that amount, Congress has provided in the Price-Anderson Act that it will "thoroughly review the particular incident and will take whatever action is deemed necessary and appropriate to protect the public from the consequences of a disaster of such magnitude."* Thus, even the $560 million limit is not absolute.

Indemnity agreements cover not only accidents occurring at the site of the contractual or licensed activity, but also accidents which might occur in the transportation of material to and from the site. The indemnity extends not only to the person or entity with whom and to whom the contract or license is extended, but any other person who may be liable to the public in connection with an incident. Liability covered by the Price-Anderson Act and related indemnity agreements remains covered even if the nuclear-related damage or injury were caused by terrorism, sabotage, or other illegal acts, unless the damage or injury resulted from illegal actions after material is diverted from the site of the indemnified activity or from the planned route of transportation.

Liability for waste facility-related nuclear accidents involving radiation contamination would be determined in accordance with applicable rules of tort law as applied by the court. Financial responsibility for such lia-

---
* 42 U.S.C, Section 2210e.

bility would be assumed by the Federal Government to the extent provided by the Price-Anderson Act.

Price-Anderson coverage for transportation of any spent fuel elements from commercial nuclear reactors to waste facilities would be afforded by indemnity agreements to be extended by the Nuclear Regulatory Commission to commercial reactor licensees. Similar coverage for the transportation of any defense-generated waste to waste facilities would be afforded by indemnity agreements entered into by the Department of Energy with its contractors.

The operation of waste facilities and all related transportation activity is expected to be performed by persons indemnified under the Price-Anderson Act; i.e., Nuclear Regulatory Commission licensees or Department of Energy contractors. As a result, any liability for damages resulting from waste facility-related accidents would be covered by the Price-Anderson Act and related indemnity agreements. Although direct involvement of DOE personnel in the handling and transportation of waste is not anticipated, any liability asserted directly against the Department of Energy would be determined in accordance with the Federal Tort Claims Act and the traditional rules of liability as established in the body of law concerning Federal torts. The $560 million limit on liability would not apply to direct Government liability under the Federal Tort Claims Act.

Long-range Planning

Planning for nuclear waste management involves much long-range forecasting and estimating. The technical design of the system and the plan to manage and maintain it must at least extend for the next 100 years, under currently proposed EPA guidelines. Thus, the analysis of potential risks of the facility and how they will be mitigated to acceptable levels involves considerable future forecasting.

Each potential risk mechanism will have to be separately examined, and suitable safeguards devised. The cost of these safeguards, now and in the future, will have to be estimated. Institutional arrangements to protect the waste repositories from natural phenomena and accidental or deliberate intrusion will have to be designed. Since most engineering structures are designed for much shorter lives—30-to-50 years—design of the nuclear repository presents unusual and difficult planning requirements.

8.
Regional Activities

Administrative and research activities for the Nuclear Waste Management Program are dispersed throughout the country. This decentralization of DOE programs is complemented and supported by 10 regional representatives (see Figure V 8-1) who represent the Secretary of DOE and who provide a mechanism to interface with States, localities, and the public. These 10 regional offices provide information to the general public and collect information relevant to public opinion on DOE policy, program, and implementation.

Regional activities differ from field activities (discussed in Chapter 4) in that the latter represent the decentralized program support offered by DOE contracted personnel and DOE personnel. Regional activities, however, represent DOE's efforts to interface with local, State, and regional governments, as well as various consumer groups, trade associations, and other sectors of the public.

This program will seek more State and local government participation in the responsibilities of decision-making regarding the waste management program.

Specifically, there has been concern over State acceptance of locating federally proposed nuclear repository sites within a State.

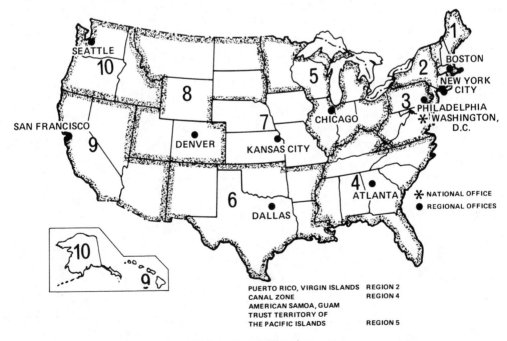

Figure V-8-1. DOE regions.

Several State and regional associations (e.g., the National Governors Association) as well as the IRG are recommending a process of "cooperative federalism" in which the States would be involved in the planning process by early reviewing of potential site characterizations.

The Nuclear Waste Management Program seeks to communicate with Governors, State legislators, and Congressmen in the site selection process. In this process, a Governor appoints an individual as a liaison between DOE and the State. The liaison has the responsibility for articulating the State's concerns regarding site selection. In addition, an Executive Planning Council composed of Governors and/or their representatives has been proposed for the following purposes:

- To establish criteria for the Nuclear Waste Management Program
- To develop procedural guidelines for States and the Federal Government for waste management and site selection.

9.
International Programs

International cooperation in radioactive waste management RD&D is an important part of ETW activities because the need to securely isolate radioactive wastes transcends national boundaries. These activities serve key U.S. foreign policy objectives and advance the U.S. energy program by the following:

- Decreasing RD&D costs through shared information and project costs
- Increasing the breadth of technical information and experience
- Reducing the time required for some phases of domestic projects through the use of facilities in other countries

Cooperation between the United States and other nations in radioactive waste management may be bilateral or multilateral, and the cooperative activities may range from information exchanges to cooperative technical projects.

PROGRAMMATIC INTEREST IN INTERNATIONAL WASTE MANAGEMENT ACTIVITIES

International waste management activities provide advantages for the United States. The European and Canadian waste management activities are generally constructed along lines similar to the U.S. program and the technical sophistication of the programs is roughly equal in particular areas, so many opportunities exist for mutually satisfactory cooperation.

For example, information on geologic disposal is of interest to the ETW defense and commercial programs. Topics of particular interest are pathway analysis, hydrologic modeling, rock properties, heat transfer, and design criteria related to granite, shale, salt, and basalt. Although the U.S. programs place less emphasis on clay as a disposal medium, basic data of isotope migration through clay formation are potentially useful. Canada, the Federal Republic of Germany, Sweden, the Netherlands, Belgium, and the United Kingdom have existing programs with known specific elements of interest.

Although the U.S. commercial activity does not involve reprocessing, defense waste does come from a reprocessing plant. A major part of the DOE defense waste activity is the selection of a suitable waste form and an appropriate production process. The European work on waste forms is relevant. Advanced programs exist in the Federal Republic of Germany, France, Belgium, and the United Kingdom. The physics and chemistry of glasses and other waste forms under long-term exposure to geologic conditions in the Federal Republic of Germany and France are of special interest because the subject is complex.

The TRU waste incineration technology in the United Kingdom and Belgium is significant for U.S. programs. A British incinerator concept has been used as a design basis. Other useful programs exist in Belgium, France, the Federal Republic of Germany, and Canada.

ETW has some interest in the European monitoring methods for burial of low-level waste. Generally, however, European programs are less extensive because sea dumping and mine disposal methods are now in use there.

The U.S. program on airborne wastes would benefit from cooperative development of trapping and disposal methods for tritium, krypton, iodine, and carbon-14 since technologies do not exist for all of these purposes. European programs are not as broad as that in the United States but can contribute to U.S. programs.

The transportation programs of Japan, the United Kingdom, France, and the Federal Republic of Germany are developing technology for package design, testing, and risk assessment which are of value to the United States. In addition, transportation safety standards should be uniform throughout the world to permit safe movement of nuclear materials.

It is probable that use of the sea bed for waste disposal can become a reality only with international support. Therefore, international cooperation in the R&D stage of seabed disposal technology will be important.

BILATERAL AGREEMENTS

As shown in Table V 9-1, the United States has four bilateral agreements relating to nuclear waste management.

The agreement with Sweden is for a specific, limited scope cooperation program at their Stripa mine. Activities include field tests and the advancement of techniques used in measuring the movement of fluids through fractures in a granite rock system. The results of these tests will provide useful data for evaluating U.S. granite as a terminal storage repository and will also provide techniques for measuring fluid movement through U.S. formations.

Other bilateral agreements provide for the mutual exchange of technical and scientific data, information, and personnel as well as collaboration on joint projects. Information on fuel reprocessing or waste management processes that require reprocessing or plutonium recycle for implementation has been withheld by the United States due to the U.S. policy on nuclear nonproliferation. The agreements cover waste management aspects of the entire fuel cycle; specifically, high-level waste solidification and packaging, treatment and packaging of intermediate- and low-level wastes, treatment of transuranic waste, trapping and storage of gases, evaluation of alternative disposal concepts, decommissioning of nuclear facilities, operating experience with low-level wastes, transportation of wastes, and retrievable surface storage.

The bilateral agreements with Canada, the Federal Republic of Germany, and the United Kingdom, as well as those pending with Belgium and Japan, are similar. The only agreement to participate in joint projects to date has been with the Federal Republic of Germany for work on geologic disposal. Information exchange under this agreement has been extensive. The Federal Republic of Ger-

Table V-9-1. Bilateral Agreements.

COUNTRY	DATE CONCLUDED	PRINCIPAL SUBJECT
Sweden	July 1, 1977	Radioactive Waste Storage
Canada	September 8, 1976	Radioactive Waste Management and Systems Analysis of Heavy Water Reactors
Federal Republic of Germany	December 20, 1974	Radioactive Waste Management
United Kingdom	September 20, 1976	Fast Breeder Reactors (FBRs)*
Belgium	Pending	Radioactive Waste Management
Japan	Pending	Fast Breeder Reactors*

*Note: These agreements on FBRs include an annex covering cooperation in nuclear waste management.

many has made extensive use of their visitation privileges at U.S. laboratories and contractors; U.S. visits to FRG facilities have been productive and well-received, although less extensive. An active exchange program is also taking place under the agreement with the United Kingdom for technologies concerned with transuranic-contaminated wastes.

MULTILATERAL AGREEMENTS

Multilateral cooperation in radioactive waste management takes place under the auspices of several international organizations of which the United States is a member or an observer. The United States participates in advisory groups, working groups, joint meetings, ad hoc projects, and studies.

Commission of the European Communities (CEC—formerly EURATOM)

CEC has established a comprehensive program of radioactive waste management technology development to be conducted in various laboratories throughout the European Economic Community (EEC). It parallels the U.S. program and supports the Eurochemic high-level waste solidification program, including geologic surveys of the suitability of disposal in various types of formations. Discussions of a possible bilateral agreement with the United States are in progress.

Current work on geologic disposal includes the use of salt formations in the Federal Republic of Germany and the Netherlands, argillaceous formations in Belgium and Italy, and crystalline rock in France and the United Kingdom. The CEC funds and operates joint research centers at Ispra, Italy; Karlsrhue, Federal Republic of Germany; Petten, Netherlands; and Geel, Belgium.

International Nuclear Fuel Cycle Evaluation (INFCE) Program

At an October 1977 organizing conference, over 50 nations and 4 international organizations agreed to join the United States in a comprehensive international evaluation of the nuclear fuel cycle. This program is a Presidential initiative which has the active support of ETW in two of the eight key working groups. Working Group 6 focuses on nuclear fuel storage; Working Group 7 focuses on nuclear waste management. The INFCE Program was completed toward the beginning of 1980.

International Atomic Energy Agency (IAEA)

The IAEA, an autonomous member of the United Nations, is devoted to the peaceful uses of atomic energy. It is actively involved in many aspects of nuclear power development, from information exchange and the sponsoring of symposia to the establishment of safety criteria for power reactors, the application of an international safeguard system for signatory nations of the Treaty on the Non-Proliferation of Nuclear Weapons, and direct technical assistance provided to developing nations.

The IAEA has published reports, sponsored scientific meetings, and prepared codes of practice on radioactive waste management. Technical guidance is offered to member countries on procedures for establishing limits for release of radioactive material into the environment, on the monitoring of actual releases from nuclear facilities, and on emergency planning.

An example activity of particular interest to ETW is the Advisory Group on Radioactive Waste Disposal into Geological Formations which was formed in 1978. The Advisory Group plans to develop guidelines for safe underground disposal over the next several years. The guidelines will address the entire range of currently conceivable options.

Organization for Economic Cooperation and Development (OECD)/International Energy Agency (IEA)/Nuclear Energy Agency (NEA)

The IEA was established by a group of developed nations in 1974 to form common policy for energy supplies. The radioactive waste management activities of the IEA are carried out through the NEA Radioactive Waste Management Coordination Committee. The NEA is the official nuclear power implementation organization of the OECD. A Steering Committee for Nuclear Energy guides the activities of several coordinating committees—one of which, together with its several subordinate groups, is devoted to radioactive waste management. The Coordinating Sub-Group on Geologic Disposal meets annually to develop methods for increasing cooperation and to exchange information on the status of national RD&D programs. A key objective of the subgroup is the periodic review of the progress achieved through national RD&D efforts in order to provide a broad and authoritative evaluation of the safety of geologic disposal.

The regular meetings of this committee have provided a useful catalyst for bilateral activities. Also, there has been considerable discussion of NEA sponsorship for multinational waste projects, although no projects have been established to date.

Part VI

Magnetic Fusion Energy

1.
U.S. Energy Policy

The National Energy Plan (NEP) and the five-part National Energy Act (NEA) were designed to produce a coherent energy policy for the United States. In addition, the Department of Energy (DOE) has recently announced its policy for fusion energy.

DOE POLICY FOR FUSION ENERGY

The DOE policy for fusion energy addresses the objectives of the NEP. During the next century, the nation will rely increasingly on three inexhaustible energy sources: solar energy, fission breeder reactors, and fusion energy. These technologies will require many years of development before they can generate large amounts of power economically. Fusion is currently the least practical economically, but its potential rewards are great. It could provide the world with an energy source whose fuel (e.g., deuterium extracted from water) is essentially unlimited and whose by-products should pose reduced environmental problems compared to coal and fission power. The present stage of knowledge suggests that fusion power stations would not have greater safety risks than ordinary fossil fuel plants and should not have significant geographical limitations.

The goal of DOE's fusion research is to develop the potential of fusion energy as a virtually inexhaustible energy source. The philosophy is to demonstrate that potential as soon as possible in an economic way that fully preserves and develops the technical base (i.e., the scientific, technological, and engineering elements).

The major decision points in the program logic are given in Figure VI 1-1. As shown in this figure, two fundamentally different approaches are being followed toward development of fusion energy by: 1) magnetic confinement, and 2) inertial confinement. In magnetic confinement, the hot ionized gases (plasmas) which will undergo fusion reactions are contained in carefully shaped magnetic fields. A combination of resistive and supplemental (e.g., neutral beam injection) heating is used to raise the plasma temperature to that required for substantial fusion reactions. In inertial confinement, the fuel which will undergo fusion reactions is contained within a small pellet. A large amount of energy is deposited into the pellet by an appropriate device, such as a laser beam, thus raising the pellet temperature high enough to cause the fusion reactions. Within each of these conceptual approaches, there are many different schemes for achieving a controlled thermonuclear reaction.

Although scientists have been able to produce conditions in the laboratory approaching those necessary for the production of fusion energy, these experiments have not yet produced more energy than was invested. In other words, net energy gain from fusion plasma reactions has not yet been attained by either approach. However, researchers are confident that this demonstration will be made within the next four to six years. While the

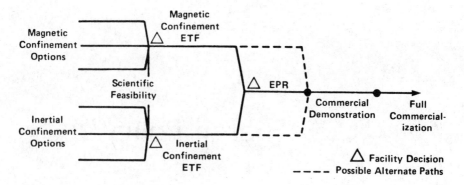

Figure VI-1-1. DOE fusion program decision logic.

highest immediate priority is to demonstrate scientific feasibility, lesser priorities are simultaneously given to projects which aim beyond that demonstration (e.g., commercialization studies).

Following the demonstration of scientific feasibility, the program will move from applied research into development. Engineering Test Facilities (ETF) will be constructed for the most promising options in both magnetic and inertial confinement fusion. These facilities will be elementary integrated energy systems, producing net energy gain, which will establish the technological requirements for each of the major components of a prototype reactor.

The next phase of the program will be to operate an Engineering Prototype Reactor (EPR). The EPR will combine the elements tested in the best ETF in a pilot plant where solutions to reactor design problems can be developed and tested. The EPR will approach, for the first time, complete energy gain, with the energy produced exceeding all energy consumed in keeping the entire plant running. The demonstration phase will be completed with the construction of one or more commercial demonstration reactors which economically produce net power output in excess of 100 megawatts. Commercialization of fusion energy will have begun when approximately one-tenth of a quad per year (about the equivalent of three 1,000 megawatt power plants) is produced.

The important near-term decisions concern the ETF scope and the choices among ETF design prospects. Until then, the fusion program will investigate many different schemes since it would be premature to narrow the field of options at this time. To complement these investigations, vigorous research in plasma physics and fusion technology and engineering will also be supported. As programs mature from applied research to engineering development, DOE will encourage a shift in responsibility for machine design and construction from DOE laboratories and universities to industry.

In summary, the near-term objectives of the DOE fusion energy policy emphasize the following:

- Demonstration of scientific feasibility
- Establishment of a sound engineering development base supporting the ETF decision
- Maintenance and development of a strong scientific and technological base
- Encouragement of research in advanced concepts, and scrutiny of their positions relative to the established major projects

The DOE laboratories will, in the near term, necessarily remain the center around which the Department organizes a large part of its research program. However, integration of more scientists and engineers from the industrial and academic communities is an important goal. DOE also hopes to expand current efforts toward international cooperation in fusion research and to include more (and larger) jointly funded projects with the Japanese, Europeans, and the Soviets. Negotiations for such increased cooperative ventures are already in progress.

DOE is committed to continuous reviews of the R&D program in fusion. The review process has been formalized with the establishment of the Fusion Review Committee (chaired by the Director of Energy Research with the Assistant Secretary for Defense Programs as member) and its supporting Fusion Working Group (representing various fusion constituencies within DOE), including advisory bodies (e.g., the Ad Hoc Experts' Group on Fusion). The program has benefited from these fusion reviews in at least two ways: a clear policy on fusion has been developed and enunciated by DOE and communication and cooperation between magnetic and inertial confinement research have improved. Perhaps most importantly, the program has taken a major step toward the establishment of the technical and managerial stature it deserves. DOE will carry on the work of the Fusion Review Committee and its Fusion Working Group and will periodically solicit advice from further Ad Hoc Experts' Groups. These reviews will assist in determining the major scientific and engineering goals of the programs and help assess progress toward those goals.

2.
FY 1980 Overview—Activities of the Office of Fusion Energy

The Office of Fusion Energy provides overall direction for the planning, development, execution, and control of the research and development program in magnetic fusion energy.

FUSION AS AN ENERGY SOURCE

This section begins with an introduction to the scientific basis of the fusion process. The practical steps required to harness this process to provide an energy source are briefly considered in the context of a conceptual fusion reactor. This background is followed by a discussion of electric and nonelectric applications. The section concludes with a report of the current status of magnetic fusion.

Introduction to the Scientific Basis of Fusion.

Fusion Reactions. Albert Einstein's special theory of relativity showed that mass can be converted to energy via his famous formula, $E=mc^2$, where m represents the change in mass from an initial state to the related final state and c represents the speed of light. The two practical methods for achieving this conversion involve the fission and fusion of atomic nuclei. In fission, heavy atoms, such as uranium, are split apart. The mass of the resulting material is less than that of the original atom, so there is a net release of energy corresponding to the change in total mass.

The second method for transforming mass into energy is fusion, in which nuclei of light atoms, such as hydrogen, are joined together to form a larger nucleus. However, the mass of the resulting nucleus is less than the sum of the masses of the light nuclei, and energy corresponding once again to the reduction in total mass is released.

Of the various possible fusion reactions, the most likely near-term fusion reaction involves two isotopes of hydrogen (H): deuterium (D) and tritium (T). The deuterium-tritium (D-T) reaction is shown in Figure VI 2-1.

The nucleus of a deuterium atom contains one proton (P) and one neutron (N); the nucleus of a tritium atom contains one proton (P) and two neutrons (N). The result of a D-T reaction is a set of particles in highly excited or energetic states. One neutron is ejected almost instantaneously (in about one-hundred-trillionth of a second) and carries away 14.1 million-electron-volts (MeV) or approximately 80 percent of the energy released; the other two neutrons and two protons form the nucleus of a helium atom (^{4}He) which carries the remaining 20 percent (3.5-MeV). In practical terms, this means that just one gram of D-T fuel can produce as much energy as 45 barrels of oil.

In addition to the D-T reaction, other promising fusion reactions use protons, a helium isotope (^{3}He), two isotopes of lithium (^{6}Li and ^{7}Li), and an isotope of boron (^{11}B) a fuel element nuclei. Table VI 2-1 lists most of the

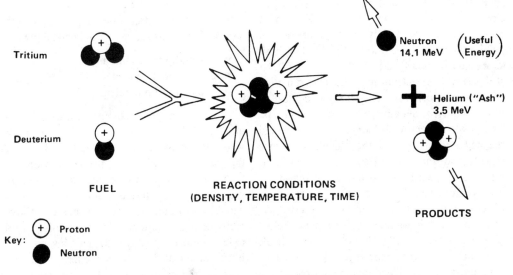

Figure VI-2-1. Deuterium-tritium (D-T) fusion reacton.

Table VI-2-1. Fusion reactions.

REACTION	REACTION ENERGY (MeV)	THRESHOLD PLASMA TEMPERATURE (keV)	MAXIMUM ENERGY GAIN PER FUSION
D + T → ^{4}He + N	17.6	10	1800
D + D → ^{3}He + N	3.2	50	70
D + D → T + P	4.0	50	80
D + ^{3}He → ^{4}He + P	18.3	100	180
^{6}Li + P → ^{3}He + ^{4}He	4.0	900	6
^{6}Li + D → ^{7}Li + P	5.0	> 900	6
^{6}Li + D → T + ^{4}He + P	2.6	> 900	3
^{6}Li + D → 2(^{4}He)	22.0	> 900	22
^{7}Li + P → 2(^{4}He)	17.5	> 900	18
^{11}B + P → 3(^{4}He)	8.7	300	30

presently considered fusion reactions. In the table, the reaction energy is the amount of energy, in MeV, released per fusion; the threshold plasma temperature, presented in energy terms in kiloelectron-volts* (keV), represents the minimum energy investment; the maximum energy gain per fusion is an approximation of the reaction energy divided by the threshold plasma temperature.

For sustained fusion reactions to take place, three conditions must be satisfied:

- The fusion fuel must be heated and maintained at a temperature near its ignition point (approximately 100,000,000 K for the D-T reaction). At this temperature, the fuel charge will exist entirely in the plasma state; that is, as a totally ionized gas.
- The energy in the plasma must be confined for a time, sufficient to ensure that, on the average, the energy (heat) from the burning fuel charge is greater than or equal to the energy supplied to heat and confine the plasma.
- The plasma must have a density, n, sufficiently high to ensure that the ions are close enough for reactions to occur. (The density is the number of ions per unit volume of plasma.) The minimum product of time and density to achieve the fusion condition can be expressed as a threshold requirement, called the Lawson Product, after its formulator, the British physicist, J.D. Lawson. The D-T reaction requires an n product of about 10^{14} ions-second per cubic centimeter (sec-cm^{-3}).

Plasma Temperature and Heating. In a fusion plasma, positively charged particles repel each other; this repulsive force comes into play long before the nuclei are close enough for a reaction to occur (this is why fusion reactions are inherently more difficult to pro-

---
*In nuclear fusion reactions, temperatures are often expressed in terms of the kilo-electron-volt, or keV (1-keV = 1,000 electron volts). 1-keV is equivalent to a temperature of 11,600,000 K (where K represents degrees Kelvin), or about 2,000 times the temperature of the sun's surface.

duce than fission reactions). Forcing the nuclei to come close enough for a reaction to occur requires extremely high temperatures (of the order of 100,000,000 K). At such high temperatures, nearly all of the atoms will be stripped of their electrons. The material then consists of positively charged nuclei, called ions, and free negatively charged electrons. The positive electrical charges of the ions and the negative charges of the electrons balance exactly, so that the plasma as a whole is electrically neutral. It behaves in many respects like a gas, but because of the presence of electrically charged particles, it conducts electricity and is affected by magnetic fields.

A temperature of 100,000,000 K is difficult to relate to ordinary experience. For comparison, most heat-resistant materials used in industry, such as ceramics, melt at about 4000 K, and the interior of the sun is estimated to be about 15,000,000 K. Plasma heating to the extraordinary temperatures required is accomplished by using several methods. A plasma can be heated to temperatures as high as a few million degrees by passing an electric current through it. This is called ohmic (or resistance) heating and depends on the electrical resistance of the plasma. As the temperature increases, the resistance of the plasma decreases and eventually becomes too low for resistance heating to be effective. Another procedure used to heat plasmas to higher temperatures is by neutral beam injection into the plasma. The neutral beams are stripped of their electrons and collide with the plasma ions and electrons, thereby raising their temperatures. Because neutral beam heating is more dependent upon mechanical resistance (i.e., ionizing collisions) than upon the electrical resistance of the plasma, it can be used as an effective supplementary technique to raise plasma temperatures higher than those achieved by ohmic heating alone. Other methods being investigated include laser, electron, and ion beams, microwave and radio frequency (RF) radiation, and magnetic compression.

Plasma Confinement and Beta. To maintain high temperatures, the plasma must be kept from hitting solid walls. Such collisions could cool the plasma below the desired temperature. There are three known physical mechanisms for confining a thermonuclear plasma:

- Gravitational confinement
- Inertial confinement
- Magnetic confinement

Gravitational confinement requires the gravitational field of an enormous body. It works well enough in the sun; so well, in fact, that the sun's energy comes primarily from the hydrogen-hydrogen fusion reaction, which requires far more stringent conditions than the deuterium-tritium reaction discussed above. However, gravitational fields of this magnitude do not exist on earth, and this approach is not practical for a fusion energy source.

In inertial confinement, the temperature and density of the plasma fuel are raised so rapidly that a significant fraction of the material reacts before it has time to disperse. One method is to heat a D-T pellet uniformly with laser, electron, or ion beams. The outer layers evaporate, and as they leave the surface they compress the remainder of the pellet, further heating it and raising its density. The inertial confinement approach is being pursued under the Assistant Secretary for Defense Programs owing to its close connection with potential military applications.

Magnetic confinement takes advantage of the fact that the plasma consists of charged particles; magnetic fields can be used to orient the motion of these particles. Figure VI 2-2 shows how charged particles react to the presence of a uniform magnetic field, which can be represented by straight, parallel lines. The charged plasma particles are constrained to follow helical paths encircling the magnetic field lines. Positive particles spiral in one

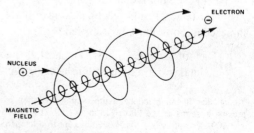

Figure VI-2-2. Effect of a uniform magnetic field on charged particles.

direction and negative particles in the opposite direction. As a result, the particles are not free to move across the magnetic field lines and can be confined long enough for a fusion reaction to take place.

A magnetic field that confines a plasma effectively exerts a pressure on the plasma. The ratio of the plasma particle pressure to the pressure from the magnetic field is called beta, and is a measure of the engineering efficiency of the use of magnetic fields. The value of beta for a confined plasma can range from 1 when the plasma pressure is equal to the magnetic pressure, down almost to zero, when the plasma pressure is very low.

Making Fusion into a Practical Energy Source

To translate the scientific basis for fusion into a practical source of fusion energy is a task whose magnitude is generally believed to be greater than the present and already difficult task of proving fusion's scientific basis. In order to achieve the requisite plasma parameters described above (i.e., T, n,), a considerable wealth of technologies has already been started to support the scientific studies; these technologies include high-intensity, large-volume magnets (both normal and superconducting), high-intensity, high-energy neutral beam injectors, radio frequency heating systems, sophisticated laser diagnostics, and data acquisition systems. Some of these are peculiar to fusion development, while others are adaptations of existing technologies.

Each of these demanding technological requirements is itself based upon state-of-the-art scientific investigations whose pursuit in the name of fusion energy extends the frontiers of basic knowledge. The large magnets require new knowledge of fundamental properties of materials at very low temperatures, the beam injectors require new extensions of ion sources, the radio frequency heating systems require new types of microwave tubes, and so forth; the radiation damage studies result in new understanding of bulk and surface materials properties.

What makes the task especially difficult is the need to apply each of these advanced technologies to the same facility in a highly integrated manner. The difficulty of turning "scientific" fusion into "practical" fusion is further increased by the growing necessity of dealing with a burning (or fusing) D-T plasma rather than the nonburning H or H-D plasma used for the past and current scientific studies. To deal with the engineering realities of handling an energy source presents the real burden of the next major period of fusion energy development. Concepts developed in the fusion program's recent period of scientific and technical achievement must now be examined for their engineering feasibilities. First, there are the engineering issues associated with handling fuel in and out and the fusion heat energy in the basic energy conversion device—the blanket. Second, one must consider these questions: How is reliability to be assured? What duty cycle can be attained? How is the economically mandated lifetime of the critical materials to be achieved under fusion conditions? What must be done to the reactors to make assembly, access, maintenance, and disassembly economic? These questions and others dealing with socioeconomic issues must be addressed and successfully resolved before the scientific basis of fusion, now believed to be nearly at hand, can be translated into a productive energy source.

To make this discussion specific, it is helpful to present in outline form, at least, the elements of a conceptual fusion reactor or energy source. Figure VI 2-3 shows a general magnetic fusion reactor. At the heart of the reactor is the vacuum chamber, where the plasma is confined by strong magnetic fields. As mentioned above, because the D-T fuel cycle is the most likely one with which to initiate fusion development, this reactor is assumed to be fueled with deuterium and tritium.

Deuterium is found in minute quantities in sea water (for every 6,500 atoms of ordinary hydrogen in water, there is one atom of deuterium), and it can be economically extracted. The oceans contain millions of years' supply of deuterium, so its availability is not a problem. Tritium does not occur naturally, but it can be produced by the reaction of lithium with a neutron. Lithium is fairly inexpensive

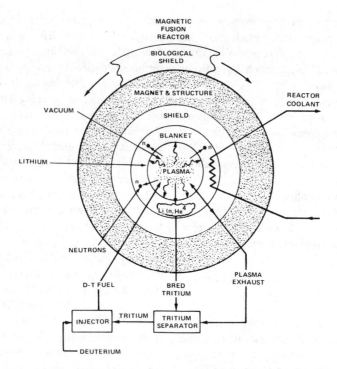

Figure VI-2-3. Schematic of a generic fusion reactor.

and is available in ample quantities as a mineral reserve.

The deuterium-tritium reaction releases a fast (14.1-MeV) neutron. In principle, this neutron can react with a lithium nucleus, and the products include tritium and a slow neutron. This slow neutron can, in turn, react with lithium to produce another atom of tritium. In short, matters can be arranged so that more tritium is produced than the amount used in the fusion reaction; i.e., the reactor "breeds" tritium.

A blanket or energy conversion device containing lithium surrounds the vacuum chamber. Fusion reaction neutrons perform two functions in the blanket: (1) they react with lithium to breed tritium, and (2) they also heat the blanket.

Radioactivity is produced in the blanket and the vacuum chamber walls. Therefore, these two elements are surrounded by local shields for personnel safety, protection of the environment, and prevention of physical damage to the reactor components. A second biological shield, completely surrounding the entire energy source, further ensures that little or no harmful radiation will reach reactor personnel and the outside environment. In the long run, fusion fuels which produce no radioactivity may be used in reactors. This would allow the reactor structure to be simplified, decrease maintenance problems, and eliminate exposure to releases of radioactive materials.

The tritium in the blanket is removed and purified. Periodically, the spent plasma is removed from the vacuum chamber, and the tritium still present is recovered. Fresh D-T fuel is either injected into the vacuum chamber for pulse systems or is continuously injected for steady-state systems.

Fusion Energy Applications

Fusion energy is released in four forms and can be used for a variety of applications as shown in Figure VI 2-4. Although the primary approach of the fusion program is the economic production of electricity, nonelectric applications of fusion energy are also being investigated.

The energetic fusion products absorbed in the reactor blanket produce heat which can be

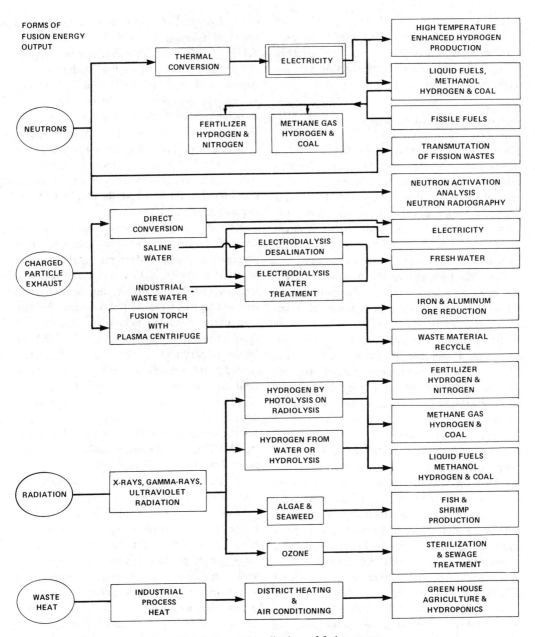

Figure VI-2-4. Potential applications of fusion energy.

used to drive conventional turbine generators. The production of electricity from this heat is a straightforward application of fusion energy. This heat could also be used for nonelectric applications such as industrial process heat and energy for the heating and cooling of buildings. In addition, two other major potential nonelectric applications of fusion energy are production of synthetic fuels and production of fissile fuel for fission reactors.

Each of the aforementioned applications requires different levels of technical and economic achievement so that some of the applications may become technically feasible before others. The plasma parameter which most directly indicates the state of technical readiness, called the Q value, is the ratio of fusion energy produced by the plasma to the energy supplied to the plasma. Q values of 2–5 are believed sufficient for fissile fuel production; Q values of 10 or more are probably required for the economic production of synthetic fuels and/or electricity.

Status of Magnetic Fusion

The history of magnetic fusion R&D shows a steady and encouraging progression of results that are drawing closer to achieving scientific feasibility for the three categories of concepts currently under development: tokamak* systems (the closest of the three to the achievement of scientific feasibility), magnetic mirror systems, and advanced concepts systems.

Scientific progress on the tokamak concept has been especially dramatic. Temperatures in the 4–7-keV range have been achieved, comparing favorably with the 5–10-keV levels probably required for breakeven** and for operation of a tokamak reactor, respectively. $n\tau$, the measure of confinement needed for successful production of net energy from fusion has been raised tenfold during the past several years (using a beam-driven plasma mode of operation) to the level of 3×10^{13} cm^{-3}-sec, a value which exceeds the theoretical minimum (approximately 10^{13} cm^{-3}-sec) needed for breakeven. Values of beta (ratio of plasma pressure to magnetic field pressure, a measure of power density and efficiency of magnetic field use) of 0.02 have been achieved, compared to a value of about 0.02 estimated for breakeven and 0.05–0.10 for operating reactors. Recently, the Princeton Large Torus (PLT) achieved an equivalent gain, Q, of 0.01, within a factor of 100 of breakeven (Q=1). However, Q increases as the square of the density and exponentially with increasing temperature; consequently, the density and temperature increases required for breakeven (factor of 3 in density and 2–3 in temperature) are more accurate measures of the difficulty of progressing to breakeven. The present status of tokamak research is summarized in Table VI 2-2. The individual parameters (n, ,T) are listed as well as the product (n, T). Advances in both the individual and product values are necessary to demonstrate understanding of the individual and integrated fusion physics processes. There is considerable confidence in the scientific community that the required parameter values for breakeven will soon be realized in a single experiment at the Tokamak Fusion Test Reactor (TFTR) to begin operation in 1982.

Although progress in tokamaks has been dramatic, other confinement concepts show promise as well. Magnetic mirrors have achieved n values of 10^{11}cm^{-3}-sec at temperatures of 13-keV. Two promising new concepts have emerged in the mirror program: the tandem mirror, and the field reversal mirror. A proof-of-principle experiment, the Tandem Mirror Experiment (TMX), has just been fabricated, and several different approaches to field reversal are being pursued. These recent results in mirror experiments have led to the development of a major scale-up experiment for this concept, the Mirror Fusion Test Facility (MFTF), now under construction at Lawrence Livermore Laboratory (LLL).

In addition to tokamaks and mirrors, several advanced concepts are being studied. Since there is now a high degree of confidence in the tokamak concept for demonstration of scientific feasibility, the primary objective of funding advanced concepts is to seek magnetic confinement configurations which may lead to more attractive fusion reactors than tokamaks. A secondary objective of the research on advanced concepts is to broaden the physics data base of the program through studies of confinement and heating in different magnetic geometries.

Technical progress and the status of each concept can be seen in Figure VI 2-5. This figure identifies the plasma operating regime for each magnetic confinement concept with respect to the two most critical plasma characteristics, namely its confinement measure ($n\tau$) and its temperature. Each contour line in the figure represents a value of energy gain, Q.

---

*Russian acronym signifying "toroidal magnetic chamber."

**Breakeven is achieved when the energy produced by the fusion reaction is equivalent to the energy required to produce and maintain the plasma.

Table VI-2-2. Status of Tokamak parameters.

PARAMETER	HIGHEST VALUES ACHIEVED (DEVICE)	ESTIMATED REQUIREMENT FOR BREAKEVEN	GOAL FOR REACTORS
CONFINEMENT MEASURE: $n\tau$ (cm^{-3}sec)	3×10^{13} (Alcator) 1.5×10^{13} (PLT)	1×10^{13}	3×10^{14}
TEMPERATURE T(keV)	7.5 (PLT) 1.0 (Alcator)	5	7-10
$n\tau T$ (cm^{-3}sec-keV)*	1×10^{12} (PLT) 3×10^{13} (Alcator)	$\sim 5 \times 10^{13}$	2.5×10^{15}
BETA	0.02 (ISX-B)	0.02	0.06-0.10
GAIN, Q**	0.02 (PLT)	1.0	10-20

*It is possible on the basis of detailed scientific understanding to consider the two experiments (PLT, Alcator) as complementary parts of a generic tokamak, based on the Alcator and PLT results. Postulating a generic tokamak would lead to $n\tau T \sim 2.3 \times 10^{14}$ cm^{-3} sec keV.

**The gains shown are calculated based on substituting a 50-50 deuterium-tritium mixture for the pure deuterium actually used in the experiments. D-T mixtures have a yield approximately 100 times higher than occurs in pure deuterium.

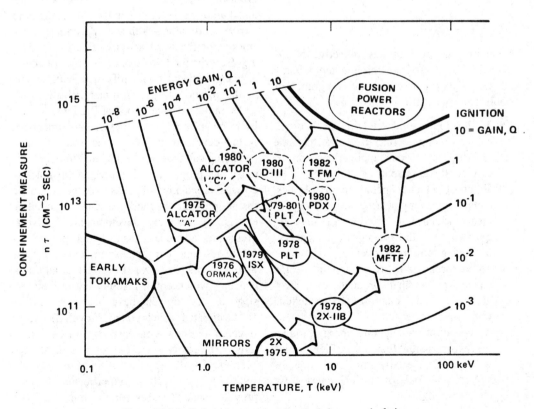

Figure VI-2-5. Technical progress and outlook in magnetic fusion.

STRATEGY FOR MAGNETIC FUSION ENERGY DEVELOPMENT

The Office of Fusion Energy, as part of its overall responsibilities for the direction of the research and development of magnetic fusion, has identified the program goal, approaches toward this goal, key objectives, and the strategy to achieve these objectives.

Program Goal and Approaches

The goal of the magnetic fusion program is to develop the potential for fusion energy as a virtually inexhaustible energy source for a variety of applications.

The primary approach of the program is the development of systems for the commercial production of central station electric power.

The secondary approach is the evaluation and development of other applications of fusion systems, including production of fissile material, synthetic fuels, industrial process heat, and heat for other end uses.

Program Objectives

To achieve the goal or key objective of the magnetic fusion program through the primary and secondary approaches, the program has established four critical objectives:

- To develop a solid scientific, technological, and engineering base so that the viability of various applications can be confidently assessed
- To construct and operate major physics/technological/engineering scaling experiments which will provide the data required for the design of energy-producing prototype reactors
- To perform parametric and point design studies of a final product with characteristics that maximize its beneficial impact on society
- To construct and operate energy-producing experimental fusion facilities, which will provide the basis for developing the fullest potential of fusion energy for commercialization

Strategy

The Office of Fusion Energy has developed a magnetic fusion program strategy designed to achieve the four critical objectives consistent with: (1) the DOE fusion energy policy, (2) the DOE research, development and commercialization (RD&C) cycle for energy technologies, (3) technical progress, and (4) availability of funds. Figure VI 2-6 demonstrates this consistency.

The top of Figure VI 2-6 shows the stages of the DOE RD&C cycle necessary to harness any energy supply option, from applied research through exploratory/technology/engineering development, demonstration, and commercialization.

The middle of Figure VI 2-6 illustrates the fusion energy RD&C process in a manner consistent with the DOE policy for fusion energy.

The bottom of Figure VI 2-6 depicts the magnetic fusion energy strategy. The program is presently centered around the first activity, Proof-of-Principle, and is making a major thrust into the second activity, Major Scaling Experiments, along with the appropriate preparation for the third activity, Energy Producing Experimental Reactors. Finally, planning is being initiated to identify the requirements of the fourth activity, Commercialization.

As indicated, the Proof-of-Principle Activity is much more than a set of experiments; it is the development of a solid scientific and technological base on which to build the applications program. The Office of Fusion Energy is broadening the base of the magnetic fusion program beyond the current mainline efforts, by fostering development of several promising advanced fusion concepts. Plans are to test one concept experimentally at the proof-of-principle level (Level 1). In addition, concepts which are at a somewhat less mature stage of development have been selected for an accelerated development (Level 2). Some research on new concepts or concepts in a more exploratory stage (Level 3) will also be supported.

The Major Scaling Experiments Activity provides sharp focuses for the transition to a

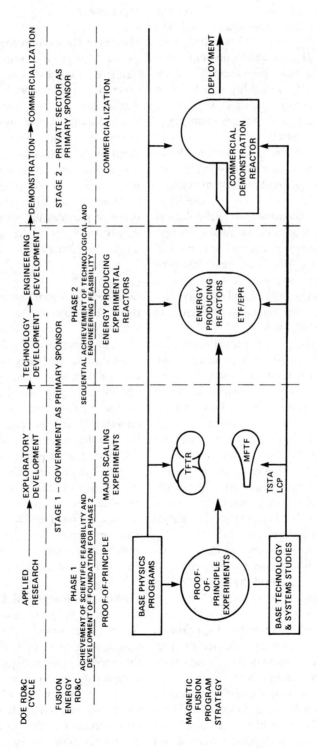

Figure VI-2-6. Magnetic fusion energy RD&C cycle and program strategy.

development emphasis through the preparation of separate, major scientific (TFTR and MFTF), technological (Large Coil Project and Tritium Systems Test Assembly), and engineering (Fusion Materials Irradiation Test) facilities. These physics and technology component-oriented facilities are expected to demonstrate the scientific feasibility of controlled fusion energy and to prepare the way for the development of complete, though elementary, fusion energy systems.

The Energy-Producing Experimental Reactors Activity provides the facility bridge between fusion energy R&D and commercialization. These facilities, the ETF and the EPR, are expected to demonstrate the technological and engineering feasibility of fusion energy. Although no energy-producing reactors will be initiated during this fiscal year, the program will continue to pursue this objective in an expeditious manner, as funding may permit, by laying the necessary design and RD&C planning groundwork.

The fourth activity, Commercialization, involves the design, construction, and operation of demonstration reactors necessary to prove the economic feasibility of commercial fusion energy systems. During this activity, significant private investment is anticipated, although a strong Federal role will be necessary to facilitate the transfer of the technology from the public to the private sector. To ensure that this future transfer will be accomplished smoothly, the magnetic fusion program planning is beginning to incorporate industrial and utility interests.

The magnetic fusion energy strategy for development of a fusion concept using the D-T reaction may be described by the following sequence of steps listed in Table VI 2-3.

Proof-of-Principle Hydrogen Experiment. This experiment determines whether the particular magnetic configuration or dynamics suggested by the concept does, in fact, confine the plasma for some significant amount of time. (The eventual goal is to obtain confinement times on the order of several minutes or hours per magnetic field pulse.) Such a test verifies whether a particular concept is workable as a confinement technique. It is usually conducted in a small- to intermediate-sized experiment.

Hydrogen/Deuterium Major Scaling Experiment. After proof-of-principle has been demonstrated, the next step is to scale-up the experiment to improve plasma performance. Again, this is a physics-type experiment carried out with either hydrogen and/or deuterium as a fuel. The primary goal is to demonstrate plasma confinement at temperature and density levels nearer to reactor level plasmas. Plasma refueling to maintain density is demonstrated. In addition, this type of experiment provides the opportunity for impurity control problems to be defined and for prototypic instrumentation and control systems to be tested.

Deuterium-Tritium Major Scaling Experiment. The objective of this experiment is to achieve the scientific breakeven point (where the energy produced by the fusion reaction is equivalent to the energy required to produce and maintain the plasma) using a D-T plasma. Specifically, the combination of plasma quality (as measured by the confinement parameter, n) and the plasma temperature will be such that a plasma energy gain (Q) of 1 is achieved. Such a plasma will produce reactor-level power densities. Owing to the large neutron yield from the D-T burning, the reactor in which the experiment is carried out will become radioactive. This experiment will provide an opportunity for testing the remote maintenance and handling systems required for commercial reactors.

Elementary Integrated Energy Source. The objective is the achievement of an elementary fusion energy source, defined as an integrated system of high Q or ignited plasma, and all the necessary technological and engineering components and aspects required to control the fusion energy. (Ignition is a state where the fusion reaction itself can provide the energy needed for its continuance; a high-gain or high-Q state is one in which there are enough fusion reactions to more than repay

Table VI-2-3. Characteristics of the magnetic fusion program strategy.

STRATEGY ACTIVITY	STEPS	PRIMARY GOAL	RELATED APPLICATIONS	PLANT AVAILABILITY	PULSE LENGTH	REMOTE MAINTENANCE
PROOF-OF-PRINCIPLE	HYDROGEN EXPERIMENT	TEST CONFINEMENT CONCEPT	PLASMA DIAGNOSTICS, PLASMA THEORY VALIDATION	N/A	VERY SHORT	NONE
MAJOR SCALING EXPERIMENTS	HYDROGEN/DEUTERIUM MAJOR SCALING EXPERIMENT	SCALED UP TEST OF CONCEPT	TEST EMPIRICAL SCALING LAWS, PLASMA MODELING	N/A	SHORT	NONE
	DEUTERIUM-TRITIUM MAJOR SCALING EXPERIMENT	DEMONSTRATE BREAKEVEN WITH D-T FUEL	MEASURE NEUTRON, He^4 PARTICLE YIELDS, COMPONENT TESTS	LOW	SHORT	PRELIMINARY
ENERGY PRODUCING EXPERIMENTAL REACTORS	ELEMENTARY INTEGRATED ENERGY SOURCE	DEMONSTRATE IGNITION OR HIGH GAIN WITH D-T FUEL	LARGE MAGNETICS (SUPERCONDUCTING), BLANKETS/BREEDING MODULES	LOW-TO-MODERATE	MODERATE	MODERATE
	ENGINEERING NET POWER REACTOR	PRODUCE POWER FROM D-T FUSION	MATERIALS TESTING MODULES, SYSTEMS INTEGRATION	MODERATE	LONG	REACTOR LEVEL
COMMERCIALIZATION	COMMERCIAL DEMONSTRATION NET POWER REACTOR & DEPLOYMENT	DEMONSTRATE SCALING TO ECONOMIC POWER	MATERIALS APPLICATION, SYSTEM OPTIMIZATION	HIGH	LONG	REACTOR LIFECYCLE LEVEL

the original energy investment needed to create the plasma.) Since either a high Q or ignition is required for a fusion reaction to be a net energy producer, completion of this phase will have achieved a major scientific goal of the magnetic fusion program. Control of the fusion energy source requires successful engineering tests of the various processes, designs, and components involved. These tests will provide elementary, but feasible, solutions to these technological and engineering problems. The facility employed for this phase is the Engineering Test Facility (ETF), at the interface between large experiments (fusion research) and the EPR (fusion reactor development); the elementary solutions developed in the ETF will establish the technological requirements and provide baseline solutions for each of the major components required by the EPR.

Engineering Net Power Production. This step will prove the engineering required for Commercial Demonstration of Net Power Production, including the generation of electricity, but at a smaller, less efficient scale. The reactor required to accomplish this objective, the EPR, is both scalable from the physics and engineering of the previous steps and scalable to the more economical characteristics of the Commercial Demonstration Reactor. Its engineering goal is to generate electrical energy with a thermal efficiency that approaches a steady-state mode and with a plant availability near that of conventional power plants. It will also demonstrate a reactor scale tritium fuel cycle, with breeding modules having a breeding ratio greater than 1, a high temperature blanket system with capability to accommodate materials testing, and reactor plant maintenance and safety systems.

Commercial Demonstration of Net Power Production. The objective of this step is to demonstrate safe and reliable electrical power generation at a near-commercial scale in a utility environment. The Demonstration Reactor will be scalable in an engineering and economic sense from the EPR and be scalable to the design and operating regime characteristic of commercial-sized reactors. This facility will provide the technical and economic groundwork to allow private industry to decide on the rate at which fusion power plants can be commercially introduced.

PROGRAM ACTIVITY AND BUDGET STRUCTURE

The budget for the Office of Fusion Energy's program in magnetic fusion is organized into four activities: Confinement Systems, Development and Technology, Applied Plasma Physics, and Reactor Projects. The function of each budget activity is as follows:

- Confinement Systems—to conduct experimental activities aimed at demonstrating and refining methods of heating and containing high-temperature plasmas in tokamak and magnetic mirror systems; the cost of MFTF construction is also included in this budget activity at this time
- Development and Technology—to provide for development of engineering and technological bases to design, construct, and operate increasingly larger and more complex fusion experiments and facilities
- Applied Plasma Physics—to pursue experimental and theoretical studies of fusion plasma phenomena needed to develop the knowledge required to predict thermonuclear plasma behavior in confinement systems and future fusion power systems; to pursue new confinement concepts which may be alternatives to the tokamak and magnetic mirror approaches
- Reactor Projects—to provide for the construction of, and project specific development for, major new facilities needed by the other activity and subactivity areas.

The actual FY 1979 BA and BO, the estimated FY 1980 BA and BO, and the FY 1981 BA and BO requests for operating expenses, capital equipment, and construction for each budget activity are shown in Table VI 2-4. The table also shows anticipated total budget requirements (i.e., required for current programs for each of the specific expenditure categories for magnetic fusion.

The current status within each of the four activities is summarized below. Table VI 2-5 shows the major milestones for each activity for FY 1978–1983.

Confinement Systems. Within the Confinement Systems Activity, support of the two primary approaches to magnetic confinement, the tokamak and the magnetic mirror, is being continued. Recent experimental results with tokamak systems have brought the program closer to the achievement of reactor-like conditions, and progress continues in five critical areas of research: heating, transport and scaling, plasma shape optimization, boundary efforts and impurity control, and fueling. Magnetic mirror efforts are also being maintained in the areas of confinement scaling, heating, and improved power balance. Construction of the MFTF is continuing.

Development and Technology. In the Development and Technology Activity, studies are continuing in six subactivities: Magnetic Systems, Plasma Engineering, Fusion Reactor Materials, Fusion Systems Engineering, Environment and Safety, and Fusion Energy Applications. Progress was made toward the development of neutral beams for heating purposes, development of large superconducting magnets, structuring of the materials program to assess the effects of high-energy neutrons on walls and structures, conceptual designs for fusion reactors, an environmental assessment of conceptual fusion power reactors, and studies of alternate fusion energy applications. Efforts will continue in each of these areas.

Applied Plasma Physics. In the Applied Plasma Physics Activity, theoretical and experimental studies continued to increase the reservoir of knowledge required to support efforts in Confinement Systems and Development and Technology. Specifically, experimental plasma studies are developing the data base in atomic and molecular physics required for impurity control, exploring promising new plasma confinement concepts, developing new diagnostic techniques to measure fusion plasma parameters, and investigating new methods of plasma production and heating.

Table VI-2-4. FY 1979-1981 Office of Fusion Energy budget activity funding (dollars in millions).

EXPENDITURE CATEGORIES	DOLLARS IN THOUSANDS					
	FY 1979		FY 1980		FY 1981	
	BA	BO	BA	BO	BA	BO
OPERATING EXPENSES						
Confinement Systems	68,975	72,921	90,090	91,490	105,100	107,630
Development & Technology	49,255	50,370	48,000	54,000	63,600	66,000
Applied Plasma Physics	54,689	52,586	49,700	51,853	65,800	65,800
Planning and Projects	36,024	37,499	42,410	42,610	48,500	56,500
Program Direction	2,410	2,410	3,089	3,089	3,117	3,117
Total Operating Expenses	211,353	215,786	233,289	243,042	286,117	299,047
CAPITAL EQUIPMENT						
Confinement Systems	12,535	11,597	15,805	14,905	18,300	16,250
Development & Technology	7,243	7,018	6,050	6,025	8,700	8,000
Applied Plasma Physics	4,362	4,949	3,675	3,515	6,700	6,500
Planning and Projects	3,060	2,207	4,270	3,855	4,400	3,000
Total Capital Equipment	27,200	25,771	29,800	28,300	38,100	33,750
CONSTRUCTION						
Confinement Systems	17,003	15,938	18,200	18,200	7,900	5,000
Development & Technology						
Large Coil Test Facility	6,654	2,253	2,000	2,000	6,500	3,500
Applied Plasma Physics	0	3,063	0	0	0	0
Planning and Projects						
Tokamak Fusion Test Reactor	42,000	57,647	30,100	56,400	30,000	48,500
Fusion Materials Irradiation Test Facility	6,900	3,778	10,000	3,000	22,600	20,600
Mirror Fusion Test Facility	40,000	24,406	28,200	20,500	7,000	46,500
General Plant Projects	4,030	3,232	4,000	4,000	5,400	7,000
Other	0	311	0	0	0	0
Total Construction	116,587	110,628	92,500	104,100	79,400	131,100
TOTAL	355,140	352,185	355,589	375,442	403,617	463,897

The plasma theory studies continue to concentrate on providing solutions to the complex equations that describe the physical characteristics of plasma. The computational capabilities of the theory group were greatly enhanced with the establishment of the National Magnetic Fusion Energy Computer Center. Research on advanced fusion concepts will be further intensified.

Reactor Projects. In Reactor Projects, construction has started on the TFTR, design has been initiated on Fusion Materials Irradiation Test (FMIT), and the High Intensity Neutron Source (HINS) has been committed to materials testing studies.

Figure VI 2-7 shows the distribution of the FY 1980 BA request among the four magnetic fusion energy program strategy activities. Proof-of-Principle R&D accounts for the largest share of the funds, approximately 55 percent. About 38 percent of the budget is related to the construction and operation of Major Scaling Experiments. Other R&D (approximately 7 percent) is directed to long leadtime technology that will be required for Energy Producing Experimental Reactors. Finally, there is a small effort (less than 1 percent) aimed at defining long-range issues associated with commercialization. Thus, while the major portion of the budget is necessarily aimed at the Proof-of-Principle and Major Scaling Experiments due to the early stage of the technology development, all areas are receiving some attention from the program.

Tables VI 2-6, VI 2-7, and VI 2-8 present the detailed FY 1980 budget structure and show the actual FY 1978 BA, the estimated FY 1979 BA, and the FY 1980 BA requests

Table VI-2-5. Major milestones.

	CONFINEMENT SYSTEMS	DEVELOPMENT & TECHNOLOGY	APPLIED PLASMA PHYSICS	PLANNING & PROJECTS
FY 1979	• RF Heating in PLT • Initiate Beta II • Evaluate Shape and Beta on ISX	• Operate HINS • Issue Revised Environmental Development Plan	• Initiate Spheromak Studies	• Begin MFTF Construction
FY 1980	• Alcator C Transport, Scaling, & Heating Exps. • Neutral Beam Heating on PDX • Evaluate Divertor on PDX • Generate Plasma Rings on Beta II • Initiate EBT-P	• Complete Fusion Reactor Safety Research Program Plan • Tokamak Commercial Reactor Design • Gyrotron RF Heating on EBT-S • Select Innovative Alloy Concepts for Study	• Operate ZT-40 Reversed Field Pinch • Evaluate Plasma Confinement from Electron Rings • Occupy new NMFECC Building • Complete Fabrication of FPRF	• Begin FMIT Construction • Decision on Upgrade of MFTF to MFTF-B • Initiate Upgrade of TFTR to TFM
FY 1981	• Decision on Future of PLT • Determine Beta Limits on PDX • Operate Impurity Control on ISX • Neutral Beam Heating on Doublet III • TMX Modification • Thermal Barrier Exps. on Phaedrus • EBT-S Plasma Exps.	• 3-Coil Tests on Large Coil Facility • Test Small Coils in High Field Test Facility • Mirror Commercial Reactor Design • Synthetic Fuel Reactor Conceptual Design • Pellet Injection on Doublet III • Test Plasma Energy Direct Recovery on TMX	• Modify ZT-40 • Begin User Exps. on Fusion Plasma Research Facility • Install a Second Class VI Computer	• Complete Facility Construction and Major Equipment Fabrication for TFTR • Complete MFTF Fabrication and Assembly
FY 1982	• Operate TFTR and MFTF • Evaluate Doublet III Beta & Heating Exps.	• 6-Coil Tests on Large Coil Facility • TFTR, MFTF Neutral Beam Systems • Tokamak ETF Conceptual Design • Tritium Operation in TSTA • Complete Generic Environmental Impact Statement	• Operate S-1	• Complete TFTR
FY 1983	• Operate TFM (TFTR Upgrade)	• Mirror ETF Conceptual Design • APIS Neutral Beam for TFTR	• Complete Surface Region Atomic Data	• Complete TFM • Complete FMIT Facility Construction • Complete MFTF-B
FY 1984	• Operate EBT-P • Operate MFTF-B	• Pellet Injection on TFTR		• Complete FMIT Installation
FY 1985		• Operate FMIT	• Install Class VII Computer	

for the Confinement Systems, Development and Technology, and Applied Plasma Physics and Reactor Projects, respectively. The last column in each table shows the increase (decrease) of the FY 1980 BA request relative to the FY 1979 BA.

MANAGEMENT APPROACH

To achieve its objectives, the Office of Fusion Energy uses a participatory management approach which tries to address the total problem of fusion in an orderly, integrated manner and to conduct the program with sound business principles. The management approach is based upon close coordination among the program participants, including Headquarters, DOE laboratories, universities, field activities, and industry. In addition, there is a growing interaction with the inertial confinement research program.

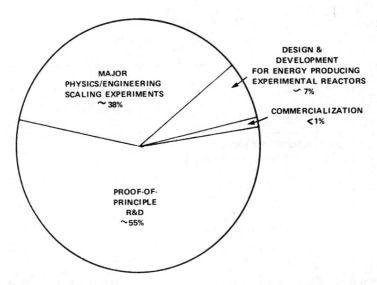

Figure VI-2-7. Distribution of FY 1980 BA request among the four magnetic fusion energy strategy activities.

Business Approach

For major projects, such as the TFTR and the MFTF, there exist formalized project management plans which document objectives and responsibilities, establish cost and technical schedules, and provide for reporting and management interface procedures. In addition, the experience, knowledge, and skills of senior technical staff within Headquarters and its major contractor installations, along with competent consultants possessing specialized backgrounds, are used in the general planning and execution of the program. Significant technical problems are analyzed collectively on a periodic basis by these groups or by special committees composed of DOE laboratory, university, industry, and Headquarters personnel. Alternatives are considered, trade-offs analyzed, milestones identified, objectives established, and resource requirements forecast. Action plans to solve the identified problems are systematically developed, subsequently reviewed by office staff, and then integrated into the office's overall activity plan. Study projects and meetings result in formal or informal reports which are widely disseminated. In this way, the judgments and knowledge of those both directly and indirectly involved with a particular problem are utilized. The technical staff responsible for activity implementation are given an opportunity to provide substantive input into decisions affecting the future direction of their particular projects.

Program Management—Headquarters

Subject to review and approval by DOE upper management, the Office of Fusion Energy is responsible for research and development on fusion power by magnetic confinement, including the formulation of strategies and the development of plans encompassing scientific and technical program content, goals, and objectives. The office also develops and defines budget requirements, defends budget requests, interprets policy, and monitors programmatic and technical progress.

The Headquarters is organized into five divisions as shown in Figure VI 2-8. The functions of the five divisions are as follows:

- Planning and Projects—responsible for strategy development, program planning, international cooperation, field/laboratory utilization, major construction projects overview, and engineering management support of other projects
- Resource Management and Acquisition—

Table VI-2-6. FY 1978-FY 1980 BA requests for the confinement systems budget activity (dollars in thousands).

ACTIVITY	SUBACTIVITY	TASK	SUBTASK	TITLE	FY 1978 BA ACTUAL ($K)	FY 1979 BA ESTIMATE ($K)	FY 1980 BA ESTIMATE ($K)	FY 1980 BA INCREASE (DECREASE) ($K)
01				**CONFINEMENT SYSTEMS**				
	10			**TOKAMAK SYSTEMS EXPERIMENTS**				
		11		Research Operations	36,612	38,100	53,500	15,400
		12		Major Device Fabrication	25,338	30,100	12,500	(17,600)
			A	ORMAK	2,121	0	0	0
			B	Princeton Large Torus (PLT)	0	0	0	0
			C	Poloidal Divertor Experiment (PDX)	3,654	600	0	(600)
			D	Doublet III	4,624	0	0	0
			E	Impurity Studies Experiment (ISX)	1,707	2,100	2,000	(100)
			F	PLT Neutral Beams (ORNL)	0	0	0	0
			G	Alcator	3,594	700	0	(700)
			H	Doublet III Neutral Beams	3,005	11,000	2,500	(8,500)
			J	PDX Neutral Beams	2,018	6,700	6,600	(100)
			K	PLT – RF	1,221	1,000	400	(600)
			L	PDX – RF	0	0	0	0
			M	Alcator RF	207	3,000	1,000	(2,000)
			N	Other	3,187	5,000	0	(5,000)
				Equipment	8,100	7,300	7,375	75
				TOTAL	70,050	75,500	73,375	(2,215)
	20			**MAGNETIC MIRROR SYSTEMS EXPERIMENTS**				
		21		Research Operations	8,344	11,720	15,100	3,380
		22		Major Device Fabrication	9,039	620	0	(620)
			A	2X	0	0	0	0
			B	Tandem Mirror Experiment (TMX)	8,503	0	0	0
			C	Phaedrus	536	620	0	(620)
			D	Elmo Bumpy Torus – Scale (EBT-S)	0	0	0	0
			E	Elmo Bumpy Torus II (EBT-II)	0	0	0	0
			F	Other	0	0	0	0
				Equipment	2,695	3,835	3,575	(300)
		23		Mirror Fusion Test Facility	17,104	48,300	44,700	(3,600)
			D	Research	0	1,308	0	(1,308)
			E	Development	3,104	2,067	11,500	9,433
			F	Auxiliary Component Fabrication	0	4,685	0	(4,685)
			G	Facility Preparation	0	240	0	(240)
				Construction	14,000	40,000	33,200	(6,800)
				TOTAL	37,182	64,475	63,375	(1,100)
				TOTAL CONFINEMENT SYSTEMS	107,232	139,975	136,750	(3,225)

formulates the budget and provides administrative, financial, and procurement support to the technical divisions
- Development and Technology—responsible for all component technologies for all development activities, and for conceptual studies of energy-producing systems, applications, environmental issues, and commercialization
- Applied Plasma Physics—responsible for conducting proof-of-principle programs for advanced fusion concepts and for all base physics, including theory and experimental plasma research

Table VI-2-7. FY 1978-FY 1980 BA requests for the development and technology budget activity (dollars in thousands).

ACTIVITY	SUBACTIVITY	TASK	SUBTASK	TITLE	FY 1978 BA ACTUAL ($K)	FY 1979 BA ESTIMATE ($K)	FY 1980 BA ESTIMATE ($K)	FY 1980 BA INCREASE (DECREASE) ($K)
02				DEVELOPMENT AND TECHNOLOGY				
	10			MAGNETIC SYSTEMS				
		11		Superconducting Magnet Development	10,197	13,700	13,200	(500)
			1	Superconducting Magnet Base Program	2,856	960	1,000	40
			2	Superconducting Magnet Conductor Development	1,587	1,810	1,600	(210)
			3	Superconducting Magnet Major Device Fabrication and Operation	5,754	10,930	10,600	(330)
		12		Energy Storage and Transfer	1,785	1,500	1,600	100
			1	Energy Storage Systems	156	0	0	0
			2	Ohmic Heating Systems	1,629	1,500	1,600	100
			3	Energy Storage Major Device Fabrication and Operation	0	0	0	0
				Equipment	2,605	2,300	2,300	0
				TOTAL	14,587	17,500	17,100	(400)
	20			PLASMA ENGINEERING				
		21		Neutral Beam Development	10,976	11,635	6,800	(4,835)
			1	Neutral Beam Base Program	9,354	7,180	4,200	(2,780)
			2	Neutral Beam Major Device Fabrication and Operation	1,622	4,455	2,600	(1,855)
		22		Alternate Heating System Development	1,505	950	1,350	400
		23		Direct Energy Conversion	509	100	150	50
		24		Vacuum Component Development	557	575	550	(25)
		25		Plasma Maintenance and Control	1,113	655	1,250	595
				Equipment	2,205	1,970	1,900	(70)
				TOTAL	16,865	15,885	12,000	(3,885)
	30			FUSION REACTOR MATERIALS				
		31		Alloy Development for Irradiation Performance	3,550	3,280	4,000	720
		32		Plasma Materials Interaction	2,173	2,365	2,600	235
		33		Special Purpose Materials Development	452	545	625	80
		34		Damage Analysis and Dosimetry	766	960	1,425	465
		35		Radiation Facilities Development and Operation	1,131	2,065	2,550	485
			1	Radiation Facilities Development	64	65	0	(65)
			2	Radiation Facilities Operation	1,067	2,000	2,550	550
				Equipment	2,030	2,000	1,950	(50)
				TOTAL	10,102	11,215	13,150	1,935
	40			FUSION SYSTEMS ENGINEERING				
		41		Conceptual Design and Engineering	2,969	2,410	2,500	90
		42		System Studies	3,037	2,500	2,500	0
		43		Blanket and Shield Engineering	772	1,135	1,000	(135)
			1	Blanket and Shield Base Program	772	1,135	1,000	(135)
			2	Blanket and Shield Major Device Fabrication and Operation	0	0	0	0
		44		Tritium Processing and Control	2,051	2,680	2,500	(180)
			1	Tritium Processing Base Program	424	380	400	20
			2	Tritium Processing System Major Device Fabrication and Operation	1,627	2,300	2,100	(200)
		45		Plasma Systems	408	460	500	40
		46		Plant Systems	418	410	500	90
				Equipment	1,365	1,130	1,100	(30)
				TOTAL	11,020	10,725	10,600	(125)
	50			ENVIRONMENT AND SAFETY				
		51		Environment Assessment	93	345	680	335
		52		Reactor Safety Research	543	630	720	90
				Equipment	120	100	100	0
				TOTAL	756	1,075	1,500	425
	60			FUSION ENERGY APPLICATIONS				
		61		Synthetic Liquid and Gaseous Fuels	758	740	1,200	460
		62		Fission Fuel Cycle Support	1,492	1,650	1,700	50
		63		Process and Space Heating and Cooling	0	0	0	0
		64		Materials Processing and Chemical Production	0	0	0	0
		65		Emergent Applications and Technology Spinoff	40	110	100	(10)
				Equipment	0	0	200	200
					2,290	2,500	3,200	700
				TOTAL DEVELOPMENT AND TECHNOLOGY	55,620	58,900	57,550	(1,616)

Table VI-2-8. FY 1978-FY 1980 BA requests for the applied plasma physics and reactor projects budget activities (dollars in thousands).

ACTIVITY	SUBACTIVITY	TASK	SUBTASK	TITLE	FY 1978 BA ACTUAL ($K)	FY 1979 BA ESTIMATE ($K)	FY 1980 BA ESTIMATE ($K)	FY 1980 BA INCREASE (DECREASE) ($K)
03				**APPLIED PLASMA PHYSICS**				
	10			ADVANCED FUSION CONCEPTS				
		11		Research Operations	9,480	21,460	20,600	(860)
				Major Device Fabrication	4,130	0	9,100	9,100
			1	ZT-40	4,130	0	0	0
			2	LLX	0	0	0	0
			3	Other	0	0	9,100	9,100
				Equipment	1,740	2,015	1,875	(140)
				TOTAL	15,350	23,475	31,575	8,100
	20			FUSION PLASMA THEORY				
				Equipment	41	0	0	0
				TOTAL	15,924	17,090	18,000	910
	30			EXPERIMENTAL PLASMA RESEARCH				
		31		Fusion Plasma Research Facility (FPRF)	2,913	2,600	3,000	400
		32		Other	7,477	9,700	10,000	300
				Equipment	2,526	2,250	1,875	(375)
				TOTAL	12,916	14,550	14,875	325
	40			NATIONAL MFE COMPUTER NETWORK				
		41		National MFE Computer Center	6,186	8,360	8,700	340
		42		User Service Center	837	1,450	1,800	350
				Equipment	1,673	400	700	300
				TOTAL	8,696	10,210	11,200	990
				TOTAL APPLIED PLASMA PHYSICS	52,886	65,325	75,650	10,325
04				**REACTOR PROJECTS**				
	01			TOKAMAK FUSION TEST REACTOR				
		01		Development	13,599	20,910	13,375	(7,535)
		02		Experiment Research	663	1,280	4,350	3,070
		03		Facility Operation	663	2,310	10,140	7,830
		04		Neutral Beam Prototype Development	6,819	0	4,735	4,735
				Equipment	2,350	2,300	3,850	1,550
				Construction	71,000	42,000	30,100	(11,900)
				TOTAL	95,094	68,800	66,550	(2,250)
	02			INTENSE NEUTRON SOURCE				
		01		Development	1,275	0	0	0
		02		Experimental Research	0	0	0	0
		03		Facility Operation	0	0	0	0
				Equipment	50	0	0	0
				TOTAL	1,325	0	0	0
	03			HIGH INTENSITY NEUTRON SOURCE				
		01		Development	1,043	0	0	0
		02		Experimental Research	0	0	0	0
		03		Facility Operation	0	0	0	0
				Equipment	100	0	0	0
				TOTAL	1,143	0	0	0
	04			FUSION MATERIALS IRRADIATION TEST FACILITY (FMIT)				
		01		Development	3,200	7,600	6,500	(1,100)
		02		Experimental Research	0	0	0	0
		03		Facility Operation	0	0	0	0
				Construction	7,500	6,900	10,000	3,100
				TOTAL	10,700	14,500	16,500	2,000
				GENERAL PURPOSE EQUIPMENT	1,950	1,600	3,000	1,400
				GENERAL PLANT PROJECTS	3,640	4,030	4,900	870
				TOTAL REACTOR PROJECTS	113,852	88,930	90,950	2,020

- Magnetic Confinement Systems—responsible for design, fabrication, and operation of tokamak and mirror proof-of-principle experiments and preparation for operation of the major scaling experiments

Project Management—Field Activities

In accordance with the policy of decentralized management of the Assistant Secretary for Energy Technology, all budget line item projects with the Office of Fusion Energy are

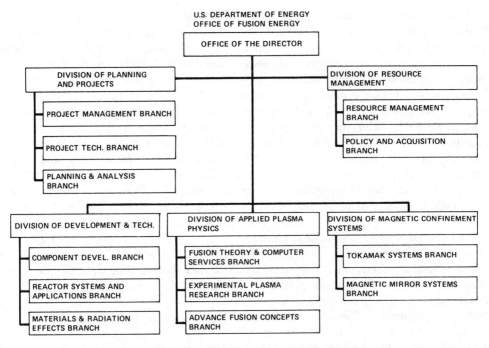

Figure VI-2-8. Headquarters organization of the office of fusion energy.

managed in the field with overall direction provided from Headquarters. The DOE field offices are responsible for the administrative management of all programs, including the control of management plans for all fabrication and construction projects. Presently, formal management plans exist for the TFTR and the MFTF with the Fusion Materials Irradiation Test (FMIT) Facility plan under development; a Memorandum of Understanding covers ongoing FMIT management arrangements.

Industry

The magnetic fusion energy activity is now funded by the Federal Government, primarily at DOE laboratories. Because of the long leadtime and financial risks associated with fusion power R&D, major support from non-Federal sources is not yet available. Utilities are currently involved in magnetic fusion R&D to the extent of several million dollars through direct contracts (primarily through the Electric Power Research Institute). Enlarging the industrial role will be required to establish a base for a fusion power industry. However, utilities and industry currently play a useful role through their assistance to overall program management.

The Fusion Power Coordinating Committee (FPCC) provides an opportunity for Headquarters and DOE laboratory personnel to meet with outside consultants to review technical progress, assess programs, and discuss management issues. The 12 outside consultants are selected for their particular expertise in scientific, engineering, and management areas.

An ETF Steering Committee is currently being formed. Its purpose will be to collect and evaluate relevant information for the current generation of experiments. The committee will provide guidance to the ETF design team for the program's next step, the ETF.

The Office of Fusion Energy will continue to use advisory bodies to assist in determining the major scientific and engineering objectives of the magnetic fusion energy programs and to review progress toward those objectives.

Cooperation with the Inertial Confinement Program

The inertial confinement approach to controlled fusion is being conducted by the Office of Laser Fusion under the auspices of the DOE Assistant Secretary for Defense Programs. Although this is a separate program, there is an exchange of appropriate information and technology. This interaction is growing as a result of the formation of the Fusion Review Committee, chaired by the Director of Energy Research, with the DOE Assistant Secretary for Energy Technology and the DOE Assistant Secretary for Defense Programs as members.

Areas in which technology transfers have been made from the Office of Laser Fusion to the Office of Fusion Energy include laser technology, pulse-power technology, electron-beam fusion concepts, and heavy-ion beam fusion techniques. Laser technology development is a key component of the overall laser fusion effort, and has been used in magnetic fusion experiments. Current experimental efforts in this area are being pursued by Princeton Plasma Physics Laboratory (PPPL), Los Alamos Scientific Laboratory (LASL), University of Washington, and Mathematical Sciences Northwest, Inc. Pulse-power technology has been sponsored by the Department of Defense and developed by private industry over the last decade. This technology may contribute to the mirror confinement research carried out at Lawrence Livermore Laboratory (LLL). Electron-beam and heavy-ion beam fusion concepts are being pursued by researchers at Sandia Laboratory, Argonne National Laboratory (ANL), Brookhaven National Laboratory (BNL), and Lawrence Berkeley Laboratory (LBL) as part of the inertial confinement program, and these concepts may be applicable to the magnetic fusion plasma preheating experiments as well.

In addition to the benefits from these transfers, the magnetic fusion research effort produces results of interest to inertial confinement researchers. Many of these contributions concern development of technology to heat fusion plasmas and handle fusion materials by-products. These developments include techniques in the areas of power supplies, tritium handling, plasma heating, and the development of superconducting magnets and reactor materials. The magnetic fusion program has also conducted reactor design studies and experiments to identify high strength, radiation-resistant materials for reactor construction. The operation of the FMIT facility will provide a data base for materials which can be utilized by both programs.

Because of a number of similarities between the magnetic and inertial confinement concepts, technology and hardware have been exchanged between the two programs. The collocation of many of the research efforts at a number of DOE laboratories, particularly LASL, LLL, Sandia, and ANL, allows researchers from each program to share information and ideas. In addition, systems studies for the Engineering Test Facility to be built for each confinement concept will provide a potential for discussion.

SIGNIFICANT ISSUES

The magnetic fusion energy program is structured so as to be responsive to the many problems associated with a technology in such an early stage of development. The program activities and subactivities are not only concerned with the technical problems of demonstrating the scientific/technological/engineering feasibility of fusion, but also with the environmental and socioeconomic issues associated with the commercialization of fusion technology. Despite the fact that fusion technology is still in the early phases of the RD&D cycle, it is not too early to begin to address the issues that will be encountered when the technology enters the later demonstration and commercialization phases. Thus, it is instructive to compare the structure of the issue areas within magnetic fusion R&D to the budget activity and subactivity structure. Such a comparison will demonstrate that there is at least one, and in many instances more than one, subactivity that addresses the key technical and environmental issues and socioeconomic issues of commercialization of fusion energy as currently perceived by the Office of Fusion Energy. These issues are defined below, and the comparison with an aggregated version of the activity structure is shown in Figure VI 2-9.

ISSUE AREAS / BUDGET SUBACTIVITY STRUCTURE	CONFINEMENT SYSTEMS				DEVELOPMENT AND TECHNOLOGY				APPLIED PLASMA PHYSICS				REACTOR PROJECTS		
	TOKAMAK SYSTEMS EXPERIMENTS	MAGNETIC MIRROR SYSTEMS EXP.	MAGNETIC SYSTEMS	PLASMA ENGINEERING	FUSION REACTOR MATERIALS	FUSION SYSTEMS ENGINEERING	ENVIRONMENT AND SAFETY	FUSION ENERGY APPLICATIONS	ADVANCED FUSION CONCEPTS	FUSION PLASMA THEORY	EXPERIMENTAL PLASMA RESEARCH	NATIONAL MFE COMPUTER NETWORK	TOKAMAK FUSION TEST REACTOR	HIGH INTENSITY NEUTRON SOURCE	FUSION MATERIALS IRRADIATION TEST FACILITY
Technical															
Plasma Behavior	X	X	X	X					X	X	X	X	X		
Scaling Laws	X	X	X							X	X	X	X		
Prototype Development	X	X	X	X		X			X	X	X		X		
Technology Development	X		X	X		X							X		
Engineering Issues						X	X	X					X		
Materials Behavior			X	X	X	X							X	X	X
Environmental															
Environmental Analysis						X									
Radioactivity Control		X			X	X	X						X	X	X
Safety Analysis and Engineering		X				X	X	X					X		
Human Health and Ecological Hazards						X									
Socioeconomic															
Cost of Fusion Energy						X									
Resource Requirements		X			X	X									
Proliferation						X	X	X							
Public Understanding						X	X	X							
Private Sector Participation	X	X	X			X		X					X		

Figure VI-2-9. Budget activity and subactivity structure versus issue areas.

Technical Issues

Technical issues related to the Magnetic Fusion Energy Program are as follows:

- Plasma behavior—The analytic and computational capability to predict plasma behavior under a wide variety of conditions needs continuing development. Of special interest are magnetic field changes, long pulses, impurities, cross-sectional area, and shape.
- Scaling laws—The ability to scale-up the present generation of experiments to larger experiments and prototype reactors must continue to be developed and validated.
- Prototype development—The various fusion power concepts (tokamak, mirror, etc.) require more experimental and theoretical development in order to evaluate their feasibility as prototypes upon which to build a more realistic, commercial-sized reactor design.
- Technology development—Some of the technological obstacles include the development and construction of large and powerful superconducting magnets, the creation and maintenance of very high vacuums, the heating of fusion fuel to ignition temperatures, and the refueling of burning plasmas.
- Engineering issues—Practical fusion power will require facilities operating at high reliability and availability. Considerable attention and ingenuity are necessary to develop designs with sufficient access, maintainability, simplicity, fabricability, etc., to reach the reliability and availability goals.
- Materials behavior—Structural materials must be developed which will have long lifetimes in a fusion reactor environment (high temperature and large flux of energetic neutrons). Some extrapolation can be made from measurements in fission reactors, but new dedicated testing facilities are required.

Environmental Issues

Environmental issues relevant to the Magnetic Fusion Energy Program include the following:

- Environmental analysis—Potential environmental impacts of magnetic fusion development and commercialization must continue to be examined.

- Radioactivity control—Methods for managing the large inventories of radioactive tritium required in large experimental devices and prototype reactors will have to be demonstrated in the Tritium Systems Test Assembly (TSTA). Low activation materials capable of functioning in a fusion reactor environment must be developed. Radioactive material management and potential accident hazards depend on the degree to which this is achieved. Nonelectric applications, such as breeding of fissile material, may present additional problems which will be examined at an appropriate time. In addition, we are investigating alternate fuel cycle fusion systems which do not use tritium as a primary fuel and, therefore, may lead to reactor plants with no greater environmental impact than present conventional plants.
- Safety analysis and engineering—Safety analyses of early conceptual fusion reactor designs are being carried out to identify potential hazards so that appropriate engineered safety features can be designed. Also, lithium, which is used to breed tritium fuel, must be carefully managed because it contains a large amount of potential energy in liquid metal form.
- Human health and ecological hazards of fusion energy as an energy supply option—The occupational health impacts of routine maintenance of fusion reactor facilities must continue to be investigated. Further research is needed on the health effects of prolonged exposure to low-level magnetic fields and electromagnetic radiation. Finally, some of the possible neutron-activated products of potential reactor materials have not yet received extensive study of their ecological and human health effects.

Socioeconomic Issues

Socioeconomic issues of the Magnetic Fusion Energy Program include the following:

- Cost of fusion power—The economic competitiveness of fusion energy with other energy sources appears to be promising but is very uncertain because existing commercial designs, based on projections from the current stage of technology, are providing general guidance and not engineering cost bases.
- Resource requirements—At the present time, the nature and magnitude of resource requirements for fusion reactors are being examined but cannot be specified. As designs are developed and progress is made, those requirements will be continually assessed. Trade-offs among resource characteristics, availability, cost, and environmental acceptability may be necessary.
- Proliferation—Pure fusion reactors do not require production or involvement with fissile material. However, the proliferation aspects of fusion-fission hybrid systems need to be further evaluated.
- Public understanding—Information for a public dialogue concerning the nature and subsequent resolution of the major technical, environmental, and socioeconomic issues will have to be developed in increasingly clearer terms.
- Private sector participation—It is difficult to obtain significant private sector funding at an early stage of fusion R&D because of high cost and risk and the long leadtime. However, as fusion R&D progresses, such financial participation will become increasingly important.
- Institutional balance—Over the next five years, the relative roles of the national laboratories and major industrial firms will likely continue to change in a significant way toward a stronger industrial presence. Preserving the valuable contributions of the laboratories while making better use of industrial strengths presents a strong challenge to program management.
- Development costs—Because of the early stage of fusion R&D coupled with the large scale of next generation experiments, appropriate funding for fusion energy research and development is difficult to judge in the context of other near-term national needs.

3.
Subactivity Descriptions

This chapter provides further descriptions of each of the four budget activity areas defined in Chapter 2:

- Confinement Systems
- Development and Technology
- Applied Plasma Physics
- Reactor Projects

Brief overviews, current status, planned milestones, and budget levels are given at the subactivity/task/subtask level for each budget activity.

The FY 1980 BA request for the four budget activity areas of the magnetic fusion program is $360.9M; the allocation of these funds among the four activities is shown in Figure VI 3-1.

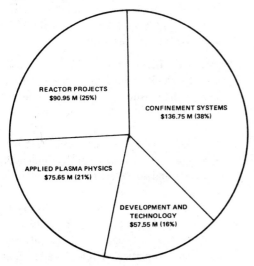

Figure VI-3-1. Office of fusion energy's FY 1980 BA request for its budget activity areas (dollars in millions).

CONFINEMENT SYSTEMS

The Confinement Systems budget activity covers the design, fabrication, and operation of proof-of-principle experiments for tokamak and mirror systems. It has two subactivities:

- Tokamak Systems Experiments
- Magnetic Mirror Systems Experiments

All tasks and subtasks within the two subactivities, with the exception of the Mirror Fusion Test Facility (MFTF) Task, are managed by the Office of Fusion Energy's Division of Magnetic Confinement Systems. The MFTF Task is managed by the Division of Planning and Projects. The MFTF will be transferred to the Reactor Projects budget activity after FY 1980.

The FY 1980 BA request for the Confinement Systems activity is $136.75M; the allocation of these funds between the two subactivities is shown in Figure VI 3-2.

Tokamak Systems Experiments

The tokamak concept, shown schematically in Figure VI 3-3, is to contain a stable high-temperature plasma in an evacuated torus (doughnut-shaped container). However, the wall of the torus is not used to confine the plasma. Confinement is accomplished by three different magnetic fields: the toroidal field, the poloidal field, and the equilibrium/stability field. The toroidal magnetic field, which is directed around the inside of the torus, is the

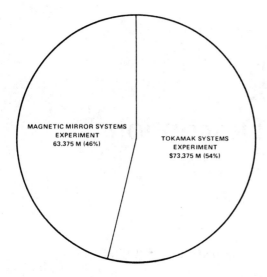

Figure VI-3-2. Confinement systems activity FY 1980 BA request (dollars in millions).

basic confining field. It is produced by coils of wire wrapped around the body of the torus. The poloidal field produces magnetic pressure, which forces the plasma toward the middle of the toroidal ring. It has a circular shape in each radial cross-section perpendicular to the plane of the ring. The poloidal field is generated by a plasma current made to flow inside the toroidal ring by increasing the electrical current in the magnetic transformer coils. The toroidal and poloidal fields combine to produce the total "resulting" magnetic field shown in the figure. A third set of magnetic fields, used to maintain plasma equilibrium and stability, is generated by smaller coils that run along the outside periphery of the torus (not shown in the figure).

Three different techniques are used to heat the plasma. Initial heating is provided by the plasma current (axial current), which also supplies the poloidal magnetic field. Further heating occurs by either the injection of high energy deuterium atoms (called neutral beam injection) or by transferring energy to plasma electrons or ions through RF or microwave electromagnetic radiation.

The objectives of tokamak development are to demonstrate the scientific feasibility of the tokamak concept and then to exploit that concept for commercial reactor development.

The program strategy is to develop a mainline of devices (moderate magnetic field, circular cross-section, and neutral-beam heated) while concurrently developing and testing variants of these designs, the results of which influence the design of the next mainline device. This approach exerts continuous pressure for attainment of the final objective while preserving an important degree of flexibility for obtaining an optimum final reactor design.

The tokamak effort has been focused on five critical areas:

- Plasma heating
- Plasma transport and scaling
- Plasma shape optimization
- Impurity control
- Fueling

Plasma heating concerns techniques for getting enough energy into the plasma to produce high plasma temperatures (e.g., ignition temperature). Plasma transport and scaling determines relationships between device parameters (e.g., plasma radius and heating power) and performance parameters (e.g., temperature, Lawson Product, and power balance, Q) that allow prediction of plasma behavior in more advanced devices. Shape optimization refers to finding the best cross-sectional plasma shape and impurity control concerns techniques to prevent plasma cooling by particles sputtered from the vacuum chamber wall. Fueling refers to the injection of additional fuel into the plasma without undue cooling.

The following experiments are within the Tokamak Systems Subactivity:

- Princeton Large Torus (PLT) to investigate tokamak scaling and provide a technology/diagnostic development base for the Tokamak Fusion Test Reactor (TFTR)
- Poloidal Divertor Experiment (PDX) to examine impurity control by magnetic methods and to investigate the effects of changing plasma shape
- Impurity Studies Experiment (ISX) for the study of nonmagnetic impurity control and the investigation of stability and transport properties of high-beta, noncircular plasmas
- I/A/Doublet III Doublet experiments to

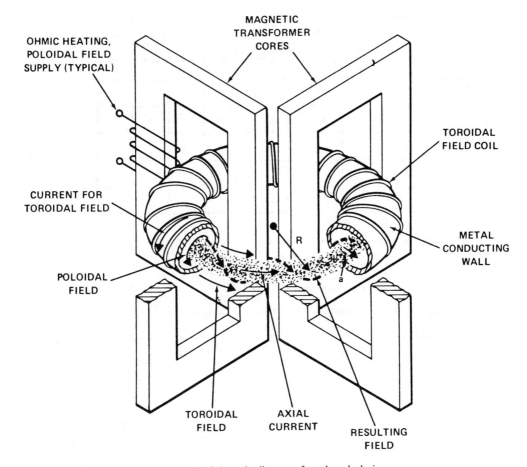

Figure VI-3-3. Schematic diagram of a tokamak device.

study properties of plasmas with different shapes
- Alcator A/C Alcator experiments to examine properties of hot, high-density plasmas

The relationship of these tokamak experiments to the problem areas is given by Figure VI 3-4, which shows the prime contractors and operational status of each experiment. Descriptions and status reports for each of these experiments follow.

Princeton Large Torus (PLT). The PLT (Figure VI 3-5, is a mainline tokamak device forming a critical intermediate step between a small-scale, exploratory tokamak device and a machine that will possess a number of commercial reactor characteristics (TFTR). The primary PLT project objective is to determine how plasma parameters (e.g., density, temperature, and confinement time) vary as machine size and heating power increase. In addition, PLT provides a test bed for high-power neutral beam and RF heating experiments, and numerous technology and instrumentation experiments needed for the TFTR.

The primary emphasis of experiments to date has been to investigate plasma stability and microscopic transport properties, to demonstrate how plasma parameters change with device parameters, and to establish a data base of plasma properties and diagnostic (measurement) techniques to help with TFTR development.

Initial experiments, using ohmic heating, showed how the Lawson Product depended on toroidal radius and plasma density. In the spring of 1978, a plasma with an ion temperature of 6-keV (to be compared with 10–15 keV needed for a reactor) was achieved using

PROBLEM AREAS \ DEVICE OR EXPERIMENT (CONTRACTOR)	PLT (PPPL)	PDX (PPPL)	ISX (ORNL)	D-IIA (GA)	D-III (GA)	ALC-A (MIT)	ALC-C (MIT)
Heating	4	1	2	4	2	–	1
Transport & Scaling	4	–	4	4	3	4	3
Shape Optimization	–	2	2	4	3	–	–
Impurity Control	–	3	4	–	–	–	–
Fueling	–	2	3	–	–	–	–

4 = Operational, Applicable Experiments in Progress

3 = Operational, Applicable Experiments in Preparation

2 = Under Design, Construction or Major Modification

1 = Proposed, Device Considered Critical for Achieving the Element Objective

Figure VI-3-4. Relationship between tokamak experiments and critical areas.

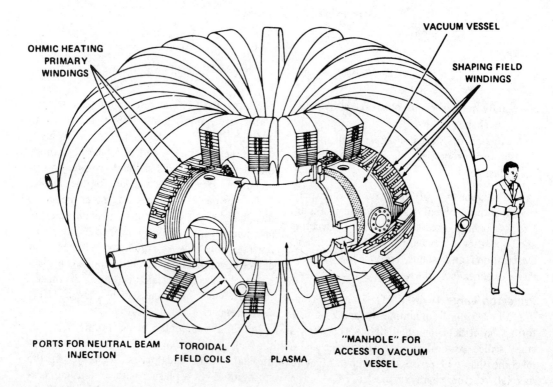

Figure VI-3-5. Schematic diagram of PLT.

four neutral-beam injectors which provided 2-MW of heating power. At these temperatures, the plasma entered the reactor regime for the first time and demonstrated better confinement than theoretically anticipated.

PLT will be used to test RF heating, a technique which may be significantly more efficient than neutral-beam injection. Installation of 2-MW of RF heating was completed in December 1978, and detailed evaluation of RF heating experiments took place in FY 1979. Additional experiments in support of TFTR were performed in FY 1979; decisions on future experiments for PLT will be made in mid-FY 1980.

Major milestones and funding for PLT are shown in Figure VI 3-6.

Poloidal Divertor Experiment (PDX). In order to attain the plasma temperatures of up to 10-keV estimated for a commercial reactor, plasma cooling must be reduced. One important way a plasma can cool is by the influx of "cold" particles (impurities) which have been sputtered from the container wall by plasma ions. This type of plasma cooling can be significantly reduced by magnetically diverting the incoming impurities into a collection chamber (separated from the plasma) before they interact with the plasma. The investigation of this magnetic diversion of impurities is the primary objective of the PDX. Other important objectives of PDX are to investigate impurity control by experimenting with different startup modes and to evaluate the stability of a D-shaped plasma (the use of a plasma with this shape could lead to a significant reduction in the amount of magnetic field needed for confinement).

The plasma in the PDX device (Figure VI 3-7) will be initially heated by the circulating current within the plasma (resistance heating). It will be necessary to add 6-MW of neutral beam power to evaluate magnetic divertor

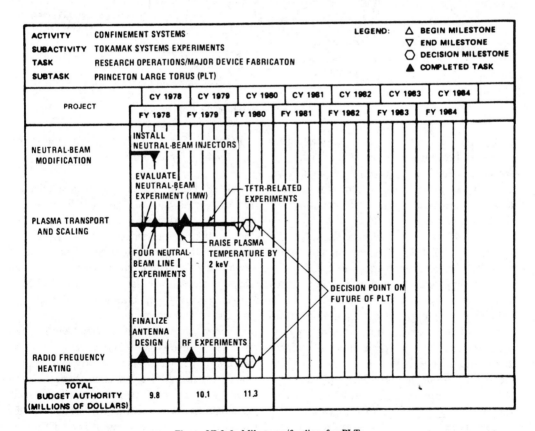

Figure VI-3-6. Milestones/funding for PLT.

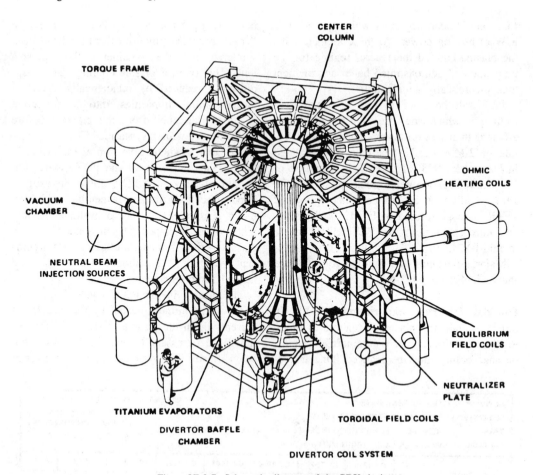

Figure VI-3-7. Schematic diagram of the PDX device.

performance on reactor-grade plasmas. This additional heating power, combined with the expected reduction in cooling by impurities, should yield plasma temperatures (at reduced densities) near the temperature required for energy breakeven.

PDX will examine many choices for the number and position of the magnetic divertor coils. The initial experiments will evaluate the divertor performance at lower temperatures using ohmically (resistance) heated plasmas. Following these experiments, neutral-beam injectors will be installed and tested in preparation for divertor experiments at higher temperatures. The higher temperature experiments will occur in FY 1980 with the dual objectives of optimizing divertor performance and evaluating the benefit of using a noncircular plasma shape. A totally optimized divertor system is expected by mid-FY 1982.

Figure VI 3-8 presents the major milestones and funding for the PDX Subtask.

Doublet. The power produced by a tokamak fusion reactor can be significantly increased by raising the density and temperature (below the ignition point) of the plasma. Large increases in these parameters occur for the same confinement magnetic field when the plasma shape is varied. The variation of plasma shape by using special magnetic fields is the basis of the Doublet experiments.

The primary objective of the Doublet subtask is to determine the best shape for the plasma by comparing the stability and equilibrium of many different plasma shapes. Doublet-IIA, currently in operation, is a small-scale tokamak device in which the shape of the magnetic field can be adjusted by coils placed around the plasma chamber. The coils may be operated "passively" by using (eddy) currents produced in the coils by the plasma, or "actively" by driving currents in the coils from an external power supply.

The Doublet-III device (Figure VI 3-9) will be used to study the effect of plasma shape on confinement time and plasma transport properties. This device, a scaled-up version of Doublet-IIA, is in the initial operating stage at General Atomic Company. The plasma temperature resulting from ohmic (resistive) heating alone should be high (about 2-keV). Additional large amounts of neutral-beam injection power (20-MW) will produce a very high-temperature plasma that should behave in a fashion similar to a reactor plasma. The resulting Lawson Product should be the same as that of a device which has reached energy breakeven.

Doublet-IIA (D-IIA), currently operating at General Atomic Company, was completed in 1974. It provided a Lawson Product roughly twice that of an equivalent circular plasma, important advances in plasma profile and position control, and techniques for increasing the efficiency or magnetic field use (beta). In addition, RF heating feasibility experiments provided evidence of good heat transfer. The D-IIA experiments continued through early 1979.

Doublet-III (D-III) will provide reactor-like conditions for the comparison of various plasma shapes. D-III began operation in February 1978 at a fabrication cost of $31.0M. Initial experiments (through the beginning of FY 1980) will evaluate different plasma shapes using ohmic heating. Near-reactor-grade plasma experiments, using up to 7-MW of neutral-beam injection power, will be completed by the end of FY 1980. A decision on whether or not to upgrade the toroidal field, plasma current, and neutral-beam heating (to 20-MW) will be made in 1980. Any upgrading will be carried out through FY 1982.

Major milestones and funding for the Doublet Subtask are given in Figure VI 3-10.

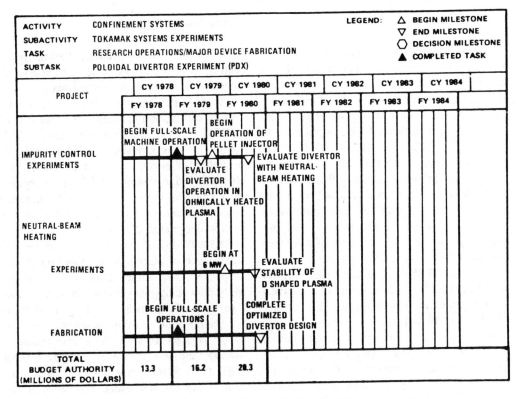

Figure VI-3-8. Milestones/funding for PDX.

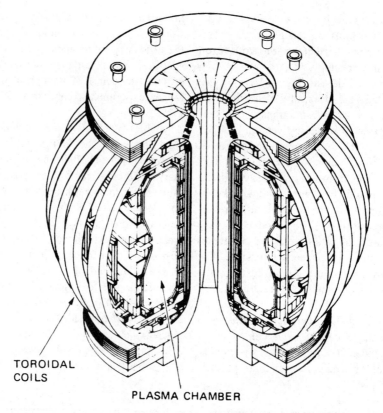

TOROIDAL COILS

PLASMA CHAMBER

Figure VI-3-9. Schematic diagram of the Doublet-III device.

Impurity Studies Experiment (ISX). The influx of "cold" impurities into a hot plasma can cause significant plasma cooling resulting in plasma instabilities. This condition makes the attainment of a reactor-type plasma considerably more difficult, if not impossible, to achieve. The ISX device (Figure VI 3-11) was originally designed to study techniques that do not use additional magnetic fields (unlike PDX) in order to control the influx of impurities. One such technique is impurity flow reversal, where the inward migration of impurities is impeded by the flow of hydrogen gas around the edge of the plasma.

The original ISX experiment was recently modified (ISX-B), and the focus of the experimental program was changed. The primary objective of ISX-B is to investigate the stability and transport properties of noncircular, high-beta (high confinement efficiency) plasmas. Additional objectives for this device include understanding impurity flow reversal control, an evaluation of ripple injection heating, a feasibility demonstration of electron-cyclotron resonance heating, testing pellet fueling, and appraising limiter materials. (A limiter is used to shape the plasma.)

The ISX device was fabricated at a total cost of $2.5M, and operated until March 1978 at ORNL. Between March and August, 1978, the device was modified to add 1.8-MW of neutral-beam power, the vacuum vessel was changed, and a new poloidal field coil system (to permit shaping of the plasma by magnetic fields) was added. This modified ISX device is now designated ISX-B.

FY 1979 ISX-B experiments include: the dependence of plasma confinement on plasma shape and beta value; determination of stability beta limits for plasmas subject to massive neutral-beam heating; the general effect of neutral-beam heating when the injected power is 5 to 10 times the ohmic (resistive) heating; and a continuation of flow-reversal impurity control experiments. Microwave heating experiments, using 200-kW of power, began in FY 1979.

In FY 1980, additional neutral-beam heat-

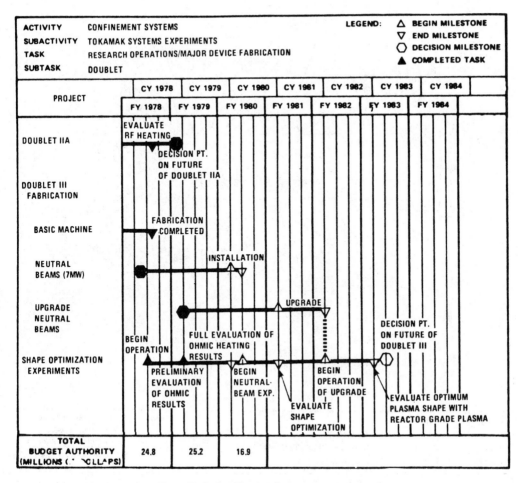

Figure VI-3-10. Milestones/funding for Doublet.

ing (for a total of 3-MW) will be used to investigate plasma properties at high values of confinement efficiency (beta about 10 percent). ORNL will design and construct a bundle divertor for ISX-B; this device may be needed to control the flux of impurities anticipated with the addition of 3-MW of neutral-beam power.

Major milestones and funding for the ISX Subtask are presented in Figure VI 3-12.

Alcator. The Alcator concept provides the basis for a smaller fusion reactor design. Reactor-grade plasmas can be obtained by using large magnetic fields and a small plasma cross-sectional area to produce high density (with corresponding large values for the Lawson Product).

The primary objective of the Alcator subtask, which is being performed at MIT, is to explore the relationship between plasma characteristics (e.g., temperature, confinement time) and plasma density. Alcator A, a small tokamak device, has proven to be capable of a broad range of operating densities (the highest densities are more than 300 times the lowest densities). A new device, Alcator C (Figure VI 3-13) has approximately twice the effective plasma radius of the existing machine, and has improved access for diagnostics (or for RF heating). With the addition of a new power supply, the device should achieve an almost 50 percent increase in maximum magnetic field over Alcator A. The resulting increase in plasma density combined with the effects of heating from a high plasma current should lead to a value of Lawson Product required for energy breakeven.

In FY 1978, Alcator A was used to explore plasma behavior in high magnetic fields.

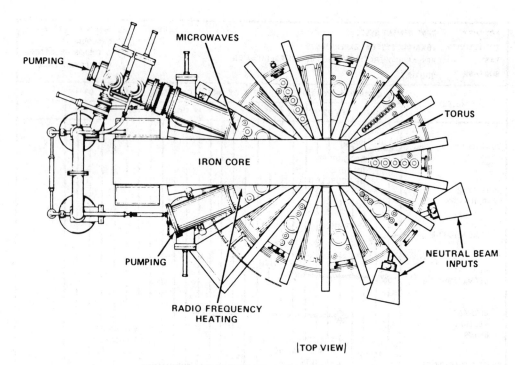

Figure VI-3-11. Schematic diagram of the ISX device.

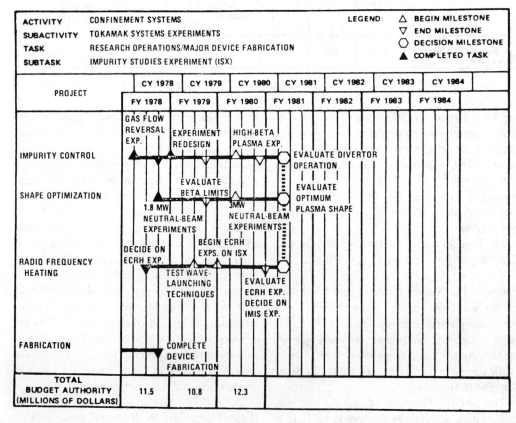

Figure VI-3-12. Milestones/funding for ISX.

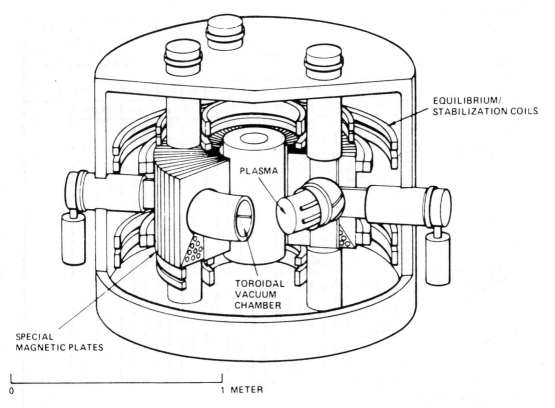

Figure VI-3-13. Schematic diagram of Alcator C.

These experiments obtained record values for a Lawson Product (3×10^{13} cm^{-3}-sec). The primary emphasis in FY 1979 is on the development of plasma diagnostics and instrumentation to aid in completing Alcator C which began operation in June 1978 at a total fabrication cost of $6.7M.

Preliminary experiments are in progress on Alcator C. The first high magnetic field experiments began in February 1979. During FY 1979, Alcator C will be routinely operated at high magnetic fields which should lead to the highest value of the Lawson Product (10^{14} sec-cm^{-3}) obtained by any tokamak experiment.

In FY 1980, two forms of RF heating will be tested on Alcator C. The first, lower hybrid heating, is expected to provide 4-MW of heating power by September 1980; the second, ion cyclotron heating, will provide 8-MW (using surplus radar equipment) and will serve as an alternate heating scheme for Alcator C. If either of these heating methods is effective, temperatures in excess of 4-keV will be obtained.

Figure VI 3-14 shows the major milestones and funding for the Alcator Subtask.

Magnetic Mirror Systems Experiments

Magnetic mirror systems rely on the property of converging magnetic field lines to reflect ("mirror") plasma particles. The increasing density (intensity) of the field lines at the ends of the device produces a force which tends to reverse the motion of the escaping particles (Figure VI 3-15). Figure VI 3-16 shows a simple magnetic mirror in which plasma is partially trapped between two coils. This simple mirror has a high leakage rate both axially through the coils for plasma particles whose motion is close to being along the axis of the device and radially outward due to instabilities associated with the outward bending of the magnetic field lines. The simple mirror by itself is not a viable reactor concept.

A better technique of mirror confinement is the minimum-B or Yin-Yang concept shown in Figure VI 3-17. This magnet creates a magnetic "well" in the center of the device (the

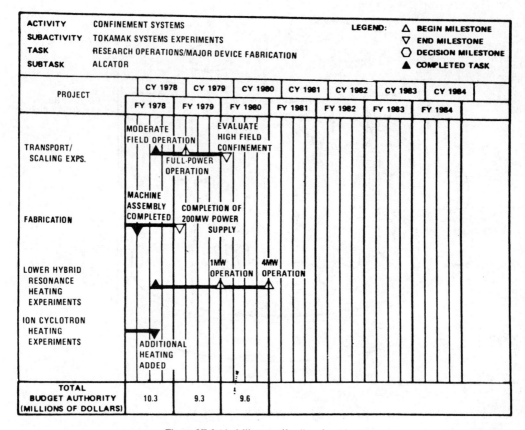

Figure VI-3-14. Milestones/funding for Alcator.

magnetic pressure is minimum in the center and the magnetic field lines are inward bending everywhere along the plasma surface), which helps to localize and stabilize the plasma. The leakage rate of plasma from the ends of the device, however, is substantial.

Two primary concepts are being pursued in order to reduce end losses in mirror systems. The first concept is the tandem mirror in which two minimum-B mirrors are placed on either end of a long solenoid (a cylindrical coil). Positively charged plasma ions are held radially in the central solenoid region by the solenoidal magnetic field and axially by electrostatic forces generated by the positive charge which builds up in the end mirrors. Heating is supplied to the mirror end cells by neutral-beam injection. In the second containment concept, field reversal, reduced plasma losses occur when the magnetic field lines deep inside the plasma are made to close on themselves (field lines effectively reverse direction). Since charged particles tend to follow magnetic field lines, the closing of lines in the plasma helps to trap the particles. Neutral-beam heating is thought to be capable of sustaining the field-reversed state once it is established.

The highest priority problem in the Mirror Subactivity is to improve the mirror power balance, Q, the ratio of energy produced during fusion to the energy required to produce the fusion. The strategy of the mirror program is to address this problem aggressively through the tandem mirror and field reversal concepts, while at the same time developing a scaled-up standard mirror device to serve as a reactor prototype for technology development. If either the tandem mirror or the field reversal concept appears to be a viable reactor concept, the scaled-up mirror device will become the key technological cornerstone for a prototype reactor design based on one or the other of these concepts.

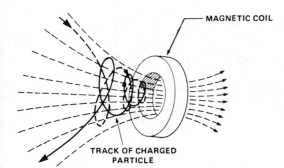

Figure VI-3-15. Particle reflection by converging magnetic field lines.

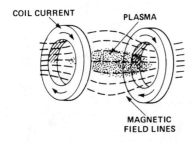

Figure VI-3-16. Schematic diagram of a simple magnetic mirror.

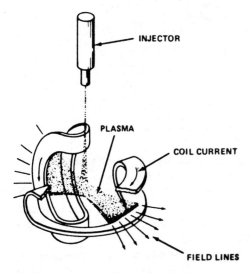

Figure VI-3-17. Schematic diagram of a minimum-B mirror.

The experiments/facilities in the Mirror Systems Subactivity are as follows:

- Beta II experiment to investigate the reversal of magnetic fields deep in the plasma
- Tandem Mirror Experiment (TMX) to investigate the scientific feasibility of the tandem mirror concept
- Phaedrus which is a tandem mirror device used to support the larger tandem mirror experiment by developing the RF heating option
- Field Reversal Experiments (FRX) to investigate the stability and confinement properties of the reversed-field plasma state
- Constance II, which is a facility to study stabilization techniques (applicable to minimum-B mirrors) and auxiliary (electron) heating methods
- MFTF to support experiments on a scaled-up minimum-B magnet; the facility could be used for large-scale experiments based on the field reversal and tandem mirror concepts

The critical study areas for the Mirror Subactivity are as follows:

- Q-enhancement
- Confinement scaling
- Heating
- Steady-state operation

Q-enhancement refers to techniques whereby a device will produce power more efficiently either through the reduction of energy losses from plasma leakage or through the attainment of plasma ignition (self-sustained burning). Confinement scaling and heating were discussed under Tokamak Systems Experiments. Steady-state operation means a device can operate in a continuous fashion, hence eliminating cyclic wall fatigue and producing continuous power as required by a utility grid. The relationship of these areas to various mirror experiments is given in Figure VI 3-18. Descriptions and current status reports for Beta II, TMX, Phaedrus, and MFTF follow.

Beta II. The Beta II experiment will use a minimum-B magnet (from the 2X-IIB experiment) to test operating procedures for the large minimum-B magnet in the MFTF and to study the reverse field concept. The Beta II device (Figure VI 3-19) consists of two Yin-Yang magnet coils. These special coils pro-

DEVICE OR EXPERIMENT	BETA II	FRX	TMX	PHAEDRUS	CONSTANCE II	MFTF
CONTRACTOR PROBLEM AREAS	LLL	LASL	LLL FY 79*	UNIV. OF WISC. FY 79*	MIT FY 79*	LLL FY 82*
Q-Enhancement	2	4	3	2	–	2
Confinement Scaling	2	4	3	2	2	2
Heating	2	–	3	2	2	2
Steady-State	–	–	–	–	–	2

4 = Operational, Applicable Experiments in Progress
3 = Operational, Applicable Experiments in Preparation
2 = Under Design, Construction or Major Modification
1 = Proposed, Device Considered Critical for Achieving the Element Objective

*Dates give beginning of operations.

Figure VI-3-18. Relationship between mirror experiments and critical problem areas.

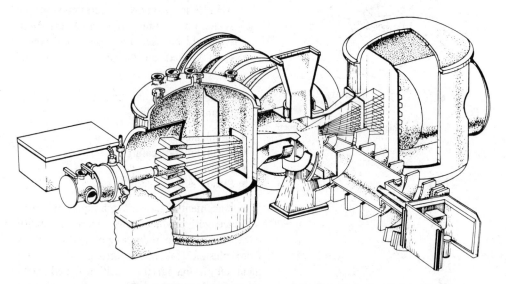

Figure VI-3-19. Schematic diagram of the Beta II device.

duce a magnetic well which helps to stabilize the hot (13-keV) plasma generated by the perpendicular injection of 12 neutral beams onto a warm (0.1-keV) plasma stream. The warm stream, injected from the ends of the device along the magnetic field lines, serves both as a neutral beam target and as a stabilizing mechanism to reduce plasma turbulence which limits the time that the plasma can be contained.

The primary Beta II objectives are to investigate single-cell (classical mirror) physics questions related to the MFTF, including start-up and stability, and to continue field reversal experiments, using a magnetized coaxial plasma gun. The field reversal confinement technique, if successful, could lead directly to designs for a small prototype commercial reactor. The Beta II technology goals directly support the new and larger Tandem Mirror Experiment (TMX) and the much larger MFTF.

The basic investigations of the minimum-B (Yin-Yang) mirror concept were carried out at LLL using the 2X-IIB experiment device. Reconfiguration of 2X-IIB into Beta II is currently in progress and was completed in early FY 1979.

During its last year of operation, 2X-IIB was used for radial size scaling, field reversal, and plasma stabilization experiments. In the radial scaling experiments, the ratio of radial plasma size to characteristic particle orbit size was increased from 2 to 6. Both the Lawson Product and the electron temperature increased for constant beta. The field reversal experiments bordered on producing the field reversed state by using 7-MW of injected neutral beam power.

Recent theoretical calculations of field reversal indicate that either the energy confinement time or the injection power will have to be substantially increased (by a factor of 3) to achieve field reversal using neutral beam injection. As a result, experiments will be pursued in Beta II to establish a reversed field plasma state which will be sustained by neutral-beam injection. The experiments based on this technique began in late FY 1979.

Major milestones and funding for the Beta II Subtask are shown in Figure VI 3-20.

Tandem Mirror Experiment (TMX). The fusion power produced in a conventional (minimum-B magnet) mirror reactor is almost equal to the neutral-beam heating power injected (i.e., Q is almost unity). In order to achieve higher reactor Q values, a tandem mirror uses as basic components two conventional mirror machines placed at either end of a solenoid (cylindrical coil). The conventional mirrors act as "plugs" to increase the plasma confinement time in the central portion of the solenoid. This function is accomplished by a positive electrostatic potential which builds up in the end mirrors as a result of an initial loss of electrons from the plasma.

The objectives of TMX are to demonstrate that the tandem mirror concept will lead to significantly larger values of energy gain Q (between 5 and 10) and will provide a basis for a technically viable fusion reactor. Once this is shown, the program objective will be to develop a tandem mirror design that can be readily scaled up to produce near-commercial, reactor-like plasma characteristics. An example of a TMX device is shown in Figure VI 3-21.

The tandem mirror concept was developed during 1976 through independent efforts in the Soviet Union and in the United States. TMX was approved, and fabrication began in 1977. The total construction cost was $11.4M. Installation of the mirror plug tanks began in

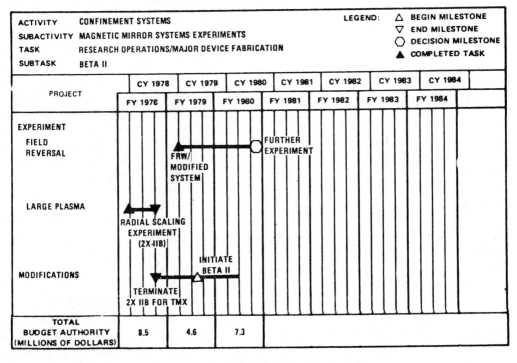

Figure VI-3-20. Milestones/funding for Beta II.

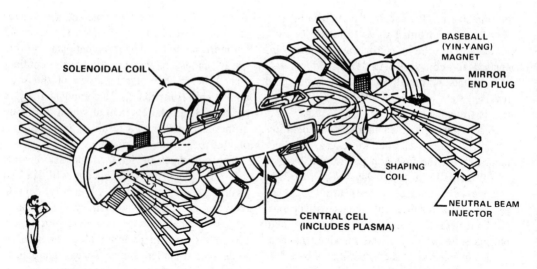

Figure VI-3-21. Schematic diagram of a TMX device.

April 1978, and some equipment was transferred from 2X-IIB (also at LLL) in June 1978. TMX became operational in early FY 1979 with initial experiments aimed at the following near-term goals:

- Demonstrating the existence of a magnetic potential well in the solenoid region between the two mirror plasmas
- Developing a magnetic geometry which can be scaled up in a larger device to confine a high-temperature, high-density, reactor-grade plasma
- Investigating the microstability of the plug-solenoid combination in order to maximize the system energy gain, Q

Figure VI 3-22 gives the major milestones and funding for the TMX Subtask.

Phaedrus. Phaedrus (Figure VI 3-23) is a tandem mirror device (similar in design to TMX but smaller) which will be used to develop RF heating for the TMX. If the RF heating experiments are successful, this technique could lead to a large reduction in the neutral-beam heating required for a tandem mirror reactor and would significantly decrease technology requirements and costs. Phaedrus will also be used to explore the trapping of plasmas (i.e., reactor refueling) by RF techniques.

The Phaedrus device at the University of Wisconsin has a 2-year fabrication schedule and a total estimated cost of $1.4M.

During the first year, one mirror end-cell and the center section were fabricated and RF trapping studies begun. Computer simulations to support the Phaedrus experiments are in progress and have already produced important results for the entire tandem mirror program. These simulations of particle motions in various magnetic field designs led to the discovery that the plasma confinement time increased when the magnetic fans in the two end cells were arranged at right angles to each other. This discovery occurred early enough to affect both the Phaedrus design and the design of the TMX at LLL.

The Office of Fusion Energy evaluated the initial results of RF heating in a single end cell in December 1978. This review preceded the commitment of funds to complete the total tandem mirror device. Initial operation began in September 1979.

Major milestones and funding for the Phaedrus Subtask are presented in Figure VI 3-24.

Mirror Fusion Test Facility (MFTF). MFTF will be a major step in exploring conventional (minimum-B) mirror characteristics under reactor-grade plasma conditions. The principal purpose of the MFTF Task is to provide a large-scale magnetic mirror device and sup-

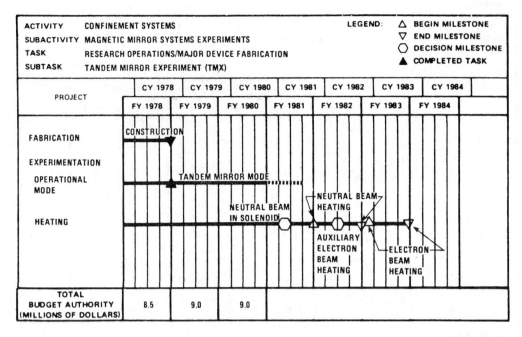

Figure VI-3-22. Milestones/funding for TMX.

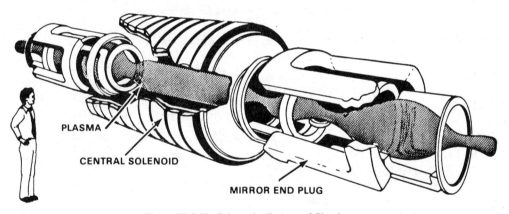

Figure VI-3-23. Schematic diagram of Phaedrus.

porting facility for performing physics experiments and technology development needed to bridge the gap between smaller existing mirror experiments (such as Beta II) and a mirror fusion ETF. MFTF will be the central facility for reactor-level mirror research in this country during the early 1980s.

MFTF will bring together the main technical elements which may be required by a mirror ETF: superconducting magnets, high-powered neutral beams, high-vacuum pumping systems, and quasi-steady-state operation. It will provide the opportunity for establishing the viability of the economically important field reversal concept and the testing of proposed solutions to the problems of end-cell operation in a large tandem mirror. If the tandem mirror concept proves feasible, MFTF could be used to perform experiments on a scaled-up version of the tandem mirror device. In addition, MFTF will provide a testbed for basic scaling studies and the exploration of plasma instability limits. MFTF should produce (by a large margin) the best values for plasma performance parameters yet achieved in a mirror machine.

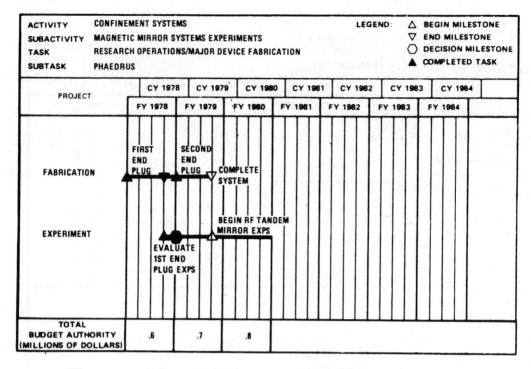

Figure VI-3-24. Milestones/funding for Phaedrus.

The main features of MFTF are shown in Figure VI 3-25. The basic mirror configuration consists of two superconducting Yin-Yang magnets with a neutral-beam injection system used both for startup and for reaction-sustaining power.

MFTF was proposed by LLL in 1976. It received favorable comments from several technical review groups and was authorized for construction at the beginning of FY 1978 at an estimated cost of $94.2M. During FY 1977, a management team for MFTF and a project office at Livermore were established. Construction of the magnet coil winder was completed in December 1977, and the practice coil winding was initiated in 1978.

The total construction time will be about 48 months, with final system checkout and initiation of plasma experiments planned for late 1981. Initially this test device will be used for the development of prototype systems and for the testing and evaluation of plasma-streaming gun, neutral-beam systems and other technical components.

Major milestones and funding for the MFTF Task are shown in Figure VI 3-26.

DEVELOPMENT AND TECHNOLOGY

The Development and Technology budget activity is managed by the Office of Fusion Energy's Division of Development and Technology. This activity has a two-fold mission:

- To provide system engineering and component development support to current

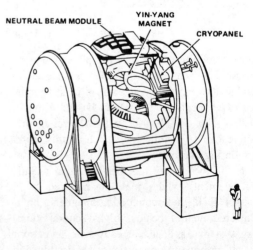

Figure VI-3-25. Schematic diagram of MFTF.

plasma devices and future fusion reactor programs
- To develop the broad technology base necessary for fusion energy to become a commercial reality

To accomplish these objectives, the Development and Technology Activity is organized into six subactivities:

- The Magnetic Systems Subactivity provides engineering design and component development of large superconducting magnets and pulsed power systems to support TFTR and major scaling experiments
- The Plasma Engineering Subactivity focuses on the design and component development of efficient plasma heating systems including neutral-beam injection and microwaves; also addresses methods of recovering energy from beams of charged particles and the development of fueling and divertor techniques for plasma maintenance and control
- The Fusion Reactor Materials Subactivity selects, tests, and evaluates candidate materials for fusion reactor components. Of particular concern is the long-term stability of the first wall which will be subject to severe neutron irradiation during reactor operation
- The Fusion Systems Engineering Subactivity develops conceptual designs for fusion reactor facilities and addresses engineering design and development of critical supporting subsystems
- The Environment and Safety Subactivity ensures that the development and ultimate operation of fusion power reactors is accomplished with minimum adverse effect on the environment and with maximum safety for both the public and operating personnel
- The Fusion Energy Applications Subactivity evaluates the role of fusion energy in nonelectrical applications

The FY 1980 BA request for the Development and Technology Activity is $57.55M; the allocation of these funds among the six subactivities is shown in Figure VI 3-27.

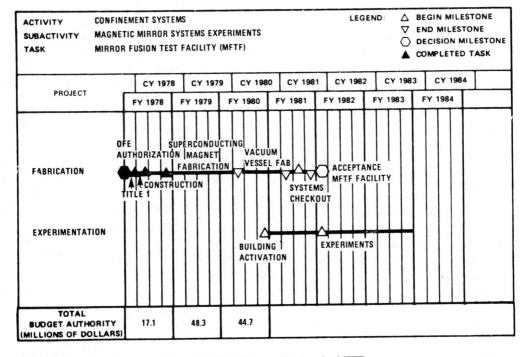

Figure VI-3-26. Milestones/funding for MFTF.

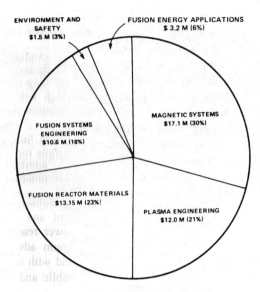

Figure VI-3-27. Development and technology activity FY 1980 BA request (dollars in millions).

Magnetic Systems

The major purpose of the Magnetic Systems Subactivity is the engineering design and component development of large superconducting magnets and pulsed power systems. The subactivity has two tasks: Superconducting Magnet Development (FY 1980 BA request—$13.2M) and Energy Storage and Transfer (FY 1980 BA request—$1.6M).

The Superconducting Magnet Development Task is developing superconducting wire and cable, magnet construction techniques, and an industrial magnet production capability to support large fusion experiments. The major effort is the tokamak-oriented Large Coil Project (LCP) at ORNL entailing the fabrication of six superconducting coils and the construction of a Large Coil Test Facility (LCTF), shown in Figure VI 3-28, for testing and evaluating these magnets (as well as future large magnets). Other activities include high-field superconductor development in support of the Mirror Systems, and a new initiative to develop high magnetic field conductors using the High Field Test Facility (HFTF) at LLL.

The Energy Storage and Transfer Task is developing the components and systems necessary to produce special magnetic or electrical fields which initiate or control plasmas. Pulsed superconducting magnets, homopolar motor generators, and vacuum breaker switches are examples of components being developed for tokamak ohmic heating and vertical field systems.

The LCP, the major project under the Superconducting Magnet Task, is proceeding on schedule. In FY 1979, final designs for the test coils were completed and short samples of the super-conducting cables were evaluated. United States industry will begin fabrication of three of the six coils (Japan, Switzerland, and Euratom will build the other three) at a cost of $20M with delivery of two coils in 1980. The LCTF at ORNL, with an estimated cost of $22M (not including the coils), will begin initial operation in late 1980 with a capability of testing three coils without a pulsed magnetic field environment. Full operation will begin in 1982.

The HFTF at LLL became operational in FY 1978 and was utilized to test the MFTF conductor. The HFTF capability will be increased to its highest magnetic field capability in 1980 and will enable high field testing of superconductors which may be used in tokamaks and mirrors.

Under the Energy Storage and Transfer Task, testing of small, pulsed superconducting energy storage coils at LASL and Argonne have demonstrated that it is time to move to larger systems to support the Engineering Test Facility needs. The homopolar motor-generator efforts at the University of Texas have produced technology spin-offs in welding and led to the invention of the compensated pulsed alternator (Compulsator) in 1978. The Compulsator may be the only practical method of efficiently producing repetitive high-energy, high-power pulses required by lasers and some alternative concepts.

Figure VI 3-29 shows the major milestones and funding for the Magnetic Systems Subactivity.

Plasma Engineering

The main thrust of the Plasma Engineering subactivity is the design and component de-

Figure VI-3-28. Schematic diagram of large coil test facility (LCTF) test chamber. (1) Vacuum vessel—11 m dia. × 13 m high, liquid nitrogen-cooled walls; (2) 6-coil Torus array—pulse magnetic field, simulated neutron heating, pool boiling & forced flow, power supplies & instrumentation; (3) large coil—2.5 m × 3.5 m bore; (4) Ormak tank—segment tests; (5) cryogenic system—liquid helium, liquid nitrogen.

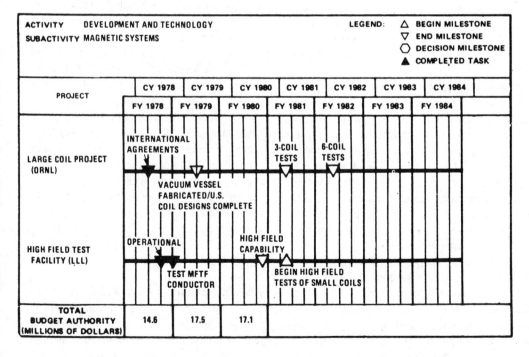

Figure VI-3-29. Milestones/funding for magnetic systems.

velopment of efficient plasma heating systems which use neutral beams and microwaves. There are five tasks under this subactivity:

- Neutral Beam Development (FY 1980 BA request—$7.45M)
- Alternate Heating System Development (FY 1980 BA request—$1.35M)
- Direct Energy Conversion (FY 1980 BA request—$0.15M)
- Vacuum Component Development (FY 1980 BA request—$0.55M)
- Plasma Maintenance and Control (FY 1980 BA request—$0.6M)

Neutral-beam heating involves the formation and injection of streams of high energy deuterium atoms into plasmas. The Neutral Beam Development Task is directed at developing higher energy, higher efficiency, and higher reliability systems which directly affect TFTR and MFTF. The program further includes design support and prototyping efforts for confinement systems users and the neutral beam test stand operation at ORNL and LLL.

Under the Alternate Heating System Development Task, the principal effort is the industrial development of gyroklystron microwave tubes. Applications include plasma bulk heating, ionization, and breakdown of plasmas during startup of tokomak and mirror devices and heating of the relativistic electron rings in Elmo Bumpy Torus (EBT).

The Direct Energy Conversion Task is developing methods and hardware for direct energy recovery from ion beams and plasma exhaust. Beam direct conversion not only recovers otherwise lost power, but reduces the equipment and operating costs of power supplies and beam dump systems. In addition, plasma direct conversion will make efficient and economic mirror fusion reactors possible and will be applied to direct recovery of tokamak divertor exhaust.

The Vacuum Component Development Task provides the engineering data and technology base for vacuum components and systems. Near-term emphasis is on selecting and qualifying the more viable vacuum pumps and investigating potential problems in large-scale helium and tritium pumping.

The Plasma Maintenance and Control Task develops fueling and divertor techniques, and hardware for longer pulsed fusion reactor systems. Near-term emphasis is on development of a solid hydrogen pellet injector for testing on PLT, Doublet III, PDX, and, if successful, on TFTR.

The PLT-style neutral beam source will be modified and demonstrated at twice its present power rating for use in ISX and PDX, and for retrofitting onto PLT. The prototype neutral-beam sources for TFTR and Doublet III have been operated at full power. Negative ion sources will be examined for later testing.

Microwave heating efforts in FY 1979 include operation of the high frequency gyroklystron on EBT-S and development work on microwave tubes for tokamaks (for bulk heating and profile control), mirrors, and EBT-II. Tests were completed for ISX in early 1979.

Plasma energy conversion efficiencies of 75 percent have been demonstrated on a test stand at LLL. Configurations oriented toward longer range applications will also be tested.

Vacuum technology efforts in FY 1979 include the development and fabrication of large compound cyropumps for testing in the Tritium Systems Test Assembly, plus a new effort at ORNL to begin development of test components for particle collection in tokamak divertors. Initial applications will be in ISX and PDX. A Vacuum Components Test Facility is being considered for FY 1980 to focus development efforts at a single location.

Plasma fueling work will be focused on the development of "pellet" and "liquid jet" fuel injector systems. Recently, a pellet injector produced pellets with speeds which may be adequate for reactor devices.

Major milestones and funding for the Plasma Engineering Subactivity are shown in Figure VI 3-30.

Fusion Reactor Materials

Successful operation of commercial fusion reactors will require the development of materials capable of withstanding the severe radiation environment resulting from the D-T fusion reaction. Critical problems are the

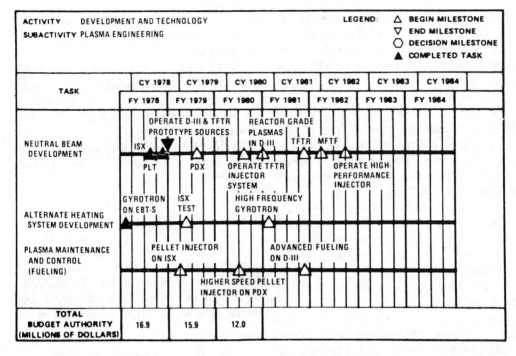

Figure VI-3-30. Milestones/funding for plasma engineering.

long-term stability of first-wall and related structural materials, and the attendant influence of those materials (as impurity sources) on the plasma. The principal objective of the Fusion Reactor Materials Subactivity is the solution of these problems using the following tasks:

- Alloy Development for Irradiation Performance (FY 1980 BA request—$4.0M)
- Plasma Materials Interaction (FY 1980 BA request—$2.6M)
- Special Purpose Materials Development (FY 1980 BA request—$0.625M)
- Damage Analysis and Dosimetry (FY 1980 BA request—$1.425M)
- Radiation Facilities Development and Operation (FY 1980 BA request—$2.55M)

The Alloy Development Task is development of radiation-resistant first-wall and structural materials. A four-path program, pursuing austenitic alloys, nickel-iron-chromium alloys, reactive-refractory metals; and innovative concepts simultaneously, is directed at the long-term objective of commercial fusion power.

The interim objective is to provide alloy selection data for the intermediate-term engineering prototype and demonstration power reactors.

The Plasma Materials Interaction Task studies the interaction of plasma irradiation with the surface of the first-wall. The objectives are to provide the data base necessary to deal with the plasma contamination problem and to minimize the detrimental effects on the first-wall material.

The Special Purpose Materials Development Task deals with the development of fusion reactor materials other than first-wall and structural materials.

Damage Analysis and Dosimetry encompasses establishment of the radiation (primarily neutron) spectra in various facilities, establishment of damage parameters and measurement of damage in materials, analysis of radiation damage resulting from different neutron energies, and extrapolation to predict effects of the fusion power spectrum.

The Radiation Facilities Development and Operations Task includes the development and operation of neutron and plasma sources to simulate the fusion reactor environment required for materials testing. Operation of the High Intensity Neutron Source (HINS) is beginning. Other work includes the conceptual design of the Fusion Materials Irradiation Test (FMIT) facility.

FY 1978 accomplishments in Fusion Reactor Materials included publication of the first Fusion Materials Program Plan and completion of the technical assessment of vanadium-base alloys for fusion reactors. It was established that wall sputtering, although introducing impurities into the plasma, was not a significant mechanism in wall erosion. Candidate materials were selected for alloy development in the austenitic, nickel-base, reactive and refractory metal systems.

In FY 1979, the first data from HINS should be available, extending high-energy neutron damage data by two orders of magnitude in exposure. The first tests of low atomic number, high-temperature coated limiters will be performed in ISX. In situ diagnostics will be developed to further define impurity transport in tokamaks, and the first mechanical property data on irradiated titanium alloys will be generated.

Plasma-Materials Interaction work in FY 1980 includes completion of a comprehensive data base for impurity injection mechanisms for early D-T devices and selection of surface coating research reference materials.

Figure VI 3-31 gives the major milestones and funding for the Fusion Reactor Materials Subactivity.

Fusion Systems Engineering

This subactivity is focused on developing conceptual designs of future fusion reactor facilities. Other efforts include the engineering design and development of critical supporting subsystems. There are six task areas:

- Conceptual Design and Engineering (FY 1980 BA request—$2.5M)
- Systems Studies (FY 1980 BA request—$2.5M)

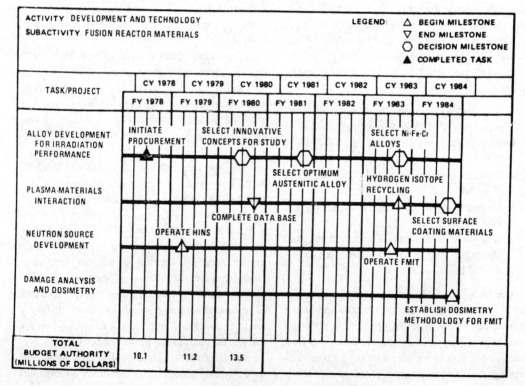

Figure VI-3-31. Milestones/funding for fusion reactor materials.

- Blanket and Shield Engineering (FY 1980 BA request—$1.0M)
- Tritium Processing and Control (FY 1980 BA request—$2.5M)
- Plasma Systems Analysis (FY 1980 BA request—$0.5M)
- Plant Systems (FY 1980 BA request—$0.5M)

The Conceptual Design and Engineering Task was established in early 1976 to define and design the ETF, the major fusion facility which will follow the TFTR and MFTF. This task provides a near-term focus for the physics and technology tasks of the magnetic fusion program. The current strategy is to develop conceptual designs for ETF which can be considered as post TFTR and MFTF devices. The ETF will provide significant advances toward fusion power and will also provide a means of obtaining necessary engineering data and technology experience. Projects evaluate these designs in terms of technical risk, schedule, and cost. Based on these evaluations, the Office of Fusion Energy will select and implement an appropriate design. Another element in the strategy is to address the need to develop an industrial base for fusion power commercialization.

The System Studies Task is assessing ongoing and near-term projects, and projections of the eventual nature of fusion power production.

Blanket and Shield Engineering is addressing the design and analysis of first-wall and blanket assemblies and is primarily oriented toward the far-term Engineering Prototype Reactor (EPR). Blanket engineering is critical to reactor success since this blanket is where the neutrons are converted into heat and tritium is produced.

Tritium Processing and Control efforts are focused on the TSTA which will be dedicated to the development, demonstration, and interfacing of technologies related to the D-T fuel cycle, including the management of radioactive tritium for first-generation fusion reactor systems. An artist's concept of TSTA is given in Figure VI 3-32.

The Plasma Systems Analysis Task develops computational engineering tools for studying the interaction of plasma-related hardware such as neutral-beam and RF heating systems, particle collection systems, and impurity control techniques. These tools are used in evaluating trade-offs among these and other plasma-related systems (such as fuel injectors)

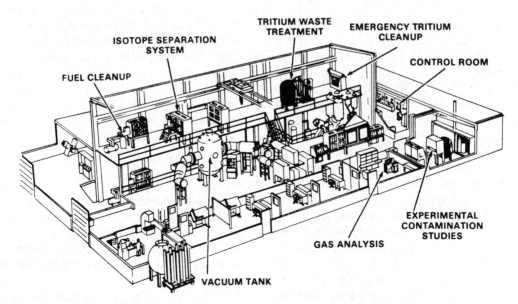

Figure VI-3-32. Artist's concept of the tritium systems test assembly (TSTA).

in order to minimize overall technological requirements.

In the Conceptual Design and Engineering Task, the "next step" tokamak design and engineering studies were initiated in FY 1976 at ORNL. In FY 1977, General Atomic Company also began to develop a "next step" design as part of a Government-funded effort. In FY 1978, PPPL and MIT initiated tokamak "next step" studies based on a TFTR upgrade study that had been completed in FY 1977. A recent shift in strategy emphasizes the design of both a tokamak and a mirror engineering test facility (ETF). Three competitive teams (ORNL, PPPL, and GA) working on the ETF design are being consolidated into a single design team which will operate at a special design center located at ORNL. A mirror ETF design effort is beginning at LLL.

The conceptual design for the tokamak ETF should be completed in FY 1980–82, and construction could start by FY 1984 with operation in FY 1990–92. For the mirror ETF, conceptual design should be completed in FY 1981–82; construction could start in FY 1984 with operation in the early 1990s. These dates are set by the physics and technology base currently available to support a particular confinement scheme.

In the Systems Studies Task, studies of potential commercial tokamak and mirror power plants are under way at the University of Wisconsin and MIT. Parametric analysis of the Elmo Bumpy Torus (EBT) alternate power reactor is continuing at ORNL. LLL is studying techniques for achieving higher energy gain and power density in tandem mirrors with less demanding requirements for neutral beam and magnet systems. Finally, an assessment of alternate confinement techniques is being conducted by LASL as a backup to the mainline tokamak and associated mirror and EBT concepts.

In the Blanket and Shield Design Task, neutron mockup experiments are under way at ORNL, with detailed design of first-wall and blanket assemblies being addressed at ORNL and ANL. Additional shielding code work and blanket and shield engineering assessments are proceeding at LASL and ANL.

The TSTA concept originated in 1976 when the Division of Development and Technology concluded that consolidation and acceleration of tritium development were essential to provide adequate input to design of any major fusion experiment in the 1980s. At the end of FY 1977, LASL was authorized to construct TSTA at an estimated total cost of $13.2M. Construction proceeded on schedule throughout FY 1978, and installation of major TSTA equipment began in FY 1979.

Plasma Systems Analysis projects are centered at ORNL, Science Applications, Inc. (SAI), and at the University of Illinois. The ORNL and SAI efforts are focused on tokamaks and constitute the major portion of this task. The University of Illinois effort is investigating plasma-related technology requirements for Reversed Field Mirror and Reversed Field Pinch configurations.

Major milestones and funding for the Fusion Systems Engineering Subactivity are shown in Figure VI 3–33.

Environment and Safety

The objective of the Environment and Safety Subactivity is to ensure that the development and ultimate operation of fusion power reactors is accomplished with minimum adverse effect on the environment and with maximum safety both for the public and for DOE and DOE contractor personnel and property. The present effort is geared strictly to D-T fusion power since it is expected that all near-term fusion power reactors will be of this type. The subactivity is divided into two tasks as shown in Figure VI 3–34:

- Environment Assessment
- Reactor Safety Research

The Environmental Assessment Task makes certain that environmental considerations are part of decisions to construct and operate major fusion facilities, and that the environmental impact of the fusion power R&D program is properly evaluated. An Environmental Impact Statement (EIS) is published for each major fusion facility.

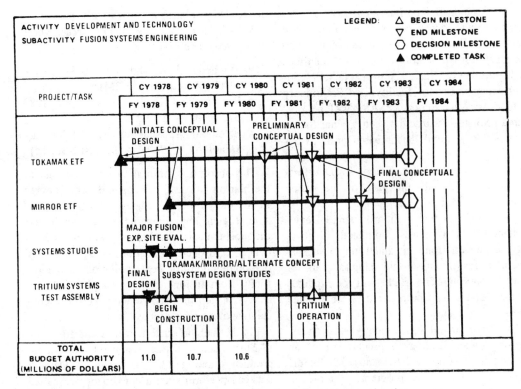

Figure VI-3-33. Milestones/funding for fusion systems engineering.

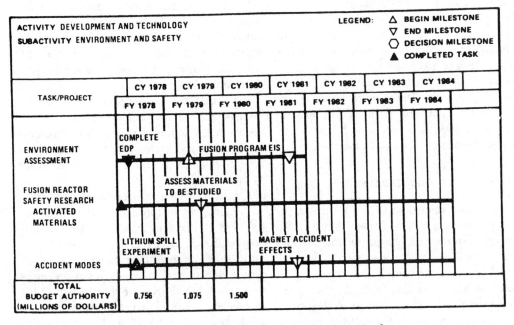

Figure VI-3-34. Milestones/funding for environment and safety.

The Reactor Safety Research Task is concerned with identification, evaluation, and control of all potential hazards associated with the operation of major fusion R&D facilities and, ultimately, fusion power reactors. For each major fusion facility, a Preliminary Safety Analysis Report (PSAR) will be prepared prior to construction, and a Final Safety Analysis Report will be prepared prior to operation.

The Environment and Safety R&D efforts are coordinated with the pace of the fusion technology program.

In FY 1978, a fusion Environmental Development Plan (EDP) was completed, identifying the key environment and safety issues in magnetic fusion and establishing a plan for their resolution. The EDP will be updated annually and should be referred to for detailed status information.

The Environment Assessment Task is focusing on studies leading to preparation of an Environmental Impact Statement (EIS) for the entire magnetic fusion program. Completion of the draft statement is expected in FY 1981. Additional studies are being initiated to consider the commercialization and public perception issues associated with fusion reactor systems.

Idaho National Engineering Laboratory (INEL) has been selected as the lead laboratory for reactor safety research studies. In addition, safety studies of conceptual fusion reactor designs are being carried out by several DOE laboratories and universities. Hanford Engineering Development Laboratory (HEDL) is conducting lithium spill experiments to provide information on lithium reactions with concrete, air, and other materials.

Fusion Energy Applications

Recent studies (including DOE's Inexhaustible Energy Resource Study) have shown that, without a substantial change in end-use energy technologies, the major energy demand after the turn of the century will in all likelihood continue to be nonelectrical. Therefore, a subactivity was established to evaluate the role of fusion energy in nonelectrical applications. This subactivity is divided into three tasks which address the major areas where fusion may contribute:

- Synthetic Liquid and Gaseous Fuels (FY 1980 BA request—$1.2M)
- Fission Fuel Cycle Support (FY 1980 BA request—$1.7M)
- Emergent Applications and Technology Spinoff (FY 1980 BA request—$0.1M)

The absorption of high-energy neutrons in the blanket can lead to very high temperatures in appropriate blanket materials. This may enable efficient high-temperature electrolytic and thermochemical synthetic fuel production cycles. Hydrogen production through a variety of combinations of heat, radiation, and electricity from fusion reactions could be utilized as a first step in any synthetic fuel or chemical manufacturing process.

Fissile fuel production can result from the 14-MeV neutron driver acting on thorium (to produce uranium 233) or uranium. Devices that can perform such a conversion are called "hybrids." By combining fusion and fission, net energy could then be made available at less stringent fusion performance levels than those required for pure fusion devices. The fissile fuels produced could be processed and delivered, in a proliferation-resistant manner, to conventional light water nuclear reactors or advanced gas-cooled reactors. (There may be some nonproliferation advantages to using fusion rather than fast breeder reactors to produce fissile fuel.)

It is likely that other opportunities for taking advantage of the unique high-temperature fusion energy source also exist. The Emergent Applications and Technology Spinoff Task has been established to seek out and foster these by-product uses.

A number of new projects to evaluate alternate fusion energy applications were begun in FY 1978. Initial results of hydrogen production studies published in October 1977 indicated that the high-temperature electrolysis process and several thermochemical processes for producing hydrogen are the most attractive approaches. Studies initiated in these areas are being continued in FY 1979. The other major emphasis is on fusion-fission fuel cycles. These fuel cycles can operate in either a combined or a tandem mode. In either mode of operation, it is the symbiotic relationship between the fusion and fission systems that results in the beneficial advantage of this configuration. An initial study of Pebble Bed Fusion-Fission Systems by General Atomic Company was completed in October 1978. Commercial reactor designs of tokamak and tandem mirror concepts will be carried out in FY 1979 and 1980.

Major milestones and funding for the Fusion Energy Applications Subactivity are given in Figure VI 3-35.

APPLIED PLASMA PHYSICS

The Applied Plasma Physics budget activity, managed by the Office of Fusion Energy's Division of Applied Plasma Physics, is responsible for conducting proof-of-principle programs for fusion concepts, other than tokamak and mirror systems, and for all base physics, including theory and computation. This activity is organized into four subactivities:

- The Advanced Fusion Concept Subactivity develops fusion concepts which represent alternatives to tokamaks and mirrors and which have the greatest promise of significant reactor advantages
- The Fusion Plasma Theory Subactivity develops and applies physics models and mathematical techniques necessary to describe the behavior of magnetically confined plasmas under real operating conditions
- The Experimental Plasma Research Subactivity provides experimental techniques and data and fundamental physics information required to operate and interpret present major confinement experiments
- The National Magnetic Fusion Energy (MFE) Computer Network Subactivity provides the computing capability essential for theoretical analyses and data reduction. The network is a highly cost-effective way of meeting the large-scale computing requirements, and it provides a central focus for cooperative efforts

The FY 1980 BA request for the Applied Plasma Physics activity is $75.6M; the alloca-

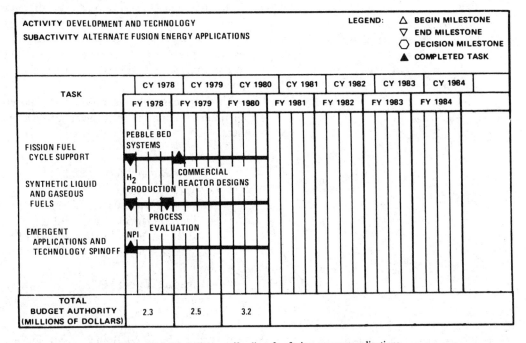

Figure VI-3-35. Milestones/funding for fusion energy applications.

tion of these funds to the four subactivities is shown in Figure VI 3-36.

Advanced Fusion Concepts

The recent scientific progress in tokamaks provides a high degree of confidence that scientific feasibility will be achieved in a tokamak device within a few years, and that alternate concepts are no longer needed to provide a scientific backup for such a demonstration. Consequently, the alternate concepts program has been restructured and redefined to include only those device concepts which have the greatest promise of significant reactor advantages.

"Significant reactor advantages" is the key idea. The Advanced Fusion Concepts Subactivity will attempt to build on the technological data base of tokamaks and mirrors but will be strongly oriented towards developing devices which have as many of the following favorable reactor characteristics as possible:

- High maintainability and accessibility
- Ignition capability (or high Q)
- Steady-state operation
- Simplicity of structure and blanket design
- Flexible reactor size
- Methods of impurity control
- Simple magnet and technology requirements
- Minimum costs of development, operation, and maintenance

For management purposes, the Advanced Fusion Concepts Subactivity has been divided into eight "concept areas," each categorized according to its level of development:

- Level 1 is a proof-of-principle test using a hydrogen plasma with most of the relevant physics parameters at or near the reactor regime
- Level 2 includes concepts in a less developed research stage than Level 1 concepts, which will be selected for accelerated development where appropriate
- Level 3 consists of concepts at the earliest stage of development; these would include new ideas or ideas lacking experimental demonstration where a small experimental device could be used to provide some early tests of the approach

The Advanced Fusion Concepts Subactivity will be managed dynamically; i.e., as definite research results become available for a concept at any level, a decision will be made either to advance that concept to the next level or to terminate funding. Funding priorities among the concepts are based upon reactor attractiveness, commercial potential, and technical and programmatic considerations.

There are currently eight advanced concepts distributed among the three levels of development:

- Level 1
  — Elmo Bumpy Torus (EBT)
- Level 2
  — Reversed Field Pinch (RFP)
  — Linear Magnetic Fusion (LMF)
  — Advanced Fuels/Multipoles (AF/MP)
  — Stellarator/Torsatron (S/T)
- Level 3
  — Compact Toroid (CT)
  — Linus
  — Tormac

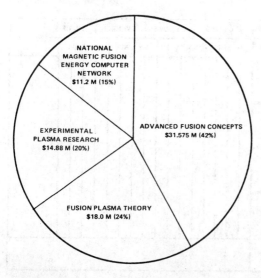

Figure VI-3-36. Applied plasma physics activity FY 1980 BA request (dollars in millions).

In level 1, the Elmo Bumpy Torus concept is that of linking many "simple" mirrors in series to form a torus so that individual mirror magnetic field lines can be joined, end losses eliminated, and plasma ignition achieved. EBT is an attractive reactor concept, with such features as steady-state operation, ignited plasma, low magnetic fields, relatively simple technology, ease of maintenance, and modular design. A schematic diagram of the EBT-I device at ORNL is shown in Figure VI 3-37.

The EBT research program, conducted at ORNL, has been highly successful. In FY 1979, EBT-S produced excellent heating and confinement results, extremely low levels of impurities (less than 1 percent), and established encouraging new scaling results, all in steady state. These results were achieved using high magnetic fields and a new electron cyclotron heating (ECH) microwave system. In FY 1980, these scaling results will be extended to higher microwave power levels (200-kW) resulting in higher temperatures.

Significant advancement of the EBT concept beyond EBT-S requires a new experimental facility designed to provide a proof-of-principle physics test. The EBT concept was selected for such a test following a public meeting held in October 1978 to review a number of potential candidates for a level 1 alternate concepts experimental program. Contractor selection was completed in 1979. Initial device operation is anticipated at the end of FY 1982. The present funding request covers all research and development work needed in support of the level 1 major device fabrication activity.

The level 2 Reversed Field Pinch (RFP) evolved from the Zeta and Z-pinch programs at Culham and LASL, respectively. The RFP concept is similar to the tokamak, but differs in that the toroidal magnetic field is reversed in the outer regions of the plasma, thereby permitting stable operation at higher plasma pressure and at lower values of the safety factor. A reactor based on this concept might be heated to ignition using only plasma current, though it would be constrained to operate in a pulsed mode.

The Linear Magnetic Fusion (LMF) approach evolved from the linear theta pinch technique researched at LASL. This concept

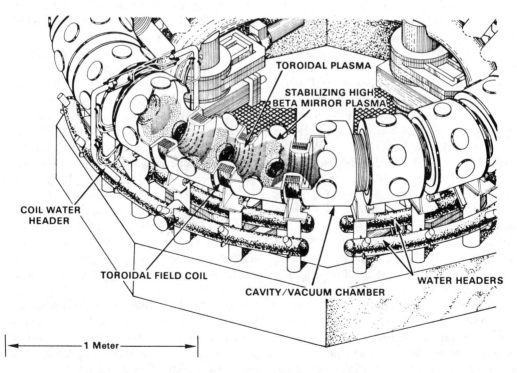

Figure VI-3-37. Schematic diagram of the Elmo Bumpy Torus.

requires a major reduction in present heat and particle loss rates to be attractive as a fusion reactor candidate.

The Advanced Fuels/Multipoles (AF/MP) program is aimed at developing fusion reactors that use neutron-lean fuel cycles. The use of such fuels could result in the elimination of the breeding blanket, of the use of tritium, and of much of the radioactivity associated with D-T fusion reactors. At the higher temperatures required for advanced fuels to burn, good plasma confinement and the absence of internal magnetic fields (to reduce synchrotron radiation losses) are desirable, leading to interest in multipole toroidal devices.

The stellarator and torsatron (S/T) are toroidal devices in which the magnetic coils are wound at an angle around the toroidal plasma vessel. The advantage of this arrangement is that a toroidal plasma current is not required and hence the devices can operate in steady state. One problem with the arrangement is the fabrication and support of the magnet windings. This effort is oriented toward design studies and theoretical work.

The RFP program, being supported primarily at LASL, is the largest Level 2 program. A major RFP device, ZT-40, costing approximately $10M began operation at LASL in FY 1979. Basic scaling results on performance versus size for the RFP approach will be obtained in FY 1980, and experiments will be extended to the long-pulse regime.

The LMF program has been evaluated and restructured, with the cessation in FY 1979 of end-loss studies on Scylla IV-P at LASL. Experiments on the LMF high-field solenoid at Mathematical Sciences Northwest, Inc., have demonstrated successful heating of a plasma in a solenoid by carbon dioxide laser irradiation over distances of 100 cm. This solenoid, which has been extended to 300 cm for studies during FY 1979, will contain a plasma that has many of the relevant physics parameters in the reactor regime.

In the AF/MP program, the emphasis in FY 1979 has been to upgrade the plasma confinement capability of SURMAC at UCLA. The upgraded device became operational in late FY 1979, and extended physics results, with higher ion temperatures and lower collisionality, will be available in FY 1980.

The only stellarator currently operating in the United States is at the University of Wisconsin and is supported by the National Science Foundation. A small-scale reactor study is being supported by DOE at MIT to assess the reactor possibilities of this concept. In addition, a stellarator advisory committee has been formed to review research progress and to promote international cooperation on stellarators by participation of two or three U.S. scientists in the stellarator experiments at foreign laboratories.

The Compact Toroid (CT) is a high-risk, but potentially high-payoff, level 3 device to achieve field reversed plasma configurations (sometimes called plasmoids or particle rings) by means such as relativistic electron beams, ion beams, neutral beams, and plasma guns.

The Linus concept uses an imploding liquid metal liner to compress a magnetic field to very high values. This high field leads to high plasma densities, and small, compact reactor designs.

The Tormac concept combines some features of the tokamak and mirror concepts. It uses poloidal cusp magnetic fields for plasma stability and a toroidal magnetic field to provide basic confinement. The advantages of tormac are steady-state operation, ignition, good impurity control, low magnetic field, and modest technological requirements.

The CT Program was reorganized and significantly strengthened during FY 1979. The desirability of providing some definitive tests of the CT concept was recognized, and programs were initiated at LLL, PPPL, and elsewhere. The CT program is a good example of a concept with attractive reactor features and high commercial potential but which is still at an early, and therefore quite speculative, stage of development. Initial physics results on this interesting approach are available.

Linus (the high field flux compression approach), being developed at the Naval Research Laboratory, has met an important program milestone with the completion of a new experimental device, LINUS-0, and its initial operation in early FY 1979. Technical data on the manipulation of liquid metal liners and their behavior under compression was obtained during FY 1979. The flux compression experiment, LINUS-1, is being designed, and fabrication will commence during FY 1980.

Tormac IV and Tormac V attained electron temperatures of 10-keV in FY 1978. A new Tormac device, P-1, was tested during FY 1979. Evaluation of the technical status of Tormac has led to the conclusion that funding for this activity will probably be redirected during FY 1980.

Figure VI 3-38 shows the major milestones and funding for the Advanced Fusion Concepts Subactivity.

Fusion Plasma Theory

Fusion Plasma Theory is responsible for developing and applying physics models and mathematical techniques necessary to describe the behavior of magnetically confined plasmas under realistic operating conditions. Plasma models are developed and used to interpret results obtained in existing confinement devices and to predict performance parameters of proposed devices. Both analytical and numerical techniques are used. However, as a general rule, the more applied and device-oriented a problem is, the greater the reliance on large-scale computing. The theoretical effort is divided into four categories: tokamak theory, mirror theory, advanced concept theory, and generic fusion theory.

The tokamak, mirror, and advanced concept theory categories provide analytical support for the Tokamak Systems Experiments, Magnetic Mirror Systems Experiments, and Advanced Fusion Concepts subactivities, respectively. The generic theory category provides broadbased support for the entire magnetic fusion program. This latter category develops computer algorithms of general applicability, provides fundamental plasma physics analyses, and supplies required atomic physics data.

During the past year, computer simulations of tokamak plasmas have provided a qualitative understanding of major plasma disruptions. These results may allow expansion of the operating range for tokamaks. Plasma shape optimization studies were aimed at achieving high power density tokamak operation to improve tokamak reactor economics. Recent theoretical studies have shown that it may be possible to use RF power to produce steady-state tokamak operation. This would further improve the economics of fusion reactor power plants. Tokamak transport codes, which assess power losses in tokamaks, were refined to the point where performance parameters for beam-injected tokamaks can now be predicted with some confidence.

A number of advances were made in the mirror theory category. The high beta operating limit of TMX was determined, and the results appear favorable for building economical TMX reactors. Successful calculations of the minimum plasma temperature and neutral-beam density required to achieve reversed field configurations explained why field reversal could not be obtained in the 2X-IIB device. Recent plasma transport calculations will be used to design future mirror reactors. Parameter windows were determined for the equilibrium and stability of ion rings used for establishing field reversed configurations.

Advanced concept theory is being realigned to conform with the recent recommendations of the Advanced Concepts Review Committee. Existing resources are being reviewed for possible reallocation to meet the needs of this program.

Recent accomplishments of the generic fusion theory category include applications of special-purpose minicomputers to a broad class of plasma problems and provisions of atomic physics data used to analyze PLT and ISX experiments.

Major milestones and funding for the Fusion Plasma Theory Subactivity are presented in Figure VI 3-39.

Experimental Plasma Research

The Experimental Plasma Research Subactivity is responsible for providing experimental techniques and data, and fundamental physics information required to operate and

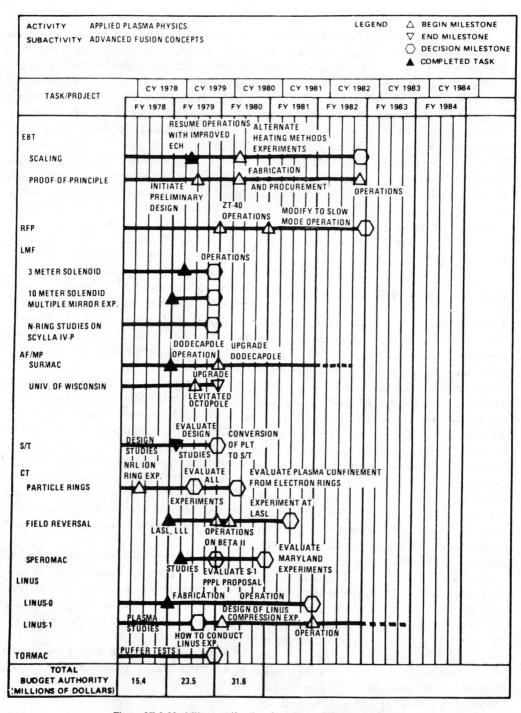

Figure VI-3-38. Milestones/funding for advanced fusion concepts.

interpret present major confinement experiments. This subactivity, which has built and supports a large number of small- and medium-scale plasma experiments, is divided into five technical areas: Plasma Properties, Plasma Heating, Diagnostics, Impurity Data and Ion Source Research, and Advanced Superconducting Materials Research. The subactivity has two tasks: the Fusion Plasma Research Facility (FY 1980 BA request—$3.0M), and the remaining base program (FY 1980 BA request—$10.0M).

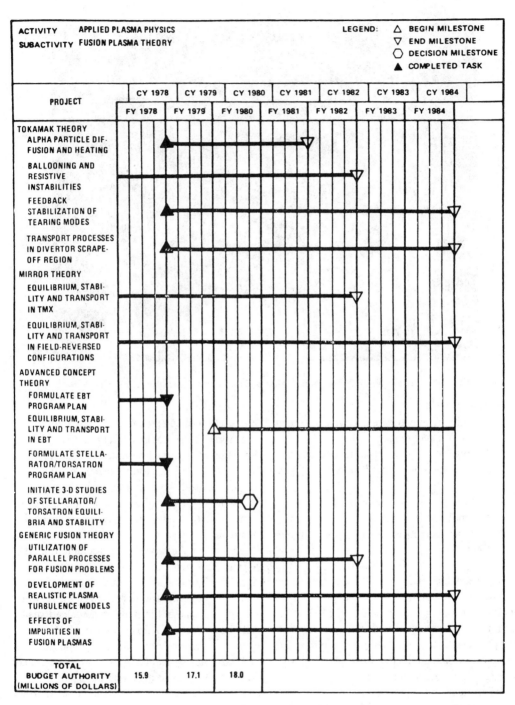

Figure VI-3-39. Milestones/funding for fusion plasma theory.

The Fusion Plasma Research Facility (FPRF) is being established at the University of Texas to provide a medium-scale fusion device (TEXT) for use by all sectors of the fusion community. A model of TEXT is shown in Figure VI 3-40. The objectives for TEXT are complementary to mainline tokamaks and include the following:

- Providing atomic and molecular data on impurity ions in tokamaks
- Testing, evaluating, and standardizing

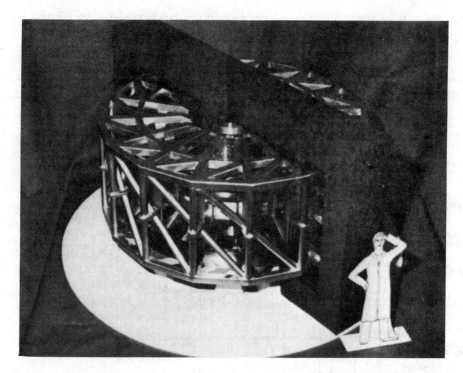

Figure VI-3-40. Model of the Texas Experimental Tokamak (TEXT).

existing diagnostics and developing new diagnostic techniques
- Testing new ideas in order to improve plasma heating, equilibrium, and stability
- Providing quantitative tests of theoretical plasma models
- Training scientists directly on confinement devices

Experimental studies in Plasma Properties provide tests, quantitative where possible, of analytic and computer models of plasma behavior. Measurements from these experiments complement data from larger machines and provide a broader information base for design of large confinement devices.

Experiments in the Plasma Heating category are directed toward exploration of alternative methods of heating plasmas to fusion temperature. Studies include development of negative ion sources for neutral beams and the entire range of RF heating techniques.

Under Diagnostics, new and improved techniques are developed for measuring plasma parameters, data handling systems are refined, and measurement techniques are standardized.

Efforts in the Impurity Data and Ion Source Research area furnish atomic data for impurity ions and describe atomic processes involved in ion sources. An important part of the activity is the operation of an atomic data center to compile and distribute data as it becomes available.

Studies in Advanced Superconducting Materials Research are focused on developing ductile, high magnetic field, superconducting materials suitable for fusion machines. Work includes a search for new or advanced materials and a detailed investigation of their mechanical and superconducting properties, as well as improvements in the processing and fabrication of existing materials.

The studies are supported at more than 25 universities and the DOE laboratories, with some support provided to industry, the National Bureau of Standards, and NRL.

Fabrication of FPRF began in FY 1978 and continued through FY 1979. Initial operation is expected to be in early FY 1980. Full-scale experiments should begin near the end of FY 1980.

Progress is continuing in each technical area. In Plasma Properties, experiments on MACROTOR have shown that the sputtering of impurities from cool regions near the walls of a tokamak can be suppressed by controlling the sheath potential near the walls. In addition, an improved discharge technique for tokamak cleaning, developed on MICROTOR, has become standard. In Plasma Heating, the construction of a gyrotron tube for generating RF radiation for plasma heating has been completed and was tested on several tokamaks in FY 1979. In the Diagnostics area, important experimental results underlying the development of a two-dimensional X-ray imaging system have been obtained. Two-dimensional imaging will provide a cross-sectional picture of the plasma and significantly help in understanding and controlling plasma behavior. Progress in Impurity Data and Ion Source Research includes developing the first experimental and theoretical data on charge exchange and ionization cross-sections for hydrogen at neutral beam energies, and obtaining basic data on plasma cooling modes. The charge exchange work is also important to the development of a negative ion neutral beam source which would provide efficient plasma heating at high beam energy.

The current program direction will be maintained through FY 1980. Plans include a test of radiation scattering in the far infrared to determine ion temperature. Alternate heating methods to be explored include the first U.S. test of lower hybrid heating in a tokamak, as well as an electron cyclotron resonance heating experiment on ISX-B using a new high-power gyrotron. A lower hybrid experiment to test generation of the electrical current required for steady, safe tokamak operation will be completed. A detailed set of selected atomic data on the impurity molybdenum (Mo) will also be completed.

Major milestones and funding for the Experimental Plasma Research Subactivity are shown in Figure VI 3-41.

National Magnetic Fusion Energy (MFE) Computer Network

The National Magnetic Fusion Energy (MFE) Computer Network subactivity has two subtasks: the National MFE Computer Center (FY 1980 BA request—$8.7M) and the User Service Center (FY 1980 BA request—$1.8M). The National MFE Computer Network (FY 1980 BA Request—$10.5M) is composed of the National MFE Computer Center (NMFECC) at LLL and five User Service Centers (at LLL, LASL, ORNL, PPPL, and General Atomic Company), which are connected to the national center by leased lines. The network (Figure VI 3-42) was conceived to meet massive computational requirements in a highly cost-effective way and to provide a central focus for cooperative efforts. In addition to the User Service Centers at the DOE laboratories and General Atomic Company, more than 30 universities and other small contractors performing magnetic fusion research are connected to the NMFECC through a dial-up capability, giving them access to the large-scale computing capability needed to solve certain problems.

The NMFECC currently houses two large scientific computers (a CDC 7600 and a CRAY-I), a CDC 6400 for handling input/output, and a CDC mass storage. One group of design engineers and systems programmers directly supports system hardward and software maintenance and development, while another group develops numerical computing procedures for the user community. The usage of available computation time is over 90 percent.

The National MFE Computer Network began operation in 1975 and presently has over 900 users. The network is now providing stable, reliable service. It handles almost all of the large-scale magnetic fusion energy computing. The sharing of large codes among the various research groups has been very successful, and cooperation promises to increase in the future. During FY 1978, a new class VI computer (CRAY-I) was added to the NMFECC and became fully operational within 3 months of acquisition. In addition, a CDC mass storage unit was installed. Construction of a new computer building for the network was started in FY 1978, and occupation began in FY 1979.

Major milestones and funding for the National MFE Computer Network Subactivity are given in Figure VI 3-43.

REACTOR PROJECTS

The Reactor Projects budget activity, managed by the Office of Fusion Energy's Division of Planning and Projects is responsible for the design and construction of large complex fusion facilities, and for the provision of project-specific technology development programs in direct support of such construction. The three principal subactivities are as follows:

- Tokamak Fusion Test Reactor (TFTR), which is a major scaling experiment which will produce fusion energy at the energy breakeven level, using a deuterium-tritium (D-T) plasma. It will be used to study the physics of burning plasmas and the engineering aspects of D-T tokamak operation with power densities near those required for a commercial reactor.
- High Intensity Neutron Source (HINS), which is the first high-energy, high-intensity neutron irradiation facility dedicated to the Fusion Reactor Materials subactivity. Its purpose is to provide an intense beam of high-energy neutrons which will be used to cause structural damage in reactor materials. The damage will be analyzed to determine ways of reducing the probability of damage, hence leading to reactor components with longer lifetimes.
- Fusion Materials Irradiation Testing (FMIT) Facility, which employs a broad spectrum, high-energy neutron source required to screen and qualify candidate fusion reactor materials. The facility will be used for testing of materials, including some accelerated life testing, and for providing information for development of alloys less likely to become radioactive or damaged due to the neutron bombardment.

The Mirror Fusion Test Facility (MFTF) is also managed by the Division of Planning and Projects, and will appear in this budget activity after FY 1980.

The FY 1980 BA request for the Reactor Projects Activity is $90.95M; the allocation of these funds among the subactivities is shown in Figure VI 3-44.

Tokamak Fusion Test Reactor

TFTR is a major scaling experiment which will produce fusion energy at the energy breakeven level, using a D-T plasma. It will be used to study the physics of burning plasmas and the engineering aspects of D-T tokamak operation with power densities near those required for a commercial reactor. It will be the first fusion device to explore the engineering requirements of remote maintenance, since the use of tritium will lead to low-level contamination of some device components. Experiments performed on TFTR will provide the critical physics and technology data base from which a Tokamak Engineering Test Facility could be developed.

A schematic diagram of TFTR is shown in Figure VI 3-45. It will be the largest U.S. tokamak device with a major radius (distance from the axis of the torus to the center of the toroidal ring) of 2.65 meters and a radius for the vacuum vessel (torus cross section) of 1.1 meters. Under optimum experimental conditions, a high-power neutral deuterium beam (20-MW) will be injected into a warm tritium plasma. The Lawson Product for the resulting plasma is estimated to be between 10^{13} and 10^{14} cm^{-3}-sec, bordering on that required for plasma ignition. TFTR will be a pulsed device (like all tokamaks) with one pulse every five minutes. It will have an extensive diagnostic system so that every important plasma property can be carefully monitored and the relationships between properties investigated.

Preliminary design work on TFTR began in February 1976, and groundbreaking ceremonies took place in October 1977. All the major design issues have been resolved and many of the key procurement contracts have been placed including the vacuum vessel, toroidal field magnet coils, and neutral beam and magnet power supplies. Total construction cost for TFTR is estimated at $239M. The facility is scheduled for completion by March 1982.

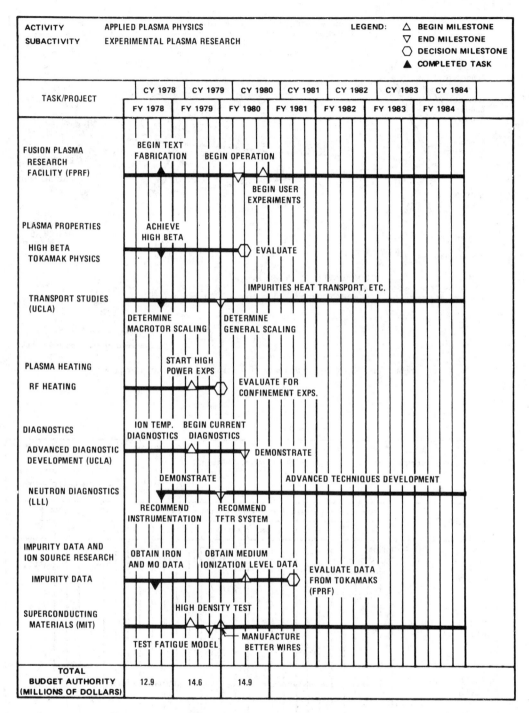

Figure VI-3-41. Milestones/funding for experimental plasma research.

Of the many challenging design problems in TFTR, two of the more difficult were the toroidal field coils and the neutral-beam injection system. High-stress levels in the first coil design led to a large redesign effort which succeeded in solving the problems. The neutral beam system has completed its final design, and a prototype beamline is nearing completion at Lawrence Berkeley Laboratory. Performance is expected to meet or exceed the

932 Magnetic Fusion Energy

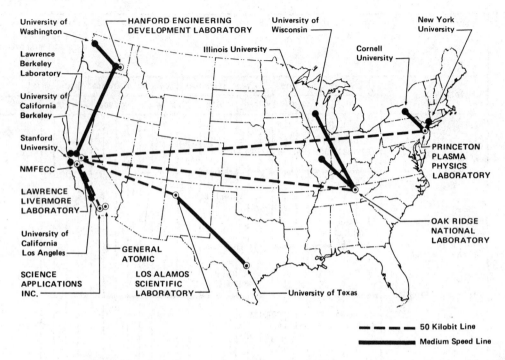

Figure VI-3-42. Major components of the national MFE computer network.

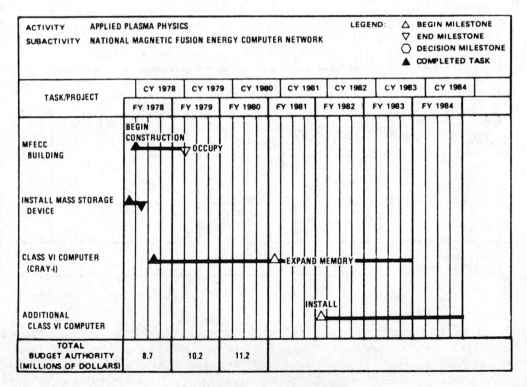

Figure VI-3-43. Milestones/funding for the national MFE computer network.

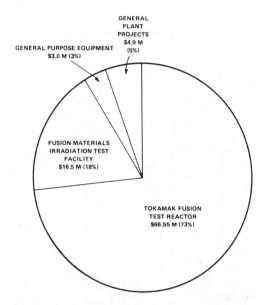

Figure VI-3-44. Reactor projects activity FY 1980 BA request (dollars in millions).

original objectives. One unresolved problem is the development of large high-vacuum seals which can be baked to 250°C and can be remotely resealed. This problem was resolved by mid-FY 1979. The Preliminary Safety Analysis Report (PSAR) was completed and published in October 1978. No significant safety problems are foreseen.

The major milestones and funding for the TFTR Subactivity are presented in Figure VI 3-46.

High Intensity Neutron Source

HINS will be the first high-energy, high-intensity neutron irradiation facility dedicated to the Fusion Reactor Materials Subactivity. Its purpose is to provide an intense beam of high-energy neutrons, which will be used to cause structural damage in reactor materials. The damage will be analyzed to determine ways of reducing the probability of damage, hence leading to reactor components with longer lifetimes.

The HINS facility has two sources. The accelerator and target room for one of these is

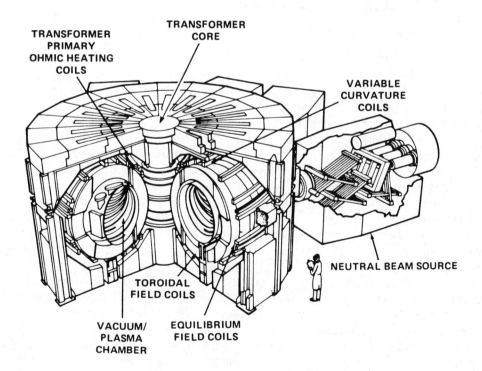

Figure VI-3-45. Schematic diagram of TFTR.

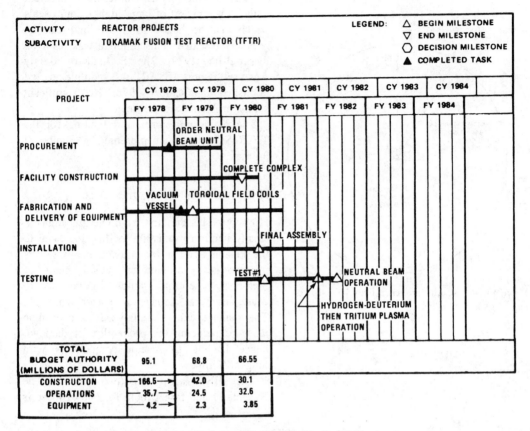

Figure VI-3-46. Milestones/funding for TFTR.

shown in Figure VI 3-47. Neutrons are generated from the fusion of deuterium and tritium—the same reaction that will probably be used in first-generation fusion reactors. A solid target coated with titanium tritide is struck by a high velocity beam of deuterium (from the accelerator). The target is 46 cm in diameter, and is rotated at 5000 revolutions per minute which limits target heating by the deuterium beam and results in a higher intensity neutron beam.

The maximum flux is 2×10^{13} n/cm^2-sec over a volume of 1.0 cm^3. The flux and the volume are not nearly high enough for life testing of materials for fusion reactors, but are sufficient to aid greatly in damage analysis. Higher fluxes and source strengths are planned for other neutron sources.

The HINS facility, located at LLL, was completed in January 1978 at a cost of $5M. Initial operation began in April 1978. Test operation and any necessary equipment modification were carried out during the remainder of FY 1979. Upon completion, the facility will be operated by the Fusion Reactor Materials subactivity. HINS will be the major high-energy irradiation testing facility until FMIT becomes operational in 1983.

Damage analysis experiments will begin in FY 1980. Stainless steels and refractory metals will be irradiated, at room and elevated temperatures, at levels 100 times higher than has previously been possible. Damage effects on the microstructure will be observable for the first time. In addition, superconductors will be irradiated at superconducting temperatures ($-269°C$), and the effect of neutron radiation on the current measured.

Major milestones and funding for the HINS Subactivity are shown in Figure VI 3-48.

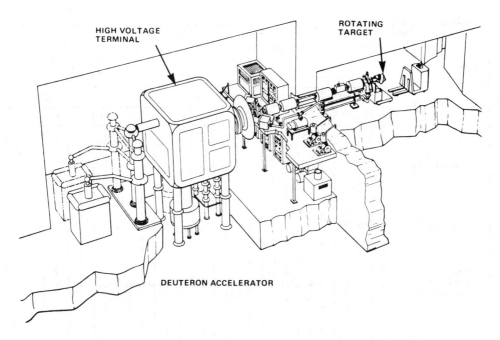

Figure VI-3-47. One of the accelerator and target rooms of the HINS facility.

Fusion Materials Irradiation Test Facility

The FMIT facility will use a broad spectrum, high energy neutron source. Both the maximum neutron flux and the test volume are greater than with HINS. The FMIT facility will provide the peak neutron flux, the test volume, and the neutron spectrum representative of fusion reactor conditions. FMIT is needed to screen and qualify candidate fusion reactor materials. The facility will be used for testing of materials, including some accelerated life testing, and for providing information for development of alloys less likely to become radioactive or damaged due to the neutron bombardment.

The neutrons for FMIT are produced by a lithium-deuterium stripping reaction. As shown in Figure VI 3-49, a deuteron (the nucleus of deuterium) beam is accelerated and aimed into a flowing stream of liquid lithium which is maintained at a temperature of about 220°C. The proton stripped from each deuteron (which consists of a proton and a neutron) is attracted to the lithium. The remaining neutrons emerge with a broad energy spectrum. The energy peak of the emerging neutrons can be adjusted by changing the energy of the incoming deuteron beam. Tritium will be produced in the reaction, and provisions for handling it are included in the FMIT facility. Heat exchangers are provided to cool the lithium.

The FMIT facility will be built at the Hanford Engineering Development Laboratory (HEDL) in Richland, Washington; facility cost is estimated at $83M.

Facility design and development have begun. R&D is proceeding on the lithium systems, the accelerator, the experimental systems, and building shielding. Design responsibility for the accelerator and development of a portion of the accelerator has been assigned to Los Alamos Scientific Laboratory (LASL).

In FY 1979, all Title I design will be completed, and actual system design began. Mockups were built and tested for various parts of the system. R&D work continued, with shielding R&D completed mid-FY 1979.

Construction and procurement will begin in FY 1980 and be completed in FY 1982. After installation and testing, the FMIT facility will become operational in mid-FY 1983; it will be

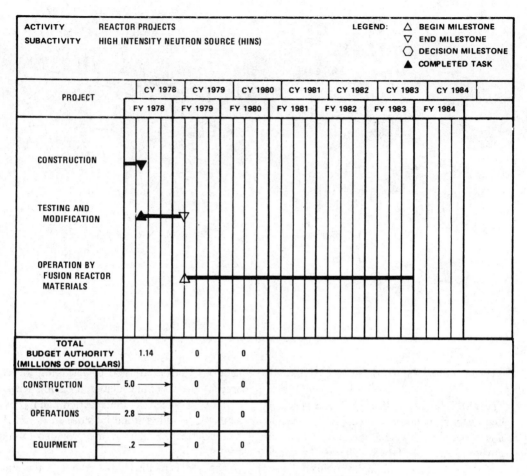

Figure VI-3-48. Milestones/funding for HINS.

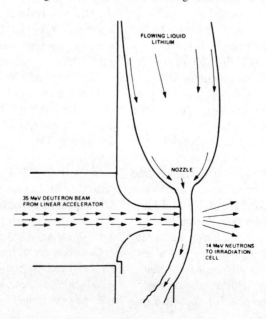

Figure VI-3-49. Schematic diagram of the neutron source for FMIT.

operated by the Fusion Reactor Materials Subactivity.

Major milestones and funding for the FMIT Facility Subactivity are presented in Figure VI 3-50.

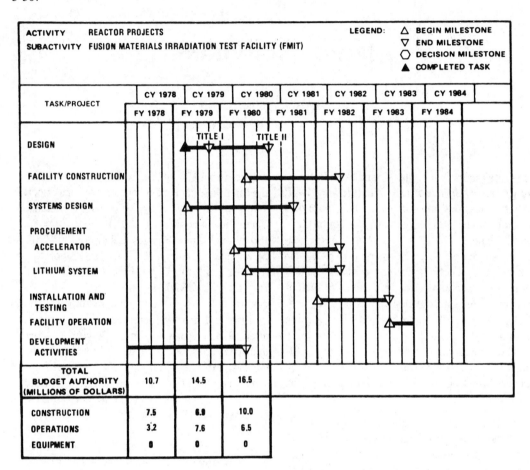

Figure VI-3-50. Milestones/funding for FMIT.

4.
Field Activities

The magnetic fusion development effort employs research talent and facilities at DOE laboratories, educational institutions, nonprofit organizations, and numerous commercial firms. These facilities contribute to the design, development, and construction of experimental magnetic fusion devices. Their personnel conduct basic research in fusion-related areas, and develop the associated technology required for construction of a working fusion energy source.

The geographic distribution of magnetic fusion research is determined by the following factors which are intrinsic to the nature of the technology:

- Existence of high-technology experimental facilities and technical personnel in educational institutions, nonprofit organizations, DOE laboratories, and commercial firms
- Necessity of concentrating Federal expenditures for hardware, operation, and follow-on research at several large research devices
- Availability of the substantial electrical power required by large confinement devices

Fusion research tends to be capital-intensive because the underlying theories and concepts necessary for constructing a commercial energy source can only be tested with increasingly large and technically sophisticated hardware. Since the commencement of the Atomic Energy Commission's fusion research program in 1951, major experimental and testing facilities have been constructed and operated at a few key sites—Los Alamos Scientific Laboratory (LASL), New Mexico; Princeton Plasma Physics Laboratory (PPPL), New Jersey; Lawrence Livermore Laboratory (LLL), California; and Oak Ridge National Laboratory (ORNL), Tennessee. After declassification of the magnetic fusion program in 1958 and greatly increased Federal support after 1973, major experiments have been conducted at these sites, and additionally at the General Atomic Company (GA), Massachusetts Institute of Technology (MIT), Hanford Engineering Development Laboratory (HEDL), University of Wisconsin, and University of Texas.

The design, construction, and operation of new experimental machines have been accompanied by extensive work in applied plasma physics research and development of new technology. These activities take place at the DOE laboratories, more than two dozen universities, and numerous private firms. This research is distributed over a wider geographical range of contractors than is major device fabrication and operation.

The majority of the magnetic fusion research budget is allocated to DOE laboratories. There is, however, significant subcontracting to private firms for design, research, or equipment fabrication. Table VI 4-1 shows the distribution of the FY 1979 research budget by activity and prime recipient. Much of the $159M (71 percent of the FY

Table VI-4-1. Distribution of FY 1979 magnetic fusion program budget to DOE laboratories and direct contracts with universities and industry (dollars in millions).

PRIME RECIPIENT	ACTIVITY BUDGET (OPERATING)				
	CONFINEMENT SYSTEMS	DEVELOPMENT/ TECHNOLOGY	APPLIED PLASMA PHYSICS	REACTOR PROJECTS	TOTAL
DOE Laboratories	56.2	42.6	32.2	28.2	159.2
Universities	8.9	2.1	14.2	0	25.2
Industry	26.1	8.1	14.4	0	48.6
Total	91.2	52.8	60.8	28.2	233.0

1979 budget) allocated to the DOE Laboratories is disbursed to architect-engineers, instrument manufacturers, engineering designers, hardware fabricators, and private-sector researchers. This decentralized contract award procedure allows the detailed equipment design and construction to be overseen by the laboratory technical staff who will use it. Headquarters staff directly authorizes and monitors prime contracts for research with broader implications for the program.

Table VI 4-1 also illustrates several other aspects of the current magnetic fusion research budget. With the major exception of the Alcator tokamak research and fabrication at MIT ($8.3M), university involvement is primarily in applied plasma physics research, which can be carried out by scientists at many U.S. universities. With the exception of the $15M allocated for construction of Doublet III by General Atomic Company, other prime industry contracts for fusion research are relatively small. Most of the industrial involvement results from DOE laboratory subcontracts.

ROLE OF THE DOE LABORATORIES

The fusion research program is centered in DOE Laboratories across the United States. Table VI 4-2 gives the FY 1979 budget distribution for the laboratories, and Figure VI 4-1 shows their locations. The major fusion research facilities are PPPL, ORNL, LASL, and LLL.

PPPL concentrates on tokamak experiments and theory and is constructing the TFTR. Other PPPL experiments include the PLT and PDX, each designed to demonstrate an important aspect of the feasibility of attaining reactor-level plasma conditions before TFTR operation.

ORNL fusion work is broad-based. Tokamak and Elmo Bumpy Torus (EBT) experiments are coupled with theory and technology work. ORNL programs include the Large Coil Testing Facility, the Impurity Studies Experiment (ISX), and the Engineering Test Facility National Design Center.

LLL continues to be the leading investigator of the mirror approach. Construction of the MFTF, a key scaling step in mirror confinement, is beginning. LLL is also the site of TMX, Beta II, and the National MFE Computer Center, the major computing facility for the magnetic fusion program.

LASL is the site of the TSTA, and is also developing a strong program in advanced concepts, highlighted by Reversed Field Pinch devices. In addition, LASL is designing the accelerator for the FMIT facility.

Other laboratory involvement includes the following seven institutions:

- Lawrence Berkeley Laboratory (LBL)— Prototype design and fabrication of the TFTR beam lines
- Hanford Engineering Development Laboratory (HEDL)—FMIT facility site
- Argonne National Laboratory (ANL)— Systems studies
- Sandia Laboratories—Materials properties
- Battelle Pacific Northwest Laboratories (PNL)—Environmental properties

Table VI-4-2. FY 1979 magnetic fusion program operating budget activity distribution to DOE laboratories (dollars in millions).

PROGRAM	ANL	BNL	HEDL	LASL	LBL	LLL	ORNL	PPPL	PNL	SANDIA	INEL
Confinement Systems				0.3	1.6	20.2	12.0	21.9			
Development/ Technology	2.4	1.8	1.0	4.7	0	9.4	20.9	0.6	0.3	1.1	0.3
Applied Plasma Physics	0.1			9.9	1.3	11.8	5.8	3.5		0.1	
Fusion Reactor Projects			7.2		4.9			16.1			
TOTAL FUSION PROGRAM	2.5	1.8	8.2	14.9	7.8	41.4	38.7	42.1	0.3	1.2	0.3

DOE Laboratory Acronyms:

ANL: Argonne National Laboratory (Illinois)
BNL: Brookhaven National Laboratory (New York)
HEDL: Hanford Engineering Development Laboratory (Washington)
LASL: Los Alamos Scientific Laboratory (New Mexico)
LBL: Lawrence Berkeley Laboratory (California)
LLL: Lawrence Livermore Laboratory (California)
ORNL: Oak Ridge National Laboratory (Tennessee)
PPPL: Princeton Plasma Physics Laboratory (New Jersey)
PNL: Battelle Pacific Northwest Laboratories (Washington)
Sandia: Sandia Laboratories (New Mexico)
INEL: Idaho National Engineering Laboratory (Idaho)

Figure VI-4-1. Locations of U.S. laboratories with major roles in the magnetic fusion program.

- Idaho National Engineering Laboratory (INEL)—Safety studies
- Brookhaven National Laboratory (BNL)—Negative ion beam and safety studies

UNIVERSITIES

University involvement in the magnetic fusion program includes extensive research and development and construction of devices. The greatest construction activity is at MIT, the site of the Alcator-C tokamak device. Table VI 4-3 shows a breakdown of university contracts for FY 1979.

Other major university contractors which operate or build devices include the University of California, the University of California at Los Angeles (UCLA), the University of Texas, the University of Maryland, and the University of Wisconsin. Texas will be the site of the Fusion Plasma Research Facility (FPRF). Wisconsin is playing a key role in the design of the Tandem Mirror Experiment at LLL; UCLA and Maryland are conducting extensive research into fusion plasma theory, computer analysis, and computer applications. Other university contractors are undertaking a wide range of fusion research projects.

Table VI-4-3. FY 1979 university contracts in magnetic fusion by budget activity (dollars in thousands).

UNIVERSITY	STATE	DEVELOPMENT AND TECHNOLOGY	APPLIED PLASMA PHYSICS	CONFINEMENT SYSTEMS	TOTAL
Arizona, Univ. of	AZ		60		60
California Inst. of Tech.	CA		210		210
California, Univ. of	CA				
Berkeley			135		135
Irvine			230		230
Los Angeles		113	2,575		2,688
Santa Barbara		110			110
Columbia Univ.	NY	50	513		563
Connecticut, Univ. of	CT		50		50
Cornell Univ.	NY		226		226
Georgia Inst. of Tech.	GA	197			197
Illinois, Univ. of	IL	120			120
Iowa, Univ. of	IA		121		121
Maryland, Univ. of	MD		1,285	150	1,435
Massachusetts Inst. of Tech.	MA	1,500	2,160	8,912	12,572
Michigan, Univ. of	MI		70		70
New York Univ.	NY		810		810
Northwestern Univ.	IL		31		31
Pennsylvania State Univ.	PA		123		123
Rensselaer Polytechnic Inst.	NY		244		244
Rochester, Univ. of	NY		65		65
Smithsonian	IL		132		132
Stanford Univ.	CA		110		110
Texas, Univ. of	TX		2,850		2,850
Virginia, Univ. of	VA	90			90
Washington, Univ. of	WA	220	330		550
Wesleyan Univ.	CT		32		32
Western Ontario, Univ. of			110		110
William and Mary, College of	VA		50		50
Wisconsin, Univ. of	WI	507	995	835	2,337
Yale Univ.	CT		110		110
TOTAL		2,907	13,627	9,897	26,431

Table VI-4-4. DOE laboratory magnetic fusion contracts and subcontracts to industry in FY 1978 (dollars in thousands).

INDUSTRY	LASL[1]	LLL[2]	ORNL[2]	PPPL[3]	DOE (PRIME)
Aerospace Corporation	$1000K				
Aerovox Corporation		263			
Airco					200
American Science & Engineering					110
Atomics International					15
Austin Research Associates					125
B. D. Malcolm Company				659	
Belden Cable Company	$100K				
Control Data Corporation				2,258	
Ebasco Services, Inc. /Grumman Aerospace Corporation				31,655	
Econ, Inc.					50
EIMAC				598	
General Atomic Company					23,312
General Dynamics Corporation			1,400		
General Electric Company			1,600	11,099	
Grumman Aerospace Corporation			200		
Hughes Aircraft Corporation		185			
Inesco					576
ITT Jennings	$100K				
McDonald-Douglas					204
Machlett Laboratories				6,350	
Magnetic Corporation of America	$500K				100
Mathematical Sciences Northwest					851
3 M Corporation	$500K				
Physics International					309
Plastoid Corporation	$250K				
RCS				566	
RTE/ASEA Corporation				829	
Science Applications, Inc.					872
Spectra Vision					40
Supercon		340			
Systems Engineering Labs				1,916	
Teledyne Readco		470			
Torr Vacuum Products		52			
Transrex				13,225	
TRW					222
United Technology					600
Varian			600		
W. P. Young		180			
Westinghouse Electric Corporation	$150K		1,800		1,497

[1] To nearest 50K only

[2] Contractors above $100,000 only

[3] Contractors above $450,000 only. PPPL listing includes total contractual commitments extending beyond FY 1978.

The breakdown of university contracts for FY 1979, given in Table VI 4-3, shows that the majority of university research is in the area of Applied Plasma Physics (APP). The universities, through their APP work, provide valuable assistance to the DOE Laboratories by expanding the theory, supporting major experiments, and performing supplemental experiments. Simultaneously, they provide a training ground for the technical personnel that will later be required by a magnetic fusion energy industry.

INDUSTRY

The capital-intensive nature of magnetic fusion experiments tends to make them too expensive to allow direct private-sector research. The only industrial site with a major experimental facility is General Atomic Company. Other major industrial participation is acquired on a subcontract basis for fabrication of major hardware items. Industrial firms are also involved with systems studies of integrated fusion energy-producing plants. This brings an industrial point of view to advanced designs and provides industry with significant learning experiences. As the fusion program progresses, more direct industrial involvement will be required.

Table VI 4-4 shows some of the major industrial involvement in the magnetic fusion program for FY 1978. This table lists a number of prime contractors and subcontractors for the four major magnetic fusion-oriented DOE laboratories. The major contractors include Ebasco Services, Inc./Grumman Aerospace Corporation, General Atomic Company, Transrex, General Electric Company, Machlett Laboratories, and Westinghouse Electric Corporation.

5. Commercialization

The DOE RD&C cycle is made up of seven phases: (1) basic research, (2) applied research, (3) exploratory development, (4) technology development, (5) engineering development, (6) demonstration, and (7) commercialization. In this framework, magnetic fusion is principally in the Applied Research phase, with programs in both the basic research and exploratory development phases. The commercialization phase is perhaps 25 to 30 years in the future. Commercialization, however, is not merely something that occurs at the end of the RD&C cycle; the requirements for successful commercialization influence efforts in the earlier phases.

Commercial introduction of a technology requires the following:

- All technical and engineering problems associated with the applications of the technology must have been resolved.
- The economics of the technology must be such that manufacturers can obtain a fair return on their investment at prices that offer users an attractive alternative to conventional technologies.
- Users must accept the technology as effective and reliable.
- The public must be satisfied that the socioeconomic and environmental effects of the new technology have been properly analyzed, and that actions are being taken to eliminate any adverse effects.

When these conditions are satisfied and commercialization proceeds, an institutional infrastructure (producers, distributors, suppliers of raw materials, industry associations, etc.) will evolve. The development of this infrastructure can be accelerated by certain Government actions such as the early establishment of industry performance standards, tax incentives, etc.

SOCIOECONOMIC ASPECTS OF FUSION ENERGY

This section examines some of the currently projected socioeconomic aspects of commercializing fusion energy. The discussion only reflects the current state of knowledge. All the socioeconomic aspects are not yet identified owing to the early stage of development of the technology.

Fusion power appears to have several potential socioeconomic advantages relative to other energy sources. These advantages include flexibility in sizing and end-use application, siting requirements, and a relatively high degree of insulation from rising fuel costs.

The anticipated socioeconomic barriers to commercializing fusion energy include its cost, materials requirements, licensing procedures, environmental and safety standards, and the dangers of nuclear proliferation with the fusion-fission hybrid systems.

Potential Socioeconomic Advantages of Fusion Energy

Earlier design studies envisioned fusion as a

large (greater than 1500-MWe), centralized, capital-intensive energy source. Based upon recent technical progress, however, it appears that fusion reactors can be designed within the range (500–1500-MWe) appropriate to projected utility system capacities. These smaller plants, either singly or possibly in aggregation, will enhance the commercialization potential by reducing capital cost requirements and the vulnerability of the total system to unscheduled outages. Given flexibility in size and allowable variations in specific design choices depending upon application, fusion reactors could be tailored to specific end-uses such as production of electricity, fissile fuels, industrial process heat, and residential/commercial space heating and cooling.

Fusion energy systems may also enjoy siting advantages over other energy sources. First, there do not appear to be significant geographic limitations. Second, the currently envisioned fusion fuel cycle will be relatively self-contained at the reactor site. Finally, at least at this early stage of the development cycle, the magnetic fusion program has the inherent flexibility to adopt approaches to engineering design which will ensure that fusion plants can meet stringent siting regulations.

Furthermore, fusion shares the same economic advantage of any virtually inexhaustible energy resource; namely, once the cost of energy from fusion plants becomes economically competitive with the energy costs of exhaustible energy technologies (e.g., coal or oil), it should become increasingly competitive because of its relative insensitivity to rising costs of scarce fuels.

Anticipated Socioeconomic Barriers to Commercializing Fusion Energy

There must be extensive analysis of the trade-off between economies of scale and smaller fusion reactor sizes. Reactor design studies are addressing the question of determining the relationship between size and cost.

Fusion systems have larger materials requirements than equivalent size fission or fossil energy systems. Because of potentially conflicting requirements concerning materials performance (e.g., mechanical integrity, resource availability, and cost), it is likely that trade-offs will have to be made among these characteristics.

The uniqueness of fusion reactors, compared with other energy sources, will require appropriate licensing procedures and environmental and safety standards to ensure public acceptance.

Although pure fusion reactors, if used as intended, do not pose a nuclear proliferation risk, hybrid fusion-fission reactors may have proliferation aspects similar in character to those of fission systems. However, it may be easier to make use of the so-called "denatured" fuel cycles in a fusion-fission system than in a pure fission system. These issues are being addressed by the present program.

To ensure that fusion energy will prove attractive to commercial users, a continuing and growing involvement of utilities and industry in the program is necessary. Such involvement is difficult to obtain at such an early stage of development because of the cost, risk, and long lead-time involved. However, utilities have currently invested several millions of dollars in fusion R&D through direct contracts (primarily through the Electric Power Research Institute). Utilities also provide advice and consultation to DOE program management. In addition, the practice of DOE Laboratory subcontracting for fabrication of components and subsystems is beginning to establish an industrial base for fusion. Also, industry is becoming involved in systems design studies, and in some cases (in particular, General Atomic Company), industry has direct responsibility for major programs. Such utility and industry participation will grow as the program progresses.

FUSION ENERGY COMMERCIALIZATION PROGRAM

Although full commercial use of fusion energy is not likely to be seen until well into the next century, the fusion program has recently begun to consider commercialization in its planning activities. In so doing, it has focused on achieving a set of commercially usable designs. This task requires resolution of the following issues both in terms of design decisions and in terms of undertaking design-specific research.

- Application—Based on current reactor technology, over 80 percent of the energy produced is in the form of high energy neutrons. This leads to many possible alternative fusion energy applications that profoundly affect the commercial system design.
- Reactor size—The reactor must be large enough to realize economies of scale. At the same time, the user must have sufficient reserve capacity to sustain power supply when the reactor cannot be operated (for example, for routine maintenance or in the event of failure). This suggests that the reactors (either individually or in a multiple grouping) should be fairly small relative to the user's total capacity. A proper balance must be found between these conflicting demands on reactor size.
- Energy storage requirements—Designs which inherently produce pulse power will require major energy storage capability to provide continuous power. A major goal is to minimize the pulsed nature of these designs.
- Complexity—The complexity of current designs must be reduced to assure easy maintenance and high reliability.
- Environmental impact and safety—This area of design concern is of major importance; therefore, design issues are discussed in Chapter 6.

The near-term objectives of the fusion energy commercialization program include the following:

- Identification of the most attractive development options by considering market demand. This task includes both careful analysis of fusion energy applications and specification of the relevant physics and technology development programs.
- Involvement of the developers, vendors, and utilities in providing program guidance. This objective includes the continuation of discussions with user groups concerning the commercialization of fusion devices and strengthening the existing relationships such as with the Electric Power Research Institute.
- Continued efforts to improve the understanding of fusion energy by the population at large.

The fusion energy commercialization program consists of two basic subprograms: Commercial Reactor Designs and Systems Studies. The commercial reactor designs serve as references against which systems studies are performed.

Commercial Reactor Designs

The near-term goal of commercial reactor designs is to establish a framework within which technological requirements, economic objectives, and environmental and safety constraints are evaluated. Emphasis is being placed on electric power and alternative applications reactor designs.

Designs for electric power are being evaluated with regard to reactor size, power cost, reactor cost, reliability, construction leadtime, and other relevant characteristics. The participation of users, vendors, and developers in this evaluation will provide valuable guidance.

The alternative applications area is focusing on two major nonelectrical energy applications: fusion-fission energy systems, and synthetic chemical fuel production reactors. In the case of fusion-fission energy, it is expected that this application may offer significant relief to the barriers facing the utilization of nuclear fission energy in the mid- to long-term. While not as well developed, synthetic chemical fuel production has the potential to reach the larger market for transportable liquid and gaseous fuels.

Systems Studies

The goal of systems studies is to address the socioeconomic aspects and the environmental implications of the fusion energy commercialization process. This task includes evaluation of economic competitiveness, market demand and penetration, costs versus benefits for various fusion power deployment schemes, institutional factors which may influence vendor/customer decisions, and political and social constraints such as environment and safety. The near-term objectives are to evaluate the cost/benefit of fusion energy, to increase the involvement of the eventual market community (i.e., both user and vendor groups), and to establish a broad base for evaluation of environmental and safety issues. Commercialization and environmental and safety studies are addressing these objectives.

The strategy of the commercialization studies is to perform preliminary market analyses, cost/benefit studies, and social perception studies to establish the critical elements of the program and then to expand these activities as the technical program progresses. The involvement of the vendor communities will be developed so that user groups can be called upon to review the anticipated commercial product and to provide recommendations on program directions. Industrial groups will be brought into the program via two mechanisms. One is to operate through the individual program elements as components and systems that can be fabricated by industry are identified. The second is to go directly to industry and award selected commercialization tasks to industrial contractors on a competitive basis.

The objective of systems studies is to ensure that the development and ultimate operation of fusion energy reactors is accomplished with minimum adverse effect on the environment and with maximum safety through: (1) identification and evaluation of potential hazards related to operation of fusion research facilities and power reactors, (2) development of appropriate safety technology and design criteria to control potential hazards, and (3) formal environmental assessments and safety analyses to guarantee that environmental impacts and plant safety implications are fully evaluated for each major facility and for the magnetic fusion program as a whole.

6.
Environmental Implications

The environmental effects of energy systems are becoming increasingly important. An Environmental Development Plan (EDP) for the magnetic fusion program was published in March 1978. This EDP, which provides an initial assessment of key environmental and safety issues, will necessarily be refined (on an annual basis) as fusion technology moves closer to commercialization and a set of mature reactor designs emerges.

The initial environmental assessment of fusion technology indicates five areas of concern:

- Tritium
- Neutron activation of structural materials
- Chemical toxicity
- Occupational hazards
- Accident risks

SURVEY OF THE MAJOR ENVIRONMENTAL IMPACTS OF FUSION

Tritium

If the D-T reaction is used, then radioactive tritium is required to fuel the reactor. An initial loading of perhaps several kilograms would be required and, thereafter, tritium would be bred from lithium in the reactor blanket. Plant inventories of up to 16 kilograms per GWe for the largest conceptual designs have been estimated. Tritium is difficult to contain because it is present in gaseous form and can readily permeate structural materials, particularly at high temperatures.

The half-life of tritium is relatively short, 12 years, and the decay products possess little penetrating power. Therefore, the hazard to man presented by tritium external to the body should not be serious. However, tritium does pose an internal hazard. It can enter humans through several paths such as inhalation and skin absorption of contaminated water vapor, and ingestion of contaminated food or drinking water in which tritium is either taken up directly or through food chain transfer. In-body residence time for tritiated water is about 12 days.

Possible releases of tritium under accident conditions are difficult to estimate quantitatively at the present stage of fusion reactor design development. The most probable accident pathways would be liquid metal fires and magnet or pressurization failures. However, projected tritium releases from fusion power plants are smaller than those currently estimated for fission fuel reprocessing plants (using current technology and normalized to an equivalent energy production), and should result in negligibly small doses to individuals living near the reactor site. The Office of Fusion Energy is presently conducting an R&D program addressing tritium handling and containment. A dedicated facility, the TSTA, is under construction and will allow extensive testing of proposed procedures for tritium handling and containment. This will include the development of techniques for decontaminating equipment and wastes containing tritium.

Neutron Activation of Structural Materials

For reactors using D-T plasmas, the high flux of neutrons from the fusion reactions will have two major effects on the structural materials in the first-wall and blanket: (1) to weaken these materials through structural damage, and (2) to make those materials radioactive. This latter effect, which is due to the absorption of fusion neutrons by the materials' nuclei, depends on several factors: the physical design and operation of the reactor, the type of structural material used, and the presence of impurities and corrosion products. The radioactivity will build up rapidly following the startup of the reactor and will reach its maximum level within a few days to a few weeks. The time required for this induced radioactivity to decay will vary greatly with the structural materials involved. For example, after a 10-year period, vanadium, titanium and aluminum have a radiation hazard to man which is lower by many factors of 10 than that posed by stainless steel.

Radioactive structural materials will require special treatment since they will have to be removed from the reactor periodically (due to structural weakness induced by the fusion neutrons) and upon decommissioning of the plant. How these materials must be handled will depend upon the nature of their radioactivity. For the lower activity materials (e.g., vanadium), it is conceivable that the waste could be stored on-site until its radioactivity decays to a harmless level and the material is recycled. For materials with longer-lived activity, such as steel, material disposal requires more involved procedures. In all cases, however, the absence of long-lived fusion products and actinides from the radioactive materials will make disposal relatively simple compared to the complexities of fission waste disposal.

When considering material requirements for fusion reactors, it is important to note that the blanket and first-wall materials must also provide structural stability. From this point of view, steel is the obvious choice at the present level of development of materials technology. However, the long-term stability of steel in a fusion reactor environment has not yet been demonstrated. Definition of the first-wall material is particularly important since it will contribute activated products into the plasma and coolant in addition to presenting a hazard during replacement and disposal. The Fusion Reactor Materials Subactivity has the goal of developing materials with the required properties.

Chemical Toxicity

Nonradioactive, but chemically toxic materials would be present in all fusion reactors based on current designs. While the quantity of air or water necessary to dilute these chemicals to innocuous levels is much less than for the radioactive substances, the level of chemical toxicity could be substantial. Taking into account the concentrations of these materials allowed in other industrial applications and the possible escape mechanisms, the greatest hazard is probably presented by the use of beryllium (if used to enhance tritium breeding), lithium, other alkali metals, lead, and perhaps mercury (if used in diffusion pumps, as in some early designs).

Lithium will be used for breeding tritium in at least early fusion reactors. Its inherent toxicity is high, and it affects the nervous system. Recent work shows that concentrations of gaseous lithium hydride above 0.025 milligrams per cubic meter cause discomfort and emphysematous changes in the lungs. The Environment and Safety Subactivity has initiated lithium spill studies which will address this issue.

Although it is extremely unlikely that substantial quantities of these toxic materials would ever be released to the environment, it is not possible at present to predict accident probabilities and consequences. The Fusion Reactor Safety Research Task will eventually lead to a calculation of the probabilities of accidental releases of chemically toxic (as well as radioactive) materials.

Occupational Hazards

A fusion reactor facility would maintain safety practices similar to those employed in industries dealing with chemically toxic and radioactive materials. Occupational hazards should not be significantly different from those for other equivalent systems, with two possible exceptions: exposure to low-level magnetic fields and to microwave radiation.

The use of very large magnetic fields within the plasma chamber of the reactor results in moderate leakage fields outside. Workers at the reactor site could therefore be exposed to moderate magnetic fields over prolonged periods of time. These magnetic fields will, however, be substantially reduced by the required shielding of ancillary electrical equipment. Very little is known about the biological effect of such fields. Current studies being supported by the DOE Assistant Secretary for Environment will increase the understanding of the biological effects of magnetic fields and of the design of magnetically shielded environments.

High-level microwave sources may be present in some concepts (for heating), and it is possible that workers could be exposed to low-level microwave radiation. It is, however, relatively simple and inexpensive to provide shielding from this radiation.

A substantial amount of radioactive material will have to be handled at the fusion reactor site during maintenance operations. This is due to the neutron activation of structural materials and to the possibility that some areas of the plant may require tritium decontamination and produce additional low-level waste streams. The volume of contaminated material that must be handled and disposed of will require special procedures to ensure the safety of workers. These procedures will probably be similar to those presently used in the commercial handling of radioactive material at various stages of the fission fuel cycle. The absence of long-lived fission products and actinides from fusion waste will reduce the relative hazards associated with these operations.

Accident Risks

A complete and detailed accident analysis cannot be performed for a technology in as early a stage of development as is fusion. However, a general analysis of the types of accidents that are possible, of the kinds of releases that might result from these accidents, and of the potential consequences can be attempted. Work is being done in these areas by the Reactor Safety Research Task. As more detailed designs become available and systems closer to commercial size and characteristics are built, this analysis will be updated and made more precise.

The most serious potential accidents would probably result from liquid metal fires and magnet or pressurization failures. Any of these could lead to the release of the tritium inventory of the reactor. The release of radioactive structural material would be much less likely because such material is bound up in the structure of the blanket and first-wall. However, it could conceivably be released in gaseous form in the event of a severe liquid metal fire. Liquid lithium may be used in the primary coolant system, and it presents a severe fire hazard if exposed to the atmosphere. Activated corrosion products might also be released in this event. Plasma instabilities, while not expected to lead to radioactive release, could cause reactor structural damage.

Because the spectrum of credible accidents for fusion reactors does not include the potentially most serious accidents possible in fission reactors, and because of the differences in the nature and distribution of radioactive materials, the envisioned accident risks in fusion reactors are currently believed to be smaller than those in fission reactors of equivalent size.

ENVIRONMENTAL AND SAFETY GOALS FOR FUSION REACTOR SYSTEMS

Based upon the currently projected environmental impacts of fusion energy, the following are goals of environmental significance to be achieved in the development of future fusion reactor systems:

- Demonstration of tritium containment and control technology under both normal and accident conditions
- Development of techniques for decommissioning equipment and materials containing tritium
- Improvement of materials performance in fusion reactor environments, and development of low activity structural materials for use in fusion reactors
- Understanding of the biological effects of prolonged exposure to moderate magnetic fields, and development of methods to mitigate any significant risk
- Detailed accident analyses of specific fusion reactor designs
- Reduction of resource requirements for fusion systems, especially for resources of anticipated scarcity
- Minimization of environmental problems caused by radioactive structural materials when a facility is decommissioned
- Possible development of more benign fuel cycles, which produce lower energy neutrons or only charged particles
- Assessment of all potential safety hazards associated with the operation of a fusion reactor

COMPARISON OF FUSION ENERGY WITH OTHER ENERGY-PRODUCING SYSTEMS

The fusion cycle, shown schematically in Figure VI 6-1, is relatively simple and does not require many of the ancillary facilities of the fossil and fission cycles. The routine emissions and overall environmental impact of the entire fuel cycle are significantly less than if fossil fuels were used to generate an equivalent amount of power. A possible exception is natural gas, a fuel which is expected to become increasingly scarce and expensive in the future. Especially important is the fact that all fossil fuels release carbon dioxide to the atmosphere when burned. There is already serious concern about the ramifications of atmospheric carbon dioxide buildup on global climate.

In comparison with fission energy systems, fusion systems will not have the severe prob-

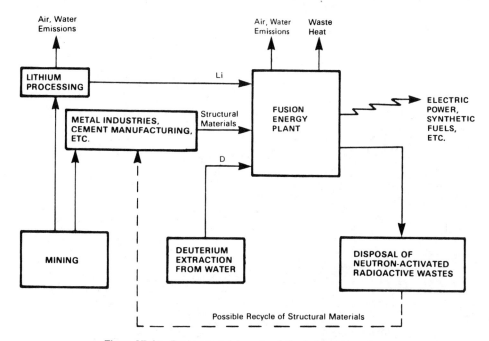

Figure VI-6-1. Environmental impacts of the fusion fuel cycle.

lem of the disposal of radioactive waste. Radioactive material generated by fusion reactors will contain none of the fission products and actinides which make long-term handling and storage such an acute problem. Plant safety hazards will be reduced because of lower operating reactor power density and lower post-shutdown afterheat levels.

As a long-term energy source, fusion should also be compared with solar energy. A meaningful comparison of fusion energy and solar energy for large-scale generation of electricity cannot be made at present because of the early stage of development of both technologies and the fundamental differences between them. However, it is possible to anticipate some important areas of comparison: use of land, materials, and other resources; the environmental impacts of producing the materials used in constructing the facility; the emissions and the safety hazards of routine operation; and the environmental trade-offs required as a result of maintenance, energy storage requirements, and overall economics. Present evidence does not give an obvious advantage to either technology when the full spectrum of such issues is analyzed.

7.
Regional Activities

Figure VI 7-1 shows the ten DOE regions and their respective regional offices. The DOE regional offices are under the jurisdiction of the Assistant Secretary for Intergovernmental and Institutional Relations. Among their primary functions are coordination of DOE activities with the public and State and local governments, planning activities, and provision of an information conduit between the regions and DOE headquarters. These offices play no role in the management or procurement of Office of Fusion Energy activities occurring within a given region; the DOE field and operations offices perform these functions.

Table VI 7-1 lists the FY 1978 Office of Fusion Energy regional funding for four types of prime contractors: DOE laboratory, non-DOE laboratory, industrial/commercial, and university. Figure VI 7-2 summarizes the total funding by region. The skew in the regional funding results from the fact that magnetic fusion research generally requires large laboratory facilities to perform the required experiments. Laboratories with strong fusion expertise are located in Region 1 (MIT), Region 2 (PPPL), Region 4 (ORNL), Region 6 (LASL), Region 9 (LLL and LBL), and Region 10 (HEDL).

Figure VI-7-1. DOE regions and regional offices.

Table VI-7-1. Office of fusion energy FY 1978 prime contractor allocations by DOE region (dollars in thousands).

REGION	STATE	DOE LABORATORIES $K	NON-DOE LABORATORIES $K	INDUSTRY $K	UNIVERSITIES $K	TOTAL $K
1	CT	0	0	0	226	226
	MA	0	0	236	11,322	11,558
	TOTAL	0	0	236	11,548	11,784
2	NJ	42,217	0	72	117	42,406
	NY	1,789	0	0	2,402	4,191
	TOTAL	44,006	0	72	2,519	46,597
3	DA	616	0	0	0	616
	DC*	0	2,818	0	0	2,818
	MD	0	0	0	1,478	1,478
	PA	0	0	1,026	185	1,211
	VA	0	0	0	143	143
	TOTAL	616	2,818	1,026	1,806	6,266
4	GA	0	0	0	113	113
	NC	0	0	0	0	0
	TN	38,695	0	0	0	38,695
	TOTAL	38,695	0	0	113	38,808
5	IL	2,442	0	0	164	2,606
	IN	0	0	0	0	0
	MI	0	0	0	72	72
	OH	21	0	0	0	21
	WI	0	0	0	1,789	1,789
	TOTAL	2,463	0	0	2,025	4,088
6	NM	16,036	0	0	0	16,036
	TX	0	0	41	3,424	3,465
	TOTAL	16,036	0	41	3,424	19,501
7	IA	0	0	0	139	139
	MO	0	0	257	0	257
	TOTAL	0	0	257	139	396
8	CO	0	370	0	0	370
	ID	251	0	0	0	251
	TOTAL	251	370	0	0	621
9	AZ	0	0	0	62	62
	CA	48,676	0	26,462	3,127	78,265
	TOTAL	48,676	0	26,462	3,189	78,327
10	WA	8,540	0	941	533	10,014

*WASHINGTON RESERVE $16,198K

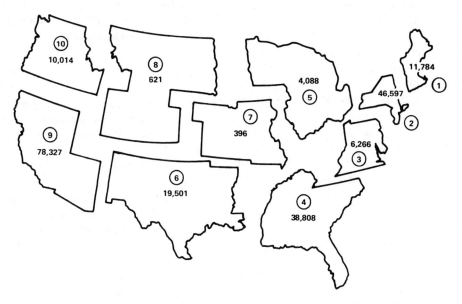

Figure VI-7-2. Office of fusion energy FY 1978 prime contractor funding for DOE regions (dollars in thousands).

8.
International Programs

For each nation developing fusion technology there is a critical trade-off between progress in technology development and protection of its technology. The United States is developing a strategy for international cooperation in magnetic fusion. This development is done in the light of increased potential for cooperation with other countries (e.g., Western Europe and Japan) and is consistent with the fundamental incentives and cautions. The incentives are twofold: (1) enhancement of progress can occur (and has occurred) through the pooling of international fusion expertise, and (2) significant reduction in technology development costs may be achieved through cost-sharing. The areas for caution are our economic security, national defense, and maintenance of sovereignty over our own national fusion program. The United States is accelerating the process of identifying and pursuing cooperative projects, supporting the development of unique foreign programs, and seeking foreign support for its programs.

NATURE OF FUSION DEVELOPMENT OUTSIDE THE UNITED STATES

The fusion technology development community consists of four primary blocs: the United States, the U.S.S.R., Japan, and Western Europe. There are also two important international agencies that provide coordination for and advice on the programs of the different blocs: the International Energy Agency (IEA), and the International Atomic Energy Agency (IAEA). In addition, the People's Republic of China has recently initiated an ambitious fusion energy research program.

U.S.S.R.

In the Soviet program, a single government agency administers both inertial and magnetic fusion. The Soviet magnetic fusion program emphasizes tokamak development, and is oriented toward a fusion-fission hybrid application. There are a number of experimental devices; each focuses on a single experimental objective with two or three small experimental devices supporting a mainline device. The Soviets appear to be attempting larger jumps in technology than the other three blocs. Results to date on the T-10 tokamak indicate that a more gradual approach might be appropriate. There are indications that the experimental programs may be suffering from the lack of adequate consolidation of supporting technology, which results from attempting too rapid an advance.

Japan

The Japanese program is aggressive and has extensive industrial involvement. Two independent, but well-coordinated, governmental agencies fund the program: the Science and Technology Agency, which funds major construction projects; and the Ministry of Education, Sciences, and Culture, which funds university research. The Japanese Atomic Energy Commission and its Nuclear Fusion Council set overall policy.

The Japanese program is broad-based technically and tokamak-oriented. Japan's major construction project (JT-60 tokamak) complements the other tokamak scaling experiments (TFTR, JET in Europe, T-20 in the U.S.S.R.) now under design or construction. Its diverse development program is investigating a number of alternative fusion concepts which are complementary to U.S. efforts.

Western Europe

Western European programs tend to be more research oriented than U.S. or Soviet programs. The Joint European Torus (JET), being designed and constructed by a Euratom team at Culham Laboratory in Great Britain, is the principal cooperative effort. Each nation also focuses its strength on one or two particular aspects as in the following examples:

- Great Britain: Bundle Divertor at Culham
- Federal Republic of Germany (FRG): Materials at KFA (Nuclear Research Center) in Juelich
- France: Neutral Beams and RF heating at Fontenay
- Italy: High Field Tokamaks at Frascati

International Atomic Energy Agency (IAEA)

The IAEA addresses fusion through its International Fusion Research Council (IFRC). This council is made up of prominent fusion experts from 13 nations and is chaired by Dr. R. S. Pease, Director, Culham Laboratory, Great Britain. It is a strong group in terms of its scientific influence and is the most powerful available force for international cooperation.

The IAEA sponsors major meetings and smaller workshops relating to fusion energy and plasma physics. The most recent biennial IAEA Conference was in August 1978, in Innsbruck, Austria; 640 individuals participated.

International Energy Agency

The IEA has a Fusion Power Coordinating Committee. This group is narrower in its base than that of the IAEA (it does not include the Soviets or French) but has mechanisms already in effect for specific project cooperation. Programs being carried out under IEA agreements include the Large Coil Project at Oak Ridge National Laboratory, cooperative neutron sources for materials testing, and plasma-surface reaction research on the TEXTOR (tokamak) device in Juelich, Germany.

U.S. INTERNATIONAL COOPERATION

The types of interaction possible for international cooperation among the major blocs are shown in Figure VI 8-1, and are discussed in the following sections.

Individual Involvement

Individual involvement has occurred via a bilateral agreement with the U.S.S.R., direct exchange of information and scientific personnel on a multinational basis, and conferences and workshops sponsored by IAEA and other organizations.

The major international individual exchange program of the Office of Fusion Energy is with the U.S.S.R. This program was initiated by a bilateral agreement signed in February 1974. In that year, the United States sent scientists to the U.S.S.R. for 68 manweeks of effort, and Soviet scientists spent 44 manweeks in the United States. These exchanges grew to about 168 manweeks in 1978, and include all aspects of magnetic fusion energy activity from basic plasma physics theory and experiments to fusion technology (e.g., neutral beam and superconducting magnet systems).

The Joint Fusion Power Coordinating Committee (JFPCC), co-chaired by Edwin E. Kintner, Director of the Office of Fusion Energy, and Evgenity P. Velikhov, Deputy Director of the Kurchatov Institute of Atomic Energy in Moscow, plans and manages the

U.S.–U.S.S.R. exchange program. The fifth annual JFPCC meeting was held at the Kurchatov Institute for Atomic Energy from May 15 to May 17, 1978, and produced the following results:

- The 1978 exchange program highlights include:
  — An impurity control technique (glow discharge cleaning with helium) was tested on the U.S.S.R. T-12 tokamak before being introduced into U.S. devices. The success of these cleaning tests will result in substantial savings in time and money for U.S. tokamak R&D.
  — Transport measurements on the T-4 tokamak provided important suggestions for the Impurity Studies Experiment to be performed at Oak Ridge National Laboratory.
  — Tandem mirror concept workshops provided helpful discussions on technical issues addressing the scientific feasibility of this device.
- The U.S.S.R. fusion program thrust continues to be in the area of fusion-fission hybrid technology emphasizing tokamak devices. The head of tokamak research at Kurchatov Institute felt that the basic physics understanding of tokamak plasma was about to undergo revolutionary progress.
- A special protocol was signed for establishing cooperation in materials research.
- Velikhov proposed that an internationally developed device on the scale of the U.S.S.R. T-20 tokamak be designed and constructed under the auspices of the IAEA. This proposal has advanced to the point at which each of the four major blocs is involved in a joint feasibility study.
- The Soviets agreed with a draft memorandum of understanding prepared by the Los Alamos Scientific Laboratory suggesting joint experiments on imploding liners.
- The U.S.S.R. agreed with a U.S. proposal for a meeting in the U.S.S.R. to review the Soviet superconductivity program.
- Two separate workshops will be organized in the U.S.S.R. to seek new directions in optimizing toroidal and linear systems for practical applications.
- The 1979 exchange schedule was negotiated and agreed on, yielding about 150 manweeks of personnel exchange from each side.

Hardware Contributions

An example of the hardware contributions category was the 1977 loan of a spare metal shell for ORMAK (Oak Ridge tokamak) to Culham Laboratory in Great Britain for use in reverse field pinch experiments. Other examples include routine and specialized commercial components

ENTITY TYPE	BILATERAL AGREEMENTS			INTERNATIONAL AGENCIES	
	USSR	JAPAN	EUROPEAN COMMUNITY	IAEA	IEA
Personnel Exchange	Formal	Formal	Informal	Workshops & Meetings	–
Coupled Experiments	Minimal	EBT/NBT Doublet III	Informal	–	Textor
Allocated Experiments	?	EBT/NBT?	RFP? Stellarator?	–	–
Technology Facilities	–	Under Discussion (FMIT, etc.)	Computers	–	LCP
Major Steps	–	–	Ignition Test (Zephyr)	Intor (Design)	–
Others	–	Joint Planning Joint Institute	Joint Institute?	–	–

Figure VI-8-1. Status of U.S. International magnetic fusion research interactions with major blocks.

such as electrical switchgears and certain conventional microwave power sources.

Coordinated and Bilateral Experiments

Coordinated experiments are those whose programs are planned in a coordinated fashion although each is separately funded and operated by its sponsoring country. Informal technical coordination through the usual workings of the fusion technical community comprise most of this type of cooperation. Bilateral experiments are those in which two countries formally share sponsorship and management of an experiment in a formal manner.

There are two major bilateral activities either in progress or planned: (1) cooperative projects under the sponsorship of IEA, and (2) projects under the proposed U.S.–Japan bilateral agreement.

IEA Programs. Cooperative projects under IEA's sponsorship are as follows:

- Materials Testing—The 1978 joint effort between Canada, Japan, Sweden, Switzerland, and the United States was based on the design and development of an intense neutron source (INS). Due to funding limitations, the U.S. INS program was cancelled for FY 1979. However, initial agreement has been reached on a material testing program using the FMIT facility currently under development at Hanford Engineering Development Laboratory.
- Large Coil Project (LCP)—Development of six large tokamak coil segments is continuing on schedule. The LCP participants are Euratom (agreement signed 6 October 1977), Japan (May 1978), Switzerland (September 1978), and the United States (three coils to be constructed by U.S. industry). Oak Ridge National Laboratory, the project operating agent, is constructing a facility where all coil segments will be tested.
- Plasma-Wall Interactions—An agreement to investigate plasma-wall interactions was signed October 6, 1977, by the United States, Euratom, Switzerland, and Turkey. It remains in force through 1985 or until completion of the experimental phase. The agreement states that the Operating Agent (Euratom/KFA, Juelich, Germany) is responsible for funding, design, construction, and operation of the TEXTOR (tokamak) device. The program consists of two phases: Phase I (3 to 4 years duration) deals with the final design and construction, and Phase II with operation. To date, the United States contribution includes participation in the inaugural meeting of the TEXTOR Executive Committee in January 1978 and the visit of two scientists to Juelich in FY 1979
- Garching Ignition Test Reactor (Germany)—An agreement for U.S. participation in the design of an ignition test reactor (to be used to examine helium nuclei heating characteristics) is in the negotiation stage. U.S. scientists are involved in the feasibility studies which will provide a technical basis for subsequent participation decisions.

U.S.–Japan Bilateral Initiative. Important negotiations are currently under way between the United States and Japanese governments. This initiative, announced on September 20, 1978, is aimed at "commercializing nuclear fusion by at least the early years of the 21st century." It proposes a large joint effort on the Doublet III peanut-shaped, noncircular, cross-section device at General Atomic Company, an extensive sharing of diagnostic techniques and data from two future experiments—TFTR and the Japanese JT-60, and a joint theoretical institute for the study of plasma physics. A formal government-to-government agreement is to be signed.

Multilateral Experiments

The most advanced level of international cooperation is a multilateral experiment equivalent to the U.S. Engineering Test Facility. This type of initiative was recently proposed by the Soviet academician Velikhov. He proposed that an internationally developed device on the scale of the U.S.S.R. T-20

Table VI-8-1. Some specific benefits associated with past international cooperative efforts in magnetic fusion research.

BLOC	AREA	SPECIFIC BENEFIT
USSR	EXPERIMENT	The Tokamak concept (technical results which influenced our program). Proof of the minimum-B principle for mirrors (1963). Evidence for stream stabilization of loss cone instabilities. Advanced communication ultimately led to the successful results on 2 X IIB in June 1975 establishing the mirror concept as the principal alternate to tokamak in the US. Parallel development and evaluation of the tandem mirror concept that permitted rapid implementation of the TMX experiment (1976). A US/USSR fusion exchange program for 1980 was negotiated (1979) to cover design and testing of elements of experimental fusion reactors.
USSR	THEORY	Neoclassical Transport Theory for Tokamaks, conceived by Galeev and Sagdeev. Neoclassical Transport Theory for Stellarators, first formulated jointly by US/USSR theorists. Anomalous Transport Theory for Tokamaks, initiated by USSR, subsequent development by US. Theories of Disruptive Instabilities in Tokamaks initiated by PPPL, important analytical contributions by USSR then detailed computational models successfully developed at both ORNL and PPPL. Development of the theory of radial transport in tandem mirrors and sharing this with us before publication (1978). A US/USSR fusion exchange program for 1980 was negotiated (1979) to cover theoretical work in the confinement and heating of plasmas in closed and open magnetic systems.
USSR	TECHNOLOGY	Insights into problems peculiar to startup and operation of a superconducting toroidal field system in a Tokamak (T-7). Concepts of conductor and coils. Results of testing promising conductors (for T-15). Data on pressure drop and heat transfer in forced flow of two-phase helium. Invention and development of the direct-extraction negative ion source that may be the basis for advanced neutral beams for mirror devices being developed at BNL (ongoing). Development of double-charge exchange negative ion source based on sodium cells: also molybdenum grids (ongoing). A US/USSR fusion exchange program for 1980 was negotiated (1979) to cover engineering problems in reactors and studies for selection of designs for fusion reactors.
EUROPE	THEORY	Cooperative development of theories of micro-instabilities in toroidal plasmas at Culham (UK). Cooperation with PPPL and LASL in a reversed-field pinch theory (Culham, UK) and support from Italian scientists on LASL ZT-40 experiment. The Errato MHD stability code was developed at Laussane, given to ORNL, and used by PPPL to validate its PEST code. Tokamak transport codes developed at Garching were used in early stages of US involvement with Tokamaks.
EUROPE	TECHNOLOGY	The Bundle divertor concept originated at Culham (UK). Neutral Beam development at Culham. A principal benefit from European cooperation has been in the area of ICRH Heating (TFR) which is the basis of the Phaedrus tandem mirror experiment at the U. of Wisconsin. A French physicist (Jacquinot) helped to design this experiment.
JAPAN	THEORY	An institutional basis for collaboration with Japanese fusion plasma theorists is being established. The objective is to focus international cooperation on fusion theory.
JAPAN	EXPERIMENT THEORY	Our cooperative effort with Japan is more recent but promises to be highly beneficial with the completion of the tandem mirror experiment Gamma 6 at Tsukuba. This is the first working tandem mirror device in the world. Its first results have demonstrated the formation of the tandem mirror bipolar potential profile that is the heart of the idea. Agreement has been reached (1979) on a 5-year Doublet III upgrade to operate with increased magnetic field, plasma current, neutral beam power, and possibly, a D-shaped vacuum vessel.

tokamak be designed and constructed under the auspices of the IAEA. The device would be a next-generation device employing superconducting magnets and allowing for hybrid operation. The four major blocs are currently participating in a feasibility study of this proposal.

The international program of the Office of Fusion Energy is currently in a transitional stage between being an academically oriented, information-exchange program and a device-technology/development-oriented program. Based upon benefits derived from past international cooperative efforts (see Table VI 8-1), international cooperation is expected to become stronger in the future.

Part VII

Reference Information

- Glossary
- Definition of RD and D Scale-up Phases
- Abbreviations and Acronyms
- References
- Energy Units and Conversion Factors
- Milestone Symbols
- Index

Glossary

Accelerator (particle accelerator). A device for imparting large kinetic energy to electrically charged elementary particles such as electrons, protons, deuterons, and helium ions through the application of electrical and/or magnetic forces. Common types of particle accelerators are direct voltage accelerators, cyclotrons, betatrons and linear accelerators.

Acid rain. Abnormally acidic rainfall, most often dilute sulfuric or nitric acid.

Acre-foot. Amount of water covering 1 acre of surface to height of 1 foot; equals 326,000 gallons.

Actinides. A group name for the series of radioactive elements from element 89 (actinium) through element 103 (lawrencium). The series includes uranium and all the man-made transuranic elements.

Aerodynamic nozzle process. A method of uranium enrichment. A gaseous mixture of UF_6 and hydrogen is forced to flow at high velocity in a semicircular path, establishing centrifugal forces which tend to separate the heavier molecules from the lighter ones.

Aerosol. A suspension of fine solid or liquid particles in gas; smoke, fog, and mist are examples.

Alluvial bottom lands. Lands where a former river bed has deposited sand, gravel, and earth.

Anaerobic. Living, active, or occurring in the absence of free oxygen.

Anthropogenic. Of, relating to, or influenced by the impact of humans or human society on nature.

Aquifer. A water-bearing stratum of permeable rock, sand, or gravel.

Base load generating costs. The cost of electricity generated by an electric facility designed to operate at constant output with little hourly or daily fluctuation.

Batch processing. Any process in which a quantity of material is handled or considered as a unit. Such processes involve intermittent, as contrasted with continuous, operation.

Binary cycle. An energy recovery system based on the transfer of heat from one fluid (e.g., hot brine from a geothermal well) to a second fluid (e.g., pure water or an organic liquid) from which the heat is ultimately extracted for use.

Bioconversion. The conversion of organic wastes into methane (equivalent to natural gas) through the action of microorganisms.

Biomass. In energy context, with potential for heat or energy generation.

Bituminous coal. An intermediate rank coal with low to high fixed carbon, intermediate to high heat content, a high percentage of volatile matter, and low percentage of moisture.

Blowdown water. A side stream of water removed from the cooling system in order to maintain an acceptable level of dissolved solids.

Blanket. A layer of fertile material such as uranium-238 or thorium 232 that is placed around the core of a fission or fusion reactor. Its major function is to produce fissile isotopes from fertile blanket material.

Boiling-water reactor (BWR). A light water reactor that employs a direct cycle; the water coolant that passes through the reactor is converted to high-pressure steam that flows directly through the turbines.

Bottoming cycle. Generating power by exploiting the (otherwise wasted in atmospheric rejection) heat content of the engine working fluid. When the working fluid is liquid boiling at a relatively low temperature, bottoming cycle generation is relatively easy.

Breeder reactor. A nuclear reactor that produces more fissile material than it consumes. In fast breeder reactors, high-energy (fast) neutrons produce most of the fissions, while in thermal breeder reactors, fissions are principally caused by low-energy (thermal) neutrons.

Breeding ratio. The ratio of the number of fissionable atoms produced in a breeder reactor to the number of fissionable atoms consumed in the reactor. The "breeding gain" is the breeding ratio minus 1.

British thermal unit (Btu). The amount of energy necessary to raise the temperature of one pound of water by one degree Fahrenheit, from 39.2 to 40.2 degrees Fahrenheit.

Burnup. The amount of thermal energy generated per unit mass of reactor core fuel load. Though burnup will vary with core management practice, it typically is one measure of reactor operating efficiency.

Bus-bar cost. The cost of electric power as it leaves the generating facility.

Candu. A nuclear reactor of Canadian design, which uses natural uranium as a fuel and heavy water as a moderator and coolant.

Capacity factor. The ratio of the amount of product (e.g., electrical energy or geothermal brine) actually produced by a given unit of time to its maximum production rate over that period. Also called "load factor."

Capital charges. The annualized costs of borrowing plus the amortization of investment and allowance for taxes.

Carrying charge. Generally the cost of carrying capital. In nuclear power context, frequently used to refer only to cost of carrying fuel inventory.

Catalysis. A modification—especially an increase—in the rate of a chemical reaction that is induced by material that remains unchanged chemically at the end of the reaction.

Centrifuge (Isotope separation). A new isotope enrichment process that separates lighter molecules containing uranium-235 from heavier molecules containing uranium-238 by means of ultra high speed centrifuges. The process is now being successfully exploited by the URENCO consortium in Western Europe.

Cesium-137. A radioactive isotope of the element cesium that is a common fission product. Its half-life is 30 years.

Chain reaction. A reaction in which one of the agents necessary to the reaction is itself produced by the reaction so as to cause similar reactions. In a nuclear reactor (or bomb) a neutron plus a fissionable atom cause a fission, resulting in a number of neutrons which in turn cause other fissions.

Clinch River Breeder Reactor (CRBR). A proposed demonstration of the liquid metal fast breeder reactor. The CRBR, which would operate on a plutonium fuel cycle and be cooled by molten sodium, would have an electrical output of about 350 MWe.

Coal conversion. Now most commonly used to denote coal gasification and liquefaction, but strictly speaking, any process that transforms coal into a different form of energy (e.g., electricity).

Coal gas. Low, intermediate, or high Btu content gases produced from coals.

Coal liquefaction. The conversion of coal into liquid hydrocarbons and related compounds by the addition of hydrogen.

Cogeneration. The generation of electricity with direct use of the residual heat for industrial process heat or for space heating.

Combined cycle. The combination, for instance, of a gas turbine followed by a steam turbine in an electrical generating plant.

Competitive price. Price of a product or service established in a competitive market.

Constant dollars. Dollar estimates from which the effects of changes in the general price level have been removed, reported in terms of a base year value.

Convection. The circulatory motion that occurs in a fluid not at a uniform temperature owing to the variation of the fluid's density and the action of gravity. The term may also denote the transfer of heat which accompanies this motion.

Conversion efficiency. The percentage of total thermal energy that is actually converted into electricity by an electric generating plant.

Conversion ratio. The ratio of the number of atoms of new fissionable material produced in a converter reactor to the number of atoms of fissionable fuel consumed. See "breeding ratio."

Converter reactor. A reactor that produces some fissionable material, but less than it consumes. In some usages, a reactor that produces a fissionable material different from the fuel burned, regardless of the ratio. In both usages the process is known as conversion.

Criticality accident. An unintended nuclear chain reaction which, depending upon the materials involved and their physical configuration, can release varying amounts of energy and radioactivity at various rates.

Curie. A measure of intensity of the radioactivity of a substance; i.e., the number of unstable nuclei that are undergoing transformation in the process of radioactive decay. One curie equals the disintegration of 3.7×10^{10} nuclei per second, which is approximately the rate of decay of one gram of radium.

Current dollars. Dollar values that have not been corrected for changes in the general price level.

Daughter products. Atoms formed as a result of the radioactive decay of other atoms. Daughter products may themselves be radioactive, continuing to decay into other daughter products.

Denature. The addition of a nonfissionable isotope to fissionable material to make it unsuitable for use in nuclear weapons without extensive processing.

Depletion allowance. A tax credit based on the permanent reduction in value of a depletable resource that results from removing or using some part of it.

Deuterium. A hydrogen isotope of mass number two (one proton and one neutron).

Devonian shale. Gas-bearing black or brown shale of the Devonian geologic age; underlies over 160,000 m^2 of the Appalachian Basin.

Discount rate. A rate used to reflect the time value of money. The discount rate is used to adjust future costs and benefits to their present day value. The effect of the discount rate (r) on the present value of a cost or benefit at time (t) in the future (C_t) is given by the expression.

$$C_{(t)} \left[\frac{1}{(1+r)^t} \right]$$

The selection of discount rates appropriate to particular situations is a matter of debate among economists, although 10 percent has been used in government calculations.

Diffusion plant. A type of plant for uranium enrichment

which uses gaseous uranium hexafluoride as its feedstock, and in which isotopic separation occurs from the differential rates at which UF_6 atoms diffuse through membranes.

District heating. A system for providing heat in commercial and residential buildings in which heat or byproduct from an electrical generating unit is distributed via an underground pipeline system.

Dose-response curve. A graph or function showing the relationship between dose or stimulus and its effect on living tissue or organism(s).

Draft environmental impact statement. A preliminary version of an environmental impact statement (EIS) which is made available for public review and comment prior to preparation of a final EIS.

Economies of scale. A production process is characterized by economies of scale when costs per unit output fall as plant size is increased.

Economic rent. Income accruing beyond that needed to bring a firm into, or keep it, in operation; arises from a variety of advantages enjoyed by the firm (e.g., location, quality of inputs, ability, changes in public policy or other external factors, often referred to as "windfalls").

Elasticities of demand. The arithmetic relations used by economists in quantifying the change in demand for a commodity in response to a change in another economic quantity.

Electrical grid. A net or system that connects a number of power-generating stations and users of electricity (load centers) to permit interchanges and more economical utilization of participating facilities.

Emission charge. A charge levied on emitters of pollutants per unit of pollutant or per unit of input of the emitting substance.

Enhanced oil recovery. Technology for lifting oil from wells beyond amount recovered through natural reservoir pressure. Commonly defined to exclude simple devices such as injecting water as a driving force.

Enriched uranium. Uranium in which the percentage of the fissionable isotope uranium-235 has been increased above that contained in natural uranium.

Enrichment. The process by which the percentage of the fissionable isotope uranium-235 is increased above that contained in natural uranium.

Entitlement. A device for allocating the higher cost of imported oil among refiners so that every refiner pays the price for crude he would pay if he purchased it in the proportion between domestic and imported oil that prevails for the industry as a whole.

Environmental Impact Statement (EIS). An analysis of environmental effects of a proposed action. Under the U.S. National Environmental Policy Act of 1969, an EIS must be prepared in conjunction with any proposed action of the federal government that would significantly affect the quality of the environment.

Epidemiology. A branch of medical science that deals with the incidence, distribution, and control of disease in a population.

Ethanol. A colorless, volatile, flammable liquid of the chemical formula C_2H_5OH, which is found in fermented and distilled liquors; also made from petroleum hydrocarbons.

Eurocurrency (Xenocurrency). A balance expressed in a currency held in a bank outside the territory of the nation issuing the currency; originally applied to deposits held in European banks, but now widely used in the generic sense.

Expected benefit. In cost-benefit analysis, the expected benefit (EB) is, broadly speaking, the weighted average of the monetary benefits of the possible outcomes (B), the weight for each being its probability of occurrence (P): $EB = P_a B_a + P_b B_b + \ldots P_n B_n$.

Expected cost. In cost-benefit analysis, the expected cost (EC) is, broadly speaking, the weighted average of the monetary costs of the possible outcomes (c), the weight for each being its probability of occurrence (P): $EC = E_a C_a + E_b C_b + \ldots E_n C_n$.

Fast breeder reactor. A breeder reactor that relies on very energetic ("fast") neutrons.

Feedstock. Any raw or processed material input stream fed into some production process.

Fertile material. A material, not itself fissionable by thermal neutrons, that can be converted into a fissile material by irradiation in a reactor. There are two basic fertile materials, uranium-238 and thorium-232. When these materials capture neutrons, they are partially converted into plutonium-239 and uranium-233, respectively.

Fissile material. Atoms such as uranium-233, uranium-235, or plutonium-239 that fission upon the absorption of a low energy neutron.

Fission. The splitting of an atomic nucleus with the release of energy.

Fission products. The medium-weight, highly radioactive daughter products of U-235 fission.

Flashing. The rapid change in state from a liquid to a vapor without visible boiling, resulting usually from a sudden reduction in the pressure maintained on a hot liquid.

Flue Gas Desulfurization (FGD). The removal of sulfur oxides from the gaseous combustion products of fuels before they are discharged into the atmosphere.

Fluidized bed. A body of finely divided particles kept separated and partially supported by gases blown through or evolved within the mass, so that the mixture flows much like a liquid.

Fluidized bed combustion. The process of burning finely divided coal in a sulfur-capturing bed of limestone-based particles, with both coal and limestone particles supported by a stream of air or other gas.

Fly ash. Fine solid particles of noncombustible ash entrained in the flue gases arising from the combustion of carbonaceous fuels. The particles of ash may be accompanied by combustible unburned fuel particles.

Focusing collector. A solar energy collector that gathers and concentrates the incoming radiation.

Food chain. An arrangement of the organisms of an ecological community according to the order of predation in which each uses the next, usually lower, member as a food source.

Front-end costs. Not as clearly defined as back-end costs; ambiguously used to cover costs arising in nuclear power generation through in-core radiation.

Fuel bank. A proposed international uranium stockpile from which nations could obtain uranium, enriched if necessary, in the event of an interruption of normal supply.

Fuel cell. A device for combining hydrogen and oxygen in an electrochemical reaction to generate electricity. Chemical energy is converted directly into electric energy without combustion.

Fuel cycle. The various processing, manufacturing, and transportation steps involved in producing fuel for a nuclear reactor, and in processing fuel discharged from the reactor. The uranium fuel cycle includes uranium mining and milling, conversion to uranium hexafluoride (UF_6), isotopic enrichment, fuel fabrication, reprocessing, recycling of recovered fissile isotopes, and disposal of radioactive wastes.

Fusion. The combining of certain light atomic nuclei to form heavier nuclei with the release of energy.

Gas-centrifuge process. A method of isotopic separation in which heavy gaseous atoms or molecules are separated from light atoms or molecules by centrifugal force.

Gas core reactor. A nuclear reactor in which the fuel is in gaseous form (uranium hexafluoride).

Gaseous diffusion. A process used to enrich uranium in the isotope uranium-235. Uranium in the form of a gas, uranium hexafluoride (UF_6), is forced through a thin porous barrier. Since the lighter gas molecules containing uranium-235 move at a higher velocity than the heavy molecules containing uranium-238, the lighter molecules pass through the barrier more frequently than do the heavy ones, producing a slight enrichment in the lighter isotope. Many stages in series are required to produce material enriched sufficiently for use in a light water reactor.

Gasifier. Equipment for converting coal to gas.

Gasohol. A program established by the state of Nebraska to aid in the development of an alternative automotive fuel containing a blend of 10 percent agriculturally derived ethyl alcohol and 90 percent unleaded gasoline. Trademark of the Agricultural Products Industrial Utilization Committee.

Genetic damage. Injury to the genetic material of a cell. Often denotes specifically damage to an organism's germ cells which can be passed on to offspring.

Geopressured brines. Formation water contained in some sedimentary rocks under abnormally high pressure; unusually hot and may be saturated with methane.

Geopressured reservoir (geothermal). A hydrothermal reservoir in which the pore fluid is confined under pressure significantly greater than normal hydrostatic pressure, developed principally by the weight of overlying rocks and sediments. Also called "overpressured" and "geopressurized" reservoirs.

Geothermal energy. The heat energy in the earth's crust which derives from the earth's molten interior. Can be tapped as steam, or by injection of water to form steam.

Geothermal gradient. The rate at which the temperature of the earth increases with depth below its surface. This varies widely from place to place, but the average or "normal" geothermal gradient is about 30 degrees Celsius per kilometer of depth (16.5 degrees Fahrenheit per thousand feet).

Global insolation. The total insolation on a horizontal surface on the earth, averaged over some specified period of time.

Greenhouse effect. A possible increase in the average temperature of the earth's atmosphere that might be caused by the release of carbon dioxide from fossil fuel combustion.

Gross National Product (GNP). The total market value of the goods and services produced in a national economy, during a given year, for final consumption, capital investment, and governmental use. (Note that GNP does not include the value of intermediate goods and services sold to producers and used in the production process itself.)

GNP Deflator. A price index that translates GNP in current dollars into GNP in constant dollars, i.e., corrects for change in GNP due to price changes.

Half-life. The period required for the distintegration of half of the atoms in a given amount of a specific radioactive substance.

Hard currency. A currency easily exchangeable anywhere at the official rate of exchange.

Heat engine. Any device that converts thermal energy (heat) into mechanical energy.

Heat pump. A device for transferring heat from a substance at one temperature to a substance at a higher temperature, by alternately vaporizing and liquefying a fluid through the use of a compressor.

Heavy oil. Crude oil of such low viscosity (usually 15° or less) that it does not flow freely enough to be lifted from a reservoir reached by the drill.

Heavy water. Water containing significantly more than the natural proportion (1 in 6500) of heavy hydrogen (deuterium) atoms to ordinary hydrogen atoms. Heavy water is used as a moderator in certain reactors because it slows down neutrons effectively and also has a low cross section for absorption of neutrons.

Heavy Water Reactor (HWR). A nuclear reactor that uses heavy water as its coolant and neutron moderator. Heavy water is water in which the hydrogen of the water molecule consists entirely of the heavy hydrogen isotope of mass number two (one proton and one neutron).

High-BTU gas. Gas that has a heating value between 900 and 1,100 Btu per cubic foot.

High-enriched uranium. Uranium sufficiently enriched in the fissionable isotope 235 to be usable as weapons material without further enrichment or processing.

High-level waste. A by-product of the operation of nuclear reactors that includes a variety of aqueous wastes from fuel reprocessing and their solidified derivatives, such as alkaline aqueous waste, calcine, crystallized salts, insoluble precipitates, salts of cesium and strontium extracts, and coating wastes from chemical decladding of fuel elements.

High-Temperature Gas-Cooled Reactor (HTGR). A graphite-moderated, helium-cooled advanced reactor that utilizes the thorium fuel cycle. The initial core is fueled with a mixture of fully enriched uranium-235 and thorium. When operated in the recycle mode, the reactor is refueled with a mixture of uranium-233 (produced from thorium) with the balance of the fissile material provided from an external source of fully enriched uranium-235.

Hot dry rock (geothermal). Naturally heated but unmelted rock sufficiently low in either permeability or pore-fluid content that wells drilled into it do not yield either hot water or steam at commercially useful rates. to be compared with "hydrothermal reservoir."

Hydrolysis. A chemical decomposition in which a compound is broken up and resolved into other compounds by reaction with water.

Hydrostatic head. The pressure created by the weight of a column of water.

Hydrothermal reservoir. A body of porous, permeable rock, gravel, or soil containing natural steam or naturally heated water at a temperature significantly above the average temperature at the earth's surface.

Income elasticity of demand. A measure of the change of consumer demand for a particular good with change in income. It is defined as the ratio of the percentage change in demand to the percentage change in consumer income. If income elasticity of demand for a commodity is low, there will be little change in consumer demand for that commodity in response to changes in income.

Indifference cost. A cost at which either of two alternatives is equally attractive, i.e., the user is indifferent as to choice.

Infrastructure. The underlying or associated foundation or basic framework required to utilize a given product or service (as of a system or organization).

Intermediation (financial). Process facilitating the flow of funds from original sources to ultimate borrowers.

Ion. An atom or group of molecularly bound atoms that carries a positive or negative electric charge is a result of having lost or gained one or more electrons.

Ionizing radiation. Radiation, including alpha, beta, or X-rays, sufficiently energetic to knock off the electrons—or "ionize"—the atoms of matter while in transit through that matter.

Insolation. The solar energy per unit time (power) crossing a unit area.

Isotope. One of two or more atoms with the same atomic number (i.e., the same chemical element) but with different atomic weights. Isotopes usually have very nearly the same chemical properties but somewhat different physical properties.

Isotope separation. The process of separating the normally and naturally commingled, and chemically identical, isotopes of a given element.

Kerogen. A solid, largely insoluble organic material, occurring in oil shale, which yields oil when it is heated in the absence of oxygen.

Kinetic energy. The energy associated with an object's motion; varies directly in proportion with the object's mass and with the square of its velocity.

Laser isotope separation. A new isotope enrichment process now in the development stage. Atoms of uranium-235, or molecules containing them, would be selectively ionized or excited by lasers, allowing physical or chemical separation of one of the isotopes.

Levelized cost. The spreading out of a cost or change in even payments over a period of time.

Life-cycle costing. Distributing one-time cost of acquisition of a piece of equipment over its estimated lifetime to calculate annual cost.

Light Water Reactor (LWR). A nuclear reactor that uses ordinary water as both a moderator and a coolant and utilizes slightly enriched uranium-235 fuel. There are two commercial light water reactor types—the boiling-water reactor (BWR) and the pressurized-water reactor (PWR).

Lignite. The lowest rank coal with low heat content and fixed carbon, high percentage of volatile matter and moisture; an early stage in the formation of coal.

Linear no-threshold model. A dose-response model which assumes that there is no threshold or "safe" dose level and that the effects are directly proportional to the size of the dose.

Liquefied Natural Gas (LNG). Natural gas cooled to -259 degrees Fahrenheit so that it forms a liquid at approximately atmospheric pressure. As natural gas becomes liquid, it reduces in volume nearly 600-fold, thus allowing economical storage and making long-distance transportation economically feasible. Natural gas in its liquid state must be regasified and introduced to the consumer at the same pressure as other natural gas. The cooling process does not alter the gas chemically, and the regasified LNG is indistinguishable from other natural gases of the same composition.

Liquefied Petroleum Gas (LPG). A gas containing certain specific hydrocarbons that are gaseous under normal atmospheric conditions but that can be liquefied under moderate pressure at normal temperatures. Propane and butane are the principal examples.

Load. The amount of energy or electric power required of an energy system during any specified period of time and at any specified location or locations.

Load factor. The annual average output of a utility system divided by its maximum potential output.

Long-wall mining. A mining method widely used outside the United States in which the entire coal seam is removed at a "long wall" and the roof allowed to cave behind the coal face.

Low BTU gas. Gas that has a heating value between 100 and 300 Btu per cubic foot.

Low-enriched uranium. Uranium that has been enriched in the fissionable isotope 235, but not sufficiently enriched for weapons use.

Low-level waste. Generally a solid by-product of special nuclear materials production, utilization, and research and development. Examples of solid low-level waste are discarded equipment and materials, filters from gaseous waste cleanup, ion-exchange resins from liquid waste cleanup, liquid wastes that have been converted to solid form by techniques such as mixing with cement and miscellaneous trash. Low-level liquid waste is generally decontaminated and released under controlled conditions.

Low sulfur coal. Coal with sulfur content generally less than 1.0 percent.

Magnetohydrodynamic power. Generation of electricity by moving hot, partially ionized gases through a magnetic field, where they are separated by electrical charge, generating an electric current that is then collected by electrodes lining the expansion chamber. Usable in combination with conventional power generating plants.

Marginal cost pricing. Charging users for all units consumed at the rate that corresponds to the cost of the final unit that needs to be supplied to meet the demand. Sometimes referred to as "incremental" pricing.

Mass number. The total number of neutrons and protons in the nucleus of an atom.

Medium BTU gas. Gas that has a heating value of between 300 and 600 Btu per cubic foot.

Megawatt. The unit by which the capacity of production of electricity is usually measured. A megawatt is a million watts or a thousand kilowatts.

Meltdown. A nuclear reactor accident in which the radioactive fuel overheats, melting the fuel cladding and threatening release of the core radioactive inventory to the environment.

Methane. The lightest compound of the paraffin series (hydrocarbons) with the formula CH_4; occurs naturally in oil and gas wells and is the principal constituent of natural gas.

Metric ton. 2,205 pounds.

Micron. One millionth (10^{-6}) of a meter.

Milling (Uranium processing). A process in the uranium fuel cycle in which ore that contains only about 0.2 percent uranium oxide (U_3O_8) is concentrated into a compound called yellowcake, which contains 80-90 percent uranium oxide.

Moderator. A material, such as ordinary water, heavy water, or graphite, used in a reactor to slow down high-velocity neutrons, thus increasing the likelihood of further fission.

Mutagen. A substance or agent that tends to increase the frequency or extent of genetic mutation.

Natural uranium. Uranium in which the naturally occurring proportions of the isotopes of mass numbers 234, 235, and 238 have not been artificially altered as, for example, through enrichment.

Neutron. An elementary particle with approximately the mass of a proton but without any electric charge. It is one of the constituents of the atomic nucleus. Frequently released during nuclear reactions and, on entering a nucleus, can cause nuclear reactions including nuclear fission.

Neutron capture. The absorption of a neutron by the nucleus of an atom. Depending on the kind of nucleus and upon the speed of the incident neutron, three outcomes are possible: The neutron can be absorbed, leaving a stable isotope of the original element; the new nucleus can decay to another of approximately the same size; or there can be fission-breakup into two smaller nuclei.

Neutron spectrum. The velocity distribution of the free neutrons in a chain-reacting assembly—for example, a reactor core.

Nominal demand. Demand expressed in current dollars.

Nontransuranic wastes. All of the constituents of nuclear waste other than the transuranic elements. The most highly radioactive of these are the so-called fission products—the approximately equal fragments into which U-235 splits.

Nuclear fuel cycle. The series of steps involved in supplying fuel for nuclear power reactors and disposing of it after use. It can include some or all of the following stages: mining, refining or uranium or thorium ore, enrichment, fabrication of fuel elements, their use in a nuclear reactor, spent fuel reprocessing, recycling, radioactive waste storage or disposal.

Nuclear waste. The radioactive products formed by fission and other nuclear processes in a reactor. Most nuclear waste is initially in the form of spent fuel. If this material is reprocessed, new categories or waste result: high-level, transuranic, low-level wastes and others.

Nucleus. The small, massive, positively charged core of an atom that comprises nearly all of the atomic mass but only a minute part of the atom's volume. The nucleus consists of neutrons and protons, in the common isotope of hydrogen, which consists of one proton only.

Ocean Thermal Energy Conversion (OTEC). Power generation by exploiting the temperature difference between surface waters and ocean depths.

Off-peak. Period of low demand, usually during 24-hour period. When applied to capacity refers to facilities preferably employed during such periods. When applied to customer rates, refers to rates charged during consumption at such periods.

Oil shale. Rock containing organic matter (kerogen) that upon being heated to 800°-1,000°F yields commercially useful oil and/or gas.

Once-through fuel cycle. Any nuclear fuel cycle in which spent fuel is disposed of without reprocessing or recycling to recover fissionable materials from the reactor.

Osmotic pressure. Pressure produced by or associated with diffusion of a solvent through a semipermeable membrane.

Overburden. Rock, soil, and other strata above a coal bed that is to be strip mined.

Ozone. A triatomic form of oxygen (O_3) that is a bluish irritating gas of pungent odor, is formed naturally in the upper atmosphere by a photochemical reaction with solar ultraviolet radiation, and that is a major agent in the formation of smog.

Ozone layer. A region in the earth's upper atmosphere containing ozone which helps shield living organisms from the sun's ultraviolet radiation.

Particulates. Microscopic pieces of solids that emanate from a range of sources and are the most widespread of all substances that are usually considered air pollutants. Those between 1 and 10 microns are most numerous in the atmosphere, stemming from mechanical processes and including industrial dusts, ash, etc.

Passive solar energy. Means of utilizing solar energy for heating and cooling, such as window shutters, eaves, or insulation, that do not consume electricity, pump fluids, or have powered mechanical components.

Payback period. The length of time required for the cumulative net revenue from an investment to equal the original investment. Often used in connection with outlays for energy conservation.

Peat. Name given to layers of dead vegetation that is sometimes regarded as the youngest member of the ranks of coal.

Permeability. The rate at which a liquid or gas can pass through a porous solid substance.

pH: A measure of acidity, calibrated logarithmically, whose values run from 0 to 14 with 7 representing neutrality, numbers less than 7 increasing acidity, and numbers greater than 7 increasing alkalinity.

Photochemical. Of, relating to, or resulting from the chemical action of radiant energy, especially light.

Photogalvanic processes. Light-induced production of electrical current.

Photon. A quantum or discrete bundle of electromagnetic radiation. According to the quantum theory underlying modern physics, all matter comes in discrete bundles, or "quanta"; "electromagnetic radiation"—X-rays, radiowave, and visible light are examples—is no exception.

Photosynthesis. The biologic process by which the containing tissues of plants use water, carbon dioxide and light energy to synthesize carbohydrates.

Photovoltaic energy. Energy obtained from devices that convert sunlight directly into electricity.

Plasma. A dilute gaseous mass of electrically charged particles.

Plutonium. A heavy, radioactive man-made metallic element with atomic number 94, created by absorption of neutrons in uranium 238. Its most important isotope is plutonium-239, which is fissionable.

Plutonium separation. The process by which plutonium is recovered from the uranium, transuranic elements, and fission products that are in spent reactor fuel.

Poison (or neutron poison). Any material (other than the deliberately introduced moderator) which absorbs neutrons within a nuclear chain reaction, slowing down or halting the rate of nuclear fission. Because of the accumulation of neutron poisons in the fuel rods, these rods must be withdrawn from a reactor core after about one year.

Polynuclear organic materials. Aromatic hydrocarbons that are important as a pollutant and possibly as carcinogens. Arise in coal conversion processes.

Present value. The current value of a future stream of costs or benefits calculated by discounting these costs or benefits to the present time.

Pressurized-Water Reactor (PWR). A light water moderated and cooled reactor that employs an indirect cycle; the cooling water that passes through the reactor is kept under high pressure to keep it from boiling, but it heats water in a secondary loop that produces steam that drives the turbine.

Price elasticity of demand. The responsiveness of demand to changes of price. It is defined as the ratio of the percentage change in the quantity demanded to the percentage change in the price of the commodity. If a given change in price results in a large change in demand, then demand is elastic; if the change in price has only a slight effect on demand, then demand is inelastic.

Primary containment. An enclosure that surrounds a nuclear reactor and associated equipment for the purpose of minimizing the release of radioactive material in the event of a serious malfunction in the operation of the reactor.

Pyrolysis. Decomposition of materials through the application of heat with insufficient oxygen for complete oxidation.

Primary energy. Energy before processing or conversion into different or more refined form.

Pumped storage. Method of generating hydroelectricity by using low cost (usually off-peak) power to pump water to an elevated storage area and releasing it at peak demand periods into a river or other body of water.

Quad. A quantity of energy equal to 10^{15} British thermal units.

Quench water. Water used to cool hot gases or solids.

Radiation activated species. Radioactive nuclei which have been produced from other stable nuclei by neutron irradiation.

Radiation damage. A general term for the adverse effects of radiation upon living and nonliving substances.

Radioactive heating. Heating caused by the decay of radioactive materials.

Radium-226. A naturally occurring radioactive isotope of radium with a half-life of 1,600 years.

Radon. A chemically inert, radioactive element which has 86 protons in its nucleus. The most common isotope is radon-222, which has a half-life of 3.8 days.

Reactor core. The central portion of a nuclear reactor, containing the fuel elements and the control rods.

Reactor year. The operation of a nuclear power reactor for one year. As the term has come to be used, unless otherwise stated, the capacity of the reactor is assumed to be 1,000 megawatts (electric).

Recycling. The reuse in a nuclear reactor of uranium, plutonium, or thorium that has been recovered from spent fuel.

Reprocessing. A generic term for the chemical and mechanical processes applied to fuel elements discharged from a nuclear reactor; the purpose is to recover fissile materials such as plutonium-239, uranium-235, and uranium-233 and to isolate the fission products.

Reserve currency. Currency in which central banks maintain foreign exchange reserves; role played by U.S. dollar for most of post-World War II period.

Reserves. Resources that are known in location, quantity, and quality and that are economically recoverable using currently available technologies.

Resource (energy). That part of the resource base believed to be recoverable using only current or near-current technology, without regard to the cost of actually recovering it. To be distinguished from both "resource base" and "reserves" (q.v.).

Resistance heat. Heat generated by an electric current as it is conducted through a substance opposing the flow of the current.

Resource base (energy). The total quantity of energy or of any given energy-producing or energy-related material that is estimated to exist in or on the earth or in its atmosphere, independent of quality, location, or the engineering or economic feasibility of recovering it.

Retort. To subject substances to heat for the purpose of distillation or decomposition; equipment for such a process.

Rolled-in pricing. Injecting a high-cost product into the market by averaging its cost with that of the bulk of the same product produced at lower cost. Typical for public utility marketing of gas and electricity.

Safeguards. Physical and institutional arrangements aimed at preventing the illegal diversion of weapons-usable materials from civilian nuclear programs.

Salt dome. A subterranean geologic formation frequently associated with occurrence of oil and/or gas; after original hydrocarbon material has been withdrawn, can be used for storage of oil or gas.

Saudi marker crude. A quality of light crude oil, with 34° viscosity, used as the basic reference material for determining differential prices for other types of oil.

Scrubber. An air pollution control device that uses a liquid spray for removing pollutants such as sulfur dioxide or particulate matter from a gas stream by absorption or chemical reaction.

Scrubbing. Technique for removing pollutants such as noxious gases and particulates from stack gas emissions.

Secondary recovery. Methods of obtaining oil and gas by the augmentation of reservoir energy, often by the injection of air, gas, or water into a production formation.

Semiconductor. Any of a class of solids, such as germanium and silicon, whose electrical conductivity is between that of a conductor and that of an insulator in being nearly metallic at high temperatures and nearly absent at low temperatures.

Semipermeable membrane. A thin, soft, pliable sheet through which some molecules (usually smaller ones) can pass while others (usually larger) cannot.

Sensitive facilities. Any facilities employed in civilian nuclear research or nuclear power programs which could conceivably be converted to weapons material production uses. Typically these are research reactors, enrichment plants, and reprocessing facilities.

Separative Work Units (SWUs). A measure of the work required to separate uranium isotopes in the enrichment process. It is used to measure the capacity of an enrichment plant independent of a particular product and tails. To put this unit of measurement in perspective, it takes about 100,000 SWUs per year to keep a 1,000 MWe LWR operating and 2,500 SWUs to make a nuclear weapon.

Short ton. 2,000 pounds.

Solar pond. An insulated pond used to store solar energy.

Solar electric cell. A device which converts sunlight directly into electricity.

Solar constant. The solar radiation falling on a unit area at the outer limits of the earth's atmosphere.

Solvent refined coal. Product of a process in which coal is dissolved in oil, treated at moderate conditions with hydrogen and filtered to produce a solid with a high heat value (approximately 16,000 Btu/lb) and an ash and sulfur content much below that of the input coal.

Somatic damage. Injury to the normal functioning of a living organism, which may or may not result in death.

Spectral-shift reactor. A reactor in which a mixture of light water and heavy water is used as the moderator and coolant. The ratio of light to heavy water is varied to change (shift) the energy spectrum of the neutrons in the reactor core. Since the probability of neutron capture varies with neutron velocity, a measure of reactor control is thus obtained.

Spent fuel. The fuel elements removed from a reactor after several years of generating power. Spent fuel contains radioactive waste materials, unburned uranium and plutonium.

Spiking. The addition of highly radioactive materials to nonradioactive weapons-usable nuclear materials. The intent is to discourage illegal diversion.

Spot price. Price formed in a market for one-time, usually quick, delivery of a specific cargo and destination. In times of stringency, spot market prices can greatly exceed prices for identical products sold under long-term contracts.

Stack gases. Gases resulting from combustion of fuels and emitted through the stack.

Strippers, stripper wells. Oil wells producing less than 10 barrels per day.

Strontium-90. A radioactive isotope of the element strontium. A common fission product, its half-life is 28 years.

Submicron. Smaller than a micron.

Sunk cost. The unrecovered or unrecoverable balance of an investment. A cost already paid or committed.

Super phenix. French commercial-scale liquid metal fast breeder reactor under construction.

Superconducting magnet. A magnet which is at very low temperature (near absolute zero).

Synergism. Combined or cooperative action of discrete agencies such that the total effect is greater than the sum of the effects taken independently.

Synthesis gas. A fuel gas containing primarily carbon monoxide and hydrogen; it can be used after careful removal of impurities, particularly sulfur compounds, for conversion to methane (high-Btu gas), methanol, liquid hydrocarbons, and a wide variety of other organic compounds.

Synthetic fuels. A term now commonly reserved for liquid and gaseous fuels that are the product of a conversion process rather than mining or drilling; most frequently applied to cover liquids or gases derived from coal, shale, tar sands, waste, biomass.

Tailings. Relected material after uranium ore is processed. Since uranium ore contains less than 1 percent uranium, essentially all of the processed ore is left as tailings near uranium mills.

Tails assay. The percentage of uranium-235 in tails. Natural uranium contains 0.71 percent uranium-235; the tails assay may be 0.3 percent or lower.

Tar sands. Hydrocarbon-bearing deposits distinguished from more conventional oil and gas reservoirs by the high viscosity of the hydrocarbon, which is not recoverable in its natural state through a well by ordinary production methods.

Tertiary recovery. Use of heat and methods other than air, gas, or water injection to augment oil recovery (presumably occurring after secondary recovery).

Therm. 100,000 British thermal units.

Thermodynamic efficiency. The highest fraction of input energy that can be converted into useful work by a heat engine. The Second Law of Thermodynamics establishes that maximum.

Thermonuclear explosive or device. A nuclear bomb, such as a hydrogen bomb, that derives its energy from the fusion of light elements into slightly heavier ones.

Thermonuclear reactor. A nuclear reactor that derives its energy from the controlled release of fusion energy.

Thorium. A radioactive element of atomic number 90; naturally occurring thorium has one main isotope—thorium-232. The absorption of a neutron by a thorium atom can result in the creation of the fissile material uranium-233.

Throwaway fuel cycle. A fuel cycle in which the spent fuel discharge from the reactor is not reprocessed to recover residual plutonium and uranium values.

Threshold. A dose level or concentration (as of a drug, toxin, or pollutant) below which there is, or it is assumed there is, no effect, especially no adverse effect, on a recipient tissue or organism.

Throwaway fuel cycle. See "Once-Through Fuel Cycle" which uranium resources and requirements are commonly calculated and reported.

Tight sands formation (gas). Gas-bearing geologic strata that hold gas too tightly for conventional extraction processes to bring it to the surface at economic rates without special stimulation.

Tokomak Fusion Test Reactor (TFTR). An experimental thermonuclear reactor, originally developed in the Soviet Union and now being developed also in the United States (for example, Princeton University). Many fusion experts believe that the Tokomak concept is the fusion reactor concept "most likely to succeed" as a commercial reactor.

Toluene. A liquid aromatic hydrocarbon (C_7H_8) that is produced commercially from coke oven gas, light oils, coal tar, and petroleum.

Transuranic elements. Radioactive nuclides generated as fission products from the fissioning of nuclear fuel during reactor operation and as induced activity from the capture of neutrons in fuel cladding, reactor structures, and reactor coolant.

Tritium. A radioactive isotope of hydrogen which has mass number three (one proton, two neutrons) and a half-life of 12.3 years.

Turnkey. A ready-to-operate industrial facility sold at an agreed price.

Ultimate disposal. Final, usually irretrievable, disposal of hazardous materials, especially nuclear wastes.

Uranium. A radioactive element of atomic number 92. Naturally occurring uranium is a mixture of 99.28 percent uranium-238, 0.71 percent uranium-235, and 0.0058 percent uranium-234. Uranium-235 is a fissile material and is the primary fuel of light water reactors. When bombarded with slow or fast neutrons, it will undergo fission. Uranium-238 is a fertile material that is transmuted to plutonium-239 upon the absorption of a neutron.

Uranium Hexafluoride (UF_6). A compound of uranium, which is used in gaseous form in the enrichment of uranium isotopes.

Uranium-232. A relatively stable isotope of uranium with a half-life of 74 years.

Uranium-233. A fissionable isotope of uranium that can be formed by bombarding thorium-232 with neutrons.

Uranium-235 (U-235). An isotope of uranium with mass number 235 that fissions when bombarded with slow neutrons. The principal fuel for light-water reactors, U-235 can, if highly enriched, be used to produce nuclear explosions.

Uranium-238 (U-238). An isotope of uranium with mass number 238 that absorbs energetic ("fast") neutrons, forming U-239 which then decays through neptunium to form plutonium-239.

Uranium hexafluoride (UF_6). A gaseous compound of uranium used in various isotope separation processes.

Uranium oxide (U_3O_8). The most common oxide of uranium that is found in typical ores.

Viscosity. Characteristic of a material determining its ability to resist flow (the lower the viscosity, the less the flowing ability).

Weapons grade material. Any mixture of uranium or plutonium isotopes that could be used to create a nuclear explosion.

Wellhead tax. A tax imposed on oil (or gas) at the top of the production well.

Working fluid. Fluid used in an electrical generator that is heated by the energy source and then expands through a turbine to produce electricity and then is recycled and reused.

Yellowcake. A uranium concentrate that results from the milling (concentrating) of uranium ore. It typically contains 80-90 percent uranium oxide (U_3O_8).

Definition of RD and D Scale-up Phases

This section provides definitions of the various research, development, and demonstration scale-up phases as they pertain to the Fossil Energy Coal Resource. Included are the distinguishing features of Process Development Units (PDU's), pilot plants, demonstration plants, and commercial demonstration plants. The key to the definition is not so much the size as the purpose—the kind of information the activity is designed to yield. In this context the relative scales can be differentiated as follows.

Process Development Unit (PDU)

The purposes of a PDU are to establish the basic technical feasibility of the process; acquire basic physical, chemical, and engineering data needed to evaluate the process; and to develop the design data necessary to allow further scale-up to a larger stage if feasible.

Other distinguishing features of the PDU are as follows:

- PDU's follow laboratory "bench" experimental work
- PDU's incorporate results of bench-scale work on key process steps to form an integrated smal-scale process to test key variables on performance. (Technology development phase of R&D)
- PDU's generally operate continuously and process the minimum amount of raw material necessary to test the process feasibility
- PDU's are never facilities in themselves, but are a component of, or contained in, an existing facility, (e.g., laboratory or plant), and can undergo considerable modification to enhance the process.

Pilot Plant

The purpose of a pilot plant is to establish the integrated process feasibility by combining commercial type (not commercial size) components into a small model plant to test and evaluate the critical parameters of scale-up, and to acquire engineering data needed to assess economic feasibility and design a larger near-commercial-size plant.

Other distinguishing features of a pilot plant are as follows:

- Pilot plants are the first scale-up facility to produce enough end-product to permit product testing and refinement
- Pilots, as experimental facilities, are subject to continuing and significant modifications to help identify candidate process and components that could be used for further scale-up to larger plant sizes (engineering phase of RD&D)
- Pilot plants are limited to three years or less of operating life and in most instances are dismantled after fact-finding is complete unless they can be modified to cost effectively test new processes
- Determining the projected economics and cost of a prospective commercial-

sized plant is a significant R&D objective of pilot plants
- Large pilot plants for some technical development may provide sufficient data about operations before pseudocommercial conditions to allow the development to proceed directly to commercial demonstration.

Demonstration Plants

The purpose of demonstration plants is to demonstrate and validate economic, environmental, and productive capacity of a near commercial-size plant by integrating and operating a single modular unit using commercial-sized components.

Other distinguishing features of a demonstration plant are as follows:

- They are still developmental in the sense that technological scale-up problems may occur and require engineering modification; however, the risk is much lower since the plant production process was developed and tested at the pilot stage
- They have a long life and are planned to be expanded to become part of the commercial plant, after their successful demonstration period, by the industrial cost-sharing partner. In this case the industrial partner will purchase the facilities at a fair market value
- They are used only to demonstrate and verify second-generation technologies (those not currently used commercially) and will demonstrate only the most feasible process surviving competition of alternatives regardless of whether previous pilot plant work was done in private industry or government
- They are neither formal systems acquisitions nor full-scale development but are the final stage in the R&D process aimed at accelerating and reducing the risks of industrial process implementation.

Commercial Demonstration Plants[1]

The purposes of commercial demonstration plants are to resolve commercial investment uncertainties by establishing the actual economic factors, environmental feasibility, socioeconomic impact, capital and resource requirements, constraints and product markets for currently available as well as newly introduced synthetic fuel products; and to encourage creation of a viable industry using these technologies.

Other distinguishing features of commercial demonstration plants are as follows:

- They demonstrate viability of a facility and/or process to produce a commercial-grade product at a rate of production considered to be commercially significant using commercial-scale equipment under commercial conditions.
- The Government's role is to provide financial incentives to stimulate industry construction and commercial production and marketing of synthetic fuel products by loan guarantees or other incentives.
- They do not constitute R&D work. Most technical problems have been solved. Their demonstration value is limited to setting industry standards and precedents for a viable synthetic fuels industry.
- They represent a scale of three to five times the productive capacity of demonstration plants by combining modular production units of the type singularly demonstrated in a demonstration plant to achieve full-scale commercial production of a product.

---

[1] Commercial demonstration projects are the responsibility of the Assistant Secretary for Resource Applications in DOE.

Abbreviations and Acronyms

A

A	Ampere
AAG	Agglomerating Ash Gasifier
AC	Alternating Current
ACS	Advanced Cogeneration Systems
A&E	Architect and Engineer
AES	Automated Extraction System
AFB	Atmospheric Fluidized-Bed
AFBC	Atmospheric Fluidized-Bed Combustion
AGA	American Gas Association
AHF	Advanced Hydraulic Fracturing
AL	Ames Laboratory
ANL	Argonne National Laboratory
API	American Petroleum Institute
APS	Advanced Power Systems
APT	Advanced Process Technology
AR&TD	Advanced Research and Technology Development
ATF	Advance Technology Face
atm	Atmosphere (of pressure)

B

BACT	Best Available Control Technology
bbl	Barrel
bbl/d	Barrels per day
Bcf	Billion Cubic Feet
BCR	Bituminous Coal Research, Inc.
BCURA	British Coal Utilization Research Association
BETC	Bartlesville Energy Technology Center
BNL	Brookhaven National Laboratory
BOM	Bureau of Mines
BSB	Blind Shaft Borer
Btu	British Thermal Unit
BTX	Benzene, Toluene, and Xylene
BWE	Bucket Wheel Excavators

C

CBS	Carbon Burn-up Cell
CC	Coal Conversion
CCC	Chemical Coal Cleaning
CCDC	Conoco Coal Development Company
CCU	Coal Conversion and Utilization
CDIF	Component Development and Integration Facility
CDTF	Component Development Test Facility
CEF	Chemical Explosive Fracturing
CEQ	Council on Environmental Quality
CFCC	Coal-Fired Combined Cycle
CFFC	Clean Fuel from Coke
CFFF	Coal-Fired Flow Facility
CFPD	Cubic Feet per Day
CG	Coal Gasifiers
CMOC	Carbondale Mining Operations Center

CMTC	Carbondale Mining Technology Center	EAS	Economic Assessment Service
CO	Carbon Monoxide	EC	Executive Committee
CO-H$_2$O	Carbon monoxide-Water	ECAS	Energy Conversion Alternative Study
CO$_2$	Carbon Dioxide	ECP	Engineering Change Proposal
COE	Crude Oil Equivalent	EDP	Environmental Development Plant
COED	Coal Oil Energy Development - Liquefaction Process	EDS	Exxon Donor Solvent
COM	Coal-Oil Mixture	EEI	Edision Electric Institute
CPCS	Cross-pit Conveying Systems	EGR	Enhanced Gas Recovery
CPU	Continuous Process Unit	EH&S	Environmental Health & Safety
CSF	Consolidation Synthetic Fuel		
CTAS	Cogeneration Technology Alternative Study	EIA	Environmental Impact Assessment
CTF	Cresap Test Facility	EIS	Environmental Impact Statement
CTIF	Component Test and Integration Facility		
CTIU	Component Test and Integration Unit	EMAC	Energy Materials Advisory Committee
		ENCON	Environmental Constraints
CW	Curtiss-Wright	EOR	Enhanced Oil Recovery
CY	Calendar Year	EPA	Environmental Protection Agency

D

		EPRI	Electric Power Research Institute
DC	Direct Current		
DCCU	Division of Coal Conversion and Utilization	ER&E	Exxon Research and Engineering Company
		ERCO	Energy Resources Company
DFFE	Division of Fossil Fuel Extraction	ERDA	Energy Research and Development Administration
DFFP	Division of Fossil Fuel Processing	ETC	Energy Technology Center
		ETF	Engineering Test Facility
DFFU	Division of Fossil Fuel Utilization		
DGE	Division of Geothermal Energy		

F

		FBC	Fluidized-Bed Combustion
DISC	Direct Injected Stratified Charge	FE	Fossil Energy
		FEARP	Fossil Energy Authorization and Review Board
DOC	Department of Commerce		
DOD	Department of Defense	FEEDS	Fossil Energy Equipment Data System
DOE	Department of Energy		
DOI	Department of Interior	RGD	Flue Gas Desulfurization
DPCS	Division of Program Control & Support	FRG	Federal Republic of Germany
		ft^3	Cubic feet
DPS	Division of Power Systems	FTP	Full-Time Permanent
DSE	Division of Systems Energy	FTT	Full-Time Temporary
		FY	Fiscal Year

E

EA	Office of Environmental Activities		

G

EACC	Eastern Associated Coal Corporation	GBF	Granular-Bed Filter
		GE	General Electric Company

GFETC	Grand Forks Energy Technology Center	**K**	
GR&DC	Gulf Research and Development Company	KC-oil	Koppers Creosote Oil
		kerogen	Organic material from which shale oil is extracted
GOCO	Government Owned, Contractor Operated	KVB	King, Vanderholl, Bocon
GPA	Geopressured Aquifer	kWh	Kilowatt Hour
GPE	Gas Producing Enterprises, Inc.		
gr	Gram	**L**	
GRI	Gas Research Institute	LASL	Los Alamos Scientific Laboratory
GURC	Gulf Universities Research Consortium	lb/hr	Pound per hour
		LBL	Lawrence Berkeley Laboratory
H		LDC	Less Developed Countries
		LETC	Laramic Energy Technology Center
H_2S	Hydrogen Sulfide		
H-Coal	Hydrogen-Fed Coal	LG	Longwall Generator
HEW	Department of Health, Education and Welfare	LI	Letter of Intent
		liquefaction	The process of being liquefied
HGMS	High-gradient Magnetic Separation	LLL	Lawrence Livermore Laboratory
HGR	Hot-Gas Recycle	LNG	Liquefied Natural Gas
hp	Horsepower	LPG	Liquefied Petroleum Gas
HPDE	High Performance Demonstration Equipment	LVW	Linked Vertical Well
HRI	Hydrocarbon Research, Incorporated	**M**	
HUD	Department of Housing and Urban Development	M	Thousand
		MA	Milliampere
Hydrogenation	The process of combining or treating with hydrogen	MAF	Moisture- and Ash-Free
		Mcf	Thousand cubic feet
		MCFC	Molten Carbonate Fuel Cells
I		METC	Morgantown Energy Technology Center
IC	Internal Combustion	MFB	Multicell Fluidized-Bed Boiler
ICE	Improved Conversion Efficiency	MFC	Methane from Coal
ICGG	Illinois Coal Gasification Group	MFCD	Multiple Face, Continuous Drivage
IEA	International Energy Agency	MHD	Magnetohydrodynamics
IGT	Institute of Gas Technology	MIT	Massachusetts Institute of Technology
IHT	Institute of High Temperatures	MIUS	Modular Integrated Utility Systems
INEL	Idaho National Energy Laboratory	ML	Mound Laboratory
In situ	In its original position or in place	MLGW	Memphis Light, Gas, and Water
IW	Improved Waterflooding	MM	Million
		MMbbl/d	Millions of Barrels per Day
J		MMscf	Millions of Standard Cubic Feet
JPL	Jet Propulsion Laboratory	MOC	Mining Operation Center

MOPPS	Market Oriented Program Planning Study	OMB	Office of Management and Budget, Executive Office of the President
MOU	Memorandum of Understanding	ONR	Office of Naval Research
MP&D	Mine Planning and Development	OPEC	Oil Producing and Exporting Countries
MPF	Micellar-Polymer Flooding	OPPA	Office of Program Planning and Analysis
MR&D	Mining Research and Development	ORBCS	Organic Rankine Bottoming Cycle System
msec	Millisecond	ORC	Occidental Research Corporation
MPS	Marine Sediment Penetrators		
MSR	Management Status Reviews	ORNL	Oak Ridge National Laboratory
MTC	Mining Technology Center		
MTCH	Mining Technology Clearing House	OS/IES	On-Site Integrated Energy Systems
MVMA	Motor Vehicles Manufacturing Association	OSHA	Occupational Safety and Health Administration
MW	Megawatt	OUA	Office of University Activities
MWe	Megawatt Electrical		
MY	Man-year		

N

P

		P&M	Pittsburg & Midway Coal Mining Company
N/A	Not applicable	PAD	Program Approval Document
NaCl	Sodium Chloride	PAF	Principal Alternate Fuels
NASA	National Aeronautics and Space Administration	PAS	Performance Assurance System
NBS	National Bureau of Standards	PB	Packed-Bed
NCB	National Coal Board	PBP	Packed-Bed Process
NEP	National Energy Plan	PCP	Physical Coal Preparation
NEPA	National Environmental Policy Act	PDS	PAS Library and Data System
		PDU	Process Development Unit
NIOSH	National Institute of Occupational Safety and Health	PERC	Pacific Energy Research Center
NL	National Laboratory	PETC	Pittsburgh Energy Technology Center
NO_x	Nitrogen Oxide		
NSF	National Science Foundation	PFB	Pressurized Fluidized-Bed
NSPS	New Source Performance Standards	PFBC	Pressurized Fluidized-Bed Combustion
NTIS	National Technical Information Service	PL	Public Law
		PMOC	Pittsburgh Mining Operations Center

O

		PNG	Petroleum and Natural Gas
		PNL	Pacific Northwest Laboratory
OCGT	Open-Cycle Gas Turbine	PON	Program Opportunity Notice
OCR	Office of Coal Research	PPD	Project Plan Documents
OCS	Outer Continental Shelf	psi	Pounds per Square Inch
OGSIST	Oil, Gas, Shale and In Situ Technology	psig	Pounds per Square Inch, Gauges
OGST	Oil, Gas, and Shale Technology	PTC	Petroleum Technology Corporation

Q

Q_e	Quads of Electric Energy
Q_f	Quads of Fossil Fuel Consumption
Q_r	Quads of Heat Rejection
Quad	10¹⁵ (Quadrillion) Btus

R

RC	Resource Characterization
R&D	Research and Development
RD&D	Research, Development, and Demonstration
retort	Distill or decompose by heat
RFP	Request for Proposal
RON	Research Octane Number
RW	Residue Fuels

S

scf	Standard Cubic Feet
scf/d	Standard Cubic Feet per Day
scf/h	Standard Cubic Feet per Hour
SCM	Superconducting Magnet
SCPE	Sublevel Caving with Pillar Extraction
SDB	Steeply Dipping Bed
SDI	Selective Dissemination of Information
SEMS	Seafloor Earthquake Measurement System
SFME	Single Face, Multiple Entry
SFMP	Solid Fuel Mining and Preparation
SGT	Small Gas Turbine
SIC	Standard Industrial Code
SL	Sandia Laboratory
SNG	Synthetic Natural Gas
SO_2	Sulfur Dioxide
SO_x	Oxides of Sulfur
SRC	Solvent Refined Coal
SRL	Solvent Refined Lignite
SSO	Source Selection Official
Syngas	Synthetic Natural Gas

T

TBD	To Be Determined
TBM	Tunnel Boring Machine
TBS	To Be Supplied
Tcf	Trillion Cubic Feet
T/D	Tons per Day
TEC	Total Estimated Cost
TETAS	Total Energy Technology Assessment
TG	Transport Gas
T/H	Tons per Hour
TIS	Technical Information Service
TMSP	Thick, Multiple, and Steeply Pitching Seams
TR	Thermal Recovery
TTU	Turbine Test Unit

U

UCG	Underground Coal Gasification
UCL	Underground Liquefaction
U.K.	United Kingdom
UOP	Universal Oil Products
USGS	United States Geological Survey
USN	United States Navy
USS	United States Steel
U.S.S.R.	Union of Soviet Socialist Republics
UTC	United Technologies Corporation

V

V	Volts

W

WEST	Westinghouse Electric Company
WGSP	Western Gas Sands Project
WMBG	Western Medium Btu Gas
WPCT	Working Party on Coal Technology
WRC	Water Resources Council

Z

$ZnCl_2$	Zinc Chloride

References

ENERGY - GENERAL

A National Plan for Energy Research, Development and Demonstration: Creating Energy Choices for the Future. Vol. 1: The Plan. Washington: Energy Research and Development Administration, 1976.

Clark, Wilson. *Energy for Survival — The Alternative to Extinction.* New York: Anchor Book, Anchor Press/Doubleday, 1975.

Considine, Douglas M. (ed.). *Van Nostrand's Scientific Encyclopedia.* 5th ed. New York: Van Nostrand Reinhold Company, 1976.

Energy — Global Prospects 1985-2000. Workshop on Alternative Energy Strategies. New York: McGraw-Hill Book Company, 1977.

Energy in America's Future. Baltimore, Maryland: The Johns Hopkins University Press, 1979.

Energy Microthesaurus. Springfield, Va.: U.S. Department of Commerce, National Technical Information Service, 1976.

Energy - The Next Twenty Years. Cambridge, Mass.: Ballinger Publishing Co., 1979.

Energy Policy Project of the Ford Foundation. *A Time to Choose: America's Energy Future.* Cambridge, Mass.: Ballinger Publishing Co., 1974.

Federal Energy Administration. *Project Independence Report.* Washington, 1974.

Hammond, Allen L., Metz, William D., and Maugh II, Thomas H. *Energy and the Future.* Second Printing. Washington: American Association for the Advancement of Science, 1973.

Hollander, Jack M. (ed.). *Annual Review of Energy.* Palo Alto, Ca.: Annual Reviews Inc.

Kahn, Herman. *The Next 200 Years.* New York: William Morrow and Company, Inc., 1976.

Maddox, John. *Beyond the Energy Crisis: A Global Perspective.* New York: McGraw Hill, 1975.

Martell, Charles L. *Batteries and Storage Systems.* New York: McGraw Hill, 1970.

The National Research Council. *Energy in Transition 1985-2020.* Final Report of the Committee on Nuclear and Alternative Energy Systems. San Francisco: W. H. Freeman and Company, 1979.

Ridgeway, James. *The Last Play: The Struggle to Monopolize the World's Energy Resources.* New York: E. P. Dutton, 1973.

Science and Public Policy Program. University of Oklahoma, Norman, Oklahoma. *Energy Alternatives, A Comparative Analysis.* Washington: Council on Environmental Quality, 1975.

Thirring, Hans. *Energy for Man — From Windmills to Nuclear Power.* Second Ed. New York: Harper Colophon Books, Harper & Row Publishers, Inc., 1979.

ENERGY CONSERVATION

Clagg, Peter. *New Low-Cost Sources of Energy for the Home.* Charlotte, Vermont: Garden Way, 1975.

Clark, Wilson. *Energy for Survival: The Alternative to Extinction.* Garden City: Anchor Press, 1974.

Dubin, Fred. Energy for Architects. Architecture Plus 1 (No. 6): 38-49, 74-75 (July 1973).

Dubin, Fred; and Long, Chalmers. *Energy Conservation Standards.* New York: McGraw-Hill, 1978.

Fond, K. W., et al. (eds.). *Efficient Use of Energy.* New York: American Institute of Physics, 1975.

Griffin, Charles William. *Energy Conservation in Buildings: Techniques for Economical Design.* Washington: Construction Specifications Institute, 1974.

Hirst, Eric. *Energy Consumption for Transportation in the U.S.,* (ORNL-NSF-EP-15). Oak Ridge: Oak Ridge National Laboratory.

Hittman Associates. Residential Energy Consumption. *Single Family Housing: Final Report.* Washington: U.S. Department of Housing and Urban Development, 1973.

Hu, David. *Industrial Conservation Handbook.* New York: Van Nostrand Reinhold, 1981.

Hunt, V. Daniel. *Handbook of Conservation and Solar Energy.* New York: Van Nostrand Reinhold, 1981.

Large, David B. (ed.). *Hidden Waste.* Conservation Foundation, 1973. *Minimum Energy Dwelling.* Washington: Energy Research and Development Administration Office of Conservation Division of Buildings and Community Systems.

Smith, Thomas W. *Household Energy Game*. University of Wisconsin: Johns Hopkins Marine Studies Center, 1974.

Yergin, Daniel, et al. (eds.). *Energy Future - Report of the Energy Project at the Harvard Business School*. Random House, New York, 1979.

SOLAR ENERGY

Anderson, Bruce. *Solar Energy — Fundamentals in Building Design*. New York: McGraw-Hill, 1977.

Adelson, E. H. *Solar Air Conditioning and Refrigeration*. Isotech Research Labs., 1975.

American Institute of Aeronautics and Astronautics. *Solar Energy for Earth*. New York: American Institute of Aeronautics and Astronautics, 1975.

Beckman, William, et. al. *Solar Heating Design — By the F-Chart Method*. New York: John Wiley & Sons, 1977.

Brinkworth, Brian Joseph. *Solar Energy for Man*. New York: John Wiley, 1973.

Chalmers, Bruce. *The Photovoltaic Generation of Electricity*. Scientific American 235 (4): 34-44 (October 1976).

Clark, Wilson. *Energy for Survival: The Alternative to Extinction*. Garden City: Anchor Press, 1977.

Corliss, William R. *Direct Conversion of Energy*. Oak Ridge: U.S. Atomic Energy Commission, Office of Information Services, 1964.

Daniels, Farrington. *Direct Use of the Sun's Energy*. New York: Ballantine Books, Division of Random House, Inc., 1964.

Davis, Albert J., and Schubert, Robert P. *Alternative Natural Energy Sources in Building Design*. Blacksburg, Va.: Passive Energy Systems, 1974.

Duffie, John A., and Beckman, William A. *Solar Energy Thermal Process*. New York: John Wiley, 1974.

Duffie, John A., and Beckman, William A. Solar Heating and Cooling. *Science*, 191: 143-149 (January 16, 1976).

Federal Energy Administration. *Buying Solar*. Washington: U.S. Government Printing Office, June 1976.

Gay, Larry. *The Complete Book of Heating with Wood*. Charlotte, Vt.: Garden Way, 1974.

Glaser, Peter E. Beyond Nuclear Power — the Large-Scale Use of Solar Energy. New York Academy of Sciences. *Transactions*, ser. 2, 31 (No. 8): 951-967 (December 1969).

Grey, J. Solar Heating and Cooling. *Astronautics and Aeronautics*: 33-37 (November 1975).

Halacy, D. S. *Earth, Water, Wind & Sun*. New York: Harper and Row, 1977.

Hoke, John. *Solar Energy*. New York: F. Watts, 1968.

Hunt, V. Daniel. *Handbook of Conservation and Solar Energy*. New York: Van Nostrand Reinhold, 1981.

Keyes, John. *Harvesting the Sun to Heat Your House*. New York: Morgan & Morgan, 1975.

Kreith, Frank, and Kreider, Jan. *Principles of Solar Engineering*. Washington, D.C.: Hemisphere Publishing Corporation, 1978.

Lucas, Ted. *How to Build a Solar Heater*. 3d ed., Pasadena, Ca.: Ward Ritchie Press, 1975.

Mazria, Edward. *The Passive Solar Energy Book*. Emmaus, Pa.: Rodale Press, 1979.

Meinel, Aden B., and Meimel, Marjorie P. *Physics Looks at Solar Energy*. Physics Today: 44-50 (February 1972).

Michels, Tim. *Solar Energy Utilization*. New York: Van Nostrand Reinhold, 1979.

Rankins II, William H., and Wilson, David A. *Practical Sun Power*. Lorien House, 1974.

Rau, Hans. *Solar Energy*. Translated by Maxim Schur. Edited and revised by D. J. Duffin. New York: Macmillan, 1964.

Russell, Charles R. Solar Energy. *Elements of Energy Conversion*. Oxford, New York: Pergamon Press, 1967.

Skurka, Norma, and Naar, John. *Living with Natural Energy. Design for a Limited Planet*. 1st ed. New York: Ballantine Books, Division of Random House, Inc., 1976.

Solar Energy for Space Heating & Hot Water. Washington: Energy Research and Development Administration, Division of Solar Energy, 1976.

Solar Energy Update. SEU 77-1 Oak Ridge: Energy Research and Development Administration, Technical Information Center, 1977.

Stoner, Carol (ed.). *Producing Your Own Power: How to Make Nature's Energy Sources Work for You*. Emmaus, Pa.: Rodale Press, Book Division, 1974.

Sun Language. Washington: Solar Energy Institute of America.

Williams, James Richard. *Solar Energy: Technology and Applications*. Ann Arbor: Ann Arbor Science Publishers, 1974.

Yanda, Bill, and Fisher, Rick. *Solar Greenhouse*, Santa Fe, New Mexico: John Muir Publications, 1976.

Yergin, Daniel, et al. (eds). *Energy Future - Report of the Energy Project at the Harvard Business School*. Random House, New York, 1979.

GEOTHERMAL ENERGY

Barnea, Joseph. Geothermal Power. *Scientific American* 226: 70-77 (January 1972).

Geothermal Energy - The Hot Prospect. *EPRI Journal* 2(3): 6-13 (1977).

Goldsmith, M. *Geothermal Resources in California: Potentials and Problems*. Pasadena: Environmental Quality Laboratory, California Institute of Technology. December 1971.

Hammond, Allen L. Geothermal Energy: An Emerging Major Resource. *Science* 177: 978-980 (September 15, 1972).

Henahan, John F. Geothermal Energy — the Prospects Get Hotter. *Popular Science* 207: 96-99 (November 1974).

Henahan, John F. Full Steam Ahead for Geothermal Energy. *New Scientist* 57 (No. 827): 16-17 (January 4, 1973).

Kruger, Paul, and Otte, C. (eds.). *Geothermal Energy — Resources, Production, Stimulation.* Stanford, Stanford University Press, 1973.

Rex, Robert. Geothermal Energy — the Neglected Energy Option. *Bulletin of the Atomic Scientists* 27 (No. 8 52-56 (October 1971).

Weaver, Kenneth F. *The Search for Tomorrow's Power.* National Geographic 142 (No. 5): 650-681 (November 1972).

WIND ENERGY

Blackwell, B. F., and Feltz, L. V. *Wind Energy — A Revitalized Pursuit* (SAND-75-0166). Livermore, California: Sandia Laboratories, March 1975.

Carter, Joe. *Windpower for the People.* Environment Action Bulletin 6 (No. 12): 4-5 (June 14, 1975).

Clark, Wilson. Energy from the Winds. *Energy for Survival.* New York: Doubleday, 1974.

Eldridge, Frank R. *Wind Machines.* Rev. 2. New York: Van Nostrand Reinhold, 1980.

Golding, Edward. *The Generation of Electricity by Wind Power.* London: E & F. N. Spon, 1955.

Hackleman, Michael A. *Wind and Windspinners: A Nuts and Bolts Approach to Wind-Electric Systems.* Sangus, California: Earthmind, 1974.

Hamilton, R. Can We Harness the Wind? National Geographic 148: 812-829 (December 1975).

Handbook of Homemade Power. By the staff of the Mother Earth News, New York: Bantam Books, 1974.

Hunt, V. Daniel. *Windpower - A Handbook on Wind Energy Conversion Systems.* New York: Van Nostrand Reinhold, 1981.

Inglis, David R. Wind Power Now. Bulletin of the Atomic Scientists 31 (8): 20-26 (October 1975).

Johnson, C. C., et al. Wind Power Development and Applications. Power Engineering 78 (No. 10): 50-53 (October 1974).

Putnam, Palmer Cosslett. *Power from the Wind.* New York: Van Nostrand Reinhold, 1974.

Putnam, Palmer Cosslett. *Energy in the Future.* New York: Van Nostrand, 1953.

Reynolds, John. *Windmills & Watermills.* New York: Praeger, 1970.

Simmons, Daniel M. *Wind Power.* Park Ridge, N.J.: Noyes Data Corp., 1975.

Sorensen, Bent. Energy and Resources. Science 189 (No. 4199): 255-260 (July 25, 1975).

Steadman, Philip. *Energy, Environment and Building.* London, New York: Cambridge University Press, 1975.

Stokhuyzen, Frederick. *The Dutch Windmill.* Translated from the Dutch by Carry Dikshourn. New York: Universe Books, 1963.

Stoner, Carol. *Producing Your Own Power; How to Make Nature's Energy Source Work For You.* Emmaus, Pa.: Rodale Press, Book Division, 1974.

Torrey, Volta. *Wind-Catchers — American Windmills of Yesterday and Tomorrow.* Brattleboro, Vermont: The Stephen Greene Press, 1976.

Wolff, Alfred R. *The Windmill as a Prime Mover.* New York: J. Wiley & Sons, 1885.

FOSSIL FUEL ENERGY

A Guide to Federal Power Commission Public Information. Enclosure No. 020137. Washington: Federal Power Commission, Office of Public Information.

Annual Review of Energy. Palo Alto, California: Annual Reviews, Inc.

Anthony Sampson. *The Seven Sisters.* New York: Viking Press, 1976.

Bakulev, G. D. *An Economic Analysis of Underground Gasification of Coal.* Washington: U.S. Dept. of the Interior, Bureau of Mines, 1962.

Coal - Bridge to the Future. Report of the World Coal Study. Cambridge, Mass.: Ballinger Publishing Company, 1980.

Hoffman, Edward Jack. *Coal Conversion and the Direct Production of Hydrocarbons from Coal-Steam Systems.* Laramie: College of Engineering, University of Wyoming, 1966.

Metz, William D. Power Gas and Combined Cycles: Clean Power from Fossil Fuels. Science 179 (No. 4068): 56ff (January 5, 1973).

Mudge, L. K., et al. *The Gasification of Coal: A Battelle Energy Program Report.* Richland, Wa.: Battelle Pacific Northwest Laboratories, 1974.

National Economic Research Associates, Inc. *Fuels for the Electric Utility Industry, 1971-1985.* New York: Edison Electric Institute, 1972.

National Research Council. Committee on Chemical Utilization of Coal. *Chemistry of Coal Utilization. Supplementary Volume.* New York: Wiley, 1963.

Pyrcioch, E. J., et al. *Production of Pipeline Gas by Hydrogasification of Coal.* Chicago: Institute of Gas Technology, 1972.

Risser, Hubert E. *Gasification and Liquefaction: Their Potential Impact on Various Aspects of the Coal Industry.* Urbana, Ill.: State Geological Survey, 1968.

Smith, Craig B. (ed.). *Efficient Electricity Use.* New York: Pergamon Press, Inc.

Thrush, Paul W., and Staff of Bureau of Mines (eds.). *A Dictionary of Mining. Mineral, and Related Terms.* Washington: U.S. Department of the Interior, Bureau of Mines, 1968.

NUCLEAR ENERGY

American Nuclear Society. *Nuclear Power and the Environment: Questions and Answers.* Hindsdale, Ill.: American Nuclear Society, 1973.

American Nuclear Society, *Nuclear Power and the Environment. Questions and Answers,* p. 50 (April 1976).

American Nuclear Society Standards Committee Subcommittee ANS-9. *American National Standard, Glossary of Terms in Nuclear Science and Technology.* Hinsdale, Ill.: American Nuclear Society, 1976.

An Economic Definition of the Nuclear Industry, prepared for Office of Planning and Analysis, U.S. Nuclear Regulatory Commission (April 1978).

Annual Report to Congress, Energy Information Administration, U.S. Department of Energy, DOE/EIA-0036/2 (1978).

Atomic Industrial Forum, INFO, No. 125 (December 1978).

Cohen B. L., The Hazards of Plutonium Disposal, *Nuclear News* 18(8), pp. 44-45 (June 1975); also *The Hazards of Plutonium Disposal*, University of Pittsburgh, Pittsburgh, Pa. (1975).

Ebbin, Steven. *Citizens Group and the Nuclear Power Controversy: Uses of Scientific and Technological Information*. Cambridge, Mass.: MIT Press, 1974.

8th Annual Review of Overall Reliability and Adequacy of the North American Bulk Power Systems, National Reliability Council (August 1978).

Eisenbud, Merril. *Environmental Radioactivity*. New York: McGraw Hill Book Company, Inc., 1963.

Electric Power Supply and Demand, 1978–87 for the Contiguous United States, DOE/ERA-0018 (July 1978).

Environmental Survey of the Uranium Fuel Cycle, WASH-1248. Washington: U.S. Atomic Energy Commission, April 1974.

Foster, Arthur R., and Wright, Robert L., Jr. *Basic Nuclear Engineering*. 3rd ed. Boston: Allyn and Bacon, Inc., 1977.

Geesey, A. H., and Schultz, M. A. *New Safety System Design for Nuclear Power Reactors*. University Park: Pennsylvania State University, College of Engineering, 1971.

Graeub, Ralph. *The Gentle Killers: Nuclear Power Stations*. Translated from the German by Peter Bostok. London: Abelard-Schuman, 1974.

Graham, John. *Fast Reactor Safety*. New York: Academic Press, 1971.

Heckman, Harry A., and Starring, Paul W. *Nuclear Physics and the Fundamental Particles*. New York: Holt, Reinhart and Winston, Inc., 1963.

Impact Assessment of 1977 New York City Black-Out, U.S. Department of Energy, DOE HCP/T5103-01 (July 1978).

Interagency Review Group on Nuclear Waste Management, *Report to the President*, TID-28817 (October 1978 draft); see also Interagency Review Group on Nuclear Waste Management, *Subgroup Report on Alternative Technology Strategies for the Isolation of Nuclear Wastes*, TID-28818 (October 1978 draft).

Komanoff, Charles, Miller, Holly, and Noyes, Sandy. *The Price of Power: Electric Utilities and the Environment*. New York: Council of Economic Priorities, 1972.

Lapp R. E., *America's Energy*, Reddy Communications, Inc., Greenwich, Conn. (1976).

Lapp, Ralph E., and Schubert, Jack. *Radiation: What It Is and How It Affects You*. New York: The Viking Press, 1957.

Leachman, Robert B., and Althoff, Phillip (eds.). *Preventing Nuclear Theft: Guidelines for Industry and Government*. New York: Praeger, 1972.

Lovett, James E. *Nuclear Materials: Accountability, Management, Safeguards*. Hinsdale, Ill.: American Nuclear Society, 1974.

McBride J.P., et. al., Radiological Impact of Airborne Effluents of Coal and Nuclear Plants, *Science 202*, p. 1945 (December 8, 1978).

McCracken S., *The War Against the Atom*, Commentary (September 1977).

Metz, William D. *Laser Fusion: A New Approach to Thermonuclear Power*. Science 177 (No. 4055): 1180 (September 29, 1972).

Minton S. A., Jr., Snakebits, *Scientific American*, pp. 114-119 (January 1957).

INFCE Working Group 8. *Improved Once-Through Fuel Cycles for Light Water Reactors*, INFCE/WG.8/USE/DOC 12-A (November 30, 1978).

Proposed Staff Response to the Jeannine Honicker Petition for Emergency and Remedial Action, U.S. Nuclear Regulatory Commission, SECY-78-560 (October 1978).

Rasmussen N., et. al., *Reactor Safety Study, An Assessment of Accident Risks in U.S. Commercial Nuclear Power Plants*, U.S. Nuclear Regulatory Commission, WASH-1400 (October 1975).

Project Management Corp., Tennessee Valley Authority, and U.S. Energy Research and Development Administration, *Final Environmental Statement Related to the Construction and Operation of Clinch River Breeder Reactor Plant*, NUREG-0139, Docket No. 50-537 (February 1977).

Proposed Final Environmental Statement, Liquid Metal Fast Breeder Reactor Program, WASH-1535 (December 1974).

Sagain, Leonard A. (ed.). *Human and Ecologic Effects of Nuclear Power Plants*. Springfield, Ill.: Thomas, 1974.

Transporting the Nation's Coal: A Preliminary Assessment, U.S. Department of Transportation (January 1978).

Lapp R. E., *America's Energy*, Reddy Communications, Inc., Greenwich, Conn. (1976).

Willrich, Mason (ed.). *International Safeguards and Nuclear Industry*. Baltimore: Johns Hopkins University Press, 1973.

U.S. Atomic Energy Commission, *Proposed Final Environmental Statement, Liquid Metal Fast Breeder Reactor Program*, Vols. I and II, WASH-1535 (December 1974).

U.S. Department of Energy, *Energy Information Administration, Annual Report to Congress. Vol. III, 1977. Statistics and Trends of Energy Supply, Demand, and Prices*, DOE/EIA-0036/3 (May 1978).

Willrich, Mason, and Taylor, T. B. *Nuclear Theft: Risks and Safeguards*. Cambridge, Mass.: Ballinger, 1974.

OCEAN ENERGY

Davey, Norman. *Studies in Tidal Power*. London: Constable & Co., Ltd., 1923.

Clegg, Peter. *New Low-Cost Sources of Energy for the Home; with Complete Illustrated Catalog*. Charlote, Vt.: Garden Way Pub., 1975.

Creager, William Pitcher, Justin, Joel D., and Hinds, Julian. *Engineering for Dams*. New York: J. Wiley & Sons; London, Chapman & Hall Ltd., 1945.

Gillett, Colin Anson. *Can the Tides be Harnessed?* Raetihi, N.Z., 1958.

International Conference on the Utilization of Tidal

Power, Nova Scotia Technical College, 1970. *Tidal Power: Proceedings.* Gray, T.J., and Gashus, O. K. (eds.). New York: Plenum Press, 1972.

Macmillan, Donald Henry. *Tides.* New York: American Elsevier Pub. Co., 1966.

Handbook of Homemade Power. By the Staff of the Mother Earth News. New York: Bantam Books, 1974.

Morton, M. Granger (ed.). Ocean Thermal Energy Conversion. *Energy and Man: Technical and Social Aspects of Energy.* New York: IEEE Press, 1975.

Paton, Thomas Angus Lyall, and Brown, J. Guthrie. *Power from Water.* London: L. Hill, 1961.

Rash, Don E., et al. *Energy Under the Ocean: A Technology Assessment.* Norman, Ok.: University of Oklahoma Press, 1973.

Reynolds, John. *Windmills and Watermills.* London: H. Evelyn, 1970.

Ross, David. *Energy From the Waves.* Oxford, England: Pergamon Press, 1979.

Wilson, Paul N. *Water Turbines.* London: H. M. Stationery Off., Palo Alto City, Calif.: Obtainable in the U.S.A. from Pendragon House, 1974.

ORGANIC (BIOCONVERSION) FUELS

Anderson, Russell E. *Biological Paths to Self-Reliance - A Guide to Biological Solar Energy Conversion.* New York: Van Nostrand Reinhold, 1979.

Bureau of Mines Circular No. 8549. *Energy Potential from Organic Wastes: A Review of the Quantities and Sources.* Washington: U.S. Bureau of Mines, 1972.

Clark, Wilson. Solar Bioconversion. *Energy for Survival.* Garden City: Doubleday, 1974.

Golueke, Clarence G., and McGauhey, P. H. Waste Materials. *Annual Review of Energy.* Jack M. Hollander (ed.). Palo Alto: Annual Review, Inc.

Horner and Shifrin, St. Louis. *Energy Recovery from Waste.* Washington: U.S. Environmental Protection Agency, 1972.

Hunt, V. Daniel. *The Gasohol Handbook.* New York, N.Y.: The Industrial Press, 1981.

Jackson, Frederick R. *Energy from Solid Waste.* Park Ridge, N.J.: Noyes Data Corp., 1974.

Lowe, Robert A. *Energy Recovery from Waste; Solid Waste as Supplementary Fuel in Power Plant Boilers.* Washington: U.S. Environmental Protection Agency, 1973.

National Research Council, Canada. Research Plans and Publications Section. *Wood and Charcoal as Fuel for Vehicles.* 3rd ed. Ottawa, 1944.

Proceedings of a Conference, March 10-12, 1976, Washington, D.C. *Capturing the Sun Through Bioconversion.* Washington: Center for Metropolitan Studies, 1976.

The Science and Public Policy Program. University of Oklahoma. Organic Farms. *Energy Alternatives: A Comparative Analysis.* Washington: U.S. Government Printing Office, May 1975.

Energy Units and Conversion Factors

MULTIPLY	BY	TO OBTAIN
atmospheres	10,333	kgs per sq meter
atmospheres	14.70	pounds per sq inch
Bars	9.870×10^{-1}	atmospheres
Bars	10^6	dynes per sq cm
Bars	1.020×10^4	kgs per square meter
Bars	1.450×10	pounds per sq inch
British thermal units	0.2530	kilogram-calories
British thermal units	777.5	foot-pounds
British thermal units	3.927×10^{-4}	horsepower-hours
British thermal units	1054	joules
British thermal units	107.5	kilogram-meters
British thermal units	2.928×10^{-4}	kilowatt-hours
Btu per min	0.02356	horsepower
Btu per min	0.01757	kilowatts
centimeters	0.3937	inches
centimeters	0.01	meters
centimeters of mercury	0.1934	pounds per sq inch
centimeters per second	1.969	feet per minute
centimeters per second	0.036	kilometers per hour
centimeters per second	0.02237	miles per hour
cubic centimeters	10^{-6}	cubic meters
cubic centimeters	10^{-3}	liters
cubic feet	1728	cubic inches
cubic feet	0.02832	cubic meters
cubic feet	7.481	gallons
cubic feet	28.32	liters
cubic feet per minute	0.1247	gallons per sec
cubic feet per minute	0.4720	liters per sec
cubic inches	5.787×10^{-4}	cubic feet
cubic meters	35.31	cubic feet
cubic meters	264.2	gallons
cubic meters	10^3	liters
decimeters	0.1	meters
degrees (angle)	60	minutes
degrees (angle)	0.01745	radians
dekameters	10	meters
ergs	9.486×10^{-11}	British thermal units

MULTIPLY	BY	TO OBTAIN
ergs	1	dyne-centimeters
ergs	7.376×10^{-8}	foot-pounds
ergs	10^{-7}	joules
ergs	2.390×10^{-11}	kilogram-calories
ergs	1.020×10^{-8}	kilogram-meters
ergs per second	1.341×10^{-10}	horsepower
ergs per second	10^{-10}	kilowatts
feet	0.3048	meters
feet per second	18.29	meters per minute
foot-pounds	1.28×10^{-3}	British thermal units
foot-pounds	1.356×10^{7}	ergs
foot-pounds	5.050×10^{-7}	horsepower-hours
foot-pounds	3.241×10^{-4}	kilogram-calories
foot-pounds	0.1383	kilogram-meters
foot-pounds	3.766×10^{-7}	kilowatt-hours
foot-pounds per minute	1.286×10^{-3}	Btu per minute
foot-pounds per minute	0.01667	foot-pounds per sec
foot-pounds per minute	3.030×10^{-5}	horsepower
foot-pounds per minute	3.241×10^{-4}	kg, calories per minute
foot-pounds per minute	2.260×10^{-5}	kilowatts
foot-pounds per second	7.712×10^{-2}	Btu per minute
foot-pounds per second	1.818×10^{-3}	horsepower
foot-pounds per second	1.945×10^{-2}	kg. calories per min
foot-pounds per second	1.356×10^{-3}	kilowatts
gallons	3785	cubic centimeters
gallons	0.1337	cubic feet
gallons	231	cubic inches
gallons	3.785×10^{-3}	cubic meters
gallons	4.951×10^{-3}	cubic yards
gallons	3.785	liters
gallons	8	pints (liq)
gallons	4	quarts (liq)
gallons per minute	2.228×10^{-3}	cubic feet per second
gallons per minute	0.06308	liters per second
grams	10^{-3}	kilograms
grams	10^{3}	milligrams
grams	0.03527	ounces
grams	0.03215	ounces (troy)
grams	2.205×10^{-3}	pounds
grams centimeters	9.297×10^{-8}	British thermal units
grams per cc	62.43	pounds per cubic foot
grams per cc	0.03613	pounds per cubic inch
grams per cc	3.405×10^{-7}	pounds per mil-foot
horse-power	42.44	Btu per min
horse-power	33,000	foot-pounds per min
horse-power	550	foot-pounds per sec
horse-power	1.014	horsepower (metric)
horse-power	10.70	kg. calories per min
horse-power	0.7457	kilowatts
horse-power	745.7	watts
horse-power-hours	2547	British thermal units
horse-power-hours	1.98×10^{6}	foot-pounds
horse-power-hours	641.7	kilogram-calories
horse-power-hours	2.737×10^{5}	kilogram-meters
horse-power-hours	0.7457	kilowatt-hours

MULTIPLY	BY	TO OBTAIN
inches of mercury	0.03342	atmospheres
inches of mercury	1.133	feet of water
inches of mercury	345.3	kgs per square meter
inches of mercury	70.73	pounds per square ft
inches of mercury	0.4912	pounds per square in
inches of water	0.002458	atmospheres
inches of water	0.07355	inches of mercury
inches of water	25.40	kgs per square meter
inches of water	0.5781	ounces per square in
inches of water	5.204	pounds per square ft
inches of water	0.03613	pounds per square in
kilograms	10^3	grams
kilograms	2.2046	pounds
kilograms	1.102×10^{-3}	tons (short)
kilogram-calories	3.968	British thermal units
kilogram-calories	3086	foot-pounds
kilogram-calories	1.558×10^{-3}	horsepower-hours
kilogram-calories	4183	joules
kilogram-calories	426.6	kilogram meters
kilometers	10^5	centimeters
kilometers	3281	feet
kilometers	10^3	meters
kilometers	0.6214	miles
kilometers	1093.6	yards
kilometers per hour	27.78	centimeters per sec
kilometers per hour	54.68	feet per minute
kilometers per hour	0.9113	feet per second
kilometers per hour	0.5396	knots per hour
kilometers per hour	16.67	meters per minute
kilometers per hour	0.6214	miles per hour
kms per hour per sec	27.78	cms per sec per sec
kms per hour per sec	0.9113	ft per sec per sec
kms per hour per sec	0.2778	meters per sec per sec
kms per hour per sec	0.6214	miles per hr per sec
kilometers per min	60	kilometers per hour
kilowatts	56.92	Btu per min
kilowatts	4.425×10^4	foot-pounds per min
kilowatts	737.6	foot-pounds per sec
kilowatts	1.341	horsepower
kilowatts	14.34	kg. calories per min
kilowatts	10^3	watts
kilowatt-hours	3,412	British thermal units
kilowatt-hours	2.655×10^6	foot-pounds
kilowatt-hours	1.341	horsepower-hours
kilowatt-hours	3.6×10^6	joules
kilowatt-hours	860.5	kilograms-calories
kilowatt-hours	3.671×10^5	kilogram-meters
meters	100	centimeters
meters	3.2808	feet
meters	39.37	inches
meters	10^{-3}	kilometers
meters	10^3	millimeters
meters	1.0936	yards
meter-kilograms	9.807×10^7	centimeter-dynes
meter-kilograms	10^5	centimeter-grams

MULTIPLY	BY	TO OBTAIN
meter-kilograms	7.233	pound-feet
meters per minute	1.667	centimeters per sec
meters per minute	3.281	feet per minute
meters per minute	0.05468	feet per second
meters per minute	0.06	kilometers per hour
meters per minute	0.03728	miles per hour
meters per second	196.8	feet per minute
meters per second	3.281	feet per second
meters per second	3.6	kilometers per hour
meters per second	0.06	kilometers per min
meters per second	2.237	miles per hour
meters per second	0.03728	miles per minute
meters per sec per sec	3.281	feet per sec per sec
meters per sec per sec	3.6	kms per hour per sec
meters per sec per sec	2.237	miles per hour per sec
miles	1.609×10^5	centimeters
miles	5280	feet
miles	1.6093	kilometers
miles	1760	yards
miles per minute	88	feet per second
miles per minute	1.6093	kilometers per minute
miles per minute	0.8684	knots per minute
miles per minute	60	miles per hour
ounces	8	drams
ounces	437.5	grains
ounces	28.35	grams
ounces	0.0625	pounds
ounces per square inch	0.0625	pounds per sq inch
pounds	444,823	dynes
pounds	7000	grains
pounds	453.6	grams
pounds of water	0.01602	cubic feet
pounds of water	27.68	cubic inches
pounds of water	0.1198	gallons
Quadrants (angle)	90	degrees
Quadrants (angle)	5400	minutes
Quadrants (angle)	1.571	radians
Radians	57.30	degrees
Radians	3438	minutes
radians per second	57.30	degrees per second
radians per second	0.1592	revolutions per second
revolutions	360	degrees
revolutions	4	quadrants
revolutions	6.283	radians
revolutions per minute	6	degrees per second
square feet	144	square inches
square feet	0.09290	square meters
square feet	3.587×10^{-8}	aquare miles
square feet	1/9	square yards
square meters	2.471×10^{-4}	acres
square meters	10.764	square feet
square meters	3.861×10^{-7}	square miles
square meters	1.196	square yards
square miles	640	acres
square miles	2.788×10^7	square feet

MULTIPLY	BY	TO OBTAIN
square miles	2.590	square kilometers
square miles	3.098×11^6	square yards
square yards	2.066×10^{-4}	acres
square yards	9	square feet
temp (degs C) + 273	1	abs temp (degs K)
temp (degs C) + 17.8	1.8	temp (degs F)
temp (degs F) + 460	1	abs temp (degs R)
temp (degs F) − 32	5/9	temp (degs C)
tons (long)	1016	kilograms
tons (long)	2240	pounds
tons (metric)	10^3	kilograms
tons (metric)	2205	pounds
tons (short)	907.2	kilograms
tons (short)	2000	pounds
tons (short) per sq ft	9765	kgs per square meter
tons (short) per sq ft	13.89	pounds per sq inch
tons (short) per sq in	1.406×10^6	kgs per square meter
tons (short) per sq in	2000	pounds per sq inch

ENERGY UNITS AND CONVERSION FACTORS

Approximate Energy Equivalents

FUEL SOURCE	UNITS	AVERAGE HEAT VALUE (Millions of Btu)
Crude Oil	Barrel	5.8
Residual Fuel Oil	Barrel	6.3
Average U.S. Coal	Short Ton	24.0
Bituminous Coal	Short Ton	26.0
Lignite Coal	Short Ton	13.5
Natural Gas (dry)	Cu.ft. $\times 10^3$	1.03
Synthetic Gas[1]: Low-Btu	Cu.ft. $\times 10^3$	0.2
Med-Btu	Cu.ft. $\times 10^3$	0.27
High-Btu	Cu.ft. $\times 10^3$	0.97
Uranium[2]	Pound U_3O_8	
a. In Thermal Reactors		215
b. In Breeder Reactors		18,500
Oil Equivalent[3]	Tonne	40
Coal Equivalent[3]	Tonne	28

[1] Low (100-500 Btu/scf), Medium (240-300 Btu/scf), High (950-1000 Btu/scf)
[2] Assuming tails assay at 0.30% and recycle of plutonium
[3] Basic tonne oil equivalent (toe) = 10×10^6 Kcal or 39.7×10^6 Btu. One tonne coal equivalent (tce)—0.7 toe (Organization for Economic Cooperation and Development)

Conversion Factors

TO OBTAIN NUMBER OF	MULTIPLY NUMBER OF	BY THE FOLLOWING CONVERSION FACTOR	
		Actual	Rule of Thumb
British Thermal Units	Kilogram-calories	3.968	4
Joules	British Thermal Units	1,055	10^3
British Thermal Units	Kilowatt hours	3415	$(1/3) \times 10^4$
Horsepower	Kilowatts	1.34	—
Gallons	Barrels of oil	42	42

Milestone Symbols

SYMBOL DEFINITIONS

◁ BEGIN MILESTONE

▷ END MILESTONE

⬡ DECISION MILESTONE

◀ BEGIN MILESTONE (TASK COMPLETED)

▶ END MILESTONE (TASK COMPLETED)

⬢ DECISION MILESTONE (DECISION MADE)

●1 RELEASE REQUEST FOR PROPOSALS OR PROGRAM OPPORTUNITY NOTICE

●2 PROPOSALS RECEIVED

●3 CONTRACTOR SELECTED

●4 CONTRACT AWARD

●5 ORDER LONG-LEAD ITEMS

●6 FINAL ENVIRONMENTAL IMPACT STATEMENT

● INFORMATION TRANSFER

SUMMARIES OF DECISIONS AND REVIEWS

SYMBOL		TASKS
⬡	ACCEPTANCE TEST	Review performance demonstration test data and other supporting data, including analyses, lower level test data, and vendor test data. Inspect facility configuration, configuration control records, test logs and pertinent documentation required to demonstrate compliance to the statement of work and performance specifications.
⬡	BID PACKAGE REVIEW	Review construction bid packages.
⬡	CONSTRUCTION STATUS REVIEW	Review (periodically) of construction against established subcontract schedules that utilize critical path methods for construction or equivalent control mechanisms.
⬡	CRITICAL DESIGN REVIEW	Review detail design drawings and specifications, analytical and experimental verification data, long lead item procurement list, bid package plan, siting and environmental impacts, final test and evaluation plan, configuration and change control procedures.
⬡	PERFORMANCE REVIEW	Establish degree of compliance with planned testing and evaluation program. Evaluate progress towards determination of commercial feasibility of process.
⬡	PILOT PLANT CONSTRUCTION	Review procurement plans, adequacy of construction go-ahead actions, construction schedules, and plans for subcontracting.
⬡	PROCESS CONCEPT REVIEW	Evaluate feasibility of proposed process concepts using the best available empirical and theoretical information. Review process system's practicality by estimating operating conditions and general economics.
⬡	PROCESS DEVELOPMENT UNIT CONSTRUCTION	Review procurement plans, adequacy of construction go-ahead actions, construction schedules, and plans for subcontracting.
⬡	PROJECT EVALUATION	Review project progress and accomplishments, compare achieved results with established objectives. Review outcome; may be used to redirect future project goals.
⬡	TECHNOLOGY ASSESSMENT	Evaluate feasibility of the proposed technology concepts, using the best available empirical and theoretical information. Review the technology's practicality by estimating operating conditions and general economics.

Index

Accident liability, 857-858
Accident risks, 938-943. See also Safety
Acid digestion, 828
ACTF. See Advanced Component Test Facility
Ad Hoc Experts' Group on Fusion, 869
Adjacent Coal Bed Mining, 95
Advanced battery development and supporting subprograms, 490
Advanced Cogeneration Systems, 24, 218-228
Advanced Combustion Technology, 250-256
Advanced Component Test Facility (ACTF), 408
Advanced environmental control technology, 23, 83, 194-211, 250
Advanced Exploratory Research (AER), 314
Advanced Fission Systems Evaluation, 43-44
Advanced Flue Gas Cleanup, 200-201
Advanced Flue Gas Desulfurization, 198-200
Advanced Fuels/Multipoles (AF/MP), 922, 924
Advanced fusion concepts and Applied Plasma Physics, 921-930
Advanced Fusion Concepts Subactivity, 921-925
Advanced gasification, 58
Advanced High Temperature Reactor, 604
Advanced Hydraulic Fracturing (AHF), 330, 331-333
Advanced Isotope Separation, 41, 570, 576, 589, 606-613
Advanced Isotope Separation Technology Program work breakdown structure, 608
Advanced Mining System, 99
Advanced nuclear systems, 43, 44, 703-736
Advanced Nuclear Systems Evaluation Program work breakdown structure, 705
Advanced Panel Mining, 99
Advanced Process Technology Program, 27, 75, 314-319
Advanced Reactor Systems Program, 41, 615-626
Advanced Research and Technology Development, 83, 172-173, 407-408, 411-413
Advanced Systems Development Activity, 408
Advanced Systems Evaluation, 703-718
Advanced Technology Systems and Assessment Program, 713-718
Advanced Water-Cooled Breeder Applications Program, 42, 670, 676-677
Advisory Committee on Reactor Safety (ACRS), 637
Advisory Group on Radioactive Waste Disposal into Geologic Formations, 863
AEF. See Atomic Energy Commission
Aerospace Corporation, 399

AF/MP. See Advanced Fuels/Multipoles
AFB Closed-Cycle Turbine Test Unit (TTU), 241-243
AFB Industrial Applications, 241
AFB 30-MW Boiler, 237, 238-240
AFR. See Away-from-reactor
AGNS-Barnwell Facility, 844
Agricultural Research Service, 254, 424
Airborne waste, 773, 801-802, 829-830
    filtration, 801
    Idaho Chemical Processing Plant, 802
    removal, 801, 803
    research and development, 802
Air conditioning. See Heating and air conditioning
Air pollution, 25
Alabama, 88
Alaska, 4, 77
Albuquerque Operations Office (ALOO), 596, 722, 769, 781, 805, 822, 828
Alcator, 894, 901, 903
Alcohol. See Ethanol
Alkali Metal Vapor Cycle, 216
ALOO. See Albuquerque Operations Office
Alternate fuel combustion, 259
Alternate Heating System Development task, 914
Alternative fuel combustion, 258-260
Alternative fuel utilization, 256-261
Alternative sources of energy, 4
American Boiler Manufacturers Association, 254
American Petroleum Institute (AIP), 467
    responsibilities, 467
American Society for Testing and Materials, 467
Ammonia turbines, 431
Anaerobic digestion, applications, 34, 438, 442
    commercialization, 438, 443-444
    environmental impact, 538
Analytical test hot cell, 700
ANL. See Argonne National Laboratory
Anthracite applications, 244-246
Anvil Points Oil Shale Facilty, 27
AP&L. See Arkansas Power & Light
API. See American Petroleum Institute
Applications, fusion energy. See Fusion Energy Applications (Subactivity)
Applied Physics Laboratory (APL), Johns Hopkins University, 733
Applied Plasma Physics Activity, 882-883, 921-930
Aquatic energy farms. See Energy Farms
Aquifer, 458, 461

Index

Aquifers, energy storage systems, 502-503, 513-514. See also Geopressured resources
Arctic, 75
Argentina, 581
Argonne National Laboratory (ANL), 171, 180, 254, 346, 399, 623, 649, 658, 660, 667-668, 701-702, 718, 738, 846, 912, 939
Arizona, photovoltaic projects, 557
Arkansas Power & Light (AP&L), 465, 598
Arms Control and Disarmament Agency (ACDA), 711
ASET. See Assistant Secretary for Energy Technology, DOE
Ash Slag Retention, 205
Assistant Secretary for Conservation and Solar Energy (CSE), 378, 388
Assistant Secretary for Defense Programs, 869, 890
Assistant Secretary for Energy Technology (ASET), DOE, 702, 888, 890
Assistant Secretary for Environment, 772
Assistant Secretary for Intergovernmental and Institutional Relations, 953
Assistant Secretary for Policy and Evaluation (PE), 388
Assistant Secretary for Resource Applications (ASRA), DOE, 772
Associated Universities of New York, 896
Atmospheric Fluidized Bed (AFB), 58, 196, 237-246
Atmospheric fluidized bed coal combustion externally fired gas turbine technology test unit, 243
Atmospheric fluidized bed combustion test and integration unit, 245
ATMX-500 rail car, 808
Atomic Energy Commission (AEC), 579, 591, 627, 638, 671, 738, 752, 756-758, 775, 787, 805, 938
Atomic Energy Commission, Environmental Impact Statement (Sept. 1974), 852
Atomic vapor laser isotope separation, 611
Atomic Vapor Process, 610-611
Auburn University, 369
Australia, 358, 364
Austria, 362
Automated extraction system, 92
Automated rail haulage, 99
Away-from-reactor (AFR), 753, 758, 840
 pilot plant, 843
 storage, 760, 842-844

Barstow, Calif., project, 404
Bartlesville Energy Technology Center (BETC), 80, 285, 312, 316, 331, 335, 342-343
Bartow, Fla., project, 438
Basalt formations, 757
Battelle Memorial Institute, 752, 757, 814
Battelle Pacific Northwest Laboratory (BPNL), 831
Batteries, 17, 31, 39, 44
Battery. See also Battery. See also Battery Storage Program
 applications, 488, 490, 494
 calcium/metal sulfide, 496
 and electric load leveling, 488
 lead/acid 490-492
 lithium/metal sulfide, 490-493, 495, 497
 metal/air, 495
 nickel/iron, 490-492
 nickel/zinc, 490-492
 reduction/oxidation, 495
 sodium/sulfur, 492, 495-496
 zinc/chlorine, 490-492
Battery Energy Storage Test (BEST) Facility, 492, 493, 500
Battery-flywheel hybrid electric car, 510
Battery Storage Program, 38, 380, 490-500
 Advanced Battery Development Subprogram, 472-494
 commercialization, 490, 534
 description, 490-500
 Dispersed Battery Applications Subprogram, 498-500
 Electrochemical Systems Research Subprogram, 495-496
 Electrolytic Technology Subprogram, 497-498
 energy saving estimate, 396
 environmental impact, 542-543
 international cooperation, 563
 Near-term Batteries Subprogram, 491-492
 objectives, 490-491
 regional activities, 525-526
 Solar Applications Subprogram, 494-496
 Supporting Research Subprogram, 496-497
 technology transfer, 534-535
BCS. See Buildings and Community Systems
Belgium, 362-363, 678, 680, 800, 861, 863
Bench-scale solvent extractor, 695
Beryllium, 949
BEST. See Battery Energy Storage Test Facility
Best Available Control Technology (BACT), 236, 261
Beta II, 905-907
Beta II device, 906
Beta value, plasma, 873, 899, 906-907
 defined, 873, 875-876
 high, 906-907, 925
Bi-Gas Pilot Plant, 132, 150
Bilateral agreements. See International Cooperation
Biofouling resistance, 33
Biological effects, 873-874, 892, 948-952
Biomass, 12, 434-446
Biomass Energy Systems Program, 376-377, 434-446
 commercialization, 33, 437, 533
 Conversion Technology Program, 440-444
 description, 33-34, 434-435
 environmental impact, 446, 538
 interagency cooperation, 400, 436-437
 international cooperation, 562
 objectives, 435
 Production Systems Program, 439-440
 regional activities, 522
 Research and Development Subprogram, 444-445
 socioeconomic aspects, 546
 strategy, 435-436
 Support and Other Programs, 445-446
 Technology Support Program, 437-439
 Thermochemical Conversion Subprogram, 440-442
Biomass Energy Systems Program Biochemical Conversion Subprogram, 442-444
Biomass resources applications, 437, 439
Biomass resources markets, 437, 439
Biomass systems. See also Anaerobic digestion; Biomass Energy Systems Program; Direct combustion; Energy farms; Fermentation
Biophotolysis, 446
Blanket reactor, 873-874, 876, 948
 and advanced concepts, 922
 and shield engineering, 917

Blind Shaft Borer, 87-88
Block Island, R.I. project, 32
BMIS. See Business Management Information System
BNL. See Brookhaven National Laboratory
Boiler fuel, 5, 9, 11
Boiling-water reactor nuclear steam supply, 592
Bonn Summit, 550
Boom-type bucket wheel excavator, 103
Boone, N. C., project, 532
Boone, North Carolina, wind energy conversion system, 428
Boron, 870
Borosilicate glass, 794
Bottoming cycle, 216
Bottoming Cycle Systems Program, 231
BPWG. See Budget and Planning Working Group
Brayton cycle, 406, 593-594
Brayton Engine, 223
Breakeven, energy, 876, 880, 897-899, 901, 930
Breeder commercialization, 571, 572
Breeder component development contractors, 648
Breeder facility projects, 661
Breeder fuels and core materials contractors, 657
Breeder materials/chemistry contractors, 659
Breeder physics activity contractors, 651
Breeder reactor. See Reactor
Breeder technology safety activity contractors, 645
British Gas/Slurry Lurgi, 155-156
Brookhaven National Laboratory (BNL), 259, 346, 399, 485, 486, 890, 941
Brown's Ferry Power Plant, 40
Budget
  advanced batteries, 494
  advanced fusion concepts, 926
  advanced isotope separation technology program, 608
  advanced nuclear systems, 703
  advanced nuclear systems economics evaluation, 714
  advanced nuclear systems evaluation, 705
  advanced nuclear systems increased siting acceptability, 719
  advanced reactor systems, 618
  advanced water breeder application, 679
  airborne waste, 803
    commercial, 830
  Albuquerque Operations Office, 782
  Alcator, 904
  applied plasma physics activity, 922
  applied plasma physics and reactor projects, 888
  Argonne National Laboratory facilities, 663
  away from reactor storage facilities, 844
  battery storage, 491
    solar applications, 495
    supporting research, 498
    thermal and mechanical storage, 489
  beta II, 907
  biochemical conversion, 445
  biomass energy systems, 438
  biomass production systems, 442
  biomass R&D, 446
  biomass technology support, 441
  breeder component development, 646
  breeder facilities, 668
  breeder fuels and core materials, 656
  breeder materials/chemistry, 659
  breeder physics, 651
  breeder reactor design studies, 641
  breeder reactor studies, 640
  breeder reactor systems, 630
  breeder technology safety, 645
  chemical/hydrogen energy storage, 510
  coal advanced environmental control technology, 197
  coal liquefaction subactivities, 113
  coal resource activities, 84
  combustion systems activity, 238
  commercial waste D&D, 836
  converter reactor systems, 590
  core flow test loop, 686
  customer storage of off-peak electricity, 507
  decontamination and decommissioning, 811
  Defense Waste Processing Facility, 797
  Dispersed battery applications, 500
  Doublet, 901
  drilling and offshore production technology, 310
  electric energy systems, 475
  electric field effects, 483
  electrochemical systems research, 496
  electrolytic technology, 499
  energy storage programs, 490
    technical/economic analysis, 516
  energy technology, 14
  energy technology centers/mining technology centers, 342-343
  enhanced gas recovery, 322
  enhanced oil recovery 283
  ETEC facilities, 668
  existing geopressured well tests, 461
  experimental plasma research, 931
  fast breeder reactor plant projects, 635
  Fast Flux Test Facility, 638
  fission energy program, 574-575
  fossil energy, 62, 63-69
    advanced research and technology development, 174
  fuel cell activity, 264
  fuel cycle proof of breeding, 700
  fusion energy applications, 921
  fusion energy environment and safety, 920
  fusion materials irradiation test facility, 937
  fusion plasma theory, 927
  fusion reactor materials, 916
  fusion systems engineering, 920
  FY 1978-1980, 15
  FY 1979 and 1980, 61
  gas cooled fast reactor, 681
    development, 684
  gas cooled thermal reactor, 615
  Gas Reactor In-Place Safety Test, 687
  gas resource activity, 321
  geochemical engineering and materials, 467
  geopressured resources, 459
    definition, 460
    reservoir engineering, 462
  geosciences, 468
  geothermal drilling and completion technology, 464
  geothermal energy, 451
    conversion systems and stimulation, 466
  geothermal technology development, 463
  Hanford Basalt Program, 819
  Hanford Engineering Development Laboratory, 664
  Hanford Reservation, 783
  heat engines and heat recovery activity, 213
  high intensity neutron source facility, 936
  high-level nuclear waste, 795
  high-level waste, 826
  high-temperature gas-cooled reactor, 606
  hot dry rock, 470

996 Index

Budget (*cont.*)
  hydrothermal engineering applications, 455
  hydrothermal facilities, 457
  hydrothermal resources, 452
  hydrothermal resource definition, 453
  Idaho National Engineering Laboratory, 785
  Impurity Studies Experiment, 902
  increased thermal energy utilization, 717
  industrial waste heat storage, 505
  in situ gasification, 165
  International Nuclear Fuel Cycle Evaluation, 712
  intermediate-scale wind energy conversion systems, 427
  international spent fuel storage, 840
  isotopes fuels, 731
  landlord responsibility (commercial), 832
  large-scale wind energy conversion systems, 429
  large solar power systems, 405
  light metal fast breeder reactor, 635
  light water breeder reactor, 676
  liquid metal fast breeder reactor safeguards, 669
  long-term waste management technology, 793
  low-level nuclear waste, 798
  low-level waste, 827
  magnetic fusion confinement systems, 886, 894
  magnetic fusion development and technology, 887
  magnetic fusion energy, 939, 940
    computer network, 932
    development and technology, 912
    reactor projects, 933
  magnetic systems, 913
  magnetohydrodynamics, 273
  mechanical energy storage, 511
  mirror Fusion Test Facility, 91
  MR&D, 85
  national waste terminal storage, 823
  near-term batteries, 493
  Nevada Test Site, 788, 821
  new geopressured drilled wells, 461
  Nonproliferation Alternative Systems Assessment Program, 710
  nuclear advanced technology systems assessments, 715
  nuclear decontamination and decommissioning, 734
  nuclear fuel cycle R&D, 690
  nuclear nonproliferation studies, 706
  nuclear power system technology and support, 729
  nuclear systems flight development, 725
  nuclear waste, 771, 772
  Oak Ridge National Laboratory, 789
  Occupational radiation dose reduction/productivity enhancement, 602
  Ocean systems, 430
    advanced R&D, 436
    engineering test and evaluation, 435
    technology development, 433
  Office of Fusion Energy, 883, 893, 954, 955
  Office of Nuclear Waste Isolation, 817
  Oil shale, 299
  petroleum advanced processes technology program, 315
  petroleum resources, 281
  Phaedrus, 910
  photovoltaic advanced R&D, 413
  photovoltaic systems engineering and standards, 418
  photovoltaic technology development, 416
  photovoltaic test and applications, 420
  plasma engineering, 915
  Poloidal Divertor Experiment, 899
  power supply load management, 478
  Princeton Large Torus, 897
  proliferation resistant concepts, 625
  proof of breeding facility, 702
  reduced enrichment research and test reactors, 623
  remote operating/maintenance demonstration area, 698
  Richland, Washington, Landlord Responsibilities, 791
  Savannah River Plant, 790
  seasonal storage in aquifers, 504
  Shippingport Atomic Power Station, 674
  small-scale wind energy conversion sytems, 426
  small solar power systems, 407
  solar electric applications, 403
  solar generic/integrations of new source technologies, 480
  solar geothermal, electric, and storage systems, 381, 382-383, 384
    research, 521
  solar heating and cooling, 506
  solar thermal advanced R&D, 409
  solar thermal power, 403
  space nuclear power system safety, 733
  space nuclear systems, 723
  space/terrestrial applications, 723
  spent fuel program management and supporting studies, 843
  storage for solar thermal power, 508
  superconducting magnetic energy, 512
  supporting commercial waste studies, 833
  surface coal gasification, 131
  Tandem Mirror Experiment, 909
  terrestrial isotope applications, 735
  thermal and mechanical storage, 501
    applications analysis, 517
    technology information systems, 518
  thermal energy storage, 503
  thermal reactor safety, 604
  thermal reactor technology program, 597
  thermochemical conversion, 443
  Tokamak Fusion Test Reactor, 934
  TRU waste, 800
  TRU waste (commercial), 829
  underground cables and compact stations, 485
  university fossil energy programs, 369
  uranium enrichment, 610
  uranium utilization, 599
  utility applications, 514
  Waste Retrieval and Treatment Facility, 801
  waste treatment technology, 825
  water cooled breeder, 671
  wind energy conversion systems, 421
    devleopment and technology, 423
  WIPP project, 807
  WIPP R&D, 805
  WIPP transportation system development, 809
Budget and Planning Working Group (BPWG), 469
Buildings and Community Systems (BCS), 380, 395
Bureau of Mines, U.S. (BOM), 244, 326, 441
Burial
  at Nevada Test Site, 757, 785-786
  at Savannah River Plant, 827
  of low-level wastes, 834

Cables. See Transmission lines
California geothermal resources, 523
Canada, 77, 361-363, 581, 677, 861, 959
Canadian Deuterium Uranium (CANDU) Reactor, 581

Candidate D&D facilities, 810
Carbon-*14*, 822
  airborne waste, 802
  immobilization technology development, 802
Carbon Dioxide Flooding, 286-289
Carbondale Mining Technology Center (CMTC), 336, 343, 346
Carter, President, 4, 43, 550, 630, 631, 688, 738, 751-753, 755, 764, 775, 808, 836
Cascade Improvement Program (CIP), 607
Catalyst Development, 151-152
Catalytic Gasification, 145-146, 151
Catalytic gasification process, 147
Caustic Flooding, 283
CCMS. See Committee on the Challenges of Modern Society
CEC. See Commission of the European Communities
Central receiver solar thermal power system, 31, 404
Central receiver solar thermal test facility (CR-STFF), 404
CEQ. See Council on Environmental Quality
Cerro Prieto, Mexico, geothermal field, 473
Cesium-*137* encapsulation, 780
Chemical Coal Cleaning, 58, 106-107
  processes, 109
Chemical explosive fracturing process, 330, 333-334
Chemical heat pump storage system, 506
Chemical/hydrogen energy storage, 508-509
Chemical ocean thermal energy conversion plant, 432
Chicago Operations Office (DOE), 81, 596, 623, 701, 722
China, 956
Civilian Nuclear Waste. See Commercial Waste Management Activity
Clayton, N. Mexico, project, 32
Clean Air Act, 105, 194, 236, 349, 356
Clinch River Breeder Reactor Plant (CRBRP), 579, 630, 638-639, 659
Closed-cycle Brayton system, 508
Closed-Cycle Gas Turbine, 208, 213
Closed-cycle magnetohydrodynamics, 277-278
Closed-Cycle Power System, 213-214, 216-217
Closed-Cycle Systems (MHD), 277-278
C0-Steam Process, 120
Coal, 21-25. See also Feedstock
  consumption of, 12
  conversion to, 83, 181-182, 188-190
  demand/reserves, 22
  International programs, 358-365
  mining R&D, 22, 99-100
  preparation, 58, 84, 105-110
  resources, U.S., 22
  slagging process, 136-137
Coal-bed methane, 260
Coal benefication, 106
Coal-cleaning, studies and support, 109
Coal Cleaning Waste—Management of, 207
Coal combustion, support and engineering evaluations, 253-256
Coal—Energy Source, 58
Coal fields (U.S.), 163
Coal-Fired Flow Facility, 276, 277
Coal—Flue Gas Cleanup, 195-203
Coal-Gas Stream Cleanup
Coal gasification, 10, 13, 23
  as a priority goal, 23
  environment and, 13
  high-Btu, 23
  in Situ, 162-172
  low-Btu, 23
  pilot plant, 23
  R&D in, 13
Coal gasifier, 203
Coal generation opportunities, 219
Coal hydrogenation, 363
Coal liquefaction, 9, 22-23, 110-131
  as a priority goal, 9
  commercialization, 111
  demonstration plants, 129-131
  products of, 9
  R&D in, 13
  strategy, 111-112
Coal Liquefaction Program, 110-131
Coal Liquids Refining, 128
Coal Mine Shaft Development, 87
Coal-oil mixture combustion, 257
Coal-Oil Mixtures, 256-258
Coal Resources Program, 20-25, 82-83
Coal Resources Strategy, 22, 58-59
COGAS Process, 153-154, 156
Cogeneration, 218-228
  concepts, 226
Cogeneration Technology Alternatives Study, 225, 227
CO_2 injection process, 290
Cold Water Pipe (CWP) systems, 33
Colorado, 26, 97
Colorado School of Mines, 370, 848
Combined closed-cycles turbine/steam turbine system, 217
Combustion System Demonstration Plants, 260-261
Combustion systems, 24, 83, 236-261
  fluidized bed, 236-237
  large-scale, 252
Combustor Design and Modification (AECT), 205
Commercial applications
  energy storage systems, 517-519, 534
  geothermal systems, 391-392, 470-472, 534
  photovoltaic systems, 409, 530-531
  solar thermal systems, 528-530
Commercial burial sites, 834
Commercial Demonstration Reactor, 881
Commercial light water reactor fuel cycle, 593
Commercial nuclear powerplants, 594
Commercial nuclear waste vitrification project, 835
Commercial Readiness Demonstration Project (CRDP), 418
Commercial Waste Management Program 761-763, 810-836
Commercial Waste Volume, 752
Commercial wind machines, 424
Commercialization, 8, 14, 849-850
  biomass systems, 533
  Commission of the European Communities, 832, 863
  electric energy systems, 534
  energy storage systems, 534-535
  fission energy, 737-739
  fossil energy development, 348-351
  fusion energy, 880, 944-947
  and reactor design, 881-882
  and socioeconomic issues, 892
  long-range issues, 890-892
  geothermal energy, 534
  hydrothermal systems, 450-451
  nuclear waste, 849-850
  ocean systems, 533
  photovoltaic systems, 530-531

Commercialization (cont.)
  solar technology, 528-539
  solar thermal systems, 528-530
  technologies and demonstrations, 350
  technology development cycle, 527
  wind energy conversion systems, 531-533
Commercialization Process, 527-528
Commercialization, Solar, Geothermal, Electric, and Storage Systems Program, 527-535
Commission of the European Communities (CEC), 832, 863
Committee on Radioactive Waste Management, NAS, 48, 762, 832
Committee on the Challenges of Modern Society (CCMS), 472
Community applications. See Electricity generation
Compact Toroid (CT), 924
Component Development and Integration Facility (CTIF), 275-276
Component Test and Integration Facility (CTIF)
Component Test and Integration Unit (CTIU), 237, 243-249
Components Handling and Cleaning Facility (CHCF), 669
Compressed Air Energy Storage (CAES), 512-514, 524, 542
Compulsator, 912
Conceptual Design and Engineering (task), 916, 918
Conceptual waste repository, 822
Conceptual waste solidification process, 794
Conceptual Design Study (CDS), 570-572, 628, 631-632, 635, 639, 648
Confinement
  gravitational, 872
  inertial, 867, 872, 889-890
  magnetic, 51-52
    concept, 872
    equilibrium/stability field, 898-899
    poloidal field, 898, 899
    toroidal field, 889, 924
Confinement systems, magnetic fusion, 882, 893-910
Conservation, energy, 4-6
Conservation and solar energy programs, 385
Conservation and Solar Energy Organization, 397
  homeowners and, 4
  legislation, 5-6
Consolidated synthetic fuel process, 126-127
Constance II, 905
Consumption, U.S. energy, 3-4, 12
  incentives for reduced, 3-4
Continuous Face Haulage, 94
Continuous Minor/Bolter, 91-92
Contour mining, 103-104
Contractors. See names of specific companies
Contractors for Industrial Boilers/heaters, 241
Converter Reactor Systems, 40-42, 575-577, 589-626, 738-739
Conveyor Systems, 104
Cooling. See Heating and air conditioning
Cooperative Power Demonstration Reactor Program 591-592
Core drilling, 820
Core Flow Test Loop (CFTL), 680, 684-686
Corona effects, 482, 487, 541
CO-steam process, 120
Council on Environmental Quality (CEQ), 753-754
CRDP. See Commercial readiness demonstration projects
Cresap Liquefaction Test Facility, 126-128

Crossridge Demonstrations, 104
CR-STTF. See Central receiver solar thermal test facility
Crude oil, 75
CRWM. See Committee on Radioactive Waste Management
CSA. See Office of the Assistant Secretary for Conservation and Solar Applications
CSE bilateral agreements, 560
CSE/IEA agreements, 556
CST. See Central Solar Technology
Culebra, Puerto Rico, project, 32
Current and projected nuclear wastes, 752
CWP. See Cold water pipe systems

D&D. See Decontamination and Decommissioning
Damage analysis, 995
Darrieus rotor, 421
Data base
  handling systems, 873, 926
  impurity control, 926-928
  materials, 890
  physics, 882-883, 930
  plasma properties, 895, 915, 928-929
  vacuum component, 914
DBER. See Division of Biomedical and Environmental Research
Decentralization of Project Management for all DOE Out Projects, 19
Decentralized applications. See Dispersed source applications
Decontamination and decommissioning, 48, 733-734, 760-762, 809-810, 834
  D&D, commercial facilities, 834
Defense nuclear waste inventory, 779
Defense Waste Management Activity, 759-761, 778-810
Defense Waste Processing Facility (DWPF), 794
Defense Waste Processing Facility reference process, 796
Delcambre No. 1 well
Denmark, 560-561
Density, plasma, 895-898
  and Alcator, 895
  and Lawson Product, 895
  and PDX, 897-898
  and PLT, 895
  required for fusion, 880
Department of Agriculture (USDA), U.S., 207, 255, 297, 422, 424, 437, 440
  biomass projects, 439-440
  wind projects, 32
Department of Defense (DOD), U.S., 397, 474, 488, 719-720, 724
Department of Energy (DOE), U.S. See also Assistant Secretary, DOE; Laboratories, DOE.
  coal strategy, 5
  contractors. See individual contractors.
  decentralization, 19, 520-521, 774
  fusion energy policy, 867-869
  nuclear waste management approach, 810
  RD&C cycle, magnetic fusion, 878
  structure, 398-399
  U.S., utilities and Department of Energy Organization Act, 396, 752
Department of Housing and Urban Development, U.S. (HUD), 241
Department of Interior, U.S., 241, 474. See also U.S. Geological Survey

Department of Labor, 741
Department of the Navy, U.S., 488
Department of State, U.S., 622, 711
Deregulation, natural gas, 6
Deuterium-tritium, 867, 914
  fusion reaction, 870-874
  and ignition point, 870-871, 880
  major scaling experiment, 880
  plasma, 871, 872, 873, 930
Development and Technology, magnetic fusion, 52, 882, 918-921
Devonian shale, 59, 324
DGE. See Division of Geothermal Energy
Direct combustion, 437-438
  applications, 34
  commercialization, 438
  environmental impact, 437
  regional activities, 522
  wood resource distribution, 437-438
Direct Energy Conversion (task), 914
Direct Hydrogenation, 112-113, 123
Direct Injected Stratified Charge, 252
Directionally Controlled Wells and Earth Fracture System Process, 334
Directly Fired Heat Cycles, 219-222
Director of Energy Research, DOE, 840
Disposable Catalyst Hydrogenation, 121-123
  process, 124
Division of Biological and Environmental Research, 487
Division of Biomedical and Environmental Research (DBER), 541-542
Division of Central Solar Technology (CST), 397, 400, 507
Division of Distributed Solar Technology, 397, 400
Division of Electric Energy Systems (EES), 29, 375, 398, 474. See also Electric Energy Systems Program
Division of Energy Storage (STOR) Systems, 398, 488
Division of Environmental Control Technologies, 487, 541
Division of Geothermal Energy (DGE), 397, 460, 539
Division of Naval Reactors, 44
Division of Nuclear Power Development (NPD), 501, 535, 596, 617
Division of Planning and Technology Transfer, 397
Division of Program Resource Management, 397
DOD. See Department of Defense, U.S.
DOE. See also Department of Energy, U.S.
  fusion program logic, 52
  regions, 859, 953
Domestic Policy Review (DPR), 35
  energy projections, 13
  of Solar Energy, 389-391
Domestic Spent Fuel Storage Activity, 763, 836, 839-894
Donor Solvent Liquefaction, 117-119
Donor solvent process, 119
Double-wall-tube steam generator, 647
Doublet experiments, 894-895, 898-899, 914
Doublet-III device, 900
Downhole Telemetry, 310-311
DPR. See Domestic Policy Review
Drilling and Offshore Technology, 27, 75, 309-314
Drilling technology, 310-312
DST. See Division of Distributed Solar Technology
D-T. See Deuterium-tritium
DWPF. See Defense Waste Processing Facility (Savannah River)

EA. See Environmental Assessment
Eastern Gas Shales, 28, 321-324
EBT. See Elmo Bumpy Torus
Ebullated-Bed (H-Coal) Pilot Plant, 112-114
Ecological systems. See Biological communities
Economic Assessment Service for Coal, 361-362
Economic Regulatory Administration, 394
Economics
  energy, U.S. growth and, 3-4
  fusion energy, 873, 875, 892
Economics Evaluation Program, 44
Edgerton, Gemeshausen, and Grier Corporation (EG&G), 826
EDP. See Environmental Development Plan
EES. See Division of Electric Energy Systems
EG&G. See Edgerton, Germeshausen, and Grier Corporation
Egypt, 473
E. I. DuPont deNemours & Co., 825, 842
EIA. See Energy Information Administration
EIA/EIS. See Environmental Impact Assessment/Statement Program
Einstein, Albert, 870
EIS. See Environmental Impact Statement
EBB, 581
El Salvador, 473, 563
Electric energy load management, 477
Electric energy systems, 36-38, 379-380, 474-487
  commercialization, 485-486, 534
  compact stations, 394
  description, 392-399
  electric field effects, 393, 481-482, 487
  emergencies, 393
  energy storage integration, 476
  environmental aspects, 393, 487, 540-542
  geothermal systems integration
  load management, 393, 476-478
  new source integration, 479
  ocean systems integration, 479
  overhead transmission, 481
  photovoltaic systems integration, 411, 479
  research, development and demonstration (RD&D), 474, 485
  solar systems integration, 481
  solar thermal systems integration, 479
  underground cables, 482-483
  wind systems integration, 479
Electric Energy Systems Program, 474-487. See also Power Delivery Program; Power Supply Integration Program
  commercialization, 485-487
  Control and Development Subprogram, 479-480
  description, 479-380
  environmental impact, 487, 540-542
  international cooperation, 563
  objectives, 37
  regional activities, 486, 523-525
  socioeconomic aspects, 486, 547
  strategy, 394
  technology transfer, 479
Electric Energy Systems, R&D, 485
Electric Field Effects, 38, 482, 487
Electric and Hybrid Vehicle Research, Development and Demonstration Act, 490, 491-492
Electric/hybrid vehicles, 488
  propulsion batteries, 488, 490-494
  regenerative braking systems, 488
Electric Power Research Institute (EPRI), 37, 114, 196,

1000 Index

Electric Power Research Institute (EPRI) (cont.).
  202, 232, 265, 276, 404, 474, 485, 493, 534,
  541, 603, 717-718, 889, 995, 996
  battery storage projects, 474, 493
  electric energy projects, 485
  solar thermal projects, 404
Electric power supply integration, 477
Electric utilities, 39, 474. See also Central power
    applications; Electric energy systems; Electricity
    generation; Load management
  battery storage systems, 474
  environmental assessment, 540-542
  land requirements, 545
  photovoltaic systems and, 417
  project management, 474
  rates, 476, 487, 544-546
Electric utilities
  and solar systems, 474
  and thermal and mechanical storage systems, 394-395
  and wind systems, 420-427
Electric Utility Powerplant, 265-266
Electricity, fusion and, 874-875, 878, 881, 945
Electricity generation
  biomass systems, 377
  energy storage systems, 488, 491, 494
  geothermal systems, 377, 450
  ocean systems, 377
  photovoltaic systems, 408-409
  solar systems, 380
  solar thermal systems, 402-403
  wind systems, 420, 427
Electrodril, 310-311
Electrolysis, 509
Electron-beam concept, 872
Electron cyclotron resonance, 929
Electropolishing, 828
Elmo Bumpy Torus (EBT), 50, 914, 918, 922-923
EMAD, 819
Encapsulated high-level waste, 46
Encapsulation, 781
Energy, supply and demand, 12
Energy Conversion Alternatives Study (ECAS), 276
Energy, Department of. See Departmen of Energy
    (DOE), U.S.
Energy farms, 438-439
Energy imports, 3-4
Energy Information Administration (EIA), 12
Energy legislation, 4-6, 374-375
Energy Materials Advisory Committee (EMAC), 203
Energy policy, 3-6
Energy Producing Experimental Reactors, 878-880
Energy Reorganization Act, 751-752, 850
Energy Research and Development Administration
    (ERDA), 118, 252, 593, 638, 757, 778, 852
Energy savings
  battery systems, 395-396
  biomass systems, 389
  compressed air energy storage, 512, 524
  hydrogen production, 508-509
  impacted, 6
  magnetic energy storage, 510-512
  mechanical energy storage, 509-510
  photovoltaic systems, 531
  solar systems, 374
  thermal energy storage, 395
  underground pumped hydroelectric storage
Energy source, elementary integrated, 880-881

Energy sources, storage systems and end uses, 381
Energy Storage and Transfer (task), 912
Energy storage delivery strategy and timetable, 396
Energy storage systems, 380-390, 394-396
  applications, 38
  aquifer, 502, 542
  chemical/hydrogen, 508-509
  commercialization, 517-519
  compressed air, 524, 542
  and electric energy systems, 395, 476, 479, 502,
    507-510
  environmental implications, 542-543
  flywheels, 414-415
  interfaces, 489
  load management and, 476-478
  magnetic, 501
  mechanical, 500-501
  and photovoltaic systems, 494
  and solar systems, 534
  and solar thermal systems, 403
  technical/economic analysis program, 515
  underground pumped hydro, 512, 542
  uses, 39
  and wind systems, 494, 500
Energy Storage Systems Program. See Battery Storage
    Program; Thermal and Mechanical Storage Pro-
    gram
  commercialization, 517-519, 534
  description, 380, 488-490
  environmental impact, 518-519, 542-543
  international cooperation, 563
  objectives, 394-395
  regional activities, 518, 525-526
  socioeconomic aspects, 519, 547-548
  strategy, 395
Energy Systems Acquisition Advisory Board (ESAAB),
    528, 770, 772, 773
Energy Tax Act (1978), 4, 5, 374
Energy technology
  decentralization, 19
  funds managed at headquarters and field, 19
  issues addressed by, 14, 16-17
  management approach, 18
  organization, 18
Energy Technology Centers, 83, 335-345
Energy Technology Engineering Center (ETEC), 645,
    646, 664-666
Energy use, 3
Engine Combustion Technology, 252-253
Engineering base, fusion, 868, 878, 881, 891, 911, 930
Engineering Development System (EDS), 248
Engineering Net Power Production, 831-882
Engineering Prototype Reactor (EPR), 51, 868, 880-881
Engineering-scale dissolver, 694
Engineering Test Facilities (ETF), 273, 868, 880-881,
    909, 917
  and tokamak, 918, 930
  status, 868, 917-918
  steering committee, 889
  system studies for, 890, 912
England, 153, 248
Enhanced Gas Recovery, 27-28, 321-334
Enhanced Gas Recovery Program, 27-28, 321-334
Enhanced oil recovery, 26, 75, 279, 280-286
Enhanced oil recovery
  cost-sharing contractors, 282
  steam displacement, 26

Entrained-Bed Gasification, 140-141
Environment, 16
  coal gasification and, 131-132
  coal liquefaction and, 128
  energy storage and, 518-519
  fluidized bed combustion and
  fossil fuel and, 16
  hydroelectric power storage and
  impact analysis, 891-892, 918-919
  increased use of coal and
  magnetic fusion and, 17, 891-892
  nuclear power and, 17
  and reactor safety, 883-884
  related fusion development, 882-886, 891-892
  solar technology and, 16
  wind technology and
Environmental Assessment (EA), 759
Environmental Development Plan (EDP), 170, 190, 305, 356, 919, 948
Environmental Development Plans,
  electric energy systems, 540
  energy storage systems, 497
  geothermal energy systems, 454, 539
Environmental impact,
  assessment (EIA), 77, 152, 171, 306, 455, 539
  biological communities, 537, 543
  biomass systems, 538
  electric energy systems, 540-542
  energy storage systems, 542-543
  fission energy, 739-744
  fossil energy, 352-355, 539
  geothermal systems, 539-540
  health and safety, 536-540, 542
  land, 536, 538-540, 542
  nuclear waste, 851-855
  ocean systems, 537-538
  photovoltaic systems, 537
  solar systems, 536-539
  solar thermal systems, 536-537
  water, 536, 538, 539, 542
  wind systems, 537
Environmental Impact Statement (EIS), 455, 539, 918-919
Environmental Impact Statement on Management of Commercially Generated Radioactive Waste (DOE 1559), 754, 852
Environmental process stream control options, 195
Environmental Protection Agency (EPA), 83, 116, 132, 142, 170-171, 194, 196-198, 202, 205, 214, 239, 261, 276, 474, 717-718, 755, 776, 781, 792, 822, 854, 858
Environmental Report (ER), 855
EPA. See Environmental Protection Agency
EPR. See Engineering Prototype Reactor
EPRI. See Electric Power Reuearch Institute
Equilibrium/stability field. See under confinement, magnetic
Equity in situ project, 304
ERA. See Economic Regulatory Administration
ERDA. See Energy Research and Development Administration
ESAAB. See Energy Systems Acquisition Advisory Board
ESR-II reactor, 652
ET. See Assistant Secretary for Energy Technology
ET. See Office of the Assistant Secretary for Energy Technology

ETEC, 665
ETF. See Engineering Test Facilities
Ethanol, 442
  production from farm products, 440
ETM. See Office of Magnetic Fusion
ETN: See Office of Nuclear Fission
ETS. See Office of Solar, Geothermal, Electric and Storage Systems
ETS. See Office of Solar Energy and Geothermal Technologies
ETW. See Office of Nuclear Waste Management
EURATOM. See Commission of the European Communities
Europe, 3, 593, 613
  fusion program in, 868, 912, 956-957, 959
EV. See Office of Environment, Assistant Secretary for Environment, DOE
Executive Planning Council, 776, 860
Experimental Breeder Reactor (EBR), 627, 651-654, 660-662
Experimental Plasma Research (subactivity), 921, 925-928
Export Promotion Program, 550
Externally fired Brayton cycle, 224
Externally fired heat cycles, 22-225
Exxon Donor Solvent (EDS), process, 23, 112

Face haulage vehicles, 93-94
Facility construction
  Near Surface Test Facility, 816
  New Waste Calcining Facility, 784
  Pacific Northwest Laboratory, 831
  shale fracture facility at Oak Ridge, 786
Facility design,
  Away-from-reactor Storage Facility, 840, 842-843
  National Waste Terminal Storage Facility, 821
  Waste Isolation Pilot Plant, 803-807
Farm applications. See Agricultural applications; Remote site applications; Residential applications
Fast breeder reactors, safety, 641-644
Fast Fluid-Bed Process (Gasification), 138-144
Fast Flux Test Facility (FFTF), 42, 579, 581, 631-632, 635-638, 648-660, 662-665
Fast Flux Test Facility driver-fuel assembly, 653
Fast Flux Test Facility drives-fuel pin, 653
Fast Flux Test Facility fuel system, 637
Feasibility, fusion energy, 867-869, 873, 876, 957-958
Federal Energy Regulatory Commission (FERC), 5, 347, 480
Federal Republic of Germany, 358-363, 365, 472-473, 605, 606, 677, 682, 861-863, 957, 959
Federal Tort Claims Act, 858
Federal Water Pollution Control Act, 194, 349
Feedstock. See also Energy farms
  chemical, 83, 501
Fenton Hill, N. Mex., project, 468-469
FERC. See Federal Energy Regulatory Commission
Fermentation, 437-438, 442-443
  applications, 437-438, 442-443
  commercialization, 438, 442-443
  ethanol production, 442-443
  feedstock production, 442-444
FFP. See Fossil Fuel Processing

Field Offices. See Operations Offices
Field Reversal Experiments (FRX), 905
Final Safety Analysis Report (FSAR), 919
Fissile fuel, fusion and, 878, 892, 919, 921
Fission energy, development, 577-582
Fission energy management program, 583
Fission energy plant work breakdown structure, 585
Fission Energy Program, 39-44, 573
  budget, 14, 15
  commercialization, 737-739
  international agreements
  management, 583-588
  national and engineering laboratories
  planning, 587-588
  socioeconomic aspects, 745-746
  strategy, 568-573
  structure, 572-573, 584-587
  relationships, 589
  supporting technology, 612-613
Fixed-bed gasification, 135-137
  stirred process, 136
Fixed-bed gasifier, slagging process, 138
Fixed-bed hydrogenation, 114
Flash Liquefaction Process, 123-128
Flight system development, 723-726
Flue gas cleanup, 23, 196-203
Flue gas desulfurization (FGD), 58, 110, 195, 207-208, 354
Fluidized bed boiler (30-MWe), 239
Fluidized bed combustion. 9-10, 109, 188, 208
  waste management of, 207
Fluidized-bed gasification, 137-144
  two-stage pressurized process, 139
Flywheel storage system, 415
Flywheel systems, 509-510
  mechanical energy, 509-510
  solar energy, 395
FMIT. See Fusion Materials Irradiation Test
Ford, President, 753
Fort Belvoir, Va., project, 411
Fort Hood, Tex., project, 406
Fort St. Vrain plant, 581, 604, 614, 740
Fossil energy
  facilities for experimental programs, 196
  issues, 20, 60-61, 75
  organization, 79
  Overview, 20, 60-81
Fossil Energy Program, 20-28, 57-370
  budget, 14, 15, 60-69
  international agreements, 358-367
  management, 78-81
  organization, 79
  technology transfer to industry, 21
  university activities, 368-370
Fossil energy resource strategy, 58-59
Fossil energy technology centers, 336-341, 344
Fossil fuel, 13. See also Fossil Energy Program; and Environment Foundation for International Technological Cooperation, 550
FPRF. See Fusion Plasma Research Facility
France, 472, 582, 595, 605, 677-678, 743, 861-863, 957
FRG. see Federal Republic of Germany
Froth flotation, 107
FRX. See Field Reversal Experiments
Fuel,
  liquid jet, 914
  pellet, 867, 900, 914
  production, biomass systems, 389
  production, solar thermal systems, 403
  reprocessing, 690-671, 778
  shearing system, 693
  synthetic, 875-876, 878, 919
Fuel cell power plant, 25
  4.8 MW, 265
Fuel cells, 25, 59, 83, 261-271
  commercialization, 83
Fuel Cells Electric Utility Power Plant (4.8 MW), 265-266
Fuel cycle, 43, 628-629, 708
  once-through, 581
Fuel Cycle R&D Program, 687-702
Fuel disassembly system, 692
  industrial, 158-160
  small-scale industrial, 160
  utility, 160
Fuel Waste Management Cycle
Fusion energy, 867-961
  applications, 874-876
Fusion Energy Applications (subactivity), 911, 919-921
Fusion energy program, 867-961
  activities, 882-884
  current status, 876
  management, 885-886, 888
Fusion-fission fuel cycle, 951-952
Fusion fuel cycle, 951
Fusion materials irradiation test source, 936
Fusion Materials Irradiation Test (FMIT) Facility, 880, 889, 916, 930, 935-937
Fusion Materials Irradiation Test
  conceptual design, 916, 935
  mangement, 889
  overview, 930, 935
  status, 935, 937
Fusion Plasma Research Facility (FPRF), 926-927
Fusion plasma theory, 921, 925
Fusion Power Coordinating Committee (FPCC), 889
Fusion reactions, 87, 870-871
Fusion Reactor Materials (subactivity), 911, 914-916
Fusion Reactor Safety Research, 919, 949
Fusion Review Committee, 869, 890
Fusion Systems Engineering (subactivity), 911, 916-918
Fusion Working Group, 869
Future fuels and engine problems, 220

Gable Mountain, Near Surface Test Facility, 818
Gas as an energy source, 27-28, 75-76
Gas-Cooled Breeder Reactor (GCBR), 42-43, 627-629, 677-687, 740-741
Gas cooled fast reactor fuel element, 682
Gas cooled fast reactor fuel fission-product vent manifold, 683
Gas cooled fast reactor fuel-rod assembly, 683
Gas cooled fast reactor program work breakdown structure, 680
Gas cooled fast reactor systems, 681
Gas cooled thermal reactor work breakdown structure, 615
Gas from oil shale, 322-324
Gaseous waste. See Airborne waste
Gas shale production, 28
Gasification, 131-135, 141-148, 440-441
  AR & RD, 175-176
  high Btu, 131-134
  low Btu, 134-135
  Test Facility, 148

Third-Generation Processes, 144-146
Gasification-Demonstration Plants, 153-162
Gasifiers in Industry, 141-144
Gasohol, 5, 442, 448, 533
Gas stream cleanup, 23, 203-207
GCTF. See Geothermal Component Test Facility
Gedser windmill, 562
GEIS. See Generic Environmental Impact Statement
General Atomic Company, 929, 938, 939, 943
  Doublet experiments, 899, 938, 959
General Atomic Company, "next-step" designs, 918
General Electric Co., 427, 519, 535
General Services Administration (GSA), 397
Generic Environmental Impact Statement (GEIS)
Generic fusion reactor, 874
Geologic formations for repositories, 812
Geologic repository, 757
  at WIPP, 804, 806
  basalt, Hanford Reservation, 757
  site evaluation, 757, 806
Geopressured aquifers, 59, 457
Geopressured energy recovery, 458
Geopressured hydrothermal energy, 35
Geopressured resources, 35, 450, 457-462
  applications, 453
  description, 457-462
  identification of, 458-459
  production tests, 455-456
  rock mechanics, 461-462
  well-production model, 461-462
Geopressured Resources Program, 457-462
  commercialization, 485-487
  description, 377-379
  environmental impact, 454-455
  Existing Well Tests Subprogram, 459-460
  New drilled wells, 460-461
  objectives, 391, 458
  regional activities, 522-523
  Reservoir Engineering Subprogram, 461-462
  Resource Definition Subprogram, 451-453
Geopressured Resources Program Reservoir Engineering Subprogram, 461-462
Georgia Institute of Technology, 408
Geosciences, 467-468
Geothermal Component Test Facility (GCTF), 456, 465
Geothermal demonstration plant, 456-457
Geothermal energy, 8, 35-36, 450-472. See also Geopressured resources; Hot dry rock resources; Hydrothermal resources
Geothermal energy
  applications, 390-391, 950
  contributions, 37
  conversion technology development, 464-466
  description, 35-36, 377
  drilling technology, 460-461
  electric energy system integration, 479
  exploration technology, 35
  geosciences, 467-468
  logging technology, 468
  materials development, 466-467
  utilization estimates, 392-393
  well stimulation, 391-392
Geothermal Energy Program. See Geopressured Resources Program; Geothermal Technology Development Program; Hydrothermal Resources Program
  commercialization, 391-392, 470-472, 534
  description, 391-392

Engineering Applications, 453-454
environmental impact, 454-455, 539
interagency cooperation, 392, 469
international cooperation, 472-473, 562-563
objectives, 36, 391
regional activities, 470-471, 522-523
socioeconomic aspects, 546-547
strategy, 391-392
technology transfer, 472, 562-563
Geothermal Energy Research Development and Demonstration Act, 550
Geothermal energy resources, 377-379
Geothermal Loop Experimental Facility (GLEF), 455-456
Geothermal power plants, 378
Geothermal Steam Act, 469
Geothermal steam field, 36
Geothermal Technology Development Program, 379, 462-469
  Drilling and Completion Technology Subprogram, 463-464
  Energy Conversion Systems and Stimulation Subprogram, 464-466
  Geochemical Engineering and Materials Subprogram, 466-467
  Geosciences Subprogram, 467-468
  Hot Dry Rock Subprogram, 468-469
  Interagency Coordination and Planning Subprogram, 469-470
  objectives, 391
Germany, Federal Republic of. See Federal Republic of Germany
Geysers, 36
  Calif., geothermal field, 35-36, 523, 539
GLEF. See Geothermal Loop Experimental Facility
GNP, 3
GNP, energy use and, 1-2
Government-owned laboratories, 848
Government-owned nuclear facilities, 848
Government role options, 34
Grand Forks ETC (GFETC), 80, 120, 136, 256, 343-344
Granite formations, 812-814
Gravitational confinement. See confinement, gravitational
Green River Formation (Colorado), 75
Greenfield, 844
GSA. See General Services Administration
Gulf Interior Salt Domes, 812, 814
Gun, plasma, 906, 910
Gyrotron, 929

Hanford Basalt Program, 816-819
Hanford Engineering Development Laboratory (HEDL), 632-633, 637-638, 654, 658, 662, 667-668, 846, 919, 934, 938-939, 953
Hanford Reservation, 756-758, 782-783, 790, 791, 810, 816
  landlord responsibilities of DOE
Haulback mining, 104
Hawaii, 424, 425
  geothermal resources, 470, 522
  Geothermal Wellhead Generator, 455
  solar resources, 522
H-Coal Process, 22, 83, 112, 114
H-D plasma, 873
HDR. See Hot dry rock resources
Health and safety,
  and biomass systems, 538

Health and safety (cont.)
  and electric energy systems, 540
  and energy storage systems, 542
  and geothermal systems, 539
  noise, 482, 487
  and nuclear waste, 856
  and ocean systems, 537
  and photovoltaic systems, 537
  and solar thermal systems, 536-537
  and transmission lines, 540
  and wind systems, 537
Heat. See also Heating and air conditioning; Process heat; Thermal energy; Thermal energy storage; Water
  and geologic disposal at Nevada Test Site
  industrial process, 24
  of spent reactor fuel at WIPP
  plasma, 911, 913, 928
    electron, 903-904
    ion-beam, 872
    levels, for fusion, 871-872, 925-928
    ohmic, 872, 895, 898-899, 912
    and pre-heating experiments, 890
    resistance, 867, 872, 897, 898
    ripple injection, 900
  recovery systems, 24
  waste. See Waste heat recovery
Heat engine and heat recovery technology, 212
Heat engines and heat recovery, 23-24, 83, 211-236
Heat Exchanger Technology, 228, 233-236
Heat exchangers, 33, 181, 465
Heat Recovery Component Technology, 228-236
Heating and air conditioning
  biomass systems, 377, 435
  energy storage systems, 395, 502
  geothermal systems, 450, 451, 453
  photovoltaic systems, 409
  solar systems, 387
  solar thermal systems, 402-403
Heavy-ion beam fusion, 890
Heavy liquids, 58
HEDL. See Hanford Engineering Development Laboratory
Helical screw expander
Heliostats, 405, 536
High-Btu Demonstration Plants, 153-155
High-Btu Gasification, 9, 131-134
High Efficiency Particulate Air (HEPA), 802
High Field Test Facility (HFTF), 92
High-grade heat recovery, 228, 230-233
High-gradient magnetic separation (HGMS, 107
High Intensity Neutron Source (HINS), 916, 930, 933-934
High Intensity Neutron Source Facility, 935
High-level liquid waste (HLLW),
  at Hanford Reservation, 783-784
  at Idaho Chemical Processing Plant
  at Savannah River Plant
High-level waste (HLW), 751, 756-758, 765-766, 778, 781, 789, 792-794, 824-825
  at Hanford Reservation, 782-783
  canister, 825
  disposal, 752, 766-767, 852
  interim storage, 782
  IRG disposal recommendations, 764-767
  research and development, 792-794
  at Savannah River Plant, 787-790
  storage tanks, 46

U.S. Government policy, 850-852
volume, current and projected, 752, 779
High Performance Demonstration Experiment (HPDE), 276
High Performance Fuel Laboratory (HPFL), 631, 663
High production advance panel mining, 95
High-production retreat panel mining, 94-95
High Temperature Gas-Cooled Reactor (HTGR), 568, 577, 581, 590, 593-596, 603-606, 614-615, 616, 688
  with direct Brayton cycle, 605
High-temperature turbines, 59
High voltage direct current transmission (HVDC), 481-484
HINS. See High Intensity Neutron Source
HLLW. See High-level liquid waste
Hoe Creek #3 Test, 166
Homopolar motor generator, 912
Horizontal drilling, 312
Hot dry rock (HDR), 35-36, 463
  extraction eperiment, 469
  resources, 379, 450
    commercialization, 468-469
    energy extraction, 468
    identification of, 468
    international cooperation, 473
    regional activities, 469
    socioeconomic aspects, 546
Hot Dry Rock Subprogram, 463, 468-469
Hot-gas cleanup, 146-149, 250
  iron oxide process, 150
  molten salt process, 149
Hot Pilot Plant (HPP), 631
Hybrid vehicles. See Electric/hybrid vehicles
Hydraulic Fracturing Process, 331-333
Hydroelectric energy,
  and energy storage systems, 509
  and ocean current systems, 434
  and wind systems, 532
Hydroelectric power
Hydrogasification, 144-145
  process, 145
Hydrogen/Deuterium Major Scaling Experiment, 880
Hydrogen Energy Coordinating Committee, 488
Hydrogen production, 508-509, 919
Hydrothermal Geothermal Commercialization Subprogram, 377
Hydrothermal-geothermal energy, 12
Hydrothermal resources, 12, 450-451
Hydrothermal Resources Program, 450-457
  commercialization, 383, 450-451
  description, 450-457
  Engineering Applications Subprogram, 453-454
  Environmental Control Subprogram, 454-455
  environmental impact, 454-455
  Facilities Subprogram, 455-457
  objectives, 451-452
  regional activities, 451, 456
  Regional Planning Subprogram, 451
  Resource Definition Subprogram, 451-453
HYGAS Demonstration Plants, 153-162
HYGAS pilot plant configuration, 157
HYGAS process, 156-157

IAEA. See International Atomic Energy Agency
ICPP. See Idaho Chemical Processing Plant
Idaho Chemical Processing Plant (ICPP), 783-784
  waste management facilities, 784

Idaho National Engineering Laboratory (INEL), 188, 346, 603, 658, 660, 666-667, 778, 783-784, 800, 829, 846-847, 919, 941
Idaho Operations Office (DOE), 81, 762, 773, 796, 802, 826, 829, 834
   airborne waste R&D, 802
Idaho Reservation, 757
Idaho State University, 848
IEA. See International Energy Agency
IGCC. See Interagency Geothermal Coordination Council
Ignition, plasma, 871, 880-881, 898
ILLW. See Intermediate-level liquid waste
Immobilization, 792. See also solidification
Imports, energy, 8
Improved thermoelectric convecter, 726
Impurities Study Experiment device, 902
Impurity control, plasma, 882, 901
   and flow reversal, 895-897, 901
   and tokamak effort, 894, 895, 898
   data base, 882, 895
Impurity Studies Experiment (ISX), 894, 900-901, 939
Incentives. See Tax credits
Incineration of TRU Waste, 797
Increased Conversion Efficiency (ICE), 252
India, 581
Indiana University, 848
Indonesia, 563
INDUS. See Industrial Energy Conservation
Industrial participation, 81
Industrial process heat, 282, 875
Industry-Government, 81
INEL. See Idaho National Engineering Laboratory
Inert gases. See Noble gases
Inertial confinement, 872, 889-890
Inexhaustible energy, 7, 12, 867, 944
INFCE. See International Nuclear Fuel Cycle Evaluation Program
INFCE working groups, 712
Initial System Evaluation Experiments (ISEE), 417-418
In-mine access, 88-89
In situ coal gasification, 162-172
Integrated Coal Conversion and Utilization Systems, 24, 213-218
Integrated Equipment Test (IET), 691, 697, 699
Integrated process demonstration area, 698
Interagency Geothermal Coordination Council (IGCC), 392, 469
   organization, 471
Interagency Nuclear Safety Review Panel, 732
Interagency Review Group (IRG) on Nuclear Waste Management, 744, 755, 764-768, 776, 808, 852
   Subgroup on Strategies for Waste Isolation, 853
Interim nuclear waste storage, 47, 759-760, 779-790
Interim Waste Operations (subactivity), 759-760, 779-790
Intermediate-Level Liquid Waste (ILLW), 784
International Atomic Energy Agency (IAEA), 863
International cooperation, 358-367
   agreements, 552
   bilateral agreements, 558-559
   biomass systems, 562
   commercialization, 552
   electric energy systems, 563
   energy storage systems, 563
   fossil energy, 358-367
   fusion energy, 956-961
   geothermal energy, 562-563
   multilateral agreements, 554-555
International cooperation, nuclear waste, 861-864
   ocean systems, 561-562
   photovoltaic systems, 557, 560
   solar systems, 549-563
   solar thermal systems, 557
   wind systems, 560-561
International Energy Agency (IEA), 234, 237, 249, 250, 359-362, 404, 466, 472, 551, 863-864, 956-959
   energy storage agreements, 504
   geothermal agreements, 466, 472
   multilateral agreements, 551-552, 554-555
   ocean systems agreements, 561-562
   solar technology, 556-562
International Energy Development Program, 550
International energy programs, 360-361
International magnetic fusion research, 958, 961
International Nuclear Fuel Cycle Evaluation (INFCE) Program, 43-44, 571, 616, 628, 631, 649, 689, 706-708, 710-711, 762, 834, 863
International Programs. See specific country
International Spent Fuel Storage Subactivity, 763, 838-839
Iodine-*129*, 781, 801-802
Iodine-*131*, 801
Ion-beam heating, 872
Ion source research, 926, 928
Iowa State University, 369
IRG. See Interagency Review Group on Nuclear Waste Management
Iron oxide process, 147-148
ISEE. See Initial system evaluation experiments
Isotope. See under specific name
Isotopes Fuels, 729-731
ISX. See Impurity Studies Experiment
Italy, 361-363, 391, 418, 466, 472, 562, 563, 863

Japan, 563, 595, 862, 956-957, 959
Jet Propulsion Laboratory, 399, 485, 486
JFK Airport natural cooling system, 503
Joint European Torus (JET), 957
Joint Fusion Power Coordinating Committee (JFPCC)

Kennedy International Airport, 502
Kenya, 563
Kilowatt Isotope Power System (KIPS), 726
KIPS, 727
Krypton, 781, 802

Labor surplus areas. See Small and disadvantaged businesses
Laboratories. See National laboratories
Laboratories, DOE. See specific facility
Land
   and biomass systems, 540
   and electric energy systems, 540
   and energy storage systems, 542
   and geothermal systems, 539
   and solar thermal systems, 536
   surface subsidence, 539, 548
   use, 536
Laramie Energy Technology Center (LETC), 80, 164, 166, 167, 292, 299, 301, 303, 306, 307, 308, 316, 344-345

Large Coil Project (LCP), 880, 912, 959
Large Coil Test Facility (LCTF), 892, 913
Large Experimental Facility (LEF), 436
Large Leak Test Rig (LLTR), 646
Large Power Systems, 402-405
Laser, 611-612, 867, 872
  technology, 867, 872
LASL. See Los Alamos Scientific Laboratory
Latent heat storage, 543
Lawrence Berkeley Laboratory (LBL), 306, 316, 346-347, 399, 445, 890, 931, 939, 953
Lawrence Livermore Laboratory (LLL), 164, 166, 167, 170-171, 252, 299, 301, 308, 316, 331, 347, 399, 847, 876, 890, 906, 908, 910, 912, 914, 918, 929, 934, 938-939, 941, 953
  and CT, 924
  and HFTF, 912
  and HINS, 934
  and MFTF, 876
  and mirror confinement research, 890, 918
  and TMX, 908, 941
Lawson Product, 871, 903
  and breakeven, 901
  and net energy production, 899
  in Beta II, 907
  values achieved, 901
LBL. See Lawrence Berkeley Laboratory
LCP. See Large Coil Project
LCTF. See Large Coil Test Facility
LEF. See Large Experimental Facility
Lewis Research Center, 399
Licensing, NRC, 751, 755
Light metal fast breeder reactor
  accident curve, 643
  heat-transport system, 629
  program contractors, 632
  work breakdown structure, 633
Light water breeder reactor blanket-rod support grid, 678
Light water breeder reactor core, 672, 676
Light water breeder reactor fuel installation, 673
Light Water Breeder Reactor (LWBR) Program, 672-674, 689
Light Water Reactor (LWR), 39, 41, 573-582, 589-602, 607, 627-629, 670, 751, 839
Light water reactor fuel assemblies, 839
Light Water Reactor, Technology Improvement Program, 569-570
Lime/limestone scrubber reliability, 198
Limiter, plasma, 898-899
Line of Assurance (LOA), 641-644
Linear Magnetic Fusion (LMF), 922-924
Linked vertical wells, 164
  process, 166
Linking sources and uses of energy, 37
Linus, 924
Liquefaction (AR&TD), 174-175
Liquefaction Demonstration Projects
Liquefied Natural Gas (LNG), 77-78
Liquid Metal Fast Breeder Reactor (LMFBR), 42, 569, 571-573, 579-582, 627-688, 737
  fuel-processing flowsheet, 691
Liquid Waste. See High-level liquid waste; Intermediate-level liquid waste
Lithium, 870, 874, 949
  as a tritium breeder, 871, 874
  in FMIT, 935
  risks, 892

LLL. See Lawrence Livermore Laboratory
LLTR, 668
LLW. See low-level waste
LMF. See Linear Magnetic Fusion
Load leveling with batteries, 499
Load management, 38, 379
Lock Kopper Valve Development Program, 151
Long-term Waste Management Technology (Subactivity), 768, 791-794, 796-802
Longwall conveyor haulage, 104
Longwall-in steeply pitching seam, 97
Longwall-in sublevel caving, 97-98
Longwall multilift mining method, 97
Los Alamos Scientific Laboratory (LASL), 171, 252, 299, 301, 307, 316, 331, 347, 399, 468-469, 472, 607, 627, 649, 683, 725-726, 730, 733, 768, 781, 805, 808, 819-820, 890, 912, 918, 923-924, 929, 935, 938-939, 953
  and FMIT, 935
  control waste collection system, 781
  RFP research, 923
  and TSTA, 918
Low-Btu Fuel Gas Demonstration, 157-158
Low-Btu fuel gas, industrial "A", 158
Low-Btu fuel gas, industrial "B", 160
Low Btu gas, 134-135
Low-Btu Gasification, 141
Low-Cost Solar Array, 913
Low-Grade Heat Recovery, 228-230
Low-Level Waste (LLW), 756, 759, 766, 778, 780, 794, 796, 825-826
Low-level waste burial sites, 758, 796, 834, 837
Low-level Waste Management Operations Subactivity, 762
Low-level waste, volume, current and projected, 758, 779
LSA. See low-cost solar array project
Lurgi Phenosolvan process, 155
LWR. See Light Water Reactor
LWR fuel cycle, 576, 577
LWR once-through fuel cycle, 575
Lyons, Kansas, 46, 757, 805

MACROTOR, 929
MAGES. See program of Research and Development of Manmade Geothermal Energy Systems
Magnet. See also Magnet, superconducting; Minimum-B magnet
  coil, 907
  superconducting, 906
  superconducting, and magnetic systems (task), 907
  superconducting, and MFTF, 906
  superconducting, development of, 882, 890
  Yin-Yang, 903, 905, 906
Magnetic confinement, 50, 872, 873
Magnetic energy storage, 39
Magnetic field effects, 487, 872
Magnetic fusion
  issues, 891
  outlook, 877
  program decision logic, 868
  program strategy, 881
  R&D, 52-53
  university contracts, 941
Magnetic Fusion Energy Program, 12, 49-53, 867-961
  environmental issues, 52, 891-892, 948-952

Index 1007

field activities, 888-889
funding, 14, 15, 882-884
objectives, 50-51
program management, 885-888
project management, 888-889
socioeconomic issues, 892, 944-945
status, 876
strategy, 878-882
Magnetic fusion energy RD&C cycle, 879
Magnetic fusion industry contracts, 942
Magnetic mirror systems, 876, 903-905. See also Minimum-B Magnet; Reversed field; Tandem mirror systems; and specific systems
budget, 883, 893-894
overview, 903-905
responsibility for, 882-890
status, 876
Magnetic Systems Subactivity, 911-912
Magnetohydrodynamics Program, 12, 25, 59, 83, 272-278
Major fuel burning installations (MFBI), 5
Major scaling experiments,
deuterium-tritium, 880
hydrogen-deterium, 880
responsibility for, 890
Management, OFE. See under Office of Fusion Energy
Management planning documents, 588
Manhattan Project, 607, 627, 756, 845
Manure. See Anaerobic digestion
Map of electric energy system projects, 524
Map of energy, storage projects, 525
Map of ETS research activities, 525
Map of geothermal projects, 524
Map of solar projects, 523
Massachusetts Institute of Technology, 445, 901
Massive hydraulic fracturing process, 332
Masuda buoys, 561
Materials Coordinating Committee, 488
Materials Supporting Technology Activity, 408
Materials test reactors, 618
Mathematical Sciences Northwest, 924
MCCC. See Mississippi County Community College
Mechanical energy
geopressured, 450
and ocean systems, 430
and solar thermal systems, 401, 402
Mechanical Energy Storage Subprogram, 509-510
Medium-Btu gas (MBG), 59
Meterological Effects of Thermal Energy Release (METER) Program, 718
Methane, 13, 28, 458
from coal, 28, 59, 321, 326-329
production by anaerobic digestion, 444
Mexico, 77, 466, 473
MFE. See Magnetic Fusion Energy
MFTF. See Mirror Fusion Test Facility
MHW RTG unit, 275
Micellar-polymer/improved waterflooding process, 283-286
Microcomputerized control interface unit, 697
MICROTOR, 929
Microwave,
heating, 873, 900, 914, 923
radiation, 894
Milestone schedules
advanced batteries, 494
advanced flue gas cleanup task, 201
advanced flue gas development task, 200

advanced fusion concepts, 926
Advanced isotope separation technology program, 609, 610
advanced nuclear systems economics evaluation, 714
advanced nuclear systems increased siting acceptability, 719
advanced research in coal gasification processes, 180
advanced water breeder application, 679
airborne waste, 803
airborne waste (commercial), 830
Albuquerque Operations Office, 782
Alcator, 904
Alternate fuels combustion task, 260
Anthracite applications, 246
Argonne National Laboratory facilities, 663
atmospheric fluidized bed (30MWe), 240
atmospheric fluidized bed and integration unit, 245
atmospheric fluidized-bed combustion demonstration plant, 262
atmospheric fluidized bed industrial applications, 242
away from reactor storage facilities, 849
battery storage supporting research, 498
BETA II, 907
biochemical conversion, 445
biomass production systems, 442
biomass R&D, 446
biomass technology support, 441
breeder component development, 646
breeder fuels and core materials, 656
breeder materials/chemistry, 659
breeder reactor studies, 640
breeder technology safety, 645
catalytic gasification process, 148
chemical coal cleaning, 110
chemical/hydrogen energy storage, 510
closed-cycle power systems task, 218
coal applied research task, 209
coal combustion, 184
coal conversion processes, 176
coal direct utilization controls, 185
coal gas process modification technology task, 206
coal heat exchanger technology, 209
coal materials and components, 191
coal materials evaluation, 189
coal mining planning and development, 102
coal-oil mixture task, 258
coal preparation and analysis, 108, 111
coal prototype systems assessment task, 228
coal systems and economic comparisons task, 210
coal waste management task, 208
CO_2 flooding subactivity, 291
combined-cycle (pressurized fluidized bed) pilot plant, 249
commercial waste D&D, 836
contour coal mining, 105
Core Flow Test Loop, 686
Cresap Test Facility, 128
customer storage of off-peak electricity, 507
decontamination and decommissioning, 811
Defense Waste Processing Facility, 797
directly-fired heat cycles task, 222
dispersed battery applications, 500
disposable catalyst hydrogenation, 125
donor solvent liquefaction, 120
doublet, 901
Eastern Gas Shales project, 323
electric field effects, 483
electric power supply load management, 478

Milestone schedules (*cont.*)
  electric utility powerplant demonstration (4.8 MW), 266
  electrochemical systems research, 496
  electrolytic technology, 499
  engine combustion technology task, 253
  enhanced gas recovery environment and support subactivity, 330
  enhanced oil recovery environmental studies, 296
  ETEC facilities, 668
  experimental plasma research, 931
  externally-fired gas turbine, 244
  externally-fired heat engines task, 225
  fast fluid bed process, 142
  Fast Flux Test Facility, 637, 638
  fixed-bed gasification, 137
  flash liquefaction, 127
  fuel cell applied research, 272
  fuel cell powerplant, 263
  fuel cycle proof of breeding, 700
  fuel gas industrial "B", 161
  fusion energy applications, 921
  fusion energy environment and safety, 920
  Fusion Materials Irradiation Test facility, 937
  fusion plasma theory, 927
  fusion reactor materials, 916
  fusion systems engineering, 920
  gas cooled fast reactor development, 684
  gasification demonstration plants, 154
  gasification test facility, 152
  gasifiers in industry, 143
  Gas Reactor In-Place Safety Test, 687
  geochemical engineering and materials, 467
  geothermal drilling and completion technology, 464
  geothermal energy conversion systems and stimulation, 466
  geopressured new drilled wells, 461
  geopressured reservoir engineering, 462
  geopressured resource definition, 460
  geopressured resources existing well tests, 461
  geosciences, 468
  Hanford Basalt Program, 819
  Hanford Engineering Development Laboratory, 664
  Hanford Reservation, 783
  H-coal process, 115
  heat exchanger technology task, 235
  high-grade (coal) heat utilization task, 233
  high intensity neutron source facility, 936
  high-level nuclear waste, 795
  high-level waste, 826
  high-temperature gas-cooled reactor, 606
  high-temperature gas-cooled reactor steam cycle, 617
  hot dry rock, 470
  hydrogasification, 146
  hydrothermal engineering applications, 455
  hydrothermal facilities, 457
  hydrothermal resource definition, 453
  Idaho National Engineering Laboratory, 785
  improved oil and gas burners task, 254
  impurities Study Experiment, 902
  increased thermal energy, utilization, 717
  industrial waste heat storage, 505
  in situ coal gasification supporting research, 173
  in situ combustion and steam flooding, 295
  intermediate-scale wind energy conversion systems, 427
  International Energy Agency, 251
  International Nuclear Fuel Cycle Evaluation, 712
  international spent fuel storage, 840
  iron oxide process, 151
  isotopes fuels, 731
  landlord responsibility (commercial), 832
  large-scale wind energy conversion systems, 429
  large solar thermal power systems, 405
  light metal fast breeder reactor, 634
  light water breeder reactor, 676
  lime/limestones reliability task, 199
  liquid metal fast breeder reactor safeguard, 669
  long-term waste management technology, 793
  low-Btu fuel gas "A", 159
  low-grade (coal) heat utilization task, 231
  low-level nuclear waste, 798
  low-level waste, 827
  magnetic fusion energy computer network, 932
  magnetic systems, 913
  magnetohydrodynamics, 274
  mechanical energy storage, 511
  methane from coal project, 328
  micellar-polymer flooding subactivity, 288-289
  mine development, 88
  mine planning and development, 90
  mirror Fusion Test Facility, 911
  molten carbonate cleanup task, 205
  molten carbonate systems development, 271
  national waste terminal storage, 823
  near-term battery program, 493
  Nevada Test Site, 788, 821
  Nonproliferation Alternative Systems Assessment Program, 710
  Northern Great Plains project, 330
  novel fossil energy systems, 96
  nuclear decontamination and decommissioning, 734
  nuclear power system technology and support, 729
  nuclear systems flight development, 728
  Oak Ridge National Laboratory, 789
  occupational radiation dose/productivity enhancement, 602
  ocean systems advanced research and development, 436
  ocean systems engineering test and evaluation, 435
  ocean systems technology development, 433
  Office of Fusion Energy, 884
  Office of Nuclear Waste Isolation, 817
  open-cast mining, 103
  open-cycle gas turbine task, 216
  onsite/integrated energy systems powerplant demonstration, 269
  petroleum advanced exploratory research subactivity, 317
  petroleum drilling subactivity, 313
  petroleum environment and support subactivity, 315
  petroleum offshore technology subactivity, 314
  petroleum product charaterization and utilization subactivity, 319
  Phaedrus, 910
  photovoltaic advanced research and development, 413
  photovoltaic energy technology development, 416
  photovoltaic systems engineering and standards, 418
  photovoltaic test and applications, 420
  plasma engineering, 915
  Polodial Divertor Experiment, 899
  pressurized fluidized-bed engineering development system plant, 251
  Princeton Large Torus, 897
  principal alternate fuels subtask, 261
  production mining, 95

proliferation resistant concepts, 625
proof of breeding facility, 702
reduced-enrichment research and test reactors, 623
remote operating/maintenance demonstration area, 698
Richland, Washington, landlord responsibilities, 791
room-and-pillar-mining, 91
Savannah River Plant, 790
slagging process, 139
seasonal storage in aquifers, 504
shale gas production, 309
shale oil production, 308
shale oil production field tests, 305
shale oil refining and utilization subactivity, 318
Shippingport Atomic Power Station, 674
small-scale industrial fuel gas, 162
small-scale wind energy conversion systems, 426
solvent refined coal, 118, 130
solar applications, 495
solar generic/integration of new source technologies, 480
solar heating and cooling, 506
solar thermal power advanced research and development, 409
solar thermal small power systems, 407
Southern Tight Basins project, 331
space nuclear power system safety, 733
spent fuel program management and supporting studies, 843
steeply dipping beds, 170
storage for solar thermal power, 508
superconducting magnetic energy, 512
supporting nuclear waste studies, 833
surface coal mining subactivity, 106
surface test facility, 101
synfuel combustion, 184
synthane pilot plant, 134
terrestrial isotopes application, 735
thermal and mechanical storage applications analysis, 517
thermal and mechanical storage technology information systems, 518
thermal reactor safety, 604
thermochemical conversion, 443
Tokamak Fusion Test Reactor, 934
TRU waste, 800
TRU waste (commercial), 829
turbine cleanup systems task, 204
two-stage pressurized process, 140
underground cables and compact stations, 485
underground coal mining subactivity, 87
underground coal mining transport 100
uranium utilization, 599
utility applications, 514
Waste Retrieval and Treatment Facility, 801
waste treatment technology, 825
Western Gas Sands project, 325
western low- and medium-Btu gas, 167
wind energy conversion systems development and technology, 423
WIPP Project, 807
WIPP R&D, 805
WIPP transportation system development, 809
zinc chloride catalyst, 123
Mine Planning and Development—Underground, 101
Mine waste, 86
Miner/bolter, 92
Minimum-B magnet, 905
  and Beta II, 905
  concept, 903, 905
  tandem mirror experiments, 904
Minimum-B mirror, 905
Mining legislation, 90
Mining planning and development, 86-90
Mining Research and Development Activity, 82-110
Mining Technology Centers, 336-341, 344-346
Mining Technology Clearinghouse, 363
Mirror experiment problem areas, 906
Mirror Fusion Test Facility (MFTF), 908, 910
  budget, 911
  project management, 910
  status, 911
Mississippi County Community College (MCCC)
Models, fusion, 921
Modified horizontal in situ mining, 303
Modified horizontal in situ process-GSG version, 304
MOD-O wind turbine, 426
Molecular laser isotope separation, 612
Molecular process, 611-612
Molten Carbonate Systems, 203-206, 260-270
Molten Salt Pressurized Process, 146-147
Molybdenum, 929
Montana Technical University
Morgantown Energy Technology Center (METC), 30, 135, 171, 254, 331, 345
Motor Vehicles Manufacturing Association (MVMA), 252
Mound Laboratory, 347, 722, 730, 734, 847
Multicell Fluidized-Bed Boiler (MFB), 238
Multilateral agreements, 554-555

NAS. See National Academy of Science
NASA. See National Aeronautics and Space Administration
NASAP. See Nonproliferation Alternative Systems Assessment Program
NASAP/INFCE tasks and interfaces, 708
NASICOU, 496
National Academy of Sciences (NAS), 775
National Aeronautics and Space Administration (NASA), 217, 232, 276, 397, 424, 488, 719-721, 723-726
National Battery Test Laboratory, 492
National Bureau of Standards, 259, 928
National Electric Reliability Council, 393
National Energy Act (NEA), 4-6, 8, 57, 374, 389, 448-449, 474-476, 480, 751-753, 773, 848, 867
National Energy Conservation Policy Act (1978), 4, 5, 374
National Energy Plan, 57, 76, 449, 567-568, 679, 751, 755, 773, 848, 867
National Environmental Policy Act (NEPA), 190, 349, 353-354, 455, 764, 767, 775-776, 791, 807, 854-855
National geothermal resource assessment update, 452
National geothermal utilization, 393
National Institute of Occupational Safety and Health, 142-143
National laboratories. See specific facility
National Magnetic Fusion Energy Computer Center (NMFECC), 921, 929-930
National magnetic fusion energy computer network, 932

National Magnetic Fusion Energy, User Service Centers, 929
National Oceanic and Atmospheric Administration (NOAA), 538
National Research Council, 770, 773, 775, 832
National Science Foundation (NSF), 276
National Uranium Resource Evaluation Program, 42, 569, 570, 572, 631
National Waste Terminal Storage (NWTS) Program, 753, 821
NATO. See North Atlantic Treaty Organization
Natural gas, 13. See also Shale
  supply of, 76
Natural Gas Policy Act, 5, 6, 374
Naval Nuclear Propulsion Program, 44-45
Naval Reactors Expended Core Facility, 670
Naval Reactors Program, 837-838
Naval Research Laboratory (NRL), 924, 928
NEA. See National Energy Act; Nuclear Energy Agency
Near Surface Test Facility, 818
Near Surface Test Facility (NSTF), Gable Mountain, 816-819
Near-term battery projects, 492
NEP. See National Energy Plan
NEPA. See National Environmental Policy Act
Net energy gain, 867-868, 876
Netherlands, 361-363, 472, 861
Neutral Beam Development (task), 914
Neutral beam injection, 867, 873
  and plasma engineering, 911
  development and technology, 882, 911, 929
  in Doublet, 899
  in ISX, 900
  in MFTF, 910
  in mirror systems, 904, 906
  in PDX device, 898
  PLT experiments, 914
Neutrons. See also High Intensity Neutron Source (HINS)
  and structural materials, 883, 929-930
  high energy, 880, 929, 930, 933, 946
Nevada Operations Office, 786
Nevada Test Site (NTS), 331, 757, 785-786, 818, 819-821
  radioactive waste area, 787
  site evaluation, 786
New Source Performance Standards (NSPS), 194, 196, 261
New Waste Calcining Facility (NWCF), 784
New York Atomic R&D Authority (NYARDA), 738
New York State Energy Research and Development Authority, 535
New Zealand, 363, 466, 472
NFS-West Valley Facility, 834, 836
Niagara-Mohawk Power Co., 535
Nicaragua, 563
NMFECC. See National Magnetic Fusion Energy (MFE)
NOAA. See National Oceanic and Atmospheric Administration
Nonproliferation. See Proliferation
  development work breakdown structure, 617
  nuclear, 43
  studies, 705-711
Nonproliferation Alternative Systems Assessment Program (NASAP), 43, 568, 616, 628, 631, 689, 705-708, 710-711
North Atlantic Treaty Organization (NATO), 472

Norway, 599
Novel Systems (Mining), 96-97
NPD. See Nuclear Power Development
NRC. See Nuclear Regulatory Commission
NSF. See National Science Foundation
NSTF. See Near Surface Test Facility
NTS. See Nevada Test Site
Nuclear/coal plant models capital costs, 713
Nuclear energy. See Fission Energy Program
Nuclear Energy Agency (NEA), 679, 751, 753-754, 832, 863-864
Nuclear Fuel Assurance Act, 738
Nuclear fuel cycle, 43, 688
Nuclear fuel cycle waste management, 833
Nuclear fuel handling, 743-744
Nuclear Materials Transportation Activity, 768-769
Nuclear Non-Proliferation Act, 550
Nuclear Power Development (NPD), 501, 535, 596, 617, 623
Nuclear power plant siting, 716-718
Nuclear radiation dose, 600-601
Nuclear Regulatory Commission (NRC), 474, 595, 602, 637, 710, 739, 751, 753, 775, 792, 811, 818, 828, 840, 850, 855, 857-858
  licensing, 751, 755, 792
Nuclear system evaluation, 44
Nuclear waste, 11-12, 751-864
  bilateral agreements, 862
  contractors, 845
  disposal, 853
  environmental implications of, 851-855
  field offices, 845
  from fuel reprocessing, 779
  headquarters interface, 846
  issues, 49, 769
  policy, 45, 756-769
  R&D, 11, 667-668
  repositories, 765
  storage, 45-47
    burial, 45
    high-level, 45-46
    low-level, 46
    packaging, 822, 826
  transportation, 760, 821-822, 807-808, 837
Nuclear Waste Management, 45-49, 751-864
  budget, 14, 15
  commercial, 48, 810-836
  commercialization, 849-850
  defense, 47-48, 778-810
  environmental impacts, 766, 851-855
  international agreements, 867-869
  public participation, 775-777
  socioeconomic issues, 756-758
  strategy, 756-769
Nuclear Waste Program Management, 774-774
Nuclear Waste Terminal Storage (NWTS), 768, 776, 821, 854-855
NWCF. See Nuclear Waste Calcining Facility

Oahu, Hawaii, project, 32, 425
Oak Ridge National Laboratory (ORNL), 47, 171, 188, 241, 347, 399, 485, 486, 506, 606, 625, 649, 668, 682, 686, 697, 701, 713, 716, 718, 730-31, 735, 757, 778, 781, 786-787, 809, 826, 847, 900, 901, 912, 814, 918, 923, 929, 938-939, 953, 959
  and EBT, 918
  and ISX, 900, 939

and LCTF, 912
shale fracture facility, 788
Oak Ridge Operations Office, 81, 639, 722, 826
Occupational hazards, 600-601, 950
Occupational health and safety. See Worker health and safety
Occupational Safety and Health Act (OSHA), 349, 356, 741
Ocean energy systems, 32-33, See also Thermal Energy Conversion (OTEC)
　applications, 428-429
　currents, 430, 434
　description, 428-429
　and electric energy systems, 479
　open cycle osmosis, 434
　platforms, 431
　salinity gradients, 428, 429
　technology development, 431-432
　testing and evaluation, 429, 433-434
　waves, 428
Ocean Systems Program, 428-434
　Advanced Research and Development Subprogram, 434
　commercialization, 533
　Definition Planning Subprogram, 430-431
　description, 400
　Engineering Test and Evaluation Subprogram, 433-434
　environmental impact, 537-538
　international cooperation, 561-562
　objectives, 388
　Project Management Subprogram, 430
　regional activities, 433-434, 522
　socioeconomic aspects, 546
　Technology Development Subprogram, 431-432
　technology transfer, 566
Ocean Thermal Energy Conversion (OTEC), 33, 377, 428-434, 448, 479
　applications, 33, 429
　commercialization, 33-533
　electric system integration, 33
　hull configurations, 431
　international cooperation, 561-562
　mist-foam, 434
　regional activities, 433-434, 522
　socioeconomic aspects, 546
　technology development, 34, 429, 433-434
Occidental oil shale process retort operation, 302
OECD. See Organization for Economic Cooperation and Development
Ogg-gas adsorption system, 696
Off-gases. See Airborne waste
Office of the Assistant Secretary for Conservation and Solar Applications (CSA), 374, 396-397
Office of the Assistant Secretary for Energy Technology (ET), 772
Office of the Assistant Secretary for Policy and Evaluation (PE), 388, 772
Office of the Assistant Secretary for Resource Applications (RA), 383, 772
Office of Energy Research, 772
Office of Environment (EV), 754
Office of Environmental Activities (EA), 356
Office of Federal Procurement Policy (OFPP)
Office of Federal Procurement Policy Act
Office of Fossil Fuels
Office of Fusion Energy, 870-892
　activities, 870
　Division of Development and Technology, 910

Division of Planning and Projects, 930
field activities, 888-889, 938-943
Headquarters, 885-886, 888
magnetic fusion energy, 51
management approach, 51
organization, 889
program objectives, 51
responsibilities, 870, 878, 885
Office of Laser Fusion, 889-890
Office of Management and Budget (OMB), 78, 192, 774
Office of Nuclear Energy Programs organization, 586
Office of Nuclear Waste Isolation (ONWI), 757, 813, 816
Office of Nuclear Waste Management. See Office of Waste Management
Office of Solar, Geothermal, Electric and Storage Systems (ETS), 381
Office of Waste Isolation (OWI), 757
Office of Waste Management (ETW), 769-775
Offshore drilling, 27
OFPP. See Office of Federal Procurement Policy
Ohmic heating, 899, 900
Oil, 57
　demand (U.S.), 22, 373
　domestic, 4, 12
　enhanced recovery of, 26
　imported, 3-4, 8
　　reduced dependence on, 7-8
　　savings of, 7-8, 13-14
　production of crude, 75
　reserves (U.S.), 22
　supply of, 4, 12
　transportation of crude, 6
Oil shale, 26, 59, 75, 280
　DOE and industry energy projects, 300
Oil in-place (U.S.), production, reserves, and residual, 281
Oil process in situ, 292
Oil Shale Program, 75, 296-298
OLF. See Office of Laser Fusion, DOE
OMB. See Office of Management and Budget, U.S.
Once-Through Fission Resources, 568-570, 573
On-site Integrated Energy System (Fuel Cells), 266-268
ONWI. See Office Of Nuclear Waste Isolation
ONWI work breakdown structure, 813
Open-cast mining, 101-103
Open-cycle coal-fired MHD system, 275
Open-cycle gas turbine/steam turbine system, 215
Open-Cycle Gas Turbine Systems, 213-216
Open cycle osmosis, 435
Open-Cycle Plasma Systems (MHD), 272-277
Operations Offices. See specific operations office
Oregon State University, 848
Organic Rankine Bottoming Cycle, 232
Organization for Economic Cooperation and Development, 472, 551, 679, 863-864
Organization of Petroleum Exporting Countries (OPEC), 556
ORMAK, 958
ORNL. See Oak Ridge National Laboratory
OTEC. See Ocean Thermal Energy Conversion
OTEC-1, 433-435, 533
　facility, 434
Outer Continental Shelf, 309
OWI. See Office of Waste Isolation

Pacific Northwest Laboratory (PNL), 347, 399, 421, 735-736, 831, 847, 936

Package testing program, 824
Pakistan, 581
Panel mining, 94-98
Paradox Basin, 812-814
PASS. See Procurement Automated Source System
PC 18 System, 267
PDU's. See Process Development Units
PDX. See Poloidal Divertor Experiment
PE. See Office of the Assistant Secretary for Policy and Evaluation
Peat gasification, 133-139
Pebble Bed Fusion-Fission Systems, 921
Pellet-Clad Interaction (PCI), 598-600
Pellet fuel, 867, 872, 900, 914
People's Republic of China. See China
Performance Assurance System, 193-194
Performance standards
  geothermal systems, 454-455, 539
  photovoltaic systems, 410, 411-412
  solar systems, 544-545
  wind systems, 545-546
Permian Basin, 812, 814
Peru, 473
Petroleum. See also Oil
  substitutes, 443-445
Petroleum extenders, 26, 442, 534
Petroleum liquids, U.S. supply, 4
Petroleum Program, 25-26
Petroleum Resource Strategy, 59
Phaedrus, 905, 908
Philadelphia Electric Company, 41
Philippines, 563
Phosphoric Acid Systems (Fuel Cells), 266-268
Photoelectric systems. See Photovoltaic systems
Photovoltaic cells, 31, 34, 409, 448
Photovoltaic device, 10, 31, 400, 409
  electricity from, 10
  pollution, 16
Photovoltaic electricity, cost, 11
Photovoltaic Energy Conversion Program, 408-418
  Advanced Research and Development Subprogram, 411-413
  commercialization, 405-407, 409, 530-531
  description, 376, 400
  energy savings estimate, 531
  environmental impact, 537
  international cooperation, 557, 560
  objectives, 388, 409-411
  regional activities, 410-411, 417-418, 522
  socioeconomic aspects, 544-545
  Systems Engineering and Standards Subprogram, 414-417
  Technology Development Subprogram, 413-414
  Test and Applications Subprogram, 411, 417-418
Photovoltaic Program Office (PVPO), 447
Photovoltaic residential system, 415
Photovoltaic systems, 416, 561
  applications, 377, 414-415, 417-418
  Balance of System (BOS) components, 409, 414-415, 417
  concentrator, 412, 414
  description, 407-409
  electric system integration, 409
  electrochemical cells, 412
  flat-panel, 411, 414
  materials development, 411-413
  technology development, 409-410, 413-414
  thin-film, 412, 414

3.5 kW, 557
  total energy, 411
  two-cell, 412
Photovoltaic Total Energy Technology, 414
Pittsburg Energy Research Center, 80, 107, 345, 441
Pittsburg Energy Technology Center (PETC), 120, 122, 132, 144, 237, 258
Plant Systems (task), 917
Plasma. See also Beta value, plasma; Density, plasma; Heating, plasma; Temperature, plasma
  behavior prediction, 891, 929
  burning, 873, 905
  circular and non-circular, 894, 898, 899
  confinement time, 871, 876, 895
  cooling, 894, 897, 900
  D-shaped, 897
  gun, 906
  ignition, 871, 880-881, 898, 905
  isotope separation technique, 613
  limiter, 900
  physics research, 898
  process, 612
  properties data research, 928
  reactor-grade, 898, 899, 901
  shape optimization, 894, 898
  spent, 872
  stabilization, 895, 898-899
  temperature and heating, 871-872
  transport, 894, 895, 899
Plasma Confinement and Beta, 872-873
Plasma Engineering (Subactivity), 911-914
Plasma Maintenance and Control (task), 914
Plasma Materials and Interaction (task), 915
PLT. See Princeton Large Torus
Plum Brook Station Project, 424
Plutonium, 577, 802
Plutonium Experimental Facility (PEF), 731
Plutonium Fuel Fabrication Facility (PuFF), 730
PNL. See Pacific Northwest Laboratories
Poland, 367
Policy (DOE), 567-582
Policy, U.S. energy, 7, 373-374
Pollution, control R&D, 13
Polodial Divertor Experiment (PDX), 894, 897-898, 914
Polodial field confinement, 893, 900
PON. See Program Opportunity Notice
Power Delivery Program, 37-38, 393-394, 481-485
  Electric Field Effects Subprogram, 393-394, 481-482
  High Efficiency Equipment and Systems Subprogram, 484-485
  High Voltage Direct Current Subprogram, 481, 483-484
  objective, 481
  Underground Transmission and Compact Stations Subprogram, 482-483
Powerplant and Industrial Fuel Use Act of 1978, 4, 5-6, 374, 474, 487
Power plants, fusion, 897, 918, 919
Power Supply Integration, 37-38, 475-481
Power Supply Integration Program Load Management Subprogram, 38, 379, 393-394, 476-478
Power Supply Integration Program Load Management Subprogram, Solar Generic/Integration of New Source Technologies Subprogram, 379, 478-479
Power Supply Integration Program Load Management Subprogram, System Control and Development Subprogram, 475-476, 479-481
Power System Technology and Support, 726-729

Index 1013

PPPL. See Princeton Plasma Physics Laboratory
PRDA. See Program Research and Development Announcement
Preliminary Safety Analysis Report (PSAR), 919, 933
Pressure water reactor, 589, 600-601
Pressurized Fluidized Bed, 188, 196, 203, 246-250, 256
Pressurized fluidized-bed combined-cycle pilot plant, 248
Pressurized-water reactor nuclear steam supply, 591
Pressurized water reactor steam generator, 601
Price-Anderson Act, 857-858
Princeton Large Torus (PLT), 51, 876, 894-897, 914
Princeton Plasma Physics Laboratory (PPPL), 890, 918, 929, 938-939, 953
Princeton I Experiment, 172
Principal Alternate Fuel, 256
Private Industry. See also Commercial Waste Management Activity
Process development units (PDU's), 111, 436, 438, 440-441
Process heat, 377, 387, 503-505
  biomass systems, 377, 389
  geothermal, 450
  ocean systems, 432
  photovoltaic systems, 394
  solar systems, 387
  solar thermal systems, 377, 388, 402, 405
Process modification technology, 205-206
Product Characterization and Utilization-Petroleum, 318-319
Production mining, 90-94, 101-104
Program of Research and Development of Manmade Geothermal Energy Systems (MAGES), 972
Program Opportunity Notice (PON), 134, 142, 245, 354
Program Reporting and Change Control System (PRCSS), 588
Program Research and Development Announcement (PRDA), 205-206
Program structure for direct utilization subactivity, 182-183
Program structure for fossil energy materials and components subactivity, 186-187
Program structure for gasification conversion processes, 178-179
Proliferation, 623-625. See also Nonproliferation, nuclear
Proof-of-Breeding Program, 701-702
Proof-of-Principle Activity, 878-880
Proof-of-Principle, EBT experiment, 925
Proof-of-Principle, hydrogen experiment, 880
Proof-of-Principle, mirror experiment, 876
Prototype fuel dissolver within hot-cell, 701
Prototype Large Breeder Reactor (PLBR), 631
Prototype reactor. See Engineering Prototype Reactor (EPR)
Prototype systems assessments, 225-228
Psychoacoustic effects, 482, 487, 541
Public, involvement in energy, 8
Public participation, nuclear waste disposal and, 775-777
Publice Service Co. of New Mexico, 456
Public Utility Regulatory Policies Act (1978) (PURPA), 5-6, 374, 474, 476, 487
Puerto Rico, 425, 532
Pulse power technology, 911-912
Pulverized coal and synfuel combustion, 177, 180-181

Pumping system, high-vacuum, 902
PVPO. See Photovoltaic Program Office
Pyrolysis, 377

Q value, 876
  enhancement, 880, 881, 905
  maximization, 907

RA. See Office of the Assistant Secretary for Resource Applications
Radiation dose levels, 739
Radiation Facilities Development and Operation (task), 915
Radio frequency (RF) heating, 872
  and Alcator experiment, 901
  development and technology, 917
  feasibility, 895
  mirror experiments, 907
  PLT experiments, 895, 897
Radio interference, 482, 487
Radioactive Waste Management Complex (RWMC), 764, 786, 799
Raft River Pilot Plant, 455, 465
Rankine cycles, 232, 406, 594, 613, 720
RD&C cycle, magnetic fusion, 889, 944
REA. See Rural Electrification Administration
Reactor. See also Nuclear Waste; Nuclear Waste Management
  breeder, 42-43, 627-702, 768
  converter, 738-739
  fuel reprocessing, 649-656
  fusion, 914-915, 945-946
    assembly and disassembly of, 873
    budget for projects, 880, 929
    conceptual, 873, 917-918
Reactor experiment complex, 836
Reactor/fuel cycle costs, 714
Reactor, fusion
  economics, 914-915, 944-945
  maintenance, 873-880
  materials, 914-916
  physical damage to, 911
  prototype, 891
  safety, 873-874, 892
Reactor, light water. See Light Water Reactor
Reactor, plutonium/highly toxic materials comparison, 742
Reactor Projects, 882-884, 893, 930-937
Reactor Safety Research (Task), 919
Redox cells, 495
Reduced Enrichment Research and Test Reactors, 618-623
Regenerative braking systems, 488, 501
Regional activities,
  biomass systems, 438, 447
  electric energy systems, 523-525
  energy storage systems, 525-526
  fusion energy, 938-943
  geothermal systems, 522-523
  nuclear waste, 859-860
  photovoltaic systems, 410-411
  solar systems, 522
  solar thermal systems, 522
  wind systems, 522
Regional energy needs, 17
Remote Controlled Continuous Miner, 92

Remote metering, 477
Remote operated, longer-than-seam-height drill, 93
Remote System Application Program, 406-407
Reprocessing and Refabrication, 738-739
Research, development, and commercialization, fusion. See RD&C cycle, magnetic fusion
Research, development and demonstration (RD&D). See Regional activities; specific programs
Research reactors using highly enriched uranium, 621
Resdential applications. See also Electricity generation; Heating and air conditioning; Water
Residential photovoltaic system, 495
Residue fuels, 258
Resistance heating, 872, 897, 898
Resource Conservation and Recovery Act, 196, 210, 349, 356
Resource Managers, 398
  geothermal/hydrothermal, 398
  program management role, 398
Retrievable Surface Storage Facility (RSSF), 757
Retrieval, stored waste
Reversed field. See Field Reversal Experiments (FRX)
Reversed field pinch (RFP) concept, 922-923, 939
RF. See Rocky Flats
RF heating. See Radio frequency heating
RFP. See Reversed Field Pinch
RHO. See Rockwell Hanford Operations
Richland Operations Office, 813-814, 816, 834
Richland, Washington (RL), 790
Right-of-access, 545
Rights-of-way, 482, 487, 547
Risk. See Safety
RL. See Richland, Washington
Rock mechanics and fragmentation, 461-462
Rockwell International Corporation, 647
Rocky Flats (RF), 847
Rocky Flats Wind Energy Test Center, 32, 422-423
Room-and-pillar-mining, 90-91
Rotary Drilling Systems, 88
RSSF. See Retrievable Surface Storage Facility
Rural applications. See Agricultural applications; Remote site applications
Rural Electrification Administration (REA), 419, 421
RWMC. See Radioactive Waste Management Complex

SA. See Solar Applications
Safe Drinking Water Act, 349
Safeguards and Security Program, Fission Energy, 666-668
Safety Analysis Report (SAR), 854
Safety Evaluation Report (SER), 637
Safety lines of assurance, 642
Safety, of personnel, 641, 918-919
Safety problems, magnetic fusion, 918-919
Safety Test Facility (STF), 661-662
Salina Basin, 812
Salinity gradients, 429
Salt formations, 812-813
  nuclear waste and, 757
San Diego Gas and Electric Co., 455
Sandia Laboratories (SL), 165, 166, 252, 292, 299, 301, 307, 308, 312, 313, 347, 399, 596, 602, 729, 733, 735, 768-769, 781, 805, 808, 819-820, 847, 890, 939
Sandusky, Ohio, wind machine, 32
SAR. See Safety Analysis Report
Saudi Arabia, 418, 556, 560, 561
Savannah River Laboratory (SRL), 701, 722, 730, 756-757, 778, 780
Savannah River Operations Office, 701, 722, 792, 825, 842
Savannah River Plant (SRP), 787-790
Savannah River solid waste burial ground, 827
Scaling. See Major scaling experiments
Schlesinger, Former DOE Secretary, 7
Science Applications, Inc. (SAI), 918
Scientific base, fusion. See under Feasibility, fusion energy
Scotland, 153, 156
SCTI, 667
SCTL, 667
Scylla, 924
Sea-base fuels plant, 432
Sea Floor Geotechnical Instrumentation, 312-313
Second generation fuel cell powerplant, 270
SERI. See Solar Energy Research Institute
Shaft Development (Underground Mine)
Shale formations, 812, 813
Shale fracturing, 781, 786
Shale, gas, 307-309
Shale, oil, 298-307
  major field tests, 298-299
  major laboratory tests, 299-303
  refining and utilization (SORU), 316-318
Shallow land burial, 758-759, 765-766
Shielding, 874, 917
Shippingport Atomic Power Station, 42, 629, 670-672, 701
Shippingport plant operating cycle, 673
Silviculture, 439-440
Simple magnetic mirror, 905
Single-circuit cable capacities, 484
Site selection
  and environmental issues, 853-854
  and IRG recommendations, 765-767
  and Nevada Test Site, 819-821
  and public opinion, 856
  for geologic repositories, 801-813
  for Hanford Reservation Basalt Program, 816-817
  for Waste Isolation Pilot Plant, 805-806
SL. See Sandia Laboratories
Slagging Lurgi Process, 153
Small Power Systems, 405-407
Snake River Plain, Idaho, project, 453
Socioeconomic aspects
  biomass systems, 546
  electric energy systems, 487
  energy storage systems, 514-517, 519
  fission energy, 745-746
  fossil energy, 355-357
  geothermal systems, 546-547
  magnetic fusion, 944-945
  nuclear waste, 856-858
  ocean systems, 546
  photovoltaic systems, 545
  solar systems, 387-389
  wind systems, 545-546
Sodium Components Test Installation (SCTI;, 645, 664-665
Sodium Loop Safety Facility (SCSF), 644
Sodium Pump Test Facility (SPTF), 647, 664-666
Solar Applications (SA), 494-496
Solar cells, 409
Solar Electric Applications Program, 29-31, 376-377, 388, 400-434. See also Ocean energy systems; Photovoltaic Energy Conversion Program; Solar Energy Thermal Power Program; Wind Energy Conversion Systems Program

description, 400-434
funding, 403
objectives, 401-402
Solar energy, 12. See also Solar, Geothermal, Electric and Storage Systems
impact, 35
international availability, 550
R&D, 381-384
Solar Energy Domestic Policy Review, International Panel, 551
Solar Energy Research, Development and Demonstration Act, 390
Solar Energy Research Institute (SERI), 390, 398, 404, 445-446, 447
biomass projects, 447
energy storage projects, 516
program management role, 396-398
project management role, 398-399
solar thermal projects, 404
wind projects, 447
tax credits, 518
Solar Generic/Integration of New Source Technologies, 38, 379, 478-479
Solar, Geothermal, Electric and Storage Systems, 28-39, 373-563
commercialization, 527, 535, 552
environment and, 536-543
funding, 14, 15, 380-393, 520
international RD&D agreements, 549-563
programs, 376-519
projects, 384
Solar, Geothermal, Electric and Storage Systems Program, regional activities, 520-526
Solar Heating and Cooling, storage, 504-505
Solar Photovoltaic Electric Energy Research, Development and Demonstration Act, 375
Solar Photovoltaics Act, 31, 389, 410
Solar photovoltaic power system, 419
Solar ponds, 434
Solar Power Corp., 531
Solar radiation network, 538
Solar systems. See Biomass Systems; Ocean energy systems; Photovoltaic Systems; Solar Technology Program; Solar Thermal Systems; Wind Systems
Solar technologies fuel displacement, 387
Solar Technology Program, 29-34, 400-449
budget, 380-381
Solvent Refined Coal (SRC), Pilot Plant, 114-117
Solvent Refined Coal, process, 22-23, 112
Source Evaluation Board, 456
Sources of energy, current and projected, 4. See also Alternative Sources of energy
South Africa, 743
South Korea, 581
Southern California Edison, 531
Southern Forest Experiment Station, 440
Soviet Union. See Union of Soviet Socialist Republics
Space and Terrestrial Systems Program, 44, 718-736
Space missions, 719
Space Nuclear Auxiliary Power (SNAP) Systems, 719
Space Nuclear Systems, 722-736
Space Power System Safety, 731-733
Space/terrestrial applications program work breakdown structure, 722
Space/terrestrial compact power systems programs, 721
Spain, 361-363, 404
Special Purpose Materials Development (task), 915
Spent fuel, 48-49, 753
handling and storage, 49, 838

Spent Fuel Storage (Activity), 763-764, 816, 836-844
international, 838-839
Spent fuel storage pool, 841
Spent fuel storage requirements, 841
Spent fuel storage work elements, 842
Solar thermal systems,
applications, 376-377
central receiver systems, 403-404, 406
description, 371, 400, 402
distributed-collectors, 404
electric system integration, 402-403
energy storage systems, 402-404
high-temperature receivers, 408
large, 529-530
materials development, 407-408
retrofitting/repowering, 403-404
small, 528-529
technology development, 407-408
total energy, 406
Solar Thermal Test Facility (STTF), 408
Solar Total Energy Systems Facility, 406
Solarex Corp., 531
Solar Technology Program
commercialization, 447-448, 997-998
description, 376-377
energy saving estimate, 387
environmental impact, 448-449, 536-539
funding, 380-381
international cooperation, 556-562
objectives, 384-388
regional activities, 447, 522
socioeconomic aspects, 449, 544-546
strategy, 34-35, 386-397, 401
Solar thermal agricultural application, 557
Solar Thermal Power Generation, 507-508
Solar Thermal Power Program, 402-408
Advanced Technology Research and Development Subprogram, 407-408
commercialization, 528-530
description, 377, 400
environmental impact, 448-449, 536-537
funding, 381
international cooperation, 557
Large Power Systems Subprogram, 402-405
objectives, 402
regional activities, 522
Small Power Systems Subprogram, 402, 405-407
socioeconomic aspects, 544-545
Solar Thermal Systems, 31
Solidification. See Immobilization
Solidification process demonstration, 48, 762, 832, 834
Solid polymer electrolytic water-splitting process, 509
Solid waste storage, 798
Solvent Extraction (Liquefaction), 114-120
Solvent Refined Coal (SRC), Demonstration Plant, 129-130
SPERS. See Storage Program Evaluation and Review System
Sputtering process, 830
SR. See Savannah River
SRC-1, 83, 112, 129-130
SRC-II, 83, 112, 130
SRC solid process (SRC-I), 116
SRC liquid process (SRC-II), 117
SRE. See System readiness experiments
SRL. See Savannah River Laboratory
SRP. See Savannah River Plant
ST. See Division of Solar Technology
Standards. See Performance standards

Stanford Univeristy, 848
State Implementation Plans, 194
Steady-state operation, 905, 909, 922
Steam displacement, 293
Steel industry heat recovery storage, 505
Steeply Dipping Beds, 164, 168-169
Steeply dipping beds, Rawlins #1 Experiment, 169
Stellarator/torsatron (S/T), 922, 924
Stirling cycle, 406
Stirling engines, 222-224
Stirling engine total energy system, 223
STOR. See Division of Energy Storage Systems
Storage, nuclear waste. See Nuclear Waste Storage
Storage Program Evaluation and Review Systems (SPERS), 517
Stratapax/Slug system, 464
Stress corrosion barrier, 599
Strontium 90, 782
Strontium and cesium capsules, 783
STTF. See Solar Thermal Text Facility
Sublevel Caving with Pillar Extraction, 98
Submarine Propulsion Reactors, 45
Substitute natural gas, 132
Superconducting cable system, 483
Superconducting Magnetic Development (task), 912
Superconducting Magnetic Energy Storage (SMES), 510-512
Supporting Studies Subactivity, nuclear waste, 48
Surface coal gasification, 131-162
Surface Coal Mining, 23, 84, 100-105
Surface Mining Control and Reclamation Act, 104
Surface ship propulsion reactors, 45
Surface Test Facility, 99-100
SURMAC, 924
Sweden, 252, 361-364, 472, 600, 861, 959
Switzerland, 364, 472, 595, 605, 677-678, 959
Syncline Ridge, 820
Synthane Pilot Plant, 132-133
Synthane Process, 133
Synthetic fuel, 12, 59, 110, 875, 919, 946
Synthetic liquids, 58
System Control and Development, 38
System readiness experiments (SRF), 418
Systems Engineering and Standards, 414-417

Taiwan, 473
Tandem Mirror Experiment (TMX), 876, 905-908
Tandem Mirror Experiment device, 908
Tandem Mirror Systems, 904-908
Tank storage, 780, 782
Tax credits, 5, 518
TEC. See Transportation Energy Conservation
Technical Information Service (TIS), 362-363
Technical progress, magnetic fusion, 875
Technology development cycle, 527, 849-850
Technology process flow chart, 529-530
Television interference, 482, 487
Temperature, plasma, 873, 894, 898
 and density, 902
 breakeven, 897
 in inertial confinement, 872
 level for fusion, 871-872
 reactor-level, 898
Tennessee Valley Authority (TVA), 196, 600, 638-639, 716, 844
Terminal Isolation (Subactivity), 48, 810-822

Terminal storage, 802-807
Terminal Storage Program, 48, 802-807
Terminal Storage Subactivity, 760, 762
Terrestrial Isotope Applications, 734-736
Terrestrial missions, 720
Test Reactor and Irradiation Test Facility, 618-619
Texas Experimental Tokamak (TEXT), 928
Texas, geopressured resources, 458, 459, 461
TEXT. See Texas Experimental Tokamak
TFTR. See Tokamak Fusion Test Reactor
Thermal and Mechanical Storage Program, 39, 380, 500-517
 applications analysis, 515-516
 chemical hydrogen storage commercialization, 501-502, 505, 506
 description, 500-517
 energy saving estimate, 510
 environmental impact, 542-543
 funding, 501
 Mechanical Storage Subprogram, 509-510
 regional activities, 486
 Superconducting Magnetic Storage Subprogram, 510-512
 technical and economic analysis subprogram, 514-517
 Technology Information Systems, 516-517
 technology transfer, 504
 Thermal Storage Subprogram, 39, 501-508
 Utility Application Subprogram, 512-514
Thermal efficiency, 715-716
Thermal energy, 377, 450, 462
Thermal energy storage load leveling, 507
Thermal Energy Storage (TES) Systems, 501-508
 applications, 501-502
 aquifer, 502-503
 chemical heat pumps, 504-506
 chemical reaction heat, 509
 industrial waste heat recovery, 503-504
 latent heat, 501
 off-peak electricity, 501, 505-507
 seasonal, 502
 sensible heat, 501
Thermal Gas-Cooled Reactor Program, 41, 568
Thermal Hydraulic Out-of-Reactor Safety (THORS) Facility, 644
Thermal reactor, technology, 41, 589, 590-606
Thermal reactor technology work breakdown structure, 596
Thermal Recovery, 289-294
Thermal Transient Facility (TTF), 648
Thermionics, 231
Thermochemical conversion, 440-442
Thin-film cadmium sulfide cells, 412-413
Third-Generation Process
 gasification, 144-146
 liquefaction 120-125
Thorium, 576
Three Mile Island, 582
Tight Sands, 59
TMX. See Tandem Mirror Experiment
Tokamak device, 895
Tokamak experiments, 896
Tokamak Fusion Test Reactor (TFTR), 876, 880, 889, 895, 897, 914, 917, 930-933
 development base, 894
 overview, 930
 project management, 889
 status, 876, 930
Tokamak parameters status, 877

Tokamak system, 891, 893-903
　budget, 894
　development and technology, 911, 914, 930
　overview, 393-395
　problem areas, 894
Topping cycles, 231
TORMAC, 922, 925
Torodial field confinement, 893, 899
Total Energy Technology Assessment Study (TETAS), 227
Toxic Substances Control Act, 349, 356
Toxicity, chemical, 948, 949
Transient Reactor Test Facility (TREAT), 644
Transmission lines
　alternating current, 5, 481-482
　compressed air, 481-482
　direct current, 481, 483-484
　environmental impact, 479, 541-542
　overhead, 481-482
　socioeconomic aspects, 547
　superconducting, 394
　underground, 481-484
Transportation Energy Conservation (TEC), 2
Transportation of nuclear waste, 48, 760, 768-769, 807-808, 821-822
Transportation Technology and Information Center, 760, 768-769, 822
Transuranic Waste (TRU), 47, 756, 760, 765-766, 778, 780-781, 786, 788-792, 796-801, 826, 842
Transuranic waste
　at Idaho Chemical Processing Plant, 783
　at Waste Retrieval and Treatment Facility, 799-800
　at WIPP, 760
　technology, 798
　volume, 779
TREAT reactor building, 662
TRIGA four-rod cluster and standard MTR plate-type element, 619
TRIGA instrumented fuel element, 620
TRIGA Mark-1 core and reflector assembly, 620
Tritium, 948-949
　radioactivity control, 917
　recovery of, 800
Tritium Systems Test Assembly (TSTA), 880, 892, 917, 948
Trojan nuclear power plant, 741
True in situ retorting, 301
TRU decontamination, 828
TRU waste, See Transuranic waste
TRU waste incinerator-conceptual layout, 799
TRU waste volume, 801
TSTA. See Tritium Systems Test Assembly
Tunnel boring machine, 89
Tunnel Boring System, 89-90
Turbine Cleanup Systems, 203-204, 206
Turkey, 959
TVA. See Tennessee Valley Authority
Two-State Liquefaction, 125, 128

Underground Cables and Compact Stations, 38
Underground coal gasification, 162-163
Underground coal mining, 84-100
Underground compressed air facility, 513
Underground pumped hydroelectric storage, 513
Underground pumped hydro storage (UPHS), 512-514, 524, 542

Union Carbide Corporation, 847
Union Oil Company of California, 456
Union of Soviet Socialist Republics, 365-367, 582, 907, 956-958
United Kingdom, 188, 252, 359, 361-365, 472, 581, 593, 644, 861-863
United States Geological Survey (USGS), 452
University of Arizona, 848
University of California, Los Angeles (UCLA), 603, 924, 941
University of Chicago, 660
University of Delaware, 369
University of Illinois, 918
University of Kentucky, 369
University of Maryland, 991
University of Miami, 848
University of New Mexico, 848
University of New York, 369
University of Pittsburgh, 369
University of Tennessee, 898
University of Texas, 938, 941
University of Utah, 848
University of Wisconsin, 908, 918, 924, 938, 941
Uranium, 573, 575-577, 589-590
　enrichment, 608-613, 738
　　studies contractors, 609
　preparation, 741-743
　recycle, 575-577
　utilization, 597-600
Uranium mill tailings, 766
U.S. bilateral agreements, 558-559
USDA. See Department of Agriculture
U.S. energy consumption, 12
U.S. forests, 439
U.S. Forest Service, 440
USGS. See United States Geological Survey
U.S Postal Service, 501
Utah, 26, 75, 718
Utilities. See Electric utilities

Vacuum Component Development (task), 914
Vacuum Components Test Facility, 914
Vacuum technology, 914
Valles, Caldera, N. Mex., projects, 468
Vanadium-base alloy, 916
Vanderbilt University, 848
Variable-geometry nuclear control concept, 677
Vehicles. See Electric/hybrid vehicles
Vertical field systems, 912
Very High Temperature Reactor (VHTR), 568, 593, 605
Virginia Polytechnic Institute, 453
Vitrification. See Immobilization; Solidification
Vitrified waste, 824
Volume. See Waste quantities

Wall, first, 915, 949
Wall, sputtering, 894, 915
Waste heat, 24
Waste heat recovery, 24
Waste Isolation Pilot Plant (WIPP), 48, 575, 760, 768, 776, 798. 804, 805-808
Waste isolation pilot project network, 761
Waste management, fossil energy, 354
Waste quantities, 752, 779

Waste Retrieval and Treatment Facility (WRTF), 799-801
Waste separation plant, 47
Waste Treatment Technology (subactivity), 48, 762
Water-Cooled Breeder Reactors, 42, 627-629, 668-677, 740
Water
 electrolysis, 535
 heating, 386, 453
 pollution, 454, 536, 538, 539, 542
 rights, 546-547
Water Pollution Control Act, 356
Wave energy, 33, 430
Wellhead generators, 455
West Valley Nuclear Center, 48, 776, 834, 836
Western Gas Sands, 28, 321, 324-326
Western Low-Btu Gasification, 164
Western Medium-Btu Gasification, 167
Western New York Nuclear Service Center (WNYNSC), 834, 837
Westinghouse Electric Corp., 328, 647
Wind Characteristics Program, 421-422
Wind energy conversion systems
 activities, 422
 objectives, 422
 products, 422
Wind Energy Conversion Systems Program, 32, 33, 418-428, 447-448
 commercialization, 447-448, 531-533
 description, 418-428
 environmental impact, 448, 537
 funding, 420
 interagency cooperation, 422, 424
 Intermediate-scale Systems Subprogram 424-427
 international cooperation, 560-561
 Large-scale Systems Subprogram, 427-428
 Multiunit Systems Subprogram, 420
 objectives, 32, 419
 Project Development and Technology Subprogram, 420-422
 regional activities, 522
 Small-scale Systems Subprogram, 32, 422-424
 socioeconomic aspects, 545-546
 technology transfer, 426
 tasks, 422
Wind Systems, 10. See also Wind Energy Conversion Systems
WIPP. See Waste Isolation Pilot Plan
WNYNSC. See Western New York Nuclear Service Center
Wood-fired spreader-stroker burner, 259
Wood resource distribution. See Direct Combustion
Wood-to-oil process development unit, 443
Worker health and safety, 600-601
WRTF. See Waste Retrieval and Treatment Facility
Wyoming, 75

X-ray imaging, 929

Yin-Yang concept, 903-906

Zero Power Plutonium Reactor (ZPPR), 649-650, 660-661
Zinc/chloride battery system, 493
Zinc chloride catalyst process, 122
Zinc Chloride Catalyst (Liquefaction), 121